NUCLEATION AND ATMOSPHERIC AEROSOLS 2000

15th International Conference

Rolla, Missouri   6–11 August 2000

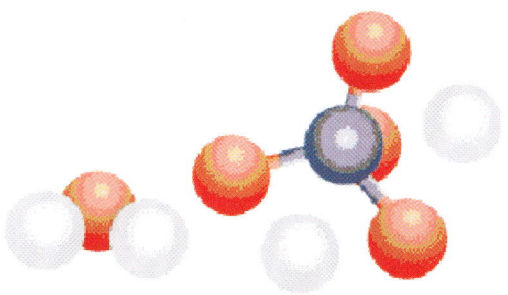

EDITORS
Barbara N. Hale
University of Missouri-Rolla

Markku Kulmala
University of Helsinki

Melville, New York, 2000
AIP CONFERENCE PROCEEDINGS ■ VOLUME 534

Editors:

Barbara N. Hale
Department of Physics
University of Missouri-Rolla
Rolla, MO 65409
USA

E-mail: bhale@umr.edu

Markku Kulmala
Department of Physics
University of Helsinki
Tallqvist Bldg.
FIN-00014 Helsinki
FINLAND

E-mail: markku.kulmala@helsinki.fi

The articles on pp. 373–385 and 783–786 were authored by U. S. Government employees and are not covered by the below mentioned copyright.

Authorization to photocopy items for internal or personal use, beyond the free copying permitted under the 1978 U.S. Copyright Law (see statement below), is granted by the American Institute of Physics for users registered with the Copyright Clearance Center (CCC) Transactional Reporting Service, provided that the base fee of $17.00 per copy is paid directly to CCC, 222 Rosewood Drive, Danvers, MA 01923. For those organizations that have been granted a photocopy license by CCC, a separate system of payment has been arranged. The fee code for users of the Transactional Reporting Service is: 1-56396-958-0/00/$17.00.

© 2000 American Institute of Physics

Individual readers of this volume and nonprofit libraries, acting for them, are permitted to make fair use of the material in it, such as copying an article for use in teaching or research. Permission is granted to quote from this volume in scientific work with the customary acknowledgment of the source. To reprint a figure, table, or other excerpt requires the consent of one of the original authors and notification to AIP. Republication or systematic or multiple reproduction of any material in this volume is permitted only under license from AIP. Address inquiries to Office of Rights and Permissions, Suite 1NO1, 2 Huntington Quadrangle, Melville, N.Y. 11747-4502; phone: 516-576-2268; fax: 516-576-2450; e-mail: rights@aip.org.

L.C. Catalog Card No. 00-105905
ISBN 1-56396-958-0
ISSN 0094-243X
Printed in the United States of America

NUCLEATION AND ATMOSPHERIC AEROSOLS 2000

Previous Proceedings in the Series of International Conferences on Nucleation and Atmospheric Aerosols (ICNAA)

	Year	Held in	Publisher	ISBN
14th	1996	Helsinki, Finland	Pergamon Press	008-0420303
13th	1992	Salt Lake City, Utah, USA	A. Deepak Publishing	0-937194-26-3
12th	1988	Vienna, Austria	Springer-Verlag	3-540-50108-8
11th	1984	Budapest, Hungary	Hungarian Meterological Service	84.392

Other Related Titles from AIP Conference Proceedings

492   Simulation and Theory of Electrostatic Interactions in Solution: Computational Chemistry, Biophysics, and Aqueous Solutions
Edited by Lawrence R. Pratt and Gerhard Hummer, November 1999, 1-56396-906-8

364   Fast Elementary Processes in Chemical and Biological Systems: 54th International Meeting of Physical Chemistry
Edited by Andre Tramer, May 1996, 1-56396-564-X

314   Advances in Plasma Physics: Thomas H. Stix Symposium
Edited by Nathaniel Fisch, 1994, 1-56396-372-8

312   Molecules and Grains in Space: Proceedings of the 50th International Meeting of Physical Chemistry
Edited by Irène Nenner, 1994, 1-56396-355-8

277   The World at Risk: Natural Hazards and Climate Change
Edited by R. L. Bras and R. G. Prinn, 1993, 1-56396-066-4

To learn more about these titles, or the AIP Conference Proceedings Series, please visit the webpage http://www.aip.org/catalog/aboutconf.html

CONTENTS

Preface xix
Committees xx
Invited Lectures xxi
Plenary Speakers xxii

HOWARD REISS SYMPOSIUM

An Application of the Kubo and Nyquist Relations to Nucleation 3
    R. McGraw, P. Schaaf, and H. Reiss
Homogeneous Nucleation Rates for Water 7
    J. Wölk, Y. Viisanen, and R. Strey
Nucleation of Water and Methanol Droplets on Ions:
The Sign Preference 11
    K. J. Oh and X. C. Zeng
Genuine Saddle Point in Binary Nucleation Kinetics 15
    J.-S. Li, I. L. Maksimov, and G. Wilemski
Nucleation of Sulfur Hexafluoride 19
    P. Ye, A. Bertelsmann, and R. H. Heist
Molecular Dynamics Studies of Evaporation and Condensation
Coefficients in Nucleation Theory 23
    P. Schaaf, B. Senger, J.-C. Voegel, R. K. Bowles, and H. Reiss
Surfactant Induced Nucleation in Supersaturated Vapors 27
    M. Rusyniak, Y. Ibrahim, V. Abedalsayed, M. Rabeony,
    and M. S. El-Shall
A Monte Carlo Discrete Sum (MCDS) Nucleation Rate Model for Water 31
    B. N. Hale and D. J. DiMattio
The Homogeneous Nucleation of a Vapor to Its Solid Phase:
Does Water Nucleation Change at 40 °C? 35
    J. L. Katz and T. P. Kole
Molecular Theory of Homogeneous Nucleation Using a More
Sophisticated Kinetic 37
    A. L. Itkin
A Theory for the Deliquescence of Small Particles 41
    P. Mirabel, H. Reiss, and R. K. Bowles
Idiosyncrasies of Nucleation in Large, Deeply Supercooled
Liquid Clusters 45
    L. S. Bartell, Y. G. Chushak, and J. Huang

NUCLEATION EXPERIMENT AND THEORY

The Homogeneous Nucleation of Water 51
    J. L. Schmitt, K. Van Brunt, and G. J. Doster

Experimental Nucleation Studies with a Laminar Flow Tube Reactor 55
    V. B. Mikheev, N. S. Laulainen, and S. E. Barlow
D_2O-H_2O Condensation in Supersonic Nozzles: I. Experiments 59
    C. H. Heath, K. A. Streletzky, J. Wölk, B. E. Wyslouzil,
    and R. Strey
D_2O-H_2O Condensation in Supersonic Nozzles: II. Modeling 63
    C. H. Heath, K. A. Streletzky, B. E. Wyslouzil, and G. Wilemski
Nucleation of Water Vapor in Methane at High Pressure 67
    P. Peeters and M. E. H. van Dongen
Laminar Flow Tube Reactor Interface with Quadrupole
Mass Spectrometer 71
    V. B. Mikheev, N. S. Laulainen, V. V. Pervukhin,
    and S. E. Barlow
Experimental Studies of the Condensation of Alkali Metals 75
    F. T. Ferguson and J. A. Nuth, III
Homogeneous Nucleation Rates from the Piston-Expansion
Tube Using a Digital Camera 79
    F. Peters and A. Graßmann
Homogeneous Nucleation of n-pentanol in a Laminar Flow
Diffusion Chamber 84
    H. Lihavainen, Y. Viisanen, and M. Kulmala
Steady State Homogeneous Nucleation Rate and Primary
Particle Size Distribution 88
    R. B. McClurg
Nucleation Rates of Lennard-Jones Clusters from Growth
and Decay Simulations 91
    H. Vehkamäki and I. J. Ford
Dynamics of Argon Clusters from Molecular Dynamics Simulations 95
    S. Wonczak and R. Strey
Propandiol Vapor Nucleation Rates 99
    M. P. Anisimov, A. G. Nasibulin, L. V. Timoshina,
    and J. A. Koropchak
Kinetics of Ion-Induced Nucleation and Ion Mobility
of Pre-critical Clusters 103
    S. P. Fisenko, D. B. Kane, and M. S. El-Shall
Gas-Liquid Nucleation in Partially Miscible Systems:
Free Energy Surface from Computer Simulation 107
    S. Yoo, K. J. Oh, and X. C. Zeng
Ternary Nucleation of H_2SO_4, NH_3 and H_2O 111
    M. Kulmala, P. Korhonen, A. Laaksonen, Y. Viisanen,
    R. McGraw, and J. H. Seinfeld
First Measurement of Prenucleation Molecular Clusters 115
    F. L. Eisele and D. R. Hanson
Experimental Determination of Molecular Content of Nuclei
by Independent Variation of Activities 119
    J. Hrubý and R. Strey
Molecular Simulation of Seed-Induced Nucleation in Vapor Phase 123
    K. Ohguchi, K. Yasuoka, and M. Matsumoto

Activation Barrier for Heterogeneous Condensation in Multicomponent Vapor Mixtures: Cases of Insoluble and Mixed Nuclei 127
    Y. S. Djikaev and D. J. Donaldson

Microscopic Effects and Kinetics of Binary Nucleation beyond the Confines of the Fokker-Planck Approximation 131
    Y. S. Djikaev, A. P. Grinin, and F. M. Kuni

The Influence of Particle Solubility on Heterogeneous Nucleation in Binary Vapor Mixtures 135
    D. Petersen, R. Ortner, A. Vrtala, A. Laaksonen, M. Kulmala, and P. E. Wagner

Use of a Turbulent Mixing CNC to Study the Influence of Composition and Vapor Properties on Heterogenous Nucleation 139
    P. K. Hopke, D. W. Lee, R. Mavliev, and H.-C. Wang

Homogeneous Nucleation in High-Pressure Multi-component Systems: Application to Mixtures of N-Alkanes 143
    J. Hrubý, P. Peeters, and M. E. H. van Dongen

Applications of the Nucleation Theorem 147
    D. Kashchiev

Molecular Simulation of Ion-Induced Nucleation in Water Vapor 151
    M. Matsumoto, K. Ohguchi, and K. Yasuoka

Reconciling Gibbs and van der Waals: A New Approach to Nucleation Theory 155
    J. W. P. Schmelzer

Comments on Kinetics and Thermodynamics of Condensations under Homogeneous Conditions 159
    S. H. Bauer, Y.-X. Zhang, and C. F. Wilcox

Equilibrium Populations of Small Stable Clusters in Nucleation Theory 163
    J. C. Barrett

Temperature-Consistent Clusters of Argon Atoms: Their Role in Free Energy and Nucleation Rate Calculations Using Monte Carlo Methods 167
    P. Pal

Nucleation Rate Determination from Measurements of Oscillatory Nucleation in Diffusion Cloud Chambers 171
    M. Rusyniak, S. P. Fisenko, and M. S. El-Shall

Statistical Theory of High Pressure Nucleation Kinetics in Vapor-Carrier Gas Mixture 175
    S. P. Fisenko

Nucleation Kinetics of Acetonitrile and Benzonitrile—Polar Molecules 178
    S. Chenthamarai, D. Jayaraman, and C. Subramanian

Critique of Molecular Theories of Nucleation 181
    H. Reiss

Dynamical Nucleation Theory 197
    S. M. Kathmann, G. K. Schenter, and B. C. Garrett

Kinetics of Cluster Evaporation and Condensation Important in Homogeneous Vapor Phase Nucleation 201
    B. C. Garrett, S. M. Kathmann, and G. K. Schenter

Nucleation Rate Changes with the Admixture Contain Grows.
I. The Case of the Cigar Like Phase State Diagrams 205
    L. Anisimova and V. Pinaev
Comparison of Modern Theories of Vapor Condensation 209
    L. Gránásy
Cluster Energetics: Models and Data 213
    I. Ford, H. Vehkamäki, and M. Knott
Translational Invariance in the Theory of Nucleation 217
    Y. Drossinos, P. G. Kevrekidis, M. Lazaridis, and P. G. Georgopoulos
Nucleation on Roads 221
    J. Kaupužs and R. Mahnke
Quantum Mechanical Treatment on the Critical Energy
in Ion-Induced Nucleation of Water Vapor 225
    K. Sakiyama, H. Takano, H. Tomida, and M. Itoh
Nucleation in Physical and Non-physical Systems 229
    R. Mahnke
Thermodynamic Properties of Molecular Clusters in Nucleation 233
    N. Fukuta and G. Guo
Nucleation Pulse Measurements for Water Vapor
at Elevated Temperatures 237
    J. Hrubý, J. Hošek, J. Blaha, and F. Maršík
Nucleation Driven by Time Dependent Diffusive Heat
and Mass Transfer 241
    I. Ford, J. Barrett, and M. Mclean
Semiempirical Cahn-Hilliard Theory of Vapor Condensation
with Triple-Parabolic Free Energy 245
    L. Gránásy, Z. Jurek, and D. W. Oxtoby
Homogeneous Nucleation of n-pentanol and Droplet Growth:
A Quantitative Comparison of Experiment and Theory 249
    T. Biet and R. Strey
MD Simulation of Heterogeneous Vapor Nucleation on a Solid Surface 253
    K. T. Kholmurodov, K. Yasuoka, and X. C. Zeng
Metastable Phase Decay on a Wide Spectrum
of the Heterogeneous Centers Activities 257
    V. Kurasov
Effective Surface Tension for Small Binary Clusters
by Monte Carlo Simulation 260
    J. Kiefer and B. N. Hale
Buoyancy-Induced Convection in the Thermal Diffusion Cloud Chamber 264
    F. T. Ferguson and J. A. Nuth, III
Molecular Dynamics Simulation of the Homogeneous Nucleation
of UF_6 Molecules 268
    S. Tanimura and K. Yasuoka
Energy Barrier Effect on Transient Nucleation Kinetics 272
    I. L. Maksimov and M. Sanada
Nucleation Rate Surface Continuity and Monotony 276
    M. P. Anisimov, S. D. Shandakov, G. V. Shandakova,
    and V. I. Poltavtsev

Thermal Diffusion Cloud Chamber—New Criteria for Proper Operation 280
    R. H. Heist, D. Martinez, Y. Chan, and A. Bertelsmann

Photoinduced Nucleation: A Novel Tool for Detecting Molecules in Air at Ultra-low Concentrations 284
    J. L. Katz, B. A. Johnson, H. Lihavainen, M. M. Rudek, and B. C. Salter

Condensation Induced Oscillations in Supersonic Expansion Flows 287
    G. Lamanna and M. E. H. van Dongen

Thermodynamic Considerations on the Homogeneous Nucleation in an Isolated System 291
    W. Vogelsberger and M. Löbbus

Nucleation at Retrograde Condensation 295
    V. G. Baidakov and G. S. Boltachev

The Thermodynamic Theory of Effects of Internal and External Electric Field in Nucleation 299
    A. K. Shchekin, V. B. Warshavsky, and M. S. Kshevetskiy

Phenomenological Model of Homogeneous Condensation 303
    R. Puzyrewski

Kinetic Theory of a Carrier Gas Effect on Nucleation in Diffusion Chambers 307
    A. L. Itkin

Improvement of the Homogeneous Nucleation Rate Measurements in a Static Diffusion Chamber with Use of a CCD Camera 311
    V. Ždímal, J. Smolík, P. K. Hopke, and J. Matas

Dynamic State Phase Diagrams for Nucleated Systems 315
    M. Anisimov, S. Shandakov, V. Pinaev, I. Shvets, and P. Hopke

Thermodynamics and Nucleation Studies of III-Nitride Materials Grown from Vapour Phase 319
    E. Varadarajan, R. Dhanasekaran, and P. Ramasamy

Evaluation of Surface Energy for Pure and Doped 4-hydroxyacetophenone from Induction Period Measurements 323
    S. Chenthamarai, D. Jayaraman, C. Subramanian, and P. Ramasamy

Thermodynamics of Cluster Ions Containing NH_3 and H_2SO_4 327
    K. D. Froyd and E. R. Lovejoy

New Particle Formation at a Rural Site in the UK 331
    P. I. Williams, H. Coe, M. W. Gallagher, K. N. Bower, K. M. Beswick, T. W. Choularton, and G. McFiggans

How Well Are Supersaturation and Temperature in Expansion Chambers Predicted by the Poisson-Equation? 335
    S. Zach, A. Vrtala, G. P. Reischl, and P. E. Wagner

Self-Consistent Binary Cluster Size Distributions of Sulfuric Acid-Water System 339
    M. Noppel

Binary Nucleation Kinetics in Size and Composition Space 343
    S. P. Fisenko and G. Wilemski

Atomic and Molecular Ions of Natural Atmosphere
as Electric System. 347
    V. V. Klingo

Water Molecules Orientation in Surface Layer 351
    V. V. Klingo

Excess Energy for the n-Octanol Critical Embryos 354
    M. P. Anisimov, I. N. Shaymordanov, Y. I. Polygalov,
    and S. A. Timoshenko

Corrected Model for Transport in Static Diffusion Chamber 358
    M. Anisimov, S. Shandakov, V. Pinaev, A. Belyshev, and R. Heist

Calculation of Vapor and Aerosol Transport in a Pipe
including Homogeneous Nucleation 362
    M. P. Kissane and I. Drosik

The Characteristic Time Scales of Condensation and Coagulation
in Ion-Induced Nucleation. 366
    L. Laakso, J. M. Mäkelä, and M. Kulmala

NUCLEATION IN CONDENSED SYSTEMS

Scaling Properties of Critical Nuclei and Nucleation Rate. 373
    R. McGraw

Nucleation in Condensed Systems 386
    K. F. Kelton

A Density Functional Approach to Nucleation in Microemulsions 398
    V. Talanquer and D. W. Oxtoby

Monte Carlo Simulation Approach to Nucleation in Microemulsions 402
    I. Kusaka and D. Oxtoby

Classical and Non-classical Descriptions of Nucleation
in an Ising Ferromagnet vs Monte Carlo Simulations 406
    V. A. Shneidman, K. A. Jackson, and K. M. Beatty

Nucleation in the Presence of Fish and Insect Ice-Growth
Inhibition ("Antifreeze") Molecules 410
    A. D. J. Haymet, A. Heneghan, P. W. Wilson,
    K. E. Zachariassen, and H. Ramløv

Bubble Nucleation Rates by Pressure Pulse Experiments 414
    B. Rathke, H. Baumgartl, and R. Strey

A Novel Method for Determining the Crystal Nucleation and Growth
Rates in Glasses Using Differential Thermal Analysis 418
    K. S. Ranasinghe, C. S. Ray, and D. E. Day

Nucleation in Liquid Solutions: The Experimental Research 420
    V. G. Baidakov and A. M. Kaverin

A Critique of Homogeneous Freezing Measurements
of Aqueous Sulfuric Acid 424
    D. J. Alofs and J. L. Vandike

Efflorescence and Ice Nucleation in Ammonium Sulfate Particles:
Analysis of Experimental Results Using Scaled Nucleation Theory 428
    T. B. Onasch, R. McGraw, A. J. Prenni, M. A. Tolbert, and D. Imre
A Universal Tool to Investigate Nucleation from Aqueous Solutions 432
    K. Tauer and K. Padtberg

ICE NUCLEATION

Nucleation of Pure and AgI Seeded Supercooled Water
Using an Automated Lag Time Apparatus 439
    A. Heneghan and A. D. J. Haymet
Measurements of Ice Nuclei at High Supersaturations 443
    D. C. Rogers, S. M. Kreidenweis, and P. J. DeMott
Atmospheric Ice Nuclei in the Arctic—Airborne Measurements
and Physico-Chemical Properties 447
    D. C. Rogers, S. M. Kreidenweis, P. J. DeMott, and K. V. Davidson
Laboratory Studies of Ice Nucleation by Aerosol Particles
in Upper Tropospheric Conditions 451
    P. J. DeMott, D. C. Rogers, S. M. Kreidenweis, and Y. Chen
Equilibrium Forms, Nucleation, Wulff's Theorem for Crystals
on Substrates-Particles with Finite Dimensions 455
    G. Miloshev
Electroscavenging of Evaporation Nuclei by Cloud Droplets
and Consequences for Contact Ice Nucleation 459
    B. A. Tinsley
Aerosol Particle Size Distribution, the Total Number
and Ice Nuclei Concentrations in Moscow Region 463
    N. O. Plaude and M. V. Vychuzhanina
Theoretical Study Formation and Development of Antarctic
Cloudiness under Different Intensity of Ice and Cloud
Droplet Nucleation 467
    S. V. Krakovskaia and A. M. Pirnach
Laboratory Studies of Ice Nucleation in Sulfate Particles:
Implications for Cirrus Clouds 471
    A. J. Prenni, M. Wise, S. Brooks, and M. A. Tolbert
Laboratory Studies on the Potential of Tropospheric Insoluble
Aerosol Components for Heterogeneous Ice Nucleation 475
    O. Möhler, H. Bunz, A. Nink, and U. Schurath
Phase Transitions in Ice Phase of Clouds under Influence
of Adsorbed Ions Electric Field 479
    V. V. Klingo
Influence of Substrate Electric Charge on Heterogeneous
Ice Phase Formation in Clouds 482
    V. V. Klingo

PARTICLE GROWTH AND AEROSOL DYNAMICS

Investigation of the Effect of Operator Splitting on the Growth
of an Aerosol Population Considering Multiphase Chemical Processes........ 487
    F. Müller
Coagulation in the Presence of Stochastic Nano-Particle Sources............. 491
    A. Vrtala and P. E. Wagner
Growth of Homogeneously Nucleated Water Droplets:
A Quantitative Comparison of Experiment and Theory 495
    A. Fladerer and R. Strey
Modelling Polydispersed Droplet Spectra in Nucleating Steam Flows......... 499
    A. J. White and M. J. Hounslow
Modeling of Homogeneous Nucleation in the Free Troposphere
and Comparison with GLOBE-2 Data 503
    K. Klein and O. B. Toon
Thermodynamics and Kinetics of Condensation on Wettable
Macroscopic Nuclei: New Results 507
    F. M. Kuni, A. K. Shchekin, and A. P. Grinin
The Characteristic Times of Establishing the Steady Rate
of Nucleation on Soluble Aerosol Particles Containing Surfactants........... 511
    A. K. Shchekin, T. M. Yakovenko, and F. M. Kuni
Asymmetric Charging of the Aerosols including
the Coagulation Mechanism 515
    S. Dhanorkar and A. K. Kamra
Sticking Probability and Uptake Coefficient—A Quantitative Comparison 519
    M. Kulmala and P. E. Wagner
Stable Droplets in Finite Volumes... 522
    A. J. H. McGaughey and C. A. Ward
Changes in the Concentration and Size Distribution
of Submicron Aerosols Accompanied with the Polar
Lows at Maitri, Antarctica 526
    A. K. Kamra and C. G. Deshpande
Simulation and Measurements of Dust Deposition in Macedonia............. 530
    S. Nickovic, V. Spiridonov, M. Andreevska, and S. Music
Numerical Simulation of Frontal Rainbands over Ukraine
under Different Mechanisms of Cloud and Precipitation Formation.......... 534
    A. M. Pirnach, S. V. Krakovskaia, and A. V. Belokobylski
Investigations of Aerosol Scavenging Efficiency by Precipitation 538
    A. A. Sinkevich, Y. A. Dovgalyuk, M. A. Ishenko,
    Y. F. Ponomarev, V. D. Stepanenko, and N. E. Veremei
Condensation of Supersaturated Vapor on Charged
Submicrometer Particles 542
    C.-C. Chen, W.-T. Tsai, and C.-J. Tao

RADIATIVE FORCING—AEROSOL CLIMATE INTERACTIONS

The Formation of Ice Clouds from Supercooled Aqueous Aerosols............ 549
   T. Koop

What Do We Know about Phase Transition Processes
Relevant to Atmospheric Aerosols?.. 561
   P. E. Wagner

Aerosol-Cloud Interactions in Global Models of Indirect
Aerosol Radiative Forcing... 565
   A. Nenes and J. H. Seinfeld

Direct and Indirect Effect of Aerosols on the Climate
of the Southeastern U.S. 569
   V. K. Saxena and S. Menon

Comparison of Simulated and Observed Aerosol Optical Depth 573
   N. Laulainen, S. Ghan, R. Easter, and R. Zaveri

Aerosol Impact on the Earth Radiation Budget with Satellite Data........... 577
   K. W. Dammann, R. Hollmann, and R. Stuhlmann

Aerosol Effects on UV Radiation.. 581
   P. Koepke, J. Reuder, and H. Schwander

Properties of Particles in the Tropopause Region........................... 585
   B. Kärcher and S. Solomon

Aerosol Optical Depth over Europe: Satellite Retrieval and Modeling 589
   C. Robles Gonzalez, G. de Leeuw, J. P. Veefkind, P. J. H. Builtjes,
   M. van Loon, and M. Schaap

Features of Aerosol Spectral Optical Depth at a Tropical
Urban Environment at Pune... 593
   G. R. Aher, N. Shantikumar Singh, and V. V. Agashe

Aerosol Concentrations and Scattering Coefficient
at Mace Head, Ireland.. 597
   C. Kleefeld, S. O'Reilly, S. G. Jennings, G. Kunz,
   G. de Leeuw, P. Aalto, E. Becker, and C. O'Dowd

Parameterization of the Optical Properties of Sulfate Aerosols 601
   J. Li, S. Schmitt, J. G. D. Wong, J. S. Dobbie, and P. Chylek

Aerosol Pollution of the Atmosphere and Its Influence
on the Direct Solar Radiation in Some Regions of Georgia.................. 605
   A. Amiranashvili, V. Amiranashvili, and K. Tavartkiladze

An Evaluation of Chemical and Size Effects on Radiative
Properties of Multi-component Aerosols.................................... 608
   S. Yu, V. K. Saxena, and B. N. Wenny

AEROSOLS AND CLOUD PROPERTIES

Organic Aerosols as Cloud Condensation Nuclei............................ 615
   T. M. Raymond and S. N. Pandis

Ultrathin Subvisible Cirrus Clouds at the Tropical Tropopause 619
    T. Peter, B. P. Luo, C. Kiemle, H. Flentje, M. Wirth,
    S. Borrmann, A. Thomas, A. Adriani, F. Cairo,
    G. Di Donfrancesco, L. Stefanutti, V. Santacesaria,
    K. S. Carslaw, and A. R. MacKenzie

Large Stratospheric Particles Observed by Lidar
during the SOLVE Mission. 623
    B. P. Luo, R.-M. Hu, T. Peter, K. S. Carslaw, C. A. Hostetler,
    L. Poole, T. J. McGee, and J. F. Burris

Lack of Closure between Dry and Wet Aerosol Measurements:
Results from ACE-2. 627
    J. R. Snider, S. Guibert, and J. L. Brenguier

Observations of the Evolution of Particulate in an Urban Plume
and Following Interaction with Cloud. 631
    K. N. Bower, M. J. Flynn, T. W. Choularton, R. A. Burgess,
    H. Coe, E. Swietlicki, B. Martinsson, J. Zhou, A. Wiedensohler,
    W. Birmili, K. Müller, and A. Berner

Aerosol Fluxes over an Urban Canopy 635
    J. R. Dorsey, E. G. Nemitz, M. Theobold, P. I. Williams,
    D. Fowler, and M. W. Gallagher

Cloud Liquid Water Content Responses to Hygroscopic
Seeding of Warm Clouds. 639
    S. S. Kandalgaonkar, G. K. Manohar, and M. I. R. Tinmaker

Variations of Atmospheric Aerosols inside and outside Cloud Air 643
    S. S. Kandalgaonkar and M. I. R. Tinmaker

Tower Based Measurements of Micrometeorological Exchange
Parameters and Heat Fluxes above a City 646
    J. R. Dorsey, E. G. Nemitz, M. W. Gallagher,
    M. Theobold, and D. Fowler

Physics and Chemistry of Atmospheric Aerosol Particles
at Zedang and Jinghong, China 650
    Y. Jun, L. Zihua, and Z. Bin

Model Studies on the Effect of Nitric Acid Vapour
on Cirrus Cloud Formation 654
    J. Hienola, M. Kulmala, and A. Laaksonen

GLOBAL ATMOSPHERIC AEROSOLS

Nucleation Properties of Aerosols in the Atmospheres
of Mars and Titan 661
    D. L. Glandorf, D. B. Curtis, T. Colaprete, O. B. Toon,
    and M. A. Tolbert

Ion and Nano-Particle Measurement in Ion-Induced Nucleation Process 665
    K. Okuyama, M. Shimada, and C. S. Kim

Modeling of Global Sulfate Aerosol Number Concentrations 677
    M. Herzog, J. E. Penner, J. J. Walton, S. M. Kreidenweis,
    and D. Y. Harrington

Modelling and Observations of Aerosol Properties in the Clean
and Polluted Marine Boundary Layer and Free Troposphere 681
    E. Vignati, F. Raes, R. Van Dingenen, and J.-P. Putaud

The Free Tropospheric Aerosol, Origin and Properties 685
    H. C. Hansson

Mineral Aerosol Production, Transport, and Removal during ACE-2:
Comparisons of an Event Model to Satellite 689
    P. R. Colarco and O. B. Toon

Plume Processing of Jet Engine Exhaust Aerosols Injected
into the Upper Troposphere and Lower Stratosphere 693
    D. E. Hagen, P. D. Whitefield, J. Paladino, and O. Schmid

MICROSCOPIC PARTICLE PROPERTIES

Novel Measurements of Atmospheric Aerosol Properties 699
    P. H. McMurry, W. D. Dick, and X. Wang

Application of Nucleation Theories to Atmospheric Aerosol Formation 711
    A. Laaksonen

Aerosol SANS: A New Method to Probe the Structure of Nanodroplets 724
    B. E. Wyslouzil, G. Wilemski, and R. Strey

Phase Changes in Internally Mixed Organic/Sulfate Aerosols 728
    S. D. Brooks, A. J. Prenni, M. E. Wise, and M. A. Tolbert

Characterization of Aerosol Particles by Their Heterogeneous Nucleation:
Activity at Low Supersaturations with Respect to Water and Various
Organic Vapors 732
    W. Holländer, W. Dunkhorst, and H. Windt

Fixation and Chemical Analysis of Single Liquid Particle 736
    M. Kasahara, S. Akashi, C.-J. Ma, and S. Tohno

Atmospheric Aerosols in the Asian Part of the Former Soviet Union 740
    R. Van Grieken, R. Jaenicke, K. P. Koutzenogii, T. V. Khodzher,
    and G. N. Kulipanov

AEROSOL CHARACTERIZATION AND PROPERTIES

Vertical Distribution Characteristics of Atmospheric Aerosols
in Liaoning, NE China 747
    Y. Jun, Z. Deping, G. Fujiu, G. Jianchun, and L. Zihua

Mass Balance of Aerosol Particles as a Function of Their Size 751
    A. Molnár, E. Mészáros, T. Feczkó, and D. Temesi

Method for Volatility Measurements on Polydisperse Aerosol 755
    O. Schmid, D. E. Hagen, P. D. Whitefield, A. R. Hopkins,
    and B. Eimer

Observational Researches on Sand Aerosol Size Distribution
in Helanshan Area 759
    N. Shengjie and Z. Chengchang

Chemical Characterization of Water Soluble Organic Compounds
in Tropospheric Fine Aerosol 761
    G. Kiss, A. Gelencsér, A. Hoffer, Z. Krivácsy, E. Mészáros,
    A. Molnár, and B. Varga

The Electric Charging of Aerosols in High Ionized Atmosphere 765
    F. Gensdarmes, D. Boulaud, and A. Renoux

Comments on the Soluble Particle Identification by the Spot
Test Technique 769
    J. Podzimek

Development of DMA-Faraday Cup Electrometer System
for Measurement of Submicron Aerosol Particles........................... 773
    M. Shimada, F. Iskandar, and K. Okuyama

Hygroscopic Properties of Ultrafine Particles in Coastal
and Forest Environments 777
    K. Hämeri and M. Väkevä

Study of the Transport of Contaminants from Norilsk Integrated
Mining Plant to the North of Western Siberia............................. 781
    V. F. Raputa, A. I. Smirnova, K. P. Koutzenogii, B. S. Smolyakov,
    and T. V. Yaroslavtseva

From Aerosol Microphysics to Geophysics Using the Method of Moments 785
    R. McGraw, D. L. Wright, C. M. Benkovitz, and S. E. Schwartz

Coating Ambient Particles for Enhanced Detection
by Mass Spectrometry... 789
    D. B. Kane, B. Oktem, and M. V. Johnston

Variations of the Weight Concentrations of Dust, Nitrogen Oxides,
Sulphur Dioxide and Ozone in the Surface Air in Tbilisi 793
    A. Amiranashvili, V. Amiranashvili, T. Gzirishvili,
    G. Gunia, L. Intskirveli, and J. Kharchilava

Temporal Variation of Atmospheric Aerosol Size Distribution
at a Tropical Station, Mysore (12 N) from Ground Based
Sunphotometer Studies 796
    B. S. N. Prasad, B. Narasimhamurthy, and N. V. Raju

Classification of Aerosol Substances by Use of Specific
Vapour Pressures 800
    J. Müller

Aerodynamic Particle Size Registration by TV-Methods..................... 804
    S. M. Kolomiets

Particle Image Formation by TV-Analyzer "ARFA" 808
    S. M. Kolomiets

AEROSOL FORMATION—NEW PARTICLES

Analysis of the Formation and Growth of Atmospheric
Aerosols Using DMPS Data 815
    M. Dal Maso, M. Kulmala, J. M. Mäkelä, P. Aalto,
    and C. D. O'Dowd

Long-Term Measurements of Events of New Particle Formation
at Hohenpeissenberg: Methods of Analysis and Climatology 819
      W. Birmili, A. Wiedensohler, C. Plass-Dülmer, and H. Berresheim

Observed H_2SO_4 and OH Concentrations and their Relation
to Particle Nucleation Events in Marine and Rural Continental Air 823
      H. Berresheim, T. Elste, C. Plass-Dülmer, W. Birmili,
      A. Wiedensohler, C. D. O'Dowd, H. C. Hansson, and J. M. Mäkelä

PARFORCE: Objectives and Achievements. 827
      C. D. O'Dowd, J. M. Mäkelä, P. Korhonen, K. Hämeri,
      M. Väkevä, L. Pirjola, H.-C. Hansson, S. G. Jennings,
      G. de Leeuw, G. Kunz, H. Berresheim, R. M. Harrison,
      A. G. Allen, Y. Viisanen, and M. Kulmala

Observations and Models of Particle Nucleation near Cloud Outflows 831
      C. Clement, I. Ford, and C. Twohy

New Particle Formation and Hygroscopical Growth
in the Lithuanian Coastal Environment 835
      V. Ulevicius, D. Sopauskiene, G. Mordas, and S. Stapcinskaite

Aerosol and Trace Gas Measurements over the Birmingham
Conurbation during PUMA 839
      K. N. Bower, K. M. Beswick, R. A. Burgess, I. M. Stromberg,
      and M. W. Gallagher

Global Transport of Gaseous Pollutants in the Troposphere
and Aerosol Particle Formation through Kinetic Processes.................. 843
      A. E. Aloyan and V. O. Arutyunyan

Particle Formation from the Oxidation of Alpha-Pinene by Ozone 847
      J. Fitzgerald, W. Hoppel, G. Frick, P. Caffrey, L. Pasternack,
      D. Hegg, S. Gao, J. Ambrusko, W. Sullivan, R. Leaitch, and C. Cantrell

CONDENSATION AND CLOUD CONDENSATION NUCLEI

Cloud Condensation Nuclei Spectral Climatology......................... 853
      J. G. Hudson, S. S. Yum, and Y. Xie

On the Distribution of Condensation Nuclei (CN)
in the Upper Troposphere/Lower Stratosphere
and the Nature of CN Sources and Sinks................................ 857
      A. G. Detwiler and L. R. Johnson

Measurements of CCN-Concentrations in the European Alpine
Aerosol Using a Newly Developed Static Thermal Diffusion Counter 861
      R. Hitzenberger, H. Giebl, A. Berner, A. Kromp, G. Reischl,
      A. Kasper-Giebl, and H. Puxbaum

Cloud Condensation Nuclei Measurement Uncertainties:
Implications for Cloud Models... 865
      J. R. Snider, W. Cantrell, G. Shaw, and D. Delene

Maritime CCN Measurement and Delayed Droplet Growth 869
      J. Gras

**Size Distribution and Critical Supersaturation Spectrum
of the Aerosol from an Electrically Heated Nichrome Wire** 873
    M. B. Trueblood, M. A. Carter, D. E. Hagen,
    P. D. Whitefield, and J. Podzimek

Performance Evaluation of Mixing Type CNC at Low Pressure 877
    K. Okuyama, M. Shimada, C. S. Km, Y. Itoh, and M. M. Lunden

Is There Aerosol/Cloud Layer near Global Tropopause? 881
    G. C. Asnani and M. K. Rama Varma Raja

**Analytical Prediction of Homogeneous Nucleation
in Rapidly Expanding Pure Vapours** 884
    J. B. Young and L. Huang

**Homogeneous Nucleation and Shock Wave Interaction
in Condensing Steam Flows** 888
    A. J. White and J. B. Young

Stable Sulfate Clusters as a Source of New Atmospheric Particles 892
    M. Kulmala, L. Pirjola, J. M. Mäkelä, and C. D. O'Dowd

**Characteristics of the Three Years Continuous Data on New Particle
Formation Events Observed at a Boreal Forest Site** 896
    J. M. Mäkelä, M. Dal Maso, A. Laaksonen, L. Pirjola,
    P. Keronen, and M. Kulmala

Author Index 901

PREFACE

This volume is a collection of papers presented at the 15th International Conference on Nucleation and Atmospheric Aerosols (ICNAA), held at the University of Missouri-Rolla, Rolla, Missouri USA, August 6-11, 2000. The first conference of this series took place in *Dublin* in 1955 and the second in *Basel and Locarno* in 1956. Thereafter conferences occurred as follows: the 3rd in *Cambridge* (1958), the 4th in *Frankfkurt am Main* and *Heidelberg* (1961), the 5th in *Clermont-Ferrand* and *Toulouse* (1963), the 6th in *Albany* and *University Park* (1966), the 7th in *Prague* and *Vienna*, the 8th in *Leningrad* (1973), the 9th in *Galway* (1977), the 10th in *Hamburg* (1981), the 11th in *Budapest* (1984), the 12th in *Vienna* (1988), the 13th in *Salt Lake City* (1992) and the 14th in *Helsinki* (1996). This series of conferences has been held jointly with the Nucleation Symposium since 1988.

At this 15th ICNAA a special session, the **Howard Reiss Symposium**, was held in honor of Professor Howard Reiss, Chemistry Department, University of Los Angeles, Californina for his many contributions to the field of nucleation. This Symposium was organized by Richard Heist (University of Rochester and Manhattan College) and included Invited Lectures by Michael E. Fisher (University of Maryland), Joel Lebowitz (Rutgers University) and Charles Knobler (University of California, Los Angeles). At the end of this symposium, Howard Reiss offered reflections on his 50 plus year career in the field of nucleation research.

Authors from 27 countries presented over 200 papers which are contained in this first electronically generated Proceedings. The International Advisory Committee provided guidance for the conference's scientific content and selected 10 Plenary Speakers to present state-of-the-art overviews of their field: K. Kelton, T. Koop, A. Laaksonen, R. McGraw P. McMurry, K. Okuyama, S. P. Pandis, H. Reiss, M. Tolbert and P. E. Wagner.

We are grateful for the support of the IAMAS, ICCP, CNAA, WMO, the National Science Foundation of the USA and the University of Missouri-Rolla in making this 15th ICNAA possible. We also thank the American Institute of Physics for making this volume available at the time of the conference.

As conference chairs we express our appreciation to the Local Organizing Committee, the staff of the University of Missouri-Rolla and the Conference Planning Consultants, S. Adams and K. Robertson. A special note of gratitude is extended to Professor J. Kiefer (St. Bonaventure University) for his invaluable assistance in preparing this Proceedings.

University of Missouri-Rolla, May 2000

Barbara N. Hale   Markku Kulmala

International Advisory Committee

D. Boulaud, France
J. Gras, Australia
B. N. Hale, USA
R. Heist, USA
J. L. Katz, USA
M. Kulmala, Finland
A. Lushnikov, Russia
E. Meszaros, Hungary
T. C. O'Connor, Ireland
K.. Okuyama, Japan

J. E. Penner, USA
Th. Peter, Switzerland
J. H. Seinfeld, US
F. Raes, Italy
H. Reiss, USA
J. Smolik, Czech Republic
R. Strey, Germany
M. Tolbert, USA
P. E. Wagner, Austria

Local Organizing Committee

D. Alofs
D. Hagen
B. Hale
J. Kiefer
J. Schmitt
P. Whitefield
G. Wilemski

Sponsors

International Association of Meteorology and Atmospheric Sciences (IAMAS)
International Commision on Clouds and Precipitation (ICCP)
Committee on Nucleation and Atmospheric Aerosols (CNAA)
World Meteorological Organization (WMO)
Cloud and Aerosols Sciences Laboratory, UMR
The National Science Foundation
University of Missouri-Rolla Physics Department
University of Missouri-Rolla

Invited Lectures

Criticality in Simple Fluids:
The Yang-Yang Anomaly and other Novelties

Michael E. Fisher

Regents Professor
Institute for Physical Science and Technology and
Department of Physics, University of Maryland
College Park, MD USA

Microscopic Origin of Macroscopic Behavior:
From Scaled Particles to Hydrodynamics

Joel Lebowitz

Department of Mathematics, Rutgers University
Piscataway, NJ 08854

Kinetics of Phase Transitions
In "Unusual Circumstances"

Charles Knobler

Department of Chemistry
University of California, Los Angeles
Los Angeles, CA USA

Plenary Speakers

1. **K. F. Kelton**, *Nucleation In Condensed Systems*

2. **T. Koop**, *The Formation Of Ice Clouds From Supercooled Aqueous Aerosols*

3. **A. Laaksonen**, *Application of Nucleation Theories to Atmospheric Aerosol Formation*

4. **R. McGraw**, *Scaling Properties of Critical Nuclei and Nucleation Rate*

5. **P. H. McMurry**, *Novel Measurements of Atmospheric Aerosol Properties*

6. **K. Okuyama**, *Ion and Nano-particle Measurement in Ion-induced Nucleation Process*

7. **S. N. Pandis**, *Organic Aerosols As Cloud Condensation Nuclei*

8. **H. Reiss**, *Critique of Molecular Theories of Nucleation*

9. **M. A. Tolbert**, *Nucleation Properties of Aerosols in the Atmospheres of Mars and Titan*

10. **P. E. Wagner**, *What Do We Know About Phase Transition Processes Relevant to Atmospheric Aerosols?*

HOWARD REISS SYMPOSIUM

AN APPLICATION OF THE KUBO AND NYQUIST RELATIONS TO NUCLEATION

Robert McGraw[1], Pierre Schaaf[2], and Howard Reiss[3]

[1]Atmospheric Sciences Division, Department of Environmental Science
Brookhaven National Laboratory, Upton NY 11973
[2]Institute Charles Sadron (C.N.R.S.-U.L.P), 6, rue Boussingault, 67083 Strasbourg Cedex, France, and Ecole Europeenne de Chimie, Polymeres et Materiaux de Strasbourg, 1, rue Blaise Pascal, BP 296F, 67008 Strasbourg Cedex, France
[3]Department of Chemistry and Biochemistry, University of California, Los Angeles, 405 Hilgard Avenue, Los Angeles, CA 90095

Abstract. We examine fluctuations in cluster evaporation and growth and express these, using the Nyquist and Kubo relations [1,2], in terms of the resistance to single-cluster motion along the coordinate of cluster size. If successful in future development, the methods introduced here will have application to the abstraction of nucleation rates from computer simulations of individual-cluster evaporation/growth events.

INTRODUCTION

In their simulation capacity, and with a consistent cluster definition, computers provide a unique source of statistical information on the molecular addition/loss steps that are responsible at equilibrium for fluctuating changes in cluster size [3-7]. Such molecular simulations can lead to reliable estimates of cluster energy, although care must be taken that the translational component is properly included [8], but not directly of nucleation rate – at least not under the levels of supersaturation encountered in nature. Specialized techniques such as umbrella sampling are needed to compensate for the fact that the frequency of appearance of clusters in the critical size range is exceedingly small [5]. Preliminary steps towards the abstraction of nucleation rates from cluster dynamics by several methods, including the present one, are reviewed in these proceedings [9]. These include application of the Bennett-Chandler scheme to compute the rate constants between two states, vapor and liquid [6], and the variational transition state approach used in the dynamical nucleation theory of Schenter et al. [7]. Here we introduce a different model – that of a Brownian particle undergoing diffusion and drift in cluster size space – that may provide an alternative route to the determination of nucleation rates from molecular-based cluster simulations.

In the following section we present a Nernst-Planck description for the total nucleation flux and an equivalent Langevin description for the Brownian-like motion of single clusters in size space. It is these latter motions of single or small groups of (rare) clusters that are amenable to computer simulation. We suggest that a fruitful analysis can be made of the random current fluctuations in the Langevin equation; leading ultimately to statistical estimation of the key transport parameters in terms of which the nucleation rate can be obtained. To illustrate the approach, we develop a shot noise model of the fluctuations and obtain Kubo and Nyquist relations for the resistance and local mobility of the cluster to changes in size.

NERNST-PLANCK AND LANGEVIN EQUATIONS

For simplicity we limit discussion to the nucleation of a single component and treat only changes in cluster size that occur through single-molecule addition/loss steps. The net current for the conversion of g-clusters (cluster containing g monomeric units) to $g+1$-clusters is:

$$J(g) = \beta(g)n(g)\left[\frac{f(g)}{n(g)} - \frac{f(g+1)}{n(g+1)}\right] \qquad (1)$$

where detailed balance has been used and the clusters are assumed not to interact. $\beta(g)$ is the rate constant for monomer addition and $n(g)$ and $f(g)$ are the constrained equilibrium and actual concentrations of g-clusters in the vapor phase. Equation 1 has the form: *current = potential difference / resistance*. This 'resistor analogy', is often a convenient way to think about nucleation currents (although certainly not an essential one), however it does not appear to have a physical basis. For example, a physical resistor will have associated with it Johnson noise through the Nyquist relation [1], and the association of $1/[\beta(g)n(g)]$ with a resistance does not naturally yield such a relation. The origin of the difficulty lies in the fact that (f/n) is not a proper conjugate force in the sense of irreversible thermodynamics [1].

The problem is solved on reformulation of Eq. 1 in terms of the Nernst-Planck equation. For this purpose we take the continuum limit of Eq.1: $J = -\beta n \nabla(f/n)$, where the gradient is with respect to cluster size. This expression, together with equilibrium population of clusters of size g, $n(g) = \exp[-W(g)/kT]$, where $W(g)$ is the work required to form the cluster [3], implies the Nernst-Planck equation:

$$J = -\beta \nabla f - \beta \frac{f}{kT}\nabla W. \qquad (2)$$

The lead term on the right side describes diffusion (in cluster size space), with size-dependent diffusion constant $D = \beta$, and the second term describes drift in the force field given by the gradient of W. The nucleation barrier, together with any applied bias potential added, forms the total potential giving rise to (linear-response) drift of the cluster in size space with the conductivity, per cluster, β/kT. It is also of interest to note that if the "equivalent" electro-chemical potential is assumed to have the standard form, $\overline{\mu}(g) = \mu_0(T,P) + kT \ln f(g) + W(g)$, the above equation for J becomes $J = -(\beta f/kT)(\partial \overline{\mu}/\partial g)_{T,P}$, which is the result demanded by irreversible thermodynamics. Key quantities in Eq. 2, specifically the nucleation current, J, and the population gradient, ∇f, cannot be easily determined from computer simulations of individual clusters. For example, the diffusion current, which is a statistical property of the full cluster distribution, cannot be determined this way.

More useful for this purpose is a Langevin equation [2] for single cluster motion in size space that is equivalent to Eq. 2:

$$\dot{g} = -\frac{\beta}{kT}\nabla W + J_r(t) \tag{3}$$

Here $\dot{g} = dg/dt$ gives the single particle current (equal to the change in the number g of molecules in the cluster with time), and β/kT is the single-particle mobility, consistent with well-known Nernst relation ($D = kT \times mobility$). $J_r(t)$ is the fluctuating current in the field-free reference system ($\nabla W = 0$) - realizable through the application of a bias potential sufficient to locally cancel the nucleation barrier gradient. The lead term on the rhs of Eq. 3 gives the drift motion in cluster space arising from the potential gradient, and the inverse mobility, $R = kT/\beta$, is the resistance to this force. We will now show that this definition of the resistance is naturally compatible with the Nyquist relation and identify the associated noise.

The analysis must first include a criterion for determining which sets of molecules form a cluster. The Stillinger criterion, in which a molecule is counted as part of the cluster if it is within some specified distance from another molecule in the cluster is a good example, although other criteria can be used [3,4,9]. Changes in cluster size take place instantaneously as single molecules (or groups of molecules, but we are excluding this case) cross the criterion boundary. Under these conditions, the fluctuation current in Eq. 3 is a series of delta functions. Additionally, we assume that molecular addition and loss are statistically independent processes [10]. Under these conditions, the fluctuating current separates into its forward (condensation) and reverse (evaporation) components: $J_r = J_r^+ - J_r^-$ with $J_r^+ = \sum \delta(t - t_i)$ where the t_i's are the random times of molecular addition events. The delta function currents may be expanded into their (white noise) frequency components [11] and with similar considerations applied to the reverse currents (cross terms vanish from the assumption of statistical independence [11]) we obtain the following result for spectral density of fluctuations in the current J_r at frequency ν_k:

$$G_{J_r}(\nu_k) = 4\beta = 4kT/R \tag{4}$$

KUBO AND NYQUIST RELATIONS AND NUCLEATION RATE

The single-particle mobility is related to the autocorrelation of the random current through the Kubo formula [1]:

$$\frac{\beta}{kT} = \frac{1}{kT}\int_0^\infty \exp(-2\pi\nu_k t)\langle J_r(0), J_r(t)\rangle dt = \frac{1}{4kT}G_{J_r}(\nu_k) \tag{5}$$

The frequency independence of the left hand side, suggesting that $J_r(t)$ has the properties of white noise, is consistent with the shot noise model [11]. From Eq. 4 and the last equality of Eq. 5 we obtain the Nyquist relation [1,11]:

$$G_{J_r}(v_k)\Delta v = \langle J_r^2 \rangle_{\Delta v} = \frac{4kT}{R}\Delta v. \qquad (6)$$

The middle expression gives the power spectrum of the current fluctuations as measured through a filter having frequency bandwidth Δv.

These results give a prescription for estimation of the nucleation rate from computer simulated sequences of cluster condensation/evaporation events. First, Eqs. 4-6 are used to determine the size-dependent transport parameter $\beta(g)$ from fluctuations (noise) in the single particle current. This is not a trivial result for nonspherical clusters where $\beta(g)$ is difficult to determine by other means. Combining this estimate for β with the average net single-particle current, $\bar{J}_r = -(\beta/kT)\nabla W$, computed by averaging Eq. 3 over fluctuations with the nucleation barrier potential present, gives the local potential gradient ∇W. Because the response is linear in force, corrections for an arbitrary bias potential – as used, for example, in umbrella sampling – are easily applied. Obtaining good statistical sampling for many different cluster sizes will be computationally expensive and effort should be expended first on those clusters nearest critical size (undoubtedly the introduction of a bias potential in this region will be required). Implementation of this prescription should provide a good estimate of the nucleation rate with minimal sampling. Multiple recrossings of the nucleation barrier are fully included in the rate estimates derived using this stochastic sampling approach.

REFERENCES

1. Wannier, G. H. Statistical Physics (Dover, 1966).
2. Kubo, R. "The fluctuation-dissipation theorem and Brownian motion", *1965 Tokyo Summer Lectures in Theoretical Physics, Part 1: Many Body Theory* (Benjamin, New York) pp 1-16.
3. Senger, B., Schaaf, P., Corti, D. S., Bowles, R., Voegel, J.-C., and Reiss, H. "A molecular theory of the homogeneous nucleation rate. I. Formulation and fundamental issues", J. Chem. Phys. 110, 6421-6437 (1999).
4. Senger, B., Schaaf, P., Corti, D. S., Bowles, R., Pointu, D., Voegel, J.-C., and Reiss, H. "A molecular theory of the homogeneous nucleation rate. II. Application to argon vapor", J. Chem. Phys. 110, 6438-6450 (1999).
5. ten Wolde P. R. and Frenkel D., "Computer simulation study of gas-liquid nucleation in a Lennard-Jones system", J. Chem. Phys. 109, 9901-9918 (1998).
6. ten Wolde P. R. and Frenkel D., "Numerical calculation of the rate of homogeneous gas-liquid nucleation in a Lennard-Jones system", J. Chem. Phys. 110, 1591 (1999).
7. Schenter G. K., Kathmann, S. M., and Garrett B. C. "Dynamical nucleation theory: A molecular approach to vapor-liquid nucleation", Phys. Rev. Lett. 82, 3484-3487 (1999).
8. Reiss H., Kegel W. K., and Katz, J. L. "Role of the model dependent translational volume scale in the classical theory of nucleation", J. Phys. Chem. A 102, 8548-8555 (1998).
9. Reiss, H. "Critique of molecular theories of nucleation", This Proceedings.
10. McGraw, R., and LaViolette, R. A., "Fluctuations, temperature, and detailed balance in classical nucleation theory", J. Chem. Phys. 102, 8983-8994 (1994).
11. Lawson J. L., and Uhlenbeck G. E., *Threshold Signals* (Dover, 1950), pg. 79.

Homogeneous Nucleation Rates for Water

Judith Wölk[*], Yrjö Viisanen[†] and Reinhard Strey[*]

[*]*Institut für Physikalische Chemie, Universität zu Köln, Luxemburger Str. 116, 50939 Köln, Germany,*
[†]*Finnish Meteorological Institute, Sahaajankatu 22E, SF-00810 Helsinki, Finland*

Abstract. Homogeneous nucleation rates of light (H_2O) and heavy (D_2O) water have been measured under identical conditions. Comparing the nucleation rates for H_2O and D_2O at the same respective vapor pressure p_v and temperature T the rates differ by a factor of 2500. However, the individual data points superimpose, if compared at the same supersaturation, i.e., they are the same within experimental scatter. Onset supersaturations S_0 for H_2O at nucleation rates of $J_0 = 10^7$ cm^{-3}s^{-1} compared with previous measurements by Viisanen et al. (Viisanen, Y., Strey, R., and Reiss, H., *J. Chem. Phys.*, **99**, 4680 (1993); Erratum: Viisanen, Y., Strey, R., and Reiss, H., *J. Chem. Phys.*, (2000) *in press*) show fair agreement. Predictions by the classical nucleation theory - using the most recent expressions for temperature-dependent vapor pressure, surface tension and density - are compared to the experimental data. While the predictions correctly yield the slope of the experimental nucleation rate curves and for one temperature (for H_2O T = 240 K) even the correct absolute rate, the temperature dependence is experimentally shown to be weaker.

INTRODUCTION

Considerable research efforts are undertaken to clarify the physico-chemical properties of pure water, both for H_2O and D_2O. For scientific purposes D_2O has become an indispensable substance, especially in small-angle neutron scattering (SANS) or nuclear magnetic resonance (NMR) applications. Very recently, the first successful experiments on condensation in supersonic nozzle flow and its detection by small-angle neutron scattering (SANS) have been conducted [2]. The condensing material for contrast reasons had to be D_2O, as usual in SANS. In nozzle flow, as well as in many other technological and environmental applications, droplet formation passes through nucleation, growth and aging stages. The size distribution and polydispersity of the precipitate depends on the initial nucleation process. The validity of quantitative theoretical models of the nucleation processes has to be checked against an experiment. Accordingly, there is considerable fundamental interest in experimental studies of homogeneous nucleation of D_2O as model system.

In the area of quantitative examinations of nucleation D_2O was considered almost as early as H_2O. Since these early investigations by Flood and Tronstad [3] in 1935 it apparently has not been investigated again. In this study we repeat the parallel investigation of H_2O and D_2O, but - due to further development of the measuring technique - we are now able to report the first nucleation *rate* study of D_2O. The nucleation of water has repeatedly been examined [4-10]. In all these works the *onset* of nucleation, but no nucleation *rates* were determined.

In order to obtain nucleation rates, nucleation and growth have to be decoupled. This is can be done in a nucleation pulse chamber. The original two piston chamber of Wagner and Strey [11], using the pressure pulse idea of Kassner et al. [12] was further developed and used by Viisanen et al. [1] to measure homogeneous nucleation rates for water in 1993. A detailed description of the chamber has been given by Strey et al. [13] in 1994. In order to see most sensitively the difference between the nucleation rates of D_2O compared to H_2O, nucleation rate measurements for H_2O have been repeated and conducted in the same chamber under identical conditions.

RESULTS

Here we report homogeneous nucleation rates of heavy and light water in argon. We confine ourselves to the use of argon as carrier gas, because Viisanen et al. [1] showed that the homogeneous nucleation of water is independent of whether helium, neon, argon, krypton or xenon was used as the carrier gas.

A. Homogeneous nucleation rates of light and heavy water

In the left diagram of Fig. 1 the homogeneous nucleation rates J of H_2O and D_2O ranging from about 10^5 to 10^9 cm^{-3}s^{-1} are plotted as function of vapor pressure p_v at constant nucleation temperature T between 220 and 260 K. This way of comparing the experimental data is as model-free as possible. It is noteworthy that the experimental nucleation rates of heavy water at the same absolute vapor pressure p_v is higher than those of light water by a factor of 2500 under otherwise identical conditions. That nucleation is governed by supersaturation is borne out by comparing the experimental results at the corresponding vapor supersaturation S at constant temperature T (Fig 1, right). The experimental nucleation rates of H_2O and of D_2O superimpose at the same T and S within experimental scatter.

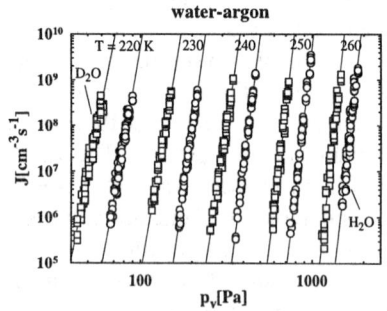
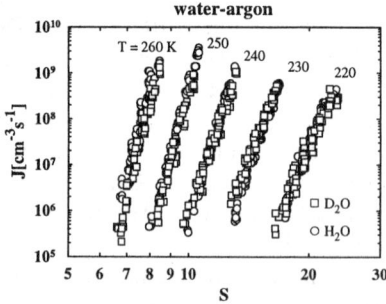

FIGURE 1. Experimental nucleation rates J of heavy (rectangles) and light water (circles) as a function of absolute vapor pressure p_v (left diagram) and as function of supersaturation S (right diagram) for five constant temperatures ranging from 220 to 260 K.

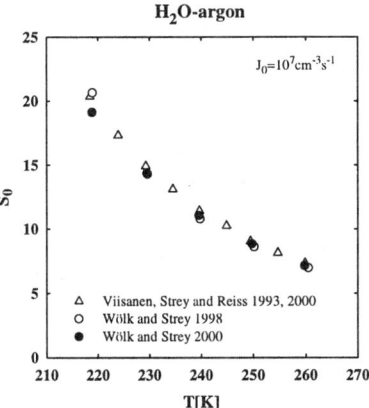

FIGURE 2. Onset supersaturation S_0 corresponding to a nucleation rate $J_0 = 10^7$ cm^{-3} s^{-1} for H_2O. Comparison between three series [1, 14, 15] recorded in different years using the same nucleation pulse chamber.

B. Comparison with previous data

A comparison between previous H_2O measurements by Viisanen et al. [1] and the new data allows checking the reproducibility of the experiment. Since the actual experimental temperatures differ somewhat, the comparison is best performed by comparing the onset supersaturation S_0 for H_2O corresponding to a nucleation rate of $J_0 = 10^7$ cm^{-3} s^{-1} as function of temperature. Fig. 2 shows, that we were able to reproduce the earlier data within experimental error. The agreement between the two most recent series is quite good. The older Viisanen, Strey and Reiss data differ somewhat but are still within experimental error. It should however be mentioned that the 1998 results [14] lead to a reanalysis of the experimental procedure and a recalculation of the previous data. A few (minor) inaccuracies were disclosed as detailed in the erratum [1, erratum].

C. Comparison with the classical nucleation theory

The classical nucleation theory of Becker and Döring [16] is the most commonly used model for quantitative prediction of nucleation phenomena. Various theoretical improvements have been reported in recent years. Due to space limitations we compare here only the measured nucleation rates for H_2O to the predictions of the classical theory. The nucleation rates have been calculated using the most recent expressions for vapor pressure, surface tension and density as function of temperature. As can be seen in Fig. 3, the discrepancy between the theoretical curves and the measurements seems to be rather small. However, the classical theory predicts a disparate temperature dependence as noted previously [1]. A nearly perfect agreement of the classical theory and experiment is found for $T = 240$ K. There both absolute rate and slope of the curves agree. A corresponding observation is made for heavy water.

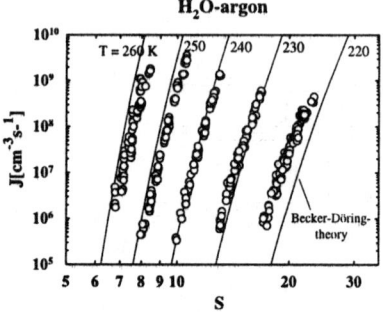

FIGURE 3. Experimental nucleation rates of light water compared to the predictions of the classical theory.

CONCLUSIONS

New measurements of homogeneous nucleation rates for H_2O and D_2O are found to be the same - within experimental scatter - for given S and T. The new nucleation rate data for H_2O show fair agreement with previous data by Viisanen et al. [1]. Classical nucleation theory quantitatively predicts the rate for $T = 240$ K, but a disparate temperature dependence.

REFERENCES

1. Viisanen, Y., Strey, R., and Reiss, H., *J. Chem. Phys.*, **99**, 4680 (1993);
   Erratum: Viisanen, Y., Strey, R., and Reiss, H., *J. Chem. Phys.*, in press. (2000).
2. Wyslouzil, B.E., Cheung, J.L., Wilemski, G., Strey, R., *Phys. Rev. Lett.*, **79**, 431 (1997).
3. Flood, H., Tronstad, L., *Z. Phys. Chem. A*, **175**, 347 (1936).
4. Wilson, C.R.T., *Phil. Trans. R. Soc. A*, London, **189**, 265 (1897).
5. Volmer, M. and Flood, H., *Z. Phys. Chem. A*, **190**, 273 (1934).
6. Sander, A. and Damköhler, G., *Naturwissenschaften*, **31**, 460 (1943).
7. Wegener, P., Lundquist, G., *J. Appl. Phys.*, **22**, 233 (1951).
8. Courtney, W.G., *J. Chem. Phys.*, **35**, 2249 (1961).
9. Katz, J.L. and Ostermier, B.J., *J. Chem. Phys.*, **47**, 478 (1967).
10. Heist, R., and Reiss, H., *J. Chem. Phys.*, **59**, 865 (1973).
11. Wagner, P.E. and Strey, R., *J. Phys. Chem.*, **85**, 2694 (1981).
12. Allard, E.F. and Kassner, J.L. Jr., *J. Chem. Phys.*, **42**, 1401 (1965).
13. Strey, R. Wagner, P.E. and Viisanen, Y., *J. Phys. Chem.*, **98**, 7748 (1994).
14. Wölk, J., Strey, R., "The Homogeneous Nucleation of Water: A Comparative Study of H_2O and D_2O", presented at the APS meeting in Los Angeles, March 16-20 (1998).
15. Wölk, J., Strey, R., *J. Chem. Phys.*, to be submitted (2000).
16. Becker, R. and Döring, W., *Ann. Phys.*, **24**, 719 (1935).

Nucleation of Water and Methanol Droplets on Ions: The Sign Preference

K.J. Oh and X.C. Zeng

Department of Chemistry, University of Nebraska – Lincoln, Lincoln, NE 68588

Abstract. The barrier to the ion-induced nucleation for water and methanol were calculated using a Monte Carlo method. The computer simulation confirms Wilson's findings in 1899 that anion is a better nucleator to produce water droplets than cation with the same magnitude of charge. The simulation also demonstrates that cation is a better nucleator to produce methanol droplets than anion. Structural and energetic analysis of prenucleation embryos offers a molecular insight into the sign preferences for hydrogen-bonding molecules in ion-induced nucleation.

INTRODUCTION

Theoretical confirmation of the sign preferences requires the determination of the free energy of formation of the prenucleation embryonic clusters and the barrier height to ion-induced nucleation. Unfortunately, the classical theory of ion-induced nucleation by Thomson (1936) and Volmer (1906) is unable to differentiate between the effects of positive and negative ions. To describe completely the nucleation on ions, we require detailed knowledge at the molecular level, and for this purpose, we need a molecular model for prenucleation embryos and molecular theory for nucleation on ions (Oxtoby, 1998; Senger *et al*, 1999). Although the density functional theory of nucleation (Kusaka *et al*, 1995; Kusaka *et al* 1995; Talanquer and Oxtoby, 1995) does predict a sign preference for simple dipolar molecules, the incorporation of molecular structure of hydrogen-bonding fluids in the density-functional approach is extremely difficult. Thus far, a molecular explanation of the sign preferences for water and methanol is still lacking. Here, we report results of the first molecular simulation to demonstrate a negative sign preference for water and a positive sign preference for methanol, and to provide general insight into the sign preferences of hydrogen-bonding fluids.

MONTE CARLO SIMULATION

We used a recently developed small-ensemble Monte Carlo (MC) method (Oh and Zeng, 2000) to compute the free energy of cluster formation and to determine the height of barrier to the ion-induced nucleation. We assumed a cluster contains one ion at most because of the extremely low density of ions used in the cloud-chamber experiments. The total number of ion-containing clusters is $N_s = \Sigma_i n^s_i$, where n^s_i is the number of clusters containing i molecules and 1 ion. The free energy of formation of an ion-containing cluster can then be computed from the cluster size distribution via the relation

$$\Delta G^s_i = k_B T \ln(P_0/P_i)$$

where k_B is the Boltzmann constant and T is the temperature of the supersaturated vapor. In the above equation, $P_i = n^s_i/N_s$ is the probability of finding the ion-containing cluster of size i, which can be evaluated by combining the grand-canonical ensemble MC method and an umbrella-sampling technique (Kusaka *et al*, 1998; Gao *et al*, 1999; Oh and Zeng, 2000). In the simulation, we used TIP4P model of water (Jorgensen *et al*, 1983) and OPLS model of methanol (Jorgensen, 1986). For model cation or anion we used a Lennard-Jones (LJ) sphere containing a point charge of $0.6e$ at the center.

RESULTS AND DISCUSSION

In Figs. 1(a) and (b) the calculated free energy of cluster formation ΔG^s_i is shown for a supersaturated water vapor at 298.15K and methanol at 275.0K. The maximum corresponds to the barrier height to the cation or anion-induced nucleation.

Figures 1(a) and (b) clearly show that the anion gives rise to a lower barrier for water but a higher barrier for methanol. As a consequence, water has a positive sign preference whereas methanol has a negative sign preference.

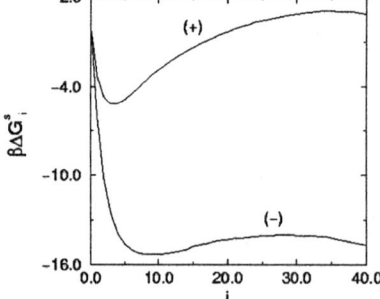
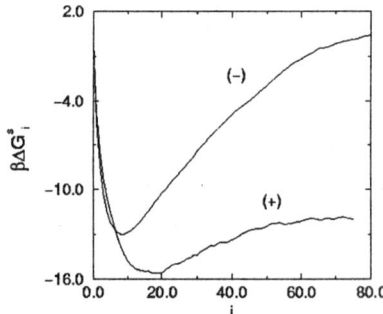

FIGURE 1. Free energy of cluster formation ΔG^s_i versus the size of cluster i for (a) ion-water cluster (left) and (b) ion-methanol cluster (right). Here (+) and (-) denotes the cation and anion-induced nucleation, respectively; $\beta = 1/k_B T$.

To shed more light on the origin of the sign preferences we examined structural characteristics of the prenucleation embryos. Figs. 2(a) - (d) display the atomic density profiles about the cation or anion for water and methanol cluster of size 30, respectively.

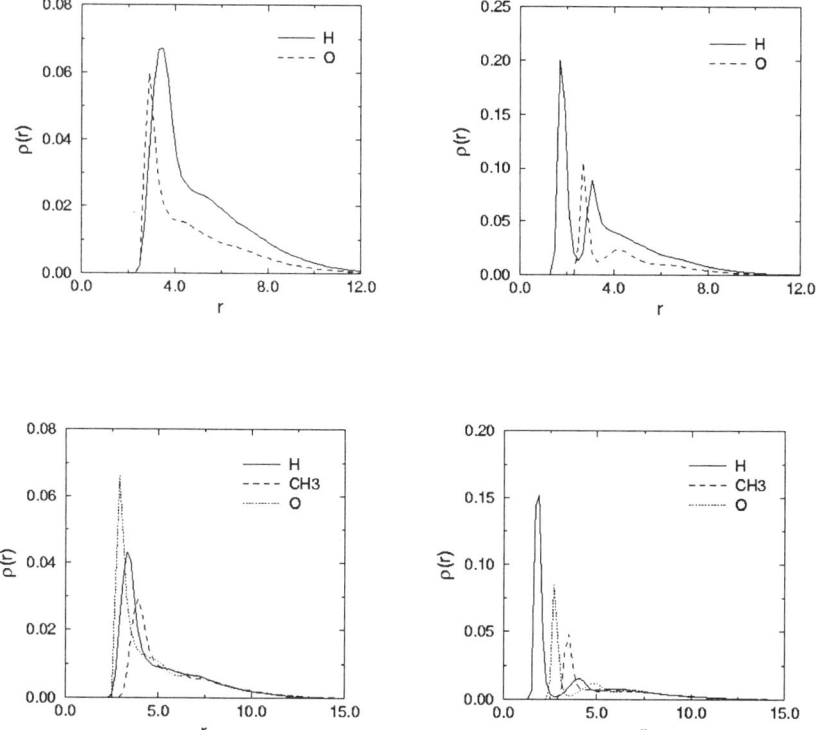

FIGURE 2. Atomic density profile about the cation or anion for the prenucleation cluster of size 30: (a) cation-water cluster (upper left), (b) anion-water cluster (upper right), (c) cation-methanol cluster (lower left), and (d) anion-methanol cluster (lower right). The units of r and ρ(r) are Å and Å^{-3}, respectively.

As shown in the figures, the average anion-H distance is considerably shorter than the averaged cation-O distance. This proximity of H to the anion results in a more favorable ion-molecule electrostatic interaction and hence leads to a negative sign preference for water. On the other hand, when a hydrogen of the water molecule is replaced by the larger methyl group, the resulting repulsive steric interaction forces the methyl group to reside at the surface of cluster, regardless the sign of ion (see the density profile of methyl group in Figs. 2(c) and 2(d)). Although the average anion-H distance is still much shorter than the averaged cation-O distance, as in the case of ion-water cluster, the favorable orientation of the methyl group results in molecule-molecule interaction that outweighs the anion-hydrogen electrostatic interaction. As a result, methanol has a sign preference opposite to that of water.

From the structural analysis of prenucleation embryos, we conclude that the opposite sign preference for water and methanol can be attributed to a subtle competition between the intermolecular steric interaction and the ion-molecular electrostatic interaction. To substantiate this view we examined the energetic characteristics of the embryos. In Figs. 3(a) and (b) we plot the potential energy difference per molecule $\Delta U = U_{anion} - U_{cation}$ versus the size of embryos i, respectively for water and methanol.

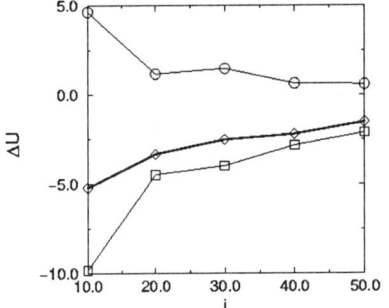
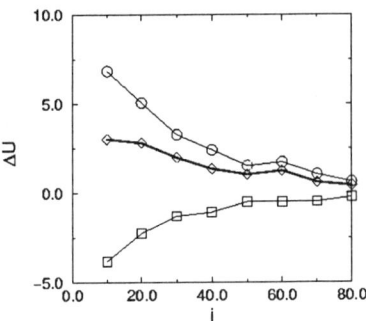

FIGURE 3. The total potential energy difference $\Delta U = U_{anion} - U_{cation}$ versus the size of cluster i for (a) water (left) and (b) methanol (right). ΔU (thick sloid line and diamonds) is a sum of molecule-molecule interaction ΔU_{mm} (solid line and circles) and ion-molecule interaction ΔU_{im} (solid line and squares). All energies have been divided by i and are in unit of kJ/mol.

One can see that ΔU is negative for water but positive for methanol, another evidence of the opposite sign preference for water and methanol. We also calculated the two contributions to ΔU, the molecule-molecule interaction (ΔU_{mm}) and the ion-molecule interaction (ΔU_{im}), which are plotted in Figs. 3(a) and 3(b) as a function of cluster size. Figures 3(a) and (b) show that for water the sign of ΔU is mainly controlled by the ion-molecule interaction ΔU_{im} but for methanol it is determined by the molecule-molecule interaction ΔU_{mm}. This difference in potential energy contribution reinforces the view that the sign preferences of hydrogen-bonding molecules stem from a subtle competition between steric and electrostatic interaction.

REFERENCES

1. J.J. Thomson, *Conduction of Electricity Through Gases*, Cambridge: Cambridge University, 1906.
2. M. Volmer, *Kinetik der phasenbildung*, Dresden : Steinkopff, 1939.
3. D.W. Oxtoby, *Acc. Chem. Res.* **31**, 91 (1998).
4. B. Senger, P. Schaaf, D.S. Corti, R. Bowles, J.-C. Voegel, and H. Reiss, *J. Chem. Phys.* **110**, 6421 (1999).
5. I. Kusaka, Z.-G. Wang, and J.H. Seinfeld, *J. Chem. Phys.* **102**, 913 (1995).
6. I. Kusaka, Z.-G. Wang, and J.H. Seinfeld, *J. Chem. Phys.* **103**, 8993 (1995).
7. V. Talanquer and D.W. Oxtoby, *J. Chem. Phys.* **103**, 3686 (1995).
8. K.J. Oh and X.C. Zeng, *J. Chem. Phys.* **112**, 294 (2000).
9. I. Kusaka, Z.-G. Wang, and J.H. Seinfeld, *J. Chem. Phys.* **108**, 3416 (1998).
10. G.T. Gao, K.J. Oh, and X.C. Zeng, *J. Chem. Phys.* **110**, 2533 (1999).
11. W.L. Jorgensen, J. Chandresekhar, J.D. Madura, R.W. Impey, and M.L. Klein, *J. Chem. Phys.* **79**, 926 (1983).
12. W.L. Jorgensen, *J. Phys. Chem.* **90**, 1276 (1986).

Genuine Saddle Point in Binary Nucleation Kinetics

Jin-Song Li,* Igor L. Maksimov, and Gerald Wilemski *

Department of Physics and Cloud and Aerosol Science Laboratory,
University of Missouri-Rolla, Rolla, MO 65409-0430
Faculty of Physics, Nizhny University, 23 Gagarin Avenue, Nizhny
Novgorod 603000, Russia

Abstract. We construct a generalized nucleation potential for binary systems, which includes both the thermodynamic and kinetic effects. We show that the major nucleation flux passes through the saddle point (termed the genuine saddle point) of this generalized nucleation potential, even if it no longer passes through the thermodynamic saddle point. The genuine saddle point concept provides a convenient way to identify systems and conditions for which the ridge crossing phenomenon occurs. Our theory agrees approximately with exact numerical results

INTRODUCTION

The problem of determining nucleation flux trajectories in systems with multiple order parameters is quite old and goes back, at least, to the early work of Reiss on binary nucleation [1]. He proposed the idea that the major nucleation flux passes through a saddle point on the surface of reversible work W^{rev} for cluster formation, and follows the path of steepest descent through the saddle point of W^{rev} (We call it the thermodynamic saddle point (TSP) to distinguish it from saddle points on other surfaces.). This assumption was followed by other authors, but Stauffer provided an important clarification by showing that the flux direction at the TSP depends on the monomer impingement rates and, in general, does not follow the path of steepest descent [2]. Following the initial suggestions of Stauffer and Kiang and Stauffer that in certain cases the major nucleation flux bypasses the TSP [2,3], Trinkaus developed an extensive theory for this phenomenon, which is referred to as ridge crossing of the W^{rev} [4]. Recent numerical results have demonstrated quite clearly that ridge crossing can occur [5-7]. Although this subject has continued to receive attention, at present it is still not easy to determine in which physical systems ridge crossing is likely to occur; nor is it simple to find the location of the major nucleation flux when it does occur.

In this paper, we describe a generalized nucleation potential that determines the nucleation pathway [8]. The potential consists of the reversible work W^{rev} plus additional terms arising from various kinetic effects. We show that the major nucleation flux passes through the saddle point of the generalized nucleation potential, which is called as the genuine saddle point. Hence, the genuine saddle point provides a convenient way to identify systems and conditions for which the ridge crossing phenomenon occurs.

GENERALIZED NUCLEATION POTENTIAL

Let us consider the process of homogenous nucleation of liquid clusters in a metastable binary vapor of species A and B at temperature T, pressure p^α and composition x_B^α of the species B. The basic equation governing the time dependent cluster concentration $f(n_A, n_B, t)$ may be written as [1]:

$$\frac{\partial f(n_A, n_B, t)}{\partial t} = -\frac{\partial J_A(n_A, n_B, t)}{\partial n_A} - \frac{\partial J_B(n_A, n_B, t)}{\partial n_B}, \qquad (1)$$

where n_i ($i = A, B$) denotes the number of i molecules in a cluster. The components J_A and J_B of the nucleation flux $\mathbf{J}$ are given by [1]:

$$J_A(n_A, n_B, t) = -F(n_A, n_B) K_A^+(n_A, n_B) \frac{\partial}{\partial n_A}\left(\frac{f(n_A, n_B, t)}{F(n_A, n_B)}\right), \qquad (2)$$

$$J_B(n_A, n_B, t) = -F(n_A, n_B) K_B^+(n_A, n_B) \frac{\partial}{\partial n_B}\left(\frac{f(n_A, n_B, t)}{F(n_A, n_B)}\right), \qquad (3)$$

where K_i^+ denotes the attachment rate of i species and $F(n_A, n_B)$ the metastable equilibrium concentration of clusters specified by (n_A, n_B) in the system.

We introduce a time-dependent orthogonal curvilinear coordinates (ξ, η), in which η denotes the contour lines of $\Phi = f/F$ (referred to elsewhere [9] as Φ lines), and ξ the lines of flow of the vector field $\mathbf{V} = -\nabla\Phi$. By transforming (n_A, n_B) into (ξ, η), the magnitude of the nucleation flux can be expressed as [10]

$$J(n_A, n_B, t) = F_0 K_A^+(n_A^*, n_B^*) \exp(-W^{GK}/kT), \qquad (4)$$

where $F_0(n_A, n_B)$ is a prefactor that depends on cluster composition, temperature or other molecular parameters [11]. In the following considerations, we neglect the composition dependence of $F_0(n_A, n_B)$. $K_A^+(n_A^*, n_B^*)$ is the value of K_A^+ at the TSP, whose location is denoted by (n_A^*, n_B^*), and W^{GK} is the generalized kinetic potential. The generalized kinetic potential consists of a force term W_0, a kinetic term W_1, a scaling term W_2, an anisotropy term W_3 and the reversible work W^{rev} [10]:

$$W^{GK} = W^{rev} + W_0 + W_1 + W_2 + W_3, \qquad (5)$$

Since nucleation involves barrier crossing kinetics, it should be possible to describe it in terms of an appropriate potential, which we will call the generalized nucleation potential W^{GN}. The generalized nucleation potential is supposed to satisfy the following requirements: (1) it includes both thermodynamic and kinetic effects; (2) it reduces to the kinetic potential of the unary system when one of the components vanishes, and (3) it has a saddle point through which the major nucleation flux passes. Hereafter, we refer to this saddle point as the genuine saddle point (GSP) as suggested by Nishioka [12].

Obviously, both the reversible work W^{rev} the kinetic potential $W^K = W^{rev} + W_1$ [13] are not this generalized nucleation potential, since they do not always satisfy condition (3). The generalized kinetic potential W^{GK} includes enough information to determine the pathway of the major nucleation flux, but it does not satisfy condition (2), since the force term W_0 and the scaling term W_2 do not vanish when one of the components vanishes.

However, we can get a satisfactory generalized nucleation potential W^{GN} by omitting the "gradient terms" (W_0 and W_2) in Eq.(5):

$$W^{GN} = W^{rev} + W_1 + W_3 \tag{6}$$

This W^{GN} obviously satisfies condition (1). The thermodynamic effect is given by the reversible work W^{rev} and the kinetic effect by the terms W_1 and W_3. The attachment kinetics gives rise to W_1, and W_3 reflects any discrepancy in the impingement rates of the two species. Equation (6) is also consistent with condition (2), since $W_3 = 0$ when one of the components vanishes. Our W^{GN} is also consistent with condition (3). The pathway of the major nucleation flux corresponds to the valley of the surface of the generalized kinetic potential, and it can be approximately determined by the equation $\partial W^{GK} / \partial \eta = 0$ [10]. Since the force term W_0 is a function that is independent of η, so that $\partial W_0 / \partial \eta = 0$. The partial derivative $\partial W_2 / \partial \eta$ may also be neglected, since Wyslouzil and Wilemski numerically found that the Φ lines were parallel for the systems that they examined [9, 14]. These numerical results imply that $\partial W_2 / \partial \eta \approx 0$. Thus the pathway of major nucleation flux can be determined by

$$\partial W^{GN} / \partial \eta = 0 \tag{7}$$

Consequently, if W^{GN} possesses a saddle point, the major nucleation flux determined by Eq.(7) will pass through it. Thus, the generalized nucleation potential W^{GN} satisfies all the conditions (1)-(3) if the Φ lines for a particular binary system are parallel.

Let us consider one example that demonstrates the value of the genuine saddle point concept. We take a model vapor-liquid system (PD2) that exhibits positive deviations from ideality. This system has been studied in detail by Wyslouzil and Wilemski [7]. Figure 1. shows the locations of the GSP and TSP for a particular set of conditions for which the gas phase activities are $a_A = 2.25$ and $a_B = 14$, respectively. We find that the GSP ($n_A = 28.5$, $n_B = 21.9$) is located far away from the TSP ($n_A = 7.7$, $n_B = 30.1$). In this case, the major nucleation flux passes through the GSP, evidently

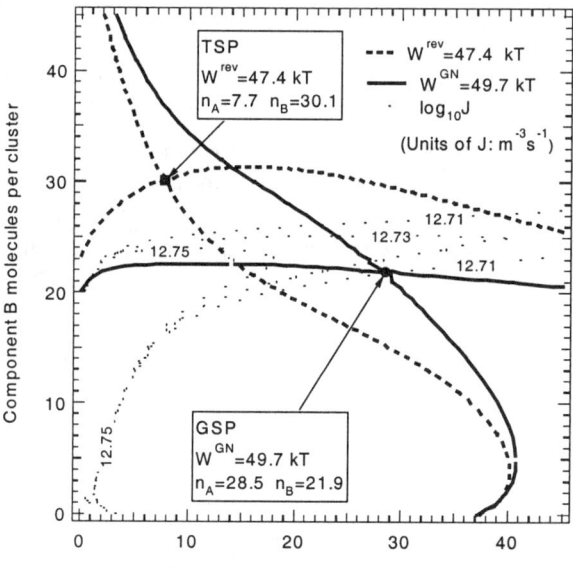

FIGURE 1. The locations of the TSP and GSP in the size space. nucleation flux for PD2 system ($a_A = 2.25$, $a_B = 14$). The physical properties of these two systems are listed in Ref.[7].

bypassing the TSP. The example is particularly important because it illustrates the power of the GSP concept as a simple means of establishing when ridge crossing is occurring.

ACKNOWLEDGMENTS

This work was supported by the Engineering Research Program of the Office of Basic Energy Sciences, U. S. Department of Energy.

REFERENCES

1. Reiss, H., *J. Chem. Phys.* **18**, 840 (1950).
2. Stauffer, D., *J. Aerosol Sci.* **7**, 319 (1976).
3. Stauffer, D., and Kiang, C. S., *Icarus* **21**, 129 (1974).
4. Trinkaus, H., *Phys. Rev. B* **27**, 7372 (1983).
5. Greer, A. L., Evans, P. V., Hamerton, R. G., Shangguan, D. K., and Kelton, K. F., *J. Cryst. Growth* **99**, 38 (1990).
6. McGraw, R., *J. Chem. Phys.* **102**, 2098 (1995).
7. Wyslouzil, B. E. and Wilemski, G., *J. Chem. Phys.* **103**, 1137 (1995).
8. Li, J.-S., Maksimov, I. L., and Wilemski, G., *Phys. Rev E* **61**, R4710 (2000).
9. Wyslouzil, B. E., and Wilemski, G., *J. Chem. Phys.* **110**, 1202 (1999).
10. Li, J.-S., Nishioka, K., and Maksimov, I. L., *Phys. Rev E* **58**, 7580 (1998).
11. Wilemski, G., and Wyslouzil, B. E., *J. Chem. Phys.* **103**, 1127 (1995).
12. Nishioka, K., private commucation.
13. Li, J.-S., Nishioka, K., and Maksimov, I. L., *J. Chem. Phys.* **107**, 460 (1997).
14. Wyslouzil, B. E., and Wilemski, G., *J. Chem. Phys.* **105**, 1090 (1996).

Nucleation of Sulfur Hexafluoride

Peng Ye, Anne Bertelsmann[†] and Richard H. Heist*

*Nucleation Laboratory, Department of Chemical Engineering
University of Rochester, Rochester, NY 14627-0166*

Abstract. We measured the homogeneous nucleation of sulfur hexafluoride using a diffusion cloud chamber under conditions that avoid the possibility of non-diffusive modes of transport within the cloud chamber thus ensuring reliable data from our measurements. We extended the critical supersaturation measurements into the critical region for SF_6 using helium as background gas.

INTRODUCTION

Based upon results of our nucleation experiments, it has become increasingly clear that homogeneous nucleation data obtained using current methods of experiment data analysis for thermal diffusion cloud chambers (TDCC) depend both on the amount and kind of background gas.[1,2] With the development of the high pressure diffusion cloud chamber (HPCC) at Rochester, it is now feasible to expand useful nucleation measurements to wide ranges of both total pressure and nucleation temperature.[1,3] In addition, it is through the increased functionality of the HPCC that we have been able to demonstrate, for the first time, the subtle and generally overlooked effects of buoyancy-driven convective flow and the importance of determining the proper ranges of the total pressure and TDCC plate temperatures before carrying out nucleation measurements.[4,5]

High Pressure Limit - The essential point in this regard pertains to maintaining the operational stability of the TDCC with respect to buoyancy-driven convective flow disturbances originating at the chamber wall.[4] Fortunately, these flows can be predicted by determining, *a priori*, the upper limit of allowable total pressure in the TDCC.[4] Results of a recent analysis have confirmed our hypothesis regarding the formation, existence, and deleterious effects of buoyancy-driven convective instabilities in the TDCC arising from density inversions at the chamber wall.[6]

Low Pressure Limit - Based on recent investigations from our laboratory, we know that there is also a lower limit of allowable total pressure within the diffusion cloud chamber.[5] This lower limit does not relate in a meaningful way to the commonly used "pressure ratio" and must be determined empirically. At too low total pressures, we have found that the cloud chamber no longer functions in a manner consistent with our current perception of the nature of the mass and energy transport processes occurring within the chamber.[5]

In this paper we report results of our investigation of the nucleation of sulfur hexafluoride obtained, for the first time, using HPCC. We used the analytic and empirical procedures referred to above to determine the proper range of operation of the HPCC <u>before</u> we made the quantitative measurements we present below. We report results of constant total pressure and constant nucleation temperature measurements with helium as the background gas. We also report preliminary measurements of SF_6 nucleation at temperatures approaching the critical point.

EXPERIMENT RESULTS

The HPCC employed in this investigation has been described in detail elsewhere.[1,3] The behavior of the vapor-gas mixture is described using the Peng-Robinson equation of state and an interaction parameter k_{12}. In our thermodynamic analysis we account for the non-ideal behavior of the gas phase, pressure effects (Poynting correction), and background gas solubility in the liquid.

Constant total pressure critical supersaturation (S_{cr}) experiments using SF_6 with helium as a background gas were conducted at 30, 40, 50, 60 bar. The results are presented in Figure 1. The important point to note is that each of these series of curves, with their corresponding envelope (not shown), shifts to larger values of the S_{cr} as the total pressure increases. The shift is more pronounced at lower temperatures.

Critical supersaturation experiments using SF_6 and helium were conducted at constant nucleation temperatures of 253.5, 261.9, 268.1, 275.8, 285.0 and 305.7K. Figure 2 shows the results of these experiments. In the analysis of this data, the value of S_{cr} was obtained using classical nucleation theory (BDZ) to identify the height of the nucleation plane. The solid lines shown on these graphs have been regressed to the data for each temperature.

We also conducted constant pressure S_{cr} studies using SF_6 in which the nucleation temperature ranged between 8 to 18K of the critical temperature (318.7K). These experiments were performed at total pressures of 60, 65 and 70 bar using helium as a background gas. The results are shown in Figure 3. For the purpose of clarity, the result of each experiment is represented as a single point (as in the constant temperature experiments) rather than a curve segment. For comparison, we have included the S_{cr} predicted by BDZ theory.

DISCUSSION OF EXPERIMENT RESULTS

Pressure Dependence - We find from our experiments involving SF_6 in helium that the S_{cr} <u>as determined for the HPCC</u> depends on total pressure. As can be seen in Figure 1, the supersaturation versus temperature curve segments shift toward higher supersaturation as the total pressure is increased. This shift is more pronounced at lower temperatures and appears to vanish the critical point is approached. This behavior can also be seen in results from our constant temperature experiments shown in Figure 2. The S_{cr} increases linearly with total pressure, and the slope of the regressed curves represents the magnitude of the total pressure effect. This observed dependence is entirely consistent with the total pressure dependence observed in our earlier experiments involving the lower molecular weight alcohols.[1,2,7]

In Figure 4, we compare selected results of the constant total pressure experiments (at 50 bar) to the results of the constant nucleation temperature experiments. The curve segments are the results from constant pressure experiments (Figure 1). The solid circles represent the values of supersaturation and temperature corresponding to the maximum calculated nucleation rate for each curve segment. The open triangles were obtained from the regressed S_{cr} versus total pressure curves (Figure 2). Agreement between the two different types of experiments is good.[2]

Experiments Near the Critical Point -The data shown in Figure 3 represent the variation of the S_{cr} of SF_6 with temperature at a variety of total pressures ranging from 60 to 70 bar. Each point corresponds to the maximum nucleation rate for each experiment determined using BDZ theory. The open region between the data shown in Figure 3 from roughly 297 K to 301 K resulted from our changing heat exchangers during our experiments and the non-overlap of their accessible ranges. The prediction of BDZ theory is included as a reference. In Figure 3, we note the coincidence of the nucleation data obtained at each of the different total pressures. This is the first experimental verification of our earlier prediction that the dependence of nucleation on total pressure would vanish as the critical point is approached.[1,2,7] We note in Figure 3 that at temperatures below approximately 301K the variation of the S_{cr} with temperature seems to decrease in the anticipated concave fashion. At approximately 301 K the variation of the data appears to abruptly change slope, become somewhat flat and then decrease in a convex fashion as the temperature increases. This is the same dependence of the S_{cr} with temperature we observed with n-hexane and helium near the critical point of n-hexane.[3] While this data is preliminary and presented here to demonstrate the apparent existence of this interesting trend, it is consistent with the predictions of the McGraw and Reiss theory.[8] Vapor to liquid nucleation measurements in the critical region are rather rare, and these data strongly suggest the need to continue these investigations.

SUMMARY AND CONCLUSIONS

We measured the homogeneous nucleation of SF_6 using the HPCC. We determined the range of proper operation of the HPCC for SF_6 - helium in order to avoid the possibility of convective transport within the cloud chamber. We observed that SF_6 nucleation depends on the total pressure. These observations are entirely consistent with results of earlier investigations from our laboratory.[1,2,7] Nucleation measurements made at constant total pressure and constant temperature are in good agreement. We extended the S_{cr} measurements into the critical region since such measurements are rare. We observed a change in the dependence of the S_{cr} with temperature (at constant pressure) in the critical region that appeared to be consistent with predictions of the theory of McGraw and Reiss.[8]

* Current address: School of Engineering, Manhattan College, Riverdale, NY 10471
† Current address: Bayer Corporation, Baytown, TX

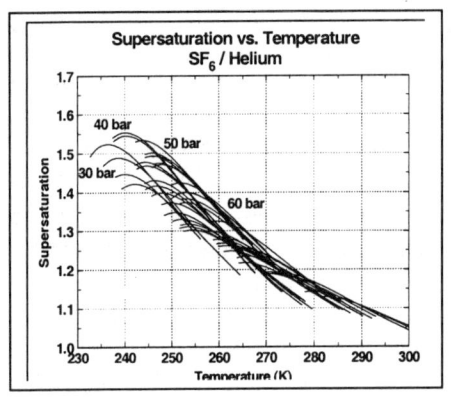 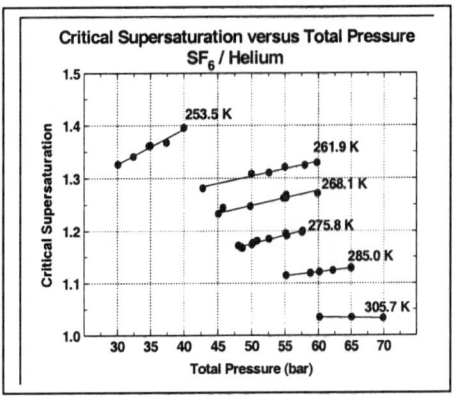

FIGURE 1. Variation of the critical supersaturation of SF_6 with temperature at 30, 40, 50 and 60 bar.

FIGURE 2. Variation of the critical supersaturation of SF_6 with total pressure.

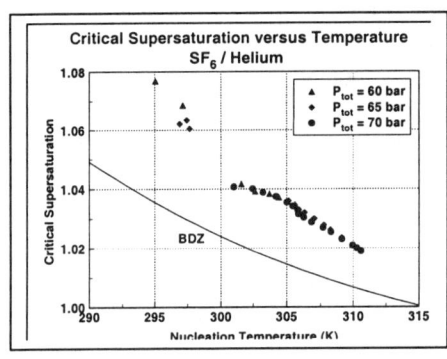 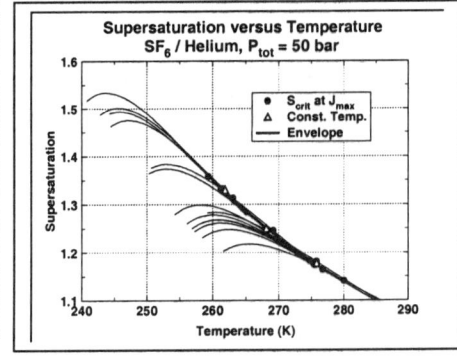

FIGURE 3. Variation of the critical supersaturation of SF_6 at with temperature at 60, 65 and 70 bar near the critical point. Prediction of the BDZ theory is included for comparison.

FIGURE 4. Comparison of constant pressure and constant temperature experiments, for SF_6 at 50 bar. The thick line represents the envelope, the curve segments the computed supersaturation profiles, and the solid dots mark the location of maximum rate on each curve. The open triangles represent the critical supersaturation obtained from constant temperature experiments.

REFERENCES

1. Heist, R. H.; Janjua, M.; Ahmed, J. *J. Phys. Chem.*, **98**, 4443 (1994).
2. Bertelsmann, A., Stuczynski, R. and Heist, R.H., *J. Phys Chem.*, **100**, 9762 (1996).
3. Heeks, M., M.S. Thesis, University of Rochester, (1983).
4. Bertelsmann, A. and Heist, R.H., *J. Chem. Phys*, **106**, 324 (1997).
5. Heist, R.H., Bertelsmann, A., Martinez, D. and Chan, Y.K., *J. Phys. Chem.* (2000), submitted.
6. Ferguson, F.T. and Nuth, J.A., III, J. Chem. Phys., **111**, 8013 (1999).
7. Bertelsmann, A. and Heist, R.H., *J. Aerosol Science and Technology*, **28**, 259 (1998).
8. (a) McGraw, R. and Reiss, H., *J. Stat. Phys.*, **20**, 385 (1979); (b) McGraw, R., *J. Chem. Phys.*, **91**, 5655 (1989).

Molecular Dynamics Studies of Evaporation and Condensation Coefficients in Nucleation Theory

P. Schaaf[a], B. Senger[b], J.-C. Voegel[b], R.K. Bowles[c] and H. Reiss[c].

[a]*Intstitue Charles Sadron, 6, rue Boussingault, 67083 Srasbourg Cedex, France.*
[b]*INSERM Unité 424, 11, rue Humann, 67085 Srasbourg Cedex, France.*
[c]*Department of Chemistry and Biochemistry, University of Claifornia, Los Angeles, CA 90095, USA.*

Abstract: A rate theory that accounts for multimolecular addition or loss of molecules from a cluster is developed for the homogeneous nucleation of a liquid from its metastable vapor. Molecular dynamics simulations of Lennard-Jones argon are used to study the evaporation coefficients of Stillinger type clusters. It is shown that steps involving the gain or loss of 2 or 3 atoms at a time make a significant contribution to the dynamics of cluster growth and that these multimolecular steps obey detailed balance. Studies of the lagtime, which to a first approximation should be related to the nucleation rate, shows that $\ln(t_{70})$ is proportional to $1/\ln^2(S)$, as predicted by classical nucleation theory.

INTRODUCTION

The move to develop a molecular theory of vapor-liquid nucleation has, in general, focused on the thermodynamic aspects of the problem, i.e. the estimation of $N(n)$, the equilibrium number of clusters of size n, through calculations of the work of formation of a cluster, or more directly, through the simulation of the probability of appearance of a cluster. On the other hand, the problem of dynamics and how clusters move over the nucleation free energy barrier has remained relatively unexplored. Progress in this direction has been made recently by ten Wolde et al[1,2], through the application of Bennett-Chandler linear response theory[3,4]. This involved the generation of the time correlation functions associated with the average flux over the barrier, by molecular dynamics, starting with critical nuclei at the top of the barrier. In this paper, we examine an alternative approach that utilizes the more familiar master equations, but includes the possibility that clusters may grow or shrink through steps involving the gain or loss of more that a single molecule, an observation already made in reference 2.

The process of nucleation has usually been assumed to occur through the addition or loss of single molecules from a cluster. A cluster of size n becomes a cluster of size $n+1$ by gaining a molecule from the vapor, or, it can lose a molecule to become a cluster of size $n-1$. The full process is then described by the set of master equations[5]

$$J(1,2) = \beta(1)f(1) - \gamma(2)f(2) \qquad (1a)$$
$$J(2,3) = \beta(2)f(2) - \gamma(3)f(3) \qquad (1b)$$
$$\ldots \ldots \ldots \ldots \ldots \ldots \ldots$$
$$J(n,n+1) = \beta(n)f(n) - \gamma(n+1)f(n+1) \qquad (1c)$$

where $J(n,n+1)$ represents the cluster flux from size n to size $n+1$, and $f(n)$ is the number of clusters of size n in a cubic centermeter. $\beta(n)$ and $\gamma(n)$ are the condensation and evaporation coefficients, respectively. In classical nucleation theory (CNT)[5]

$$\beta(n) = \frac{PS(n)}{\sqrt{2\pi mkT}} \qquad (2)$$

where P is the vapor pressure, $S(n)$ is the surface area of the cluster, m is the mass of a molecule, kT is the thermal energy and it is assumed that the sticking coefficient is equal to unity. Under constrained equilibrium, the fluxes $J(n,n+1)$ are zero. The determination of $N(n)$ from simulation and the estimation of $\beta(n)$ then allows the calculation of $\gamma(n)$. However if one is interested in experimental conditions under which the nucleation rate is small, steady state conditions can be assumed. In this case the fluxes $J(n,n+1)$ are no longer zero but are all constant and equal to the nucleation rate J. The nucleation rate can then be estimated directly from Eqs 1 and 2.

A refined rate theory based on a set of master equations equivalent to the Eq. 1, where multistep processes (a cluster gains or loses more than one molecule at a time) are formally available[6]. Assuming steady state conditions

$$\beta(n,n+i) = \frac{N(n+i)}{N(n)} \gamma(n+i,n) \qquad (3)$$

where $\beta(n,n+i)$ and $\gamma(n+i,n)$ are the condensation and evaporation coefficients for the gain or loss of i molecules respectively, $N(n+i)$ and $N(n)$ are the number of clusters of size $n+1$ and n under constrained equilibrium.

Equation 3 is true for any cluster model. Although $\beta(n)$ usually is treated as the primary quantity from which $\gamma(n)$ is then obtained, we have chosen to regard $\gamma(n)$ as primary, since unlike $\beta(n)$, it is almost independent of both pressure and time. We have performed an initial study of $\gamma(n+i,n)$ for Stillinger clusters of Lennard-Jones (argon) atoms. These clusters, along with its variations, have been the focus of a number of recent nucleation studies[1,2,7-10]. Stillinger clusters are characterized by a connectivity distance, d_c, so that atoms within d_c of each other belong to the same cluster. The evaporation of an atom from the cluster occurs when it moves a distance greater than d_c away from all other atoms in the cluster. Conversely, a vapor molecule condenses onto the cluster when it comes closer than d_c to any of the atoms already in the cluster.

Determination of the Evaporation Coefficients.

Using molecular dynamics simulations that utilize a Verlet algorithm[11], the $\gamma(n+i,n)$ are determined in the following manner. N_p particles are placed in a cubic box, with periodic boundary conditions, with their center of mass at the center of the box and with initial velocities selected according to a Maxwell-Boltzmann distribution corresponding to the chosen temperature (e.g. $T = 85K$). After each cycle, which consists of moving all the particles through a time δt in accordance with the Verlet algorithm, the system is examined for clusters according to the Stillinger criterion with a given d_c. Only the largest cluster is considered to be a cluster while all smaller clusters are regarded as part of the vapor. Transforming the coordinate system of the atoms relative to the center of mass of the cluster, placed at the center of the cell, avoids complications associated with the connectivity criteria at the periodic boundaries.

During the time step $\delta t = 10$ fs, a cluster can disappear *a priori* through a number of different pathways: (1) it can gain one or more atoms, (2) it can lose more than one atom, or (3), it can lose exactly one atom. At the end of each cycle, the cluster is removed from the vapor. This is the starting configuration for the calculation of the $\gamma(n+i,n)$ (i.e. $t = 0$). The isolated cluster is allowed to evolve according to the usual molecular dynamics until an evaporation event takes. The time between the $t = 0$ configuration and the evaporation event is used to determine $\gamma(n+i,n)$. The main simulation is then resumed, starting from the point prior to the extraction of the cluster.

Lagtime and Detailed Balance

Molecular dynamics simulations also allow us to follow a system as it evolves dynamically. Starting with 216 argon atoms placed randomly in a box with periodic boundaries, at a temperature $T = 85K$, the system is allowed to continuously evolve. At the end of each time step, the size n of the largest cluster is recorded. Initially n fluctuates around some small value before increasing dramatically and eventually settling at some much larger (equilibrium) value, at long times. In the steady state, this relaxation time (lagtime) is well approximated by the reciprocal of the steady state rate[12]. Operationally, we define the lagtime t_{70}, as the last time the cluster was size $n = 70$, before it grows to a size $n = 120$ (fig 1). Once the cluster has more than 120 particles it is never seen to shrink again, and is assumed to be over the nucleation barrier. Fig 2a shows that $\ln(t_{70})$ varies linearly with $1/\ln^2(S)$, where S is the supersaturation. This relation is predicted by CNT when the lagtime is equated to the inverse of the steady rate. The simulation was repeated three times at each degree of supersaturation to obtain an estimate of the fluctuation in t_{70}. This simulation also allowed us to determine the frequency of transitions between clusters of size n and $n+i$. We found that, to a high degree of accuracy, the transitions followed detailed balance. Fig 2b shows the case of the transitions starting with $n = 160$ and ending in n'. Moreover, it should be noted that the $(n, n \pm 2)$ and $(n, n \pm 3)$ transitions are far from negligible, so that they need to be taken into account in the set of master equation describing the kinetic evolution of the system.

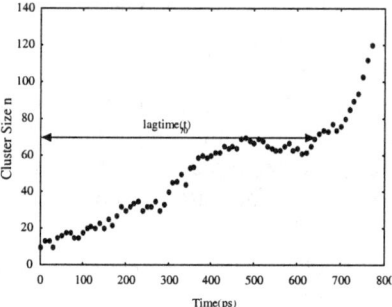

Figure 1. The size of the largest Stillinger Cluster n as a function of time during Molecular dynamics simulation. n is averaged over 10ps intervals.

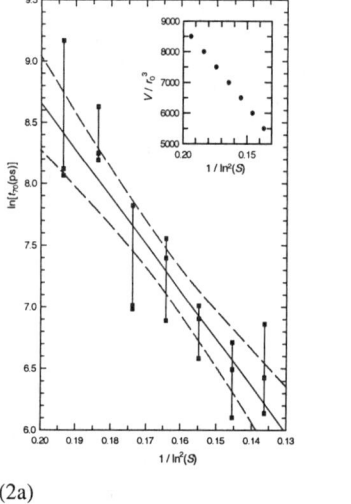

(2a)

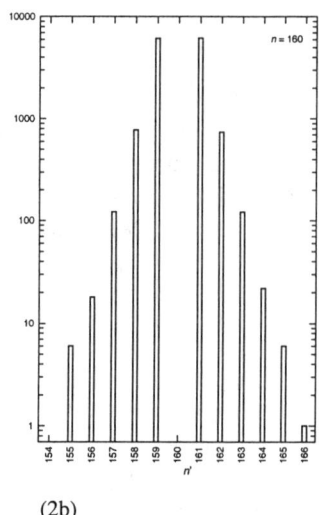

(2b)

Figure 2. (a)$\ln(t_{70})$ vs $1/\ln^2(s)$. Solid line is the least squared fit. Dashed line is the 95% confidence interval. (b) Frequency of n to n' transitions for a cluster of $n=160$.

1. ten Wolde, P.R. and Frenkel, D., *J. Chem. Phys.*, **109**, 9901 (1998).
2. ten Wolde, P.R. Ruiz-Montero,, M.J. and Frenkel, D., *J. Chem. Phys.*, **110**, 1591 (1999).
3. Bennett, C.H., in *Algorithms for Chemical Computations*, Ed. Christofferson, R.E., (Am. Chem. Soc. Washington D.C. 1977).
4. Chandler, D. *J. Chem. Phys.* **68**, 2959 (1978).
5. Oxtoby, D.W., *j. Phys. :Condens. Matter*, **4**, 7627 (1992).
6. Katz, J.L., Saltsburg, H., and Reiss, H., *J. Coll. Interface Sci.*, **21**, 560 (1966).
7. Senger, B., Schaaf, P., Corti. D.S., Bowles, R.K., Voegel, J.-C., and Reiss, H., *J. Chem. Phys.*, **110**, 6241 (1999); Senger, B., Schaaf, P., Corti. D.S., Bowles, R.K., Pointu, D., Voegel, J.-C., and Reiss, H., *J. Chem. Phys.*, **110**, 6438 (1999).
8. Yasuoka, K. and Matsumoyo, M., *J. Chem. Phys.* **109**, 8451 (1998).
9. Yasuoka, K. and Matsumoyo, M., *J. Chem. Phys.* **109**, 8463 (1998).
10. Oh, K.J., and Zeng, X.C., *J. Chem. Phys.*, **110**, 4471 (1999).
11. Allen, M.P. and Tildesley, D.J. in *Computer Simulation of Liquids*.(Clarendon Press, Oxford, 1987).
12. Shugard, W.J., and Reiss, H., *J. Chem. Phys.*, **65**, 2827 (1976).

Surfactant Induced Nucleation in Supersaturated Vapors

M. Rusyniak, Y. Ibrahim, V. Abedalsayed, M. Rabeony and M. S. El-Shall[*]

*Department of Chemistry, Virginia Commonwealth University
Richmond, VA 23284-2006*

Abstract. The roles of fluoroalcohols in enhancing the binary clusters with water as compared to aliphatic alcohols, and in lowering the barrier to homogeneous nucleation have been investigated. The results suggest that fluorocarbons in the gas phase can be used to lower the surface tension of the condensation nuclei in supersaturated vapors, and therefore enhance the rate of homogeneous nucleation.

INTRODUCTION

Recent experiments have shown that the vapors of fluorocarbons at low pressures can greatly change the surface tension of many liquids and produce some new surface phenomena [1]. The experiments suggest fluorocarbons at low concentrations in the gas phase exhibit properties of surfactants by essentially reduce the surface tension of organic and inorganic liquids. This surprising result appears to contradict the known fact that gases produce minimal reduction in the surface tension of liquids at normal low-pressure conditions since any gas affects the liquid surface tension to a degree proportional to its density [2].

In this work, we address the question of whether the presence of fluorocarbons in the gas phase can influence the rate of homogeneous nucleation from supersaturated vapors. In other words, does the addition of fluorocarbons at low concentrations to a supersaturated vapor change the composition of the condensation nuclei in a way that enhances the nucleation process by reducing the surface tension of the growing droplets? Our approach to investigate these questions involves two experimental methods. The first is to compare the extent of clustering produced from a supersonic expansion of water vapor containing a small concentration of the fluoroalcohol with that produced by a similar expansion of water vapor containing the corresponding aliphatic alcohol. The second method involves measuring the homogeneous nucleation rate of a supersaturated vapor of an aliphatic alcohol in the presence of a small concentration of the fluoroalcohol within a thermal diffusion cloud chamber (DCC).

In this paper, we report preliminary results from the two experimental methods. These results provide evidence for the enhancement of the water-fluoroalcohol binary clusters as compared to water-aliphatic alcohol clusters, and for the lowering of the barrier to homogenous nucleation of a supersaturated vapor in the presence of a low concentration of fluoroalcohol.

EXPERIMENTAL RESULTS

Binary ethanol-water and trifluoroethanol-water (TFE-W) clusters were generated by pulsed adiabatic expansion in a supersonic cluster beam apparatus [3]. The essential elements of the apparatus are jet and beam chambers coupled to a coaxial quadrupole mass spectrometer as shown in Figure 1. During operation a saturated ethanol-water vapor mixture is formed by flowing ultrahigh-purity He at a pressure of 2-4 atmosphere through two separate reservoirs filled with ethanol and water both at 24 °C. The vapor mixture is then expanded through a conical nozzle in pulses of 100-200 µs duration at repetition rates of 6-10 Hz. The jet is skimmed and passed into a high vacuum chamber (10^{-7} torr). The cluster beam is ionized by electron impact and the amplified signal from a particle multiplier is processed using a boxcar integrator set to sample at arrival times appropriate for the detected ions.

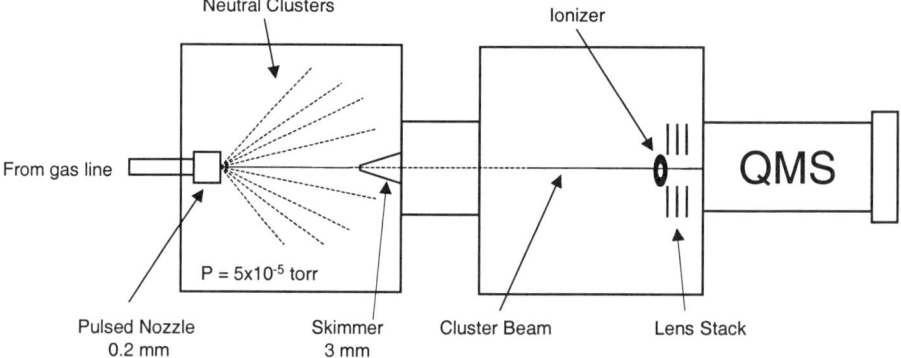

FIGURE 1. Set-up of a cluster beam apparatus coupled with a quadrupole mass spectrometer.

Figure 2 displays the mass spectra of the resulting ethanol-water (EW) and TFE-water (TW) clusters. In the ethanol-water system, protonated clusters of the form $H^+E_mW_n$ are observed only at m≥7. In contrast, in the TFE-water system, protonated clusters of the form $H^+T_mW_n$ are observed starting from T=1 and with n=1-12. This indicates that the extent of clustering of water and TFE molecules is much higher than that of water and ethanol molecules. This behavior is illustrated in Figure 3, which exhibits a comparison of the normalized relative intensities of $H^+T_mW_n$ and $H^+E_mW_n$ clusters. It should be noted that while the addition of several water molecules is only observed on large ethanol clusters (m≥7), the monomer TFE and its small as well as large clusters tend to add a significant number of water molecules (up to 7 or 8). This is probably due to the stronger TFE-water interaction as compared to ethanol-water interaction, owing to the electronegativity and the electron withdrawing effects of the fluorine atoms [4].

In order to investigate the effect of TFE on enhancing the rate of homogeneous nucleation, the following experiment was designed. Isopropanol was selected as a convenient source of a supersaturated vapor in the DCC. Detailed description of the DCC and the principles of its operation are given elsewhere [5,6]. The nucleation was observed by measuring the forward-scattered light from a He-Ne laser transmitting across the center of the DCC. A photomultiplier positioned to detect the forward-scattered light and a counting circuit were used to measure the rate of nucleation within a well-defined volume.

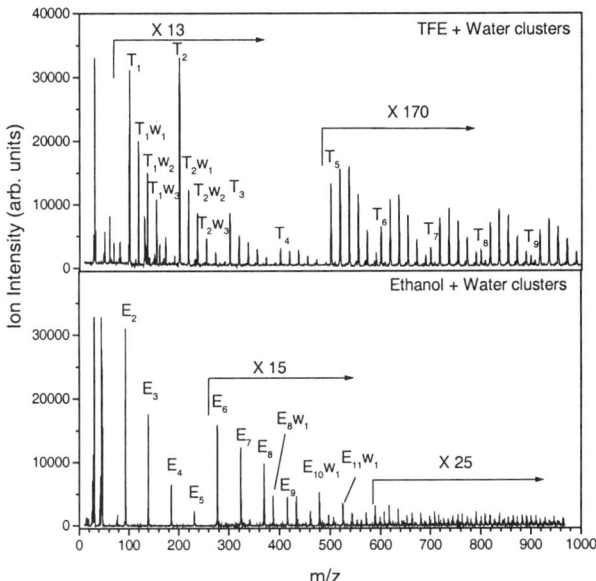

FIGURE 2. Mass spectra of TFE-Water (T_mW_n) and Ethanol-Water (E_mW_n) clusters.

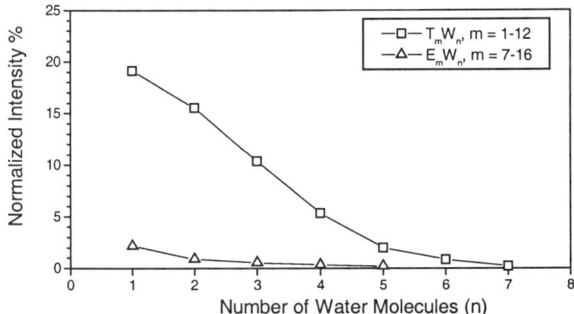

FIGURE 3. Normalized intensity of TFE-Water (T_mW_n) and Ethanol-Water (E_mW_n) clusters as a function of the number of water molecules (n) in the cluster.

The temperature of the chamber plates were adjusted to produce a steady nucleation rate of about 1.3 drops/cm^3/s. The nucleation count was measured every 10^3 s and the measurements were repeated three times yielding an average count of 1.3±0.1 drops/s. The total pressure was lowered in the DCC and a mixture containing 1 mL of TFE and 5 mL of isopropature was slightly lower than the original pressure. The chamber was allowed to recover and the temperatures of both plates were similar to their values before the addition of the mixture. The nucleation count was measured five times over an interval of one hour yielding an average count of 4.7±0.1 drops/s. The chamber parameters (p_{tot}, total pressure in torr; T_{lower} and T_{upper}, temperatures of lower and upper plates in Kelvin) and the corresponding nucleation count before and after the addition of TFE are given in Table 1.

TABLE 1. Droplet count in supersaturated isopropanol vapor before and after the addition of TFE.

P_{tot}, torr	T_{lowe} K	T_{upper} K	Count (drops/s)
(A) Before addition of TFE			
329.1	300.0	260.4	1.3
B) After addition of TFE			
315.2	300.0	260.4	4.7

It is clear that the addition of TFE at low concentration causes the homogeneous nucleation rate of isopropanol to increase more than three times the original rate. Experiments are in progress to investigate the temperature dependence and the mechanism of such enhancement.

ACKNOWLEDGMENTS

The authors gratefully acknowledge support from NASA Microgravity Materials Science Program (Grant NAG8-1484).

REFERENCES

1. Stoilov, Y. Y. *Langmuir* **14**, 5685-5690 (1998).
2. Abramson, A. A.; Schykin, E. D., Eds. *"Surface Phenomena and Surfactants"*; Chemistry: Leningrad (1984).
3. El-Shall, M. S.; Marks, C.; Sieck, L. W.; Meot-Ner, M. *J. Phys. Chem.* **96**, 2045-2051 (1992).
4. Kinugawa, K.; Nakanishi, K. *J. Chem. Phys.* **89**, 5834-5842 (1988).
5. Kane, D. B.; Fisenko, S. P.; Rusyniak, M.; El-Shall, M. S. *J. Chem. Phys.* **111**, 8496-8502 (1999).
6. Kane, D. B.; El-Shall, M. S. *J. Chem. Phys.* **105**, 7617 (1996).

A Monte Carlo Discrete Sum (MCDS) Nucleation Rate Model for Water

Barbara N. Hale and David J. DiMattio

*Physics Department and Cloud and Aerosol Sciences Laboratory,
University of Missouri – Rolla
Rolla, MO 65401 USA*

Abstract. The experimental homogeneous nucleation rate data, $J_{exp}(T,S)$, for water are compared with a steady state Monte Carlo nucleation rate model, $J_{MC}(T,S)$, in which the small cluster energies of formation are determined from a discrete sum of Monte Carlo generated Helmholtz free energy differences, $\delta f_c(n)$. This formalism includes the $\tau \ln(n)$ term of Fisher, where for the critical water clusters ($n^* > 20$ molecules) we use $\tau = 2.2$. It is shown that the $\delta f_c(n)$ scale like $[T_c/T -1]$ and that the discrete summation over $n = 2,3,...8$ produces a term that cancels the temperature dependence of the monomer flux factor. The result is a scaled form for the nucleation rate, $J_{scaled}(T,S) = J_o \exp[-(16\pi/3)[\Omega(T_c./T-1)]^3 /(\ln S)^2$, where $J_o = 10^{26}$ cm^{-3} s^{-1} and $\Omega = 1.47$ (close to the value predicted by liquid water bulk surface tension). This form has the same T dependence as the data of Viisanen, Strey and Reiss and gives an improved prediction for the data of Miller, Kassner and Anderson.

HOMOGENEOUS NUCLEATION DATA FOR WATER

The homogeneous nucleation rate data for water of Viisanen et al. [1] and Miller et al. [2], $J_{exp}(S,T)$, are plotted versus the classical model prediction for the nucleation rate, $J_{class}(S,T)$, in Fig. 1. One can observe lines of approximately constant temperature falling roughly parallel to the dashed line (with T increasing left to right). With a perfect model prediction all the points would collapse onto the dashed line. We have suggested [3] that the spreading of the lines of constant T in the steady state classical model arises from the monomer flux factor, whose T dependence is roughly, $e^{-W[Tc/T-1]}$ where $W \approx 6$. In fact, this is easy to

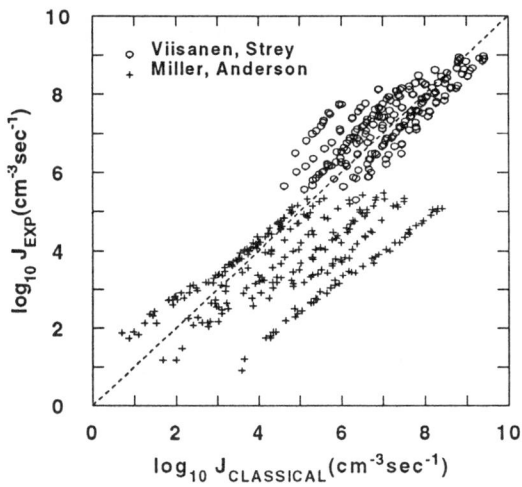

FIGURE 1. Comparison of the experimental nucleation rate data [1-3] with the classical model prediction.

verify as when one sets the classical prefactor equal to a constant, the above data of Viisanen et al. do indeed fall on a line. This is exactly what the scaled model predicts [3,4],

$$J_{scaled} = J_o \exp [-(16\pi/3)\Omega^3 [T_c/T-1]^3 / (\ln S)^2]. \tag{1}$$

Figure 2 shows the experimental data plotted vs. the scaled model, with $J_o = 10^{26}$ cm^{-3} s^{-1} and $\Omega = 1.47$. In comparison the scaled model provides a much improved prediction of the Viisanen et al. experimental nucleation rate temperature dependence. Even the Miller et al. data appear more closely clustered about the dashed line, though still indicating a spread at the higher temperatures

Since the monomer flux prefactor itself is not in question, one looks for additional terms in the n-cluster energy of formation to cancel the monomer flux T dependence. In our first attempt we made a simple summation over (classical) surface free energy differences, $\delta f(n) \approx (2/3)An^{-1/3}$, and determined the n-cluster free energy of formation from $\Delta f(n) = \sum_{n'=2,3,...n} [(2/3)An'^{-1/3}$

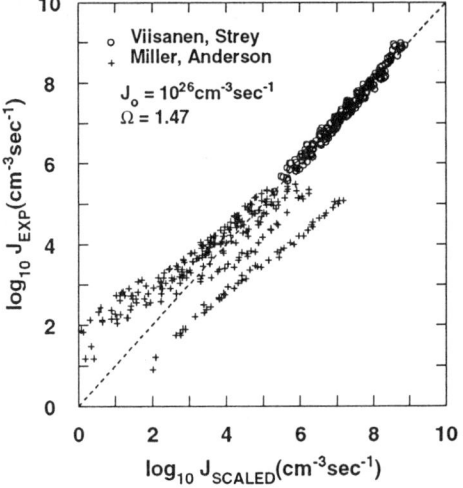

FIGURE 2. Comparison of the scaled model for J with the experimental data

$- \ln S] \approx An^{2/3}[1 + 1/(3n)] - (4/3)A - (n-1)\ln S$.[3] The $(4/3)A$ term cancelled the T dependence of the monomer concentration $[N(1) = SN_{eqb}(1)]$ in the flux as well as in $N(n^*) = N(1)\exp[-\Delta f(n^*)]$. But the extra term gave an overall nucleation rate which was three to four orders of magnitude too large. At this time Dillmann and Meier [5] reported their remarkable small cluster energies of formation which used the $\tau \ln(n)$ term of Fisher [6]. This latter term had negligible temperature and the correct sign and magnitude to restore agreement [3] with the toluene experimental data [Schmitt et al.[7]] and the experimental nonane data [Adams et al. [8]]. For water, however, this simple approach was less successful and we turned to a molecular model for the small cluster free energy differences. [9,10]

A MONTE CARLO MODEL FOR SMALL WATER CLUSTERS

The present work assumes a similar summation process over small cluster sizes but uses the Bennett Monte Carlo technique [11] to determine configurational Helmholtz free energy differences, $\delta f_c(n)$, for small water clusters interacting via the TIP4P water-water potentials[12]. The general procedure is described in references [9,10], and here we report TIP4P free energy differences for clusters ranging in size from n = 2 to n = 192 molecules at T = 260 K, 280 K, and 300 K The $-\delta f_c(n)$ are

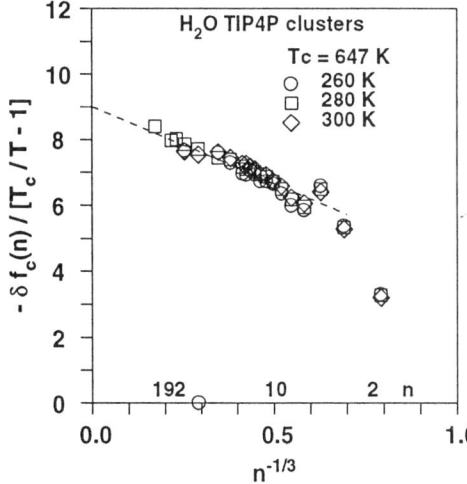

found to scale roughly like [$T_c/T-1$], where T_c is the critical temperature for water, 647 K. To demonstrate this we have plotted $-\delta f(n)/[T_c/T-1]$ vs. $n^{-1/3}$ in Fig. 3. One can take advantage of this temperature dependence to generate free energy differences for arbitrary T within the range noted. This circumvents the prohibitive task of simulating clusters at a large number of temperatures.

FIGURE 3. The configurational Helmholtz free energy differences for neighboring sized clusters (divided by kT) scaled with [$T_c/T-1$] for T = 260 K, 280 K and 300 K.

DISCRETE SUMMATION OVER SMALL CLUSTER SIZES

In this approach we have assumed a classical statistical mechanical vapor-cluster system consisting of a non-interacting mixture of ideal gases with each cluster size, n, constituting an ideal gas of N(n) clusters in equilibrium with N(1) monomers in a total volume, V. Each n-cluster is described by a classical Hamiltonian with classical effective atom-atom potentials dependent only on separation distance. For such a system the law of mass action gives N(n) in terms of the cluster configurational partition functions, Q(n), as follows: [9, 10]

$$N(n) = N(1) \exp \sum_{n'=2,n} \ln[Q(n')/[Q(n'-1)Q(1) v_{n'}/V]] - \ln[(n'/v_{n'})/(N(1)/V)]$$
$$= N(1)\exp\sum_{n'=2,3,...n} [\quad -\delta f_c(n') \quad - I_o + \ln S] \quad (2)$$

The $I_o = \ln[\rho_{liquid}/\rho^1{}_{vapor}]$ where ρ_{liquid} and $\rho^1{}_{vapor}$ are the liquid and equilibrium vapor monomer number densities, respectively. In most cases $\rho^1{}_{vapor} \approx P^o/kT$, where P^o is the equilibrium vapor pressure. Using this approximation $I_{o\,exp} = 9[T_c/T-1]$ for water. The method of Lee, Barker and Abraham is used to define the n-cluster and $v_n = \alpha n/\rho_{liquid}$, where α is of the order of 5. Equation (2) is formally independent of α.

One can see from Fig. 3 that the free energy differences behave like $n^{-1/3}$ for n larger than about 8. So, for n > 8, we fit the $-\delta f_c(n)$ data to a line passing through the experimentally predicted intercept, $I_{o\,exp}$, as follows [10]:

$$-\delta f_c(n) = I_{o\,exp} - (2/3)(36\pi)^{1/3}\Omega_{fit}[T_c/T-1] n^{-1/3} - \delta[\tau\ln(n)] \quad (3)$$
$$\approx I_{o\,exp} - (2/3)(36\pi)^{1/3}\Omega_{fit}[T_c/T-1] n^{-1/3} \quad \text{for large n.}$$

In Eq. (3) we use the form $An^{2/3} = (36\pi)^{1/3} \Omega [T_c/T-1] n^{2/3}$ from the scaled model [4]. A discrete summation of $-\delta f_c(n)$ is carried out from n = 2 to n = 8. While the

$\delta[\tau\ln(n)]$ term is *included* in the $-\delta f_c(n)$, it is too small to be extracted for large n and too uncertain in functional form to be identifiable for small n = 2,3,...8. The summation approach however gives $\Sigma_2^n \delta[\tau\ln(n')] = \tau\ln(n) - \tau\ln(n')|_{n' \to 1}$. We assume $\tau\ln(n')|_{n' \to 1} = 0$ and for $n \geq 20$ $\tau \approx 2.2$ as proposed by Dillmann [5].

The steady state nucleation rate is calculated from $J_{MC} = J_{o\,class}(T,S) N(n^*)/V$, where $J_{o\,class}(T,S) = [8kT/(\pi m)]^{1/2} 4\pi r_{n^*}^2 \gamma S N_{eqb}(1)/V$ and γ is the Zeldovitch factor [14]. The critical cluster size is found from $d\ln N(n)/dn |_{n=n^*} = 0$ and includes the $\tau\ln(n)$ contribution. In Fig. 4, J_{MC}, is plotted versus the experimental data.

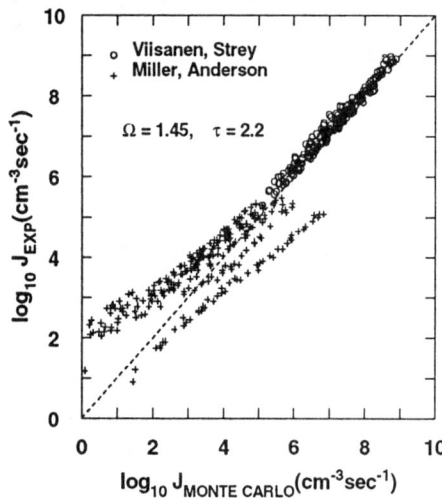

FIGURE 4. The molecular Monte Carlo nucleation rate, J_{MC}, plotted *vs.* the experimental data for water.

CONCLUSIONS AND COMMENTS

One can see from Figs. 3 and 4 that our MCDS model approach produces nucleation rates for water which are effectively scaled. Cancellation of the monomer flux T dependence via the discrete summation over n = 2,3,..8 in Eq. (2) and the scaling of $-\delta f_c(n)$ with $[T_c/T -1]$ give rise to the scaling of J. Our intent here has been to show how the scaling can emerge from a simple steady state model by including small cluster discrete size effects. And we note that the *validity* of the MCDS model for water otherwise rests on a number of factors including the assumed value of τ, the accuracy of the classical TIP4P potentials, and the simulation cluster definition.

REFERENCES

1. Y. Viisanen, R. Strey and H. Reiss, *J. Chem. Phys.* **99**, 4680 (1993); **112**, 8205 (2000)
2. R. C. Miller, R. J. Anderson, J. L. Kassner and D. E. Hagen, *J. Chem. Phys.* **78**, 3204 (1983)
3. B. N. Hale, *Metallurgical Trans.* **23A**, 1863 (1992).
4. B. N. Hale, *Phys. Rev. A* **33**, 4156 (1986); *Lecture Notes in Physics* **309**, 323 (1988).
5. A. Dillmann, Ph.D. thesis, Gottingen, 1989; A. Dillmann and G. E. A. Meier, *Chem. Phys. Lett.* **160**, 71 (1989); *J. Chem. Phys* **94**, 3872 (1990).
6. M. E. Fisher, Physics **3**, 255 (1967).
7. J. L. Schmitt, R. A. Zalabsky and G. W. Adams, *J. Chem. Phys* **79**, 4496 (1983)
8. G. W. Adams, J. L. Schmitt and R. A. Zalabsky, *J. Chem. Phys.* **81**, 5074 (1984).
9. B. N. Hale, Aust. J. Phys. **49**, 425 (1996); B. N. Hale and R. C. Ward, J. Stat. Phys. **28**, 487 (1982).
10. D. J. DiMattio, *Calculation of Scaled Nucleation Rates for Water Using Monte Carlo Generated Cluster Free Energy Differences*, Ph.D Thesis, University of Missouri-Rolla, August 1999.
11. C. H. Bennett, *J. Computat. Phys.* **22**, 245 (1976).
12. W. L. Jorgensen and J. D. Madura, *Mol. Phys* **56**, 1381 (1986).
13. J. K. Lee, J. A. Barker and F. F. Abraham, *J. Chem. Phys.* **58**, 3166 (1973).
14. Y. B. Zeldovich, *Acta Physicochim.* **18**, 1 (1943).

The Homogeneous Nucleation of a Vapor to its Solid Phase: Does Water Nucleation Change at 40°C?

Joseph L. Katz*, and Thomas P. Kole*

Department of Chemical Engineering, The Johns Hopkins University, Baltimore, MD 21218, USA

Abstract. The homogeneous nucleation of water is investigated in a downward diffusion cloud chamber. Nucleation rates are measured as a function of supersaturation to determine whether homogeneous nucleation of water below -40°C occurs via vapor to liquid or vapor to solid nucleation.

This paper addresses the question of whether or not it is possible to homogeneously nucleate a vapor into its crystalline phase. Nucleation of condensed phases from the gas phase occurs via the formation of clusters of molecules. Molecules are added to and removed from the cluster one at a time. In vapor to liquid nucleation, molecules cluster in somewhat random orientations and shape themselves into typical liquid forms in a length of time, which is almost instantaneous when compared to the frequency of arrival or departure of another molecule. However, for direct vapor to crystal nucleation to be possible, molecules must cluster in very specific orientations. Thus, the probability that a group of molecules will arrive simultaneously, oriented correctly to form a crystalline cluster, is vanishingly small. The only possibility to form a crystalline cluster is for the molecules to arrive sequentially and stay together while they move about until a group of them finds a crystalline orientation. What occurs instead is the homogeneous nucleation of these more random clusters, which then commence to grow. The probability that a small group of molecules in these growing clusters (now called droplets since they are larger than the critical size for homogeneous nucleation of the liquid) will orient and space themselves into the required crystalline structure increases proportionally to the number of molecules in the droplets. Thus, when large enough, ice nucleation occurs within the droplet. For water, it is well known that small liquid droplets can be subcooled to approximately -40°C.[1] Below these temperatures ice nucleation occurs very rapidly.

Lets consider the possibility of direct crystal to ice nucleation at -43°C. At this temperature the vapor pressure of water in equilibrium with ice is approximately 50% smaller than the vapor pressure of water in equilibrium with water. Thus, the driving force (the supersaturation) to form a crystal is 50% larger than the driving force to form a liquid cluster. The rate of nucleation is an extremely strong function of supersaturation, and at these conditions increases about one order of magnitude for every 10% increase in supersaturation. It is unlikely that if vapor to ice nucleation actually were occurring, it would do so at a supersaturation exactly equal to that of vapor to liquid nucleation. Even if such an unlikely event were to occur at -43°C, it certainly would not occur at other temperatures. Thus, if measurements were to show that the critical supersaturations measured below -40°C were a smooth continuation of those above -40°C we would assert that this would be strong evidence that nucleation occurs via homogeneous nucleation of liquid water, then rapid growth of the thus formed water droplets, followed by ice crystals homogeneously nucleating in these small water droplets.

EXPERIMENTAL PROCEDURE

The homogeneous nucleation of water in the desired temperature range is being studied using a downward thermal diffusion cloud chamber. In the downward mode, the top plate is covered by a thin layer of ice (previously condensed onto it) and is maintained at a higher temperature than the bottom plate. Water evaporates from this solid layer and diffuses downward through a noncondensible gas, and condenses on to the bottom plate. In contrast to an upward diffusion cloud chamber, with increasing partial pressure of the noncondensible gas, the downward chamber becomes increasingly stable with respect to convection. Operating the chamber at higher total pressures also results in smaller mass fluxes and thus smaller evaporation rates of water from the top plate.

A layer of solid water is condensed on the top plate by cooling the plate temperature below 0°C and maintaining the bottom plate at room temperature. After some time a sufficiently thick layer of ice forms on the top plate (approximately 0.1 mm). The temperature gradient across the chamber is then reversed, i.e.; the bottom plate is cooled to a temperature lower than that of the top plate. The plate temperatures are adjusted until the temperature at the predicted chamber height of the maximum rate plane is at the desired temperature. Nucleation rates are determined by a counting system which consists of a 1.5 mW He-Ne laser, beam shaping optics, a photomultiplier tube, and a computer-based counting program. The laser and its beam shaping optics produce a thin horizontal ribbon of light situated below the middle of the chamber (a reduced height, Z, of 0.35). A photomultiplier tube and lens assembly detects light scattered by each droplet which falls through the intersection area of the ribbon of light and the photomultipliers line of vision. This information is passed on to a computer where it processed through a counting algorithm which calculates the nucleation rate.

The special experimental difficulties encountered due to the extremely low temperatures involved have been solved. Nucleation rates are being measured as a function of supersaturation and temperature. Results will be presented for several maximum rate plane temperatures within the range of –60°C to -25°C and will be interpreted to show whether the homogenous nucleation of water below -40°C occurs via vapor to liquid or vapor to crystalline nucleation.

REFERENCES

1. Broto, F., and Clausse, D., *J. Phys. Chem.* **9**, 4251-4257 (1976).

Molecular Theory of Homogeneous Nucleation Using a More Sophisticated Kinetic

Andrey L. Itkin[1]

Institute of High-Performance Computing and Databases, Russia

Abstract. We use a recent *non ad hoc* model of microcluster (MM) proposed by Prof. Howard Reiss and co-authors [1,2] together with a more rigorous theory of the nucleation kinetic equation. It is shown that this model satisfactory predicts at least qualitative features of nucleation and the nucleation rate is insensitive to the connectivity distance d_c.

INTRODUCTION

Recently a new *non ad hoc* model of microcluster (MM) has been proposed by Prof. Reiss and co-authors [1,2] based on generalization of n/v-Stillinger's model. This model has been selected on the basis that it offered the best chance for constructing a non ad hoc theory of the nucleation rate in the sense that predicted rates would be insensitive to the magnitudes of the parameters, especially the "connectivity distance". What further the authors of MM reasonably decided to do was just to guess how they could utilize this comprehensive cluster model to get a proof of MM. At the present time unless a precise experimental technique is available to measure the cluster distribution of size, and perhaps exclusive volume as well as to register clusters' isomers, apparently the only way to carry out such verification could be as follows. One has to choose some appropriate kinetic theory, introduce the MM cluster definition in this theory and then try to predict available experimental data on nucleation rate based on such a combined approach.

The authors of [1,2] chose a simplest and so far the most popular classical nucleation theory (CNT) as *it is* to derive a nucleation rate. Thus, they used MM mostly to derive an equilibrium distribution of the clusters and also to calculate an averaged formation rate constant of the cluster containing *j* molecules and having an arbitrary volume between minimum and maximum admissible ones. In other words, they did not do any modification of CNT to make it more appropriate to the considered situation when clusters in addition to the number of molecules have also some additional characteristics, in particular cluster's volume. In such a form all this approach faces two problems. The first one is that now an *ad hoc* element present in the *combined* model in sense that nobody made efforts to derive kinetic equations of nucleation for MM clusters. And consequently, nobody proved that the original form of CNT can be used to make a proof of MM. Second, it turned out that nucleation rate

---

[1] Current position: IBES International, One World Trade Center, 18 fl., New York, NY 10048-1818, itkin@chem.ucla.edu

obtained as a result of such amalgamation more or less depends on d_c - a connectivity distance which is not a fundamental parameter. The authors of MM argued qualitatively that the final expression obtained for the nucleation rate contains several terms depending on d_c that could compensate each other, but they did not show it explicitly. Finally they expressed a hope that an account of more sophisticated cross-sections describing an interaction between two molecules could improve the situation. And later they made some modifications of MM [3] and introduce reasonable arguments why the nucleation rate could be less sensitive to the magnitude of d_c but they did not overcome this inconsistency completely.

This report aims to show that this inconsistency is not a problem of MM, but rather a problem of CNT which can not be used, at least in its original form, to derive a nucleation rate of the MM clusters. For instance, MM is developed for a *finite* system with N molecules placed in volume V at temperature T while CNT has been originally developed for an *open* system. It means that CNT considers number density of monomers n_1 as an external parameter while for the finite system it has to be determined from the conservation condition that the total number of monomers (both free and inside all clusters) is constant.

At present more comprehensive theories have been elaborated, for instance [4], that are able to describe closed systems in a proper way and allow one to overcome many problems of CNT. In particular, refer to our recent paper [5] where such an approach is discussed in detail. Thus, the idea of this work is just to eliminate using CNT and instead put the MM definition of the cluster into the basis of a more comprehensive kinetic theory and then to *derive* appropriate macroscopic kinetic equations from more general ones.

THEORY

Certainly, this is a very cumbersome task because as mentioned either in [1] based on a general form of cluster's partition function inherent to the MM one can not expect to derive an equilibrium cluster distribution of sizes in an explicit form. Therefore, instead we use a simpler model of the microcristalline cluster, but what is important this model does not contradict MM and even more, it is a particular case of MM under some assumptions made. But for this model all integration can be explicitly carry out and all assumptions made within our formalism to derive proper macroscopic kinetic equations can be verified. Thus, there is an *ad hoc* element in our theory but i) in a considerably less degree than in CNT and ii) this ad hoc element has no relation at all to the MM because we use an original *non ad hoc* MM just within the frame of our approach.

We consider our system as a canonical ensemble while in [1,2] the authors considered only a constant pressure ensemble. Nevertheless, their approach can be adopted also in case of a canonical ensemble as, for instance, they showed for a simple model in [6]. Therefore we use an equilibrium distribution of the n/v-Stillinger cluster in a canonical ensemble N_{iv} which gives the equilibrium number of clusters having i molecules and volume between v and $v+dv$.

Usually in view of a great complexity the Helmholtz free energy A_{iv} of the n/v-Stillinger cluster is evaluated by means of MC simulation under a given potential function $U(\mathbf{r})$. Then, for instance in [2] the equilibrium distribution of clusters have been substituted to the usual CNT expression in order to get the nucleation rate. Therefore, this approach doesn't touch the kinetic scheme of nucleation, but mainly takes care about the form of the equilibrium distribution of clusters with allowance of the excluded volume. Even in this sense this work is not completed because for a *closed* system the equilibrium number of monomers is not a parameter of the theory known in advance. Instead, it has to be found from the conservative law which for a closed system can be written in the form $\Sigma_i \Sigma_v i N_{iv} = N$. As shown in [4] a general method to find equilibrium density of monomers is to substitute expression for N_i into this equation and then solve the obtained equation regarding N_1 further restoring a *real* equilibrium distribution of clusters substituting this found value of N_1 into the expression for N_{iv}.

Unfortunately, even if this procedure could be directly applied to N_{iv} represented in a general form, the final distribution can be obtained only numerically so further it is very difficult to analyze it within the frame of a certain nucleation theory. That is why in this work we give a simple example of the cluster for which this procedure can be fulfilled analytically. We consider clusters located in nodes of some structure. In other words, such a model is more appropriate to explore nucleation in liquids. Nevertheless, with the help of such an example we are able to examine the model [1] within the frame of a more comprehensive kinetic scheme and just to explore how much the "connectivity distance" influences the nucleation rate.

Equlibrium Density of Monomers

For such a system one can apply a well-known method called "normal mode approximation". Applying further a method of [4] one can obtain the equilibrium number density of monomers that in the thermodynamic limit at $N \, \Delta T$, $V \, \Delta T$, $n = const$ coincides with thermodynamic equilibrium number density $n_{1e}(T) = N_{1e}/V$.

Quasisteady Distribution of Clusters

The equilibrium distribution of clusters given in [1] with N_1 found above can be now utilized to determine the vapor nucleation rate in the considered system. However, at this point we have to attract already kinetics of nucleation while so far we operated only with thermodynamics. In this report we demonstrate that the initial equations that are a subject of consideration in CNT have to be somewhat modified in the case of the clusters characterizing not only by the number of molecules in the cluster but the cluster's volume as well. There is a clear analogy between this case and the case of clusters with nonequilibrium distribution of the internal energy described and explored in detail in [4] because in the cited literature the concentrations of the clusters also were considered as a function of two variables - the number of molecules and the average internal energy. Therefore, our method is based on the physical ideas and mathematical methods developed in that book. The solution is the following. Nonequilibrium quasisteady concentrations of clusters coincide with that given in [4] where the rate constant of clusters' formation K_j^+ depend upon the minimum possible volume v_0 which can have a cluster with j molecules inside. Same is true for the nucleation rate I. The main difference is that while v is a function of d_c, v_0 does not depend on d_c. Indeed, imagine that we increase d_c that is the *maximum* distance for the molecule which is a part of the cluster from the another one. But it does not influence the configuration with the *minimum* volume when the average distance between the centers of molecules in the cluster is about the diameter of the molecule, i.e. about a parameter of the potential.

DISCUSSION

How does our I depend upon d_c. The only terms that could pretend this role if we associate d_c with the parameter of the potential σ_0 are the frequencies of intercluster vibrations ω. It can be shown that in such a case $d_c \propto \sigma_0$ and $\omega \propto 1/d_c$. Therefore with the d_c increase the nucleation rate stops to depend on d_c at all. It is also possible to argue why it is not correct to associate d_c with the parameter of the potential. If this is true, the nucleation rate does not depend on d_c at all. Thus, in both cases the present approach gives much more satisfactory results that the theory which uses a very perspective MM model [1] combined with the CNT.

ACKNOWLEDGMENTS

This work was sponsored by the National Science Foundation and fulfilled in UCLA under supervision of Prof. Reiss to whom the authors express his deep gratitude.

REFERENCES

1. Senger, B., Schaaf, D., and et al., *J. Chem. Phys.* **110**, 6421-6437 (1999).
2. Senger, B., Schaaf, P., and et al., *J. Chem. Phys.* **110**, 6438-6450 (1999).
3. Senger, B., Schaaf, P., and et al., *Phys. Rev. E* **60**, 771-777(1999).
4. Itkin, A., and Kolesnichenko, E., *Microscopic theory of condensation in gases and plasma*, New York: World Scientific, 1997.
5. Itkin, A., *J. Chem. Phys.* **108**, 3660-3677 (1998).
6. Ellerby, H., Weakliem, C., and Reiss, H., *J. Chem. Phys.* **95**, 9209-9218 (1991).

A Theory for the Deliquescence of Small Particles

Philippe Mirabel[a], Howard Reiss[b] and Richard K. Bowles[b]

[a] *Department of Chemistry, Université Louis Pasteur, Strasbourg, France.*
[b] *Department of Chemistry and Biochemistry, University of California, Los Angeles, CA 90095, USA.*

Abstract. A theory is developed for the deliquescence of small particles that incorporates the effects of surface phenomena on the deliquescence process. The theory is based on rudimentary surface thermodynamics since we are primarily interested in the orders of magnitude of the various effects. Applying the theory to a generic crystal with properties similar to NaCl, in water vapor, we find that the surface tension has a significant effect for crystals with a radius smaller than 10^{-5}cm. We suggest an experiment for the determination of the surface tension of soluble crystals that utilizes the deliquescence of these small particles. Such an experiment will need to be accompanied by a further development of the rigorous thermodynamic theory of solid surfaces.

DELIQUESCENCE IN SMALL SYSTEMS

In a macroscopic system deliquescence implies the coexistence between the solid solute and saturated solution divided by an interface. The interface contributes a positive free energy that the system would like to remove by developing completely into one or the other phase. However, the driving force behind this process is small because the excess free energy is small relative to the bulk free energy. In contrast, when the total system consists of a small drop the relative free energy of the surface is large, so the system fluctuates between solid and liquid states in a process more like a chemical isomerization for which the rate is determined by a free energy of activation related to the interfacial free energy. These fluctuations have already been observed in computer simulations of the "melting/freezing" of small argon clusters and several theories have been advanced to describe this process[1,2].

Under these circumstances the deliquescence pressure cannot be determined in the conventional manner, i.e. by the equating of the chemical potentials of the two phases, because there is no coexistence equilibrium. Instead, we derive an expression for the free energy G_A, of the system consisting of the dry particle and the surrounding vapor, which takes into account the surface tension of the particle and the pressure of the vapor, P_{vap}. Similarly, we derive an expression for G_B, of the system consisting of the vapor and the drop resulting from deliquescence, in which the surface tension of the drop now appears. The deliquescence pressure, P_{vap}^{del}, is then determined by

$$G_A\left(P_{vap}^{del}\right) = G_B\left(P_{vap}^{del}\right), \tag{1}$$

i.e. by the pressure at which the system can fluctuate to either state with equal probability. G_B is a function of the pressure and the composition of the drop at P_{vap}^{del} can be obtained from the Kelvin Relation[3]. Since it is almost impossible to fine tune the pressure to the point where the fluctuation can be observed, the alternative is to locate the pressure of the deliquescence transition.

To quantify, magnitudewise, the dependence of deliquescence pressure on particle size we develop our expressions for G_A and G_B using only the most elementary formulation of interfacial thermodynamics. For simplicity we assume that the crystal is spherical with a radius r^c, so we can write

$$G_A = N\mu_1^v + n_2\mu_2^c + \sigma^c a^c, \quad (2)$$

$$G_B + n_1\mu_1 + n_2\mu_2 + (N - n_1)\mu_1^v + \sigma a, \quad (3)$$

where N is the total number of solvent molecules in the system and n_1 is the number of solvent molecules in the resulting drop. n_2 is the number of solute molecules in the crystal and the drop, since it is involatile. The surface tension of the drop, σ, will depend on the drop composition, as will its area, a. μ_1 and μ_2 are the chemical potentials of the solvent and the solute in the drop respectively, while μ_1^v and μ_2^c are the chemical potentials of the vapor and the crystal. All the chemical potentials are taken at the pressure P_{vap}^{del} outside the drop[4]. Finally the surface tension of the crystal σ^c, is assumed to be independent of the of the crystal plane and a^c is the surface area of the dry crystal. Substitution of Eqs. 2 and 3 into Eq. 1 yields

$$n_1\mu_1^v + n_2\mu_2^c + \sigma^c a^c = n_1\mu_1 + n_2\mu_2 + \sigma a \quad (4)$$

If it is assumed that the vapor is ideal and that the solution is ideal and dilute then the following standard expressions for the chemical potentials can be used:

$$\mu_1^v = \mu_1^o + kT \ln S = \mu_1^o + kT \ln(P_{vap}/P_\infty^o), \quad (5)$$

$$\mu_1 = \mu_1^o + kT \ln x_1, \quad (6)$$

$$\mu_2 = \mu_c^o + kT \ln(x_2/\bar{x}_2), \quad (7)$$

where P_∞^o is the vapor pressure of the bulk solvent and $S = P_{vap}/P_\infty^o$ is the supersaturation. μ_1^o and μ_c^o are the chemical potentials of the pure bulk solvent and crystal respectively, while $\bar{x}_2$ is the mole fraction of the solute in a bulk saturated solution and x_1 and x_2 are the mole fractions of the solvent and solute in the drop. Substitution of Eqs. 6 through 8 into Eq. 4 yield

$$n_1 \ln(S/[x_1 - x_2]) - \ln(x_2/\bar{x}_2) = (\sigma a - \sigma^c a^c)/kT. \quad (8)$$

The vapor pressure of the volatile component of the vapor-drop system, is given by the Kelvin relation[3,5,6]

$$\frac{P_{vap}}{P_\infty^o} = x_1 \exp\left\{\frac{2\sigma v_1}{rkT}\right\}, \quad (9)$$

where r is the radius and v_1 is the partial molar volume of the solvent, of the drop. Eq. 9 requires that the vapor be ideal and that the involatile solute be present in dilute proportion so we may use x_1 as the thermodynamic activity. The geometric quantities of Eqs. 8 and 9, r, a and a^c, can be expressed in terms of the partial molar volumes, v_1, v_2 and v^c of the solvent and solute in the drop, and of the crystal respectively, and, n_1 and n_2. Using these along with Eq. 9, in Eq. 8 we obtain an expression in which n_2 is the only unknown. Solving this equation and using the result in Eq. 9 we obtain the pressure of deliquescence, P_{vap}^{del} for a particle of size n_2. The vapor pressure of the bulk saturated solution, $\overline{P}_{vap}^{del}$, can be determined by substituting $\overline{x}_1 = 1 - \overline{x}_2$ in to Eq. 9, so that we may use the ratio $R = P_{vap}^{del}/\overline{P}_{vap}^{del}$ as a measure of the surface effects on the deliquescence process.

Deliquescence of a Generic NaCl Crystal.

To examine the general features and orders of magnitude, of the surface effects on the deliquescence process, we calculated R as a function of n_2 for a generic crystal and drop with simplified thermodynamic quantities[7] (fig 1). In particular we assume the surface tension of the drop to be independent of the composition. However, where possible we have selected the thermodynamic parameters to be similar to a NaCl-H_2O system. For very small particles the surface tension of the crystal has a significant effect on the deliquescence pressure. Smaller values of σ^c produce an increase in the pressure required for a particular size crystal to deliquesce. As the crystal becomes larger than 10^7 particles the surface effects have all but disappeared. The maximum that appears for the smallest surface tension is due to competing effects of Raoults law

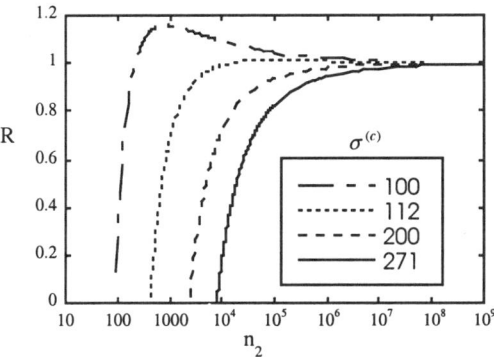

Figure 1. R versus n_2 for a generic crystal with different values of the crystal surface tension σ^c.

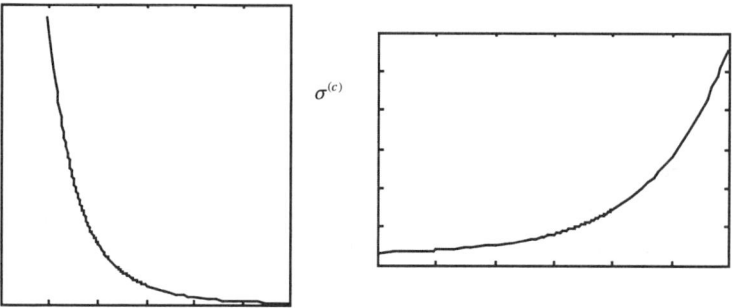

Figure 2. (a) Mole fraction of the solute in the drop at deliquescence, x_2, vs. the size of the crystal n_2. (b) The largest crystal with a given σ^c that always deliquesce in a vapor of vanishing density vs. n_2.

at high concentrations of solute but small values of n_2, and the Kelvin relation which dominates when the drop becomes dilute (fig 2a). As the surface tension of the crystal increases along with the concentration of the solute, the Kelvin relation can no longer compete with Raoults law and the maximum disappears. Fig 1 also shows that R goes to zero at some crystal size for each value of σ^c. Particles smaller that this will always deliquesce, even in a vapor of vanishing density (fig 2b). This is simply a result of the enormously increased solubility of the crystal due to its surface tension.

The size of the surface effects on the deliquescence pressure of these small crystals suggests that they may be utilized in methods for determining the surface tension of soluble crystals such as NaCl. One possibility is to suspend an aerosol of small particles over a bulk solution master solution of known vapor pressure and to observe Raleigh scattering. The intensity of the scattering should dramatically increase at the deliquescence pressure, where the particles abruptly change size. If this occurs for extremely small particles, soft x-ray scattering would be necessary. An alternative method involves the measurement of particle mobility, which is likely to change dramatically upon deliquescence. The use of fig. 2b provides an immediate upper bound to σ^c if a particle of a known size is seen to remain undeliquesced in a vapor of vanishing density. Still another possible experiment involves the comparison of deposited NaCl aerosols that have and have not been exposed to water vapor, by electron microscopy[8].

Acknowledgements

This work was supported by the National Science Foundation.

1. Berry, R.S., Jellinek, J., and Natanson, G., *J. Phys. Rev.*, **A30**, 919 (1984).
2. Reiss, H., Mirabel, P., and Whetten, R.L., *J. Phys, Chem.*, **92**, 7241 (1988).
3. Köhler, H., *Geofys. Pub. Krestiana*, **1**, 1 (1921); *ibid*, **2**, 6 (1922).
4. Reiss, H., *Methods of Thermodynamics*, (Dover, Mineloa, N.Y. 1996).
5. Mirabel, P., Reiss, H., and Bowles, R.K., *J. Chem. Phys*, In Press
6. Reiss, H., and Koper, G.J.M., *J. Chem. Phys.*, **99**, 7837 (1995).
7. Mirabel, P., Reiss, H., and Bowles, R.K., *J. Chem. Phys*, In Press.
8. Espenscheid, W.F., Matijevic, E., and Kerker, M., *J. Phys. Chem.*, **68**, 2831 (1964).

Idiosyncrasies Of Nucleation In Large, Deeply Supercooled Liquid Clusters

Lawrence S. Bartell, Yaroslav G. Chushak, and Jinfan Huang

Department of Chemistry, University of Michigan, Ann Arbor, Michigan 48109, USA

Abstract. Characteristic differences between the nucleation of solids in bulk liquids and in liquid clusters are identified in computer simulations, and the reasons for these differences are discussed.

INTRODUCTION

What is idiosyncratic about the nucleation in deeply supercooled clusters? To answer the question we first have to consider what is considered normal nucleation. Until quite recently, a picture of what happens on a molecular scale during a phase change appeared only in the imagination of the theorist. Experiments were blind to the details, and theories of the complex process had to be simplified in order to make progress. The classical theory of homogeneous nucleation (CNT) was based upon the idea proposed by Gibbs according to which a (nominally spherical) particle of the new phase appears in the old phase by structural fluctuations at the cost of the free energy to produce the interface between the old and new phases. Direct measurements of the interfacial free energy between the two condensed phases could only be performed at the equilibrium temperature of the phase change[1], and nucleation experiments were feasible only at modest degrees of supercooling where critical nuclei were large and presumably bulk-like in properties. In our investigation the nucleation process is studied by carrying out computer experiments using the molecular dynamics (MD) technique based on realistic intermolecular interaction potential functions[2]. In the computer simulations, molecules in liquid clusters can be watched as their unbiased trajectories carry the system into whatever regions of phase space are decided by chance. Phase changes take place spontaneously at different times for different members in a set of supercooled clusters. Deviations from what might be considered to be conventional nucleation behavior can come from two sources. (a) Clusters in this study are minuscule in comparison with the droplets in conventional experiments and this, alone, can lead to altered physical properties of the system. (b) The supercooling is enormously deeper, leading to considerably smaller critical nuclei. Some of the consequences are sketched below.

SIMULATIONS

The clusters whose behavior we report typically contain 100 to 2000 molecules. We first elected to chose clusters with free boundaries in order to avoid the possible interferences that can occur when periodic boundary conditions are imposed as in simulations of the bulk. Later, clusters proved to be worthy subjects in their own right.

Their typical supercoolings are in the vicinity of half the melting temperature. Rates of spontaneous transitions at much shallower supercooling would be too slow to observe within reasonable CPU times. Although it has been proposed that such deep supercooling as we examine can be expected to lead to spinodal decomposition instead of nucleation[3], we have seen no evidence of such decomposition in our simulations. The kinetics of transformation of an ensemble of clusters has always been first-order, to within the limits of our ability to determine order, and the nuclei initiating the transition can be watched as they form and grow. Moreover, their kinetics of growth is in good agreement with the kinetics expected from a modified form of the Kolmogorov-Johnson-Mehl-Avrami equations[4] based on first-order kinetics.

Some theorists protest that aggregates of matter as small as our clusters cannot be considered to exist in what can legitimately be called a thermodynamic phase. Likewise, they dispute that structural changes in such small systems can be viewed as phase changes. It turns out that there is no difficulty in recognizing the state of the clusters, as a rule. Liquids are readily discriminated from crystalline forms, both by their haphazard structure and their facile self-diffusion. Crystalline clusters can be uniquely identified with bulk crystal structures by their periodic translational and orientational orders. Sometimes, at very fast cooling rates to deep supercooling, the solid clusters formed may consist of mixtures of several different crystalline structures, and this may confuse the identification of the structure.

Although the crystalline phases encountered can be identified, they are not necessarily those obtained when the bulk material is cooled. This is because kinetics, rather than thermodynamics, controls what happens in the nanosecond time scale of the processes.

RESULTS

Differences Between Various Clusters

Clusters investigated in this report include molecular, ionic, and metallic clusters. These behave quite differently from each other when their liquid forms are deeply cooled or when their crystalline forms are melted. When molten salt clusters are cooled, however rapidly, they have an extremely strong tendency to crystallize. If the cooling is not unphysically rapid, they tend to freeze to *single* crystals. Clusters of molecular liquids are easily frozen to glasses if cooled sufficiently rapidly. They often freeze to single crystals, however, at cooling rates of ~ 10^{11} K/s, a rate not normally considered to be slow! Metals are intermediate. Generally, both molecular and ionic crystals retain their form while heated until they melt, even if they begin with metastable structures. Crystalline clusters of metals, however, may partially transform to a mixture of crystalline forms before melting.

Differences Between Clusters And Bulk Matter

When solid clusters are heated, they invariably start melting first at their surfaces. Alternatively stated, at a given temperature, molecules in surface layers of crystals tend to be more disordered and mobile, i.e., more liquid-like than interior molecules. Therefore, we had originally supposed that when a liquid cluster froze, the freezing would begin in the interior . Surfaces seemed to be less hospitable sites for ordered arrays. Our conjecture was wrong. For all of the kinds of supercooled liquid clusters we have studied, molecular, ionic, and metallic, crystal nuclei materialize preferentially at or very near the surface. The reasons for this are not entirely clear. This is one of the idiosyncrasies seen in clusters. A consequence is that nucleation rates in clusters tend to be higher than in the bulk because clusters have a much higher surface-to-volume ratio.

Is Nucleation In Clusters Homogeneous?

A virtue of computational studies of clusters is that the subjects in the simulations are perfectly pure with no traces of the submicroscopic contaminants that are believed to initiate heterogeneous nucleation in, and thereby subvert, studies of truly homogeneous nucleation in bulk liquids. Still, can the nucleation observed in liquid clusters be considered as truly homogeneous? Since nucleation tends to occur in surface layers, it could be argued that the nucleation is heterogeneous, catalyzed by the surface. Of interest in this respect are the solid state transitions seen in some of the simulations. For example, in the case of our clusters of hexafluorides, if freezing is carried out at temperatures not too deeply supercooled, the solid phase first formed is body-centered cubic, the same as that observed in the bulk substances. Then, if the temperature is low enough, the solid transforms to monoclinic, the stable low-temperature phase for SF_6 and SeF_6. MD simulations clearly show that the bcc to monoclinic change is facilitated by grain boundaries or solid-liquid interfaces. A cluster consisting of a single bcc crystal at the same temperature is much less likely to transform in a given time than a polycrystalline cluster or one that is partially liquid. Therefore, as in prior experimental studies of solid state transformations, the nucleation can be considered heterogeneous.

Size Of Critical Nuclei In Deep Supercooling

The number of molecules, n*, in critical nuclei found in the simulations has been much larger than the value inferred from the CNT using the interfacial free energy parameter deduced from the nucleation rate via the CNT. A similar disagreement arises when Gránásy's diffuse interface theory[5] (DIT) is applied, even though, in some respects, the DIT is superior to the CNT. In deeply supercooled liquid clusters, well before the concerted growth of the new phase signals that a critical nucleus has been formed, there exist large numbers of molecules passing the Voronoi and other tests for molecules with solid-like connectivity. They number far in excess of the n* inferred via the CNT and exist in fluctuating filaments and sheets of molecules. In view of this phenomenon, which does not correspond at all closely with the picture of precritical embryos associated with the CNT model, our criterion for the onset of nucleation is not based on the number of contiguous molecules judged to have the local structure of the new phase. It has proven to be more appropriate to monitor those molecules which not only pass the Voronoi and other order parameter tests but which can also be considered to be "bulk-like solid" in nature. In transitions to a bcc phase these are molecules which are surrounded by at least 12 other "bcc" molecules within the distance to the first minimum of the pair correlation function. Molecules in the sheets and filaments do not pass this test. When the temporal evolution of the *total* number of bcc molecules is followed, the onset of nucleation is poorly defined but when only the "bulk-like solid" molecules are monitored, the onset of nucleation is usually fairly sharp and definite. Note that the value of n* based on the CNT at our characteristic supercoolings is typically a mere half-dozen. So our number of molecules qualifying to be considered in the critical nucleus (at least 13) is already larger than the CNT n*, and the total number of contiguous molecules passing the tests of solid-like connectivity before onset considerably exceeds the CNT n*. This is another of the idiosyncrasies of deep supercooling.

Why are the critical nuclei larger than the n* of the CNT even when the sheets and filaments are disregarded? It is worth noting that a single layer of molecules surrounding a CNT critical nucleus would complete an aggregate of roughly the size of the critical nuclei we find. Now, the CNT critical nucleus is based on a theory that completely neglects any diffuse interface between the old and new phases. Therefore, it might be speculated that the discrepancy results from a disregard in the CNT of a possible difference between the equimolar surface and the surface expressing the interfacial free energy. Other reasons may be more important. In the CNT, the cost of forming a nucleus of the new phase in the old is the product of the interfacial free energy and the area of the boundary, and the driving force to lower the free energy ultimately is $N \Delta G$, where N is the number of moles in the nucleus and ΔG is the free energy of freezing per mole of the *bulk*. For very large critical nuclei formed at shallow supercooling this is probably not a bad approximation. For the extremely small nuclei formed at deep supercooling, the absolute magnitude of $N \Delta G$ is probably a substantial overestimate. It is also quite possible that our present criterion for identifying crystalline nuclei using a structural order parameter is somewhat too liberal. In any event, nucleation at deep supercooling differs appreciably from the nucleation encountered in conventional studies at shallow supercooling. Specific examples of the behavior discussed in the foregoing will be illustrated in the presentation.

REFERENCES

1. Jones, D. R. H. *J. Mater. Sci.* **9**, 1 - 17 (1974).
2. Bartell, L. S., *Annu. Rev. Phys. Chem.* **49**, 43 - 72 (1998).
3. ten Wolde, P. R., Ruiz-Montereo, M. J., Frenkel, D. *J. Chem. Phys.* **104**, 9932 - 9947 (1996).
4. Avrami. M., *J. Chem. Phys.* **7**, 1103 (1939); **8**, 212 (1940); **9**, 177 (1941).
5. Gránásy, L., and Iglói, F., *J. Chem. Phys.* **107**, 3634 - 3644 (1997).

NUCLEATION EXPERIMENT AND THEORY

The Homogeneous Nucleation of Water

John L. Schmitt, Kari Van Brunt and G. Jay Doster

Department of Physics and Cloud and Aerosol Sciences Laboratory, University of Missouri-Rolla, Rolla, MO, 65401 USA

Abstract. We present new measurements with an expansion cloud chamber of the homogeneous nucleation rate for water at approximately 265 and 285K. The new data show nucleation at higher supersaturation ratios than those previously reported. We discuss possible sources of error in the measurements and conclude that the data has been correctly measured.

INTRODUCTION

The measurements made by Miller, Anderson, Kassner and Hagen [1], of the homogeneous nucleation of water were the first extensive experimental study of the homogenous nucleation of a single material over a wide range of temperature, nucleation rate and supersaturation ratio. The measurements were made with a Wilson expansion chamber specifically developed for the measurement of homogeneous nucleation. Water was selected as the material because of its interest to atmospheric science.

The expansion chamber used by Miller and Anderson was replaced with a chamber constructed completely of stainless steel, glass and fluorocarbon seals so that materials other than water could be measured, Schmitt [2]. Recently (binary) nucleation of the octane isomers, n-octane and i-octane, was measured with this chamber, Doster *et al.* [3]. In the course of the octane measurements, it was discovered that long-term cleaning by repeated cycling of the chamber to supersaturations higher, in the octane case, than the supersaturation of the nucleation measurements was efficacious. Repeated cleaning over time produced the result that much higher supersaturations were needed to produce a given number of drops than had been previously measured in the course of the octane work. Our general conclusion based on the observation that the chamber cleans itself over time, is that we represent our experimental data as the highest supersaturation that we were able to measure for a given number of drops rather than as true "homogeneous" nucleation data.

The work on octane raised this question: Would the same cleaning technique produce nucleation at higher supersaturation ratios if we re-measured the homogeneous nucleation of water in our chamber?

EXPERIMENTS

Our new measurements of the homogeneous nucleation of water are exhibited in Fig. 1. The new measurements were done at initial chamber temperatures of 25C and 45C, producing nucleation at approximately 265 and 285K, respectively. The solid square data points labeled "Old" are from Miller et al. [1] and were taken at 45, and 24.5C initial chamber temperatures. The crosses are our data points and are labeled "New". It is obvious that the new data is at higher supersaturation ratios than the old. In the new 25C data, one clearly sees data points at lower supersaturations probably due to impurities in the system. The new 25C data may also show further evidence of impurities at the lower nucleation rates. The difference between the old and new data is approximately a factor of 1000 in nucleation rate.

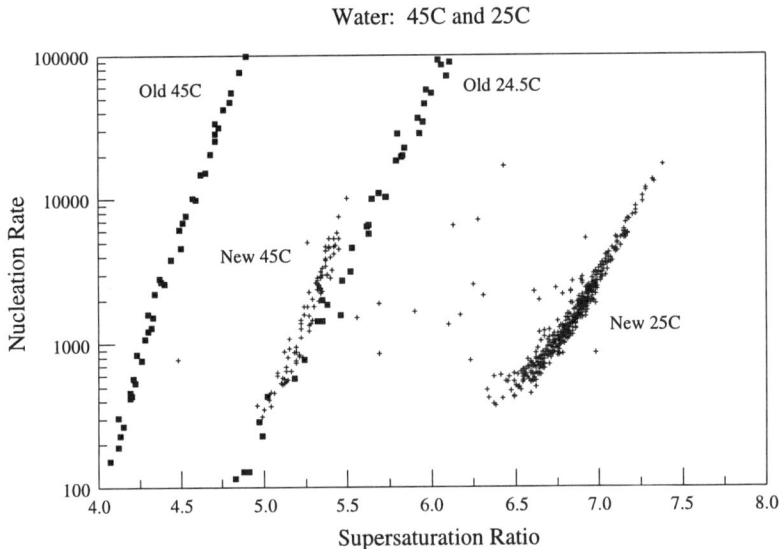

FIGURE 1. The data from this work and the data from Miller, Anderson, Kassner and Hagen [1].

DISCUSSION

The octane data were obtained by cleaning the chamber by expanding, usually twice, to a supersaturation above that required for homogenous nucleation and subsequently expanding the chamber to a slightly lower supersaturation for the data. This proceedure produced the lowest drop count for a given supersaturation. With water as the material, we found, this technique produced the same or higher drop counts than the previously established technique, also used by Miller and Anderson, of cleaning with supersaturations slightly below those required for homogenous nucleation and then expanding to higher supersaturations for measurements. We believe the primary experimental difference between the two materials is,when water droplets evaporate they leave re-evaporation nuclei, Anderson *et al.* [4], Wilson [5], Smith *et al.* [6] and octanes do not. Our first conclusion is that the (octane) technique does not work with water. However, Fig. 1 shows nucleation at higher supersaturation ratios than previous data therefore it is therefore important that we carefully examine possible sources of error.

The temperature of the chamber is controlled with reference to several temperature sensors, both transistor thermometers constructed in house and commercially purchased thermocouples and thermistors. These internally agree to within their accuracies; therefore we believe our temperatures are correct.

We believe the temperature and vapor fields are re-established in the 4minute wait time (this is the minimum time that was used) between expansions of the chamber. We placed a 0.002-inch thermocouple in the center of the chamber and measured the temperature during an expansion. Condensation in the thermocouple prevents accurate measurements during and immediately after the expansion, but we were able to verify that after 4 minutes the gas temperature was back to the pre-expansion temperature. We calculated the diffusive transport time for the chamber and found that 4 minutes was adequate to re-establish vapor equilibrium in the chamber. The real chamber however typically exhibits turbulence after we photograph the drops and turbulence will help to re-establish the vapor field. Also only a very small amount of the vapor is actually condensed into droplets: in octane, *e. g.*, 100 drop/cm^3 contain less than 1% of the vapor. Finally, we have made an independent measurement of the vapor content with water in the chamber. We used a commercial dew point meter, using the cooled mirror technique, with a stated accuracy of 0.1C. The argon and water vapor in the chamber were allowed to come to thermal equilibrium and then were pushed through the meter by raising the piston. The meter agreed with the controlled temperature of the chamber, 25C, within 0.1C.

What other effects might play a role in the chamber? We believe that the following process takes place. Impurities in the system are nucleated on and the drops rain into the liquid pool where the impurities are trapped. The impurities evaporate back into the vapor and the measurements are made in a dynamic system. If the impurity level is high, many evaporate back into the vapor before the next expansion. We do observe that if the chamber has not been cycled, new measurements many produce higher droplet counts. A possible source of impurities in the Miller and Anderson measurements could be the rubber o-ring seals and the rubber piston seal in that chamber. Finally, re-evaporation nuclei may be important (and interesting) and need to be extensively investigated.

CONCLUSIONS

We conclude that our measurements of the nucleation of water have been performed correctly and that our data is correct. We indeed measure the nucleation of water at higher supersaturations than previous investigators. Furthermore, we suggest that experimental homogeneous nucleation data should be characterized as the fewest number of drops that were measured at the highest supersaturation under the stated conditions. True homogenous nucleation may occur at a higher supersaturation, but cannot occur at a lower supersaturation.

REFERENCES

1. R. C. Miller, R. J. Anderson, J. L. Kassner, Jr., and D. E. Hagen, J. Chem. Phys. **78**, 3204 (1983).
2. J. L. Schmitt, Rev. Sci. Instrum. **52**, 1749 (1981).
3. G. J. Doster, J.L. Schmitt, and G. Bertrand, submitted to J. Chem. Phys. (2000).
4. R. J. Anderson, R. C. Miller, J. L. Kassner, Jr., and D. E. Hagen, J. Atmos. Sci. **37**, 2508 (1980).
5. J. G. Wilson, *The Principles of Cloud-Chamber Technique* (Cambridge Univ. Press, Cambridge, 1951), p. 13
6. J. G. Smith, J. L. Kassner, Jr., and A. J. Biermann, J. Rech. Atmos. **3**, 63 (1968).

Experimental Nucleation Studies With A Laminar Flow Tube Reactor

Vladimir B. Mikheev, Nels S. Laulainen, and Stephan E. Barlow

Pacific Northwest National Laboratory, P.O. Box 999, Richland, WA 99352, USA

Abstract. A Laminar Flow Tube Reactor (LFTR) has been used as a quantitative tool for nucleation measurements. It has been specially designed and constructed in order to minimize the potential experimental inaccuracies. Careful attention has been paid to the temperature conditions inside the LFTR. Other sources of experimental uncertainties, such as initial vapor concentration conditions, calculations of the parameters of the nucleation zone, and particle concentration measurements have been thoroughly analyzed as well. The nucleation rate dependence as a function of supersaturation and temperature has been measured for the dibutyl phthalate, ethylene glycol, and glycerol. The experimental results have been compared with the other experimental data and with the theoretical predictions.

INTRODUCTION

Among the various types of the experimental techniques used for the study of nucleation [1], currently only two of them, namely the thermal diffusion cloud chamber [2] and the expansion cloud chamber [3], can provide reliable quantitative nucleation rate measurements as a function of supersaturation and temperature over relatively wide range of nucleation rates. Other experimental techniques are also contending to be in such a category and one of them is the laminar flow technique [4-6]. As we will show, it appears very promising as a precise tool for nucleation measurements. However, the current state of the measurement of dibutyl phthalate (DBP) nucleation [4-6], which can be considered as a traditional standard substance for the laminar flow technique, shows that additional work is certainly required to understand the reasons for the large experimental discrepancies observed. A main objectives of this paper is to quantify the experimental uncertainties of the LFTR.

EXPERIMENTAL SETUP

The principle of operation of the LFTR is relatively simple. Typically (Fig.1), it consists of two thermally separated parts. A "hot" part called the saturator and a "cold" part called the condenser. In our case both saturator and condenser are made from aluminum cylinders of 76 mm external diameter. The total length of the saturator is 225 mm and the length of the condenser can be varied from 150 up to 650 mm. The diameter of the internal channel is 8 mm. Both the saturator and condenser are constructed with jackets for liquid circulation to support the required temperatures.

NESLAB RTE Refrigerated Circulators are used for temperature control and have a temperature stability of ±0.05°C. A teflon gasket (5 mm thick) serves for the thermo-insulation between saturator and condenser. However, because of the cone-shaped interior protrusion on both faces of the saturator and condenser and consequently the concave cone shape for the teflon gasket, we have been able to attain a 1 mm thick insulation between the internal channels of the saturator and of the condenser. Several narrow (less than 1 mm diameter) wells have been drilled at different places in both saturator and condenser to monitor the temperature at any part of the reactor. In addition, special temperature measurements directly inside the internal channel can be made. K-type thermocouples are used for the temperature measurements.

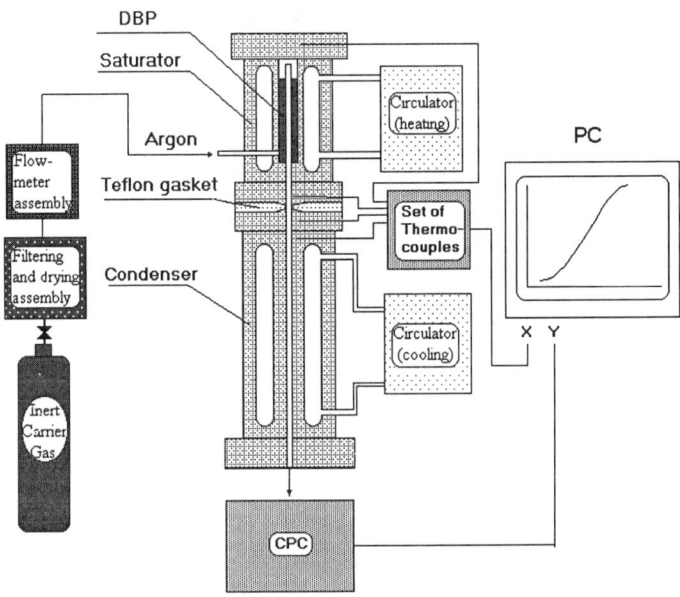

FIGURE 1. General schematic of the experimental set up.

The operation of the LFTR is as follows. First, an inert carrier gas (ultra high purity argon, helium, nitrogen) is injected into the saturator, where, while passing through molecular sieves (1 mm grain diameter) impregnated by some substance of interest (DBP, ethylene glycol, glycerol), the carrier gas becomes saturated by this vapor at the temperature of the saturator. The vapor-gas flow then enters the internal channel of the reactor and after passing through the length of the saturator section, it enters the internal channel of the condenser. Here, through cooling, a supersaturated state is achieved and particles begin to nucleate. Once the gas flow leaves the condenser, the concentration of the particles formed in the LFTR can be determined by an appropriate particle counter (*e.g.*, a TSI-3025A CPC). The entire system is interfaced to a PC, so the particle concentration dependence upon the temperatures and flow rate can be recorded and displayed graphically. Because the flow is laminar (volume flow rate is 5 cm^3/s), one can calculate the spatial distributions of temperature, vapor concentration and supersaturation using the known initial conditions (vapor concentration and temperature) of the flow at the inlet to the channel and also using the known boundary conditions at the walls of the channel (temperature and equilibrium vapor concentration). This allows us to define the nucleation zone, as well as to determine the nucleation volume and the residence time, after which the experimental value of the nucleation rate can be determined as a function of temperature, vapor concentration and/or supersaturation.

ANALYSIS OF EXPERIMENTAL UNCERTAINTIES

In spite of its simple appearance of construction and operation, the comparison of experimental data obtained by different groups of experimentalists [4-6] for the same substance (DBP) shows serious discrepancies. Thus, the evaluation of the possible inaccuracies that may lead to the observed discrepancies is imperative.

Question 1: Is there sufficient residence time in the saturator for the inert gas to become completely saturated by the vapor of the substance under study? It is difficult to calculate residence time for the internal geometry of the saturator used in our study and for the constructions used by other investigators [5,6]. We tried different internal geometries of the saturator and found empirically that if the diameter of the internal column of the saturator is several times larger than the diameter of the outlet, then the inert gas has enough time to become saturated by vapor of substance under study. For more details, the reader is referred to our recent paper [7].

Question 2: Are the correct temperatures measured in the reactor, especially in the region of the thermo-insulation between the saturator and the condenser? In order to provide uniform temperature conditions for the whole saturator, as well as for the whole condenser (or at least to diminish the temperature gradients as much as possible), all parts of the reactor (excluding the thermo-insulating teflon gasket between saturator and condenser) have been made from aluminum. The temperature was measured at different locations inside the aluminum body of both the saturator and the condenser and along the entire internal channel of the reactor as well. These measured data have been used as input into the boundary conditions of the calculations, so we can nearly avoid any errors and/or uncertainties in the temperature boundary conditions within an accuracy ± 0.1C.

Question 3: What are the sources of computational errors? Because the temperature is well known, the only likely source is the inaccuracy of the physicochemical properties of the vapor and the carrier gas data used in the study (*e.g.*, the thermal diffusivity and the molecular diffusion coefficient). A sensitivity analysis has been performed and the results of this analysis do not differ from the results reported by other experimentalists [6].

Question 4: How accurate are the particle concentration measurements? Contrary to expectation, we found that the commercial instruments do not necessarily provide accurate measurement for all types of particles. We have tested the ability of TSI-

3025A to measure the concentration of the particles freshly nucleated from water, pentanol, pentadecane, ethylene glycol, glycerol, and dibutylphthalate over a wide range of nucleation temperatures (from about −30 C up to +20 C). The counter only worked well for DBP over the entire range of temperatures studied. It worked with some restrictions for ethylene glycol and for glycerene, although some problems arose at room temperatures at high concentration levels. It does not work at all for pentanol, pentadecane and water.

DISCUSSION AND CONCLUSIONS

Figure 2 shows DBP experimental data and the comparison with the theoretical predictions based on the SCC theory. It is easy to see that the slopes of the experimental curves are consistent with the theoretical predictions. At the same time the temperature dependence of nucleation rate does not follow the theoretical predictions. One can see that the discrepancies with the theoretical predictions increase with decreasing temperature and, therefore, with decreasing critical cluster size. Generally speaking, this conclusion is reasonable, because, the critical cluster size is only six molecules at a temperature of −30.2 C. Classical considerations should break down under these conditions.

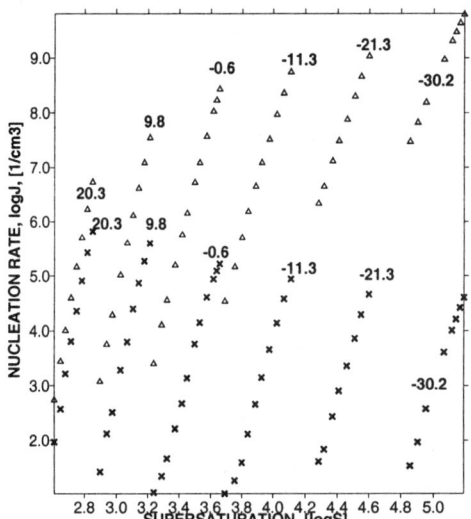

ACKNOWLEDGMENTS

This work has been supported by the Atmospheric Chemistry Program of the U. S. Department of Energy (DOE) under Contract DE-AC06-76RLO 1830. Pacific Northwest National Laboratory is operated for DOE by Battelle Memorial Institute.

FIGURE 2. Nucleation rate of DBP as a function of supersaturation. Crosses - experiment. Triangles - SCC theory. Temperature at every curve is given in Celsium.

REFERENCES

1. Heist, R. and H. He, *J. Phys. Chem. Ref. Data*, **23**, 781-805 (1994)
2. Fisk, J. A. and J. L. Katz, *J. Chem. Phys.*, **104**, 8649-8656, (1996)
3. Strey, R., P. E. Wagner, and Y. Viisanen, *J. Chem. Phys.*, **98**, 7748-7758 (1994)
4. Mikheev, V. B. and N. S. Laulainen, *J. Aerosol Sci.* **28**, Suppl. 1, S167-S168 (1997)
5. Bedanov, V. M., V. S. Vaganov, G. V. Gadiyak, G. G. Kodenyov, Khim. Fiz. **7**, 555-563 (1988)
6. Hämeri, K., M. Kulmala, E. Krissenel', and G. Kodenyov, *J. Chem. Phys.* **105**, 7683-7695 (1996)
7. Mikheev, V. B., N. S. Laulainen, S. E. Barlow, M. Knott, and I. J. Ford, *J. Chem. Phys.* (submitted).

D_2O-H_2O Condensation in Supersonic Nozzles: I. Experiments

Christopher H. Heath[*], Kiril A. Streletzky[*], Judith Wölk[†], Barbara E. Wyslouzil[*], and Reinhard Strey[†]

[*]*Department of Chemical Engineering, Worcester Polytechnic Institute, Worcester MA 01609, USA*
[†] *Institute for Physical Chemistry, University of Cologne, D-50939 Cologne, Germany*

Abstract. Pressure trace measurements and small angle neutron scattering (SANS) were used to probe the binary condensation of D_2O-H_2O mixtures in a supersonic nozzle. Each expansion started from the same initial pressure (carrier gas and condensible vapor) and temperature. The partial pressures of D_2O and H_2O were adjusted so that the onset of condensation always occurred at the same position in the nozzle. Under these conditions, the total pressure of condensible at onset varied linearly between the pressures of the pure components. Furthermore, the partial pressure at onset for pure H_2O was 29-34% higher than that for pure D_2O. The SANS scattering signals also varied systematically with the mixture composition. As the mixtures became progressively richer in H_2O, the scattering intensity dropped rapidly because the scattering length of H_2O is much smaller in magnitude and of opposite sign to that of D_2O. Further analysis of the scattering spectra showed that the particle size of the aerosol was increasing as the mixtures became more water rich. The increase in particle size was consistent with the increase in condensible molar flowrate needed to maintain onset at a fixed position in the nozzle as the mixture becomes richer in H_2O.

INTRODUCTION

Water is the most important fluid for life. Water condensation occurs in many natural and industrial processes including cloud formation, power generation and turbomechanical flows. Condensation of water in nozzles has been studied for over 50 years[1], and experiments range from those studying pure stream condensation[2,3] to the more dilute systems[3,4] described here. Another way to enhance our understanding of the behavior of this important substance is to conduct experiments using light water, H_2O, heavy water, D_2O, and their intermediate mixtures. In this paper we describe the results of our D_2O-H_2O binary condensation studies in which we combine conventional pressure trace measurements[5] with aerosol SANS measurements[6,7].

EXPERIMENTAL

A schematic diagram of the equipment used to conduct the pressure trace and the SANS experiments is illustrated in Fig. 1. The key components of the apparatus include[5] the carrier gas generator, the condensible vapor generators and the supersonic nozzle. In the nozzle, a mixed gas stream consisting of $N_2(g)$ with ~2 mol%

condensible expands and cools. The supersaturation of the condensible vapor increases until a point is reached where rapid particle formation and growth deplete the vapor and reduce the supersaturation to close to 1. Under typical operating conditions[6,7] the resulting aerosol has a number concentration N on the order of 10^{11} cm^{-3}, an average particle radius $<r>$ in the range of 5 to 12 nm, and a relative polydispersity of about 20-25%. Conventional investigations of particle formation in the nozzle have focused on measuring the deviation of a state variable (pressure or density) for the expansion containing a condensible vapor from that observed for the expansion of the carrier gas alone. In our experiments we measure the static pressure as a function of position, and define the onset of condensation as that point in the nozzle where the temperature of the condensing flow curve is 0.5 K higher than the temperature of the (hypothetical) isentropic expansion of the mixed gas stream. Once the onset behavior of the aerosol is well characterized, we can use SANS to directly measure the parameters of the corresponding size distributions.

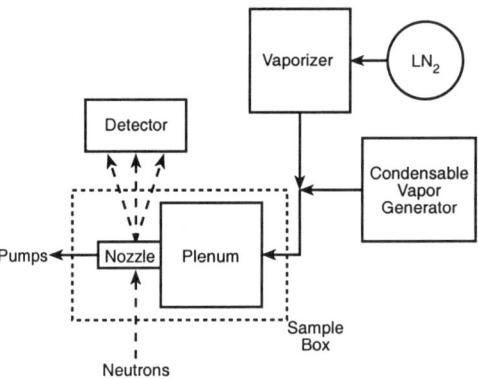

FIGURE 1. The experimental setup as it is used at NIST. To measure the axial pressure profile, a thin static pressure probe traverses the nozzle from 2-3 cm upstream to about 8-9 cm downstream of the throat. For the SANS experiments the pressure probe is removed.

To examine binary condensation in the nozzle, we first fix the total pressure p_0 and the temperature T_0 in the plenum, and then measure the static pressure profiles for different flowrates of one binary mixture. For each mixture, the values of pressure and temperature at onset define a straight line on a plot of $\ln p$ versus T. The mixture composition is then changed, and the process is repeated for other mixture compositions including the pure components. From the onset lines ($\ln p$ vs T) generated this way we are able to define the partial pressure of each species required to maintain onset at constant temperature. When the experiments begin with the same value of T_0, constant onset temperature corresponds to a fixed position in the nozzle. Figure 2 summarizes the pressures required to maintain onset at 231 K for D_2O, H_2O, and four intermediate mixtures.

The corresponding SANS scattering spectra are illustrated in Fig. 3. Because the onset position is fixed, each gas mixture experiences the same gasdynamic history up to onset. In this almost degenerate system, one might expect that the droplets comprising the aerosol are internally well mixed, their composition is independent of size, and the particle size distributions are essentially the same. If these assumptions are valid, the droplet composition should reflect the overall composition of the condensible mixture, the background subtracted curves should have the same shape, and the intensity should scale directly with the contrast factor corresponding to the droplet composition. In Fig. 3(b) we tested this hypothesis by finding the best fit parameters for the pure D_2O aerosol assuming a Gaussian size distribution of droplets moving[8] at 435 m/s and fitting the data available at both sample-to-detector distances (2.9 and 1.0 m). The best fit parameters are given in Fig. 3(b). For the other mixtures we scaled the D_2O curve by the ratio of the contrast factors of the mixtures. At first glance, the data follow the expected trend reasonably well.

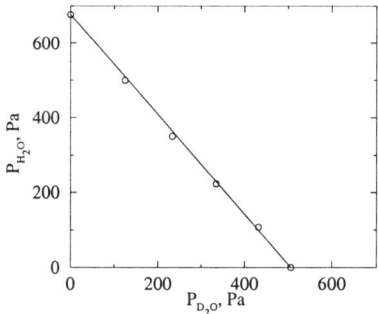

FIGURE 2. The pressures of D_2O and H_2O required to maintain onset at $T=231$K vary linearly between the endpoints. For these experiments $T_0 = 26°C$ and $p_0 = 60$ kPa.

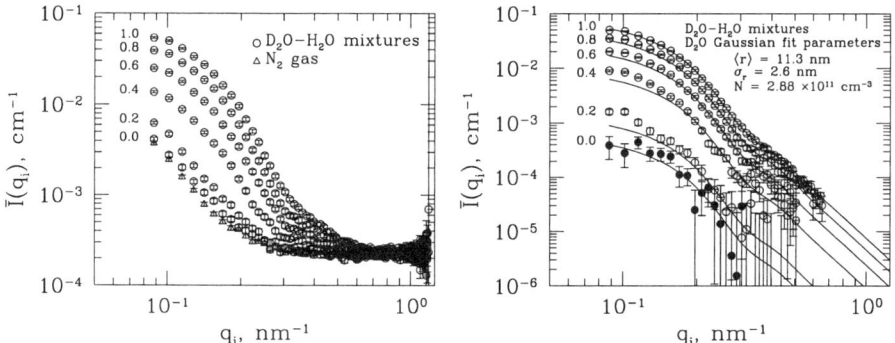

FIGURE 3. (a) The raw scattering spectra for the D_2O-H_2O aerosols and from the N_2 alone. (b) The background subtracted spectra. The corresponding mole fraction D_2O is indicated to the left of each curve. Although we were able to measure scattering from pure H_2O aerosol, the data are discounted in the remaining analysis because signal is too weak.

In the limit of $q \to 0$, the scattering intensity I_0 varies as $N \langle r^6 \rangle (b_1 x + b_2 (1-x))^2$ where b_i is the scattering length of component i and x is the mole fraction of component 1. Thus, if the assumptions stated above are correct, $\sqrt{I_0}$ should vary linearly with x. Figure 4, however, illustrates that for the mixtures $\sqrt{I_0}$ lies significantly above the solid line. Further analysis showed that the radii of gyration R_g for the mixtures were higher

(3-11%) than that for the pure D_2O aerosol. To see if the difference in droplet size can explain the observed behavior of I_0, we scaled the values of $\sqrt{I_0}$ by $(R_g/R_{g0})^3$. The corrected curve agrees with the measured intercepts almost perfectly, suggesting that the particle formation process for D_2O and H_2O under these conditions is not very different. Furthermore, the increase in particle size is consistent with the increased molar flowrate needed to maintain onset at the desired location for the H_2O rich mixtures.

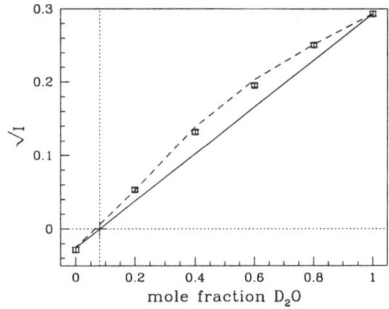

FIGURE 4. $\sqrt{I_0}$ obtained from Guinier fits to the data as a function of D_2O mole fraction. The solid line is the value of $\sqrt{I_0}$ for the pure D_2O aerosol scaled by the mixture scattering length. The short dashed lines show the location of the D_2O-H_2O null mixture. The long dashed line corrects $\sqrt{I_0}$ for the change in particle size.

In summary, the formation of D_2O-H_2O binary aerosols has been studied in a supersonic nozzle using pressure trace measurements and SANS. At this level of analysis, the composition of the droplets appears to equal that of the initial condensible vapor.

ACKNOWLEDGMENTS

We thank G. Wilemski for many helpful discussions. This work was supported by the National Science Foundation, the Donors of the Petroleum Research Fund administered by the American Chemical Society, and by a NATO Travel Grant. The work was based on activities supported by the National Science Foundation under agreement No. DMR-9423101.

REFERENCES

1. Oswatitsch, K. *Z. Angew. Math. Mech.* **22**, 1 (1942).
2. Moses, C.A. and Stein, G.D., *J. Fluids Eng.* **100**, 311 (1978).
3. Barschdorff, D., *Phys. Fluids* **18**, 529 (1975).
4. Stein, G.D., Ph.D. Thesis, Yale University, New Haven, CT (1967).
5. Wyslouzil, B. E., Heath, C. H., Cheung, J. L., and Wilemski, G. submitted, (2000)
6. Wyslouzil, B.E., Cheung, J.L., Wilemski, G., and Strey, R., *Phys. Rev. Letters* **79**, 431 (1997).
7. Wyslouzil, B.E., Wilemski, G., Cheung, J.L., Strey, R., Barker, J., *Phys. Rev. E* **60**, 4330 (1999).
8. The particle velocity is derived from pressure trace measurements.

D₂O-H₂O condensation in supersonic nozzles. II. Modeling

Christopher H. Heath[*], Kiril A. Streletzky[*], Barbara E. Wyslouzil[*], and Gerald Wilemski[†]

[*]Department of Chemical Engineering, Worcester Polytechnic Institute, Worcester MA 01609, USA,
[†]Department of Physics, University of Missouri, Rolla, MO 65409, USA

Abstract. An integral steady state model of nucleation and condensation was used to examine the formation and growth of D₂O droplets in a supersonic nozzle. The classical nucleation rate expression was used together with isothermal and nonisothermal droplet growth laws. For each experiment, the nucleation rate expression was multiplied by a temperature independent parameter in order to match the experimentally observed onset of condensation. In all cases, the predicted pressure traces lie above the measured ones. For one of the condensation experiments, the corresponding neutron scattering spectrum was also available. Modeling showed that once the rate expression was adjusted to match onset, the predicted scattering spectrum was a strong function of the growth law. Furthermore, the match between the measured and predicted scattering spectra was much better for the isothermal growth law than for the nonisothermal growth law.

INTRODUCTION

Modeling of condensation in supersonic nozzles has often proceeded concurrently with experiments[1] as a means of inferring nucleation rates, particle concentrations, and particle size distributions. The main problem with this approach is that there is usually no unique pair of nucleation and growth models that gives a best fit to the measured flow pressure[2] and density[3] profiles. Light scattering experiments[4] provide only limited help in resolving this lack of uniqueness because in the Rayleigh regime the angle-independent scattering signal gives no information about the differential particle size distribution. In our investigations of condensation in supersonic nozzles, we combine conventional pressure trace measurements[5] with small angle neutron scattering (SANS) experiments[6,7]. Since the typical wavelength (0.5-1.5 nm) of cold neutrons is smaller than the mean particle radius, SANS spectra are sensitive to the particle size distribution. Thus, independent estimates of the average particle size, $<r>$, the width of the size distribution, σ, and the droplet number concentration, N, are obtained using only a weak assumption about the general shape of the distribution (e.g. Gaussian or log-normal). A brief description of the experimental setup and procedures used for this work is presented in an accompanying paper[8]. Here we describe the model applied to predict condensation in our nozzle, and summarize our first attempts to reconcile the pressure trace and SANS data using this model. In this paper only D₂O results are presented.

MODELING

Our model uses the integral steady state approach first developed by Oswatitsch[1] and applied extensively by others[2-4]. The model simulates particle nucleation and growth and includes the effects of heat addition to the flow through the diabatic gasdynamics equations[2]. Cluster sizes are assumed to change only by monomer addition and evaporation. The change in the number density, ΔN, of new particles formed at each step is computed from the steady state nucleation rate, J, and the conservation law, $\Delta N=(J/u)\Delta x$, where u is the local flow velocity. The condensate mass fraction, g, can be calculated as a function of position using an appropriate droplet growth law. From g and the latent heat of condensation, ΔH_{vap}, the change in the flow properties can be obtained by integrating the diabatic flow equations using the measured nozzle profile.

In our calculations we used the classical nucleation rate expression, J_{cl}, for J. Following conventional practice, a temperature independent, multiplicative adjustment factor, Γ, is used to bring the calculated and measured values of the onset temperature into agreement; i.e. $J = \Gamma J_{cl}$. The following expression was used for J_{cl},

$$J_{cl} = \left(\frac{2\gamma\mu_c}{\pi N_A}\right)^{1/2}\left(\frac{p_v}{k_B T}\right)\frac{1}{\rho_c}\exp\left[-\frac{16\pi}{3}\left(\frac{\gamma}{k_B T}\right)^3\left(\frac{v_c}{\ln S}\right)^2\right].$$

Here γ, μ_c, v_c, and ρ_c are the surface tension, molecular weight, molecular volume, and density of the condensate, respectively; p_v is the partial pressure of the condensible vapor, N_A is Avogadro's number, T is temperature, S is supersaturation, and k_B is the Boltzmann constant. The physical property correlations for D_2O were as follows: vapor pressure comes from Hill et al.[9], ΔH_{vap} is derived from the vapor pressure correlation, surface tension and density are those developed by Wölk and Strey[10]. The isothermal droplet growth law is that developed by Wegener et al.[11] with the droplet temperature equal to the local gas temperature. The nonisothermal droplet growth law is that of Peters and Paikert[12]. In both cases the mass accommodation coefficient equals one.

Figure 1 illustrates the conventional modeling results. Once the growth law is selected and Γ is adjusted to match onset, there is only a slight difference between the pressure traces for the two growth rate models. Both models overpredict the final pressure ratio, because each predicts that about 90% of the material condenses by the end of the nozzle. In contrast, by integrating the diabatic flow equations[5] using the measured pressure traces we find that only ~75% of the material condenses. The values of Γ required to make the growth models match the data are included in the legend. These values are not constant, since the temperature dependence of the nucleation rate is overpredicted by classical theory. The predicted pressure traces do not fit the data very well beyond the onset of condensation, although more of the observed pressure trace could be matched when the isothermal growth law was used.

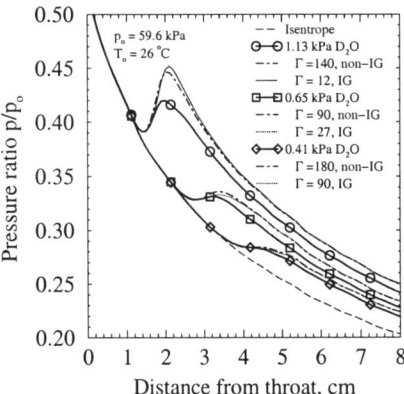

FIGURE 1. Three experimentally measured pressure traces are compared to isothermal (IG) and nonisothermal (non-IG) growth models. Once a reasonable value of Γ is found, there is no significant difference between predictions of the two growth models; the onset criteria of 0.5 K difference between the condensing and isentropic expansion is predicted within 0.7 mm from where it is observed.

The more interesting results are illustrated in Fig. 2 where we compare the predicted and measured (SANS) scattering spectra and their respective underlying size distributions. As illustrated in Fig. 2(a), the scattering spectrum predicted using the isothermal growth law has a shape similar to that of the measured spectrum. In contrast, the scattering spectrum predicted by invoking nonisothermal growth clearly fails to describe the data. This result is surprising because nonisothermal calculations predict that the growing droplets are significantly hotter than the background gas and, thus, nonisothermal effects should be important.

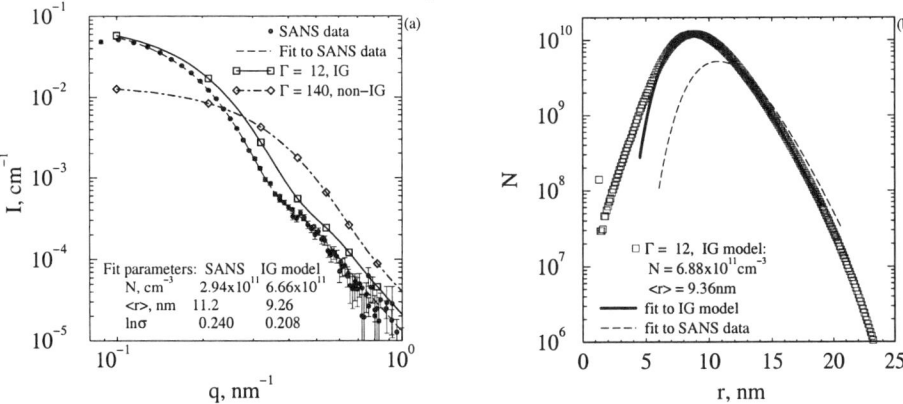

FIGURE 2. 1.13 kPa D_2O results: (a) The measured scattering spectrum, which has been Doppler corrected for a particle velocity of 435 m/s, is compared to the model generated spectra assuming isothermal (IG) or nonisothermal (non-IG) growth. The size distribution parameters for the fit to the SANS and to the IG spectra are given in the legend. (b) The size distribution for the IG model is compared to the log-normal fits to the model generated and measured (SANS) data.

To further quantify the differences between the measured and predicted scattering spectra we first treated the I vs. q data generated by the model using our standard fitting procedures. We found that this synthetic spectrum was far more consistent with a log-normal particle size distribution than a Gaussian one. As illustrated in Fig. 2(b), the full size distribution predicted by the model is accurately described by a log-normal size distribution for $r > 6$ nm. Both are quite different from the size distribution obtained from the measured SANS spectra. The difference between the SANS and model size distributions indicate that experimentally we observe a smaller number of small particles but roughly the same number of large particles. The larger mean radius of the measured size distribution may indicate that our model growth law underpredicts the actual droplet growth rate. To match the onset of condensation, the modeling appears to make up for the predicted smaller particles by boosting the nucleation rate. As a result, experimentally we observe a narrower size distribution of droplets with larger $<r>$ than is modeled. We also compared the experimental and modeled volume fractions of the condensed material. From SANS spectra the deduced condensed volume fraction is $\phi = 2.1 \times 10^{-6}$. The pressure trace data give the same result to two significant figures. The isothermal growth model predicts ϕ of 2.8×10^{-6}, while the fit to the generated $I(q)$ spectra leads to $\phi = 2.9 \times 10^{-6}$. The values of ϕ are consistent with the earlier conclusion that less material actually condenses than is predicted by the model.

ACKNOWLEDGEMENTS

This work was supported by the National Science Foundation, the Donors of the Petroleum Research Fund of the ACS, the Engineering Research Program of the Office of Basic Energy Sciences at the US DOE, and by a NATO Travel Grant. The work was based on activities supported by the National Science foundation under agreement No. DMR-9423101.

REFERENCES

1. Oswatitsch, K. *Z. Angew. Math. Mech.* **22**, 1 (1942).
2. Hill, P.G., *J. Fluid Mech.* **25**, 593 (1966).
3. Wyslouzil, B.E., Wilemski, G., Beals, M.G., and Frish, M.B., *Phys. Fluids* **6**, 2845 (1994).
4. Moses, C.A. and Stein, G.D., *J. Fluids Eng.* **100**, 311 (1978).
5. Wyslouzil, B. E., Heath, C. H., Cheung, J. L., Wilemski, submitted, (2000).
6. Wyslouzil, B.E., Cheung, J.L., Wilemski, G., and Strey, R., *Phys. Rev. Lett.* **79**, 431 (1997).
7. Wyslouzil, B.E., Wilemski, G., Cheung, J.L., Strey, R., Barker, J., *Phys. Rev. E* **60**, 4330 (1999).
8. Heath, C.H, Streletzky, K.A., Wölk, J., Wyslouzil, B.E., and Strey, R., ICNAA (2000).
9. Hill, P.G. and MacMillan, R.D.C., *Ind. Eng. Chem. Fundam.* **18**, 412 (1979).
10. Wölk, J. and Strey, R., "The Homogeneous Nucleation of Water: A Comparative Study of H$_2$O and D$_2$O", presented at the APS meeting in Los Angeles, March 16-20 (1998).
11. Wegener, P.P., Clumpner, J.A., and Wu, B.J.C., *Phys. Fluids* **15**, 1869 (1972).
12. Peters, F. and Paikert, B., *Int. J. Heat Mass Transfer.* **37**, 293 (1994).

Nucleation of Water Vapor in Methane at High Pressure

P. Peeters and M.E.H. van Dongen

Eindhoven University of Technology, Department of Applied Physics, P.O. Box 513, 5600 MB Eindhoven, The Netherlands

Abstract. Preliminary experimental results of nucleation of water vapor in methane at high pressure are presented. The results are obtained by applying the nucleation pulse method, using a modified shock tube. The nucleation conditions were taken at two different temperatures (250 K and 240 K) and three different pressures (10 bar, 25 bar and 40 bar).

INTRODUCTION

In natural gas research the methane-water system is of great interest. The main component of natural gas, as it comes out of a well, is methane. Depending on the well conditions, the methane is often saturated with water and other (heavy) hydro-carbons. Before further processing the natural gas, the vapor components have to be separated, thereby lowering the dew point. One of the key processes in separating the vapors is (homogeneous) nucleation. Nucleation experiments of water in methane at high pressure were performed with the final aim of modeling nucleation in natural gas.

EXPERIMENTAL SETUP

The experiments were performed in a pulse expansion wave tube, which is basically a modified shock tube. The set-up is described in detail by Looijmans and Van Dongen (1) and Luijten *et al.* (2).

The gas-vapor mixture is prepared using a bubbler set-up, which is submerged in a thermostatic bath. Different (final) compositions in the test section of the wave tube are obtained by changing the ratio of dry to wetted gas let into the test section and/or changing the temperature of the thermostatic bath. Before each experiment, the mixture in the test section is homogenized by mixing it at constant pressure in a recycling loop. The final vapor fraction is determined from a specially calibrated humidity sensor. The calibration procedure is described in detail by Luijten *et al.* (3). Nucleation rates were determined at two different temperatures (250 K and 240 K) and three different pressures (10, 25 and 40 bar).

RESULTS

For each experimental condition the saturation ratio $S=y/y^{eq}$ was evaluated, S being defined as the ratio of the actual vapor fraction to the equilibrium vapor fraction. The equilibrium vapor fraction can be expressed as a function of measurable quantities, $y^{eq}=fe(p,T)p^s(T)/p$, where p is the total pressure, p^s the pure component saturated vapor pressure, and fe the enhancement factor. The enhancement factor incorporates real gas effects. It is calculated from an equation of state, and fitted as a function of pressure and temperature (4). The resulting expression is

$$\ln(fe) = \left(b_0 + \frac{b_1}{T} + \frac{b_2}{T^2}\right)(p - p^s) + \left(c_0 + \frac{c_1}{T} + \frac{c_2}{T^2}\right)(p - p^s)^2, \qquad (1)$$

where,

$b_0 = -7.087E\text{-}3 \text{ bar}^{-1}$, $\qquad b_1 = 3.413 \text{ Kbar}^{-1}$, $\qquad b_2 = 1.400 \text{ K}^2\text{bar}^{-1}$,
$c_0 = -1.207E\text{-}4 \text{ bar}^{-2}$, $\qquad c_1 = 2.673E\text{-}2 \text{ Kbar}^{-2}$, $\qquad c_2 = 2.358 \text{ K}^2\text{bar}^{-2}$.

The experimental results will be compared to the theoretical predictions of the ICCT model. The expression for the nucleation rate can then be written in the form (5)

$$J = \frac{y p f e p^s}{(Z k_B T)^2}\left(\frac{2\sigma M}{p N_A}\right)^{1/2} \frac{1}{\rho_l} \exp\left[-\frac{16\pi M^2}{3 N_A^2 \rho_l^2 k_B^3} \frac{\sigma^3}{T^3 (\ln(S))^2}\right], \qquad (2)$$

where Z is the compressibility of the mixture, k_B is Boltzmann's constant, N_A is Avogadro's constant, σ is the surface tension, M is the molar mass of the vapor, and ρ_l is the liquid density. The surface tension of a liquid in a binary system depends both on the pressure and temperature. For the methane-water system different data series are available (6, 7, 8, 9). The data of Slowinski et al. (9) were found not to agree with the other data and were therefore discarded. The remaining data were fitted with an expression based on the Langmuir adsorption model,

$$\sigma = \sigma_0 - n_a k_B T \ln\left(\frac{p + p_L}{p_L}\right). \qquad (3)$$

Here, σ_0 is the surface tension of the pure vapor component. For supercooled water it is given by (10)

$$\sigma_0 = 0.127245 - 1.89845 \cdot 10^{-4} T \text{ [N/m]}, \qquad (4)$$

where T is in kelvin. The number of adsorption sites n_a is taken equal to $5.4E18 \text{ m}^{-2}$ (11). The Langmuir pressure p_L is used as a temperature dependent fit parameter. Since the data are at three different temperatures, p_L is fitted as a linear function of

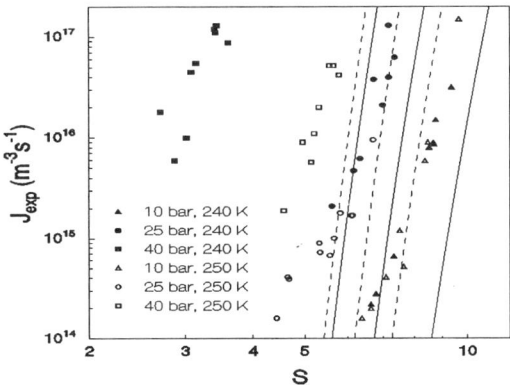

FIGURE 1. Nucleation rate J as a function of the saturation ratio S, for the system methane-water. The markers indicate the experimental results. The solid and broken lines indicate the results calculated from the ICCT model at 240 K and 250 K, respectively

temperature. When the pressure range is taken from 0 to 100 bar the result is

$$p_L = -481.95 + 2.1211T \text{ [bar]}. \qquad (5)$$

Although the nucleation rate data are obtained at a temperature outside the domain of the fit function, the values for the surface tension still seem reasonable.

Preliminary results of a series of experiments are shown in Fig. 1, together with experimental predictions. There is still some uncertainty about the absolute values of the supersaturation, so that care should be taken in interpreting the data. There is some evidence for a systematic error, which we are currently investigating. Still, the results can be used for observing trends and temperature and pressure dependencies.

DISCUSSION

At nucleation pressures of 10 and 25 bar, the experimentally found nucleation rates at 240 K and 250 K coincide, although the theory predicts a different rate. A similar behavior was found for the methane-n-nonane system. The experimental nucleation rates at 230, 240, and 250 K and 25 bar appeared to be temperature independent (2). Quite peculiar is the nucleation behavior of the methane-water system at 240 K and 40 bar. While at 250 K and 40 bar the experiments still form a smooth curve, at 240 K and 40 bar the results are much more scattered, and strongly deviate from the results at 250 K. A similar behavior was again found for the methane-n-nonane system, at 240 K and 40 bar (2,12), and at 230 K and 30 bar (12).

By applying the nucleation theorem (2) to the new experimental data, the size and composition of the critical cluster have been determined at each condition. The results are shown in table 1 and 2. At 40 bar and 240 K the numbers of gas and vapor molecules in the critical cluster are of equal magnitude. Again, similar experimental results were found by Luijten (2) for the methane-n-nonane system at this pressure and temperature.

TABLE 1. The number of water molecules in the critical cluster.

	10 bar	25 bar	40 bar
240 K	14±1	13±2	18±3
250 K	15±2	13±3	14±2

TABLE 2. The number of methane molecules in the critical cluster.

	10 to 25 bar	25 to 40 bar
240 K	4±2	24±5
250 K	4±2	7±2

Clearly there is considerable difference between the experimental results and the theoretical predictions, which can be partly due to the uncertainties in the absolute values of the supersaturation. The theory predicts larger critical clusters (steeper slope in the J-S-plot) and a pronounced temperature dependence. The fact that this temperature dependence is not supported by these and other measurements (2), could mean that at these conditions the surface tension increases with temperature. The origin of the anomalous behavior of the methane-water and methane-n-nonane system at certain rather extreme conditions remains to be investigated.

CONCLUSION

From these preliminary results we can conclude that the methane-water system has similar features as the methane-n-nonane system. The nucleation rate appears to be temperature independent, indicating that at these conditions the decrease in temperature is compensated by a simultaneous decrease in surface tension. Furthermore, at 240 K and 40 bar anomalous nucleation behavior is observed, which remains to be investigated.

REFERENCES

1. Looijmans, K. N. H, and Van Dongen, M. E. H., *Exp. Fluids* **23**, 54-63 (1997).
2. Luijten, C. C. M., Peeters, P., and Van Dongen, M. E. H., *J. Chem. Phys.* **111**, 8535-8544 (1999).
3. Luijten, C. C. M., Stormbom, L. E.,and Van Dongen, M. E. H., *Sens. Act.: B. Chem.* **49**, 279-282 (1998).
4. Data provided by the Shell labaratory in Amsterdam.
5. Luijten, C. C. M.,and Van Dongen, M. E. H., *J. Chem. Phys.* **111**, 8524-8534 (1999).
6. Sachs, W., and Meyn, V., *Colloids Surfaces A: Physicochem. Eng. Aspects* **94**, 291-301 (1995).
7. Jho, C., Nealan, D., Shogbola, S, and King Jr., D., *J. Colloid Interface Sci.* **65**, 141-154 (1978).
8. Massoudi, R., and King Jr., A. D., *J. Phys. Chem.* **78**, 2262-2266 (1974).
9. Slowinski Jr., E. J., Gates, E. E., and Waring, C. E., *J. Phys. Chem.* **61**, 808-810 (1957).
10. Hacker, P. T., *Experimental values of the surface tension of supercooled water,* National Advisory Committee for Aeronautics, 1951, technical note 2510.
11. Eriksson, J. C., *Acta Chem. Scand.* **16**, 2199-2211 (1962).
12. Looijmans, K. N. H., *Homogeneous nucleation and droplet growth in the coexistence region of n-alkanes/methane mixtures at high pressure,* PhD-thesis, Eindhoven University of Technology, Department of Applied Physics, 1995.

Laminar Flow Tube Reactor Interface With Quadrupole Mass Spectrometer

Vladimir B. Mikheev, Nels S. Laulainen, Viktor V. Pervukhin, and Stephan E. Barlow

Pacific Northwest National Laboratory, P.O. Box 999, Richland, WA 99352, USA

Abstract. Nucleation of supersaturated water vapor was studied using the Laminar Flow Tube Reactor (LFTR) technique. In order to check the presence of contaminants in the freshly nucleated water particles, the LFTR has been connected to a mass spectrometer. Trace amounts contaminants arising from the substances used as a circulating liquid to maintain temperatures of the LFTR have been detected. The results of the mass spectroscopic analysis are in full agreement with the observed values of nucleation rate.

INTRODUCTION

The potential influence of unknown contaminants on the results of any nucleation rate measurement is always in question. Nucleation can be distorted by such a small amount of a contaminant that it can not be detected by any of the traditional analytical techniques [1]. This is particularly important for the measurement of water nucleation because of the universal solubility of water. Thus, the analysis of the chemical content of the freshly nucleated water particles becomes crucial during the interpretation of the results of water nucleation measurements. This work describes an attempt to provide such an analysis using the interface between the LFTR and a quadrupole mass spectrometer.

MEASUREMENT OF THE WATER NUCLEATION RATE

The nucleation rate of supersaturated water vapor has been measured using the LFTR technique. Briefly, the operation of the LFTR is as follows. The LFTR consists of two thermally separated parts (Fig.1). A "hot" part called the saturator and a "cold" part called the condenser. An inert carrier gas (ultra high purity 99.999% helium) flows through the saturator. The saturator is filled with water impregnated molecular sieves. The carrier gas becomes saturated by water vapor at the temperature of the saturator. The vapor-gas flow then enters the internal channel of the reactor and after passing through the length of saturator section it enters the internal channel of the condenser. Here, through cooling, a supersaturated state is achieved and particles begin to nucleate. Once the gas flow leaves the condenser, the concentration of the particles formed in the LFTR can be determined by an appropriate particle counter.

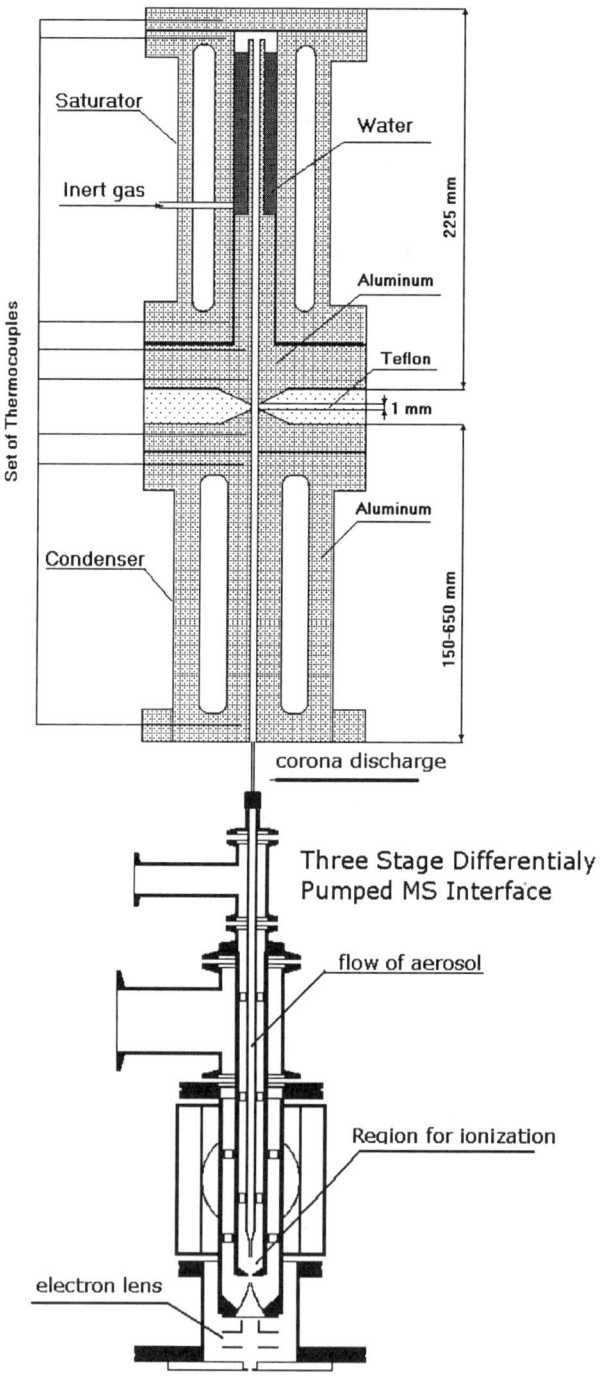

FIGURE 1. LFTR connected to Mass Spectrometer Inlet

The entire system is interfaced to a PC, so the particle concentration dependence upon the temperatures and flow rate can be recorded and displayed graphically. The nucleation zone can be defined by calculations. Consequently the nucleation rate can be measured as a function of supersaturation and temperature. For more details, the reader is referred to our recent paper [2].

Some problems were encountered during the measurements. First was the measurement of particle concentrations. The main problem is to count the particles before they evaporate. It was solved by using only the optical counting chamber of the TSI CPC. We have found that for pure water particles (sizes > 1 μm) this approach is more appropriate. The second problem was to provide enough time for measurement, because the walls of the internal channel of the condenser section were kept at temperatures below 0 C, which led to rapid icing of the channel. This problem was resolved in two ways. First, the condenser temperature was decreased to about – 40 C. That allowed the saturator temperature to be about 0 C to reach nucleation conditions. As a result, the concentration of water vapor introduced into the internal channel of the reactor was sufficiently low to avoid the rapid icing. Second, the internal channel was dried using a high flow of dry filtered pure nitrogen.

Reproducible results were obtained over a wide range of nucleation rate values (about 4 orders of magnitude). Comparison with theoretical predictions, however, has shown that measured nucleation rates are of several orders of magnitude higher than the theoretical calculations at lower supersaturations and then decreases with increasing supersaturation. The results were checked by careful temperature measurements of the entire reactor. Also measurements were made for different flow rates and a sensitivity analysis of the calculations of the nucleation zone parameters was made as well. Once we were convinced that the measurements were correct, the only plausible cause for deviations with theory is the influence of contaminants.

Interface of the LFTR with a Mass Spectrometer

To determine the composition of the water particles, the LFTR was interfaced with a mass spectrometer(MS). The MS system consists of an atmospheric pressure sampling inlet with three stages of differential pumping (Fig.1), a quadrupole mass filter, and pulse counting detection. Ionization is accomplished in the first stage by striking a current limited corona discharge between the tip of the inlet tube and the wall. This type of ionization gives the mass spectrometer rather limited sensitivity to gas-phase molecules, but seems to be ideal for particle work. This differential sensitivity occurs because charge - either positive or negative - will preferentially move to larger particles. Further, we find that as particles evaporate, the charge preferentially stays with contaminants that may be present. Figure 2 provides a striking example of this effect. Mass spectra were obtained at fixed saturator temperature (+2.5 C) by gradually lowering the temperature of the condenser.

For condenser temperatures as low as -30 C no mass-spectrum signals were detected. Below that point, peaks of ethanol, methanol, ethyl acetate, and methyl iso-butyl ketone (present in the denatured ethanol used for cooling the condenser) appeared (Fig.2) and increased with further decrease of the condenser temperature (Fig. 2 was obtained at -40 C). It is clear the results imply that without particles, no signal is observed, but once the critical supersaturation is reached then the contaminants are detected. We conclude that denatured alcohol leaked into the system and participated in nucleation. Thus, instead of pure water nucleation, a multicomponent nucleation rate was measured. Obviously this rate is significantly higher than the nucleation rate of pure water. The level of contamination was extremely low, because before we reached nucleation conditions, the mass spectrum did not indicate any presence of it. The combination of LFTR and MS gives a new tool for the detection of contaminants at extremely low concentrations.

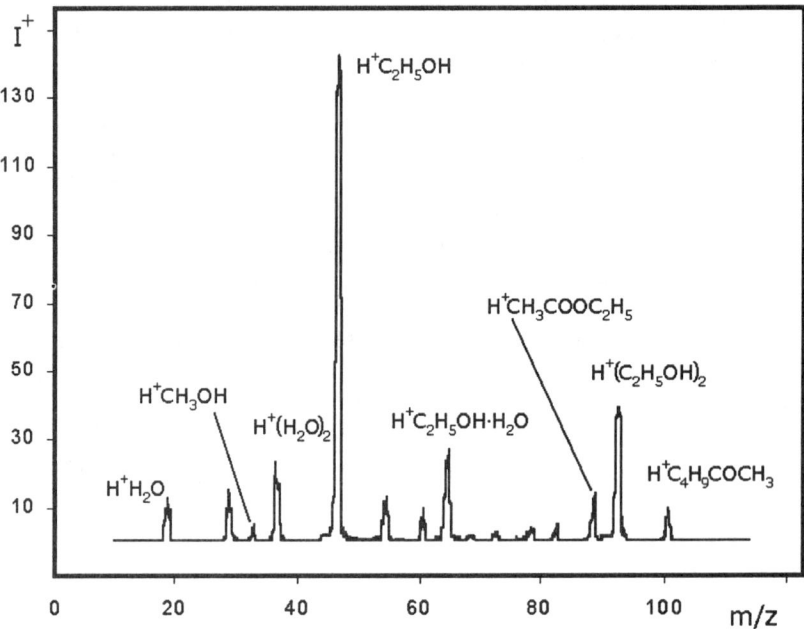

FIGURE 2. Mass spectrum from contaminated water particles. No signal was observed without the presence of particles.

ACKNOWLEDGMENTS

This work has been supported by the Atmospheric Chemistry Program of the U. S. Department of Energy (DOE) under Contract DE-AC06-76RLO 1830. Pacific Northwest National Laboratory is operated for DOE by Battelle Memorial Institute.

REFERENCES

1. Kogan Y., Molecular condensation nuclei, *Dokl. Akad. Nauk USSR*, **161**, pp. 388-391 (1965).
2. Mikheev, V. B., N. S. Laulainen, S. E. Barlow, M. Knott, and I. J. Ford, *J. Chem. Phys.* (submitted)

Experimental Studies of the Condensation of Alkali Metals

Frank T. Ferguson[a] and Joseph A. Nuth, III[b]

[a]*Department of Chemistry, Catholic University of America, Washington, DC 20064*
[b]*Code 691, NASA Goddard Space Flight Center, Greenbelt, MD 20771*

Abstract. In this paper we give a brief overview of a series of experiments focused on the condensation of the alkali metals. The condensation of sodium is currently in progress and results are given for recent work with lithium. Supersaturation values for lithium range from approximately 300 at 830 K to 7 at 1100 K. The measured supersaturations tend to be much higher than the predictions of Classical Nucleation Theory and seem to be more closely aligned with the predictions of Scaled Nucleation Theory.

INTRODUCTION

Small dust grains play an extremely important role in the life cycle of stars and in the chemistry and physics of the interstellar medium. The exact composition of these grains is unknown (1) but they must contain a significant fraction of refractory species (primarily Fe, Mg & Si) observed in the gaseous outflows of supernovae or around late-type, AGB stars. The nucleation of these species is an extremely important astrophysical process and is key to understanding both how these grains form and grow and to correlating this information with observational evidence of grain composition and crystal structure.

Unfortunately, there have only been a handful of studies on the nucleation of higher-temperature, refractory species (2). Most of these studies are characterized by high supersaturations and fluxes, low critical cluster sizes, and little agreement between data and Classical Nucleation Theory (CNT) (3). One exception is recent work on cesium. Experiments have been performed in a thermal diffusion cloud chamber over a fairly large temperature range and the data agree with the internally consistent classical theory in a manner similar to that seen with volatile species (4).

Because there is a lack of experimental data and a suitable comparison between data and theory cannot be made, our goal is to develop a larger database on the nucleation of refractory elements. Although our interest lies in the highly refractory species such as carbon, SiO and iron, (because of their astrophysical significance), we are currently studying the more volatile, alkali metals. Sufficient vapors of these materials can be generated at reasonable temperatures and our goal is to develop the experimental techniques and equipment to transition to the more refractory species. One interesting property of the alkali metals is that the vapors of these species have a signifcant population of dimers. Therefore, it should be interesting to see the effect these associated vapors have on the nucleation behavior.

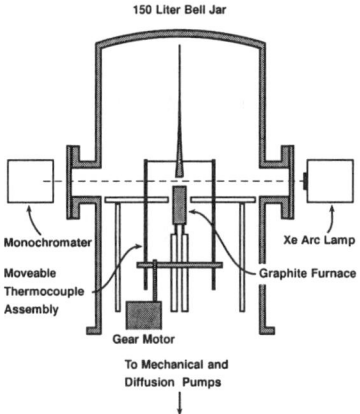 

Figure 1. Cross sectional diagram of the gas evaporation apparatus and an example of a lithium condensation plume. Above the heater there is a strong fluoresence due to the lithium dimer and a sharp transition from the vapor region to the particle rich region.

EXPERIMENTAL

The apparatus used to study alkali metal condensation is shown in Figure 1 and is similar to the system used in our laboratory for previous gas evaporation studies. Material is placed within a furnace assembly and heated under an approximately 200 torr, inert gas atmosphere until a buoyant plume of condensed material forms above the heater. There is a sharp transition between the vapor region and the particle-rich region in this plume. An example of a plume of lithium smoke particles is also shown in Figure 1. The temperature at the crucible source is measured with a fixed thermocouple while the temperature at the smoke cloud interface is measured using a thin-wire thermocouple stretched over the furnace assembly. This thermocouple is positioned so that the thermocouple junction is centered above the cylindrical furnace assembly and is mounted on a moveable assembly so that it can be raised or lowered to various locations above the furnace assembly.

When heating is first initiated, a faint orange fluoresence due to the 670.8 nm resonance line of lithium is visible under the illumination of the xenon arc lamp. Somewhat later (when the crucible temperature is high enough to yield sufficient vapor) a plume of condensed smoke particles forms as shown in Figure 1. When this plume forms there is a region of green fluorescence just above the crucible that extends to the base of the condensing plume. This glow is due to the fluoresence of the lithium dimer: similar behavior is also seen for sodium. This fluoresence is probably indicative of the region of highest vapor concentration. The ability to view the dimer fluoresence gave us some insight as to the behavior of the vapor transport process.

Previously, supersaturation values were estimated by assuming that the strong, upward convective currents curtailed the spread of the vapor and confined it to a column approximately the diameter of the crucible. It was also assumed that the vapor concentration at the condensation point was essentially identical to that at its crucible source. Now the supersaturation data is collected using a similar, but slightly modified model for the vapor transport based on observations of the smoke plume.

During a run, the temperature at the vapor source and condensation point is measured along with the width of the condensing plume. The concentration of the vapor at the source is assumed to be given by the equilibrium vapor pressure. If the width of the plume is comparable to the opening of the vapor source, then the vapor pressure at the condensation point should be approximately equal to the concentration at the source. Otherwise, a correction factor based on the ratio of the cross-sectional area of the plume to the cross-sectional area of the crucible is applied to the calculated pressure at the condensation point.

In order to test this hypothesis, the vapor concentration is measured using a second procedure. The alkali metals have strong absorption lines and these can be utilized to estimate the vapor concentration. In the bell jar apparatus a monochromater has been added to make such a concentration measurement. The monochromater is placed just below the point of typical condensation and it is assumed that the path length for absorption is equal to the width of the condensing plume. The absorption signal is measured and used to calculate the concentration of alkali metal just before the condensation point and this information is then used, in turn, to calculate the supersaturation.

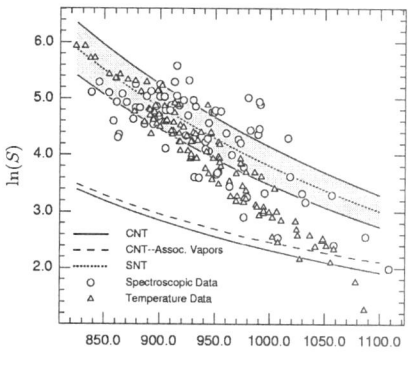

Figure 2. Supersaturation Data for Lithium. Shown in the plot are supersaturation data calculated via the spectroscopic method, (circles), and the temperature/model method, (triangles). Also shown in the plots are the predictions of Classical Nucleation Theory, Classical Nucleation Theory modified to account for associating species, and Scaled Nucleation Theory. The gray shaded area is the prediction of Scaled Nucleation Theory based on a critical temperature for lithium of 2700 ± 100 K.

RESULTS

Experiments with lithium have been performed and we are currently working with sodium. Shown in Figure 2 are supersaturation values for lithium ranging from approximately 300 at 830 K to 7 at 1100 K. In this figure, the open circles represent data taken using the spectroscopic measurements while the triangles are data taken by using the crucible temperature to estimate the initial vapor concentration and measurement of the plume width to calculate a vapor dilution correction factor.

As shown in the figure there is considerable scatter in the experimental results and we are currently working on ways of reducing these errors. The two sets of data are reasonably consistent, especially at the lower temperatures, but the temperature/model data tends to have a steeper drop with temperature than the spectroscopic data.

Using an estimated value of the flux, (10^9 cm^{-3}·s^{-1}), and the value of the surface tension for lithium at the melting point, the predictions of CNT are shown in Figure 2. With the exception of the very highest values, all of the supersaturation values lie well above the predictions of Classical Nucleation Theory.

Lithium vapor contains a small fraction of dimer in its equilibrium vapor. Katz, Saltsburg, and Reiss developed a modficiation to CNT to account for vapor-assocation and this theory predicts higher critical supersaturations in the absence of these associated species than if the monomer alone were present (5). To test whether the magnitude of this effect could account for the discrepancy between data and theory we have also plotted the results for this theory in Figure 2. Although the supersaturation values are indeed higher, they are certainly not as high as the data and this does not appear to be the source of the discrepancy between the data and CNT.

There has been some favorable agreement between refractory nucleation data and Scaled Nucleation Theory (SNT) (2,6). SNT depends upon the critical temperature for the material, which, in the case of many of the refractories, is unknown. Nevertheless, these critical temperatures can be estimated and the predictions of SNT with a critical temperature of 2700 K are shown in Figure 2. The gray shaded region represents a ±100 K variation in this critical temperature estimate. Although the scatter in the data prohibits making definitive statement, it does appear the experimental data are more closely aligned with the predictions of Scaled rather than Classical Nucleation Theory.

CONCLUSIONS

We are studying the condensation of alkali metals and earths in order to provide more information on the nucleation of higher temperature, refractory species. We are beginning to study the nucleation of sodium and have presented recent work on lithium. There is a considerable amount of scatter in the presented data and we are currently working to improve the quality of the dataset. However, the current results quite clearly differ from the predictions of CNT or modfications to CNT that account for association of the vapor. In contrast, much of the data at the lower temperature range tends to coincide with the predictions of SNT.

REFERENCES

1. Tielens et al., *Ap. & SS*, **255**, 415, 1998
2. Nuth, J.A., and Ferguson, F.T., *Ceramic Trans.*, **30**, 23, 1993.
3. Frurip, D.J., and Bauer, S.H., *J. Phys. Chem.*, **81**, 1001, 1977.
4. Rudek, M.M., Katz, J.L., and Uchtmann, H., *J. Chem. Phys.*, **110**, 11505, 1999.
5. Katz, J.L., Saltsburg, H., and Reiss, H., J. of Colloid and Interface Science, **21**, 560, 1966.
6. Hale, B. Kemper, P., Nuth, III, J.A., *J. Chem. Phys.*, **91**, 4314, 1989.

Homogeneous nucleation rates from the piston-expansion tube using a digital camera

Franz Peters* and Arne Graßmann

*Department of Mechanical Engineering,
University of Essen, 45127 Essen, Germany*

Abstract. Homogeneous nucleation rates of n-pentanol in nitrogen are obtained from a piston-expansion tube (pex-tube) involving the nucleation pulse method which generates a limited number of nuclei that grow into droplets. The detection of the droplets is achieved by a new counting method developed on the basis of a CCD camera in combination with a laser light sheet. Nucleation rates between 10^4 and 10^9 $cm^{-3}s^{-1}$ are covered for the nucleation temperature 260 K. The rates are plotted as isotherms versus supersaturation. An influence of the initial expansion temperature on the nucleation rate is identified. Literature data from other expansion experiments agree with our finding.

INTRODUCTION

Experimental techniques for the investigation of homogeneous nucleation from the supersaturated vapor phase are either based on rapid adiabatic gas expansions or on rather slow diffusion processes. Various research groups, as reviewed by Heist and He [1], have succeeded in tuning experimental set-ups such that the rate becomes measurable at which nuclei emerge from a well defined supersaturated state which is the nucleation rate. The most prominent among the expansion devices are the shock tube [2] the piston-expansion tube (pex-tube) [3] and the expansion chambers [4]. Shock tubes provide the fastest expansions (e.g. 2 ms) because their time response relies on pure gas dynamics. The benefit of the high expansion rate is that the assumption of an adiabatic process entering the determination of the nucleation state is most likely satisfied. The pex-tube combines a piston motion with 1-D expansion gas dynamics. As compared to the shock tube the piston slows the expansion down to about 5 ms. Adiabatic conditions can nevertheless be verified. The valve controlled expansion chambers release the high pressure chamber gas against a low pressure reservoir on about the time scale of the pex-tube. The nucleation state relies on adiabatic expansion and pressure measurement. All of the methods achieve a short nucleation state at the tail of the expansion appearing as a pressure dip of less than a ms. After the dip there is sufficient supersaturation of the vapor left such that the born nuclei grow into the μm range. The detection of the droplets is generally achieved by constant angle Mie-scattering in which the scattering amplitude yields the nucleation rate on basis of a rather difficult calibration. In this paper we show how the pex-tube method is applied with two innovations. We secure the nucleation state by piston travel eliminating the pressure transducer calibration. And we replace the Mie-scattering method by a direct visualization of the droplets introducing a light sheet technique in combination with a video camera. The data refer to n-pentanol in nitrogen.

EXPERIMENTAL METHOD

The working principle of the pex-tube is, in short, to subject a vapor diluted in a carrier gas to a fast expansion from an initially undersaturated to a short-lived supersaturation peak where nucleation occurs followed by droplet growth at a reduced saturation level. Realization of this principle is through the set-up shown in Fig.1.

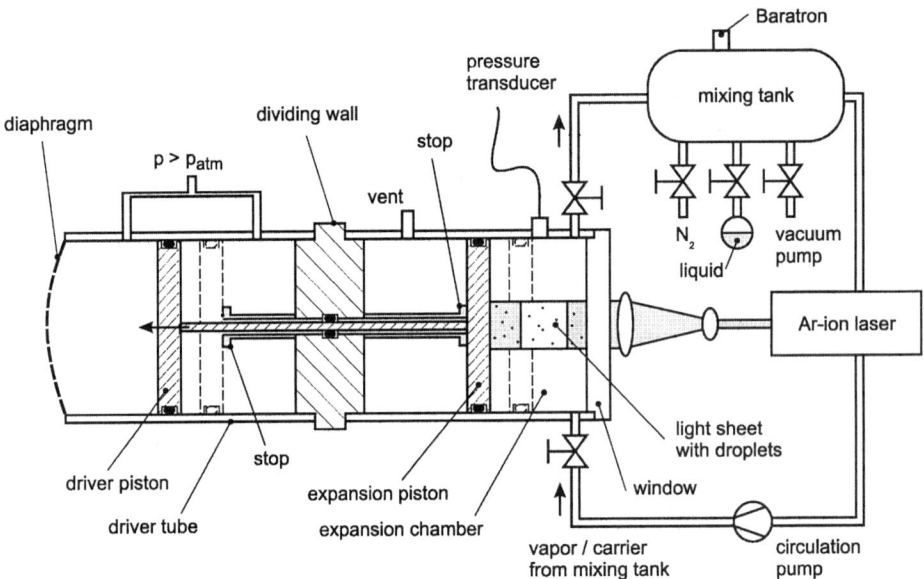

FIGURE 1. Experimental set-up of the piston-expansion tube

The instrument is, to a large extent, identical to earlier versions a detailed discussion of which is found in [3] . On the right hand side the tube (i.d. 70 mm), the expansion piston and the end wall window form the expansion chamber. A piston rod links the expansion to the driver piston which operates in the driver tube at the left hand side. Initially, the driver piston faces equal air pressure above atmospheric on either side. When the diaphragm is ruptured a pressure difference builts up moving both pistons. The expansion chamber enlarges at a dropping internal pressure. Everything is balanced such that the expansion piston barely makes it to the stop where it bounces off a little bit and rests.

The expansion pressure is monitored by a dynamic pressure transducer (Kistler 6031). Fig.2 displays a typical signal with initial pressure, the expansion and the pressure dip of less than 1 ms which is the nucleation period.

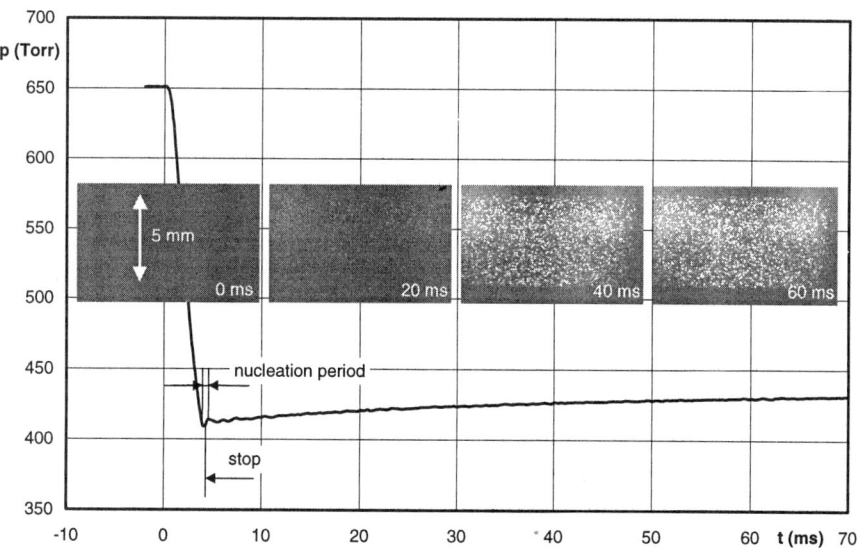

FIGURE 2. Pressure signal with video frames of condensing droplets.

When such a signal has been established the initial state and the nucleation state have to be secured. As to the initial state the pex-tube is temperature controlled by a feed-back heating system. The same control is applied to the mixing tank together with the connecting lines and the circulation pump (membrane type). The vapor/gas samples are prepared by partial pressure measurement in mixing tank and expansion chamber at open chamber valves. During this filling process the circulation pump maintains a flow which guarantees a homogeneous vapor/carrier gas composition. The partial pressures are measured by an accurate Baratron (390HA) transducer. When the preparation of the sample is finished the chamber valves are closed for a run. However, between the runs they are open for further circulation.

When the initial state is secured the nucleation state is inferred assuming an adiabatic expansion free of dissipation. In such an isentropic expansion the temperature T, the total pressure p and the vapor pressure p^v all depend on the volume expansion in the following way

$$\left[\frac{T}{T_0}\right]^{1/(1-\kappa)} = \left[\frac{p}{p_0}\right]^{-1/\kappa} = \left[\frac{p^v}{p_0^v}\right]^{-1/\kappa} = \frac{L_0 + \Delta L}{L_0} \qquad (1)$$

Subscript 0 denotes the initial state and κ the ratio of specific heats. At the nucleation state the ratio of the vapor pressure and the equilibrium vapor pressure $p_e(T)$ defines the supersaturation

$$S = \frac{p^v}{p_e(T)} \qquad (2)$$

Due to the fixed tube diameter, the volume expansion ratio may be expressed as a length ratio in which ΔL is the piston displacement and L_0 the initial length of the expansion chamber. The straightforward way to get T and p^v is to measure the pressure drop $\Delta p = p_0 - p$ and proceed from there which, of course, requires calibration of the transducer. Our experience has shown that, despite the most careful procedure, calibration is not sufficiently stable. That is why we switched over to the volume expansion ratio expressed as a length ratio. To be sure for each run that the piston has in fact travelled the distance ΔL the contact of the piston with the stop is monitored.

A serious problem always encountered in expansion experiments is heat transfer from the warm walls to the expanding cool gas questioning the validity of Eq.(1). To make sure that there is no interference with our nucleation state we have conducted comprehensive calculations based on the non-steady 2-D heat transfer equation.

The nuclei formed within the nucleation period grow rapidly into the μm-range where they become countable through visualization. The latter is achieved by the beam of an argon-ion laser which is spread to a light sheet by means of cylinder lenses and is directed through the end wall window into the nucleation chamber. A standard CCD-video camera (not shown) observes the droplets within the sheet from a direction normal to the sheet through a small side wall window. The camera operates at the video frequency of 25 Hz providing a single frame every 20 ms. The first frame is triggered simultaneously with the diaphragm rupture at 0 ms where the expansion begins. An example sequence of four images is inserted into Fig. 2. We see that the droplets start to become visible at 20 ms.

Evaluation of the droplet number is by counting by eye when there are only a few ones. When the number is high like in the frames of Fig.2 an image processing software is adapted as it is used in fluid mechanics for the measurement of the velocity of particles or droplets carried in a streaming gas. The droplet number divided by the observation volume which is determined by the sheet depth and the observation window, yields the number concentration. This, divided by the nucleation time read off the pressure trace gives the nucleation rate. Compared to the Mie-method this method extends the measurable range of nucleation rates by two orders of magnitude at the lower end so that this method covers five orders of magnitude.

RESULTS

The nucleation rate vs. S was measured for n-pentanol in nitrogen at 260 K . The n-pentanol was provided from the lot referred to by Rudek et al. [5]. Fig.3 presents the results in three series #1, #3, #5. The data is spread over roughly two orders of magnitude in nucleation rate equivalent to about two counts in supersaturation. The question is why #5 differs so much from #1 and #3. The main experimental difference between the series is the initial temperature of expansion which is 297.7 K for #1 and #3 and 310.1 for #5. Therefore, the higher initial temperature corresponds to the lower nucleation rate. In terms of the classical nucleation theory this is not to be expected. Yet, the effect is there and interestingly the literature data fit into the picture quite well. The series #4 is from an expansion chamber [4] with an intermediate initial temperature of 305.8 K. and the series #2 is from a shock tube [2] experiment starting at 293.7 K. Although these results were obtained with Ar and He, respectively, adding another parameter, we think that the initial temperature plays a role among the expansion experiments awaiting further investigation and explanation. The two curves in Fig.3 represent the classical nucleation theory for 260 and 270 K calculated following the paper by Rudek et al. [5].

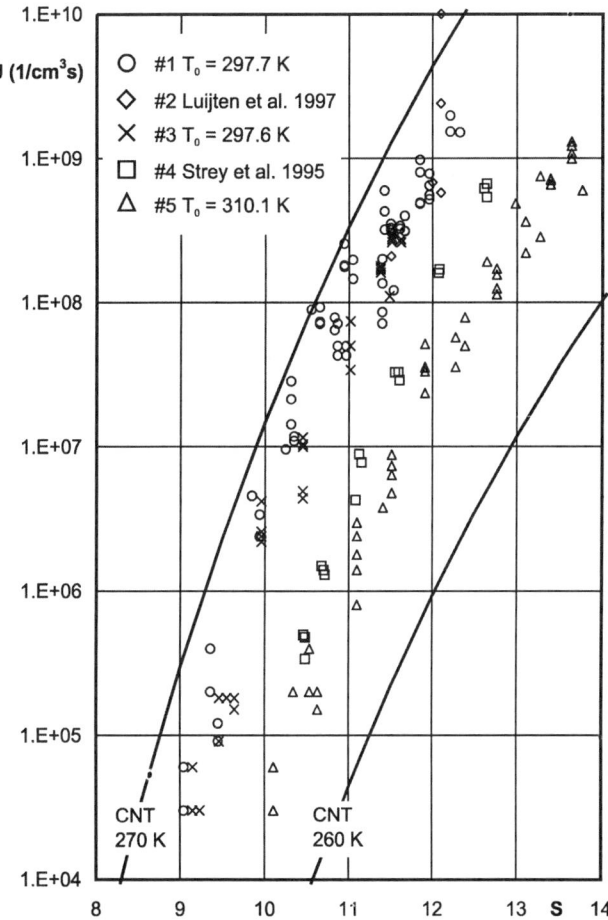

FIGURE 3. Nucleation rates for n-pentanol in nitrogen as function of supersaturation for the nucleation temperature 260 K.

REFERENCES

1. Heist,R.H., and He,H., *J.Phys.Chem.Ref.Data*, **23** (5), 781-805 (1994)
2. Luijten,C.C.M., Baas,O.D.E., and van Dongen,M.E.H., *J.Chem.Phys.* **106** (10), 4152-4156 (1997)
3. Peters,F., and Rodemann,T., *Exp.Fluids* **24**, 300-307 (1998)
4. Strey,R., Viisanen,Y., and Wagner,P.E., *J.Chem.Phys.* **103** (10) 4333–4345, (1995)
5. Rudek,M.M., Katz,J.L., Vidensky;I.V., Zdimal,V. and Smolik,J., *J.Chem.Phys.*,**111** (8), 3623-3629 (1999)

Homogeneous Nucleation of *n*-pentanol in a Laminar Flow Diffusion Chamber

H. Lihavainen, Y. Viisanen and M. Kulmala[*]

Finnish Meteorological Institute, Air Quality Research, Sahaajankatu 20 E, FIN 00810 Helsinki, Finland
Department of Physics, University of Helsinki, P.O. Box 9, FIN-00014 Helsinki, Finland

Abstract. As a contribution to the Joint Experiment on Homogenous Nucleation, a version of a laminar flow diffusion chamber was developed for nucleation rate measurements. The design and operational characteristics of the chamber will be presented. Homogenous nucleation rates of n-pentanol were measured as a function of saturation ratio in the temperature range between 260 K and 290 K. The results were compared to the classical nucleation theory. The experimental results were three orders of magnitudes higher than the theoretical predictions. The difference was almost constant over the whole temperature range. The results were compared with results from other experimental devices, they were in good agreement at lower temperatures.

INTRODUCTION

Nucleation can be studied experimentally using various methods based on different measuring principles. However, the comparison of experimental results from different devices is usually difficult because the different experimental methods are limited often to different substances, temperatures, pressures and nucleation rates. In order to make it possible to quantitatively compare measurements, in the workshop on Nucleation Experiments – State of the Art and Future Developments, Praque, 1995, a decision was made to start a series of Joint Experiment on Homogenous Nucleation. Its goal is that different research groups would perform measurements with the same substances and at the same conditions. As a part of this Joint Experiment, a version of a laminar flow diffusion chamber was developed to measure homogeneous nucleation rates as a function of saturation ratio. The nucleation rate range of the laminar flow diffusion chamber falls between the two commonly used devices in nucleation studies; static diffusion chamber and expansion cloud chamber.

EXPERIMENTAL TECHNIQUE

The developed laminar flow diffusion chamber (from now on referred as LFDC) consists from three main parts; saturator, preheater and condenser. In the saturator the carrier gas is fully saturated with the vapor under investigation by evaporation from the surface of a liquid pool. The stream of the mixture of the carrier gas and the vapor achieves steady and laminar flow profile in the preheater. The mixture is suddenly cooled in the condenser by heat exchange with the wall of the condenser. Because the equilibrium vapor pressure is a exponentially rising function of temperature, supersaturation is achieved. Because the heat transfer is faster than the mass transfer the nucleation zone occurs at the center of the tube, inside a volume where the saturation ratio and the temperature are quite constant. After the nucleation zone the vapor is still supersaturated and the particles can grow to a size where they are optically detectable. Particles are counted with the optical head of TSI's Condensation Particle Counter 3010.

Reliable measurement of flow, temperature and saturation ratio profiles from laminar flow is an impracticable task. They have to be calculated. It was assumed that the flow is steady, laminar and incompressible. The Reynols number in the measurements was typically about 140 so the tube flow was clearly laminar. The effect of radial velocity was assumed to be negligible and was set to zero. With these assumptions the equations of motion reduces to a form, which have been analytically solved resulting in a parabolic velocity profile (1). The temperature and vapor pressure profiles were solved numerically because satisfactory analytical solution does not exist. The experimental temperature and nucleation rate range was chosen so that the effects of aerosol dynamics on saturation ratio and temperature could be left out from the calculations. In the temperature profile calculations convective heat transfer in axial direction and heat transfer in axial and radial directions by conduction were taken into account. In vapor pressure profile calculation convective mass transfer in axial direction, mass transfer by diffusion and thermal diffusion in the axial and radial directions were taken into account. Because the heat and mass transfer equations are coupled, they were solved iteratively. The iterative solutions of finite element method calculations converged quite fast, after four iteration the difference between iterations was negligible. Figure 1 presents the calculated temperature, equilibrium vapor pressure, vapor pressure, saturation ratio and nucleation rate (according to the classical nucleation theory) profiles with boundary conditions typical for the measurements at the center of the tube. It can be seen that the temperature and therefore also the equilibrium vapor pressure drops much faster than the actual vapor pressure. This leads to supersaturation of the vapor. Maximum nucleation rate occurs slightly before the saturation ratio maximum because the nucleation is both function of saturation ratio and temperature. Inside the nucleation volume, where 90 % of the particles are formed the, the temperature changes about 1.5 K and saturation ratio about 3 %.

During the measurement of nucleation rate isotherm the temperatures of the preheater and the condenser were kept constant. The saturation ratio was controlled with the temperature of the saturator.

RESULTS

Experimental results were compared to the classical nucleation theory. Experimental output particle concentrations were plotted as a function of theoretical predictions (both axes on a logarithmic scale). A straight line was fitted to each particle concentration isotherm. If the theory predicts the experimental results correctly, the intercepts should be zero and the slopes of the lines should be one. The

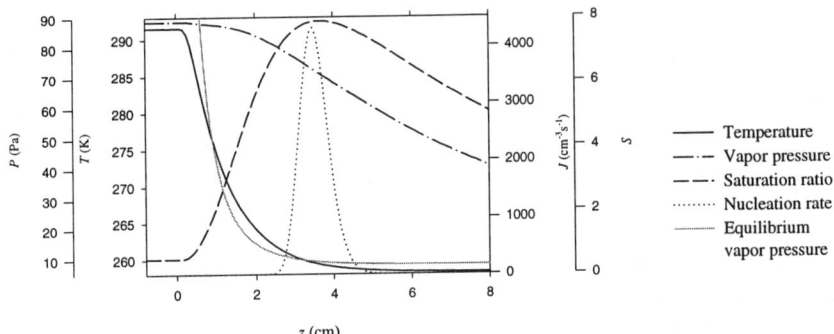

FIGURE 1. Temperature, T, equilibrium vapor pressure and vapor pressure, P, saturation ratio, S, and nucleation rate, N, profiles at the center of the condenser tube as a function of the tube length z.

intercepts and slopes from line fitting are presented in figure 2. The slopes are very close to unity. This implies that the results and the theory have the same saturation ratio dependency. The intercepts have nonzero value around three. Even though the absolute value is incorrect the intercepts do not have noticeable temperature dependency.

The results have to be converted to nucleation rates ($cm^{-3}s^{-1}$) as a function of saturation ratio in order to compare the results with other measurements. This was done using the same method that is used to convert the static diffusion chamber data (2). It gives the experimental nucleation rate as a function saturation ratio and temperature. The temperature and the saturation ratio are those in the thepretical nucleation rate maximum. Comparison was made with results from five different experimental apparatus; a two piston expansion chamber (3), nucleation pulse chamber (4), expansion wave tube (5) and two different static diffusion chambers (6). They all agree with this study that the saturation ratio dependence of the classical nucleation theory is well predicted. The temperature dependence is however different with different devices, see figure 3. Results from this study, the nucleation pulse chamber (4) and from the expansion wave tube (5) show that even though the absolute values of the nucleation rates are differ from the theoretical values, the deviations is quite constant over the measured temperature range. The results from the two-piston expansion chamber (3) and from the static diffusion chamber (6) show different temperature dependence, the dependence of the measured nucleation rates on temperature is weaker than predicted by theory. It is quite surprising that the results from this study and the results from the static diffusion chamber differ even though they rely on the same operation principle.

SUMMARY

A version of a laminar flow diffusion chamber was designed. A mathematical model was developed to analyze the events inside the chamber. Nucleation rates of n-pentanol were measured as a function of saturation ratio at temperature range from 260 K to 290 K. The results were compared to the classical nucleation theory. The saturation ratio dependency of the measured nucleation rates was the same that of the theory. This was consistent with the results from five different experimental devices. Data from different devices and the data from this study were consistent at lower temperatures.

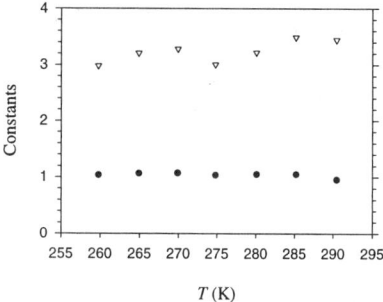

FIGURE 2. The constants from line fitting to $\log N_{exp}$ vs. $\log N_{theor}$ curves as a function of the temperature at the theoretical nucleation maximum. The white symbols are the intercepts and the black symbols are the slopes.

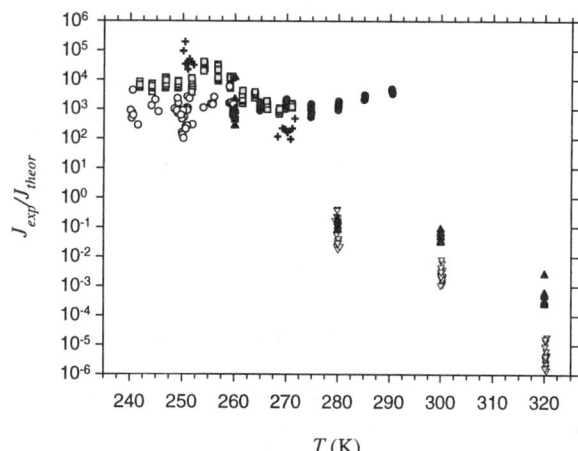

FIGURE 3. Experimental nucleation rates divided by theoretical predictions plotted as function of nucleation temperature. The black circles = this study, the crosshairs = two piston expansion chamber (3), the grey squares = nucleation pulse chamber (4), the white circles = expansion wave tube (5), the black upward triangles = static diffusion chamber I (6), the grey downward triangles = static diffusion chamber II (6).

REFERENCES

1. Bird, R. B., Stewart, W. E. Stewart, and Lightfoot, E. N, *Transport Phenomena*, New York, John Wiley & Sons, 1960, ch. 2, pp. 42- 47.
2. Hung, C., Krasnopoler M. J., and J.L. Katz, *J. Chem. Phys.* **90**, 1856-1865 (1989)
3. Strey R., Wagner P.E., Scheming T., *J. Chem. Phys* **84**, 2325-2335 (1986)
4. Hrubý, J., Viisanen, Y., and Strey, R:, *J. Chem. Phys.* **104**, 5181-5187 (1996)
5. Lujiten, C. C. M., Baas, O. D. E., and van Dongen, M. E. H., *J. Chem. Phys.* **106**, 4152-4156 (1997)
6. Rudek, M. M., Katz, J. L., Vidensky, I. V., Zdímal, V., and Smolík, J., *J. Chem. Phys.* **111**, 3623-3629 (1999)

Steady State Homogeneous Nucleation Rate and Primary Particle Size Distribution

Richard B. McClurg

Department of Chemical Engineering and Materials Science, University of Minnesota, Minneapolis, MN 55455-0132

Abstract. A self-consistent, analytic model for the steady state homogeneous nucleation rate and primary particle size distribution is presented. Classical homogeneous nucleation models give the rate of particle production, but are silent on the resulting primary particle size distribution. This size distribution is important for determining the dynamics of the resulting particles in the environment or for industrial applications. When the evolution of the particle size distribution has been considered, nucleation has generally been decoupled from subsequent particle dynamics. Therefore, a size distribution for the primary particles had to be assumed. The present model is an analytic solution of the steady state cluster concentration evolution equations, including terms for particle transport from the region where nucleation is occurring. Therefore, both the nucleation rate and primary particle size distribution are calculated on the same basis. The classical homogeneous nucleation rate equations are a special case of this more general model with the particle transport terms set to zero.

INTRODUCTION

Classical homogeneous nucleation models give the rate of particle production, but are silent on the resulting primary particle size distribution. This size distribution is important for understanding the dynamics of the resulting particles in the environment [1,2] and in industrial applications [3]. When the evolution of the particle size distribution has been considered, nucleation has generally been decoupled from the subsequent particle dynamics. Therefore, a size distribution for the primary particles had to be assumed [2]. In this work, the cluster balance equations contain terms for particle transport that allow for the primary particle size distribution to be estimated on an equal basis with the nucleation rate.

MODEL EQUATIONS

The species balance for clusters of size i is

$$\frac{dC_i}{dt} = k_{1,i-1}^a C_1 C_{i-1} - k_{1,i-1}^f C_i - k_{1,i}^a C_1 C_i + k_{1,i}^f C_{i+1} - h_i C_i \qquad (1)$$

where C_i is the concentration of clusters of size i, t is the time, $k^a_{n,m}$ is the rate constant for aggregation of clusters of size n and m, $k^f_{n,m}$ is the rate constant for fission of a larger cluster into clusters of size n and m, and h_i is the rate constant for transport of clusters of size i from the region where nucleation is occurring. As for most nucleation calculations, Equation (1) is based on a Markovian growth mechanism wherein clusters grow and decay in steps of one monomer each. The inclusion of h_i makes the present model different from classical nucleation models that do not account for particle transport via gravitational settling, diffusion, electro-magnetic fields, or phoresis. Equations for calculating h_i for each of these mechanisms can be found elsewhere [4]. Clusters that leave the nucleation region are called primary particles. The total flux of primary particles is the nucleation rate (J).

$$J = \sum_i h_i C_i \qquad (2)$$

MODEL SOLUTION AND RESULTS

To solve Equations (1) and (2) for the steady state cluster concentrations and nucleation rate, it proves useful to define three functions,

$$\theta_i = k^a_{1,i} C_1^{eq} C_i^{eq} S^{i+1} \qquad (3)$$

$$\varphi_i = h_i C_i^{eq} S^i \qquad (4)$$

$$\gamma_i = C_i / (C_i^{eq} S^i) \qquad (5)$$

to simplify subsequent results. Here S, is the supersaturation ratio which is the ideal gas monomer activity relative to full thermodynamic equilibrium at saturation conditions.

$$S = C_1 / C_1^{eq} \qquad (6)$$

The equilibrium cluster concentrations (C_i^{eq}) can be estimated using any of a wide variety of cluster models [5,6], or simulation techniques [7] described elsewhere.

Using the principle of microscopic reversibility, Equations (1) through (6), and a bit of algebraic manipulation, the steady state nucleation rate is expressed as a continued fraction.

$$J = \varphi_1 + \cfrac{1}{\left(\cfrac{1}{\theta_1}\right) + \cfrac{1}{\varphi_2 + \cfrac{1}{\left(\cfrac{1}{\theta_2}\right) + \cfrac{1}{\varphi_3 + \cfrac{1}{\ddots}}}}} \qquad (7)$$

The cluster concentrations in the nucleation region are then calculated iteratively using Equation (5) and

$$\gamma_{i+1} = \gamma_i - R_i / \theta_i \tag{8}$$

$$R_{i+1} = R_i - \varphi_{i+1} / \gamma_{i+1} \tag{9}$$

starting with
$$\gamma_1 = 1 \tag{10}$$

$$R_1 = J - \varphi_1 \tag{11}$$

Finally, the flux of primary particles of size i from the nucleation region is

$$J_i = h_i \gamma_i C_i^{eq} S^i \tag{12}$$

DISCUSSION

Details of the solution method, a continuous function approximation to Equations (7) and (12), example calculations, and comparisons with classical nucleation rate predictions will be published elsewhere.

The particle size distribution affects transport and aggregation kinetics as well as particle properties. It is my hope that the ability to self-consistently calculate both the nucleation rate and the primary particle size distribution will aid the modeling, simulation, and design of nucleation processes in the environment and for industrial applications.

ACKNOWLEDGMENTS

I gratefully acknowledge support for this research from the Shell Oil Foundation as a Shell Faculty Fellow.

REFERENCES

1. Ping, J. I., and Harrison, R. M., *Env. Sci. Tech.* **33**, 3730-3736 (1999).
2. Pirjola, L., Kulmala, M., Wilck, M., Bischoff, A., Stratmann, F., and Otto, E., *J. Aerosol Sci.* **30**, 1079-1094 (1999).
3. Wu, C. Y., and Biswas, P., *Env. Eng. Sci.* **17**, 41 (2000).
4. Seinfeld, J. H., and Pandis, S. N., *Atmospheric Chemistry and Physics*, New York: Wiley, 1998, pp. 465-484.
5. McClurg, R. B., and Flagan, R. C., *J. Chem. Phys.* **201**, 194-199 (1998).
6. Zeng, X. C., and Oxtoby, D. W., *J. Chem. Phys.* **94**, 4472-4478 (1991).
7. Senger, B., Schaaf, P., Corti, D.S., Bowles, R., Voegel, J.C., and Reiss, H., *J. Chem. Phys.* **110**, 6421-6437 (1999).

Nucleation Rates Of Lennard-Jones Clusters From Growth And Decay Simulations

Hanna Vehkamäki[1] and Ian J. Ford [2]

[1] Department of Physics, P.O. Box 9 00014 University of Helsinki, Finland
[2] Dept. of Physics and Astronomy, University College London, Gower Street, London WC1E 6BT, U.K.

Abstract. We have studied singles clusters of Lennard-Jones atoms using a novel Monte Carlo simulation technique. We computed canonical ensemble averages of the grand canonical growth and decay probabilities of the cluster as a function of the cluster size. The critical size is identified as the one for which growth and decay are equally probable. The size and average internal energy the critical cluster was found for different temperatures and vapour chemical potentials. We used this information together with nucleation theorems to predict the behaviour of the nucleation rate as function of the two external parameters. Our results are in line with the results found in the literature, and roughly correspond to the predictions of classical theory.

INTRODUCTION

A vapour can be described as a collection of free molecules and quasi-bound molecular clusters which are gaining and losing molecules at various rates. Small clusters are more likely to decay than grow, and this makes it possible to understand how a supersaturated vapour, one that is thermodynamically metastable with respect to a condensed phase, can be maintained in existence. The phase transformation is impeded since it must proceed through the formation of these relatively unstable small molecular clusters. However, the ratio of growth to decay probability, per unit time, increases with cluster size. Viewing the clusters as tiny versions of continuum droplets in thermal equilibrium, the size dependence of the growth and decay probabilities is easily understood. There is competition between the free energy cost of creating the droplet-vapour interface, and the bulk reduction in free energy afforded by the phase transformation. For the so-called critical size, the probabilities of growth and decay are equal. Since growth and decay are stochastic, an individual cluster can reach the critical size through improbable sequences of molecular acquisitions. The formation of critical clusters results is key to the phenomenon of nucleation, where droplets appear from a supersaturated vapour. This common but inadequately understood phenomenon has been a subject of numerous theoretical studies, aimed at interpreting the growing body experimental data.

THEORY

We have developed a novel Monte Carlo simulation technique to obtain the critical cluster information. Having tested the method for the case of new phase nucleation in the Ising model of interacting spins [1], we have gone on to calculate the averaged growth and decay probabilities for clusters of Lennard-Jones atoms in a grand canonical ensemble [2]. To enhance statistics, we actually obtain the averages of grand canonical growth and decay probabilities within a canonical scheme. The decay probability of a certain configuration is the average annihilation probability over all the atoms in the cluster. The growth probability is the average creation probability over N_{max} randomly chosen sites around the cluster, where the number density of these sites is around 100 times the liquid density of Lennard-Jones fluid. During simulation we obtain the canonical ensemble averages of these decay and growth probabilities. We also evaluate the canonical ensemble average of the cluster binding energy for each cluster size. We identify the critical cluster as the size for which the growth and decay probabilities are equal. We exploit the nucleation theorems to determine the variation in nucleation rate as the conditions are changed. The first nucleation theorem gives the number of molecules in the critical cluster, n^c as

$$n^c = (\partial J/\partial S)_T - (\partial J_0/\partial S)_T$$

Where $J = J_0 \exp(-W_c/kT)$ is the nucleation rate, W^c is the formation free energy of the critical cluster, S is the saturation ratio, T is the temperature and k is the Boltzmann constant. The second nucleation theorem gives the excess energy (compared to the energy the constituent molecules would have in the bulk liquid at the same temperature) of the critical cluster E_x^c as

$$E_x^c = [(\partial J/\partial T)_S - (\partial J_0/\partial T)_S]kT^2$$

We get n^c and E_x^c from simulations; J_0 can be evaluated using the know thermodynamic properties of Lennard-Jones fluid. Also chemical potential μ which besides the temperature is the other parameter entering our simulations, can be related to the chemical potential difference $\Delta\mu = \mu_v(P_v) - \mu_l(P_v)$ and further to saturation ration using the equation of state of Lennard-Jones vapour. $\mu_v(P_v)$ and $\mu_l(P_v)$ are the chemical potentials of vapour and liquid, respectively, at the pressure of the vapour P_v. We performed two sets of simulations: one to get the critical size as a function of chemical potential difference at a constant temperature, and the other to get the critical size as a function of temperature with constant $\Delta\mu/kT$. To allow comparison with literature values, we have used truncated and shifted Lennard-Jones potential with cut off distance 2.5σ for the first set and full Lennard-Jones potential for the second set. Lennard-Jones parameters defining the length scale and the energy scale are denoted by σ and ε, respectively. To obtain the actual nucleation rates rather than only the derivatives of them we need a reference rate at one temperature and supersaturation. We take these reference nucleation rates from literature. We have defined the cluster as a connected network of neighbours, where two atoms are neighbours if their centers are less than 1.5σ apart.

RESULTS

Figure 1 shows the results of a set of simulations for $\Delta\mu/kT=0.702$ for the truncated and shifted potential. The smooth curve fitted to the growth probability over the decay probability data indicates that the critical size is about 37. The energy starts to fluctuate at larger sizes due to in sufficient statistics. If a more accurate estimate for critical cluster energy is needed, it can be obtained with extended simulations for this size only. Figure 2 shows the critical size and critical cluster energy as a function of the chemical potential difference. The uncertainty in the critical size is less or equal to ±1. The results of our simulation are compared with the results of ten Wolde et al., as well as with classical prediction. We obtain a fairly good agreement with the lowest critical sizes studied by ten Wolde et al. [3], bearing in mind that we define the cluster in a slightly different way. The classical theory agrees reasonably well with both sets of simulations. Figure 3 shows the comparison between nucleation rates obtained using nucleation theorem with a result of ten Wolde et al. [4] as a reference rate and classical nucleation theory. The lowest chemical potential we studied is $\Delta\mu/kT=0.6$ and so we have to extrapolate the nucleation rate from the reference point up to this point using classical predictions for the critical cluster size. Ten Wolde et al. [3] have shown the the classical critical cluster size is valid up to about $\Delta\mu/kT=0.65$ and so your extrapolation is justified. We also show that using classical critical cluster size for the whole chemical potential range results in a nucleation rate curve with the same slope as the classical curve, as it should. Ten Wolde et al. [3,4] use the truncated and shifted potential, and report the thermodynamic properties of the truncated and shifted Lennard-Jones fluid at $kT/\varepsilon=0.741$.

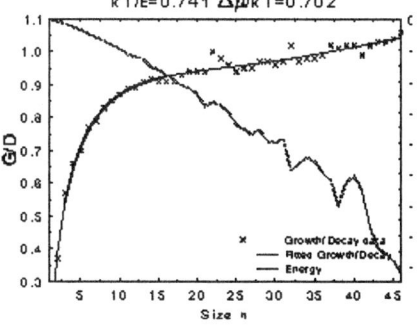
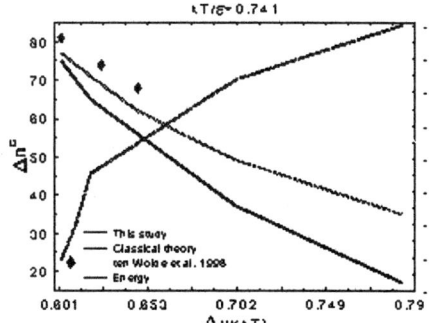

FIGURE 1. The ratio of average growth rate (G) to decay rate (D), and the average energy as a function of size for Lennard-Jones clusters.

FIGURE 2. The critical size and the average internal energy of the critical cluster as a function of chemical potential difference.

This data is used in classical predictions and in converting chemical potential to chemical potential difference and saturation ratio.. Figure 4 shows the nucleation rate as a function of temperature when $\Delta\mu/kT = 1-39305$ is kept constant. The corresponding saturation ratio is also shown. Thermodynamic data found in the literature for temperatures other than $kT/\varepsilon = 0.741$ refer to the full potential [5,6], and for

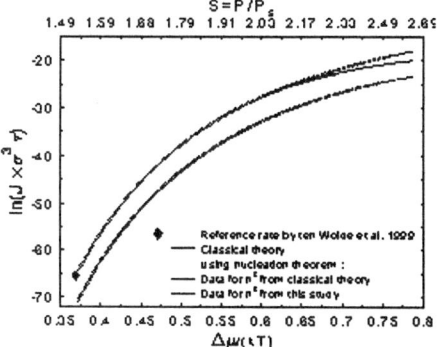
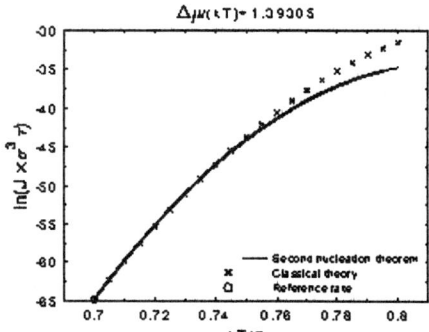

FIGURE 3. The nucleation rate as function of chemical potential difference. The temperature is $kT/\varepsilon = 0.741$.

FIGURE 4. The nucleation rate as a function of the temperature. The classical nucleation rate at is $kT/\varepsilon = 0.7$ is used as the reference rate in the method based on nucleation theorems

this reason the full potential is used in the second set of simulations. This leads to the fact that we can not use the reference rate from ten Wolde et al.[4], but we have chosen to use the classical result for the nucleation rate at $kT/\varepsilon = 0.7$ as the reference rate. At lower temperatures, where the critical size agrees well with the classical results, also the classical nucleation rate follows the same slope as the results from the nucleation theorem. At high temperatures the deviation becomes significant.

CONCLUSIONS

We have presented a new technique for obtaining the relative probabilities for the growth and decay of molecular clusters in a vapour. The size for which growth and decay probabilities are equal is identified as the critical size. For Lennard-Jones atoms the critical sizes obtained using our method are consistent with literature values and classical theory predictions. We use the critical cluster size and energy, together with two nucleation theorems, to determine the behaviour of the droplet nucleation rate when the temperature and vapour supersaturation are changed. We can focus our simulation only on sizes around the critical size and thus the simulations are computationally effective. The price paid is that we then only know the derivatives of the nucleation rate, and need a reference case to predict the absolute value of the nucleation rate. We use literature values for the reference rate. The simulation method and analysis can be extended to more complicated systems, for example molecules such as water, and multicomponent clusters. Also, the cluster definition we used can be easily modified. Our method has proved to be an effective way to gather information about nucleating clusters. We have therefore demonstrated the power of nucleation theorems in the analysis of molecular simulations.

ACKNOWLEDGMENTS

The authors thank I. Kusaka for providing a grand canonical Monte Carlo code used as a starting point yin the work. Thanks are due to P.R. ten Wolde for giving us the additional pressure-density data for Lennard-Jones vapour. We thank the Center for Scientific Computing (CSC) in Espoo, Finland for computer time. The work was funded by the U.K. Engineering and Physical Sciences Research Council under grant GR/L78499 and by the Academy of Finland under project number 64314.

REFERENCES

1. Vehkamäki, H., and Ford, I. J., Phys. Rev. E 59, 6483-6488 (1999).
2. Vehkamäki, H., and Ford, I. J., J. Chem. Phys. 112, 4193-4202 (2000).
3. ten Wolde, P. R. and Frenkel, D., J. Chem. Phys. 109, 9901-9918 (1998).
4. ten Wolde, P.R., Ruiz-Montero, M. J. and Frenkel, D., J. Chem. Phys. 110, 1591-1599 (1999).
5. Lotfi, A., Vrabec, J. and Fischer,.J., Mol. Phys. 76, 1319-1333 (1992).
6. Mecke, M. and Winkelmann, W., J. Chem. Phys. 107, 9264-9270 (1997).

Dynamics of Argon Clusters from Molecular Dynamics Simulations

Stephan Wonczak and Reinhard Strey

Institute for Physical Chemistry, University of Cologne
Luxemburger Straße 116, D-50939 Köln, Germany

Abstract. The understanding of nucleation processes is important for many natural phenomena and industrial processes. Real experiments are limited when it comes to investigation of the early stages of nucleation, because current methods cannot detect the very small clusters involved there. Here molecular dynamics simulations provide a useful tool for examining this part of the process. Small clusters and subsequently larger droplets can be seen developing from a uniform gas. The condensation and evaporation rates of the small clusters, which play an important role in many nucleation theories, can also be gained from molecular dynamics simulations.

INTRODUCTION

It is important to understand the physics of the nucleation process. Many experimental [1,2] and theoretical [3 - 6] studies, to name a few, are performed to this end. Current experiments are unable to observe the nucleation process directly due to the small size of the nuclei, which usually comprise 10 to 100 monomers. Nucleation theories are limited by certain idealizing assumptions. Here molecular dynamics simulations provides a useful tool for obtaining information about the nucleation process. In addition it gives data on the dynamics of the early stages of the growth of the particles formed during the nucleation process.

It is advisable to study a simple substance such as argon first before tackling more complicated compounds like water. With current computers it is possible to simulate systems with 10^4 - 10^5 argon atoms on time scales of several nanoseconds without getting to excessively long simulation times.

SIMULATIONS

We model argon by a Lennard-Jones gas with pair-wise interactions. The parameters for the atoms [7] were set to $\sigma = 0.3405$ nm and $\varepsilon/k_B = 120$ K. The simulations were performed with 4096 atoms in a box with a side length of 20 nm and periodic boundary conditions. The system was thermostated by simple velocity scaling.

One goal of the simulations was to mimic a real experiment in an expansion chamber [2]. Here the gas is subjected to an adiabatic expansion which leads to a temperature drop. The result is a supersaturated vapor which then nucleates and generates small droplets which are subsequently detected by light scattering. After a small interval the nucleation process is quenched by a small recompression. The droplets so obtained are subsequently detected by light scattering.

There are some problems reconstructing this experiment in simulation, however. The time scale of the real experiment in expansion chambers is on the order of milliseconds. The interval for measurements of the nucleation rate in real systems is between 10^2 cm^{-3}s^{-1} and 10^{10} cm^{-3}s^{-1}. The fastest expansions occur in supersonic nozzles where nucleation takes place in microseconds and rates of the order of 10^{18} cm^{-3}s^{-1} are reached. To actually see a nucleation event in simulation with these parameters one would have to simulate exceedingly large systems for very long times, which is clearly beyond the capacity of current computers. One solution for this problem is to go to very high supersaturations and thus very high nucleation rates. To do this in the simulation, homogeneous argon vapor was prepared at 150 K, which was then cooled down to 80 K instantaneously by rescaling the atomic velocities to the desired value. This results in supersaturated argon vapor at a pressure of 5.15 bar, while the experimental equilibrium vapor pressure at 80 K is 0.434 bar giving a supersaturation of 11.87. The system then was allowed to relax to its new equilibrium, and the nucleation and subsequent growth of small droplets was observed. Figure 1 shows a visualization of such an experiment.

The nucleation rate can be gained from the number of droplets that developed in the simulation box and the time elapsed before they became discernible. Figure 2 shows the size and growth of biggest cluster in the system, from which the nucleation time is estimated to be on the order of 1 ns, while from Figure 1c it can be seen that four clusters eventually grow to larger size. From these data the nucleation rate is estimated to 10^{26} cm^{-3}s^{-1}. This rate is much higher than those of present experimental work.

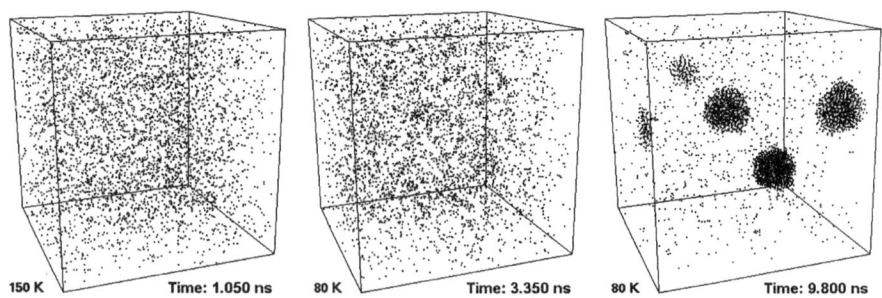

FIGURE 1. Visualizations of nucleating argon vapor. a) homogeneous gas at 150 K, b) density fluctuations and nuclei after cooling down to 80 K, c) growing droplets

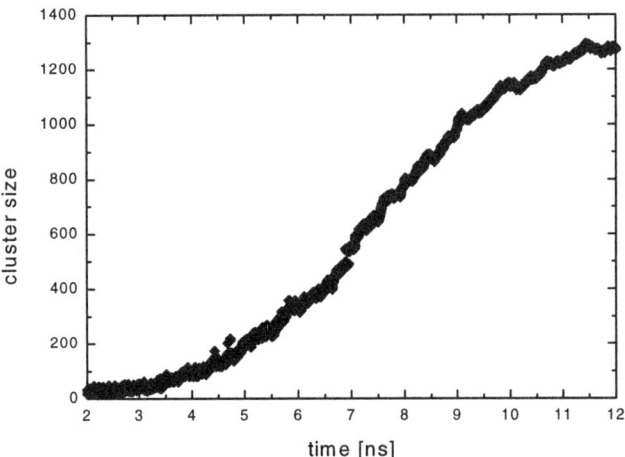

FIGURE 2. Size of the biggest cluster vs. time. Only the time after the jump to 80 K is shown. The jump in the curve at 7 ns denotes a coagulation event: A smaller cluster impinged upon a larger cluster and was incorporated into it.

A unique feature of simulations is the complete reversibility with respect to time. This way, one can easily identify a well-defined cluster, then go backwards in time and identify the critical cluster and examine the actual process that leads to its formation.

For the conditions of simulation the Gibbs-Thomson equation predicts a critical cluster size of $n^* = 10.5$ particles. As can be readily verified from the simulation data, a cluster this small has a very high probability of evaporating again. Only when a size of about 20 particles has been reached, the cluster has an even chance of growing to macroscopic size.

Another interesting feature from the simulation is the fact that the condensation and evaporation rates of the clusters, which play an important role in many molecular theories of nucleation, are readily accessible at all times. Figure 3 shows a cluster exchanging atoms with the surrounding gas. At the start of the simulation, all atoms belonging to the cluster are marked in black, while the particles belonging to the gas are shown in white. With progressing time, more and more atoms are exchanged between gas and cluster until an equilibrium distribution is reached, as can be seen in Figure 4a as well. The rate with which the atoms are evaporating and condensing on the cluster remains constant over time, however; i.e. the cluster size fluctuates around 3680. The fluctuating nature of the process becomes obvious if one considers the number of atoms actually exchanged in time intervals of 5 picoseconds (c.f. Fig. 4b).

FIGURE 3. Visualization of the exchange of cluster atoms with the surrounding gas. a) Situation at the start of the simulation run. The cluster atoms are marked black, the surrounding gas is shown in white. b) Situation after 150 ps. Some gas atoms have been incorporated into the cluster, and some cluster atoms have evaporated. c) Situation after 1 ns. Almost all of the gas atoms have been incorporated into the cluster; the gas is now composed almost exclusively from former cluster atoms.

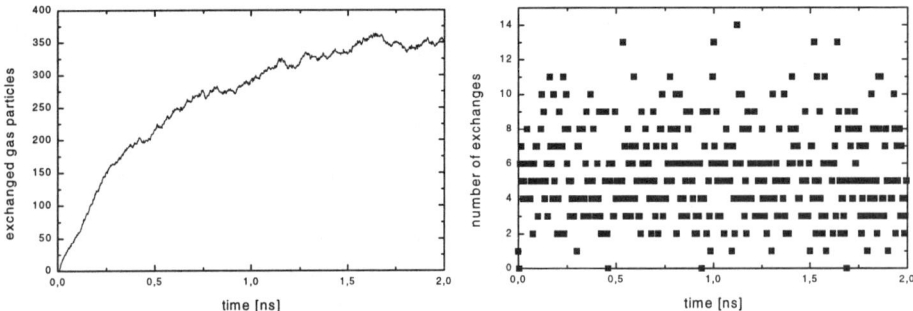

FIGURE 4. a) Number of atoms originally part of the cluster exchanged between cluster and gas. An equilibrium is reached after about 2 ns. b) Number of atoms exchanged in 5 ps intervals beween cluster and gas. The average rate is about one atom per picosecond.

REFERENCES

[1] Heist, R.H., He, H., *J. Phys. Chem. Ref. Data* **23**, 781 (1994) and references cited therein
[2] Strey, R., Wagner, P.E., Visanen, Y., *J. Phys. Chem.* **98**, 7748 (1994)
[3] Becker, R., Döring, W., *Ann. Phys.* **24**, 719 (1935)
[4] Oxtoby, D.W., *J. Phys.: Condens. Matter* **4** 7627 (1992)
[5] Senger, B., Schaaf, P., Corti, D.S., Bowles, R., Voegel, J.-C., Reiss, H., *J. Chem. Phys.* **110**, 6421 (1999); Senger, B., Schaaf, P., Corti, D.S., Bowles, R., Pointu, D., Voegel, J.-C., Reiss, H., *J. Chem. Phys.* **110**, 6438 (1999)
[6] Schaaf, P., Senger, B., Voegel, J.-C., Reiss, H., *Phys. Rev. E* **60**, 771, 1999
[7] Haile, J.M., *Molecular Dynamics Simulation*, New York: John Wiley & Sons, 1992, p. 462

Propandiol Vapor Nucleation Rates

M. P. Anisimov, A. G. Nasibulin, L. V. Timoshina, and J. A. Koropchak[+]

Nucleation Laboratory at Kemerovo, Institute of Catalysis SB RAS
41A-119 Moskovskiy Prospect, 650065 Kemerovo, Russia.
[+]*Southern Illinois University, Carbondale, Illinois 62901-4409, USA*

Abstract. Consideration of vapor-gas nucleation as binary vapor nucleation (instead widely used the one component approximation for nucleation of this system now) may lead the progress in the development of nucleation theory. Observations of phase transitions initiated by the carrier gas in the critical embryos of condensate can be a sufficiently convincing argument in this discussion. In order to confirm the role of the carrier gases received in the recent research [1], in present study 1,2-propanediol and 1,3-propanediol vapor nucleation rates were measured. Carbon dioxide (T_c=304.2 K, P_c=7.39 MPa) and sulfur hexafluoride (T_c = 318.7 K, P_c= 3.75 MPa) were chosen as the carrier gases, because of their low and convenient critical temperatures, T_c, and critical pressures, P_c. Analysis of the experimental data shows that gas-carrier molecules are involved in new phase embryo formation. Vapor nucleation of investigated substances in a carrier gas atmosphere can be considered as nucleation of binary system

INTRODUCTION

Vapor condensation, boiling, crystallization, or haze formation, polymerization, etc. – all of these events are examples of nucleation, i.e. new phase generation. Nucleation is a subject of multiple theoretical and experimental studies. Currently, the most significant progress in the theory and experiment has been for vapor nucleation. Classical nucleation theory is a significant step in the development of new phase formation description. But the agreement between experimental rates and theory predictions are satisfactory only for some experimental results on nucleation. This suggests that the theory have significant axiomatic inconsistencies. It is well known that any vapor or gas has a critical point. The critical line connects the critical points of two individual components. Obviously, the condition of a nucleation must be below the critical conditions, because there are no heterogeneous states at and above the critical conditions. Consideration of vapor-gas nucleation as binary vapor nucleation (instead widely used the one component approximation for nucleation of this system now) may lead to progress in the development of nucleation theory.

In order to alleviate these contradictions, it is necessary to carry out experiments under conditions where the influence of the carrier-gas is obvious. Observations of phase transitions initiated by the carrier gas in the critical embryos of condensate can be a sufficiently convincing argument in this discussion. It is known that at the first order phase transition temperature, the chemical potential of a condensed phase has a singularity. The free energy of critical embryo formation and therefore, the vapor nucleation rate, reflect this singularity. In other words, the chemical potential singularity is the basic reason to find experimentally a singularity of the nucleation rate surface at the condition of critical embryo phase transition. The vapor nucleation rate should have an experimentally detectable response to the non-monotone temperature behavior of the free energy of new phase critical embryos. As can be seen in the example of glycerin-carbon dioxide system nucleation, a nucleation rate singularity can be experimentally observed [1]. This singularity can be detected in the measurement of the vapor nucleation rates with high resolution in the nucleation rate and temperature.

In order to confirm the role of the carrier gases received in previous research [1], in present study 1,2-propanediol and 1,3-propanediol vapor nucleation rates were measured. Carbon dioxide (T_c=304.2 K, P_c=7.39 MPa) and sulfur hexafluoride (T_c = 318.7 K, P_c= 3.75 MPa) were chosen as the carrier gases, because of their low and convenient critical temperatures, T_c, and critical pressures, P_c. These parameters allowed variation of experimental conditions in the nucleation temperature from below to above the critical temperatures of the gases. It is known that these gases have no singularities of heat-mass transfer constants over the pressure interval $P = 0.1 \div 0.3$ MPa.

II. EXPERIMENTAL SET UP

Aerosol formation was experimentally studied in a flow diffusion chamber (FDC), which is suitable to measure the nucleation rate at different total pressures and nucleation temperatures of system under investigation. Description of FDC can be found, for example, in paper by Anisimov at al.[1] and briefly repeated here.

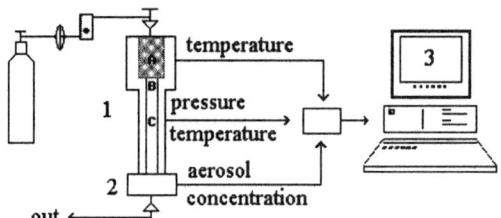

Fig. 1. The experimental setup scheme
1 - Aerosol generator, containing hot thermostat with vapor saturation volume, A; Laminator, B; and Cooler, C; 2 - Laser aerosol counter; 3 - System for computer collection of experimental parameters.

The experimental setup (Fig. 1) contains an aerosol generator (1), which consists of hot (A, B) and cold thermostats (C). A high-pressure cylinder provided the inert carrier gas, which maintained the flow via a gas pressure reducer, filter, flow meter, and pressure regulator to the hot thermostat. The gas than passed through a chromatographic stationary phase, alumina (A), where the gas become saturated in the substance under investigation. Inside the second thermostat (B), a steady state laminar flow was established. The vapor-gas mixture becomes supersaturated in the cooler (C) and aerosol particles are generated. Temperatures were measured with copper-constantan thermocouples, which were calibrated in the appropriate temperature range by using mercury thermometers with an accuracy of 0.1^0C.

The aerosol concentration was measured by a laser photoelectric aerosol counter (2). The sensitivity of particle detection was sufficient to count particles such as 0.1 µm in diameter and bigger. Computer control of the aerosol concentration allowed collecting a sufficient statistical data in the complete range of the experimental aerosol nucleation rates achievable in these experiments. The standard deviation for measured aerosol concentration did not exceed 3% in these experiments. The sensor's signals were digitized and passed to a personal computer (3) for processing.

The carrier gas (with purity not less than 99.99 vol. %) passed through the chromatographic stationary phase soaked by 1,2-propanediol or 1,3-propanediol, where the carrier gas becomes vapor saturated in those vapors. The Reynolds number was around 100. The maximum rate of droplet formation was calculated using the algorithm developed for the flow diffusion chamber by Wagner and Anisimov[2].

III. RESULTS AND DISCUSSION

A. Nucleation of 1,2-proranediol and 1,3-proranediol vapors in the sulfur hexafluoride atmosphere

Investigations of the influence of conditions in the vicinity of the critical line of the binary systems on the vapor nucleation of 1,2- and 1,3-propanediols in sulfur hexafluoride (SF_6) were carried out at total system pressures $P = 0.10$; 0.20; and 0.30 MPa.

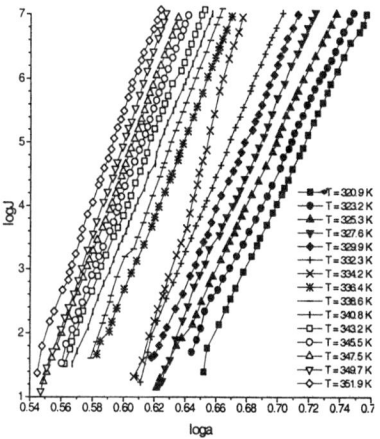

Fig. 2. Sulfur hexafluoride – 1,3 propandiol nucleation rate, J, on vapor activity, a, at total pressure $P=0.30$Mpa

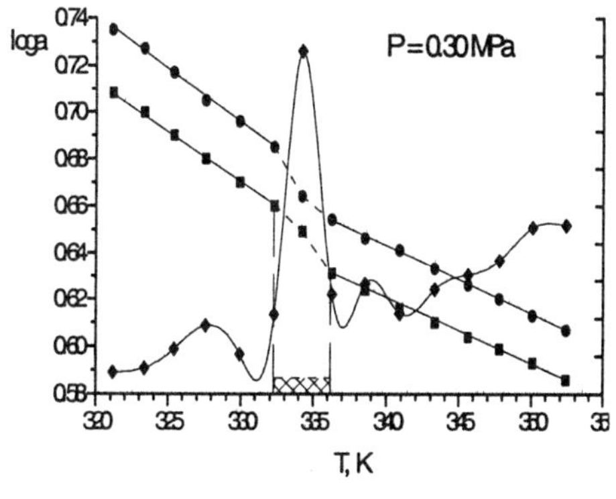

Fig. 3. Sulfur hexafluoride – 1,3 propandiol vapor activity , a, on nucleation temperature, T, at total pressure $P=0.30$Mpa and the critical embryo size, n^* (diamond-shaped points)

Figure 2 displays the experimental nucleation rates, $logJ$, versus the 1,3-propanediols vapor activities, $loga$, at different nucleation temperatures and total pressure 0.3 MPa. As example of the temperature dependencies [3] of $loga$ on nucleation temperatures of 1,3-propanediol vapors in SF_6 is shown in fig.3 at two sections of nucleation rates ($logJ = 4.5$ and $logJ = 6.0$) and total pressure in the system at $P = 0.30$ MPa. As one can see, the logarithms of activity at different binary system pressures exhibited breaks, which do not coincide with critical temperature of pure SF_6. The temperature of the phase transition is shifted to lower temperatures as the carrier gas pressure grows. In other words, increasing the sulfur hexafluoride content shifts the phase transition temperature to the critical temperature of pure SF_6.

B. Isothermal nucleation of 1,2-proranediol and 1,3-proranediol vapors in carbon dioxide atmosphere

Experimental investigations of 1,2- and 1,3-propanediol vapor nucleation in carbon dioxide in the vicinity of its critical temperature were carried out at total pressures $P = 0.10$; 0.20; and 0.30 MPa. The experimental 1,2- and 1,3-propanediol vapors activities, $loga$, on nucleation temperature, T, (at constant nucleation rates, i.e. for the 1,2-propanediol – carbon dioxide at $logJ = 5.0$ and $logJ = 6.5$ and for the 1,3-propanediol – carbon dioxide at $logJ = 4.5$ and $logJ = 6.0$) are presented in our paper [3].

One can see nonmonotone temperature behavior for activity logarithms [3]. With increasing the nucleation temperature, growth the $log\ a(T)$ values falls monotonous, then breakup initiated by first order phase transition in critical embryos is appeared. These are phase transitions of the first order because the second order phase transitions have continuos first derivatives for chemical potential with respect to temperature, therefore the nucleation rate surface can not have the broken first derivative as the result of second order phase transitions. Further nucleation temperature growth leads the regular, monotonous changing of $log\ a(T)$. It is remarkable that the temperatures of phase transitions in critical embryos are decreasing with increasing the total pressure in these systems. The system behavior can be explained by approximation of a binary system. The increasing of carbon dioxide contents in the system leads the shift of the phase transition temperature toward the critical temperature of pure CO_2.

V. CONCLUSIONS

A phase transition initiated by critical lines of binary systems was found experimentally. The nucleation rate surface singularity and a gap in the number of molecules in critical embryos reflect the influence of the critical line on nucleation. Shifts of the phase transition temperatures were revealed by increasing the pressure (or concentrations) of the carrier gases. This behavior is peculiar to binary systems. Analysis of the experimental data shows that gas-carrier molecules are involved in new phase embryo formation. Vapor nucleation of investigated substances in a carrier gas atmosphere can be considered as nucleation of binary system[4].

ACKNOWLEDGEMENT

Authors acknowledge Dr. S.D. Shandakov and Dr. V.A. Pinaev for the help in the heat- mass transfer problem solution.

REFERENCES

1. Anisimov, M. P., Koropchak, J. A., Nasibulin, A. G., and Timoshina, L. V. *J.Chem.Phys.* **109** (21), 9979 (1998).
2. Wagner, P.E. and Anisimov, M.P. *J.Aerosol Sci.* **24**:103 (1993).
3. Anisimov, M. P., Koropchak, J. A., Nasibulin, A. G., and Timoshina, L. V. *J.Chem.Phys.* (2000) (in print)
4. Critical embryo compositions and comparison of the experimental and calculated temperatures of phase transitions are presented in Web Page:
   http://www.geocities.com/ResearchTriangle/System/7762/1.html

Kinetics of Ion-Induced Nucleation and Ion Mobility of Pre-Critical Clusters

S. P. Fisenko, D. B. Kane and M. S. El-Shall[*]

Department of Chemistry, Virginia Commonwealth University, Richmond, VA 23284-2006

Abstract. Two limiting regimes of ion induced nucleation kinetics are discussed. Analytical expressions for the nucleation rate are obtained. The influence of the molecular nature of the ion on the nucleation rate is shown. Experimental evidence of the limiting regimes is presented.

INTRODUCTION

Ion-induced nucleation is a phenomenon of great importance not only as a subject for basic scientific inquiries of complex phenomena in chemical physics but also for many implications such as chemical reactions in the ionosphere, condensation of interstellar dust, and for other applications in radiation chemistry and combustion processes. The history of ion-induced nucleation research dates back about 100 years to the original experiments by Wilson [1]. While research since that time has greatly increased our understanding of this phenomenon, nevertheless there remain many unsolved problems. In particular the area of ion nucleation kinetics has been neglected by researchers. Most efforts have been devoted to understanding the thermodynamics of ion nucleation and little attention has focused on the kinetics.

KINETICS OF ION NUCLEATION

For illustration, the thermodynamic barriers for ion- induced nucleation and for homogeneous nucleation are shown in Figure 1. It is clear that the free energy of cluster formation, according to Thomson's electrostatic model for ion nucleation, exhibits both a local minimum and a maximum [2]. The minimum's position corresponds to a stable solvated ion g_s and the maximum's position corresponds to the size of critical cluster g_c in a supersaturated vapor.

The kinetic equation for the distribution function of the nucleated clusters can be written as:

$$\frac{df(g,t)}{dt} = \partial_g \left[f(g,t) L_{11} \partial_g \left(\ln f(g,t) + \beta \Phi(g,q) \right) \right] \quad (1)$$

where $\beta = 1/kT$, is the kinetic coefficient (k is the Boltzmann constant and T is temperature), L_{11}, is the total flux of the vapor molecules on the critical cluster and is given by:

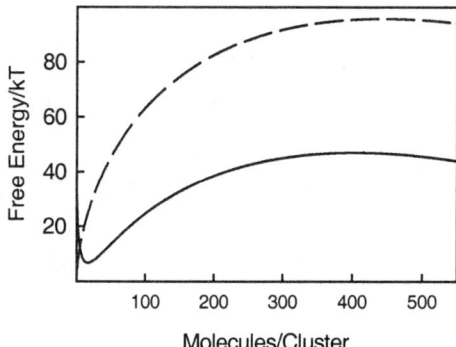

FIGURE 1. Free energy of cluster formation versus the number molecules per cluster. Supersaturation = 1.54 and temperature T = 275 K. Top curve is for the homogeneous nucleation of methanol, and the bottom curve for the ion-induced nucleation of methanol induced by benzene positive ions (ion radius = 4.54 10^{-10} m).

$$L_{11} = \frac{nkT\Sigma}{\sqrt{2\pi mkT}} \qquad (2)$$

where n is the vapor density, m is the molecular mass and Σ is the surface of a sphere with radius R_k. Equation (2) is similar to the flux on the critical cluster in the kinetic equation for homogeneous nucleation expect that the radius, R_k, reflects the dynamics of collisions between vapor molecules and the charged cluster. For vapor molecules with permanent dipole moment, d_0, the expression for R_k can be written as:

$$R_k \cong \sqrt{\frac{qd_0}{6\pi\varepsilon_0 kT}} \qquad (3)$$

where ε_0 is the dielectric permeability of vacuum. It follows from expressions (2) and (3) that the kinetics of ion nucleation depends on the dipole moment of the condensing vapor molecules. The lower the temperature of the vapor the larger the contribution of this effect.

Steady-State Nucleation Kinetics

When the barrier height, the difference between the free energy minimum and maximum is much greater kT (at low supersaturations), we can consider ion-induced nucleation as Brownian motion in cluster size space. The typical boundary conditions for the nucleation kinetic equations in this case are:

$$f(g,t) = 0, \quad g \to \infty \qquad (4)$$
$$f(g,t) = n_i, \quad g \to 0 \qquad (5)$$

where n_i is the equilibrium cluster ion distribution. Using standard mathematical methods to solve the Fokker-Plank equation, the following expression for the ion-induced nucleation rate, I, can be obtained:

$$I = \frac{1}{\pi} N_i \sqrt{|\alpha|\gamma} L_{11} \exp(-\Delta\Phi\beta) \qquad (6)$$

where N_i is the ion density, $\Delta\Phi = \Phi(g_c,q) - \Phi(g_s,q)$ is the nucleation barrier height and $|\alpha|$ defines the characteristic width of the thermodynamic barrier near its maximum $\Phi(g_c,q)$ and can be expressed as:

$$\alpha = \frac{1}{2}\partial^2_{gg}\Phi(g_c,q)\beta \qquad (7)$$

and γ is the characteristic width near the local minimum g_s, and is given by:

$$\gamma = \frac{1}{2}\partial^2_{gg}\Phi(g_s,q)\beta \qquad (8)$$

From (6), it is evident that the steady-state ion nucleation rate is directly proportional to the ion density. Also the exponential term in the expression (6) is independent of the ion radius and depends on the square of ion charge q (according to Thomson's model). However, the ion radius affects the characteristic widths of the minimum and maximum of the free energy. This leads to a dependence of the nucleation kinetics on the ion radius. Another asymptote of ion-induced nucleation exists for relatively high supersaturations. For methanol vapor at temperature 275K, this behavior takes place when the supersaturation is above 1.9. In this case, the second term on the right hand side of Eq. (1) disappears and the nucleation rate can be expressed as:

$$I \cong \frac{N_i}{g_c - g_s}\sqrt{\frac{\gamma}{4\pi}}L_{11} \qquad (9)$$

It is clear that at high supersaturations all solvated ions will lead to nuclei of the new phase. Numerical results for intermediate regime of ion-induced nucleation will also be presented.

COMPARISON WITH EXPERIMENT

As mentioned above, the minimum in the free energy of Thomson's model corresponds to the formation of a solvated cluster ion g_s. Further growth occurs with an increase in free energy until the cluster reaches a size g_c, after which it grows with decreasing free energy to form a liquid droplet. With increasing the vapor supersaturation, the size of the solvated cluster ion increases and the size of the critical cluster decreases. The verification of the trend predicted by Thomson's model has been recently provided by measuring the ion mobility of pre-critical clusters as a function of vapor supersaturation [3]. The method is based on REMPI (Resonance Enhanced Multiphoton Ionization) nucleation technique, which allows the generation of selected ions within a supersaturated host vapor held in a diffusion cloud chamber [4-7]. The ions are generated below the maximum supersaturation region of the chamber; methanol molecules cluster around the ions, and the cluster ions drift under the influence of the uniform field toward the maximum supersaturation region where nucleation occurs. The resulting condensation nuclei rapidly grow into macroscopic liquid droplets, and the droplets fall by gravity into a He-Ne laser beam. The light signals scattered by the droplets are detected by a photomultiplier and recorded using a computer interface. The arrival time distribution is measured as a function of the drift field across the chamber. The data is used to calculate the mobility of the cluster ions, and from the dependence of the measured mobility on the vapor supersaturation, we can relate the size distribution of the pre-critical cluster ions to the nucleation parameters. Figure 2a shows the supersaturation dependence of the size of the solvated ion as predicted by Thomson's model and Figure 2b illustrates the supersaturation dependence of the measured ion mobility. The similarity between the two trends is evident. Note that from the measured mobility, the size of the pre-critical clusters can be estimated through the mobility-size relationships based on either the continuum model or on kinetic theories. The observation of this trend provides the first direct experimental evidence for the dependence of the size of the pre-critical clusters on the vapor supersaturation as predicted by Thomson's model.

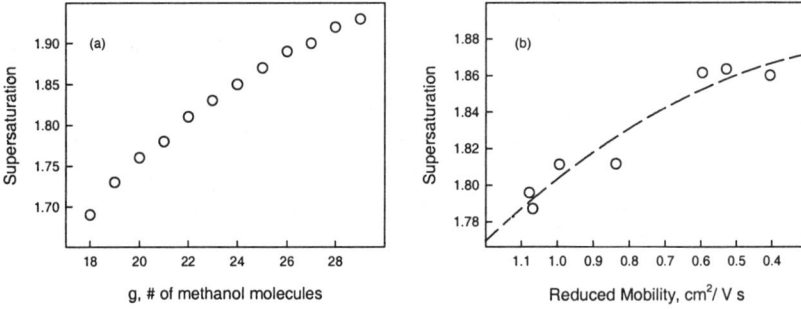

FIGURE 2. (a) Dependence of the size of the solvated cluster ion, toluene$^+$(methanol)$_g$, on the supersaturation of methanol vapor at T = 276K according to Thomson's model. (b) Dependence of the measured reduced mobility of the toluene$^+$(methanol)$_g$ pre-critical clusters on the supersaturation of methanol vapor within the temperature range of 274-276 K.

ACKNOWLEDGMENTS

The authors gratefully acknowledge support from NASA Microgravity Materials Science Program (Grant NAG8-1484).

REFERENCES

1. Wilson, C. T. R., *Phil. Trans. R. Soc. London* **189**, 265 (1897).
2. Thomson J. J., and Thomson, G. P., *Conduction of Electricity through Gases*, 3rd Ed. Vol. 1, Cambridge University, 1928.
3. Kane, D.; Rusyniak, M.; Fisenko, S. P.; El-Shall, M. S. *J. Phys. Chem.* **104**, *in press* (2000).
4. Kane, D.; Daly, G. M.; El-Shall, M. S. *J. Phys. Chem.* **99**, 7867 (1995).
6. Kane, D.; El-Shall, M. S. *Chem. Phys. Letters* **259**, 482 (1996).
6. Kane, D.; Fisenko, S.; El-Shall, M. S. *Chem. Phys. Letters* **277**, 6, 12 (1997).
7. Kane, D. *"Application of Resonance Enhanced Multiphoton Ionization to the Study of Ion Nucleation in Supersaturated Vapors"*, Ph.D. dissertation, Virginia Commonwealth University, Richmond, VA, 1997.

Gas-liquid nucleation in partially miscible systems: Free energy surface from computer simulation

S. Yoo, K.J. Oh, and X.C. Zeng

Department of Chemistry
University of Nebraska-Lincoln
Lincoln NE 68588 USA

Abstract. The formation free energy surface of binary clusters in a non-ideal mixture is obtained using a constrained-ensemble Monte Carlo approach. The special feature of the surface is that two minima appear at cluster sizes greater than the size of critical cluster. The results of this large-scale Monte Carlo simulation are consistent with those of ten Wolde and Frenkel who used an umbrella-sampling Monte Carlo approach.

INTRODUCTION

The formation free energy $\Delta G_{i,j}$ of (i, j)-cluster composed of i molecules of species 1 and j molecules of species 2 provides valuable information for understanding nucleation behavior of binary fluids. By locating a saddle point on the free energy surface, one can determine the formation free energy of the critical cluster $\Delta G_{i,j}^*$ and estimate the rate of nucleation J from the equation

$$J = K \exp(-\Delta G_{i,j}^* / k_B T) \quad (1)$$

where k_B is the Boltzmann constant, T is temperature, and K is the kinetic prefactor.

To evaluate $\Delta G_{i,j}^*$, classical nucleation theory has rested on Gibbsian surface thermodynamics which describes a cluster as a macroscopic liquid drop. Recent experimental studies[2-4], however, have revealed that classical nucleation theory has certain limitations in describing the experimental observations. For non-ideal mixtures, for example, it even yields thermodynamic inconsistencies, that is, it predicts a positive slope in the plot of the onset activity of one component as a function of the activity of the other at constant nucleation barrier height. This inadequacy of the classical nucleation theory points to the need for molecular-based approaches such as Monte Carlo simulation[5-7] or density functional theory[8-11] to the evaluation of $\Delta G_{i,j}^*$. In this study, we employed a recently developed constrained-ensemble Monte Carlo approach to evaluate the formation free energy of cluster for a non-ideal mixture.

METHOD

The system chosen is the Lennard-Jones (12, 6) fluid, for which the pairwise intermolecular potential is given by

$$u(r) = 4\ \varepsilon_{ij}\ [\ (\sigma_{ij}/r)^{12} - (\sigma_{ij}/r)^{6}\ ], \qquad (2)$$

where r is the intermolecular distance, ε_{ij} is the Lennard-Jones well depth corresponding to the interaction between species i and j, and σ_{ij} is the corresponding Lennard-Jones diameter. For potential parameters, we take $\sigma_{11} = \sigma_{22} = \sigma_{12} = \sigma$, $\varepsilon_{22} = \varepsilon_{11} = \varepsilon$, and $\varepsilon_{12} = (1 - \Lambda/2)\ \varepsilon$ where $\Lambda = 0.7$.

Canonical ensemble Monte Carlo simulation was performed and the conventional Metropolis algorithm was adopted to generate configurations of a supersaturated vapor. A special feature in the constrained-ensemble Monte Carlo approach is that a Monte Carlo move will be rejected if it leads to the size of a cluster larger than the upper bound value i_{max}[6]. Here i_{max} is used as a constraint to maintain the supersaturated vapor in metastable equilibrium. In this study, we used $i_{max} = 50$.

To identify clusters in the supersaturated vapor, we adopted the Stillinger criterion[12,13] that a group of Lennard-Jones particles can be considered as a cluster if every particle has at least one nearest neighbor within a distance less than 1.5σ (a value close to the first minimum of the pair correlation function of the Lennard-Jones liquid near the triple point). By simply counting the number of (i, j)-clusters in the supersaturated vapor and then taking the ensemble average of the number over the entire Monte Carlo run, one can obtain the formation free energy via the equation[1]

$$\Delta G_{i,j} / k_B T = -\ln(\ N_{i,j} / N_1\ ), \qquad (3)$$

where $N_{i,j}$ is the average number of (i, j)-clusters, $N_1 = N_{1,0} + N_{0,1}$ is the total number of monomers.

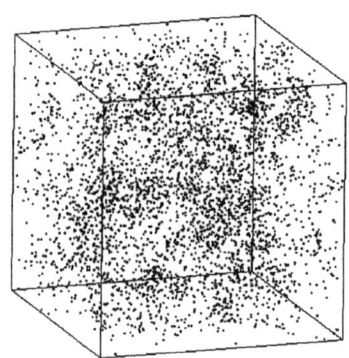

FIGURE 1. A snapshot of the supersaturated vapor from simulation.

RESULT AND DISCUSSION

In the Monte Carlo simulation, the temperature of the system is set at $T = 0.670\ \varepsilon/k_B$ and the volume is $V = 70000\sigma^3$. The number of molecules of species 1 and species 2 is 2000. Periodic boundary condition was implemented in the Monte Carlo simulation.

Fig. 1 shows a typical snapshot from the simulation. Fig. 2(a) shows the obtained formation free energy surface spanned by the number of molecules of species 1 and species 2. The critical cluster size is about 30. To find out the shape of the free energy surface near the critical size, we plot in Fig. 2(b) the formation free energy at the constant cluster size $i+j = 25$, 30, and 40 as a function of mole fraction $i/(i+j)$ of species 1 in the cluster. As can be seen from Figs. 2(b), one minimum appears in the size range of precritical cluster whereas there are two minima showing up in the size range of postcritical cluster. The two minima have nearly the same value of free energy. As the cluster size increases, the two minima are lowered and the height of barrier separating the two minima increases. These features of the free energy surface have also been observed by ten Wolde and Frenkel[11] from an umbrella-sampling Monte Carlo computer simulation.

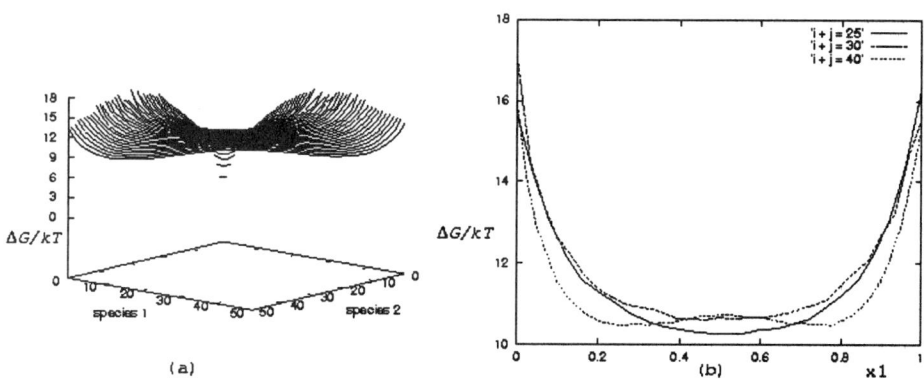

FIGURE 2. (a) The formation free energy surfaces for the non-ideal mixture $\Lambda = 0.7$, and (b) values of formation free energy for all i and j, but $i + j = 25$, 30 or 40, as a function of the mole fraction of species 1.

The present study is only semiquantitative in a sense that a larger number of molecules in the system is needed to achieve better statistics. Nevertheless, the basic physical features of binary nucleation of the non-ideal mixture are well captured in the obtained free energy surface.

In summary, we have performed a large-scale Monte Carlo simulation to evaluate the free energy surface of cluster formation for a non-ideal mixture. The preliminary results are consistent with the previous one from a small-ensemble Monte Carlo simulation. In passing, we point out a recent study on the binary mixture of spherical

Lennard-Jones atoms (monomers) and dumbbell molecules composed of two Lennard-Jones atoms (dimers) by Napari and Lakksonen[14]. These workers found that an increased interaction anisotropy between the dimer sites and the monomers can result in mutual enhancement of nucleation. They also found that the critical cluster has a lamellar structure at high dimer activities. A computational study on this system using the present approach is underway.

REFERENCES

1. Reiss, H., *J. Chem. Phys.* **18**, 840—848 (1950).
2. Strey, R., Viisanen, Y., and Wagner, P. E., *J. Chem. Phys.* **99**, 4680—4704 (1993).
3. Viisanen, Y., Strey, R., Laaksonen, A., and Kulmala, M., *J. Chem. Phys.* **100**, 6062—6672 (1994).
4. Strey, R., Viisanen, Y., and Wagner, P. E., *J. Chem. Phys.* **103**, 4333—4345 (1995).
5. ten Wolde, P. R., and Frenkel, D., *J. Chem. Phys* **109**, 9901—9918 (1998).
6. Oh, K. J., and Zeng, X.C., *J. Chem. Phys.* **110**, 4471—4476 (1999).
7. ten Wolde, P. R., and Frenkel, D., *J. Chem. Phys.* **109**, 9919—9927 (1998).
8. Zeng, X. C., and Oxtoby, D. W., *J. Chem. Phys.* **95**, 5940—5947 (1991).
9. Laaksonen, A., and Oxtoby, D. W., *J. Chem. Phys.* **102**, 5803—5810 (1995).
10. Talanquer, V., and Oxtoby, D. W., *J. Chem. Phys.* **104**, 1993—1999 (1996).
11. Napari, I., and Laaksonen, A., *J. Chem. Phys.* **111**, 5485—5490 (1999).
12. Stillinger, F. H., *J. Chem. Phys.* **38**, 1486 (1963).
13. Senger, B., Corti, D. S., Bowles, R., Pointu, D., Voegel, J.-C., and Reiss, H., *J. Chem. Phys.* **110**, 6438—6450 (1999).
14. Napari, I., and Laaksonen, A., *Phys. Rev. Lett.* **84**, 2184—2187 (2000).

Ternary Nucleation of H_2SO_4, NH_3 and H_2O

M. Kulmala[1], P. Korhonen[1,2], A. Laaksonen[3], Y. Viisanen[2], R. McGraw[4] and J.H. Seinfeld[5]

[1]*Department of Physics, University of Helsinki, Finland*
[2]*Air Quality Research, Finnish Meteorological Institute, Helsinki, Finland*
[3]*Department of Applied Physics, University of Kuopio, Finland*
[4]*Environmental Chemistry Division, Brookhaven National Laboratory, Upton, New York, USA*
[5]*Division of Engineering and Applied Science and Department of Chemical Engineering, California Institute of Technology, USA*

Abstract. A classical theory of the ternary homogeneous nucleation of sulfuric acid – ammonia – water is presented. For NH_3 mixing ratios exceeding 1 ppt, the presence of ammonia enhances the binary (sulfuric acid –water) nucleation rate by several orders of magnitude. However, the limiting component for ternary nucleation – as for binary nucleation – is sulfuric acid. The sulfuric acid concentration needed for significant ternary nucleation is several orders of magnitude below that required in binary case.

INTRODUCTION

Water vapor is the most abundant gaseous species in the atmosphere that may take part on the new particle formation. However, for the homogenous nucleation process, in which new particles form from a single species only, its typical concentrations are far too small. Thus it is generally accepted that the new particle formation in the atmosphere occurs via homogenous heteromolecular nucleation process, in which two or more vapor species form new stable particles. In the past the atmospheric new particle formation was almost always assumed to take place via binary nucleation of water (H_2O) and sulfuric acid (H_2SO_4) vapors. Field measurements have been shown that there exists situations where the new particle formation cannot be explained with this nucleation route alone [1-3]. Although meteorological factors (e.g. mixing) may play an important role in some cases, researchers have begun to search for additional mechanisms that may launch nucleation in the atmospheric conditions. Because the presence of ammonia in the aerosol particles considerably decreases the vapor pressure of sulfuric acid above the solution surface [4], it has been suggested, that ammonia (NH_3) forms new particles with sulfuric acid [4] or with sulfuric acid and water [5].

We have recently shown theoretically [6] that ternary nucleation is important phenomena in the atmosphere. Its importance in aerosol dynamics point of view have also been studied [7]. In this paper we apply the ternary nucleation model on atmospheric conditions. We also summarize the ternary nucleation theory and also the thermodynamics used.

THEORY FOR THE TERNARY NUCLEATION

In the following we focus on the formation of new particles via nucleation of stable H_2O - NH_3 - H_2SO_4 clusters. The ternary water - ammonia - sulfuric acid solution nuclei are assumed to be in the liquid phase. The nucleation rate of stable water-ammonia- sulfuric acid clusters (J) is obtained from

$$J = C\exp(\frac{-\Delta G^*}{kT}) \tag{1}$$

Here C is a kinetic factor. In this study the minimum work for the critical nucleus formation is determined using the so called revised classical theory. The minimized Gibbs free energy change (in the limits of capillarity approximation) is obtained from

$$\Delta G^* = \frac{4}{3}\pi r^{*2} \sigma_{s/a} \tag{2}$$

where r^* is the critical radius of the cluster and $\sigma_{s/a}$ is the surface tension. The critical nucleus composition is obtained from the following equations (Arstila et al., 1998):

$$-kT\ln\left(\frac{P_1}{P_{s,1}}\right) + \frac{2\sigma_{s/a}v_1}{r} = -kT\ln\left(\frac{P_2}{P_{s,2}}\right) + \frac{2\sigma_{s/a}v_2}{r} = \ldots = -kT\ln\left(\frac{P_i}{P_{s,i}}\right) + \frac{2\sigma_{s/a}v_i}{r} = 0 \tag{3}$$

where P_i is the ambient partial vapor pressure of species i, $P_{s,i}$ is the equilibrium vapor pressure of species i above the flat solution surface, r is the radius of the cluster and v_i is the partial molecular volume of species i.

When the critical cluster is formed from water, ammonia and sulfuric acid equation (3) becomes:

$$v_{h2so4}\ln\left(\frac{P_w}{P_{s,w}}\right) - v_w\ln\left(\frac{P_{h2so4}}{P_{s,h2so4}}\right) = 0$$

$$v_{nh3}\ln\left(\frac{P_w}{P_{s,w}}\right) - v_w\ln\left(\frac{P_{nh3}}{P_{s,nh3}}\right) = 0 \tag{4}$$

Here subscript w refers to water, h2so4 to sulfuric acid and nh3 to ammonia. From equations (4) one can solve the composition of the critical cluster by numerical iteration. When the composition of the cluster is known, the critical radius is obtained from the Kelvin equation

$$r^* = \frac{2\sigma_{s/a}v_i}{kT\ln\left(\frac{P_i}{P_{s,i}}\right)} \tag{5}$$

In this context *i* refers either water or ammonia or sulfuric acid. We have used recently presented rigorous kinetic factor for the ternary system [8]. In order to solve the radius and composition of the critical cluster for the systems presented above, one needs surface tension (surface free energy), density of the solution and equilibrium vapor pressures of the various species above the flat solution surface. When the classical nucleation theory is used, the thermodynamical properties of the nucleus are assumed to be those of bulk substance in question. Our recent paper [6] summarizes the thermodynamical model.

RESULTS AND DISCUSSION

According to the present knowledge, the nucleation of new aerosol particles in the atmospheric conditions in significant extent can occur The results of this model study suggest that nucleation of water – ammonia - sulfuric acid clusters occurs easier than the nucleation of water – sulfuric acid clusters in the atmospheric conditions. The results also suggest that the composition of the critical clusters in the atmospheric conditions is typically that of water - ammonium bisulfate solution. However, the formed H_2O - NH_3 - H_2SO_4 clusters are very small and highly concentrated: in this study they appear to be supersaturated with respect to ammonium sulfate almost in every case we have studied [6].

The nucleation rate is a strong function of sulfuric acid concentration, ammonia concentration and temperature. In Figure 1 the ternary nucleation rate as a function of temperature is presented. Sulfuric acid concentration is 1×10^6 cm^{-3} and ammonia concentration 25 pptv. In Figure 2 the ternary nucleation rate as a function of ammonia concentration is presented. Sulfuric acid concentration is 1×10^5 cm^{-3} and temperature 278.15 K.

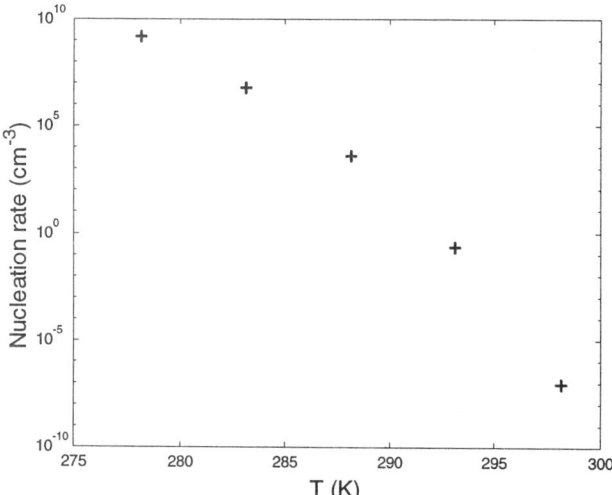

FIGURE 1. Ternary nucleation rate as a function of temperature. Sulfuric acid concentration is 1×10^6 cm^{-3} and ammonia concentration 25 pptv.

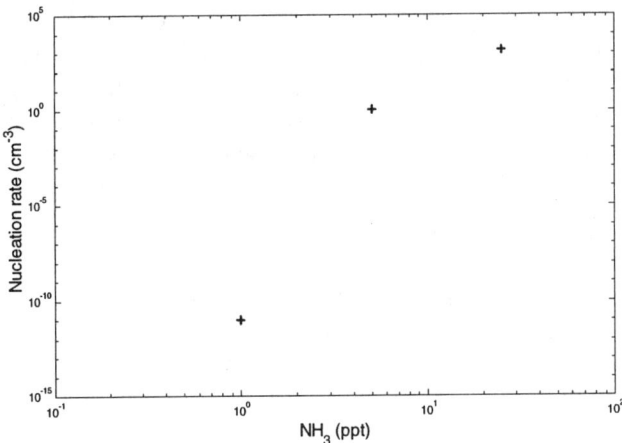

FIGURE 2. Nucleation rate as a function of ammonia concentration. Sulfuric acid concentration is 1×10^5 cm^{-3} and temperature 278.15 K.

Although the ternary nucleation is able to predict the formation of new 1 nm particles, it is not able to predict the formation of 3 nm particles. This is due to the fact that the sulfuric acid concentration needed in ternary nucleation is too small to be able to explain the growth of 1 nm particles to 3 nm size [7], and some other vapours are needed for condensation growth.

REFERENCES

1. Covert D.S., Kapustin V.N., Quinn P.K. and Bates T.S., New Particle Formation in the Marine Boundary Layer. J. Geophys. Res., 97, 20581-20589 (1992).
2. Hoppel W.A., Frick G.M. and Fitzgerald J.W., Marine boundary layer measurements of new particle formation and the effects nonprecipating clouds have on aerosol size distribution. J. Geophys. Res., 99, 14443-14495 (1994).
3. Mäkelä J.M., Aalto P., Jokinen V., Pohja T., Nissinen A., Palmroth S., Markkanen T., Seitsonen K., Lihavainen H., Kulmala M., Observations of ultrafine aerosol particle formation and growth in boreal forest. Geophys. Res. Lett., 1219-1222 (1997).
4. Scott W.D. and Cattell F.C.R., Vapor pressure of ammonium sulfates. Atmos. Env., 13, 307-317 (1979).
5. Coffmann D.J. and Hegg. D.A., A Preliminary study of the effect of ammonia on particle nucleation in the marine boundary layer. J. Geophys. Res., 100, 7147-7160 (1995).
6. KorhonenP., Kulmala M., Laaksonen A., Viisanen Y., McGraw R., Seinfeld J.H., Ternary nucleation of H_2SO_4, NH_3 and H_2O in the atmosphere. J. Geophys. Res., 104, 26349-26353 (1999).
7. Kulmala M., Pirjola L., Mäkelä J.M., Stable sulphate clusters as a source of new atmsopheric particles. Nature, 404, 66-69 (2000).
8. Arstila H., Korhonen P. and Kulmala M., Ternary Nucleation: Kinetics and application to water - ammonia - hydrochlorid acid system. Technical note, J. Aerosol Sci. 30, 131-138 (1999).

First Measurement of Prenucleation Molecular Clusters

F. L. Eisele[*] and D. R. Hanson

Atmospheric Chemistry Division, NCAR, Boulder, CO 80303
[*] *also School of Earth and Atmospheric Sciences, Georgia Institute of Technology, Atlanta, GA 30332*

Abstract. The molecular cluster ions $HSO_4^-(H_2SO_4)_{n-1}$ corresponding to the neutral species $(H_2SO_4)_n$ for n = 3 to 8 have been observed using a transverse chemical ionization scheme located inside a cooled flow tube. The contribution of ion-molecule clustering reactions was ascertained and readily separated from the ionization of the neutral clusters. The presence of the clusters were strongly dependent on temperature and humidity. Ratios of successive clusters thought to be representative of steady-state are reported.

INTRODUCTION

There is growing interest in atmospheric aerosols because of the large uncertainties in their influence on global radiative forcing, their potential health hazards in urban and industrial areas, and their largely unexplored role in tropospheric chemistry. A major source for atmospheric aerosol particles is the gas-to-particle nucleation process which is probably very nonlinear and is not well understood. Other sources not normally included in the area of nucleation, for example from combustion engines or wild fires, may involve a gas-to-particle nucleation process in the early stages. Thus, an improved understanding of the rather elusive gas-to-particle nucleation process is central to quantifying both natural particle production and anticipated increases in these production rates resulting from human activity.

The important first steps to at least a portion of new particle formation events in the atmosphere likely involve the sulfuric acid molecule, H_2SO_4.[1] For example, two H_2SO_4 molecules may combine to give the H_2SO_4 dimer, another H_2SO_4 molecule may collide with the dimer resulting in a H_2SO_4 trimer, etc. $K_2, K_3, ...K_n$ denote the equilibrium constants for the successive addition of H_2SO_4 to these clusters:

$$H_2SO_4 + H_2SO_4 \rightleftarrows (H_2SO_4)_2 \quad (1a)$$

$$H_2SO_4 + (H_2SO_4)_2 \rightleftarrows (H_2SO_4)_3 \quad (1b)$$

$$H_2SO_4 + (H_2SO_4)_{n-1} \rightleftarrows (H_2SO_4)_n \quad (1n)$$

There are other species in the atmosphere that may play a role in particle nucleation, notably H_2O and NH_3.[2,3] Reactions (1a-n) represent a simplification of the system in these important cases.

This paper describes the measurement of molecular clusters of sulfuric acid under quiescent particle-growth conditions. It may be the first report of the detection of neutral H_2SO_4 clusters containing 2 to 8 H_2SO_4 molecules; the large clusters being comparable to the critical cluster size found in laboratory experiments.[4] Although the measurements cover only a small range of concentration, the results provide a test for theory as well as reveal the power of the experimental technique. This technique provides a new window through which particle nucleation can be observed: each step of the process can now be studied. Finally, rough estimates of the steady state ratios of successive clusters are derived from the measured ion cluster distributions.

APPARATUS, AND MEASUREMENT AND ANALYSIS TECHNIQUES

The measurement of H_2SO_4 and its clusters were made in a thermostatted 9.5 cm ID flow tube using chemical ionization mass spectrometry (CIMS). The CIMS ion source and mass spectrometer inlet extend radially into the flow tube (3-to-6 and ~1 cm, respectively, from the wall) to detect species *in situ*. The ion-molecule reaction time was usually varied by changing the voltage applied to the ion-source; the distance between source and inlet could also be varied. A variable ion-molecule reaction time provided a means to distinguish between ionization of neutral H_2SO_4 clusters and ion-molecule clusters produced by stepwise addition of H_2SO_4 molecules to HSO_4^- ions. The H_2SO_4 concentration was kept low (~2×10^9 cm^{-3}) to minimize ion-molecule clustering while the flow tube temperature, T_f, was held at ~240 K to induce formation of neutral clusters of H_2SO_4.

Flow and Temperature

The temperature of the gas in the flow tube during these initial studies was not uniform owing to the large cooling that must occur. The gas temperature a few cm from the flow tube wall was measured with a thermocouple probe. The aluminum shower head cools the gas from 298 K to a temperature of ~ T_f +12 K while it and the flow tube are at T_f. The gas cools to a temperature of ~ T_f + 5 K at a position of 25 cm into the flow tube (from the top) and to T_{gas} ~ T_f + 2 K after 40 cm of travel into the flow tube. These were the two positions of the ion detection region. The data presented here was generally taken at the 40 cm distance because of the longer residence time. Despite the non-uniform temperature of the gas, the distributions determined from the signal ratios for some of the clusters could be representative of steady state values.

How well the measurements reflect the steady-state distributions of the clusters is dependent on how fast each cluster equilibrates. If the forward rate coefficients for (1) are close to the collision rate and the dissociation rates are fast (comparable to or faster than the pseudo-first-order forward reaction rate coefficients), our measurements may reflect steady-state or even equilibrium distributions. Results for the 25 cm and 40 cm distances (i.e., different neutral reaction times) are comparable and suggest that these assumptions are valid at least for the smaller clusters (n=2-4).

Determining equilibrium constants from the measured distributions requires more stringent criteria.

Ionization of clusters:

The principal difference between the two main pathways for producing cluster ions, ionization of a pre-existing neutral cluster versus product ion clustering with H_2SO_4, is their dependence on time. The production of ions from the $(H_2SO_4)_n$ neutral clusters has a linear time dependence for all n; thus, the ratio of any two of the ions produced from neutral clusters would be independent of ion reaction time. On the other hand, the $HSO_4^-(H_2SO_4)_{n-1}$ product of ion induced clustering depends on time to the nth power and the ratio of any two ion clusters would have a dependence on time equal to the difference in the number of H_2SO_4 molecules in the clusters. Therefore, the ion drift time dependence of the signals can be used to distinguish between neutral clusters and ion-induced clusters. By going to short reaction times and low $[H_2SO_4]$ such that ion induced clustering is suppressed, particularly in the production of large clusters, the observed ions can be attributed to the ionization of neutral clusters.

RESULTS

Figure 1a shows a plot of $ln([HSO_4^-]/[NO_3^-] + 1)$ as a function of ion drift time for water partial pressures of 0.12 and 0.04 torr at ~ 236 K. Ion drift time was set by varying the voltage on the ion source while maintaining a constant source-inlet distance of 4 cm. Since the ordinate should be equal to $k_1t[H_2SO_4]$ a linear time dependence is expected and $[H_2SO_4]$ is estimated to be 1.2 and 2.4×10^9 molecule cm^{-3}, respectively. Figure 1b-d show the ratios of successive ion clusters of H_2SO_4 to that of the monomer. A time dependence is expected for any such ratio if the ion cluster is due to successive H_2SO_4 monomer addition to HSO_4^- ions, and this is clearly exhibited for the dimer (linear) and for the trimer (quadratic) when the relative humidity is low, ~20% RH. On the other hand, the ion clusters derived from a proton exchange between NO_3^- and a pre-existing neutral cluster should be time independent and this is clearly evident for the $HSO_4^-(H_2SO_4)_{n-1}$ clusters for $n \geq 3$ at 63 % RH and for $n \geq 4$ at 20% RH. Further evidence for the detection of the neutral clusters is exhibited in the magnitudes of the signals for the trimer and higher clusters at high RH versus those at low RH. The signals at low RH can be taken to be an upper limit to the effect of ion-induced clustering at high RH (indeed $[H_2SO_4]$ is lower at high RH) assuming no water dependence for this process. Thus the observed signals cannot be due to ion clustering processes and are attributed to the presence of the neutral clusters $(H_2SO_4)_3$, $(H_2SO_4)_4$ and $(H_2SO_4)_5$.

We also observed larger clusters, up to the octamer, and it is likely the n=7 and 8 clusters were larger than the size of a critical nucleus for that temperature and relative humidity. Thus, in addition to the first steps of the nucleation process, some of the growth steps may now be observable.

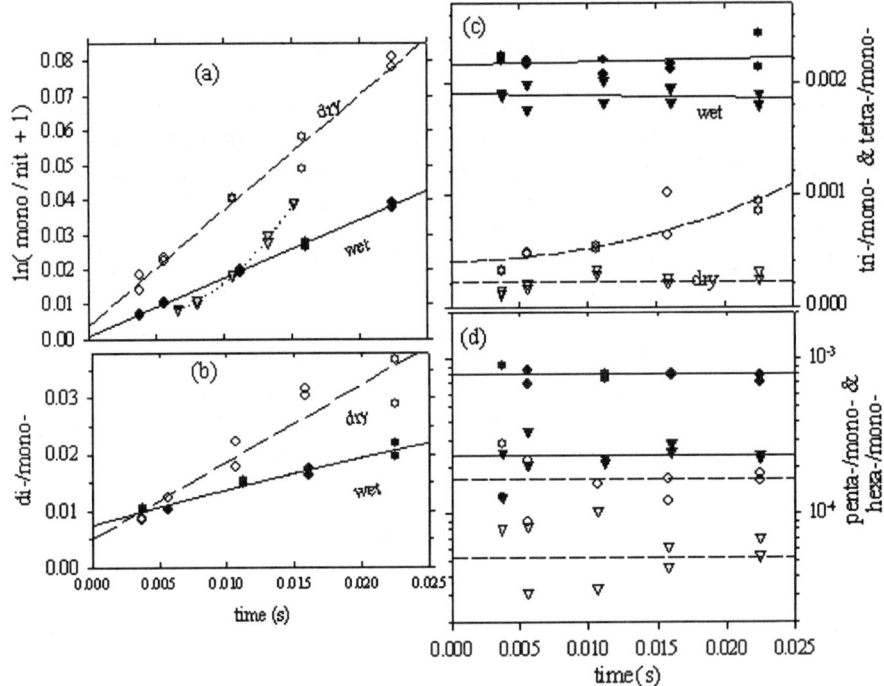

FIGURE 1. Ion ratios versus ion-molecule reaction time for two experiments at 236 K with water vapor at 0.12 torr (filled symbols, solid lines) and 0.04 torr (open symbols, dashed lines). (a) The monomer to nitric ratios are shown along with an experiment at the high RH where the source-inlet distance was changed (shaded triangles), (b) the dimer to monomer signal ratios, (c) the trimer and tetramer to monomer ratios, and (d) the pentamer and hexamer to monomer ratios are plotted on a log axis versus time.

REFERENCES

1. Weber, R.J.; McMurry, P.H.; Mauldin, R.L.; Tanner, D.J.; Eisele, F.L.; Clarke, A.D.; Kapustin, V.N. *Geophys. Res. Lett.*, **26, 1999**, 307.
2. Weber, R.J.; Marti, J.J.; McMurry, P.H.; Eisele, F.L.; Tanner, D.J.; Jefferson, A. *Chem. Eng. Comm.*, **1996**, *157*, 53.
3. Coffman, D.J.; Hegg, D.A. *J. Geophys. Res.*, **1995**, *100*, 7147.
4. Ball, S.M.; Hanson, D.R.; Eisele, F.L.; McMurry, P.H. *J. Geophys. Res.*, **1999**, *104*, 23709.

Experimental Determination of Molecular Content of Nuclei by Independent Variation of Activities

J. Hrubý* and R. Strey[+]

*Institute of Thermomechanics, Academy of Sciences of the Czech Republic, Dolejškova 5, CZ-182 00 Prague 8, Czech Republic
[+]Institute for Physical Chemistry, University of Köln, Luxemburger Str. 116, D-50939 Köln, Germany

Abstract. An expansion cloud chamber has been equipped with a new mixing unit, based on continuous mixing and evaporation of two liquid flows and a dry gas flow. The most remarkable feature of the new setup is that activities of the vapor components can be varied independently. This development provides a new view on the nucleation in binary (two-vapor) systems. The experimental results, obtained during tests of the set-up with water-n-propanol-argon system, lead us to the suggestion that Henry's and Raoult's laws can be generalized for the case of finite nucleation rate, i.e. for supersaturated systems. The ordinary laws are just a special case for zero nucleation rate (equilibrium).

INTRODUCTION

The expansion cloud chamber is an effective tool to measure homogeneous nucleation rates. Mixtures containing up to three vapor components and a background gas have been investigated (1) using the present device with another mixing unit. However, it was not possible to vary experimental activities independently. This made it difficult to deduce size and composition of the critical clusters using the Nucleation Theorem (2,3). The slopes of the nucleation rate surface were found based on analysis of several data series, obtained with separate vapor carrier-gas mixtures (4). Imperfect reproducibility of the mixture composition, although of minor importance when considering the overall accuracy of the nucleation rates, had a major effect on the scatter of the derivatives. In the limiting case where one of the activities approaches zero, the previous procedure was too uncertain to observe the proper shape of the nucleation rate surface. With the new device, these limitations are overcome and application of the Nucleation Theorem is greatly facilitated.

THEORY

We consider the nucleation rate J (number of nuclei per unit of volume and unit of time) in a mixture of two vapors and an inert carrier gas. J is measured as a function of the activities a_1, a_2 of the vapor components and temperature T. The activities are defined via the identity $RT \ln a_i \equiv \mu_i - \mu_{pure,i}^{sat}(T)$. At low pressures, the activities are usualy well approximated as $a_i \approx p_i / p_{pure,i}^{sat}(T)$. Strey and Viisanen (4) had written the Nucleation Theorem, originally suggested by Kaschiev (2) and later generalized by Oxtoby and Kashchiev (3), in a form

$$n_i^* = \partial \ln J(a_1, a_2, T) / \partial \ln a_i - c_i, \quad i = 1,2. \tag{1}$$

The small term c_i is of kinetic origin (it stems from the week activity dependency of the kinetic pre-factor of the nucleation rate expression). Here we realize that c_1 (c_2 similarly) must satisfy

$$\lim_{a_1 \to 0} c_1 = 0 \quad \text{and} \quad \lim_{a_2 \to 0} c_1 \approx 1 \tag{2}$$

in order to match the relations proven for the single-component case (5). If the vapor component 1 is diluted, the experiments show limiting behavior of the following kind:

$$\ln(J / J_{pure2}) = k_1 a_1, \quad a_2 = \text{const.}, \quad a_1 \to 0. \tag{3}$$

Straightforward application of the Nucleation Theorem (eq.1) gives

$$n_1^* = k_1(a_2, T) a_1, \quad a_1 \to 0. \tag{4}$$

In words, the number of molecules of species 1 in the critical cluster is proportional to the activity of this component. The proportionality factor k_1 is (in the limit) only function of activity of the other vapor component and temperature. Expressing a_1 from the last equation we find

$$a_1 = (n_{pure2}^* / k_1) \times [n_1^* / (n_1^* + n_2^*)] \times [(n_1^* + n_2^*) / n_{pure2}^*] \equiv K_1 x_1 b_1, \tag{5}$$

where x_1 is the overall (i.e. including both bulk and surface molecules) molar fraction of component 1 in the critical cluster. The factor b_1 approaches unity for $a_1 \to 0$. One might understand this relation as a generalization of Henry's rule to the kinetic case, K_1 being a dimensionless Henry's constant. The equilibrium Henry's rule is obtained for the limit of an infinitely large clusters (or $J=0$).

We define the *onset nucleation rate* J_o as nucleation rate of $10^7 \text{cm}^{-3}\text{s}^{-1}$. In the unary case $a_1=0$, the *onset activity* $a_{o,pure2}$ is activity of the component 2 for which $J=J_o$. For small a_1, the nucleation rate surface can be described as

$$\ln(J / J_o) = (n_{2,pure}^* + 1) \ln(a_2 / a_{2o,pure}) + k_1 a_1, \tag{6}$$

which is a plane in ($\ln J$, a_1, $\ln a_2$) coordinates. We approximate this expression as

$$\ln(J / J_o) = (n_{2,pure}^* + 1)(a_2 / a_{2o,pure} - 1) + k_1 a_1. \tag{7}$$

Making a horizontal cut at $J=J_o$, we find

$$a_{2o} \approx a_{2o,pure}[1 - k_1 a_1 / (n_{2,pure}^* + 1)] \approx a_{2o,pure} x_2 - a_{2o,pure} x_1 / n_{2,pure}^*. \tag{8}$$

The last equation can be viewed as a kinetic generalization of Raoult's law. The small second term is of kinetic origin and it vanishes in the equilibrium limit.

EXPERIMENT

The expansion cloud chamber, described elsewhere (6), has been equipped with a new mixture-preparation unit. Noteworthy features of the new device are: (i)

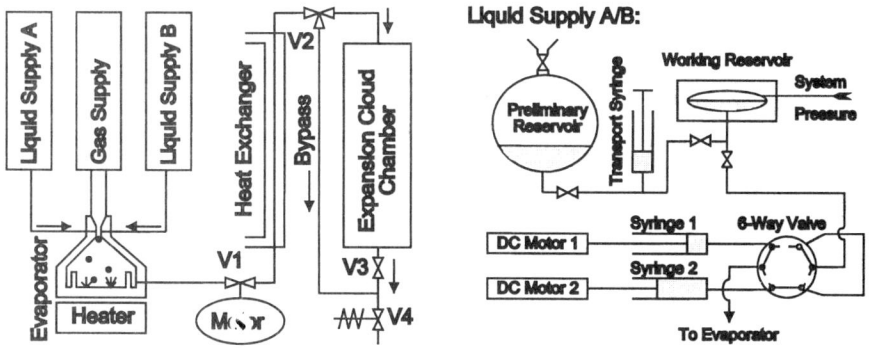

FIGURE 1. Scheme of the new mixing unit connected to the expansion cloud chamber.

activity of each vapor component and the total pressure can be varied independently[1], (ii) the mixture of vapors and carrier gas can be supplied continuously, (iii) binary systems in which the vapor pressure of one component is very low can be handled, (iv) systems where the liquid components are immiscible can be handled. The mixing unit includes two liquid supplies (A and B in Fig.1), a gas supply, and evaporator. Each liquid supply includes of two syringes (10 μl or 100 μl capacity), driven by two DC motors with magnetic encoders, connected to a micrometer screws. The velocities of the syringe plungers can be varied in the practical range of about 0.001 to 0.5 mm/s. The internal diameters of the syringes are calibrated by weighing. The accuracy of the liquid flow rates are within 0.5%. Both syringes, the capillary leading to the evaporator, and the capillary leading to the working reservoir are connected to ports of an automated 6-way valve (one for each liquid supply). The valve has two positions: in the first position, syringe 1 is connected to the evaporator and syringe 2 to the liquid reservoir, in the second position the connections are interchanged. Thus, one syringe injects the liquid to the evaporator, whereas the other one sucks from the reservoir. The valve needs about 0.3s to switch from between positions. In order to prevent a disturbance due to the switching, the following procedure is used: The velocity of the sucking syringe is set higher than velocity of the injecting one. After being filled, it starts to injects *back to the reservoir*. In the moment of switching, both syringes are injecting. In this way, the effect of mechanical clearances is removed. Second, the pressure in the liquid reservoir is kept equal to the pressure in the evaporator in order to eliminate the effects of compressibility of the liquid and of the dilatation of elastic parts. A flexible diaphragm in the working reservoir is used to equalize the pressures. The working reservoirs can be re-filled with liquids degassed in the preliminary reservoirs. The dry gas flow is controlled using 5 selectable calibrated critical orifices of diameters 20 to 100 μm and an accurate pressure control system. Capillaries of 200 μm inner diameter bring the two liquids into a nozzle in top of the evaporator. There they are dragged by rapidly flowing gas. Tiny droplets of the liquids hit the heated (up to 200°C) lower plate of the evaporator. The gas jet from the nozzle provides a stirring action. After leaving the evaporator, the pressure of the mixture is

---

[1] For binary cases. In multi-component cases, two selected activities can be independently varied, the others remaining constant.

reduced using valve V1 to the level required as the initial pressure for the expansion, and the mixture is cooled in the heat exchanger to the required initial temperature. Before expansion, the mixture flows at a slow rate through the expansion cloud chamber, in order to saturate its walls. During expansion, the chamber is bypassed. The pressure in the chamber is controlled using a proportioning electromagnetic valve V4. The movement of the four syringes and the action of the 6-way valve are PC-controlled. The software enables direct programming of the required composition.

RESULTS

The expansion cloud chamber equipped with the new mixing unit was tested with n-propanol-water-argon system. Figure 2 shows the most interesting measurements, in the limit of small water activity. The results strongly support equation 3, which served as a starting point to derive the kinetic generalizations of Henry's law (Eq.5) and Raoult's law (Eq.8). Besides the new thermodynamic insight just described, the new method provides an experimentally tractable way of making well-defined vapor-trace gas mixtures, which are difficult to prepare due low volatility of one of the substances, and because of depletion effects due to adsorption on the chamber walls. In that case, one of the injected liquids is a dilute solution of the trace, and the other liquid will be the pure major component.

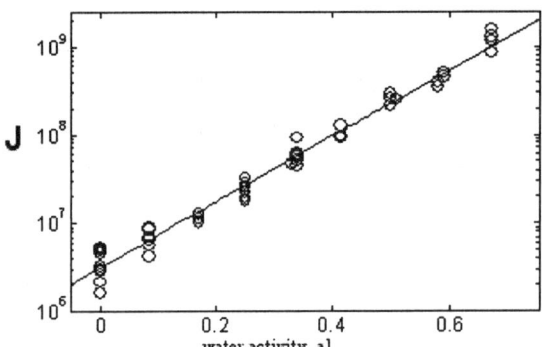

FIGURE 2. Nucleation rate, J (cm^{-3} sec^{-1}), in the water-n-propanol-argon system at 240 K, 63 kPa, and n-propanol activity a_2=5.95.

ACKNOWLEDGMENTS

J.H. acknowledges the support by Deustcher Akademischer Austauschdienst (DAAD) and grants GA AVČR A2076703 and GAČR 101/00/1282.

REFERENCES

1. Viisanen, Y. and Strey, R. *J. Chem. Phys.* **105**, 8293-8300 (1996).
2. Kashchiev, D., *J. Chem. Phys.* **76**, 5098-5102 (1982).
3. Oxtoby, D.W. and Kashchiev, D., *J. Chem. Phys.* **100**, 7665-7670 (1994).
4. Strey, R. and Viisanen, Y., *J. Chem. Phys.* **99**, 4693-4704 (1993).
5. Hrubý, J., Viisanen, Y., and Strey, R., *J. Chem. Phys.* **104**, 5181-5187 (1996).
6. Strey, R., Wagner, P. E, and Viisanen, Y., *J. Chem. Phys.* **98**, 7748-7758 (1994).

Molecular Simulation of Seed-Induced Nucleation in Vapor Phase

Koji Ohguchi,* Kenji Yasuoka** and Mitsuhiro Matsumoto*

*Department of Engineering Physics and Mechanics, Kyoto University,
Kyoto 606-8501, JAPAN
**Department of Mechanical Engineering, Keio University
Yokohama 223-8522, JAPAN

Abstract. Molecular dynamics (MD) simulations were executed to investigate the molecular mechanism of nucleation around a seed in Lennard-Jones vapor. It was found that the seed-induced nucleation takes place at a supersaturation ratio that is too low to cause homogeneous nucleation. At low supersaturation ratio, no clusters larger than a certain size (15~20) appear around the seed. We also executed grand canonical Monte Carlo (GCMC) simulations to estimate the cluster formation free energy under the same conditions. The free energy curves estimated from GCMC simulations are consistent with the results of the MD simulations.

INTRODUCTION

We had executed molecular dynamics (MD) simulations for homogeneous nucleation in Lennaed-Jones vapor [1] and water vapor [2], and had found that the nucleation rate obtained from our MD simulation is by several order of magnitude different from a prediction of a classical nucleation theory. In this paper, we report a similar model simulation for seed-induced nucleation.

In recent years, several Monte Carlo studies for estimation of cluster formation free energy have been published [3-5]. We estimated the free energy under the same conditions by grand canonical Monte Carlo (GCMC) simulations.

MOLECULAR DYNAMICS SIMULATION

Simulation Method

The system we used contains 1,000 condensable particles ('condensables' for short), only one seed particle and 1,027 carrier gas particles. The interaction potential between condensables and the seed or condensables each other is a Lennard-Jones (LJ) type. In this paper physical properties are expressed in units reduced by LJ parameters: length in σ, energy in ε, and time in $[m\varepsilon^2/\sigma]^{1/2}$. The energy parameter between condensables and the seed is ten times larger than that between condensables. Interaction potential between condensables and carrier gas, the seed and carrier gas or

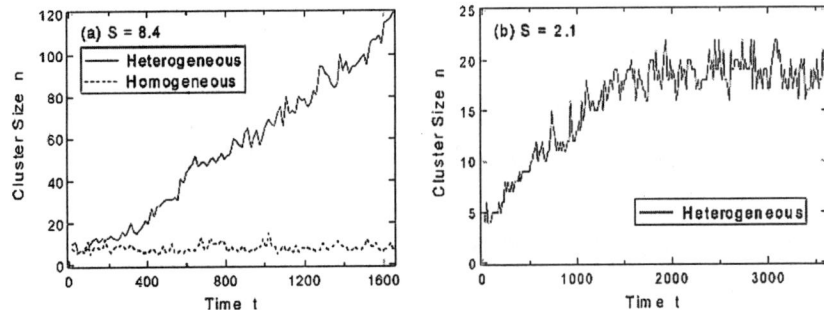

FIGURE 1. The size change of seed-including cluster. (a) Supersaturation ratio $S = 8.4$: with seed (solid line), a cluster grows large. However, no homogeneous nucleation occurs at this supersaturation ratio (dashed line). (b) $S = 2.1$: no large cluster appears although the system contains a seed.

carrier gas each other is a soft core type. Reduced temperature T_c^* of carrier gas was kept at 0.67 by Nosé-Hoover method [6] to remove the condensation heat from cluster surfaces.

Simulations were carried out at two supersaturation ratio ($S = 8.4, 2.1$) by changing the volume of the simulation cell. The 'supersaturation ratio' in this paper means initial supersaturation ratio without seed. Each supersaturation ratio we used is too small to cause homogeneous nucleation within our simulation time.

Results

The definition of cluster we used is Stillinger type [7]: two particles whose distance is less than 1.5σ is 'connected', and 'cluster' is a group of 'connected' particles.

Figure 1 shows the size change of the cluster including the seed under each condition. At higher supersaturation ratio ($S = 8.4$), which is too low to cause homogeneous nucleation, the cluster size increases with time. This means that a seed-induced nucleation takes place. At lower supersaturation ratio ($S = 2.1$), however, the cluster can not grow larger than a certain size (~ 20). No nucleation occurs in spite of existence of the seed.

GRAND CANONICAL MONTE CARLO SIMULATION

Simulation Method

We executed GCMC simulations [8] to estimate the cluster formation free energy for each condition. This method includes two more operations, creation and annihilation, in addition to particle move in normal (canonical) Monte Carlo method. The operation ratio (move) : (creation) : (annihilation) was chosen to be 8 : 1 : 1.

In addition, we employed an umbrella sampling method [9]: suppose the transition probability is proportional to $W\exp(-U/k_B T)$, where W is an arbitrary weighting function, U is potential energy, k_B Boltzmann constant and T temperature of the system. With this method, we can estimate large free energy difference from fewer

Monte Carlo steps. The weighting function we used was a function of the cluster size n, and was improved by trial and error. Probability $c(n)$ with which a cluster of size n appears in the system was calculated by following equation:

$$c(n) = \frac{c_w(n)/W(n)}{\langle 1/W(n)\rangle_w}, \qquad (1)$$

where $c_w(n)$ is the probability under the weighting function $W(n)$. The cluster formation free energy $\Delta G(n)$ was calculated from

$$\Delta G(n) = T^* \ln[c(n)/c(1)], \qquad (2)$$

where T^* is reduced temperature of the system. We set T^* to 0.80 because the cluster temperature obtained from MD simulations was around 0.8.

The cell size used in the GCMC simulation was smaller than that used in the MD simulation, but much larger than the diameter of the cluster appearing in the GCMC simulation. Thus, the system contained vapor particles, differing from Ref. [3]. The definition of cluster was the same as the MD simulation. The seed was fixed at the center of the cell.

Results

Grand canonical ensemble needs chemical potential μ of the system. Thus, GCMC simulation was carried out without a seed with changing μ until vapor density obtained from the GCMC simulation is equal to that from MD simulation. We obtained μ for each supersaturation ratio as -9.6 ($S = 8.4$) and -10.4 ($S = 2.1$).

The calculated cluster formation free energy at each supersaturation ratio is shown in Fig. 2. At lower supersaturation ratio ($S = 2.1$), the free energy curve has a deep and narrow minimum around $n \sim 20$. This means that a cluster of size ~ 20 is stable. At higher supersaturation ratio ($S = 8.4$), the minimum around $n \sim 40$ is shallow and wide, and the cluster can easily grow larger. Thus, the estimated free energy curves are consistent with the results of the MD simulations.

CONCLUSION

We executed MD and GCMC simulation to investigate heterogeneous (seed-induced) nucleation. From the MD simulations, we found that the seed-induced nucleation takes place at the supersaturation ratio that is too low to cause homogeneous nucleation. At lower supersaturation ratio, no clusters larger than a certain size (15~20) appear around the seed. From the GCMC simulations, we calculated the cluster formation free energy. The results of the GCMC simulations are consistent with the results of the MD simulations.

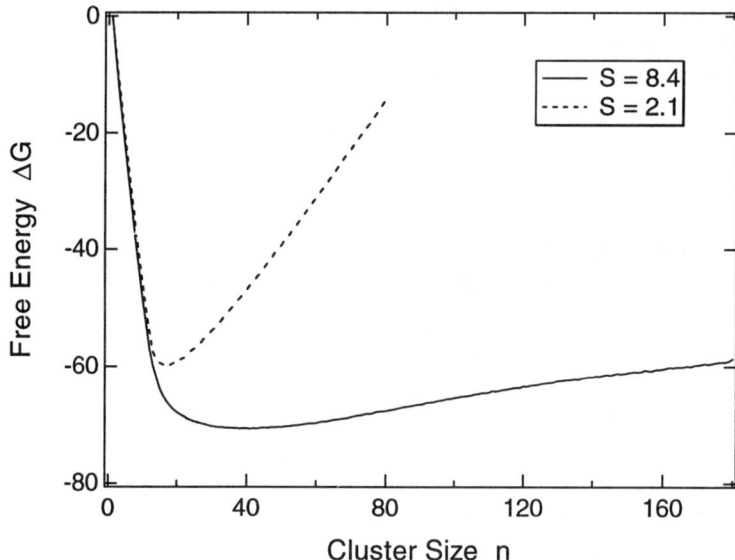

FIGURE 2. Cluster formation free energy. At lower supersaturation ratio (dashed line), free energy curve has a deep and narrow minimum around $n \sim 20$. At higher supersaturation ratio (solid line), the minimum around $n \sim 40$ is shallow and wide.

REFERENCES

1. Yasuoka, K., and Matsumoto, M., *J. Chem. Phys.* **109**, 8451 (1998).
2. Yasuoka, K., and Matsumoto, M., *J. Chem. Phys.* **109**, 8463 (1998).
3. Kusaka I., Wang, Z.-G., and Seinfeld, J.H., *J. Chem. Phys.* **108**, 3416 (1998).
4. ten Wolde, P. R., and Frenkel, D., *J. Chem. Phys.* **109**, 9901 (1998).
5. Oh, K. J., and Zeng, X. C., *J. Chem. Phys.* **110**, 4471 (1999).
6. Hoover, W. G., *Phys. Rev.* **A31**, 1695 (1985).
7. Stillinger, F. H., *J. Chem. Phys.* **38**, 1486 (1963).
8. Yao, J., Greenkorn, R. A., and Chao, K. C., *Mol. Phys.* **46**, 587 (1982).
9. Torrie, G. M., and Valleau, J. P., *J. Comput. Phys.* **23**, 187 (1977).

Activation Barrier for Heterogeneous Condensation in Multicomponent Vapor Mixtures: Cases of Insoluble and Mixed Nuclei

Y. S. Djikaev and D. J. Donaldson

Department of Chemistry, University of Toronto
Toronto, Ontario, M5S 3H6, Canada

Abstract. We develop a thermodynamic treatment of heterogeneous condensation in multicomponent vapor mixtures in the cases where nucleating centers are either insoluble but completely wettable or are partially soluble. This method allows one to find all the main features of the free energy surface without explicitly knowing the free energy itself. The mathematical relations which serve as criteria for whether heterogeneous condensation is barrierless or it occurs in a fluctuational way or does not occur at all are obtained. The theoretical results are illustrated by numerical evaluations for atmospherically relevant systems.

CONDITIONS OF BARRIERLESS CONDENSATION

The formation of cloud droplets is thought to occur by the condensation of water and various other atmospheric gases onto pre-existing nucleating centers, which are ubiquitous in the atmosphere.

Developing our previous work on heterogeneous binary condensation on insoluble nuclei[1], we present here a general approach to the thermodynamics of heterogeneous condensation on insoluble and mixed nuclei. The approach proposed is based on the generalization of the method developed by Kuni et al.[2] for the thermodynamics of heterogeneous unary condensation.

An insoluble nucleus. Let us consider a nucleating center (nucleus) in a vapor mixture of N components. The nucleus is assumed insoluble but completely wettable in a liquid solution of N components. We also assume that it has a spherical shape with the radius R. Considering multicomponent condensation as the formation of a wetting film of liquid solution on the nucleus, we impose no restriction on the thickness h of the film. Clearly, at the beginning of its growth the film is thin (its surface layers may overlap), and important information is lost by assuming that the film is thick (its surface layers do not overlap). We will use the Greek superscripts α and β for marking values in the film (α) and vapor mixture (β). χ denotes the set $\{\chi_1,...,\chi_{N-1}\}$ of mole fractions in the droplet defined as $\chi_i = v_i/(v_1+...+v_N)$ ($i=1,...,N$) where v_i is the number of molecules of component i in a droplet. If we choose h and χ as independent variables of state of the droplet, the partial derivative of the free energy F with respect to h is given by the expression

$$\partial F / \partial h = f(h,\chi)(\mu^\alpha - \mu^\beta), \qquad (1)$$

where $f(h,\chi)$ is a positive, continuous function with continuous partial derivatives of any order, and

$$\mu^\alpha = \tfrac{1}{k_BT}\sum_{i=1}^{N}\chi_i(\mu_i^\alpha - \mu_{i\infty}^\alpha), \quad \mu^\beta = \tfrac{1}{k_BT}\sum_{i=1}^{N}\chi_i(\mu_i^\beta - \mu_{i\infty}^\beta), \quad (2)$$

represent the average deviations of the chemical potentials, μ_i ($i=1,...,N$), from their equilibrium values, $\mu_{i\infty}$, in the droplet(α) and vapor(β), respectively, expressed in units k_BT (k_B is Boltzmann's constant and T is the temperature of the system).

Let us denote by μ^α_∞ the effective chemical potential of the bulk solution: $\mu^\alpha \to \mu^\alpha_\infty$ as $h \to \infty$. One can show that $\mu^\alpha \to -\infty$ as $h \to 0$ at any fixed χ. On the other hand, one can show that at any fixed χ the effective chemical potential in the droplet, μ^α, as a function of h has a maximum, μ^α_{max}, at some $h_*\equiv h_*(\chi)$, i.e., $\mu^\alpha_{max}\equiv\mu^\alpha_{max}(\chi)=\mu^\alpha(h_*(\chi),\chi)$. The existence of this maximum is due to the concurrence of two contributions to μ^α which arise from the capillary and disjoining pressures within the droplet (film)[1]. Thus the equality $\mu^\beta_* = \mu^\alpha_{max}$ determines the threshold value μ^β_* of the effective chemical potential of the vapor mixture as a function of χ.

A partially soluble nucleus.[3] Let us consider a multicomponent vapor mixture in which there is a nucleating center consisting of an insoluble core and M soluble non-volatile species. A droplet is formed as a result of condensation of molecules from the vapor mixture onto the nucleus. As a consequence, a particle of $(N+M)$-component solution is formed, in which the insoluble core is immersed.

Let us label attributes of the components which are present initially only in the vapor mixture with the subscript i ($i=1,...,N$). When in the droplet, the component i will be referred to as "condensate i".

Attributes of the nonvolatile soluble species of the nucleus are labelled with the subscript j ($j=1,...,M$). We will use q_j to denote the number of molecules of nonvolatile soluble species j of the nucleus. When in the droplet, the nonvolatile soluble species j will be called "solute j".

Once a droplet has been formed on a nucleus, the insoluble core is immersed in a liquid solution formed by $q_1+...+q_M$ solute molecules and by $v_1+...+v_N$ condensate molecules. All the constituents of the droplet are assumed to be miscible and non-surfactant, so that the liquid part of the droplet consists of one single phase. The soluble part of the nucleus is completely dissolved. The dissociation of nonvolatile species is also assumed to be complete if it occurs. Again, we use the Greek superscripts α and β for marking values in the droplet (α) and vapor mixture (β).

Let us introduce the variable v as $v=v_1+...+v_N$ and the variable χ as the set $\{\chi_1,...,\chi_{N-1}\}$ of nominal mole fractions of condensates in a droplet defined as $\chi_i=v_i/(v_1+...+v_N)$ ($i=1,...,N$), where v_i is the number of molecules of condensate i in the droplet, and let us choose v and χ as independent variables of state of the droplet. The real mole fraction of condensate i in the droplet is $x_i=v_i/(v+q)$, where $q=q_1+...+q_M$. Usually, the condition $q/v \ll 1$ is perfectly fulfilled for nuclei and droplets of interest in the atmosphere. Therefore, $x_i=\chi_i(1-q/v)$.

The first derivative of the free energy F of formation of a droplet on a mixed nucleus with respect to v is

$$\partial F / \partial v = \mu^\alpha - \mu^\beta, \quad (3)$$

where, once again, μ^α and μ^β are defined by eq.(2) and represent the average deviations of the chemical potentials, μ_i ($i=1,...,N$), from their equilibrium values, $\mu_{i\infty}$, in the droplet(α) and vapor(β), respectively, expressed in units k_BT.

As above, let us denote by μ^α_∞ the effective chemical potential of the bulk solution: $\mu^\alpha \to \mu^\alpha_\infty$ as $v \to \infty$. One can show that $\mu^\alpha \to -\infty$ as $v \to 0$ at any fixed χ. On the other hand, one can show that at any fixed χ the effective chemical potential in the droplet, μ^α, as a function of v has a maximum, μ^α_{max}, at some $v_* \equiv v_*(\chi)$, i.e., $\mu^\alpha_{max} \equiv \mu^\alpha_{max}(\chi) = \mu^\alpha(v_*(\chi),\chi)$. The existence of this maximum is due to the concurrence of two contributions to μ^α which arise from the presence of solute molecules in the droplet and from the capillary pressure in the droplet[3]. Thus, the equality $\mu^\beta_* = \mu^\alpha_{max}$ determines the threshold magnitude μ^β_* of the effective chemical potential of the vapor mixture as a function of χ.

Criteria for barrierless condensation. Taking account of eqs.(3) and (5), one can affirm that heterogeneous condensation occurs in a barrierless way if for a given set of saturation ratios, $\zeta_1,...,\zeta_N$, there exists at least one set of mole fractions such that

$$\mu^\beta \geq \mu^\beta_*. \qquad (4)$$

If at given $\zeta_1,...,\zeta_N$ condition (4) is not satisfied for any χ but yet

$$\mu^\alpha_\infty < \mu^\beta \leq \mu^\beta_* \qquad (5)$$

for some χ or manifold of χ, then heterogeneous condensation occurs in a usual way demanding droplets to overcome the activation barrier of condensation.

If at given $\zeta_1,...,\zeta_N$ there is no χ at which conditions (4) or (5) are satisfied, i.e.,

$$\mu^\beta \leq \mu^\alpha_\infty \qquad (6)$$

for all χ, then formation of droplets growing regularly and irreversibly is impossible, i.e., condensation does not occur at all.

ACTIVATION BARRIER OF CONDENSATION

Since the following arguments are identical for both insoluble and mixed nuclei, we will use the notation y for the variables h or v.

Using the properties of μ^α, one can show that under condition (5) for any fixed χ the equation $\mu^\alpha - \mu^\beta = 0$ has two solutions, $y'_e = y'_e(\chi)$ and $y'_c = y'_c(\chi)$, such that

$$(\partial F/\partial y)_e = 0, \quad (\partial F/\partial y)_c = 0, \quad (\partial^2 F/\partial y^2)_e > 0, \quad (\partial^2 F/\partial y^2)_c < 0. \qquad (7)$$

These mean that y'_e and y'_c are the points of minimum and maximum, respectively, of F as a function of y at fixed χ satisfying conditions (5).

Let us introduce $\chi^{(i)}$ as the set χ without χ_i and consider the behaviour of F as a function of χ_i at fixed y and $\chi^{(i)}$. One can show that $\partial F/\partial \chi_1 \to -\infty$ as $\chi_1 \to 0$ and $\partial F/\partial \chi_1 \to \infty$ as $\chi_1 \to 1$. Thus there exists at least one χ_i which provides the solution for the equation $\partial F/\partial \chi_i = 0$ at any fixed y and $\chi^{(i)}$ (F attains its minimum at this point).

Thus the free energy surface, defined by the function $F = F(y,\chi)$, has <u>at least</u> two extrema (a "well" point and a "saddle" point) whose coordinates are determined by

$$\partial F/\partial y = 0, \quad \partial F/\partial \chi_i = 0 \quad (i=1,...,N-1). \qquad (8)$$

If some of the rightmost equations in eq.(8) have multiple roots, the coordinates of the saddle and well points are selected by the conditions:

$$\det F_c'' < 0, \quad \det F_e'' > 0, \quad (\partial^2 F/\partial y^2)_e > 0, \quad (\partial^2 F/\partial \chi_i^2)_e > 0 \quad (i = 1,...,N-1), \quad (9)$$

where F'' is the matrix of second derivative of F and the subscripts "e" and "c" label values at the well and saddle points, respectively.

NUMERICAL CALCULATIONS AND CONCLUSIONS

Numerical calculations. The theoretical results were illustrated by numerical calculations for heterogeneous condensation on quartz particles and NaCl containing mixed nuclei. We considered several multicomponent vapor mixtures which are of great interest for atmospheric cloud formation (e.g., sulfuric acid-nitric acid-water).

Conclusions. We have developed a general approach to the thermodynamics of heterogeneous condensation on insoluble and partially soluble in a multicomponent vapor mixture. Based on the differential relations for the free energy of formation F, that approach allows one to find out all main features of the free energy surface without knowing the function F itself.

Using that method, we have shown that the disjoining pressure within the droplet (in the case of an insoluble nucleus) and the presence of the solute molecules in the droplet (in the case of a partially soluble nucleus) play a key role in the thermodynamics of heterogeneous condensation. In respective cases, the contributions of the disjoining pressure or solute molecules to the free energy of formation greatly lower the height F_c-F_e of the activation barrier of heterogeneous condensation and can cause the disappearance of the barrier at very low metastability of the vapor mixture.

Using this approach, we have found the inequalities which, at given metastability of the vapor mixture, allow us to definitely predict whether heterogeneous multicomponent condensation is barrierless or it occurs in a usual (fluctuational) way or does not take place at all.

In the case where heterogeneous condensation occurs in a fluctuational way, we have shown that the free energy surface has at least one well point and one saddle point and we have found explicit equations for determining their coordinates.

ACKNOWLEDGMENTS

This work was financially supported by NSERC.

REFERENCES

1. Djikaev, Y. S., and Donaldson, D. J., *J. Geophys. Res.* **104**, 14,283-14,292 (1999).
2. Kuni, F. M., Shchekin, A. K, and Rusanov, A. I., *Colloid J. Russ. Acad. Sci.*, Engl. Transl., **55**(2), 174-183 (1993).
3. Djikaev, Y. S., and Donaldson, D. J., submitted to *J. Geophys. Res.* in March, 2000.

Microscopic Effects and Kinetics of Binary Nucleation Beyond the Confines of the Fokker-Planck Approximation

Y. S. Djikaev[a], A. P. Grinin[b], and F. M. Kuni[b]

[a] Department of Chemistry, University of Toronto, Toronto, Ontario, M5S 3H6, Canada
[b] Department of Statistical Physics, St-Petersburg State University, St-Petersburg, 198094, Russia

Abstract. The kinetic equation of isothermal binary nucleation is derived from the discrete equation of balance. It is shown that under some circumstances the Fokker-Planck approximation in the kinetic equation is not applicable. For such a case, we establish the hierarchy of the time scales of binary nucleation. This hierarchy allows one to separate and analytically describe the stage of concentration relaxation of binary nucleation during which the distribution of nuclei with respect to the concentration of solution in them approaches the quasi-equilibrium Gaussian distribution. During the subsequent evolution of nuclei the kinetic equation is solved by using the method of Enskog and Chapman, and all the main characteristics of the kinetics of binary nucleation are found. We also study two microscopic effects of nucleation, namely, the effect of fluctuations of the nucleus composition and the effect of material quasi-isolation of a nucleus. It is shown that the contributions of these two effects to the kinetics of binary nucleation compensate.

MICROSCOPIC EFFECTS OF BINARY NUCLEATION

Binary condensation is a very widespread first order phase transition and hence is of great interest in many fields. Though the heterogeneous mechanism of condensation is more favorable compared to the homogeneous one, in many cases homogeneous binary condensation remains of great importance as a source for the formation of secondary aerosols[1] which can serve as cloud condensation nuclei (CCN).

Let us denote by v_i $(i=1,2)$ the number of molecules of component i in a nucleus. The concentration of the solution in the nucleus can be characterized by the mole fraction χ of the first component: $\chi = v_1/(v_1+v_2)$. The temperature of the nuclei is assumed to be constant and equal to that of the vapor mixture, T.

Let us consider the vapor mixture where the number density n_2 of molecules of the undersaturated vapor of component 2 is much greater than the number density n_1 of molecules of the supersaturated vapor of component 1 (the degree of supersaturation does not matter). Then the quasi-equilibrium is quickly established between the nucleus and the vapor with respect to component 2 at any given v_1. We denote by $v_2^* \equiv v_2^*(v_1)$ the value of v_2 which ensures the equilibrium concentration χ_c of the solution in the nucleus: $\chi_c = v_1/(v_1+v_2^*)$.

In developing the kinetic theory of binary nucleation two microscopic effects should be taken into account.

First, there is[2] the effect of fluctuations of the solution concentration in a nucleus: though at any v_1 the nucleus is in quasi-equilibrium with respect to v_2 and it contains the solution of equilibrium concentration χ_c, the fluctuations of v_2 about its equilibrium value v_2^* cause the fluctuations of χ about χ_c. These fluctuations influence the ability of the nucleus to emit the molecules of component *1*.

Secondly, there is[3] the effect of material quasi-isolation of a nucleus. Let us denote by δ_i (*i=1,2*) the change in the variable χ of the nucleus when it absorbs or emits a molecule of component *i*. Since the act of emission of a molecule of component *i* is not an instantaneous event and during this event the solution concentration in the nucleus changes from χ to $\chi_c - \delta_i$, it remains unclear by what value of concentration from this interval the ability of the nucleus to emit the molecules of component *i* is determined. Note that in our case one can neglect the effect of material quasi-isolation of a nucleus with respect to component *2* as well as the influence of the fluctuations of the nucleus composition on the nucleus ability to emit molecules of component *2*.

One can show[3] that, in main order of magnitude of the small parameter of the theory, the joint consideration of the effect of fluctuations of the nucleus composition and of the effect of material quasi-isolation of a nucleus leads to their compensation.

KINETIC EQUATION BEYOND THE FOKKER – PLANCK APPROXIMATION

The state of the nucleus can be determined by the independent variables v_1 and χ. Let us denote by $\rho(v_1,\chi,t)$ the two-dimensional distribution of nuclei with respect to v_1 and χ at time *t*. The time evolution of this distribution is governed by the discrete equation of balance. Reducing that equation to a differential form in the vicinity of the saddle point of the free energy surface, one can obtain the kinetic equation of binary nucleation; usually it is a two-dimensional Fokker – Planck equation.[4]

Let us introduce the variable ξ as

$$\xi = (\chi - \chi_c)/2^{1/2}\Delta\chi \quad (1)$$

and define the function $P(v_1,\xi,t)$ by

$$\rho(v_1,\chi,t) \equiv [2\pi\Delta\chi]^{-1/2} e^{-\xi^2} P(v_1,\xi,t), \quad (2)$$

where $\Delta\chi$ is the equilibrium rms fluctuation of the variable χ due to the fluctuations of v_2. Introducing the parameters α_1 and α_2 by

$$\alpha_i = \delta_i/2^{1/2}\Delta\chi \quad (i=1,2) \quad (3)$$

and assuming them to satisfy the conditions

$$2^{1/2}\alpha_1 \leq 1, \quad 2^{1/2}\alpha_2 \ll 1, \quad (4)$$

one can obtain[5] the kinetic equation governing the time evolution of the function $P(v_1,\xi,t)$ (from the following equation on we omit the arguments $v_1,\xi,$ and *t* of all quantities and the free energy *F* of nucleus formation is expressed in units $k_B T$):

$$\frac{\partial P}{\partial t} = -\frac{\partial}{\partial v_1}\left[\hat{L} - W_1 \sum_{m=1}^{\infty}\frac{\alpha_1^m}{m!}\frac{\partial^m}{\partial \xi^m}\right]P + \sum_{m=1}^{\infty}\frac{(-1)^m \alpha_1^m}{m!}\hat{L}\left(\frac{\partial}{\partial \xi} - 2\xi\right)^m P$$

$$- W_1 \sum_{m \neq m'=1}^{\infty}\frac{(-1)^{m'}\alpha_1^{m+m'}}{m!m'!}\left(\frac{\partial}{\partial \xi} - 2\xi\right)^{m'}\frac{\partial^m}{\partial \xi^m}P \qquad (5)$$

$$W_1\left[\frac{q+1}{q}\alpha_1^2\left(\frac{\partial}{\partial \xi} - 2\xi\right)\frac{\partial}{\partial \xi} - \sum_{m=2}^{\infty}\frac{(-1)^m \alpha_1^{2m}}{m!m!}\left(\frac{\partial}{\partial \xi} - 2\xi\right)^{m'}\frac{\partial^m}{\partial \xi^m}\right]P,$$

where

$$q = (W_1/W_2)(1-\chi_c)^2/\chi_c^2, \quad \hat{L} = -W_1 F_1' - W_1 \frac{\partial}{\partial v_1}, \qquad (6)$$

W_i (i=1,2) is the number of molecules of component i being absorbed by the critical nucleus; F_1' is the first derivative (with respect to v_1) of the free energy of formation of a nucleus (v_1, χ_c). The fact that equation (5) goes beyond the framework of the Fokker-Planck approximation is clearly due to the leftmost condition in eq.(4).

STAGE OF CONCENTRATION RELAXATION OF NUCLEI

Assuming the inequality $\alpha_1 q/2(q+1) << 1$ to be fulfilled, the analysis of eq.(5) allows one to separate the stage of concentration relaxation in the evolution of the function P. During this stage the distribution of nuclei with respect to the concentration of solution in them approaches the quasi-equilibrium Gaussian distribution. At this stage the solution of eq.(5) is given by

$$P = g + \sum_{i=1}^{\infty}\exp(-2\alpha_1^2 i \lambda_i W_1 t) g_i H_i, \qquad (7)$$

where $H_i \equiv H_i(\xi)$ (i=0,1,2,...) is the Hermit polynomial,

$$\lambda_1 = (q+1)/q, \quad \lambda_i = (q+1)/q + (i-1)! \sum_{m=2}^{i}\frac{(2\alpha_1^2)^{m-1}}{m!m!(i-m)!} \quad (i \geq m), \qquad (8)$$

g and g_i are independent of ξ and t and are determined through the initial two-dimensional distribution $P_0 \equiv P|_{t=0}$ as

$$g = (H_0, P) = (H_0, P_0), \quad g_i = (2^i i!)^{-1}(H_i, P_0), \qquad (9)$$

with the scalar product (Φ, Ψ) of the real functions Φ and Ψ of ξ being defined by

$$(\Phi, \Psi) = \pi^{-1/2}\int_{-\infty}^{\infty} d\xi\, e^{-\xi^2} \Phi\Psi. \qquad (10)$$

The function $g \equiv g(v_1)$ is the distribution of nuclei, having $\chi = \chi_c$, with respect to v_1.

At the end of the stage of concentration relaxation we have

$$P \approx g \quad (t \geq t_\xi), \quad t_\xi = 1/2\alpha_1^2 \lambda_1 W_1 = q/(2(q+1)2\alpha_1^2 W_1,$$
$$t_v \sim (\Delta v_1)^2/W_1, \quad t_\xi/t_v \sim q/2(q+1)(\Delta v_1)^2 \alpha_1^2 << 1, \qquad (11)$$

where t_ξ represents the duration of the stage of concentration relaxation, Δv_1 is the half width of the vicinity of the saddle point along the v_1 axis, and t_v is the characteristic time of change of the one-dimensional distribution g with respect to v_1. The inequality $t_\xi/t_V \ll 1$ expresses the hierarchy of the time scales of the evolution of nuclei.

STAGE FOLLOWING CONCENTRATION RELAXATION

After the distribution of nuclei with respect to the variable ξ approaches the Gaussian (i.e., P approaches g), the solution of eq.(5) can be constructed with the help of the Enskog-Chapman method, the quasi-equilibrium distribution being taken as the zeroth approximation. One can find that

$$P = g + \sum_{i=1}^{\infty}(2\alpha_1)^{-i}x_i(g)H_i, \quad x_i(g) = \left(\delta_{i1}\frac{q}{q+1} + \frac{1}{q+1}b_i\right)\frac{1}{W_1}\hat{L}g, \quad (12)$$

where δ_{i1} is the Kroneker symbol and $b_1, b_2, b_3, \ldots$ are found by solving the equations

$$b_i = \sum_{j=1}^{\infty}\Gamma_{ij}b_j + (1-\delta_{i1})\frac{(2\alpha_1^2)^{i-1}}{i!i\lambda_i} \quad (13)$$

with the coefficients Γ_{ij} defined as follows: if $i=j$, then $\Gamma_{ij}=0$; and if $i \neq j$, then

$$\Gamma_{ij} = \frac{j!}{i\lambda_i}\sum_{m=\gamma_{ij}}^{i}\frac{(2\alpha_1^2)^{i-j+m-1}}{m!(j-m)!(i-j+m)!} \quad [\gamma_{ij}=1 \ (i>j); \ \gamma_{ij}=1-i+j \ (i>j)]. (14)$$

The time evolution of the distribution g itself is governed by the equation

$$\frac{\partial g}{\partial t} = -\frac{\partial}{\partial v_1}J, \quad J = (1+q)^{-1}(1+q^{-1}b_1)\hat{L}g, \quad (15)$$

where J is the nucleation rate and the function g is subject to the usual boundary conditions

$$g/g_e = 1 \ (v_1 \leq v_{1c} - \Delta v_1), \quad g/g_e = 0 \ (v_1 \leq v_{1c} + \Delta v_1) \quad (16)$$

(g_e is the equilibrium distribution g, and "c" labels attributes of the critical droplet).

Thus the problem of finding the solution of eq.(5) is reduced to the well-studied problem of solving one-dimensional kinetic equation (15) with boundary conditions (16). Therefore, all the main characteristics of binary nucleation can be found.

ACKNOWLEDGMENTS

This work was partially (Y.S.D) supported by NSERC.

REFERENCES

1. Kerminen, V.-M., Wexler, A. S., and Potukuchi, S., *J. Geophys. Res.* **102**/D3, 3715-3724 (1997).
2. Grinin, A. P., Kuni, F. M., and Dzhikaev, Y. S., *Vestn. LGU, Fiz. Khim.* (in Russian) 2(11), 93-96 (1989).
3. Grinin, A. P., Dzhikaev, Y. S., and Kuni, F. M., *Vestn. LGU, Fiz. Khim.* (in Russian) 3(18), 79-81 (1990).
4. Reiss, H., *J. Chem. Phys.* **18**, 840-848 (1950).
5. Grinin, A. P., Dzhikaev, Y. S., and Kuni, F. M., *Sov. Phys. Tech. Phys.* **37**(6), 589-593 (1992).

The Influence of Particle Solubility on Heterogeneous Nucleation in Binary Vapor Mixtures

D. Petersen[1], R. Ortner[1], A. Vrtala[1], A. Laaksonen[2], M. Kulmala[3], P. E. Wagner[1]

[1]*Institut für Experimentalphysik, Universität Wien*
Boltzmanngasse 5, A-1090 Wien, Austria

[2]*Department of Physics, University of Kuopio*
P.O. Box 1627, FIN-70211 Kuopio, Finland

[3]*Department of Physics, University of Helsinki*
P.O.Box 9, FIN-00014 Helsinki, Finland

Abstract. Heterogeneous nucleation of binary n-propanol - water vapor mixtures on non-soluble or partially soluble 8nm Ag and NaCl particles has been investigated experimentally. At constant temperature the vapor activities have been stepwise increased and the number of particles activated to condensational growth shows a comparatively steep increase until all particles have been activated. Onset activities were obtained for various mixing ratios of the binary vapor mixtures. In contrast to homogeneous nucleation, for binary heterogeneous nucleation only a conucleation without significant mutual enhancement was observed. Heterogeneous nucleation calculations based on Fletcher theory using experimentally obtained contact angles have been performed. Satisfactory agreement of the slopes of the nucleation probability curves was observed. However, considerable deviations between experimental and theoretical onset activities indicate problems of the binary nucleation theory based on the capillarity approximation. For NaCl particles and water rich mixtures the Köhler theory is in good agreement with the experimental data. On the other hand, for n-propanol rich mixtures and NaCl particles the heterogeneous nucleation calculations based on Fletcher theory provide a reasonable approximation of the experimental data. It appears that neither Fletcher nor Köhler theory are applicable to NaCl particles in the region of transition from water rich to n-propanol rich mixtures.

INTRODUCTION

Heterogeneous nucleation in multicomponent vapor mixtures is a common atmospheric process. However, unfortunately there is only limited information available on heterogeneous nucleation of unary vapors [1,3,4]. Evidence on investigations of heterogeneous nucleation of *binary* vapor mixtures can hardly be found in the literature. In this paper the influence of particle surface properties and solubility on the nucleation properties is presented. Non-soluble particles with either non-oxidized or oxidized surfaces, as well as particles soluble in one compound of the binary liquid mixture were considered. Binary vapor mixtures with well known physico-chemical properties and various mixing ratios have been studied in these experiments. Measurements were performed at a constant nucleation temperature.

METHOD

Heterogeneous nucleation in binary vapor mixtures has been investigated experimentally under well defined laboratory conditions. Particles were activated to condensational growth and enlarged to optically detectable droplet sizes. The growing droplets are illuminated by a laser beam and the Constant Angle Mie Scattering (CAMS) method [6] is used for evaluation of the number concentration of the activated particles. The fraction of particles activated depends on the physico-chemical properties of the particle surface as well as on the prevailing vapor activities. Well defined Ag and NaCl particles and n-propanol-water vapor mixtures with various mixing ratios were considered. The vapor phase activities were increased during each series of measurements. Beyond a certain vapor activity all of the particles were activated to condensational growth. The values of the vapor activities at which 50 % of the particles were activated to condensational growth are considered as onset vapor activities for heterogeneous nucleation.

The Ag or NaCl particles were generated in a high temperature furnace tube by evaporation and subsequent condensation of the substance either in an air or a nitrogen flow. Using nitrogen possible changes of the Ag particle surface due to oxidation were prevented. The flows were precisely controlled by means of a critical orifice flow control unit. A monodispersed fraction of the particles generated was obtained by electrostatic classification.

The vapor mixtures considered were generated by means of the evaporation of liquid mixtures injected by a syringe pump and diluted by particle-free dry clean air. Increasing the liquid flow rate from the syringe pump leads to an increase of the vapor activities. The mixing ratio of the compounds in the liquid mixture has been varied to observe the influence of the concentration of each compound on the heterogeneous nucleation behavior. Subsequently, the monodispersed condensation nuclei aerosol was mixed with the binary vapor mixture so that a humidified aerosol with well known vapor activities of each compound of the vapor mixture was obtained.

The heterogeneous nucleation in binary vapor mixtures has been investigated by means of the Size Analyzing Nuclei Counter – SANC. In this pressure defined fast expansion chamber a pressure drop leads to a temperature drop and supersaturation and particles acting as condensation nuclei start growing to detectable sizes. For unary vapors the agreement between the calculated supersaturation and the supersaturation actually prevailing in the expansion chamber has been verified by means of droplet growth measurements, as droplet growth is a very sensitive indicator for the supersaturation. Varying the vapor activities by variation of the liquid flow from the syringe pump allows to obtain nucleation probability curves at constant temperature after expansion. Beyond certain activities a steep increase of the particle number concentration of the activated particles was observed. Eight mixing ratios of water and n-propanol were considered. The onset vapor activities for NaCl particles were obtained at a temperature 288K and the onset vapor activities for Ag particles were obtained at a temperature 285K.

RESULTS AND DISCUSSION

Nucleation probabilities and onset activities at constant temperature have been investigated for heterogeneous nucleation of 8nm Ag and NaCl particles in water- n-propanol vapor mixtures. NaCl is soluble in water and non-soluble in n-propanol. On the other hand, Ag is non-soluble in both of those substances. Accordingly, the influence of the solubility on the heterogeneous nucleation behavior can be studied. For Ag as well as for NaCl particles nucleation probability curves have been experimentally determined. They indicate a steep increase of the normalized particle number concentration of the activated particles beyond a certain value of water or n-propanol activity, respectively. With increasing concentration of n-propanol in the n-propanol-water vapor mixture a decrease of the water onset activity was observed. On the other hand increasing concentration of water in the n-propanol-water vapor mixture a decrease of the n-propanol onset activity was found.

As shown Figure 1 the onset activity curves for Ag and NaCl particles reveal a qualitatively similar behavior showing a rather convex shape of the onset activity curve which indicates a conucleation of n-propanol and water vapor but no mutual enhancement. This feature is different from the behavior of binary *homogeneous* nucleation of n-propanol – water, where a concave shape of the onset activity curve and thus significant mutual enhancement of the homogeneous nucleation was perceived[5]. As expected due to the solubility of NaCl in water, the onset water activity for NaCl particles is much lower than for Ag particles. On the other hand, the n-propanol onset activity for NaCl is higher than for Ag particles. Oxidation of the Ag particle surface reduces the value of the water onset activity in experiments with unary water vapor and increases the value of the n-propanol onset activity in experiments with unary n-propanol vapor.

Theoretical calculations[2] based on the capillarity approximation were performed. For these calculations the values of the contact angles are needed. The macroscopic contact angles have been measured for various mixing ratios by the Wilhelmy plate method. The slopes of the experimental nucleation probability curves were found to be in satisfactory agreement with theory, however, deviations of the actual nucleation probabilities and correspondingly onset activities usually occur, which may be connected with the macroscopic assumptions of the capillarity approximation. Even though discrepancies were also reported for the case of binary homogeneous nucleation, for homogeneous nucleation in *unary* vapors usually reasonable agreement between calculations and experimental data was observed[5]. In contrast, in the case of heterogeneous nucleation on Ag particles even for *unary* water vapor already quite a large disagreement between calculations and experimental data has been observed. This disagreement could tentatively be explained by differences between macroscopic and microscopic contact angles. The important role of the surface properties in the heterogeneous nucleation behavior is emphasized. The deviations for non oxidized Ag particles were found to be slightly smaller than for oxidized Ag particles. For NaCl particles on the water rich side a quite good agreement with Köhler theory is consistent with the solubility of NaCl in water. As NaCl is non-soluble in n-propanol, on the n-propanol rich side the heterogeneous nucleation theory has been applied resulting in a reasonable approximation. These calculations were performed using the experimentally obtained contact angle of 0° between NaCl and n-propanol. A quantitative description of the transition from the water rich side to the n-propanol rich side requires an extended theoretical approach, which considers the solubility in one compound and the non-solubility in the other compound of the binary mixture.

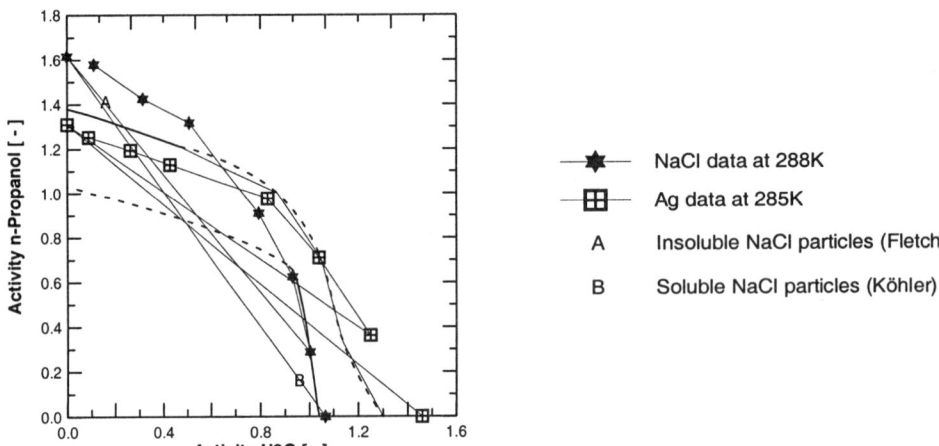

FIGURE 1. Onset Activities for the Heterogeneous Nucleation of 8 nm NaCl and Ag Particles in n-Propanol - Water Vapor Mixtures. The Curves A, B Show Calculations for NaCl Particles Based on Fletcher Theory and on Köhler Theory, respectively.

REFERENCES

1. Kotzick, R., Panne, U., Niessner, R. *J. Aerosol Sci.* **28**, 725 (1997).
2. Lazaridis, M., Kulmala, M., Gorbunov, B. Z. *J. Aerosol Sci.* **23**, 457 (1992).
3. Porstendörfer, J., Scheibel, H. G., Pohl, F. G., Preining, O., Reischl, G., Wagner, P. E., *Aerosol Sci. Technol.* **4**, 65 (1985).
4. Smolik, J., Schwarz, J., *J. Colloid Interface Sci.* **185**, 382 (1997).
5. Strey, R., Viisanen, Y., Wagner, P. E., *J. Chem. Phys.* **103**, 4333 (1995).
6. Wagner, P. E., *J. Colloid Interface Sci.* **105**, 456 (1985).

Use of a Turbulent Mixing CNC to Study the Influence of Composition and Vapor Properties on Heterogenous Nucleation

Philip K. Hopke[1], D.W.Lee[1], Rashid Mavliev[2], Hwa-Chi Wang[2,3]

1) Department of Chemistry, Clarkson University, Potsdam NY 13699 USA; 2) Department of Chemical and Engineering, Illinois Institute of Technology, Chicago, IL 60616 USA; 3) Air Liquide, Chicago Research Center, Countryside, IL USA

Abstract. A new method for changing the supersaturation in the Turbulent Mixing CNC has been developed and used to study the influence of composition and vapor properties on heterogeneous nucleation. Supersaturation was controlled by changing the condensing vapor pressure in nozzle flow by saturating only a predetermined part of the flow while the total flow and temperature remain constant. This approach allows changing the initial vapor pressure while keeping the flow structure and temperature field unchanged. Experimental results for transitions from heterogeneous nucleation to homogeneous nucleation are presented for NaCl and WOx particles at various DBP vapor pressures. With increasing of the DBP vapor pressure, the concentration of enlarged particles increases until it reaches a plateau. At higher initial values of DBP pressure, homogeneous nucleation prevails and the number concentration of particles follows a curve typical for homogeneous nucleation recorded in the absence of nuclei. Nuclei with different mobility diameters were activated at different values of vapor pressure. There are significant differences in the slopes of particle activation curves for NaCl and WOx particles. The reasons for such differences are the subject for continuing research.

INTRODUCTION

Recently in the field of nanometer sized particles, attention has been paid to homogeneous nucleation, binary nucleation, ion-induced nucleation and simultaneous heterogeneous and homogeneous nucleation because of the importance of these processes in atmospheric particle formation (Kim et al, 1998). In addition, discrepancies between theory and experiment have stimulated new research in this field. Several new techniques have been developed to more accurately characterize heavy ions and particles in the nanometer size range.

The Condensation Nuclei Counter (CNC), that grows primary particles (nuclei) to more easily detectable sizes, is one of the most widely used devices for studying particles below 0.1 µm. Several types of CNC's are used in aerosol research. The main difference among these CNC designs is the way they produce supersaturation that leads to particle growth up to a predetermined size for subsequent detection.

Although a CNC is primarily devoted to measuring the number concentration of particles, in recent years it has been shown that a CNC can be used measuring the size distribution of nanometer particles. The size distribution of nuclei can be measured by means of changing the CNC's sensitivity (McDermont et al. 1991) and by means of measuring the size of grown particles (Rebours et al., 1996; Saros et al., 1996). The last approach is based on the fact that the growth of smaller particles is delayed because of the Kelvin effect that results in the final size of particles being dependent on initial nucleus size.

In this work, a new approach for changing the sensitivity of turbulent mixing (TM) CNC is described. The approach is based on changing the vapor pressure of working fluid by changing the ratio of saturated and by-pass flows in the vapor generator. This approach was used to investigate the transition from heterogeneous to homogeneous nucleation for two types of initial nuclei.

CNC OPERATION PRINCIPLE

In TMCNC the aerosol flow ("cold" flow) mixed with gas flow saturated by working fluid vapor ("hot" flow). The mixing of "cold" and "hot" flows causes supersaturation because the saturation pressure corresponding to the temperature after mixing is lower than the vapor pressure in the mixed flow. The common technique to vary the degree of supersaturation in a CNC, and therefore, the equivalent Kelvin diameter, is adjusting the temperature of the vapor generator. A disadvantage of this method is that the temperature of the mixing zone depends on the on temperature of the vapor generator, and therefore the effect of supersaturation is convoluted with that of the temperature field in the mixing zone (Adachi et al., 1992). In the present research, a different approach was developed to control supersaturation without changing the saturator temperature and flow rate and thus preserving the temperature and flow field.

Two modifications of TM CNC were used in these experiments. The design and performance of the first version of TM CNC is described elsewhere (Mavliev and Wang 2000). The scheme of the modified TMCNC is shown in Figure 1. The modified version of TM CNC consists from three major parts – saturator, mixer and condenser. All three parts of TM CNC are machined from a block of aluminum alloy. The flow channels are machined by drilling through the aluminum block. Additional holes are drilled to install the heating elements and a solid-state temperature sensors. This approach allows uniform temperature distribution within the block.

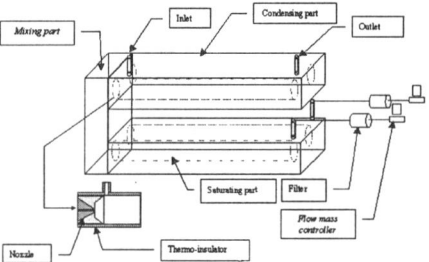

FIGURE 1. The scheme of the modified Turbulent Mixing CNC.

The vapor generator (saturator) and mixer are placed in an outer Teflon shell for insulation up to 150 C. Temperatures of these two parts are controlled separately by an circuit board and Data Acquisition System (5500MF from ADAC corp.) with an accuracy of ±0.1 C. Two chambers of equal size approximately 80 mL of volume are machined in saturator block. Aluminum alloy crucible, which can be filled with a working fluid of 15-20 cm^3 by volume, can be placed in one of the chambers to saturate flow with working fluid vapor. Another chamber is kept empty and dry. Flows from these two chambers are connected and mixed in mixer part (see figure 1) before passage through nozzle to the condenser part. The degree of supersaturation is adjusted by varying the ratio of flows passing through these two chambers, which are equilibrated at the same temperature. The saturated vapor flow is directed to condenser through a 1 mm diameter nozzle. The aerosol flow is directed into the condenser through a circular opening. These two coaxial flows are mixed and directed through outlet connector to a particle detection system. The residence time of particles in the condensation part is about 1.9 seconds, assuming a fully developed flow. The actual residence time may be somewhat shorter because of the turbulent jet flow structure at the entrance of the growth tube. However, it is sufficient for particle growth if the saturator temperature is set at 110 C or above (Mavliev and Wang, 2000). The condenser is cooled by a water flow from a circulating bath at a temperature of 18-20 C to prevent the condensation of DBP vapor on optical elements downstream. The concentration of particles after condensation growth is measured using a TSI CNC 3760, an optical particle counter (PMS OPC-501) and a laser aerosol spectrometer (PMS LAS-X). The optical counter (OPC-501) has a detection range of 0.5 μ and 5 μ and is designed for a flow rate of 2.8 lpm. The laser aerosol spectrometer (LAS-X) has four measuring ranges with detection limits of 1.5 μm, 0.3 μm, 0.17 μm, and 0.12 μm respectively. The third instrument used to measure number concentration of grown particles, TSI CNC model 3760, has a detection limit of 14 nm and was used at a standard flow rate of 1.5 lpm.

As was shown above, the activation of particle growth and CNC detection limits depend on supersaturation. The increase of supersaturation allows starting heterogeneous nucleation for smaller particles and lowering the detection limit. As supersaturation increases to a certain level, homogeneous nucleation starts to compete with heterogeneous nucleation. Therefore, homogeneous nucleation is the terminal limiting factor for this method since the operating conditions have to avoid the initiation of homogeneous nucleation. An important question is how far from this point should conditions be set and what criteria can be used to choose the operating parameters.

These criteria can be determined by scanning the supersaturation over the range of values up to the onset of homogeneous nucleation. Scaling parameters can be applied to obtain proper working conditions. Homogeneous nucleation causes a very characteristic exponential growth in particle number and can be easily identified. The onset point of homogeneous nucleation can be used to determine the supersaturation value and the corresponding Kelvin diameter under the specific working conditions.

Supersaturation scanning in the TM CNC is achieved by changing the initial vapor pressure of DBP in the "hot" flow. The carrier gas to the vapor generator (0.8 lpm) is split into two flows at the entrance. One fraction of flow is directed to the vapor generator chamber containing DBP ("bottom") and saturated with DBP vapor. Another part of the vapor generator flow is directed to the chamber without DBP ("top") but equilibrated at the same temperature. These two flows are recombined inside the mixing part before being directed to the nozzle. This approach permits changing the initial DBP vapor pressure in the vapor generator flow by changing the ratio of flows through "top" and "bottom" chambers while keeping the flow structure and the temperature field unchanged.

The DBP concentration in the outlet of the vapor generator was measured gravimetrically. Porous metal cups were placed in the air stream at the nozzle outlet. The weight of the cups was measured before and after exposure. The DBP accumulation is directly proportional to the exposure time indicating that the measurement procedure is correct. To measure the penetration of vapor through the porous cups, two cups are placed sequentially. The DBP amount on the second cup is negligible indicating complete condensation of vapor on the first cup. The measurement results show that the dependence of DBP vapor concentration in the outlet of the saturator chamber is in good agreement with the calculated values.

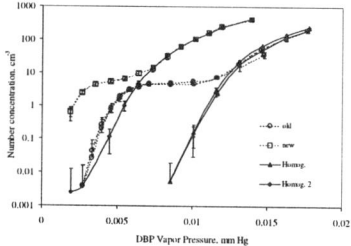

FIGURE 2. The dependence of particle concentration on DBP vapor pressure for heterogeneous and for homogeneous nucleation.

RESULTS AND DISCUSSION

The data presented in Figure 2 shows a typical example of the transition from heterogeneous to homogeneous nucleation for NaCl particles of initial DMA size of 1.7 nm for both designs of TM CNC (marked "old" and "new" in Figure 2). In the case of homogeneous nucleation, the DMA voltage was turned to zero and FCE current was at background levels. The "old" results for 1.7 nm and for homogeneous nucleation (Homog.) consist of two sets of data obtained by decreasing and increasing the DBP vapor pressure. Both of the heterogeneous nucleation curves in Figure 2 have similar patterns, but shifted in abscissa scale probably due to difference in condensation temperature. As the vapor concentration increases, the number of detected particles starts to increase, the increasing concentration reaches a "plateau" before starting to increase again. The second increase of particle concentration is caused by homogeneous nucleation of DBP vapor as indicated by the curves for homogeneous nucleation in Figure 2. Note that curves for homogeneous nucleation have shifts similar to heterogeneous nucleation curves. Measuring the onset of heterogeneous nucleation from according homogeneous nucleation curve allows correcting the possible temperature effects and to measure precisely the Kelvin diameters of introduced nuclei.

Simultaneous measurements of FCE current and particle concentration detected by CNC allow comparison between these two concentrations in order to determine the detection efficiency of the CNC. The FCE current is recalculated to particle concentration assuming each particle carries a single negative charge (the fraction of multiple charged particles for a particle size below 20 nm is negligible). The concentration is reduced by the dilution factor in the outlet of the DMA system and in the CNC. Diffusion losses in the tube connecting the DMA system and CNC inlet are estimated and taken into account. CNC/FCE ratios measured for different particle sizes and compositions at different DBP vapor pressures are presented in Figure 3. The impact of homogeneous nucleation caused by the increase in particle concentration at DBP pressures above 0.012 mm Hg (see Figure 2) has been subtracted.

Experiments were performed with two types of nuclei. WOx and NaCl particles were generated simultaneously and were used in equal experimental conditions. There are significant differences between these two types of nuclei. The major difference in data for WOx and NaCl is in the slope of the rising part of the detection efficiency curves. The curves for different particle sizes of the same substance are essentially parallel. When curves are reaching a "plateau", the detection efficiency also varies with particle size and substance.

The CNC detection efficiency for WOx particles is substantially lower than that for NaCl particles for DBP vapor pressure values of 0.01 mm Hg and above. As observed from Figure 3, the situation reverses at lower DBP pressures where WOx particles have higher detection efficiency. This fact cannot be explained with available theory. Further experimental and theoretical research on this subject is required.

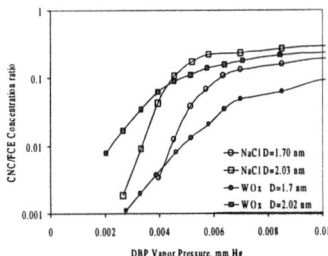

FIGURE 3. Dependence of detection efficiency on vapor pressure for different sizes and compositions of nuclei.

ACKNOWLEDGMENTS

This work is supported by US Environmental Protection Agency under grant R826654.

REFERENCES

1. Adachi, M., Okuyama, K., Seinfeld, J., H., (1992). *J. Aerosol Sci.*, **23**, 4:327-337.
2. Kim, T., O., Ishida, T., Adachi, M., Okuyama, K., Seinfeld, J., H., (1998) *Aerosol Sci. Technol.*, **29**:112-125
3. Mavliev R., Wang H.-C.(2000). *J. Aerosol Sci.*, (in press).
4. McDermont W.T., Ockovic R.C. and Stolzenburg M.R. (1991) *Aerosol Sci. Technol.*, **14**:278-287
5. Rebours A., Bouland D., Renoux A. (1996) *J. Aerosol Sci.*, **27**: 1227-1242
6. Saros, M.T, Weber, R.J.,. Marti, J.J , McMurry, P.H.. (1996) *Aerosol Sci. Technol.* **25**:200-213.

Homogeneous Nucleation In High-Pressure Multi-Component Systems: Application To Mixtures Of N-Alkanes

J. Hrubý, P. Peeters and M. E. H. van Dongen

*Eindhoven University of Technology, Department of Applied Physics,
P.O. Box 513, 5600 MB Eindhoven, The Netherlands*

Abstract. We present a new method of determining the formation energies of critical and near-critical clusters in multi-component systems and discuss the application to mixtures of n-alkanes, serving as a model of natural gas and derived mixtures.

INTRODUCTION

Homogeneous nucleation of droplets in natural gas and similar mixtures occurs in a number of industrial processes. These processes are characterized by high pressure and high nucleation rates. The mixtures include a whole spectrum of hydrocarbons, carbon dioxide, nitrogen and water. The nucleation behavior of water-free natural gas is governed by heavy hydrocarbons and their interaction with the prevailing light component, methane. For the present purposes, such mixture can be replaced with an *effective* mixture of n-alkanes. Determination of the *formation energy* of the critical and near-critical clusters appears the key problem in resolving the nucleation kinetics. The particular objective of this work is to find, based on the capillarity approximation, the formation energy of clusters in mixtures of n-alkanes.

AN OUTLINE OF THE THEORY

We consider a droplet of a liquid (β-phase) in a gaseous environment (α-phase). Let r be the number of components in the system. The chemical potentials in the core of the droplet may differ from the environmental values by $\Delta\mu_i^\beta = \mu_i^\beta - \mu_i^\alpha$. The chemical potential in the interface region is not considered an independent quantity, but rather a function of the bulk potentials: $\mu_i^\sigma(\underline{\mu}^\alpha, \underline{\mu}^\beta)$ [1]. Arguments can be given proving that the coefficients $\alpha_{ij} \equiv \partial\mu_i^\sigma / \partial\mu_j^\alpha$ and $\beta_{ij} \equiv \partial\mu_i^\sigma / \partial\mu_j^\beta$ satisfy the relation $\alpha_{ij} + \beta_{ij} = \delta_{ij}$ (the Kronecker δ) at near-critical conditions. Additional arguments show

---

[1] The underlined notation means a vector.

that the non-diagonal coefficients ($i \neq j$) must vanish. The capillarity approximation yields then the formation energy of the droplet in the form

$$\Delta E = A\sigma/3 + \sum_{1 \leq i \leq r}(V\rho_i^\beta + A\beta_{ii}\Gamma_i)\Delta\mu_i^\beta, \tag{1}$$

where A is the surface area and V the volume of the sphere of tension of radius R, which is related to the pressure difference Δp by the Laplace equation $R=2\sigma/\Delta p$. The surface tension σ is related to the chemical potentials through the non-critical adsorption equation

$$d\sigma = -\sum_{1 \leq i \leq r}\Gamma_i d\mu_i^\sigma = -\sum_{1 \leq i \leq r}(\alpha_{ii}\Gamma_i d\mu_i^\alpha + \beta_{ii}\Gamma_i d\mu_i^\beta). \tag{2}$$

This formulation of surface thermodynamics is similar to that developed by Nishioka and Kusaka (1). Knowing the adsorptions Γ_i and the coefficients, β_{ii} ($=1-\alpha_{ii}$), the surface tension can be extrapolated from an equilibrium coexistence state (where it is known) and the formation energy can be computed. In this section we focus on the determination of adsorptions, assuming that the coefficients β_{ii} can be estimated: for non-volatile components we have $\beta_{ii}=1$, for typical gases $\beta_{ii}=0$, and components appreciably present in both phases carry intermediate values.

The critical surface tension σ^* is a special case of σ for $\mu^\alpha = \mu^\beta$, and the equilibrium coexistence surface tension σ^{coex} is a special case of σ^* for $\Delta p=0$. For critical cases, equation 2 reduces do the Gibbs adsorption equation, $d\sigma^* = -\sum\Gamma_i d\mu_i$. At constant temperature, $r-1$ state variables z_j (e.g. molar fractions x_j^β) characterize a state of equilibrium coexistence of the two phases, and the set $\{z, \Delta p\}$ characterizes the critical coexistence. Let us express the chemical potentials in terms of z_j's and Δp:

$$d\mu_i = \sum_{1 \leq j \leq r-1} M_{ij}dz_j + M_{ir}\Delta p. \tag{3}$$

The coefficients M_{ij} can be found from an equation of state. Let us substitute this expression into the Gibbs adsorption equation. We find

$$d\sigma^* = -\sum_{1 \leq j \leq r-1} Z_j dz_j - Z_r d\Delta p, \tag{4}$$

where $Z_j \equiv \sum_{1 \leq i \leq r}\Gamma_i M_{ij}$, $j=1,...,r$. $\tag{5}$

We further write the adsorptions in the form

$$\Gamma_i = t_o \Delta\rho_i + \Delta\Gamma_i, \tag{6}$$

where $\Delta\rho_i \equiv \rho_i^\beta - \rho_i^\alpha$ and t_o is an offset length. We insert expression 6 into equation 5. Using the Gibbs-Duhem equation written for both phases we find

$$Z_j = \sum_{1 \leq i \leq r}\Delta\Gamma_i M_{ij}, \quad j=1,...,r-1, \tag{7}$$

$$Z_r = t_o + \sum_{1 \leq i \leq r}\Delta\Gamma_i M_{ir}. \tag{8}$$

So far, the offset length t_o does not have a definite meaning. We fix it by postulating

$$\sum_{1 \leq i \leq r}\Delta\Gamma_i M_{ir} = 0. \tag{9}$$

We focus now on the equilibrium coexistence. Since the surface tension σ^{coex} can be measured using standard techniques, it can be assumed a known function of the $r-1$

variables z_j. Hence, the coefficients $Z_1^{coex},\ldots,Z_{r-1}^{coex}$ are known: $Z_j^{coex} = -\partial \sigma^{coex}/\partial z_j$. Equations 7 and 9, with coefficients Z_j evaluated at equilibrium coexistence, can be solved to obtain r unknowns $\underline{\Delta\Gamma}^{coex}$. The missing piece of information needed to compute the adsorptions $\underline{\Gamma}^{coex}$ is the offset length $t_o^{coex} = -\partial \sigma^* /\partial \Delta p|^{coex}$.

The droplet and its environment are characterized by $2r$ variables: say the molar fractions and pressures. For changes maintaining criticality, only r variables can be varied independently. For example, changes $d\underline{x}^\beta$ and $d\Delta p$ must be accompanied by appropriate changes $d\underline{x}^\alpha$ and dp^α. Let us vary the pressure difference Δp at constant $\underline{z}$. The resulting variation of the critical surface tension σ^* will be a result of the following two groups of effects:

I. intrinsically size-dependent effects, arising from the fact that change of radius is directly coupled to change of Δp via the Laplace equation,

II. thermodynamic effects due to the changes of composition and pressure.

Group I includes, in particular, effects arising when the surface layer thickness and/or the liquid correlation length become comparable with the droplet radius. For unary systems well bellow their critical point we only have group I effects. For multi-component systems, however, group II effects can become decisive. In certain cases it is possible to find a set of variables $\underline{z}$, which eliminates almost fully the group II effects. In particular, if the gaseous phase is of low pressure (say, <1 bar), it is sufficient to take $\underline{z} = \underline{x}^\beta$. The second example is less trivial. Let the system contain s components $i = 1,\ldots,s$ rich in the liquid phase and almost absent in the gaseous phase, and $r-s$ gases $i = s+1,\ldots,r$, almost absent in the liquid phase. Let x_s^β be the highest liquid molar fraction and x_r^α be the highest gas molar fraction. An appropriate set of variables is then $\underline{z} = \{x_1^\beta,\ldots,x_{s-1}^\beta,x_{s+1}^\alpha,\ldots,x_{r-1}^\alpha,p^\alpha\}$, enabling to treat the influence of the adsorbed gases properly. A similar method of classification of components was introduced by Koenig (2). In the two cases described the offset length t_o^{coex} has a meaning analogous to the Tolman length δ^{coex} (3) of an unary system. In a *general case* it is not possible to eliminate the group II effects by choosing an appropriate set of $r-1$ variables. In that case we accept any set of variables, practical for description of the equilibrium coexistence surface tension σ^{coex}, and determine $\underline{\Delta\Gamma}^{coex}$.

To develop a new general procedure to determine t_o^{coex}, we consider a process characterized by *constant molar fractions in both phases and constant pressure of the gaseous phase*. This change necessarily breaks the criticality condition (the final and/or initial states are non-critical). For this process the non-critical adsorption equation 2 gives

$$d\sigma = -t_{III}\,d\Delta p = -2t_{III}\sigma\,dq/(1+2t_{III}q), \qquad (10)$$

where $t_{III} \equiv \sum_{1 \le i \le r} \Gamma_i \beta_{ii} \bar{v}_i^\beta$, (11)

$q \equiv 1/R$ is the surface curvature and $\bar{v}_i^\beta$ is the liquid partial molar volume. Evaluating equation 11 at equilibrium coexistence with adsorptions expressed in form 6, we find

$$t_o^{coex} = \left[\left(t_{III} - \sum_{1 \leq i \leq r}\Delta\Gamma_i\,\beta_{ii}\,\bar{v}_i^\beta\right) / \sum_{1 \leq i \leq r}\Delta\rho_i\,\beta_{ii}\,\bar{v}_i^\beta\right]^{coex}. \qquad (12)$$

For unary cases we find

$$t_o^{coex} = \delta^{coex} = \left[t_{III}\,\rho^\beta / (\Delta\rho\,\beta)\right]^{coex}. \qquad (13)$$

Available studies discussing the numerical value of the Tolman length only focus on the unary case. Unlike the original estimates by Tolman (3), the contemporary studies show that δ^{coex} is negative. Koga and Zeng (4) suggest an empirical relation

$$\delta^{coex} \approx \left\{-\sigma\left[(\rho^\beta)^2\chi^\beta - (\rho^\alpha)^2\chi^\alpha\right]/(\Delta\rho)^2\right\}^{coex}, \qquad (14)$$

where χ^α and χ^β are isothermal compressibilities of both phases. For multi-component cases, we suggest that this expression can also be used together with equation 13 to find t_{III}^{coex}, which, in turn, is used to determine t_o^{coex} through equation 12. The total densities and compressibilities of the bulk phases are used in equations 13 and 14. In equation 13 we assume an effective coefficient $\beta = \sum x_i^\beta \beta_{ii}$. Our suggestions are justified by the fact that the change characterized by constant $\underline{x}^\alpha$, $\underline{x}^\beta$ and p^α mimics a change in a unary fluid.

APPLICATION TO MIXTURES OF N-ALKANES

In order to find the formation energy of critical and near-critical clusters (eq. 1), it is necessary to estimate the coefficients β_{ii}. For pressurized natural gas-like mixtures at laboratory temperature and below, the methane liquid molar fraction is not negligible so that a value $\beta_{11} \approx 0.1$ might be assumed. Starting about from pentane it seems reasonable to assume $\beta_{ii} = 1$. The coefficients for the intermediate alkanes can be interpolated (in a β vs. r plot). We tested the methods described here for binary methane-nonane mixtures, for which experimental nucleation data were obtained in our laboratory (5). However, it turned out that the equations of state (cubic Peng-Robinson or Readlich-Kwong-Soave equations) do not provide an adequate description of the behavior of nonane in the gaseous phase. This problem stems from the fact that experimental data for nonane solubilities in the gaseous phase, to which the interaction coefficients k_{ij} are fitted, are only available at higher temperatures.

REFERENCES

1. Nishioka, K., Kusaka I., *J. Chem. Phys.* **96**, 5370-5376 (1991)
2. Koenig, F. O., K., *J. Chem. Phys.* **18**, 449-549 (1950)
3. Tolman, R. C., *J. Chem. Phys.* **17**, 333-337 (1949)
4. Koga, K., and Zeng, X. C., K., *J. Chem. Phys.* **110**, 3466-3471 (1999).
5. Luijten, C. C., Peeters, P., and van Dongen M. E. H., *J. Chem. Phys.* **111**, 8535-8544 (1999).

Applications of the Nucleation Theorem

Dimo Kashchiev

*Institute of Physical Chemistry, Bulgarian Academy of Sciences,
ul. Acad. G. Bonchev 11, Sofia 1113, Bulgaria*

Abstract. It is demonstrated how the nucleation theorem can be applied to analyze experimental data for quantities which characterize one-component nucleation of condensed phases or other processes involving this kind of nucleation.

The nucleation theorem is a most general relationship between the nucleation work W^*, the number n^* of molecules or atoms constituting the nucleus and the supersaturation $\Delta\mu \equiv \mu_{old} - \mu_{new}$, μ_{old} and μ_{new} being the chemical potentials of a molecule or atom in the old and the new phase. As shown by Kashchiev (1, 2), for the one-component condensed-phase nucleus defined by the equimolecular dividing surface the nucleation theorem reads

$$dW^*/d\Delta\mu = -n^* \tag{1}$$

provided the density of the nucleus is considered as equal to that of the bulk condensed phase.

The considerations to follow are aimed at demonstrating how the nucleation theorem can be used for a model-independent determination of the nucleus size n^* with the help of four different types of experimental data for quantities which characterize one-component nucleation of condensed phases or other processes involving this kind of nucleation.

1. We first consider data for the stationary nucleation rate J as a function of the supersaturation $\Delta\mu$. In this case, if $\Delta\mu$ is varied at constant absolute temperature T, the dependence of n^* on $\Delta\mu$ is obtainable by using the formula (1, 2) (k is the Boltzmann constant)

$$n^* = kT d(\ln J)/d\Delta\mu - 1. \tag{2}$$

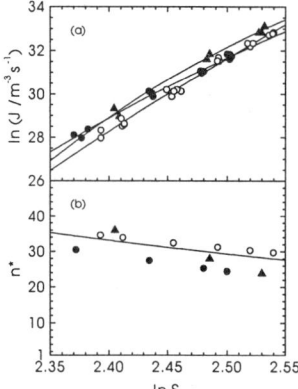 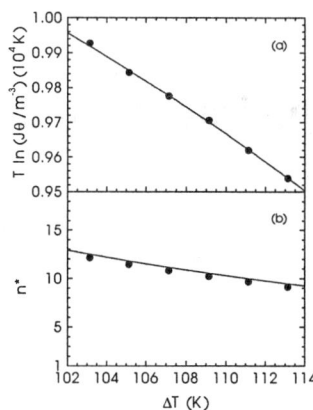

FIGURE 1. Dependence of J and n^* on S in homogeneous nucleation of n-butanol droplets in vapors.
FIGURE 2. Dependence of $J\theta$ and n^* on ΔT in homogeneous nucleation of ice in amorphous water.

With $\Delta\mu = kT \ln S$ where S is the supersaturation ratio, this formula is employed for analysis of the experimental data of Viisanen and Strey (3) in Fig. 1a for homogeneous nucleation of n-butanol droplets in vapors at $T=240$ K in the presence of argon (the solid circles and triangles) or xenon (the open circles). The lines in this figure represent the function

$$y = a/x + b + cx \tag{3}$$

fitting the three sets of data with $x=\ln S$, $y=\ln J$ and the best-fit values of the mathematical parameters a, b and c. The symbols in Fig. 2b depict the $n^*(\Delta\mu)$ dependence calculated from Eq. (2) with the help of the $\ln J$ derivative resulting from Eq. (3) for each of the data sets. This model-independent result for n^* as a function of $\Delta\mu$ allows testing the validity of the classical Gibbs-Thomson equation for three-dimensional nucleation

$$n^* = A/\Delta\mu^3 \tag{4}$$

where A is a constant related to the nucleus specific surface energy. In the case studied by Viisanen and Strey (3) the value of A is known independently and the line in Fig. 1b displays Eq. (4) with no free parameter. We observe that despite the smallness of the n-butanol nucleus droplet the classical nucleation theory predicts well the n^* values obtained with the help of Eq. (2) without using any concrete model of nucleation.

2. The second to consider are data for the $\Delta\mu$ dependence of the product $J\theta$ of the stationary nucleation rate J and the delay time θ in nonstationary nucleation. In this case the $n^*(\Delta\mu)$ dependence can be found with the aid of the formulae (2)

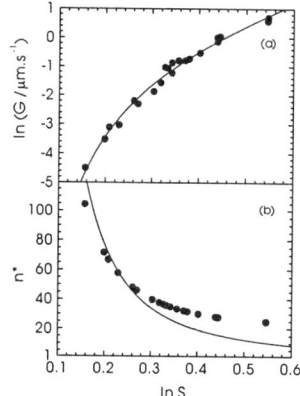
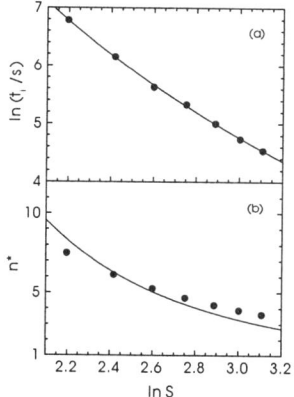

FIGURE 3. Dependence of G and n^* on S in nucleation-mediated growth of paraffin in solution.
FIGURE 4. Dependence of t_i and n^* on S in precipitation of $CaCO_3$ in aqueous solution.

$$n^* = kTd[\ln(J\theta)]/d\Delta\mu \tag{5}$$

$$n^* = kd[T\ln(J\theta)]/d\Delta\mu - k(dT/d\Delta\mu)\ln C_0 \tag{6}$$

when $\Delta\mu$ is varied isothermally or by means of T, respectively. In Eq. (6) C_0 is the concentration of nucleation sites in the system. The usage of this equation with $\Delta\mu=\Delta s\Delta T$ (Δs is the entropy of melting, ΔT is the undercooling) is demonstrated in Fig. 2. In Fig. 2a the circles present the $J\theta$ product determined by Koverda et al. (4) for nucleation of ice in amorphous water, and the line visualizes the best-fit function (3) with $x=\Delta T$ and $y=T\ln(J\theta)$. The n^* values calculated from Eq. (6) with the aid of this function and $C_0=3.1\times10^{28}$ m^{-3} corresponding to homogeneous nucleation are shown by the circles in Fig. 2b. The line in this figure is drawn according to Eq. (4) with no free parameter, since A is known independently (5). As seen, the classical Gibbs-Thomson equation is in force although the ice nucleus comprises only about 10 water molecules.

3. The third to consider are data for the nucleation-mediated growth rate G of a crystal face as a function of the supersaturation $\Delta\mu$. In this case, when $\Delta\mu$ is varied isothermally, the following formula yields the $n^*(\Delta\mu)$ dependence (2):

$$n^* = (1+2\nu)kTd(\ln G)/d\Delta\mu - (1+2\nu) - 2\nu/(e^{\Delta\mu/kT} - 1) \tag{7}$$

where $\nu\leq1$ is the exponent in the growth law $r\propto t^\nu$ of the two-dimensional (2D) supernucleus clusters on the crystal face (r is the cluster radius, and t is time). Shown by the circles in Fig. 3a are the $G(S)$ data of Simon et al. (6) for nucleation-mediated growth of the (100) face of paraffin $C_{36}H_{74}$ crystal in solution of petroleum ether. The line in this figure illustrates the best fit to the data, provided by Eq. (3) with $x=\ln S$ and

$y=\ln G$. Using Eq. (7) with $\Delta\mu=kT\ln S$ and the $\ln G$ derivative calculated from the best-fit function (3) yields the n^* values represented by the circles in Fig. 3b. This model-independent experimental finding for the nucleus size can be used for verifying the validity of the classical Gibbs-Thomson equation for 2D nucleation

$$n^* = B/\Delta\mu^2 \qquad (8)$$

where B is a constant related to the nucleus specific edge energy. In this case the B value is unknown and the line in Fig. 3b displays the best fit resulting from Eq. (8) with B as a free parameter. One sees that the classical nucleation theory is not in good agreement with experiment, particularly when the 2D nucleus on the face of the paraffin crystal contains less than about 40 molecules.

4. Finally, we consider data for the $\Delta\mu$ dependence of the induction time t_i in precipitation. In this case, under conditions of isothermal variation of $\Delta\mu$, n^* can be determined as a function of $\Delta\mu$ with the help of the expression (2)

$$n^* = -(1+vd)kTd(\ln t_i)/d\Delta\mu - 1 - vdkTd(\ln G)/d\Delta\mu. \qquad (9)$$

Here $v \leq 1$ is the exponent in the growth law $r=(Gt)^v$ of the precipitating crystallites, G is their growth constant (or rate when $v=1$), and d is the dimensionality of their growth ($d=1$, 2 or 3, e.g., for cylinders of length $2r$ and fixed diameter, disks of radius r or spheres also of radius r, respectively). Equation (9) is applicable when t_i is determined by measuring the volume or the mass of the precipitated phase. This equation is employed for analysis of the $t_i(S)$ data (the circles in Fig. 4a) of Verdoes et al. (7) for precipitation of $CaCO_3$ in aqueous solution. The data are fitted with the help of $y(x)$ from Eq. (3) with $x=\ln S$ and $y=\ln t_i$ (the best fit is visualized by the line in Fig. 4a). Accordingly, the circles in Fig. 4b represent the experimental n^* values determined from Eq. (9) upon setting $\Delta\mu=kT\ln S$, using the best-fit function (3) for evaluation of the $\ln t_i$ derivative and accounting for the $G(S)$ dependence reported by Verdoes et al. (7). The line in Fig. 4b illustrates the best fit provided by the Gibbs-Thomson equation (4) with A as a free parameter (in this case A is unknown). It is seen that the classical $n^*(\Delta\mu)$ dependence follows closely the experimental one despite that the $CaCO_3$ nucleus is constituted of a few molecules only.

REFERENCES

1. Kashchiev, D., *J. Chem. Phys.* **76**, 5098 (1982).
2. Kashchiev, D., *Nucleation: Basic Theory with Applications*, Oxford: Butterworth-Heinemann, 2000.
3. Viisanen, Y., and Strey, R., *J. Chem. Phys.* **101**, 7835 (1994).
4. Koverda, V. P., Skripov, V. P., and Bogdanov, N. M., *Kristallografiya* **19**, 613 (1974).
5. Butorin, G. T., and Skripov, V. P., *Kristallografiya* **17**, 379 (1972).
6. Simon, B., Grassi, A., and Boistelle, R., *J. Cryst. Growth* **26**, 77 (1974).
7. Verdoes, D., Kashchiev, D., and van Rosmalen, G. M., *J. Cryst. Growth* **118**, 401 (1992).

Molecular Simulation of Ion-induced Nucleation in Water Vapor

Mitsuhiro Matsumoto*, Koji Ohguchi*, and Kenji Yasuoka[†]

*Department of Engineering Physics and Mechanics, Kyoto University, Kyoto 606-8501, Japan
†Department of Mechanical Engineering, Keio University, Yokohama 223-8522, Japan

Abstract. Molecular dynamics simulation is carried out to investigate ion-induced nucleation in water vapor. Comparison is made between ions with similar size and opposite electric charge, a potassium ion and a chloride ion. At large supersaturation ratios, no qualitative difference is discernible in clustering dynamics. At a small supersaturation ratio, clustering around the cation stops at a size ~12, while clustering around the anion continues. The molecular structure of ion clusters explains the difference.

INTRODUCTION

Using molecular dynamics (MD) computer simulation, we have examined homogeneous nucleation processes at a molecular level for Lennard-Jones (LJ) fluid (1) and water (2). In these studies, we have estimated the nucleation rate from the number of generated clusters in supersaturated vapor; it is found that quantitative disagreement between the simulation and the prediction with a classical nucleation model can be very large (several orders of magnitude), depending on the conditions. In general, the observed nucleation rate is slower than the prediction for LJ fluid, and faster for water. Size dependence of the cluster formation free energy has been also estimated from the MD simulation data (1,2); the obtained free energy barrier is consistent with the clustering dynamics. The structure of small clusters is very different for LJ fluid and water, which is the main reason for the difference in the nucleation rate.

In this paper, we carry out a similar MD simulation to investigate ion-induced nucleation processes in water vapor. This kind of seed-induced nucleation is important and relevant in most cases of fluid phase change. Classical continuum models are well established, but molecular models are not fully developed yet, in which proper cluster structures should be taken into account. The detailed cluster structure is especially important for the case of ion-induced nucleation because the long-ranged Coulombic interaction affects the solvation around the ion. As a first MD approach to the problem, we study the nucleation process in a (water vapor + a small inorganic ion) system at several supersaturation conditions. As the seed ion, we choose potassium ion (K^+) and chloride ion (Cl^-), which have a similar ionic radius but an opposite sign of the charge, in order to investigate how the ion charge affects the clustering rate and the cluster structure.

SIMULATION METHOD

We use 1,000 water molecules, one ion particle, and 1,000 carrier gas particles. Jorgensen's TIP4P rigid rotor model (3) is adopted for water-water interaction. The carrier gas is to remove the condensation heat; the gas particles have soft-core ($\propto r^{-12}$) repulsive interaction with each other and with water molecules, and are temperature-controlled by velocity scaling method. As the model ion, we choose a simple "LJ + Coulombic charge" models; "argon + positive e"(elementary charge unit) roughly corresponds to a potassium ion and "argon + negative e" roughly corresponds to a chloride ion. Thus, the ionic radius of both species in this study is exactly the same because the main purpose is to investigate the charge effect on the clustering process.

The temperature of the carrier gas is controlled to be 350K, as in the homogeneous nucleation case (2). We carry out the simulation with three different cell volumes. The supersaturation ratio S (the ratio of the vapor density to the saturation vapor density) is 14.6, 7.3, and 2.9; the condition S=14.6 is the same as in Ref. (2).

After equilibrating the system at 1000K, the particle velocities are re-scaled so that the system is quenched to be 350K, from which we the main calculation. After the quench, only the carrier gas particles are temperature-controlled.

SIMULATION RESULTS

Water Cluster: Definition

Two molecules are defined to be 'bonded' when the potential energy between them is less than a threshold value; this energy criterion is often used for defining a 'hydrogen bond', but in this study, we also adopt this definition for ion-water pairs. As the threshold value, we use -10 kJ/mol, which is a typical value for a hydrogen bond.

Cluster Growth

Nucleation (cluster formation) starts as soon as the system is quenched to 350K. In most cases, the generated clusters include the ion, i.e., the ion-induced nucleation takes place; note that it requires several hundred picoseconds as the induction time for homogeneous nucleation to start even at the highest supersaturation condition (2).

The size of the generated ion-water cluster is plotted against time in Fig.1. At S=14.6 and 7.3, no qualitative difference is observed between the systems of potassium ion and chloride ion. The sudden size change observed in Cl⁻ system (S=7.3, ~230ps) was caused by collision of two clusters. Overall clustering seems slightly faster for K⁺ system than Cl⁻ system, but much more statistics is required for further quantitative discussion.

The difference between K⁺ and Cl⁻ is observed at low supersaturation condition, S=2.9. Clustering stops at the size ~12 for the cation case, while gradual clustering proceeds for the anion case.

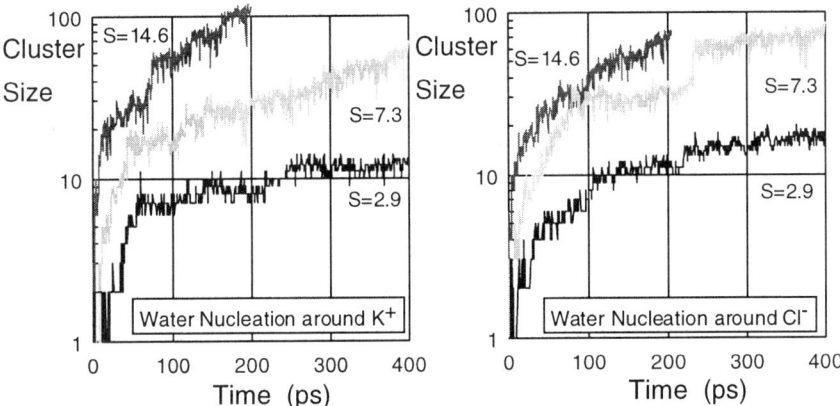

FIGURE 1. Cluster size change during the MD simulation.

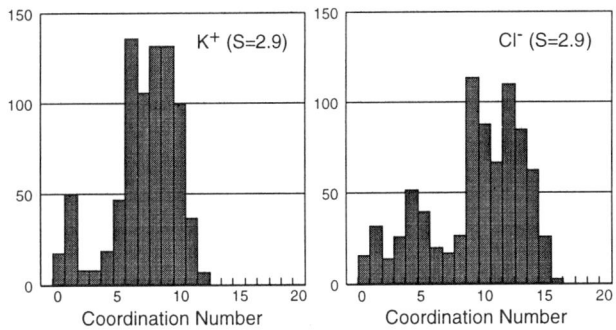

FIGURE 2. Histogram of the number of water molecules surrounding the ion.

Cluster Structure

The observed effect of ionic charge is beyond the classical (continuum) nucleation model because the model deals with the water layer around the ion only as dielectric continuum. To explain the charge sign effect, molecular structure of the generated ion-water clusters should be taken into account. Detailed analysis from this microscopic viewpoint is still under way, but we can point out the following observations:

1. The number of water molecules surrounding the central ion is larger for Cl^- than for K^+; as shown in Fig.2, K^+ is surrounded typically 6-10 water molecules, while Cl^- is surrounded by 9-14. The definition of this number is slightly different from the usual coordination number of ions in aqueous solutions (4), but the concept is similar.

2. From a structural viewpoint, Cl^- clusters seem more compact than K^+ clusters, as seen in Fig.3. This is partly due to the difference in the coordination number described above.

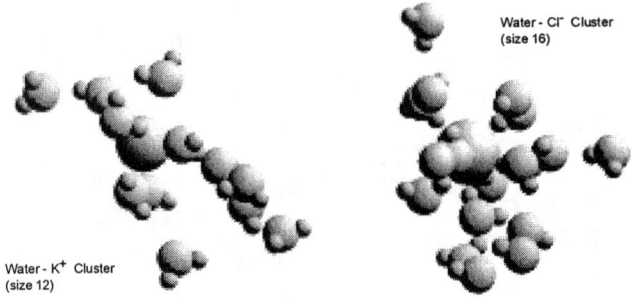

FIGURE 3. Typical structure of generated clusters.

CONCLUSION

Ion-induced cluster formation in water vapor was studied with molecular dynamics simulation. At low supersaturation conditions, clustering behavior was found to depend on the sign of the ion. Microscopic structure of ion-water clusters should be taken account.

The relaxation, energetic as well as structural, inside the cluster is considered to be rather fast; for example, the life time of hydrogen bonds in water is order of 1-10 ps (5). Thus, clusters are expected to be in almost equilibrium. In this case, cluster formation free energy should be the most relevant to the nucleation rate; the estimation of the free energy is our next target.

ACKNOWLEDGMENTS

This study is financially supported in part by Japan Society for the Promotion of Science (Grant-in-Aid for Scientific Research, No. 10750147). We are grateful to the Supercomputer Laboratory (Institute for Chemical Research, Kyoto University) and the Center for Promotion of Computational Science and Engineering (Japan Atomic Energy Research Institute) for the allowance of their computer resources.

REFERENCES

1. Yasuoka, K., and Matsumoto, M., *J. Chem. Phys.*, **109**, 8451-8462 (1998).
2. Yasuoka, K., and Matsumoto, M., *J. Chem. Phys.*, **109**, 8463-8470 (1998).
3. Jorgensen, W.L., Chandrasekhar, J., Madura, J.D., Impey, R.W., and Klein, M.L., *J. Chem. Phys.*, **79**, 926-935 (1983).
4. Robinson, G.W., Zhu, S-B., Singh, S., and Evans, M.W., *Water in Biology, Chemistry and Physics*, Singapore: World Scientific, 1996, ch. 7.
5. Bertolini, D., Cassettari, M., Ferrario, M., Grigolini, P., and Salvetti, G., *Adv. Chem., Phys.*, **62**, 277-320 (1985).

Reconciling Gibbs and van der Waals: A New Approach to Nucleation Theory

Jürn W. P. Schmelzer

Department of Physics, University of Rostock, 18051 Rostock, Germany

Abstract: A new general method for the determination of the work of critical cluster formation in nucleation theory is developed. The method consists in a generalization of Gibbs' approach retaining its advantages and avoiding some of its shortcomings. For small supersaturations, the results are in agreement with the classical Gibbs' approach (when in Gibbs' method, in addition, the capillarity approximation is employed). However, in contrast to the classical and in agreement with van der Waals' type approaches to the determination of this quantity, for initial states near the spinodal curve the work of critical cluster formation is shown to tend to zero.

INTRODUCTION

Despite the prolonged history of research, aimed at the proper understanding of the kinetics of first-order phase transitions, diverse problems in the formulation of the theory as well as in the interpretation of experimental results remain unsolved till now. One of these problems consists in the appropriate determination of the work of critical cluster formation in nucleation theory. The present contribution is devoted to an analysis of this problem.

GIBBS' APPROACH: BASIC ASSUMPTIONS AND RESULTS

To be definite, we consider first-order phase transformations proceeding at constant external pressure, p, and constant temperature, T. The characteristic thermodynamic potential, appropriate for the description of nucleation under such conditions, is then the Gibbs free energy, G. The change of the Gibbs free energy, ΔG, due to the formation of a cluster of size R can be written - according to Gibbs' thermodynamic treatment (1) - generally as

$$\Delta G = \sigma A + (p - p_\alpha) V_\alpha + \sum n_{j\alpha} (\mu_{j\alpha} - \mu_{j\beta}) + \sum n_{j0} (\mu_{j\beta} - \mu_{j0}) . \tag{1}$$

Here σ is the surface tension, μ the chemical potential, A the surface area of the cluster, V_α the volume and $n_{j\alpha}$ the number of moles or particles of the k different components in the cluster ($j=1, 2, \ldots, k$). The subscript α specifies hereby the parameters of the cluster phase, while β refers to the actual state of the ambient phase. The values of the different quantities in the initial metastable state are denoted by the subscript 0.

Following Gibbs' approach, the volume of the cluster and the number of particles of the different components in the cluster have to be considered as independent variables. Thus, we get from Equation (1) quite generally (cf. also (2))

$$d\Delta G = [(p-p_\alpha) + \sigma \, dA/dV_\alpha] \, dV_\alpha + \sum (\mu_{j\alpha} - \mu_{j\beta}) \, d\, n_{j\alpha} \,. \tag{2}$$

The necessary equilibrium conditions for a cluster of critical size are determined thus - in Gibbs' approach employing the surface of tension as the dividing surface - by

$$(p - p_\alpha) + \sigma \, dA/dV_\alpha = 0 \quad \text{or} \quad (p_\alpha - p) = 2\sigma/R, \tag{3}$$

$$\mu_{j\alpha}(T, \rho_{1\alpha}, \rho_{2\alpha}, \ldots, \rho_{k\alpha}) = \mu_{j\beta}(T, \rho_{1\beta}, \rho_{2\beta}, \ldots, \rho_{k\beta}). \tag{4}$$

The composition of the critical cluster - and thus the pressure in the bulk state chosen as the reference state for the description of the properties of the cluster - is determined via Equation (4).

In Gibbs' approach, for the specification of the reference states for the description of the bulk properties of the critical clusters, the properties of the newly evolving macroscopic phases are chosen. Thus, the bulk properties of the clusters are qualitatively similar to the respective macroscopic phases, quantitatively somewhat different due to the increased pressure in the critical cluster. While for sufficiently large critical clusters this choice of the reference state corresponds widely to the actual properties of the cluster, this is not the case for small cluster sizes. By this reason, for small clusters, an appropriately chosen curvature dependence of the surface tension (3) has to be incorporated into the description to accurately determine the work of critical cluster formation. However, remaining in the framework of Gibbs' approach such determination cannot be carried out.

Summarizing we note that

- the actual state of the critical clusters - as determined via Gibbs' approach - does not describe, in general, properly the bulk properties of the cluster;
- the necessity of incorporation of a curvature dependence of the surface tension into the picture is thus intimately connected with deviations of the actual state of the clusters from the reference state;
- the curvature dependence of the surface tension is introduced in Gibbs approach primarily via the size dependence of the properties of the coexisting phases (cf. (3)).

Since these properties depend on the cluster size via Equations (3) and (4), equivalently a curvature dependence of the surface tension enters the description;
- the work of critical cluster formation may be described correctly via the introduction of an appropriately defined curvature dependence of the surface tension.

A NEW METHOD OF DETERMINATION OF THE WORK OF CRITICAL CLUSTER FORMATION

Recently, we developed a new thermodynamic method of determination of the work of critical cluster formation (4). In this approach, we start with Equation (1) and analyze the question, how - for a given state of the cluster - the Gibbs free energy depends on its size. Hereby it is assumed that changes of the state of the ambient phase can be neglected (this way, $(\mu_{j\beta} - \mu_{j0})=0$ holds) and that the surface tension - as in Gibbs' approach - depends primarily on the parameters both of cluster and ambient phases. The critical cluster size and the work of critical cluster formation are obtained then as

$$R_c = 2\sigma/[(p-p_\alpha) + \sum \rho_{j\alpha}(\mu_{j\alpha} - \mu_{j0})], \quad \Delta G_c = 1/3 \, \sigma A_c, \quad A_c = 4\pi R_c^2. \tag{5}$$

Both quantities, R_c and ΔG_c, depend of course on the state of the cluster assumed. Now, in the next step, the state of the cluster may be varied to find those compositions or densities, which refer to a minimum of ΔG_c as compared with all other possible cluster states. Evidently, the critical cluster corresponds thus - similarly to Gibbs' approach - to a saddle point of the Gibbs free energy (to a maximum with respect to variations of the cluster size and to a minimum in dependence on cluster composition).

In the newly developed approach, the possible states of the clusters - chosen as reference states for the description of their bulk properties - are not restricted to those structures and compositions, which are characteristic for the newly evolving macroscopic phases. In other words, the bulk contributions are chosen to fit in the best possible way the actual properties of clusters (which may change of course in dependence on the initial supersaturation in the system). By this reason, the macroscopic values of the surface tension for a coexistence of the two considered states of the system near planar interfaces describe the interfacial properties of the cluster in the ambient phase with sufficiently high precision. Consequently, the introduction of an (unknown, in general) specific expression for the curvature dependence of the surface tension - as in Gibbs' approach - is not required.

APPLICATION TO SEGREGATION IN SOLUTIONS

In Ref. (4), the newly developed method was employed for a description of nucleation in a regular solution. The interfacial tension was described, according to a theoretical analysis performed by Becker, via the relation $\sigma \propto (x_1 - x_2)^2$. Here x_1 and x_2 are the molar fractions of one of the components in the both coexisting phases. Employing the approach, as characterized above, the composition of the critical clusters (corresponding to the minimum of the work of critical cluster formation for a given initial composition, x) can be calculated in dependence on x. For initial states near the binodal curve, the composition of the critical clusters corresponds to the state of the newly evolving macroscopic phase (alternative branch of the binodal curve). For initial compositions, approaching the spinodal curve, the composition of the critical cluster becomes nearly the same as the concentrations in the ambient phase. As the result, the work of critical cluster formation tends to zero for initial states near the spinodal curve.

DISCUSSION

The newly developed approach, sketched here briefly, is to some extent similar to van der Waals (5) or density functional (6) approaches to the determination of the work of critical cluster formation. In each of the mentioned methods, first the work of critical cluster formation is formulated for any arbitrary states of the cluster and then those compositions are searched for corresponding to a minimum of ΔG_c. The newly developed method leads to, at least qualitatively, similar results. A more detailed quantitative comparison is in progress (7).

As an advantage, the newly proposed method requires exclusively the knowledge of the bulk as well as of the interfacial properties of the systems for the different possible states under consideration and for a coexistence at planar interfaces. These properties can be established experimentally or theoretically in a relatively simple way. The approach can be extended straightforwardly to any other cases of phase formation of interest including phase formation in multi-component systems. By these reasons, the proposed method is considered as a valuable new tool in the solution of the complicated problem of the determination of the work of critical cluster formation in nucleation theory.

REFERENCES

1. Gibbs, J. W.: *The Collected Works*, vol. 1, *Thermodynamics*, Longmans & Green, New York, 1928
2. Schmelzer, J. W. P., Röpke, G., Priezzhev, V. B.: *Nucleation Theory and Applications*, JINR Publishing House, Dubna, Russia, 1999
3. Schmelzer, J. W. P., Gutzow, I., Schmelzer, J. Jr., J. Colloid Interface Science **178**, 657 (1996)
4. Schmelzer, J. W. P., Schmelzer, J. Jr., Gutzow, I., J. Chem. Physics **112**, 3820 (2000)
5. Rowlinson, J. S., Widom, B.: *Molecular Theory of Capillarity*, Clarendon Press, Oxford, 1982
6. Oxtoby, D. W., Accounts of Chemical Research **31**, 91 (1998)
7. Baidakov, V. G., Boltashev, G. S., Schmelzer, J. W. P., *Work of Critical Cluster Formation in Segregation Processes in Solutions: Comparison of Different Approaches*, in preparation

Comments on Kinetics and Thermodynamics of Condensations under Homogenous Conditions

S.H. Bauer, Y-.X. Zhang and C.F. Wilcox

Department of Chemistry and Chemical Biology, Cornell University

For over eight decades an elusive goal in statistical mechanics has been the development of a self-consistent, parameter-free model for avalanche condensation from supersaturated vapors under strictly homogeneous conditions. As is well known, the classical theory (CNT) permits calculation of condensation flux levels [J=f(css;T)] that deviate from experimental observations by many orders of magnitude. This theory, as initially formulated and subsequently embellished, appears to be internally inconsistent because it asserts that the free energy of a not well defined cluster is equal to a surface free energy term minus a (T∗entropy) term. The conceptual focus is on a "critical size condensation nucleus" that initiates cluster growth. That concept appears to be a carry-over from the generally accepted model for condensation on ions or foreign particles. There is no experimental evidence that critical size nuclei (however defined) play comparable roles under strictly homogeneous conditions. Cluster growth may be best described as a smooth sequence of accretions (via monomers, dimers, etc) balanced by evaporations, the net being controlled by the difference between an enthalpy ($-\Delta H^\circ_{u-1,u}$) minus an entropy ($T\Delta S^\circ_{u-1,u}$) for successive steps. This difference controls the ratio of rate constants for conversion of (u-1) to (u); i.e. the ratio of rate constants, k(accretion)/k(evaporation).

An internally consistent model of condensation under homogeneous conditions has been formulated in chemical kinetics terms (KMM). As such, molecular parameters are introduced, so that there is no pretence at providing predictions of flux levels for specified supersaturations. The objective is to explain the data that have been derived from carefully performed, painstaking experiments. The range of adjustable parameters is rather narrow and well within established magnitudes for interaction potentials, collision diameters, etc. A "critical size nucleus" does not appear. Extended cluster growth begins to dominate when for some size u,

$$|\Delta H^\circ_{u-1,u}| > T \Delta S^\circ_{u-1,u}$$

i.e. when the enthalpy for accretion outstrips the corresponding loss of entropy. This model has evolved into a user friendly computer program. Thus, by adjusting the appropriate thermochemical parameters the computed cluster growth rates reproduce the observed flux levels. In turn, one may then estimate values for entropies of clusters, as a function of size, for specified temperatures.

The following features of KMM contrast with those of CNT.
(a) The surface free energy of very small collections of monomers does not appear.

(b) Clusters are defined by their standard state thermochemical parameters --- size dependent enthalpy of formation and entropy.
(c) Condensation takes place under "steady state" conditions.
(d) The basic kinetics relations allow for reduction of rates of accretion in the presence of high pressures of inert diluents via corrections for limited diffusion rates.
(e) Accretion via dimers, trimers, etc. are readily incorporated in the program.

There is merit in calling attention to a fundamental assumption. In CNT flux levels, however measured, are of macroscopic entities, but are identified with numbers of "critical size nuclei" that consist of fewer than 100 monomers. This presumes that coalescence is negligible during the time of observation. In KMM, J(exp) is set equal to numbers/cm3-s of microclusters that flow through the growth sequence at steady state.

The basic relations introduced in the kinetics formulation are briefly summarized below. Their applications to four cases wherein dimers as well as monomers participate in accretion illustrate the utility of the program.

Condensation occurs by sequential accretion (balanced by evaporation).

$$A_{u-1} + A_1 \underset{k_{-u}}{\overset{k_u}{\Leftrightarrow}} C_{u1}$$

$$Cu_1 + X \underset{k_{-us}}{\overset{k_{us}}{\Leftrightarrow}} A_u + X$$

$$A_{u-2} + A_2 \underset{k_{-us2}}{\overset{k_{us2}}{\Leftrightarrow}} C_{u2}$$

The "hot" species C_u are either stabilized via collisions with X or they dissociate. X designates a composite de-excitation collision partner. It consists of all A_u plus the (inert) diluent G. The de-excitation efficiency of G relative to A's is lower by a factor $\lambda \, (\approx 0.1)$.

$$[X] = \sum_{u-1}^{m} [A_u]^{eq} + \lambda [G]$$

In the absence of any indications to the contrary, we assumed that $C_{u1} \Leftrightarrow C_{u2}$ rapidly equilibrate. As the associations progress, two types of steady states develop. Very rapidly $[C_u]$ reach levels that are about two orders of magnitude lower than for the adjacent $[A_u]$'s. The dynamic concentrations of the latter initially rise rapidly, then slow down to attain their steady states. Finally, their rates of production become equal to their rates of dissociation; i.e. the "constrained equilibrium" sets in.

Expressions for the rate constants depend on the molecular parameters of the species involved. Standard magnitudes are inserted for collision diameters, reduced masses, etc. with rational dependencies on cluster sizes. The following rate constants were derived for particles that interact via a Lennard-Jones attractive potential plus hard core repulsion:

$$k_u = \pi \sigma^2_{u-1,u} \{8k_bT/\pi\mu_{u-1,u}\}^{1/2} * \{8(\varepsilon_{u-1,u}/k_bT\}^{1/3} \, \Gamma(5/3) \, Z_1$$

$$k_{us} = \pi \sigma^2_{u,x} \{8k_bT/\pi\mu_{u,x}\}^{1/2} * \{8(\varepsilon_{u,x}/k_bT\}^{1/3} \, \Gamma(5/3) \, Z_2$$

$$k_{u2} = \pi\sigma^2_{u-2,u}\{8k_bT/\pi\mu_{u-2,u}\}^{1/2} *\{8(\varepsilon_{u-2,u}/k_bT\}^{1/3}\ \Gamma(5/3)\ Z_3$$
$$k_{-u} = (k_bT/h)\ Z_1 \qquad \text{maintains detailed balance}$$
$$k_{-us} = (k_{-u}/k_{-u})*(k_{-us}/K_{u-1,u})\ \text{and}\ k_{-u2} = (k_{us}/k_{-us})*(k_{u2}/K^c_{u-2,-u})$$

The magnitudes of the potential well depths (ε_{-u}) were set equal to ≈ 0.2 of the corresponding energies for association. For the nominal steps; $A_{u-1} + A_1 \Leftrightarrow A_u$, the ratios of association to dissociation rate constants were set equal to the corresponding equilibrium constants:

$$K^{(c)}_{u-1,u} = \{[A_u]/[A_{u-1}]*[A_1]\}_{eq} = RT\ \exp\{-(\Delta H^o_{u-1,u}/RT\}\ \exp\{(\Delta S^o_{u-1,u}/R\}$$

There are four significant adjustable parameters a, b, α and β in the expressions for the enthalpy and entropy decrements:

$$\Delta H^o_{u-1,u} = -Q(v)\ [1 - a(1-b)\ u^{-b}]$$

$$\Delta S^o_{u-1,u} = -\delta S^o_l\ [1 - \alpha u^{-\beta}\ ]$$

where $Q(v)$ is the heat of vaporization of the liquid at T. Let p(atm) be the equilibrium vapor pressure of the liquid (at T), then

$$\delta S^o_l = Q(v)/T + R\ \ln\ p(atm)$$

The "best" measured (or calculated) values for the enthalpies and entropies of accretion, per monomer, for cluster sizes $2 \leq u \leq 6$, determine the initial steps of the sequence of associations. The magnitudes of the four parameters are derived by successive trials for matching the computed flux levels to those measured. The computed net reaction rates appear in the printouts of the program. Consider times when the concentrations of the various species nearly attain steady state. Then the rate of production of any cluster of size u (via monomer or dimer accretion) MINUS the rate of loss of a monomer or dimer from species u, is identified as the computed flux, because at steady state, the net loss of any $[A_u]$ is the spill-over to $[A_{u+1}]$. The dependence of enthalpy increments and of entropies on cluster size are thus derived from experimental data.

When the entropy of the monomer is available, $S^o_1(T)$, the entropy of clusters can be evaluated for the temperature range covered in the condensation data.

$$S^o(u;T) = u*S^o_1(T) + \sum_{u=2}^{u} \delta S^o(u-1,u).$$

The results of applying the program developed for KMM to four sets of measured flux levels {water, n-pentanol, Cs and methyl alcohol} are briefly summarized below in tabular form. The temperatures, critical supersaturations, J(obs) vs J(cal) are cited, along with the fitted values of the four critical parameters. Attention is directed to the narrow range of magnitudes required to fit the measured flux levels (within factors of 2).

TABLE 1. Temperature dependent condensation flux levels for water --- Observed vs calculated values, derived with the listed parameters

T(K)	css	a	b	α	β	J(exp)	J(cal)
259.1	7.22	0.950	0.333	-0.170	0.050	1.9E5	(1.3-1.6)E5
259.1	8.73	0.950	0.333	-0.180	0.050	9.2E8	(5 - 8)E8
233.5	14.84	0.950	0.333	-0.185	0.050	3.7E7	(1.6 -3)E7
217.1	22.16	0.950	0.340	-0.200	0.050	3.1E6	2.4 E6
217.1	24.47	0.950	0.340	-0.200	0.050	5.5E7	8.5 E7

TABLE 2. Temperature dependent condensation flux levels for n-pentanol --- Observed vs computed magnitudes, using the listed parameters.

T(K)	css	J(exp)	J(cal)	a	b	α	β
260	8.212	23.1	32 - 15	0.960	0.333	-0.230	0.040
260	7.16	0.315	0.27-0.21	0.960	0.333	-0.230	0.040
280	6.226	25.4	35 - 47	0.960	0.333	-0.210	0.040
300	4.29	0.631	0.33-0.35	1.000	0.333	-0.210	0.055
320	3.442	0.117	0.10-0.05	1.000	0.333	-0.200	0.065
320	3.633	10.5	10	1.000	0.333	-0.200	0.061

TABLE 3. Fitting Parameters for $J \approx$ unity(obs) --- for Cs Vapor

T(K)	a	b	α	β	J(calc)	τ(s) to s.s.
554	1.000	0.320	-0.175	0.050	$1.82 \to 0.46$	10^{-3}-10^{-2}
466	1.000	0.320	-0.2000	0.050	$0.13 \to 0.28$	~ 1
377	1.000	0.333	-0.196	0.050	$5.58 \to 1.23$	~ 1
289	1.000	0.337	-0.220	0.050	$0.80 \to 0.20$	10^{-2}-10^{-3}

TABLE 4. Temperature dependent condensation flux levels: CH_3OH

T(K)	css	J(exp)	J(cal)	a	b	α	β
229.3	4.07	1.9E9	2.8E9	1.000	0.333	-0.270	0.040
270.9	2.71	8.4E8	6.9E8	1.000	0.333	-0.225	0.040

Equilibrium Populations of Small Stable Clusters in Nucleation Theory

Jonathan C. Barrett

Department of Nuclear Science and Technology
HMS SULTAN, Military Road, Gosport PO12 3BY, U.K.

Abstract. Small clusters containing molecules with high relative momenta are likely to decay rapidly and so such arrangements should be excluded from consideration in nucleation theory. The effect of limiting the range of the momentum integrations when calculating the partition function of clusters containing 2, 3, and 4 molecules interacting via a Lennard-Jones potential is investigated. The resulting values are larger than those calculated using the harmonic-oscillator /rigid-rotor approximation, the relative difference increasing with cluster size. However, results using parameters and experimental values representative of Argon show that internally consistent classical theory predicts the populations of these small clusters surprisingly accurately.

INTRODUCTION

Statistical mechanical studies of nucleation usually begin with the partition function (or a related quantity) for clusters containing i molecules, q_i, which involves integrations over the positions and momenta of the i particles in the cluster. However, the exact definition of such a cluster has long been a subject of controversy. Much of this controversy relates to the spatial boundaries of the cluster but here we consider another aspect: the boundaries on the allowable momenta of the molecules in the cluster. If the molecules have large momenta relative to the cluster's center of mass, the cluster is likely to decay rapidly by the emission of molecules, and so cannot take part in the nucleation process. We shall term such clusters unstable, and those that are relatively long-lived as stable. Except for dimers, it is difficult to specify a stability criterion precisely. Here we adopt a suggestion by Hill[1,2] who extended the idea of negative energy states in dimers to higher i-mers.

DEFINITION OF STABLE CLUSTERS

Hill[1] proposed a definition of a stable dimer that involved limiting the integrations over particle momenta to values for which the sum of the potential and relative kinetic energies of the two monomers is always negative. The partition function for such a dimer can be written $q_2 = \frac{1}{2} V \Lambda^{-6} \int d\mathbf{r}_{12} \exp(-u_b(r_{12})/kT)$ where Λ is the de Broglie wavelength, V is the system volume, and $u_b(r)$ is related to the pair potential $u(r)$ by,

$$u_b(r) = \infty, \qquad u(r) > 0$$
$$= u(r) - kT \ln(P(\tfrac{3}{2}, -u(r)/kT)), \quad u(r) < 0 \qquad (1)$$

where $P(a,x)$ is the incomplete gamma function. Hill further suggested splitting up the partition function for larger clusters into regions representing bound and unbound groups. Defining $u_u(r)$ by $\exp(-u_u(r)/kT) = \exp(-u(r)/kT) - \exp(-u_b(r)/kT)$, then the partition functions for i-mers with every molecule bound to at least one other in the group can be related to integrals involving $u_b(r)$ and $u_u(r)$. For $i = 4$, these integrals are of the form,

$$I_{bbwxyz} = 4\pi^2 \int_0^\infty dr_{12}\, r_{12} f_b(\sigma r_{12}) \int_0^\infty dr_{13}\, r_{13} f_b(\sigma r_{13}) \int_{|r_{12}-r_{13}|}^{r_{12}+r_{13}} dr_{23}\, r_{23} f_w(\sigma r_{23})$$
$$\left[2\int_0^\infty dr_{14}\, r_{14} f_x(\sigma r_{14}) \int_{|r_{12}-r_{14}|}^{r_{12}+r_{14}} dr_{24}\, r_{24} f_y(\sigma r_{24}) \int_0^\pi d\phi\, f_z(\sigma r_{34}) \frac{1}{r_{12}} \right] \qquad (2)$$

where $f_w(\sigma r) = \exp(-u_w(\sigma r)/kT)$ and $w,x,y,z = b$ or u (only six of the 16 possible combinations represent distinct arrangements of bound molecules). σ is an (as yet) arbitrary scaling length and ϕ is the angle between the planes containing the points labeled 1,2,3 and those labeled 1,2,4. From geometric considerations we have $r_{34}^2 = r_{13}^2 + r_{14}^2 - 2r_{13}r_{14}\left(\sqrt{(1-\mu_{23}^2)(1-\mu_{24}^2)}\cos\phi + \mu_{23}\mu_{24}\right)$, where $\mu_{jk} = \mathbf{r}_{1j} \cdot \mathbf{r}_{1k}/r_{1j}r_{1k}$. For $i = 3$, the terms in square brackets are omitted and $q_3 = V\sigma^6 \Lambda^{-9}(I_{bbb} + 3I_{bbu})/3!$.

The cluster definition described above is not entirely appropriate for use in nucleation theory for the following reasons. Firstly, it is possible that transfer of kinetic energy from one molecule to another may lead to emission of that molecule, even if both molecules are "bound" to a third. Such states are likely to be short-lived and so should also be omitted from the nucleation process. Secondly, there are positive energy rotational states that are not included in this cluster definition but that may be long-lived. In the case of a dimer, these states have been examined by Lushnikov and Kulmala.[3] For molecules interacting by a Lennard-Jones potential, $V(r) = 4\varepsilon\left([\sigma/r]^{12} - [\sigma/r]^6\right)$, they showed that states with positive energies contribute up to about 26% of the total partition function. Unfortunately, their analysis cannot be readily extended to higher i-mers. We do not consider these limitations further here. Instead, we examine the consequences of Hill's cluster definition in nucleation theory, extending our earlier work on consistency.[4]

Given the cluster partition function, the population of i-mers in volume V, N_i, can be found from the law of mass action: $N_i = (N_1/q_1)^i q_i$. On the other hand, classical theory (with the $1/S$ correction factor) uses $N_i = N_1 \exp\left((i-1)\ln S - \gamma A_1 i^{2/3}/kT\right)$ where S is the vapor saturation, γ the liquid surface tension and A_1 the surface area per monomer (extrapolated from the bulk liquid density). Since $N_1 = n_{eq}VS$, where n_{eq} is the equilibrium vapor number density at temperature T, the saturation dependence of these expressions is the same. However, the classical expression must be multiplied by a factor F_i to be consistent with the statistical mechanical result, where

$$F_i = \frac{\hat{q}_i}{\hat{q}_1^i} \exp\left((i-1)\ln(\sigma^3 n_{eq}) + \frac{\gamma A_1}{kT} i^{2/3} \right) \qquad (3)$$

In equation (3), the scaled partition functions are defined by, $\hat{q}_i = q_i \Lambda^{3i}/(V\sigma^{3(i-1)})$. For structureless monomers, $\hat{q}_1 = 1$. For $i = 1$, equation (3) gives the correction factor which, together with the $1/S$ factor already included, is used to obtain "Internally Consistent Classical Theory" (ICCT).

RESULTS AND DISCUSSION

We have performed calculations for molecules interacting via a Lennard-Jones potential. The integrals in equation (2) must be performed numerically. Since the integrand is small for large values of r_{12} or r_{13} we truncated the range of integration (at 10σ) and used Gauss-Legendre quadrature. Typically, 30 quadrature points in each dimension gave results accurate to better than 5% with a few hours computation time (for $i = 4$) on a desktop PC. Figure 1 shows values of the scaled partition functions $\hat{q}_2$, $\hat{q}_3$ and $\hat{q}_4$ for dimensionless temperature $kT/\varepsilon = 0.25$ and 0.42. Also shown are values derived from the results of McGinty,[5] who used the harmonic-oscillator/rigid-rotor approximation. These values are significantly less than the ones we calculate, despite the fact that there is no limit to the momenta in McGinty's calculations. This is because the approximate harmonic-oscillator potential rises much more rapidly with separation than the actual Lennard-Jones potential.

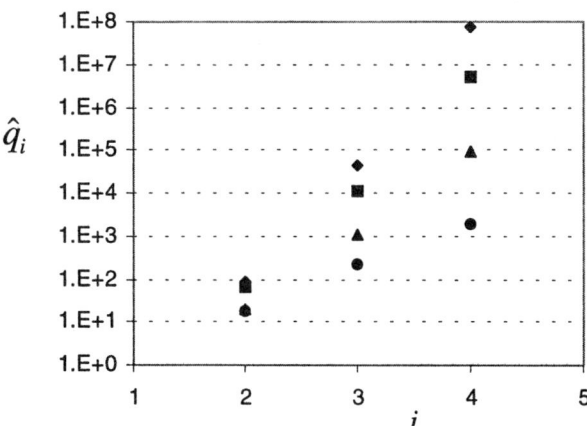

FIGURE 1. Scaled partition functions $\hat{q}_i$ for $i = 2, 3$, and 4. Bound clusters at $kT/\varepsilon = 0.25$ (diamonds) and 0.42 (triangles). Also shown are McGinty's values at $kT/\varepsilon = 0.25$ (squares) and 0.42 (circles).

Figure 2 shows the effect of requiring consistency between classical theory and our statistical mechanical predictions for the populations of small Lennard-Jones clusters. To represent Argon, we used the parameters $\varepsilon/k = 119.8$ K, $\sigma = 0.3405$ nm in the potential, and experimental values for the surface tension and liquid density of Argon in classical theory (it should be noted that there are considerable uncertainties in these experimental values). Remarkably, given the difference between the classical and stable cluster models, the correction factors for different values of i are almost identical and have the same temperature dependence. In other words, ICCT seems to predict the equilibrium populations of these small stable clusters quite accurately.

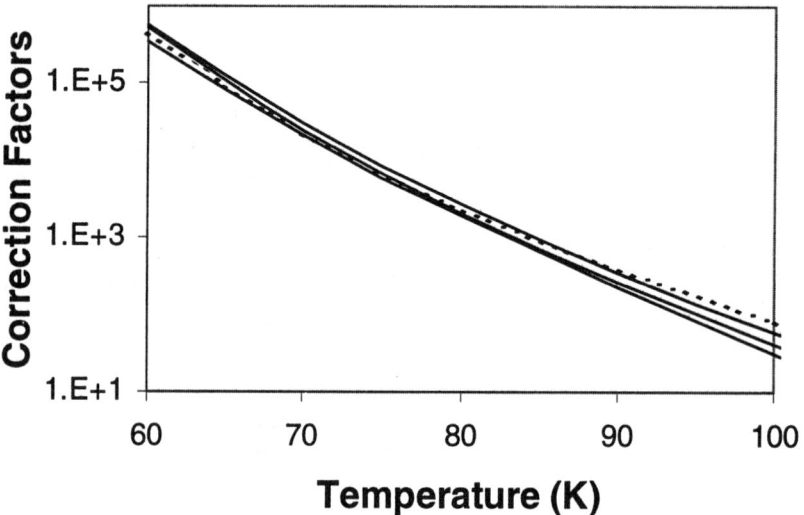

FIGURE 2. Correction factors F_i for Lennard-Jones Argon as a function of temperature. The dashed line is F_1, used in ICCT. The solid lines show F_2 (top line), F_3 (middle) and F_4 (bottom).

We have also investigated the application of this approach to more complex fluids by fitting the second virial coefficient of the Lennard-Jones fluid to experimental values. For nonane, we find a reasonable fit in the temperature range 213-313 K by using $\varepsilon/k = 1530$ K, $\sigma = 0.35$ nm. Using bulk properties in the classical expression, we again find that the correction factors have a temperature dependence very similar to the ICCT correction, although in this case their magnitudes increase with i.

REFERENCES

1. Hill, T. L., *J. Chem. Phys.* **23**, 617-622 (1955).
2. Hill, T. L., *Statistical Mechanics Principles and Selected Applications*, New York: Dover, 1987.
3. Lushnikov, A. A. and Kulmala, M., *Phys. Rev. E* **58**, 3157-3167 (1998).
4. Barrett, J. C., *J. Aerosol Sci.* **29**, S47-S48 (1998).
5. McGinty, D. J., *J. Chem. Phys.* **55**, 580-588 (1971).

TEMPERATURE-CONSISTENT CLUSTERS OF ARGON ATOMS: THEIR ROLE IN FREE ENERGY AND NUCLEATION RATE CALCULATIONS USING MONTE CARLO METHODS

P. PAL

Royal Holloway, University of London,
Egham Hill, Egham, Surrey TW20 0EX
UNITED KINGDOM

Abstract. In this study the pre-nucleation phenomenon is modelled as a canonical ensemble of N-body systems and the necessary thermodynamic states are determined using the conventional Monte Carlo simulation method. A distinct approach has been adopted towards building the essential, *a priori*, knowledge of the initial configurations of the N-body systems by developing the temperature-consistent clusters from the derivatives of the interaction potential. The results, including the Helmholtz free energy and the thermodynamic states at incipient nucleation, are presented for comparison with experimental and theoretical investigations on Argon clusters.

THE TEMPERATURE-CONSISTENT CLUSTER MODEL

At a temperature $T = 0K$ the pursuit of equilibration of a cluster with N mass units is reduced to attaining the absolute minimum potential energy. At $T > 0K$ the problem is reduced to the evaluation of the required thermodynamic states corresponding to the most likely values of their position and momentum coordinates that are consistent with the temperature T. We consider the variation in r, where r is the separation or bond length between a pair of atoms in a cluster, from its equilibrium value r_{eq} at 0K. This variation is denoted by $\Delta r = r - r_{eq}$. A general two-body scalar potential can be expanded in a power series in Δr as follows:

$$U = U''(\Delta r)^2 + U'''(\Delta r)^3 + U''''(\Delta r)^4 + \ldots \quad (1)$$

In the above equation U'', U''', U'''' refer to the 2nd, 3rd and 4th derivatives of the potential. The 3rd and 4th derivatives represent the asymmetry of mutual repulsion in a pair of atoms, and the general softening effect of the potential due to large displacements. These are the basic anharmonic terms responsible for the dilation of the pair separation, r, as a function of temperature. The thermal average of Δr as a function of temperature, T, denoted by $\overline{\Delta r}(T)$, can be evaluated using the Boltzmann distribution function, which weights all possible Δr values commensurate with the thermodynamic probability. This is given by

$$\overline{\Delta r}(T) = \frac{\int_{-\infty}^{\infty} \Delta r \, e^{-\beta U(\Delta r)} d(\Delta r)}{\int_{-\infty}^{\infty} e^{-\beta U(\Delta r)} d(\Delta r)} \quad (2)$$

For low anharmonic energy, the integrands in the *Gaussian* integrals appearing in Eqn (2) can be expanded further. This expansion gives

$$\cong \frac{\int e^{-\beta U''(\Delta r)^2}[\Delta(r) + \beta U'''\Delta(r)^4 + \beta U''''\Delta(r)^5]d(\Delta(r))}{\int e^{-\beta U''\Delta(r)^2}d(\Delta(r))} \quad (3)$$

$$= \frac{(U'''\beta)(U''/\beta)^{5/2}(3\pi^{1/2}/4)}{(\pi\beta/U'')^{1/2}} = \frac{3U'''}{4\beta(U'')^2} \quad (4)$$

where $\beta = 1/kT$; and kT represents the classical thermal energy of the pair. This is the key equation, which gives the linear expansion in terms of the potential derivatives and forms the basis of our temperature-consistent cluster model.

For the basic 2-body interaction potential, we adopt the Lennard-Jones function which is given by:

$$U(r) = 4\epsilon[(\sigma/r)^{12} - (\sigma/r)^6] \quad (5)$$

For Argon clusters; $\epsilon/k = 119.8K$ and $\sigma = 0.3405$ nm. For a primary dimer (N = 2), at $T = 0K$, $\Delta r = 0$; therefore $r = r_{eq} = 2^{1/6}\sigma$, and its static energy reaches the minimum value of the potential function. At $T > 0K$ the pair separation is expanded by $\overline{\Delta r(T)}$ as given by Eqn (4). Obviously this expansion results in the readjustment of the coordinates of the two atoms, consistent with T. Extending the idea to an N-atom cluster, all its independent bond lengths are dilated by the amount $\overline{\Delta r(T)}$ uniformly by suitable readjustment of the 3N coordinates of its constituent atoms. In essence, this specifies a homogeneously dilated ("warm") cluster at T, albeit in an approximated form. The expansion may be applied to clusters of any geometry, crystalline or non-crystalline, within a temperature range up to the bulk melting temperature of test species. However, it is possible to select a set of so-called "cold" clusters with known high stability and dilate them homogeneously as a function of T, using Eqn (4). These clusters are likely to emerge with configurational energy values within the proximity region of the *expectation value* at T. For this study, we have selected non-linear clusters of known geometrical form in the size range $3 \leq N \leq 12$ and members of the icosahedral family in the range $12 \leq N \leq 115$. In this range, clusters of the icosahedral family are known[1] to possess minimal potential energy.

The Monte Carlo Search

Each "cold" cluster of the stated size range was warmed up consistent with T and used as the initial configuration of the N-atom system in the Monte Carlo search. During this search, the coupling between all pairs of atoms was switched off; and the cluster was no longer treated as a solid. This allowed the atoms to move randomly in the configuration space. The exploration involved a machine simulation of a directed random walk by each atom in accordance with the Metropolis[2] algorithm. The random

TABLE 1. Helmholtz Free Energies of a selection of Argon clusters.

N	U(N) x 10^{12} ergs	$<A(N,T)>$ x 10^{12} ergs			
	0K	10K	30K	50K	65K
3	-0.049	-0.067	-0.074	-0.077	-0.081
7	-0.273	-0.320	-0.360	-0.376	-0.390
12	-0.628	-0.730	-0.781	-0.825	-0.836
13	-0.733	-0.742	-0.890	-0..950	-0.981
43	-3.230	-3.120	-3.540	-4.120	-4.290
55	-4.310	-4.150	-5.270	-5.820	-6.020
80	-6.700	-6.420	-7.650	-8.150	-8.420
90	-8.220	-8.010	-8.150	-8.660	-9.020
100	-8.740	-8.580	-8.810	-9.180	-11.510

walk was emulated by varying each coordinate in sequence which generated a chain of Boltzmann weighted energy states. Thus 3N × 2000 discrete moves were performed on an N-atom system for calculating the canonical averages. The computed results gave the derivative of the Helmholtz free energy (internal energy) between two temperatures. The absolute $<A(N,T)>$ was determined by direct integration of this derivative with respect to a reference state. In the present study, this state was chosen at $T = 1K$. Some of the computed $<A(N,T)>$ results are tabulated in Table 1 as a function of temperature and N. The static energies of the "cold" clusters (0K) which have been used as the generic systems are also shown in the same table for reference purposes. The free energy values for the clusters with N>12 show a qualitative agreement with the results of other independent[3,4,5] studies.

Thermodynamic States at Incipient Nucleation

Since each N-atom cluster was formed of N single atoms, the next step was to determine the corresponding Gibbs free energy of formation from its classical Helmholtz free energy function. The Gibbs energy of formation is given by

$$\Delta G(N,T,p) = G_0(N,T) - NG_0(1,T) + (1-N)\log(p/p_0) \quad (6)$$

The cluster free energy $G_0(N,T)$ appearing in Eqn (6) includes the classical Helmholtz free energy $<A(N,T)>$ of the N-atom system (computed from the ensemble) together with the corporate translational component and the reversible work[6] done in the formation process. $G_0(1,T)$ includes the translation of the single atoms; p and p_0 refer to the pressure of the nucleation chamber and the standard atmospheric pressure respectively. The $\Delta G(N,T,p)$ profile obtained from this investigation has exhibited a maximum[7] at $N = N^*$ (critical nucleus) for a range of temperatures T and pressures p, similar to that of the liquid drop model. These p and T values at the ΔG maxima define the thresholds of thermodynamic states in which nucleation occurs. In this context, the nucleation limit represents the deepest penetration of a system into the domain of the meta-stable state. At the onset of nucleation, if p is constant, it is the highest temperature T below the critical point that

TABLE 2. Argon nuclei of critical size, N*, at a range of $\Delta G(N,T,p)$ maxima, with the associated temperature T and pressure p. The related values from nucleation experiments are also included.

N*	p(torr) Pal	p(torr) Expt	T(K) Pal	T(K) Expt
35	0.8	0.8	45	43
45	8.0	8.2	52	50
50	25.0	30.0	55	53
60	80.3	82.0	65	63

a system can sustain without undergoing phase transition; analogously it is the lowest pressure p, if T is constant.

The characteristic steep rise in the rate of steady state nucleation was found from our calculations at different p and T as expected. Our investigation was further tested through comparison of the theoretical p and T values, corresponding to different ΔG maxima, with the experimental results. The experimental results [8,9,10] were recorded by the Yale University team on the nucleation of Argon using supersonic nozzles at low Mach numbers. The comparison between the theoretical and experimental results is shown in Table 2. The agreement indicates that the physical process involved at the nucleation onset is qualitatively well predicted by our treatment.

Discussion

The main thrust of this study lies in the development of temperature-consistent clusters using the potential derivatives. This provides an N-body system in which the constituents are brought to their approximate equilibrium positions consistent with the temperature T. At this stage, the mutual coupling among the atoms of the system is switched off and the stochastic exploration of the potential energy surface is initiated using the MC techniques. Thus by combining the ideas of solid state and statistical physics, we achieve efficient convergence to the *expectation values* of the target thermodynamic properties, leading to the evaluation of critical states at the onset of homogeneous nucleation. Apart from nucleation, the study of these temperature-consistent N-body systems provides valuable insight into the structural and thermodynamic properties that are mathematically tractable. These investigations are in progress.

REFERENCES

1. Northby J. A., *J Chem Phys* **87**, 6166-6177 (1987).
2. Metropolis N., Rosenbluth A.W., Rosenbluth M.N., Teller A.H., and Teller E., *J. Chem Phys,* **21**, 1087-1092 (1953).
3. Lee Jong, Barker J.A., and Abraham F.F., *J. Chem Phys,* **58**, 3166-3180 (1973).
4. McGinty D.J., *J.Chem Phys,* **58**, 4733-4742 (1973).
5. McClurg R.B., Flagan R.C., and Goddard W.A., *J. Chem Phys,* **105**, 7648-7663 (1996).
6. Reiss H., Tabazadeh A., and Talbot J., *J. Chem Phys,* **92**, 1266-1274 (1990).
7. Hoare M.R., Pal P., and Wegener P.P., *J Colloid and Interface Sci,* **75**, 126-137 (1980).
8. Steinwandel J., and Bucholtz T., Aerosol Sci. Tech. **3**, 71 (1984).
9. Matthew M.W., and Steinwandel J., J Aerosol Sci. *14*, 755-763 (1983).
10. Wegener P.P.and Mirabel P., *Naturweiss,* **74**, 111-119 (1987).

Nucleation Rate Determination from Measurements of Oscillatory Nucleation in Diffusion Cloud Chambers

M. Rusyniak, S. P. Fisenko and M. S. El-Shall[*]

*Department of Chemistry, Virginia Commonwealth University
Richmond, VA 23284-2006*

Abstract. The oscillatory nucleation in supersaturated vapors of the higher alkanes: dodecane, hexadecane and octadecane has been measured using the diffusion cloud chamber technique. The frequency of oscillation shows a dependence on the physical properties of the condensing vapor. Measurements of the frequency of oscillatory nucleation can be used to measure higher nucleation rates above the typical limits of the diffusion cloud chamber.

INTRODUCTION

The development of the thermal diffusion cloud chamber (DCC) has provided the foundation for significant advances in the study of nucleation phenomenon from a supersaturated vapor [1]. The DCC is a reliable tool for the production of supersaturated vapors, and is applicable to a wide variety of nucleation studies including homogeneous, photo- induced and ion-induced nucleation mechanisms. There are, however, some limitations to the use of the DCC. One of the major limitations is that the system must be maintained in a steady state mode of operation. At relatively high supersaturation, corresponding to high rates of nucleation, non-linear effects can result in the breakdown of the steady state condition. The system undergoes a transition from steady state behavior to stable oscillatory behavior [2]. In practice the use of the DCC is limited to low supersaturation systems, corresponding to rates of nucleation less than ~10 drops cm^{-3} s^{-1}.

Oscillatory nucleation in the DCC has been previously observed in the homogeneous nucleation study of supersaturated methanol vapor [2]. However, the correlation of the frequency of oscillation with properties of the condensing vapor has not been investigated. In this work, we use the DCC to study the oscillatory nucleation in supersaturated vapors of dodecane, hexadecane and octadecane. The objective of the present work is twofold. First, to investigate the dependence of the measured frequency of the oscillatory nucleation on the physical properties of the condensing vapors. The second objective is to show that the nucleation rate can be determined from the measurements of the oscillatory nucleation. Our approach is to use the measured oscillation frequency in connection with a non-steady state mathematical model of the heat and mass transfer in the DCC, which includes condensation of vapor on the falling droplets.

EXPERIMENTAL RESULTS

Detailed description of the DCC and its principles of operation are given elsewhere [1]. The nucleation is observed by measuring the forward-scattered light from a He-Ne laser transmitting across the center of the chamber. A photomultiplier positioned to detect the forward-scattered light is used to measure the intensity of the nucleation signal. The signal from the photomultiplier is sampled as a function of time at 10,000 Hz by an analog to digital converter and recorded by a computer. Due to the high nucleation rates in the oscillatory regime single droplet counting is not possible.

Figure 1 displays the scattered light intensity from the nucleated droplets as a function of time and the corresponding Fourier transformation of the data for several temperature gradients of the oscillatory nucleation of dodecane. It is clear that as the temperature difference between the top and bottom surfaces of the cloud chamber increases, well-organized collections of nucleation pulses are observed where each pulse contains a number of individual droplets. The measured oscillation has a fundamental frequency of about 4.2 Hz and higher harmonics start to develop as the temperature gradient increases.

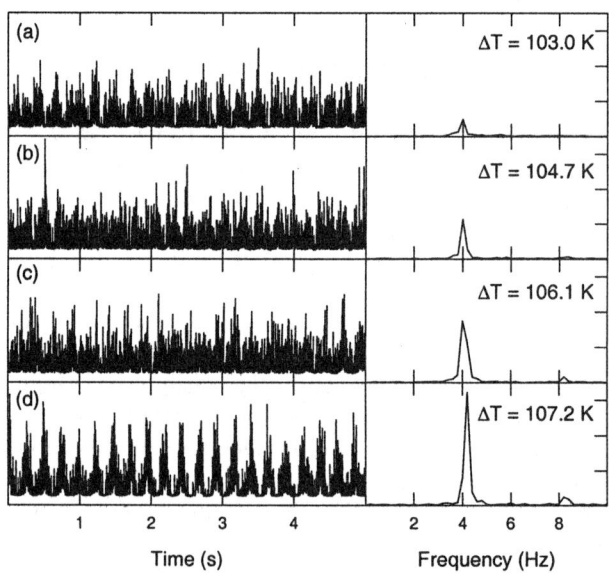

FIGURE 1. Evolution of the oscillatory nucleation of dodecane as a function of temperature gradient.

Figure 2 displays the scattered light intensity versus time along with the Fourier transform power spectra for the studied higher alkanes at the experimental conditions corresponding to the oscillatory regime in the DCC. It is clear that the fundamental frequency of the oscillatory nucleation decreases from dodecane to hexadecane to octadecane. This is also the order of increasing molecular weight, normal boiling point, critical temperature and the enthalpy of vaporization among the three alkanes.

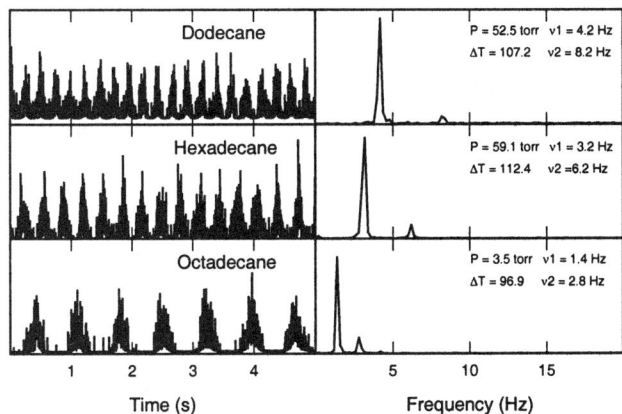

FIGURE 2. Oscillatory nucleation of dodecane, hexadecane and octadecane.

MATHEMATICAL MODEL OF OSCILLATORY NUCLEATION

The nucleated droplets regulate their own formation by depleting, through condensation, the concentration of the monomer species until a sufficient number of droplets are removed to allow the monomer concentration to build up again. The nonlinear effects in nucleation and growth are general manifestation of the inhibitory feedback control [3]. The mathematical model used here includes the diffusion equation of the vapor molecules in the DCC and three equations that describe the growth and motion of the droplets and the steady-state equation for the temperature profile. The diffusion equation can be written as

$$\partial_t n(x,t) = \partial_x [D(x)\partial_x n(x,t)] - I(R(z(t)), <n(z(t))>), \quad (1)$$

where n(x,t) is the vapor density, $<n(z,t)>$ is the average vapor density in a spatial domain occupied by droplets near the center of mass of the droplet cloud z(t), D(x) is the vapor diffusion coefficient at point x, I is a function which describes the condensation of vapor on a moving droplet, R(z(t)) is the average radius of droplet. The moving source is written as a result of spatial averaging over the droplet cloud. Galerkin's method was used for solving Eq. (1) [4]. The evolution of the amplitude A_1 of the first mode defines the fundamental frequency of oscillations in the nucleation rate (see Figure 3). This mode can be written as: $A_1(t)\sin(\pi x/H)$, where H is the chamber height. Our calculations show that the period of the oscillation is equal to the sum of the time required for the droplet cloud to grow and fall and the time needed for the original vapor profile to be restored by diffusion.

INVERSE PROBLEM OF OSCILLATORY NUCLEATION

There are a number of important parameters needed for the mathematical model under investigation. These include the number density of the droplets (N) and the amplitudes of the various modes during oscillatory nucleation. The periodic process only exists for a limited range of N. For small N ($N<7\times10^5$ droplet/m^3), there is only a steady-state solution. At high droplet densities ($N>5\times10^6$ droplet/m^3) the vapor depletion is so large, that the nucleation time is longer than the oscillation period. Oscillatory nucleation can only occur within these extremes for a supersaturated dodecane vapor given the thermodynamic parameters for the system.

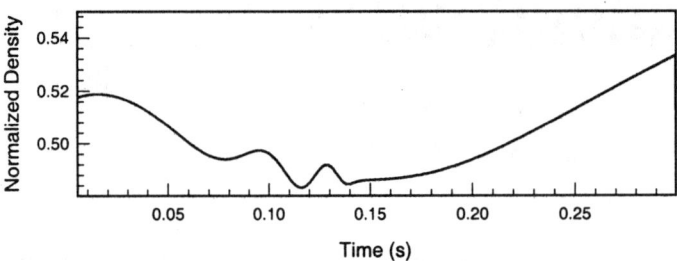

FIGURE 3. Evolution of normalized density at the nucleation zone for the data shown in Figure 1.

To solve the inverse problem, i.e. determining the nucleation rate, the following algorithm based on the available experimental data is used. For fixed thermodynamic conditions and value of N, the value of A_1 during nucleation is chosen to fit the experimentally measured fundamental frequency (the time required to restore the vapor density in nucleation zone). The amplitudes of other modes are assumed to be the steady-state values during nucleation. From the amplitudes of the various modes the vapor density can be found and the nucleation rate can be calculated using the classical nucleation theory. For the data shown in Figure 1-d the average nucleation rate is equal to approximately 7×10^7 droplets/m^{3}s, and the nucleation time is about 0.06s. For the data shown in Figure 1-a, the vapor depletion is less than 1% (for $N\sim7\times10^5$ droplets/m^3). In this case the DCC exhibits a modulated steady-state nucleation rate.

ACKNOWLEDGMENTS

The authors gratefully acknowledge support from NASA Microgravity Materials Science Program (Grant NAG8-1484).

REFERENCES

1. Kane, D. B.; Fisenko, S. P.; Rusyniak, M.; El-Shall, M. S. *J. Chem. Phys.* **111**, 8496- 8502 (1999).
2. Brito, J.; Heist, R. H. *Chem. Eng. Commun.* **15**, 133-149 (1982).
3. McGraw, R.; Saunders, J. H. *Aerosol Sci. Technol.* 367-380 (1984).
4. Fletcher, C. A. J., *Computational Galerkin Methods*, New York: Springer- Verlag (1984).

Statistical Theory Of High Pressure Nucleation Kinetics In Vapor-Carrier Gas Mixture

Sergey P. Fisenko

A.V. Luikov Heat & Mass Transfer Institute of National Academy of Sciences
P. Brovka Str., 15 ,220072, Minsk, Belarus

Abstract. The statistical derivation of the system of the equations of high pressure nucleation kinetics is presented. For kinetic coefficients the representation in Green- Kubo form is obtained. It was shown the nucleation coefficient is inversely proportional carrier gas pressure. Some features of the high pressure nucleation kinetics qualitatively are discussed.

INTRODUCTION

The resent experimental studies of nucleation kinetics in high pressure (about several MPa) vapor- carrier mixture have shown deviation of results from predictions of classical nucleation theory [1 and references there]. It is not surprise, because, according to its derivation procedure, the equations of classical nucleation theory in gases are limited the conditions of free molecular regime. The free molecular regime means that Knudsen number Kn is much greater than 1, Kn = λ / R_c, here R_c is the radius of a critical embryo and λ is the mean free path of vapor molecule. When Kn $\leq$ 0.1 nucleation kinetics will be in the diffusion regime. Using the diffusion regime of nucleation as the zeroth order approximation this work seeks corrections that are proportional Kn. The aims of this paper are to derive the equations for nucleation kinetics in high pressure environment using the methods of non-equilibrium statistical thermodynamics [2], and to analyze some features of these kinetics.

MATHEMATICAL MODEL OF NUCLEATION KINETICS

Dynamic Equations of Nucleation Kinetics

The total Hamiltonian H of the system under investigation (mixture of vapor, carrier gas and new phase cluster) is the sum

$$H = H_1 + H_2 + V. \tag{1}$$

where H_1, the Hamiltonian of a molecules, which can condense, H_2, the Hamiltonian of carrier gas and V, the Hamiltonian of interaction between of the cluster and carrier gas. The numerical density of molecules, which can condense (vapor and cluster), n(x) can be defined following to general methods of statistical mechanics as

$$n(x) = \sum_i \delta(x - x_i). \qquad (2)$$

where the summation is doing over all these molecules. The dynamic equation of n(x) is obtained by means of the Poisson bracket with the Hamiltonian H [2]. For density n(g)=δ(g-N) in cluster size space we have dynamic equation

$$\partial_t n(g) = -\partial_g \left(n(g) \dot{N} \right) \qquad (3)$$

where N is the dynamic number of molecules in cluster of volume Ω and $\dot{N}$ is the dynamic rates of change of the number of molecules in the cluster [3]. We consider the number of molecules as continuos variables for convenience of use methods of the nonequilibrium statistical thermodynamics. The macroscopic equations are the result of averaging dynamic equations with nonequilibrium statistical operator.

Non-Equilibrium Statistical Operator

The nonequilibrium statistical operator (distribution function for all dynamic variables) for averaging dynamic equations is obtained by means of Zubarev's method [2]. The averaging with this statistical operator means integration over Gibbs's phase space and the cluster size space. Other words for nucleation problem the phase space is the direct product of Gibbs's phase space and the cluster size space, its dimension is odd. Zubarev's operator ρ can be written as

$$\rho(t) = \rho_l - \int_{-\infty}^{0} dt_1 \, \exp(\varepsilon + L) t_1 L \rho_l. \qquad (4)$$

Where L is the Liouville operator, ρ_l is locally –equilibrium statistical operator, $\varepsilon \to 0$ after integration over time [2,3]. Note that a quantum-mechanical description of dynamic processes of nucleation was used in work [3]. In this we use classical mechanics description and expanding phase space, therefore Liouville operator takes the form

$$L... = \{..., H\} + \partial_g (... \dot{N}) \qquad (5)$$

where {...,H} is the Poisson bracket with the Hamiltonian (1). Free energy of cluster formation $\Delta\Phi$ at zero approximation relatively $\dot{N}$ and gradient of vapor chemical potential can be written as

$$\Delta\Phi(g) = F_c(g) + F_v(N-g) - F_v(N). \qquad (6)$$

where the first term is free energy of cluster from g-molecules, the second one is free energy of equilibrium vapor from N-g molecules and the last term is the free energy of non-equilibrium vapor from N molecules in the same volume.

Use nonequilibrium statistical thermodynamics allow to obtain kinetic equations by averaging dynamic equations (2-3) with nonequilibrium statistical operator (4).

CONCLUSIONS

In the so –called Markovian approximation of thermodynamic parameters, kinetic equation for distribution function of new phase clusters takes the form

$$\partial_t f(g,t) = \partial_g \left\{ f(g,t) \middle| \partial_g [L_{11}(\ln f + \beta \Delta \Phi(g))] + + \beta \int \nabla(\mu) J_{11}(x) d^3 x \right\} \quad (7)$$

and the equation for determining of vapor density near cluster

$$\partial_t n(x,t) = \text{div} \left\{ D_{11} \beta \nabla \mu(x) + J_{11}(x) \beta \int f(g,t) \partial_g \Delta \Phi(g) dg \right\} \quad (8)$$

The terms with J_{11} take into account the influence of Knudsen's boundary layer near the surface of new phase cluster. This influence can be described by means of effective free energy cluster formation which depends on Kn number for Kn>0.1.

The representation in the form integral over temporal correlation functions (Green-Kubo formula) of dynamic flows for kinetic coefficients is obtained:

$$L_{11} = \int_{-\infty}^{0} dt_1 \exp[(\varepsilon + L)t_1] < \dot{N}\dot{N}(t_1) >_0. \quad (9)$$

$$J_{11} = \int_{-\infty}^{0} dt_1 \exp[(\varepsilon + L)t_1] < \dot{N}j(x,t_1) >_0. \quad (10)$$

$$D_{11} = \int d^3x' \int_{-\infty}^{0} dt_1 \exp[(\varepsilon + L)t_1] < j(x)j(x',t_1) >_0. \quad (11)$$

Index "0" means here averaging with equilibrium Gibb's distribution function.

Using obtained above the Green - Kubo relations for kinetic coefficients it easy to obtained the exact relationship between $J_{11}(x)$ and L_{11}

$$L_{11} = \int \text{div} J_{11}(x) d^3x. \quad (12)$$

Analysis of the evolution of the first moment of distribution function f(g,t) by means kinetic equation (9) leads to the formula

$$L_{11} = 4\pi D R_c n. \quad (13)$$

where R_c, radius of critical cluster, D, diffusion coefficient, n, numerical vapor density. As follows from Eqs. (11-12), kinetic coefficients L_{11} and J_{11} depend on the pressure of carrier gas. As well known D~1/P, where P is carrier gas pressure. It's principal deviation our model of nucleation kinetics from classical nucleation kinetics theory which does not depend carrier gas pressure. As follows from (9-10), interesting feature high pressure nucleation kinetics arise in the case of high supersaturation: diffusion interaction between new phase clusters can lead to interesting effects at nucleation kinetics. Expressions (6-9) could be used for interpretation of MD simulation of nucleation kinetics. The results of statistical thermodynamics investigation and MD simulation, as in [4], should be complementary to each others.

REFERENCES

1. Luijten, C. C. M. and van Dongen, M. E. H., *J. Chem. Phys.* **111**, 8524- 8534 (1999).
2. Zubarev, D.N., *The Nonequilibrium Statistical Thermodynamics*, New York: Plenum, 1974
3. Bashkirov, A. G. and Fisenko, S. P., *Theoretical and Mathematical Physics*, **48**, 636-640 (1982)
4. ten Wolde, P.R., Ruiz-Montero, M.J. and Frenkel, D., *J. Chem. Phys.* **110**, p.1591 (1999)

NUCLEATION KINETICS OF ACETONITRILE AND BENZONITRILE - POLAR MOLECULES.

S. Chenthamarai, D. Jayaraman and C. Subramanian

Presidency College, University of Madras, Chennai 600 005, India
*Crystal Growth Centre, Anna University, Chennai 600 025, India

The nucleation study of polar molecules is quite interesting because of dipole-dipole interactions in the nucleation kinetics. The classical nucleation theory does not take proper account of the dipole-dipole interactions in calculating the free energy of the clusters of polar molecules. Polar molecules exhibit a preferential orientation at the droplet surface leading to a surface tension that differs from the bulk surface tension assumed by classical nucleation theory. By considering the effects of dipole-dipole interactions on the surface free energy of the embryos, Abraham [1] derived an expression for a size dependent surface tension for clusters with oriented surface dipoles. The predictions of this model depend upon the size of the dipole moment, the size of the cluster and the separation between surface molecules in the cluster. The consequence of dipole orientations is that for highly polar molecules the embryonic droplets may become somewhat elongated due to energetically favourable head to tail and antiparallel alignments of the dipoles. This would tend to give the clusters a hyperboloidal shape resulting in an increase in surface to volume ratio and thus an increase in surface free energy.

In the present work, a new expression for the curvature dependent surface energy by assigning a hyperboloidal shape to the nucleus to favour anti parallel alignment of the dipoles has been derived. The density of the nucleus is also assumed as a function of size. After incorporating the dipole-dipole interactions and density, a relation between surface energy and size of the cluster has been derived based on the concept of thermodynamics.

$$\sigma(a) = \sigma_o \left[1 - \frac{\delta_1^2}{a^2} - \frac{2\delta_1}{(2\chi + \delta^2\theta^2)a}\left(1 - \frac{\delta_1}{a}\right) \right]$$

where $\chi = \delta^2 \left[\frac{\sinh^{-1} e/\delta}{4e} + \frac{1}{4\delta}\left(1 + e^2/\delta^2\right)^{1/2} \right]$

$\delta = b/a = (e^2 - 1)^{1/2}$

$a_1 = b\theta$

a and b are the semi-widths of the nucleus along the x-axis and y-axis respectively, e is the eccentricity of the hyperbola used to generate the hyperboloidal shape about the y-axis and θ is the angle made by the line joining the center of the nucleus to the periphery of the top or bottom surface of the nucleus and its value is considered to be 10°. A study has been made on the critical supersaturation and nucleation parameters for Acetonitrile and Benzonitrile and the theoretical predictions are compared with the experimental results. The microscopic surface energy and density values are used to calculate the critical size and free energy change for different temperatures using the following expressions.

$$a^* = \frac{2\delta_1}{3(2\chi+\delta^2\theta^2)}\left[1+\frac{\pi\sigma_o(2\chi+\delta^2\theta^2)}{\alpha\delta_1}\right]$$

$$+\left\{\left[\frac{2\delta_1}{3(2\chi+\delta^2\theta^2)}\frac{1+\pi\sigma_o(2\chi+\delta^2\theta^2)}{\alpha\delta_1}\right]^2 - \frac{4\pi\sigma_o\delta_1}{3\alpha}\left[1-\frac{\alpha\delta_1}{4\pi\sigma_o}\left(1-\frac{2}{2\chi+\delta^2\theta^2}\right)\right]\right\}^{1/2}$$

$$\Delta G^* = 2\pi a^{*2}(2\chi+\delta^2\theta^2)\sigma_o\beta - \alpha\beta a^{*3}$$

where

$$\alpha = 8\pi k T d_o N\delta \ln S / 3M$$

$$\beta = \left[1 - \frac{\delta_1^2}{a^{*2}} - \frac{2\delta_1}{(2\chi+\delta^2\theta^2)a^*}\left(1-\frac{\delta_1}{a^*}\right)\right]$$

The nucleation parameters of Acetonitrile and Benzonitrile are presented in Tables 1 and 2. The nucleation barrier is decreased due to the impact of size dependent surface energy. The plots of the critical supersaturation against temperature for the above systems of molecules taken for the present study are shown in Figures 1 and 2 along with the experimental results of Wright and El-Shall [2]. The theoretical predications for a hyperboloidal shaped nucleus are found to be in better agreement with the experimental results than those for a spherical shaped nucleus.

Table 1 Nucleation Parameters of Acetonitrile

Temp. (K)	a^* (nm)	B^* (nm)	$\Delta G^*/kT$
240	0.4487	8.0641	63.75
250	0.4581	8.2332	63.47
260	0.4684	8.4173	63.24
270	0.4781	8.5918	62.89
280	0.4889	8.7859	62.81
290	0.4998	8.9816	62.54
300	0.5109	9.1820	62.16

Table 2 Nucleation Parameters of Benzonitrile

Temp. (K)	a^* (nm)	B^* (nm)	$\Delta G^*/kT$
270	0.5053	9.5880	67.51
280	0.5121	9.7160	67.15
290	0.5185	9.8370	66.33
300	0.5258	9.9770	66.05
310	0.5325	10.1030	65.23

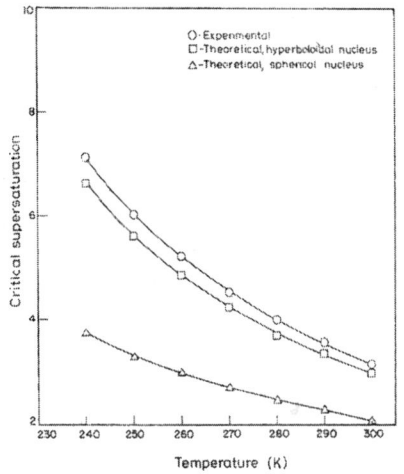 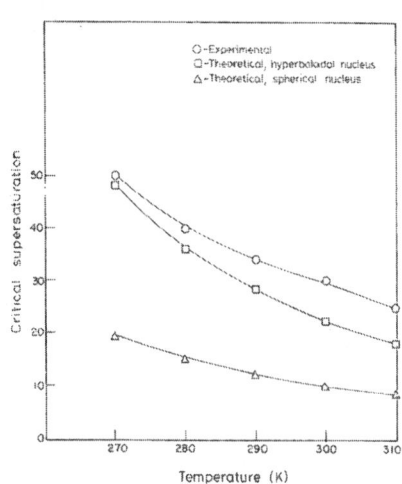

FIGURE 1. Variation of critical supersaturation with temperature for condensation of Acetonitrile vapour

FIGURE 2. Variation of critical supersaturation with temperature for condensation of Benzonitrile vapour

REFERENCES

[1]. F.F. Abraham, Academic Press, New york, 1974.
[2]. D.Wright and M.S. El-Shall, J.Chem. Phys., 98(1993)3369.

Critique of Molecular Theories of Nucleation

Howard Reiss

Department of Chemistry and Biochemistry, University of California, Los Angeles

Recent developments of molecular theories of nucleation, involving both analysis and simulation are considered. Subtle issues that arise in connection with these theories are analyzed and discussed with an emphasis on exactness. The focus is on three topics. These are (i) mapping of fluctuations simulated in a small cell onto the macrosystem, (ii) kinetics and dynamics of the nucleation rate, and (iii) the "nucleation theorem".

Exactness

During the last decade, impressive advances have been made in the development of molecular theories of nucleation (starting with the intermolecular potential). The bulk of this effort has been concerned with small molecule systems, e.g. Lennard-Jones (argon) fluids and water, but the importance of the studies should not be underestimated. Results from a completely molecular theory can provide guidance in the development of the more phenomenological theories that must be used in connection with the great variety of more complicated systems of practical interest. This suggests that molecular theories constructed for the simpler systems should be as exact as possible. The increased focus on such theories and their successful development has been due, in no small part, to use of the computer in both analytical and simulative modes. Thus, molecular approaches have ranged from those that are almost purely analytical[1-13], to those that are a mix of analysis and simulation (but more analytical than simulative)[14-17], to those that are more simulative than analytical[18-25], and finally to those that are so fully simulative[26-29] as to almost constitute a computer experiment rather than a theory.

The occurrence of a nucleus in a metastable initial phase is usually such a rare event that the direct simulation of both the nucleus and the phenomena surrounding it is often beyond the capacity of the computer. As a result, computational tricks, many of them brilliant, have been devised to circumvent this difficulty. In the process, however, subtle features of inaccuracy have crept into the methodology. These inaccuracies have rarely been serious enough to threaten the value of a particular theory or procedure but, in view of the goal of achieving maximum exactness, it is desirable to eliminate them when possible or at the very least to recognize them.

Although the title of this paper is ambitiously broad, the practical limits on length force me to focus my remarks on a small number of topics and to avoid mathematical derivations, but I will supply adequate references. Among the few topics addressed, I will spend a disproportionate amount of time on the issue concerning the careful separation of the translational and internal degrees of freedom of the physical clusters involved in a nucleation process, because it highlights the subtleties that abound in a molecular theory of nucleation. This issue has already plagued the phenomenological classical theory of nucleation (CNT) in the form of the "replacement free energy controversy" [30-33] that lasted for more than thirty years. Unfortunately, to a degree, it has reappeared in molecular theories, especially in those that rely more on simulation than analysis. Another topic concerns the dynamics of the nucleation process. Here, progress has been slow (but still remarkable), and those few who have fashioned molecular theories of rate have been well aware of subtleties, but I will attempt a brief discussion in this area. A third topic has received thermodynamic as well as molecular treatment, and involves the so-called *"nucleation theorem"* [34-37] which, although useful, may or may not be "remarkable".

Statistical Treatment of Rare Clusters

A physical cluster must usually be defined by a set of criteria that distinguishes it from the rest of the molecular system of which it forms a part. An early and still very viable example is the Stillinger cluster[2] in which a molecule, in order to belong to the cluster, must be closer than some assigned distance to at least one other molecule of the cluster. A second example is the classic LBA cluster[38,39] in which n molecules are confined to a rigid spherical container, of assigned volume v, whose center coincides with the center of mass of the molecules. A third example is the so-called n/v-Stillinger cluster[15,16] in which the n molecules of a Stillinger cluster are confined to a nonrigid spherical container centered on the molecules' center of mass and in which the radius of the container is defined as the distance between that center of mass and the cluster molecule furthest from it. Still another example might be called the Kusaka cluster[21,22,40], after its inventor, and in which n molecules are enclosed in a cell of volume v with one of them fixed to the center of the cell. Other models have been suggested but these four should be sufficient to make the point. As it is probably well known to the reader, the use of such clusters in a nucleation theory generally requires consideration of both their equilibrium (equilibrium size distribution) and kinetic properties. It is not unreasonable to hope that, if both types of property are treated consistently, the calculated rate of nucleation will be relatively insensitive to the particular model involved.

As far as the equilibrium distribution of clusters is concerned, a special simplification is available when the cluster is rare (as is the case with most nuclei involved in nucleation theory). This simplification consists of being able to equate Π_n the probability that the macrosystem contains *at least* one n-cluster (one cluster containing n molecules) with $\langle N_n \rangle$ the average number of such clusters in the system[9]. In turn

$$\Pi_n = \frac{Q_n(N,V,T)}{Q(N,V,T)} = \langle N_n \rangle \tag{1}$$

$Q(N,V,T)$ is the canonical ensemble partition function for the macrosystem at temperature T, containing N molecules in the volume V. $Q_n(N,V,T)$ is the partition function of the system restricted to states in which *at least* one n-cluster is present. It can be shown[9] that $Q_n(N,V,T)$ is given, exactly and quite generally, by

$$Q_n(N,V,T) = Q(N-n,V,T)q_n^*(V,T)\langle e^{-\beta U_\sigma}\rangle_0 \qquad (2)$$

where $\beta = 1/kT$ and $q_n^*(V,T)$ is the partition function of the *isolated* n-cluster, including its translation through the volume V, and $Q(N-n,V,T)$ is the partition function of the N-n molecules in the surrounding phase. In the integrations involved in both $q_n^*(V,T)$ and $Q(N-n,V,T)$, the conditions that define the cluster are enforced. Finally,

$$\langle e^{-\beta U_\sigma}\rangle_0 = \int_V d\mathbf{R}\int_c..\int_c dr_2'..dr_n'.\int_V..\int_V dr_{n+1}'..dr_N' \phi_n^o e^{-\beta U_\sigma} \phi_{N-n}^o \qquad (3)$$

In this equation U_σ is the interaction potential between the n molecules in the cluster and the N-n molecules outside of it. **R** is the coordinate of the center of mass of the cluster, and r_j' is the coordinate of the jth molecule relative to the center of mass. The limit c on some integrals indicates that those integrals must conform to the cluster requirement. If necessary, the cluster requirement may be included in U_σ. Finally, ϕ_n^o is the n particle distribution function (satisfying the cluster condition) for an isolated (decoupled from the environment) cluster while ϕ_{N-n}^o is the similar quantity for the N-n molecules outside of the cluster, but not required to satisfy the cluster condition. Eq. (2) remains valid if center of mass coordinates are replaced by laboratory coordinates, provided that the proper Jacobians for the transformation are used.

Eqs.(1)-(3) include, exactly, possible interactions between the cluster with molecules in the surrounding system and with other clusters. However, if these equations are specialized to a system in which the surrounding system is an ideal gas, and the cluster condition c includes a volume v from which the N-n surrounding molecules are excluded, it can be proved[9] that

$$\langle N_n\rangle = N\exp\{-\beta\tilde{W}_n\} \quad \text{or} \quad P_n^{int} = \langle N_n\rangle/N = \exp\{-\beta\tilde{W}_n\} \quad \text{or} \quad \tilde{W}_n = -kT\ln P_n^{int} \quad (4)$$

In these equations $\tilde{W}_n$ is the Gibbs free energy of formation of the n-cluster, provided that its center of mass is limited to $\tilde{v}$ the volume per molecule in the vapor. P_n^{int} has been called the *intensive free energy of formation* of the n-cluster[19].

We note, at this point, that the first of the three equations in Eq(4) has appeared commonly in the literature[41] for many years, but that the exact nature of $\tilde{W}_n$ does not seem to have been interpreted. It is also worth noting that $\tilde{W}_n$ contains a large component of translational free energy. This is evident from the translational component of q_n^* in Eq.(2). In fact, for argon vapor near its triple point, the translational contribution is of the order of 20kT out of a total free energy lying between 50 and 70 kT[9]. This emphasizes the importance of dealing carefully with the separation of translational and internal degrees of freedom.

The last of Eqs. (4) is particularly interesting. In the more analytical of the recent molecular theories of nucleation[15,16], any simulative work is usually aimed at a direct evaluation of $\tilde{W}_n$. This has proved to be a daunting task requiring large computer capacity and extensive computer time. Furthermore, because of the unusually large values of $\tilde{W}_n$ involved in situations of interest, the first two of Eqs(4) show that $\tilde{W}_n$ must be computed with high precision, since a small fractional error can then result in very large errors (orders of magnitude) in either $\langle N_n \rangle$ or P_n^{int}. On the other hand, some of the approaches that appear to involve more simulation and less analysis[18,19,21,22] have introduced a clever idea. This consists of simulating P_n^{int} directly and then using the third of Eqs(4) to evaluate $\tilde{W}_n$. It turns out that far less simulation is required in this approach than in the direct simulation of $\tilde{W}_n$. Furthermore, since the extremely sensitive P_n^{int} is the direct target of simulation and its logarithm is taken in order to arrive at $\tilde{W}_n$, that value of the free energy must be far more accurate than the value obtained via direct simulation. Thus, paradoxically, the approach that could be characterized as more analytical and less simulative demands more simulation than the one that appears to be less analytical and more simulative. On the other hand, as we shall see, the subtle problems involving the separation of translational and internal degrees of freedom do not arise in the more analytical approach.

In closing this section, it is useful to repeat the argument of the preceding paragraph in another way. This is to say that, if it were possible to simulate a partition function, Eq(1) could be used to evaluate $\langle N_n \rangle$. However, it is well known[42] that partition functions cannot be simulated so that one is forced to simulate probabilities. Finally, if a cluster probability is simulated in a small simulation cell, then a way must be found to map this probability onto the macrosystem within which the cluster is immersed so as to generate $\langle N_n \rangle$ a quantity fundamental to the evaluation of the *rate* of nucleation. It is in this mapping process that several of the subtle problems, mentioned above, appear. Furthermore, it is the mapping process that involves the separation of translational and internal degrees of freedom, although sometimes this may not be apparent.

Mapping and the Simulation of Probability: Two Examples

In analytical theories[9,43] where the translational free energy of a cluster is evaluated in a clear and distinct manner, the numerical specification of $\langle N_n \rangle$ is usually accomplished without paradox. A possible exception occurs in the density functional approach[7] where the density fluctuation that models the cluster is not centered on its center of mass. However, because of the limitations of space we will not analyze this situation in this paper. Also, because of the limitation on space and the fact that theory is incomplete, problems in condensed systems will not be analyzed. The focus will be on nucleation in vapors.

Very large scale simulations[26-27] that are almost computer experiments can avoid the separation problem implicitly, but in order to address probabilities large enough to permit the observation of a cluster, they must be performed at levels of metastability that are unrealistic for most laboratory experiments. Furthermore, as soon as an "experiment" is attempted with a system of intermediate size, models must be introduced in order to identify clusters and to evaluate the nucleation rate. As a result, the methods of preference seem to have been reduced to those that simulate the probability of cluster formation in a small simulation cell coupled to a mapping of these probabilities onto the surrounding macrosystem in order to generate $\langle N_n \rangle$ [19-22]. In these simulations, "tricks" are used to overcome the problem of contending with dramatically small probabilities, so that levels of metastability consistent with laboratory experiments can be addressed.

In the author's opinion, the special problem of simulating a fluctuation in a small cell transcends the phenomenon of nucleation and extends to the simulation of nanoscale fluctuations in general. The problem of simulating and mapping inhomogeneities is distinct from that of simulating a uniform bulk property of the system.

To render these ideas more concrete we will consider two effective but different examples of such local simulation and the mapping procedures adopted in each case. In the first example, due to Kusaka and coworkers[21,22], the cluster effectively consists of all of the molecules in the simulation cell, while in the second example, due to ten Wolde and coworkers[19-20], it is made up of only a small fraction of the molecules in the cell, but it is identified as a Stillinger cluster with an assigned connectivity distance (and, in addition, each cluster molecule is coordinated to at least 5 nearest neighbors).

We begin our discussion with the Kusaka cluster and confine our remarks to a low density vapor, one of the systems considered by Kusaka et al[21,22,40]. Figure 1 is a schematic of the total volume V of the gas, spanned by a lattice of cells, each of volume v. Thus the number of cells is M=V/v. The filled circles indicate molecules. Now, if the gas is rarefied and a physical cluster forms, it will have to be *compact* since the gain in binding potential energy will have to overcome the large gain of system entropy that would attend the evaporation of the cluster. Such a compact cluster is shown in the cell marked A, and a similar cluster is shown straddling cells B and C.

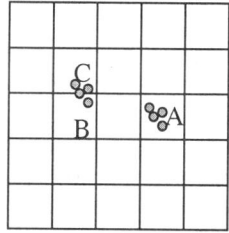

Figure 1

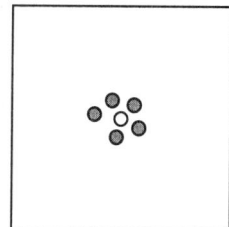

Figure 2

The cell volume v is chosen so that it is considerably larger than the extent of compact clusters in the range of sizes of interest, but yet small enough so that the vast majority of cells in the attenuated vapor are vacant, i.e so that the rare clusters will not, on the average, interact.....a cell occupied by a cluster will in general be surrounded by many empty cells. These two criteria are in some opposition and cannot be satisfied for every gas or vapor, e.g. it is not possible to satisfy them for argon vapor above the triple point[44]

When the requirements can be met, a particular cell is regarded as a small grand ensemble and since n molecules within the cell can only be a compact cluster, one is motivated to equate the grand probability that n molecules occupy the cell with the probability of formation of an n-cluster within that cell. Then it seems perfectly natural to evaluate the average number of n-clusters in the macrovolume V as

$$\langle N_n \rangle = \frac{V}{v} p_n = M p_n \qquad (5)$$

where p_n is the grand probability that n molecules will occupy v. The probability p_n then becomes the target of simulation. In eq(5), V/v is the factor that "maps" the probability p_n onto the macrovolume V in order to generate $\langle N_n \rangle$. It should be reemphasized that the ultimate goal is the determination of $\langle N_n \rangle$ and that, in principle, this quantity could be determined directly from a partition function of the system containing the various kinds of clusters[9]. We shall consider two different cluster models, and it is convenient for the purpose of discussion to label them. Thus we shall denote the model in Figure 1 (n molecules...including n=0... in an open cell of volume v) as a cluster of type A. The second cluster..."type B"..., related to type A, will be considered below and, as an illustration of one of the above mentioned subtleties, we will examine which of the two clusters permits the most rigorous and expeditious evaluation of $\langle N_n \rangle$.

In the small grand ensemble used in the representation of the A cluster, there is supposed to be no interfacial interaction between the molecules in the cell and those in the surrounding medium, i.e. in the M-1 remaining cells of the system[21]. There are several problems with this approach[11,12,45]. For example, consider the cluster straddling cells B and C in Figure 1, and suppose that cell B is the one involved in the small grand ensemble. If the cluster contains n molecules, then only n' < n molecules will be in cell B and, under the rules, we will count cell B as an n'-cluster. But at the same time we will violate the assumption that there are no interactions between the molecules in cell B and those in cell C which now lie in the surrounding medium. Thus the appropriate grand ensemble would have to exhibit an interfacial term in the corresponding grand potential, and the method loses precision because of the possibility of clusters straddling the interface between two cells.

A second problem concerns the magnitude of p_n. In the attenuated vapor this will be so small that, in any feasible simulation, occupation of the cell by n molecules will almost never be observed.

Finally, there is a conceptual problem that is highlighted by the possibility of having clusters in a "straddling" position. The lattice of cells in Figure 1 is an abstraction imposed upon the system; it is not part of the physical system. As such, it must remain "passive" in the determination of the physical properties of the system, and, *in general*, it makes little sense to define a cluster in terms of the occupation of the cell. What is needed is a cluster model which, if a cell is involved, links the cell to the degrees of freedom of the group of molecules constituting the cluster.

In references 21 and 22, an attempt is made to remedy these problems by introducing the model cluster (type B) shown in Figure 2. This consists of a container or cell of volume v at whose center a molecule is fixed. This "open" container, immersed in the rarefied gas, is then used as a grand ensemble and when occupied by n-1 molecules, additional to the one at the center, is regarded as an n-cluster. This n-cluster will still be compact and will surround the central molecule (denoted by an open circle in Figure 2) so that all of its molecules will lie far from the container boundary and there will be little concern that any of them can interact with vapor molecules outside of the container. Furthermore, since the container boundary is *far* from the cluster molecules, it might be assumed that the properties of the cluster are independent of the container. This cannot be entirely true since if v is made large enough the cluster will evaporate within it[46-49], although it is known[38,39] that a range of volumes exists within which the cluster properties are relatively insensitive to its size. However, the container cannot be detached from the cluster for another reason. This is related to the fact that the molecules in it are indistinguishable so that it is not possible to unambiguously say which of them should be chosen as the central one. The choice of different molecules results in a set of different locations for the container. What *is* unique is the *location of the container*, i.e the location of the center of the container. By attaching a molecule to this center we make the container part of the cluster and indeed couple it to the degrees of freedom of this molecule. Thus, in effect, the container participates in the degrees of freedom of the molecules within it, and its center defines the location of the cluster.

However, the attachment of the container center to the molecule *marks* that molecule and distinguishes it from the remaining n-1 molecules. One might try to use this marked molecule in a proof[40] of an equation that resembles eq(1), but which refers to the probability that an n-cluster of type B lies within a volume v (now v is not necessarily a cell of the lattice illustrated in Figure 1).The argument is that the probability $p_c(n)$ that the marked molecule, and therefore a cluster of size n, lies within v is clearly given by

$$p_c(n) = \langle N_n^* \rangle \frac{v}{V} \qquad (6)$$

where $\langle N_n^* \rangle$ is the number of B type n-clusters in V. This equation can be rearranged to

$$\langle N_n^* \rangle = \frac{V}{v} p_c(n) \qquad (7)$$

which resembles eq(1). The problem with this argument is that the actual cluster consists of a group of mutually close indistinguishable molecules so that unless the molecule chosen to be central is *permanently* marked the statement that the cluster lies within v is ambiguous. For example, if some molecules of the group, defined as noncentral, lie outside of v, even though the central one lies within v, the cluster could very well lie outside of v if the mark is removed and assigned to an exterior molecule. In Figure 3 we illustrate this situation for a 3-cluster, a trimer. If molecule 1 is considered to be central, then the cluster lies within v, but if molecule 3 is chosen as central the cluster lies outside of v. Since, in reality, there are *no* marks on the molecules, we cannot locate the cluster unambiguously. Note that for simplicity, in Figure 3, the cells are depicted as spherical since they no longer need to tile a close packed lattice. Also, the volume that is supposed to be occupied by the cluster is

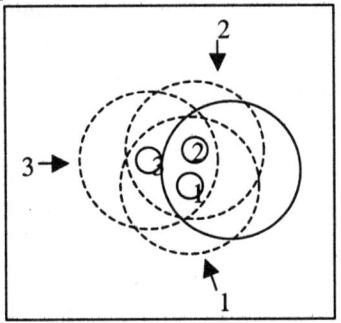
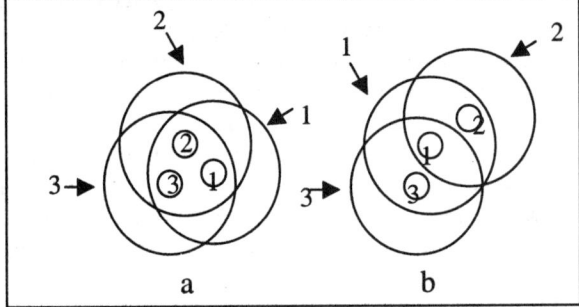

Figure 3   Figure 4

identified by having a solid boundary, while the volumes that bound the various clusters, distinguished by the choice of a central molecule, have dashed boundaries.

The logic underlying eq(6) involves some further assumptions. For example, it is assumed that the center of the cluster container, always associated with a molecule, has a uniform probability density of being located anywhere within v and, for that matter, anywhere within V. This is the origin of the factor v/V which is supposed to represent the probability that one particular cluster, among the $\langle N_n^* \rangle$ of them, will be found in v. However, it can be shown[12] that this probability is not uniform, at least for a cluster of type B. Thus eq(6) is not exact. The nonuniformness of the probability density is due to the internal degrees of freedom of the cluster, including the effects of ambiguity[12], especially near the boundaries of v.

The type B cluster in Figure 2 eases the problem of observing a cluster within v, i.e. within the container, during a simulation. This follows from the fact that at least one molecule is always attached to the container center so that its strong potential can invite the entry of other molecules. The probability that v is occupied by n-1 molecules, additional to the one at the center can then be denoted by p_n^{**}.

The grand ensemble probability that n-1 molecules, additional to the one at the center, occupy the container is thus p_n^{**} where $p_n^{**} \gg p_n$. Thus, it is feasible to attempt a simulation of p_n^{**}, when it is almost impossible to simulate p_n. The application of the Kusaka cluster has been limited to compact versions, because, on a qualitative basis, compact clusters of both types should satisfy the relation, $p_c(n) \approx p_n$. This reasoning is based on the fact that if a compact type B cluster lies within a cell of volume v (i.e. its central molecule is in v) whose breadth is much greater than the breadth of the cluster, then all n molecules of the cluster are likely to lie within v. But the probability $p_c(n)$ that a B cluster is in v is then equivalent to the probability that n molecules occupy v and this is just p_n. Hence the two probabilities converge on one another. Then, if one can simulate $p_c(n)$, either eq(1) or (7) can be used to generate $\langle N_n^* \rangle \approx \langle N_n \rangle$. The mapping factor V/v is not exact but, as we discuss below, it may be a good approximation when the clusters are compact enough.

The *compact* B cluster is useful in other ways. For example, in spite of the ambiguity in the choice of the central molecule, very little change in cluster location will occur upon the choice of different centers. Furthermore, in most locations throughout v, all of the molecules of the cluster will lie within v. Thus, since the possible nonuniformness of the probability density of location within v or V is connected to ambiguity (involving the internal degrees of freedom of the cluster) the nonuniformity will be minimized. Also, situations involving straddled clusters like the one shown in Figure 1 will be less probable, since the breadth of a compact cluster is small compared to the breadth of v.

Although, as a mapping factor, V/v is not exact, it turns out that the correct factor for A clusters can not only be rigorously derived[11,12], but it can also has been shown that it remains correct for *noncompact* as well as compact clusters.

In view of the several points made in the above discussion, a reasonable strategy, in the case of compact clusters, for the evaluation of $\langle N_n \rangle$ is the following. First simulate p_n^{**}, and then derive a relation between p_n^{**} and $p_c(n)$ so that the latter can be used in either Eqs(5) or (7) to generate $\langle N_n \rangle$. This is the strategy followed in references 21 and 22 where $\langle N_n \rangle$ is viewed as the number of B clusters.

In references 21 and 22 a relation between $p_c(n)$ or p_n and p_n^{**} is derived on the basis of several approximations that are reasonable for compact clusters. Actually, what is simulated is not p_n^{**}, but rather p_n^{**}/n, and the major approximation is that the B cluster, in any configuration, can be *rigidly* translated (by translating its central molecule) through a cell of volume v without concern for the cell boundaries. The relation derived in this manner is

$$p_n = p_c(n) = \frac{v\rho p_n^{**}/n}{p_1^{**} + v\rho} \tag{8}$$

in which ρ is the density of the vapor (equal to the thermodynamic activity in the attenuated gas). More recently an exact relation, taking the cell boundaries into account, has been derived[12]. It is

$$p_n = \left[\frac{\phi_1^{**}(0)\{p_n^{**}/n\}}{\phi_n^{**}(0)p_1^{**}} \right] p_1 \tag{9}$$

in which $\phi_n^{**}(0)$ is the probability density that a molecule will be found at the center of v when the cell contains n molecules with none of them fixed at the center. This relation is valid for a noncompact as well as a compact cluster. For a compact cluster where the breadth of the cell is much larger than the breadth of the cluster, $\phi_n^{**}(0) \approx 1/v$, so that the ϕ's cancel out of the ratio in Eq(9). This, coupled with the fact that, in Eq(8), $v\rho \ll p_1^{**}$ because the vapor is of low density, and that, in the attenuated vapor $v\rho \approx p_1$, leads to the reduction of Eqs(8) and (9) to identical expressions. However, the simulation of the ϕ's can now provide a quantitative criterion for the accuracy of the approximation in Eq(8). We note that the problem of ambiguity regarding the central molecule of a B cluster has no relevance for an A cluster since the latter *has* no central molecule. Thus even when the cluster is not compact, Eq(9) not only remains exact, but the p_n that it describes is free of ambiguity.

Now p_1^{**} in Eq(8) is essentially unity, since for a B cluster, by definition, it is certain that there is one molecule in v, i.e. the central one, and because ρv is so small it is virtually certain that a second molecule will not occupy v. Using this fact, setting $v\rho$ in the denominator of Eq(8) equal to zero, and substituting the result into Eq(5) or (7), then yields

$$\langle N_n \rangle = \frac{N p_n^{**}}{n} \qquad (10)$$

where again N is the total number of molecules in the system. We can provide a physical interpretation of this equation. First consider the numerator in Eq(10). The probability p_n^{**} can be validly interpreted as the chance that any molecule in the system forms the central molecule of an n-cluster of type B. Thus, if for the moment, we were to ignore the n in the denominator in Eq(10) or set it equal to unity, Eq(10) would prescribe the number of n clusters in the system as a fraction p_n^{**} of all the molecules in the system. Unfortunately, some of these molecules would be so close together (themselves forming an n-cluster) that the n-clusters that they bear would overlap. This simply means that the several n-clusters share some of the same molecules. In Figure 4a, this situation is illustrated for a compact 3-cluster. Notice that if we select any one of the three molecules as central, the other two molecules will lie within the sphere centered on it. Thus, for such a compact cluster, the three molecules can be thought of as constituting a single *physical* cluster even though the assembly consists of three *mathematical* clusters which Eq(10)...without the denominator... would count as three separate clusters. However, inclusion of the denominator rescues the situation since division by n, in this case 3, leaves us with only a single cluster, just as the physical situation demands. Now, consider the less compact 3-cluster illustrated in Figure 4b. Here we see that the sharing of molecules results in only one 3-cluster, accompanied by two 2-clusters! Division by 3, as required by Eq(10), now gives us only one third of a 3-cluster and two thirds of a 2-cluster, whatever this means? What we are seeing is a failure of both Eq(1) and its mapping factor V/v, as well as the failure of Eq(8), when the cluster is not sufficiently compact.

The use of Eq(9), on the other hand, together with a mapping factor that remains valid even for a noncompact cluster, should avoid the difficulties and uncertainties discussed in the preceding paragraph, although then $\langle N_n \rangle$ would then refer to A clusters.

It is now appropriate to discuss the correct mapping factor. For this purpose we confine attention to A clusters (the more difficult factor for B clusters has not yet been derived). Multiplying p_n in Eq(9) by the factor for the A cluster yields $\langle N_n \rangle$ for both compact and noncompact clusters. In the absence of a lattice, as an aid in defining an A cluster, one can envision a molecular microscope with a spherical aperture of volume v that can be used to search V for groups of n molecules, compact enough to fit within the aperture which should be large enough to encompass the widest nonuniform group. Molecules outside of the group should be far enough removed so that the group or cluster does not interact with them. In the formal analysis[11,12], noninteraction between the cluster and the surrounding medium can be implemented through the use of Eq(1) with this restriction, among others, built into $Q_n(N,V,T)$. The main caution involves the avoidance of "positional redundancy" which, if not avoided would result in an overcounting of configurations in $Q_n(N,V,T)$. Such redundancy is illustrated in Figure 5. Figure 5a shows the spherical cell (aperture) of volume v in an initial position and containing n molecules (open circles). The N-n molecules (filled circles) in the surrounding vapor of volume V-v are also shown. Figure 5b shows the cell (solid circle) translated from its original position through the vector distance **ds**. The cell in its initial position is shown as the dashed circle. The molecules within the cluster are now all within the cell in its initial position as well as in its new position (see the lens shaped region of overlap). Clearly, the cluster has not been shifted even though the cell (aperture) has. If we regarded the configuration in Figure 5b as a shifted cluster we would be overcounting. Such redundant counting would lead to a spurious $Q_n(N,V,T)$. It is possible to avoid this redundancy[11,12]. Then, using the correct $Q_n(N,V,T)$ in Eq(1), we arrive at

$$\langle N_n \rangle = \frac{(P_n^w + P)V}{kT} p_n = \frac{(P_n^w + P)V}{kT} \left[\left[\frac{\phi_1^{**}(0)\{p_n^{**}/n\}}{\phi_n^{**}(0)p_1^{**}} \right] p_1 \right] \quad (11)$$

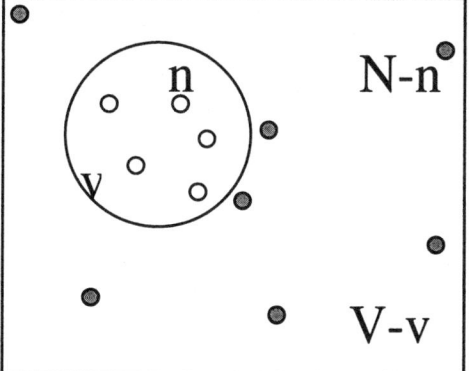
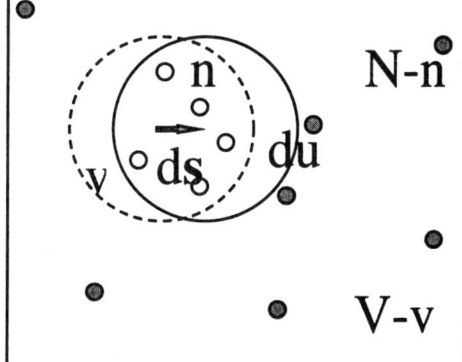

Figure 5a Figure 5 b

where we have used Eq(8), and where P_n^w is the pressure exerted by the n molecules on the interior wall of the cell while P is the pressure of the surrounding vapor. As indicated earlier, this expression is correct for type A clusters (compact or noncompact) surrounded by a vapor of low enough density so that there is no interaction between it and the cluster molecules.

It is clear from Eq(10) that the mapping factor is $(P_n^w + P)V/kT$. If, as will be the case for a low density vapor, $P_n^w \gg P$, P can be neglected, and if the cluster is so compact that it can be considered to constitute an ideal gas consisting of a single molecule, then $P_n^w V/kT \approx V/v$, so that the mapping factor in Eqs(5) and (7) is recovered. Also, the second factor on the right of Eq(11) reduces to the expression in Eq(10) so that for compact clusters the less rigorous scheme is approximated.

However, the conclusion to be drawn from the foregoing is that B clusters play a useful role because they can be simulated and can provide values of p_n through Eq(9), so that it is $\langle N_n \rangle$ for the A cluster that is ultimately evaluated. As advertised, the argument is subtle. In this short paper there is not room to cover all of the subtleties of the mapping problem associated with the Kusaka cluster but, hopefully, enough ground has been covered to demonstrate that the mapping is not straightforward.

We now turn to the ten Wolde approach[19]. In this approach, as many as 5000 molecules are simulated within a periodic cell of volume v. The system is simulated as an NPT ensemble in order to imitate a nucleation experiment in which both pressure P and chemical potential μ are maintained constant as a cluster forms. The cluster itself is identified as a Stillinger cluster. In this case, the very small probability of cluster formation is overcome by the use of umbrella sampling[42]. Since, neglecting boundary effects, the frequency of cluster observation includes its formation anywhere within v, the appropriate mapping factor is indeed V/v, and for the *intensive* probability of occurrence the factor is $\tilde{v}/v$. The logarithm of this probability is then used in Eq(4) for the determination of $\tilde{W}_n$.

A subtle aspect of this approach is the following. In the thermodynamic limit, all of the system's intensive variables are determined when any two of them are fixed. However, in a small system in which P and T are fixed, as in the NPT ensemble, μ can fluctuate or, alternatively, in a grand ensemble with μ fixed, P can fluctuate. For example, it can be estimated[50] that for a cell containing 5000 argon molecules at T=85K and at the estimated critical pressure[20] where the nucleus has $n \approx 300$, the fluctuation in μ will be of the order of 0.022 kT. Multiplied by a factor of 300, this enters into $\tilde{W}_n$ to result in a change in the nucleation barrier height of about 6.7 kT. Because $\tilde{W}_n$ is located in an exponent, this results in a change in the nucleation rate by a factor of the order of 8.8×10^2. In the physical measurement, this means that the rate could differ by this much from that obtained via simulation.

Rate of Nucleation: Dynamics

A molecular theory offers the opportunity to develop a rather exact theory of the rate of nucleation. Indeed, impressive progress in this direction has recently been made by ten Wolde et al[20] through the application of Bennett-Chandler linear response theory[51,52] to the barrier crossing process. In this work a barrier free energy surface in an order parameter (reaction coordinate) space, for a defined cluster, is first generated by Monte Carlo simulation. Then molecular dynamics simulation is used to generate time correlation functions associated with the average flux over the barrier top, starting at that top. In this approach the flux over the barrier is expressed as

$$J^* = \left\{ \frac{\langle \delta(\Phi - \Phi^*) \rangle}{\langle \theta(\Phi^* - \Phi) \rangle} \right\} \left\{ \frac{\langle (d\Phi/dt)\delta(\Phi - \Phi^*)\theta[\Phi(t) - \Phi^*] \rangle}{\langle \delta(\Phi - \Phi^*) \rangle} \right\} \tag{12}$$

Φ is the reaction coordinate describing the path on the free energy surface connecting the vapor and liquid states (the asterisk refers to the top of the barrier), δ is a delta function, and θ is a unit step function, while the angle brackets denote canonical averaging. With these definitions, the quantity in the first curly brackets is the (equilibrium) probability of finding the system at the top of the barrier when it is known that the system is in the vapor state while that in the second curly brackets, the time correlation function, is the average flux over the top of the barrier for a system known to be at the top. The molecular dynamics results for argon vapor not only provide a credible value for the nucleation rate, but also clearly show that the barrier crossing is diffusive in the Kramers sense[53] and the process involves more than single step additions and losses of molecules.

Since accurate rate experiments involving simple systems like argon are extremely difficult to perform, an additional check on the accuracy of a dynamical molecular rate theory is afforded by a comparison of the results of more than one such theory, each developed with the same degree of care.

For example, an expression for the rate involving multimolecular steps, more reminiscent of the conventional master equation of nucleation theory, is[54]

$$J_{i-(1/2)}(t) = \sum_{m=1}^{\infty} \sum_{j=1}^{m} m \{ \beta_{m,i-j}(t) f_{i-j}(t) - \gamma_{m,i+m-j}(t) f_{i+m-j}(t) \} \quad i \geq 3 \tag{13}$$

where $\beta_{m,i-j}$ is the rate at which a cluster of size i-j grows via an m molecule step and $\gamma_{m,i+m-j}$ is the rate at which a cluster of size i+m-j diminishes via an m molecule step. β and γ are evaporation and condensation coefficients which, in principle, can be evaluated by simulation, while the f's are the concentrations of clusters at the time t. This expression is exact and can be used in a molecular theory. A simplification (not necessarily recommended) is obtained by passing to the continuum[54] and using detailed balance. Then the steady state rate of nucleation is easily found to be[54]

$$J^* = \sum_{m=1}^{\infty} m^2 \beta_{m,i} \cdot \left\{ \int_1^{\infty} di/N(i) \right\}^{-1} \tag{14}$$

The discrete set of master equations, Eq(13) can be solved[55] when the β's and γ's are simulated. This process is facilitated by the simulation of γ, which is almost time and pressure independent, and from which β can be determined through an application of detailed balance. Initial molecular dynamics simulations for argon[56] show that detailed balance is well satisfied and confirm the findings of reference[20] that multimolecular steps are important.

As part of this critique, mention should be made of promising approaches that, like linear response theory are related to Onsager regression[57] and Kubo theory[58]. Thus, for the simple single step version of barrier crossing theory, in which both detailed balance and the continuum limit is used, the cluster flux can be represented as $J = -\beta(i)[\partial f/\partial i] - \{\beta(i)f(i)/kT\}[dW(i)di]$ where $W(i)$ is the barrier height as a function of cluster size i. This equation can be regarded as a Nernst-Planck equation[59] in which $\beta(i)/kT$ plays the role of a "conductivity" at position i. Indeed, if the "equivalent" electro-chemical potential is assumed to have the standard form, $\bar{\mu}(i) = \mu_o(T,P) + kT \ln f(i) + W(i)$, the above equation for J becomes $J = -[\beta(i)f(i)/kT](\partial\bar{\mu}/\partial i)_{T,P}$ which is the result demanded by irreversible thermodynamics[60]. The identification of the "conductivity" for the nucleation flux allows us to write an equivalent Nyquist relation[61], $\langle J^2(i)\rangle = 4\beta(i)f(i)\Delta\upsilon$ for the canonical average of the square of the noise current, where $\Delta\upsilon$ is the bandwidth over which the noise is observed[62]. In the Nernst-Planck equation for J, only the second term affects the behaviors of the *individual* (generally rare) clusters whose motions are tracked in a computer simulation. The corresponding nucleation current, $J(i) = -[\beta(i)/kT](dW/di)$, is thus measurable and can be used with a separate measurement of the "diffusivity" $\beta(i)$ to determine dW/di.

Finally, among the several developing molecular theories of rate that can be compared to assess the validities of all of them, the interesting approach of Garrett and coworkers[8], who are developing a theory based on variational transition state theory[63], must be included. It should be remarked that these authors have also elected to calculate the evaporation coefficient as a primary quantity rather than the condensation coefficient.

The Nucleation Theorem

The *nucleation theorem*[34-37], in its most current form, may be expressed as

$$\left[\partial\tilde{W}_{n^*}/\partial\mu\right]_{V,T} = -(n^* - \bar{n} - 1) \tag{15}$$

where $\tilde{W}_{n^*}$ is the intensive free energy of formation of the nucleus, μ is the chemical potential of the system, and $\bar{n}$ is the average number of molecules that would occupy a volume v centered on a nucleus if that volume contained vapor at the uniform average density, while n^* is the actual number of molecules in v when the nucleus is present. Since the density within the nucleus converges on $\bar{n}$ far from its center, the excess number of molecules $n^* - \bar{n}$ is well defined and, if v is large enough, will be independent of v and can be used as a measure of the number of molecules in the nucleus. If the approximation is made that the derivative in Eq(15) is equal to $kT(\partial \ln J/\partial\mu)_{V,T}$ where J is the nucleation rate, Eq(15) can clearly be used to measure the number of molecules in the nucleus through a measurement of rate. This seems a remarkable result, but it has been confirmed by a number of experiments, both physical and simulative[35,19,64]. For later reference, it should be added that all of these confirmations have been performed in vapors that were almost ideal, i.e. $\bar{n} \approx 0$.

The original theories that led to the theorem were phenomenological (CNT)[34], molecular and based on statistical mechanics[35], or thermodynamic[36,37]. These theories failed to consider translational degrees of freedom. The statistical mechanical theory assumed the vapor to be ideal, but the thermodynamic theories did not.

Recently the theorem has been derived as a manifestation of the law of mass action, but again, for the case of an ideal vapor[13], for which $\bar{n}$ in Eq(15) is set to zero. If it is a consequence of the law of mass action it remains useful but not so remarkable.

In that same recent paper[13], the theorem was derived exactly, including translation, beginning with Eq(1), but also for the case of an ideal vapor. From the structure of the derivation, it appears that it would require a mathematical miracle for the theorem to be valid for an imperfect vapor, but miracles are not ruled out. In any event, it is important to determine whether the theorem is indeed applicable to nonideal systems, especially because there is a beginning stream of papers[65-68] in which the applicability of the theorem to nonideal systems is called upon to validate important predicted results.

It is fitting to end this critique of molecular theories of nucleation on this note, because it is evident that a relation such as Eq(1), based on molecular considerations, has been instrumental in calling into question the broad range of conditions assigned to the nucleation theorem, and could be equally instrumental in validating them. Thus we have a compelling example of a molecular theory providing guidance to theories that are more phenomenological.

1. S..J. Reed, J. Chem. Phys. **20**, 208 (1952).
2. F. H. Stillinger, J. Chem. Phys. **38**, 1486 (1963).
3. H. P. Gillis, C.C. Marvin, and H. Reiss, J.Chem. Phys. **66**, 214 (1977).
4. .B. N. Hale and S. M. Kathmann, *Nucleation and Atmospheric Aerosols*, Ed. M. Kulmala and P. E. Wagner, p. 30 (Pergamon , Elsevier Science, Inc. Tarrytown, NY, 1996); B.N. Hale, Aust. J. Phys. **49**, 425 (1996).
5. H. M. Ellerby, C. L. Weakliem, and H. Reiss, J. Chem. Phys. **95**, 9209 (1991).
6. K. J. Oh, X.C. Zeng, and H. Reiss, J. Chem. Reiss, **107**, 1242 (1997).
7. V. Talanquer, and D. W. Oxtoby, J. Chem. Phys. **111**, 5190 (1994).
8. G. K. Schenter, S.M. Kathmann, and B.C. Garrett, J. Chem. Phys. **110**, 7951 (1999); S.M. Kathmann, G.K. Schenter, and B.C. Garrett, ibid, **111**, 4688 (1999).
9. H. Reiss and R. K. Bowles, J. Chem. Phys. **111**, 7501 (1999).
10. H. Reiss and R. K. Bowles, J. chem. Phys. **112**, 1390 (2000).
11. H. Reiss, J. Mol. Struct. **485-86**, 465 (1999).
12. H. Reiss and R.K. Bowles, Submitted to J. Chem. Phys.
13. R. K. Bowles, R . McGraw, P. Schaaf, B. Senger, J.-C. Voegel, and H. Reiss, submitted to J. Chem. Phys.
14. Z. Li and H. A. Scheraga, J. Chem. Phys. **92**, 5499 (1990).
15. B. Senger, P. Schaaf, D. S. Corti, R. K. Bowles, J.-C. Voegel, and H. Reiss, J. Chem. Phys. **110**, 6241 (1999).
16. B. Senger, P. Schaaf, D. S. Corti, R. K. Bowles, D. Pointu, J.-C. Voegel, and H. Reiss, J. Chem. Phys. **110**, 6438 (1999).
17. K.. J. Oh and X. C. Zeng, , submitted to J. Chem. Phys.
18. P. R. ten Wolde, M.J. Ruiz-Montero, and D. Frenkel, Faraday Discuss. **104**, 93 (1996).
19. P. R. ten Wolde and D. Frenkel, J. Chem. Phys. **109**, 9901 (1998).
20. P. R. ten Wolde, M.J. Ruiz-Montero, and D. Frenkel, J. Chem. Phys, **110**, 1591 (1999).
21. I. Kusaka, J. H. Seinfeld, and Z,-G. Wang, J. Chem. Phys. **108**, 3416 (1998).
22. I. Kusaka and D. W. Oxtoby, J. Chem. Phys. **110**, 5249 (1999).
23. I. Kusaka and D. W. Oxtoby, J. Chem. Phys, **111**, 1104 (1999).
24. D. I. Zhukhovitskii, J. Chem. Phys.. **103**, 9401 (1995).
25. K. J. Oh and X, C. Zeng, J. Chem. Phys, **110**, 4471 (1999).

26. W. C. Swope and H. C. Andersen, Phys. Rev. B. **41**, 7042 (1990).
27. K. Yasuoka and M. Matsumoyo, J. Chem. Phys. **109**, 8451 (1998)
28. K. Yasuoka and M. Matsumoyo, J. Chem. Phys. **109**, 8463 (1998).
29. R. K. Bowles, J. Chem. Phys. **112**, 1122 (2000).
30. J. Lothe and G, M. J. Pound, J. Chem. Phys. **36**, 2080 (1960).
31. J. Lothe and G, M. J. Pound, in *Nucleation*, A. C. Zettlemoyer, Ed. (Marcel Dekker, New York, 1969).
32. H. Reiss, J. L. Katz, and E. R. Cohen, J. Chem. Phys. **48**, 5553 (1968).
33. H. Reiss, W. K. Kegel, and J, L. Katz, J. Phys. Chem. A **102**, 8548 (1998).
34. D. Kaschiev, J. Chem. Phys. **76**, 5098 (1982).
35. Y. Viisanan, R. Strey, and H. Reiss, J. Chem. Phys. **99**. 4680 (1993).
36. D. W. Oxtoby and D. Kaschiev, J. Chem. Phys. **100**, 7665 (1944).
37. I. J. Ford, J. Chem. Phys. **105**, 8324 (1996).
38. J. K. Lee, J. A. Barker, and F.f. Abraham, J. Chem. Phys. **61**, 1221 (1999).
39. N. G. Garcia, and J. M. S. Torroja, Phys. Rev. lett. **47**, 186 (1981).
40. I. Kusaka, D. W. Oxtoby, and Z.-G. Wang, J. Chem. Phys. **111**, 9958 (1999)
41. J. Frenkel, *Kinetic Theory of Liquids*, Chapter VII (Clarendon Press, Oxford, 1946).
42. D. Frenkel and B. Smit, *Understanding Molecular Simulation*, (Academic Press, Boston, 1996).
43. P. Schaaf, B. Senger, and H. Reiss, J. Phys. Chem. **101**, 8740 (1997).
44. K. J. Oh and X. C. Zeng, J. Chem. Phys.**108**, 4683 (1998).
45. H. Reiss and R. K. Bowles, J. Chem. Phys. **111**, 9965 (1999).
46. A. J. Yang, J. Chem. Phys. **82**, 2082 (1985).
47. M. Rao, B. J. Berne, and M. H. Kalos, J. Chem. Phys. **68**, 1325 (1978).
48. C.L. Weakliem and H. Reiss, J. Chem. Phys. **99**, 5374 (1993).
49. V. Talanquer and D. W. Oxtoby, J. Chem. Phys. **100**, 5190 (1994).
50. Estimate by author
51. C. H. Bennett, in *Algorithms for Chemical Computations*, E.d R. E. Christofferson (Am. Chem. Soc. Washington D. C. 1977)
52. D. Chandler, J. Chem. Phys **68**, 2959 (1978).
53. H. A. Kramers, Physica (Utrecht) **7**, 284 (1940).
54. J. L. Katz, H. Saltsburg, and H. Reiss, J. Coll. Interface Sci. **21**, 560 (1966).
55. D. T. Wu, J. Chem. Phys. **97**, 2644 (1992).
56. P. Schaaf. N1.6C, This Proceedings
57. H. S. Robertson, *Statistical Thermophysics*, pp. 413-421 (Prentice Hall, Englewood Cliffs NJ, 1993).
58. G. H. Wannier, *Statistical Physics*, pp. 492-500 (Dover, Mineola, NY, 1987).
59. A. J. Bard and L. R. Faulkner, *Electrochemical Methods*, p. 27 (John Wiley & Sons, New York, 1980).
60. K.S. Forland, T. Forland, and S.K. Ratkje, *Irreversible Thermodynamics, Theory and Applications* (John Wiley and Sons, New York, 1988).
61. G. H. Wannier, *Statistical Physics*, pp. 487-492 (Dover, Mineola, NY, 1987).
62. R. McGraw, N1.1 This Proceedings.
63. B. C. Garrett and D. G. Truhlar, J. Phys. Chem. **83**, 1052 (1979).
64. Private Communication, Professor R. Strey.
65. D.W. Oxtoby and A. Laaksonen, J. Chem. Phys, **102**, 6846 (1995).
66. D. Kashchiev, J. Chem. Phys. **104**, 8671 (1995).
67. C. C. M. Luijten and M. E. H. van Dongen, J. Chem. Phys. **111**, 8524 (1999).
68. C. C. M. Luijten, P. Peeters, and M. E. H. van Dongen, J. Chem. Phys. **111**, 8535 (1999).

Dynamical Nucleation Theory

Shawn M. Kathmann, Gregory K. Schenter, and Bruce C. Garrett

Environmental Molecular Sciences Laboratory, Pacific Northwest National Laboratory, Richland, Washington 99352, USA

Abstract. The nucleation rate is governed by the corresponding evaporation and condensation rate constants for the multi-step kinetics for populations of individual clusters. In our recent publications on Dynamical Nucleation Theory (DNT) a formalism was presented which allows the evaporation and condensation rate constants to be determined *directly*. Given these rate constants the transient and steady state behavior of the nucleation rate can be investigated by solving the pseudo-first order kinetic equations. The use of variational transition state theory to estimate the monomer evaporation rate constant provides a unique and physically consistent value for the cluster constraining volume. Application of DNT to small polarizable water clusters and future work on multi-component systems will be discussed.

INTRODUCTION

Nucleation is a process by which embryos of a new phase are produced. In general, these embryos are small clusters of *i*-molecules which, after initial formation by the nucleation process, grow into droplets, crystals, or bubbles of the new phase. Vapor phase nucleation can be viewed as a series of gas-phase association and dissociation reactions. The ultimate goal is to develop a molecular-level theory of nucleation that can elucidate the factors governing the extreme sensitivity of nucleation phenomena. A remaining challenge has been the development of a consistent general theory that provides a molecular level understanding of the individual reaction mechanisms, reaction rates, and the resulting nucleation rates. Dynamical Nucleation Theory (DNT)[1] utilizes variational transition state theory (VTST)[2] to provide an expression for the monomer evaporation rate constant. Using VTST the thermal rate constant is approximated by the equilibrium one-way flux through a dividing surface separating reactants from products. We choose a spherical dividing surface (for simplicity and to make a connection with previous work on clusters) with its center placed at the center-of-mass of the *i*-cluster to separate reactant and product regions in configuration space. The radius of the sphere, r_{cut}, determines the location of the dividing surface. The fundamental dynamical assumption of VTST provides an optimal dividing surface located in a region of configuration space that minimizes the reactive flux. DNT thus allows the identification of those clusters that are most stable with respect to evaporation and hence contribute the most to the overall nucleation rate.

THEORY

The generalized TST (GT) expression for the evaporation rate constant is given by

$$\alpha_i^{GT} = \frac{k_B T}{h} \frac{q_i^{GT}(r_{cut}, T)}{q_i^{R}(r_{cut}, T)}. \tag{1}$$

The zero of energy for the two partition functions, q_i^{GT} and q_i^{R}, has been chosen to be identical so that the Boltzmann factor that is normally present in the traditional TST expression is omitted. The reactant partition function is given by

$$q_i^{R}(r_{cut}, T) = \frac{\gamma^i}{i!} \int d\mathbf{r}^i \exp\left[-\frac{U(\mathbf{r}^i)}{k_B T}\right] \prod_{j=1}^{i} \theta(r_{cut} - |\mathbf{r}_j - \mathbf{R}_i|), \tag{2}$$

where $U(\mathbf{r}^i)$ is the interaction potential, $\gamma = (2\pi m k_B T)^{3/2} / h^3$, k_B is Boltzmann's constant, $\theta(x)$ is the Heaviside step function, $\mathbf{R}_i = (1/i) \sum_{j=1}^{i} \mathbf{r}_j$ is the i-cluster center-of-mass, and T is the temperature. The generalized transition state partition function can be expressed in terms of the reactant partition function as

$$q_i^{GT}(r_{cut}, T) = \left[\frac{h^2}{2\pi m k_B T}\right]^{\frac{1}{2}} \frac{dq_i^{R}(r_{cut}, T)}{dr_{cut}}. \tag{3}$$

The TST approximation to the reactive flux, $\alpha_i^{GT} q_i^{R}(r_{cut}, T) = (k_B T/h) q_i^{GT}(r_{cut}, T)$, is an upper bound to the exact classical reactive flux. The minimum in the reactive flux is found by variationally optimizing the location (note: $r_{cut}^{\ddagger} = r_i^{\ddagger}$) of the dividing surface as

$$q_i^{R}(r_i^{\ddagger}, T) \alpha_i^{\ddagger} = \min_{r_i}\left[\frac{k_B T}{h} q_i^{GT}(r_i, T)\right] = \frac{k_B T}{h} q_i^{GT}(r_i^{\ddagger}, T) \tag{4}$$

The Helmholtz free energy for an i-cluster is given by

$$A_i(r_i, T) = -k_B T \ln[q_i^{R}(r_i, T)]. \tag{5}$$

Using Eqs.(3)-(5) allows the variationally optimized evaporation rate constant to be expressed as

$$\alpha_i^{\ddagger} = -\frac{1}{\sqrt{2\pi m k_B T}} \frac{dA_i(r_i, T)}{dr_i}\bigg|_{r_i = r_i^{\ddagger}}. \tag{6}$$

The surprising result is that the evaporation rate constant is proportional to the derivative of the Helmholtz free energy with respect to the radius of the spherical dividing surface. Thus, although previous molecular theories have assumed the Helmholtz free energy to be independent of the constraining radius, we have shown that the evaporation rate constant is determined by its rate of change. In addition, the optimum value of $r_{cut} = r_i^{\ddagger}$ is uniquely determined by VTST. Therefore, DNT provides the first physically justified procedure for selecting a unique volume for an i-cluster. The evaporation rate can be expressed in a form reminiscent of the gas-kinetic collision rate

$$\alpha_i^{\ddagger} = \frac{\bar{c}}{4}\left[4\pi(r_i^{\ddagger})^2\right]\frac{p_i^{int}}{k_B T}, \tag{7}$$

where the average molecular speed $\bar{c} = \sqrt{8k_B T / \pi m}$, and the internal pressure of the i-cluster is given by $P_i^{int} = -dA_i(r_i^\ddagger, T)/dv\big|_{r_i = r_i^\ddagger}$. This expression points to the importance of knowing and exploiting the dependence of the Helmholtz free energy on the constraining volume $v_i = 4\pi r_i^3/3$.

NUCLEATION KINETICS

The kinetic mechanism that describes monomer addition and loss is given by

$$A_1 + A_{i-1} \underset{\beta_{i-1}}{\overset{\alpha_i}{\longleftrightarrow}} A_i \tag{8}$$

where the time dependence of the cluster distribution function is

$$\frac{dN_i}{dt} = \beta_{i-1} N_{i-1} - \alpha_i N_i - \beta_i N_i + \alpha_{i+1} N_{i+1}, \tag{9}$$

which at equilibrium gives the detailed balance condition

$$\frac{\beta_{i-1}}{\alpha_i} = \frac{N_i^{EQ}}{N_{i-1}^{EQ}} = K_{i,i-1}^{EQ}. \tag{10}$$

The equilibrium constants $K_{i,i-1}^{EQ}$ are determined by the Helmholtz free energy differences between adjacent-sized clusters using an extended Bennett technique[1]. Using a statistical mechanical treatment of the nucleating vapor, the i-cluster distribution function is

$$N_i^{EQ} = \exp\left\{-\beta\left[A_i^\ddagger + pv_i^\ddagger - k_B T \ln(i^{3/2} \gamma V) - i\mu_1\right]\right\}, \tag{11}$$

where $A_i^\ddagger$ is the internal Helmholtz free energy for an i-cluster sampling the configuration space constrained to the DNT volume $v_i^\ddagger$, the monomer chemical potential $\mu_1 = k_B T \ln[N_1/\gamma V]$, V is the mother phase volume, and $\beta = 1/k_B T$. Thus, given the evaporation rate constants and equilibrium constants, the corresponding condensation rate constants are determined via detailed balance.

RESULTS, COMMENTS AND CONCLUSIONS

To demonstrate the feasibility of DNT, the volume dependent Helmholtz free energies for water cluster sizes from $i = 2$ to 30 were calculated at 243K using the Dang-Chang[3] polarizable water model. From these results, the corresponding rate constants were determined and are shown in FIG. 1. Work is in progress involving simulations of the largest clusters. Given these rate constants, the pseudo-first order kinetic equations can be solved numerically to obtain homogeneous nucleation rates for water for comparison with new experiments being performed at PNNL. Inevitably, almost all experimental conditions are such that the nucleating vapor is characterized by some degree of contamination (unless immaculately arranged and the resulting aerosol analyzed for impurities e.g. by mass spectrometry). In conclusion, DNT provides a theoretical formalism in which the relevant dynamical quantities can be calculated.

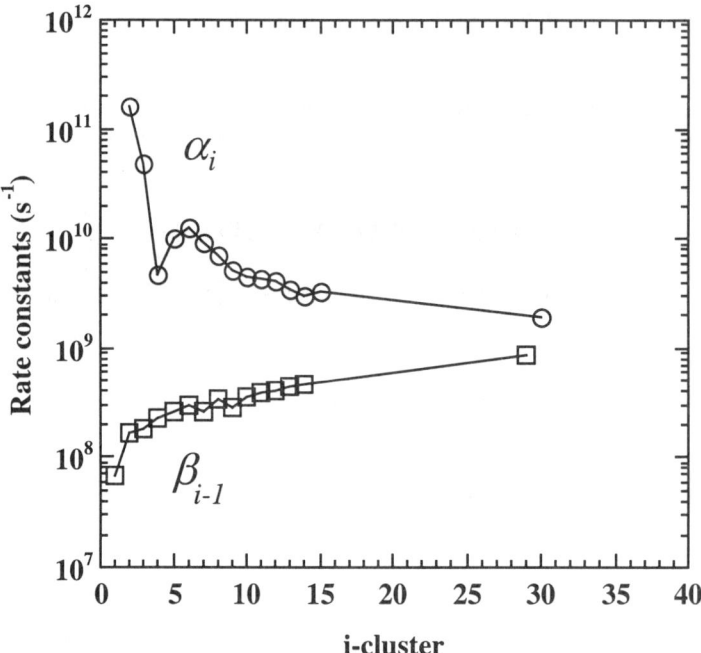

FIGURE 1. Rate constants for water clusters at 243K and supersaturation S = 10.

ACKNOWLEDGMENTS

This work was supported by the Division of Environmental Science, Office of Biological and Environmental Research, (S.M.K. and B.C.G.) and the Division of Chemical Sciences, Office of Basic Energy Sciences, (G.K.S.) of the U.S. Department of Energy. This research was performed in part using the Molecular Science Computing Facility (MSCF) in the William R. Wiley Environmental Molecular Sciences Laboratory a national scientific user facility sponsored by the Department of Energy's Office of Biological and Environmental Research and located at Pacific Northwest National Laboratory (PNNL). Battelle operates PNNL for the Department of Energy.

REFERENCES

1. Schenter, G.K., Kathmann, S.M., and Garrett, B.C., *J. Chem. Phys.*, **110**, 7951 (1999); Schenter, G.K., Kathmann, S.M., and Garrett, B.C., *Phys. Rev. Lett.*, **82**, 3484 (1999); Kathmann, S.M., Schenter, G.K., and Garrett, B.C., *J. Chem. Phys.*, **111**, 4688 (1999).
2. Wigner, E., *J. Chem. Phys.*, **5**, 720 (1937); Wigner, E., *Trans. Faraday Soc.*, **34**, 29 (1938).
3. Dang, L.X., and Chang, T.M., *J. Chem. Phys.*, **106**, 8149 (1997).

Kinetics of Cluster Evaporation and Condensation Important in Homogeneous Vapor Phase Nucleation

Bruce C. Garrett, Shawn M. Kathmann, and Gregory K. Schenter

Environmental Molecular Sciences Laboratory, Pacific Northwest National Laboratory, Richland, Washington 99352, USA

Abstract. Variational transition state theory (VTST) is used to estimate rate constants for the processes of evaporation and condensation that are important in homogeneous vapor-phase nucleation. We review the approximations inherent in the calculations of the rate constants, and indicate how VTST provides a framework in which systematic improvements to the approximations can be made.

INTRODUCTION

Although molecular theories of nucleation go back over 30 years to the pioneering work of Howard Reiss and his coworkers,[1] advances in molecular approaches are still being made today. This is a testament to the difficulty of treating nucleation in a rigorous manner. The extreme sensitivity of the nucleation phenomena to physical conditions requires a high level of accuracy in molecular theories for them to be truly predictive. This requirement of accuracy in turn calls for exactness of molecular theories of nucleation, which has been a goal of Reiss' research over the years.[2] We have recently presented a new molecular approach, Dynamical Nucleation Theory (DNT),[3] in which we do not achieve exactness, but we aspire to clearly understand the validity of approximations inherent in our approach.

We focus our attention on homogeneous vapor-phase nucleation. A common starting point for many theories of vapor-to-liquid nucleation is that addition and removal of monomers from clusters adequately describes the nucleation kinetics, e.g.,

$$N_1 + N_{i-1} \underset{\alpha_i}{\overset{\beta_{i-1}}{\rightleftarrows}} N_i, \; i=1,2,\ldots.$$

where N_1 is the number of monomers, N_i ($i>1$) are numbers of clusters containing i monomers, and α_i and β_i are evaporation and condensation rate constants. The assumption that monomer addition and loss control cluster growth is generally made so that the nucleation rate can be obtained by solving the pseudo-first order kinetic equations (assuming N_1 is large and constant). With the advent of fast computers, solution of the nonlinear differential equations can be obtained numerically so that this approximation is not necessary. In the following we will discuss the theory for monomer addition and loss, but our approach is general and can include condensation of two clusters to form a new cluster or evaporation of a daughter cluster from a larger parent cluster. The key quantities for the kinetic mechanism are the evaporation and condensation rate constants. In this presentation we focus on the approximations inherent in the calculation of the evaporation and condensation rate constants.

VARIATIONAL TRANSITION STATE THEORY APPROACH TO EVAOPRATION AND CONDENSATION RATE CONSTANTS

In our approach we view the processes of evaporation and condensation as gas-phase dissociation and association reactions. This view of the nucleation process as a gas-phase reaction mechanism composed of dissociation and association reactions is different than the liquid-phase perspective of density fluctuations in a supersaturated fluid. For nucleation of water vapor, the view of nucleation as a gas-phase reaction mechanism is quite reasonable since the average distance between water monomers and clusters in the vapor is quite large (e.g., greater than 10^4 Å for water at 298 K and a supersaturation of 10); the vapor looks more like a rarified gas than a liquid.

We take an approach in which the evaporation and condensation rate constants are calculated *directly* from knowledge of the intermolecular potential energy surface for the system. This is distinct from most molecular theories that approximate condensation rate constants by collision rates of monomers with the cluster, where collision cross sections are approximated by surface areas of a spherical liquid drop, with unit sticking probability. Once this approximation is made for the condensation rate, the evaporation rate constant can be evaluated using detailed balance, which requires calculations of equilibrium constants (or cluster free energies).

In DNT we do not employ the simple approximation of a gas kinetic rate constant for condensation, but instead use transition state theory (TST)[4,5] to evaluate equilibrium rate constants for evaporation and condensation. An exceptionally clear exposition of the approximations inherent in TST were given in a seminal paper by Wigner.[5] As stated there, TST for thermal rate constants yields equilibrium rate constants (ones in which an equilibrium of reactant states is maintained). One key question is whether equilibrium rate constants are the proper parameters for use in nucleation kinetics. For unimolecular dissociation reactions, reactant clusters will maintain equilibrium energy distributions if collision rates of buffer gas molecules with the clusters are much faster than dissociation rates for the clusters. For small clusters (several molecules) this may not be the case, but the approximation becomes better as the clusters become larger (e.g., tens of molecules typical of critical clusters that control the nucleation rate). The validity of this approximation can be easily checked by comparing evaporation rates computed from TST with gas collision rates of buffer gas molecules with the cluster. If needed, this approximation can be systematically improved by using microcanonical TST rate constants in an RRKM formalism[6] to compute the pressure dependent rate constants.

Wigner went on with his exposition by stating "that the transition state method is based, in addition to well-established principles of statistical mechanics, on only three assumptions, two of which are generally accepted." The first two assumptions were those that Wigner categorized as 'generally accepted': (1) electronic adiabaticity of the reaction and (2) adequacy of classical mechanics to treat the motion of the nuclei. The third assumption has become know as the fundamental assumption or the no-recrossing assumption of TST. The assumption of electronic adiabaticity is a good one for weak interactions between closed shell molecules treated in the nucleation process. The adequacy of classical mechanics is more suspect, and we will return to it after a discussion of the fundamental assumption of TST.

Classically, the rate constant is proportional to the flux of reactive trajectories through a dividing surface separating reactants from products through which all reactive trajectories must pass. The fundamental dynamical assumption is stated as follows: a reactive trajectory originating in reactants must cross the dividing surface only once and proceed to products. With this approximation the TST rate constant is expressed using quasiequilibrium statistical mechanics without the need to calculate classical trajectories. TST correctly accounts for all reactive trajectories, but also can count as reactive trajectories that recross the dividing surface and do not react. Recrossing trajectories cause TST to overestimate the exact classical rate constant. A corollary to this upper bound principle of TST is that the best estimate of the rate constant can be obtained by optimizing the dividing surface to minimize the rate constant. This is the basis for variational transition state theory (VTST).[7,8]

For evaporation of a monomer from an i-molecule cluster, reactants are the i-molecule cluster and products are a monomer infinitely separated from an $(i-1)$-molecule cluster. In our previous studies,[2] we chose the dividing surface to be a sphere with its center at the center of mass of the i-molecule cluster; the radius of the sphere then determines the location of the dividing surface. VTST gives a prescription for defining a unique location of the dividing surface (or radius of the sphere) for each dissociation reaction. The VTST estimate of the rate constant can be systematically improved by using more flexible definitions of the dividing surface (see 8 and references therein). Alternatively, the effects of recrossing can be estimated from ensembles of trajectories started at the optimized dividing surface.

For the dissociation reaction considered here the energy profile from reactants to products is monotonically increasing, so that there is no barrier maximum for quantum mechanical tunneling to occur. In this case, quantum mechanical effects will only be important for bound motions of the cluster. We treat the molecules as rigid so that the high frequency internal modes of each molecule, which require a quantum mechanical treatment, are not explicitly treated. A classical statistical mechanics description of lower frequency motions between molecules is a reasonable first-order approximation for the system considered here. This approximation can be systematically improved using Feynman path integral methods to include quantum mechanical effects in equilibrium properties such as free energies.

For gas-phase chemical reactions, the largest uncertainties in VTST rate constants are typically not from approximations in VTST, but arise from uncertainties in the underlying potential energy surface. This is also true for the evaporation rate constants important in homogeneous vapor-to-liquid nucleation of molecular systems. For studies of water nucleation there are many potential energy surfaces for the water-water interactions to choose from. The majority of these have been parameterized to reproduce bulk properties of liquid water. A smaller number have been fitted to reproduce ab initio electronic structure data for small water cluster (up to about 6 waters). To date there have been no potentials for water that have been parameterized for the clusters that control nucleation rates (i.e., critical clusters of tens of molecules). The development of potentials that accurately describe molecular interactions in clusters of 10's to 100's of molecules is a major challenge for the application of Dynamical Nucleation Theory (and other molecular theories of nucleation) to molecular systems such as water. We are currently exploring methods to use ab initio electronic structure to obtain accurate information about the interaction potentials that can be used directly in calculations of the evaporation and association reactions of nucleation, without the need to fit the data to global potential energy surfaces.

SUMMARY

Variational transition state theory provides a convenient framework for calculating the elementary evaporation and condensation rate constants that are important in nucleation mechanisms. VTST provides a framework for systematically improving the quality of the calculations by including important dynamical effects. The largest errors in the computed rates will likely be from uncertainties in the interaction potentials used in the rate constant calculations. This systematic improvement will be necessary to make molecular theories of nucleation consistent with experimental results.

ACKNOWLEDGMENTS

This work was supported by the Division of Environmental Science, Office of Biological and Environmental Research, and the Division of Chemical Sciences, Office of Basic Energy Sciences, of the U.S. Department of Energy. This research was performed in part using the Molecular Science Computing Facility (MSCF) in the William R. Wiley Environmental Molecular Sciences Laboratory a national scientific user facility sponsored by the Department of Energy's Office of Biological and Environmental Research and located at Pacific Northwest National Laboratory (PNNL). Battelle operates PNNL for the Department of Energy.

REFERENCES

1. Reiss, H., Katz, J. L., and Cohen, E. T., *J. Chem. Phys.* **48**, 5553-5560 (1968).
2. For recent examples see Reiss, H. and Bowles, R. K., *J. Chem. Phys.*, **111**, 7501 (1999); *ibid*, **112**, 1390 (2000).
3. Schenter, G. K., Kathmann, S. M., and Garrett, B. C., *J. Chem. Phys.*, **110**, 7951 (1999); Schenter, G. K., Kathmann, S. M., and Garrett, B. C., *Phys. Rev. Lett.*, **82**, 3484 (1999); Kathmann, S. M., Schenter, G. K., and Garrett, B. C., *J. Chem. Phys.*, **111**, 4688 (1999).
4. Eyring, H. (1935) *J. Chem. Phys.* **3**, 107 (1935).
5. Wigner, E., *Trans. Faraday Soc.*, **34**, 29 (1938).
6. Robinson, P. J. and Holbrook, K. A., *Unimolecular Reactions*, Wiley-Interscience: London, 1972.
7. Wigner, E., *J. Chem. Phys.*, **5**, 720 (1937).
8. Truhlar, D. G., Garrett, B. C., Klippenstein, S. J., *J. Phys. Chem.* **100** 12771 (1996).

Nucleation Rate Changes with the Admixture Contain Grows. I. The Case of the Cigar Like Phase State Diagrams

Lyubov Anisimova and Victor Pinaev

Kemerovo Institute of Commerce, 37 Kuznetskiy Prospect, 650099 Kemerovo, Russia

Abstract. Using the obvious relation for the nucleation rate surfaces and the diagrams for phase equilibrium is the main idea of the present consideration. The nucleation rate surfaces for the cigar like phase state diagrams are constructed. It is concluded: 1) Increasing of admixture contains leads the vapor nucleation rate growth for substances with lower boiling temperature and the nucleation rate goes down for the higher boiling temperature compounds. 2) With growing fraction of the second component (as admixture) the nucleation rate has the relatively bigger changes for substances with lower boiling temperature and smaller changes for substances with higher boiling temperature at the grows of the evaporation enthalpy of components.

INTRODUCTION

Currently, the requirements for the purity of the nucleated vapor are being studied, but they are not known enough. If one has, for example, a vapor sample with high purity, then this sample contains not less than trillion of the admixture molecules per cc. Is it too mach or no? In any case it means that all of the investigated vapors are the multicomponent systems. Fortunately the current experiment illustrates the relatively low sensitivity of the nucleation rate to the admixture contains. For example, Strey et al. (1995) found experimentally that 10 percents of second component shift nucleation rate less than 10 orders of magnitude. For the experimentally investigated water-alcohol systems one percent of second component,

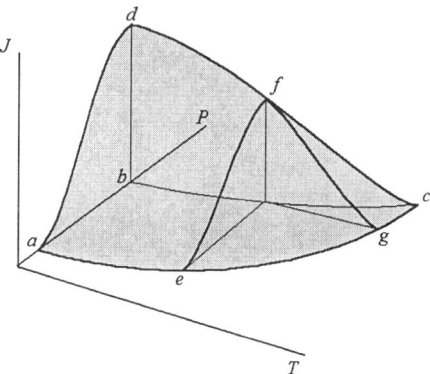

Fig.1. The nucleation rate surface topology for one component vapor.

added as admixture, produce practically undetectable influence on the nucleation rate. Seemingly the current state of the nucle- ation experiment accuracy allows the admixture contains on the level 0.1-0.3 percent. Nevertheless, it is obvious that the scientific knowledge of the admixture influence on the nucleation rate should be developed to get at least the qualitatively estimations the admixture influence on the vapor nucleation rate.

RESULTS AND DISCUSSION

The recent idea for the semi-empirical construction of nucleation rate surfaces for the binary systems, suggested by Anisimov et al., 1998, looks as the promising tool for this consideration. Using the obvious relation for the nucleation rate surfaces and the diagrams for phase equilibrium is the main idea of this consideration. This idea is applied for the analysis the admixture influence on the ideal pure substance nucleation rate. For n - component system nucleation at the constant of n-1 component partial vapor pressures the nucleation rate surface should have topology similar to the one component system. The part of this nucleation rate surface, involved the critical conditions, is shown in Fig. 1. Here J, P, and T are nucleation rate, total pressure, and temperature respectively; c is critical point; lines ac and bc are the vapor-liquid equilibrium line and spinodal line respectively. It is assumed that the highest nucleation rates (line dc) are realized over the spinodal line. Lines ef and gf represent the nucleation under the constant nucleation temperature and the constant nucleation pressure respectively. It was shown, for example by Anisimov and Cherevko (1985), that slope for lines ef and fc are determined by the numbers of the molecules and by enthalpy formation of critical embryos. It is obvious that any deviations from these line directions make interpretation heavier and to avoid mistakes, the experimental results should be recalculated to the one of these two nucleation conditions.

For the small admixture amount one can consider system in the ideal solution approximation. Let us to consider the case of the binary system with the cigar like phase state diagram, which describe the phase states for the ideal system. To self-help the Figure 2 is drawn. It represents the PTx diagram near the critical conditions c and c_1 for two component system. Here x is composition; lines afc and $a_1 c_1$ are the vapor -liquid equilibrium lines for

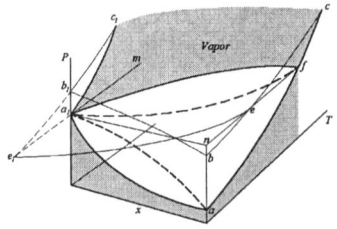

Fig. 2. The volume of the binary vapor metastable states as well as Tx and Px diagrams of the binary vapor – liquid equilibrium.

pure components; *bec* and $e_1b_1c_1$ are spinodal lines; cigars $a_1 1f$ and aa_1 are phase diagrams for the constant total pressure and constant temperature respectively; the flat surface ma_1 *nef* represents the constant level for pressure; $e_1\ a_1\ m$ is the straight line; ee_1 and $b_1\ b$ are spinodal lines for binary system.

One can consider the nucleation rate surface at the constant total pressure (P_{tot} = const). In Fig. 3 *JTx* diagram is shown, where J is the supersaturated vapor nucleation rate; T is the system temperature; and x is mole fraction. Description of the lines is referred to Fig. 2. One has the cigar like phase diagram. The vapor with composition x_o can be under cooling. At the condition, close to the equilibrium, system produces the condensed phase with composition x_1. At the fast cooling the

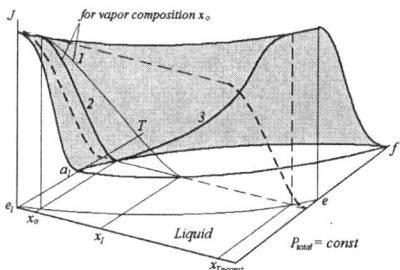

Fig. 3. The nucleation rate surface for isobaric nucleation. Lines 1 and 2 refer to nucleation rates in the vapor with initial composition x_o. Line 3 illustrates the isothermal nucleation and different initial vapor compositions at the constant total pressure. The nucleation rate shifting is shown by dotted lines for both components.

system can become the spinodal decomposition with the initial composition of condensed phase, x_o. These limiting points could be connected by line numbered by number 1. This line contains the knowledge about the condensed phase compositions. Usually these compositions are unknown and all nucleation rates referred to the initial composition, x_o (curve 2). Both lines are referred to the nucleation at the constant total pressures and the constant initial vapor composition. For the constant temperature nucleation and constant total pressure one can get nucleation rates, shown schematically by curve 3. For the real nucleation at constant temperature one has variation of the total pressure and continuum of figures like Fig. 3 should be considered. For lower boiling substances one has grows (dotted line) of the nucleation rate as for the shown example for composition x_o. For higher boiling compound, one can see the lower (dotted) nucleation line.

The more common case is the nucleation at the constant temperature and different contains of vapor (at the constant carrier gas pressure). In Fig. 4 the nucleation rate surface is shown for the constant nucleation temperature, $T=const$, and at variation total pressure. Description of the lines can be referred to Fig. 3. One has the cigar like phase diagram for $T=const$. The vapor with composition x_o can be under compressing. At condition close to equilibrium in point d (Fig. 4) system produces the condensed phase with composition x_1. At the fast compressing the system can become the spinodal decomposition (point h) with the initial composition of condensed phase,

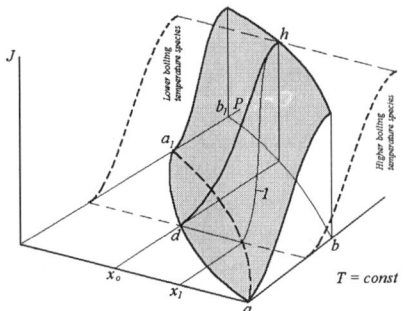

Fig. 4. The nucleation rate surface for isotermal nucleation rate. The nucleation rate shifting is shown by dotted lines for both components.

x_o. These limiting points could be connected by line numbered by 1. This line contains knowledge of the condensed phase compositions. Usually these compositions are unknown and all nucleation rates referred to the initial composition, x_o (curve dh). Both lines are referred to the nucleation at the constant nucleation temperature and the constant initial vapor composition, x_o. One has the same result, i.e. for the lower boiling substances one has grows (dotted line) of the nucleation rate as for the shown example for composition x_o. For higher boiling substances, one can see the schematically lower (dotted) nucleation line.

CONCLUSION

The relations of the phase equilibrium diagram shape and the condensation enthalpies of the system components are known (for example, Elliott, 1965, Hansen, 1958). The components' melting or evaporation enthalpy grows leads grows of the width of the cigar like diagram. If evaporation enthalpies for substances are approximately (not more than 20% difference) equal, than follow the present results one can easily conclude:

1. With increasing of admixture contains the vapor nucleation rate grows for substances with lower boiling temperature and goes down for the higher boiling temperature substances.

2. With growing of the second component fraction the nucleation rate has the relative bigger changes for substances with lower boiling temperature and smaller changes for substances with higher boiling temperature at the grows of the evaporation enthalpy of components.

REFERENCES

Anisimov, M., and Cherevko, A. *J. Aerosol Sci.* **16**(2): 97 (1985).
Anisimov, M., Hopke, P., Rasmussen, D., Shandakov, S., and Pinaev, V. *J. Chem. Phys.* **109**:1435 (1998).
Elliot, R. *Constitution of Binary Alloys.* Suppl. 1. McGraw-Hill, New York (1965).
Hansen, M. *Constitution of Binary Alloys.* McGraw-Hill, New York (1958).
Strey, R., Viisanen, Y., and Wagner, P. *J. Chem. Phys.* **103**: 4333 (1995).

Comparison of Modern Theories of Vapor Condensation

L. Gránásy

Research Institute for Solid State Physics and Optics, H-1525 Budapest, P. O. B. 49, Hungary

Abstract. Recent theories of homogeneous vapor condensation (density functional theory, self-consistent classical theory, phenomenological diffuse interface theory, and the scaling theory by McGraw, Laaksonen, and Talanquer) are compared with experiments on non-polar, weakly polar, and metallic liquids. In accord with a theoretical analysis, remarkable agreement between predictions of the density functional theory and the diffuse interface theory is observed.

INTRODUCTION

During the past decade, experiments and theoretical investigations made clear that the classical nucleation theory cannot be accepted as a fully satisfactory model of homogeneous vapor condensation. The main problem, not yet fully resolved, is the evaluation of the free energy of liquid-like vapor fluctuations. The tiny size of such fluctuations conflicts the bulk physical properties, one needs to assume in most theoretical approaches. Another aspect of the problem is the contribution of rotational/translational degrees of freedom to the free energy of such fluctuations. While a recent analysis in terms of a detailed statistical mechanical treatment establishes that the nucleation prefactor of the classical approach is almost indistinguishable from the non-classical result [1], the excess free energy of small droplets remains elusive with the exception of the simplest cases.

The attempts to improve the description of small droplets range from simple phenomenological approaches to molecular theories based on first principles. Although these approaches have been confronted with experiments, a systematic comparison on equal footing, with the same experiments, is not yet available. Such a comparison is attempted here for a selection of the latest cluster models.

NON-CLASSICAL MODELS

Approaches that do not contain adjustable parameters are considered: The density functional theory (DFT) [2,3], a DFT related scaling theory (MLT) [4,5], the self-consistent classical theory (SCCT) [6], and a phenomenological diffuse interface theory (DIT) [7]. The successful phenomenological approach of Dillman and Meier [8] is not incorporated into the analysis as it relies on excess information (2nd virial coefficient) unavailable for some of the substances used here. Similarly, the van der

Waals/Cahn-Hilliard type gradient theories that use density as the order parameter are excluded, since according to previous investigations [9,10], they yield no significant improvement with respect to the CNT.

Let us briefly recall the essential features of these non-classical models. The SCCT corrects for the nonzero free energy of the monomers via subtracting the free energy of the monomer from all cluster sizes [6]. In the DIT a strongly curvature dependent surface tension is introduced [7], which is related to a characteristic interface thickness expressed in terms of measurable quantities. The most appealing model is the DFT [2,3] that incorporates molecular interactions explicitly. The radial density profile of the critical fluctuation is determined by finding the extremum of the free energy functional via variational methods. The scaling theory MLT relates the curvature dependence of the surface tension to the number of molecules in the classical droplet at the spinodal [4,5] (the free energy – order parameter relationship of a hard sphere fluid with Yukawa type perturbative attraction has been adopted).

The nucleation rate is calculated as $J = J_0 \exp\{- W^*/kT\}$, where W^* is the free energy of critical fluctuations, and the nucleation prefactor J_0 of the CNT is used.

RESULTS AND DISCUSSION

The nucleation rates predicted by the non-classical theories for nonane and dibutyl-phthalate are compared in Figs. 1(a) and 1(b). In the practically accessible range, $J \sim 0.01 - 10^9$ cm^{-3}s^{-1}, the curves from the DFT, DIT, and MLT are fairly close to each other, and are enveloped by those from the CNT and the SCCT. Note the differences in the slopes, which suggest that the vertical sequence of the curves varies with supersaturation (they intersect each other).

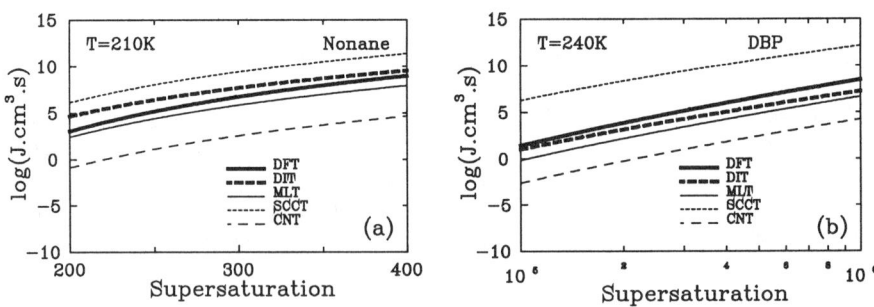

FIGURE 1. Nucleation rate vs. supersaturation as predicted by the density functional theory (DFT), the phenomenological diffuse interface theory (DIT), the scaling theory by McGraw, Laaksonen, and Talanquer (MLT), the self-consistent classical theory (SCCT), and the classical nucleation theory (CNT) for nonane and dibutyl-phthalate. In the typical range of experiments, usually the lowest rate is predicted by the CNT, and the highest by the SCCT. [$\Delta\mu/\Delta\mu_s \sim 0.3$ and ~ 0.65 for (a) and (b), where $\Delta\mu_s$ is the relative chemical potential at the spinodal].

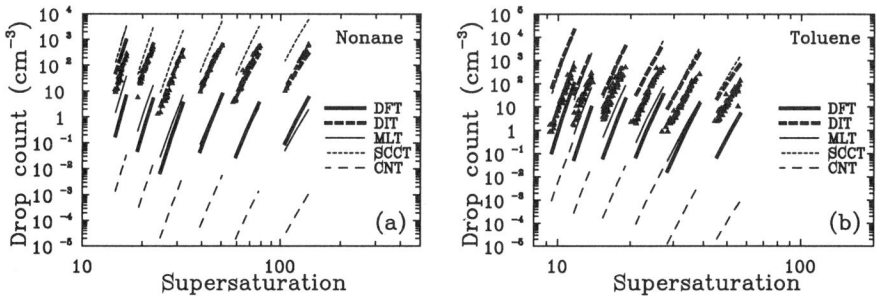

FIGURE 2. Comparison of selected theories with expansion cloud chamber experiments on the homogeneous condensation of nonane [11] and toluene [12]. (The drop count is proportional with the nucleation rate. The data sets from left to right correspond to pre-expansion temperatures 45, 35, 25, 15, 5, and −5 °C, respectively.) The properties on nonane were taken from [13].

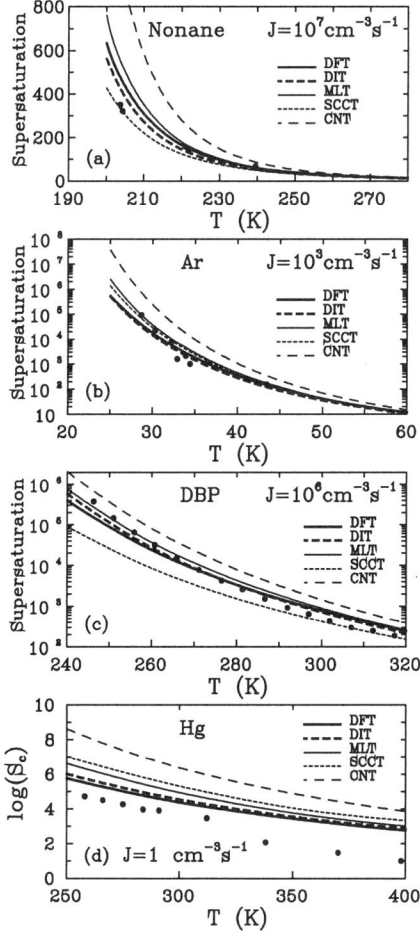

FIGURE 3. Critical supersaturation vs. temperature as predicted by non-classical theories. For comparison, experimental results (nonane [14], Ar [15], dibutyl-phthalate [16], Hg [17]) are also shown. Note the closeness of the DFT and DIT, predictions.

211

The predicted drop counts (a quantity proportional with nucleation rate) are compared to experiments in Fig. 2. Considering that the uncertainty of the calculated rates, associated with the experimental error of the surface tension and other input parameters, is about an order of magnitude, it is difficult to decide which of the models should be preferred. Extension of the analysis to other compositions may give a general impression on the reliability of the models.

The theoretical and experimental results for the temperature dependence of the critical supersaturation S_c are presented in Fig. 3 for non-polar, weakly polar, and metallic substances. It appears, that the DFT and the DIT yield fairly similar $S_c(T)$ relationships for all these liquids. In the case of nonane, argon, and dibutyl-phthalate, they are also in a fair agreement with experiments. In contrast, for mercury, experiment and theory lead to conflicting results, a finding that might be attributed to a metal – non-metal transition expected in the case of small Hg clusters [17]. The similarity of the DFT and DIT predictions has been expected on the basis of a theoretical analysis [18] that found a remarkable agreement between the nucleation barrier heights of the two approaches. The comparison of the scaling approach and the self-consistent classical theory is rather inconclusive; it is difficult to decide which is superior to the other. While the former is usually closer to the DFT result, the SCCT is often closer to the experiment [Figs. 2(a), 3(a), 3(b)].

Work is under way to extend this analysis to further substances and models.

ACKNOWLEDGMENT

This work was supported by the National Scientific Research Fund (Hungary, Grant No. OTKA-T-025139).

REFERENCES

1. Talanquer, V., and Oxtoby, D. W., *J. Chem. Phys.* **100**, 5190-5200 (1994).
2. Oxtoby, D. W., and Evans, R., *J. Chem. Phys.* **89**, 7521-7530 (1988).
3. Nyquist, R. M., Talanquer, V., and Oxtoby, D. W., *J. Chem. Phys.* **103**, 1175-1179 (1995).
4. Talanquer, V., *J. Chem. Phys.* **106**, 9957-9960 (1997).
5. McGraw, R., and Laaksonen, A., *Phys. Rev. Letters* **76**, 2574-2577 (1996).
6. Girshick, S. L., and Chiu, C.-P., *J. Chem. Phys.* **93**, 1273-1277 (1990).
7. Gránásy, L., *J.Chem. Phys.* **104**, 5188-5198 (1996).
8. Dillmann, A., and Meier, G. E. A. *J. Chem. Phys.* **94**, 3872-3884 (1991).
9. Gránásy, L., *J. Non-Cryst. Solids* **219**, 49-56 (1996).
10. Barrett, J. C., *J. Phys.: Condens. Matter* **9**, L19-L26 (1997).
11. Adams, G. W., Schmitt, J. L., and Zalabsky, R. A., *J. Chem. Phys.* **81**, 5074-5078 (1984).
12. Schmitt, J. L., Zalabsky, R. A., and Adams, G. W., *J. Chem. Phys.* **79**, 4496-4501 (1983).
13. Hung, C.-H., Krasnopoler, M. J., and Katz, J. L., *J. Chem. Phys.* **90**, 1856-1865 (1989).
14. Wagner, P. E., and Strey, R., *J.Chem. Phys.* **80**, 5266-5275 (1996).
15. Wu, B. C., Wegener, P. P., and Stein, G. D., *J. Chem. Phys.* **69**, 1776-1777 (1978).
16. Hämeri, K., Kulmala, M., Krissinel', E., and Kodenyov, G., *J. Chem. Phys.* **105**, 7683-7695 (1996).
17. Martens, J., Uchtmann, H., and Hensel, F., *J. Phys.Chem.* **91**, 2489-2492 (1987).
18. Talanquer, V., and Oxtoby, D. W., *J. Phys. Chem.* **99**, 2865-2874 (1995).

Cluster Energetics: Models and Data

Ian Ford, Hanna Vehkamäki and Michael Knott

Department of Physics and Astronomy, University College London, Gower Street, London WC1E 6BT, United Kingdom.

Abstract. A wealth of information about the energies of clusters of several tens of molecules has been obtained recently through the analysis of nucleation rate data using the nucleation theorems. We describe here studies of n-pentanol and dibutylphthalate clusters, and using multicomponent nucleation theorems, we present data for binary clusters of water and ethanol, and n- and i-octane. We then attempt a synthesis of alkane cluster information obtained to date, and find that the cluster excess energies show a rough proportionality to the molecular size raised to the 2/3 power.

INTRODUCTION

The temperature dependence of the rate of droplet nucleation is related to the excess energy of the critical cluster, and the dependence on vapour supersaturation is related to the size of the critical cluster in molecules. These results are the nucleation theorems (1,2). By analysing experimental nucleation rates, therefore, we can determine the energies of the molecular clusters that served as the critical clusters in the given experimental circumstances (2). The experimental data can, if detailed enough, provide information about a wide range of sizes of clusters, and we can hope to discover the size dependence of the cluster energy.

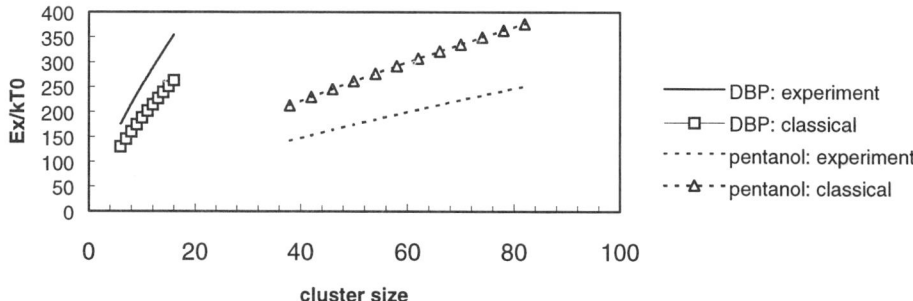

FIGURE 1. Excess energies of DBP and pentanol clusters, in units of kT_0, where T_0=273.15K.

SINGLE COMPONENT CLUSTERS

Our work to date has focussed on single component droplet nucleation. Our latest results, shown in Figure 1, are for n-pentanol and dibutylphthalate (DBP) (3,4,5). The excess energies take the form of curves since we fit a function to the measured data. We compare these results with the excess energies predicted by the capillarity approximation (3,5), and observe a rough order of magnitude agreement. However, we suspect that this is just coincidental: we do not expect clusters of such sizes to behave like scaled-down macroscopic droplets.

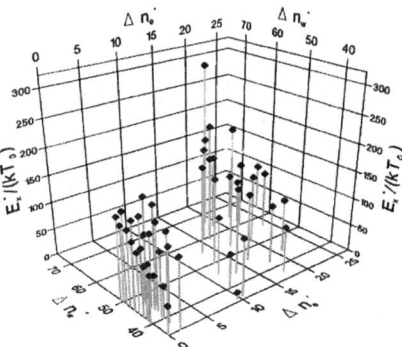

FIGURE 2. Excess energies of binary clusters containing Δn_w molecules of water and Δn_e molecules of ethanol. These clusters are rich in water: ethanol-rich clusters were measured but are not shown.

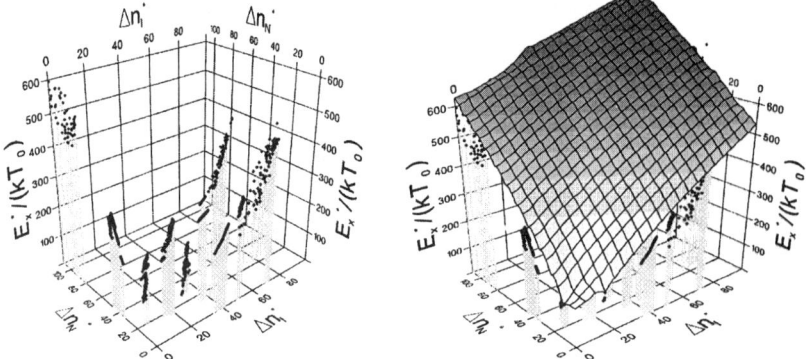

FIGURE 3. Excess energies of clusters containing Δn_n molecules of n-octane and Δn_i molecules of i-octane. The right hand plot suggests ease of mixing between the two components.

BINARY CLUSTERS

Nucleation theorems for multicomponent nucleation have recently been derived (6) allowing us to analyse data for water-ethanol droplet nucleation (7), and very recently obtained measurements of n- and i-octane nucleation (8). The excess energies for a range of cluster compositions for these binary systems are shown in Figures 2 and 3.

For water-ethanol, the data is sparse, and the resulting coverage of cluster composition is relatively poor. For the mixture of octane isomers, the data are more plentiful, and a wider range of cluster compositions can be explored (9). It appears that the data favour a fairly smooth excess energy surface. The capillarity approximation predicts a surface lying somewhat above these data, as shown on the right hand side.

THERMALLY ADJUSTED ENERGY PLOTS

Each excess energy displayed in the above plots refers to a cluster at the temperature of the nucleation experiment for which it happened to be the critical cluster. The excess energy of a cluster will be temperature-dependent, and so in order to create a genuine plot of cluster excess energy against size, we must adjust the results according to the formula $E_x(T_0) = E_x(T) + C(T_0 - T)$, where $E_x(T)$ is the excess energy of the cluster at the experimental temperature T, C is the heat capacity of the cluster, and T_0 is the reference temperature, which we take to be equal to 273.15 K. The heat capacity is unknown, but we can estimate it to be $3kN$, where N is the number of molecules in the cluster. The extracted values of E_x/kT_0 should therefore be adjusted by approximately $3N(T_0 - T)/T_0$, which turns out to be in the region of 10-20; small but not negligible. Earlier data for a variety of alkanes (2) have been reanalysed in this way, to produce the plot of excess energy against size in Figure 4. In an attempt to produce a universal plot, each dataset has been scaled by the latent heat, per molecule, for each substance at T_0. This is a measure of the strength of bonding in each case. The plot is compared with excess energies that arise from the rigid spherical cluster model described by Lee *et al* (10), and also with a power law expression $E_x \propto N^{2/3}$. The data seem compatible with both models. Of course, the capillarity approximation also produces this dependence.

CONCLUSIONS

The information about small molecular clusters we have extracted from nucleation data is becoming detailed enough for us to seek general features that might help us to understand their structures. The available information now extends to binary clusters, and we can compare substances that appear to mix easily, like isomers of octane, and those like water and ethanol that appear to segregate. Universal behaviour for the excess energy can be seen if we present the excess energies of clusters of alkanes, correcting for their different temperatures, and normalising by the latent heat. We see a dependence on cluster size that corresponds to the very simple rigid spherical cluster model of Lee *et al* (10).

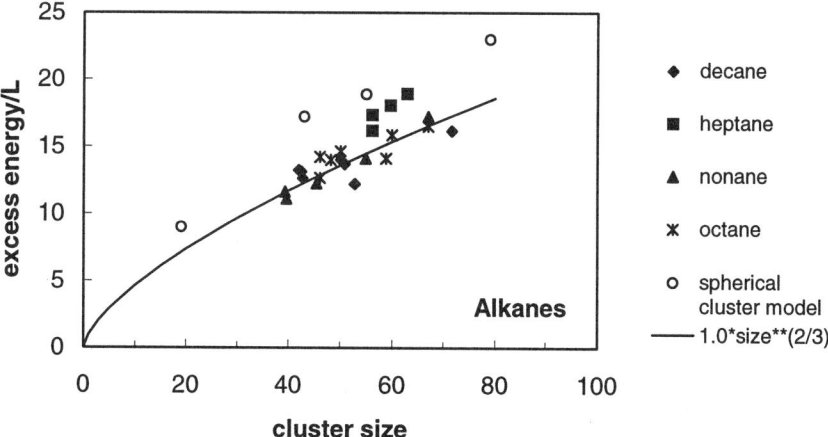

FIGURE 4. Excess energies of alkane clusters, adjusted approximately to a temperature of 273.15K, and normalised by the molecular latent heat for each substance at this temperature. The open circles are predicted excess energies according to a rigid spherical cluster model (Lee et al 1973).

ACKNOWLEDGEMENTS

We thank Drs V. Mikheev, N. Laulainen and S. Barlow for giving us access to their data for DBP; Drs V. Zdimal and M. Rudek for making available data for n-pentanol, and Dr J.L. Schmitt for his data on n- and i-octane. We thank the U.K. Engineering and Physical Science Research Council and the Academy of Finland for support.

REFERENCES

1. Oxtoby, D.W., and Kashchiev, D., *J. Chem. Phys.* **100**, 7665 (1994).
2. Ford, I.J., *J. Chem. Phys.* **105**, 8324 (1996); *Phys. Rev. E* **56**, 5615 (1997).
3. Knott, M., Vehkamäki, H., and Ford, I. J., "Energetics of small n-pentanol clusters from droplet nucleation rate data", to appear in *J. Chem. Phys.*
4. Zdimal, V., and Smolik, J., *Atmos. Res.* **46**, 391 (1998); Rudek, M. M., Katz, J. L., Vidensky, I.V., Zdimal, V., and Smolik, J., *J. Chem. Phys.*, **111**, 3623 (1999).
5. Mikheev, V.B., Laulainen, N.S., Barlow, S.E., Knott, M., and Ford, I.J., "The laminar flow tube reactor as a quantitative tool for nucleation studies: experimental results and theoretical analysis of homogeneous nucleation of dibutylphthalate", submitted for publication.
6. Vehkamäki, H., and Ford, I.J., "Analysis of water-ethanol nucleation rate data with two component nucleation theorems", submitted for publication.
7. Schmitt, J.L., Whitten, J., Adams, G.W., and Zalabsky, R.A., *J. Chem. Phys.* **92**, 3693 (1990).
8. Schmitt, J.L., Doster, J., and Bertrand, G., submitted for publication.
9. Vehkamäki, H., and Ford, I.J., "Excess energies of n- and i-octane molecular clusters", submitted for publication.
10. Lee, J.K., Abraham, F.F., and Pound, G.M., *Surface Sci.* **34**, 745 (1973).

Translational Invariance in the Theory of Nucleation

Y. Drossinos[1], P.G. Kevrekidis[2], M. Lazaridis[3] and
P.G. Georgopoulos[2]

[1] *European Commission, Joint Research Centre, I-21020 Ispra (Va), Italy*
[2] *EOHSI, Rutgers University and UMDNJ, 170 Frelinghuysen Rd., Piscataway, NJ 08854*
[3] *Norwegian Institute for Air Research, Instittutvein 18, P.O. Box 100, N-2007 Kjeller, Norway*

Abstract. A connection between field-theoretic descriptions of condensation and density-functional theories of nucleation is established. The time-independent differential equation for the density profile of a liquid droplet that mediates the transition from liquid to gas phase is derived based on a variational principle. The droplet profile depends crucially on an attractive interaction term. A physically motivated form for the one-dimensional droplet profile is postulated, the droplet translational eigenmode for an infinite system is identified, and its contribution to the nucleation-rate prefactor is evaluated.

INTRODUCTION

The discrepancy between classical nucleation theory predictions and experimental results has triggered a large controversy and a long-standing dispute. Recently, Oxtoby and co-workers (see, for example, [1]), as well as Barrett [2], used density-functional theory to alleviate these discrepancies. An alternative, field-theoretic, approach to the theory of condensation consists of Langer's [3] semi-phenomenological and essentially one-dimensional analysis.

Herein, we adopt the field-theoretic approach to obtain a connection with density-functional theories. The motivation for our work is the statistical mechanics of Bose-Einstein condensation where the wavefunction of the macroscopically-occupied ground state of a Bose gas becomes the superfluid order parameter. The explicit analog (appropriate order parameter) is between the condensed bosons in the (quantum mechanical) ground state and, given Langer's field theoretic description of condensation, a cluster of molecules in the condensed phase that forms a liquid droplet.

ORDER PARAMETER

Consider the many-body Hamiltonian of a system of N interacting particles in second-quantized notation [4]

$$\hat{H} = \int d\mathbf{r}\hat{\Psi}\dagger(\mathbf{r})\left[-\frac{\hbar^2}{2m}\nabla^2\right]\hat{\Psi}(\mathbf{r}) + \frac{1}{2}\int d\mathbf{r}d\mathbf{r}'\hat{\Psi}\dagger(\mathbf{r})\hat{\Psi}\dagger(\mathbf{r}')V_{int}(\mathbf{r}-\mathbf{r}')\hat{\Psi}(\mathbf{r}')\hat{\Psi}(\mathbf{r}), \quad (1)$$

where $\hat{\Psi}$, $\hat{\Psi}\dagger$ are boson annihilation and creation field operators and V_{int} is the two-particle interaction potential. The basic idea, dating back to Bogoliubov [5], is to separate out the condensate part from the total wavefunction by writing $\hat{\Psi}(\mathbf{r}, t) = \Phi(\mathbf{r}, t) + \hat{\Psi}'(\mathbf{r}, t)$ where the function Φ is the Bose macroscopic wavefunction (a classical field rather than an operator). The field Φ plays the role of the order parameter for the superfluid transition, and the local density is its magnitude squared. In the mean-field approximation we will be neglecting fluctuations, retaining only the condensate part of the wavefunction. The interaction potential is decomposed as follows

$$V_{int}(\mathbf{r}-\mathbf{r}') = V_{att}(|\mathbf{r}-\mathbf{r}'|) + g\,\delta(\mathbf{r}-\mathbf{r}'), \quad (2)$$

where the first term is the attractive potential, assumed to be central, and the second term the repulsive one. The repulsive potential has been written in the simplest approximation arising only from binary, hard-sphere collisions. The attractive potential will be determined self-consistently.

The evolution equation for Φ can be derived from a variational principle on the free-energy functional E

$$E[\Phi, \Phi^*] = \int d\mathbf{r}'\left[\frac{\hbar^2}{2m}|\nabla\Phi(\mathbf{r}',t)|^2 + \frac{1}{2}F(\mathbf{r}',t)|\Phi(\mathbf{r}',t)|^2 + \frac{g}{2}|\Phi(\mathbf{r}',t)|^4\right], \quad (3)$$

where the attractive term is

$$F(\mathbf{r}',t) = \int d\mathbf{r}''V_{att}(|\mathbf{r}'-\mathbf{r}''|)|\Phi(\mathbf{r}'',t)|^2. \quad (4)$$

The Euler-Lagrange equation for the time-independent spatial profile of the droplet wavefunction $\Phi(\mathbf{r})$ that spatially mediates the transition from the liquid to the vapor phase is obtained by considering a time-dependent solution of the form $\Phi(\mathbf{r}, t) = \varphi(\mathbf{r})\exp(-i\mu t/\hbar)$ where μ is the thermodynamic chemical potential of a state with N particles, and ϕ is real and normalized to the total number of particles, $N = \int d\mathbf{r}\,\phi^2$. Then, the droplet-profile equation, after appropriate rescalings of the chemical potential and the interaction-potential parameters, becomes

$$\nabla^2\phi + [\mu - f(\mathbf{r})]\phi - g\phi^3 = 0, \quad (5)$$

where we have defined the attractive term to be

$$f(\mathbf{r}) = \int d\mathbf{r}' V_{att}(|\mathbf{r} - \mathbf{r}'|)\phi^2(\mathbf{r}') < 0. \tag{6}$$

This equation provides the desired connection: it has been derived from a quantum mechanical Hamiltonian, and is valid in three spatial dimensions. The attractive term $f(\mathbf{r})$ acts as a symmetry-breaking term, a term that appears naturally in the equation and does not necessitate *ad hoc* assumptions. The connection to density functional theories becomes apparent since the attractive term $f(\mathbf{r})$ is the term referred to as $\varphi_{\text{eff}}(\mathbf{r})$ in [1]. Furthermore, for a uniform fluid of density ρ with a finite-range attractive potential the coefficient of the linear term becomes $\mu + \alpha\rho$, where α is the (positive) integrated strength of the potential.

TRANSLATIONAL MODES AND NUCLEATION RATE

The eigenvectors with zero eigenvalue may be obtained by linearizing the appropriate equation. We consider the one-dimensional version of Eq. 5: the three dimensional extension will be analyzed in future investigations. The order parameter field is decomposed into a mean-field part and a fluctuations term as follows

$$\phi(x) = \phi_0(x) + \varepsilon\phi_1(x). \tag{7}$$

The zeroth-order equation is

$$\frac{d^2}{dx^2}\phi_0(x) + [\mu - f(x)]\phi_0(x) - g\phi_0^3(x) = 0, \tag{8}$$

where $f(x)$ is the one-dimensional version of Eq. 6. Of course, the zeroth-order equation is just the Euler-Lagrange equation for a one-dimensional density profile dependent on the spatial variable x. It is easy to show that $\varphi_1(x)$ is an eigenvector with zero eigenvalue. Furthermore, translational invariance, an exact symmetry only for infinite systems, can be shown to imply that the translational mode is

$$\phi_1(x) = \frac{d\phi_0(x)}{dx}. \tag{9}$$

The translational modes make an important contribution to the nucleation rate expression because they ensure that the nucleation rate is extensive, a consequence of the invariance of the free energy functional under translations of the center of mass of the droplet. Following Langer [3] the contribution of the translation eigenmode to the nucleation rate prefactor is

$$A_{tran} = 2L\left[\int_{-L}^{L} dx \left(\frac{d\phi_0}{dx}\right)^2\right]^{1/2}, \tag{10}$$

where $2L$ is the size (volume) of the system.

Our analysis up to this point has been general. The nucleation-rate prefactor may be evaluated by choosing a molecular interaction potential and subsequently solving the Euler-Lagrange equation numerically. Alternatively, generic properties of the solution may be obtained by postulating an *ansatz* for the droplet profile. We choose a functional form that is physically motivated and is based on the exact solution of the one-dimensional equation without the attractive term,

$$\phi_0(x) = \frac{(\sqrt{\rho_v} - \sqrt{\rho_l})}{2} \tanh\left(\frac{x - R_c}{\xi}\right) + \frac{(\sqrt{\rho_l} + \sqrt{\rho_v})}{2}, \qquad (11)$$

where ρ_l is the liquid density and ρ_v the vapour density. The variable R_c is the radius of the critical droplet, a quantity that may be calculated from the classical theory of nucleation, and ξ is the interfacial correlation length. The droplet profile has the expected limits close to the origin (for large R_c/ξ) and far way from it. Note that these limits impose constraints on the asymptotic behaviour of the attractive term. The requirement that the proposed functional form solve Eq. 8 imposes self-consistency conditions that may be used to determine the attractive potential and the correlation length [6]. Using Eq. 11 in Eq. 10, the prefactor A_{tran} for a one-dimensional system, under the previously described constraints, becomes (up to exponentially small L-dependent corrections to extensivity)

$$A_{tran} = \frac{2\sqrt{3}}{3\sqrt{\xi}}\left(\sqrt{\rho_l} - \sqrt{\rho_v}\right) L \ . \qquad (12)$$

The parameters in Eq. 12 are determined by imposing the normalization condition, the self-consistency requirements that the *ansatz* solve the droplet-profile equation, and by considering the limit of the attractive term at infinity.

YD and ML acknowledge partial support from the European Commission (Fifth Framework Programme, Environment and Sustainable Development, project SUB-AERO, contract EVK2-CT-1999-00052). PGK and PGG acknowledge support from the "Alexander S. Onasis" Public Benefit Foundation, the NJDEP funded Ozone Research Center, U.S. DOE Cooperative Agreement DE-FC01-95EW55084 (CRESP), U.S. EPA Cooperative Agreement CR827033, and U.S. EPA Grant R826768-01.

REFERENCES

1. Oxtoby, D. W. and Evans, R., *J. Chem. Phys.* **89**, 7521 (1988).
2. Barrett, J., *J. Chem. Phys.* **107**, 7989 (1997).
3. Langer, J. S., *Ann. Phys.* (NY) **41**, 108 (1967); Langer, J. S., *ibid.* **54**, 158 (1969).
4. Dalfovo, F., Giorgini, S., Pitaevskii, L. P. and Stringari, S., *Rev. Mod. Phys.* **71**, 473 (1999).
5. Bogoliubov, N., *J. Phys.* (Moscow) **11**, 23 (1947).
6. Drossinos, Y., Kevrekidis, P. G. and Georgopoulos, P. G. (in preparation).

Nucleation on Roads

J. Kaupužs* and R. Mahnke[†]

*Institute of Mathematics and Computer Science, University of Latvia
29 Rainja Boulevard, LV-1459 Riga, Latvia
†Universität Rostock, Fachbereich Physik, D-18051 Rostock, Germany

Abstract. A stochastic approach based on Master equation is proposed to describe the process of formation and growth of car clusters in traffic flow in analogy to usual aggregation phenomena such as the formation of liquid droplets in supersaturated vapour, especially a cloud formation in the earth atmosphere. By this method a coexistence of many clusters on a one–lane circular road has been investigated. Results have been summarized in the flux–density or fundamental diagram of traffic flow and have been compared to the measured vehicular experimental data on German highways.

INTRODUCTION

The formation and growth of clusters is a widely known phenomenon in physics. We mention the formation of liquid droplets in a supersaturated vapour [1]. The formation of car clusters (jams) at overcritical densities in traffic flow is an analogous phenomenon in the sense that cars can be considered as interacting particles [2], and the clustering process can be described by similar equations. In particular, the probability that the system has a given cluster distribution at a time t in both cases can be described by the stochastic Master equation. The transition probabilities depend on the specific physical model under consideration. Our purpose is to describe the congested traffic on a one–lane circular road (freeway) allowing a coexistence of many car clusters. The proposed approach enables us to describe and interpret the experimental traffic data reflected in the fundamental flux–density diagram.

THE MODEL

Here we consider a model of traffic flow on a one–lane circular road of length L. The total number of cars is N. The distance between the front bumpers of two neighbouring cars, in general, is $\ell + \Delta x$, where ℓ is the effective length of a car. Car clusters (jams) are characterised by a constant headway Δx_{clust} between "effective" cars. In free flow the headway is $\Delta x_{free} > \Delta x_{clust}$. The situation without congested cars is illustrated in Fig. 1. In this case equally–spaced cars are moving along the circular road as indicated by an arrow. In Fig. 2 the coexistence of two car clusters and free traffic flow is shown. In general, the total number of clusters is denoted by N_{cl} and the total number of congested cars by n. The details of the model are given elswhere [4,5]. The traffic flow is described as a stochastic process where adding a vehicle to a given car cluster (any of N_{cl} clusters) is characterized by a transition frequency (attachment probability per time unit) $w_+(n, N_{cl})$ and the opposite process by a frequency $w_-(n, N_{cl})$. The stochastic variables are numbers of car clusters of different sizes N_k with $k = 1, 2, ..., N$, whereas the transition frequencies depend on n and N_{cl}, as discussed below. We have assumed that the free cars are distributed uniformly over the spacings between clusters, i. e., all these parts of the free road are characterised by the same mean headway $\Delta x_{free}(n, N_{cl})$. This allows us to formulate an appropriate ansatz for transition frequencies based on physically motivated assumptions about the behaviour of individual drivers depending on the distance ($\Delta x_{free}(n, N_{cl})$ in free flow and Δx_{clust} in a cluster) to the next car (cf. [4,5]). Some stochastic event or perturbation of the free traffic

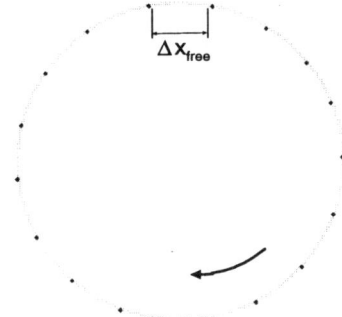

FIGURE 1. Free traffic flow on a one–lane circular road.

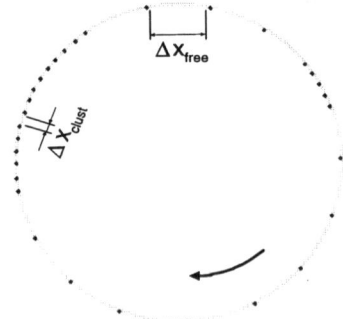

FIGURE 2. Congested traffic flow on a one–lane circular road. There are two car clusters of different length coexisting with free flow.

flow is necessary to initiate formation of a new cluster. Such stochastic events are simulated assuming that any car belonging to free flow can randomly reduce its velocity, i. e., become a single congested car or a cluster of size $k = 1$, called also a pre–cluster. The probability of such an event per time for a given free car is w_+^*. We have excluded any merging and splitting of clusters which is usually also done to describe aggregation in supersaturated systems like droplets [1].

The stochastic variables N_k may be considered as components of an N-dimensional vector $\mathbf{N} = (N_1, N_2, \ldots, N_N)$ or $\mathbf{q} = \sum_k N_k \mathbf{q}_k$ where $\mathbf{q}_k$ is a unit vector the i-th component of which is $\delta_{i,k}$. In such a notation, the stochastic Master equation describing the evolution of the probability distribution function $P(\mathbf{q}, T)$ with the dimensionless time $T = t/\tau$ reads

$$\frac{1}{\tau} \frac{dP(\mathbf{q}, T)}{dT} = (N - n + 1) w_+^* P(\mathbf{q} - \mathbf{q}_1, T)$$
$$+ \sum_{k=2}^{N} (N_{k-1} + 1) w_+(n - 1, N_{cl}) P(\mathbf{q} + \mathbf{q}_{k-1} - \mathbf{q}_k, T)$$
$$+ (N_1 + 1) w_-(n + 1, N_{cl} + 1) P(\mathbf{q} + \mathbf{q}_1, T) \qquad (1)$$
$$+ \sum_{k=1}^{N-1} (N_{k+1} + 1) w_-(n + 1, N_{cl}) P(\mathbf{q} + \mathbf{q}_{k+1} - \mathbf{q}_k, T)$$
$$- \left[(N - n) w_+^* + N_{cl} \left(w_+(n, N_{cl}) + w_-(n, N_{cl}) \right) \right] P(\mathbf{q}, T) .$$

The stochastic process has several reaction channels, written on r. h. s. of Eq. (1), which are transitions changing the cluster distribution from $\mathbf{N} = (N_1, \ldots, N_{k-1}, N_k, \ldots, N_N)$ to $\mathbf{N'} = (N_1 + 1, \ldots, N_{k-1} + 1, N_k - 1, \ldots, N_N)$ and vice versa (valid for $k > 2$). The formation and dissolution of a pre–cluster (jam of size $k = 1$) and a dimer (jam of two vehicles) have also to be considered. Generally Eq. (1) describes condensation and evaporation of car clusters due to stochastic one–step processes as attachment or detachment of one car only.

RESULTS AND DISCUSSION

Based on the Master equation (1), stochastic trajectories for the time evolution of the cluster distribution can be simulated by Monte–Carlo method. An interesting example of such a stochastic trajectory for the total number of congested cars n vs time T is shown in Fig. 3. A switching from free traffic flow to a congested state occurs at some time moment $T \approx 25\,000$.

One of the most important characteristics of the traffic flow is the so–called fundamental diagram or flux–density relation. The fundamental diagram calculated from our model is shown in Fig. 4 by smooth relatively thicker solid line. At any density c, this line represents the dimensionless flux j averaged over time. Analytical solutions for an infinitely long road with free and congested traffic are shown by smooth thin solid line and dashed line, respectively. As it is evident from

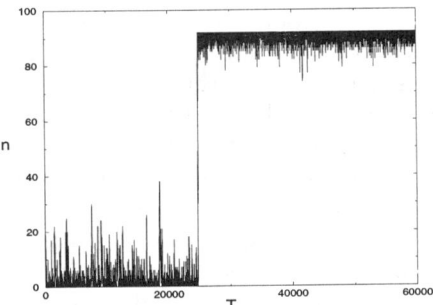

FIGURE 3. A stochastic trajectory showing the total number of congested cars n vs dimensionless time T.

Fig. 4, our theoretical results are comparable with the experimental data from German highways [3] depicted by points.

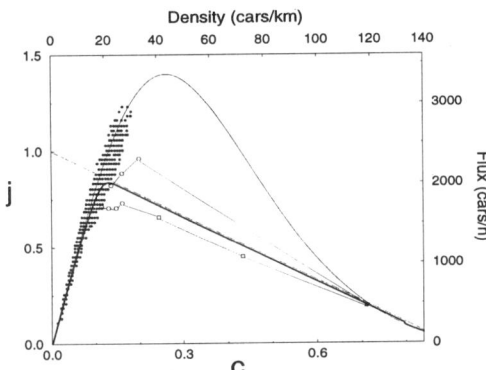

FIGURE 4. Fundamental diagram of traffic flow: comparison between theory (smooth lines) and vehicular experiment (points).

REFERENCES

1. Schmelzer J., Röpke G., and Mahnke R., *Aggregation Phenomena in Complex Systems*, Weinheim: Wiley–VCH, 1999.
2. Prigogine I. and Herman R., *Kinematic Theory of Vehicular Traffic*, New York: Elsevier, 1971.
3. Kerner B. S. and Rehborn H., Phys. Rev. E **53**, R4275 (1996).
4. Kaupužs J. and Mahnke R., Europ. Phys. J. B **14**, 793 (2000).
5. Mahnke R. and Kaupužs J., submitted to J. Networks and Spatial Theory (2000).

Quantum Mechanical Treatment on the Critical Energy in Ion-induced Nucleation of Water Vapor

Kaname Sakiyama, Hiroshi Takano, Hiroyuki Tomida and Masayuki Itoh

Department of Chemical Engineering and Materials Science,
Faculty of Engineering,
Doshisha University
Kyoto, Japan

Abstract. The properties of ion-induced nucleation regarding super-saturation ratio (S = 7,9) were investigated with the combination of Monte-Calro simulation and *ab initio* calculation method. The free energy differences were calculated by *ab initio* method according to the arrangement of molecules in an embryo ($[Na(H_2O)_N]^+$, N=3-16) estimated from Monte-Calro simulation. The meta-stable state and the critical state of ΔG were investigated, and ion cluster of meta-stable was confirmed quantitatively both for classical and quantum mechanical models. It was concluded that the energetic properties showed a good qualitative agreement between the models, still far from quantitative ones.

INTRODUCTION

The formation process such as a nucleation is very important to estimate the properties of nano-phase particles. Sometimes such a process is strongly dependent on the presence of minute particles like as an ion. Theoretical analysis of the phenomena of ion-induced nucleation is an important part for understanding the atmospheric condensation. Because of complexes of growth it is very difficult to predict properties of actual system of embryo, though, it is worthwhile to make clear the properties of ion-induced nucleation.

In this work, the molecular configuration of embryo in ion-induced nucleation and the free energy difference, ΔG were investigated with the combination of Monte-Calro simulation and *ab initio* calculation. After a briefly introduction of some results on the general framework of ion-induced nucleation theory, both ΔGs derived by current classical nucleation theory and by our *ab initio* method were compared qualitatively and quantitatively.

GENRAL FRAMEWORK OF CLASSICAL NUCLEATION

The Gibbs free energy is known to play an important role determining the growth of embryo. A change in the free energy resulted from the embryo formation is between the free energy of the final state G and the initial state G_0,

$$\Delta G = G - G_0. \tag{1}$$

This change of free energy can be expressed as a sum of two parts representing the volume and the surface of the embryo.

$$\Delta G = G_v - G_s \tag{2}$$

First let us consider the formation for homogeneous nucleation. Expression (2) can be written as

$$\Delta G = -\frac{4}{3}\pi r^3 \frac{kTN_a \rho_L}{M} \ln S + 4\pi r^2 \sigma, \tag{3}$$

where r is the radius of embryo; k, T, N_a, ρ_L, S, M and σ are the Boltzmann constant, the temperature, the Avogadro constant, the density of liquid, the saturation ratio, the molecular weight and the surface tension, respectively. The expression (3) can be described by differentiation with respect to radius r,

$$\frac{\partial \Delta G}{\partial r} = -\frac{4\pi r^2 kTN_a \rho_L}{M} \ln S + 8\pi r \sigma. \tag{4}$$

The critical radius r^* and critical change of free energy ΔG corresponding to the critical radius r^* with the condition of $\frac{\partial \Delta G}{\partial r}$,

$$r^* = -\frac{\sigma M}{kTN a \rho_L \ln S}, \quad \Delta G^* = \frac{4\pi r^{*2} \sigma}{3}.$$

Analogous to the homogeneous nucleation, the change of free energy can be derived as an equation in the case of ion-induced nucleation[1],

$$\Delta G = -\frac{4}{3}\pi r^3 \frac{kTNa\rho_L}{M} \ln S + 4\pi r^2 \sigma - \frac{q^2}{8\pi}\left(\frac{1}{\varepsilon_0} - \frac{1}{\varepsilon_0 \varepsilon_r}\right)\left(\frac{1}{r_i} - \frac{1}{r}\right), \tag{7}$$

where $\varepsilon_0, \varepsilon_r, q$ and r_i are the dielectric constant, the relative dielectric constant, the ion charge, the radius of ion cluster in meta-stable state. The equilibrium state is obtained by the differentiation of the equation (7) with respect to radius r,

$$\frac{\partial \Delta G}{\partial r} = -4\pi r^2 \frac{N a \rho kT}{M} \ln S + 8\pi r \sigma - \frac{q^2}{2r^2}\left(\frac{1}{\varepsilon_0} - \frac{1}{\varepsilon_0 \varepsilon_r}\right) \tag{8}$$

The critical radius r^* and the critical change of free energy ΔG^* corresponding to the critical radius r^* can be derived with the same manner to the homogenous nucleation.

DETAILS OF THE SIMULATION

Let us consider the system consists of a sodium ion and some water molecules. The Monte-Carlo method was usually applied to determine the molecular in embryo. The potential energy U can be divided into two parts; the potential regarding water and water interaction and the potential between ion and water. The potential energy on the interactions between water molecules was described by the following function, i.e., L-J type potential[2].

$$U_{H2O-H2O} = 4\varepsilon_{00}\left[\left(\frac{\sigma_{00}}{r_{00}}\right)^{12} - \left(\frac{\sigma_{00}}{r_{00}}\right)^{6}\right] - \frac{1}{4\pi\varepsilon_0}\sum_{a=1}^{3}\sum_{b=1}^{3}\frac{q_a q_b}{r_{ab}}, \quad (9)$$

where the subscription of a or b stands for the number of molecule and σ_{00}, r_{00} and ε_{00} are the parameter. The model for water molecules is shown in Fig.1.

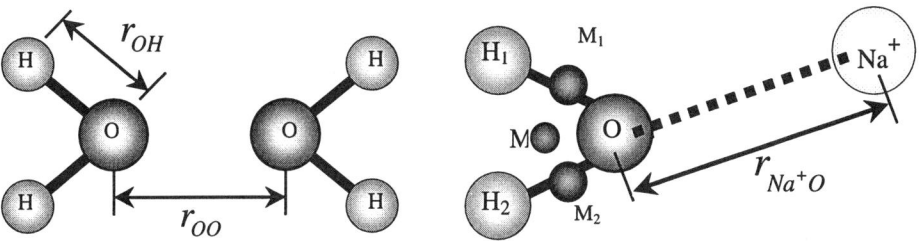

Fig.1 Schematic models for the water and water (left image) and the water and sodium ion (right image).

The potential function describing the interaction between the sodium ion and water was proposed by Kistenmacher[3],

$$U_{Na^+ - H_2O} = Q\left(\frac{1}{r_{Na^+H_1}} + \frac{1}{r_{Na^+H_2}} - \frac{2}{r_{Na^+M}}\right) - Q_M\left(\frac{1}{r_{Na^+M_1}} + \frac{1}{r_{Na^+M_2}} - \frac{1}{r_{Na^+O}}\right)$$
$$+ a_1[\exp(-a_3 r_{Na^+H_1}) + \exp(-a_3 r_{Na^+H_1})] + a_2 \exp(-a_4 r_{Na^+O}), \quad (10)$$

where each Q, r and a is the constant depending on sodium and water.

After determining the molecular configuration in the embryo, *ab initio* calculations of the free energy of the embryo regarding both volume and surface in equation (2) were calculated with GAMESS (Ref.4). The critical super saturation ratio for the classical ion-induced nucleation can be derived by the equation (8). The temperature was kept at 288K in this simulation.

RESULTS AND DISCUSSION

The changes of free energy for ion-induced nucleation ($[Na(H_2O)_n]^+$, n=3-16) were shown in Fig. 2 for saturation ratios of 7 and 9. The first minimum and maximum of the curves in Fig 2 agreed qualitatively with the classical one. The meta-stable state and the critical state of ΔG were investigated, and ion cluster of meta-stable was confirmed quantitatively both for classical and quantum mechanical models. It was concluded that a combination of Monte-Calro simulation and *ab initio* calculation method is applicable for sophisticated understanding on ion-induced nucleation.

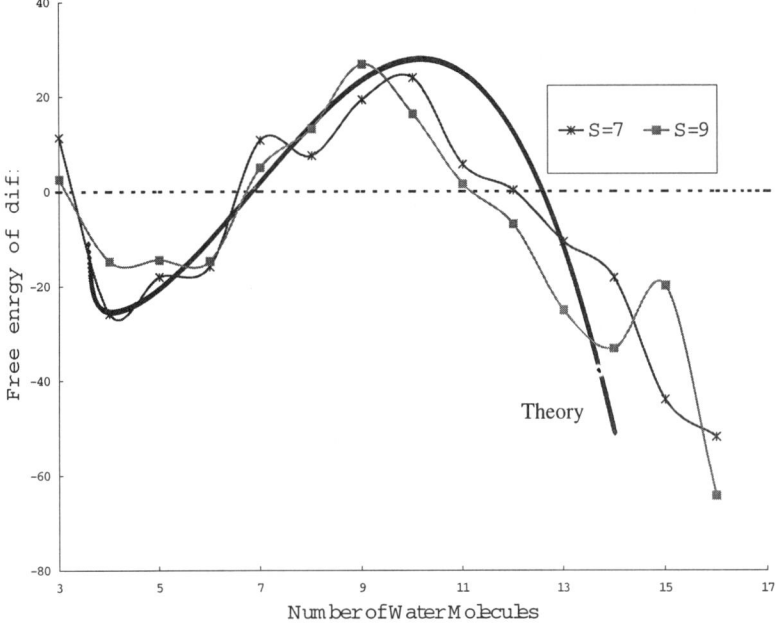

Fig 2. The changes of free energy for ion-induced nucleation.

ACKNOWLEDGMENTS

This study is funded by a part of the Research Project of the Ministry of Education, Science, Sports and Culture of Japan for 1996-2000, and by a part of the Research Promotion Fund of Aid at Doshisha University for 1996-2000.

REFERENCES

1. J.J. Thomsn, *Conduction of Electricity thorough Gases* (Cambridge University Press, Cambridge, 1906)
2. H.J.C.Berendsen, et al., *J.Chem.Phys*., 91, 6269 (1987)
3. H.Kistenmacher, H.Pokie, E.Clementi,j, *J.Chem.Phys.*, 59, 5842 (1973)
4. M.W.Schmidt *et al.*•"GAMESS User's Guide",•http: // www.classic.chem.msu.su/gran/games / index.html (1998)

Nucleation in Physical and Non-Physical Systems

Reinhard Mahnke[*1]

Universität Rostock, Fachbereich Physik D-18051 Rostock, Germany

Abstract. The aggregation of particles out of an initially homogeneous situation is well known in physics. Depending on the system under consideration and its control parameters the cluster formation in a supersaturated (metastable or unstable) situation has been observed in nucleation physics as well as in other branches. We investigate the well--known example of condensation (formation of liquid droplets) in an undercooled (water) vapour. The formation of bound states as a stochastic process is related to transportation science. We present a comparative analysis of nucleation to the probabilistic description of traffic congestion.

INTRODUCTION

The term *cluster* as a *group of similar things* is used in science in many different fields. It is common on all scales from subatomic physics up to astronomy. The existence of bound states of atoms or molecules as clusters with different sizes in gases, liquids and solid matter is an intermediate step between the elementary isolated particle and the macroscopic condensed state. In general the properties of clusters, especially the binding energy, depend on the cluster size and shape.

The formation of clusters takes place on different scales in physical (nucleation of droplets, condensation of clouds) and non--physical systems (insect societies, socioconfigurations, transport communities). Spin glasses are one of the most investigated complex systems to understand the clustering of states. In chemically reacting systems the domain formation is simulated by Monte Carlo experiments and by lattice—gas automata.

The clustering behaviour can be considered as a phase transition between different states of matter. In dependence on the boundary conditions a quenched system as for instance an undercooled vapour or an overcrowded road undergoes a relaxation process to the new phase. The theoretical understanding of the physical mechanism for the dynamics of the aggregation phenomenon has important implications for the more general task to realize relaxation phenomena in equilibrium and nonequilibrium complex systems. [1].

NUCLEATION IN SUPERSATURATED VAPOURS

If we consider a vapour at equilibrium then a certain change of the thermodynamic parameters enable us to remove the system into a nonequilibrium state. The vapour becomes supersaturated [1]. The basic quantity describing the situation is the cluster distribution **N** at time t

$$\mathbf{N}(t) = (N_1, N_2, ..., N_n, ..., N_N) \tag{1}$$

which gives the number of clusters N_n of size n. The free particles (molecules) are called monomers of size $n=1$. The bound states are clusters of size $n \geq 2$. Investigating a finite system the overall number of particles is fixed. The boundary condition

$$M_0 = \sum_{n=1}^{N} n N_n = const \tag{2}$$

takes into account that the particles are either free or bounded in clusters.

From the canonical partition function we can calculate the thermodynamic quantities. The state function Free energy $F(T,V,\mathbf{N})$ reads

$$F(T,V,\mathbf{N}) = kT \sum_{n=1}^{N} N_n \left[\ln(\lambda_n(T)^3 N_n / V) - 1\right] + \sum_{n=1}^{N} N_n f_n(T) \tag{3}$$

with the so-called de Broglie wave length defined as $\lambda_n(T) = n^{-1/2} \lambda_1(T) = n^{-1/2} (h^2/2\pi mkT))^{1/2}$. The negative binding energies $f_n(T)$ for each cluster size n have to be introduced empirically. Using the well-known Bethe-Weizsäcker ansatz $f_n(T) = \mu_\infty(T) n + \sigma A_n$ with the surface of a size–n–cluster $A_n = (c_{clust} 4\pi/3)^{-3/2} n^{2/3}$, indicating volume and surface contributions, we take this formula valid for cluster sizes $n \geq 2$ and $f_1(T)=0$ for isolated monomers. The chemical potential of a monomer over a flat surface $\mu_\infty(T) = kT \ln(\lambda_1(T)^3 c_{eq}(T))$. The equilibrium concentration $c_{eq}(T)$ is connected with the equilibrium vapour pressure $p_{eq}(T)$ by the ideal gas law $c_{eq}(T) = p_{eq}(T)/(kT)$. In the framework of thermodynamics the question of nucleation is connected with the calculation of extrema of the thermodynamic potential. The search for these states of stable or unstable equilibrium, for example of the free energy $F = F(T,V,\mathbf{N})$, can be done by changing the cluster distribution $\mathbf{N}$ via reactive collisions. The monomer - cluster – reactions $N_1 + N_n \Leftrightarrow N_{n+1}$ (formation and dissolution of clusters by a single monomer) are the most important reaction channels. The state of equilibrium in this system corresponds to the minimum of the Free energy $F(T,V,\mathbf{N})$ with fixed temperature T, volume V amd total particle number M_0.

CAR CLUSTER MODEL AND ITS STOCHASTIC DESCRIPTION

In analogy we consider a model of traffic flow on a one-lane road according to which N cars are moving along a circle of length L. If a road is crowded by cars, each car requires some minimal space or length which, obviously, is larger than the real length of a car. We call this the effective length ℓ of a car. The maximal velocity of each car is v_{max}. The desired (optimal) velocity v_{opt}, depending on the distance between two cars Δx, is given in dimensionless variables $w_{opt} = v_{opt}/v_{max}$ and $\Delta y = \Delta x/\ell$ by the formula [2,3]

$$w_{opt}(\Delta y) = \frac{(\Delta y)^2}{d^2 + (\Delta y)^2}, \qquad (4)$$

where the parameter $d = D/\ell$ is the interaction distance. D is the distance between two cars corresponding to the velocity value $v_{max}/2$.

Measurements on highways have been shown that the density of cars in congested traffic $\equiv \rho_{clust}$ is independent of the size of the dense phase (jam). As a consequence the distance between jammed cars, the spacing $\Delta x_{clust} = \ell \Delta y_{clust} = \text{const} \geq 0$, is well known and has to be treated as a given measured quantity.

The length of the cluster (jam) depending on the number of congested cars n is defined by $L_{clust} = \ell n + (n-1)\Delta x_{clust}$. According to this, the average distance $\Delta x_{free} = \ell \Delta y_{free}$ between two cars outside the jam (or free cars) distributed over the free part of the road with length $L_{free} = L - L_{clust}$ is given by

$$\Delta y_{free}(n) = \frac{L/\ell - N - (n-1)\Delta y_{clust}}{N - n + 1}, \qquad (5)$$

where N is the total number of cars on the road (circle of length L).

The traffic flow is described as a stochastic process where adding a vehicle to a car cluster of size n is characterized by a transition frequency (attachment probability per time unit) $w_+(n)$ and the opposite process by a frequency $w_-(n)$. The number n of cars in the cluster is the stochastic variable which may have values from 1 to N. According to our model only one car cluster does exist at any time. The basic equation for the evolution of the probability distribution $P(n,t)$ to find a cluster of size n at time t with probability P is known as Master equation. The one-dimensional stochastic equation reads

$$\frac{1}{\tau}\frac{dP(n,T)}{dT} = w_+(n-1)P(n-1,T) - w_+(n)$$
$$- w_-(n)P(n,T) + w_-(n+1)P(n+1,T). \qquad (6)$$

where $T = t/\tau$ is the dimensionless time. The time constant τ will be specified below.

The main task is to formulate expressions for both transition probabilities w_+ and w_-. We have assumed that the detachment frequency $w_-(n)$ or the average number of cars leaving the cluster per time unit is a constant independent of cluster size n. The ansatz for $w_+(n)$ is now formulated allowing for Δx_{clust} to be nonzero. Our general assumption is that a vehicle changes the velocity from $v_{opt}(\Delta x_{free})$ in free flow to $v_{opt}(\Delta x_{clust})$ in jam and approaches the cluster as soon as the distance to the next car (the last car in the cluster) reduces from Δx_{free} to Δx_{clust}. This assumption allows one to calculate the average number of cars joining the cluster per time unit or the attachment frequency $w_+(n)$. Thus, we have the ansatz which in dimensionless quantities reads

$$w_+(n) = \frac{b}{\tau} \frac{w_{opt}(\Delta y_{free}(n)) - w_{opt}(\Delta y_{clust})}{\Delta y_{free}(n) - \Delta y_{clust}} \qquad (7)$$

$$w_-(n) = 1/\tau = const, \qquad (8)$$

where $b = v_{max}\tau / \ell$ denotes a dimensionless parameter. The parameter τ is a time constant, which can be understood as the waiting time for the escape (detachment) of the first car out of the jam into free flow [2].

Our purpose is to solve the Master equation (6) and to extract from this solution information about the formation of traffic jams and about the various possible regimes of traffic flow depending on the parameters of the system, as well as to calculate the flux-density or the fundamental diagram of traffic flow. Finally, our aim is to compare the results of calculation with experimental data [2,3].

DISCUSSION

A stochastic approach based on Master equation has been discussed to describe the process of formation and growth of car clusters in traffic flow in analogy to usual aggregation phenomena such as the formation of liquid droplets in supersaturated vapour. In particular, there are three different regimes in traffic flow (free jet of cars, coexisting phase of jams and isolated cars, highly viscous heavy traffic) and two phase transitions between them. In distinction to the aggregation in supersaturated systems, there is no critical cluster size due to the absence of cluster surface curvature. In essence, the traffic flow is one--dimensional, therefore such effect cannot be, in principle, observed. As a result, in the supersaturated vapour the final stage of aggregation is one large cluster, whereas in traffic flow there are many smaller clusters.

REFERENCES

1. Schmelzer, J., Röpke, G. and Mahnke, R., *Aggregation Phenomena in Complex Systems*, Weinheim: Wiley-VCH, 1999.
2. Mahnke, R. and Kaupuzs, J.," Stochastic theory of freeway traffic" *Phys. Rev. E* **59**, *117 (1999)*.
3. Mahnke, R. and Pieret, N., Stochastic master--equation approach to aggregation in freeway traffic, *Phys. Rev. E* **56**, 2666 (1997).
4. R. Mahnke, J. Schmelzer and G. Röpke: *Nichtlineare Pheanomene und Selbstorganisation*, Teubner, Stuttgart, 1992

Thermodynamic Properties of Molecular Clusters in Nucleation

Norihiko Fukuta* and Guang Guo*

Department of Meteorology, University of Utah, Salt Lake City, Utah 84112-0110, USA

Abstract. Applying a pair-wise Lennard-Jones type intermolecular potential and finding empirical relationships for the normal components as well as surface molecular areas, the size dependence of the surface free energy, σ, has been re-evaluated. σ for monomer and dimer are found as 0 and 2.8% of the flat surface value, respectively. The cluster vapor pressure based on the new σ with Ono-Kondo factor resulted in discovery of a maximum at about 13-mer which quantitatively matches the vapor pressure of homogeneous condensation nucleation at 0°C and resolves the Second Law violation of the classical nucleation theory which permits spontaneous vapor flow from low to high pressure in subcritical region. The cluster at the maximum is found to possess a true critical thermodynamic characteristic and unlike the critical embryo of classical theory, it is independent of the supersaturation of the nucleating environment. Use of other size-dependent surface potentials results in totally different maximum vapor pressures.

INTRODUCTION

Among the fundamental properties of clusters in nucleation, the surface free energy, σ, the size dependence in particular, is most crucial but controversial. Applying the essential behaviors of heat and work in thermodynamics to the microclusters with the help of Sutherland type intermolecular potential, we obtained the size dependence of σ for the first time in clarity [1,2]. While Sutherland potential made a number of successes in gas kinetics, the use of rigid sphere left some doubt about the accuracy of obtained σ and its further application to the cluster properties. It is the purpose of this paper to report σ and the related properties of microclusters through application of more realistic, compressible Lennard-Jones (L-J) type intermolecular potentials and comparison and verification of the results with the experimental data.

LENNARD-JONES TYPE POTENTIALS AND MOLECULAR PACKING

The L-J intermolecular potential for relative evaluation of size dependent σ takes the form

$$u(r) = r^{-12} - Cr^{-n}, \qquad (1)$$

where n is a parameter describing the nature of the attractive potential and C a positive constant to be determined by the equilibrium condition of the bulk. For a non-polar substance, $n = 6$, and a polar substance, $n < 6$. The molecules, water for the present study at 0°C, are assumed to be spherical and take a hexagonal closest packing. The assumption is needed in order to sum up the pair-wise potentials since an integral approach is found to be in large error [1,2] even if the radial distribution function were used. i-cluster is assumed to be a sphere of volume iv_0, where v_0 is the volume of single molecule in the bulk. Under the assumed packing and in equilibrium, any molecule in the bulk is surrounded by 12 molecules on average, and a force balance holds among them. The molecule in question receives force through the nearest neighbors dividing the solid angle of the entire space into 12 or 6 pairs in the form of rhombic dodecahedron. The summation of all the attractive forces in the differential form of Eq. (1) between the molecule in question and between molecules within the paired solid angles (rubber band effect) determines C.

SIZE DEPENDENCE OF SPECIFIC SURFACE ENERGIES

The normal potential, imbalanced to the outside, was determined for each surface molecule of a cluster as before [1,2], except that it carried $\cos^2\theta$ factor since it is a work or product between the displacement and the resisting force, both carrying $\cos\theta$ factor, where θ is the angle between the direction of interaction and the normal. The ratio of the interaction factor for the nearest neighbor (nn) interaction is obtained in empirical form as

$$R_{i,n} = \frac{f_{i,nn}}{f_{\infty,nn}} = \frac{i-1}{i+2}, \tag{2}$$

where i is the number of molecules in the i-mer cluster and ∞ the flat surface (see Fig. 1).
The empirical equation of best fit for the surface molecular area is found as

$$R_{a,i} = \frac{A_{1,i}}{A_{1,\infty}} = 1 + 50.77 i^{-.04} - 47.34 i^{-0.3967}, \tag{3}$$

(see Fig. 2). $A_{1,\infty} = 1.0514 \times 10^{-19}\ m^2$.

The surface free energy ratio of nn interaction is given by the quotient

$$\left(\frac{\sigma_i}{\sigma_\infty}\right)_{nn} = \frac{R_{i,nn}}{R_{a,i}} \tag{4}$$

It was found that small clusters like dimer, trimer and tetramer all consist surface molecules, and their interactions are among the nearest neighbors only. For larger clusters including 13-mer and flat surface, each surface molecule has interactive molecules beyond the nearest neighbors, and their contribution must be counted. The ratio

$$\gamma = \frac{f_{\infty,nn}}{f_{\infty,lr}}, \tag{5}$$

where $f_{\infty,nn}$ and $f_{\infty,lr}$ are the nearest neighbor and the long-range (lr) interaction factors of flat surface, respectively, reduces $f_{\infty,nn}$ of small clusters when they are applied to cluster systems of long-range interaction, so that

$$\left(\frac{\sigma_i}{\sigma_\infty}\right)_{lr} = \frac{i+2}{i - i^\dagger + \dfrac{i-1}{\gamma R_{i^\dagger}}} \left(\frac{\sigma_i}{\sigma_\infty}\right)_{nn}, \tag{6}$$

where superscript (†) refers to the cluster through which Eq. (2) passes.

Table 1 lists estimated C and γ, and Fig. 3 describes σ_i / σ_∞ for L-J type potential in comparison with those for nearest neighbor interaction, Sutherland type potential and Tolman [3]. Except for Tolman, all three Values pass through $I = 0$, but L-J type is the lowest. Since σ is based on the interaction energy and a monomer has no molecule associated, this relationship of new results should be reasonable.

VAPOR PRESSURE OF MICROCLUSTERS

Under the equilibrium, the chemical potential of single water molecule in the vapor phase matches the free energy an evaporating molecule takes away, or

$$kT \ln(\frac{e_i}{e_\infty}) = \left(\frac{\partial \Delta F_s}{\partial V}\right)v_L = \left(\frac{2\sigma}{r} + \frac{\partial \sigma}{\partial r}\right)v_L, \qquad (7)$$

where k is the Boltzmann constant, T the temperature, e_i the vapor pressure of i-mer, e_∞ that over the flat surface, v_L the molecular volume in the liquid phase, V the volume, r the cluster radius and $\Delta F_s = \sigma(r)A(r)$. The first term in the bracket is due to Kelvin if $\sigma=\sigma_\infty$ and the second Ono and Kondo[4]. Figure 4 compares the saturation ratio, $S=e_i/e_\infty$, estimated with various surface energies in place of σ. It is clear that all of S show maxima, a new finding in this study, but except for the normal component, or σ, they do not reach $S=5$ for homogeneous condensation nucleation at 0°C[5].

In Fig. 5, contributions of $2\sigma/r$ and $\partial\sigma/\partial r$ terms to S are compares. Both terms cause maxima, the latter earlier, but neither of them takes the maximum up to $S=5$; only their combination does. Figure 6 describes the effect of n on the maximum of S or S_{max}. Smaller n results in slightly lower S_{max} and larger i_{max}. Nevertheless, the difference is small and $n = 5.75$ appears to give $S_{max}=5$ at about 13-mer position in contrast to about 60-mer of the classical nucleation theory [6]. It is reasonable from the viewpoint of polar water molecules. Unlike the critical embryo of classical theory which varies with the condition of the environment, S_{max} thus discovered suggests that it is a property of the given substance independent of environment S.

Appearance of the maximum also resolves the 2nd Law violation of the classical theory in the subcritical zone where the low-pressure environmental vapor has to flow towards the high-pressure vapor of clusters. Under the condition, the assumed fluctuation current cannot take place.

ACKNOWLEDGEMENT

This work was partially supported by NSF under Grant ATM 8626600.

REFERENCES

1. Wiśniewska, M., *Surface Free Energy of the Molecular Clusters*, M.S. Thesis, University of Utah,, 1995, 66pp.
2. Fukuta, N., and Wiśniewska, M., "Molecular-Theory of Ultramicro Clusters and Nucleation", in 14*ICNAA Proceedings*, edited by M. Kulmala and P.E. Wagner, Pergamon, Oxford, U.K., 1996, pp. 212-215
3. Tolman, R.C., *J. Chem. Phys.*, **17**, 333-337 (1949).
4. Ono, S., and Kondo S., *Molecular Theory of Surface Tension in Liquids*, Springer-Verlag, Berlin, 1960, 280pp.
5. Miller, R.C., Anderson, R.J., Kassner, Jr, J.L., and Hagen, D.E., *J. Chem. Phys.*, **78**, 3204-3211, (1983).
6. Becker, R., and Döring, W., *Ann. Phys.*, **24**, 719-725 (1935).

Table I. C and γ for L-J Type Potential.

n	C (C.G.S.Unit)	γ
5.0	2.9748×10^{52}	0.5527
5.5	5.5450×10^{48}	0.6350
5.75	7.4744×10^{46}	0.6790
6.0	9.9856×10^{44}	0.6977

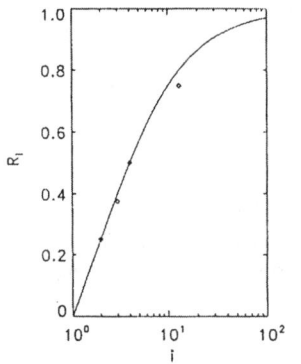

FIGURE 1. R_i for the *nn* interaction model plotted as a function of i.

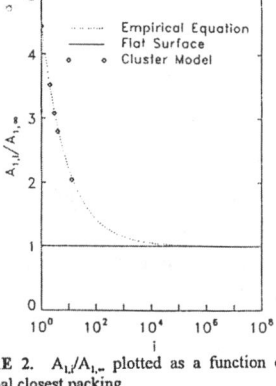

FIGURE 2. $A_{1,i}/A_{1,\infty}$ plotted as a function of i under hexagonal closest packing.

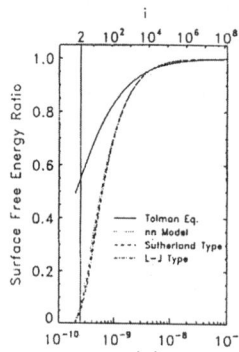

FIGURE 3. Surface free energy ratio for *nn* and *lr* interaction models ($i^\dagger=4$, $n=6.0$) plotted with r.

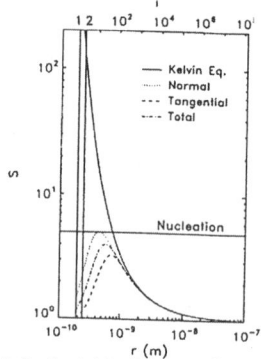

FIGURE 4. S estimated for different surface energies with L-J type potential ($i^\dagger=4$, $n=5.75$) plotted with r.

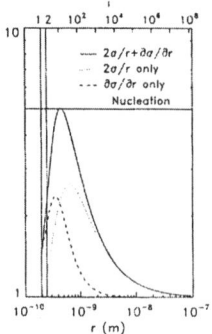

FIGURE 5. Contribution of $2\sigma/r$ and $\partial\sigma/\partial r$ in S for L-J type potential ($i^\dagger=4$, $n=5.75$) as a function of r.

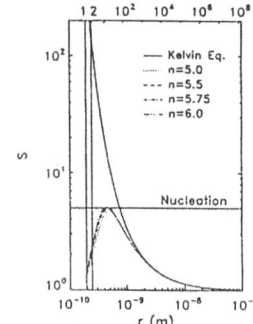

FIGURE 6. S as a function of r for L-J type, *lr* interaction model ($i^\dagger=4$) for different n.

Nucleation Pulse Measurements for Water Vapor at Elevated Temperatures

J. Hrubý, J. Hošek, J. Blaha, and F. Maršík

Institute of Thermomechanics, Academy of Sciences of the Czech Republic, Dolejškova 5, CZ-182 00 Prague 8, Czech Republic

Abstract. At elevated temperatures, droplet growth becomes very fast. We present a new method to generate a nucleation pulse using a shock tube. This method is particularly suitable for nucleation and growth studies at high-temperature conditions, because the falling edge of the nucleation pulse is very sharp. A comparison of the wave-based methods to generate a nucleation pulse is given. Preliminary results for water-argon measurements are discussed.

INTRODUCTION

The nucleation pulse technique is perhaps the most fundamental experimental approach to study droplet nucleation and growth. In general, the nucleation pulse is a short period of time, during which the system is so supersaturated that significant homogeneous nucleation takes place. However, the newly born nuclei are too small to be observed directly. Therefore, the nucleation pulse is followed by a long plateau, during which the system is still somewhat supersaturated, so that existing nuclei can grow, but the supersaturation is too low for the nucleation process to continue. The allowable duration of the nucleation pulse and the slope and roundness of its limits (in a supersaturation vs. time plot) are determined by the characteristic times of the growth. If the pulse is too long, the droplet distribution is too broad (polydisperse). Moreover, thermodynamic conditions can vary during the pulse (depletion and heating due to condensation of earlier nucleated droplets). Expansion cloud chambers use adiabatic expansion followed by a small re-compression to generate the nucleation pulse. At elevated temperatures (we consider temperatures about between 0 and 100°C), the growth of water droplets becomes very fast. The main reason is the exponentially increasing saturated vapor pressure. Under such conditions, the nucleation pulse must be rather short and have sharp limits. Devices generating the nucleation pulse based on gas-dynamical principles (like the shock tubes) avoid the necessarily slow moving mechanical parts and are, therefore, more suitable for the higher temperature range.

A NEW METHOD TO GENERATE THE NUCLEATION PULSE

For cloud chambers, the nucleation pulse was implemented by Schmitt (1) and Wagner and Strey (2). The last device has been continuously improved (2,3). The cloud chamber method can be considered quasi-static, since a sound wave caused by action of a piston or opening/closing a valve travels many times through the chamber and always brings only a very small change of pressure and temperature. The gas-dynamic based methods, on the other hand, rely on a combination of solitary waves propagating through a quasi-one dimensional tube after a thin diaphragm ruptures. The speed, magnitude and shape of these waves can be predicted with a good accuracy mathematically, starting from the initial conditions of the experiment and the tube geometry.

The first method to generate the nucleation pulse using a shock tube was invented by Peters (4, see Fig.1a). The vapor-background gas mixture is located in the high-pressure section (HPS) of a shock tube. Before the expansion, a thin diaphragm separates the high and low pressure sections (HPS and LPS). After the diaphragm ruptures, an expansion wave propagates into the HPS, causing adiabatic cooling of the mixture and establishing supersaturated conditions. The observation place (O) is located near the bottom plate of the shock tube. Here the expansion wave reflects. The shock wave runs from the diaphragm place into the low-pressure section (LPS) and reflects partially on a constraint C. The partially reflected shock wave (PRSW), and its reflection (RPRSW), form the right edge of the nucleation pulse.

The Eindhoven group (5, see Fig.1b), working in the high-pressure range, developed a different technique. Here both the start and the end of the nucleation pulse are generated by reflections of the shock wave on a local widening of the LPS.

In Prague we have developed still another technique (see Fig.1c). The LPS has somewhat smaller cross section than the HPS. The nucleation is terminated by a weak compression wave (WCW, not a shock wave). WCW originates as a partial reflection of the reflected expansion wave (REW) at the place of broadening.

THE APPARATUS

The high-pressure section (HPS) of our shock-tube is very short, some 180 mm, and it has an internal diameter of 46 mm. Yet the flow pattern is to a good approximation one-dimensional and it can be predicted using a simple mathematical model. The advantage of the shortness is that the cooling rate is very high (cooling of 50 K is achieved within about 0.3 ms with argon and even quicker with helium). The shock tube is oriented vertically (see Fig.2). This enables that a shallow liquid pool can be located at the bottom of the HPS. The HPS is thermostated using a circulation thermostat. The LPS is thermostated using electric heaters to prevent condensation of the liquid near the junction of the LPS and HPS. These facts enable that vapor-liquid equilibrium can establish in the HPS. After this equilibrium is established, the thin plastic diaphragm is punctured and it opens fully within a few microseconds. Near the bottom of the HPS, three windows are located, enabling illumination with a linearly polarized diode laser (red, 35 mW) and measurement of the transmitted and 90° scattered light using a photodiode and photomultiplier, respectively.

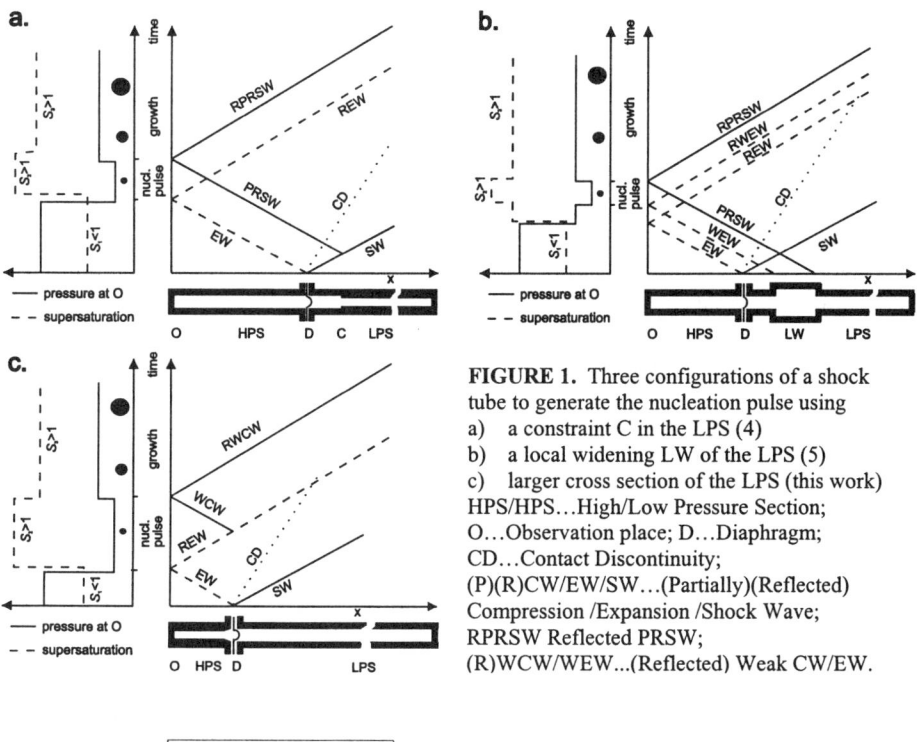

FIGURE 1. Three configurations of a shock tube to generate the nucleation pulse using
a) a constraint C in the LPS (4)
b) a local widening LW of the LPS (5)
c) larger cross section of the LPS (this work)
HPS/HPS...High/Low Pressure Section;
O...Observation place; D...Diaphragm;
CD...Contact Discontinuity;
(P)(R)CW/EW/SW...(Partially)(Reflected) Compression /Expansion /Shock Wave;
RPRSW Reflected PRSW;
(R)WCW/WEW...(Reflected) Weak CW/EW.

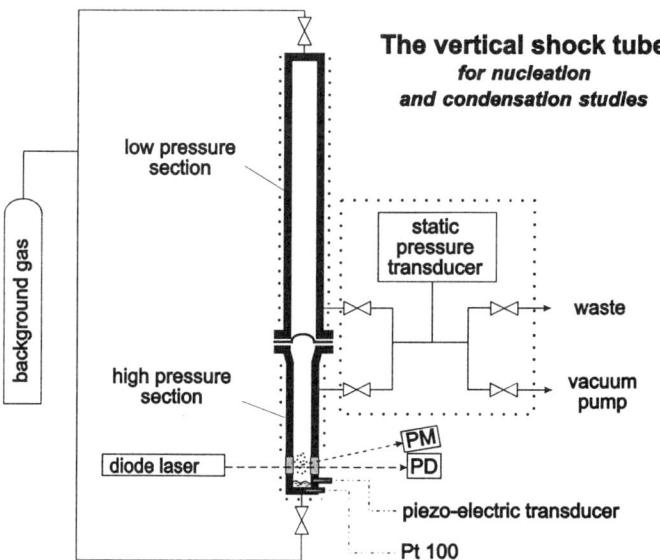

FIGURE 2. Scheme of the vertical shock tube developed for nucleation and condensation studies.

RESULTS

Figure 3 shows an example of the signals recorded for an experiment with the water-argon system. The pressure signal shows peaks stemming from the reflections of the shock wave on the LPS end. Although the temperature is only about 265 K, the peaks and valleys of the scattered light signal become shallow and smooth, showing that the droplet cloud is significantly polydisperse. In fact, the 90° scattering was not found particularly suitable for higher temperatures. The reason is that the Mie-peaks at this angle are narrow-spaced and the pattern is very sensitive to polydispersity.

Preliminary results show that rather high cooling rates and sharp falling edge of the nucleation pulse can be achieved with the present method, enabling a well-defined start of the nucleation process. Although the nucleation pulse can be shortened down to a few tenth of millisecond, it is necessary to count with a polydisperse cloud of droplets. Therefore, we are currently improving the optical measuring methods in order to overcome this problem.

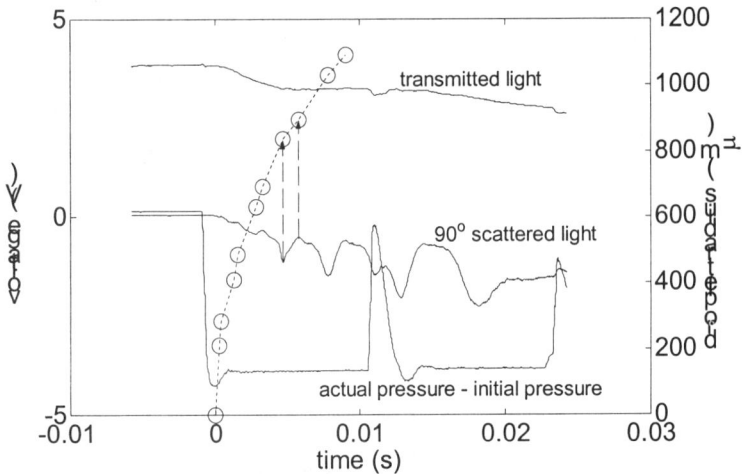

FIGURE 3. An example of pressure scattered and transmitted light signals in electrical units (water-argon, 265 K / 200 kPa nucleation temperature/pressure), combined with time-resolved mean radius of the droplets, evaluated using the Mie-Lorenz theory of light scattering. The first Mie-peak and first valley are almost unreadable because of the polydispersity.

ACKNOWLEDGMENTS

We acknowledge the support by grants GA AVČR A2076703, GAČR 101/00/1282 and S2076003.

REFERENCES

1. Schmitt, J. L., *Rev. Sci. Instrum.* **52**, 1749-1754 (1981).
2. Wagner, P. E., and Strey R., *J. Phys. Chem.* **85**, 2694-2698 (1981).
3. Strey, R., Wagner, P. E., and Vissanen Y., *J. Phys. Chem.* **98**, 7748-7758 (1994).
4. Hrubý, J., and Strey, R., *ibid.*
5. Peters, F., *Experiments in Fluids* **1**, 143-148 (1983).
6. Looijmans, K. N. H., and van Dongen, M. E. H., *Experiments in Fluids* **23**, 54-63 (1997).

Nucleation Driven by Time Dependent Diffusive Heat and Mass Transfer

Ian Ford[1], Jonathan Barrett[2] and Michael Mclean[1]

[1]*Department of Physics and Astronomy, University College London, Gower Street, London WC1E 6BT, United Kingdom.*
[2]*Department of Nuclear Science and Technology, HMS Sultan, Gosport, Hampshire, PO12 3BY, United Kingdom.*

Abstract. Nucleation of fresh aerosol particles in the atmosphere is driven by the prevailing vapour supersaturation and temperature, which often vary with time. We consider the diffusion of heat and vapour into a parcel of gas and the resulting potential for particle nucleation. We focus on the case where a nucleation burst occurs which is self-terminated by the condensation of vapour onto the particles created. Simple consideration of the timescales for the increase and decrease of the nucleation rate at various points within the parcel is sufficient to characterise the productivity of the burst, in agreement with more elaborate numerical simulations.

INTRODUCTION

The creation of new aerosol particles from a supersaturated, or metastable, vapour phase is responsible for maintaining the atmospheric particle number density, roughly balancing the removal rate by fallout or the rate of particle coagulation. It must be emphasized, however, that few situations in atmospheric physics are in a steady state, and that in practice it is believed that particle nucleation is intermittent, occurring in bursts at particular locations, and that thermodynamic conditions (namely the supersaturation and temperature) are often not suitable for new particle production.

This time dependence in a nucleation burst is due in part to the coupled dynamics involving the nucleated particles and the parent phase: condensation depletes the vapour and reduces the driving force for continued nucleation. A second factor is that temperatures and supersaturations are determined by heat and mass transfer processes in the atmosphere, and that these can also drive the nucleation in a time-dependent manner.

In this work, we first examine the different regimes of nucleation burst dynamics that result from a one-dimensional vapour subject to heat and mass flows from a boundary. We model the generation of particles numerically, and then for the vapour depletion regime, apply a simplified model of burst dynamics that successfully accounts for the features we observe. This model could be used in more general circumstances.

BOUNDARY LAYER HEAT AND MASS TRANSFER

The diffusion of heat and vapour into a space initially at a lower temperature and vapour concentration can lead to the development of a vapour supersaturation. Observers at an individual point in the space would perceive an increase followed by a decrease in the saturation ratio S. For suitable S and temperature T, droplets can nucleate from the vapour. There are then two regimes to consider, depending on whether the droplets created do or do not affect the global heat and mass transfer rates. If they do *not*, then the concentration of particles produced may be calculated simply from knowledge of the spatial and temporal pattern of S and T, which is obtained from the solution to two uncoupled heat and mass diffusion equations. But when condensation upon the particles, and the associated latent heat release, begin to affect the global heat and mass transfer, then more detailed calculations are necessary. This can readily be achieved with numerical modelling of various degrees of sophistication. Of particular interest, however, is whether simple models of particle yield can be developed by applying recent ideas of the dynamics of nucleation bursts.

ANALYTIC MODEL

The simple model we employ here has been developed by Barrett [1] to describe bursts of nucleation driven by vapour production or cooling. Our extension of this work is to apply it to a spatially inhomogeneous system. The ideas are very simple. We assume that a nucleation rate J is known as a function of T and S. If T and S are given as functions of time and space, then so, therefore, is J. Assuming no coupling between the global heat and mass transfer through condensation on the droplets, T is given by

$$\frac{\partial T}{\partial t} = \kappa \frac{\partial^2 T}{\partial x^2}$$

where κ is the thermal diffusivity, and x and t are space and time respectively. The vapour mole fraction in the gas mixture c is determined by a similar equation (involving the vapour diffusivity) and hence the supersaturation $S=c/c_s(T)$ is defined, where $c_s(T)$ is the equilibrium vapour mole fraction at temperature T. Thus for all positions, the nucleation rate J is defined as a function of time, as long as the heat and mass transfer are decoupled. We can work out the local value of $d\ln J/dt$, which we denote $1/t_R$, which defines a timescale $t_R(x,t)$ for the increase in nucleation rate by a factor of e.

For a system where the condensation strongly affects the heat and mass transfer, the equations that control the evolution of T and c are quite different. The temperature rises according to

$$\frac{\partial T}{\partial t} = \frac{L\dot{m}}{\rho c_p}$$

and the vapour concentration falls according to

$$\frac{\partial c}{\partial t} = -\frac{(1-c)\dot{m}}{\rho}$$

where L is the latent heat of condensation, $\dot{m}$ is the mass condensation rate, c_p is the mixture specific heat capacity and ρ is the vapour density. These equations describe a different variation of T and S in time and space, and hence a different field of nucleation rates $J(x,t)$. The decrease of J with time may be modelled by $d\ln J/dt = -1/t_C$, introducing a timescale for the decrease of the nucleation rate by a factor of e in this regime.

The timescales t_R and t_C are functions of time and space, and we seek to use them to identify the position and time of the nucleation burst. The peak of the nucleation burst will be characterised by the condition $t_R = t_C$, such that there is a balance between the forces driving the evolution of J. This condition will be met at position x at a time $t_N(x)$, and it remains to be decided where and when the burst actually takes place. But a reasonable additional requirement is to demand that the condition $t_N \geq t_R$ should hold as well as $t_R = t_C$. This would allow time for the development of the proposed burst. The location x_N of the burst is formally obtained from the equation $t_N = t_R(x_N, t_N) = t_C(x_N, t_N)$ These requirements are sketched in Figure 1.

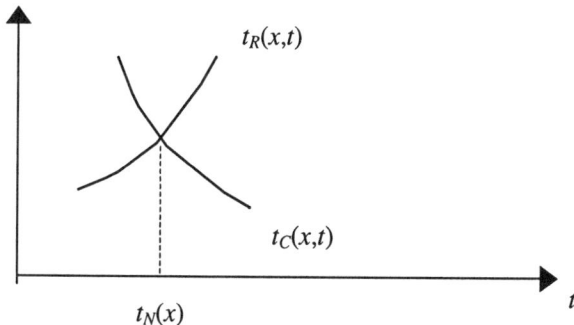

FIGURE 1. The variation in the timescales for the growth (t_R) and decay (t_C) of the nucleation rate at a particular position x, for various times t. The condition for the peak in the burst is $t_R = t_C$, and in addition we require the elapsed time t_N to be not less than these timescales.

Once the position and time of the burst are determined, the total number of particles generated in the burst can be evaluated, by integrating the nucleation rate over a period $2t_N$ for conditions appropriate to the position of the burst. We have explored the sensitivity in the model to the exact relationship between t_R and t_N and there seems to be little dependence.

COMPARISON WITH NUMERICAL TREATMENT

We have also performed numerical simulations of the nucleation burst [2], using a simplified treatment of the size distribution [3], and have confirmed that the approach we have used is well able to account for the number of particles generated in the burst, as well as the position and timing. Thus we have an analytic model of the nucleation of particles in a self-terminated burst driven by global diffusive heat and mass transfer. The requirements for the model are the fields of vapour concentration and temperature as functions of time, together with a model nucleation rate. We are now seeking to extend our work to treat diffusion-controlled bursts in cylindrical geometries, which might be applicable to particle production due to heat and mass transfer from hot wires into a surrounding environment.

CONCLUSIONS

The diffusion of heat and mass into a parcel of air can lead to the development of vapour supersaturation and hence new particle production. A simple burst would be driven passively by the evolution of the supersaturation and temperature due to the heat and mass diffusive currents. A more complex burst would result if the condensation on the nucleated droplets and the release of the latent heat of condensation competes with the diffusive currents. We have focussed on the latter process here, and have shown that simple considerations of the dynamics controlling the nucleation rate in various regimes can provide good estimates of the particles nucleated, given the initial conditions and a characterisation of the diffusive processes.

REFERENCES

1. Barrett, J.C., Timescales for nucleation and growth in supersaturated vapour-gas mixtures, *J. Aerosol Sci.* **31**, 51 (2000).
2. Ford, I.J., Barrett, J.C., and Mclean, M., unpublished work.
3. Warren D.R., and Seinfeld, J.H., Predictions of aerosol concentrations resulting from a burst of nucleation, *J. Coll. Int. Sci.* **105**, 136 (1985).

Semiempirical Cahn-Hilliard Theory of Vapor Condensation with Triple-Parabolic Free Energy

L. Gránásy,[*] Z. Jurek,[*] D. W. Oxtoby[◊]

[*]*Research Institute for Solid State Physics and Optics, H-1525 Budapest, P. O. B. 49, Hungary*
[◊]*James Franck Institute, The University of Chicago, 5640 S. Ellis Avenue, Chicago, Illinois 60637*

Abstract. A highly accurate analytical approximation for the density functional theory of vapor condensation is presented which is based on the square-gradient transcription proposed by Iwamatsu and Horii, and a triple-parabolic free energy – order parameter relationship.

INTRODUCTION

The density functional theory (DFT) of vapor condensation [1,2] takes the attractive part of the molecular interaction into account as a perturbation relative to a hard sphere contribution. The density profile for condensation droplets is found by searching for an unstable solution of the integral Euler equation that corresponds to the extremum of the free energy functional. In a semiempirical version of the DFT [3], the model parameters [inverse range of potential (λ), integrated interaction strength (α), and hard sphere radius (d)] are fixed so that the measured liquid density, equilibrium vapor pressure, and the surface tension are reproduced. This approach proved far superior to the classical theory in predicting homogeneous nucleation rates in non-polar and weakly polar vapors [3]. Simplification of the associated mathematics would reduce the difficulties in obtaining reliable estimates for the nucleation rates.

In a previous work, Iwamatsu and Horii have cast the DFT into the form of a square-gradient theory akin to that by Cahn and Hilliard, however, with the chemical potential of the HS fluid as the order parameter [4]. This reduces the mathematical task to the solution of an ordinary differential equation. Recently, the possibility of approximating such gradient theories via a triple-parabolic expansion of the free energy – order parameter relationship has been explored [5,6]. This approach proved a powerful tool in describing ice nucleation in undercooled water [6]. Here we use a similar approach to develop an accurate analytical approximation for condensation.

TRIPLE-PARABOLIC GRADIENT THEORY

In the gradient form of the DFT, the free energy of a spherical fluctuation reads as

$$W = \int_0^\infty 4\pi r^2 dr \left\{ \frac{1}{2} A \left(\frac{d\mu_h}{dr} \right)^2 + p_0 - V(\mu_h) \right\}, \quad (1)$$

where A is the coefficient of square-gradient term, the order parameter μ_h is the chemical potential of the HS fluid (related to the number density via the highly accurate Carnahan-Starling expression), p_0 the pressure of the supersaturated vapor, $-V$ is a function of μ_h [4] and have a double-well shape [Fig. 1(a)]. Being in an unstable equilibrium, the critical fluctuation corresponds to a saddle point of the functional W. Hence the $\mu_h(r)$ profile can be obtained by solving the Euler-Lagrange equation $\mu_h'' + (2/r)\mu_h' - \lambda^2 \Delta\mu = 0$, where $\Delta\mu$ is the local chemical potential difference relative to the supersaturated vapor [see Fig. 1(b)].

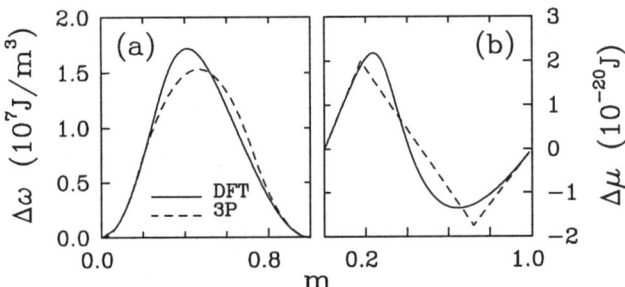

FIGURE 1. Equilibrium thermodynamics for nonane at $T = 315$ K in the density functional theory (DFT) and the triple-parabolic approximation (3P): (a) The grand potential density $\Delta\omega = p_0 - V$ and (b) the chemical potential $\Delta\mu$ of the homogeneous system relative to the saturated vapor as a function of the order parameter $m = (\mu_h - \mu_{h,v})/(\mu_{h,l} - \mu_{h,v})$, related to the hard sphere chemical potential μ_h.

Adopting the dimensionless order parameter $m = (\mu_h - \mu_{h,v})/(\mu_{h,l} - \mu_{h,v})$, where $\mu_{h,l}$ and $\mu_{h,v}$ are the HS chemical potentials for the equilibrium liquid and vapor phases, we approximate the grand potential density as $\omega = \frac{1}{2} \lambda_i (m - m_i)^2 - \beta m + \delta_{i,1} \omega_b$, where λ_i are the curvatures of the parabolas, m_i the positions of the parabola tips in equilibrium, β a parameter chosen so that the free energy difference between the minima corresponds to $\rho_L kT \ln(S_p)$, ρ_L the number density of the liquid, S_p the supersaturation. Here subscripts $i = 0, 2$ and 1 denote parabolas for the vapor and liquid minima and the maximum in between, respectively. $\delta_{i,j}$ is Kronecker's delta, while ω_b is the equilibrium height of the free energy barrier. With this notation, $\Delta\omega = \omega - \omega[m(r \to \infty)] = p_0 - V$. Introducing reduced quantities $M_i = m - m_i - \beta/\lambda_i$, the Euler-Lagrange equation transforms to

$$M_i'' + (2/r) M_i' \pm \Gamma_i^2 M_i = 0, \quad (2)$$

where $\Gamma_i = (\lambda_i/2c)^{1/2}$ for $i = 0$ and 2, $\Gamma_1 = (|\lambda_1|/2c)^{1/2}$, $c = A(\mu_{h,l} - \mu_{h,v})^2/2$ while the negative sign of the third term applies for $i = 0$ and 2, and the positive for $i = 1$.

This transcription enables us to develop *analytic* solutions for the order parameter distribution, and the free energy of critical fluctuations along the lines described in [6]. The semiempirical version has been implemented as follows. While α, and d are determined as in [3], in the gradient formalism λ is related to the surface tension via a simple integral handled numerically. Once these parameters are defined, $\Delta\mu(\mu_h)$ and A can be readily evaluated. [A is related to the surface tension via the planar version of Equation (1).] We use here the exact value of A. The curvatures λ_0 and λ_2 are fixed by $d\mu/d\mu_h$ in the vapor and liquid minima of the grand potential density in equilibrium, while λ_1 is chosen so that the experimental surface tension is reproduced. The triple-parabolic approximant of $\Delta\mu(\mu_h)$ is compared with the exact relationship in Fig. 1(b). For small deviations from equilibrium, the shape of the free energy wells and the respective portions of the chemical potential curve are reproduced accurately. Deviation from the exact (numerical) solution is only expected at large supersaturations (near the spinodal). Since the condensation experiments are usually performed well away from the spinodal point, a fairly accurate reproduction of the DFT results is expected under typical conditions.

RESULTS AND DISCUSSION

The close agreement between the DFT and its analytic triple-parabolic approximant (3P) is demonstrated in Figs. 2 and 3 that compare the respective radial density profiles and nucleation rates. It appears that below a critical supersaturation, the 3P model reproduces the density profiles of the DFT with a high accuracy (almost exactly), although, it fails in the spinodal region. The latter follows from the fact that the spinodal point of the 3P model is somewhat below that of the density functional theory, yielding a faster reduction of the nucleation barrier with supersaturation than in the DFT. Nevertheless, this region is usually out of the reach of experiments. Even for the high nucleation rates (low nucleation barriers) achieved in expansion cloud chambers, the nucleation rates predicted by the triple-parabolic approximant are very close to those from the exact theory. Similar results were obtained for mercury, dodecane, hexadecane, octadecane, and dibutyl phtalate. In the light of these, we believe that the triple-parabolic approximation can be safely applied for estimating the nucleation rate and density profile for non-polar and weakly polar substances.

Work is under way to extend this analysis to the condensation of further substances.

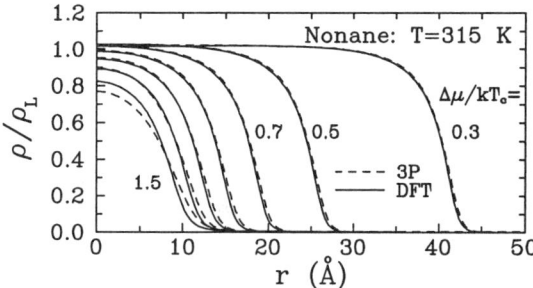

FIGURE 2. Comparison of radial density profiles predicted by the density functional theory (DFT) and the present analytic approximation (3P). The calculations were performed using the physical properties of nonane [7]. Note the close agreement between the predictions up to $\Delta\mu/kT_c = 1.3$, where T_c is the critical temperature.

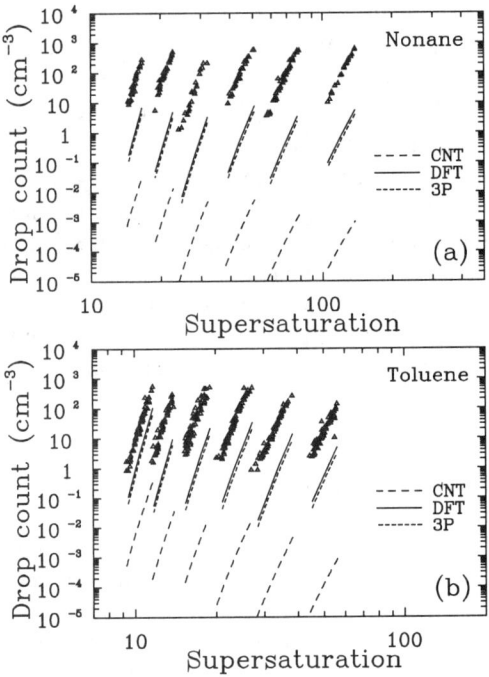

FIGURE 3. Drop counts (a quantity proportional with the nucleation rate) predicted by the density functional theory (DFT) and the triple-parabolic approximation (3P) for (a) nonane and (b) toluene. For comparison, experimental results from expansion cloud chamber measurements (triangles; nonane [8], toluene [9]) and predictions of the classical nucleation theory (CNT) are also presented. The data sets from left to right correspond to pre-expansion temperatures 45, 35, 25, 15, 5, and −5 °C, respectively. Note the similarity of the results of the DFT and the analytical approximation.

ACKNOWLEDGMENTS

This work was supported by the National Scientific Research Fund (Hungary, Grant No. OTKA-T-025139) and the National Science Foundation (Grant No. CHE-9800074).

REFERENCES

1. Oxtoby, D. W., and Evans, R., *J. Chem. Phys.* **89**, 7521-7530 (1988).
2. Zeng, X. C., and Oxtoby, D. W., *J. Chem. Phys.* **94**, 4472-4478 (1991).
3. Nyquist, R. M., Talanquer, V., and Oxtoby, D. W., *J. Chem. Phys.* **103**, 1175-1179 (1995).
4. Iwamatsu, M., and Horii, K., *J. Non-Cryst. Solids.* **205-207**, 919-923 (1996).
5. Barett, J. C., *J. Phys.: Condens. Matter* **9**, L19-L26 (1997).
6. Gránásy, L., and Oxtoby, D. W., *J. Chem. Phys.* **112**, 2399-2409 (2000).
7. Hung, C.-H., Krasnopoler, M. J., and Katz, J. L., *J. Chem. Phys.* **90**, 1856-1865 (1989).
8. Adams, G. W., Schmitt, J. L., and Zalabsky, R. A., *J. Chem. Phys.* **81**, 5074-5078 (1984).
9. Schmitt, J. L., Zalabsky, R. A., and Adams, G. W., *J. Chem. Phys.* **79**, 4496-4501 (1983).

Homogeneous Nucleation of *n*-pentanol and Droplet Growth: A Quantitative Comparison of Experiment and Theory

Thorsten Biet and Reinhard Strey

Institut für Physikalische Chemie, Universität zu Köln,
Luxemburger Str. 116, D-50939 Köln, Germany

Abstract. Homogeneous nucleation of *n*-pentanol droplets and their growth was measured. Quantitative comparison between the measured experimental nucleation rates and classical nucleation theory yields a disparate temperature dependence as previously reported for other systems. Also, the experimental rates are always much higher (within the temperature range studied) than theory predicts. Droplet growth calculations according to the *Fuchs* and *Sutugin* approach were found to accurately describe the experimental growth patterns without adjustable parameters..

I. INTRODUCTION

Nucleation theories easily predict the variation of the homogeneous nucleation rate to vary from 10^{-5} to 10^{25} cm^{-3}s^{-1} as function of supersaturation S at constant temperature T.

The experimental situation is more difficult. Nevertheless, in the past decades experiments have been developed to test the theoretical predictions. Although most devices were not able to provide nucleation rates as function of supersaturation at constant temperature, still S and T may be specified when a certain (approximate) nucleation rate J is observed. For instance, Volume and Flood [1] in 1934 studying water found $S = 4.2$ at $T = 275$ K for $J = 10^3$ cm^{-3}s^{-1}; or Wegener and Wu [2] in 1977 report $S = 14$ and $T = 213$ K to yield $J = 2 \cdot 10^{19}$ cm^{-3}s^{-1} in a supersonic nozzle for ethanol.

In recent years a number of experiments have been developed to measure nucleation *rates* over several orders of magnitude [3]. For instance, the diffusions cloud chamber can be used to obtain rates from 10^{-5} to 10^{-2} cm^{-3}s^{-1}, the piston expansion chamber (10^1 to 10^5 cm^{-3}s^{-1}), the nucleation pulse (two-valve) chamber (10^5 to 10^9 cm^{-3}s^{-1}) and the pex-tube (10^7 to 10^{10} cm^{-3}s^{-1}). New development of a pulse version of the super sonic nozzle will add nucleation rates in the range 10^{17} to 10^{19} cm^{-3}s^{-1}. In all these devices temperature and from this supersaturation is calculated not measured. Therefore it was decided a while ago to use a single common substance, *n*-

pentanol to measure the nucleation rates at prescribed 240, 260, 280 K in order to be able to compare the various techniques.

II. RESULTS

We report here the homogeneous nucleation rates for n-pentanol and the growth patterns of the droplets formed as obtain using the nucleation pulse chamber [4].

Homogeneous Nucleation

In Fig. 1 the nucleation rates of n-pentanol in argon as carrier gas are shown. As can be seen steep series of data points are obtained. The higher T the steeper the curves and the lower S. The experimental window is limited towards high temperature by too fast growth of the droplets and towards low temperatures by the low total vapor pressure.

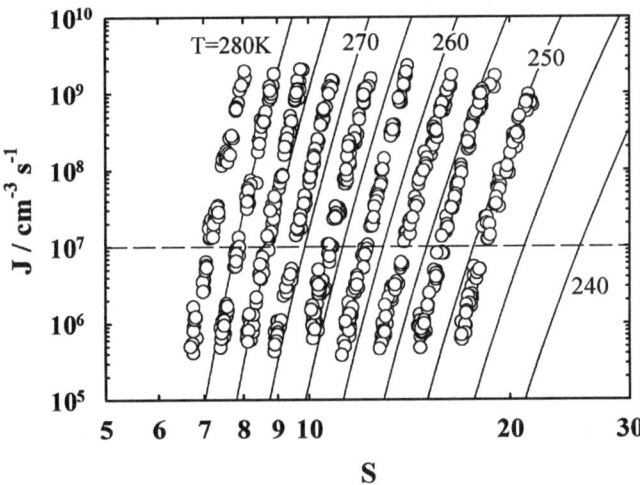

FIGURE 1. Experimental nucleation rates vs. supersaturation (circles) from T=280K to 240K in 5K steps. The full lines are calculated according to the classical nucleation theory of Becker-Döring. Note that experimental rates are higher by 2 orders of magnitude at T=280K and 4 orders at 240K.

For comparison the nucleation rate predicted by classical nucleation theory of Becker and Döring is shown as full lines in Fig. 1. The predicted rates are lower by a factor 10^4 at low T and by 10^2 at high T. Accordingly, a temperature-dependent correction to the classical theory is needed. The same conclusion had previously been reached for the homologous series of alcohols [5].

In Fig. 2 we compare the onset supersaturations S_0 at $J = 10^7$ cm^{-3}s^{-1} (c.f. Fig. 1, dashed line) to previous measurements using the same chamber. As can be seen the agreement is fair. It should be noted that the present experimental procedure includes calibration of the piezo-electric pressure transducer immediately before and after the measurements.

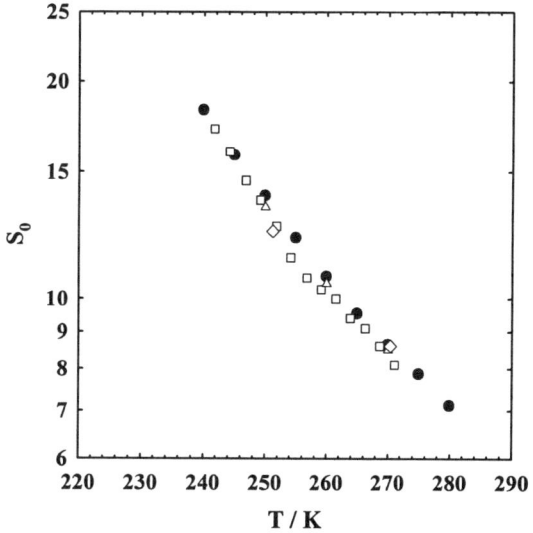

FIGURE 2. Experimental onset supersaturations S_0 (full circles) from T=240K to 280K in 5K steps. For comparison the results of previous measurements are shown. Open diamonds: Strey et al. [5]. Open squares: Hruby et al. [6]. Open triangles: van Remoortere et al. [7].

Droplet growth

In Fig. 3 the growth of the droplets homogeneously nucleated is shown. As detailed in previous publications the nucleation pulse chamber detects the growing droplets by CAMS. Wagner has developed this technique into a practical tool for measuring the radius as function of time [8,9]. To this end the experimentally discernible maxima in the scattered light vs. time recordings are identified with the corresponding maxima of theoretical Mie-calculations for the scattered intensity as function of radius. It should be mentioned that *n*-pentanol is a particularly suited substance for growth measurements. Due to the smooth and slow growth high resolution scattering curves are obtained sometimes permitting a wealth of maxima being identified leading to a high data density seen on some of the curves in Fig. 3. For comparison the growth functions predicted by the Fuchs-Sutugin [10] formalism is shown. As can be seen the agreement is quite satisfactory.

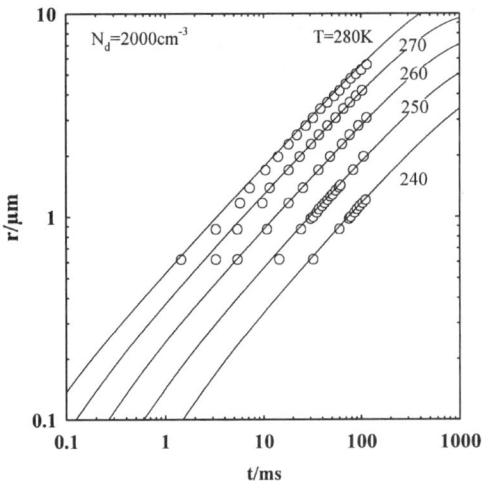

FIGURE 3. Experimentally determined radii of n-pentanol droplets as function of time (circles) from T=240K to 280K. The full lines are calculated according to *Fuchs* and *Sutugin* [10].

III. CONCLUSIONS

We examined nucleation and growth of n-pentanol droplets from the vapor phase. The classical nucleation theory shows a disparate temperature dependence. The approach of *Fuchs* and *Sutugin* is suitable for describing the growth of homogeneously nucleated n-pentanol droplets. It will be interesting to compare the results of this study to those by other techniques. This will be done elsewhere.

REFERENCES

1. Volmer, M. "Kinetik der Phasenbildung", Steinkopff, Leipzig (1939)
2. Wegener, P.P., and Wu, B.J.C. in "Nucleation Phenomena", Zettelmoyer, A.C., ed., Elsevier Publ., , p.325 (1977)
3. for review see: Heist, R.H. and He, H. J. Phys. Chem. Ref. Data 23, 781 (1994).
4. Strey, R., Wagner, P. E., and Viisanen, Y., *J. Phys. Chem.* **98**, 7748 (1994).
5. Strey, R., Wagner, P. E., and Schmeling, T. *J. Chem. Phys.* **84**, 2325 (1986).
6. Hruby, J., Viisanen, Y., and Strey, R., *J. Chem. Phys.* **104**, 5181 (1996).
7. van Remoortere, P., Heath, C., Wagner, P.E., and Strey, R., Proceedings of the 4th International Conference on Nucleation and Atmospheric Aerosols, Helsinki (1996)
8. Wagner, P. E., *J. Colloid Interface Sci.* **44**, 181 (1973).
9. Wagner, P. E., "Aerosol Growth by Condensation" in *Topics in Current Physics* **29**: *Aerosol Microphysics II*, edited by W. H. Marlow, Springer-Verlag, Berlin, 1982, pp. 129.
10. Fuchs, N. A., and Sutugin, A. G., *Highly Dispersed Aerosols*, Ann Arbor Science Publications, Ann Arbor (1970).

MD Simulation of Heterogeneous Vapor Nucleation on a Solid Surface

K. T. Kholmurodov[a,+], K. Yasuoka[b] and X. C. Zeng[c]

[a]*Computational Science Division, Advanced Computing Center, The Institute of Physical and Chemical Research (RIKEN), Hirosawa 2-1, Wako, Saitama, 351-0198, Japan tel:+81-48-467-9415, fax:+81-48-467-4078, e-mail: mirzo@atlas.riken.go.jp*
[b]*Department of Mechanical Engineering, Keio University, Yokohama 223-8522, Japan*
[c]*Department of Chemistry, University of Nebraska-Lincoln, Lincoln, NE 68588, USA*
[+]*Permanent address: Laboratory of Computing Techniques and Automation, Joint Institute for Nuclear Research, Dubna, Moscow region, 141980, Russia*

Abstract. Molecular dynamics simulations of nucleation of a supersaturated Lennard-Jones vapor confined between parallel solid walls have been performed. The walls in the system, unlike the structure-less one [K.Yasuoka, G. T. Gao, and X. C. Zeng, J. Chem. Phys. 112, pp.4279-4285 (2000)], are simulated in a more realistic fashion. The particles in the vapor and the solid walls are exactly the same and are described by the same Lennard-Jones potential. The walls are fixed in place in an fcc lattice structure by a combination of the restoring tethering forces and constraint mechanism. The restoring potential confines wall atoms to their lattice sites at the equilibrium state. The walls during simulations are thermostatted, so the heat being dissipated through conduction to the walls, as it is in an experiment [S. T. Cui, P. T. Cummings, and H. D. Cochran, J. Chem. Phys. 111, pp.1273-1280 (1999)]. Condensation of a supersaturated vapor on a solid surface is investigated with the increasing strength of the wall-vapor interaction. The wall surfaces are the same (100) surface, but with the different strengths of the attraction of vapor atoms: strongly adsorbing wall (with $\varepsilon_1=\varepsilon$) and weakly adsorbing wall (with $\varepsilon_2=0.1\varepsilon$). Transition from a metastable supersaturated vapor phase to a stable condensed state is studied by monitoring the molecular dynamics configurations of the system in real time and identifying the short-lived droplet nucleus.

HETEROGENEOUS VAPOR NUCLEATION ON A SOLID SURFACE

In the last years there has been a growing interest for the investigation of fluids confined between two surfaces, which are separated few nanometers. First of all, we note importance of the understanding of physical properties of confined fluids in nanotribology. The interest is related here to both industrial applications and fundamental research study. It is well established for a number of molecular systems that the properties of confined fluids differ from those of the bulk. A fluid, confined within two solid walls of a very small spaces of a few molecular diameters, emerges behavior from continuum to molecular one. Both experimentally and theoretically it is found that a confined fluid shows a strong anisotropy of many dynamical properties (molecular ordering, oscillatory and non-uniform layered density distribution, liquid to solid-like phase transition, non-trivial diffusion properties, etc.) [1-5]. One of the striking features here is a large anisotropy of the atomic diffusion, which contrasts with the modification of the effective activation barriers due to the nanopores. This fundamental behavior should also be taken into account for the surface diffusion of gases, which are confined between solid walls. The surface diffusion of adsorbed molecules plays an important role in the nucleus formation [6]. Understanding the properties of the vapor and gases in nanopores (in the gas-liquid transition region) is very important, for example, in materials science, in designing of porous materials, for many physical, chemical, and biological cleaning processes.

In this work we are mainly interested in the nucleation of a supersaturated vapor, which is confined between two solid walls. Supersaturated vapor represents itself as an essentially metastable phase, and its dynamics in the confined area may strongly be affected by the presence of solid walls. Transitions from a metastable supersaturated vapor, into thermodynamically stable state, such a liquid, goes through a droplet nucleation, where the cluster critical size defines the dynamics of the system in the coexisting region. Capillary condensation is a gas-liquid phase transition shifted by confinement. Capillary condensation is one of the processes (viz., condensation and evaporation, wetting and drying, etc.), showing a huge metastability of the coexisting phases of the liquid-vapor transition. Nucleation, the precursor of the first-order phase transition, is also expected to be strongly influenced by the presence of the walls. Because nucleation is a thermally activated process, the rate of nucleation depends exponentially on the height of the barrier characterized by the formation of free energy of the critical nucleus. The latter is very sensitive to the small changes in environment, such a nanoscale confinement by two planar walls [1-7].

In our previous work [6], through the molecular dynamics simulations, the mechanism of nucleus formation inside the nanopores has been investigated as we varied the strength of attraction between the wall and the vapor molecules. It was shown that this attraction strongly affects the process of nucleus formation: if the attraction is weak (a drying wall), nuclei tend to form in the middle of the pore, whereas if the attraction is strong (a wetting wall), the nucleus formation originates from two sources, the surface diffusion of adsorbed molecules and deposition of clusters formed in the middle of the pore. The aim of this work is to study the nucleation and condensation of Lennard-Jones (LJ) supersaturated vapor, confined between two planar solid walls. The walls in the system, unlike the structure-less one [6], are simulated in a more realistic fashion. The walls are fixed in place in an fcc lattice structure by a combination of the restoring tethering forces and constraint mechanism. The wall surfaces are the same (100) surface, but with the different strengths of the attraction of vapor atoms: strongly adsorbing wall and weakly adsorbing wall. Condensation of a supersaturated vapor on a solid surface is investigated with the increasing strength of the wall-vapor interaction.

We have performed MD simulations on the Fujitsu VPP700 vector computer using DL_POLY_2.11 code [8] to study vapor condensations on a solid surface. (The package was not designed specifically to run on vector machines, so the efficiency of the code on the vector computer have preliminarily been enhanced. Some optimizations of DL_POLY on a vector computer, providing significant enhancements in performance, have been reported previously [9]).

The vapor mixture consists of 1024 target particles and 1024 carrier gas particles. The carrier gas is used here to avoid unnatural energy exchanges and to release the latent heat generated from cluster formation. The target-target potential is a Lennard-Jones type, the potential between target-carrier and carrier-carrier is a soft-core (no attraction), and the potential parameters for the argon we have considered [10]. The vapor phase is separately equilibrated at $T = 180$ K using leapfrog algorithm for numerical integration of classical equations of motion. The box size is 18×18×18 for the vapor mixture, so the number density of target particles at initial state is the same as that in homogeneous vapor [10]. The time step is chosen to be 5 fs and cutoff radius between particles is 4.5 in LJ-distance parameter unit.

The solid walls are four layers of a face-centered cubic (fcc) lattice of atoms of the same LJ potential parameters as those are for the vapor phase. These four layers are equally divided into two parts: wall 1 and wall 2, each consisting of two layers (800 atoms per layer). We have used the same technique as it is used in works [11] to simulate the solid walls. The walls are fixed in place in an fcc lattice structure by a combination of the two mechanisms: (1) tethering (harmonic spring) forces; (2) constraint mechanism. The walls were separately equilibrated at $T = 80$ K, using slab boundary conditions in plain (x,y) (there is no periodic boundary condition in z-direction perpendicular walls).

After separately equilibrating, the supersaturated vapor and the solid walls are joined to evolute together. The vapor equilibrated at $T = 180$ K is quenched to a state at $T = 80$ K and then it was positioned at a well-separated distance on a solid wall. The periodic boundary conditions are applied in all three directions, so one of the walls is simply, as the periodic image, plays a role of a second confining one. The whole system is thermalized at setting temperature for about 100 steps. In Fig.1 four sequential MD configurations of the evolution of the system are displayed. The first snapshot corresponds to the initial state, when the vapor and the solid walls are joined together. The following three configurations of Fig.1 show us the dynamics of the system for the moments of $t = N\tau$, $3N\tau$, and $4N\tau$, respectively ($\tau = 5$ fs is time step; $N = 50000$ is the number of steps). As mentioned above the solid walls during simulations are thermostatted, so the heat being dissipated through conduction to the walls, as it is in an experiment [3]. It is worth to note that the heat, being pumped away from the system, includes the latent heat generated by nucleus formation too. The latent heat in the simulations is mainly transferred to the environment via the carrier gas, rather than target particles. The particles of the carrier gas cannot exist inside the nucleus. We employed a thermostat [10] to let carrier gas connect to a hypothetical heat bath, so the temperature of carrier is kept constant.

By monitoring the MD configurations of the system in real time and identifying the short-lived droplet nucleus, transition from a metastable supersaturated vapor phase to a condensed state is studied. The rate of the transition (the number of droplets appearing per unit time and per volume from a supersaturated vapor) is calculated. The nucleation rate allows one to estimate the dynamics of the clusters, affected by the presence of the walls. The vapor condensation is mainly affected by the strongly adsorbing wall 1, and the local density near wall 1 is much higher than the overall mean density. The monitoring of the molecular dynamics configurations showed a heterogeneous nature of the nucleation process near the strongly adsorbing wall.

A detailed analysis of the mechanism of heterogeneous nucleation, the peculiarities of the cluster surface diffusion and its contribution to the heterogeneous nucleation process is under way.

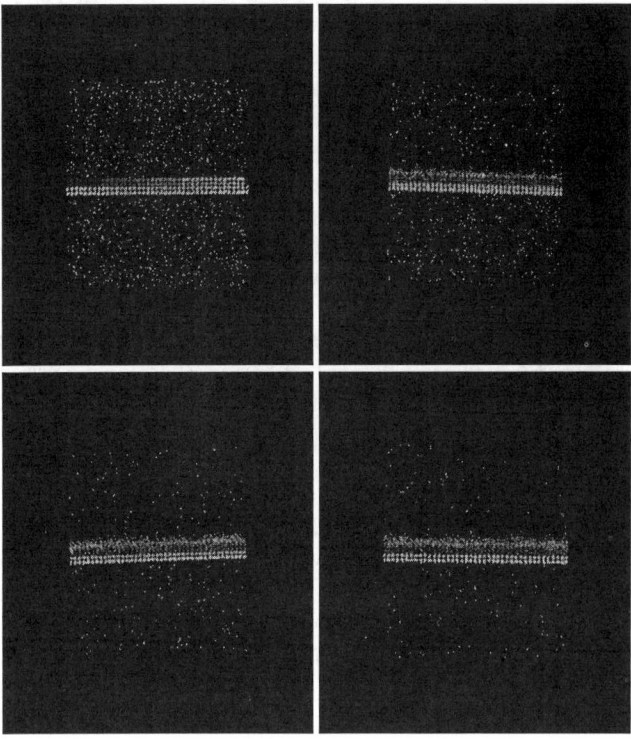

FIGURE 1. Four sequential snapshots of the MD simulation. The carrier gas particles are not shown. The periodic boundary conditions are used in all three dimensions.

REFERENCES

1. Bhushan, B., Israelachvili, J. N., and Landman, U., Nature (London) 374, 607 (1995).
2. Gee, M. L., McGuiggan, P. M., Israelachvili, J. N., and Homola, A. M., J. Chem. Phys. 93, 1895 (1990).
3. Cui, S. T., Cummings, P. T., and Cochran, H. D., J. Chem. Phys. 111, 1273 (1999).
4. Restagno, F., Bocquet, L., and Biben, T., Phys. Rev. Lett., v. 84, No. 11, 2000, pp. 2433-2436.
5. Oxtoby, W., Acc. Chem. Res. 31, 91 (1998).
6. Yasuoka, K., Gao, G. T., and Zeng, X. C., J. Chem. Phys. 112, 4279 (2000).
7. Knott, M., Vehkamaki, H., and Ford, I. J., J. Chem. Phys. 112, 5393 (2000).
8. Smith, W., Forester, T. R., Molecular Graphics, 14, 136 (1996).
9. Kholmurodov, K., Smith, W., Yasuoka, K., and Ebisuzaki, T., Comp. Phys. Comm.,125, 2000, pp. 167-192.
10. Yasuoka, K., and Matsumoto, M., J. Chem. Phys. 109, 8451 (1998).
11. Todd, B. D., Evans, D. J., and Davis, P. J., Phys. Rev. E, v. 52, No. 2, 1995, pp. 1627-1639.

Metastable Phase Decay on a Wide Spectrum of the Heterogeneous Centers Activities

Victor Kurasov

St.Petersburg State University
Department of Physics

One of the typical examples of the system evolution in the first order phase transition is the metastable phase decay in the closed system. After the instantaneous creation of initial supersaturation Φ the external influence doesn't take place and the evolution of the supersaturation ζ in the system occurs only due to the nucleation kinetics.

The supercritical embryos of a new liquid phase, i.e. the droplets, appear mainly on heterogeneous centers inevitably existing in any real system. The number of heterogeneous centers slightly changes in time, but their activation barriers can attain rather arbitrary values. Then one can speak about the distribution η_{tot} of the heterogeneous center numbers over heights ΔF of activation barriers, i.e over activities w of heterogeneous centers.

It is natural to suppose that the mentioned distribution is rather smooth and covers some region of the activation barriers heights.

The subject of current publication is to give the simplest variant for the nucleation description (the full theory is given in [1]). The property of activity spectrum to be the wide one will be seen a few moments later. We consider the homogeneous system of a unit volume. The regime of the droplets growth is supposed to be the free molecular one, the process will be an isothermal one.

The rate of nucleation I is the intensity of the droplet formation. To characterize the droplet with ν molecules it is covenient to introduce the linear size $\rho = \nu^{1/3}$. The rate of ρ growth is ζ/τ with some characteristic constant τ.

One can easily extract the main dependencies in the nucleation rate I behavior over the supersaturation ζ and over the free (unoccupied by droplets) number η of the heterogeneous centers: $I \propto \exp(-\Delta F)\eta$, where ΔF is the height of the activation barrier taken in thermal units.

The end of nucleation corresponds to the relatively small fall of supersaturation from initial value Φ. Then one can linearize ΔF as a function of ζ at the nucleation period and come to the following approximation:

$I(\zeta,\eta) = I(\Phi,\eta_{tot})(\eta/\eta_{tot})\exp(\Gamma(\zeta-\Phi)/\Phi)$ where $\Gamma = -\Phi d\Delta F/d\zeta \mid_{\zeta=\Phi}$.

One can also see that during the nucleation the rate of droplet growth ζ/τ is approximately constant in time Φ/τ.

To get the behavior of the supersaturation one has to know $\zeta(t)$. It can be gotten by the balance equation $\dot\zeta = \Phi - G/n_\infty$, where G is the number of the molecules in the liquid phase and n_∞ is the number of the molecules in the saturated vapor.

For G one can get $G = \int dw\, g(w)$, where the integral is taken over the whole spectrum of activities and $g(w)$ is the number of molecules in the droplets on the centers with activity w. The last value is equal to

$$g(w) = \int_0^t (t-t')^3 (\frac{\Phi}{\tau})^3 I(t',w)\,dt'$$

Having mentioned that the main dependence of I over the activity w is $\propto \exp(-w)$ one can come to

$$G = \int dw \int_0^t (t-t')^3 \exp(-w+w_*)\frac{\eta(w,t')}{\eta_{tot}(w_*)}(\frac{\Phi}{\tau})^3 I(w_*,\Phi,\eta_{tot}(w_*))\exp(\Gamma\frac{\varsigma(t')-\Phi}{\Phi})dt'$$

with some parameter w_*.

In the last equation there exists $\eta(w_*,t')$. It can be given by the following expression

$$\eta(w,t) = \eta_{tot}(w)\exp(-\int_0^t \exp(-w+w_*)\frac{I(w_*,\Phi,\eta_{tot}(w_*))}{\eta_{tot}(w_*)}\exp\Gamma\frac{\varsigma(t')-\Phi}{\Phi})dt'$$

The system of condensation equations is closed now.

The choice of w_* is rather arbitrary. From the last equation one can see that if w_* is chosen to satisfy condition $\eta(w_*,t=\infty) = \eta_{tot}(w_*)/2$ then all centers with $w > w_* +1$ remain practically unexhausted and all centers with $w < w_* -1$ are practically exhausted.

So, the interesting region $[w_* -1, w_* +1]$ is rather narrow. One can put in $[w_* -1, w_* +1]$ the value $\eta_{tot}(w)$ to $\eta_{tot}(w_*)$.

Distribution $\eta_{tot}(w)$ has to be limited from below to ensure the finite value of integral nucleation rate. Denote by w_0 the minimum of w. If $w_* - w_0 \geq 3$ then one can say that the relatively "wide spectrum" of activities takes place in nucleation. In the opposite situation one can consider nucleation as taking place on the similar heterogeneous centers.

Consider the situation of wide spectrum. We need to know the behavior of supersaturation at the end of nucleation when the fall of supersaturation stops the nucleation. All centers with $w > w_+ \equiv w_* +2$ will be exhausted until the first quarter of the nucleation period duration. The droplets formed on these centers can be regarded as formed at the initial moment of time. Then one can write for G:

$G = G_> + G_<$

$$G_> = \int_{-\infty}^{w_+} \eta_{tot}(w)\,dw\, t^3 (\frac{\Phi}{\tau})^3$$

$$G_< = \int_{w_+}^{\infty} dw \int_0^t dt'(t-t')^3 \exp(-w+w_*)\frac{\eta(w,t')}{\eta_{tot}(w_*)}(\frac{\Phi}{\tau})^3 I(w_*,\Phi,\eta_{tot})\exp(\Gamma\frac{\varsigma(t')-\Phi}{\Phi})$$

The second term is negligible in comparison with the first one. Then $G = G_>$ and the behavior of supersaturation is determined as $\zeta = \Phi - G_>$.

The last step to do is to determine w_*. For w_* we have

$$\frac{1}{2} = \exp(-\int_0^\infty \frac{I(w_*, \Phi, \eta_{tot}(w_*))}{\eta_{tot}(w_*)} \exp(\Gamma \frac{\zeta - \Phi}{\Phi}) dt')$$

or

$$\frac{1}{2} = \exp(-\int_0^\infty \frac{I(w_*, \Phi, \eta_{tot}(w_*))}{\eta_{tot}(w_*)} \exp(-\frac{\Gamma}{\Phi n_\infty} \int_{-\infty}^{w_*} dw \eta_{tot}(w) t'^3 (\frac{\Phi}{\tau})^3) dt')$$

Having calculated the integral one can come to

$$\frac{1}{2} = \exp(-\frac{I(w_*, \Phi, \eta_{tot}(w_*))}{\eta_{tot}} [\frac{\Gamma}{n_\infty \Phi} \int_{-\infty}^{w_*} dw \eta_{tot}(w) (\frac{\Phi}{\tau})^3]^{-1/3} 0.9)$$

The last equation together with an explicit expression for I given by the classical theory of nucleation is an ordinary algebraic equation.

The number $N_\infty(w)$ of the droplets formed on the centers with activity w can be calculated

$$N_\infty(w) = \eta_{tot}(w)(1 - \exp(-\exp(-w + w_*)) \frac{I(w_*, \Phi, \eta_{tot}(w_*))}{\eta_{tot}(w_*)} [\frac{\Gamma}{\Phi n_\infty} \int_{-\infty}^{w_*} \eta_{tot} dw]^{-1/3} \frac{\tau}{\Phi} 0.9))$$

The total number of droplets can be calculated by the integration over all w.

Consider the opposite situation. The spectrum of activities participating in the nucleation process is so narrow the one can consider all centers as the similar ones. Then the system of equations will be the following

$$\eta(t) = \eta_{tot} \exp(-\frac{I(\Phi, \eta_{tot})}{\eta_{tot}} \int_0^t dt' \exp(-\frac{\Gamma g(t')}{n_\infty \Phi})$$

$$g(t) = I(\Phi, \eta_{tot}) \int_0^t dt'(t-t')^3 (\frac{\Phi}{\tau})^3 \exp(-\frac{\Gamma g(t')}{n_\infty \Phi}) \frac{\eta}{\eta_{tot}}$$

One can use iterations defined as $\eta_{i+1} = Q(g_i)$, $g_{i+1} = P(g_i, \eta_i)$ with initial approximations $g_0 = 0, \eta_0 = 0$. Here P, Q define the r.h.s. of the two previous equations. The second iteration gives already suitable results. Then one can get with a relatively high precision for the total number of droplets

$$N_\infty = \eta_{tot}[1 - \exp(-0.9 \frac{I(\Phi, \eta_{tot})^{3/4}}{\eta_{tot}} (\frac{4 n_\infty \tau^3}{\Gamma \Phi^2})^{1/4})]$$

The last relation solves the problem.

REFERENCES

1. Kurasov, V., *Universality in kinetics of the first order phase transitions*, St.Petersburg: Chemistry Research Institute of St.Petersburg University, 1997, pp.300-350.

Effective Surface Tension for Small Binary Clusters by Monte Carlo Simulation

J. Kiefer[1] and B. N. Hale[2]

[1]Physics Dept., St. Bonaventure University
St. Bonaventure, NY 14778 USA
[2]Physics Dept. and C.A.S.L., University of Missouri-Rolla
Rolla, MO 65401 USA

Abstract. Configurational Helmholtz free energy differences between n and $n-1$ Lennard-Jones unary and binary clusters are calculated using the Bennett-Metropolis Monte Carlo technique. When plotted versus $n^{-1/3}$ the slope of the energy differences yields an effective surface tension, σ. When plotted in a scaled form, the slopes for both the unary and binary clusters display the same excess entropy/k_B per atom, Ω, as well as the same $\ln(\rho_{liquid}/\rho_{vapor})$. Clusters as small as 5 atoms show bulk surface free energy properties.

INTRODUCTION

The present work focuses on free energy differences between neighboring sized clusters of atoms and on the scaling of these free energy differences with a universal temperature dependence, $(T_c/T - 1)$. Previous work has addressed unary Argon clusters and water clusters.[1-4] The intent here is to test the feasibility of applying the same scaling to binary clusters.

Theory

Previously, it has been shown that the difference in free energy between the n cluster and the $n-1$ cluster can be written as[4]

$$-\delta F_c(n) \equiv \ln\left(\frac{Q(n)}{Q(n-1)Q_n(1)}\right), \qquad (1)$$

where $Q(n)$ is the unitless configurational integral for a cluster of n atoms and $Q_n(1)$ that for n monomers. In the context of the classical liquid drop model for nucleation,[5]

$$-\delta F_c(n) \approx \Gamma'_0 - \frac{2}{3}An^{-\frac{1}{3}} \qquad (2)$$

where the A is related to the liquid surface tension. The $\Gamma'_0 = \ln(\rho_{liquid}/\rho_{vapor})$. As the $\delta F_c(n)$ can be evaluated through a Bennett-Metropolis Monte Carlo[6,7] simulation of clusters of differing sizes, it is possible to estimate an effective surface tension for small n.

It has been proposed that A be rewritten in a scaled form[1]

$$A = (36\pi)^{\frac{1}{3}} \Omega \left(\frac{T_c}{T} - 1 \right), \qquad (3)$$

where T_c is the critical temperature, and T is the cluster temperature. The Ω is the excess surface entropy/k_B per atom. When $-\delta F_c(n)$ is plotted versus $n^{-1/3}$, for n greater than or equal to about 5, a straight line is obtained, whose slope is $-(2/3)A$ and whose intercept on the vertical axis is I'_0. According to the proposed scaling rule, both the Ω and the scaled $I_0 = I'_0(T_c/T-1)^{-1}$ should be independent of the composition and temperature of the clusters, for binary as well as unary clusters.

Model And Calculations

A cluster consists of n atoms interacting via the Lennard-Jones (6-12) potential energy function and confined in a spherical volume. The volume is chosen to be $5n/\rho_{liquid}$, where ρ_{liquid} is the bulk liquid number density; this volume is chosen to minimize the effect of the confining surface on the free energy.[8] The cluster center of mass is fixed. Unary clusters of two species (Argon and Krypton) are simulated, as well as two varieties of binary clusters, having the ratio of Argon to Krypton atoms of 3:1 and of 1:1. The Lennard-Jones parameters used for the Argon-Argon interaction are $\epsilon/k = 119$K and $\sigma = 3.4$Å. For the Krypton-Krypton interaction the parameters are $\epsilon/k = 171$K and $\sigma = 3.6$Å. The usual combination rules are used to evaluate the parameters for the Argon-Krypton interaction, giving $\epsilon/k = 143$K and $\sigma = 3.5$Å.[9]

For each unary cluster, the difference in free energy between a cluster of n atoms (ensemble B) and a cluster of n-1 atoms plus a monomer (ensemble A), $\delta F_c(n)$, is computed using the Bennett-Metropolis Monte Carlo technique. In the cases of the binary clusters, a "monomer" means a combination of one Krypton and k Argon atoms, while $n = m(k+1)$ is the total number of atoms. Therefore, $\delta F_c(n)$ is the difference between the free energy of a cluster of n atoms and the free energy of a cluster of n-k-1 atoms. However, the atoms of a binary "monomer" are not required to remain together as a unit. Snap shots of a binary cluster in the two ensembles are shown in Figure 1. The initial configuration of a cluster is chosen randomly, restricted only in that two atoms are not allowed to start too close to one another, and that all atoms lie within the constraining volume. Normally, the $\delta F_c(n)$ is averaged over $2n$ million Monte Carlo steps. The free energy differences are scaled; that is, divided by the quantity $(T_c/T - 1)$. For the binary clusters, $T_c = (kT_{cAr}+T_{cKr})/(k+1)$, where $T_{cAr}(T_{cKr})$ is the critical temperature of Ar(Kr). Now, the Lennard-Jones critical temperature is about $1.3\epsilon/k_B$ and that is used in this work rather than the experimental critical temperatures for Argon and Krypton.

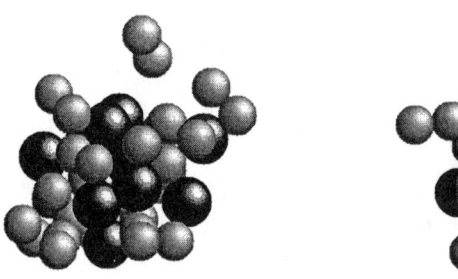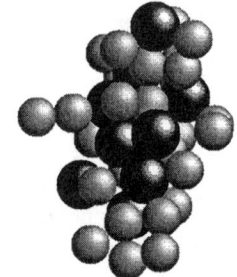

FIGURE 1. Snap shots of a binary cluster consisting of 10 Krypton(blue) and 30 Argon(green) atoms. The left frame is ensemble A, while the right frame is ensemble B.

RESULTS

The results of the simulations are shown in Figure 2. Although the clusters are at different temperatures and have different compositions, the plotted points for the Krypton, Argon, and binary clusters all fall on the same curve. For clusters of size n equal to 5 and above, the points lie along a straight line. It should be possible, therefore, to predict Ω and I_0 for any unary or binary cluster of noble gas atoms without computing $\delta F_c(n)$ afresh for every species of unary cluster, or mole fraction of binary cluster.

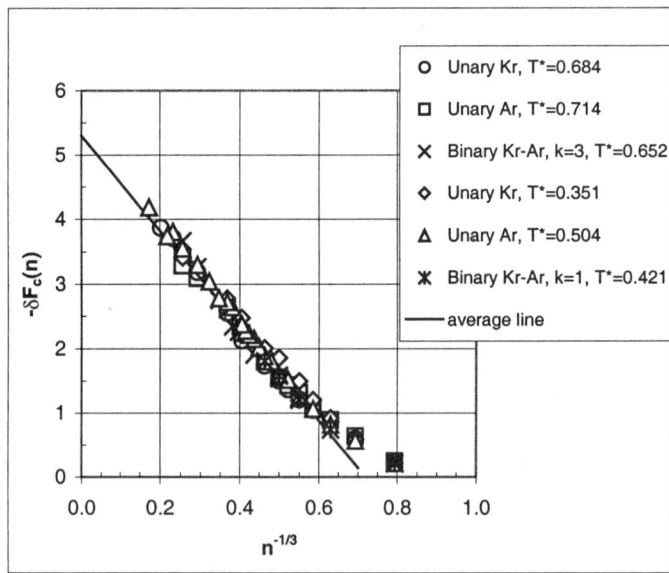

FIGURE 2. Configurational Helmholtz free energy differences between n and $n - 1$ Lennard-Jones clusters versus $n^{-1/3}$, scaled by $T_c/T - 1$. T^* is the reduced temperature; k is the number of Argon atoms in the cluster for each Krypton atom.

Least squares fits to a straight line were done for the linear portions of the six curves in Figure 2. The intercepts and slopes (Ω) obtained are shown in Table 1. The average I_0 is 5.3 ± 0.2 and the average Ω is 2.3 ± 0.2. A dependence of Ω on the composition of the clusters is not evident in Table 1.

TABLE 1.

Cluster species	T^*	I_0	Ω
Unary Krypton	0.684	5.4	2.40
Unary Argon	0.714	5.1	2.18
Binary Kr & Ar, $k=3$	0.652	5.9	2.74
Unary Krypton	0.351	5.2	2.09
Unary Argon	0.504	5.3	2.24
Binary Kr & Ar, $k=1$	0.421	4.9	2.07

CONCLUSION

The scaling of free energy differences of neighboring sized clusters, $\delta F_c(n)$, has been demonstrated through Monte Carlo simulations of unary and binary Lennard-Jones clusters of atoms. When scaled by a factor $(T_c/T-1)$, plots of $-\delta F_c(n)$ vs. $n^{-1/3}$ collapse to a single curve, which approaches a straight line for $n > 5$. The slope of that line yields an estimate for the excess surface free energy of the clusters, Ω, while the $\ln(\rho_{liquid}/\rho_{vapor})$ can be estimated by extrapolation to $n = \infty$. The motivation for this work has been to examine the scaling and mole fraction dependence of simple binary cluster free energy differences and the feasibility of estimating from them scaled model nucleation rates of binary noble gas clusters.[10,11]

REFERENCES

1. Hale, B. N., *Phys. Rev. A*, **33**, 4156 (1986).
2. Hale, B. N. and Ward, R. C., *J. Stat. Phys.* **28**, 487 (1982).
3. Hale, B. N. and Kemper, P., *J. Chem. Phys.* **91**, 4314 (1989).
4. Hale, B. N., *Aust. J. Phys.* **49**, 425-434 (1996).
5. Fisher, M. E., *Physics* **3**, 255 (1967).
6. Bennett, C. H., *J. Comutat. Phys.* **22**, 245 (1976).
7. Metropolis, M., Rosenbluth, A., Rosenbluth, M., Teller, A., and Teller, E. *J. Chem. Phys.* **21**, 1087 (1953).
8. Lee, J. K., Barker, J. A., and Abraham, F. F., *J. Chem. Phys.* **58**, 3166 (1973).
9. Hirschfelder, J. O., Curtiss, C. F., and Bird, R. B., *Molecular Theory of Gases and Liquids*, New York: John Wiley & Sons, 1963, p. 168.
10. Hale, B. N., *Metallurg. Trans.* **23A**, 1863 (1992)
11. Hale, B. N. and Wilemski, G., *Chem. Phys. Lett.* **305**, 263-268 (1999).

Buoyancy-induced Convection In The Thermal Diffusion Cloud Chamber

Frank T. Ferguson[a] and Joseph A. Nuth, III[b]

[a]*Department of Chemistry, Catholic University of America, Washington, DC 20064*
[b]*Code 691, NASA Goddard Space Flight Center, Greenbelt, MD 20771*

Abstract. An important question currently concerning thermal diffusion cloud chamber operation is what effect, if any, does buoyancy-induced convection have on measured supersaturations. In this paper we highlight some of the important results from a model of the chamber which includes such convective effects. The results indicate that the reduction in the maximum attainable supersaturation is likely to be small, but a general method of determining when these effects will become significant is needed.

INTRODUCTION

There is increasing concern regarding the limits of stable operation of the thermal diffusion cloud chamber (TDCC). This is especially true in view of the recent effort to develop a reliable and extensive database on the nucleation of certain substances. These high-quality sets of data can then be used in the testing of new and existing theories of nucleation.

One problem currently plaguing the TDCC is that the supersaturation results obtained with the TDCC appear to be dependent on the pressure of the background gas whereas no background gas effect is seen with typical expansion studies (1). This pressure effect has been studied extensively by Heist et al.(2-4). Another important consideration in TDCC work is the stability of chamber operation with respect to buoyancy-induced convection. Walls of the chamber are typically heated to prevent condensation and if the walls are heated too much, strong convective currents can develop within the chamber. Evidence of these currents are clearly visible due to the movement of droplets entrained in the flow. A more insidious problem may occur at lower heating rates. It is possible that very subtle convective currents can systematically alter the maximum attainable supersaturation within the chamber and go undetected. The purpose of this paper is to highlight some results of recent convective modeling of the TDCC aimed at determining the possible magnitude of these flows and their possible effects on the measured supersaturations.

MODEL OF THE CHAMBER

To examine the magnitudes of the flow possible with the TDCC we have developed a model that includes buoyancy-induced convection arising from both thermal and solutal gradients. The model is similar in concept to the work by Bertelsmann and Heist's 2D model of the TDCC (5), but includes the additional convective terms. In addition to the equations for the transport of mass and energy that are typically used to describe the chamber, the momentum and continuity equations for the system are also included. A full description of the model is given in (6) and only important highlights will be presented here.

Boundary conditions for the model are similar to those for the typical 1D modeling. Important differences are additional boundary conditions for the velocity components and sidewall boundary conditions. For the momentum components, all boundary velocities are assumed to be zero except at the centerline—in this case symmetry conditions apply and there is no gradient in the axial velocity.

For the sidewall, two different types of boundaries are considered in this work— wet and dry operation. Under dry operation, the sidewall is heated just enough to prevent condensation. In this case we adopt the best possible heating arrangement, namely that the supersaturation is equal to 1.0 at all points along the wall. In this critically heated arrangement there is no overheating of the walls, and the mole fraction at the wall is given by P_{eq}/P_{total} where P_{eq} is the equilibrium vapor pressure of the material and P_{total} the total chamber pressure.

In the wet wall case, the condensing vapors are allowed to wet the walls and it is assumed that the vapor is in equilibrium with the liquid. Therefore, the same concentration boundary condition is used for the wet wall case as in the dry case. For the temperature field it assumed that there is a minimal disturbance to the temperature field so that there is no thermal gradient at the sidewall.

Unlike previous modeling of TDCC's, the results will depend upon the size of the chamber as well as the aspect ratio. A chamber inside diameter of 10.38 cm with a diameter to height ratio of 7.5 is taken as a baseline case. These dimensions correspond to the high pressure cloud chamber (HPCC) used by Heist to study 1-propanol in helium (2). Results will be given for 1-propanol under a 1.18 bar atmosphere of helium at a lower plate temperature of 302.9 K and an upper plate temperature of 256.5 K. This corresponds to the same test case used by Bertelsmann and Heist in their 2D modeling of the TDCC (5).

The equations are solved using finite differences and the SIMPLER method (6). Results are computed on a 60x60 grid and iterations on this grid are continued until the continuity and energy equations are sufficiently satisfied at every control volume within the solution domain.

RESULTS

Contour plots for the dry wall case are shown in Figure 1. Shown in the figure are a). the temperature profile, b). the stream function, c). the mass fraction of propanol, and d). the supersaturation profile. In the case of the temperature profile, the isotherms are fairly flat and layered except at the sidewall where there is a sharp increase in temperature due to heating. As shown in the stream function plot, under these conditions there is a single convective cell with a high velocity along the sidewall balanced by a broad, weaker flow downward at the centerline. The mass

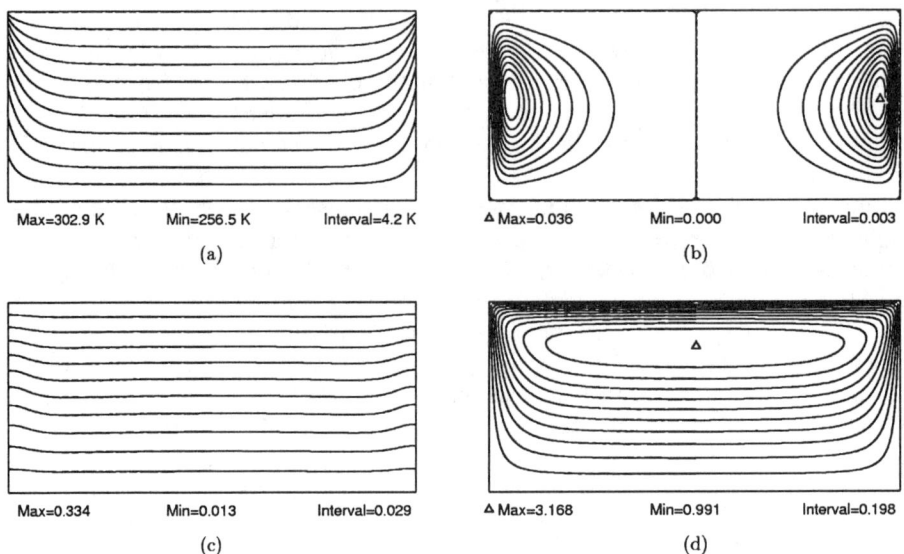

Figure 1. Contour plots of the results for the critically heated, dry wall case. Shown in the figure are plots of a). the temperature field, b). the stream function, c). mass fraction of propanol, and d). supersaturation profile.

fraction profile is also layered, but the upward convective flow along the sidewall tends to pull the concentration profile up slightly at the edges. The effects due to buoyancy alone can easily be distinguished by running the model with a gravitational level of zero. In this case, the maximum supersaturation ratio is 3.224. In the critically heated, dry wall case, this maximum supersaturation has been reduced just slightly to 3.168. Overheating would tend to decrease this value further.

In the wet wall case, the effect of convection is similar and the flow and supersaturation profiles look very similar to those of the dry wall case. In contrast to the dry wall case, the disturbance to the temperature profile in the wet wall case is minimal while the concentration gradient at the wall is high. In this case the convective flows are somewhat stronger than in the dry wall case and the maximum supersaturation is reduced to 3.068.

An important point to emphasize is that unlike the typical modeling of the TDCC, buoyancy-induced flows within the chamber will depend upon the size of the chamber as well as the aspect ratio. To highlight this, various runs were made for the propanol case at various chamber diameters and aspect ratios. The results of these calculations are shown in Figure 2. Plotted in the figure are the reduced maximum supersaturations, defined as the maximum attainable supersaturation with buoyancy effects included to the same quantity without these effects. Therefore, a reduced supersaturation of 1.0 indicates no buoyancy-induced influence on the maximum supersaturation. As shown in the figure, as the chamber size becomes smaller, the

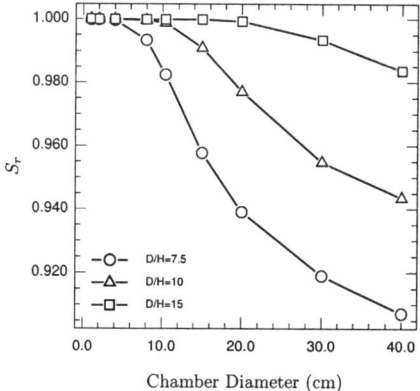

Figure 2. The effect of the chamber size on the maximum attainable supersaturation within the chamber. The reduced supersaturation, Scr, is defined as the ratio of the maximum supersaturation with buoyancy effects included to the same value when these effects are neglected. Results are shown for various chamber diameter to height ratios

effects of buoyancy are diminished as viscous drag with the chamber walls impedes such flows. As the chamber size is increased, buoyancy effects become more pronounced and there is a noticeable decrease in the reduced supersaturation. These effects are also greater for smaller aspect ratios.

CONCLUSIONS

In this paper, the results for only one particular example have been shown. In these cases, flows exist but their effects on the maximum attainable supersaturation are rather small. Unfortunately it is not currently possible to predict when these minute flows will have a significant influence on the data collected using the thermal diffusion cloud chamber without resorting to such modeling. Future work will involve developing a method to help predict when such flows will become significant.

REFERENCES

1. Viisanen, Y., Strey, R., and Reiss, H., *J. Chem. Phys.*, **99**, 4680, 1993.
2. Heist, R.H., Janjua, M., and Ahmed, J., *J. Phys, Chem.*, **98**, 4443, 1994.
3. Heist, R.H., Ahmed, J., and Janjua, M., *J. Phys. Chem.*, **99**, 375, 1995.
4. Bertelsmann, A., Stuczynski, and Heist, R.H., *J. Phys. Chem.*, **100**, 9762, 1996.
5. Bertelsmann, A., and Heist, R.H., *J. Chem. Phys.*, **106**, 610, 1997.
6. Ferguson, F.T., and Nuth III, J.A., *J. Chem. Phys.*, **111**, 8013, 1999.

Molecular Dynamics Simulation of the Homogeneous Nucleation of UF$_6$ Molecules

Shinobu Tanimura[a] and Kenji Yasuoka[b]

[a]*RIKEN (The Institute of Physical and Chemical Research), Hirosawa 2-1, Wako, Saitama 351-0198, Japan*
[b]*Department of Mechanical Engineering, Keio University, Hiyoshi 3-14-1, Kohoku-ku, Yokohama 223-8522, Japan*

Abstract. Molecular dynamics simulations of nucleation of UF$_6$ molecules have been performed. We observed two phenomena which demonstrate that the nucleation process occurs in the state far from thermal equilibrium. First, the excited hot clusters were produced and continued to exist during the nucleation process. Second, the relationship between the potential energy and temperature of the clusters depends on the monomer temperature, that is, the potential energy at a temperature decreases with the increase in monomer temperature. In the simulations, various types of cluster configurations were observed: prolate, oblate, sphere-like, and confeito-like. The confeito-like cluster is composed of one core and a few horns, and it was found predominantly in the hotter clusters. As a result, we found that the spectra measured in the supercooled state can be attributed to the excited hot clusters, the configurations of which are confeito-like.

INTRODUCTION

The thermal equilibrium state, which is assumed in the classical nucleation theory (CNT), is not necessarily reached in the nucleation process of the supersaturated (supercooled) vapor.[1-3] Molecular dynamics (MD) simulations of the homogeneous nucleation for UF$_6$ molecule,[1] Lennard–Jones fluid,[2] and water[3] showed that the temperature of the clusters are higher than that of the monomer, and tend to be higher for larger clusters. Tanimura et al.[4] investigated the configurations of UF$_6$ clusters in supercooled and near-equilibrium states by using Fourier transform infrared (FTIR) spectroscopy. They found that the spectra of the clusters in the supercooled state are completely different from those of the clusters in the near-equilibrium state. The measured spectra were compared to the calculated ones of the model clusters with tetrahedral or octahedral configurations. As a result, the spectra in the near-equilibrium state were attributed to the configurations of low potential energy. In contrast, the spectra in the supercooled state were attributed to the configurations of high potential energy, though the temperature of the monomer was lower than that in the near-equilibrium state. This result coincide with the finding in the MD simulation of the UF$_6$ molecule described above, that is, the "exited hot clusters", whose temperature is higher than that of the monomer, exist in the nucleation process of the supersaturated UF$_6$ vapor. However, the potential energy and configurations of the clusters were not investigated in the previous MD simulation.[1] Since, the direct measurements of the temperature, potential energy, and configuration of the clusters were impossible in the previous experiment,[4] the quantitative determination of those values by MD simulation is desired for the detailed understanding of the exited hot clusters in the nucleation process.

In the present study, the temperature, potential energy, and configurations of the UF_6 clusters produced in the supercooled states are determined by using MD simulation technique. The IR spectra of the clusters in the simulation are calculated and compared with those measured in the previous experiment.[4]

SIMULATION METHOD

The same technique as in a previous study[1] was used. A cubic cell was filled with 4000 UF_6 molecules and 4000 Ar atoms. The periodic boundary conditions were imposed for all three dimensions of the cell, and the leap-frog method was adopted for time integration of the classical equations of motion. The system was initially equilibrated at a relatively high temperature and quenched to a lower temperature in order to start the nucleation. After the start of nucleation, only Ar was connected to a hypothetical heat bath using the Nosé-Hoover method to maintain the temperature of Ar at the initial temperature. We used the same model as used in the previous study[4] to calculate the IR spectrum of the UF_6 cluster. The more detail conditions of the simulation are described in Ref. 5.

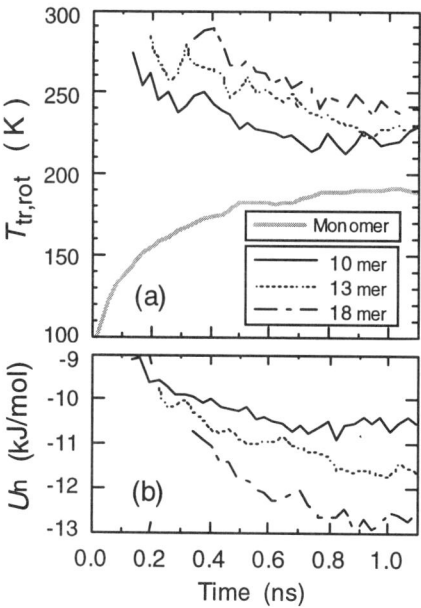

FIGURE 1. (a) Time development of the temperature $T_{tr,rot}$ for monomer and clusters at the initial temperature 100 K. $T_{tr,rot}$ denotes the average of translational and rotational temperatures. (b) Time development of the potential energy U_n for clusters. (Ref. 5)

SIMULATION RESULTS

Figure 1(a) shows the time development of temperature for monomer and clusters. As shown in the figure, the exited hot clusters, the temperature of which is much higher than that of the monomer, are formed and continue to exist because of the slow relaxation to the thermal equilibrium. In Fig. 1(b) the potential energy of the clusters decrease with time as similarly as the temperature of those in Fig. 1(a). These figures demonstrate that the temperature of monomer increase and the potential energy of clusters decrease according as the system relax to the thermal equilibrium. This finding coincide with the result of the previous study[4] described in Introduction, that is, the spectra in the supercooled state were attributed to the configurations of high potential energy, though the temperature of the monomer was lower than that in the near-equilibrium state.

Typical configurations of the 13 mer sampled in the simulation and the corresponding calculated IR spectra are shown in Fig. 2. Figures. 2(a) and 2(b) indicate the configurations, calculated spectra of which are similar to the experimental one measured in the supercooled state. The shapes of those clusters are composed from one core and some horns. The dark spheres in those snapshots denote the ends of the horns. These shapes are similar to that of the *confeito*, which is a type of small candy made by crystallizing sugar around a core. More than ten horns of the crystal of sugar are grown from the core. Therefore, we call those type of the clusters in Figs.2(a) and 2(b) as confeito-like cluster. The confeito-like clusters with one horn or four horns were also observed. As a result of the simulation, the confeito-like clusters were found to be dominant in the cluster of high temperature at the beginning of the nucleation. With the decrease in the temperature of the cluster, the population of the confeito-like cluster decreased and those of the prolate and sphere-like clusters in Figs. 2(c) and 2(d) increased. The prolate cluster was dominantly populated in the clusters of low temperature. The calculated spectrum of the prolate cluster in Fig.2(C) is similar to the measured one at the near-equilibrium state. Therefore, the change of the cluster configuration from the confeito-like to the oblate and sphere-like can explain the difference between the measured spectra at the supercooled and near-equilibrium states.

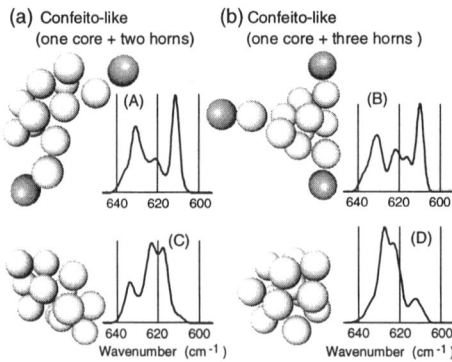

FIGURE 2. Typical configurations of the 13 mer sampled in the simulations. Snapshots are shown in (a)–(d), in which UF_6 molecule is represented by a sphere. Calculated spectra corresponding to the configurations (a)–(d) are shown in (A)–(D), respectively. (Ref. 5)

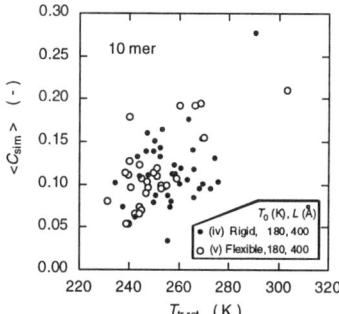

FIGURE 3. Relationships between the similarity coefficients $\langle C_{sim} \rangle$, which is defined at Ref. 5, and temperature $T_{tr,rot}$ for 10-mer.

Figure 3 shows the relationships between the similarity coefficients $\langle C_{sim} \rangle$ and the temperature $T_{tr,rot}$ for the rigid molecule model and the flexible one. The similarity coefficients was defined in Ref. 5. In the previous MD simulation study,[4] we demonstrated that the nucleation rate in the flexible model of UF_6 molecules was about twice as large as that in the rigid model of UF_6 molecules. This acceleration in the nucleation rate was attributed to the flow of condensation heat into the intramolecular vibrations. The relationships between $\langle C_{sim} \rangle$ and $T_{tr,rot}$ in the two models were compared, and the differences were found to be negligibly small. Therefore, the intramolecular vibrations were confirmed to have little effect on the configuration. The intramolecular vibrations of the UF_6 molecules in the cluster seem to act as a heat bath, absorbing the condensation heat. As a consequence, the findings in the MD simulation of the rigid model, as well as in the case of the flexible model, were confirmed to be correct.

CONCLUSIONS

In the simulations of the nucleation of the supersaturated UF_6 vapor, we found that the exited hot clusters, the temperature of which are much higher than that of the monomer, were produced and continued to exist in the nucleation process. This phenomenon demonstrates that the nucleation process goes on in the state far from the thermal equilibrium. The confeito-like clusters were observed in the simulations, whose population were larger in the hotter cluster. The IR spectra measured in the supercooled state were found to be attributed to the exited hot cluster, the configurations of which are confeito-like.

REFERENCES

1. Tanimura, S., Yasuoka, K., and Ebisuzaki, T., *J. Chem. Phys.* **109**, 4492-4497 (1998).
2. Yasuoka, K., and Matsumoto, M., *J. Chem. Phys.* **109**, 8451-8462 (1998).
3. Yasuoka, K., and Matsumoto, M., *J. Chem. Phys.* **109**, 8463-8470 (1998).
4. Tanimura, S., Okada, Y., and Takeuchi, K., *J. Chem. Phys.* **107**, 7096-7105 (1997).
5. Tanimura, S., Yasuoka, K., and Ebisuzaki, T., *J. Chem. Phys.* **112**, 3812-3819 (2000).

Energy Barrier Effect on Transient Nucleation Kinetics

Igor L. Maksimov* and Masaaki Sanada**

*Nizhny Novgorod University, 23 Gagarin Ave., Nizhny Novgorod 603000, Russia
**University of Tokushima, 2-1 Minamijosanjima, Tokushima 770-8506, Japan

Abstract. Energy barrier effect on the transient nucleation kinetics is studied. For high barrier case an advanced expression of the 'boundary-layer' type is suggested that is valid for the entire time interval. For low barrier case a new 'similarity' solution is reported.

INTRODUCTION

Transient nucleation (TN) is an important stage of the first-order phase transition during which the system undergoes a transition from the metastable state to that of a greater stability. The time necessary for the nucleation to be completed provides the information about the parameters of the nucleating phase. As it is typical for the first-order phase transition the competition between the bulk term and the surface term in a Gibbs free energy determines the actual barrier height that controls the rate of the process. In the majority of research devoted to the transient nucleation problem a high-barrier limit was employed which is adequate for a near equilibrium condition (see e.g. Ref.1). However for the condition of far-from-equilibrium nucleation (high supersaturation and/or high temperature) when the barrier height drastically decreases the problem of the TN has to be analysed on the basis of different approach.

MODEL

According to Zeldovich[2] the equation for the cluster distribution function $c(n,t)$ governing the nucleation kinetics in the limit $n \gg 1$ reads

$$\frac{\partial c(n,t)}{\partial t} = \frac{\partial}{\partial n}\left[K^+(n)c_0(n)\frac{\partial}{\partial n}\left(\frac{c(n)}{c_0(n)}\right)\right]. \quad (1)$$

Here the equilibrium cluster distribution $c_0(n)$ is given by[3]

$$c_0(n) = N\exp\left(-\frac{W^{rev}(n)}{kT}\right), \quad (2)$$

where $W^{rev}(n)$ is thermodynamic reversible work, N is a normalization factor, k denotes the Boltzmann constant and T is the absolute temperature. Eq.(1) is obtained by expressing the detachment rates $K^-(n)$ through the attachment rates $K^+(n)$ in the master equation (1) with the help of the detailed balance condition.[3] In what follows we employ conventional boundary conditions

$$c(1,t)/c_0(1) = 1, \qquad c(n,t)/c(\infty) = 0, \qquad (3)$$

supplemented by the traditional initial condition[1]

$$c(n,0) = c_0(n)H(1-n), \qquad (4)$$

where $H(x)$ is the Heaviside step function.

NUCLEATION BARRIER

The measure of the barrier height is controlled by the dimensionless parameter

$$w^* = W^{rev}(n^*)/kT = 0.5\Delta\mu n^*/kT, \qquad (5)$$

where $n=n^*$ represents thermodinamical critical nucleus[1] and $\Delta\mu$ is the difference of the chemical potentials between the parent phase and the bulk nucleating phase. The conventional Becker-Doring-Zeldovich theory,[2,3] taking advantage of the high barrier limit ($w^* >> 1$) makes it possible to use aquadratic expansion of $W^{rev}(n)$ near n^* inside the so-called critical region (CR) $n_c^- \leq n \leq n_c^+$ defined by the condition

$$W^{rev}(n_c^\pm) = W^{rev}(n^*) - kT. \qquad (6)$$

HIGH-BARRIER LIMIT

The governing equation for the transient nucleation stage deserved great attention among theorists who attempted to solve it by many different approaches. The majority of the advanced recent solutions employes quite complex singular perturbation method which is applicable not only in the vicinity of n^* but also well outside the CR. Here we focus our attention on the study of the so-called boundary-layer (BL) type solution.[4-7]

In particular, we will suggest a constructive generalization of the previous BL theory which results in a self-consistent erfc-type solution (see below) valid for arbitrary cluster size within the entire time interval. General expression of $c(n,t)$ for the erfc-type BL solution reads

$$\frac{c(n,t)}{c_0(n)} = \frac{1}{2}erfc\left[\frac{A + B\exp(t/t_i)}{\sqrt{\exp(2t/t_i) - \zeta}}\right], \qquad (7)$$

where parameter A is a model-dependent quantity and the controlling parameter ζ takes two values: $\zeta = 1$ for the majority of the models except that of Shi et al[5] for which $\zeta = 0$. The function $B(n)$ satisfies the condition $B(n^*)=0$ for any BL model. The main deficiency of the majority of reported BL-type solutions[4-7] attempting to describe TN kinetics is the so-called asymptotics catastrophe' which reflects the fact that the asymptotical value of the nucleation flux $J(n,t)$ at $t \to \infty$ does not saturate at its steady state value J_s for arbitrary cluster size n.

In order to eliminate an obvious inconsistency we suggest a 'salvation' procedure which treats the function $B(n)$ as a trial one for the BL - like distribution (7). The parameter A is unambiguously defined as $A=-B(1)$ (for the case of no preexisting nuclei) in order to satisfy initial condition (4) for $c(n,t)$. Furthermore, we select $B(n)$

in such a way that the steady-state normalization condition $J(n,t)/J_s=1$ for $t\to\infty$ is satisfied. From the above requirement one arrives at the closed-form expression for the function $B(n)$

$$B(n) = erf^{-1}\left[2Z_D \int \exp[\Delta W^{rev}(n')/kT]\left(n^*/n'\right)^{2/3} dn'\right], \quad (8)$$

where $\Delta W^{rev}(n) = W^{rev}(n) - W^{rev}(n^*)$ and $erf(z)$ is an error function. While deriving Eq.(8) the condition $B(n^*)=0$ was taken into account (the details are presented in Ref.8). It can be shown that the ansatz (10) makes the BL solution (9) asymptotically equivalent (at $t\to\infty$) to the conventional[2,3] steady-state solution $c_s(n)$ given by

$$c_s(n) = J_s c_0(n) \int_u^\infty \frac{dn'}{c_0(n')K^+(n')}, \quad (9)$$

where u according to (3) is: $u(n) \equiv n$. The correct expression for the nucleation flux $J(n, t)$ now reads

$$\frac{J(n,t)}{J_s} = \frac{1}{\sqrt{1-\zeta\exp(-2t/t_i)}} \exp\left[B^2 - \frac{[A+B\exp(t/t_i)]^2}{\exp(2t/t_i)-\zeta}\right]. \quad (10)$$

We should emphasize that Eq.(10) is valid for ARBITRARY cluster size. For the case $\zeta=1$ Eq.(10) is formally equivalent to the expression given by Trinkaus and Yoo theory.[4] For the case $\zeta=0$ Eq.(10) represents a modified version of the "double-exponential" asymptotics reported earlier.[5,7] The time-dependent integrated flux,

$$I(n,t)/J_s \approx t - t_L.$$

for the distribution (7) is presented in a standard asymptotical ($t \gg t_i$) form

$$I(n,t) = \int_0^t J(n,t')dt'$$

The lag-time t_L can be straightforwardly expressed through the parameter A and the function $B(n)$ of Eq.(8). Thus, the expressions (7) and (10) with an appropriate selection of the parameters (see Eq.(8)) provide a useful interpolation which embraces both the growth region and the near-critical region of cluster sizes and provides an adequate description of the nucleation kinetics within the entire time interval ($0<t<\infty$). Since the expression (10) (with $\zeta=1$) is generically close to that by Trinkaus and Yoo[4] obtained within the parabolic expansion scheme, it could be referred to as the advanced parabolic interpolation (API). Note that for the later stage of nucleation $t>t_i$ the API is practically equivalent to the results given by the most advanced transient nucleation theory[7] with the only reservation that the latter approach does not apply at the early nucleation stage.

LOW-BARRIER NUCLEATION KINETICS

In case of high supersaturation and/or temperature the barrier parameter $w^* \approx \Delta\mu^2(kT)^{-1}$ may be reduced appreciably to become of the order of 1 (and even less). This situation may take place, for example, near the critical point where n^* tends to decrease drastically with temperature. Under such conditions, conventional theory of the TN is not formally valid and obviously requires an adequate modification. Below we present the analysis performed in the extreme case of low barrier ($w^* \ll 1$). The limit $w^* \ll 1$ suggests one to neglect the barrier while solving the governing equation. By employing the expression (9) and introducing the cluster-size-dependent non-stationary variable $u=u(n,t)$ the equation (1) is reduced to the nonlinear-diffusion one with respect to the unknown function u. The solution to this equation can be obtained with the help of a self-similar ansatz

$$u(n,t) = n\left[1 + \phi(n/t^\beta)\right] \tag{11}$$

For the case of three-dimensional nucleation one finds (for details see Ref.9) the exponent β=3 and the function $\phi(x)$ expressed in a closed form as

$$\phi(x) = \frac{1}{16x}\left(x^{1/3} - 1\right)^3 \left[3x^{1/3} + 1\right] H(x-1). \tag{12}$$

It is very important that solution (11, 12) holds as long as

$$t \leq \tilde{t}_{ind}(n) = \tau n^{1/3}. \tag{13}$$

For $t > \tilde{t}_{ind}(n)$, when $u(n) \equiv n$ (since $\phi(x) \equiv 0$) the steady-state distribution $c_s(n)$ (see Eq. (9)) is established. Thus, $t_{ind}(n_0)$ from Eq.(13) represents the runaway-nucleation induction time necessary to reach steady state within the cluster size domain $n > n_0$.

SUMMARY

We have studied the effect of the energy barrier on the transient nucleation kinetics. We have obtained analytical expressions for the cluster distribution function and for the time-dependent nucleation flux.
1. For a sufficiently high barrier case an advanced parabolic interpolation is suggested of the 'boundary-layer' type which provides an adequte description of the TN kinetics for arbitrary cluster size n.
2. For a low barrier case a new 'similarity' solution is reported, that is characterized by a power-law time evolution of the nucleation flux.
3. The transient nucleation kinetics exhibits a crossover from the barrier-controlled nucleation at low supersaturation, which is characterized by the exponentially-sharp cluster size distribution, to the runaway regime (for high supersaturation) characterized by the power-law evolution in time.

REFERENCES

1. Binder, K., and Stauffer, D., *Adv. Phys.* **25**, 343 (1976).
2. Zeldovich, Ya. B., *Acta Physicochim (USSR)* **18**, 1 (1943).
3. Becker, R., and Doring, W., *Ann. Phys.* **24**, 719 (1935).
4. Trinkaus, H., and Yoo, M. H., *Phil. Mag. A* **55**, 269 (1987).
5. Shi, G., Seinfeld, J. H., and Okuyama, K., *Phys. Rev. A* **41**, 2101 (1990).
6. Demo, P., and Kozisek, Z., *Phys. Rev. B* **48**, 3620 (1993); *Phil. Mag. B* **70**, 49 (1994).
7. Shneidman, V. A., *Zh. Tekh. Fiz.* **57**, 131 (1987); *Phys. Rev. A* **44**, 2609 (1991).
8. Maksimov, I. L., Sanada, M., and Nishioka, K., *J. Chem. Phys.* (to be published).
9. Maksimov, I. L., and Nishioka, K., *Phys. Lett. A* **264**, 51 (1996).

Nucleation Rate Surface Continuity and Monotony

M.P. Anisimov, S.D. Shandakov, G.V. Shandakova, and V.I. Poltavtsev

Nucleation Laboratory at Kemerovo, Institute of Catalysis, Siberian Branch of the Russian Academy of Science, 41A-119 Moscowskiy Prospect, 650065 Kemerovo, Russia.

Abstract. The condition for surface continuity and monotony is applied as criteria for analysis of the experimental results on vapor nucleation to detect the phase transitions in the condensed states. Criteria is tested with broad spectrum of the available experimental results on vapor nucleation. It is shown that criteria could be helpful to detect the singularities for any data set.

I. INTRODUCTION

As a rule, the modern theories of nucleation represent only modifications of classical theory, and they still remain unsuitable for quantitative prediction of experimental results. In order to create the universal description of supersaturated vapor nucleation, as well as evaluate other theories, we need to get the reliable experimental results. Now experimental skills are high enough for precision measurements of the vapor nucleation rates as well as detect and evaluate, for example, temperatures of phase transitions in the new phase critical embryos. That is, one can find experimentally the singularity of nucleation rate surface at conditions for the critical embryo phase transitions. Visual analysis of the experimental results has the low potential to detect the nucleation rate surface singularity and the surface singularity can be missed sometime. One major problem for singularity detection is the noise of experimental results. This problem still exists even if one has experimental results of high accuracy. The condition for surface continuity and monotony could be applied as criteria for analysis of experimental data for the vapor nucleation rates. Criteria are tested for the broad spectrum of the experimental results in the vapor nucleation. As an example of the criteria using, the experimental results for the n-pentanol - water binary system of nucleation have been analyzed. A peculiarity of this system is the existence of a miscibility gap between the components. It was found that criteria as function of n-pentanol mole fraction has the singularity in the vicinity of this gap. This singularity was smoothed in the experimental result presentation and omitted in the original research. The proposed criteria can be used to search the experimental data set singularity as function of their variable(s).

II. CONDITION FOR THE SURFACE CONTINUITY AND MONOTONY IN APPLICATION FOR EXPERIMENTAL RESULTS ANALYSIS

To increase the sensitivity of singularity detection of a nucleation rate surface, a sensitive formal criterion can be applied. As mentioned above, it is attractive to use mathematical conditions for surface continuity and monotony. If one has a function, $F(x_1, x_2, ..., x_n)$, where $x_1, x_2, ..., x_n$ are independent variables. When function F has continuous partial derivatives, the surface, described by the function $F(x_1, x_2, ..., x_n)$, is continuous and monotonous. For condition $F = constant$ function $F(x_1, x_2, ..., x_n)$ has $(n-1)$ independent variables. For one independent variable and a continuous and monotonous surface a full derivative of F with respect of x_1 at $F = constant$ and $n = 2$ can be written as

$$\left(\frac{dF}{dx_1}\right)_{x_2} = \left(\frac{\partial F}{\partial x_1}\right)_{x_2} + \left(\frac{\partial F}{\partial x_2}\right)_{x_1}\left(\frac{\partial x_2}{\partial x_1}\right)_F = 0. \tag{1}$$

Subscripts show the constant values of variables as a condition for derivative evaluation. Eq. 1 can be rewritten as

$$\left(\frac{dF}{dx_1}\right)_{x_2} = \left(\frac{\partial F}{\partial x_1}\right)_{x_2} + \left(\frac{\partial F}{\partial x_2}\right)_{x_1}\left(\frac{\partial x_2}{\partial x_1}\right)_F = A, \tag{2}$$

where A has nonzero value in points of the surface, $F(x_1, x_2)$, singularity. For two independent variables, at $F = constant$, the condition for surface continuity and monotony can be written in the next form:

$$\left(\frac{dF}{dx_1}\right)_{x_2,x_3} = \left(\frac{\partial F}{\partial x_1}\right)_{x_2,x_3} + \left(\frac{\partial F}{\partial x_2}\right)_{x_2,x_3}\left(\frac{\partial x_2}{\partial x_1}\right)_{x_2,x_3,F} = S_1;$$

$$\left(\frac{dF}{dx_2}\right)_{x_1,x_3} = \left(\frac{\partial F}{\partial x_2}\right)_{x_1,x_3} + \left(\frac{\partial F}{\partial x_1}\right)_{x_1,x_3}\left(\frac{\partial x_1}{\partial x_2}\right)_{x_1,x_3,F} = S_2. \tag{3}$$

In common case, for function, $F(x_1, x_2, ..., x_n)$, with independent variables: $x_1, x_2, ..., x_n$, the complete derivative F with respect of x_1 could be written in form such as:

$$\frac{dF}{dx_1} = \frac{\partial F}{\partial x_1} + \frac{\partial F}{\partial x_2}\frac{\partial x_2}{\partial x_1} + \frac{\partial F}{\partial x_3}\frac{\partial x_3}{\partial x_1} + ... + \frac{\partial F}{\partial x_n}\frac{\partial x_n}{\partial x_1}. \tag{4}$$

For full set of complete derivatives the matrix equation could be written such as:

$$\begin{vmatrix} \frac{\partial F_1}{\partial x_1} & \frac{\partial F_1}{\partial x_2} & \cdots & \frac{\partial F_1}{\partial x_n} \\ \frac{\partial F_2}{\partial x_1} & \frac{\partial F_2}{\partial x_2} & \cdots & \frac{\partial F_2}{\partial x_n} \\ \cdot & \cdot & & \cdot \\ \cdot & \cdot & & \cdot \\ \frac{\partial F_n}{\partial x_1} & \frac{\partial F_n}{\partial x_2} & \cdots & \frac{\partial F_n}{\partial x_n} \end{vmatrix} * \begin{vmatrix} 1 & \frac{\partial x_1}{\partial x_2} & \cdots & \frac{\partial x_1}{\partial x_n} \\ \frac{\partial x_2}{\partial x_1} & 1 & \cdots & \frac{\partial x_2}{\partial x_n} \\ \cdot & \cdot & & \cdot \\ \cdot & \cdot & & \cdot \\ \frac{\partial x_n}{\partial x_1} & \frac{\partial x_n}{\partial x_2} & \cdots & 1 \end{vmatrix} = \begin{vmatrix} a_{11} & a_{12} & \cdots & a_{1n} \\ a_{21} & a_{22} & \cdots & a_{2n} \\ \cdot & \cdot & & \cdot \\ \cdot & \cdot & & \cdot \\ a_{n1} & a_{n2} & \cdots & a_{nn} \end{vmatrix} = M \tag{5}$$

One can see easily, that each diagonal term a_{ii} of matrix M is equal to full derivative of F with respect of x_i; $i = 1, 2, ..., n$, i.e. $a_{ii} = \frac{dF_i}{dx_i}$. If Eq. (5) has the condition $F_i = constant$, $i = 1, 2, ..., n$, then $\frac{dF_i}{dx_i} = 0$ for a continuous and monotonous surface, and all diagonal terms are equal to zero. It is known, that the matrix determinant with zero diagonal terms is equal to zero, i.e. $M=0$ at these conditions. The condition $F = constant$ at different levels $F_1, F_2, ..., F_n$ is included in Eq. (5). It is remarkable that it can be used up $(n-1)$ independent conditions only, because $F_1, F_2, ..., F_n$ could not be independent from each others for the constant level of the continuous and monotonous surface. Hence, the determinant of matrix M has zero value, excluding the case if one (or more) derivatives of $F_1, F_2, ..., F_n$ is (are) not exist or if function F has at least one independent variable more than n, i.e. F is under the influence of uncontrolled parameter(s). A determinant value, M, could be applied as criteria for any set of the experimental results to test a condition for continuous and monotonous surface. It is reasonable to use standard deviation of M as a nondimensional characteristic of experimental data accuracy. For set of experimental results the nonmonotony of M values cold be used for singularity or free parameter detection.

For the model of nucleation rate, based on the application of two variables, such as temperature, T, and vapor supersaturation, S, with all other variables $(P_1...P_n)$ constant. Thus, a nucleation rate is the function of state, $J = f(T,S)$. It is customary to present a nucleation rate experimental data in the form $J=J(T)$ at $S=$ *constant* or $J = J(S)$ at $T =$ *constant*. An application of theory of partial derivatives leads a simple relation for nucleation rate, as function of two variables along section $J =$ *constant*, such as:

$$\left(\frac{\partial \ln J}{\partial T}\right)_{S,P_1...P_n} + \left(\frac{\partial \ln J}{\partial \ln S}\right)_{T,P_1...P_n} \cdot \left(\frac{\partial \ln S}{\partial T}\right)_{J,P_1...P_n} = A. \qquad (6)$$

If the vapor nucleation rate surface $J = f(T,S)$ has no singularity, A in Eq. (6) should be equal to zero, i.e., function, $f(T,S)$, represents a continuous and monotonous surface. Due to the noise of the experiment, A can have nonzero value which can be described by standard deviation. Deviation of A from zero value along the level $J =$ *constant* characterizes the quality of experimental results (if set of experimental values for J represents a continuous and monotonous surface). Singularity of the nucleation rate surface (or surface for any other set of

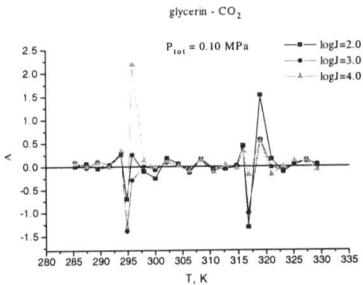 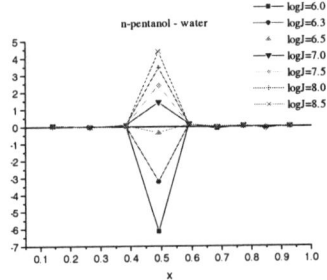

Figure 1. The criteria A distribution on nucleation Temperature[1], T. for log$J$=2.0: <A> = 0.04 and σ=0.16; for log$J$=3.0: <A> = 0.03, and σ=0.15

Figure 2. Criteria A values on embryo composition, x, for data by Strey and Wagner [2].

experimental results) breaks the local condition for surface continuity and monotony and, as a result, the local values of A under the influence of independent variable should be far enough from zero.

In common case, Equations (2,3,5) may be used to characterize the quality of any experimental results and/or to search any experimental data set singularity as function of their variable(s). Criteria usage allows detecting a presence of uncontrolled parameter(s) of the experiment or inconsistency in the mathematical model used. Regular criteria deviation from zero line or noise higher than the possible standard deviation calculated from known quality of used equipment could indicate these problems.

III. RESULTS AND DISCUSSION

Eq. (6) is applied for the experimental nucleation rate data, which have been collected recently. Results of calculations are shown in Figures 1-2. The most number of the experimental nucleation rates have the average value for A around <A>= 0.1. The standard deviation σ, has values around 0.2 and less. Seemingly this relation of the average value for A and for the standard deviation σ, is the level for current state of the nucleation experiments. The big values of <A> could be explained by the nucleation rate surface singularities (or experiments have the problems). For glycerin – carbon dioxide nucleation [1] one can find on Fig. 1 two phase transitions in critical embryos. Bulk glycerin has no phase transitions between melting and critical points. Nevertheless critical embryos have a phase transition of first order[1]. Criteria A detect these phase transitions. The accuracy of criteria could be better with application of the smaller step in experimental nucleation temperature and/or pressure. In Fig. 2 results for binary nucleation are shown for n-pentanol-water nucleation [2]. One can see a beautiful, close to zero, experimental values for criteria A, except for one point at critical embryo composition: $x \approx 0.5$. This jump is clear evidence of a singularity in the nucleation rate surface at constant temperature, omitted in the original paper [2]. Criteria have positive and negative values in point of singularity, as well as in case shown on Fig. 1. It means that there is the nucleation rate, J, where criteria can has zero value. Evaluation of A at different constant levels of J (or other constant variable) should avoid the possibility to miss the singularity. This jump is clear evidence of a singularity existence in the nucleation rate surface at constant temperature. The jump of criteria A is a result of the n-pentanol-water miscibility gap in critical embryos. Hopefully, more careful experimental research of n-alcohol - water series will detect two-channel nucleation, produced by the partially miscible liquids.

IV. CONCLUSIONS

As has been demonstrated, the condition for surface continuity and monotony can be applied as criteria for analysis of the experimental results on vapor nucleation to detect the phase transitions in the condensed states. Criteria were tested with broad spectrum of the available experimental results on vapor nucleation[1]. It was shown that criteria could be helpful to detect the singularities for any nucleation rate surfaces. For example, using criteria, the experimental results for the n-pentanol - water binary system nucleation [2] have been analyzed. The peculiarity of this system is the existence of the miscibility gap between the components. It was found that the nucleation rate surface for the n-penatanol - water binary system has the singularity in the vicinity of this gap. This singularity was smoothed in the experimental result presentation and omitted in original paper by Strey et al.[2] It should be mentioned that the proposed criteria could be used for any experimental data to search the experimental data set singularity as function of their variable(s). In the most cases the singularity of the experimental data surface has the direct relation with the phase transitions in the system under investigation[3].

ACKNOWLEDGEMENT

Authors acknowledge Dr. A. Nasibulin for the useful discussion and technical help in calculations.

REFERENCES

1. Anisimov, M.P., Koropchak, *et al. J.Chem.Phys.* **109**, 10004 (1998).
2. Strey, R., Viisanen, Y., and Wagner, P.E. *J. Chem. Phys.* **103**, 4333 (1995).
3. Anisimov, M.P., Nasibulin, A.G., and Shandakov, S.D. *J.Chem.Phys.* **112**(5), 2348 (2000).

Thermal Diffusion Cloud Chamber - New Criteria for Proper Operation

Richard H. Heist*, Daniel Martinez, Yuk Chan, and Anne Bertelsmann[†]

*Nucleation Laboratory, Department of Chemical Engineering
University of Rochester, Rochester, NY 14627-0166*

Abstract. We report results of new nucleation experiments involving 1-pentanol with hydrogen as the background gas utilizing the high-pressure diffusion cloud chamber (HPCC). We discuss the important issue of buoyancy-driven convective motion and cloud chamber operation, and we focus on the lower total pressure limit required for stable chamber operation. We provide, for the first time, an empirical procedure for determining the lower total pressure limit.

INTRODUCTION

The thermal diffusion cloud chamber (TDCC) has been used for nucleation research for more than four decades. It has been responsible for approximately 50% of all critical supersaturation and nucleation rate data published in the literature.[1] The mass and energy transport that occur within the TDCC during operation have routinely been modeled as 1-D transport through a stagnant gas. However, results of recent investigations clearly indicate the need for a careful re-examination of these assumptions.[2-7] In this paper we address the important issue of the lower limit of total pressure for the TDCC. We employ our HPCC with 1-pentanol as our working fluid and hydrogen as our background gas.

EXPERIMENTAL RESULTS

Our recent experiments on the pentanol/hydrogen system span a total pressure range of 10 to 3000 kPa and a nucleation temperature range from 280 K to 380 K. Here, we present only select results of these studies.[8] In Figure 1 we show S_{cr} (critical supersaturation) data as a function of total pressure obtained at constant nucleation temperatures of 331K, 306K, and 280K. The solid line running though each set of data is a best-fit line to the solid data symbols. All open symbols in Figure 1 refer to operational regions of the HPCC that are expected to give rise to unstable behavior and, hence, unreliable data. These data points are discussed below.

In Figure 2 we plot the upper plate temperature for each experiment versus the total pressure corresponding to that same experiment. The upper solid curve is the upper total pressure stability limit for our experiments.[7] The lower solid curve is an empirically determined locus of points that represents the lower total pressure stability limit and will be discussed in more detail below. The solid lines drawn through the nearly vertical data sets are best-fit lines to indicate trends.

REGIONS FOR DIFFUSION CLOUD CHAMBER OPERATION

The data presented in Figure 1 represented as solid symbols are limited to those points that were obtained in what we define to be the "proper" operating range for the TDCC. In Figure 2 we see that there are three regions that can be identified that pertain to the reliability of the nucleation data shown in the figures. These three regions are marked I, II and III.

Region III: This region corresponds to a range of undesirable total pressures. When the total pressure in the diffusion cloud chamber is increased above a certain limit, local inversions of the density gradient occur at the wall of the chamber. These inversions can lead to a buoyancy-driven convective flow that is extremely difficult to detect and can seriously affect the reliability of the nucleation data.[3-5,7]

Region II: This region corresponds to the range of chamber operation that we define to be "proper." It corresponds to the range of data represented by the solid points in Figure 1. It is in this region that mass and energy transport are presumed to occur as diffusion and conduction through a (supposed) stagnant background gas and to be accurately described by the routinely used heat and mass transport equations.[2-7]

Region I: As seen in Figure 1, at lower total pressures the measured supersaturation tends to deviate from that expected based upon extrapolation of the data in region II. As with previous results,[3,7] we are led to the conclusion that this deviation is caused by a change in the transport process within the cloud chamber. If we consider the constant nucleation temperature data plotted in Figure 2 and linked by solid lines, we note a sudden change in slope at the lower total pressure end of each series. Based upon our experience with operational stability of the TDCC, we conclude that this change in slope is caused by the onset of an additional mode of transport not considered in our analysis of cloud chamber conditions and for this reason we consider data obtained over this range of low total pressures to be unreliable. We have indicated these data points with open symbols on the graph. We have connected these data points using a separate line, and the lower solid line in Figure 2 connects the intercepts of these lines and thus represents the empirically determined boundary between the regions of "proper" chamber operation (i.e. region II) and the (to be avoided) low pressure region of operation (i.e. region I).

We examined the lower total pressure limit of TDCC operation in a more quantitative fashion by examining the conditions that could give rise to an inversion in the density gradient at the central portion of the TDCC in the region just above the lower plate. We have determined that all the data discussed in this paper are stable in this regard.[8] We have also considered the possibility that "double-diffusive" phenomena may be responsible for this behavior.[9] Application of the criteria for bounds of double-diffusive phenomena to our experimental data, however, did not support this explanation. It may also be that Bénard or Marangoni instabilities in the liquid pool give rise to turbulent flow or mixing in the region just above the surface of the pool that affects the overall transport through the HPCC.[10]

If we examine the stability diagram shown in Figure 2, we make the important observation that at lower temperatures total pressures, the allowable, stable range of operation (region II) becomes seriously limited, and it is increasingly difficult to use the TDCC and obtain reliable nucleation data under these conditions.

We do not yet have an analytical expression to predict this lower total pressure limit, but we can determine this lower limit empirically.[8] The supersaturation is not the only quantity that shows a sudden change in behavior at the lower total pressure boundary. In fact, the heat flux, the mass flux and the plate temperatures all display a similar behavior. We propose that a combination of all these quantities be used to identify empirically the location of the lower total pressure limit of operation in order to create the stability diagram necessary to determine the limits of "proper" operation for the TDCC.[8]

DETERMINING THE "PROPER" RANGE OF OPERATION

In the past, ratios in excess of two or three of the total pressure to the equilibrium vapor pressure of the working fluid at the lower plate temperature have been used to define stable TDCC operation at lower total pressures.[8] In our study, we found no correlation between the pressure ratio and our empirically determined lower total pressure limit. In fact, we have obtained data points with pressure ratios of approximately 1.5 which lie above the lower pressure limit[11] and data points with pressure ratios of approximately 8 to 10 which fall below the lower limit.[8]

When using the TDCC for nucleation measurements the first step in determining the range of operation is to calculate the upper pressure limit.[7] Next, we recommend a series of S_{cr} measurements at constant nucleation temperature, spanning the largest possible total pressure range consistent with the upper total pressure limit. Extrapolation of the pressure dependent data to lower pressures in conjunction with the heat and mass flux and plate temperature criteria described earlier should be used to empirically determine the lower total pressure limit. The region between the upper and lower total pressure limits gives the range of "proper" TDCC operation.

SUMMARY AND CONCLUSIONS

We report results of new nucleation experiments involving 1-pentanol with hydrogen as the background gas. We discuss the existence of a lower total pressure limit for operating the TDCC, and we describe how violating this limit is manifested by the experimental data. We also provide, for the first time, an empirical procedure for determining the lower total pressure limit. Since the upper and lower total pressure limits together impose substantial limitations on the allowable operational range of the TDCC - especially at lower nucleation temperatures and total pressures - there appears to be a serious question concerning the reliability of a significant amount of the nucleation data in the literature resulting from past TDCC studies. Finally, we suggest that the TDCC is best suited for nucleation measurements at higher temperatures and total pressures where the region of stable operation (e.g., region II) is widest.

* Current address: School of Engineering, Manhattan College, Riverdale, NY 10471
† Current address: Bayer Corporation, Baytown, TX

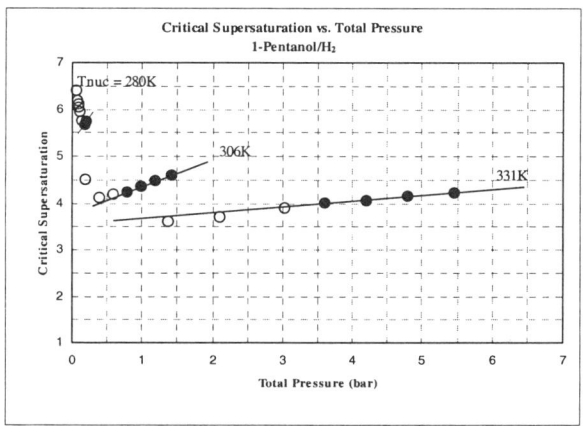

FIGURE 1. Experimentally determined variation of the critical supersaturation of 1-pentanol with total pressure using hydrogen as a background gas. The nucleation temperatures are indicated. The solid lines are fit to the data indicated by solid symbols. See the text for a discussion of the open symbols

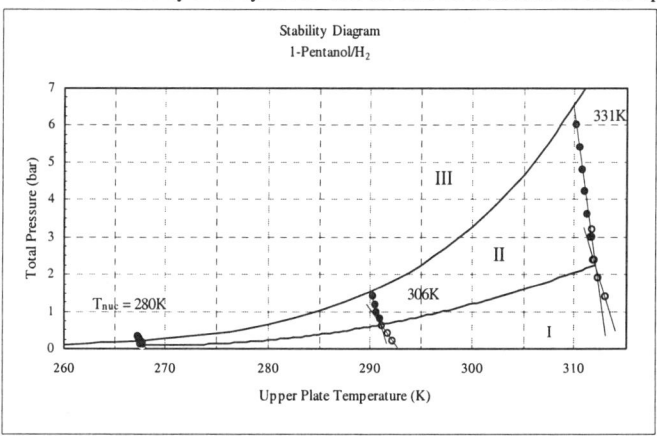

FIGURE 2. Stability Diagram for 1-Pentanol - Hydrogen. The upper and lower total pressure limits of stability are indicated by the solid curves. Data points obtained in the region of "proper" chamber operation (II) are indicated by solid symbols and data points obtained while operating below the lower limit (I) by open symbols. The straight lines are added to illustrate the change in slope of the constant nucleation temperature data.

REFERENCES

1. Heist, R.H. and He, H., *J. Phys. Chem. Ref. Data* **23**, 781 (1994)
2. Bertelsmann, A., Stuczynski, R. and Heist, R.H., *J. Phys. Chem.* **100**, 9762 (1996)
3. Bertelsmann, A. and Heist, R.H., *Atmospheric Research*, **46**, 195 (1998)
4. Ferguson, F.T., Heist, R.H. and Nuth, III, J.A., *J. Chem. Phys.* submitted (2000)
5. Ferguson, F.T. and Nuth, III, J.A., *J. Chem. Phys.* **111**, 8013 (1999)
6. Bertelsmann, A. and Heist, R.H., *J. Chem. Phys.* **106**, 610 (1997)
7. Bertelsmann, A. and Heist, R.H., *J. Chem. Phys.* **106**, 624 (1997)
8. Heist, R.H., Martinez, D., Chan, Y.F and Bertelsmann, A., *J. Phys. Chem.* submitted (2000)
9. Turner, J.S., *Ann. Rev. Fluid Mech.* **17**, 11 (1985); Veronis, G., *J. Mar. Res.* **23**, 1 (1965)
10. Palmer, H.J., private communication.
11. Ye, P., Bertelsmann, A. and Heist, R.H., submitted to *J. Phys. Chem.* (2000)

Photoinduced Nucleation: A Novel Tool for Detecting Molecules in Air at Ultra-low Concentrations

Joseph L. Katz[†], Brent A. Johnson[†], Heikki Lihavainen[‡], Markus M. Rudek[†], Brian C. Salter[†]

[†]Department of Chemical Engineering, The Johns Hopkins University, Baltimore, MD 21218.
[‡]Visiting research scientist at The Johns Hopkins University Department of Chemical Engineering from the Finnish Meteorological Institute, Sahaajankatu 20E, 00810 Helsinki, Finland.

Abstract. This work presents the development of a novel detection method named Photoinduced Nucleation Detection (PIND). PIND is capable of detecting substances at extremely low concentrations in ambient air. Methods of obtaining substance specificity including the substance's UV light absorption spectrum, and gas chromatography are also presented.

Photoinduced Nucleation Detection (PIND) consists of illuminating an air stream containing molecules of a contaminant with UV light in a fused silica cell. The photoproducts formed in the cell then flow into a saturator, which is saturated with a nucleating vapor. The stream is then cooled to make the vapor supersaturated. The nucleating vapor then condenses onto the photoproducts formed in the silica cell to a size that is easily detected by light scattering.

2,4,6 trinitrotoluene and 2,6 dinitrotoluene were studied. The saturation concentrations of these substances at room temperature are approximately 10 ppb and 700 ppb respectively. Concentrations as small as about 5 ppt for TNT and 10 ppt for 2,6 DNT were detectable (see figure 1). Note that these concentrations are 2000 and 70000 fold *smaller* respectively than their saturation concentrations; a result which suggests that PIND can be developed into a tool for detecting unwanted substances in air including explosives, drugs, etc. Since these results will be published well before the meeting[1] they will not be further described here.

However, in addition to detection sensitivity, it is important to identify the substance causing the photoinduced nucleation. One way of accomplishing this is by measuring the dependence of the nucleation rate on the wavelength of the UV light. Figure 2 (also taken from reference 1) shows the nucleation rate dependence on wavelength (called "the nucleation rate spectrum") for a flowing stream of air containing 100 ppm of OTA illuminated by light

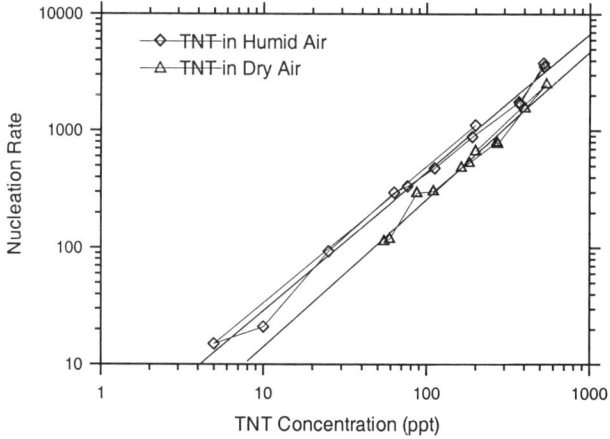

Figure 1. The dependence of nucleation rate on TNT concentration in both humid and dry air.

from a 150 W Xenon arc lamp - monochromator combination. This figure also shows its vapor phase light absorption spectrum. One sees that the light absorption spectrum exactly matches the variation with wavelength of the nucleation rate spectrum. Thus, the nucleation rate spectrum enables one to identify the substance. One also can determine the concentration by calibration.

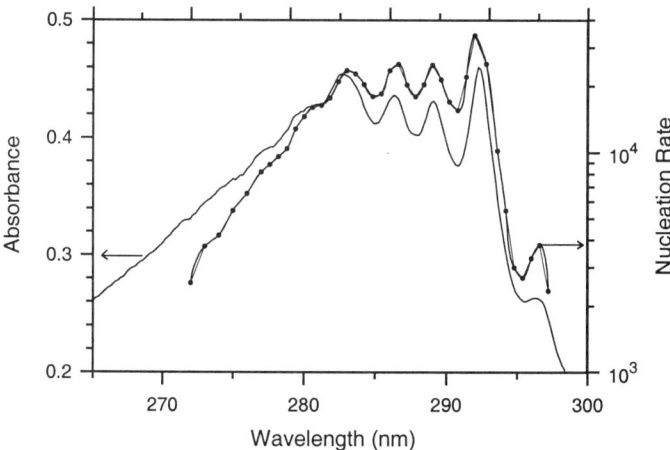

Figure 2. Comparison of the nucleation rate spectrum of OTA (the line with the data points) with its vapor phase light absorption spectrum.

A special difficulty which arises in measurements at extremely low concentrations is that adsorption of the molecules to be detected onto surfaces (e.g. tubing and the fused silica cell) will give misleading results. In the earlier studies, to prevent adsorption the tubing and fused silica cell in the experimental setup were wrapped with insulated heating tape and heated to 200°C. However, it is possible that at this temperature some of the substance of interest (e.g. the TNT) may decompose and thus give misleading results. Studies are underway to determine the affects.

As another means of obtaining substance specificity, PIND was combined with gas chromatography. Separations using packed columns were possible for OTA and 2,6 DNT at 4 ppb (see figure 3), however the high background rate due to bleeding of the stationary phase resulted in a detection limit for these substances approximately three orders of magnitude larger than those obtained from previous experiments. A 1 m *uncoated* multicapillary column consisting of 900 fused silica tubes in parallel, each tube 20 μm in diameter was obtained from Alltech. This column is now being investigated and the results will be reported at the meeting.

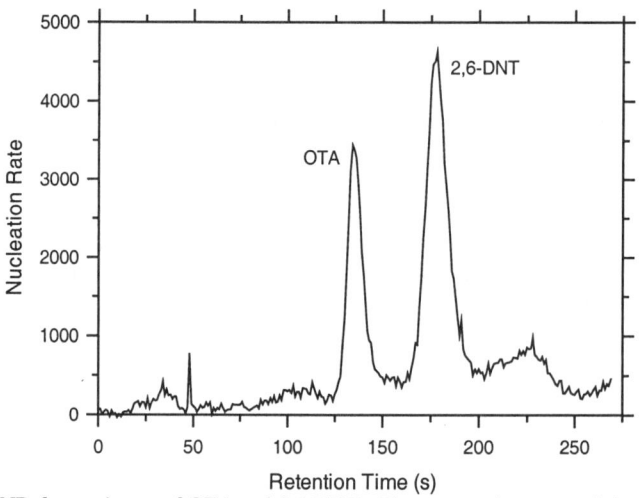

Figure 3. GC/PIND for a mixture of OTA and 2,6 DNT. The vapor mixture was injected into a GC, separated into individual substances, and detected using PIND.

References

1. Katz, J.L., Lihavainen, H., Rudek, M.M., Salter, B.C., *Journal of Chemical Physics*, **112**, No.19, 2000

Condensation Induced Oscillations in Supersonic Expansion Flows

G. Lamanna and M.E.H. van Dongen

*Eindhoven University of Technology, Dept. of Applied Physics,
P.O. Box 513, 5600 MB Eindhoven, The Netherlands*

Abstract. In supersonic expansion flows of humid nitrogen, at relatively low cooling rates, self-sustained oscillations may occur due to the combined effects of nucleation, droplet growth, and latent heat release. Different modes of oscillation are found depending on the initial humidity. It is shown numerically that the stability limit and the oscillation frequencies of the different modes strongly depend on the nucleation rate. Numerical results are compared with experiments of Adam and Schnerr and own experiments for nozzles of different shapes. The ICCT nucleation model yields correct results for both series of experiments, when corrected with a factor of the order of 0.01.

INTRODUCTION

During the expansion of humid nitrogen in a supersonic nozzle, a state of high supersaturation may be reached which is indicative of a radical departure from the condition of equilibrium phase distribution. Upon the occurrence of both nucleation and droplet growth, latent heat is released to the flow, which reacts by re-adjusting itself to the new thermodynamic state. Depending on the initial stagnation conditions, different flow regimes can be envisaged. If the amount of heat released exceeds a critical value, the flow becomes thermally choked and a steady shock appears embedded in the nucleation zone. This flow situation is depicted in Fig.1.

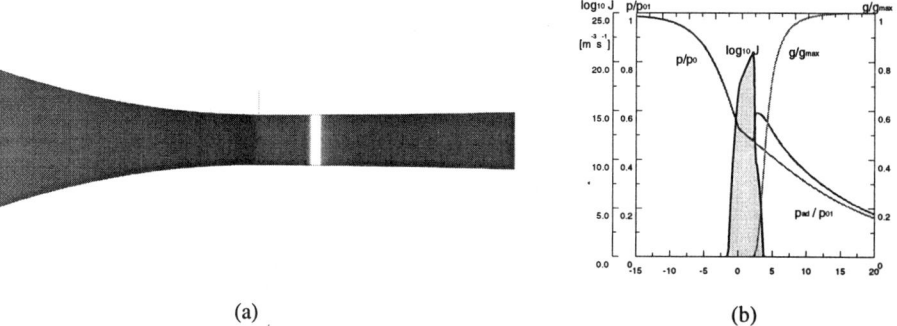

(a) (b)

FIGURE 1. Steady flow for nozzle G2 (a) - Stagnation conditions: $P_0 = 8.67 \cdot 10^4$ Pa, $T_0 = 296.6$ K, $S_0 = 0.48$. Part (a): Numerical schlieren picture of a steady normal shock. Part (b): Axial distribution of non-dimensional pressure, condensate mass fraction, and nucleation rate.

From the picture the details of the condensation process can be inferred immediately. First nucleation sets in and no perturbation of the flow properties is observed. As soon as the growth process starts, heat is released to the flow and a shock appears which interrupts the nucleation process. As a result the nucleation rate is a sharply peaked function of x. In fact, nucleation effectively occurs at one specific thermodynamic state.

If the amount of heat released increases further, a steady solution cannot be obtained and self sustained oscillations appear. A detailed description of the different oscillation modes can be found in Adam & Schnerr [1]. Here the influence on the nucleation rates is stressed, as shown in Fig. 2, which depicts profiles of temperature and nucleation rate at different times in one cycle.

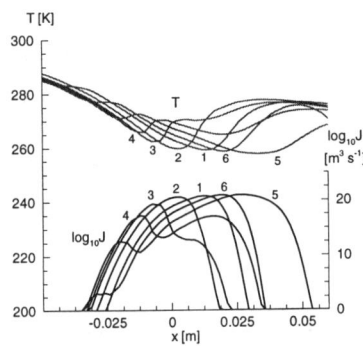

FIGURE 2. Unsteady flow - Stagnation conditions: $P_0 = 8.63 \cdot 10^4$ Pa, $T_0 = 296.9$ K, $S_0 = 1.30$. Time evolution of nucleation rate and temperature along the nozzle axis. $\Delta t = 0.86$ ms.

Contrary to the steady case, now the effects of condensation are spread along a much wider zone. This is due to the interaction between the nucleation process and upstream moving disturbances: as the temperature perturbation propagates upstream, it quenches the nucleation at different positions along the nozzle axis (curves 6, 1, 2, 3, 4). As the disturbance moves away, the nucleation process re-acquires again its maximum strength (curve 5). Due to this interplay mechanism, nucleation takes place at different temperatures in the range [260÷270] K. Unsteady flow regimes, thus, may constitute a valid tool to test nucleation rates for a range of temperatures and supersaturations. It is the intention of this paper to validate a particular nucleation model on the basis of different sets of experiments for two nozzle shapes.

MODELLING

Modelling of homogeneous condensation can be separated in a nucleation and droplet growth model. On the basis of Luijten's experiments on water-nitrogen systems [2, 3], the *Internally Consistent Classical Theory* (ICCT) was chosen to model the nucleation process. For the droplet growth process, a generalized transitional growth model, encompassing both the continuum and free molecular growth regimes, was adopted. A detailed description of the growth model can be found in [4]. It should be explicitly noted that for the correct simulation of condensation effects in Laval nozzle flow, it is fundamental to predict exactly the rate of heat addition at the very beginning of the growth process. In this phase, growth is controlled by the Hertz-Knudsen law, with the droplet temperature calculated explicitly via the wet-bulb equation. This issue is discussed in a detailed manner in [5]. The two-phase flow is assumed inviscid and two-dimensional; for the vapor mixture, ideal gas behavior is adopted. The time dependent Euler equations are solved in conservation form, plus 4 additional equations describing the formation of condensate. For a detailed description of the numerical scheme and related accuracy, the reader is referred to Prast's Ph.D. thesis [4].

EFFECT OF NUCLEATION MODEL ON OSCILLATORY BEHAVIOUR

As pointed out in the previous section, it is essential to predict correctly the rate of heat addition in correspondence of maximum nucleation rates. This, in fact, determines the shock strength, its relative speed, and thus its upstream travelling time, which ultimately defines the frequency of the oscillating regime. On the other hand, the specific expression for the nucleation rate is also very important. In fact, the Mach number, at which nucleation is maximum, influences drastically the transition from steady to unsteady flow regime or from one mode of oscillation to another. This effect is shown clearly in Fig. 3. By expressing the nucleation rate as $J_{ICCT} = \xi \cdot J_{ICCT}$, four different conditions were considered, corresponding to values of the parameter ξ of 1, 0.1, and 0.01 respectively.

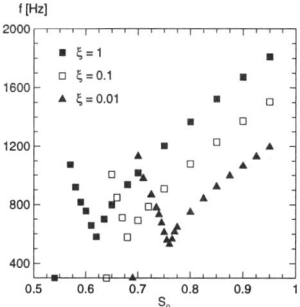

FIGURE 3. Frequency plot of the different oscillation modes as function of the relative humidity. Nozzle: S2 – Adam & Schnerr [1]. Stagnation Conditions: $P_0 = 1.00 \cdot 10^5$ Pa, $T_0 = 295.0$ K.

The points on the horizontal axis are representative of the steady limit, an additional increase in flow humidity causes the flow to become unsteady. For higher values of the parameter ξ, the stability limit shifts towards lower values of S_0: stronger nucleation rates shift the onset of condensation upstream towards the nozzle throat and thus the stability limit is reached earlier in correspondence of lower supersaturations.

The switching from Mode I to Mode II of oscillations can be also identified very easily: it corresponds with the frequency minimum in the plot. Similar considerations hold also for this limiting point.

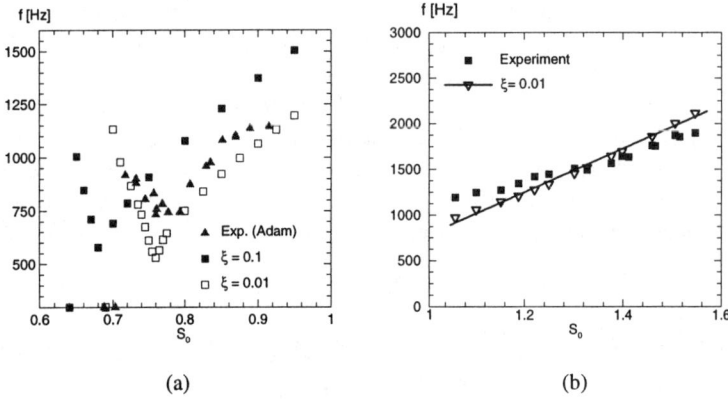

FIGURE 4. Experimental and numerical frequencies versus initial supersaturations. Part (a): Nozzle S2 - $P_0 = 1.00 \cdot 10^5$ Pa, $T_0 = 295.0$ K. Part (b): Nozzle G2 - $P_0 = 8.78 \cdot 10^4$ Pa, $T_0 = 296.6$ K.

Figure 4 compares the numerical results with the experimentally determined frequencies for two different nozzles: nozzle S2 [5] and G2 [4]. The best agreement is found for values of the parameter ξ close to 0.01, thus indicating that the ICCT nucleation theory slightly overestimates the actual nucleation rates.

CONCLUSIONS

It has been shown that supersonic nozzle flows can be a useful tool in assessing the quality of a certain nucleation model, since they provide the possibility to test nucleation rates for a range of supersaturations and temperatures. Especially the unsteady experiments are particular useful in this respect. In this paper, the quality of the ICCT nucleation theory has been checked for a range of temperatures between [240-300] K and a stagnation pressure of 1 bar. Comparison with experimental results shows that the best agreement is found with a correction factor of about 0.01.

ACKNOWLEDGMENTS

The authors would like to express their gratitude to the Netherlands Foundation for Fundamental Research on Matter for supporting this work with grant no.97.1293.

REFERENCES

1. Adam, S., and Schnerr, G., *J. Fluid Mechanics* **348**, 1-28 (1997).
2. Luijten, C.C.M., and van Dongen, M.E.H., *J. Chem. Phys.* **111**, 8524-8534 (1999).
3. Luijten, C.C.M., and van Dongen, M.E.H., *J. Chem. Phys.* **111**, 8535-8544 (1999).
4. Prast, B., Ph.D. Thesis, Eindhoven University of Technology, (to be published).
5. van Dongen, M.E.H., Lamanna, G., and Prast, B., *ZAMM*, (to be published).
6. Adam, S., Ph.D. Thesis, Universität Karlsruhe (TH) (1996).

Thermodynamic Considerations on the Homogeneous Nucleation in an Isolated System

Wolfram Vogelsberger and Mario Löbbus

Institute of Physical Chemistry, Chemistry and Earth Science Faculty, Friedrich-Schiller-University Jena, Lessingstrasse 10, D-07743 Jena, Germany

Abstract. The nucleation of a pure vapor is investigated in an isolated system. It is assumed that the amount of vapor is reduced by the number and size of the droplets formed. The entropy as the natural thermodynamic potential is calculated dependent on both of these variables droplet radius and number of droplets. Possible stable states of the system are relative maxima of the entropy. The most stable state is given by the Clausius-Clapeyron equation.

INTRODUCTION

Supersaturated states of a vapor are commonly generated by isentropic expansion. The supersaturated state generated by this way is considered as an isolated system, i.e. a box of volume, v, containing an amount of vapor molecules, n_0, having a pressure p_0, and a temperature, T_0. Within this system nucleation occurs by the formation of a number of droplets, Z, of spherical shape and uniform radius, R. Therefore, the vapor pressure in the system is reduced during nucleation according to the size and the number density of the droplets. Throughout the droplet formation the heat of condensation, $\Delta_K u$, is released which results in a temperature increase of the system. The appropriate thermodynamic potential of an isolated system is the entropy, S. It is therefore necessary to calculate the entropy of the system for all combinations of droplet size and droplet number density which can be formed from the initial amount of vapor molecules. The initial supersaturated vapor can be used as the reference state. Stable states of the system are given by possible maxima of the entropy.

THERMODYNAMIC CALCULATIONS

To calculate the change of entropy of the system during the transition from the initial supersaturated vapor to the system containing Z droplets of size R we have to look for a reversible way between these states. The reversible way consists of the following steps

i. Isothermal expansion of the vapor from the initial state to the saturated state, p_0^s.
ii. Isobaric heating of the system from temperature T_0 to temperature T.
iii. Isothermal compression of the vapor from p_0^s to a new equilibrium pressure p_1^s.
iv. Formation of an amount of liquid, n_l.
v. Formation of Z droplets of size R from the amount of liquid, n_l.
vi. Isothermal expansion or compression of the system to the initial volume, v.

The total amount of substance, n_0, the volume, v, and the total energy of the system have to be maintained since we are concerned with an isolated system. Hence, the actual temperature, T, of the system is given by the following equation

$$\frac{T}{T_0} = \frac{1 - \left(\frac{\Delta_K U(T_0) - \Delta_K C_V T_0}{C_V^g T_0}\right)\frac{n_l}{n_0} - \frac{\sigma}{n_0} \frac{o_l}{C_V^g T_0}}{1 + \frac{\Delta_K C_V}{C_V^g} \frac{n_l}{n_0}} \quad . \tag{1}$$

$\Delta_K U$ is the molar energy of condensation, $\Delta_K C_V$ is the change in heat capacity at constant volume during the condensation, and C_V^g is the molar heat capacity at constant volume of the vapor, σ is the surface tension and o_l is the surface area of the droplets.

The actual pressure in the system is given by

$$p = p_0 \left(1 - \frac{n_l}{n_0}\right) \frac{T}{T_0} \quad . \tag{2}$$

Calculation of the entropy of the system results in the following expression

$$S = \left(\frac{C_V^g}{R_G} + \frac{\Delta_K C_V}{R_G} \frac{n_l}{n_0}\right) \ln\left(\frac{T}{T_0}\right) + \frac{n_l}{n_0}\left(\frac{\Delta_K U(T_0)}{R_G T_0} - 1\right) + \ln y$$
$$- \left(1 - \frac{n_l}{n_0}\right) \ln\left\{y\left(1 - \frac{n_l}{n_0}\right)\right\} - \frac{\sigma}{n_0 R_G T} \frac{o_l}{} \tag{3}$$

The entropy, S, in eq. (3) is a dimensionless quantity obtained by dividing the entropy by the product, $n_0 R_G$. Furthermore, y is the supersaturation p_0/p_0^s and R_G is the gas constant. First the surface term (last term) in eq. (3) is set equal to zero, i.e. the investigation of a bulk liquid. Derivation of eq. (3) with respect to n_l/n_0 and setting $\partial S/\partial(n_l/n_0)=0$ leads to the familiar Clausius-Clapeyron-equation with the second Ulich approximation.

$$\ln\left(\frac{p_1^s}{p_0^s}\right) = \left(\frac{\Delta_K U(T_0) - \Delta_K C_V T_0}{R_G}\right)\left(\frac{1}{T} - \frac{1}{T_0}\right) - \left(\frac{\Delta_K C_V}{R_G} - 1\right) \ln\left(\frac{T}{T_0}\right) \tag{4}$$

In eq. (4) p_1^s is the saturation pressure at the temperature, T. The further discussion is appreciably simplified if the following dimensionless quantities are introduced

$$\frac{n_l}{n_0} = \frac{4\pi}{3V_l} \frac{Z}{n_0} R^3 = zr^3, \quad z = \frac{Z}{N_0}, \quad r = R\left(\frac{4\pi N_A}{3V_l}\right)^{\frac{1}{3}} = \frac{R}{R_0}, \tag{5}$$

N_0 is the initial number of molecules in the system and N_A is the Avogadro constant and R_0 is the radius of a molecule. The entropy of the isolated system will be discussed with these quantities.

DISCUSSION

The entropy is a function of the number and the size of the droplets formed in the system, S(r,z). This is shown in figure 1. The representation and all calculations are carried out with Mathematica [1]. The lower part of this figure shows the surface itself. In the upper part of the figure a contour plot of the entropy surface is depicted. By way of illustration S(r,z) is calculated for water at 298 K and a supersaturation of 1.75. This surface covers all possible states of the corresponding isolated system. The first state is the initial state where all molecules are present in the vapor phase. On the other hand, all the molecules may condense in liquid droplets, which is, of course, very unlikely. In this case a large number of states is possible since the molecules may be contained in a different number of droplets. These states are given by the relation $zr^3=1$. The representation of this relation limits the range of definition of S(r,z). The entropy surface therefore covers an infinite number of states characterized by different distributions of the molecules on vapor and droplets with a certain number and size.

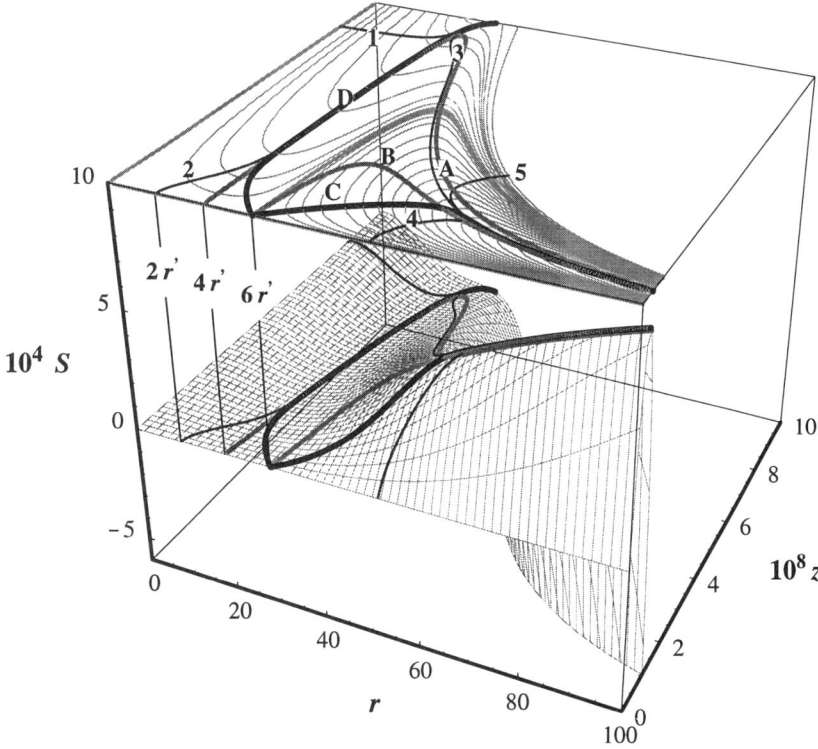

FIGURE 1. Surface of the entropy of droplet formation, S, as a function of number of droplets, z, and radius of droplets, r.

Here, the question arises, whether there are some states given by special properties in a mathematical sense. In principle states of high entropy are preferred in the system. The presented approach tries to give some answers in this direction. The contour line for S(r,z)=0 starts at 6r' where r' is given by r'=$\sigma V_l/(R_G T_0 \ln(y) R_0)$. Extrema of the entropy can be found by derivation of eq. (3) with respect to r and z. For it the corresponding partial derivatives have to be set equal to zero. The resulting transcendental expressions give a relation between z and r. These equations can only be solved numerically. The expression which results from $(\partial S(r,z)/\partial r)_z=0$ is depicted in figure 1. by the gray line labeled A. In the range $0<z<9.01\ 10^{-8}$ for a given value of z two radii can be found which correspond to two extrema of S. The smaller radius belongs to a minimum and the larger one to a maximum. An analogous situation can be found for the equation $(\partial S(r,z)/\partial z)_r=0$ in the range $0<z<2.67\ 10^{-8}$. In this case the gray line labeled B results. The starting points of both of these curves may be determined by

$$a)\quad \lim_{z\to 0}\left(\frac{\partial S(r,z)}{\partial r}\right)_z = 0, \quad\text{and}\quad b)\quad \lim_{z\to 0}\left(\frac{\partial S(r,z)}{\partial z}\right)_r = 0 \qquad (6)$$

which yields

$$a) \quad r = \frac{4}{R_0} \frac{V_l \sigma}{RT_0} \frac{1}{\ln y} \quad \text{and} \quad b) \quad r = \frac{6}{R_0} \frac{V_l \sigma}{RT_0} \frac{1}{\ln y} \qquad (7)$$

These expressions are similar to those well known from the familiar Kelvin equation. However, in case a) twice and in case b) three times the Kelvin value for droplets is found. Both curves A and B approach each other for large radii and small z values.

Since a real system tends to approach the most stable state the question arises what happens if an arbitrary point (r,z) on the surface is selected. The higher the entropy the higher the stability is. The gradient gives the direction of the steepest ascent. Therefore, the system evolves in this direction. Examples for this evolution of the system are given by the graphs labeled with numbers 1 to 5. The development of the system along this paths proceeds in the direction of increasing entropy. Two groups of curves may be distinguished. Graphs 1 and 2 lead to final states where either r or z are equal to zero. In this case the system tends to have its initial state, i.e. the pure supersaturated vapor. The other group graphs 3 to 5 approach the line labeled C. The line C starts at zero entropy and the droplet radius given by eq. (7b). Obviously, this path represents the crest of the entropy surface. The end of this crest is the state of maximal entropy in the system and, therefore, the most stable state. All the curves A, B, and C approach each other and meet at this state. For the most stable state of the system eq. (4) holds (Clausius-Clapeyron equation) which is the limes of entropy for $z \to 0$ and $r \to \infty$. Numerical evaluation leads to $n_l/n_0 = 6.312 \cdot 10^{-3}$ and $S = 1.736 \cdot 10^{-3}$ for the most stable state of the given system. The two groups of graphs are separated from each other by the curve labeled D. Curve D starts also at r=6r'. It rapidly approaches line A and coincides with A for a while before reaching the maximum of A. A system characterized by a z-r combination on the left hand side of curve D tends to decompose in molecules. On the other hand a system having a combination z-r on the right hand side of curve D tends to approach the most stable state. The graphs discussed in the contour plot are also depicted in the representation of the entropy surface in the lower part of figure 1.

CONCLUSIONS

The nucleation process in an isolated system has been investigated by the calculation of the entropy dependent on the number of droplets and droplet size. Temperature and pressure change during this process. Possible states of the system, liquid droplets-vapor, form an entropy surface S(r,z). This surface is subdivided into two parts by the graph D in figure 1. States on the left hand side of graph D (small droplets) tend to decompose in the molecular species (pure vapor). States on the right hand side (larger droplets) refer to a system consisting of droplets and vapor. Some of these states are more stable than neighboring states due to their higher entropy. The most stable state of the system is determined by the Clausius-Clapeyron equation which gives the equilibrium between bulk liquid and saturated vapor. Graph C represents the most probable path of a system consisting of small droplets and vapor to achieve the stable equilibrium state. This refers to the bulk liquid under saturation conditions. Graph C represents a crest of the entropy surface.

REFERENCES

1. Wolfram, S., *Mathematica 3. Auflage*, Bonn et. al.: Addison-Wesley, 1997.

Nucleation at Retrograde Condensation

Vladimir G. Baidakov and Grey Sh. Boltachev

*Institute of Thermal Physics, Ural Branch of the Russian Academy of Sciences,
GSP-828, Ekaterinburg, 620219, Russia*

Abstract. A binary Van-der-Waals solution is considered. Within the framework of the Gibbs and Van-der-Waals capillary theories under the conditions of constancy of the temperature and the composition the size dependences of surface tension $\sigma(R)$ are calculated for critical nuclei of liquid and vapor. The phenomenon of "closure" of size dependences $\sigma(R)$ and the presence anomalously of high maximums on them are established.

The homogeneous nucleation rate is largely determined by value of critical nucleus formation work W_*. According to Gibbs dividing surfaces approach [1]:

$$W_* = \frac{16\pi}{3} \frac{\sigma^3}{(p_*'' - p')^2}, \qquad (1)$$

where σ is the surface tension at boundary of a critical nucleus, p is the pressure. Hereinafter one prime indicates an initial, metastable phase and two a prime indicate a nucleus. In a binary system the dissolution of fugitive component in solvent liquid is subject to accompaniment by large adsorption at liquid-vapor boundary. Last leads to a substantial variation of surface tension.

The thermodynamic and structural characteristics of binary solution interface can be described within the framework of the Van-der-Waals capillary theory [2]. Then the Helmholtz free energy of an inhomogeneous system is noted as

$$F\{\rho_1(r), \rho_2(r)\} = \int \left(f(\rho_1, \rho_2) + \sum_{i,j=1}^{2} \kappa_{ij} \nabla \rho_1 \nabla \rho_2 \right) dr, \qquad (2)$$

where f is the free energy per unit volume of a homogeneous system, κ_{ij} is the matrix of influence parameters. The equilibrium density distributions of components at interface $\rho_1(r)$, $\rho_2(r)$ correspond to a saddle point of the grand potential $\Omega = F - \sum \mu_i N_i$ (μ_i is a chemical potential, N_i is number of particles of ith component) and are deduced from a solution of Euler-Lagrange equation system for this functional:

$$\frac{d^2\rho_0}{dr^2} + \frac{2}{r}\frac{d\rho_0}{dr} = \frac{\mu_2(\rho_1,\rho_2) - \mu_2(\rho_1',\rho_2')}{2\kappa_{22}}, \qquad \rho_0 = (\kappa_{11}\kappa_{22})^{1/2}\rho_1 + \rho_2, \qquad (3)$$

$$(\kappa_{22}/\kappa_{11})^{1/2}[\mu_1(\rho_1,\rho_2) - \mu_1(\rho_1',\rho_2')] = [\mu_2(\rho_1,\rho_2) - \mu_2(\rho_1',\rho_2')]. \qquad (4)$$

The boundary conditions are $\rho_i \to \rho_i'$ at $r \to \infty$ and $d\rho_i/dr \to 0$ at $r \to 0$. The critical nucleus formation work is

$$W_* = \min \max \Delta\Omega\{\rho_1,\rho_2\}, \qquad (5)$$

the surface tension is determined by Eq. (1) and the radius of the tension surface R is determined by Young-Laplace equation:

$$p_*'' - p' = 2\sigma/R. \qquad (6)$$

The pressure in a critical nucleus p_*'' and the composition of it x_*'' are determined by requirements of substantial equilibrium [1]

$$\mu_1(p',x') = \mu_1(p_*'',x_*''), \qquad \mu_2(p',x') = \mu_2(p_*'',x_*''). \qquad (7)$$

States of the binary solution were described by Van-der-Waals equation of state with one-fluid approximation:

$$\tilde{p} = \frac{8\tilde{\rho}\tilde{T}}{3 - \tilde{b}\tilde{\rho}} - 3\tilde{a}\tilde{\rho}^2, \qquad (8)$$

where $\tilde{p} = p/p_{c,2}$, $\tilde{T} = T/T_{c,2}$, $\tilde{\rho} = \rho/\rho_{c,2}$; $p_{c,2}$, $T_{c,2}$, $\rho_{c,2}$ are the pressure, the temperature and the density at critical point of the second component;

$$\begin{aligned}\tilde{a} &= (1 - x + \xi_a x)^2, & \xi_a &= \sqrt{a_{11}/a_{22}}, \\ \tilde{b} &= (1 - x + \xi_b x)^2, & \xi_b &= \sqrt{b_{11}/b_{22}},\end{aligned} \qquad (9)$$

$x = \rho_1/(\rho_1 + \rho_2)$ is the concentration, a_{ii} and b_{ii} are pure component parameters. For influence coefficients in the Eq. (2) within the framework of the mean-field theory we have [3]:

$$\kappa_{ij} = \frac{13}{6}a_{ij}\left(\frac{3b_{ij}}{2\pi}\right)^{2/3}. \qquad (10)$$

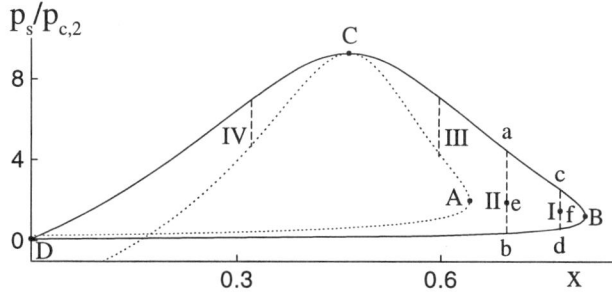

FIGURE 1. The phase diagram of explored solution at $T=0.6$. The solid line is binodal, the dotted line is spinodal; I, II, III and IV are lines of penetration into metastable region.

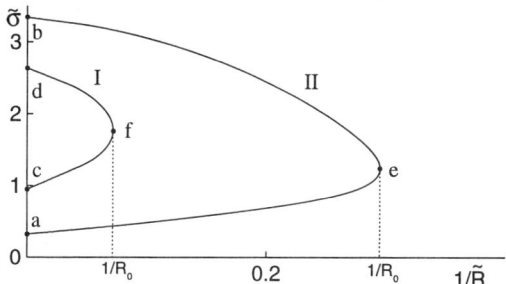

FIGURE 2. The size dependences of surface tension $\tilde{\sigma}=\sigma\rho_{c,2}^{1/3}/p_{c,2}$ at $T=0.6$ along lines I and II (see Fig. 1); $\tilde{R}=R\rho_{c,2}^{1/3}$; R_0 is the minimum size of a critical nucleus along a selected line.

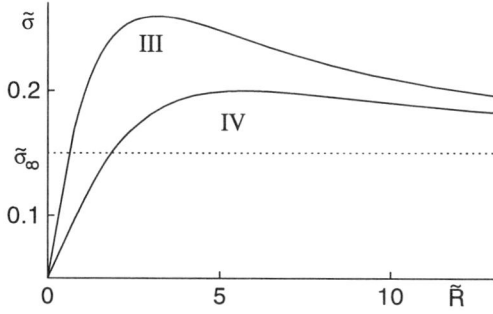

FIGURE 3. The size dependences of surface tension at $T=0.6$ along lines III and IV (see Fig. 1)

The equation of state (8) contains two individual parameters ξ_a and ξ_b, which specify a type of the solution.

Fig. 1 demonstrates the phase diagram of solution with $\xi_a = 0.4$ and $\xi_b = 1$. This solution is characterized by presence of retrograde condensation region. The critical point C is a point of tangency of binodal and spinodal. The branch of binodal at higher content of the fugitive component describes the vapor phase. The presence at fixed temperature and composition of two values of saturation pressure causes an opportunity of passing here of peculiar processes of phase formation which are not characteristic for one-component system. As follows from the Fig. 1 if the initial concentration of mixture is more than the concentration at the point A of spinodal then the isothermal process of vapor compression from the curve DB or vapor expansion from the curve CB can continue uninterruptedly without a phase transition.

The energy barrier of nucleation W_* conserves finite value different from zero on all interval (the line ab). It leads to peculiar size dependences of nuclei surface tension $\sigma(R)$ (Fig. 2). The expansion of vapor phase from the point a corresponds to increase of surface tension. As this takes place, the radius of a critical nucleus decreases down to some minimal size R_0. Thereafter further expansion leads to growth of the critical nucleus size without disruption of stability.

If the expansion of vapor phase at $x = $ const happens more to the left of the point A then the process of phase formation undergoes qualitative changes. Here the regions of retrograde and normal condensation are separated by region of solution instability. Going toward the spinodal, R and $\sigma \rightarrow 0$. At retrograde condensation for critical drops of liquid on the curve of dependence $\sigma(R)$ there is a maximum and the surface tension at the point of a maximum σ_{max} can more than twice exceed the flat limit σ_∞ (Fig. 3). For comparison, in solutions with a complete mixing of components the quantity $(\sigma_{max} - \sigma_\infty)/\sigma_\infty$ tends to be a part of percent [3].

ACKNOWLEDGMENTS

The authors thank RFBR (Grant No. 99-02-16377) for support of this work.

REFERENCES

1. Gibbs, J. W., Collected Works, New Haven, Connecticut: Yale University Press, 1948.
2. Van der Waals, J. D., Kohnstamm, Ph., Lehrbuch der Thermostatik, Leipzig: Verlag Von Johann Ambrosius Barth, 1927.
3. Baidakov, V. G., Boltachev, G. Sh., J. Chem. Phys. **71**, 1965-1970 (1997) (in Russian).

The Thermodynamic Theory of Effects of Internal and External Electric Field in Nucleation

Alexander K. Shchekin, Vadim B. Warshavsky and Mikhail S. Kshevetskiy

Department of Statistical Physics, Research Institute of Physics, St Petersburg State University, Ulyanovskaya 1, Petrodvoretz, 198904 Russia

Abstract. The procedure for finding the profile and nucleation characteristics of a dielectric droplet in an arbitrary axisymmetric electric field is illustrated in the case of external uniform electric field and the case of non-uniform internal field of the heterogeneous nucleus with an electric dipole moment. The results of analytical theory (perturbation theory assuming small deformation of the droplet shape in the electric field) for the chemical potential and free energy of droplet formation are shown to be valid even for rather strong fields due to mutual compensation in the surface and electric contributions.

INTRODUCTION

The effects of the electric field in nucleation reveal through the chemical potential of condensate in nucleating droplet or the work of droplet formation [1,2]. In order to obtain these thermodynamic characteristics of nucleation, we need a reliable procedure allowing simultaneous solving of the coupled non-linear equations for the equilibrium droplet shape and the electric field potential. A numerical approach to this problem in the case of uniform external electric field was proposed in Refs. [3,4]. Nevertheless, neither the thermodynamic characteristics relevant for nucleation theory nor the effects of the internal non-uniform non-spherical electric field were considered. In this communication we compare the results of our numerical approach applied for an arbitrary axisymmetric electric field with the analytical ones obtained previously [2,5,6] for the cases of the external uniform electric field and the non-uniform internal field of the heterogeneous condensation nucleus with an electric dipole moment.

1. BASIC EQUATIONS

Let us consider a droplet condensing out of the vapor-gas environment in the electric field of the axial symmetry. The source of the electric field may be located outside or inside the droplet (as in the case of the droplet formed on heterogeneous condensation nucleus). Hereafter, the indices α and β denote the quantities referred to the dielectric liquid and the vapor phase, correspondingly.

We choose the coordinate system with the center at the droplet mass center and the polar axis directed along the vector of the electric field intensity $\vec{E}$, and introduce $x \equiv \cos\theta$ where θ is the polar angle between the polar axis and the radius vector r' of the observation point. In this system the equation for the droplet profile $r(x)$ can be written as the pressure balance equation at the droplet surface in the form

$$\frac{\gamma}{\sqrt{r^2 + (1-x^2)r_x^2}}\left[2 + \frac{(1-x^2)(r_x^2 - rr_{xx}) + xrr_x}{r^2 + (1-x^2)r_x^2} + \frac{xr_x}{r}\right] - \frac{\varepsilon^\alpha - \varepsilon^\beta}{8\pi} \times$$

$$\times \left\{ \frac{\left[r^2\Phi_r^\alpha - (1-x^2)r_x\Phi_x^\alpha\right]\left[r^2\Phi_r^\beta - (1-x^2)r_x\Phi_x^\beta\right]}{r^2\left[r^2 + (1-x^2)r_x^2\right]} + \frac{(1-x^2)\left(\Phi_x^\alpha + r_x\Phi_r^\alpha\right)\left(\Phi_x^\beta + r_x\Phi_r^\beta\right)}{r^2 + (1-x^2)r_x^2} \right\} - P_0^\alpha + P_0^\beta = 0 \quad (1.1)$$

where γ is the scalar surface tension of the droplet, $\varepsilon \equiv \varepsilon(\mu)$ is the dielectric permittivity determined as a function of chemical potential μ, $r_x \equiv dr/dx$, $r_{xx} \equiv d^2r/dx^2$, $\Phi_r \equiv \partial\Phi/\partial r|_{r'=r(x)}$, $\Phi_x \equiv \partial\Phi/\partial x|_{r'=r(x)}$, the scalar pressure P_0 is determined in absence of the electric field at the same chemical potential μ as in presence of the electric field. In order to determine the electric potentials $\Phi^\alpha(r',x)$ и $\Phi^\beta(r',x)$, we need to solve the Laplace equations

$$\Delta\Phi^\alpha = 0, \quad \Delta\Phi^\beta = 0 \quad (1.2)$$

with the standard boundary conditions at the droplet surface $\Phi^\alpha\big|_{r'=r(x)} = \Phi^\beta\big|_{r'=r(x)}$, $\varepsilon^\alpha(\nabla\Phi^\alpha, \vec{n})\big|_{r'=r(x)} = \varepsilon^\beta(\nabla\Phi^\beta, \vec{n})\big|_{r'=r(x)}$ and the specific boundary conditions at the locations of field sources.

In view of the axial symmetry of the electric field, the droplet profile and electric potentials can be expanded in the Legendre polynomials $P_m(x)$ (m is the order of polynomial). In this way eqs.(1.1) and (1.2) are transformed into set of non-linear algebraic equations for coefficients of the expansion that can be solved by the Newton iteration procedure. Specifying the volume of the droplet with the profile found, one is able to reveal the dependence of the chemical potential of condensate in the droplet on droplet size. The droplet profile and the electric potentials also give the surface and electric contributions to the work of droplet formation.

2. EXTERNAL UNIFORM ELECTRIC FIELD

The effects of external uniform electric field on shape and stability of dielectric droplet are most worked out [2-5,8,9]. At small deviations of a droplet shape from sphere, parameter $\epsilon^2 = 9(\varepsilon-1)^2 \varepsilon^\beta RE_\infty^2 / 16\pi(\varepsilon+2)^2 \gamma$ (where $\varepsilon \equiv \varepsilon^\alpha/\varepsilon^\beta$, R is the radius of sphere having the same volume as the droplet, E_∞ the electric field intensity far from the droplet) has a geometrical meaning of the eccentricity of a prolate spheroid. The droplet profile in the uniform external electric field looks like prolate (along the field direction) spheroid for all values of parameter ϵ^2 until its critical value ϵ^2_{crit}, which depends on ε.

Figure 1 shows the dependence on ϵ^2 for the ratio G of the chemical potentials of droplet substance in presence and absence of the external field at $\varepsilon = 76$ ($\epsilon^2_{crit} = 0.4776$). Figure 2 depicts the behavior of sum $\overline{W}$ of the surface and electric contributions to the work of droplet formation as a function of ϵ^2. In Figs.1, 2 curve 1 represents the results of our numerical approach, curve 2 and curve 3 correspond to the analytic results from [5] and [2], respectively. As one can see, agreement between numerical and analytical results is rather good.

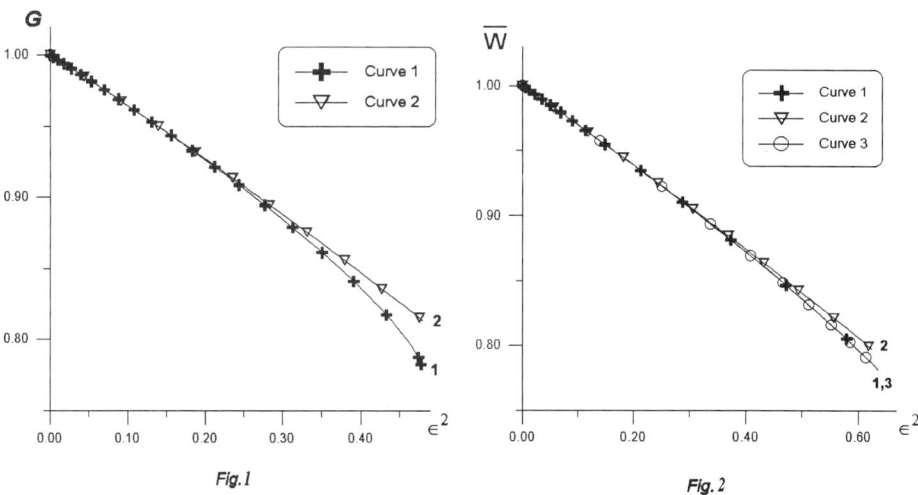

Fig.1 Fig.2

3. INTERNAL NON-UNIFORM FIELD OF HETEROGENEOUS NUCLEUS WITH AN ELECTRIC DIPOLE MOMENT

The droplet profile in the electric field of the dipole of the heterogeneous nucleus looks like oblate (along the dipole moment direction) spheroid only for small values of parameter $\in^2 = 9(\varepsilon-4)(\varepsilon-1)p^2 / 16\pi\varepsilon^\alpha(\varepsilon+2)^2 \gamma R^5$ (where p is the electric dipole moment of the nucleus, R is the radius of sphere having the same volume as the droplet) and transforms to the apple-like shape at large $\in^2$.

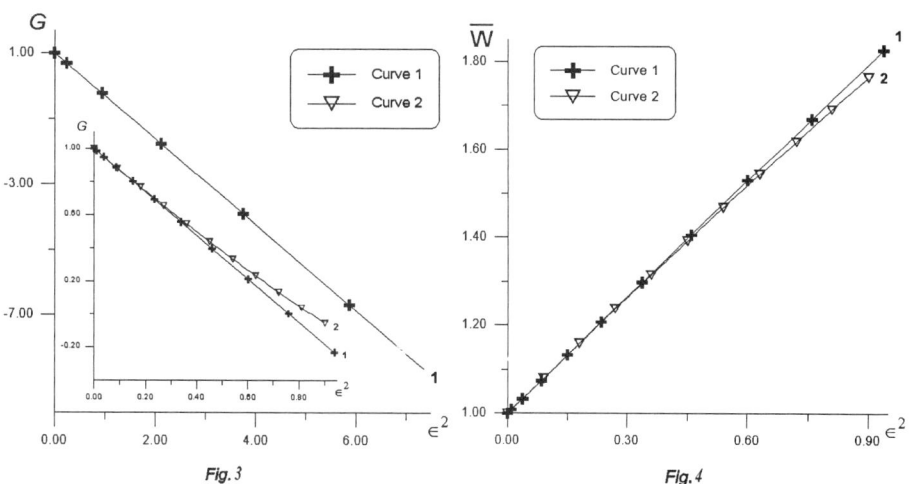

Fig.3 Fig.4

Figure 3 shows the dependence of the chemical potential ratio G (in presence and absence of the electric field of nucleus) on ϵ^2. Figure 4 shows the behavior of sum $\overline{W}$ of the surface and electric contributions (excluding the energy of wetting of the nucleus in the bulk liquid that does not depend on droplet size) to the work of droplet formation as a function of ϵ^2. In Fig. 3 and 4 curve 1 represents the results of our numerical approach, curve 2 corresponds to the analytical results [6]. As one can see, the agreement of both curves is good even for high values of ϵ^2.

CONCLUSIONS

While the validity of the droplet profiles derived analytically [5,6] is really limited by small deviations from the spherical shape, the region of validity of the formulas for the chemical potential and the free energy of droplet formation is much wider. The main reason for that is provided by the mutual compensation in the surface and electric contributions. The partial compensation of these contributions should take place even in general case of an arbitrary electric field. The distortion of the spherical shape of the droplet in the axisymmetric electric field leads to decreasing of the electric contribution to the free energy of droplet formation. Because the surface energy contribution in minimal for sphere (with a fixed volume), the distortion of the spherical shape increases this contribution. In view of generating properties of the work of droplet formation in nucleation theory, the same conclusion stays valid for other nucleation characteristics.

ACKNOWLEDGMENTS

Alexander Shchekin thanks the Program "Universities of Russia" (project no. 2145). Vadim Warshavsky and Mikhail Kshevetskiy thank the Euler Program (Freie Universität Berlin) for the financial support of this work.

REFERENCES

1. Kuni, F.M., Shchekin, and A.K., Rusanov, A.I., *Colloid J. of the USSR* **45**, 598 (1984).
2. Cheng, K.J., *Phys. Lett. A* **106**, 403 (1984).
3. Basaran, O.A., and Scriven, L.E., *Phys. Fluids A* . **1**, 799 (1989).
4. Wohlhuter, F.K., and Basaran, O.A., *J. Magn. & Magn. Mater.* **122**, 259 (1993).
5. Warshavsky, V.B., and Shchekin, A.K., *Colloids and Surfaces A*. **148**, 283 (1999).
6. Shchekin, A.K., and Varshavskii, V.B., *Colloid J.* **58**, 538 (1996).
7. Varshavskii, V.B., and Shchekin, A.K., *Colloid J.* **61**, 579 (1999).
8. Oh, K.J., Gao, G.T., and Zeng, X.C., *J. Chem. Phys.* **108**, 4683 (1998).
9. Gao, G.T., Oh, K.J., and Zeng, X.C., *J. Chem. Phys.* **110**, 2533 (1999).

Phenomenological Model of Homogenous Condensation

Romuald Puzyrewski

Technical University of Gdańsk, Mechanical Dept.
PL-80-952 Gdańsk, Poland

Abstract. Molecular effects govern nucleation and condensation phenomena. Due to the high complexity of treating such a phenomena on molecular level [1], the phenomenological approach seems to bridge the gap between the real scale of nucleation and condensation and technical needs for a simpler model. Accuracy in comparisons with experimental data can be achieved within the frame of phenomenological models by means of proper definition and reliable values of coefficients, which describe the process of nucleus creation, rate of nucleation and growth of nuclei up to the droplets of macroscopic size. The paper presents the rearrangement of classical Becker-Döring (B-D) equation into the form of Fokker-Planck-Kolmogorow equation (F-P-K) where nucleus growth plays an important role. The set of equations for the nuclei growth is discussed. The well known Volmer-Frenkel (V-F) nucleation rate can be derived from the master F-P-K equation for isothermal nucleus growth. It has been shown that non-isothermal growth of nucleus deviates from isothermal one.

REARRANGEMENT OF B-D EQUATION INTO F-P-K EQUATION

The number of clusters f_g, containing g molecules and having phenomenological surface s_g, is to evaporating and condensing molecules exposed. The rates α for evaporating and β for condensing molecules have the dimensions $1/(m^2s)$. The stream of clusters between classes f_{g-1} and f_g can be expressed as

$$J_g = f_{g-1} s_{g-1} \beta - f_g s_g \alpha \qquad (1)$$

The change of number of clusters in the class f_g is given by the difference

$$\Delta f_g = J_g - J_{g+1} \qquad (2)$$

It leads simply to the form of Becker-Döring equation

$$\frac{\partial f}{\partial t} = -\frac{\partial J}{\partial g} \qquad (3)$$

The relation (1) can be rearranged [2]

$$J_g = f_{g-1} s_{g-1} \beta - f_g s_g \beta + f_g s_g \beta - f_g s_g \alpha \qquad (4)$$

The first part of right hand side can be rewritten into the form of diffusion rate

$$J_d = -\beta(f_g s_g - f_{g-1} s_{g-1}) = -\beta \frac{\partial fs}{\partial g} = -\frac{\partial}{\partial g}(Df) \qquad (5)$$

here indexes are omitted. The diffusion coefficient is in the form $D = \beta s$ with units $1/s$. The second part of right hand side of (4) can be interpreted as driven term of the character

$$J_v = f_g s_g \beta - f_g s_g \alpha = f(s\beta - s\alpha) = f\frac{dg}{dt} \tag{6}$$

where $\frac{dg}{dt}$ means the growth of single clusters of a surface s exposed to the evaporating and condensing molecules. Now if one inserts (5) and (6) into (3) the following form arises

$$\frac{\partial f}{\partial t} + \frac{\partial}{\partial g}(f\frac{dg}{dt}) - \frac{\partial^2}{\partial g^2}(Df) = 0 \tag{7}$$

which is of the type of Fokker-Planck -Kolmogorow (F-P-K) equation. The position of two parameters involved in the above equation D and $\frac{dg}{dt}$ classifies the equation as prospective one. The stream of clusters (4), after omitting the indexes, can be written as

$$J = -\frac{\partial}{\partial g}(Df) + f\frac{dg}{dt} \tag{8}$$

There are two parameters, which essentially govern the solution, namely the growth of nuclei $\frac{dg}{dt}$ and diffusion coefficient D. Models for nucleus growth form the foundation for further investigation of nuclei governed equation (7).

GROWTH OF NUCLEI

In the relation (6) the growth of nuclei is introduced in the form

$$\frac{dg}{dt} = s\beta - s\alpha \tag{9}$$

For metastable critical size of droplets one gets $\alpha = \beta$ and hence for $g = g_*$ we have $\frac{dg}{dt} = 0$. From Knudsen formula the number of striking molecules per unit surface and time can be evaluated as

$$b = \frac{p}{\sqrt{2\pi kmT}} \tag{10}$$

Treating the surface of clusters as a macroscopic value one can introduce the condensing stream of molecules as [3]

$$\beta = \frac{2\alpha_c}{2 - \alpha_c}\frac{p}{\sqrt{2\pi kmT}} \tag{11}$$

where the condensation coefficient appears as α_c. The parameters pressure p and temperature T of surrounding condensing gas need not necessarily be equal to pressure inside the nuclei treated as macroscopic droplets. Number of molecules g of the volume v_m and mass m determines the volume and mass of droplet with radius r. Metastable droplet's radius (critical droplet) determines the pressure inside the droplet p_d at the temperature T_d. Thus one can get droplet pressure in the form

$$p_d(r, T_d) = p_{do} e^{\frac{1}{kT_d}(\frac{2\sigma}{\rho_d r} + \frac{L}{T_{do}}(T_d - T_{do}))} \tag{12}$$

where the references of pressure and temperature are p_{do}, T_{do}, latent heat L, surface tension σ, density of condensed gas ρ_d, gas constant R. The equation (9) can be rearranged into

$$\frac{dr}{dt} = \frac{2\alpha_c}{2-\alpha_c} \frac{m}{\rho_d \sqrt{2\pi km}} \left(\frac{p}{\sqrt{T_d^i}} - \frac{p_{do}}{\sqrt{T_d}} e^{\frac{1}{RT_d}(\frac{2\sigma}{\rho_d r} + \frac{L}{T_{do}}(T_d - T_{do}))} \right) \qquad (13)$$

here T_d^i means the temperature at droplet surface which differs from droplet temperature T_d and far field temperature T_∞ due to heat transfer. Temperature T_d^i close to the droplet surface can be estimated by means of temperature jump coefficient α_T and mean free path l_o as follows

$$T_d^i = \frac{T_d + \alpha_T \frac{l_o}{r} T_\infty}{1 + \alpha_T \frac{l_o}{r}} \qquad (14)$$

For the droplet temperature T one can derive the equation from energy balance

$$\frac{dT_d}{dt} = \frac{3}{rc} \frac{dr}{dt} L - \frac{3\lambda}{\rho_d c r^2} \frac{T_d - T_\infty}{1 + \alpha_T \frac{l_o}{r}} \qquad (15)$$

Far field temperature is governed by the expansion rate Θ and ratio of droplet mass to the mass of condensed gas M/M_∞. The equation for far field temperature is

$$\frac{dT_\infty}{dt} = \frac{\kappa - 1}{\kappa} T_\infty \Theta + \frac{M_d}{M_\infty} \frac{L}{c_p} \frac{3r^2}{r_{do}^3} \frac{dr}{dt} \qquad (16)$$

The equations (13), (15) and (16) are closed with respect to r, T and T_∞. The initial conditions are given for critical droplet at $r = r^*$ with number of molecules $g = g_*$. The growth of nuclei in time can be investigated by means of the above set of equations.

CLASSICAL SOLUTION V-F FROM F-P-K EQUATION

It is worth noting that classic Volmer-Frenkel (V-F) solution for the nucleation rate follows from the model described above as the solution under very strong additional assumptions. Let us assume the isothermal growth of nuclei, so the equations (15) and (16) can be dropped. If the linear procedure is applied to eq. (13), in terms of molecule numbers for the neighborhood of critical droplet with g_* molecules, one gets

$$\frac{dg}{dt} = a(g) = a_1(g - g_*) = \frac{2}{9} \frac{\beta s_s^2 \sigma}{kT} g_*^{-\frac{2}{3}} (g - g_*) \qquad (17)$$

Depending on the sign of difference $(g - g_*)$ one has either growth or collapse the near-critical droplet. For the diffusion coefficient D at critical size of the droplet one can assume constant value of D

$$D = \beta s = \beta s_s g_*^{\frac{2}{3}} = const \qquad (18)$$

Equation (7) reduces then to the simpler form

$$\frac{\partial f}{\partial t} + \frac{\partial}{\partial g}(a(g)f) - D\frac{\partial^2 f}{\partial g^2} = 0 \qquad (19)$$

If one assumes the stationary state of f, then an ordinary differential eq. follows from (19)

$$D\frac{d^2 f}{dg^2} - \frac{d}{dg}(a(g)f) = 0 \qquad (20)$$

The second order ordinary equation for f includes two constants. The first one is equal to C_1=-J according to (8), the second one C_2=0. The evaluation of integral

$$\int_{-\infty}^{+\infty} e^{-\frac{a_1}{2D}(g-g_*)^2} dg = \sqrt{\frac{2D\pi}{a_1}} \qquad (21)$$

leads to the F-V solution for critical droplets rate (n is the concentration of single molecules),

$$J = n\sqrt{\frac{a_1 D}{2\pi}} e^{-\frac{s_g \sigma}{3kT}g_*^{\frac{2}{3}}} = \frac{p}{kT}\sqrt{\frac{a_1 D}{2\pi}} e^{-\frac{s_g \sigma}{3kT}g_*^{\frac{2}{3}}} . \qquad (22)$$

NON-ISOTHERMAL DROPLETS GROWTH

The assumption of isothermal droplet growth used in procedure of V-F solution is very far from the non-isothermal growth. Let us consider the following set of parameters for the solution of (13), (15), (16) set of equations:

$$p = 10^5 \, Pa, \; \lambda = 2.48 \, 10^{-1} W/(mK), l_0 = .3 \, 10^{-7} m, c_p = 1.846 \, 10^{-3} J/(kgK),$$

$$k = 1.38 \, 10^{23} m^2/(s^2 K), m = 3 \, 10^{-26} kg, \sigma = 6 \, 10^{-2} kg/s^2, \alpha_c = 1.$$

The critical droplet was chosen to have $g_* = g_0 = 102$ molecules. The initial jump to grow the nuclei as $\Delta g = 1$ was chosen. Figures 1a and 1b show the resolutions for the short time t=10^{-7}s for isothermal growth T=const and non-isothermal growth T=var.

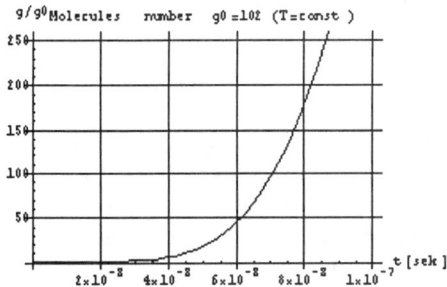
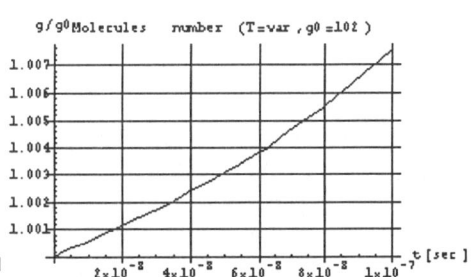

FIGURE 1. Isothermal and non-isothermal nuclei growth.

Long time solution for t=10^{-3}s has even more pronounceable difference of order of 10^8 higher, between numbers of molecules for isothermal growth to non-isothermal. The nucleation rate given by V-F solution seems to need verification due to the above mentioned differences.

REFERENCES

1. T. Ytrehus, Molecular Effects in Evaporation and Condensation at Interfaces" MultiphaseScience and Technology, 1997 Vol.9, No. 3.
2. R. Puzyrewski „Condensation of steam in de Laval nozzle" 1969 PWN (in polish)
3. R.W. Schrage „A Theoretical Study of Interphase MassTransfer" Columbia University Press, New York 1953

Kinetic Theory of a Carrier Gas Effect on Nucleation in Diffusion Chambers

Andrey L. Itkin[1]

Institute of High-Performance Computing and Databases, Russia

Abstract. Kinetic theory is presented which proposes an explanation of a carrier gas effect on nucleation in DCC. The main idea of this theory is that for DCC transport of condensing molecules to the cluster surface is determined by their diffusion through a carrier gas that results in new kinetic equations or, in other words, in usual kinetic equations where rate constants of the cluster formation depends upon carrier gas pressure. Analytical dependencies of critical supersaturation S_* and derivative $\partial S_*/\partial T$ on a carrier gas pressure P_0, temperature T and the nature of the carrier gas are derived that reproduce experimental data of Heist and co-authors.

MAIN IDEA OF THE THEORY

First we have to discuss the main physical idea of our theory that in more detail is given in [1]. We consider nucleation process when a certain molecule attaches a cluster. In the existing theories usually rate constant of this process is not supposed to depend on any characteristic (for instance, pressure) of background gas. This is true unless transport of this molecule to the surface of the cluster is regulated by diffusion of the molecule through the background gas, and the rate of chemical reaction (attachment of the molecule to the cluster) is much more than the rate of diffusion. As shown in [1] typical conditions in DCC just meet these requirements and hence this is the case when nucleation kinetics is a diffusion-limited one.

In a bit more detail, if a certain cluster with j molecules transits to another sort with $j+1$ molecules by attaching a monomer the rate of this process K_j^+ in a usual nucleation theory is proportional to the free-molecular flux of monomers on the surface of this cluster, and this flux is determined by the numerical concentration of monomers. Under conditions of DCC this assumption stops to be valid because the real concentration of monomers at the external boundary of the Knudsen layer over the cluster surface differs from the concentration n_1 determined far from the cluster because of the presence of the background gas. It is caused by the existence of a concentration jump in the vicinity of the cluster surface, and the value of this jump becomes considerable when the mean free path of the condensing molecules in the carrier gas is of order of the cluster radius.

As for DCC the number density of the background gas n_0 is few orders more that the number density of the vapor n_1, the mean free path of the vapor molecule in the own vapor λ_v is about two orders more in length than the mean free path of this

---

[1] Currrent position: IBES International, One World Trade Center, 18 fl., New York, NY 10048-1818, itkin@chem.ucla.edu

molecule in the carrier gas λ_{vg}. That is why in the vicinity of the cluster surface where we should determine K_j^+ vapor molecules collide only with the carrier gas ones and therefore there is reason to expect vapor concentration n_v to be equal to n_l if the number of such collisions is high. Taking this effect into account we could obtain the following new representation for K_j^+ [1]

$$K_j^+ = v_j^+ \frac{S\Theta_j + 1}{S(\Theta_j + 1)} \approx v_j^+ \frac{S\Theta_j + 1}{S}, \quad \Theta_j \, ? \, 1$$

$$\Theta_j \approx \Theta j^{-1/3}, \quad \Theta = \frac{3}{2\pi^2(1 + C/T)} \frac{k_B T}{P_0 \sigma_{vg}^2} \left(\frac{4\pi \rho_l}{3 m_1} \right)^{1/3}$$

Here S is supersaturation, C is the Sutherland constant, m_1 is a vapor molecular mass, ρ_l is liquid density, k_B is the Boltzmann constant, $\pi\sigma_{vg}^2/4$ is a collision cross-section of the vapor and gas molecules, therefore σ_{vg} is approximately a half of the sum of effective diameters of the vapor and gas molecules. We also introduced notation v_j^+ for the traditional (kinetic) rate constant that is proportional to square of the droplet radius and Θ_j is our correction for the case of the diffusion-limited kinetics. As follows from analysis of [1] $\Theta_j < 1$ and is inversely proportional to P_0 - the pressure of the carrier gas because the diffusion coefficient is inversely proportional to P_0. In contrast the rate constant of the inverse process K_j^- does not depend on the external environment of the cluster and is determined only by intracluster processes.

SOLUTION OF KINETIC EQUATIONS

Despite we derived a new representation for the cluster formation rate constant which now depends on the background gas pressure we can not proceed just substituting this expression into the nucleation rate, for instance, provided by classical nucleation theory. Rather, we need to reconsider kinetic equations describing transport and nucleation of vapor in DCC and try to find their solutions. According to our microscopic nucleation theory (MNT) [2] a studying mixture of gases and clusters is treated as a mixture of ideal gases each of that is characterized by the size of the identical clusters composing it. Continuity and diffusion equations for the mixture components obtained in the Navier-Stokes approximation are used, that based on the analysis of the DCC operation specificity and our asymptotic method [1,3] we managed to analytically solve, obtaining the following representation for concentrations n_j of clusters with j molecules

$$n_j \approx n_{ye} S \prod_{i=1}^{j-1} \frac{i^{1/3} + S\Theta}{i^{1/3} + \Theta}, \quad j \le y$$

$$n_j \approx n_{je} S^y \prod_{i=1}^{y-1} \Theta_i \prod_{k=y}^{j-1} \frac{\frac{\partial \ln K_{j+1}^+}{\partial j} + \frac{\partial \ln n_{je}/n_{j+1,e}}{\partial j}}{\frac{\partial \ln K_{j+1}^+}{\partial j} + \frac{\partial \ln \Theta_j}{\partial j}}, \quad j > y$$

Here n_{je} is the equilibrium concentration [2], parameter y is introduced as a cluster size being a boundary between two kinds of solution. In [1] an equation for y is derived which takes a simple form if S exceeds approximately 1.5 (as it usually takes place in a zone of active nucleation in experiments in DCC)

$$\frac{2b}{3y^{1/3}} - \ln\left(\frac{S\Theta_j + 1}{1 + \Theta_j}\right), \quad b = \frac{4\pi\sigma}{k_B T}\left(\frac{3m_1}{4\pi\rho_l}\right)^{2/3}$$

where σ is surface tension. At $\Theta_j \cong 1$, y coincides with the critical size.

DERIVING OF NUCLEATION RATE MEASURED IN DCC

To understand what is the nucleation rate measured in the experiment we use the following consideration. At the initial stage of the DCC work a certain quasisteady distribution of clusters is formed corresponding to the established distribution of S and T over the height of DCC. After that clusters (droplets) start to move due to gravity, the drag and thermophoretic forces that results in a violation of this quasisteady distribution. A certain amount of clusters reaches zone D where they are detected. Thus all clusters capable to fall down or to reach the top wall leave the zone of active nucleation.

If nucleation stopped, only these falling droplets would be registered during a short period of time while after this period no droplets should be visible (all of them reach either the bottom or top plates). The flux of droplets per unit square per unit time is $v_g n_j$, where n_j is their concentration and v_g - velocity of the clusters' precipitation. However, experimentalists present another value - the number of droplets per unit volume per unit time. For instance, it could be find by counting the falling drops within a certain (apparently rather long) period of time and then divide this number of droplets by the time, the square of the chamber and the width of the laser beam. If the period of detection is more than the time necessary for droplets to reach the bottom plate but the droplets are still detected, it means that due to transport of vapor molecules to clusters (diffusion) and chemical reactions between them (nucleation) the quasisteady distribution of clusters is restored and anew formed droplets fall down following the previous ones. In what follows the process is frequently repeated.

Based on this consideration we obtained for the nucleation rate J [3]

$$J \approx A(T_*)\left[\exp(-by^{2/3})\prod_{j=1}^{y}\frac{S\Theta + j^{1/3}}{\Theta + j^{1/3}}\right]^2 j_d^{2/3} S^2 K_{y+2}^{-} \frac{2b}{9y^{7/3}} \quad (1)$$

where $A(T)$ is a preexponent of the equilibrium cluster distribution [2], and expression for j_d is given in [3]. For some regimes $j_j < y$.

EFFECT OF A CARRIER GAS

Now we can put $J = 1$ drop/sec/cm^2 as say in experiments of Heist [4,5] and differentiate both sides of (1) on T at $P_0 = const$ or on P_0 at $T = const$ to reveal how maximum supersaturation depends on T or P_0 or even on the nature of the carrier gas.

The results are the following [3] (we mark parameters in maximum nucleation plane by subscript *).

- S_* linearly increases as P_0 increases at $T_* = const$ that reproduces the available experimental data of [4-6];
- Critical supersaturation S_* increases as T_* decreases at $P_0 = const$ (also in accordance with the experiments);
- If under conditions of DCC $S\Theta \cong 1$ then $S\Theta = c(T)$ and y depends only on T and does not depend on P_0 and S. In such a case function $\partial \ln S_*/\partial T$ does not depend on P_0. Moreover, as $S_* \propto P_0$ one has $\partial S_*/\partial T \propto P_0$.
- The slopes of the straight lines $S_*(P_0)$ decrease as the temperature T_* increases.
- The heavier the carrier gas the more is the efficient diameter of its molecule and hence the more pronounce dependence of S_* on P_0 occurs as well as the slopes of the lines $S_*(P_0)$ increase. This fact is also in good qualitative agreement with the experimental data of [4].
- As the molecular mass of the molecule of the condensing vapor decreases usually it gives rise to the decrease of the carrier gas effect. Such a situation has also been observed in many experiments.

The main conclusion which is made here based on the results obtained is that the dependence of S_* on P_0 observed in the experiments in DCC has no connection with the nucleation kinetics itself and is determined by the peculiarities of the transport processes in DCC.

ACKNOWLEDGMENTS

This work was supported in part by the Russian Ministry of Science and Technologies and Russian Foundation for Basic Researches grant 97-03-32434.

REFERENCES

1. Itkin, A., *Aerosol Sci. and Tech.* (in press) (2000).
2. Itkin, A., and Kolesnichenko, E., *Microscopic theory of condensation in gases and plasma*, New York: World Scientific, 1997.
3. Itkin, A., *Chem. Phys.* (in press) (2000).
4. Heist, R., Janjua, M., and Ahmed, J., *J. Phys. Chem.* **98**, 4443-4453 (1994).
5. ertelsmann, A., Stuczynski, R., and Heist R., *J. Phys. Chem.* **100**, 9762-9773 (1996)
6. Kane, D., and El-Shall, M., *J. Chem. Phys.* **105**, 7617-7631 (1996).

Improvement of the Homogeneous Nucleation Rate Measurements in a Static Diffusion Chamber with Use of a CCD Camera

Vladimír Ždímal[*,1], Jiří Smolík[#], Philip K. Hopke[*], and Jiří Matas[&]

[*]*Department of Chemistry, Clarkson University, Box 5810, Potsdam NY 13699-5810, U.S.A.*
[#]*Institute of Chemical Process Fundamentals AS CR, Rozvojová 135, 16502 Praha 6, Czech Republic*
[&]*Centre for Machine Perception, Czech Technical University, Karlovo n. 13, 121 31 Czech Republic*

Abstract. A photographic approach introduced recently to study nucleation kinetics in a static diffusion chamber, has been further improved by using a sensitive CCD camera instead of a photographic one. The method is based on illuminating the chamber from side by a flattened vertical laser beam. Trajectories of droplets, formed by nucleation and growing inside this beam, are recorded by a camera positioned perpendicular to this beam. Evaluation of starting points of droplet trajectories leads to vertical distribution of nucleation rate, which can then be related to calculated local values of temperature and supersaturation. This approach is independent on any nucleation theory. The main drawback of the photographic method, tedious production and evaluation of photographs, is replaced here by an automated procedure. Digitized pictures are downloaded to a PC and evaluated by standard methods of image analysis. The starting points of individual droplets are sought using a normalized correlation with a template. The position of liquid films is found using a method of maximizing sum of intensities on a line. The approach seems to keep the advantages of the photographic one, but is much more effective.

STATE OF THE ART

Since the pioneering experiments by Wilson (1), many experimental techniques have been developed to study homogeneous nucleation in supersaturated vapors. In past three decades some of these techniques reached the ability to determine nucleation rate quantitatively (2). The thermal diffusion cloud chamber, invented by Langsdorf (3), and first used for condensation studies by Franck and Hertz (4), became a standard tool for studying nucleation in seventies (5).

Principle of the chamber operation

The chamber is a closed flat cylindrical vessel using nonisothermal diffusion to produce supersaturation of vapor. It consists of two metal plates separated by a glass cylinder. The bottom plate with a film of liquid is heated, vapor diffuses through an inert gas, and condenses on the cooler top plate. The top plate is slightly conical, so that the condensate flows to its edge and along the glass wall back to the pool. If the

---

[1] On leave from Institute of Chemical Process Fundamentals, Academy of Sciences of the Czech Republic, Rozvojová 135, 16502 Praha 6, Czech Republic

sort and amount of the inert gas is properly selected, so that convection is avoided and the chamber operates at the steady state, temperature and partial vapor pressure decrease almost linearly from the bottom to the top. Since the equilibrium vapor pressure decreases with temperature more rapidly than the actual one, the vapor in the chamber becomes supersaturated. By increasing the temperature difference between the plates one can increase supersaturation gradually and arrange the state in which supersaturation is sufficient for homogeneous nucleation to begin. The droplets, once formed by nucleation, grow rapidly to visible size and fall down to the liquid pool. Calibrated thermocouples are used to measure temperature of both plates, pressure transducer allows to measure total pressure inside the chamber. A mathematical model of transport processes inside the chamber is used to find profiles of vapor-gas mixture properties in the chamber.

Determination of the nucleation rate

Two experimental approaches have been developed to determine nucleation rate in the chamber. Marvin and Reiss (6) illuminated the chamber by a He-Ne laser beam, shaped by cylindrical optics to a horizontal ribbon, passing the chamber close to the bottom. The droplets, formed by nucleation, fall through the beam, scatter light and are counted using a photomultiplier and connected electronics. This approach was initially used to study kinetics of homogeneous nucleation on isotherms by Katz et al. (7), who introduced the conversion of measured droplet flux to nucleation rate at the "maximum rate plane" via theory.

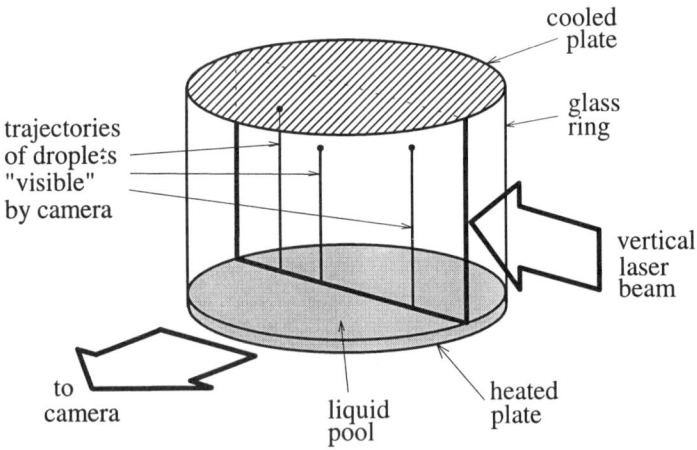

FIGURE 1 Principle of the photographic approach - a schematic sketch

Another approach to determine nucleation kinetics was developed by two of us (8). The chamber was also illuminated from side by a flattened laser beam, but positioned vertically as in (9), so that the beam illuminates the whole height of the chamber. It is shown schematically on Fig. 1. Droplets formed inside the beam scatter light and are photographed by a camera, positioned perpendicular to the beam. Each recorded droplet can be characterized by its trajectory and starting point. Under conditions, when thermo- and diffusiophoretic forces can be considered negligible (10), this starting point corresponds to the point of droplet origin by nucleation. By evaluating enough starting points we obtain vertical distribution of nucleation rate. It is then related to previously calculated local values of temperature and supersaturation.

Method was tested in vapors of dioctylphthalate and other substances, most recently in the system n-pentanol –He, where it was directly compared with other experimental methods during an international study „Joint Experiments on Homogeneous Nucleation" (11). The main advantage of the last approach is its independence on theory, and also the possibility to check quality of the experiment visually on droplet trajectories. On the other hand the approach is very tedious and time consuming, due to the necessity to process films and manually evaluate lot of pictures.

Recent improvements

In order to keep the advantages of the photographic approach and at the same time, minimise its drawbacks, the photographic camera was replaced by a digital CCD camera. A 16 bit camera ST-7 (SBIG, U.S.A.) widely used in astronomy for imaging of faint objects was chosen for its sensitivity and acceptable resolution (512x756 pixels). A standard camera lens was connected to it via adapter. Recorded images are downloaded to a PC through its parallel port. The camera is software controlled, and it makes possible to take a series of pictures with predetermined parameters. The noise caused by camera electronics, is usually negligible in comparison with the optical noise from the experiment (reflections). The result is that quality of pictures taken by this camera is comparable with quality of photographs, but the pictures can be evaluated automatically using image analysis.

The starting points of droplet trajectories are detected as local maxima of normalised correlation with a template (12). On a Pentium PC under Linux the processing is completed within seconds. The position of liquid films, defining the coordinate system, is found by a randomised optimisation approach. The criterion function - the sum of intensities along a line - is approximately evaluated from a set of samples (typically 10% of pixels along a line). Exhaustive search is performed in the space of nearly horizontal lines (a pair of lines in the case of the ceiling). Since the optimisation is completed in a fraction of second, more sophisticated methods suitable for the task, e.g. the Hough transform (13) were not needed.

A typical CCD image is shown in Fig. 2. The black horizontal lines at the bottom and top are the positions of the liquid films as found by the algorithm. White vertical lines denote droplet trajectories, black crosses are the assigned positions of the starting points. The estimated error in determining the vertical position is less than 1 pixel. It means that this approach slightly surpasses in precision former visual evaluation of photographs. The ability to process a series of pictures automatically makes it much more effective.

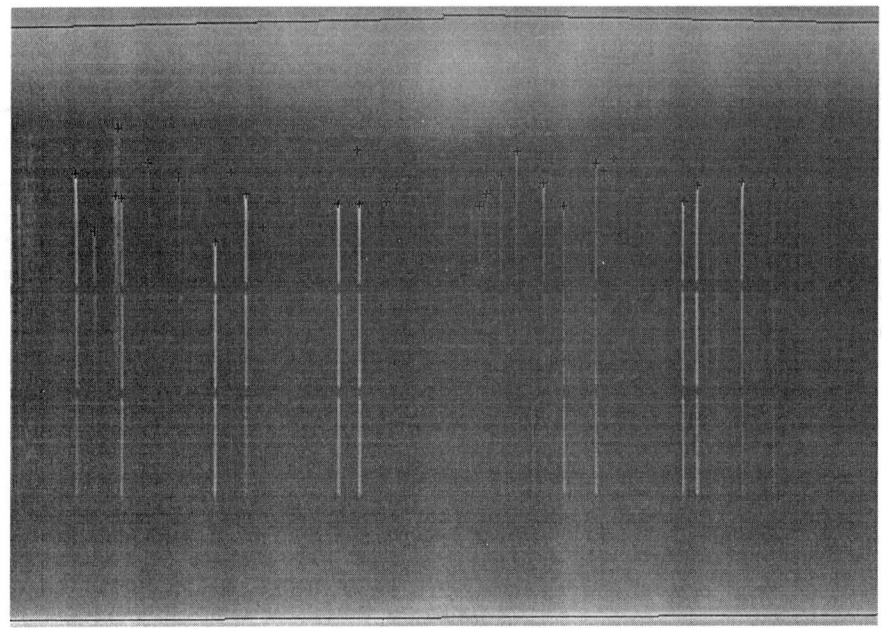

FIGURE 2. CCD image of the droplets nucleated in a static diffusion chamber. Black lines at the bottom and top are positions of liquid films as found by the algorithm. White vertical lines denote droplet trajectories, black crosses are found positions of starting points (n-pentanol, isotherm 280 K, 30 s exposure, full resolution).

ACKNOWLEDGMENTS

The authors gratefully acknowledge support of this work partly by the Grant Agency of the Czech Republic under a grant No. 104/97/1198 (V.Ž. and J.S.) and partly by NSF-NATO grant No. DGE-98-04534, awarded in 1998 (V.Ž. and P.K.H.).

REFERENCES

1. Wilson, C.T.R., Phil. Trans. Roy. Soc. (London) A 189, 265 (1897).
2. Heist, R.H., and He, H., J. Phys. Chem. Ref. Data 23, 781 (1994).
3. Langsdorf, A., Rev.Sci Inst. 10, 91 (1939).
4. Franck, J.P., and Hertz, H.G., Zeitschrift für Physik 143, 559 (1956).
5. Katz, J.L., J. Chem. Phys. 52, 4733 (1970).
6. Marvin, D.C., and Reiss, H., J. Chem. Phys. 69, 1897 (1978).
7. Katz, J.L., Hung, C.-H., and Krasnopoler, M., „The homogeneous nucleation of nonane" in Lecture Notes in Physics 309, edited by P.E.Wagner and G.Vali, Proc. of the 12th ICNAA, Springer, Berlin 1988, pp. 356-359.
8. Smolík, J., and Ždímal, V., Aerosol Sci.Technol. 20, 127 (1994).
9. Flageollet-Daniel, C., Ehrhard, P., and Mirabel, P., J. Chem. Phys. 75, 4615 (1981).
10. Ždímal, V., Tříska, B., and Smolík, J., Colloids Surfaces A: Physicochem. Eng. Aspects 106, 119 (1996).
11. Rudek M.M., Katz J.L., Vidensky I.V., Ždímal V., and Smolík J., J. Chem. Phys., 111, 3623 (1999).
12. Šonka, M., Hlaváč, V., and Boyle, R.D., Image Processing, Analysis and Machine Vision, Boston: PWS, 1998.
13. Matas, J., Computer Vision and Image Understanding 78, 1077 (2000).

Dynamic State Phase Diagrams For Nucleated Systems

M. Anisimov[1], S. Shandakov[2], V. Pinaev[3], I. Shvets[2], and P. Hopke[1]

[1] *Division of Chemical and Physical Sciences, Clarkson University, Potsdam, NY 13699-5810*
[2] *Kemerovo State University, 6 Krasnaya Str., 650043 Kemerovo, Russia,*
[3] *Kemerovo Institute of Commerce, 37 Kuznetskiy Prospect, 650099 Kemerovo, Russia*

Abstract. In the present research the topological analysis of nucleation rates surfaces based on the nucleation rate surface and the phase diagram relationship is used. The qualitative relation of phase (stable) state diagram behavior with intermolecular interaction is well known in theory of solutions. The dynamic state phase diagrams are introduced and discussed here. Transformation of these diagrams with deviation of system from equilibrium states is considered. It is found in agreement with results of non-equilibrium thermodynamics, that for system moving away from the equilibrium state the molecules are losing individuality and intermolecular interaction becomes interaction similar to the unimolecular system.

The gas influence on vapor nucleation is more evident, for example in the case of gas hydrates forming. This effect is not described in the framework of classical nucleation theory, in which interaction between molecules of condensed phase and gas is reduced to thermostat effect only. Earlier offered [1,2] topological analysis of nucleation rates surfaces based on the nucleation rate surface and the phase diagram relationship showed that nucleation in vapor-gas system may be more complicated than it is predicted by classical nucleation theory. The qualitative relationship of phase (stable) state diagram behavior with intermolecular interaction is well known in theory of solutions. For example the positive (or negative) deviation of real solution behavior from ideal one may be explained by, the first, increasing (or decreasing) of intermolecular interaction between uniform molecules with respect to different sorts (binary interactions) of molecules, and, the second, breaking (or growing) of molecular associations. The intermolecular interaction in metastable state medium is practically unknown in comparison with stable one. In present research the dynamic state phase diagrams are discussed and transformation of diagrams with deviation of system from equilibrium states is considered.

In the case of the non-equilibrium phase transition (nucleation in the supersaturated vapor) a metastable phase can be characterized by stationary flux of new phase nuclei[3]. In approach of the relatively small hydrodynamic flows inside of the studied systems the value of a new phase nuclei flux, J, (nucleation rate) may be used as a measure of deviation of nucleated system from equilibrium. This allows to consider nucleation isorates, $J=const$, in corresponding coordinates as dynamic phase diagrams.

On the basis of steady-state diagram and experimental nucleation rates for water-n-alcohol system dynamic P-x phase diagrams have been drown. The dynamic phase diagrams transformation for water-ethanol system with changing nucleation rate from zero to 10^{10} drop/cm^{3}s is presented in figure 1. Here we used experimental data by Kogan et al.[4], Mirabel and Katz[5], the Schmtt et al.[6], Strey et al.[7], Zahoransky and Wittig[8] and calculated data by Laaksonen et al.[9]. These data were received at temperature of 293 K, excluded results by Kogan et al[4]. and Strey et al[7]. which are at 313 and 260 K, respectively. To compare experimental data at different temperatures and pressures, which are differ in a few orders the $P-x$ diagram presentation is more convenient in dimensionless form. To illustrate the effects of real solution behavior the pressure was normalized with respect to pressure of ideal solution at the same mole fraction of alcohol. Pressure of hypothetical ideal solution is calculated according to Raul low

$$P_{id.sol.} = P_{01}x_1 + P_{02}x_2, \qquad (1)$$

where P_{01} and P_{02} are vapor pressures for pure liquids "1" and "2", respectively; x_1 and x_2 are mole fraction of components "1" and "2" in considered phase (liquid or gas) respectively. Pressure of saturated vapors was calculated with using of Antuan's equation. Line for liquid for dynamic phase diagram at $J = 10^7$ drop/cm^{3}s was build on the base of nuclei composition presented by Strey et al.[7]. As seen in fig. 1. the positive deviation from Raul low, appeared in the line of liquid for static $(J = 0)$ phase diagram, is changed to negative one for dynamic $(J = 10^7$ drop/cm^{3}s$)$ phase diagram. It is worth to mentioned that calculated dynamic phase diagram

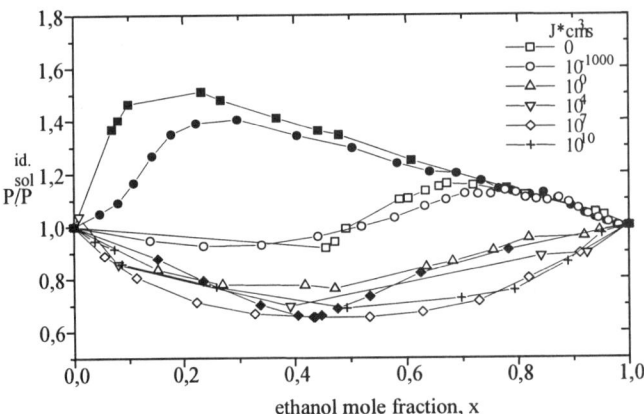

Fig. 1. The ratio of vapor pressure of real binary water-ethanol system to pressure of ideal solution vs. ethanol mole fraction in condensed phase (solid symbols) and in vapor phase (open symbols) at different values of vapor nucleation rate, J. Kogan et al.[4]; Laaksonen et al.[9]; Mirabel et al.[5]; Schmtt et al.[6]; Strey et al.[7]; Zahoransky et al.[8] works are at J= 0; 10^{-1000}; 1; 10^4; 10^7; 10^{10} drop/cm^{3}s, respectively.

with a small deviation $(J = 10^{-1000}$ drop/cm^{3}s$)$ from equilibrium state is closely to the static one. As for line of gas (open symbols in fig.1) in sight of consideration of gas as solution that it's transformation had the same tendency as for line of liquid. Thus attraction energy is modifying of binary interaction between the same sort molecules with moving away from equilibrium state or with increasing of nucleation rate, J. Other mechanism of solution deviation from ideal system can be the molecule associations. In the framework of this consideration the associations of different sorts molecules in metastable phase (in liquid or gas) with increasing of nucleation rate are assumed even if such associations are not existed in stable phase.

The transformation of dynamic state phase diagrams for water-n-propanol system has more large deviation from ideal system in the comparison with the water-ethanol system. To illustrate the relationship of experimental dynamic phase state

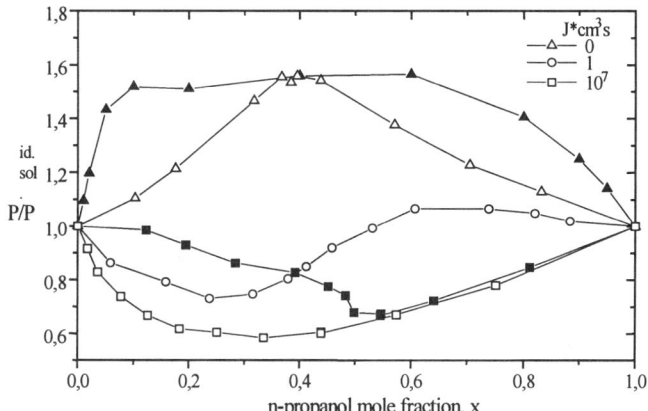

Fig. 2. The ratio of vapor pressure of real binary water-n- propanol system to pressure of ideal solution vs. propanol mole fraction in condensed phase (solid symbols) and in vapor phase (open symbols) at different values of vapor nucleation rate, J. Kogan et al. [4]; Flageollet-Daniel et al. [10]; Strey et al.er [7] works are at J= 0; 1; and 10 [7] drop/cm^{3}s, respectively.

diagrams with line of spinodal decomposition, these lines for water-n-propanol system are schematicallypresented in figure 3 with retaining of curves shape. To compensate difference in temperatures for used experimental data in fig. 3 the pressure is reduced to saturated n-propanol vapor pressure at appropriate temperatures.

Experimental nucleation rate data for $J = 1$ drop/cm^{3}s and $J = 10^7$ drop/cm^{3}s are taken from works of Flageollet-Daniel et al.[10] and Strey et al.[7], respectively. Static phase state diagram is building on the base of experimental data at $T = 298$ K presented in work of Kogan et al.[4]. Fig. 3 incorporates the static phase diagram in dynamic one, namely aseotrop point confirm the relationship of experimental nucleation rate surfaces with static phase state diagrams. Line of spinodal decomposition defined as nucleation rate maximum and it is calculated by method[11] with using of Van-der-Vaals state equation for binary mixture where binary interaction factor, k_{12}.is introduced. The shape of schematically presented line of spinodal decomposition in fig. 3 corresponds to factor $k_{12} \sim 4$ at temperature 260 K. A value of k_{12} determines the curvature of this line and consequently the shape of dynamic phase state diagram at enough nucleation rates. The effect of spinodal line *lbd* asymmetry the dynamic phase state diagram transformation from static one must be appeared in the first place in composition of water solution. It is appeared in experiment. Mutual

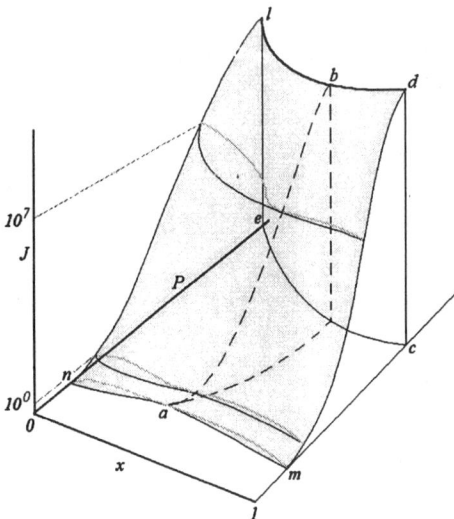

Fig. 3. Schematic nucleation rate surface for water-n-propanol system.

disposition of static phase diagram and spinodal decomposition lines may lead to qualitatively different dynamic phase diagram transformation.

Using the dynamic state phase diagrams leads to conclusion that the binary interaction of water molecules with ethanol or n-propanol molecules becomes stronger with increasing of nucleation rate. It is in agreement with results of non-equilibrium thermodynamics[12] where for system moving away from equilibrium state the molecules are losing individuality and intermolecular interaction behave as uniform system. This result means that for an uncondensed component a binary intermolecular interaction may become essential with the system moving away from equilibrium state if even its negligible in equilibrium.

REFERENCES

1. Anisimov M.P. *J.Aerosol Sci.* **24** (Suppl.1): 23 (1990).
2. Anisimov M.P., Hopke P.K., Rasmussen D.H., Shandakov S.D., Pinaev V.A. *J.Chem.Phys.* **109** (4): 1435 (1998).
3. Skripov V.P. in: *Termodinamika i kinetika fazovyh perehodov.* Nauka, Ekaterinburg, Russia, 1992.
4. Kogan V.B., Fridman V.M., and Kafarov V.V *Ravnovesie Mezhdu Zhidkostyu i Parom.* Nauka, Moscow, Russia, 1966.
5. Mirabel P. and Katz J. *J.Chem.Phys.* **67**: 1697 (1977).
6. Schmtt J.L., Whitten J., Adams G.W., and Zalabsky R.A. *J.Chem.Phys.* **92** (6): 3693 (1990).
7. Strey R., Viisanen Y., and Wagner P.E. *J.Chem.Phys.* **103** (10): 4333 (1995).
8. Zahoransky R.A. and Wittig S., in *Proceedings of thr 13th International Symposium on Shock Tubes and Waves: Niagara Falls, 1981,* edited by C.E. Treanor and J.G. Hall (SUNY Press, Albany, 1982), p. 682.
9. Laaksonen A., Kulmala M., and Wagner P.E. *J.Chem.Phys.* **99** (9): 6832 (1993).
10. Flageollet-Daniel C., Garnier J.P., and Mirabel P. *J.Chem.Phys.* **78** (5): 2600 (1983).
11. Ushenin I.Y.and Baydakov V.G., in: *Termodinamika i Kinetika Fazovyh Perehodov.* Nauka, Ekaterinburg, Russia , 1992.
12. Prigogin I., *From Being to Becoming: Time and Comlexity in the Physical Sciences.* W.H. Freeman and Company, 1980.

Thermodynamics and Nucleation Studies Of III-Nitride Materials Grown From Vapour Phase

E.Varadarajan, R.Dhanasekaran and P.Ramasamy

Crystal Growth Centre, Anna University, Chennai-600 025, India

Abstract. The initial stage of growth of III-Nitrides ternary compound on binary substrate were analysed using the classical heterogeneous nucleation theory, incorporating lattice mismatch between the grown alloy and the substrate. The explicit expression for the lattice mismatch induced supercooling for the growth of the chosen system was established, and it was used to evaluate the nucleation parameters. It has been theoretically shown that the nucleation barrier for the formation of ternary alloy on binary substrate depends on the composition of the alloy.

1. Introduction

III-Nitrides are semiconducting compounds of large energy gap. GaN and its ternaries with AlN and InN are the most promising material for light emitting devices in the spectral region of blue, violet and ultra violet because of its direct energy band gap [1]. These compounds are high temperature stability and low electrical leakage. Much efforts has been made in the recent year to examine various characteristics and parameters of the GaN, AlN, InN and their ternary compounds. Currently, Molecular Beam Epitaxy (MBE) and Metal Organic Chemical Vapour Deposition (MOCVD) methods are applied to grow thin layers of these compounds on foreign substrate. Recently light emitting diodes and lasers using InGaN/GaN and AlGaN/GaN heterostructures have been demonstrated. GaN based metal semiconductor field effect transistor (MESFET) and short gate-length modulation doped field effect transistor based on AlGaN/GaN heterostructures [2] would be a promising application. Nucleation kinetics and the mechanism of the initial stages of the growth of these ternaries and quaternaries have yet to be studied in detail in order to get the high quality device materials.

2. Stress Induced Supercooling

In this paper we make an attempt to investigate the thermodynamic and nucleation kinetics of AlGaN/GaN and InGaN/GaN using the reported thermodynamic data. However, the epitxial growth of In containing nitrides is more problematic than that of GaN and AlGaN. The investigation of the initial stage nucleation of ternary alloys on binary substrate is essential to understand the phenomena of the initial growth of ternary alloys and hence to control the composition, lattice mismatch and crystalline quality of the layers. The interface between ternary layers and binary substrate is strained due to difference in the lattice parameter of these two[3]. The difference in lattice constant leads to the stress at the interface between the nucleus and the substrate. The stress at the interface created a certain amount of supercooling. The expression for lattice mismatch, stress induced supercooling due to mismatch for the growth of chosen system have been evaluated numerically. In order to evaluate the amount of stress induced supercooling due to lattice mismatch between the ternary layers on binary substrate the work of Bolkhovityanov [4] was followed. The expression for stress induced supercooling is

$$\Delta T = \frac{TYM(a_s - a_l)^2}{H(1-\upsilon)a_s^2 f} \quad (1)$$

$$f = 1 + \frac{2YM}{1-\upsilon}\left(\frac{da_l}{dx}\right)^2 \left[a_s^2 \left(\frac{d^2\Delta F}{dx^2}\right)_{x=x_0}\right]^{-1} \qquad (2)$$

$$\Delta F = xRT \ln\left(\frac{x}{x_0}\right) + (1-x)RT \ln\left(\frac{1-x}{1-x_0}\right) - \beta(x-x_0)^2 \qquad (3)$$

$$H_{AlGaN} = xH_{GaN} + (1-x)H_{AlN} - 2\beta \qquad (4)$$

$$H_{InGaN} = xH_{InN} + (1-x)H_{GaN} - 2\beta \qquad (5)$$

where T is the equilibrium temperature, Y is the Young's modulus, υ is the Poisson's ratio, M is the molar weight, H is the effective enthalpy for the formation of layers from its two binaries, a_s is the lattice parameter of the substrate, a_l is the lattice parameter of the alloy, f is the composition stabilisation coefficient, R is the universal gas constant, x is the composition and β is the interaction parameter of the two binaries. Using the numerical values for the chosen system, the stress induced super cooling as the function of composition of the AlGaN and InGaN alloys have been calculated and the results are shown in Figures 1and 2. It is noted that in the fig.2 ΔT is positive for almost all the composition, so the superheating is produced at the interface due to the lattice mismatch.

3. Nucleation Kinetics Of Ternary System

Using regular solution model [5] the interfacial tension between the nucleus and substrate and hence the interfacial tension between substrate and mother face and thermodynamical potential of the constituent components of the compounds have been calculated. The interfacial energy between the ternary layer and binary substrate has been calculated by determining the excess energy associated with the interface. The lattice planes numbered positively, beginning with 0, into the ternary alloy, and negatively with -1, into the substrate so that the interface between the ternary layer and binary substrate is located between -1 to 0. If the alloy were uniform at the composition x then the energy between the planes -1 to 0 would be

$$\sum\nolimits_{alloy} = gP\left[x^2 \varepsilon_{AA} + 2x(1-x)\varepsilon_{AB} + (1-x)^2 \varepsilon_{BB}\right] \qquad (6)$$

where g is the number of bonds per atom between planes, P is the number of atoms per plane and ε_{AA}, ε_{BB}, and ε_{AB} are the bonding energies between A-A, B-B, and A-B respectively. (A=AlN, B=GaN for AlGaN and A=InN, B=GaN for InGaN system) Similarly for the substrate is given by

$$\sum\nolimits_{sub} = gP[\varepsilon_{AA}] \qquad (7a)$$

and for the alloy and substrate

$$\sum\nolimits_{as} = gP[x\varepsilon_{AA} + (1-x)\varepsilon_{AB}] \qquad (7b)$$

The change in this energy process is

$$\sum = 2\sum\nolimits_{as} - \sum\nolimits_{alloy} - \sum\nolimits_{sub} \qquad (8)$$

If the area per atom in the surface is A, then the total interfacial area of the system is $\Omega = 2PA$. Then the surface energy per unit area is

$$\sigma_{12} = \frac{\sum}{\Omega} = -\left(\frac{g\phi}{2A}\right)(1-x)^2 \qquad \text{(where } \phi = \varepsilon_{AA} + \varepsilon_{BB} - 2\varepsilon_{AB}) \qquad (9)$$

Substituting the values we get the interfacial tension between the substrate and nucleus (σ_{12}) and hence the interfacial tension between the nucleus and mother face (σ_{13}) can be calculated. These values are plotted for different values of composition (x) of the GaN in the alloy in figure 3 and 4. It can be seen that the composition of the alloy plays a dominant role in determining the interfacial tension between the nucleus and substrate.

The chemical potentials of the components of the compound not only depend on the degree of supercooling but also on the composition of the alloy. The amount of driving force available for the nucleation has been explicitly determined for different composition and degree of supercooling. These values have been used to evaluate the nucleation parameters like radius of critical nucleus, total number of molecules and free energy barrier of the nucleus. Figure 5 shows the numerical value of the radius of the AlGaN nucleus corresponding to the saddle point as a function of actual supercooling between the substrate and layer for various values of composition (x). The radius of critical nucleus decreases exponentially with in increase in supercooling, but the variation depends strongly on the composition of the alloy. Figure 6 represents the total number of AlN and GaN molecules in the critical AlGaN nucleus as function of supercooling for various compositions of the ternary alloy. From our numerical analysis on the nucleation data corresponding to the growth of ternary alloy on binary substrate, it is shown that the nucleation barrier for the formation of ternary nucleus (AlGaN,InGaN) on GaN substrate depends strongly on the composition of the alloy.

REFERENCES:

1. S.Nakamura, M. Senoh, S. Nagahama, N. Iwasa, T. Yamada, T. Matsushita, Y. Sugimoto and H. Kiyoku, Appl. Phys. Lett. 70(1997) 1135.
2. M.A. Khan, A.Bhattarai, J.N. Kuznia and D.T. Olsen, Appl.Phys. Lett. 63(9) (1993)1214.
3. Yu. S.Karpov, MRS.Internet J. Nitride Semicond. Res. 3 16(1998)
4. Yu. B. Bolkhovityanov, J.Cryst. Growth 57 (1982) 84.
5. H. Resis and M. Shugad, J. Chemical Physics 65 (1976) 5280.

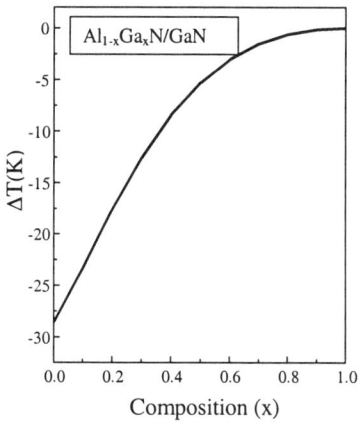

Figure. 1

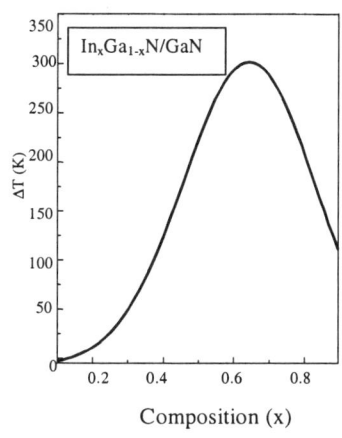

Figure. 2

FIGURES 1 and 2. The numerical values of stress induced supercooling produced at the interface due to lattice mismatch as function of composition (x)

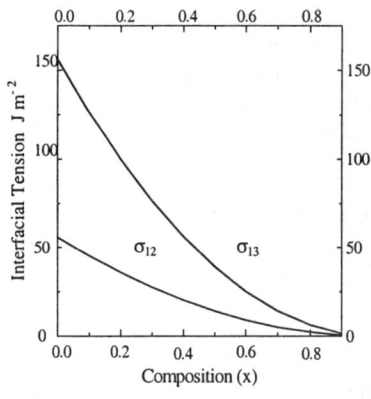

Figure 3 [Al$_{1-x}$Ga$_x$N/GaN]

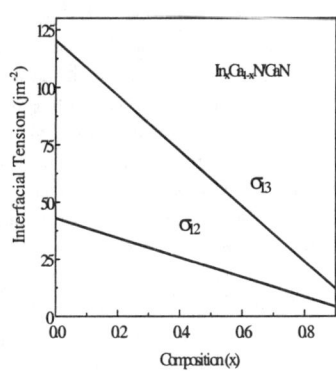

Figure 4 [In$_x$Ga$_{1-x}$N/GaN]

FIGURE 3 and 4. The numerical values of interfacial tension between the substrate and layer((σ_{12}) and the interfacial tension between the nucleus and mother phase (σ_{13}) as function of compsition (x)

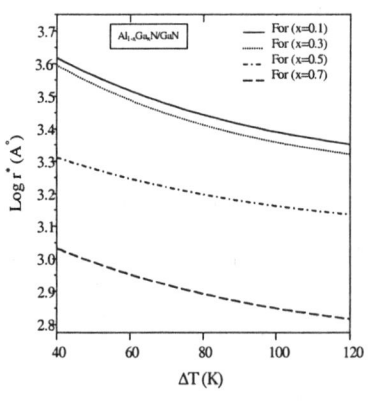

Figure 5

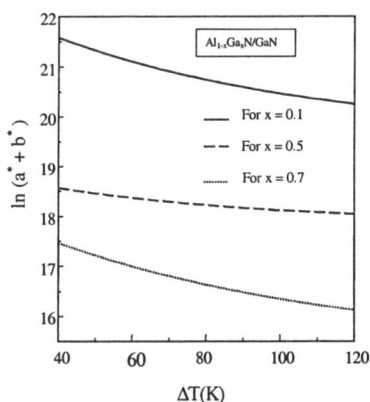

Figure 6

FIGURE 5. Dependence on the radius of the saddle point nucleus of AlGaN for different supercooling for various values of composition

FIGURE 6. Total number of GaN and AlN molecules in the critical nucleus of GaAlN as a function of supercooling for different values of Composition (x).

EVALUATION OF SURFACE ENERGY FOR PURE AND DOPED 4-hydroxyacetophenone FROM INDUCTION PERIOD MEASUREMENTS

S. Chenthamarai, D. Jayaraman,
*C. Subramanian and P. Ramasamy

Presidency College, University of Madras, Chennai 600 005, India
*Crystal Growth Centre, Anna University, Chennai 600 025, India

Crystal growth of nonlinear optical materials has become most indispensable in the field of materials science and engineering. In recent years molecular engineering has led to the development of organic crystals possessing better nonlinear optical properties. The material 4-hydroxyacetophenone is one of the organic crystals exhibiting NLO property to find applications in telecommunications and possibilities for optical information storage, frequency conversion, etc. It crystallizes in the orthorhombic system with space group $P2_12_12_1$ and lattice parameters a = 9.48 Å, b = 24.15 Å and c = 6.09 Å [1]. An attempt has been made to grow crystals of 4-hydroxyacetophenone by slow cooling technique. The material is then doped with nitro group to improve the NLO property.

When a supersaturated condition is achieved, embryos are formed by single molecular addition starting from the monomer at the beginning. Thus it takes some time for the formation of critical nucleus from the monomers. The time taken between the achievement of supersaturation or supercooling and the appearance of the crystal nucleus in a supersaturated system is known as induction period. Depending upon the solubility of the material one can choose the appropriate technique available for determining the induction period. The induction periods of pure and doped solutions were measured by visual observation method for various supersaturations. From the solubility data the supersaturated solution of 1.18 was prepared at 32 °C and placed in a constant temperature bath of controlling accuracy ±0.01 °C. Normally, the time required for the growth of crystal nucleus to a detectable size is negligibly small compared with the time needed between the achievement of supersaturation and the appearance of crystal nucleus. Therefore the total time elapsed between the achievement of supersaturation and the appearance of nucleus of detectable size can be measured as induction period. The appearance of the first speck of nucleus was seen at the bottom of the container. Induction periods were recorded for different supersaturation values of 1.19, 1.22, 1.28 and 1.33.

The interfacial energy can be estimated by performing nucleation experiments[2]. However, there are certain difficulties in conducting the nucleation experiments which include

(i) The requirement of the solution which is mostly free from foreign particles to enable homogeneous nucleation and
(ii) The laborious and time consuming nature of the experiment.

Hence in order to arrive at a quick estimate of surface tension, a theoretical approach using the physico-chemical data of the solution has gained considerable interest. Many empirical expressions have proposed and all these expressions have a common feature of predicting a linear dependence of Υ_0 on $\ln C$ where C is a measure of concentration.

The nucleation is treated as heterogeneous type due to the appearance of the first sparkling speck of the nucleus at the bottom of the container. From nucleation theory the induction period is written as

$$\tau \alpha \exp(\Delta G^*_{het} / kT) \tag{1}$$

where ΔG^*_{het} is the critical free energy change of the cap shaped nucleus, k is the Boltzmann constant and T is the constant temperature of the solution.

$$\ln \tau = \ln B + \Delta G^*_{het} / kT \tag{2}$$

$$\Delta G^*_{het} = \frac{16 \pi \Upsilon_0^3 \varphi(\theta)}{3 \Delta G_v^2} \tag{3}$$

and $\varphi(\theta) = (2 - 3\cos\theta + \cos^3\theta) / 4 \tag{4}$

where B is a constant, Υ_0 is the bulk value of interfacial energy, ΔG_v is the bulk energy change per unit volume and θ is the angle of contact which is acute for crystal-glass interface.

The bulk energy change per unit volume is given as
$$\Delta G_v = \frac{-(kT \ln S)}{v} \tag{5}$$

where v is the specific volume and S is the supersaturation. Therefore

$$\ln \tau = \ln B + \frac{16\pi \Upsilon_0^3 \varphi(\theta)}{3k^3 T^3 (\ln S)^2} \tag{6}$$

A plot of $1/(\ln S)^2$ against $\ln \tau$ is a straight line (Fig.1). The intercept on y axis gives the value of $\ln B$. From the slope of the straight line, the value of Υ_0 has been evaluated, assuming the angle of contact to be 60° for the crystal - glass interface. This value of Υ_0 is compared with the value predicted by Bennema et al [3].

$$\Upsilon_0 = (kT/a^2) \{0.173 - 0.248 \ln(x_m)\} \tag{7}$$

where a is the interionic distance and x_m is the mole fraction of the solute. The value of a is estimated using the expression[4] a=(abc)/2 where a,b,c are lattice parameters of the crystal.

After incorporating the size dependent interfacial energy $\Upsilon=\Upsilon_o(1-\delta/r)$, the nucleation parameters of the 4-hydroxyacetophenone nucleus have been calculated using the following expressions

$$r^* = \frac{-\Upsilon_0 \pm \sqrt{\Upsilon_0^2 + 8\delta\Delta G_v \Upsilon_0}}{\Delta G_v} \qquad (8)$$

$$\Delta G^*_{het} = 4\pi r^* \{ \Upsilon_0(r^*-\delta) + r^{*2} \Delta G_v / 3 \} \varphi(\theta) \qquad (9)$$

$$\text{and } J = A \exp(-\Delta G^*_{het}/kT) \qquad (10)$$

where δ is the radius of the monomer and A is the pre-exponential factor. The value of A is taken as 10^{25}.

The experimentally evaluated values of Υ_0 for pure and nitro doped 4-hydroxyacetophenone from the induction period measurement have been found to be in good agreement with the theoretically predicted values (Table.1). A range of supersaturation ratio taken in the present work is narrow so that the interfacial energy evaluated is assumed to be constant. Table2 show the nucleation parameters of pure and doped crystals. It is found that the values of nucleation parameters are increased after doping the materials.

REFERENCES

[1]. B.K.Vaishtein, G.M.Lobanova and G.V.Guruskaya, Sov.Phys.Crystallog.,19(1974)329.
[2].A.Mersmann, J.Crystal Growth,102(1990)841.
[3].P.Bennema and Sohnel, J.Crystal Growth, 102(1990)547.
[4].O.Sohnel, J.Crystal Growth, 57(1982)101.
[5].D.Jayaraman, C.Subramanian and P.Ramasamy, J.Mater.Sci.Lett., 8(1989)1399.

Table1 Interfacial energy of the pure and nitrodoped 4-hydroxyacetophenone single crystals.

Sample	σ_0 (mJ/m^{-2})	
	Experiment	Theory
Pure	9.97	9.00
Doped	11.37	10.47

Table 2. Nucleation parameters of pure and nitro doped 4-hydroxyacetophenone

S	r^* (nm)		$\Delta G^*_{het}/kT$		J ($m^{-2}s^{-1}$)	
	Pure	Doped	Pure	Doped	Pure	Doped
1.33	75.9	89.1	0.64	3.40	5.26×10^{24}	3.34×10^{23}
1.28	88.8	104.1	0.92	4.84	3.96×10^{24}	7.94×10^{22}
1.22	111.3	140.2	2.24	7.76	1.06×10^{24}	4.25×10^{21}
1.19	128.0	159.5	6.58	10.54	1.39×10^{22}	2.64×10^{20}
1.18	134.7	167.3	7.35	11.77	6.44×10^{21}	7.76×10^{19}

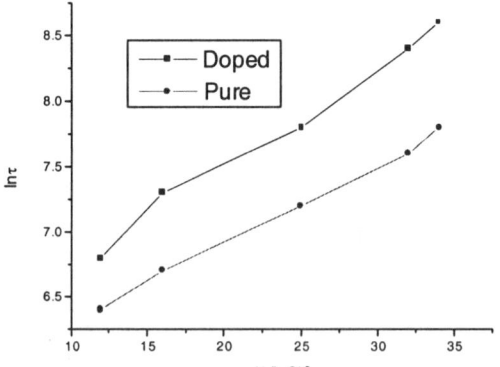

Fig. 1 Plot of $(1/\ln S)^2$ Vs. $\ln \tau$

Thermodynamics of Cluster Ions Containing NH_3 and H_2SO_4

Karl D. Froyd[†] and Edward R. Lovejoy

NOAA Aeronomy Laboratory, Boulder, Colorado 80303

[†] *Also affiliated with the Department of Chemistry and Biochemistry, and The Cooperative Institute for Research in the Environmental Sciences, University of Colorado, Boulder, Colorado 80309*

Abstract. The stability of cluster ions of the form NH_4^+ $(NH_3)_X$ $(H_2SO_4)_Y$ and HSO_4^- $(NH_3)_X$ $(H_2SO_4)_Y$ was studied with an ion-molecule flow reactor coupled to a quadrupole mass spectrometer. Equilibrium constants for the association of NH_3 with these species were measured and Gibbs free energy changes were derived. The binding of NH_3 appears to be enhanced by the presence H_2SO_4, and vice versa. These stable cluster ions may be important precursors for atmospheric aerosols.

INTRODUCTION

The mechanisms and rates of homogeneous gas phase nucleation in the atmosphere are poorly understood. Sulfuric acid and water are implicated as necessary components for nucleation (1,2). Recent work has shown that NH_3 may also play an important role in stabilizing H_2SO_4/H_2O clusters and enhancing nucleation rates (3-5). Additionally, modeling and experimental studies indicate that ion-induced nucleation may provide a more efficient means of particle production in the atmosphere than neutral nucleation mechanisms (6,7).

Naturally produced atmospheric ions have lifetimes of 100-1000 seconds, long enough to undergo ionic interconversion and chemical reaction. Positive ions generated in the atmosphere quickly convert to stable protonated species such as H_3O^+ and NH_4^+. Much of the negative charge becomes incorporated into NO_3^- and HSO_4^- (8). Association of water and other neutral molecules to these core ions in the atmosphere is efficient, and at equilibrium the stable core ions are clustered to several ligands.

The thermodynamics of $NH_3/H_2SO_4/H_2O$ cluster ions have not been previously reported. Free energy values for these species are required in order to calculate ion-induced nucleation rates. We have investigated both the positive and negative cluster ions of the NH_3/H_2SO_4 binary system. Equilibrium constants and ΔG° values for NH_3 association reactions were determined at room temperature.

EXPERIMENT

Positive and negative cluster ions were studied using an ion-molecule flow reactor coupled to a quadrupole mass spectrometer. Ions were generated by adding NH_3 and H_2SO_4 to a main He flow downstream of an electron-emitting filament to produce cluster ions of the form, $NH_4^+ (NH_3)_X (H_2SO_4)_Y$ and $HSO_4^- (NH_3)_X (H_2SO_4)_Y$. Ions were sampled from the flow reactor into a high vacuum region through a 50 μm orifice. The ions were then collimated and transported by electrostatic lenses to the quadrupole mass filter for mass analysis.

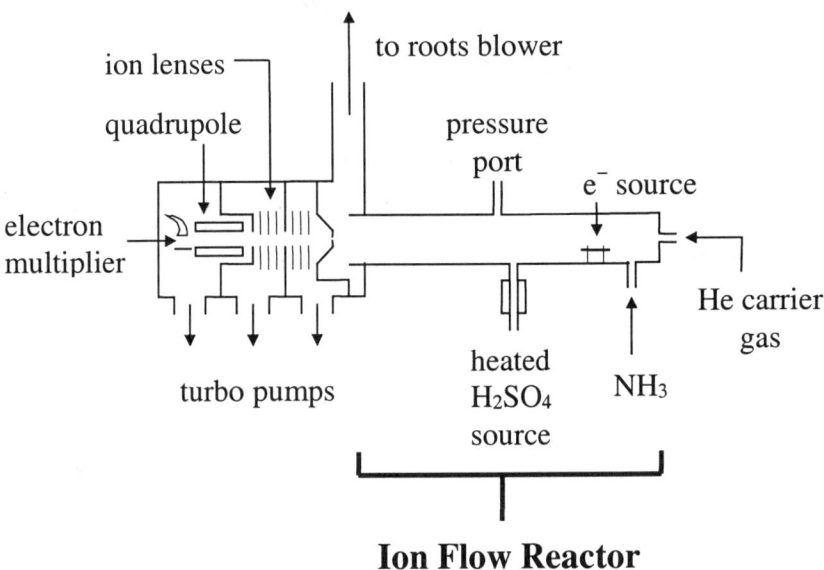

FIGURE 1. The ion-molecule flow reactor system used to study NH_3/H_2SO_4 cluster ions. The pressure inside the flow reactor was 0.5 – 3 Torr, and ion residence times ranged from 10 – 100 ms.

OBSERVATIONS AND RESULTS

Positive Ions

As shown in Figure 2, the entire 500 amu range of the positive ion spectrum is dominated by species of the form $NH_4^+ (NH_3)_X (H_2SO_4)_Y$, with $X = 2$-5 and $Y = 0$-4. The cluster ions equilibrated with NH_3 at NH_3 concentrations of $(5 - 100) \times 10^{12}$ molecules cm^{-3} inside the flow reactor. Equilibrium constants for NH_3 association reactions were calculated using the ratio of reactant and product cluster ion signal intensities and the concentration of NH_3, measured by UV absorption prior to entering the flow reactor.

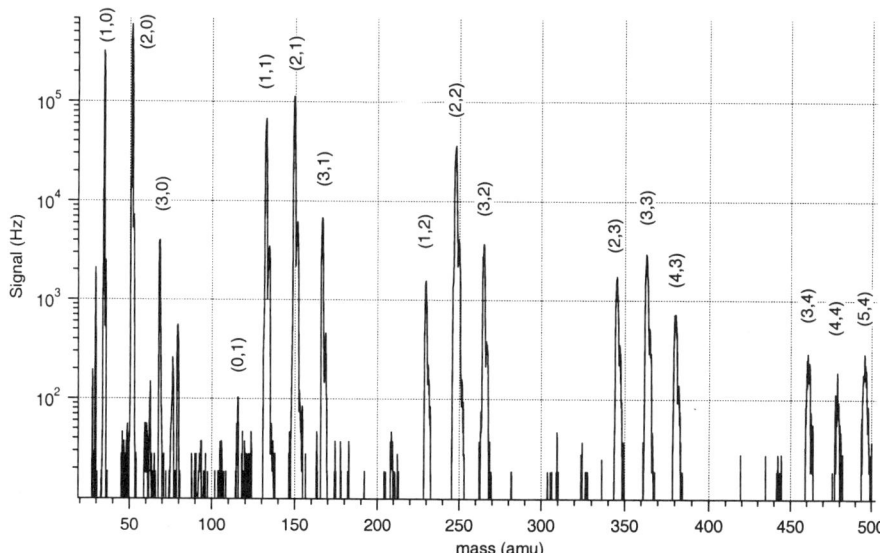

FIGURE 2. A positive ion spectrum showing NH_4^+ $(NH_3)_X$ $(H_2SO_4)_Y$ cluster ions, where (X,Y) denotes the number of NH_3 and H_2SO_4 molecules, respectively, in the cluster. Concentrations in the flow reactor are $[NH_3] = 5.0 \times 10^{12}$ and $[H_2SO_4] < 4 \times 10^{11}$ molecules cm^{-3}.

NH_3 is particularly stable in NH_4^+ $(NH_3)_X$ $(H_2SO_4)_Y$ cluster ions where X = Y. However, attachment of one, two, and three additional NH_3 molecules onto these stable species is increasingly less favorable, indicated by rising values of the free energy of association, $\Delta G°_{(X-1,X)}$, for each successive NH_3 molecule. This diminishing stability of NH_3 with increasing X is also observed in the well known NH_4^+ $(NH_3)_X$ system. In contrast, with the number of ammonia ligands held constant, $\Delta G°_{(X-1,X)}$ values show that ammonia binding is enhanced by the addition of H_2SO_4.

Sulfuric acid concentrations in the flow reactor were estimated to be $[H_2SO_4] < 5 \times 10^{11}$ molecules cm^{-3}, and consequently, cluster ions were probably not in equilibrium with H_2SO_4. Therefore, equilibrium constants for H_2SO_4 addition to the cluster ions could not be accurately measured. However, upper limits for $\Delta G°_{(Y-1,Y)}$ indicate that H_2SO_4 is more strongly bound than NH_3 in NH_4^+ $(NH_3)_X$ $(H_2SO_4)_Y$ clusters when X ≥ Y, suggesting that H_2SO_4 is likewise stabilized by the presence of NH_3.

To validate the experimental technique for studying cluster ion equilibria, protonated ammonia clusters were also examined in the absence of H_2SO_4. These results are in good agreement with $\Delta G°_{(X-1,X)}$ literature values (9).

Negative Ions

The negative ion system is built on the very stable HSO_4^- $(H_2SO_4)_y$ family of clusters, y = 1-4. Attachment of $(NH_3)_x$, x = 1-3, was observed for y ≥ 2. $\Delta G°_{(x,x-1)}$ for NH_3 association decreased successively with the number of H_2SO_4 molecules in the cluster. Hence, as with the ammonium-based positive clusters, addition of H_2SO_4

molecules enhanced the NH_3 binding in the negative clusters. However, in contrast to the NH_4^+ $(NH_3)_X$ $(H_2SO_4)_Y$ system, ammonia binding did not decrease as more ammonia molecules were added to the HSO_4^- $(NH_3)_x$ $(H_2SO_4)_y$ clusters. There also does not appear to be an unusually stable set of negative ion species, as seen with the X = Y positive cluster ions.

CONCLUSION AND IMPLICATIONS

This work demonstrates a synergistic enhancement in ligand stability between ammonia and sulfuric acid in both the positive and negative binary cluster ion systems. The thermodynamic results for NH_3 association suggest that several of the species studied may exist in the atmosphere, likely in a hydrated form. Furthermore, the strong NH_3 and H_2SO_4 association to some of the cluster ions may be essentially irreversible, providing a possible mechanism for growth of small clusters in the atmosphere.

ACKNOWLEDGMENTS

The authors would like to thank C. J. Howard and A. R. Ravishankara for their insight and helpful suggestions. This work is supported by NOAA's Climate and Global Change Program.

REFERENCES

1. Weber, R. J., McMurry, P. H., Eisele, F. L., and Tanner, D. J., *J. Atmos. Sci.*, **52**, 2242-2257 (1994).
2. Kulmala, M., Kerminen, M., and Laaksonen, A., *Atmos. Environ.*, **29**, 377-382 (1995).
3. Coffman, D. J., and Hegg, A. D., *J. Geophys. Res.*, **100**, 7147-7160 (1995).
4. Korhonen, P., Kulmala, M., Laaksonen, A., Viisanen, Y., McGraw, R., and Seinfeld, J. H., *J. Geophys. Res.*, **104**, 26349-26353 (1999).
5. Ball, S. M., Hanson, D. R., and Eisele, F. L., *J. Geophys. Res.*, **104**, 23709-23718 (1999).
6. Yu, F., and Turco, R. P., *Geophys Res. Lett.*, **27**, 883-886 (2000).
7. Kim, T. O., Adachi, M., Okyuama, K., and Seinfeld, J. H., *Aerosol Sci. Tech.*, **26**, 527-543 (1997).
8. Viggiano, A. A., *Mass. Spect. Rev.*, **12**, 115-137 (1993).
9. Keesee, R. G., and Castleman, A. W., *J. Phys. Chem. Ref. Data*, **15**, 1011-1071 (1986).

New Particle Formation at a Rural Site in the UK

P. I. Williams[1], H. Coe[1], M. W. Gallagher[1], K. N. Bower[1], K. M. Beswick[1], T. W. Choularton[1] and G., McFiggans[1,2]

Physics Department, UMIST[1]
P. O. Box 88
Manchester M60 1QD
UK

School of Environmental Sciences[2]
University of East Anglia
Norwich NR4 7TJ
UK

Abstract. Particle size distribution measurements were made at a coastal site in the UK. These are presented and the behaviour of recently formed ultra-fine particles is discussed. No ultra-fine particles were observed in maritime air masses, however 3 to 7 nm particles were frequently observed at enhanced concentrations when the wind direction was from the land. Their formation was favoured at lower temperatures, when 1 ppbv or more of SO_2 was present and in air masses that had not been aged extensively. On days when enhanced ultra-fine particle concentrations were observed, 3 nm particles increased sharply in the morning, approximately 30 to 90 minutes after the UV solar flux first increased. By early afternoon the ultra-fine particle concentration had returned to background levels. Rapid measurements of 5 nm particles showed no correlation with turbulence parameters, although the boundary layer mixing scales were similar to growth times of freshly nucleated particles to 5 nm diameter. However, ultra-fine particle concentrations do correlate with the availability of sulphuric acid vapour. A delay of approximately an hour between the increase of H_2SO_4 in the morning and a large increase in ultra-fine particle concentrations is due to the growth of particles to observable sizes, not the nucleation process itself. An analysis of the timescales for growth showed that coagulation may be important immediately the particles have nucleated but its effectiveness reduces as number concentration falls. Conversely, growth by condensation is initially slow due to the Kelvin effect but increases in importance as the particles reach observable sizes.

INTRODUCTION

Ultra-fine aerosol particles, between 3 and 10 nm diameter, provide an important source of new aerosol particles in the atmosphere; however, much uncertainty surrounds their formation and growth. Observations of these particles have now been made in a number of different environments including the remote marine boundary layer[1], coastal locations[2,3], forests[4], remote continental sites[5] and the free troposphere[6]. In all cases the presence of these particles is sporadic, occurring in events or bursts lasting from several minutes to a few hours, and only occurs during daylight, implying the gas precursors are photochemically produced.

BACKGROUND TO THE EXPERIMENT

The experiment took place at the Weybourne Atmospheric Observatory[7] on the north coast of Norfolk, UK (52.951 N, 1.125 E) from 14th June to the 30th June 1998. The observatory lies within 100 m of the coastline to the north at a height of 15 m above sea level on grassland sloping down to a pebble sea defence. In northerly wind directions air, often from north of the Arctic Circle, advects over the North Sea directly to the station. To the south and west, Weybourne is surrounded by flat arable land and is well-removed from immediate local sources. When sampling air from these sectors north Norfolk is characteristic of much of the rural UK and is several hundred kilometres downwind of the large conurbations of London (190 km) and Birmingham (250 km). Most of the largest point sources of SO_2 in the UK are located approximately 150 km to the west and north west of the site.

Particle size distributions in the range 3 nm to 450 nm diameter were measured using a Differential Mobility Particle Sizing (DMPS) system, comprising two differential mobility analysers (DMAs). Number concentrations of larger aerosol particles (100 nm to 3 μm diameter) were measured as a function of their size using an optical scattering probe (ASASP-X). Full size distributions were retrieved every 10 minutes throughout the experiment. However, ultra-fine particle concentrations can vary rapidly over very short timescales as their lifetimes are very short. To investigate these fluctuations an additional DMA was set up to continuously measure 5 nm diameter particles with a time response of 1 second.

Important chemical precursors and physical parameters relevant to new particle formation were also measured including SO_2, photolysis rates and O_3 at a time resolution of, at most 1 minute so that processes influencing the behaviour of new particles could be correlated with the fast 5 nm particle measurements. In addition a full range of meteorological parameters was measured using an automatic weather station mounted at the top of the 10 m sampling tower and a sonic anemometer measured the three wind components at 20 Hz.

SUMMARY OF RESULTS

General

During periods of northerly winds no evidence of ultra fine aerosol (< 10 nm) was observed in marine air at Weybourne. In contrast ultra-fine particles were observed regularly in rural air. Their presence was inhibited at higher temperatures, reduced concentrations of SO_2 and large pre-existing aerosol populations. New particles were not observed in air advected from continental Europe. These periods were marked by high temperatures (>20 C) and very high aerosol surface areas (2 x 10^5 cm^2 cm^{-3}), which were caused by long range transport of aged anthropogenic aerosol and also Saharan dust. On the remaining days of the experiment the air arriving at Weybourne was cyclonic, rapidly advecting cold, unstable air across the UK from the Atlantic Ocean. On these days new particle concentrations were only observed to be substantially elevated when the air was from a west or north-westerly direction and not when it was from the south westerly sector. The SO_2 mixing ratios observed in the former sector were generally above 1 ppbv and showed structure typical of a plume from a source tens of kilometres distant. Those in the latter sector were generally less than 1 ppbv and often at the detection limit of the instrument. Large point sources of SO_2 exist in an area approximately 100 km to the west and north west of the site and it is likely that these sources are responsible for the enhanced SO_2 concentrations observed.

On days when new particles were formed they grew rapidly from background levels between 30 minutes and 1.5 hours after the initial rise in UV radiation in the morning, implying a photolytic source of precursors. By early afternoon new particle concentrations had again returned to background levels and the particles formed during the morning had appeared to grow into the Aitken mode and add to its number concentration.

Rapid Measurements of 5 nm Particles, the Effects of Turbulence and Precursors

Fast measurements of 5 nm aerosol particles showed no clear correlation between rapid temperature fluctuations and new particle concentrations on timescales of minutes, indicating that 5 nm particles are well mixed throughout the boundary layer. These data do not appear to show any evidence for enhancements of new particle formation by mixing previously[8,9], although the effects of small scale turbulence could not be investigated using this approach. However, as the growth times of freshly nucleated particles to 5 nm diameter are of the same order as mixing time scales throughout the boundary layer it is also possible that vertical variations in the nucleation rate will be obscured by mixing if larger particles are investigated.

The fast measurements of 5 nm particles, N_5, are closely related to the availability of sulphuric acid vapour, A_A. This parameter is calculated from a steady state analysis of the oxidation of SO_2 by OH and is given by

$$[H_2SO_4] \propto k_1 j_{O(^1D)} [O_3][SO_2] k_{loss}^{-1} = A_A$$

where k_1 is rate of SO_2 oxidation by OH, $j_{O(1D)}$ is the photolysis rate of ozone to yield $O(^1D)$ and k_{loss} is the loss rate of H_2SO_4 to pre-existing particle surfaces. The correlation between H_2SO_4 and 5 nm particles is very similar to that obtained previously in descending air at a remote location in the US[5], a very different environment to the rural environment at Weybourne. Daily averaged values of N_5 and the availability of sulphuric acid show that an increase in new particle concentrations by approximately an order of magnitude occur on days when there is a greater than three fold increase in the amount of acid available. The mechanism of particle formation remains unclear but a likely candidate is a ternary process involving NH_3, in addition to H_2SO_4 and water.

The observations show a delay of between 30 minutes and 1.5 hours from the initial increase in the availability of H_2SO_4 vapour and the presence of 5 nm particles. If, as expected, ternary nucleation is rapid then the observed delay is due to the growth of stabilised particles below observable size to 5 nm diameter.

Growth of Nucleated Particles to Observable Sizes

We have considered the timescales for this growth due to both coagulation and condensation. The former process will be significant if the particle concentration is greater than 10^6 cm^{-3}, these levels are never observed for particles greater than 3 nm. It is apparent therefore that growth of the newly formed particles by coagulation is not occurring at observable sizes. However, sufficient mass exists in the observed ultra-fine aerosol for initial number concentrations of 1 nm diameter particles to have been at least 10^6 particles cm^{-3} in the absence of condensation.

Condensation of H_2SO_4 onto particles of 1 nm diameter is restricted by the Kelvin effect and initial growth will be slow, although the calculated growth times are of similar order to the observed delay times. In the model calculation around 1 pptv H_2SO_4 is required for grow a particle from a nucleated cluster of 1 nm diameter to a 5 nm particle in around 40 minutes, in good agreement with the observations.

It seems likely that as the timescales associated with both coagulation and condensation are approximately the same, they will both play a part in growing the nucleated particles. Coagulation will be important initially when the number concentrations of the small particles are high and condensation is limited by the Kelvin effect. After a short time in the growth of these particles the number of particles will have decreased, reducing the effectiveness of coagulation, the Kelvin effect will lessen and condensation will dominate as the particles reach observable sizes.

ACKNOWLEDGMENTS

This work was supported by UK NERC grants GR9/03438 and GR3/R9686.

REFERENCES

1. Covert D. S., V. N. Kasputin, P. K. Quinn and T. S. Bates, New particle formation in the marine boundary layer, *J. Geophys. Res.*, *97*, 20581-20589, 1992.
2. Allen A.G., J. L. Grennfell, R. M. Harrison, J. James and M. J. Evans, Nanoparticle formation in marine airmasses: contrasting behaviour of the open ocean and coastal environments, *Atmos. Res.*, *51*, 1-14, 1999.
3. O'Dowd C., G. McFiggans, D. J. Creasey, L. Pirjola, C. Hoell, M. H. Smith, B. J. Allan, J. M. C. Plane, D. E. Heard, J. D. Lee, M. J. Pilling and M. Kulmala, On the photochemical production of new particles in the coastal boundary layer, *Geophys. Res. Letts.*, *26*, 1707-1710, 1999.
4. Makela J. M., P. Aalto, V. Jokinen, T. Pohja, A. Nissinen, S. Palmroth, T. Markkanen, K. Seitsonen, H. Lihavainen, M. Kulmala, Observations of ultrafine aerosol particle formation and growth in boreal forest, *Geophys. Res. Letts.*, *24*, 1219-1222, 1997.
5. Weber R. J., J. J. Marti, P. H. McMurry, F. L. Eisele, D. J. Tanner and A. Jefferson., Measurements of new particle formation and ultrafine particle growth rates at a clean continental site, *J. Geophys. Res.*, *102*, 4375-4385, 1997.
6. Clarke A. D., Atmospheric nuclei in the remote free troposphere, *J. Atmos. Chem.*, *14*, 479-488, 1992.
7. Penkett S. A., J. M. C. Plane, F. J. Comes, K. C. Clemitshaw, H. Coe, The Weybourne Atmospheric Observatory, *J. Atmos. Chem.*, *33*, 107-110, 1999
8. Easter R. C. and L. K. Peters. Binary nucleation: Temperature and relative humidity fluctuations, nonlinearity, and aspects of new particle production in the atmosphere, *J. Appl. Meteorol.*, *33*, 775-784, 1994.
9. Nilsson E. D. and M. Kulmala, The potential for atmospheric mixing processes to enhance the binary nucleation rate, *J. Geophys. Res.*, *103*, 1381-1389, 1998.

How Well Are Supersaturation and Temperature in Expansion Chambers Predicted by the Poisson-Equation?

S. Zach, A. Vrtala, G. P. Reischl, P. E. Wagner

Institut für Experimentalphysik
Universität Wien
Boltzmanngasse 5, A-1090 Wien, Austria

Abstract. Various experimental studies of nucleation processes have been reported, which were performed using expansion chambers. In these studies the gas temperatures and vapor supersaturations at the end of the adiabatic expansions considered have usually been determined by means of the Poisson equation. Recently, deviations from the Poisson equation were reported and an empirical correction has been suggested [2]. In the present study gas temperatures and vapor supersaturations at the end of adiabatic expansions were experimentally determined from quantitative measurements of condensational drop growth rates. Except for one expansion at an extremly high expansion ratio no significant deviations from the Poisson equation were found. The trend predicted by the empirical relation [2] has not been observed experimentally.

INTRODUCTION

Experimental studies of nucleation and condensation processes in the gas phase have been reported by many authors (see, e. g., Heist and He [1] for review). The crucial condition for well-defined nucleation experiments is the generation of condensable vapors with accurately determined supersaturations. Mainly adiabatic expansion, nonisothermal diffusion and turbulent mixing have been applied for this purpose. Direct experimental determination of temperature and supersaturation in supersaturated vapors is difficult, however, as vapor condensation will generally occur on the surface of a temperature sensor and thus the temperature of the condensing liquid rather than the gas temperature is measured. Accordingly, gas temperature and vapor supersaturation are usually *calculated*. Particularly for the case of adiabatic expansions of dilute vapor-carrier gas mixtures the gas temperature is easily calculated using the Poisson equation. Recent nucleation experiments [2], however, seem to indicate systematic deviations from these calculated temperatures and an empirical correction has been suggested [2].

In the present study the thermodynamic conditions at the end of adiabatic expansions have been *experimentally* determined from measurement of condensational drop growth rate, which can be used as a sensitive indicator for vapor supersaturation and temperature. The results are compared to corresponding predictions.

MEASUREMENT PROCEDURE

We studied adiabatic expansions of vapor – carrier gas mixtures in an expansion cloud chamber. The expansion times did not exceed 10 ms. The subsequently occuring condensational drop growth was measured by means of the Constant-Angle Mie Scattering (CAMS) method [3]. Comparatively small supersaturations were chosen as the correspondingly slow drop growth is most sensitive to changes of supersaturation and temperature.

While keeping the vapor fraction constant, we performed series of different expansions starting from different initial total gas pressures in the expansion chamber and leading to the identical final total gas pressure and thus the identical vapor pressure. The initial (chamber) temperatures were chosen in such a way that the final temperature after expansion, calculated according to the Poisson equation

$$T_f = \frac{T_i}{\beta^{\frac{\kappa-1}{\kappa}}} \qquad (1)$$

was kept constant. Here T_i and T_f are the gas temperatures before and after adiabatic expansion, $\beta = p_i/p_f$ is the ratio of initial and final total gas pressure (pressure expansion ratio), and κ is the adiabatic index for the vapor - carrier gas mixture, which can be determined according to Richarz [4]. Thus, according to the Poisson equation, the identical thermodynamic conditions at the end of the different expansions and correspondingly the same drop growth rates can be expected.

In first measurement series comparatively small expansion ratios were considered. Within experimental scatter actually the same drop growth rates were found indicating that the Poisson equation is consistent with experiment at small expansion ratios. Accordingly, at small expansion ratios the actual temperature after expansion is known thus allowing calibration of the observed drop growth rates with respect to the gas temperature for the considered conditions at the end of the expansion.

Subsequently, measurements were performed over an extended range of expansion ratios and from the measured drop growth rates the actual temperatures after expansion can be determined. It should be emphasized that the temperature measurement method applied in this study does not involve drop growth calculations and thus the results obtained are independent of drop growth theory.

EXPERIMENTAL RESULTS

Temperature and supersaturation at the end of adiabatic expansions were experimentally determined from drop growth measurements. Here we report experimentally determined vapor supersaturations in n-propanol - air mixtures.

First measurement series have shown that the accuracy of the results is critically dependent on the adsorption equilibrium at the walls of the measuring system. In order to establish adsorption equilibrium, extensive flushing of the measuring system *at the selected initial thermodynamic conditions before expansion* was performed prior to all measurements.

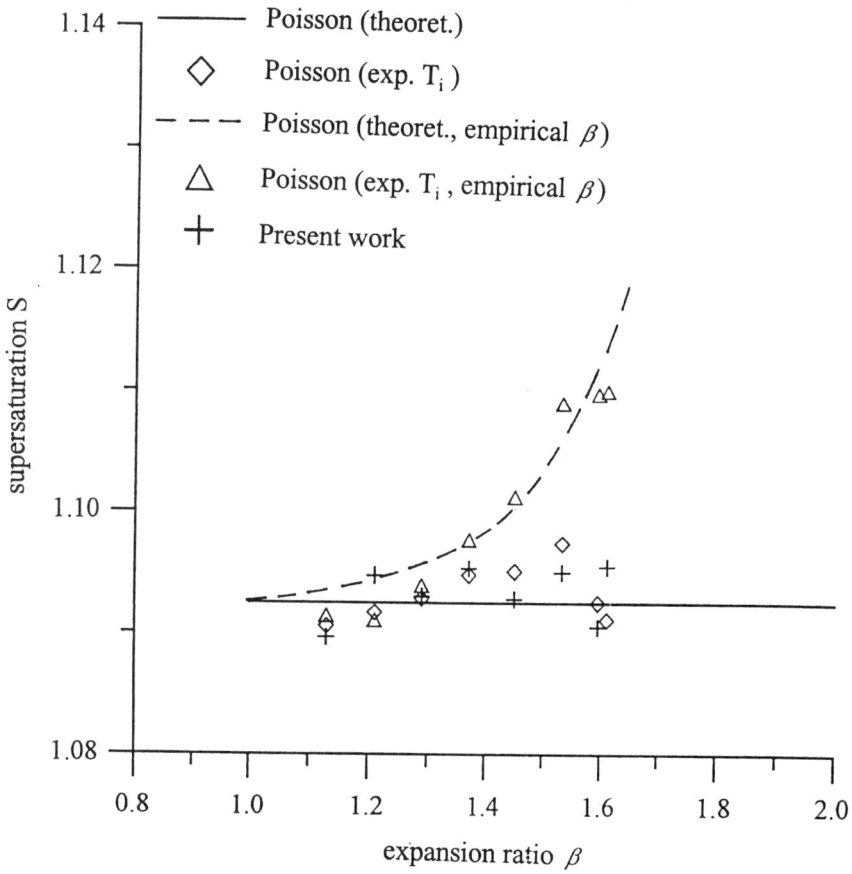

FIGURE 1. Final Vapor Supersaturation after Adiabatic Expansion of n-Propanol - Air Mixtures.

Expansion ratios between about 1.1 and 1.6 and corresponding temperature drops ranging from about 10 K up to as much as 40 K were considered. The constant final vapor saturation ratio at the end of expansion was chosen to be around 1.09 resulting in comparatively slow drop growth rates. The corresponding initial vapor saturation ratios before expansion were varied from about 0.65 down to 0.15. As mentioned above, during each series of experiments the final temperature and supersaturation after expansion, calculated according to the Poisson equation 1, were kept constant. In Figure 1 the experimental final supersaturation after expansion, as obtained in the present study, is shown vs. the expansion ratio (crosses). It can be seen that the Poisson equation (solid line) is consistent with the experimental final supersaturation. Accordingly, for the experimental conditions considered in the present study, supersaturation and temperature in an expansion chamber appear to be quite well predicted by the Poisson equation.

We have also performed a comparison of the experimental results obtained in this study to the empirical relation

$$\frac{1}{\tilde{\beta}} = \frac{1}{\beta} - 0.2915 \left(1 - \frac{1}{\beta}\right)^{5.036} \qquad (2)$$

according to Viisanen et al.[2], where β denotes the *actual* experimental pressure expansion ratio, while $\tilde{\beta}$ is a modified value for the pressure expansion ratio to be used in the Poisson equation 1. As can be seen from Fig. 1, the trend predicted by this empirical relation (broken line) has not been observed experimentally.

For one experiment, performed with water vapor at the particularly high expansion ratio of 1.9, we did observe a significant deviation from the prediction by the Poisson equation. This appears to be connected with extreme temperature gradients in the measuring chamber and with turbulence. This observed deviation is in fact opposite to the trend predicted by the empirical relation[2] shown in equation 2.

ACKNOWLEDGMENTS

This work has been supported by the Fonds zur Förderung der Wissenschaftlichen Forschung, Projekt Nr. P 9421 - GEO, and by the Hochschuljubiläumsstiftung der Stadt Wien.

REFERENCES

1. Heist, R. H., and He, H., *J. Phys. Chem. Ref. Data* **23**, 781 (1994).
2. Viisanen, Y., Strey, R., and Reiss, H., *J. Chem. Phys.* **99**, 4680 (1993).
3. Wagner, P. E., *J. Colloid Interface Sci.* **105**, 456 (1985).
4. Richarz, F., *Ann. Phys.* **19**, 639 (1906).

Self-Consistent Binary Cluster Size Distributions of Sulfuric Acid-Water System

Madis Noppel

Institute of Environmental Physics, University of Tartu,
18 Ülikooli St., 50090 Tartu, Estonia

Abstract. The new self-consistent and partially self-consistent forms of binary distribution function are proposed. The activities (number concentrations) of sulfuric acid and water calculated by these distributions at unit nucleation rate are compared between themselves and with experimental data

INTRODUCTION

Recent study by Wilemski et al. [1] of nucleation kinetics have revealed the importance of issues of self-consistency (SC) of equilibrium distributions. It was shown that an acceptable equilibrium binary cluster distribution should obey the law of mass action, should reduce to appropriate forms for the unary distributions and should be equal the monomer concentration when distributions are evaluated for a single monomer unit. In this paper, in relation with self-consistency, the incorporation of additional information on free energies of small clusters in a cluster distribution of capillarity approximation is considered. The possible source of additional information may be, for instance, *ab initio* calculations or experimental data. The new SC and partially SC forms of binary distribution function are proposed. Using these distributions the vapor number concentrations of sulfuric acid and water, required to obtain unit nucleation rate, are calculated and compared between themselves and with experimental concentrations.

Mass Action and A Cluster Distribution

The law of mass action requires that $P(i,j)$, the equilibrium partial pressure of clusters containing i monomers of type A and j monomers of type B, be expressible in the form

$$P(i,j) = P_A^i P_B^j \exp(-\frac{\Delta G_{i,j}}{kT}) = \exp(-\frac{\Delta G_{i,j} - ikT \ln P_A - jkT \ln P_B}{kT}), \quad (1)$$

where P_v is the partial pressure of monomers of species v ($= A$ or B) in atmospheres, k is the Boltzmann constant, T is the absolute temperature, and $\Delta G_{i,j}$ is the standard free energy of formation of a cluster of size (i, j), i.e. the free energy of formation of a cluster from vapors of species v at normal pressure. The free energy of formation of a monomer from a monomer is equal to zero, $\Delta G_{0,1} = \Delta G_{1,0} = 0$. If we knew the standard free energy $\Delta G_{i,j}$ Eq. (1) would give us a SC distribution. In the classical nucleation theory the capillarity approximation is used to estimate the values of $\Delta G_{i,j}$ and this is the point where issues of self-consistency arise. The common expression of capillarity approximation for $\Delta G_{i,j}$ (in the following $\Delta G_{i,j}$ denotes the standard free energy value of capllarity approximation) does not give zero value when evaluated for a single monomer unit. Let us suppose that we have free energy estimates $\Delta G^S_{i,j}$, for some small clusters, obtained from a source that is more reliable than classical approximation. The question is how to join the different estimates, $\Delta G^S_{i,j}$ and $\Delta G_{i,j}$ into the united free energy surface? How to deal with possible unrealistic big jumps at the junction line between free energy surfaces $\Delta G^S_{i,j}$ and $\Delta G_{i,j}$.

Relative position of clusters (except monomers) on the free energy surface $\Delta G_{i,j}$ does not change when we add a constant value to free energies. The free energy surface remains smooth. In the case of unary distribution the free energy values ΔG^S_g and ΔG_g are components of free energy curves, and it seems natural to add a suitable constant to ΔG_g to get smooth united free energy curve from these components. For instance, if to take that $\Delta G^S_2 = \Delta G_2 - \Delta G_1$ is a more reliable value for a dimer than the value ΔG_2, then relation $\Delta G_g - \Delta G_1$ gives us a smooth united free energy curve. Substitution of this relation into Eq. (1) gives us the SC classical unary distribution

$$N(g) = N^\infty \left(N(1)/N^\infty\right)^g \exp\left[-\frac{\sigma s_1}{kT}\left(g^{2/3} - 1\right)\right], \qquad (2)$$

where $N(g)$ is number density of clusters containing g monomers, N^∞ is the equilibrium monomer vapor concentration over flat surface of liquid, σ is the surface tension, s_1 is surface area of a monomer.

Partially Self-Consistent Equilibrium Binary Distribution

In the case of binary distribution the situation is more complex. Now the procedure of a constant addition permits us to get smooth transition (the term is used rather loosely here) from the surface of $\Delta G^S_{i,j}$ to the surface of $\Delta G_{i,j}$ at one point of the conjunction line. For instance, if we have values of $\Delta G^S_{1,1}$ and $\Delta G^S_{1,2}$ then we can use the value of $\Delta G^S_{1,2} - \Delta G_{1,2}$ as a correction constant. It is assumed that values of $\Delta G_{i,j} - \Delta G_{2,1}$ are closer to the corresponding real values than values of $\Delta G_{i,j} - \Delta G_{1,1}$ or $\Delta G_{i,j}$, i.e. capillarity approximation fails more for smaller clusters. For the clusters of size (1,1) and (1,2) the values $\Delta G^S_{1,1}$ and $\Delta G^S_{1,2}$ are used, respectively. For clusters of other sizes the values of capillarity approximation, $\Delta G_{i,j} - \Delta G_{1,2} + \Delta G^S_{1,2}$, are used. Free energies for monomers estimated by the above expression are not zero, but have to be defined zero. In this sense the obtained distribution do not satisfy I type limiting consistency [1]. This distribution can be expressed as

$$N(i,j) = N_A P_B^2 K_{1,2} \exp(-\frac{W_{i,j} - W_{1,2}}{kT}), \qquad (3)$$

where $N(i,j)$ is the number concentration of the clusters of size (i,j), $K_{1,2}$ is the equilibrium constant ($K_{1,2} = \exp(-\Delta G^S{}_{1,2})$), $W_{i,j}$ is the free energy of formation of the cluster of size (i,j) estimated by capillarity approximation ($\Delta G_{i,j}$ denotes the standard free energy) and $N(1,1)$ is determined by $N_A P_B K_{1,1}$. In the above approach minimal changes are made in the initial values of free energy $\Delta G_{i,j}$ and it is not applicable to obtain a binary distribution that reduces to the form of Eq. (2) for both species of A and B as one of them vanishes. In other words, the standard free energies $\Delta G^S{}_{i,0} = \Delta G_{i,0} - \Delta G_{1,0}$ and $\Delta G^S{}_{0,j} = \Delta G_{0,j} - \Delta G_{0,1}$ for species A and B, respectively, are considered more reliable to keep.

Self-Consistent Equilibrium Binary Distribution

To obtain a binary distribution that reduces to the form of Eq. (2) for both unary distributions we use the general form of binary distribution given in the paper [1] and consider a cluster of size (i,j) to consist of $i+j$ fictitious average monomers with the molecular volume equal to the average molecular volume of the solution with composition corresponding to the interior of the cluster. We insist that the binary distribution with general form must coincide by form with the unary distribution of fictitious average monomers described by Eq. (2) to obtain

$$N(i,j) = \left(N_A^\infty\right)^{\frac{i}{i+j}} \left(N_B^\infty\right)^{\frac{j}{i+j}} \left(\frac{N_A}{N_A^\infty}\right)^i \left(\frac{N_B}{N_B^\infty}\right)^j \exp\left[-\frac{\sigma s_1}{kT}\left((i+j)^{2/3} - 1\right)\right], \quad (4)$$

where N_A^∞, N_B^∞, σ, s_1 are the functions of x, which is the mole fraction attributed to the interior of the cluster of size (i,j). The distribution (4) is an analogue of the SC distribution proposed by Wilemski [1].

Nucleation Calculations of Sulfuric Acid-Water System

We use the above distributions to calculate nucleation rates of sulfuric acid-water system. To obtain the free energy of formation of noncritical clusters we assume that a noncritical cluster and its surrounding vapor are separately in equilibrium. The Gibbsian formulation of Nishioka et al [2] is used. The surface tension of a cluster is considered to be independent of cluster size. The bulk mole fraction attributed to the interior of a cluster with given numbers of molecules is determined from the Gibbs adsorption isotherm and from the properties of equimolar surface. The reversible work to form a noncritical cluster is calculated on the basis of the expression given by Nishioka *et al.* [2]. The effect of sulfuric acid hydrates on the work of formation is taken into account as described in [3]. The necessary equilibrium constants of hydrate formation are estimated by the above distributions. The value of $\Delta G^S{}_{1,2}$ (A – acid, B – water) are estimated by the values of enthalpy change $\Delta H_1 = -44.3$ kJ/mol and entropy change $\Delta S_1 = -94.6$ J/mol in the gas-phase addition of a water molecule to a sulfuric acid molecule and by corresponding values, $\Delta H_2 = -54.3$ kJ/mol and $\Delta S_2 = -145$ J/mol, in hydration of a sulfuric acid monohydrate. These values are obtained by fitting [4] to the experimental pressure data. The kinetic part of nucleation rate is calculated by the theory of Stauffer (formulas of paper [3] are used). The results are presented in Fig. 1.

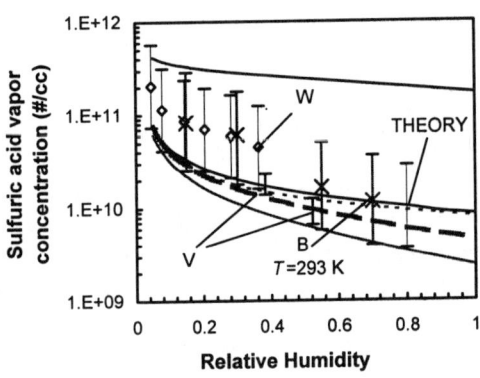

FIGURE 1. Total sulfuric acid number concentrations required to obtain nucleation rate $J = 1$ cm^{-3}s^{-1} according to the equilibrium distributions given by Eqs. (3) and (4) at $T= 298$ K. The upper solid curve correspond to Eq. (4). The middle solid curve corresponds to Eq. (3). The lower solid curve corresponds to Eq. (4), where free energies of critical cluster formation are calculated at free acid concentrations obtained from total acid concentrations according to Eq. (3). The dotted curve corresponds to the distribution of Eq. (4) corrected for mono- and dihydrates as in Eq. (3). The bold dashed curve corresponds to the classical non-self-consistent distribution by Reiss, where free acid concentrations are determined from the balance equation of hydrates by the same distribution. W refers to the experiments of Wyslouzil et al. [5]. V refers to the experiments of Viisanen et al. [6] and B refers to the experiments of Bouland et al. [7]. The middle values (diamonds) of W experiments and theory are calculated using pure H_2SO_4 saturation vapor pressure of $1.88 \cdot 10^{-3}$ Pa. The upper and lower limits for W experiments and theory refer to the uncertainty of pure acid saturation pressure ($6.7 \cdot 10^{-4}$ Pa...$5.2 \cdot 10^{-3}$Pa).

The distribution by Eq. (3) as well as the distribution by Reiss agree the best with experimental data, but the latter involves unphysical behavior [1]. The large difference between upper and lower solid curves reflects the large effect that hydration of acid molecules has on nucleation. The SC distribution by Eq. (4) fails to describe the hydration. When the distribution by Eq. (4) is corrected for mono- and dihydrates analogically with Eq, (3), the result curve is close to the curve of Eq.(3).

ACKNOWLEDGMENTS

This research has been supported by the Estonian Science Foundation grant 3903.

REFERENCES

1. Wilemski, G., and Wyslouzil, B. E., *J. Chem. Phys.* **103**, 1127-1136 (1995).
2. Nishioka K., and Kusaka, I., *J. Chem. Phys.* **96**, 5370-5376 (1992).
3. Noppel, M., *J. Chem. Phys.* **109**, 9052-9056 (1998).
4. Noppel, M., accepted by *J. Geophys. Res.*
5. Wyslouzil, B. E., Seinfeld, J. H., Flagan, R. C., and Okuyama, K., *J. Chem. Phys.*, **94**, 6842-6850 (1991).
6. Viisanen, Y., Kulmala, M., and Laaksonen, A., *J. Chem. Phys.* **107**, 920-926 (1997).
7. Boulaud, D., Madelaine, G., Vigla, D., and Bricard, J., *J. Chem. Phys.* **66**, 4854 (1977).

Binary Nucleation Kinetics In Size And Composition Space

Sergey P. Fisenko* and Gerald Wilemski

*Physics Department and Cloud and Aerosol Sciences Laboratory,
University of Missouri-Rolla, Rolla, MO 65409-0640*

Abstract. The main features of binary nucleation kinetics in size and composition space are presented. An approximate expression for the total nucleation rate of binary nucleation is discussed for the special case of passage through a narrow barrier.

INTRODUCTION

Multicomponent nucleation plays a significant role in many areas of science and technology, ranging from astrophysics to nanoparticle production. Binary nucleation is a special case of multicomponent nucleation, but one that is very important scientifically. The classical kinetics equations of binary nucleation, obtained by H. Reiss [1], involve the distribution function $f(g_a, g_b)$ of new phase clusters, where g_a and g_b are the numbers of molecules of kind a and b, respectively. In this paper we investigate the possibility of developing the theory of binary nucleation kinetics in terms of the new variables g, x. Here, $g = g_a + g_b$ is the total number of molecules in the cluster, and x ($= g_b/g$), is the composition of cluster. The generalization of our results to the multicomponent nucleation is obvious and will be presented later.

MATHEMATICAL MODEL

The distribution function in the new variables $\varphi(g,x,t)$ should be defined as [2]

$$f(g_a, g_b, t) dg_a dg_b = \varphi(g, x, t) dg dx. \tag{1}$$

The saddle point on the surface of the free energy of cluster formation, $\Delta\Phi(g,x)$, can be found as the simultaneous solution of the two equations [3]

$$\frac{\partial \Delta\Phi}{\partial g} = \frac{\partial \Delta\Phi}{\partial x} = 0. \tag{2}$$

The solution to these equations will be denoted as g^* and x^*. If we treat g and x as continuous variables, the full set of kinetics equations for $\varphi(g,x,t)$ can be written as

$$\partial_t \varphi(g,x,t) = -[\partial_g J_g + \partial_x J_x] \tag{3}$$

---

*Permanent Address: A. V. Luikov Heat & Mass Transfer Inst., National Academy of Sciences of Belarus, Minsk, Belarus, CIS where the flux components are defined as

$$J_g = (L_{aa} + L_{bb})F_g + a(g,x)F_x, \quad J_x = a(g,x)F_g + b(g,x)F_x \qquad (4)$$

and the "force" components are

$$F_g = -(\partial_g + \beta \partial_g \Delta\Phi - g^{-1})\varphi, \quad F_x = -(\partial_x + \beta \partial_x \Delta\Phi)\varphi. \qquad (5)$$

Also we have $\beta=1/kT$, where T is the absolute temperature of the gas-vapor mixture and k is Boltzmann's constant. The kinetic coefficients in Eq. (4) are obtained from the kinetic coefficients of the classical kinetics equations of binary nucleation

$$L_{aa} = \frac{p_a \Sigma}{\sqrt{2\pi m_a kT}}, \qquad (6)$$

where p_a and m_a are, respectively, the partial pressure and molecular mass of species "a", Σ is the surface area of the binary cluster. The explicit form of L_{bb} is similar to Eq. (6), and

$$a(g,x) = \frac{L_{bb}(1-x) - xL_{aa}}{g}, \quad b(g,x) = \frac{L_{bb}(1-x)^2 + L_{aa}x^2}{g^2}. \qquad (7)$$

As follows from the interpretation of the kinetic equation (3) as describing a random walk in cluster size and composition space, the sum $(L_{aa} + L_{bb})$ is the coefficient of Brownian diffusion in the cluster size space, and $b(g,x)$ is the coefficient of Brownian diffusion in cluster composition space. The latter is always positive and inversely proportional to g^2.

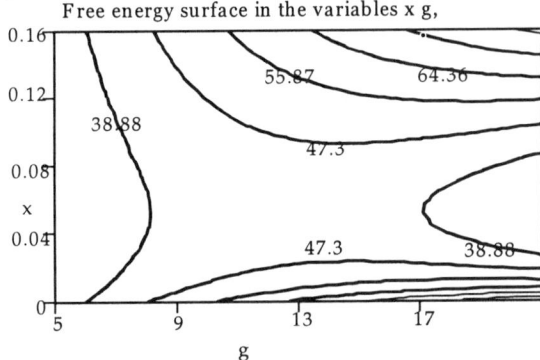

FIGURE 1. Free Energy for Cluster Formation for Binary Nucleation of Mixture of m-o xylene. The saddle point parameters are: $\beta\Delta\Phi(g^*_a, g^*_b)=42.57$, $g^*= 12.544$, $x^*=0.082$; $x_k=0.486$

It is easy to show that there are two characteristic compositions for binary nucleation: One of these is x^*, the composition of the cluster at the saddle point, and the other is x_k, the kinetic composition for which the coefficient $a(g,x)$ equals zero. The value x_k is defined by purely kinetic factors (partial pressures and molecular masses). The composition x_k defines the asymptotic composition of relatively large new phase clusters well after overcoming the thermodynamic barrier.

Total Nucleation Rate

As follows from Eq. (3), the total nucleation rate I can be expressed as

$$I = \int_0^1 J_g(x)dx. \tag{8}$$

Next we show results for a special, but interesting case involving an approximate form for the free energy of cluster formation.

Very Narrow Passage Through Saddle Point

In this section we consider the steady–state approximate solution of (3), assuming that the distribution function can be represented as

$$\varphi(g,x) = f(g)\delta(x-x^*). \tag{9}$$

The expression (9) should be a good approximation at relatively low activity of one of the components when the pass over the free energy surface in the vicinity of the saddle point is very narrow. Then the total nucleation rate I at this case can be written as

$$I = C_1(L_{aa} + L_{bb})\exp[-\beta\Delta\Phi(g^*,x^*)]\sqrt{\frac{\alpha}{\pi}}, \tag{10}$$

where $\alpha = -0.5\partial_{gg}\Delta\Phi(g^*,x^*)$. The parameter α defines the half-width of thermodynamic barrier along the g-axis for the composition x*. The characteristic time lag τ for binary nucleation in this approximation can be estimated as

$$\tau \sim \frac{1}{\alpha(L_{aa} + L_{bb})}. \tag{11}$$

Note that if $p_a \to 0$ or $p_b \to 0$ (implying x→0 or x→1) expression (10) reduces to the classical rate expression for unary nucleation kinetics.

CONCLUSIONS

A transformed kinetic equation of classical binary nucleation was found using a new set of independent variables $\{g, x\}$, where g is the total number of molecules in the cluster and x is the concentration of one of the components. This transformation provides new opportunities to obtain approximate expressions for the total nucleation rate for binary nucleation. An approximate solution for the total nucleation rate was obtained for a special case. In this case $I \sim (L_{aa} + L_{bb})\exp[-\beta\Delta\Phi(g^*,x^*)]$, and it is possible that a similar structure will be found in the expressions for the total rate of multicomponent nucleation [4].

Comparison of experimental and theoretical data in composition space is practically impossible to do at present because the composition of nanodroplets is very difficult to measure. Only small angle neutron scattering currently provides this possibility [5].

ACKNOWLEDGMENTS

This work was supported by the Engineering Research Program of the Office of Basic Energy Sciences, U. S. Department of Energy.

REFERENCES

1. Reiss, H., *J. Chem. Phys.* **18**, 840 (1950)
2. Risken, H., *The Fokker –Plank Equation*, Berlin: Springer, 1984.
3. Hale, B. N. and Wilemski, G., *Chem. Phys. Lett.* **305**, 263 (1999).
4. Kalikmanov, V. I. and van Dongen, M. E. H., "Theory of Multicomponent Nucleation" in *Nucleation and Atmospheric Aerosols 1996*, edited by M. Kulmala and P. E. Wagner, Proceedings of 14 ICNAA Conference, Helsinki: Pergamon, 1996, pp. 188-192.
5. Wyslouzil, B. E., Cheung, J., Wilemski, G., and Strey, R., *Phys. Rev.Lett*, **79**, 431 (1997).

Atomic and Molecular Ions of Natural Atmosphere as Electric System

V. V. Klingo

Voeikov MGO RCARS, St. Petersburg, Russia

Abstract. The calculations of ions electric interaction with various Earth's atmosphere aerosols as well as ion induced nucleation suppose ions electric structure. The electrostatic parameters of the atomic and molecular atmosphere ions have been determined on the basis of the simplest quantum-mechanical approach. The ion geometric dimension identifies with a radius of the most remote sphere with maximum probability of electron residence. The charge distribution on the sphere (along latitude) depends on an angular part of an electron wave function. The classical analogues of two-atom molecules ions electric structure have been found taking into account basic principles of molecular structures calculations in quantum chemistry. The overlap of bonding orbitals integrals have been obtained by means of an artificial mathematical method. The two-atom molecules ions: $H_2^+, N_2^+, O_2^+, and\, O_2^-$ are tree centers of definite electric charges: two at atomic nuclei and one between nuclei with experimental value of a distance between nuclei.

INTRODUCTION

The interactions of hard aerosol particles and cloudy droplets with varied ions in Earth's atmosphere are electrical interactions first of all. This interactions include the ions electrical structure. Hence, it is reasonably to present the ions as electrical systems.

The initial principles of such ions presentations have been set forth in (1). However, there it has been made the assumption that simplifies a problem strongly. The electrical charge distribution in the ions has a spherical symmetry. The further consideration of the electrical ions structure demands a refusal from this assumption that is take into account the angular dependenc of ions charge space distribution. The angular dependence of charge distribution in the ions considers in this investigation.

THEORY

Electrostatic parameters of atomic and molecular ions are calculated on a basis of the simplest quantum-mechanical approach. Electrostatic model of a water molecule have been constructed from such approach (2). Electron wave functions of atomic orbitals ψ_{nlm} are taken in approximate but analytical Slater-Zener form. They are in spherical coordinate system

$$\Psi_{nlm} = \frac{N_n}{a_0^{3/2}} \left(\frac{r}{a_0}\right)^{n-1} \exp\left\{-\varsigma \frac{r}{a_0}\right\} Y_{lm}(\vartheta,\varphi),$$

$$N_n = (2\varsigma)^{\frac{2N+1}{2}} \left[(2n)!\right]^{-1/2}, \quad Y_{lm} = \theta_{lm}(\vartheta)\Phi_m(\varphi), \quad \varsigma = \frac{Z-S_{nl}}{n} \qquad (1)$$

where a_0 is the Bohr radius, Z is the atomic number, S_{nl} is the electron screening value of a given electron, $Y_{lm}(\vartheta, \varphi)$ is the spherical function; n, 1, an dm are correspondingly principal, azimuthal, and magnetic quantum numbers.

An ion geometric dimension is the electron orbit radius a_{nl} with the quantum numbers n and l, which identifies with the radius of the most remote sphere with the maximum probability of electron residence. Radius a_{nl} takes a form from the expression (1):

$$a_{nl} = \frac{n^2 a_0}{Z-S_{nl}} \qquad (2)$$

where S_{nl} is established on a basis of the Slater rules. Thus, the atomic ion electrostatic model of definite chemical composition includes the point ion charge, which is localized in spherical volume center with radius a_{nl}, and angular charge distribution of external nonfilled shell electrons with $1 \geq 1$. The ions radii calculations with the indication of the external electrons configuration are given in Table 1. It is known that summary electron density of three electrons of nonfilled electronic shell is spherically symmetrical. This Table shows that ions N_7^+ and O_8^- only have not spherically symmetrical electronic charge distribution.

TABLE 1.

ion	H_1^-	Li_3^+	N_7^+	O_8^+	O_8^-
configuration	$1s^2$	$1s^2$	$2s^2$	$2p^2$	$2p^3$	$2p^5$
$a_{nl}, m \cdot 10^{10}$	0.76	0.20	2.23	0.50	0.45	0.50

ion	Na_{11}^+	Na_{11}^-	K_{19}^+	K_{19}^-	Rb_{37}^+	Rb_{37}^-
configuration	$2p^6$	$3s^2$	$3p^6$	$4s^2$	$4p^6$	$5s^2$
$a_{nl}, m \cdot 10^{10}$	0.31	2.57	0.61	3.91	9.30	4.58

Modern method features of the electron structure molecules calculation in a quantum chemistry far deviate from the purpose of a present investigation. Basic principles only will be used. The molecule complex structure will be reduced to an electrical field source of sufficiently simple form: two positive charges with a center in atomic nuclei and a negative charge with a center at a middle of distance between nuclei. The molecule electrical structure is calculated at immovable atomic nuclei (the Born-Oppenheimer adiabatic approach). Molecular ions electron wave functions are linear combinations of atomic wave functions. Each molecular ion consists of neutral atom and atomic ion. Atomic electron shells subdivide into electron closed shells and valent shells. The wave functions of the closed shells conserve electron configuration of isolated atoms to a considerable extent. All electrons of a covalent bond is ascribed to two nuclei. Constructed electronic wave functions of the ions satisfy basic principles both quantum-mechanical description of electron systems and a formation of stable molecules with the covalent bonds.

$$\Psi_{nlm} = \alpha \Psi_{\alpha,nlm}(r,\vartheta,\varphi) + \beta \psi_{\beta,nlm}(r',\vartheta',\varphi') \qquad (3)$$

where the electron wave functions are taken from (1), α and β are constant coefficients, they are equal by symmetry. The normalizing condition of the wave function ψ_{nlm} gives

$$\alpha = \beta = \frac{1}{\sqrt{2(1+I_{nlm})}}, \quad I_{nlm} = \int \psi_\chi \psi_\alpha d \xrightarrow{r} \alpha \qquad (4)$$

where I_{nlm} is the integral of overlap of bonding orbitals which is calculated in coordinate system with center in the atomic nucleus of α index.

The angular coordinates are eliminated from the expressions under integral sing (4) with use of second mean value theorem. Then integrals I_{nlm} is found by means of the Fourier transformation.

$$\frac{e^{-\gamma r}}{r} = \frac{1}{2\pi^2} \int \frac{e^{i\vec{k}\vec{r}}}{k^2+\gamma^2} d\vec{k}. \qquad (5)$$

Differentiating both part of (5) with respect to γ necessary number of times, we find that integrals I_{nlm} reduce to combinations of form integrals

$$I(c) = \int \frac{e^{i\vec{k}\vec{R}}}{(k^2+\gamma^2)^c} d\vec{k}, \quad \vec{R} = \vec{r}_\alpha - \vec{r}_\beta. \qquad (6)$$

Here it is taken into account the equality

$$\int e^{i(\vec{k}_1+\vec{k}_2)\vec{r}} d\vec{r} = 8\pi^3 \delta(\vec{k}_1+\vec{k}_2) \tag{7}$$

where δ is delta-function. Differentiating both part of (5) with respect to γ^2 corresponding number of times, we obtain all $I(c)$ from (6) and hence all I_{nlm} from (4) after very bulky calculations.

Electron <<parts>> in a probabilistic meaning from (3) and (4) are $\alpha^2 = \beta^2 = \frac{1}{2}(1+ I_{nlm})$ between the nuclei. Thus, electrostatic structure of two-atomic molecules are tree point charges. This classical scheme gives upper limit of electrical field strength.

Calculation results of total charges (in charge electron units e) at nuclei Q_n^+ and between nuclei Q_{ov}^- on account of all common valent electrons are represented in Table 2. Q_n^+ includes a nuclei charge Ze^+, closed shells $1s^2 2s^2$ charge of N and O atoms with $I = 1$, and charge part of the valent electrons. Table 2 shows that spherical nonsymmetrical wave function in form (1) with $l = 1$ considerably specifies the values of the charges Q_n^+ and Q_{ov}^- in comparison with previous obtained values (1).

TABLE 2.

Ion	H_2^+	N_2^+	O_2^+	O_2^-
$R \cdot 10^{10}$ m	1.06	1.26	1.34	1.34
Q_n^+ in e	0.684	1.22	1.18	0.50
Q_{ov}^- in e	0.368	1.44	1.36	2.00

Values of the total charges Q_n^+ at nuclei and between nuclei Q_{ov}^-, the experimental distances between nuclei R.

REFERENCES

Klingo, V.V., *Natural atmospheric ions as electrostatic systems*, in Proceedings of the 11th International Conference on Atmospheric Electricity, Huntsville, USA, 1999.

Eisenberg, D., Kauzmann, W., *The structure and properties of water*, Oxford, 1969.

Water Molecules Orientation In Surface Layer

V. V. Klingo

Voeikov MGO RCARS, St. Petersburg, Russia

Abstract. The water molecules orientation has been investigated theoretically in the water surface layer. The surface molecule orientation is determined by the direction of a molecule dipole moment in relation to outward normal to the water surface. Entropy expressions of the superficial molecules in statistical meaning and from thermodynamical approach to a liquid surface tension have been found. The molecules share directed opposite to the outward normal that is hydrogen protons inside is equal 51.6%. 48.4% water molecules are directed along to surface outward normal that is by oxygen inside. A potential jump at the water surface layer amounts about 0.2 volts.

INTRODUCTION

Physical adsorption of charged aerosol particles by water drops, penetration these particles into a depth of the drops are depended on a electrical structure of the water surface layer, which is bound up with the molecules – dipoles orientation in this layer. The surface molecule orientation is determined by the direction of the molecule dipole moment in relation to a outward normal to the water surface.

Langmuir (1927) attempted to elaborate a molecules orientation theory and a double electrical layer formation at the water surface. Superficial entropy values have been considered for various liquids by Good (1957). The liquids with strong hydrogen bonds have appreciably less the superficial entropy value. Hence, such liquids must have the molecules surface orientation. However, the hydrogen bonds influence by no means explain on the superficial entropy value.

THEORY

Theoretical calculations of the water molecules surface orientation first have been performed by Fletcher (1962, 1968). The molecules are directed opposite to the outward normal in the first work and along this normal in the second work. These calculations include a great number of qualitative assumptions, which cannot be grounded strictly.

The direction and magnitude of the surface molecules orientation are determined on a base of strict physical data about water only. It is supposed like Fletcher that all molecules of the surface layer are directed along or opposite to the outward normal to the surface. This assumption is impossible to derive because a liquid theory does not exist at present. The molecules orientation magnitude, a order magnitude of the surface layer system can be expressed through its entropy. The entropy expression is written taking into account of the water molecule structure (Fletcher, 1970). A geometric water model includes: two directions of electronic orbitals to hydrogen atomic nuclei, two directions of oxygen electronic orbitals that is lone pair orbitals. It is considered local Cartesian system: origin of coordinates is atom oxygen nucleus; the electronic orbitals directions to the hydrogen nuclei arranges in a plane H-O-H; the direction of the water molecule dipole moment coincides with angle bisectrix between these directions. The lone pair directions are found perpendicularly of the plane H-O-H.

It is estimated a hydrogen bond probability in water. Experimental data about the hydrogen bonds energy in water and a value of a ice heat of fusion indicate that 20% of ice hydrogen bonds are broken in water. Thus, the hydrogen bond formation probability in water is equal 0.8.

It is calculated the entropy of superficial molecules mole N. Surface molecules relative number with orientation inside of surface is α and molecules orientation outside is $1 - \alpha$. Number of surface molecules microscopic states, which realizes macroscopic state, includes $C_N^{\alpha N}$ states. Number of combinations for the hydrogen bonds those aN molecules includes $2^{1.6aN}$ states. Each from two bonds realizes with probability 0.8 independently from other bonds by two modes. The expression of molar entropy in statistical meaning can be written in the from

$$S_N \; k \ln \left\{ 2^{1.6aN} \frac{N!}{(\alpha N)! [(1-\alpha)N]!} \right\} \tag{1}$$

where k is Boltzmann's constant. Replacing factorials by Stirling's formula and neglecting small terms (without coefficient N) we obtain

$$S_N = kN[-\alpha \ln \alpha - (1-\alpha)\ln(1-\alpha) + 1.6\alpha \ln 2]. \tag{1'}$$

The first two terms in right part have maximum value at $\alpha = \frac{1}{2}$. This corresponds equal number of the molecules directions.

Entropy of the water surface layer S_N can be found on the base of thermodynamical approach about the liquid surface tension.

$$\sigma_N = U - TS_N \tag{2}$$

where σ_N is the surface tension of one mole liquid that is free energy of the liquid surface layer, U is internal energy of the liquid surface layer. It is assumed that U very weakly depends on temperature T.

$$\frac{d\sigma_N}{dT} = -S_N. \tag{3}$$

Inserting S_N from (3) into (1') it is got a equation about α. Mean value of experimental data about a dependence of surface tension coefficient for unit surface area from positive temperature $\sigma_1(T)$ gives

$$-\frac{d\sigma_1(T)}{dT} \cong 0.150 \frac{erg}{cm^2}.$$

One mole of the water molecules occupies the surface area A_N.

$$A_N = (\pi N)^{1/3} \left(\frac{3M}{4d}\right)^{2/3} = 7.01 \bullet 10^8 cm^2 \qquad (4)$$

where M is a water molecular weight, d is a water density.
On account of (1'), (3) and (4) a equation for a takes the form

$$kN\left[-\alpha \ln\alpha - (1-\alpha)\ln[1-\alpha] + 1.6\alpha \ln 2\right] = 0.150 \cdot 7.01 \cdot 10^8. \qquad (5)$$

$\alpha = 0.516$ from (5). Thus, 51.6% the water surface layer molecules have the direction opposite to the outward normal that is the hydrogen protons inside and 48.4% molecules are directed along to the surface outward normal that is oxygen inside.

The potential jump φ at the water surface, double electrical layer, conditioned by 3.2% non-compensative molecules directed the hydrogen protons inside is expressed

$$\varphi = 4\pi \cdot 0.032 \, un \cong 0.2 volts \qquad (6)$$

where the water molecule dipole moment $\mu = 1.84D$, the number molecules on $1cm^2$ of the surface area $n = 8.59 \cdot 10^{14} cm^{-2}$. Usually it is supposed that $\varphi \cong 0.1 - 0.2 volts$.

REFERENCES

Langmuir, I. (1927), *Coll. Science Monograph*.
Good, R.J. (1957), *J. Phys. Chem.*, v.61, N6, 810-813.
Fletcher, N.H. (1962), *Phyl. Mag.*, v.7, N74, 255-269.
Fletcher, N.H. (1968), *Phyl. Mag.*, v.18, N156, 1287-1300.
Fletcher, N.H. (1970), *The Chemical Physics of Ice*.

Excess Energy for the n-Octanol Critical Embryos

M.P.Anisimov, I.N. Shaymordanov, Yu.I. Polygalov [+], S.A Timoshenko[+]

*Aerosol Nucleation Lab at Kemerovo, Institute of Catalysis SB RAS,
41A-119 Moskovskiy Prospect, 650065 Kemerovo, Russia*
[+]*Kemerovo State University, 6 Krasnaja Ave., 650043 Kemerovo, Russia*

Abstract. In the framework of the thermodynamic approach we suggested the new algorithm for determination of excess energy for critical embryos. Offered algorithm was implemented for determination of excess energy for the critical embryos in the n-octanol - sulfur hexafluoride system nucleation. We found that the experimental excess energy of n-octanol critical embryos are below the droplet model prediction. In the limit of the big embryo size the experimental results agree with the droplet model. It have been shown the linear dependence of the excess energy on the critical embryo volume. We concluded that the critical embryos might form the chain-like particles.

INTRODUCTION

Aerosol formation plays an important role in many fields of human activities. The first theoretical investigations of new phase formation from supersaturated vapor appeared in the Gibbs work [1]. Volmer and Weber [2], Becker and Döring [3] and others developed the nucleation theory. In complete version the theory was named Classical Nucleation Theory (CNT). In the first experiments CNT was tested by measurements of the vapor critical supersaturation [4, 5]. Sometime theoretical predictions are in a good agreement with experimental results [6, 7 and others.]. The last time investigations illustrate the influence of carrier gas on nucleation in opposite the predictions of the CNT [8-12 and others].

It needs to say that the theory does not postulate correctly macroscopic properties such as a surface tension etc. for critical embryos. For example, surface tension described by Tolman's equation [13] is parametric only and it is true for narrow intervals of parameters. Investigators have founded that there is very important information receivable from the nucleation rates [14-18]. It was shown that from the nucleation rate measurements the embryo size and excess surface energy could be calculated.

Recently in the framework of the thermodynamic approach we suggested the new algorithm for determination of excess energy for critical embryos [19, 20]. Offered algorithm was implemented for determination of excess energy for the critical embryos in the n-octanol - sulfur hexafluoride system nucleation.

THEORY

Here we describe briefly the idea and equation assumptions made to excess critical embryo measurements. Exponential form of nucleation rate (J) certainly is such as $J = J_0 \exp(-\Delta G/kT)$. The pre-exponential factor can be written as follows:

$$J_0 = \frac{1}{\rho}\sqrt{\frac{2\sigma m}{\pi}}\left(\frac{P_0}{kT}\right)^2 S^2, \qquad (1)$$

where ρ and σ are bulk density and surface tension of embryo, respectively; m - molecular mass; P_0 and S are the equilibrium vapor pressure and vapor supersaturation. The Gibbs free energy, ΔG, for critical embryo, containing $i(\Delta\mu, P)$ unary molecules, can be written as:

$$\Delta G(i, \Delta\mu, P) = i\Delta\mu + \Theta(i, \Delta\mu, P), \quad (2)$$

where P, $\Delta\mu$, and Θ are total pressure, difference in chemical potentials of macroscopic bulk liquid (or solid) and gaseous phases, and the excess energy for critical embryos. In experiments the total pressure can be supported at the constant level, i.e. $P=const$. In the approximation of the ideal gas one could write $\Delta\mu = -kT\ln S$. The equation (2) can be changed to another independent variables such as temperature, T, and vapor supersaturation, S. The embryo size and the excess energy for critical embryos can be expressed as function of these variables, $i=i^*(T,S)$ and $\Theta^*(T,S)$. In this case

$$\Delta G^*(T, S) = -i^* kT \ln S + \Theta^*(T, S). \quad (3)$$

Now we can rewrite the equation for critical embryo excess energy. By differentiating (1) and (3) we can get, respectively:

$$\left(\frac{\partial \ln J_0}{\partial T}\right)_S = -\frac{\partial \ln \rho}{\partial T} + \frac{1}{2}\frac{\partial \ln \sigma}{\partial T} + 2\frac{\partial \ln P_0}{\partial T} - \frac{2}{T}. \quad (4)$$

and

$$\frac{\partial}{\partial T}\left(\frac{\Delta G^*}{kT}\right)_S = -\left(\frac{\partial i^*}{\partial T}\right)_S \ln S + \frac{1}{kT}\left(\frac{\partial \Theta^*}{\partial T}\right)_S - \frac{\Theta^*}{kT^2} \quad (5)$$

We neglected of the first and second terms in the right hand of the equation (4) because they are much smaller in comparison with the third term. The third term could be found by Clausius-Clapeyron equation application as $\frac{\partial \ln P_0}{\partial T} = \frac{\Delta H}{kT^2}$. Here ΔH is evaporation heat per one molecule.

The differential equation for nucleation rate will be as follows:

$$\left(\frac{\partial \ln J}{\partial T}\right)_S = \frac{2(\Delta H - kT) + \Theta^*}{kT^2} + \left(\frac{\partial i^*}{\partial T}\right)_S \ln S - \frac{1}{kT}\left(\frac{\partial \Theta^*}{\partial T}\right)_S. \quad (6)$$

In the linear approximation the nucleation rate can be continued up to the vapor-to-liquid spinodal conditions. Equation (6) can be considered as the first order linear differential equation [21] such as $y' + f(x)y = g(x)$. Solution of this equation is presented below

$$y = \exp(-F)\left(\eta + \int_\xi^x g(x)\exp(F)dx\right) \quad (7)$$

$$F(x) = \int_\xi^x f(x)dx$$

assuming that: (i) $f(x)$ and $g(x)$ are continuos, (ii) integral curve has a known point (ξ, η), and (iii) $y = \Theta^*$ and $x = T$, we can find the excess surface energy Θ. The nucleation rates at the constant vapor supersaturation are taken up to the vapor-liquid spinodal condition. It is well known that spinodal is line where the non-barrier transformation to new phase appeared [22] (i.e., $\Delta G^*_{sp} = 0$, and $\Theta^*_{sp} = i^*_{sp} kT_{sp} \ln S$, here subscript sp refers to spinodal conditions). Finally we have the next equation

$$\frac{\Theta^*}{kT} = i^* \ln S - \frac{2\Delta H}{kT}\left(1 - \frac{T}{T_{sp}}\right) - 2\ln\frac{T}{T_{sp}} - \ln\frac{J}{J_{sp}}. \quad (8)$$

In order to evaluate the critical embryo excess energy from the nucleation experiments one needs to know the experimental critical embryo size, i^*, and the condition at the vapor spinodal decomposition, such as temperature, T_{sp}, and nucleation rate, J_{sp}. It should be noted that we avoided questionable problem such as the embryo size and surface energy at spinodal condition description. Spinodal decomposition temperature can be found from the state equation (here used the. van der Waals equation) at the isotherm maximum where condition $(\partial P/\partial V)_T = 0$ exists. Nucleation rate at temperature T_{sp} might be determined from suitable extrapolation of $J = J(T)$.

EXPERIMENTAL RESULTS

The experimental results for n-octanol vapor nucleation rates in sulfur hexafluoride atmosphere [23] are used for excess energy evaluations. For the details of the experimental set-up and descriptions of the heat-mass transfer problem solution we refer to our previous work [8].

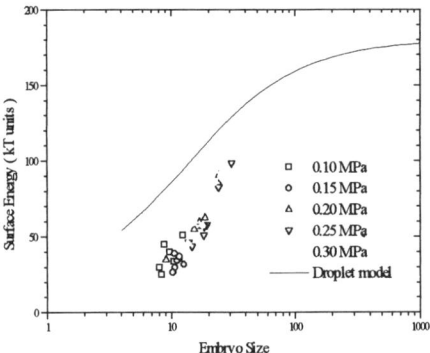

Fig.1. Surface energy versus n-octanol number of molecules. Points are experimental data at different total pressures. The droplet approximation is shown by the solid line.

Figure 1 illustrates the excess energy versus embryo size experimentally determined by equation (8) using at total pressures from 0.10 to 0.30 MPa with step 0.05 MPa. Critical embryo size is measured by using algorithm [15]

$$\left(\frac{\partial \ln J}{\partial \ln S}\right)_T = i^* + 2. \qquad (9)$$

Droplet approximation is shown by solid line. Here the surface energy and the droplet size had been evaluated, respectively as

$$\frac{\Theta^*_{class}}{kT} = 16\pi \left(\frac{\sigma}{kT}\right)^3 \left(\frac{m}{\rho \ln S}\right)^2 \qquad (10)$$

and

$$i^*_{class} = \frac{32\pi}{3} \left(\frac{\sigma}{kT \ln S}\right)^3 \left(\frac{m}{\rho}\right)^2. \qquad (11)$$

One can see that experimental values are below the droplet model results. The largest difference (about one third part from the droplet model predictions) observed for the small embryos, and the discrepancy diminishes when particles become sufficiently large size. One can say that droplet approximation agrees better in case of large droplets (100 molecules and more).

Experimental data can be approximated by linear excess energy versus embryo size. In the double logarithmic scale we found that the slope of this approximation is nearly to unit while the surface energy depends on the particle number in power 2/3. This fact may be explained from known result that the embryo surface (excess) energy relates to whole volume [24, 25]. Experiment shows the reasonable agreement with the computer-simulation prediction for polar fluid [26], where it is shown that critical embryos might be chain-like clusters up to 100-200 molecules.

CONCLUSION

Description of the method for determination of the critical embryo excess energy is presented. Our algorithm had been applied to the experimental results on the n-octanol supersaturated vapor nucleation in the sulfur hexafluoride atmosphere. We found that the experimental excess energy of n-octanol critical embryos are below the droplet model prediction. In the limit of the big embryo `size experimental results agree with the droplet model. It have been shown the linear dependence of the excess energy on the critical embryo volume. We concluded that the critical embryos might form the chain-like particles.

REFERENCES

1. Gibbs, J.W. *The Scientific Papers of J.W.Gibbs* (Dover, New York, 1961), vol.1
2. Volmer, M. and Weber, A. *Z.Phys.Chem.*, **119**, 77 (1926)
3. Becker, R. and Döring, W. *Ann.Phys.*, **24**, 719 (1935)
4. Volmer, M. and Flood, H. *Z.Phys.Chem.*, **170**, 273 (1934)
5. Langsdorf, A. *Rev.Sci.Instr.*, **10**, 91 (1939)
6. Strey, R. Wagner, P.E. and Schmeling, T. *J.Chem.Phys.*, **84**, 2325 (1986)
7. Katz, J.L., Fisk, J. and Chakarov, V. in *Nucleation and Atmospheric Aerosols* (ed. by N.Fukuta and P.E.Wagner, Deepak Publishing, Hampton, Virginia, 1992), p.11
8. Anisimov, M., Koropchak, J., Nasibulin, A., and Timoshina, L. *J.Chem.Phys.*, **109**, 10004 (1998)
9. Anisimov, M.P. and Nasibulin, A.G. *Reports of Academy of Science of Russian Federation* (DAN RF), Russia), **356**, 261 (1997)
10. Anisimov, M.P., Nasibulin, A.G., Timoshina, L.V. and Polygalov, Yu.I. *Abstracts of AAAR'96. 15th Annual Conference* (Kemper Wood Center, Orlando, 1996), p.159
11. Heist, R.H., Janjua, M. and Ahmed, J. *J.Chem.Phys.*, **98**, 4443 (1994)
12. Luijten, C.C., Bosschaart, K.J., van Dongen, M.E.H. *J.Chem.Phys.*, **106**, 8116 (1997)
13. Tolman, R.C. *J.Chem.Phys.*, **17**, 333 (1949)
14. Allen, J.B. and Kassner Jr., J.L. *J.Colloid and Interface Sci.*, **30**, 81 (1969)
15. Anisimov, M.P. and Cherevko, A.G. *Izd.Akad.Nauk USSR* (Siberian Div.), Ser.Khim.Nauk, **4**, 15 (1982)
16. Kashchiev, D. *J.Chem.Phys.*, **76**, 5098 (1982)
17. Ford, I.J. *Phys.Rev.E*, **56**, 5615 (1997)
18. Ford, I.J. *J.Chem.Phys.*, **105**, 8324 (1996)
19. Anisimov, M.P. Nasibulin, A.G. and Shaimordanov, I.N. *Abstracts of AAAR'99. 18th Annual Conference* (Tacoma, Washington, 1999), p.318
20. Anisimov, M.P. Nasibulin, A.G. and Shaimordanov, I.N. *J.Aerosol Sci.*, **30,Suppl.1**, S317 (1999)
21. Kamke, Dr.E. *Lösungsmethoden und Lösungen*, Leipzig, 1959
22. Unger, C., and Klein, W. *Phys.Rev.B*, **29**, 2698 (1984)
23. Anisimov, M.P. Nasibulin, A.G. and Shaimordanov, I.N. *J. Chem. Phys.*, (Submitted in 2000)
24. Scherbakov, L.M. *Colloid Zhurnal* (Russia), **14**, 379 (1952)
25. Scherbakov, L.M., Ryazantsev, P.P. and Phylippov, N.P. *Colloid Zhurnal* (Russia), **23**, 338 (1961)
26. Rein ten Wolde, P., Oxtoby, D.W., and Frenkel, D. *J.Chem.Phys.*, **111**, 4762 (1999)

Corrected Model for Transport in Static Diffusion Chamber

M. Anisimov[1], S. Shandakov[1], V. Pinaev[2], A. Belyshev[1], and R. Heist[3]

[1]*Kemerovo State University, 6 Krasnaya Str., 650043 Kemerovo, Russia.*
[2]*Kemerovo Institute of Commerce, 34 Kuznetskiy Prospect, 650099 Kemerovo, Russia*
[3]*Nucleation Laboratory, Manhattan College, Riverdale, NY, USA*

Abastract. In present consideration for the static diffusion chamber the model of mass and heat transfer and its analytical solution are presented for the pseudo-open in one direction system.

Experimental research of homogeneous nucleation kinetics of supersaturated vapors is based on the determination of relation between nucleation rate and, as a rule, vapor supersaturation (activity) under the constant other parameters. The static diffusion chamber (SDC) allowed direct measurements of nucleation rates could be a powerful instrument for nucleation study. The vapor supersaturation is calculated by using the heat and mass transfer models. Recently the effect of the radial vapor flow to sidewall [1,2] has been calculated. A good agreement between one-dimensional and two-dimensional models (with zero mass average velocities at boundary) was received. It is customary to accept that the diffusion flux is zero at the chamber boundary where vapor is in equilibrium with wet wall. But SDC can be considered as an open system in one direction because of the mass transport of vapor from one boundary (hot plate) to other (cold plate). For this reason idea of a zero vapor flow to boundary is not correct. The model with zero vapor flow may be realized in the case of zero vapor concentration only. At nonzero vapor concentration one have for this model three boundary conditions (first and second are respectively equilibrium concentration values on hot and cold plates, third is zero value of the vapor diffusion flux) for one differential equation of 2-d order with two integration constants only. Recently the nonzero diffusion vapor flux to boundary was accounted in the initial transport equations [3]. Presented numerical solution [3] is in a good agreement with an existing one - dimension model based on numerical solution of the Stefan-Maxwell and heat transfer equations. In present consideration the model of mass and heat transfer and its analytical solution are presented for the open in one direction system. Calculated values for vapor supersaturation are compared with results of Czech group by Rudek et al.[4] on the example of n-pentanol nucleation rates.

According to Hirschfelder et al.[5] system of mass - heat transfer equations in the case of axial-symmetry for the binary mixture stationary flow at the relatively small mass average velocities may be presented as

$$\frac{d(\rho u)}{dz}=0 \qquad \frac{dp}{dz}=0, \qquad \rho u \frac{dc}{dz}=-\frac{dJ_z}{dz} \qquad \rho u \frac{dh}{dz}=-\frac{dq_z}{dz}, \qquad (1)$$

$$J_z=-\rho D_{12}\left(\frac{dc}{dz}+k_{T2}\frac{M_1 M_2}{M^2}\frac{d\ln T}{dz}\right), \quad q_z=-\lambda\frac{dT}{dz}+\left(h_2-h_1+k_{T2}\frac{p}{(1-c)c\rho}\right)J_z. \qquad (2)$$

Here p, ρ and u are the total pressure, density and the average mass velocity, respectively; c is the vapor mass fraction; h, h_1 and h_2 are specific enthalpies of mixture, gas and vapor, respectively; J_z and q_z are the axial mass flux of vapor and heat flux, respectively; T is temperature; λ, D_{12} and k_{T2} are heat conductivity, binary diffusion coefficient of mixture and thermal diffusion factor of vapor, respectively; M_1, M_2 and $M = [(1-c)/M_1+c/M_2]^{-1}$ are molar weights of gas, vapor and mixture, respectively.

It is assumed that at given temperatures on hot plate, T_0, and cold plate, T_1, the vapor concentrations, as boundary conditions, are equal to the equilibrium vapor concentrations, c_0 and c_1, respectively. The boundary average mass velocity is determined from condition of not penetrating gas through boundaries. At $z=0$ this condition is

$$(1-c_0)\rho_0 u_0 - J_{z0} = 0. \qquad (3)$$

Accepted that specific enthalpy $h = (1-c)h_1 + ch_2$. Equation (1) after integration with the taken into account the boundary condition (3) may be expressed as

$$\rho u = \rho_0 u_0 = \frac{J_{z0}}{1-c}, \quad p = p_0, \tag{4}$$

$$\frac{J_{z0}}{1-c_0}(c - c_0) = -(J_z - J_{z0}), \tag{5}$$

$$\frac{J_{z0}}{1-c_0}(h_1 - h_{10} + c(h_2 - h_1) - c_0(h_{20} - h_{10})) = -(q_z - q_{z0}). \tag{6}$$

Here and further values at reference plate (at $z=0$) are indicated by index "zero". After transformation the equation (5) can be written as

$$J_z = \frac{1-c}{1-c_0} J_{z0}. \tag{7}$$

Substituting (7) into (2) we have

$$\frac{dc}{dz} = -\frac{1-c}{1-c_0} \frac{J_{z0}}{\rho D_{12}} \left(1 + \frac{k_{T2}}{T} \frac{M_1 M_2}{M^2} \frac{dT}{dc}\right)^{-1}, \tag{8}$$

$$q_z = -\lambda \frac{dT}{dz} + \left(h_2 - h_1 + k_{T2} \frac{p_0}{(1-c)c\rho}\right) \frac{1-c}{1-c_0} J_{z0}. \tag{9}$$

After substitution of heat fluxes, q_z and q_{z0}, in the form (9) into (6) and after the result transformation we receive the next equation:

$$\lambda \frac{dT}{dz} - \lambda_0 \left(\frac{dT}{dz}\right)_0 + \left(h_{20} - h_2 + k_{T2}^0 \frac{p_0}{c_0 \rho_0} - k_{T2} \frac{p_0}{c\rho}\right) \frac{J_{z0}}{1-c_0} = 0. \tag{10}$$

Specific heat capacity, c_{p2}, of vapor at constant pressure and temperature is constant. Using state equation for mixture, $P/\rho = RT/M$, equation (10) can be rewritten to form

$$\lambda \frac{dT}{dc}\frac{dc}{dz} - \lambda_0 \left(\frac{dT}{dc}\right)_0 \left(\frac{dc}{dz}\right)_0 = \frac{J_{z0}}{1-c_0} c_{p2}\left(T - T_0 + \frac{\gamma_2 - 1}{\gamma_2}\left(\frac{k_{T2} M_2 T}{cM} - \frac{k_{T2}^0 M_2 T_0}{c_0 M_0}\right)\right). \tag{11}$$

Here gas constant is presented through adiabatic ratio for the vapor as $R = c_{p2}(\gamma_2 - 1)/\gamma_2$.
One can substitute expressions (8) for dc/dz and $(dc/dz)_0$ into equation (11). After transforming we come to

$$J_{z0} \cdot \left[\frac{Le(1-c)}{K + dc/dT} - \frac{Le_0(1-c_0)}{K_0 + (dc/dT)_0} - \left(T - T_0 + \frac{\gamma_2 - 1}{\gamma_2}\left(\frac{k_{T2} M_2 T}{cM} - \frac{k_{T2}^0 M_2 T_0}{c_0 M_0}\right)\right)\right] = 0 \tag{12}$$

$$K_0 = \frac{k_{T2}^0 M_1 M_2}{M_0^2}, \quad K = \frac{k_{T2} M_1 M_2}{M_0},$$

where Lewis numbers, respectively Le and Le_0, are expressed as

$$Le = \frac{\lambda}{\rho D_{12} c_{12}}, \quad Le_0 = \frac{\lambda_0}{\rho_0 D_{12}^0 c_{p2}}.$$

The equation (12) has two solutions, namely, the first, for zero diffusion flux, J_{z0}, and the second, for nonzero vapor flux, J_{z0}. One can get the dimensionless temperature, $t = T/T_0$, and designations

$$A = \left(\frac{dc}{dt}\right)_0 + K_0, \quad B(\tau, c) = \frac{1}{Le_0(1-c_0)}\left[1 - \tau + \frac{\gamma_2 - 1}{\gamma_2}\frac{M_2}{M_0}\left(\frac{k_{T2}^0}{c_0} - \frac{M_0}{M}\frac{\tau \cdot k_{T2}}{c}\right)\right], \tag{13}$$

we rewrite equation (12) for the case of nonzero diffusion flux to form

$$\frac{dc}{dt} = \frac{A}{1+A\cdot B(\tau,c)}\frac{1-c}{1-c_0}\frac{Le}{Le_0} - K. \qquad (14)$$

Thus we received a first order differential equation, where integrating constant A is defined from one of two boundary conditions for vapor concentration at plates with different temperatures, T_0 and T_1. Le, K and $B(\tau,c)$ are not simple functions of vapor concentration and temperature in common case.

Usually experiments in the static diffusion chamber are doing in atmosphere of light gas (helium). In this case the gas molar weight is much smaller than vapor one ($M_1<<M_2$) and vapor mass fraction is not closely to unit. In the frame of presentation by Hirschfelder et al.[5] work the heat conductivity of mixture in the form of Vassiljev's equation[6] may be presented as

$$\lambda = \lambda_1(1-c)\frac{M}{M_1}. \qquad (15)$$

In the same approximation, with account the equation (15), the Lewis numbers can be expressed through interaction integrals, Ω,[5] as

$$\frac{Le}{Le_0} = \frac{1-c}{1-c_0}, \qquad (16)$$

$$Le_0 = \frac{25}{8}\cdot\frac{\gamma_2-1}{\gamma_2}\left(\frac{4}{15}\frac{c_{V1}M_1}{R}+\frac{3}{5}\right)\sqrt{\frac{2M_2}{M_1+M_2}}\cdot\frac{M_2}{M_1}\left(\frac{\sigma_{12}}{\sigma_1}\right)^2\frac{\Omega^{(1,1)}(kT_0/\varepsilon_{12})}{\Omega^{(2,2)}(kT_0/\varepsilon_1)},$$

where c_{V1} is a specific heat capacity of gas at constant volume; s and e/kT are force constants for the Lennard-Jones (6-12) potential. According Hirschfelder et al.[5], thermal diffusion factor of vapor can be present as

$$k_{T2} = c(1-c)f\frac{M^2}{M_1 M_2}, \quad f \approx \frac{6\gamma_2 Le_0}{5(\gamma_2-1)}\frac{M_1}{M_2}(c_{12}^*-5/6), \qquad (17)$$

where c_{12}^* is function of $T_{12} = kT/e_{12}$, expressed through the interaction integrals, Ω. For n-pentanol-helium system in present work is used extrapolation of thermal diffusion factor as in work[4]

$$f = \left(\left(-0.7272-\frac{T}{16.36-0.2882\cdot T}\right)\cdot\left(1.12281-c\frac{M}{M_2}\right)+0.089303\right)^{-1}.$$

With account of equations (16) and (17) the equation (14) may be represented as

$$\frac{d}{d\tau}\left(\frac{1-c_0}{1-c}\right) = \frac{A_0}{1+A_0 B_0} - c\cdot\frac{f}{\tau}\cdot\frac{1-c_0}{1-c}, \qquad (18)$$

$$A_0 = \frac{(dc/dt)_0 + c_0 f}{1-c_0}, \quad B_0 = \frac{1-\tau}{Le_0}\left[1+f\frac{\gamma_2-1}{\gamma_2}\frac{M_0}{M_1}\frac{1-c_0}{1-\tau}\left(1-\tau\cdot\frac{M}{M_0}\frac{1-c}{1-c_0}\right)\right]. \qquad (19)$$

Solution for equation (18) is

$$c^{(1)} = 1-(1-c_0)\left(1-Le_0\ln\left(1+\frac{1-\tau}{1-\tau_1}\left(\exp\left(\frac{c_0-c_1}{1-c_1}Le_0^{-1}\right)-1\right)\right)\right)^{-1}, \qquad (20)$$

$$y^{(n)} = \left(G^{(n)}(1,\tau_1)-\frac{1-c_0}{1-c_1}\right)\frac{F^{(n)}(1,\tau)}{F^{(n)}(1,\tau_1)}-G^{(n)}(1,\tau)+1, \quad c^{(n)} = 1-\frac{1-c_0}{y^{(n)}}, \qquad (21)$$

$$F^{(n)}(1,\tau) = \int_1^\tau (1+A_0^{(n-1)}B_0^{(n)})^{-1}d\tau, \quad G^{(n)}(1,\tau) = \int_1^\tau c^{(n-1)}\frac{f}{\tau}\frac{1-c_0}{1-c^{(n-1)}}d\tau,$$

$$A_0^{(1)} = \frac{Le_0}{1-\tau_1}\left(\exp\left(\frac{c_0-c_1}{1-c_1}Le_0^{-1}\right)-1\right), \quad A_0^{(n)} = \left(G^{(n)}(1,\tau_1)-\frac{1-c_0}{1-c_1}\right)/F^{(n)}(1,\tau),$$

$$B_0^{(n)} = \frac{1-\tau}{Le_0}\left[1 + f\frac{\gamma_2 - 1}{\gamma_2}\frac{M_0}{M_1}\frac{1-c_0}{1-\tau}\left(1 - \tau \cdot \frac{M^{(n-1)}}{M_0}\frac{1-c^{(n-1)}}{1-c_0}\right)\right], \quad n = 2,3,\ldots$$

The solution (20) of equation (18) is for zero thermal diffusion factor (f=0). The solution (21) gives the n-order iteration ones ($n = 2,3,4,\ldots$) with account of thermal diffusion effect. The present calculations were performed for 4-th order iteration ($n = 4$).

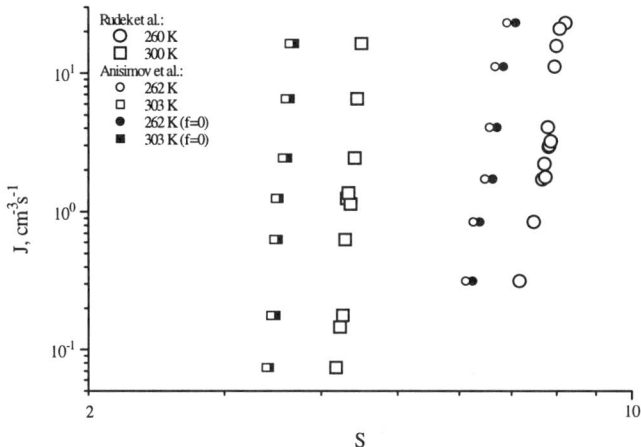

Fig. 1. Nucleation rates, J, versus vapor supersaturation, S, (Rudek et al.) and the same data calculated with the present research assumptions.

To evaluate difference between two treatments, namely with account of the interplate vapor flux and without it, the calculations of vapor supersaturations based on the experimental n-pentanol nucleation results were done. Present calculations was done at the same boundary conditions as by Rudek et al.[4] The value of supersaturation was corresponded the maximum of theoretical nucleation rate calculated by using the classical nucleation theory:

$$J_{theor} = \frac{V}{(kT)^2}\sqrt{\frac{2\sigma}{\pi \cdot m}} \cdot P^2 \cdot \exp\left(-16\pi \cdot \sigma^3 \cdot V^2 / 3(kT)^3 \cdot (\ln S)^2\right),$$

where σ is the surface tension; m, V are molecular mass and volume, and S is the vapor supersaturation ratio (or vapor activity), P is the partial vapor pressure, T is the nucleation temperature, k is the Boltzmann's constant. As seen in fig. 1 this difference is very large even for relatively small mass vapor fraction at temperature 260 K. Herewith the calculated nucleation temperatures are larger as ones in Rudek et al.[4] work. Our calculations are done with and without the thermal diffusion influence. The thermal diffusion influence on transport process is not too essential.

REFERENCES

1. Bertelsmann, A. and Heist, R.H. *J.Chem. Phys.*, **106**, 610; (1997).
2. F. T. Ferguson and J.A. Nuth, J. Phys. Chem. **111**, 8013 (1999).
3. Stratmann, F., Zdimal, V., Wilck, M., and Smolik, J. *J.Aerosol Sci.*, **30, Suppl.1**, S75-S76. (1999)
4. Rudek, M., Katz, J.L., Vidensky, I.Y., Zdimal, V., and Smolik, J. *J Chem Phys.* **111**(8), 3623. (1999)
5. Hirschfelder, J.O., Curtiss, C.F., and Bird, R.B. *Molecular Theory of Gases and Liquids*. Wiley, NY. (1954)
6. Reid R.C., Prausnitz J.M., and Sherwood T.K. *The properties of gases and liquids*. 3rd ed. McGrow-Hill, New York, p. 508 (1977)

Calculation of Vapor and Aerosol Transport in a Pipe Including Homogeneous Nucleation

M.P. Kissane and I. Drosik

Institut de Protection et de Sûreté Nucléaire,
Département de Recherches en Sécurité,
CEA/Cadarache, France

Abstract. In the analysis of hypothetical severe accidents in nuclear power plants the behavior of hot vapors and aerosols as they flow through metal pipes is studied. Computer codes have been developed enabling calculations to be made and the present article describes application of the SOPHAEROS code emphasizing the treatment and impact of homogenous nucleation.

INTRODUCTION

Nuclear power plant safety studies include very low probability events which may lead to severe accidents. Accident scenarios, involving breach of the reactor coolant system (RCS) and loss of safety systems, are analyzed for potential radionuclide emissions. One step is evaluation of radionuclide masses and physico-chemical forms released from the RCS. Results and analyses of experiments simulating accident conditions, e.g. the small-scale Falcon[1] and large-scale Phebus FP tests[2], show releases to be sensitive to chemical and aerosol phenomena. However, adequate treatment of the important phenomena must avoid excessive calculation times.

SOPHAEROS is a computer code[3] which models (equilibrium) vapor-phase chemistry, homo- and heterogeneous nucleation and aerosol agglomeration, deposition and resuspension. SOPHAEROS version 2.0 deals with 32 elements of which 23 can form compound species in the vapour-phase (work is in progress extending this database based on identification of 65 elements needed to cover the full range of applications). Thermodynamic data for version 2.0 were mainly determined using the Thermodata database and the COACH tool[4]. This note describes the modeling and consequences of homogeneous nucleation of the vapors formed.

HOMOGENEOUS NUCLEATION

Nucleation Rate

The volumetric rate of homogeneously nucleated mass, J_m, for a vapor is calculated with the Girshick, Chiu and McMurray model[5]:

$$J_m = \bar{J}.i^*.m_L = \frac{1}{S}\exp\left[\frac{4\pi\sigma\left(\frac{3v_L}{4\pi}\right)^{\frac{2}{3}}}{kT_f}\right]\frac{1}{\sqrt{2\pi m_L kT_f}} 4\pi R^{*2} n_L \exp\left[\frac{-\Delta G^*}{kT_f}\right].i^*.m_L$$

where

$\bar{J}$ is the volumetric number rate of particle formation ($m^{-3}.s^{-1}$),

$i^* = \frac{4\pi R^{*3}}{3v_L}$ is the number of molecules in a critical-size particle,

m_L is the molecular mass of the nucleating vapor (kg),

$R^* = \frac{2\sigma v_L}{kT_f \ln S}$ is the critical particle radius (m), k is Boltzman's constant,

$\Delta G^* = \frac{4}{3}\pi R^{*2}\sigma$ is the enthalpy of formation variation for a critical-size particle (J),

S is the vapor saturation ratio, T_f is the flow temperature (K),

n_L is the vapor-phase molecular number concentration (m^{-3}),

v_L is the liquid-phase molecular volume (m^3) and

σ is the liquid surface tension ($N.m^{-1}$).

This model was chosen since in an assessment of approaches for the intended applications it was found to provide the best compromise between accuracy and simplicity[6]. In particular, it was shown to give results similar to microscopic approaches simulating detailed combination and disintegration of molecular clusters.

Surface tension is defined as a linear function of temperature requiring inclusion of two coefficients in the database with the thermodynamic data for the vapor species. If the coefficients are unknown for a liquid, values for UO_2 are used since these aerosols will be among the first to form as the flow cools. Currently the surface tensions of the species Ag, Ba, Cd, In, Mo, Ru, Sb, Si, Sn, Te, UO_2, CsI and CsOH are covered while those of many more are being included in the new SOPHAEROS database.

Particle Size

In the SOPHAEROS code the nucleation model is only used to determine the mass rate of formation of aerosols, particle size not being calculated. This is because the code applications usually involve rapid cooling of a flow as it travels along a pipe. It can thus be assumed that nucleation rapidly generates high concentrations of seed particles followed by fast agglomeration due to Brownian motion. The nucleated mass can thus be introduced into the lowest size class of the sectional aerosol size scheme used for calculating agglomeration and deposition. This simplification has been born out by showing that progressive reduction of the minimum aerosol size of the discretized scheme eventually has no effect on calculations since agglomeration of the smaller and smaller seed particles becomes so rapid as to render the effect of further size reduction negligible. Nevertheless, in any calculation where homogeneous nucleation is significant, sensitivity to the chosen minimum size must be checked.

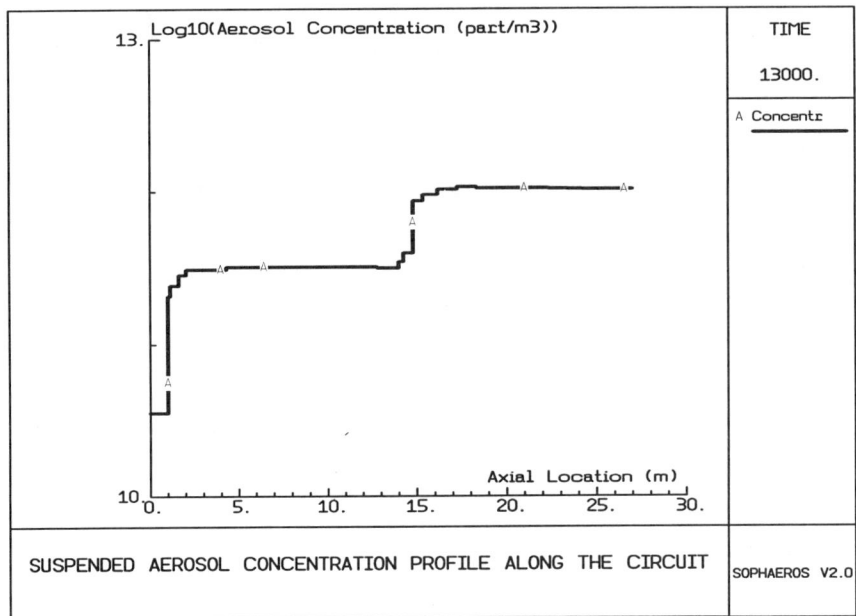

FIGURE 1. Particle number concentration as a function of distance.

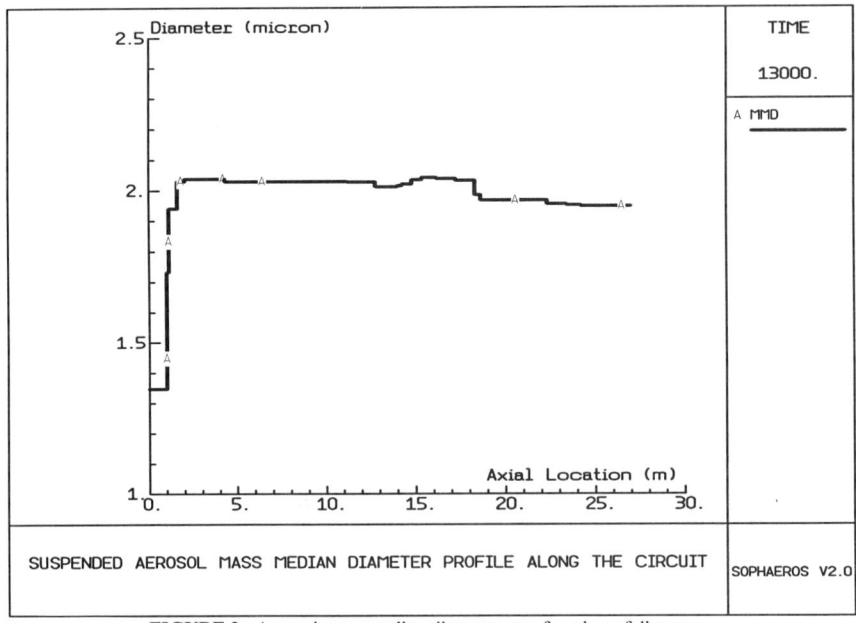

FIGURE 2. Aerosol mass median diameter as a function of distance.

EXAMPLE CALCULATION

The impact of homogeneous nucleation is seen in analysis of test Phebus FPT1. The test involves a 28m-long pipe where the gas at the entrance is at around 1600K and 420K at the exit. There are two zones of steep temperature gradient: at the entrance and between 12 and 16m. In addition to the hydrogen and oxygen of the carrier gas, the analysis accounts for 11 elements introduced into the pipe. Two calculations were performed, one with homogeneous nucleation and one without, all other phenomena being active. Figures 1 and 2 show the evolution of aerosols along the pipe when nucleation is active. A number of vapors nucleate in the first few meters of the pipe where temperatures cool to 1000K, particle concentration and size increasing, then in a second zone due to condensation of cadmium about 13m from the entrance. Decrease in average particle size is seen where inertial deposition (bends and turbulence) is significant. Deposition for most elements is about 50%, only hydrogen, oxygen and molybdenum deposit less since they remain mainly in the vapor-phase at 420K. The difference with the calculation with no homogeneous nucleation is striking: in this case deposition is greater than 98% for all elements except again for H, O and Mo.

CONCLUSIONS

It appears that adequate account can be taken of homogeneous nucleation for the applications considered here by using the Girshick-Chiu-McMurray model. Neglecting it leads to substantial overprediction of deposition due to vapor condensation on surfaces. Alternatively, modeling homogeneous nucleation obviates the need to inject aerosols artificially into the flow (with associated assumptions) to induce heterogeneous nucleation and avoid unphysical behavior. Lastly, taking adequate account of the important phenomena seems possible without excessive calculation times.

REFERENCES

1. Beard, A.M., Beattie, I.R., Benson, C.G., Bennett, P.J., Bowsher, B.R., Brunning, J., Codron, L., Freemantle, N.E., Mignanelli, M.A., Newland, M.S., and Stansbury, A.R., CEC report EUR 15766/1 EN, 1994.
2. Schwarz, M., Arnaud, A., Clément, B., and Von der Hardt, P., "The severe accident Phebus FP programme: results from FPT0 and FPT1 and future experiments", Int. Conf. on Severe Accident Risk Management, Piest'any, Slovakia, June 1997.
3. Missirlian, M., Kissane, M.P., and Schmitz, B.M., "Sophaeros v2.0 : development and validation status of the IPSN reactor coolant system code for fission product transport ", 3rd OECD Specialist Meeting on Nuclear Aerosols in Reactor Safety, 15-18 June 1998, Cologne, Germany.
4. Cheynet, B., Rivet, A., and Fisher, E., "COACH Version 1.0, computer aided chemistry", THERMODATA/INPG/CNRS, Grenoble, France, 1992.
5. Girshick, S.L., Chiu, C.-P., and McMurry, P.H., *Aerosol Sci. Technol.* **13**, pp. 465-477 (1990).
6. Martin, F., "La nucléation homogène: étude des interactions vapeurs-aérosols dans le circuit primaire d'un réacteur nucléaire lors d'un accident grave", *Doctoral Thesis*, Université de Provence / Aix-Marseille I, 1997.

The Characteristic Time Scales of Condensation and Coagulation in Ion-induced Nucleation

Lauri Laakso, Jyrki M. Mäkelä and Markku Kulmala

Department of Physics, PB 9, FIN-00014
University of Helsinki, Finland

Abstract. The characteristic time scales of the condensation, coagulation and recombination during ion-induced nucleation have been investigated. The classical theory for ion-induced nucleation of water and sulfuric acid has been used in these calculations. The effect of the Coulomb potential on coagulation and condensation has also been taken into account. The results show that in certain conditions, with low pre-existing particle concentrations and high atmospheric temperatures, the new particles can grow to the sizes they are stable even without electric charge. This system is however very sensitive to changes in pre-existing aerosol particle concentration and temperature.

INTRODUCTION

Several studies have shown nucleation events in the atmosphere; Covert et al. observed new particles in the Arctic Ocean [1], Hõrrak et al. [2] and Mäkelä et al. [3] in the continental sites of Tähkuse, Estonia and Hyytiälä, Finland, respectively. There are also observations of nucleation events in the tropopause made for example by Reus [4].

One of the possible routes for particle formation in the atmosphere is the ion-induced nucleation. Laboratory measurements with relative high ion concentrations have shown that the presence of the ions does enhance the nucleation rate. Also the classical and density functional nucleation theories predict this enhancement. Despite this, the direct measurements and observations in atmospheric conditions are still lacking.

Compared to the homogeneous nucleation, ion-induced nucleation is less investigated and it gives a unique opportunity both for theoretical and experimental studies. Recent theoretical hypothesis like thermodynamically stable clusters [5] and ion mediated nucleation [6] increase the relevance of the studies related to ion-induced nucleation and the properties of ions as a possible method for the investigations.

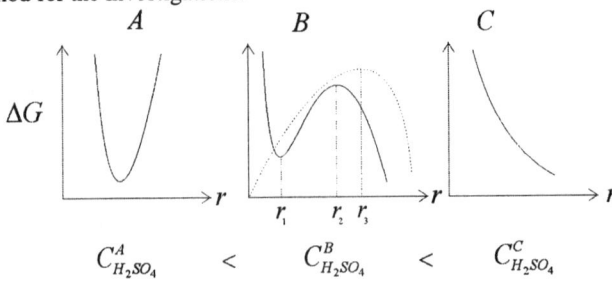

FIGURE 1. The projection of Gibbs free energy surface for ion-induced nucleation with different sulfuric acid concentrations. In the fig. (B) upper, dashed curve represents homogeneous nucleation. r_1 and r_2 are the critical radii for ion-induced and r_3 for homogeneous nucleation. The sulfuric acid concentration increases from the left to the right.

One important phenomenon related to ion-induced nucleation is the survival of the freshly formed nanoparticles during the growth. The classical nucleation theory separates three different cases for Gibbs free energy as a function of the sulfuric acid concentration as presented in Figure 1. In the case (A) there is no nucleation, the other end is the case (C) when all ions will nucleate. The most interesting case is however (B), when there exists a barrier that limits the nucleation. In this situation the stable ions of size r_1 will nucleate to charged particles of size r_2. Before the particles have grown to the size r_3, there exists particles that are stable only because of the charge. This means that if they are electrically neutralized before they grow over the critical size of homogeneous nucleation, r_3 they will be evaporated [7]. These small particles can also coagulate with the pre-existing aerosol particles during their growth.

In this study we have investigated the growth between sizes r_2 and r_3 and the corresponding time scales of coagulation and recombination. This size range is studied because it is critical for the ability of ion-induced particle formation to change the number concentration of the particles in measurable sizes (>3 nm diameter).

THE MODEL DESCRIPTION

The theory for classical binary water-sulfuric acid ion-induced nucleation is based on the work of Yue and Chan [8]. In this approach the Gibbs free energy for ion-induced particle formation and the corresponding critical radii has been calculated using the concept of virtual monomer [9].

The coagulation calculations are based on the theory by Fuchs which has been modified with enhancement factors due to the Coulomb interactions. The theory used in the calculations of enhancement factors is taken from Kerminen [10], who has based his studies on the work done by Amadon and Marlow [11]. The coagulation loss is calculated between charged particles of the radius r_2 and neutral aerosol particles between 0.5 and 1000 nm. This assumption of neutral particles as a second partner of the coagulation is based on the fact that the effect of charges of the same and the opposite signs in particles would approximately cancel each other.

The theory for condensation used here is taken from Kulmala [12] and it is based on the work done by Fuchs and Sutugin [13]. This theory assumes particles to be all the time in the equilibrium with respect to water vapour because of the high concentration of water vapour compared to the concentration of sulfuric acid. The enhancement in condensation due to the electric potential proposed by Turco [6] is based on calculations presented by Hoppel, who used it in the calculations of the attachment of ions on aerosol particles [14], [15]. This theory, when used on condensation is quite rude but there are no better theories available at the moment. The model presented above will be used as a part of the comprehensive ion and aerosol dynamical model AEROCHARGE.

INITIAL VALUES

The pre-existing particle distribution used in this study is calculated from the typical size distribution measured in Hyytiälä, Finland. The total particle concentration between 0.5 and 1000 nm is 2400 1/cm^3 which is divided logarithmically in 27 size classes. The ion concentration is assumed to be 1000 1/cm^3 and the ambient pressure 1 atm.

THE RESULTS

The critical radii for ion-induced and homogeneous nucleation are shown in the Figure 2. The critical sulfuric acid concentration for nucleation rate of 0.1 #/cm^{3}s in ion-induced nucleation is indicated by a vertical line.

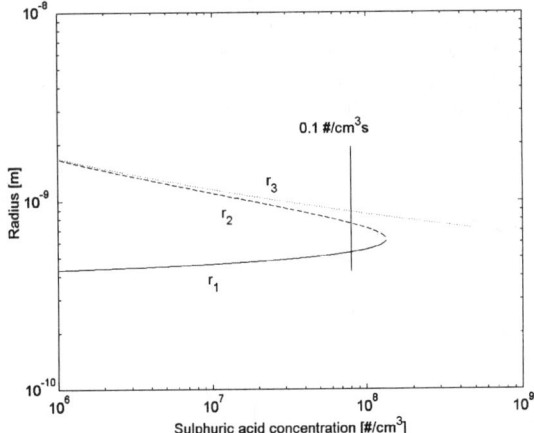

FIGURE 2. The solutions for critical radii according the classical nucleation theory for ion-induced and homogeneous nucleation in a binary water-sulfuric acid nucleation. The vertical line presents the nucleation rate of 0.1 # /cm^{3}s corresponding the case (B) in Figure 1. RH is 60 % and the temperature 293 K

The calculated time scales for condensation, enhanced condensation and coagulation as a function of temperature are presented in Figure 3. These characteristic time scales are averages for nucleation rates over 0.1 #/cm^{3}s i.e. between the vertical line and the point where r_1 and r_2 join (see Figure 2). It can be seen that the enhanced condensation rate exceeds the combined recombination and coagulation rate at the temperatures over 275 K and the normal condensation rate at the temperature of 287 K.

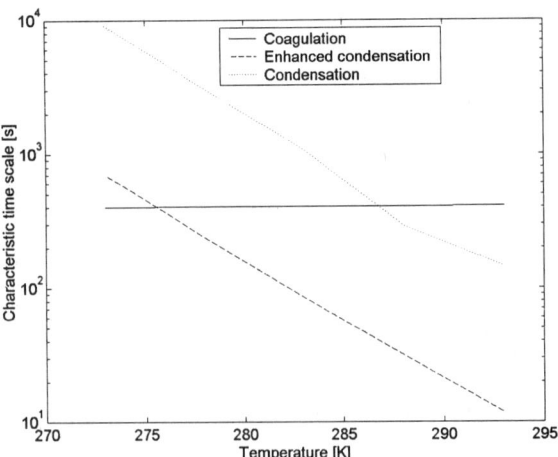

FIGURE 3. The characteristic survival time during the growth from r_2 to r_3 in ion-induced nucleation. The position of the curve representing coagulation will be elevated if the ion and particle concentrations are lower and vice versa.

The more interesting situation rises when the temperature is lower, under 275 K. In this case the radii obtained from the classical nucleation theory are smaller and the radial distance the freshly formed charged particles have to grow is larger. This changes the order of time scales so that the coagulation and recombination dominate. This behavior depends on the concentration of the pre-existing particles so if the particle concentration is lower than the one used here, the characteristic time for coagulation is larger. Even if the background concentration is low, the coagulation sink sufficient to change the order of the time scales can be reached at some point during the nucleation event. The more detailed analysis will be done when this part of the model is used as a part of model AEROCHARGE.

These results, even preliminary, raise an interesting question about the ion-induced nucleation to be an important phenomenon in the atmosphere. The possibility of ion-induced nucleation to contribute to the particle number concentration depends on different variables like ion production rate and pre-existing particle concentration. In some cases it is possible that ion-induced nucleation may change the growth rates or mass of the pre-existing particles rather than the number concentration.

ACKNOWLEDGMENTS

I wish to thank Dr. Liisa Pirjola and Dr. Veli-Matti Kerminen for their help with the computer code and Prof. Ari Laaksonen for discussions as well as Prof. A. Lushnikov for clarifying Fuchs theories.

REFERENCES

1. D.S.Covert et al., J.Geophys.Res 97, 20581-20587 (1992)
2. U. Horrak et al., J.Geophys.Res. 103, 13909-13916 (1998)
3. J.M. Mäkelä et al., Geophys.Res.Lett. 24, 1219-1222 (1997)
4. M.de Reus et al., J.Geophys.Res. 103, 31255-31263 (1998)
5. M. Kulmala et al. Nature 404, 66-69 (2000)
6. F.Yu and R.P.Turco, Geophys.Res.Lett. (in press)
7. F.Raes and A. Janssens, J. Aerosol Sci. 16, 217 (1985)
8. G.K. Yue and L.Y.Chan, J. Colloid and Interface Sci. 68, 501 (1979)
9. M.Kulmala and Y.Viisanen, J. Aerosol Sci. 22, Suppl. (1991)
10. V-M.Kerminen, Report Series in Aerosol Sci. 22, 1993
11. A.S.Amadon and W.H.Marlow Phys.Rew. A. 43, 5483-5492 (1991)
12. M.Kulmala and A.Laaksonen, J.Chem.Phys. 93, 696-701 (1990)
13. N.A. Fuchs, Mechanics of Aerosols, Pergamon, New York (1964)
14. W.A.Hoppel and G.M.Frick, Aerosol Sci.Tech. 5, 1-21 (1986)
15. W.A.Hoppel, In Electrical Processes in Atmospheres, edited by H.Dolazalek and R.Reiter, Steinkopff Verlag, Darmstadt, pp. 60-69 (1977)

NUCLEATION IN CONDENSED SYSTEMS

SCALING PROPERTIES OF CRITICAL NUCLEI AND NUCLEATION RATE

Robert McGraw

Atmospheric Sciences Division, Department of Environmental Science
Brookhaven National Laboratory, Upton NY 11973

Abstract. Scaling approaches to nucleation are reviewed first for molecular properties of the critical nucleus and then for the nucleation rate. Scaling properties of critical nuclei are based on the Kelvin relation. These are reexamined in light of more recent density functional and molecular dynamics calculations. For the nucleation rate scaling we derive thermodynamic phase space paths of constant reduced nucleation barrier height, $W*/kT$. These paths, expressed in terms of the conjugate variables T and lnS, where T is temperature and S is saturation ratio, approximate conditions at the nucleation threshold and contours of constant nucleation rate. Finally we show that both scaling approaches (molecular and rate) are fundamentally connected through nucleation theorems and may therefore be combined. The combination of Kelvin-based molecular scaling and scaled nucleation theory for the rate is derived and shown to give a temperature dependence that is in agreement with experiment.

INTRODUCTION

The have been two central motivations for introducing scaling concepts into nucleation theory. First, as shown below, there has been remarkable success with the recent development of scaling relations for molecular properties of clusters containing as few as 50-100 molecules. These developments have given new significance to the Kelvin relation through which the number of molecules in the critical nucleus is estimated from measurable bulk thermodynamic properties. The new feature of the recent work has been the derivation of a number of other scaling relations for critical cluster properties that are equivalent to the Kelvin relation. The resulting collection of equivalent, Kelvin-based, scaling relations forms the foundation for the molecular-properties scaling used in this paper.

The second motivation for the introduction of scaling derives from the need for efficient thermodynamic parameterizations and correlations of nucleation threshold conditions and nucleation rates. Scaling approaches to nucleation rate were developed during the eighties (1-5), prior to the widespread use of nucleation theorems (6-12), and development of more advanced phenomenological (13-16) and molecular based nucleation theories (17-20). Here scaling refers to the correlation of nucleation rates over wide classes of materials, and wide ranges of environmental conditions, using reduced (nondimensional) thermodynamic coordinates. The greatest success of the early scaling theory was in the prediction of nucleation threshold conditions and not of nucleation rate. Because nucleation threshold conditions typically correspond to a range of about nine orders of magnitude uncertainty in absolute nucleation rate (1), agreement with experiment is not as stringent a test of the theory. Nevertheless the theory was sufficient as most laboratory measurements were of nucleation threshold conditions, although significant experimental advances were beginning to yield accurate measurements of nucleation rate (21,22,7).

More recently there has been increasing realization, due largely to development of "nucleation theorems", that nucleation rate measurements correlate directly with molecular properties of clusters of critical size. Specifically, the nucleation theorems provide a direct, model free, and quantitative link between laboratory rate measurements and molecular properties, such as the number of molecules in the critical nucleus, energy of the critical nucleus, etc.. Thus the development of a phenomenological scaling theory for molecular cluster properties provides, at once, a parallel theory for prediction of nucleation rate. Through this productive linkage, which is demonstrated below, the comparison of accurate rate measurements with classical and molecular-based predictions can lead to both a quantitative description of the observed departures from classical nucleation theory (CNT), and development of phenomenological approaches that are in better agreement with both molecular theory and experiment. The present study will re-examine scaling approaches to nucleation in light of these advances, especially with respect to the nucleation theorems, and suggest potentially productive directions for their future development. For simplicity we limit discussion to nucleation of a single component daughter phase.

SCALING PROPERTIES OF CRITICAL NUCLEI IMPLIED BY THE KELVIN RELATION

Simple scaling relations have been derived under the ansatz that the Kelvin relation is preserved even for clusters of critical size (13-15). The Kelvin relation gives the number of molecules present in the critical cluster, g^*, as a homogeneous function of the driving free energy for nucleation, $\nabla \mu = kT \ln S$, where S is the saturation ratio:

$$g^* = C(T)(\nabla \mu)^{-3} = \left[32\pi \gamma_\infty^3 /(3\rho_c^2)\right](\nabla \mu)^{-3}. \qquad (1a)$$

The prefactor $C(T)$ is determined in the last equality by the fact that the Kelvin relation is expected to be valid along the coexistence curve ($\Delta \mu \to 0$) where the nucleus is large and the capillarity drop model provides a reliable estimate for g^*. As noted previously, Eq. 1a gives a prediction for nucleus size in terms of measurable bulk thermodynamic properties.

The combination of Eq. 1a with the isothermal nucleation theorem (6-8),

$$\left(\frac{\partial W^*/kT}{\partial \ln S}\right)_T = -g^*, \qquad (2a)$$

where we neglect the number of vapor molecules in the exclusion volume of the cluster, which is negligible when the vapor is dilute, yields additional scaling forms for the nucleation barrier height (13):

and

$$W_{CNT}^* - W_{SNT}^* = D(T) \tag{1b}$$

$$\frac{W_{SNT}^*}{g^* \Delta\mu} = \frac{1}{2} - \frac{D(T)}{C(T)}(\Delta\mu)^2. \tag{1c}$$

W_{SNT}^* is the nucleation barrier height in the scale model based on the Kelvin relation and W_{CNT}^* is the barrier height in the CNT. The function $D(T)$ gives the displacement of the barrier height from its classical value. That $D(T)$ is a function of temperature alone is of utmost significance to what follows.

Talanquer (16) evaluated $D(T)$ using equation of state parameters under the assumption that Eq. 1c remains valid deep within the coexistence region and at the spinodal where, according to density functional theory, the height of the nucleation barrier is expected to go to zero. Thus the location of the spinodal, as determined from the equation of state, determines $D(T)$ and there are no adjustable parameters in the theory. Comparisons of the resulting scaled nucleation theory with experimental rate measurements for nonpolar and weakly polar vapors, including measurements from Ref. 21, showed significant improvement over CNT.

Equations 1a- 1c are equivalent manifestations of the Kelvin relation, which is now seen to imply a generalization of CNT, for which $D(T) = 0$. This Kelvin-based scaling theory gives results in much better agreement with density functional calculations (13-14) and with the temperature-dependent trends seen in experimental measurements of nucleation rate (16,22). Furthermore, a molecular basis for persistence of the Kelvin relation to small clusters has recently been established using the nonuniform spherical drop model, and Gibbs dividing surface methods (15). These results yield an interpretation of $D(T)$ in terms of the interfacial curvature free energy. Specifically, we have shown the Kelvin relation implies additional scaling relations, which within the context of the Gibbs dividing surface model are equivalent to Eqs. 1a-1c. These are (15):

$$R_e = R^* \tag{1d}$$

$$\gamma_e = \gamma_\infty + \frac{k_s}{R_e^2} \tag{1e}$$

$$\frac{\gamma_\infty}{R_e} = \frac{\gamma_s}{R_s} \tag{1f}$$

were R^* is the radius of the critical nucleus in the capillary drop model of CNT, R_e is the radius of the equimolecular dividing surface, R_s is the radius of the surface of tension, γ_e is the surface tension at the equimolecular dividing surface, γ_s is the surface tension at the surface of tension, and γ_∞ is the bulk surface tension for a flat interface. In the notation of Eq. 1e, we find:

$$D(T) = -4\pi k_s \qquad (3)$$

where k_s is the rigidity coefficient and $4\pi k_s$ is the interfacial curvature free energy (15). It is interesting that Talanquer also obtains a result for the surface tension that follows Eq. 1e, however in his model k_s is interpreted as proportional to the number of molecules in the critical nucleus at the spinodal.

Preservation of the Kelvin relation, under this curvature correction to the surface tension, provides additional justification for its use to predict critical cluster size. Nevertheless, it would be unrealistic to assume that Eqs. 1 remain valid under all conditions. Additional insight can be gained from independent studies of free-energy expansions in density and higher-order density gradients for nonuniform spherical drops. The usual curvature-independent surface tension arises from the square-gradient term. The interfacial curvature free energy accounts for higher-order corrections up through the square-Laplacian term (see references cited in Ref. 15). It is remarkable that the Kelvin relation, Eq. 1a, and its equivalent forms, Eqs. 1b-1f, are each preserved under the interfacial curvature free energy correction. Indeed the Kelvin relation provides a closed-form, analytic prediction for all higher-order terms in the free-energy expansion (15). Analytic evaluation of the next higher-order terms in the nonuniform drop model, for comparison with the scaling prediction, requires a considerable algebraic effort that apparently has not been carried out.

Alternatively, the density functional calculations of Koga and Zeng provide a numerical assessment of corrections to the Kelvin relation for clusters near coexistence (23). These authors found that, although Eqs. 1 describe the largest departures observed from the capillary drop model, additional smaller corrections are needed to more accurately match the density functional results. A particularly interesting finding suggests that the symmetry between liquid drop and bubble nucleation implied by Eq. 1c, which is symmetric in the sign of $\Delta\mu$, is broken by odd-order expansion terms not included in the scaling relations (23). Nevertheless, the asymmetric correction terms are small compared to the largest correction term already included in the scaling theory, which preserves this symmetry. Overall, the findings of Ref. 23 suggest that the Kelvin relation provides an accurate description of the largest curvature-dependent discrepancies between the density functional and classical capillary drop models. Nevertheless, departure from the Kelvin relation for these density-functional clusters is clearly seen.

In concluding this section, we call attention to the independent confirmation of molecular scaling obtained by molecular dynamics and Monte Carlo cluster simulation methods applied to gas-liquid nucleation in a Lennard-Jones system (19). This study found excellent agreement with the Kelvin relation (Eq. 1a) and with the scaling prediction of constant ($\Delta\mu$ independent) departure of the barrier height from CNT given by Eq. 1b. The observed difference $\delta = R_e - R_s$ between the equimolecular dividing surface and the surface of tension as a function of the excess number of molecules in the critical nucleus also appears to be well described by the scaling theory (19).

SCALING OF NUCLEATION THRESHOLD CONDITIONS AND NUCLEATION RATE

This section will review some of the early research in development of scaling approaches for prediction of nucleation threshold conditions (1-5) within the context of the nucleation theorems. The potential for future extensions of these approaches to yield more accurate prediction of nucleation rates will then be examined.

Scaling approaches have led to development of remarkably simple, yet highly effective, rules for predicting the thermodynamic phase space paths of constant reduced nucleation barrier height. Corresponding states correlations first showed the universality behavior of these phase space paths for widely different classes of materials undergoing vapor-liquid nucleation (1). Reported values for the logarithm of critical supersaturation, $\ln S_c$, were studied as a function of reduced temperature T/T_c, where T_c is the critical temperature, along paths of constant $W*/kT$ (1). To the extent that variations in the kinetic prefactor to the nucleation rate can be neglected, these paths approximate contours of constant nucleation rate. Universal behavior was found for the homogeneous nucleation thresholds of the vapors of simple fluids and correlated using CNT together with Guggenheim's empirical fits to surface tension and density. This studies, augmented by the work of Rasmussen and Babu who introduced the Eotvos constant into such correlations (2), motivated development of Hale's scaled nucleation theory (3-5). Scaled nucleation theory has proven to be powerful tool for correlating nucleation threshold behavior for diverse materials ranging from simple fluids, to metallic vapors (4-5), and vapors of refractory materials (24). As in the corresponding states theory, scaled nucleation theory yields paths of constant $W*/kT$ in the thermodynamic phase space coordinates $\ln S$ and T. As shown by Hale, these paths take the form (4):

$$\ln S_c \cong \Gamma \Omega^{3/2} \left(\frac{T_C}{T} - 1 \right)^{3/2} \qquad (4)$$

where Γ is a weak function of T and S, and will include a nucleus shape factor for homogeneous crystallization. $\Omega = \gamma_0 \rho_c^{-2/3}/k$ is close in relation and numerical value to the Eotvos constant, and has the physical interpretation of excess surface entropy per molecule in units of k. γ_0 is the prefactor in the empirical linear relation between temperature and surface tension $[\gamma_\infty \cong \gamma_0(T_c - T)]$ and ρ_c is the nucleus density. Equation 4 is also preserved when the quantity $\gamma_\infty/\rho_c^{2/3}$ is linear in T as found empirically by Eotvos (28) (see also Eq. 8c below). It is remarkable, to mention only one application of Eq. 4, that good estimates can be make for the critical temperatures of such materials as silver and SiO from fits to nucleation data acquired at temperatures well below T_c (4).

The development of nucleation theorems over the past decade can shed new light on thermodynamic phase space paths of constant reduced nucleation barrier height. Figure 1 shows how such paths can be analyzed using the nucleation theorems. With

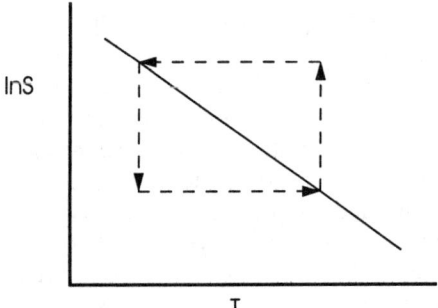

FIGURE 1. Differentials along a path (solid curve) of constant reduced nucleation barrier height. Vertical and horizontal and components are given by the isothermal and nonisothermal nucleation theorems, respectively, and correspond to the first and second terms on the rhs of Eq. 5a. Note that the path is not linear in temperature and only appears to be so on this magnified differential scale.

$\ln S$ and T as coordinates we have along such a path:

$$d(W^*/kT) = \left(\frac{\partial W^*/kT}{\partial \ln S}\right)_T d\ln S + \left(\frac{\partial W^*/kT}{\partial T}\right)_{\ln S} dT = 0 \quad (5a)$$

or, equivalently,

$$\left(\frac{d\ln S}{dT}\right)_{W^*/kT} = -\frac{\left(\frac{\partial W^*/kT}{\partial T}\right)_{\ln S}}{\left(\frac{\partial W^*/kT}{\partial \ln S}\right)_T} \quad (5b)$$

The denominator on the right hand side (rhs) of Eq. 5b is given by the isothermal nucleation theorem (Eq. 2a) and the numerator is given by the nonisothermal nucleation theorem (8), obtained for these coordinates by Ford who uses the designation "second nucleation theorem" (9):

$$\left(\frac{\partial W^*/kT}{\partial T}\right)_{\ln S} = -\frac{E_{g^*} - g^* E_1^b}{kT^2}. \quad (2b)$$

Substitution of these results into the rhs of Eq. 5b yields an expression for the paths of constant W^*/kT directly in terms of the molecular properties of the critical nucleus itself:

$$\left(\frac{d\ln S}{dT}\right)_{W^*/kT} = -\frac{1}{g^*}\frac{E_{g^*} - g^* E_1^b}{kT^2} \xrightarrow{CNT} -\frac{1}{kT^2}\frac{A_{g^*}}{g^*}\left(\gamma_\infty - T\frac{d\gamma_\infty}{dT}\right). \quad (6)$$

E_{g*} is the critical cluster energy and E_1^b is the energy per molecule in the bulk liquid phase. The arrow shows the specialization of this general result to the capillary drop model of CNT under the additional assumption of temperature-independent density, ρ_c. Here the quantity in parenthesis is the surface energy per unit area, also equal to the surface enthalpy per unit area because the interface has zero volume in the Gibbs dividing surface model, and A_{g*} is the nucleus surface area. The rhs of Eq. 6 contains the surface enthalpy per molecule for the critical cluster. Under the empirical linear relation between surface tension and temperature, the quantity in parenthesis is constant, and integration of Eq. 6 yields Eq. 4. Paths having values of $W*/kT$ in the 50-70 range are characteristic of nucleation threshold conditions and thus yield a predicted temperature dependence for the critical supersaturation, S_c (1).

Paths of constant $W*/kT$ can also be derived by more conventional means by combining various expressions for the nucleation barrier height. The results are equivalent to the general first equality of Eq. 6, but are in a form that is closer to the original scaling theory. Additional variants of the nucleation theorems can also be derived this way. For this analysis we begin with three expressions for $W*$ that are generally valid in the Gibb's dividing surface model. These are: (i) $W* = (1/2)n_f * \Delta\mu$ where $n_f *$ is the number of molecules within the region of the critical nucleus bounded by the surface of tension (13), (ii) the Gibbs relation, $W* = (1/3)A_s \gamma_s$, and (iii) $W* = (1/2)V_f * \Delta P$ where $V_f *$ is the volume bounded by the surface of tension and $\Delta P = P_f - P_v$ is the Laplace pressure across the dividing surface. Expressions (ii) and (iii) are connected through the Laplace relation:

$$\Delta P = \frac{2}{3}\frac{A_s}{V_f *}\gamma_s. \quad (7)$$

At this point it is necessary to have some ansatz relating nucleus surface area and volume. For this purpose we invoke the assumption of preservation of similarity of nucleus shape with size. This holds trivially for a spherical nucleus, for which we have the volume-area scaling relation: $d\ln V_f * = (3/2)d\ln A_s$. This similarity equation also holds for crystalline nuclei if one assumes the minimum free-energy crystal geometry dictated by the Wulff relations. This volume-area scaling relation accounts for the ubiquitous 3/2 term in scaled nucleation theory. Combining expressions (ii), (iii), volume-area scaling (Eq. 7) gives the following general result for the evolution of the Laplace pressure along a path of constant reduced nucleation barrier height:

$$\left(\frac{d\Delta P}{dT}\right)_{W*/kT} = \frac{3}{2}\Delta P\left(\frac{1}{\gamma_s}\frac{d\gamma_s}{dT}\right) - \frac{\Delta P}{2T}. \quad (8a)$$

Introducing the assumption that the nucleus is incompressible, as in CNT, we can utilize the thermodynamic relation $\Delta P = \rho_c \Delta\mu$ to obtain a similar equation in $\Delta\mu$:

$$\left(\frac{d\rho_c \Delta\mu}{dT}\right)_{W*/kT} = \frac{3}{2}\rho_c \Delta\mu \left(\frac{1}{\gamma_s}\frac{d\gamma_s}{dT}\right) - \frac{\rho_c \Delta\mu}{2T}. \quad (8b)$$

Transformation to the lnS coordinate gives:

$$\left(\frac{d\ln S}{dT}\right)_{W*/kT} = \frac{3}{2}\ln S\left[\frac{d\ln(\gamma_s/\rho_c^{2/3})}{dT}\right] - \frac{3}{2}\frac{\ln S}{T}. \quad (8c)$$

Equations 8 provide remarkably simple and elegant expressions for paths of constant reduced nucleation barrier height. Their main disadvantage is that the surface tension defined for the surface of tension, γ_s, remains to be determined. In classical nucleation theory $\gamma_s = \gamma_\infty$, and the quantity in parenthesis ($\gamma_\infty^{-1} d\gamma_\infty / dt$) is the relative slope of the Eotvos line, describing the empirical linear relation between surface tension and temperature at temperatures well below T_c (27). Under these conditions Eq. 8c is readily integrated to obtain the scale nucleation theory result of Eq. 4. Combining Eq. 8c, for constant ρ_c, and the barrier height expression (i), from above, results in a generalization of the CNT result given on the rhs of Eq. 6:

$$\left(\frac{d\ln S}{dT}\right)_{W*/kT} = -\frac{1}{kT^2}\frac{1}{g*}\left\{\frac{g*}{n_f*}A_3\left(\gamma_s - T\frac{d\gamma_s}{dT}\right)\right\}. \quad (9)$$

Comparison of this result with the middle term in Eq. 6 shows that the quantity in curly brackets is the excess cluster energy, $E_{g*} - g*E_1^b$, in the Gibbs dividing surface model. In the classical theory $n_f* = g*$, $\gamma_s = \gamma_\infty$, and the rhs of Eq. 9 reduces to the CNT result (Eq. 6). The CNT limiting forms of Eqs. 6 and 8a are used elsewhere in this Proceedings for analysis of laboratory measurements of salt efflorescence and ice nucleation from ammonium sulfate solutions using scaled nucleation theory (26).

In the following section it is shown that within the framework of the scaled nucleation cluster theory of Eqs. 1, because $g*$ is a known function temperature, a determination of the cluster energy E_{g*} is sufficient to obtain paths of constant reduced barrier height (or rate). Conversely, if such paths are known, the most important of these being the path giving S_c as a function of T, E_{g*} may be determined. Finally it is clear from Eqs. 5b and 6 that paths of constant reduced barrier height (or rate) map directly, in a model-free way through the nucleation theorems, to specific molecular properties of the critical cluster itself.

<u>Remarks on Nucleation Theorems:</u> Considering the important role played by the nucleation theorems in scaling, it is worthwhile mentioning a few perspectives on their interpretation gained from recent research (10-12). The first of these concerns the connection between nucleation theorems and the law of mass action. If we restrict consideration to an ideal cluster mixture, so that even the exclusion volume of the cluster is neglected, it is clear that the nucleation theorem of Eq. 2a is an immediate consequence of the law of mass action (10). This can be seen by considering the constrained equilibrium association reaction $gn_1 \Leftrightarrow n_g$, where $n_g = \exp[-W(g)/kT]$ is

the number of clusters per unit volume, and recalling that the equilibrium constant, which for this reaction is $K_g(T) = n_g/(n_1)^g$, is a function of temperature alone. Equation 2b follows in similar fashion from the temperature derivative of $\ln K_g(T)$ given by the Gibbs-Helmholtz relation, and for an ideal mixture of clusters Eq. 2b is an immediate consequence of that relation (11). These results hold for any fixed size cluster and not just for clusters of critical size. On the other hand, the critical cluster size is not fixed, changing as it does with changes in either S or T. To be formally correct, evaluation of the derivatives in Eqs. 2 must take account of the additional term in the "chain rule" that represents this change in critical size. However, that term vanishes by the property, specific to critical clusters, that the size derivative of the work of formation, $dW(g)/dg$, is zero at the critical size. This last point provides some justification for the label "nucleation theorems", even though Eqs. 2 are valid for clusters of any size.

The second remark is related to the paths of constant $W*/kT$. One would ideally like to take into account the nucleation kinetics and express results in terms of paths of constant nucleation rate (J). In this regard, Ford (25) has evaluated the derivatives appearing in Eq. 2 with $\ln J$ replacing $W*/kT$. Surprisingly the resulting expressions are not that much more complicated. The main difference is that the critical cluster size and critical cluster energy appearing on the right hand side are replaced by certain averages of size and energy over the constrained equilibrium distribution, with the heaviest weighting, as expected, for clusters nearest the critical size. Similar kinetic extensions of the nucleation theorems have been derived for an ideal mixture of clusters through a combination of the laws of mass action and detailed balance (11). Future developments of scaled nucleation theory might well begin by replacing the reduced barrier height appearing in Eqs. 5a and 5b by $\ln J$.

The final remark concerns the effect of background subtraction due to cluster exclusion volume, which is included in rigorous derivations of the nucleation theorem for vapor-liquid nucleation (7,8). We have neglected this effect here by assuming that, for the cases under study, the vapor is ideal and sufficiently dilute. However the scaled nucleation theory of Hale applies equally well to homogeneous crystallization from a liquid parent phase (3, 26). Conventional, CNT-based, treatments of ice nucleation from supercooled solutions, important in atmospheric models of cirrus cloud formation, would be in gross error if it were required that the displaced water in the parent phase be subtracted from $g*$. This is because CNT makes no such subtraction. To address this question, we have applied the Gibbs-Duhem approach of Ref. 8, accounting also for the equilibrium between the parent phase (e.g. a supercooled salt-water solution) and the water vapor in equilibrium with it. When the analysis is carried through for ice nucleation, we find that the displaced liquid phase water cancels, with the net effect that, just as in the case of vapor-liquid nucleation, it is the much smaller number of water molecules in the *vapor* phase exclusion volume, which to be rigorously correct should be subtracted from $g*$. Thus there is no need to subtract (or add) large corrections to account for parent-phase displacement when applying the nucleation theorem to crystallization from atmospheric-pressure solutions. For nucleation studies at much higher pressures, this conclusion must be modified to take into account the volume work contribution to changes in the condensed phase chemical potential, which can lead to significant correction.

COMBINED APPROACHES

In the previous section we obtained paths of constant reduced nucleation barrier height in the Gibbs dividing surface model. This is a full generalization of the CNT to this model; but it may be too general to be immediately useful simply because the quantities g^*, n_f^*, A_s, and γ_s are all unknown. On the other hand there is good evidence in support of the Kelvin scaling theory, including the density functional and molecular dynamics calculations cited above. The strongest evidence for Kelvin scaling derives from widespread observations that the ratio, J_{exp}/J_{CNT}, where J_{exp} is the experimental rate, and J_{CNT} is the rate predicted by the classical theory, is a function of temperature alone (21, 22). To the extent that changes in kinetic prefactor can be neglected, this behavior follows directly from Eq. 1b, and by studying this experimentally measurable ratio we can hope to get a handle on $D(T)$: $\ln(J_{exp}/J_{CNT}) = D(T)/kT$. Conversely, if J_{exp}/J_{CNT} is a function of temperature alone, all of the equivalent formulations of scaling (Eqs. 1a-1f) apply. Specifically, the Kelvin relation itself (Eq. 1a) is obeyed. Thus the observation that J_{exp}/J_{CNT} can be expressed as a function of temperature alone supports the validity of the molecular scaling relations given by Eqs. 1.

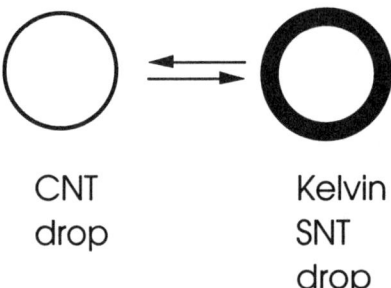

CNT drop   Kelvin SNT drop

FIGURE 2. Schematic depiction of the isomerization equilibrium between clusters modeled in the classical nucleation theory (CNT) and scaled nucleation theory (SNT) based on the Kelvin relation.

Unlike the much more general Gibbs model, Kelvin-based scaling remains simple enough that a useful phenomenological approach to the prediction of thermodynamic phase space paths of constant nucleation rate can still be derived. To obtain such a model we begin by investigating the difference in energy between the CNT and Kelvin-scaled (SNT) clusters. This difference was previously obtained by Ford (9) by applying the second nucleation theorem (Eq. 2b) to our model. The same result follows directly from the Gibbs-Helmholz equation. Consider the hypothetical constrained equilibrium between CNT and SNT critical clusters. Because g^* is the same in both models this can be view as an isomerization equilibrium, as depicted schematically in Fig. 2. The corresponding equilibrium constant is:

$$K_{iso}(T) = n_{g*}(SNT)/n_{g*}(CNT) = \exp[-(W_{SNT}* - W_{CNT}*)/kT] = \exp[D(T)/kT] \quad (10)$$

where the last equality follows from Eq. 1b. The Gibbs-Helmholtz relation is:

$$\frac{d\ln K_{iso}}{dT} = \frac{d[D(T)/kT]}{dT} = \frac{E_{SNT}* - E_{CNT}*}{kT^2} \quad (11)$$

or

$$E_{SNT}* - E_{CNT}* = kT^2 \frac{d[D(T)/kT]}{dT} = -\left(D - T\frac{dD}{dT}\right), \quad (12)$$

which is identical to the result obtained from the second nucleation theorem (Eq. 2b). Combining Eqs. 6 and 12 gives the path of constant reduced barrier height in the Kelvin-scaled nucleation theory

$$\left(\frac{d\ln S}{dT}\right)_{W*/kT} = \frac{3}{2}\ln S\left[\frac{d\ln(\gamma_\infty/\rho_c^{2/3})}{dT}\right] - \frac{3}{2}\frac{\ln S}{T} + \frac{1}{kT^2}\frac{1}{g*}\left(D - T\frac{dD}{dT}\right) \quad (13)$$

where we have written the first two terms on the rhs using Eq. 8c so as to include the temperature dependent density as in CNT. $g*$ is given by Eq. 1a. Thus all terms on the rhs of Eq. 13 are readily obtained from bulk measurements with the exception of $D(T)$. Theoretical guidance on the behavior of $D(T)$ is presently unclear. Density functional calculations from different research groups appear to be consistent as to the overall size of the correction but not for its slope. Molecular dynamics (MD) simulations (19), although in excellent agreement with the predictions of Eqs. 1, even for critical nuclei consisting of only 50-100 particles, yield a smaller value for $D(T)$ than does density functional theory. Perhaps, as suggested by the authors of Ref. 19, this due to the use of truncated potentials in the MD simulations. One possibility, suggested by analysis of experimental nucleation rates for n-alcohol vapors (22), is that $D(T)$ tends to be linear in T. This remark is supported by the data analysis plots of Ref. 22 that show generally strong linear behavior in $\ln(J_{exp}/J_{CNT})$ vs. $1/T$. From our previous analysis, this would require that $D(T)/kT$ itself be linear in $1/T$ or, equivalently, that the last term in parentheses in Eq. 13 is constant! The rate analysis of Ref. 22 also suggests strong systematic behavior throughout the studied homologous n-alcohol vapor series. These findings are reminiscent of the corresponding states correlations seen in the earliest developmental stages of scaled nucleation theory. If history repeats itself, a much clearer understanding of the problem may soon be at hand through the identification of new Eotvos-like parameter correlations, for $D(T)$, and parallel extensions of scaled nucleation theory based on the Kelvin relation.

REFERENCES

1. McGraw R. "A corresponding states correlation of the homogeneous nucleation thresholds of supercooled vapors", J. Chem. Phys. 75, 5514-5520 (1981).
2. Rasmussen D. H. and Babu S. V. "A corresponding states correlation for nucleation from the vapor", Chem. Phys. Lett. 108, 449-451 (1984).
3. Hale B. N. "Scaled models for nucleation", in Lecture Notes in Physics 309, 323-349 (1988).
4. Hale B. N., Kemper P., and Nuth J. A. "Analysis of experimental nucleation data for silver and SiO using scaled nucleation theory", J. Chem. Phys. 91, 4314-4317 (1989).
5. Hale B. N. "Application of a scaled homogeneous nucleation-rate formalism to experimental data at $T \ll T_c$", Phys. Rev. A 33, 4156-4163 (1986).
6. Kashchiev D. "On the relation between nucleation work, nucleus size, and nucleation rate", J. Chem. Phys. 76, 5098-5103 (1982).
7. Viisanen Y., Strey R., and Reiss H. "Homogeneous nucleation rates for water", J. Chem. Phys. 99, 4680-4692 (1993).
8. Oxtoby D. W. and Kashchiev D. "A general relation between the nucleation work and the size of the nucleus in multicomponent nucleation", J. Chem. Phys. 100, 7665-7671 (1994).
9. Ford I. J. "Thermodynamic properties of critical clusters from measurements of vapour-liquid homogeneous nucleation rates", J. Chem. Phys. 105, 8324-8332 (1996).
10. Bowles R. K., McGraw R., Schaaf P., Senger B., Voegel J.-C., and Reiss H. "A molecular-based derivation of the nucleation theorem", Submitted for publication (2000).
11. McGraw R. and Wu D. "Kinetic extension of the nucleation theorem", Submitted for publication (2000).
12. Reiss, H. "Critique of molecular theories of nucleation", This Proceedings.
13. McGraw R. and Laaksonen A. "Scaling properties of the critical nucleus in classical and molecular-based theories of vapor-liquid nucleation", Phys. Rev. Lett. 76, 2754-2757 (1996).
14. Laaksonen A. and McGraw R. "Thermodynamics, gas-liquid nucleation, and size-dependent surface tension", Europhys. Lett. 35, 367-372 (1996).
15. McGraw R. and Laaksonen A. "Interfacial curvature free energy, the Kelvin relation, and vapor-liquid nucleation rate", J. Chem. Phys. 106, 5284-5287 (1997).
16. Talanquer V. "A new phenomenological approach to gas-liquid nucleation based on the scaling properties of the critical nucleus", J. Chem. Phys. 106, 9957-9960 (1997).
17. Zeng X. C. and Oxtoby D. W., "Gas-liquid nucleation in Lennard-Jones fluids", J. Chem. Phys. 94, 4472-4478 (1991).
18. Senger, B., Schaaf, P., Corti, D. S., Bowles, R., Voegel, J.-C., and Reiss, H. "A molecular theory of the homogeneous nucleation rate. I. Formulation and fundamental issues", J. Chem. Phys. 110, 6421-6437 (1999).
19. ten Wolde P. R. and Frenkel D., "Computer simulation study of gas-liquid nucleation in a Lennard-Jones system", J. Chem. Phys. 109, 9901-9918 (1998).
20. Schenter G. K., Kathmann, S. M., and Garrett B. C. "Dynamical nucleation theory: A molecular approach to vapor-liquid nucleation", Phys. Rev. Lett. 82, 3484-3487 (1999).
21. Adams G. W., Schmitt J. L., and Zalabsky R. A. "The homogeneous nucleation of nonane", J. Chem. Phys. 81, 5074-5078 (1984).
22. Strey R., Wagner P. E., and Schmeling T. "Homogeneous nucleation rates for n-alcohol vapors measured in a two-piston expansion chamber", J. Chem. Phys. 84, 2325-2335 (1986).
23. Koga K., and Zeng X. C. "Thermodynamic expansion of the nucleation free-energy barrier and size of the nucleus near the vapor-liquid coexistence", J. Chem. Phys. 110, 3466-3471 (1999).
24. Ferguson F. T., Nuth J. A, and Lilleleht L. U. "Experimental studies of the vapor phase nucleation of refractory compounds. IV. The condensation of magnesium", J. Chem. Phys. 104, 3205-3210 (1996).
25. Ford I. J. "Nucleation theorems, the statistical mechanics of molecular clusters, and a revision of classical nucleation theory", Phys. Rev. E 56, 5615-5629 (1997).
26. Onasch T. B., McGraw R., Prenni A. J., Tolbert M. A., and Imre D. "Efflorescence and ice nucleation in ammonium sulfate particles: Analysis of experimental results using scaled nucleation theory", This Proceedings.

27. Pippard A. B., in Classical Thermodynamics (Cambridge University Press, Cambridge, 1957) pg. 86.
28. Hale B. N., "The scaling of nucleation rates", Metallurgical Trans. A <u>23</u>, 1863-1868 (1992).

NUCLEATION IN CONDENSED SYSTEMS

K. F. Kelton

Department of Physics, Washington University, St. Louis, MO 63130

Abstract. Studies in condensed phases allow the exploration of some aspects of nucleation that are weakly manifest in gas phase condensation. Time-dependent rates that depend strongly on thermal history, competition between long-range diffusion and interfacial processes, and the influence of the structures of the initial and final phases, are examples. These are discussed and illustrated with representative experimental studies. A new model for time-dependent homogeneous nucleation that takes account of the coupled fluxes of interfacial attachment and long-range diffusion is developed and key predictions are discussed.

INTRODUCTION

In 1724, Fahrenheit performed the first recorded undercooling experiment,[1] noting the tendency of water to resist the formation of ice as the temperature was lowered below the freezing point, thus providing the first evidence for the nucleation barrier. Studies of nucleation in condensed systems continue to provide new information about the kinetic processes underlying phase transformations, and the structural and compositional relations between the initial and final phases. Here, selected nucleation studies for liquids and glasses are reviewed.

CLASSICAL THEORY OF NUCLEATION

Homogeneous nucleation data in condensed phases are typically analyzed using the classical theory of nucleation, a phenomenological model that describes the time-evolution of a size distribution of clusters of the new phase.[2,3] The nucleation rate is determined by the work of formation of a cluster of size n, W_n. Assuming spherical clusters, negligible stress effects (which should be true for crystallization from liquids and glasses), and a sharp interface between the cluster and the parent phase, W_n is approximately

$$W_n = n\delta\mu + (36\pi)^{1/3} \bar{v}^{2/3} n^{2/3} \sigma \ . \qquad (1)$$

Here, $\delta\mu$ is the Gibbs free energy per molecule (or atom) of the new phase less that of the initial phase, $\bar{v}$ is the molecular (or atomic) volume, and σ is the interfacial free energy per unit area.

The competition between the volume free energy favoring cluster formation and the surface free energy opposing it for small sizes leads to a maximum in the work of formation, W*, for a critical cluster size, n* (fig. 1.a),

$$W^* = \frac{16\pi}{3} \frac{\sigma^3}{|\delta\mu/\overline{v}|^2}, \quad n^* = \frac{32\pi}{3\overline{v}} \frac{\sigma^3}{|\delta\mu/\overline{v}|^3}. \quad (2)$$

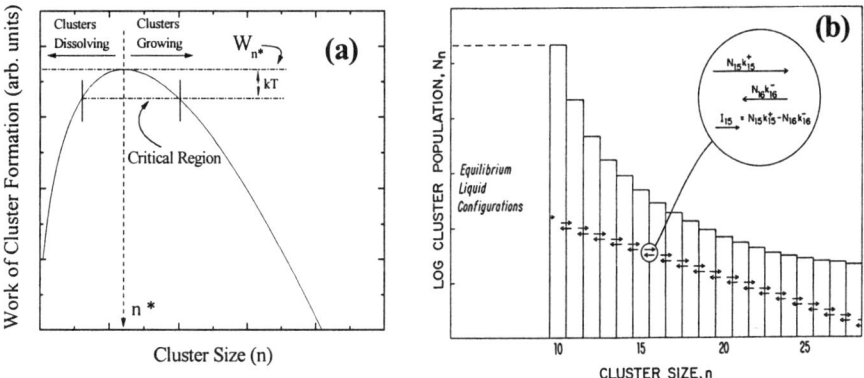

FIGURE 1. (a) Schematic diagram showing the work of cluster formation as a function of cluster size. The critical size and the critical region (the region within k_BT of W_{n^*}) are indicated. (b) Cluster population as a function of cluster size, n, showing the cluster flux underlying nucleation and growth. Below some size, here n=10, clusters of the new phase are indistinguishable from equilibrium fluctuations in the liquid (from Kelton[3]).

Clusters grow or shrink by the addition or loss of a single molecule at time, governed by the forward, k_n^+, and backward, k_n^-, rate constants (fig. 1.b). The time-dependent cluster density as a function of size, n, $N_{n,t}$, is determined by solving a system of coupled differential equations,

$$\frac{dN_{n,t}}{dt} = N_{n-1,t} k_{n-1}^+ - [N_{n,t} k_n^- + N_{n,t} k_n^+] + N_{n+1,t} k_{n+1}^- \quad (3)$$

The time-dependent nucleation rate for a cluster size n, $I_{n,t}$, is the flux of clusters past that size and is given by

$$I_{n,t} = N_{n,t} k_n^+ - N_{n+1,t} k_{n+1}^-. \quad (4)$$

The nucleation rate is then generally a function of the time and the cluster size at which it is measured. For a steady-state cluster distribution, however, the rate is constant (steady-state nucleation rate) for all times and cluster sizes. It is given by

$$I^s = \frac{24Dn^{*2/3}N_o}{\lambda^2}\left(\frac{|\delta\mu|}{6\pi k_B Tn^*}\right)^{1/2}\exp\left(-\frac{W_{n*}}{k_B T}\right) = A^*\exp\left(-\frac{W_{n*}}{k_B T}\right), \quad (5)$$

where D is the diffusion coefficient, λ is the atomic jump distance, k_B is the Boltzmann constant, T is the temperature (K), W_{n*} is the work of formation for a critical cluster, and N_o is the number of atoms. For liquids and glasses, the increasing driving free energy causes I^s to increase sharply with decreasing temperature below the melting temperature.

The thermodynamics of cluster formation are likely incorrect for small clusters, largely due to the assumption of a sharp division between volume and surface contributions. Density functional methods and thermodynamic studies have led to more correct formulations.[4]

NUCLEATION IN LIQUIDS

The maximum undercooling of a liquid below its melting temperature ($\Delta T_{max} = T_{min} - T_m$) can be used to estimate the nucleation rate of the crystal phase. The nucleation rate rises sharply with undercooling in a temperature range where the growth velocity is still large, so that the time scale for crystallization is dominated by the time required to form a nucleus.[3]

Maximum Undercooling

Generally, undercooling limits are set by heterogeneous rather than homogeneous nucleation, catalyzed by the container walls, the surface of the liquid, and structural impurities within the liquid. Fig. 2 illustrates some of the methods used to minimize the effects of heterogeneous nucleation. In the emulsion[5,6] and substrate[7] techniques (fig. 2.a and 2.b) the liquid is subdivided into a dispersion of small droplets. To suppress droplet coalescence and interactions with the container in the emulsion technique, the droplets are frequently coated with organic or inorganic acid surfactants or molten salt mixtures.[8] If the dispersions are sufficiently fine ($\approx$ 100 μm in diameter), a significant number of the droplets will contain no heterogeneous sites; the maximum undercooling observed is then taken to be the homogeneous limit. The fluxing technique (fig. 2.c) can be used for large liquid samples. The liquid is coated with a material (typically amorphous) that isolates it from the container walls and the atmosphere and dissolves the heterogeneous impurities (fig. 2.c).

Because nucleation is typically catalyzed by interactions of the sample with the container walls, there is a growing interest in containerless solidification of bulk samples (fig. 2.d). The maximum undercooling is determined from the recalescence temperature, arising from a rapid heat evolution upon crystallization. In the drop-tube method (fig. 2.d-top), the sample is melted and allowed to solidify while falling down a long, evacuated, or inert-gas-filled tube.[9] In electromagnetic levitation (fig. 2.d-middle) metallic droplets are levitated and melted using an rf-field, potentially allowing an accurate control of the sample position and the temperature.[10] For electrostatic levitation,[11,12] charged droplets are held between two charged plates (fig. 2.d-bottom). Since the droplets are typically melted by laser heating, the power required for levitation and heating are decoupled, allowing the investigation of a wider range of undercooling. To eliminate gravity-induced segregation and convective stirring, which might mask diffusion effects on nucleation and growth, rf-levitation studies have been successfully carried out on space shuttle missions, using TEMPUS, a German facility.[13,14] In addition to the maximum undercooling, the specific heat, viscosity and surface tension of the liquid have been measured with this technique, providing fundamental data for the analysis of the nucleation data.

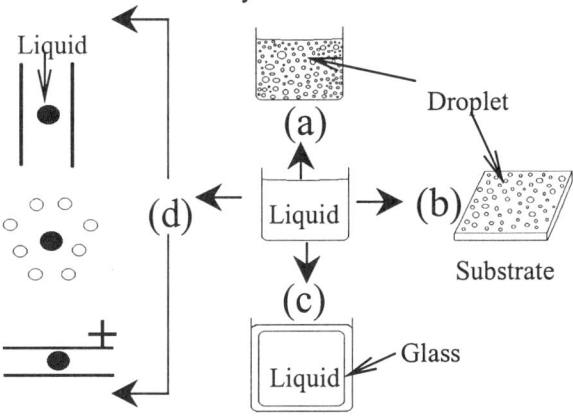

FIGURE 2. Schematic illustration of common methods for effectively minimizing the role of heterogeneous sites in the melt.

Selected maximum undercooling data for metals are listed in Table 1. The reduced undercooling values, $\Delta T_r = \Delta T_{max}/T_m$, tend to lie near 0.2. From a more complete set of data, the reversible work of cluster formation, W^*, is estimated using eq. (5) as $(60\pm2)k_BT$ (see reference 3).

TABLE 1. Maximum undercooling obtained for selected metals.

Metal	T_m (K)	ΔT_{max} (K)	$\Delta T_{max}/T_m$	Ref
Hf	2423	450	0.19	15
Ir	2725	440	0.16	16
Mo	2890	520	0.18	9
Nb	2740	525	0.19	17
Pt	2042	380	0.19	9
Re	3450	975	0.28	18
Rh	2233	450	0.20	9
Ta	3269	625	0.19	16
Ti	1948	350	0.18	9
Zr	2125	430	0.17	9
W	3690	560	0.15	16

Icosahedral order in condensed phases

To explain the large undercoolings observed, Frank[19] first proposed that the undercooled liquid contains a large amount of polytetrahedral order (taken to be icosahedral), which is incompatible with the translational periodicity of crystalline phases. Recently, a new phase of condensed matter that is characterized by a non-crystallographic orientational symmetry, and extended quasiperiodic translational order, was discovered.[20] The most common of these quasicrystals, the icosahedral phase, or i-phase, has the rotational symmetry of the icosahedral point group ($m\bar{3}\bar{5}$).[21]

Numerous studies have shown that the nucleation rates of quasicrystals from liquids and glasses are high, consistent with a similar local order in initial and final phases. Maximum undercooling studies by rf-levitation on metallic liquids that form quasicrystals and related crystal phases, called crystal approximants, reveal a trend consistent with this conclusion. The results of these studies are summarized in Table 2.

TABLE 2. Maximum undercooling by rf-levitation for phases of decreasing icosahedral order. (from Holland-Moritz[22])

Phase Type	Alloy	$\Delta T_{max}/T_m$
Icosahedral Phase	$Al_{60}Cu_{34}Fe_6$	0.09
" "	$Al_{58}Cu_{34}Fe_8$	0.09
" "	$Al_{72}Pd_{21}Mn_{17}$	0.11
Crystal Approximant	λ-phase - $A_{13}Fe_4$	0.12
" "	λ-phase – $Al_{62}Cu_{25.5}Fe_{12.5}$	0.14
" "	μ-phase - Al_5Fe_2	0.14
CsCl	β-phase - $Al_{65}Cu_{20}Co_{15}$	0.25

The reduced undercooling decreases with increasing icosahedral order in the condensed phase. The simple crystal phase having the CsCl structure shows the greatest undercooling, consistent with Frank's hypothesis that the nucleation barrier arises from the breaking of icosahedral order in the undercooled liquid. Based on these results, the work of cluster formation can be ordered,

$$W_I^* < W_\lambda^* < W_\mu^* < W_D^* < W_\beta^* \ . \tag{6}$$

NUCLEATION IN GLASSES

Silicate Glasses

Silicate glasses have been studied extensively, due to their ease of glass formation, their slow devitrification kinetics, and their technological interest. For many silicate glasses, the peak in the steady-state nucleation rate occurs at a sufficiently low temperature that growth of the nuclei is slow. Nucleation rates can therefore be determined quantitatively by first annealing the samples at a temperature, T_N, where the nucleation rate is large but the growth velocity is small, to develop a population of nuclei. The samples are then annealed at a higher temperature, T_G, where the nucleation rate is small, to grow all clusters larger than the critical size at T_G to visible size, allowing $I(n_{T_G}^*,t)$ to be measured directly.[23,24]

For illustration, fig. 3.a shows the measured steady-state homogeneous nucleation rates of the stable crystal phases in several silicate glasses that crystallize to phases of the same composition. The data are in quantitative agreement with the behavior expected from the classical theory (eq. 5).

For a steady-state nucleation rate, a plot of the number of nuclei, N_v, as a function of annealing time should give a straight line, with a slope equal to the steady-state nucleation rate, I^s. During glass formation the melt is cooled on a time scale that is rapid compared with the time required for atomic rearrangements. A time-dependent nucleation rate is, therefore, often observed, as the inherited cluster distribution evolves toward the steady-state one at the annealing temperature. This is shown in figure 3.b for the nucleation of the cubic crystal phase in $Li_2O.2SiO_2$ glass. The nucleation rate (given by the local slope) is initially low, but increases with time to the steady-state value. For long annealing times, N_v, is approximately

$$N_v = I^s(t-\theta), \quad for\ t >> \theta \quad (7)$$

Here θ is an effective time lag for nucleation, often called an induction time, which is obtained by extrapolating the number of nuclei produced as a function of time in the steady-state regime to the time axis.

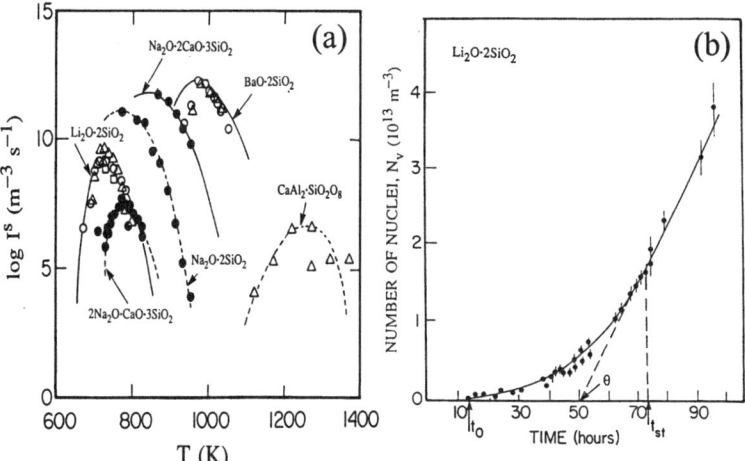

Figure 3. (a) Measured and fit (solid lines) steady-state homogeneous nucleation rates as a function of temperature in a variety of silicate glasses (from ref. 3). The interfacial free energy was assumed to increase with temperature, consistent with enhanced ordering in the liquid (or glass) near the cluster interface; (b) the number of nuclei produced as a function of time at 703K in lithium disilicate glass (from ref. 25).

Detailed discussions of the numerical and analytical solutions for the time-dependent nucleation rate can be found in references 3 and 26. Numerical calculations have the advantage that they are free from the many, largely untested, approximations made to obtain the analytical solutions and they can be extended readily to include arbitrary cluster distributions, nonisothermal annealing treatments,[27] glass formation,[27] nucleation of a crystal of composition different from that of the liquid or glass[28] and heterogeneous nucleation.[29] The numerical treatments demonstrate that time-dependent nucleation in these cases is understood. They also demonstrate that while there are difficulties with the thermodynamic model for nucleation, the kinetic model is quantitatively correct.[30,31]

Few quantitative studies of compositional effects on the nucleation rate exist. Changes in the undercooling as a function of composition for metallic alloys are primarily determined by changes in $\delta\mu$.[32] This is also true for nucleation in many pseudo-binary glasses, including Na_2O-BaO-SiO_2,[33] Li_2O-BaO-SiO_2,[33] and $Li_2O \cdot 2SiO_2$-$BaO \cdot 2SiO_2$.[33] The nucleation rates peak at a composition near the stoichiometric composition, decreasing slightly on either side of this composition as the liquidus temperature decreases. Recent measurements in $Na_2O.CaO.3SiO_2$ glasses reveal a strong dependence of the steady-state nucleation rate on $[SiO_2]$, with only minor changes in the induction times, likely originating from a composition dependence of σ.[28,34]

Nucleation in Metallic Glasses

The nucleation and growth curves are often not well separated in metallic glasses, making the results from two-step annealing studies difficult to interpret. Nucleation rates have typically been estimated from fits to the isothermal and nonisothermal devitrification kinetics or TEM measurements of the crystal size distributions.[35,36,37] Homogeneous steady-state and time-dependent nucleation studies in rapidly-quenched metallic glasses by these methods show no new features over those found (and better studied) in the silicate glasses.

Recently, crystallization nano-structures were found in several families of metallic glasses, including the Zr/Ti-(late transition metal) bulk glasses, and the rapidly-quenched Al-Fe,Ni-(rare earth) and Fe-Si-B glasses, indicating enormously high nucleation rates, sometimes approaching $10^{23}/m^3.s$. Key devitrification features include: (i) a fine uniform microstructure with grain sizes between 5 and 20 nm[38,39,40] indicating a very high nucleation rate and a low growth velocity; (ii) an anomalously small activation energy for growth;[41] and (iii) a strong size dependence for grain growth, with small grains growing rapidly until their diameters reach a few nm, after which growth slows sharply;[42] (iv) significant time-dependent nucleation behavior in the early stages of crystallization, followed by a dramatic decrease in the nucleation rate.[42] The resultant nano-structured materials frequently have significantly enhanced properties of technological interest.

The complicated free energy surfaces of five component bulk metallic glasses that show this nucleation behavior could lead to phase separation into two or more thermodynamically favored undercooled melts or glasses on a nano-scale.[43,39] If nucleation occurred readily in the phase separated regions, and if growth outside the region were slow, due to the required long-range diffusion, the size scale for the devitrification product would be set by the wavelength of the decomposition. Spatially rapid composition fluctuations have been inferred from transmission electron microscopy (TEM), atom-probe field-ion microscopy (FIM), secondary ion mass spectrometry (SIMS), and small angle neutron scattering (SANS) studies of several bulk metallic glasses, supporting this model for those systems (see review in ref. 44).

Recent measurements have produced evidence for phase separation on a similar length scale in Al-RE-LTM glasses, which crystallize to a high density of nanocrystals (10^{20}- 10^{23} m^{-3}) of α-Al.[45,46] This is surprising in an alloy containing 88-92 at.% Al, a conclusion supported by thermodynamic calculations. It may reflect a kinetic rather than a thermodynamic process, however, resulting from a coupling of interfacial attachment and long-range diffusion for a new type of nucleation.

COUPLED DIFFUSION /INTERFACIAL ATTACHMENT NUCLEATION

The formation of the new phase frequently requires diffusion of atoms of the proper type to the interface. This flux is competitive with the interfacial attachment in setting the time-scale for nucleation. A proper account of the coupling between these two fluxes, however, is extremely difficult. In a lowest-order model, first proposed by K. Russell,[47] the complete diffusion field around each cluster is replaced by a shell of variable composition (fig. 4). This was recently developed further by Kelton.[48,49]

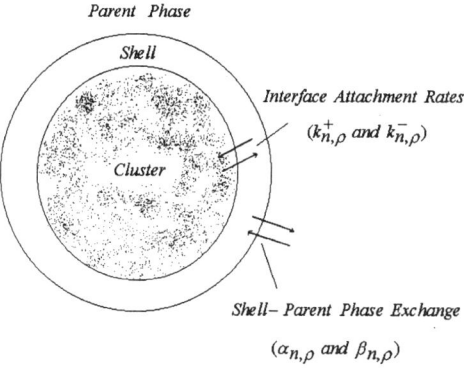

Figure 4. Schematic illustration of the linked-flux model for nucleation when the parent and daughter phases have a different composition.

In that model, atoms are transported to and from the shell by diffusion in the parent phase. At the interface, the normal attachment/detachment kinetics assumed in the classical theory are assumed. The cluster distribution is now a function of both the cluster size, n, and the number of solute atoms in the nearest neighbor shell to the cluster, ρ. Cluster growth is based on the relative rates of exchange of solute atoms between the parent phase and cluster neighborhood, and between the neighborhood and the cluster. This leads to a new set of coupled linear differential equations that describe the time-dependent population of clusters, $N(n,\rho,t)$ (the rates are indicated on fig. 4)

$$\frac{\partial N(n,\rho)}{\partial t} = \alpha(n,\rho-1)*N(n,\rho-1) - [\alpha(n,\rho) + \beta(n,\rho)]*N(n,\rho)$$
$$+ \beta(n,\rho+1)*N(n,\rho+1) + k^+(n-1,\rho+1)*N(n-1,\rho+1) \quad (8)$$
$$+ k^-(n+1,\rho-1)*N(n+1,\rho-1) - [k^+(n,\rho) + k^-(n,\rho)]*N(n,\rho) .$$

These equations have been solved by iteration, using parameters appropriate to precipitation transformations in solids (see ref. 48,49 and 50). Those studies show that it is impossible to model nucleation in partitioning systems where diffusion is important without taking into account the linking of the interfacial and diffusive fluxes. In many ways the predicted time-dependent nucleation behavior for coupled-flux nucleation is similar to that calculated from the classical theory and observed in transformations where the initial and final phases have the same composition. For example, the time-dependent nucleation rate and the number of nuclei generated as a function of time have the same form in the two models. The nucleation rate and the number of nuclei are small initially, reflecting the small density at the measurement cluster size. After a long annealing time, the number of nuclei grows linearly with time, consistent with the steady-state-rate, I^s. An extrapolation of the linear portion of the curve for the number of nuclei produced as a function of time to the time axis, gives the induction time for nucleation, θ. As in the classical theory, θ is a strong function of the cluster size at which it is measured.

The time-dependent nucleation rate scales with the relevant mobility, *i.e.* interfacial for rapid diffusion processes and the diffusion mobility for cases where long-range diffusion is the rate-limiting step. As long-range diffusion becomes more important, the nucleation rate becomes smaller and the induction time becomes larger than calculated from the classical theory, sometimes by several orders of magnitude. As the diffusion rate is increased, or the concentration of the initial phase becomes more similar to that of the new phase, the nucleation rate and the induction time take on their classical theory values.

Surprisingly, the solute concentration predicted from the coupled-flux model is always enhanced near sub-critical clusters, rising above the mean composition of the parent phase, and falling below the mean matrix composition for large clusters - as expected for diffusion-limited growth. To see this, the average composition near the cluster neighborhood as a function of cluster size, n, is computed

$$<c_n> = \frac{<\rho_n>}{\text{shell volume}} = \left(\frac{\sum_{\rho=0}^{\rho_n^{\max}} \rho N(n,\rho)}{\sum_{\rho=0}^{\rho_n^{\max}} N(n,\rho)} \right) \left(\frac{1}{(4\pi)^{1/3}(3n\overline{v})^{2/3}\lambda} \right), \quad (9)$$

where $\rho_n^{\max}$ is the maximum number of sites in the cluster neighborhood. As shown in figure 5, the concentration is enhanced for small clusters. The neighborhood composition equals the mean sample composition near the critical size, n* = 8 here. This behavior is reasonable from physical arguments. Atoms that diffuse to the cluster interface will attach and detach from the cluster interface many times before they are either permanently incorporated into the cluster or diffuse away from it. For clusters smaller than the critical size, detachment is favored (fig. 1.a). If diffusion in the parent phase is then slow compared with the interfacial kinetics, these detached atoms will tend to remain in the cluster neighborhood, raising the local solute concentration. Because attachment is biased for clusters larger than the critical size, the neighborhood becomes depleted in solute, filled only by atoms diffusing within the original phase to the region of the cluster interface.

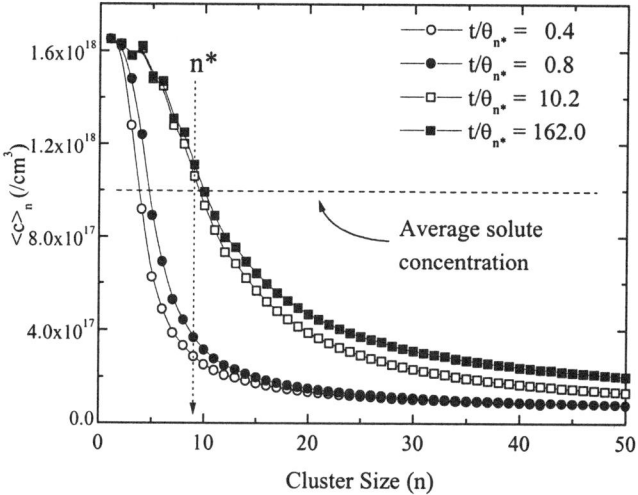

Figure 5. Effective concentration, $<c>_n$, of the parent phase in the nearest-neighborhood shell of clusters of size n and different annealing times (scaled relative to the induction time at the critical size). The neighborhood for small clusters near n* becomes enhanced above the average sample concentration (shown as a dotted line) (from Kelton[49]).

The coupled-flux approach predicts a stronger compositional dependence than can be argued within the classical theory. Partially due to this, recent calculations have produced much better agreement with the precipitation of oxygen in single crystal silicon[51] than has been possible within the classical theory.[52] Coupled-flux nucleation may also explain the devitrification nano-structure in the Al-TM-RE metallic glasses discussed in the last section. Most of the glasses that form nano-structured materials contain atoms of significantly different size, making long-range diffusion important. Since the cluster distribution in as-quenched glasses is characteristic of a high temperature (with a large critical size),[53,27] the coupled-flux calculations indicate that most of these clusters will be surrounded by a glass matrix that is closer in composition to the transformation phase. Depending on free energy and kinetic considerations at the annealing temperature, these clusters might either grow directly to nano-crystal α-Al, or serve as preferential sites for phase separation by nucleation in the liquid. In either case, growth will occur quickly until the nearby composition approaches that of the bulk glass; subsequently slower growth will be limited by long-range diffusion, giving the observed microstructures.

CONCLUSIONS

Studies of nucleation in condensed phases are important for basic and practical reasons. The smaller values for the maximum undercooling of quasicrystals and related phases argue for a local icosahedral order in liquids and glasses. The kinetic model for the classical theory of nucleation can quantitatively explain the time-dependent nucleation rates typically observed in glasses that crystallize to phases of the same composition. In carefully selected cases, it can also be used to study crystallization to phases of different composition. Finally, a coupling of the interfacial attachment processes with long range diffusion gives rise to new nucleation behavior. This is important for understanding solid state-precipitation processes and may be one cause for the phase separation and devitrification nano-structure observed in some metallic glasses.

ACKNOWLEDGMENTS

The partial support of this work by NASA under Cooperative Agreement NCC 8-85 is gratefully acknowledged.

REFERENCES

1. Fahrenheit, D.B., *Phil. Trans. Royal Soc.*, **33**, 78-84 (1724)
2. Turnbull, D., "Phase Changes," in *Solid State Physics*, edited by F. Seitz and D. Turnbull, New York, Academic Press, 1956, pp. 225-306.
3. Kelton, K.F., "Nucleation," in *Solid State Physics*, edited by H. Ehrenreich and D. Turnbull, Boston, Academic Press, 1991, pp. 75-177.
4. Oxtoby, D.W., *J. Phys. Condens. Matter*, **4**, 7627-7650 (1992).
5. Vonnegut, B., *J. Colloid Sci.*, **3**, 563-68 (1948).
6. Turnbull, D., *J. Chem. Phys.*, **20**, 411-24 (1952).
7. Turnbull, D. and Cech, R.E., *J. Appl. Phys.*, **21**, 804-810 (1950).
8. Paik, J.S. and Perepezko, J.H., *J. Non-Cryst. Solids*, **56**, 405-410 (1983).
9. Hoffmeister, W.H., Robinson, M.B. and Bayuzick, R.J., *Appl. Phys. Lett.*, **49**, 1342-44 (1986).
10. Willnecker, R., Herlach, D.M., and Feuerbacher, B. *Phys. Rev. Lett.*, **62**, 2707-2710 (1989).
11. Rhim, W.-K., Chung, S. K., and Elleman, D., *Mater. Res. Soc. Symp. Proc.*, **9**, 115-19 (1982).
12. Rulison, A.P.J. and Rhim, B.K., *Rev. Sci. Instr.*, **65**, 695-700 (1994).
13. Piller, J., Knauf, R., Lohöfer, G., Herlach, D. M. and Preu, P., *Proc. 6th Europ. Symp. on Materials and Fluid Sciences under Microgravity,* Bordeaux, ESA SP-256, 437-44 (1986).
14. Herlach, D.M., Cochrane, R. F., Egry, I., Fecht, H. J., Greer, A. L., Int. Mat. Rev., **38**, 273-347 (1993).
15. Hofmeister, W.H., Robinson, M.B., and Bayuzick, R.J. *Appl. Phys. Lett.*, **49**, 1342-1344 (1986).
16. Cortella, L. and Vinet, B., *Phil. Mag. B.*, **71**, 11-21 (1995).
17. Lacy, L.L., Robinson, M.B., and Rathz, T.J., *J. Cryst. Growth*, **51**, 47-60 (1981).

18. Vinet, B., Cortella, L., and Favier, J.J., *Appl. Phys. Lett.*, **58**, 97-99 (1991).
19. Frank, F.C., *Proc. Royal Soc.*, **215A**, 43-46 (1952).
20. Shechtman, D., Blech, I., Gratias, D., and Cahn, J. W., *Phys. Rev. Lett.*, **53**, 1951-53 (1984).
21. Kelton, K.F., "Quasicrystsals and Related Stuctures," in *Intermetallic Compounds: Principles and Practice*, edited by J.H. Westbrook and R.L. Fleischer, West Sussex, England, John Wiley & Sons 1987, pp. 453-491.
22. Holland-Moritz, D., *Int. J. of Non-Eq. Processing*, **11**, 169-199 (1998).
23. James, P.F., *Phys. Chem. Glasses*, **15**, 95-105 (1974).
24. Kelton, K.F. and A.L. Greer, *Rapidly Quenched Metals*, edited by S. Steeb and H. Warlimont, 223-26 (1985).
25. Fokin, V.M., Kalinina, A.M., and Filipovich, V.N., *J. Cryst. Growth*, **52**, 115-21 (1981).
26. Wu, D.T., "Nucleation Theory," in *Solid State Physics*, edited by H. Ehrenreich and F. Spaepen, Boston, Academic Press, 1997, pp. 37-187.
27. Kelton, K.F., Lakshmi Narayan, K., Levine, L. E., Cull, T. C., and Ray, C. S., *J. Non-Cryst. Solids*, **204**, 13-31 (1996).
28. Narayan, K.L. and Kelton, K.F., *J. Non-Cryst. Solids*, **220**, 222-30 (1997).
29. Narayan, K.L., Kelton, K.F., and Ray, C.S., *J. Non-Cryst. Solids*, **195**, 148-57 (1996).
30. Kelton, K.F. and Greer, A.L., *Phy. Rev.*, **B38** 10089-92 (1988).
31. Greer, A.L. and Kelton, K.F., *J. Am. Cer. Soc.*, **74**, 1015-22 (1991).
32. Thompson, C.V. and Spaepen, F., *Acta Metall.*, **31**, 2021-27 (1983).
33. Burnett, D.G. and Douglas, R.W., *Phys. Chem. Glasses*, **12**, 117-24 (1971).
34. Narayan, K.L. and Kelton, K.F., *Acta Mater.*, **46**, 3159-64 (1998).
35. Morris, D.G., *Acta Metall.*, **29**, 1213-20 (1981).
36. Köster, U. and Blank-Bewersdorff, M., *Mater. Res. Soc. Symp. Proc.*, **57**, 115-27 (1987).
37. Buchwitz, M., Adlwarth-Dieball, R. and Ryder, P.L., *Acta Metall. Mater.*, **41**, 1885-92 (1993)
38. Kim, Y.H., Inoue, A., and Masumoto, T., *Mater. Trans. JIM*, **31**, 747-49 (1990).
39. Schneider, S., Thiyagarajan, P. and Johnson, W. L., *Appl. Phys. Lett.*, **68**, 493-95 (1996).
40. Allen, D.R., Foley, J.C., and Perepezko, J.H., *Acta Mater.*, **46**, 431-40 (1998).
41. Omata, S., Tanaka, Y., Ishida, T., Sato, A., and Inoue, A., *Phil. Mag. A.*, **76**, 397-412 (1997).
42. Calin, M. and Köster, U. in *Mater. Sci. Forum*, **262-272**, 749-54 (1998).
43. Busch, R., Schneider, S., Peker, A., and Johnson, W. L., *Appl. Phys. Lett.*, **67**, 1544-46 (1995).
44. Kelton, K.F, *Int. J. Non-Eq. Processing*, **11**, 141-68 (1998).
45. Gangopadhyay, A., K. Croat, T. K., and Kelton, K. F., "The Effect of Phase Separation on Subsequent Crystallization in $Al_{88}Gd_6La_2Ni_4$," *Acta Mater.*, (submitted).
46. Croat, T. K., Gangopadhyay, A. K., Kelton, K. F. "Time Dependnt Nucleation in Al-based Metallic Glasses," *Mater. Res. Soc. Symp. Proc*, Symposium E, (in press).
47. Russell, K.C., *Acta Metall.*, **16**, 761-69 (1968).
48. Kelton, K.F., Phil. Mag. Lett., **77**, 337-43 (1998).
49. Kelton, K.F., "Time-Dependent Nucleation in Partitioning Transformations," *Acta Mater.* (in press).
50. Kelton, K.F., "Kinetic Model for Nucleation in Partitioning Systems," *J. Non-Cryst. Solids* (in press).
51. Wei, P.F., K.F. Kelton, and R. Falster, "Coupled-Flux Nucleation Modeling of Oxygen Precipitation in Silicon," *J. Appl. Phys.* (submitted).
52. Kelton, K. F., Falster, R., Gambaro, D., Olmo, M., Cornara, M., and Wei, P. F., *J. Appl.Phys.*, **85**, 8097-8111 (1999).
53. Kelton, K.F. and A.L. Greer, J. Non-Cryst. Solids, **79**, 295-309 (1986).

A Density Functional Approach to Nucleation in Microemulsions

Vicente Talanquer* and David W. Oxtoby[†]

*Department of Chemistry, University of Arizona, Tucson, Arizona 85721 and
[†]James Franck Institute, University of Chicago, Chicago, Illinois 60637

Abstract. We develop a microscopic approach to nucleation in microemulsions based on density functional theory. Using a simple free energy functional to describe amphiphiles (consisting of fused hydrophilic and hydrophobic spheres) interacting with water molecules, we explore the phase diagram and calculate the free energies of small clusters of amphiphiles. Our calculation reveals the existence of stable micelles and vesicles, as well as the potential critical nuclei (free energy maxima) that occur along pathways to their formation.

Microemulsions are states of matter that exhibit phase separation on a microscopic, not a macroscopic, scale. They are characterized by the presence of two components that do not readily mix (oil and water) and a bifunctional amphiphilic component (the surfactant) which has one end strongly attracted to water and the other to oil. In this paper we describe recent statistical mechanical approaches to phase transitions and nucleation in amphiphilic mixtures, using the techniques of density functional theory.

Several years ago, we explored gas-to-liquid nucleation in a ternary vapor that contained water, an alkane, and an alcohol (1), and showed that the amphiphilic alcohol molecule could act as a surfactant in inducing the alkane to nucleate in a droplet around the water, with a layer of alcohol in between. Here we discuss ternary liquid mixtures, in which liquid-liquid phase separation is prevented by the presence of an amphiphile. The approaches we describe are based on density functional theory, a powerful method of studying phase transitions and nucleation (2).

Earlier work has explored two free energy functionals for ternary amphiphiles. The first treated the three species as spherical force centers, with a point dipolar interaction on the amphiphile, causing one end to be attracted to water molecules and the other to oil. This was applied by Telo da Gama and coworkers (3, 4) to microemulsions and micelles, and is the same type of free energy used by us for nucleation in ternary vapors (1). We later introduced a second approach (5) in which a specific, short-ranged attractive interaction between the water and amphiphile species (giving rise to association) was added to a van der Waals free energy functional. The Wertheim theory of associated liquids was employed to calculate the effects of this interaction on the ternary phase diagram and on interfacial properties. This approach gives realistic ternary phase diagrams, but is suited only for weak amphiphiles.

In this paper, we use a third approach to amphiphile mixtures. The amphiphile (C) consists of a pair of fused hard spheres, of which one sphere is more strongly attracted to nearby water molecules (A) and the other to oil molecules (B). The water and oil molecules are spherical, although it would be straightforward to make the hydrocarbon a chain of spheres (a more realistic model that we are pursuing). For simplicity, all particles have the same hard-sphere radius. The results presented here are for the simpler case of a binary water-surfactant (A-C) mixture in the absence of oil.

The Helmholtz free energy F has the form

$$\beta F[\rho(\mathbf{r})] = \int d\mathbf{r} \rho_0(\mathbf{r})[\ln \rho_0(\mathbf{r}) - 1] + \sum_{i=1}^{2} \int d\mathbf{r} \rho_i(\mathbf{r}) \ln f_i(\mathbf{r}) - \iint d\mathbf{r} d\mathbf{r}' s(|\mathbf{r} - \mathbf{r}'|) f_1(\mathbf{r}) f_2(\mathbf{r}')$$

$$+ \int d\mathbf{r} \Psi(\bar{\eta}(\mathbf{r})) \rho_s(\mathbf{r}) + \frac{\beta}{2} \sum_{i,j=0}^{2} \iint d\mathbf{r} d\mathbf{r}' \phi_{att}^{ij}(|\mathbf{r} - \mathbf{r}'|) \rho_i(\mathbf{r}) \rho_j(\mathbf{r}')$$

Here $\beta = 1/kT$ and the subscript 0 designates water, with subscripts 1 and 2 referring to the hydrophilic and hydrophobic spheres making up the amphiphile. The total local density (the sum of the three individual densities) is $\rho_s(\mathbf{r})$; $s(r)$ is a delta-function bond constraint on the two spheres in the amphiphile; $f_i(\mathbf{r})$ is a local site activity for species i; $\Psi(\eta)$ is the free energy density for a hard sphere fluid evaluated at a weighted packing fraction η; ϕ_{att}^{ij} is the attractive potential between species i and j, taken to be an inverse sixth power outside the hard sphere radius. Ref. 6 gives additional details.

Once a series of potential parameters is chosen, the phase diagram can be calculated for this binary mixture by setting the densities to constant values and minimizing the Helmholtz free energy. More interesting, however, are the inhomogeneous states in which the densities vary with position through the system. In a planar geometry, we examined the structure and free energy of a bilayer membrane, in which two layers of amphiphile molecules (with hydrophilic ends pointing outward) are sandwiched between outer layers of water in a planar geometry. This free energy functional gives the type of narrow interface expected for a strong amphiphile. Although we looked only at a single bilayer interface, it would be straightforward to examine a lamellar phase in which a series of such bilayers repeat in a periodic fashion.

Next, we turned to spherical geometries for the amphiphile clusters surrounded by water, in order to look for micelles and vesicles, as well as the critical nuclei that may occur on the pathway to the formation of these structures from a uniformly dispersed solution. Figure 1 shows a sample result from such a calculation. It displays the work of formation to create a cluster containing i_e excess molecules of amphiphile relative to the original phase, which is a dilute solution (mole fraction less than 0.01) of amphiphile in water. As shown in that figure, the free energy to create such a cluster first rises, then falls, then rises again, and then falls again. Let us explore the origins of this seemingly complex behavior.

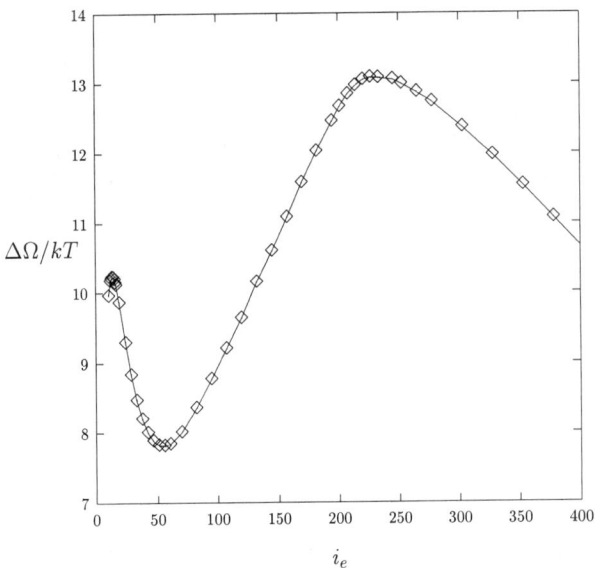

FIGURE 1. Dependence of work of formation of a cluster on the number of amphiphile molecules it contains, i_e.

The minimum that occurs in the free energy (work of formation) in the vicinity of 50 amphiphile molecules is a stable micelle; its free energy rises if molecules are removed or added to the cluster. Such a micelle is in thermodynamic equilibrium with the surrounding solution of amphiphiles in water, and in fact under some conditions its actual excess free energy is negative. An examination of the density profiles that give rise to this micelle show the expected form: the hydrophobic spheres cluster near the center while the hydrophilic spheres are on the outside to increase their contact with water. The peak in free energy to the left of this (near $i_e = 12$) is a free energy barrier to formation of the micelle, which is a thermally activated process. Putting together the first few amphiphile molecules costs free energy (because of loss of translational entropy), before the energetic savings take effect and the free energy falls.

An increase in the number of amphiphile molecules beyond the first minimum also causes the free energy of the cluster to go up. As this occurs it becomes increasingly difficult to lower the energy through hydrophobic and hydrophilic interactions. Instead, a transition takes place in which water enters the center of the sphere and a bilayer surfactant interface appears; this corresponds to a vesicle. This observation is confirmed by examining the particle density profiles that emerge for larger clusters. The fact that the free energy falls as the size of the vesicle increases in size suggests that in this case the curved vesicle is metastable with respect to a planar lamellar phase, although we have not yet carried out a detailed calculation of the full phase diagram. Although a critical nucleus that is spherical on average is plausible for formation of a micelle, it seems unlikely that a micelle-to-vesicle transition would proceed through a spherical transition state. More likely is a non-spherical intermediate state in which a fluctuation causes the micelle to open up on one side and let water enter in as the interface simultaneously changes over to a bilayer structure.

These results demonstrate that a simple free energy functional can give rise to a range of structures characteristic of complex fluids: bilayer membranes, micelles, and vesicles. Much more work remains to be done. First, we need to add the third component (oil) to study true oil-in-water micelles as well as water-in-oil reverse micelles. We need to calculate the elastic constants from our free energy functional in order to make contact with phenomenological approaches to amphiphile solutions. We need to explore the full equilibrium phase diagram to locate lamellar phases and phases of cylindrical symmetry. Subsequent work can explore nucleation dynamics for the formation of stable and metastable phases in these systems.

We are also extending our calculation to a more realistic model for the water-alkane-surfactant mixtures studied experimentally. Specifically, we are developing a free energy model for chain-like alkane and surfactant molecules, which consist of series of spherical force centers joined together through bonding constraints (7). This will enable us to study more systematically the effect of chain length on phase behavior and surface thermodynamics, and make closer contact with the experimental studies of Strey and coworkers on polymethyl-polyether amphiphiles (8).

ACKNOWLEDGMENTS

This work was supported by the National Science Foundation (through grant CHE 9800074 and through the NSF Materials Research Science and Engineering Center at the University of Chicago). We acknowledge useful discussions with Dr. Chaok Seok.

REFERENCES

1. Talanquer, V., and Oxtoby, D. W., *J. Chem. Phys.* **106**, 3673-3680 (1997).
2. Oxtoby, D. W., *Accts. of Chem. Res.* **31**, 91-97 (1998).
3. Telo da Gama, M. M., and Gubbins, K., *Mol. Phys.* **59**, 227-239 (1986).
4. Guerra, C., Somoza, A. M., and Telo da Gama, M. M., *J. Chem. Phys.* **109**, 1152-1161 (1998).
5. Talanquer, V., and Oxtoby, D. W., *Faraday Disc.* **112**, 91-101 (1999).
6. Talanquer, V., and Oxtoby, D. W., *J. Chem. Phys.* submitted.
7. Seok, C., and Oxtoby, D. W., unpublished work.
8. Strey, R., *Curr. Opin. Colloid Interface Sci.* **1**, 402-415 (1996).

A Monte Carlo Simulation Approach to Nucleation in Microemulsions

Isamu Kusaka and David W. Oxtoby

The James Franck Institute, The University of Chicago, 5640 South Ellis Avenue, Chicago, Illinois 60637

Abstract. We study nucleation of amphiphilic molecules in a solution by Monte Carlo simulation aided by the umbrella sampling technique. The method is free of any ad hoc molecular level cluster definition and is also very efficient in constructing the free energy surface relevant to nucleation.

Recently, we presented a new approach to cluster simulation within the context of nucleation.[1-4] The method has been applied successfully to a multitude of nucleation phenomena, including cavitation,[3] and has provided a unified viewpoint in understanding nucleation at a molecular level. In this paper, we apply this method to nucleation of amphiphiles in a binary solvent-surfactant system.

Our basic idea is to follow the stochastic evolution of the system by means of a Monte Carlo simulation. Under a proper choice of the system size, the fluctuation we observe can be identified as a single cluster, which we characterize by a set of global order parameters.[1-4] In this way, we can avoid any ad hoc molecular level cluster definitions introduced in a conventional cluster simulation. The method yields directly the equilibrium cluster size distribution.

In the case of fluid to fluid nucleation in a single component system, two order parameters have been identified: the molecular content in the system and the potential energy. When nucleation occurs in an attenuated vapor phase, the emerging clusters are compact and the two order parameters correlate strongly with each other. In this case the cluster can be characterized by the molecular content alone. In the case of crystal nucleation, van Duijneveldt and Frenkel[5] identified the bond-orientation order as the relevant quantity.

The clusters identified in this manner and their free energies are attributes of an equilibrium ensemble. Such "equilibrium" clusters are nonetheless relevant in the dynamics of nucleation. This is a direct consequence of the fact that these order parameters all exhibit a separation of time scales, one characterizing the rapid fluctuation of these quantities and the other showing their systematic change during the course of nucleation.[2]

Let us now turn to the binary solvent-surfactant system. Following Refs. 1-4 and

6, the system is taken to be a small part of the macroscopic solution and satisfies the following two conditions. (1) The system is large enough to be regarded as statistically independent of the surroundings. (2) It is at the same time small enough that the probability of finding more than one uncorrelated fluctuation at any instant is negligible.

In simulating a condensed phase, it is convenient to work with a closed system in which the total number of molecules N remains fixed. An increase in the number of surfactant molecules N_C then necessarily accompanies an equal decrease in the number of solvent molecules, indicating that N_C rather than N is a relevant order parameter. We take as the second order parameter the potential energy U_C due to the interaction among the surfactant molecules. This choice appears natural if we note that a large portion of the solvent remains unaffected by nucleation of the surfactant molecules.

Under the present choice of the system, its surroundings can be regarded as a reservoir imposing the pressure P and the difference between the fugacity z_C of the surfactant and that of the solvent z_A. Thus, the probability p of finding the system in a macrostate characterized by (N_C, U_C) is

$$p(N_C, U_C) = \frac{B e^{(z_C - z_A) N_C}}{(N - N_C)! N_C!} \int dV e^{\beta PV} \int d\{N\} e^{-\beta U_N}, \qquad (1)$$

where V is the system volume, $\{N\}$ collectively denotes the configuration of N molecules, $\beta = (k_B T)^{-1}$ is the reciprocal of the temperature, and U_N is the total potential energy of the system. B is a constant ensuring the normalization of p. The probability p can be sampled by a usual Isothermal-Isobaric Monte Carlo simulation[7] combined with an identity exchange between a solvent molecule and a surfactant molecule.

Under periodic boundary conditions, p is related to the equilibrium cluster size distribution n in a very simple manner.[4] In fact, it can be readily shown that

$$n(N_C, U_C) \approx \frac{\langle \rho \rangle}{N} p(N_C, U_C), \qquad (2)$$

where ρ is the number density of the molecules and the angular bracket indicates that a thermal average is taken. A key observation in arriving at Eq. (2) is that, for a given value of V, the cluster can be found anywhere in V with uniform probability. In an ensemble in which V is constant, $\langle \rho \rangle / N$ reduces to $1/V$ and Eq. (2) is an exact equality, thereby recovering our earlier result.[1,4]

For an efficient and accurate evaluation of p, we employ the umbrella sampling technique.[8] One possible implementation of this technique is to perform simulations while confining (N_C, U_C) to small subintervals so that p is comparable within each of the subintervals.[1-4] Once p is calculated by simulation for various values of (N_C, U_C) in each of the subintervals, p is determined by requiring it to be normalized and continuous over the entire range of (N_C, U_C) of interest. When p changes rapidly

TABLE 1. Well-depth ϵ of the Lennard-Jones inteaction

	site1	site2	site3	site4
site1	1	1	1.36	0.34
site2	1	1	1.36	0.34
site3	1.36	1.36	1.2	0.8
site4	0.34	0.34	0.8	1.2

with N_C or U_C, this approach becomes inefficient. Instead, we introduce a function f that is chosen to render fp nearly independent of (N_C, U_C), and evaluate

$$p_f(N_C, U_C) = \frac{B' f}{(N - N_C)! N_C!} \int dV e^{\beta PV} \int d\{N\} e^{-\beta U_N}, \qquad (3)$$

from which p follows immediately. B' is the normalization constant. Obviously, the most suitable choice of the function f is p^{-1}. Of course, p is the very quantity we wish to evaluate by the simulation, while f has to be supplied prior to the simulation. This does not present any difficulty since, in practice, a rough estimate of f from a relatively short simulation was found to be sufficient for our purpose. This technique of rescaling the probability p in a Monte Carlo simulation is due to Lyubartsev et al. and is referred to as the method of expanded ensembles.[9]

In this work, both solvent and surfactant molecules are taken to be diatomic molecules, each consisting of two Lennard-Jones interaction sites separated by the Lennard-Jones radius $\sigma = 1$. The well-depth ϵ between the pairs of the interaction sites are given in Table 1. Sites 1 and 2 belong to the solvent molecules and are chosen to be identical. Sites 3 and 4 represent the hydrophilic end and the hydrophobic end of a surfactant molecule, respectively. Aside from the fact that the solvent molecule is diatomic and that the repulsive part of the potential is not a hard-core, our choice of the model potentials corresponds to a particular case of various model systems studied by Talanquer and Oxtoby using a density functional approach. In this work, we have deliberately chosen the same geometry for solvent and surfactant molecules so that the distribution over N_C can be sampled accurately with a relatively small computational effort.

As a preliminary calculation, we dropped U_C and worked only with a single order parameter N_C, which is found to be acceptable for large enough clusters at sufficiently low temperature. A typical density profile of a cluster is shown in Fig. 1, which exhibits a characteristics micelle structure. We are currently investigating the free energy surface using both N_C and U_C as the order parameters to obtain the complete pathway of nucleation.

ACKNOWLEDGMENTS

This work was supported by the National Science Foundation (Grant No. CHE-9800074).

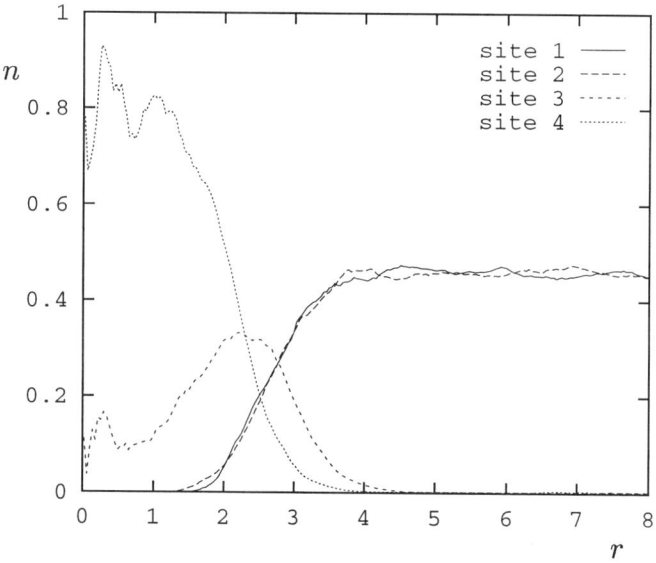

FIGURE 1. The number densities of each interaction site. $T = 0.65$, $P = 0$, $N = 2048$, and N_C is restricted to [48, 54]. The center of the cluster ($r = 0$) is conventionally taken as the center of mass of the largest of the aggregates, each identified using Stillinger's criteria[10] with the cutoff distance 3.5.

REFERENCES

1. I. Kusaka, Z.-G. Wang, and J. H. Seinfeld, J. Chem. Phys. **108**, 3416 (1998).
2. I. Kusaka and D. W. Oxtoby, J. Chem. Phys. **110**, 5249 (1999).
3. I. Kusaka and D. W. Oxtoby, J. Chem. Phys. **111**, 1104 (1999).
4. I. Kusaka, D. W. Oxtoby, and Z.-G. Wang, J. Chem. Phys. **111**, 9958 (1999).
5. J. S. van Duijneveldt and D. Frenkel, J. Chem. Phys. **96**, 4655 (1992).
6. K. Nishioka and G. M. Pound, Adv. Colloid and Interface Sci. **7**, 205 (1977).
7. M. P. Allen and D. J. Tildesley, *Computer Simulation of Liquids* (Oxford University Press, New York, 1987).
8. D. Chandler, *Introduction to Modern Statistical Mechanics* (Oxford University Press, New York, 1987).
9. A. P. Lyubartsev, A. A. Martsinovski, S. V. Shevkunov, and P. N. Vorontsov-Velyaminov, J. Chem. Phys. **96**, 1776 (1992).
10. F. H. Stillinger, Jr., J. Chem. Phys. **38**, 1486 (1963).

Classical and Non-Classical Descriptions of Nucleation in an Ising Ferromagnet vs Monte Carlo Simulations

V. A. Shneidman, K. A. Jackson and K. M. Beatty

Department of Physics, New Jersey Institute of Technology, Newark, NJ 07102
Department of Materials Science and Engineering, University of Arizona, Tucson AZ 85721

Abstract. Large scale Monte Carlo simulations and theoretical analysis of thermodynamic properties of small clusters and large nuclei on two-dimensional Ising lattices were performed in order to test the kinetic and thermodynamic aspects of several mainstream nucleation approaches. The kinetics predicted by the classical theory (i.e. the time evolution of cluster distributions) is accurate at intermediate temperatures as long as the description relies on measured (not calculated!) equilibrium distributions and as long as coagulation between clusters can be neglected. However, the classical theory seriously overestimates the equilibrium cluster populations – this can be unambiguously demonstrated since the interfacial tensions for two-dimensional Ising systems are known exactly from the Onsager solution and from later developments. Alternatives to classical description are also considered, and ways to improve the correspondence between theory and simulations are discussed.

INTRODUCTION

The standard expression for the nucleation rate can be written as

$$I = A\exp(-W^*/kT)$$

The exponential part was obtained by Volmer and Weber (1926), while the work W^* required to form a critical nucleus was earlier calculated by Gibbs. The pre-factor A, however, remains undetermined.

There are two aspects in determination of A, namely its relation to the kinetics and the thermodynamics of a given system. Thermodynamics allows one to determine the (quasi)equilibrium cluster distributions. Kinetics, on the other hand, is determined by the specific mass exchange mechanism between the nucleus and the metastable phase; kinetic contributions to A were considered by Farkas (1927), Becker and Doring (1935), Zeldovich (1942) and Frenkel (1946), forming the basis of what is now known as classical nucleation theory.

Experimental verification of the classical theory and distinguishing it from any of the competing approaches is often hindered by the fact that thermodynamics and kinetics are not known with sufficient accuracy, and only averaged information about cluster distributions can be obtained from measurements. Otherwise, time dependencies of such distributions can be observed only for selected systems (glasses) for which little is known about the key thermodynamic parameter, the interfacial tension.

Two-dimensional Ising systems provide an important complement to real-life experiments in several ways. First, not only the location of phase equilibrium for those systems is known exactly, but also the interfacial tension is often either known or can be deduced from known expressions; some rigorous results on cluster kinetics are also available in mathematical literature. Second, large scale Monte Carlo simulations allow one to trace the evolution of cluster distributions in time to any desirable detail. Finally, the input parameters can be modified in a controlled manner in order to test one or the other aspect of a given theory.

In Refs. [1(a)] and [1(b)] we considered dynamic Ising models on a triangular and square lattices, respectively. The triangular lattice is less studied analytically, and we treated it is an "experimental" system, placing the main emphasis on the kinetic studies (growth and decay of individual clusters, "measurement" of the nucleation rate, time-dependent effects, onset of coagulation, etc. – see next section). For a square lattice more analytical results are available, allowing one to gain some insight into the thermodynamics of the problem and to attempt a first-principle description of the nucleation kinetics.

KINETICS

Kinetics was studied in two stages [1(a)]. First, one examines an individual nucleus with n spins. At equilibrium (zero field, h) a nucleus decays with n changing linearly with time. This allows one to determine the time scale τ which enters the expression for the nucleation rate (see below). In the undercooled region, $h>0$, a nucleus will grow or decay depending on the relation of n to the critical size, n^*; a special case is the flat interface which corresponds to $n \to \infty$. Kinetics can be fully predicted from equilibrium measurements, as long as one can neglect coagulation (coagulation also can be switched-off by an artificial modification of the spin-flip dynamics).

In the multicluster kinetics of main interest is the value of I, the nucleation rate. Three methods to extract this value from simulations can be used depending on the size of the system. (Size-dependence of the nucleation process was also studied in Ref.[2]). In a small system I approximately corresponds to the inverse of the waiting time to detect this first nucleus. For a larger system (equivalently, in larger fields) many nuclei can be formed. Here I can be obtained from the observed number of clusters. [This second method is the closest to the actual experimental studies of isothermal nucleation in glasses, as reported, e.g., by James (1974) or Kalinina et al. (1976)].

The third way to obtain I is to use the Kolmogorov-Avrami expression for the transformed area together with the known (measured or calculated) growth rate of a flat interface. This method is less accurate than direct cluster counting but is also simpler for realization and corresponds to approaches used by experimentalists when analyzing crystallization data. It is also the only method which can be applied, at least formally, in the high field region when coagulation between clusters starts practically simultaneously with nucleation. The Kolmogorov-Avrami curves turn out to be remarkably accurate when describing the *shapes* of the phase transformation curves [1,3].

Once reliable methods to measure the nucleation and growth rates are established, one can test the kinetic prediction of the classical theory

$$I = \frac{\Delta}{2\tau\sqrt{\pi}} f(n^*)$$

In the above expression Δ is the width of the near-critical region, defined in a standard manner, while $f(n^*)$ is the quasiequilibrium cluster distribution at the critical size. This formula was found reasonably accurate [1(a)] for a triangular lattice, provided the values of I, τ and $f(n^*)$ were deduced from measurements rather than calculated. Similarly, the above formula holds for a square lattice [1(b)] with analytical estimations of τ and $f(n^*)$, although the latter deviates from classical predictions as discussed in the "Thermodynamics" section.

Time-Dependent Nucleation Effects

Even if the goal of a study is to extract the steady-state rate, I, an account for time-dependent nucleation effects can be important for the following reason. Due to relatively low values of the nucleation barrier W^* (about *10 kT*, in contrast to *30-40 kT* for real experiments) coagulation will inevitably enter the scene if one waits for the steady-state cluster distribution to be established up to sufficiently large sizes. To avoid coagulation, one is restricted to smaller times with pronounced transient effects. For example, the distribution of clusters above the critical size is

$$f(n,t) = j(n,t)/v(n)$$

where $v(n)$ is the deterministic growth rate. The transient nucleation flux $j(n,t)$ is given by I times a "double-exponential" function [4] of a dimensionless parameter $x = [t - t_i(n)]/\tau$, where $t_i(n)$ is the incubation time. Comparing the above expression for $f(n,t)$ with measured distributions one can deduce the values of I. Similarly, the number of nuclei larger than n is given by $I\tau E_1[\exp(-x)]$ (where E_1 is the first exponential integral) [4(b)], which gives a family of t-dependent curves for different values n. Once the value of τ is known from independent growth/decay measurements, comparison with simulation data provides an accurate way to measure I.

Low-Temperature Kinetics

At small T is the square shape of a nucleus with nearly perfect flat faces. Here additional nucleation of a step is required for a nucleus to grow [5]. Strictly speaking, the conventional Becker-Doring description is not applicable here since the attachment rate strongly depends on whether a nucleus is a square or not. Nevertheless, one-dimensional nucleation of steps can be treated exactly, providing a unique possibility of a first-principle kinetic description.

THERMODYNAMICS

Suppose one knows from simulations the distribution of clusters, $F(n)$, at equilibrium with $h = 0$. Due to an extremely simple field-dependence of the cluster free energy in the Ising model, the above distribution allows one to predict the quasiequilibrium distribution of subcritical clusters

$$f(n,h) = F(n)\exp(2nh/kT)$$

This can be used to evaluate $f(n^*)$ in the expression for the nucleation rate. A first-principle evaluation of $F(n)$, however, is much more complex. For example, the classical theory assumes

$$F(n) \approx N\exp[-W(n)/kT]$$

with N being the total number of spins in a system and $W(n)$ the minimal work required to form a given nucleus. For large n this work can be calculated exactly for both square [1(b)] and triangular [6] Ising lattices due to available expression for anisotropic interfacial tension. Theories of Fisher (1963) and Langer (1971) suggest additional power-law factors in the expression for $F(n)$. On the other hand, non-classical nucleation approaches of the type developed by Oxtoby (1992,1996) propose finite-n corrections to $W(n)$.

Comparison with simulation results shows that the classical theory does not predict a correct pre-exponential of the cluster distribution in the high-temperature region $(> 0.8\ T_c)$ either for the square or the triangular Ising lattices. The likely reason is coagulation between clusters which are abundant in this region, although alternative explanations are also possible [1(b)].

The pre-exponential factor of M. Fisher is in agreement with observations in the immediate vicinity of T_c which was already noted in the early simulations by Binder and Stauffer (1976). In the intermediate high-temperature regime Langer's approach is in better agreement (see also Refs.[2,3]), although there does not exist a single theory which would span a sufficiently large temperature region. Further lowering the temperature seems to favor the classical description except for the absolute value of the pre-factor in the expression for the cluster distribution.

In the cold temperature region simulations are hindered by the fact that the distribution rapidly decays with size n and only the smallest clusters can be examined. On the other hand, a powerful additional feature here is the possibility of exact analytical evaluation of populations of such clusters. This allows a more consistent evaluation of the pre-factor, substantially improving the accuracy of the classical predictions [1(b)] .

REFERENCES

1. Shneidman, V. A., Jackson K. A., and Beatty K. M. (a) Phys. Rev. **B 59**, 3579 (1999); (b) J. Chem. Phys. **111**, 6932 (1999).
2. Rikvold, P. A., Tomita H., Miyashita S., and Sides S. W. , Phys. Rev. **E 49**, 5080 (1994).
3. Ramos R.A., Rikvold P.A., and Novotny M. A. Phys. Rev. **B 59**, 3579 (1999).
4. Shneidman, V. A., Sov. Phys. Tech. Phys. (a) **32**, 76 (1987); (b) **33**, 1338 (1988).
5. Shneidman, V. A., Jackson K. A., and Beatty K. M., J. Cryst. Growth **212**, 564 (2000).
6. Shneidman, V. A., and Zia R. K. P. *Wulff Shapes and the Critical Nucleus on a Triangular Ising Lattice* , in preparation.

Nucleation in the Presence of Fish and Insect Ice-Growth Inhibition ("Antifreeze") Molecules

A.D.J. Haymet, Aaron Heneghan, Peter W. Wilson[*], Karl Erik Zachariassen[†] and Hans Ramløv[‡]

Department of Chemistry, University of Houston, Houston TX 77204 USA
[*]*Physiology Dept. Medical School, University of Otago, Dunedin, New Zealand*
[†]*Norwegian University Science & Technology, Department of Zoology, N-7034 Trondheim, Norway*
[‡]*Department of Life Sciences and Chemistry, Roskilde University, DK-4000, Roskilde, Denmark*

Abstract. Temperature plays an essential role in the survival of all organisms. For example, some organisms have adapted to situations that would otherwise be unfavorable to the survival of that organism. The fish winter flounder from the Arctic and the bark beetle from Norway are capable of withstanding temperatures well below the equilibrium freezing temperature of their inner fluids. This avoidance of freezing is attained through the ability to suppress crystal growth. The reason for this avoidance stems from a class of "antifreeze" molecules, but it is not fully understood at the molecular level why or by what mechanism these organisms are able to exhibit this property. A separate but equally interesting question is the nucleation properties of these molecules. By using the automated lag time apparatus (ALTA) for collecting statistics of nucleation on a single sample, we study this phenomenon.

BACKGROUND

In most systems, the phenomenon of lower freezing temperature is explained using colligative properties. The idea is the system is dependent upon the number of solute particles present. The more solute particles present, the lower the freezing temperature. This is the mechanism by which ethylene glycol functions in automobiles. Nonetheless, these organisms are able to withstand temperatures which are 300-500 times lower than what would be expected on the number of solute particles present [1, 2]. Hence, these molecules are termed kinetic ice growth inhibition molecules. Small ice crystals are indeed formed within the organism, but these "antifreeze" molecules act to prevent the growth which would result in damage to the cell membranes [3-7].

An important fingerprint for these molecules is thermal hysteresis [8-11]. Simply stated, this is the difference between the equilibrium freezing temperature and the ice growth temperature. As expected, this difference in pure water is zero °C, and any seed crystal inserted into the system below zero °C will result in ice growth. One way to test this hysteresis is to use a Clifton nanoliter osmometer. This osmometer is a thermoelectric cooling module with a precise temperature control. The apparatus reads a value in milliOsmoles. By calibrating the osmometer with various standards, one can directly associate a change in freezing temperature to the value of the solution in milliOsmoles. The data from the osmometer provide a good starting point for the temperature at which to run samples in the automatic lag time apparatus (ALTA).

Collection of Samples and Experimental Procedure

Larvae of the bark beetle *Rhagium inquisitor* [12, 13] were collected from stumps of trees in Norway in January of 1999. These larvae were kept overnight at 5 °C before haemolymph was collected. The actual haemolymph was collected by puncturing of the cuticle near the anterior. The larvae were then squeezed gently while the haemolymph was collected in 1.5 mL eppendorf tubes. This haemolymph was then stored at -80 °C until it was used in experiment. The collection of samples from the fish winter flounder *Pleuronectus americanus* has been documented elsewhere [14].

The pure haemolymph collected was spun down using a 0.22μM filter. This spun down solution was diluted 1:3 with distilled water producing an approximate 25% solution. This original sample was placed in ALTA for collection of statistics; however, long times greater than 10,000s were observed without a freezing event. As a result, several dilutions were made on the sample. Dilutions were made by taking 100μL of the parent solution and diluting with 100μL of distilled water. A final concentration of approximately 3.2% was found to give a good distribution of times to nucleation.

These samples were run in a new generation automatic lag time apparatus(ALTA). This new apparatus differs from that described by Barlow [15] by employing a solid state cooling of a smaller sample volume to produce rapid and reproducible supercooling temperatures. The sample, usually 100-200μL, is placed in a shortened NMR tube which sits inside a block of aluminum. This block is wedged between two 100 watt peltier devices. Temperature is monitored using a thermocouple as close to the sample as possible but still outside of the sample cell. Liquid samples permit light to reach the photodiode, while decreases in the intensity at the photodiode are due to scattering of the laser by a frozen sample. Upon freezing, the current to the peltiers is automatically reversed resulting in heating of the sample to +10 °C for four minutes while ensuring the melting of all the ice crystals. After these four minutes, the current is once again reversed and supercooling is again achieved. The timer collects this lag time again and again.

Analysis of Data

In this experiment we collect the time it takes a single sample to freeze. This time to nucleation is collected many times on a single sample with computer simulations giving a good statistical distribution when $N_o \approx 500$ runs. From these data, it is possible to determine the fraction unfrozen as a function of time. It is this fraction unfrozen which is plotted as a function of time. As expected, the fraction unfrozen at time zero should be 100% while the fraction unfrozen should go to zero at some finite time when all of the samples have frozen. Figure 1a and Figure 1b, shown below, help to describe what is going on in the course of the experiment.

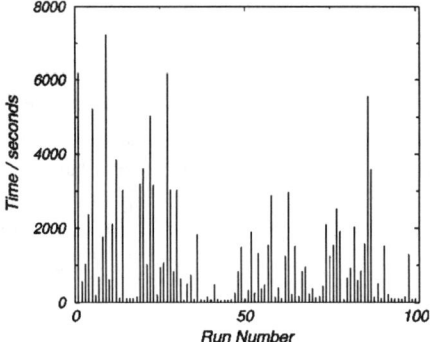 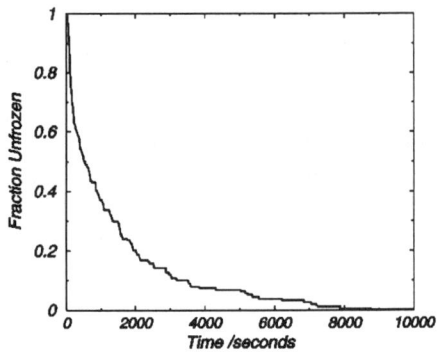

Figure 1 (a) Time to nucleation of a single sample containing ice growth inhibition molecules (b) Fraction unfrozen for this sample data set as a function of time.

It is from this fraction unfrozen curve that we begin to analyze these data. The first approach is to assume macroscopic kinetics govern this nucleation process. From this assumption, it is possible to derive the kinetic rate equations for cases where first order and second order kinetics are used, for example. This analysis helps in the determination of what is going on macroscopically. The goal of this work is to gain understanding of the microscopic processes involved in the overall nucleation process. Hence, a unified theory of nucleation which meshes the microscopic regime with experimental evidence and macroscopic understanding is sought. The combination of the above methods will aid in examining such a microscopic theory as well as provide insight into this ice growth inhibition mechanism.

ACKNOWLEDGMENTS

This research is supported in part by ACS-PRF Grant AC 33707-AC9 and UH Environmental Institute, to whom grateful acknowledgement is made.

REFERENCES

1. Haymet, A.D.J., L.G. Ward, and M.M. Harding, *Winter flounder "antifreeze" proteins: Synthesis and ice growth inhibition of analogues that probe the relative importance of hydrophobic and hydrogen-bonding interactions.* Journal of the American Chemical Society, 1999. **121**(5): p. 941-948.
2. Eastman, J.T. and A.L. DeVries, *Antarctic Fishes.* Scientific American, 1986. **November**: p. 49-58.
3. Davies, P.L. and B.D. Sykes, *Antifreeze Proteins.* Current Opinion in Structural Biology, 1997. **7**(6): p. 828-834.

4. Yeh, Y. and R.E. Feeney, *Antifreeze Proteins - Structures and Mechanisms of Function.* Chemical Reviews, 1996. **96**(2): p. 601-617.
5. Chapsky, L. and B. Rubinsky, *Kinetics of Antifreeze Protein-Induced Ice Growth Inhibition.* FEBS Letters, 1997. **412**(1): p. 241-244.
6. Chen, L.B., A.L. Devries, and C.H.C. Cheng, *Convergent Evolution of Antifreeze Glycoproteins in Antarctic Notothenioid Fish and Arctic Cod.* Proceedings of the National Academy of Sciences of the United States of America, 1997. **94**(8): p. 3817-3822.
7. Graham, L.A., et al., *Hyperactive Antifreeze Protein from Beetles.* Nature, 1997. **388**(6644): p. 727-728.
8. Haymet, A.D.J., et al., *Valine substituted winter flounder antifreeze - preservation of ice growth hysteresis.* FEBS Letters, 1998. **430**(3): p. 301-306.
9. Harding, M.M., L.G. Ward, and A.D.J. Haymet, *Type I "antifreeze" proteins: Structure activity studies and mechanisms of ice growth inhibition.* European Journal of Biochemistry, 1999. **264**: p. 653-665.
10. Loewen, M.C., et al., *Alternative roles for putative ice-binding residues in type I antifreeze protein.* Biochemistry, 1999. **38**(15): p. 4743-4749.
11. Meyer, K., M. Keil, and M.J. Naldrett, *A leucine-rich repeat protein of carrot that exhibits antifreeze activity.* FEBS Letters, 1999. **447**(2-3): p. 171-178.
12. Zachariassen, K.E. and J.A. Husby, *Antifreeze effect of thermal hysteresis agents protects highly supercooled insects.* Nature, 1982. **298**: p. 865-867.
13. Kristiansen, E., et al., *Antifreeze activity in the cerambycid beetle Rhagium inquisitor.* Journal of Comparative Physiology B, 1999. **169**(1): p. 55-60.
14. DeVries, A.L. and Y. Lin, *Structure of a Peptide Antifreeze and Mechanism of Adsorption to Ice.* Biochimica et Biophysica Acta, 1977. **495**: p. 388-392.
15. Barlow, T.W. and A.D.J. Haymet, *ALTA: an automated lag-time apparatus for studying the nucleation of supercooled liquids.* Rev. Sci. Instruments, 1995. **66**: p. 2996-3007.

Bubble Nucleation Rates By Pressure Pulse Experiments

B. Rathke[2], H. Baumgartl[1], R. Strey[2]

(1) BASF AG, D – 67056 Ludwigshafen, Germany
(2) Institute of Physical Chemistry 1, University of Cologne, Luxemburger Strasse 116, D – 50939 Köln, Germany

Abstract. We designed a precision method for observation of liquid-gas first order phase-transitions. At first we prepare a gas-saturated liquid of well known composition. Then we induce the phase transition by a rapid pressure quench under simultaneous detection of pressure and scattered light. To this end the proceeding phase separation is detected by time resolved light-scattering at constant angle (CAMS). Specifically, we obtain detailed information about the microscopic characteristics of the phase separation process (nucleation / spinodal decomposition), the maximal pressure undershoot, the expansion rate and the preselected growth of the newly evolving phase as well as the nucleation rate. It is the aim of the study to measure homogeneous nucleation rates as function of supersaturation in gas-saturated liquids at constant temperature and to compare the results with classical and other nucleation theories.

INTRODUCTION

Kinetics of first order liquid - gas phase transitions are of scientific and technical interest. Bubble formation takes place in technical processes dealing with degassing of liquid systems (foaming, pool boiling, sparkling drinks) influenced by heterogeneous or homogeneous nucleation. The main design of industrial processes is usually based on empirical results, but principles are up to now not fully understood; there is a controversy between classical and non classical nucleation theories how to describe nucleation quantitatively. Experimental results usually show a large discrepancy with theoretically predicted nucleation rates of some orders of magnitude. Investigations up to now were limited to the determination of critical supersaturations. Nucleation rates were not directly measured.

The design of an experimental procedure which allows direct measurements of nucleation rates as function of supersaturation as well as following the growth of the nucleated bubbles is the aim of the present work.

As shown during the last two decades the nucleation - pulse method is one of the most powerful experiments to investigate gas-liquid phase transitions [1]. The method yields as direct result model-independent nucleation rates as function of supersaturation and permits following the growth process as well.

The present work applies this method to the study of bubble nucleation of gas-saturated liquids. The principle idea is to prepare a large amount of a binary gas-liquid mixture of known composition and to bring a small aliquot by a rapid pressure quench into a supercritically supersaturated state. Bubble nucleation occurs under these supersaturated conditions. After a certain time interval (some ms) the supersaturation is lowered by recompression to a subcritical value to stop nucleation. Nucleated bubbles grow to visible size; nucleation and growth are decoupled. Detection is done by constant angle Mie-scattering (CAMS) [2]. The features of such a pressure pulse are demonstrated in Fig. 1.

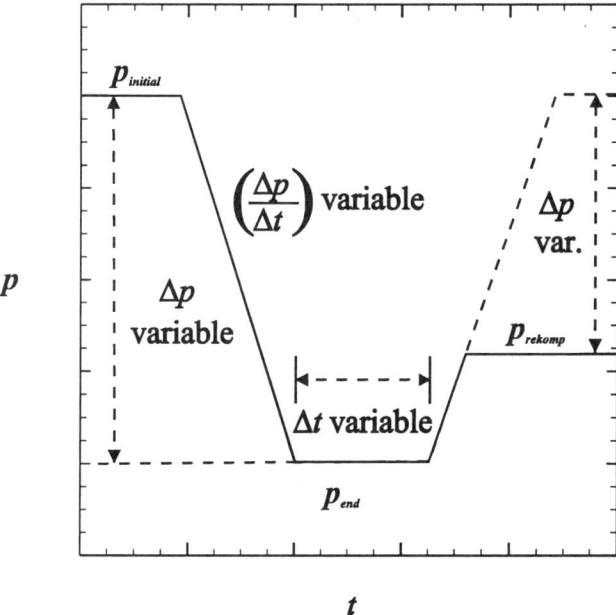

FIGURE 1. Schematic presentation of a typical pressure pulse.

EXPERIMENT

The basic requirement is the reproducible preparation of large samples of known composition as well as the exact knowledge of the phase diagram and additional thermodynamic properties of the system. We developed a mixing-device which allows the handling of mixtures (V=300ml) of a low-volatile compound (long-chain alkane e.g. $C_{16}H_{34}$) with gas (e.g. CO_2) under p and T control (1-200 bar; 10-120 °C). The basic idea is to prepare two coexisting phases; their compositions are fixed by pressure and temperature, only their volumes - according to the lever rule - vary with the total amounts of the components.

The set-up includes a network of tubes which in connection with a piston pump allows
1. circulation and fast mixing of the components,
2. quasi-isobaric sampling of volumes of 20 ml,
3. determination of the samples composition analytically by weight,
4. on-line determination of density (and isothermal compressibility) for each sample by using an oscillating tube densimeter and
5. fast and simple filling and refilling of the nucleation chamber,

all under preselected and maintained pressure p.

Figure 2 shows as an example the phase diagram determined using the newly constructed device at $T = 40$ °C for the system $C_{16}H_{34}/CO_2$.

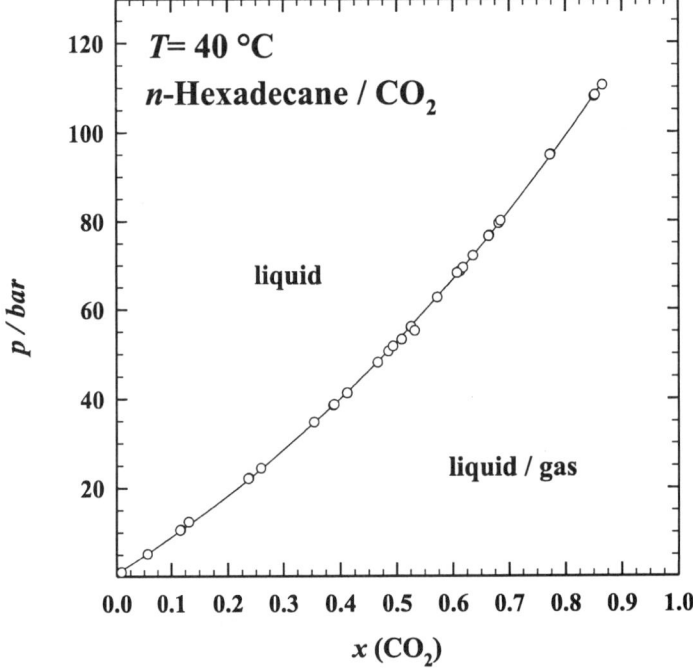

FIGURE 2. Isothermal phase diagram at $T = 40$ °C of the system n-Hexadecane/CO_2.

Nucleation measurements are performed inside a modified high-pressure light-scattering cell. The basic design of the light-scattering apparatus follows the classical set-up for measuring angle-dependent scattering intensities ($\theta = 2\text{-}178°$). Only the sample tube is a sapphire cylinder which allows time-resolved CAMS [2]. Connected with a pressure pulse unit we are able to produce rapid pressure changes, the rate being variable up to 20000 bar/s. Intensities of scattered light, transmitted light and pressure profiles are recorded simultaneously with high time resolution. Comparison between the measured signals and theoretical predictions of the Lorentz-Mie theory yields number concentration of the bubbles along with their radius. Knowledge of the time period of nucleation allows determination of the nucleation rate for conditions of stationary nucleation.

ACKNOWLEDGMENTS

We acknowledge a cooperation with Dr. J. Schmelzer, University of Rostock and helpful discussions with Prof. U.K. Deiters.

REFERENCES

1. R. Strey, P.E. Wagner, Y. Viisanen, *J. Phys. Chem* **98**, 7748-7758 (1994).
2. P.E. Wagner, *J. Colloid Interface Sci* **53**, 439-446 (1975).

A Novel Method for Determining the Crystal Nucleation and Growth Rates in Glasses Using Differential Thermal Analysis

Kisa S Ranasinghe, Chandra S Ray and Delbert E Day

Department of Physics and Graduate Center for Material Research, University of Missouri-Rolla

Abstract. The nucleation rate (I), the crystal growth rate (U), and the concentration of quenched in nuclei (N_q) in lithium disilicate ($Li_2O \cdot 2SiO_2$) and soda-lime-silica ($Na_2O \cdot 2CaO \cdot 3SiO_2$) glasses have been determined using a newly developed differential thermal analysis (DTA) method. This new DTA technique which is faster, less tedious, and requires a smaller amount of sample than in the conventional method of measuring I, U, and N_q, yields results that are in excellent agreement with those determined by conventional method.

INTRODUCTION

Determining the rates of nucleation (I) and crystal growth rate (U) as a function of temperature and the concentration of quenched-in nuclei (N_q) in glass is important to understand the stability of glasses in practical applications as well as to develop the glass-ceramics with desired microstructure and properties. The existing method of determining I, U, and N_q is time consuming and tedious. An alternate method using differential thermal analyzer (DTA) has been developed recently [1] which is much faster, less tedious and requires a smaller amount of sample than the conventional method. This newly developed DTA method is reported [1] to yield values of I, U and N_q for a lithium disilicate, $Li_2O \cdot 2SiO_2$ (LS$_2$) glass that are in excellent agreement with the values of I, U and N_q determined by the conventional method.

In the DTA method, a small amount of powdered glass sample (40-50mg) is first given a nucleation heat treatment at temperature T_N for time t_N and then a crystal growth heat treatment at temperature T_G for a time t_G. This is followed by a DTA scan at a fixed heating rate (10°C/min). Two such DTA scans are taken where all the experimental procedures are the same except the crystal growth heat treatment time at T_G is covered out with two different times, t_{G1} and t_{G2}. Using the two DTA scans the number of quenched-in nuclei N_q and the nucleation rate I at T_N can be calculated using the following equation.

$$\frac{A_1}{A_2} = \frac{M_1\left[1-\frac{\pi}{3}(It_N + N_q)Ut_{G1}^3\right]}{M_2\left[1-\frac{\pi}{3}(It_N + N_q)Ut_{G2}^3\right]} \quad (1)$$

where A_1 and A_2 are the areas of the two DTA peaks after crystal growth at T_G for time t_{G1} and t_{G2} respectively and M_1 and M_2 are the masses of the glass samples used in the two DTA runs, and U is the growth rate at temperature T_G.

However when this method is applied to determine I, U, and N_q, the (NC_2S_3) glass the DTA method appeared to yield erroneous results. For example, the number of quenched-in nuclei, N_q, determined using equation (1) was the same for different crystal growth time t_G in LS_2 glass, while for the NC_2S_3 glass N_q depends strongly on the crystal growth time t_G as shown in the Table.1. A likely reason for this deviation in

TABLE 1. Number of quenched-in nuclei calculated using equation (1) for LS_2 NC_2S_3 glasses

t_{G1} (min)	t_{G2} (min)	N_q (cm^{-3}) - LS_2	N_q (cm^{-3}) - NC_2S_3
10	15	4.05 X 10^6	63566
10	20	4.57 X 10^6	29101
10	30	4.51 X 10^6	9853.5

N_q for NC_2S_3 glass, may be due to the overlap of I and U(as a function of temperature) where both rates are large where simultaneous nucleation and crystal growth rates cannot be avoided. Equation (1) assumes that the I and U curves are separated such as in the case of LS_2 glass, which therefore cannot be applied to determined I, U and N_q for the glasses where I and U have a considerable overlap.

Present Work

The discussion and the results in Table.1 suggest that a more generalized equation that equation (1) needed to developed to accommodate the overlap of I and U curves. In the present work a new equation which takes account the overlap between I and U curves has been developed. The new equation reduced to equation (1) under the special case where I and U curves do not overlap. The values of I, U and N_q for the NC_2S_3 glass are now being determined using DTA and the revised equation.

Preliminary calculations show that the values for nucleation rate (I) and number of quenched-in nuclei (N_q) in NC_2S_3 glass, using the present technique are in excellent agreement with those of the conventional method[2].

ACKNOWLEDGMENTS

This work is supported by National Aeronautics and Space Administration, contract NAG8-1465.

REFERENCES

1. Ray, C. S, Fang X, and Day, D.E, *J. Ame.Ceram.Soc.* **83** 865-72 (2000).
2. Gonzalez-Oliver, C.J.R and James, P.F, *J. Non-Cryst.* **38&39**, 699-704 (1980).

Nucleation in Liquid Solutions. The Experimental Research

Vladimir G. Baidakov, Alexei M. Kaverin

Institute of Thermal Physics, 620219, Ekaterinburg, Russia

Abstract: Nucleation in binary solutions of cryogenic liquids has been investigated. Experiments have been made in range of nucleation rates from 10^4 to 10^8 m^{-3}s^{-1} at temperatures close to the critical line. The results of experiments are compared with the homogeneous nucleation theory, which takes into account the main factors limiting the nucleus growth. The effect of the curvature of the nucleus separating surface is allowed for in the nucleation work with the help of the Van-der-Waals - Cahn & Hilliard method. It is shown that the allowance for the curvature effect in the homogeneous nucleation theory improves its agreement with experiment.

INTRODUCTION

In many technological and natural processes phase transitions proceed in a multicomponent medium. Intensification of such processes results in deviations from thermodynamic equilibrium and realization of metastable phase states. The metastable state decay begins with formation of a new-phase nucleus. In pure conditions this process is described by the homogeneous nucleation theory [1].

THEORY OF NUCLEATION IN SOLUTIONS

A critical nucleus in solution is characterized by several separated variables. In a binary solution such variables may be, for instance, the volume of a bubble, the pressure in it and the concentration of one of the components in a bubble. Then nucleation is a multidimensional diffusion process in the space of separated variables in a field of thermodynamic forces, which are prescribed by the relief of the system themodynamic potential in the vicinity of its saddle point. The movement of nuclei over the passing is edscribed by a multidimensional kinetic equation of the Kramers–Zeldovich type [2]. A stationary solution of this equation for the nucleation rate J applicable to the boiling up of binary liquid solutions in the most general case has been obtained [3] in the form

$$J = C \, |\lambda_0| \, \frac{R_*^2}{v_*} \left(\frac{kT}{\sigma} \right)^{\frac{1}{2}} \exp(-W_*/k_B T), \qquad (1)$$

where k_B is the Boltzman constant, T is the absolute temperature, W_* is the work of formation of a critical nucleus, λ_0 is the decrement of increase of an unstable variable (bubble volume), σ is the surface tension at the vapor bubble – liquid boundary, R_* and v_* are the radius and the volume of a critical nucleus, and C is the normalization constant of the equilibrium distribution function. The volume of λ_0 is determined by the dynamics of gross of new phase nuclei. For finding λ_0 in [3] a cubic algebraic equation has been obtained. In deducing this equation account was taken of the viscosity of a liquid solution, the volatility of components in a mixture, the diffusional supply of a substances to a growing bubble. In the simplest case $c \approx \rho_1' + \rho_2'$, where ρ_1', ρ_2' are the numerical densities of components in a metastable solution.

WORK OF NUCLEUS FORMATION

The determining contribution to the nucleation rate (1) is made by the exponential factor, which contains in the exponent the work of formation of a critical nucleus W_*. Traditionally the work W_* is calculated in the framework of the Gibbs method of separating surfaces [4]

$$W_* = \frac{16}{3}\pi \frac{\sigma^3}{(p_*'' - p_0')^2}, \qquad (2)$$

p_*'', p_0' are the pressures in the bubble and the liquid, respectively. Expression (2) is written as applied to the surface of tension.

As the dependence of σ on the separating surface curvature is unknown, in the nucleation theory use is traditionally made of a capillary approximation – the surface tension is considered to be independent of the separating surface curvature and equal to its value of σ_∞ at a flat interface. In this case the value of p_*'' is expressed in terms of the pressure of saturated vapors with the help of approximate relations [1].

Another way of determining W_* is the Van-der-Waals capillarity theory [5], which was first applied to the problem of nucleation by Cahn and Hilliard [6]. In this case

$$W_* = min\,max \int \left[\Delta f + \sum_{ij} k_{ij} (\nabla \rho_i)^2 \right] dV, \qquad (3)$$

$$\Delta f = f - f_0 - \sum_i (\rho_i - \rho_{i0}) \mu_{i0}. \qquad (4)$$

Here f is the Helmholtz free energy density of a homogeneous solution, μ is the chemical potential, k_{ij} is the matrix of the influence parameters.

EXPERIMENT

The method of measuring the life time of a liquid solution in a metastable state was used to study the nucleation kinetics [7]. A liquid was transferred to a metastable state by pressure rejection at fixed values of temperature and concentration, and the solution time of expectation of boiling-up (life time) τ was measured. The results of measuring from 30 to 100 values of τ were used to determine the mean life time $\bar{\tau}$ and calculate the nucleation rate $J = (\bar{\tau} V)^{-1}$, where $V \approx 68$ mm^3 is the volume of the superheated liquid. Experiment traced the temperature, the concentration and the baric dependences of the nucleation rate under its changes from 10^4 to 10^8 sec^{-1}m^{-3}.

RESULTS

The systems under investigation were: argon - krypton, nitrogen - oxygen and helium - oxygen. The first two solutions belong to a class of solutions with a complete solubility of components, the last one is a gas-filled solution. Solutions of both the classes are characterized by the identifi of T, p, x dependences of J and similarity of the temperature and the baric dependences to one-component systems. On the curves $J = J(T, x=\text{const}, p=\text{const})$ one can differentiate two sections: a section with a weak temperature dependence of J ($J<J_*$) and a section with a strong temperature dependence ($J>J_*$). The first section is connected with heterogeneous and initiated nucleation, the second one with homogeneous nucleation. In the region of heterogeneous and initiated nucleation:
- an increase in the concentration of krypton in solution results in a monotonic decrease, by approximately an order of magnitude, of the rate J_*, which may be connected with the enhanced radiation resistance of krypton with respect to argon;
- dissolution of 0.1 mole % helium in liquid oxygen reduces the solution mean life time by a factor of 3 to 4.

COMPARISON OF THEORY AND EXPERIMENT

Comparison of experimental data in the region of homogeneous nucleation with theory in a capillary approximation has shown that:
- the discrepancy between theory and experiment in the superheat temperature for argon - krypton solutions does not exceed 0.6 K. It is maximum for pure substances and minimum for solutions with concentration $x \approx 0.5$;
- experimental values of superheat temperatures are always smaller than theoretical ones;
- dissolution of 0.1 mole % helium in liquid oxygen reduces the limiting superheat $\Delta T = T_n - T_S$ by approximately 10 %.
- temperature and concentration dependences of J are close to theoretical;
- a temperature shift of 1 K results in a change of J by 6 to 12 orders of magnitude;

- at a pressure $p = 1$ Mpa and $x = 0.5$ the superheat of argon – krypton solution achieved by experiment was $\Delta T_n \approx 27.3$ K. The superheat $\Delta T_{sp} \approx 36.2$ K correspond to the solution spinodal.

The dependence of the surface tension of critical bubbles in the investigated solutions on the interface curvature has been determined in the framework of the Van-der-Waals capillarity theory. The mean results obtained in this case are:
- the surface tension of vapor – gas bubbles of pure components is a monotonically decreasing function of the surface tension curvature;
- addition of a substance to be dissolved into a liquid results in disturbance of the monotonic character of the dependence $\sigma(R)$. At large radii of curvature, considerably larger than bubble radii in the region of spontaneous nucleation, in the vicinity of $x \approx 0.5$ the function $\sigma(R)$ has a characteristic maximum;
- the value of the surface tension in the region of the $\sigma(R)$ maximum exceeds the value of σ at a flat interface by approximately 1 %.

Comparison of the homogeneous nucleation theory with allowance for the dependence of the properties of new-phase nuclei on their size and experimental data has shown that:
- allowance for the dependence $\sigma(R)$ in the work of formation of a critical nucleus improves the agreement between theory and experiment not only for pure substances, but also for solutions. In the whole range of concentrations discrepancies in the superheat temperature do not exceed 0.3 K;
- the nonmonotonic character of the dependence $\sigma(R)$ in the vicinity of $x \approx 0.5$ leads to drawing nearer the values of $\sigma(R)$ for critical bubbles and σ at a flat interface.

ACKNOWLEDGMENTS

The work was done with a financial support of the Russian Foundation of Basic Researches, project 99-02-16377.

REFERENCES

1. Skripov, V.P., *Metastable liquids*, New York: Wiley, 1974.
2. Zeldovich, Ya.B., *JETP* **12**, 525-538 (1942)
3. Baidakov, V.G., *J. Chem. Phys.* **110**, 3955-3960 (1999)
4. Gibbs, J.W., *The Collected Works, Vol.1, Thermodynamics*, New York: Longmans & Green, 1928.
5. Van der Waals, J.D., Kohnstamm, Ph, *Lehrbuch der Thermodynamik*, Leipzig und Amsterdam, 1908.
6. Cahn, J.W., Hilliard, J.E., *J. Chem. Phys.* **28**, 258 -267 (1958); **31**, 688 -699 (1959).
7. Baidakov, V.G., Kaverin, A.M., Boltachev, G.Sh., *J. Chem. Phys.* **106**, 5648-5657 (1997)

A Critique of Homogeneous Freezing Measurements of Aqueous Sulfuric Acid

Darryl J. Alofs and John L Vandike

University of Missouri-Rolla,
Department of Mechanical and Aerospace Engineering and Engineering Mechanics
and Cloud and Aerosol Sciences Lab

Abstract. Two laboratory measurements of homogeneous freezing of aqueous sulfuric acid particles are critiqued: The first by Bertram *et al*, 1996, J. Phys. Chem., vol. 100, pp.2376-2383: the second by Koop *et al*, 1998, J. Phys. Chem. A, vol.102, pp.8924-8931. Calculations for a proposed experimental artifact are inconclusive for Bertram *et al*. A proposed artifact for Koop *et al* is shown to be insignificant.

INTRODUCTION

Homogeneous freezing of aqueous ammonium sulfate droplets is a possible source of first ice in cirrus clouds (ref. 1-5). Data for ammonium sulfate are lacking so data for sulfuric acid are of interest. There is a discrepancy between laboratory measurements of homogeneous freezing of aqueous sulfuric acid. The Bertram *et al* (1996) experiments give higher freezing temperatures than those reported by Koop *et al* (1998). Bertram *et al* pass an aqueous sulfuric aerosol through a cold tube. At the downstream end of the cold tube spectroscopy is used to detect freezing and to measure acid concentration. Koop *et al* deposit drops on a quartz plate coated with silane to avoid vapor deposition. They cool the quartz plate and watch the drops under a microscope to observe freezing temperatures. When all the drops are frozen, the quartz plate is heated and the melting temperature of each drop is used to determine the composition of each drop. For each experiment we propose an experimental artifact which might explain the discrepancy and then we perform calculations for the proposed artifact.

CALCULATIONS FOR THE BERTRAM *ET AL* EXPERIMENTS

Bertram *et al* (1996) pass sulfuric acid particles through a plenum chamber kept at –40C and then into a colder flow tube about 5 cm diameter and 150 cm long (dimensions from J.J. Sloan, personal communication) with a mean flow velocity of about 1 cm/s. At the exit end of the flow tube, FTIR extinction spectroscopy is used to measure the weight percent sulfuric acid of the particles (wt %) and to detect freezing. This wt % is adjusted by varying the conditions in the aerosol generator. For a given wt %, extinction spectra are measured at a variety of flow tube temperatures. One possible experimental artifact is that if the particle composition varies with axial position (z) in the flow tube, the freezing nucleation rate (J) could be a decreasing function of z such that the composition at z = 150 cm is not the effective one.

To test for this artifact, the bulk mean temperature and water vapor pressure in the flow tube were computed as a function of z. To simplify the computation, condensation on the particles was treated as a perturbation quantity having only a small influence. Thus the Nusselt number for heat transfer as well as for vapor transfer were approximated to be the same function of Graetz number as for the case with zero condensation (Shah and London, 1978). The growth rates of the particles were computed, and J values were calculated using the theory reported by Khvorostyanov and Sassen (1998), but freezing was not allowed to occur. If this analysis shows the effect of condensation is small, and that the calculated J field decreases with z, the proposed artifact is shown to be significant. If the effect of condensation is large, the simplifying assumption is violated and the results are inconclusive.

The saturation ratio with respect to ice (SRI) is assumed to be unity at the entrance and on the wall. The entrance temperature is $-40C$, and the wall temperature is 208K. The particle radius at the inlet is 0.3 μm, and the inlet wt % is taken to be 70 %. Let N denote the number of aerosol particles per cc of gas. Fig 1 shows J (number per cc per sec) versus z for various N values. Starting from the top curve, the N values are 1.2, 1.7, 2.26, and 3 million particles/cc, respectively. These values are about an order of magnitude higher than the estimate of Bertram *et al*, which the authors say is only a crude order of magnitude estimate. The top curve in Fig. 1 has about the right mean J for 50% of the drops to freeze. The peak J for the top curve is at about z = 45 cm. At this point the SRI is also at its peak value (SRI = 1.315) and the wt % is 28.8. The effective wt % (that most of the drops freeze at) is thus 28.8%. At the exit of the chamber, the SRI is at 1.18 and the wt % is 32.1%. Since the wt % is measured at the exit, the effective wt % is thus overestimated. This error would be in the right direction to explain the deviation with Koop *et al*.

The shape of the bottom two curves is ideal in that the J is rising or flat, so that measuring wt % at the exit would give a good measure of the drop composition at freezing. But the bottom two curves have such low nucleation rates that freezing would not be observed. Because the four curves of Fig.2 are so different, The N values are too large to treat condensation as a perturbation quantity, and thus the computation is in error by unknown amounts. For values of N sufficiently low that condensation is a perturbation quantity, the analysis shows J decreasing strongly as z increases. These results are computationally accurate, but do not apply to Bertram *et al* because the N values are well below their estimates.

In summary, these calculations are inconclusive in proving the proposed artifact, but they show that condensation and evaporation have the following tendency: Near the entrance, condensation lowers the peak SRI and moves it downstream. In the downstream region, evaporation keeps the SRI value bigger than unity, in that the aerosol temperature has reached the wall temperature, but the vapor pressure has not reached the wall vapor pressure. The effect of condensation and evaporation is thus in the right direction to validate the Bertram *et al* experiments.

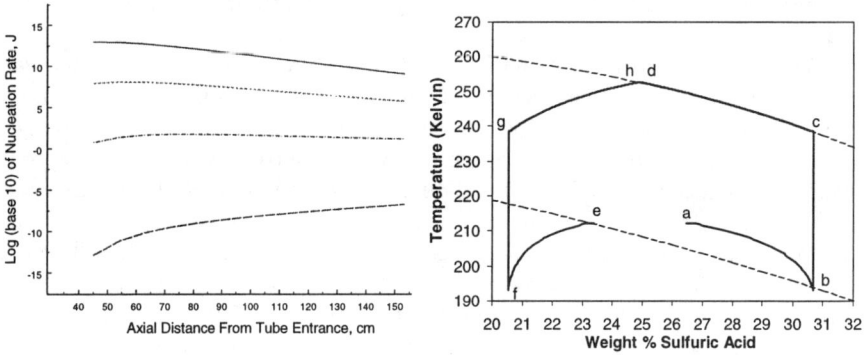

Fig.1 *Bertram et al*   Fig. 2 *Koop et al*

CALCULATIONS FOR THE KOOP *ET AL* EXPERIMENTS

Koop *et al* (1998) describe an experimental artifact which takes place for cooling rates slower than they used for their final measurements; viz., the first drops to freeze act as sinks for water vapor coming from the unfrozen drops. They show the effect of varying the cooling rate for the case of almost pure water, but it seemed worthwhile to explore this artifact for particles with higher acid content. A simple deterministic calculation is to assume that 50 % of the particles have a slightly different composition than the other 50%. Alternatively, one could assume that 50% of the drops are slightly warmer than the other 50%. We have calculated the water diffusion rates for both types of perturbations, and find that the perturbation does grow during the cooling process so that significant composition changes result by the time all drops have frozen. However, these changes reverse during the heating process, such that the final error in measured composition for the first-frozen drops is about the same magnitude as the perturbation assumed. The last-frozen drops in all cases have the same composition on melting as they do on freezing.

Fig. 2 shows the computed behavior. The top dotted curve is the melting curve by Gable *et al* (1950). The bottom dotted curve is an assumed median freezing curve (the one by Bertram *et al* is chosen). Point e represents drops with 23.37 wt % which have just frozen. Point a, at the same temperature as e, represents 26.43 wt % drops. During the cooling process (10K/min), the frozen drops move from e to f, and the unfrozen drops move from a to b, where they freeze. Then the drops are heated at the standard (reported) rate of 1 K/min. For the paths from b to c and f to g, both drop populations have the vapor pressure of ice so there is no composition change. At point c, the last-frozen drops melt. Thereafter, they move from c to d. The calculated diffusion rate is such that the melted drops follow the melting curve in composition, thereby supplying vapor to the frozen drops, which follow the path from g to h, where they melt, and where their composition equals that for the other population of drops.

Fig. 2 is for 7.8 μm radius particles at point a and at point e. The composition at d and h is 24.8%. The reciprocal of the wt % at d and h is the average of the reciprocals of the weight percents at a and e. When a heating rate of 100 times the standard heating rate is assumed, the path from c to d at first travels into the region above the melting curve, but the points d and h still have the same composition.

The particle size for Fig. 2 is near the middle of the range stated in the paper. For smaller sizes, the changes in composition during cooling (difference between a and b or e and f) are larger, but the points d and h still have the same composition. The large change in composition (between a and b or e and f) explain why Koop *et al* found it necessary to do drop-by-drop measurement of composition and freezing temperature.

In conclusion, the final error is about a quarter of the assumed perturbation. The perturbation is divided in half twice: first because for half the population the freezing composition is measured correctly; second because for the other population the freezing composition is overestimated by half the assumed perturbation. This error is not large enough, and is in the wrong direction, to explain the deviation with the Bertram *et al* experiments.

ACKNOWLEDGMENTS

This work was support by the Division of Atmospheric Sciences, National Science Foundation under Grant SGER ATM 9528465.

REFERENCES

1. Sassen, K., and Dodd, G. C., *J. Atmos. Sci.* **46**, 3005 (1989).
2. Heymsfield, A. J., and Sabin, R. M., *J. Atmos. Sci.*, **46**, 2252 (1989).
3. Jenson, E. J., and Toon, O. B., *Geophys. Res. Ltrs.* **21**, 2019 (1994).
4. DeMott, P. J., Meyers, M. P., and Cotton, W. R., *J. Atmos. Sci.* **51**, 77 (1994).
5. Jensen, E. J., Toon, O. B., Westphal, D. L., Kinne, S., and Heymsfield, A. J., *J. Geophs. Res.*, **99**, 10443 (1994).
6. Bertram, A. K., Patterson, D. D., and Sloan, J. J., *J. Phys. Chem.*, **100**, 2376-2383 (1996).
7. Koop, T., Ng, H.P., Molina, L.T., and Molina, M. J. *J. Phys. Chem. A.*, **102**, 8924-8931 (1998).
8. Khvorostyanov, V., and Sassen, K.,*Geophys. Res. Ltrs.*, **25**, 3155-3158 (1998).
9. Shah, R. K., and London, A. L., *Laminar Flow Convection in Ducts*, New York, Academic Press, 1978, pp.99-109.
10. Gable, C. M., Betz, H. F., and Maron, S. H., *J. Am. Chem. Soc.*,**72**, 1445-1448 (1950).

Efflorescence and Ice Nucleation in Ammonium Sulfate Particles: Analysis of Experimental Results Using Scaled Nucleation Theory

T. B. Onasch[1], R. McGraw[1], A. J. Prenni[2], M. A. Tolbert[2] and D. Imre[1]

[1]*Brookhaven National Laboratory, Atmospheric Sciences Division, Environmental Sciences Department, Upton, NY 11973*
[2]*University of Colorado, Department of Chemistry and Biochemistry and CIRES, Boulder, CO 80309*

Abstract. Temperature-dependent efflorescence and ice nucleation results of ammonium sulfate aerosol obtained from flow tube studies are analyzed using a newly developed model based on scaled nucleation theory. The presented thermodynamic model defines trajectories of constant reduced nucleation barrier height as a function of temperature and yields simple analytic expressions for the liquid-solid nucleation boundaries. Within the context of the model the experimental results are interpreted in terms of surface enthalpy, entropy and nucleus size. These thermodynamic properties, along with a full thermodynamic model, can be then used to predict changes in the composition and phase of ammonium sulfate particles over the wide range of relative humidity and temperature conditions found in the atmosphere.

INTRODUCTION

In this paper we describe a thermodynamic model of homogeneous liquid-solid nucleation based on the Gibbs dividing surface model of a nucleus and apply it to experimental ammonium sulfate aerosol data. This model relies on the use of scaling forms for such physical quantities as surface tension, density, and temperature to nondimensionalize the nucleation barrier height [1, 2, 3]. The model yields compact analytical expressions for trajectories of constant reduced barrier height, and therefore, because the effect of the kinetic prefactor is less, essentially constant nucleation rate [2]. Some advantages of this approach are that the scaled quantities are fewer in number and tend to show considerably less variability than, for example, the cube of the surface tension that appears in the barrier height in the classical nucleation theory.

Ammonium sulfate aerosol was selected for study because it is a major component of the atmospheric aerosol and its phase transformation properties are of considerable importance to atmospheric chemistry and climate. The equilibrium and metastability phase diagrams have been rigorously investigated [4, 5, 6, 7] and thermodynamic models developed [8]. Thus the ammonium sulfate aerosol system provides an excellent model for the development and testing of new approaches to liquid-solid nucleation.

FLOW TUBE EXPERIMENTS

Deliquescence, efflorescence and ice nucleation experiments have been carried out on small (submicrometer) ammonium sulfate aerosol in a temperature-controlled flow tube system [4, 5]. Large supercoolings are readily and reproducibly achieved in these freely floating aerosol droplets. Reproducibility of a well defined supercooling limit, together with the absence of heterogeneous nucleation sites, suggests that the limit to supercooling is determined by homogeneous nucleation. The temperature-controlled flow tube system has been described elsewhere [4, 5]. An FTIR spectrometer is used for estimating the size distribution and the phase transitions of the ammonium sulfate particles. The relative humidity in the flow tube was controlled by conditioning the aerosol flow using sulfuric acid solutions and ice coated walls. The relative humidity was directly measured by FTIR and by a dew point hygrometer for the efflorescence study, and calculated from the temperature of the ice-coated walls for the ice nucleation experiments. The experiments yielded accurate determinations of the deliquescence and efflorescence points, and ice nucleation thresholds, as a functions of aerosol temperature and relative humidity (RH).

THERMODYNAMICS

Following the standard thermodynamic treatment of binary solutions [9] we obtain the following equation relating the vapor pressure of water $p1$, the mole fractions of water and salt in solution (x_1 and x_2, respectively), and the driving free energy for salt crystallization ($\Delta\mu_2 = kT\ln S_2$, where S_2 is the saturation ratio for salt in solution relative to the crystalline phase):

$$\frac{d\ln p_1}{dT} + \frac{x_2}{x_1}\frac{d(\Delta\mu_2/kT)}{dT} = -\frac{\Delta H_v}{kT^2} - \frac{x_2}{x_1}\frac{\Delta H_s}{kT^2}. \quad (1)$$

For pure water, $x_2=0$ and Eq.1 reduces to the familiar Clausius-Clapeyron relation where $\Delta H v$ is the heat of vaporization. For a saturated salt solution, $\Delta\mu_2=0$ and Eq. 1 describes the vapor pressure along the (equilibrium) deliquescence line where ΔH_s is the integral heat of solution [7].

In terms of the driving free energy for ice crystallization ($\Delta\mu_1 = kT\ln S_1$, where S_1 is the saturation ratio of water vapor relative to ice), we obtain:

$$\frac{d\ln p_1}{dT} + \frac{d(\Delta\mu_1/kT)}{dT} = -\frac{\Delta H_v}{kT^2} - \frac{\Delta H_{melt}}{kT^2}, \quad (2)$$

where ΔH_{melt} is the heat of melting. For ice in equilibrium with water vapor, $\Delta\mu_1=0$, and Eq. 2 relates the equilibrium water vapor pressure over ice to the heat of sublimation.

Figure 1 shows $\ln p_1$ versus $1/T$ curves for pure water and ice [8], and experimental data for deliquescence and efflorescence of salt, and ice nucleation. Lines through the data points are linear fits. The two upper curves on the left side of the figure correspond to the two special cases of Eq.1 described above. The dashed curve corresponds to the special ice sublimation case of Eq. 2. The lower (efflorescence) curve on the left also corresponds to a special case of Eq. 1 - that for which $\Delta\mu_2$ assumes values marking the salt nucleation threshold. In similar fashion, the ice nucleation curve (upper right curve) corresponds to a special case of Eq. 2 -that for which $\Delta\mu_1$ assumes values marking the ice nucleation threshold. In the following section we describe trajectories for $\Delta\mu_1$ and $\Delta\mu_2$ at the homogeneous nucleation thresholds for ice nucleation and salt nucleation, respectively. We obtain these trajectories solely in terms of the molecular properties of the critical nucleus using the Gibbs dividing surface model and scaled nucleation theory of Hale [3].

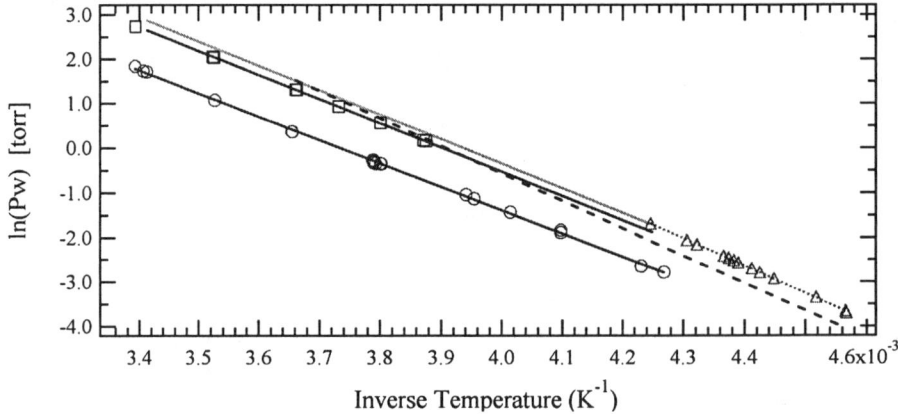

FIGURE 1. Plot of the equilibrium and metastable phase transitions for ammonium sulfate. The solid line is the vapor pressure of water over pure liquid water and the dashed line is the vapor pressure of water over ice, both derived from Clegg et al. [8]. The squares are the deliquescence data, the circles are the efflorescence transition [4] and the triangles are the ice nucleation data [5].

SCALED NUCLEATION THEORY

Throughout this section $\Delta\mu$ and S will refer to either component and γ_s is the interfacial free energy, per unit area, between the nucleus (either salt or ice) and the parent solution. We begin with the following trajectory equation for $\Delta\mu$, which defines a path of fixed nucleation barrier height (or approximately constant nucleation rate):

$$\frac{d\Delta\mu}{dT} = \frac{3}{2}\Delta\mu\left(\frac{1}{\gamma_s}\frac{d\gamma_s}{dT}\right) - \frac{\Delta\mu}{2T}. \qquad (3)$$

Eq. 3 assumes that the nucleus is incompressible and has a temperature-independent density. Transforming to the $\ln S$ coordinate gives:

$$\frac{d(\Delta\mu/kT)}{dT} = \frac{d\ln S}{dT} = -\frac{1}{kT^2}\frac{A_s}{n_f^*}\left(\gamma_s - T\frac{d\gamma_s}{dT}\right) = -\frac{1}{kT^2}\frac{\Delta H_{surface}}{n_f^*} \quad (4)$$

where $\Delta H_{surface}$ is the total surface enthalpy and n_f^* is the number of molecules within the critical nucleus boundary defined by the surface of tension (surface area A_s). Eqs. 3 and 4 are not new and can be found in papers of Ford [10] and earlier, in very similar form, in the scaled nucleation theory of Hale [3]. Their derivation using the Gibbs-dividing surface model, is described in this proceedings [1].

Substitution of Eq. 4 into Eqs. 1 and 2 yields thermodynamic conditions along a phase-space trajectory of constant nucleation rate. Note that the right hand side of Eq. 4 contains the total surface enthalpy divided by the number of molecules in the nucleus. The closeness in slopes of the linear fits in Fig. 1 suggest that this quantity is small compared to the enthalpies per molecule of vaporization and sublimation.

Eq. 3 is a remarkably simple and elegant expression for the nucleation threshold conditions. It implies that an entire nucleation threshold trajectory can be generated using only bulk solution properties and an estimate for $\gamma_s^{-1} d\gamma_s/dT$. The latter quantity is the relative slope of the Eötvös line, describing the empirical linear relation between surface tension and temperature [11]. It is clearly much less sensitive to uncertainty in γ_s then is the nucleation rate itself, or even the exponent of the nucleation rate.

As an example, the surface tension between an ice nucleus and the ammonium sulfate solution has been estimated using Antonoff's rule. The derived per-unit-area surface entropy and surface enthalpy are found to be independent of temperature, in agreement with Eötvös rule [11]. Nucleation trajectories based on the derived surface tension are compared with the experimentally determined ice nucleation data.

Acknowledgments. This work was supported by the U.S. Department of Energy (Contract DE-AC02-98CH10886). TBO acknowledges a DOE Alexander Hollaender Distinguished Postdoctoral Fellow.

REFERENCES

1. R. McGraw, Scaling properties of critical nuclei and nucleation rates, 15th International Conference on Nucleation and Atmospheric Aerosols, University of Missouri - Rolla (2000).
2. R. McGraw, Journal of Chemical Physics 75, 5514 (1981).
3. B. N. Hale, Lecture Notes in Physics, P. E. Wagner, G. Vali, Eds., 12th International Conference on Nucleation and Atmospheric Aerosols, University of Vienna - Austria (1988).
4. T. B. Onasch, et al., Journal of Geophysical Research 104, 21,317-21,326 (1999).
5. A. J. Prenni, M. E. Wise, S. D. Brooks, M. A. Tolbert, Submitted to Journal of Geophysical Research (2000).
6. J. Xu, D. Imre, R. McGraw, I. Tang, The Journal of Physical Chemistry B. 102, 7462-7469 (1998).
7. I. N. Tang, H. R. Munkelwitz, Atmospheric Environment 27A, 467-483 (1993).
8. S. L. Clegg, P. Brimblecombe, A. S. Wexler, Journal of Physical Chemistry 102A, 2137-2154 (1998).
9. E. A. Guggenheim, in Thermodynamics . (North-Holland, Amsterdam, 1959) pp. 296.
10. J. Ford, Journal of Chemical Physics 105, 8324-8332 (1996).
11. A. B. Pippard, in Elements of classical thermodynamics for advanced students of physics . (Cambridge University Press, Cambridge, 1957) pp. 86.

A Universal Tool to Investigate Nucleation from Aqueous Solutions

Klaus Tauer and Klaus Padtberg

Max Planck Institute of Colloids and Interfaces
Am Mühlenberg, D-14476 Golm, Germany

Abstract. It has turned out that on-line monitoring of conductivity and turbidity is a universal tool to study nucleation from aqueous solutions. In this paper, we describe an experimental setup allowing to determine the onset of nucleation from aqueous phases. Examples are presented for emulsion polymerization of different monomers and precipitation reactions of different salts.

INTRODUCTION

It is a matter of fact that the preparation and characterization of both organic and inorganic nanoparticles or nanodroplets which can be either polymeric or monomeric in nature have become a field of great activities. The basic theoretical ideas of nucleation and growth from supersaturated solutions can be applied for precipitation or crystallization processes as well as polymerization reactions, for instance emulsion polymerization [1]. From the experimental point of view, the exact determination of the onset of nucleation is a sophisticated problem. Unfortunately, optical techniques which can be easily applied are no absolute methods. Note, the onset of nucleation means the time when during an unseeded emulsion polymerization the first crop of particles is formed. The maximum or final number of particles is governed by the stabilizing ability of the particular emulsion polymerization system and has to do only little with the primary nucleation event. For emulsion polymerization of styrene we were able to show that the combination of on-line conductivity and on-line turbidity measurements allows the determination of the onset of nucleation and moreover the on-line calculation of the development of particle number and size [1, 2]. A schematic drawing of the experimental setup shows Figure 1. In this contribution we report the application of this experimental setup to study firstly, the nucleation in emulsion (homo)polymerizations of styrene (STY), methyl methacrylate (MMA), and vinyl acetate (VAC) and secondly, the precipitation of barium sulfate, calcium oxalate, and calcium carbonate. During emulsion polymerization the nucleating species are formed *in-situ* by aqueous phase polymerization started with peroxodisulfate radicals. Thus, these systems are reactive or dynamic with regard to the chemical composition. The water soluble oligomers begin to nucleate when a critical supersaturation is reached. To investigate the precipitation of inorganic salts under comparable dynamic conditions corresponding complementary water soluble precursor salts were continuously fed into the reactor. Note, this is a different approach than the frequently used conductivity measurements to determine the inhibition period in the case of supersaturated solutions without any feed streams [3, 4].

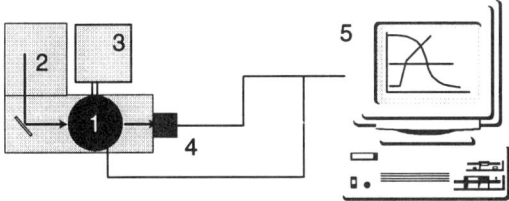

FIGURE 1. Schematic drawing of the experimental setup; 1 – reactor (500 ml volume) with stirrer, conductivity and temperature probe, 2 – spectrophotometer (365 nm), 3 – syringe pump with two feed streams, 4 – photomultiplier, 5 – data acquisition and monitoring

EXPERIMENTAL PART

All chemicals were from Sigma-Aldrich and used as received except for the monomers. Deionized water was taken from a Seral purification system (PURELAB Plus™) with a conductivity of 0,06 µS cm^{-1} and degassed prior to use. The water was filled into the reactor with a higher temperature than reaction temperature. The monomers were distilled prior to use under reduced pressure and stored in a refrigerator. Only oligomer free monomers were used. For the polymerization 390 g of water and 5 g of monomer were allowed to equilibrate in the reactor at 70 °C. The polymerization was started by injecting 10 ml of a 4,15 mM (MMA, VAC) or 20,75 mM (STY) potassium peroxodisulfate (KPS) solution. The precipitation reactions were carried out at 35 °C by feeding (syringe pump model 22, Havard Apparatus) equimolar solutions of 0,5 M $CaCl_2$ and Na_2CO_3, 0,5 M $Ba(CH_3COO)_2$ and $(NH_4)_2SO_4$, 0,1 M $CaCl_2$ and NaC_2O_4 for calcium carbonate, barium sulfate, and calcium oxalate precipitation, respectively. The feed rate was 0,5 ml h^{-1} ($BaSO_4$), 1 ml h^{-1} ($CaCO_3$), and 2 ml h^{-1} (CaC_2O_4). For both the polymerization and the precipitation transmission (T) (Spekol, Carl Zeiss Jena), conductivity (κ) (CDM 92, Radiometer), and temperature were monitored online [2].

RESULTS

Emulsion Polymerization

It is assumed that the monomers investigated behave quite different in emulsion polymerization [5, 6] as their water solubilities differ about three orders of magnitude (STY<<MMA<VAC) [7, 8]. However, in the nucleation experiments all monomers apart from minor differences behave similar thus indicating a generally valid behavior. The common feature is a sharp bend in the conductivity curves from one second to the other (cf. Figure 2A) which is directly related to the onset of nucleation. The conductivity under the particular conditions is governed by protons which are generated during the decomposition of peroxodisulfate [9] and as soon as particles are

formed protons will be either specifically adsorbed or incorporated in the electrical bilayer and loose mobility. In the consequence, the increase in conductivity becomes lower. The κ-values in Fig. 2A are at the one hand caused by different KPS concentrations (for STY five-fold higher than for MMA and VAC) and on the other hand by hydrolysis of VAC and MMA. During the equilibration κ increases for VAC much stronger than for MMA due to hydrolysis but remains constant for STY. In contrast to κ, the transmission curves look at the first glance quite different. However, this has nothing to do with the primary nucleation step but with the development of the particle number and average particle size in the course of the polymerization.

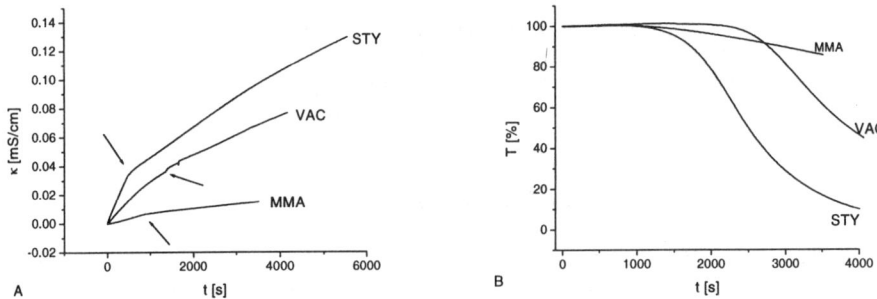

FIGURE 2. Conductivity (A) and transmission (B) time curves during emulsion polymerization of STY, MMA, and VAC; the arrows indicating the bend and the onset of particle nucleation

Precipitation Reactions

The results summarized in Figure 3A show that the experimental setup is suited to distinguish between different salts. The curves represent average values from at least five repetitions. In contrast to emulsion polymerization no sharp bend of the conductivity is observed but a rather smooth transition. This transition region is of special interest as an analysis of the data leads to some quantitative conclusions regarding the precipitation kinetics. Besides the precipitation curve (κ) for $CaCO_3$ Figure 3B shows also the conductivities of the starting compounds ($κ_{st}$) and the remaining NaCl ($κ_{sa}$) measured in separate experiments. The κ curve shows 4 characteristic points with the whole precipitation period between t_1 and t_4. Note, that at $t < t_1$ κ and $κ_{st}$ coincide and that at $t > t_4$ the slope of the κ curve is clearly even if only a little higher than that of the $κ_{sa}$ curve. The latter indicates that not only NaCl contributes in that region to κ but also some $CaCO_3$ dissolved in water. At t_1, t_2, t_3, and t_4 the amounts of $CaCO_3$ fed into the reactor are $1,426 \cdot 10^{-7}$, $6,788 \cdot 10^{-7}$, $9,513 \cdot 10^{-7}$, and $1,192 \cdot 10^{-6}$ mol ml^{-1}, respectively. Disparity exists in the published data concerning the solubility of $CaCO_3$ in water and hence it is only possible to say that the concentration at t_1 inside the reactor is still lower than the lowest solubility ($1,59 \cdot 10^{-7}$ mol ml^{-1} at 35 °C) value published [10]. A further analysis of the data is possible if the conductivity $κ_{sa}$ is subtracted from both κ and $κ_{st}$. The difference $κ_{st} - κ_{sa}$ and $κ - κ_{sa}$ correspond to the theoretical conductivity of a non-precipitating $CaCO_3$ solution ($κ_{th}$) and to the real conductivity change during $CaCO_3$ precipitation ($κ_{pr}$). If we assume that (1) clusters or aggregates formed during the precipitation do not contribute to the conductivity, which is on the basis of their larger size very reasonable, and (2) the equivalent conductivities of the ions remain unchanged $d(\kappa_{th} - \kappa_{pr})/dt$ corresponds to an apparent cluster formation rate. Cluster formation begins at the time of t_1 with a low apparent rate.

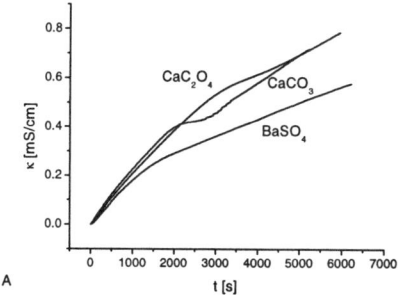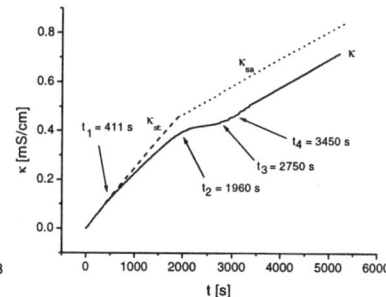

FIGURE 3. Conductivity change during precipitation reactions of BaSO4, CaCO3, and CaC2O4 (A) and a more detailed contemplation of CaCO3 precipitation (B)

Between t_2 and t_3 the apparent rate steadily increases and reaches a maximum between t_3 and t_4 at 2447 seconds. The apparent rate reaches an almost constant value at $t > t_4$ corresponding to an equilibrium situation. Furthermore, the data analysis revealed that during this period the concentration of dissolved $CaCO_3$ increases proportional to the amount of precipitate formed. However, the latter increases stronger than the dissolved carbonate in dependence on the total amount fed into the reactor.

ACKNOWLEDGMENT

The authors gratefully acknowledge the Max Planck Society for financial support.

REFERENCES

1. Tauer, K., and Kühn, I. "Particle nucleation at the beginning of emulsion polymerization" in *Polymeric Dispersions: Principles and Applications*, edited by J. M. Asua, Kluwer, Dordrecht, The Netherlands, 1997, pp. 49-65.
2. Kühn, I., and Tauer, K., *Macromolecules* **28**, 8122-8128 (1995)
3. Aoun, M., Plasari, E., David, R., and Villermaux, J., Chem. Engng. Sci. 54, 1161-1180 (1999)
4. Söhnel, O., and Garside, J., *Precipitation: Basic principles and industrial applications*, Butterworth-Heinemann, Oxford, 1992, pp. 72-73.
5. Gilbert, R. G., *Emulsion Polymerization A Mechanistic Approach*, Academic Press, London, 1995, pp. 292-342.
6. Fitch, R. M., *Polymer Colloids: A Comprehensive Introduction*, Academic Press, San Diego, 1997, pp. 6-47.
7. Gardon, J. L., *J. Polym. Sci.* **6**, 2859-2879 (1968)
8. Vijayendran, B. R., *J. Appl. Polym. Sci.* **23**, 733-742 (1979)
9. Tauer, K., Deckwer, R., Kühn, I., and Schellenberg, C., *Coll. Polym. Sci.* **277**, 607-626 (1999)
10. D'Ans Lax *Taschenbuch für Chemiker und Physiker*, 4th Edition, edited by R. Blachnik, Springer, Heidelberg, 1998, pp.1210 – 1229.

ICE NUCLEATION

Nucleation of Pure and AgI Seeded Supercooled Water using an Automated Lag Time Apparatus

Aaron Heneghan and A.D.J. Haymet

Department of Chemistry, University of Houston, Houston TX 77204 USA

Abstract. The evolution of non-equilibrium systems into equilibrium has been studied for many years. However, there is still a need for a unified theory which describes this evolution in time. It is known that all systems reach equilibrium eventually, but difficulties arise in determining how long eventually is. The Automatic Lag Time Apparatus(ALTA) was constructed to collect statistics of nucleation on a single sample. The ergodic hypothesis states that the time average is equal to the ensemble average. Hence, running one sample 500 times is equivalent to trying to prepare and monitor 500 identical samples. ALTA repeatedly supercools a sample to a set supercooling point, measures the time until nucleation, heats the sample for four minutes to ensure all of the ice crystals have melted, and repeats the above steps as many times as possible in order to gain statistics of nucleation. By analyzing these statistics of nucleation, various theories of nucleation can be tested.

INTRODUCTION

Nucleation may be defined as the first irreversible formation of a nucleus that leads to an equilibrium phase. In the case considered here, a nucleus of the solid phase would be formed within the supercooled fluid leading to the formation of the equilibrium solid phase. However, difficulties arise in trying to explain the underlying rates and mechanisms involved in the nucleation process. In this work, we study the repeated nucleation of samples of supercooled solutions.

In "classical" nucleation theory [1-5], clusters of the equilibrium phase are treated as large spherical clusters which have contributions to the free energy from separate volume and surface terms. For liquid to solid nucleation, Talanquer and Oxtoby [6] reaffirmed recently that the assumptions of the "classical" theory may not be true. Density Functional (DF) theory [7-10] yields two important results. The first result is that the nucleus which forms in nucleation may be only a few to a few hundred molecular diameters, and hence a decomposition into terms proportional to surface area and volume may not be possible. Secondly, the "classical" nucleation theory assumes the existence of a sharp interface between the newly formed nucleus and the surrounding supercooled phase, and uses a single surface free energy for a "flat" interface. The same papers show that the interface may be diffuse over several molecular diameters, again contradicting the "classical" description. Density Functional theory is able to add such non-classical effects as curvature dependant surface tensions and finite interfacial widths [11, 12].

Our goal is to provide experimental data which may test candidate theories of nucleation. Shown below in Figure 1 is a plot of two modes in which ALTA may be run. The difference between the two modes will be explained below.

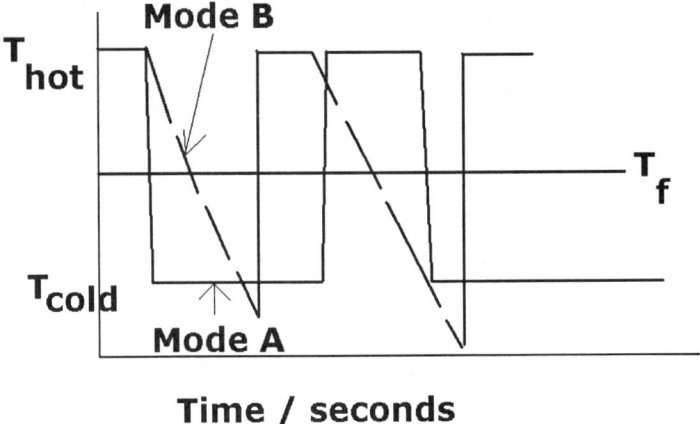

Figure 1 Temperature as a function of time for the two modes of operation

Supercooling at a Fixed Temperature (Mode A)

Currently, ALTA is running in Mode A, defined in Figure 1. This mode repeatedly cools a single sample to a fixed supercooling point and holds the sample at this temperature until a freeze occurs. For a detailed reference of the apparatus see reference [13]. The amount of time between reaching the set supercooling point and freezing is known as the lag time, τ, also called the induction time. Using a single sample, this cycle is performed as many times as necessary, N_0, to generate unambiguous statistics for nucleation. Simulations show that $N_0 \approx 500$ is desirable. From the distribution of lag times as a function of Run Number, one can calculate the fraction unfrozen at each instant in time, $N(t) / N_0$. Upon obtaining the fraction unfrozen as a function of time, it is possible to plot this fraction unfrozen as a function of time. It is this fraction unfrozen plot which serves as the central tool for analysis of these data. By introducing an AgI crystal, it is possible to make the supercooling temperature higher [13-15]. That is, the AgI crystal provides sites for nucleation to occur. Shown below in Figure 2a is a sample of the fraction unfrozen as a function of time for a sample of 200μL of distilled water with no AgI crystal. An alternate analysis of this kind of data, which assumes classical theory, is given by Hale [1].

Water both unseeded and seeded with an AgI crystal will be discussed. Ultimately, this experiment yields the average lag time, $<\tau>$, as a function of the amount of supercooling, ΔT. However, slight changes in the supercooling temperature may produce large changes in the average lag time. Hence, experiments which collect 500 runs on a single sample may take weeks to get one point on this curve. We address this problem below.

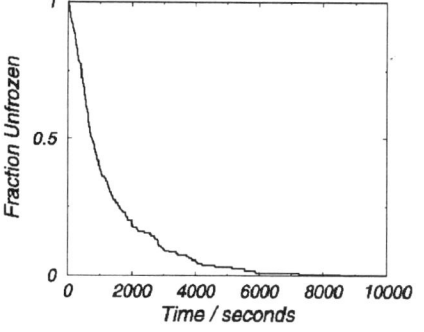 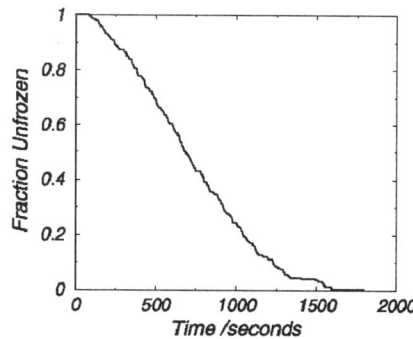

Figure 2. (a) The fraction unfrozen as a function of time for 175 runs on a single sample. (b) Fraction Unfrozen in Linear Cooling Mode B from computer simulation with $N_0=500$ runs.

Simulations and Linear Supercooling (Mode B)

A new mode, linear supercooling (Mode B), has been developed through the aid of computer simulations of kinetic equations for supercooled systems. In preparing the simulation, two key assumptions are made. The first assumption is that the system of interest undergoes first order kinetics, namely the fraction unfrozen $N(t) / N_0$ yields a single exponential decay of form $\exp(-t/\tau)$ at a fixed supercooling temperature. Secondly, a special functional form is assumed for the lag time, τ, as a function of supercooling temperature, $\tau(\Delta T)$. For simplicity, here we assume $\tau(\Delta T) \sim 1/\Delta T$. In the linear cooling mode, shown above in Figure 1, $\Delta T \sim t$, and hence $\tau(T(t)) = K / t$, where K is a constant. Computer simulations on the resultant kinetic equations yield the fraction unfrozen as a function of time. The number of runs can be made as large as necessary to obtain a good statistical distribution.

A sample simulation for $N_0=500$ is shown in Figure 2(b). With the above assumptions, fraction unfrozen can be shown to be a Gaussian distribution, which is well resolved in the simulation.

In the future, the apparatus will be in this mode and the fraction unfrozen measured, thus testing the assumed functional form of $\tau(\Delta T)$. This mode will provide data much more quickly, since the fraction of the statistically correct long run at fixed supercooling are naturally truncated by the ever-decreasing temperature. A set of 500 runs may be collected in a day or so, instead of the weeks it may take in mode A. If a candidate functional form is found, the apparatus can be run in mode A (above) to test this functional form $\tau(\Delta T)$. Simulations show that the measured fraction unfrozen is sensitive to the actual functional form $\tau(\Delta T)$ and the kinetic laws. For example, a given system may exhibit second order kinetics with a cubic dependence of lag time on supercooling temperature. By examining various combinations, the relationships between lag time, temperature and time may be deduced.

ACKNOWLEDGMENTS

This research is support in part by ACS-PRF Grant AC 33707-AC9 and the UH Environmental Institute, to whom grateful acknowledgement is made.

REFERENCES

1. Hale, B.N., *Scaled Models of nucleation*, in *Atmospheric Aerosols and Nucleation*, P.E. Wagner and G. Vali, Editors. 1988, Spriner-Verlag: Berlin. p. 323-349.
2. Shneidman, V.A., K.A. Jackson, and K.M. Beatty, *On the applicability of the classical nucleation theory in an Ising system*. Journal of Chemical Physics, 1999. **111**(15): p. 6932-6941.
3. Sen, S. and T. Mukerji, *A generalized classical nucleation theory for rough interfaces: application in the analysis of homogeneous nucleation in silicate liquids*. Journal of Non Crystalline Solids, 1999. **246**(3): p. 229-239.
4. McGraw, R. and R.A. Laviolette, *Fluctuations, temperature, and detailed balance in classical nucleation theory*. Journal of Chemical Physics, 1995. **102**(22): p. 8983-8994.
5. Laaksonen, A., M. Kulmala, and P.E. Wagner, *On the cluster compositions in the classical binary nucleation theory*. Journal of Chemical Physics, 1993. **99**(9): p. 6832-6835.
6. Talanquer, V. and D.W. Oxtoby, *Density Functional Analysis of Phenomenological Theories of Gas-Liquid Nucleation*. Journal of Physical Chemistry, 1995. **99**(9): p. 2865-2874.
7. Oxtoby, D.W., *Nucleation of First-Order Phase Transitions*. Accounts of Chemical Research, 1998. **31**(2): p. 91-97.
8. Haymet, A.D.J., *Theory of the Equilibrium Liquid-Solid Transition*. Annual Review of Physical Chemistry, 1987. **38**: p. 89-108.
9. Laaksonen, A., V. Talanquer, and D.W. Oxtoby, *Nucleation - Measurements, Theory, and Atmospheric Applications*. Annual Review of Physical Chemistry, 1995. **46**: p. 489-524.
10. Shen, Y.C. and D.W. Oxtoby, *Density Functional Theory of Crystal Growth - Lennard-Jones Fluids*. Journal of Chemical Physics, 1996. **104**(11): p. 4233-4242.
11. Oxtoby, D.W. and Y.C. Shen, *Density Functional Approaches to the Dynamics of Phase Transitions*. Journal of Physics-Condensed Matter, 1996. **8**(47): p. 9657-9661.
12. Shen, Y.C. and D.W. Oxtoby, *Nucleation of Lennard-Jones Fluids - a Density Functional Approach*. Journal of Chemical Physics, 1996. **105**(15): p. 6517-6524.
13. Barlow, T.W. and A.D.J. Haymet, *ALTA: an automated lag-time apparatus for studying the nucleation of supercooled liquids*. Rev. Sci. Instruments, 1995. **66**: p. 2996-3007.
14. Chen, S.L. and C.L. Chen, *Effect of nucleation agents on the freezing probability of supercooled water inside capsules*. High Vacuum Research, 1999. **5**(4): p. 339-351.
15. Smorodin, V.Y., *Endothermal wetting effect and ice nucleation mechanisms of silver iodide*. Journal of Aerosol Science, 1994. **25**(1).

Measurements of Ice Nuclei at High Supersaturations

David C. Rogers, Sonia M. Kreidenweis, Paul J. DeMott

Department of Atmospheric Science, Colorado State University, Fort Collins. CO USA

Abstract. Airborne measurements in clouds at −10 to −30°C often show concentrations of ice crystals much greater than would be predicted from ice nuclei (IN) measurements made at or below 2% supersaturation with respect to liquid water (SSw). There may be small regions of such clouds where cloud droplets and CCN are removed by scavenging and precipitation processes, and large SSw (>10%) can develop. It has been speculated that in such regions, high concentrations of IN may activate, thereby explaining the high ice concentration. We explored this idea by obtaining a few measurements of natural IN with a continuous flow thermal gradient diffusion (CFD) chamber at high SSw, up to 40%. However, competition for vapor by activated and growing particles limit the supersaturation that can be achieved in this instrument. To examine this limit, numerical simulations and laboratory experiments were performed.

INTRODUCTION

An extensive body of measurements shows that atmospheric concentrations of ice nucleating aerosol particles (IN) increase rapidly with water vapor supersaturation [1-5]. The usual range of humidity for IN measurements is from ice saturation to a few percent water supersaturation (SSw). Instrumented aircraft studies show ice crystal concentrations (ICC) in natural clouds are often much larger than would be predicted from IN measurements from about -10 to -30°C [6,7]. It has been hypothesized [8] that high ICC sometimes occur in clouds where conditions favor the localized production of SSw much higher than the 1 or 2% typically associated with clouds. These two factors, (1) IN increasing rapidly with supersaturation and (2) observations of high numbers of ice crystals, led to speculations that large numbers of ice crystals may result from IN activating in small regions of high SSw. The mechanisms leading to high SSw have been studied through microphysical modeling simulations [9-13]. They suggest that high SSw may be produced in regions where processes that concentrate or remove droplets and newly activated CCN (coalescence, accretion and turbulence) continue over relatively long time periods (~30 min). We explored part of this question by attempting to measure natural IN at high supersaturations.

INSTRUMENTATION

A continuous flow thermal gradient diffusion (CFD) chamber was used for these measurements [14]. The chamber is the annular space between two vertically oriented concentric cylinders that are coated with ice and held at different temperatures, thus creating a vapor supersaturation between them. The sample air is ~10% of the total flow and is sandwiched between two particle-free sheath flows. Ice crystals nucleate and grow to diameters of ~3 to 10 µm diameter in ~5 s residence time. Upstream of the chamber, all particles in the sample air larger than 2 µm diameter are removed with inertial impactors. At the outlet of the chamber, an optical particle counter detects all particles larger than ~0.8 µm. Those particles larger than 3 µm are assumed to be the newly formed ice crystals and comprise the ice nucleus count. This technique detects ice nuclei that activate through deposition or condensation-freezing mechanisms. Nucleation mechanisms that require times longer than ~1 s are not detected (contact-freezing and immersion-freezing).

Two factors introduce errors in this technique for measuring IN at high supersaturations: (1) the reduction of supersaturation due to vapor sinks and (2) erroneously identifying as ice crystals those droplets that grow >3 µm. The sample SSw in the CFD chamber is calculated from wall temperatures and flow rates. In the absence of vapor sinks, SSw can reach 60% and higher. Droplets and ice crystals are vapor sinks and limit the attainable supersaturation. To evaluate the first error, numerical simulations were performed using equations from [15,16]. The calculations assumed up to 1000 cm^{-3} of CCN particles, -20°C, ambient pressure (850 mb), ice particle numbers as predicted by [17], and SSw of 5, 10, 20, 30, 40, 50 and 60%. The results indicated that the concentrations of ice crystals were too low to affect SSw significantly. The effect of droplets was estimated by comparing SSw with and without droplets present. The effect was slight at low SSw (e.g., reducing SSw from 5.00% to 4.97%) and increased with SSw (40% without drops decreased to 36% with drops). We concluded that vapor competition would not significantly affect IN measurements for SSw <~40% at -20°C and 850 mb.

To evaluate the second error (droplets growing >3 µm), droplet sizes in the simulations were analyzed, and laboratory experiments were performed. The simulations indicated that droplets could reach >3 µm for SSw ≥18%. Conceptually, ice crystals would reach sizes larger than droplets, since ice supersaturation is greater than water supersaturation. However, differentiating crystals from droplets solely on the basis of size becomes more difficult at higher SSw where the size difference is less pronounced. Unknowns in the timing of ice nucleation and growth rates at small sizes make this size differentiation difficult to calculate with certainty.

The laboratory tests used 0.2 µm diameter ammonium sulfate particles, which are active CCN but poor IN at temperatures above -45°C [18]. Sample temperature was -20°C, and particle size spectra were measured as SSw increased from -3% to more than 40%, as shown in Figure 1. At SSw below 25%, a few particles >3 µm were detected. As supersaturation increased, the droplet sizes increased dramatically and consistently. For the supersaturations shown, the number of particles >3 µm was 0, 15, 104, 540 and 3446. The results from this experiment were consistent with the numerical simulations and indicated that droplet size can exceed our 3 µm "crystal threshold" when SSw ≥20%. With natural CCN particles instead of these laboratory aerosols, the values of SSw where droplets exceed 3 µm are unknown, but a value ~20% is a reasonable estimate. Therefore, until other methods are developed for differentiating droplets from ice crystals, it appears that peak SSw ~20% is an upper limit for this approach.

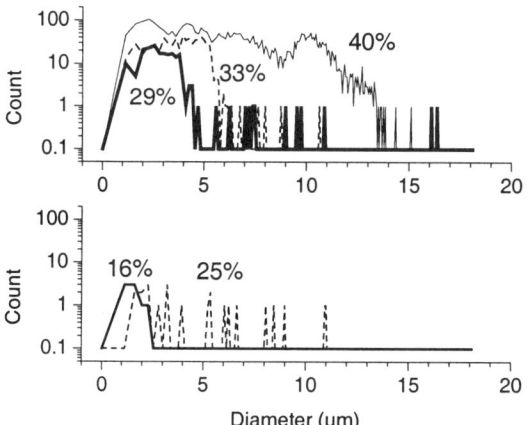

Figure 1. Size distributions of 0.2 μm (dry) ammonium sulfate particles growing in CFD chamber at -20°C and extremely high SSw (%). For convenience, zero counts are plotted 0.1 on logarithmic scale.

AIRBORNE MEASUREMENTS OF ICE NUCLEI

An example of IN measurements from airborne data is shown in Figure 2. Note the general increase of particles exceeding 3 μm (ice nuclei) where SSw is greater. There is a particularly strong peak at 19:19 when SSw ≈20%. Average particle size distributions were grouped by SSw from 19:15 to 19:27, through the SSw peak period, and are shown in Figure 3. A dramatic increase of particles >3 μm is evident as SSw increases. Based on the results of our simulation studies and laboratory experiments, it seems likely that the particles at 19:19 are not all ice crystals (ice nuclei), but a combination of crystals and droplets.

ACKNOWLEDGMENTS

This material is based upon work supported by the National Science Foundation under Grants No. ATM93-11606, ATM96-32917 and ATM97-14177 and by NASA grant NAG1 2063. Any opinions, findings, and conclusions or recommendations expressed in this material are those of the authors and do not necessarily reflect the views of the National Science Foundation.

REFERENCES

1. Huffman, P.J., *J. Appl. Meteor.* **12**, pp. 1080-1082 (1973).
2. Berezinskiy, N.A. and G.V. Stepanov, *Izvestiya, Atmos. Oceanic Physics* **22**, 722-727 (1986).
3. Vali, G., *J. Rech. Atmos.* **19**, 105-115 (1985).
4. Rogers, D.C., *Atmos. Res.* **29**, 209-228 (1993).
5. Mizuno, H. and Fukuta, N., *J. Meteor. Soc. Japan* **73**, 1115-1122 (1995).
6. Mossop, S.C., *Bull. Amer. Meteor. Soc.* **66**, 264-273 (1985).
7. Cooper, W.A., *Meteor. Monographs* **21**, Amer. Meteor. Soc., Boston, 29-32 (1986).
8. Hobbs, P.V. and Rangno, A.L., *J. Atmos. Sci.* **47**, 2710-2722 (1990).
9. Young, K.C., *Conf. on Cloud Physics*, Tucson, Amer. Meteor. Soc., 95-98 (1974).
10. Hall, W.D. *J. Atmos. Sci.*, **37**, 2486-2507 (1980).

11. Baker, B.A., *J. Atmos. Sci* **48**, 1904-1907 (1991).
12. Rogers, D.C., DeMott, P.J. and Grant, L.O., *Atmos. Res.* **33**, 151-168 (1994).
13. Shaw; R.A., *J. Atmos. Sci.,* (in press).
14. Rogers, D.C., DeMott, P.J., Kreidenweis, S.M. and Chen, Y., *J. Atmos. Ocean. Technol.* (2000, in press)
15. Rogers, D.C., *Atmos. Res.* **22**, 149-181 (1988).
16. Plooster, M.N., Report 5-31701, Denver Research Inst, Univ. Denver, Denver, CO, 78pp. (1985)
17. Meyers, M.P., P.J. DeMott, and W.R. Cotton, *J. Appl. Meteor.* **31**, 708-721 (1992).
18. Chen, Y., DeMott, P.J., Kreidenweis, S.M., Rogers, D.C. and Sherman, D.E., *J. Atmos. Sci.*, (2000; in press).

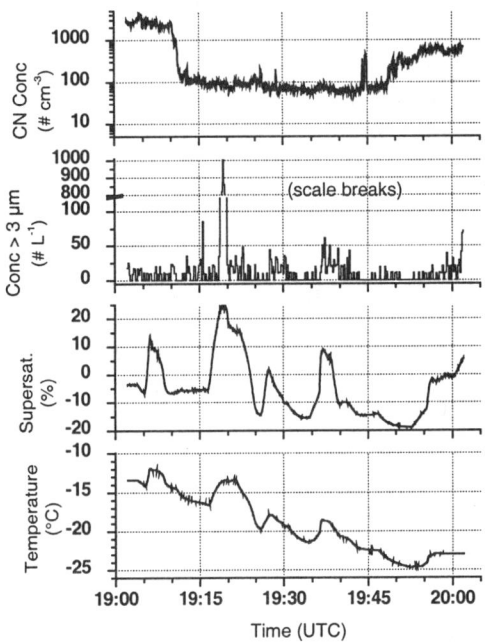

Figure 2. One hour of airborne IN measurements from May 27, 1998. CN (top), particles >3 μm at CFD chamber outlet, sampling temperature and supersaturation (bottom two panels).

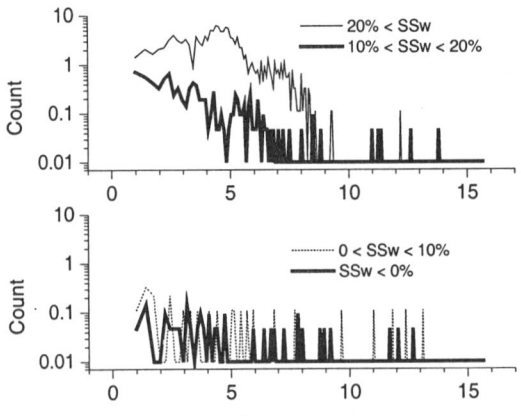

Figure 3. Size distributions from 19:15 to 19:27, during period of extremely high SSw.

Atmospheric Ice Nuclei in the Arctic - Airborne Measurements and Physico-Chemical Properties

D.C. Rogers, S.M. Kreidenweis, P.J. DeMott and K.V. Davidson

Department of Atmospheric Science, Colorado State University, Fort Collins. CO USA

Abstract. Measurements of the number concentration of ice nucleating aerosol particles (IN) were made over the Arctic Ocean and pack ice with a continuous flow diffusion (CFD) chamber. The measurements ranged from -10 to -34°C and from ice saturation to ~20% water supersaturation. The IN concentrations ranged from <1 to hundreds L^{-1}. Occasionally, small regions of high IN concentrations were detected near the surface. During selected time periods, the ice crystals from the chamber were sampled onto electron microscope (EM) grids for analyses of the IN; samples of total aerosol were also collected for comparison. The IN were a few tenths micrometer in size and often contained carbonaceous and crustal materials but little sulfur; they had widely varying morphology. In contrast, for the total aerosol, sulfur was a dominant component. Many IN particles produced weak or no x-ray signatures, most likely due to the dominance of a low-molecular-weight component not detected by the EDX system, probably carbonaceous. This lack of x-ray signature was rarely seen in the total aerosol samples.

INTRODUCTION

Earlier studies of Arctic stratus clouds [1] indicated that ice crystal concentrations were generally very low (~0.005 to 0.3 L^{-1} at -5 to -20°C). There was general agreement between the concentrations of ice crystals and IN measurements from a settling cloud chamber or from filter samples [2]. Observations [3] of "diamond dust" (clear sky ice crystal precipitation) along the coast of northern Alaska in March showed ice crystals forming at low altitudes in concentrations of ~1-30L^{-1} at ~-25°C over regions of open water at humidities between ice and water saturation. Crystals were small (<100 μm), suggesting they formed first by condensation into droplets that subsequently froze. Arctic flights [4] showed extensive ice cloud regions were often encountered at low altitudes (> 80% of flight time in April).

IN filter measurements in Alaska [5] yielded ~0.1-0.5 L^{-1} at -20°C. Shipboard IN filter samples of ~60 minute duration were made during the summer-fall transition [6] and were processed at -12.5, -15, and -17.5°C. They showed generally low concentrations, ~0.01 L^{-1} mid-summer, decreasing to ~0.001 L^{-1} by early fall. Air trajectory analyses indicated that higher IN concentrations were associated with passage over nearby open water leads, suggesting that the ocean surface might be a source of IN.

Measurements of IN are crucial to understanding ice formation in Arctic cloud systems and for investigating whether anthropogenic effects have the potential to modify natural Arctic cloud processes. This paper summarizes IN measurements from eight flights over the Arctic Ocean north of Alaska during in May 1998. These flights were part of the Arctic Cloud Experiment (NASA FIRE program) and SHEBA (Surface Heat Budget of the Arctic, NSF and ONR) [7].

INSTRUMENTATION

The NCAR C-130 aircraft was used for these studies. A large variety of measurements were obtained. Instruments for measuring IN, condensation nuclei (CN) and for collecting ambient particles were inside the aircraft cabin. They shared a common air inlet that was flushed at ~1000 L min^{-1}. A continuous flow thermal gradient diffusion (CFD) chamber was used for the IN measurements [8]. The chamber is of concentric cylinder design. In this chamber, ice crystals nucleate and grow to diameters of ~3 to 10 µm diameter in ~5 s residence time. At the outlet of the chamber, an optical particle counter (OPC) detects all particles larger than ~0.8 µm. Those particles larger than 3 µm are assumed to be the newly formed ice crystals and comprise the IN count. This technique detects deposition or condensation-freezing IN. Nucleation mechanisms that require times >~1 s are not detected (contact-freezing and immersion-freezing).

An inertial impactor immediately downstream of the OPC was used during selected time periods (20-60 minutes) to collect crystals (containing IN) on transmission electron microscope (TEM) grids. The grids were copper with a Formvar backing and thin carbon coating. Single particle analysis and energy dispersive x-ray (EDX) microprobe techniques were used to determine the number, size, morphology and elemental composition of the nuclei [9], [10]. The minimum particle size for EDX analysis is affected by detector sensitivity, atomic number and mixed composition of particles [11]. The size limit is ~0.1 µm, which is within the range expected for IN particles [12], [10]. Several hundred particles were usually collected.

Samples of ambient aerosol particles >0.06 µm diameter (AP) were also collected onto TEM grids with a separate impactor. TEM analyses were the same way as for IN particles. CN concentration was measured with a butanol type instrument (TSI-3076).

Over a period of time, ice nuclei measurements were made both above and below water saturation, with the maximum water supersaturation (SSw) typically about +5%, and the minimum typically -10%. The general strategy was to maintain constant (T, SSw) conditions during vertical profiles and when impactor samples were collected. Occasionally, samples were made at very high SSw (>10%) for a few minutes.

Counts of IN (crystals) were measured at 10 Hz and then averaged to 10 s to increase the number of non-zero values while preserving temporal resolution. Since the C-130 typically flies at ~100 m s^{-1}, the 10 s averages correspond to ~1 km distance and ~0.17 L volume.

RESULTS

Cumulative probability distributions of the ice nuclei and CN measurements for all eight flights are shown in Figure 1 when the aircraft was within 100 km of the SHEBA ice camp. The IN measurements covered wide ranges of temperature and supersaturation, from -10 to -34°C and from ice saturation to 20% SSw. Those measurements with SSw >5% are not included here because such large supersaturations are not representative of typical conditions in Arctic clouds. The calibration and interpretation of the CFD technique for high supersaturations is under investigation [13].

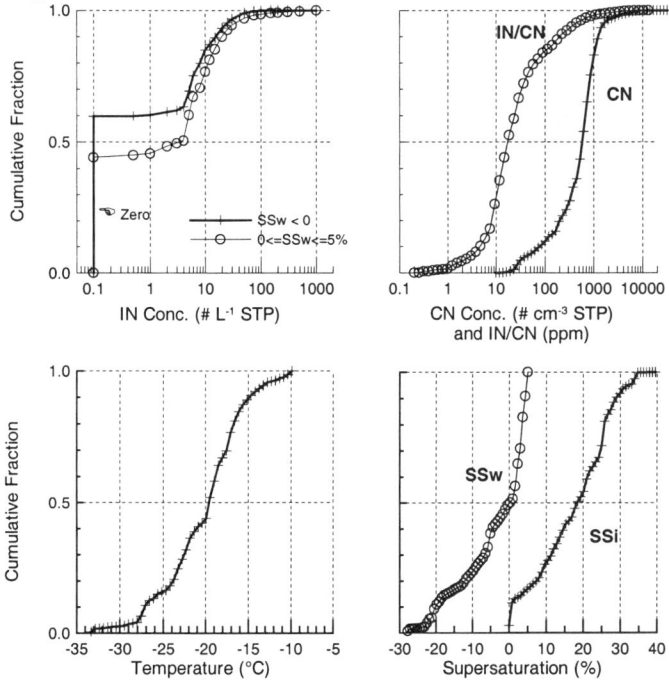

FIGURE 1. Cumulative frequency distributions of 10 s average IN and CN concentrations and sampling conditions for eight Arctic flights in May 1998. For plotting convenience on logarithmic scale, zero IN concentration was plotted at 0.1. Total number of points 41196.

The fraction of all particles active as IN is the ratio of IN and CN concentrations. Figure 1 shows that the IN/CN ratio varied from zero to ~0.02, with a median ~20 ice nuclei per million CN. Median CN was ~600 cm^{-3}. CN values associated with extremely clean air (<100 cm^{-3}) occurred ~10% of the time.

Because the IN concentrations covered a very wide range, the probability distribution was calculated using geometrically increasing bins of concentration. A notable characteristic of the IN curves is that they are highly skewed. Note that concentrations of zero IN comprised ~60% of the 10 s measurements below water saturation, and ~45% of those above water saturation. On average, higher SSw was associated with greater IN concentrations, as expected. Nevertheless, some very high concentrations were measured below water saturation.

An unexpected and surprising observation was the rare occurrence of small regions of high IN concentrations (100's per liter) near the surface of the ice pack. The small extent of high IN suggests the source may be local or that higher IN might be confined within thin stable layers. It is the topic of further investigation.

SINGLE PARTICLE ANALYSES

The EM analyses are summarized in Table I. Generally, IN particles were a few tenths micrometer in size, and their shapes were non-spherical. Elements often detected in the IN samples included Si, S and some metals (Zn, Al, Fe). S and Na were more common in low altitude IN. Carbonaceous material may have been present in many IN particles with weak or no x-ray signatures, due to small mass or low atomic numbers (e.g., elemental or organic carbon). By comparison, a high proportion of the total particles (AP) contained primarily sulfur or S and Si. These results are in general agreement with [10] who reported that the elemental composition of IN is systematically different from the total ambient aerosol.

TABLE I. Electron microscope analysis summary of IN and all particles (AP). Values are percent of particles containing indicated element (rounded to nearest whole percent). Altitude of collection high (700-3000m) or low (50-170m).

ident	# ptls.	dia. (μm)	shape Osphere 1other	Al	Ca	Cl	Co	Cr	Fe	K	Mg	Na	S	Sb	Si	Sn	Ti	V	Zn	no x-ray
IN high	110	0.46	0.82	3	2	0	0	0	2	2	0	2	7	0	38	1	0	0	5	38
IN low	152	0.36	0.68	1	2	0	0	1	5	0	1	15	18	0	37	0	2	0	0	14
AP high	80	0.41	0.06	1	0	0	0	0	1	2	1	1	87	0	6	0	0	0	0	1
AP low	123	0.50	0.04	1	1	0	0	1	1	3	1	3	59	0	29	2	0	0	0	2

ACKNOWLEDGMENTS

This material is based on work supported by NASA grant NAG-1-2063, and NSF grant ATM97-14177. Any opinions, findings, and conclusions or recommendations expressed in this material are those of the authors and do not necessarily reflect the views of the National Science Foundation. We wish to acknowledge the technical support from the Research Aviation Facility at NCAR for helping us install the instruments and conduct the flights. We are grateful for the dedicated work of other FIRE – ACE and SHEBA investigators whose observations are helping our studies.

REFERENCES

1. Jayaweera, K.O.L.F. and T. Ohtake, T., *J. Res. Atmos.*, **7**, 199-207, (1973).
2. Bigg, E.K. and Stevenson, C.M., *J. Rech. Atmos.*, **2**, 41-58, (1970).
3. Ohtake, T., Jayaweera, K. and Sakurai, K-I., *J. Atmos. Sci.*, **39**, 2898-2904, (1982).
4. Curry, J.A., Meyer, F.G., Radke, L.F., Brock, C.A. and Ebert, E.E., *Intl. J. Clim.*, **10**, 749-764, (1990).
5. Fountain, A.G. and Ohtake, T., *J. Clim. Appl. Meteor.*, **24**, 377-382, (1985).
6. Bigg, E.K., *Tellus*, **48B**, 223-233, (1996).
7. Curry, J.A. et al., *Bull. Amer. Meteor. Soc.* **81**, 5-29 (2000).
8. Rogers, D.C., DeMott, P.J., Kreidenweis, S.M. and Chen, Y., *J. Atmos. Ocean. Technol.* (2000, in press)
9. Kreidenweis, S.M., Chen, Y., Rogers, D.C. and DeMott, P.J., *Atmos. Res.*, **46**, 263-278, (1998).
10. Chen, Y., Kreidenweis, S.M., McInnes, L.M., Rogers, D.C. and DeMott, P.J., *Geophys. Res. Lett.*, 25, 1391-1394, (1998).
11. Markowitz, A., Raeymaekers, B., van Grieken, R., and Adams, F., "Analytical electron microscopy of single particles," in *Physical Chem. Char. Individ. Airborne Particles*, John Wiley and Sons, NY, 173-197, (1986).
12. Rosinski, J., Haagenson, P.L., Nagamoto, C.T. and Parungo, F., *J. Aerosol Sci.*, 18, 291-309, (1987).
13. Rogers, D.C., Kreidenweis, S.M. and DeMott, P.J., *15th Int.l Conf Nucl. Atmos. Aer.*, Rolla (2000).

Laboratory Studies of Ice Nucleation by Aerosol Particles in Upper Tropospheric Conditions

Paul J. DeMott, David C. Rogers, Sonia M. Kreidenweis and Yalei Chen

Department of Atmospheric Science, Colorado State University, Fort Collins, CO 80526

Abstract. Ice formation in sulfate, sulfuric acid and black carbon/sulfate aerosol particles under upper tropospheric conditions was studied using a continuous flow thermal diffusion chamber. No clear difference in the homogeneous freezing conditions (temperature, relative humidity) as a function of degree of liquid sulfate neutralization was found, consistent with most other studies. Results support that homogeneous freezing nucleation cannot alone explain observed conditions for cirrus cloud formation. Some types of black carbon (soot) associated with sulfates in mixed particles will induce freezing in preference to the homogeneous process, but only at quite large particle sizes. Small soot produced from burning a particular jet fuel did not show heterogeneous ice nucleation activity until water saturation conditions were exceeded at upper tropospheric temperatures.

INTRODUCTION

In cirrus clouds, purely soluble particles may freeze as haze particles by a process that is much like the homogeneous freezing of pure water, but modified by solution effects. The temperature and relative humidity (RH) at which this process ensues is expected to depend on the size of particles, their thermo-chemical properties, and their kinetic adjustment to cirrus conditions. Insoluble particles or mixed soluble/insoluble particles may nucleate ice heterogeneously under less stringent thermodynamic conditions. These nucleation processes have been studied for particles of relevant sizes and compositions at cirrus temperatures using a continuous flow ice-thermal diffusion chamber.

EXPERIMENTAL

Aerosol generation systems were designed for sulfate, sulfuric acid, soot and combustion aerosols. Aerosols were characterized for size, concentration, and soluble species composition. A water vapor saturator and sample air precooler were used to assure that particles were liquid-phase prior to entering a continuous flow diffusion (CFD) chamber, where particles adjusted to the low temperature water vapor pressure and the conditions of ice formation were determined. The CFD chamber is described in [2] and [3]. Aerosols were isolated by particle-free flows between ice surfaces held at different temperatures. Particles were exposed to set humidities above ice saturation at temperatures down to -60°C. Sample residence times were > 12 s at room pressure

(840 hPa). Haze and ice particle growth calculations showed these times to be sufficient to nearly achieve Köhler equilibrium between the submicron aerosols and the ambient water vapor concentration and to grow nucleated ice crystals to sizes above a few µm in most cases. The size distribution of particles above 0.3 µm diameter was measured with an optical counter. Fractional freezing conditions were measured as a function of RH. Liquid particle composition was inferred using available water activity data and Köhler theory.

RESULTS

Data on the homogeneous freezing of $(NH_4)_2SO_4/H_2O$ and H_2SO_4/H_2O aerosol particles are presented and compared to other recent studies in Figure 1. The data from the current study [2] and from other recent studies [4-8] were used to determine homogeneous freezing point depressions (ΔT_{hf}), given by,

$$\Delta T_{hf} = T_{hf(water)} - T_{hf(solution)} = \lambda \Delta T_m, \qquad (1)$$

$T_{hf(water)}$ is the freezing temperature of pure water drops of the same size and at the same nucleation rate as observed at $T_{hf(solution)}$, ΔT_m is the equilibrium melting point depression, and λ is a constant. Depressions are positively valued in this equation. This presentation permits comparison of data sets using different methods and particle sizes and is useful for applying results to numerical cloud models. Other studies on particles with fixed composition include observations of freezing of supermicron drops on a chilled surface [7] and in emulsions [4], and FTIR studies of submicron drops in chilled flow tubes [5,6,8]. The greatest discrepancy occurs within the latter method of study. Otherwise, a value of λ equal to 2 could be selected for either ammonium sulfate or sulfuric acid with limited uncertainty. This result indicates much deeper supercooling of these solution drops from equilibrium than for pure water drops. Consequently, cirrus formation conditions are not well explained by homogeneous freezing of pure solution drops ([2,4, 6-8]). Results on the effect of sulfate phase state and particle size are described in [2].

Some discrepancies of freezing conditions at unit compositions in Figure 1 appear to be true reflections of nucleation rate differences in the studies and true variation in the constant λ with nucleation rate. A first analysis of estimated homogeneous freezing nucleation rates is given in Figure 2. The different studies indicate a consistent lowering of the slope of nucleation rate versus temperature for solutions compared to pure water. Reasons for this behavior are under study. The data in this form should be valuable for constraining classical theory and providing more fundamental understanding.

Figure 3 summarizes studies of ice formation on various black carbon (BC) aerosol particles at upper tropospheric temperatures. Some of these results were summarized in [1]. Figure 3 includes additional data on ice formation by soot particles produced from burning jet fuel (Jet Fuel A with PRIST). The data, in most cases, indicate freezing at ice supersaturation conditions that align with those required to homogeneously freeze ice in the liquid portion of the soot aerosols. Only the larger BC particles with larger acid coatings indicated distinct heterogeneous nucleation activity lower than -53°C. The poor heterogeneous nucleation activity of simulated jet exhaust aerosols is consistent with field observations of contrail formation conditions [10]. Studies are continuing to identify more effective ice nuclei in the upper troposphere and to determine the role of organic carbon particles.

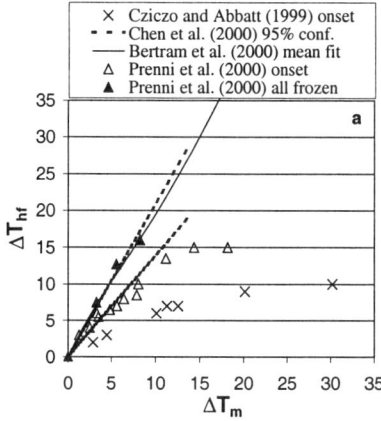
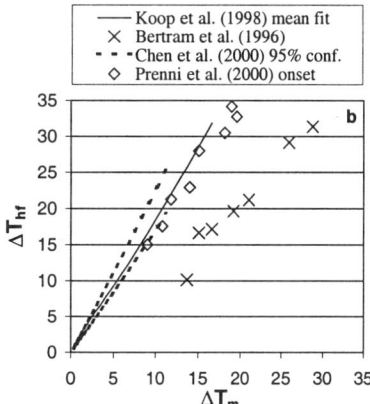

Figure 1. Summary of homogeneous "freezing point" depression versus melting point depression inferred from recent investigations [2-8] of low temperature freezing of sulfates and sulfuric acid aerosols. Data sources are indicated in the legends for a) $(NH_4)_2SO_4$-H_2O, and b) H_2SO_4-H_2O particles. These freezing conditions are relevant for equilibrium droplet conditions, but the slope of the results is useful for making computations of the freezing rates of haze droplets as a function of temperature and instantaneous composition (see text). Adapted from [2].

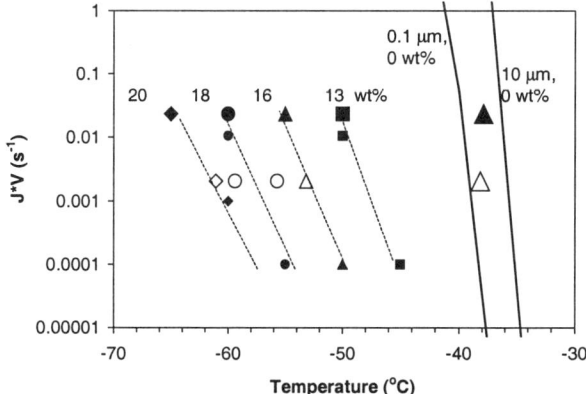

Figure 2. Ice formation rates (particle^{-1} s^{-1}) as a function of temperature and H_2SO_4/H_2O composition estimated from selected data taken in drop freezing (medium and large filled data points: [6]), flow tube (open data points: [8]), and continuous flow diffusion chamber (small filled data points - this study: [2]) studies. Triangles (large), squares, triangles (medium or small), circles and diamonds are data for 0,13,16, 18, and 20 (±1) weight percent (wt%) particles, respectively. Solid lines show the expected freezing rates of pure water droplets based on previous studies, while dashed lines are suggested trend lines of constant wt%. Typical drop sizes were <0.1 μm in [2], 0.3 μm in [8] and 7 μm in [6].

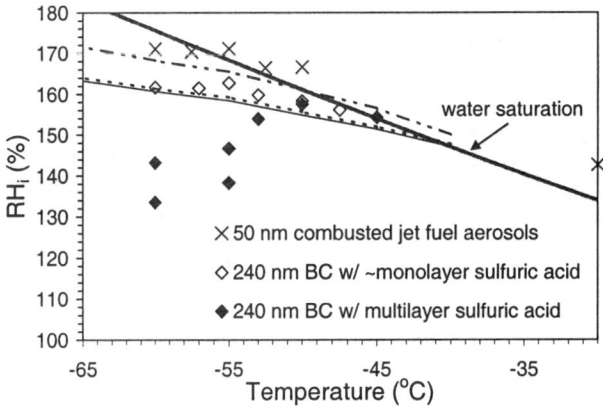

Figure 3. Comparison of the ice relative humidity observed to be required for the formation of ice in 1% of BC based aerosols during the CFD chamber residence time (~12 s here). The thin solid line is the calculated condition for homogeneously freezing the H_2SO_4-H_2O solution on multilayer coated commercial BC particles (mean particle size noted). The dashed and dashed-dotted lines give the calculated homogeneous freezing conditions for monolayer coated BC and combusted jet fuel aerosol particles, respectively. Water saturated conditions are shown by the thick solid line. The data point at -35°C represents the maximum fraction of jet fuel aerosols that would freeze -30°C. Partially adapted from [1].

ACKNOWLEDGMENTS

This work was supported by NSF-ATM9632917, and the NASA Atmospheric Effects of Aviation Program (Jet Propulsion Laboratory Contract 961353). Any findings and conclusions are those of the authors. Y.C. was supported by NASA Earth System Science Fellowship (NGT5-30001). We thank A. Middlebrook, S. L. Clegg, T. Koop, S. Martin, T. Prenni, A. Bertram and D. Cziczo for helpful discussions and/or assistance with experimental design issues

REFERENCES

1. DeMott, P.J., Chen, Y., Kreidenweis, S.M., Rogers, D.C., and Sherman, D.E., *Geophys. Res. Lett.* **26**, 2429-2432 (1999).
2. Chen, Y, DeMott, P.J., Kreidenweis, S.M., Rogers, D.C., Sherman, D.E., *J. Atmos. Sci.*, (2000; in press).
3. DeMott, P.J., Kreidenweis, S.M., Rogers, D.C., Chen, Y., Sherman, D.E., "Laboratory studies of aerosol effects on ice formation in cirrus clouds," in *Proceedings, Intnl. Conf. on Clouds and Precip.* (2000)
4. Bertram, A.K., Koop, T., Molina, L.T., and Molina, M.J., *J. Phys. Chem. A.* **104**, 584-588 (1999).
5. Bertram, A.K., Patterson, D.D., and Sloan, J.J., *J. Phys. Chem.* **100**, 2376-2383 (1996).
6. Cziczo, D.J. and Abbatt, J.P.D., *J. Geophys. Res.* **104**, 13781-13790 (1999).
7. Koop, T, Ng, H.P., Molina, L.T., and Molina, M.J., *J. Phys. Chem. A* **102**, 8924-8931 (1998).
8. Prenni, A.J., Wise, M.E., Brooks, S.D., and Tolbert, M.A., Submitted to *J. Geophys. Res.* (2000).
9. Rogers, D.C., P.J. DeMott, S.M. Kreidenweis, and Y. Chen: A continuous flow diffusion chamber for airborne measurements of ice nuclei. *J. Atmos. Ocean. Technol.* (2000; in press)
10. Jensen, E.J. et al., ., *J. Geophys. Res.* **103**, 3929-3936 (1998).

Equilibrium Forms, Nucleation, Wulff's Theorem for Crystals on Substrates-Particles with Finite Dimensions

George Miloshev

Bulgarian Academy of Sciences, Geophysical Institute
Acad.G.Bonchev Str. bl.31113 Sofia, Bulgaria

1. During last years the investigations of small phases formation became deeply connected with the processes of nucleation and their application in nanotechnologies. It was shown in [1,2,3] that the formation of cubic nuclei on cubic particles is accomplished under two possible mechanisms: growing of the crystal nuclei on three of the particle's walls, i.e. on a convex top, or on one of the particle walls. The free energy for nucleus formation under these mechanisms, compared to the homogeneous case $\Delta G_0 = n_1^2 \psi$ are respectively:

$$\frac{\Delta G_1}{\Delta G_0} = 1 + 2\frac{n_0^3}{n_1^3} - 3\frac{n_0^2}{n_1^2}\frac{\psi'}{\psi} \qquad \frac{\Delta G_2}{\Delta G_0} = 1 - \frac{n_0^2}{n_1^2}\frac{\psi'}{\psi} \qquad (1)-(2)$$

The realization of each of these mechanisms depends on the ratios $n_0/n_1 \lessgtr \psi'/\psi$ when respectively $\Delta G_1/\Delta G_2 \lessgtr \Delta G_2/\Delta G_0$. Here ψ and ψ' are the works of separation of an element from own crystal and from a substrate. $\psi'/\psi = \beta/2\sigma$ defines the wettability of the substrate - particle (β is the adhesive work nucleus - particle and σ the surface tension of the nucleus). In the above mentioned two possible mechanisms however the influence of the substrate - particle on the equilibrium form of the crystal, its size and the free energy of its formation, are not taken into account.

2. In the case on Fig.1b, since the particle is symmetrically placed, the nucleus will decrease in three directions [4]. So the equilibrium form will rest a cub, but with a decreased edge n_1'. The relation n_1'/n_1 of the equilibrium forms of the crystals from Fig.1a and 1b, since they are under one and the same supersaturation, will be determined from the Stranski-Kaishev equation [5] and the equality of the mean separation works $\overline{\varphi}_1 = \overline{\varphi}_2''$.

$$KTS = \varphi_{1/2} - \overline{\varphi}_1 = \varphi_{1/2} - \overline{\varphi}_2''; \quad \frac{2\psi}{n_1} = \frac{2n_1'\psi - 2n_0\psi'}{n_1'^2 - n_0^2};$$

$$\frac{n_1'^2}{n_1^2} - \frac{n_1'}{n_1} + \left(\frac{n_0}{n_1}\frac{\psi'}{\psi} - \frac{n_0^2}{n_1^2}\right) = 0; \quad \frac{n_1'}{n_1} = \frac{1}{2} + \frac{1}{2}\sqrt{1 - 4\left(\frac{n_0}{n_1}\frac{\psi'}{\psi} - \frac{n_0^2}{n_1^2}\right)} \qquad (3)$$

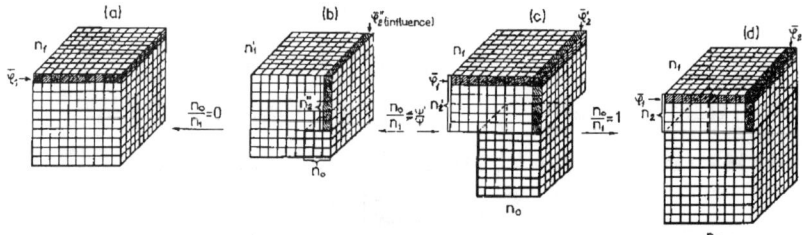

FIGURE 1.

It is seen that the size of the nucleus (n_1'/n_1) depends not only on the wettability ψ'/ψ but also on the size of the particle (n_0/n_1). The relation n_1'/n_1 has it sense only for $0<n_0/n_1\leq\psi'/\psi$ i.e. for the configuration of Fig 1b. For $n_0/n_1=\psi'/\psi$, $n_1'/n_1=1$, i.e. $n_1'=n_1$. For $1\geq n_0/n_1\geq\psi'/\psi$, $n_1'/n_1>1$, i.e. $n_1'>n_1$ which has no physical sense. The relation n_1'/n_1 is shown on Fig.2. The edge of the nucleus decreases with increasing of the particle size reaching a minimum after which start to increase up to its unaffected value when $n_0/n_1=\psi'/\psi$. This is due to the fact that the decreasing of the nucleus is conditioned by the adhesive forces (wettability) of the substrate-particle, which are directed against the capillary pressure forces, which depend on the nucleus size and could not be decreased infinitely. The competition between these two effects was shown by B.Mutafchiev [6]. The two forces are competitive only for $0<n_0/n_1\leq\psi'/\psi$. The edge of the decreased crystal could be presented as sum of the crystal thickness n_2'' and the size of the particle n_0 i.e. $n_1'=n_2''+n_0$. So for n_2'' we obtain:

$$\frac{n_2''}{n_1} = \frac{1}{2} + \frac{1}{2}\sqrt{1-4\left(\frac{n_0}{n_1}\frac{\psi'}{\psi} - \frac{n_0^2}{n_1^2}\right)} - \frac{n_0}{n_1}. \tag{4}$$

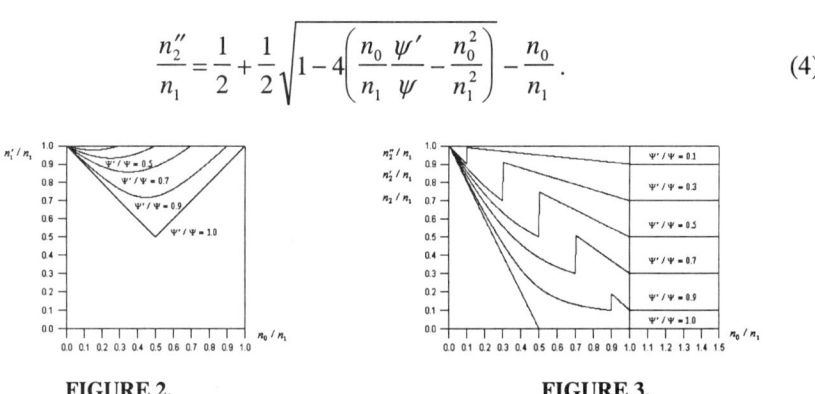

FIGURE 2. **FIGURE 3.**

When $n_0/n_1 \gtrless \psi'/\psi$ a change in the growing mechanism is provoked and the nucleus starts growing on one of the particle's wall - Fig.1c. From the equality of the mean separation works $\overline{\varphi}_1 = \overline{\varphi}_2'$ for the thickness of the crystal we obtain:

$$\frac{2\psi}{n_1} = \frac{n_2'\psi + n_1\psi - n_0\psi'}{n_1 n_2'}, \quad \text{i.e.} \quad \frac{n_2'}{n_1} = 1 - \frac{n_0}{n_1}\frac{\psi'}{\psi} \tag{5}$$

It is easily seen that the equilibrium form in this case is a parallelepiped, i.e. the crystal's decreasing is only by one side.

The relations (4) and (5) are shown on Fig.3. Under the mechanism of growing on a convex top, the thickness of the crystal n_2'' decreases with increasing of the particle size and wettability up to values of $n_0/n_1=\psi'/\psi$. For bigger particles i.e. $n_0/n_1 \gtreqless \psi'/\psi$ the growing mechanism changes - Fig.1c and the crystal thickness n_2' with a jump increases followed by next decreasing up to values of $n_0/n_1=1$, when it reaches the crystal thickness on infinitely large substrate n_2. Of special interest are crystal thicknesses under complete wettability, i.e. $\psi'/\psi =1$. In this case the nucleus fully envelopes the particle. The influence on the crystal's edges will be over all six walls with one and the same power. It is observed that for particle sizes equal or greater than half nucleus size, i.e. $n_0/n_1>1/2$ the crystal thickness is nullified which means that a monomolecular crystal layer is spontaneously formed.

3. The free energy for nucleus formation - Fig.1b, also decreases [7]. Its formula will be the same as (1) but with decreased edges consisting of n_1' building elements. Compared to ΔG_0 we obtain:

$$\frac{\Delta G'_{1(affected)}}{\Delta G_0} = \frac{n_1'^2 \psi}{n_1^2 \psi}\left(1+2\frac{n_0^3}{n_1'^3}-3\frac{n_0^2}{n_1'^2}\frac{\psi'}{\psi}\right) \quad (6)$$

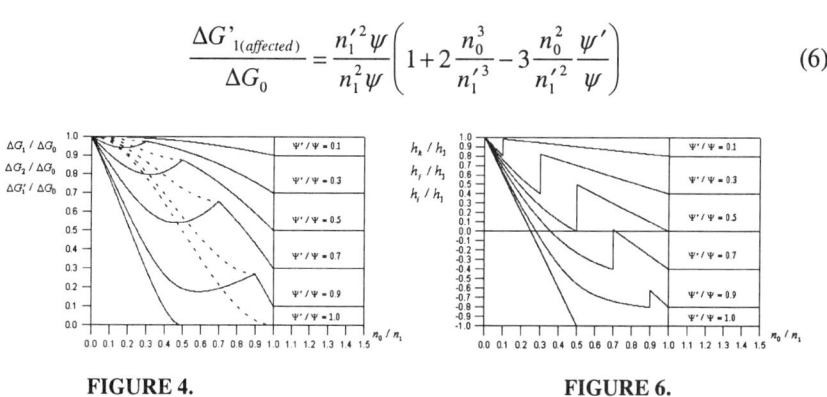

FIGURE 4.   FIGURE 6.

The ratios (1), (2) and (6) are presented on Fig.4. It is seen that the free energy for nucleus formation deeply decreases following the run of the competitive forces from Fig.2 up to $n_0/n_1=\psi'/\psi$. Under further increasing of the particle's size the free energy does not change and is given by (2). Only the equilibrium form changes. Here also on completely wettable particles ($\psi'/\psi=1$) the free energy is deeply reduced up to annulation for particle sizes ($n_0/n_1 \gtreqless 1/2; n_0 \gtreqless 1/2 n_1$), which turned to be ready nuclei for new crystals growth.

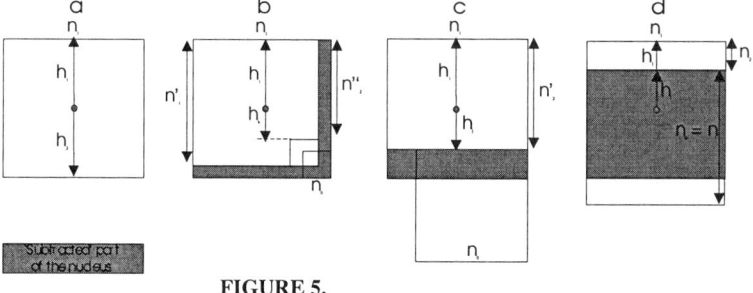

FIGURE 5.

4. After all examinations up to here we obtain a possibility to define an extension of the Wulff's theorem for the case of nucleation on substances-particles with finite dimensions.

For the cases on Fig.1c and 5c replacing n_2'/n_1 from (5) in the relation h_j/h_1, we see that the distance h_j to the wall bordering the foreign phase - the particle, is proportional to the difference between the specific surface energy of this wall and the adhesive energy between the wall and the particle, multiplied by the relation n_0/n_1, i.e. the particle size under given supersaturation.

$$h_1:h_2:\ldots:h_j = \sigma_1:\sigma_2:\ldots:(\sigma_j - \beta_j \frac{n_0}{n_1}). \tag{7}$$

The elongation to the wall bordering the particle here is decreased with the ratio $\dfrac{\sigma_j - \beta_j x n_0 / n_1}{\sigma_j}$ compared to the elongations to the free walls. For the cases on Fig.1b and 5b replacing n_2''/n_1 from (4) in the ratios h_k / h_1 and after transformations we obtain:

$$h_1:h_2:\ldots:h_k = \sigma_1:\sigma_2:\ldots:\sqrt{\sigma^2 - 2\frac{n_0}{n_1}\sigma\beta + 4\frac{n_0^2}{n_1^2}\sigma^2 - 2\frac{n_0}{n_1}\sigma} \tag{8}$$

here the elongation h_k to the wall bordering the foreign phase - the particle, is given with (8) and is reduced compared to the distances to the free walls with the ratio (8)/σ. For all possible cases the Wulff's theorem could be presented with the relation

$$h_1:h_2:\ldots:h_i:h_j:h_1':h_k = \sigma_1:\sigma_2:\ldots:(\sigma_i - \beta_i):(\sigma_j - \beta_j \frac{n_0}{n_1}):$$

$$\left(\sqrt{\sigma_k^2 - 2\frac{n_0}{n_1}\sigma_k\beta_k + 4\frac{n_0^2}{n_1^2}\sigma_k^2 - 2\frac{n_0}{n_1}\sigma_k}\right) \tag{9}$$

The ratios (7) and (8) are extensions of the Wulff's theorem for the cases of nucleus formation on substrates - particles with finite dimensions. When $n_0/n_1=1$ from (7) we obtain the Wulff's theorem (extended by Kaishev) for infinitely large substances. When $n_0=0$ from (8) we obtain the original Wulff's theorem. In addition we obtain possibility to present on Fig.6 the changes of the elongation from the Wulff's point in the crystal to the crystal walls bordering the foreign substrate - the particle (under the two mechanisms of growing) and on infinitely large substrate.

REFERENCES

1. Miloshev G., *C.R.Acad.Bulg.Sci.*, v.16, No.5 (1963).
2. Miloshev G., *C.R.Acad.Bulg.Sci.*, v.16, No.6 (1963).
3. Miloshev G., *C.R.Acad.Bulg.Sci.*, v.16, No.7 (1963).
4. Miloshev N., *C.R.Acad.Bulg.Sci.*, v.52, No.9 (1999).
5. Kaishev R., *Arbeitstatung Festkorperphisik - Dresden, Berlin, Deutsch.-Vlg.Wiss.*, 81-83 (1953).
6. Mutaftschiev B., *Bulg.ChemicalCommunication*, v.27, No.1 (1994).
7. Miloshev N., *C.R.Acad.Bulg.Sci.*, v.53, No.2 (2000).

Electroscavenging of Evaporation Nuclei by Cloud Droplets and Consequences for Contact Ice Nucleation

Brian A. Tinsley

Center for Space Science, FO22, University of Texas at Dallas, Richardson, TX, 75083-0688

Abstract. Droplets in clouds collect aerosol particles that are charged due to the (hitherto largely neglected) action of image charge forces between the conducting surface of the droplet and the particle. This process (electroscavenging) is in addition to phoretic scavenging. We have made new computations of scavenging to include electrical, phoretic, and gravitational forces in a self consistent way. We show that for only a few tens of elementary charges on particles of radii from 0.1 µm to more than 1.0 µm, and for droplets of radii more than 15 µm and for 98% relative humidity, the collision efficiency is dominated by electroscavenging. Charges on the droplets (in comparison to charges on the particles) have little effect on the collection efficiencies, even when the charges are one or two orders of magnitude larger and of the same or opposite sign as those on the particles. Electroscavenging provides a pathway for contact ice nucleation in supercooled clouds, when aerosol particles with charges of order 10^2 elementary charges result from evaporation of droplets, with typical droplet charges, that are subject to mixing processes at cloud tops. Contact ice nucleation of remaining unevaporated droplets is favored not only by collection rates enhanced by electroscavenging, but also because the sulfate and organic material that is adsorbed onto droplets before evaporation remains as coatings on the residual charged evaporation nuclei. The coatings are thought to have good ice nucleating properties. The retention of the coatings and the charges, although temporary, is long enough (with continued mixing) to greatly increase the contact ice nucleation rate compared to that from other aerosol particles that have not been just previously processed through droplets.

INTRODUCTION

Droplets at cloud tops of non-thunderstorm clouds have typically more than a hundred elementary charges on them (1, sect. 18.4; 2). In the presence of mixing and entrainment processes a fraction of them evaporate, and almost all of the charge is retained on the residual evaporation nucleus (3). The approximate decay time constant for the charge is typically 10^3 s, owing to the low conductivity in clouds (4). The mixing processes also bring the charged particles (evaporation nuclei) into proximity with unevaporated droplets. The electrical force between the charged aerosol particle and its image charge induced on the conducting droplet can be large, compared to the phoretic forces due to the gradients of temperature and water vapor concentration near the droplet (5).

The numerical trajectory calculations made by Tinsley et al. (4) considered the electrical forces between droplets and particle separately from the other forces. Here we include in a self consistent treatment the thermophoretic and diffusiophoretic forces, and the gravitational force on the particle as well as on the droplet. We also evaluate the effect of Brownian motion.

APPROACH TO TRAJECTORY CALCULATIONS

For the droplet sizes and aerosol particle sizes of interest here, it is a reasonable approximation, and greatly simplifies the numerical integration, to consider the particles as point charges and the inertial forces negligible. The basic theory of particle collection and phoretic effects is given in Wang et al. (6) and reviewed in Pruppacher and Klett (1, sect. 17.4.2). The treatment of the electrical force resulting from the point charge near the conducting droplet, in terms of a long range Coulomb force and a short range image charge force, is given by Tinsley et al., (4) as:

$$F_e = (q^2/4\pi\varepsilon_o A^2)[(Qr/q + 1)/r^3 - r/(r^2 - 1)^2] \qquad \dots\dots (1)$$

where q is the charge on the particle; ε_o is the permittivity of free space, A is the radius and Q is the charge of the droplet; r is the normalized radial distance of the particle, i.e. $r = b/A$, where b is the distance of the particle from the origin at the center of the droplet. The first term in the square brackets is a long range repulsive force if Q and q have like signs and is attractive for unlike signs; the second is a short range attractive force that becomes very large as r tends to 1, i. e., as the particle approaches the droplet surface.

The numerical integration of trajectories for a given set of input parameters proceeded as in Tinsley et al. (4). The impact parameter measured perpendicular to the vertical axis of flow symmetry (the Y axis) was increased until the last trajectory for a particle to be collected was determined with impact parameter $xmax$. We define collection efficiency E as the ratio of the volume containing aerosol particles collected in a given time to the volume swept out by the falling droplet. For particle radius a small compared to A it is sufficient to take $E = (xmax)^2$ where $xmax$ is in units of $r = b/A$.

RESULTS

Figure 1 shows plots of collection efficiencies $(xmax)^2$ for three of the nine droplet sizes modeled in the range 10 μm to 100 μm radius; for droplet charges Q of $500e$ and $0e$, (where e is the elementary charge of 1.6×10^{-19} C); for particle charges q of $1e$ to $100e$ and for particle radii a from 0.01 μm to 2 μm. The atmospheric pressure, temperature, and humidity were 540 hPa, -17°C, and 98%. Other models (not shown) were run for $Q = 100e$, with results intermediate between those for Q of $500e$ and $0e$, and for $Q = -100e$, with results similar to those for $Q = 0e$. The collection efficiencies for Q of $500e$ and particle radius < 0.1 μm decrease because the long range repulsion prevents the particles from being swept close enough for short range attraction to take over.

The collection rates (for unit concentration of droplets and nuclei) depend on the collection efficiencies multiplied by the volume swept out by a droplet in unit time, which depends on the droplet fall velocity (approximately proportional to A^2) and cross section (also proportional to A^2). This A^4 dependence results in significant collection rates only for droplets with radii greater than about 15 μm.

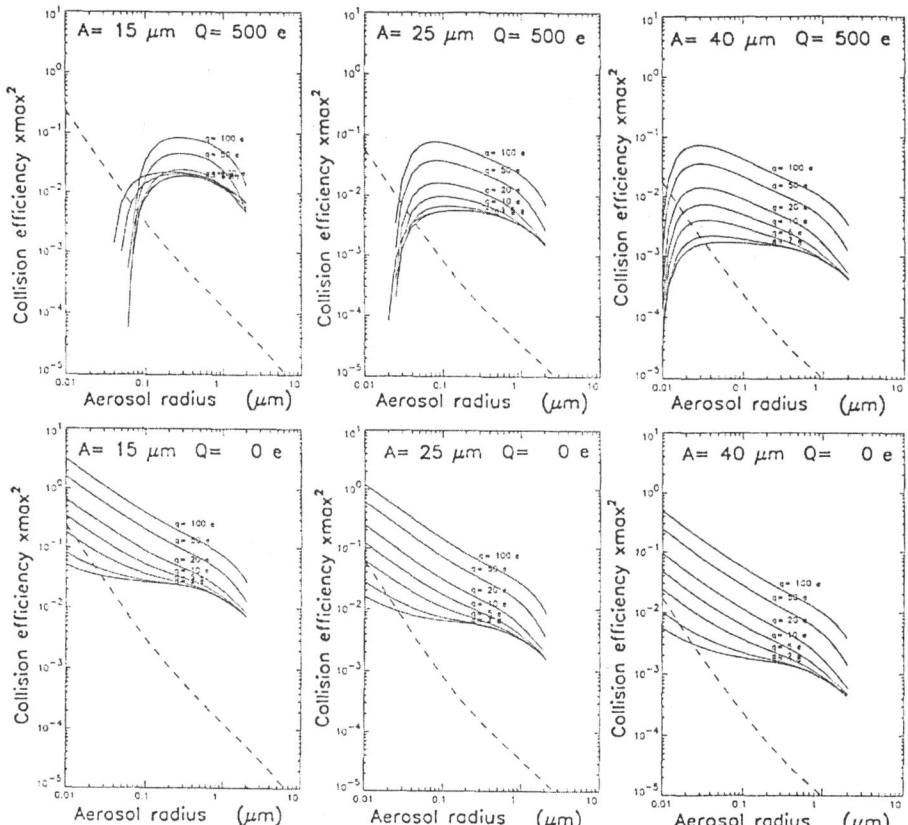

FIGURE 1. Variations of collision efficiency with droplet radius A and charge Q, for aerosol particle charges q of 1 to 100 elementary charges, as a function of aerosol radius. The Brownian collection efficiency is shown as a dashed line.

CONSEQUENCES FOR CONTACT ICE NUCLEATION

In his analysis of the properties of evaporation nuclei (collected outside of clouds and after decay of charge) Rosinski (6) found them to be effective ice forming nuclei in the deposition and condensation-followed-by freezing modes. Thus it is reasonable to expect that they would also be effective as ice forming nuclei in the contact mode. The evaporation nuclei have surface properties that enhance the probability of ice formation on them, evidently due to soluble sulfate compounds and insoluble organic compounds that were previously scavenged from the air by the droplet (8, 9, 5, 7). The ice nucleating activity of the evaporation nuclei was found to decrease after the temperature was cycled through higher values, or when the humidity was cycled through lower values.

Observations of concentrations of atmospheric nuclei effective as ice forming nuclei in different modes at different temperatures were made by Cooper (10). The same equipment was used to obtain similar data on AgI nuclei created in the laboratory. At temperatures from -20°C to -15°C the concentrations of both the atmospheric and the

AgI particles active as contact nuclei were an order of magnitude greater than the concentrations active as deposition or immersion nuclei. At -10°C and warmer the concentrations were 30 to 100 times larger.

Pruppacher and Klett (1, p. 340 - 341) discuss possible explanations for this behavior, including that of Fukuta (11, 12), who suggested that the difference is associated with the movement of the water-air interface relative to the surface of the contacting particle during contact. Fukuta pointed out that the rapid spreading of the water along the hydrophobic solid surface forces its local wetting, and thereby temporarily creates high interface-energy zones that can increase the likelihood of ice nucleation. The experiments of Abbas and Latham (13) showed that freezing could be triggered in supercooled droplets merely by sudden disruption of them by impacting objects.

Uncertainties in the above effects prevent an accurate determination of contact nucleation rates at present. Other uncertainties concern whether the additional impact velocity and electric field at impact in electroscavenging enhance nucleation probability. Even without such enhancement, estimates suggest that for clouds with droplet size distributions extending appreciably beyond 15 µm radius, and temperatures above -20°C, contact nucleation rates exceed deposition nucleation rates for the same populations of nuclei. This could explain the Hobbs and Rangno (14) observations of high ice nucleation rates in clouds with similar broad droplet size distributions.

ACKNOWLEDGEMENTS

I wish to thank K. Beard for many stimulating discussions. This work has been supported by NSF grant ATM 9903424.

REFERENCES

1. Pruppacher, H. R., and Klett, J. D., *Microphysics of Clouds and Precipitation*, 2nd. ed., Kluwer, 1997.
2. MacGorman, D. R., and Rust, W. D., *The Electrical Nature of Storms*, Oxford Univ. Press, 1998.
3. Robertson, J. A., *Interactions Between a Highly Charged Aerosol Droplet and the Surrounding Gas, Ph. Dissertation,* Univ. of Illinois, 1969.
4. Tinsley, B. A., *J. Atmos. Sci.*, in press, (2000).
5. Beard, K. V., *Atmos. Res.*, **28**, 125-152 (1992).
6. Wang, P. K., Grover, S. N., and Pruppacher H. R., *J. Atmos. Sci.*, **35**, 1735-1743 (1978).
7. Rosinski, J., *Atmos. Res.*, **38**, 1351-359 (1995).
8. Garrett, W. D., *Pure Appl. Geophys.*, **116**, 316-334 (1978).
9. Rosinski, J., and Morgan, G., *J. Aerosol Sc.*, **22**, 122-133 (1991)
10. Cooper, W. A., A Method of Detecting Contact Ice Nuclei Using Filter Samples, *8th. Intern. Conf. Cloud Phys.*, Clermont-Ferand, France, 15-19 July, 665-668, 1980.
11. Fukuta, N., *J. Atmos. Sci.*, **32**,1597-1603 (1975).
12. Fukuta, N., *J. Atmos. Sci.*, **32**, 2371-2373 (1975).
13. Abbas, M. A., and Latham, J., *J. Fluid Mech.*, **30**, 663-670 (1967)
14. Hobbs, P. V., and Rangno, A. L., *J. Atmos. Sci.*, **42**, 2523-2549 (1985).

AEROSOL PARTICLE SIZE DISTRIBUTION, THE TOTAL NUMBER AND ICE NUCLEI CONCENTRATIONS IN MOSCOW REGION

N.O. Plaude and M.V. Vychuzhanina

Central Aerological Observatory
Dolgoprudny, Moscow Region
141700 Russia

Abstract. The results of measuring the characteristics of atmospheric aerosol in the period of 1987 through 1999 are given. It is shown that no trend in the variation of the high values of total aerosol concentration and ice nuclei concentration was observed in these years. The data collected on the synchronous influence of industrial pollution sources on the total aerosol concentration and the concentration of ice nuclei, along with the previously obtained results, allow one to expect the number of ice nuclei to increase in the atmosphere with the growing man-made pollution.

INTRODUCTION

Since 1987, periodic measurements of atmospheric aerosol characteristics have been carried out on the territory of the Central Aerological Observatory, 20 km north of Moscow. The concentration and size distribution of sub-micron particles (SP) within a $0.01 - 1$ μm diameter range (electric aerosol size analyzer TSI 3030), the concentration and size distribution of large particles (LP) $0.3 - 10$ μm in diameter (photoelectric counter), and the concentration of ice nuclei (IN) active at -20^0C (cloud chamber described by Vychuzhanina et al. (1)) are measured. Atmospheric air samples are taken from a 12-m height above the ground. The measurements are made between 9:00 and 10:00 LT on working days in monthly series. In 1987, 1994-95, and 1997-99, they were carried out during a half-year from January through June. From the data collected one can infer the mean characteristics of atmospheric aerosol in this region.

RESULTS

Fig. 1 shows the mean semi-annual concentrations of sub-micron and large particles of atmospheric aerosol as well as ice nuclei concentrations observed in 1987-

99. No clearly observable trend in the variation of the concentrations of the three atmospheric aerosol constituents was detected in the years when the measurements were carried out. A certain increase in IN and LP concentrations has been observed during the last two years. In Moscow Region, the mean aerosol and ice nuclei concentrations have been measured to be 33000 cm^{-3} and 24 liter^{-1}, respectively.

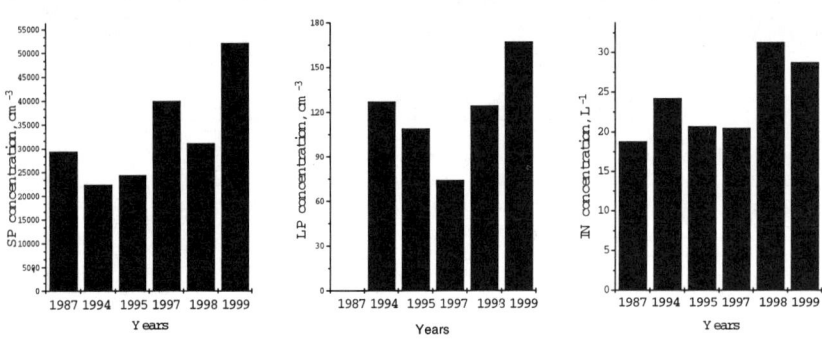

FIGURE 1. Mean semi-annual SP, LP, and IN concentrations in 1987-1999

Figure2 presents the mean monthly changes in SP, LP, and IN concentration within a half-year period. The seasonal changes in these concentrations reveal considerable inter-annual variations mainly resulting from weather variability. In particular, this refers to a sub-micron aerosol fraction. Very frequently observed is an increase in LP and IN concentration in spring that is the period of vegetation growth and flowering.

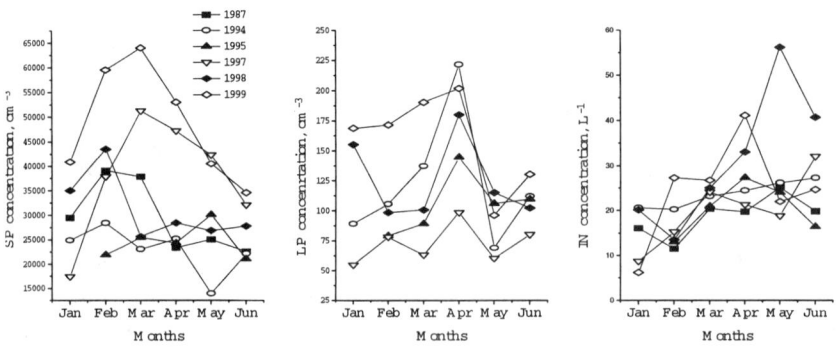

FIGURE 2. Seasonal variations of mean monthly aerosol particle and ice nuclei concentrations.

The configuration of the mean monthly size distribution shows an extremely low variability in different years. Three maxima at 0.02, 0.75, and 2 μm persistently occur in mean size distributions. Changes in the state of the underlying surface in spring and summer months cause an increase in the concentration of particles larger than 1 μm. In its integral number concentration and size distribution shape, the aerosol in the measurement area corresponds to the type that is transitional between aerosols characteristic of industrial and rural areas, as defined by Jaenicke (2).

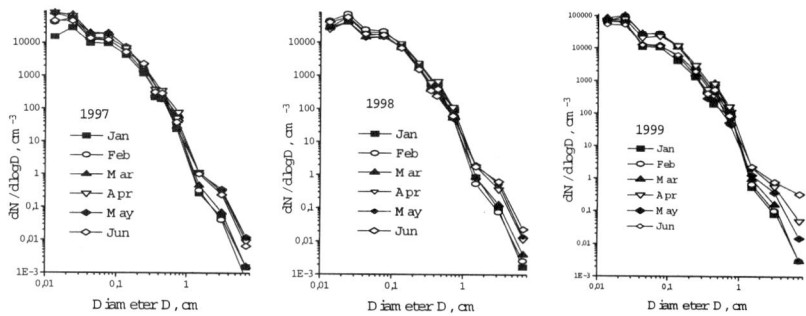

FIGURE 3. Mean monthly aerosol particle size distribution in 1997, 1998, and 1999.

The influence of local pollution sources shows in the growth of the concentration of all aerosol constituents at winds blowing from the direction of industrial Moscow districts (south -east-south) (Fig. 4).

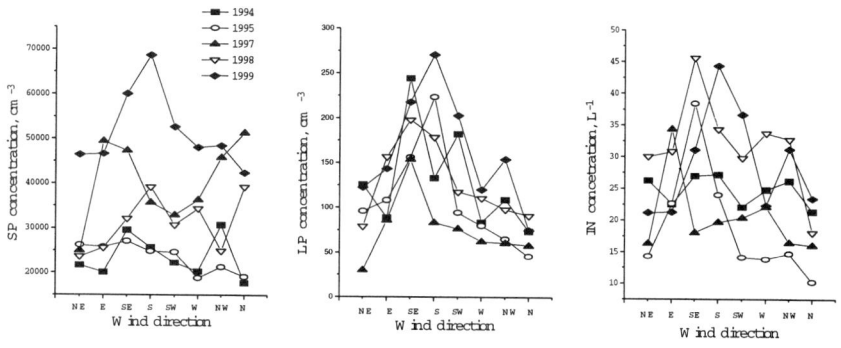

FIGURE 4. SP, LP, and IN concentrations at different wind directions during the periods of January-June.

As previously pointed out by the authors (3,4), the growth of the total concentration of atmospheric aerosol resulting from man-made atmospheric pollution is accompanied by increasing IN concentration.

Figure 2 and 4 show that the changes in ice nuclei concentration most closely follow the changes in large particle concentration. This proves the well-known fact that ice nuclei are predominantly large atmospheric aerosol particles. Note that the daily LP and IN concentrations in the measurement period did not show any clearly discernible correlation (5).

CONCLUSION

Atmospheric aerosol in Moscow Region is characterized by a high total aerosol concentration and a high concentration of ice nuclei. No trend is detected in the variation of aerosol and ice particle concentrations during the period of 1987-1999, which could be expected considering the decline of manufacturing industry in Moscow area during this period. The monthly averaged size distribution is a conservative characteristic of regional aerosol which is of an intermediate type between that of urban and rural areas. The enhanced concentration of atmospheric aerosol, caused by industrial air pollution, entails the growth of ice nuclei concentration. Thus the influence of the total particle concentration enhancement prevails over the possible effect of ice particle deactivators, and under conditions of the growing man-made air pollution, an increase of ice nuclei content in the atmosphere can be expected.

REFERENCES

1. Vychuzhanina,, M. V., Miroshnichenko, V. I., and Plaude, N. O., *Optics of atmosphere and ocean*, **10** (6), 1-6 (1997) (in Russian).
2. Jaenicke, R., *Russ. Chem. Rev*, **59**, issue 10, 1651-1675 (1990) (in Russian).
3. Vychuzhanina, M. V., Grishina, N. P., Parshutkina, I. P., Plaude, N.O., and Potapov, Ye. I., *J.Aeros. Sci.*, **20** (8),1237-1240 (1989).
4. Vychuzhanina, M. V., and Plaude, N. O., *Optics of atmosphere and ocean*, **9** (6), 858-861 (1996) (in Russian).
5. Plaude, N.O., and Vychuzhanina, M.V., *Optics of atmosphere and ocean*, **11** (10) 1139-1142 (1998) (in Russian).

Theoretical Study Formation and Development of Antarctic Cloudiness under Different Intensity of Ice and Cloud Droplet Nucleation

Svetlana V. Krakovskaia and Anne M. Pirnach

Ukrainian Hydrometeorological Research Institute
37 Nauki Avenue, Kiev – 03028, Ukraine

Abstract. 1-D numerical model with a detailed description of the evolution of cloud particles (CP) (drops, crystals, cloud nuclei, etc.) is used to study formation and development of suppercooled mixed frontal clouds over the Antarctic Peninsula on 02.04.98. Outputs of 3-D nowacasting limited area model (LAM) based on rawinsonde data of Bellingshausen station are used as initial data for 1-D forecasting microphysical model. Thermodynamical conditions in the troposphere are continuously updated as the system moved over the initial point in 1-D simulation. A set of equations is used to simulate the evolution of the processes of condensation, nucleation, freezing, sedimentation, accretion, collection, etc. Frontal clouds are simulated under different intensity of CP generation on CCN and IN. Relationships between these two processes and optimal parameters for realization of the whole cloud moisture were found.

INTRODUCTION

The spatial distribution and quantity of precipitation in the Antarctic are very poorly understood because of the small number of synoptic observing stations especially over the large data sparse ocean areas. The uncertainties of making in situ measurements of cloud microphysics parameters and the lack of reliable remote sensing techniques for precipitation measurement from the space are also a problem for the ice continent. On the other hand, with the increasing use of general circulation models (GCM's) of the atmosphere for predictions and other aims it is essential to ensure that these models are representing correctly the cloudiness and precipitation in the modelled domain, particularly over the Antarctic. For the above problems numerical models with detailed cloud microphysics can help in obtaining needed characteristics for parameterization of cloud and precipitation formation processes.

This present study is focused on (1) the adaptation of previously worked out numerical models to the specific region with limited initial data for construction and verification of the model's outputs; and (2) the investigation the effect of CCN and IN concentrations on the cloud and precipitation development. Frontal rainbands associated with a deep depression of explosive cyclone type, moved over the Antarctic Peninsula from the South Pacific and caused severe weather and heavy precipitation on 02.04.98 are chosen for simulations as an extreme but characteristic event. Note, 80% of precipitation reports in that area are associated with cyclonic disturbances and 51% of lows are from mid-latitudes (Turner et al., 1995).

A SHORT DESCRIPTION OF THE NUMERICAL MODELS

Only the essential details and principles of 1-D microphysical and 3-D mesoscale models construction required for a full understanding of the presented results are described. The reader is referred to Pirnach et al. (1994) and Krakovskaia et al. (1998) for a detailed presentation of the methodology, equations and numerics.

The spectrum of liquid cloud drops is divided into two parts: cloud droplets (radii < 20 µm) and raindrops (radii > 20 µm). The spectrum of cloud droplets is the result of condensation, turbulent diffusion and dynamical motion. Additionally, the spectrum of raindrops is determined by the collection of droplets. The size distribution function of the ice particles is assumed to result from sublimation, turbulent diffusion, motion, riming, glaciation (heterogeneous freezing) and accretion. Coefficients for droplet gravitational collection by raindrops and ice crystals are variable and depend on CP radii. For $T < 243$ °K (T is the temperature of the air) spherical ice crystals transform to plates fell with v_2=30 cm/s (v_i is the fall speed of CP, hereinafter, i=1 is for drops, i=2 is for cloud ice particles) and collision processes are stopped. The kinetic equation of CP size distribution functions f_i is used in the following form:

$$\frac{df_i}{dt} + \frac{\partial}{\partial r}(R_i f_i) - v_i \frac{\partial f_i}{\partial Z} = I_{\alpha i} \pm I_{fi} - \delta f_i + \Delta f_i, \qquad (1)$$

$$I_{\alpha 1} = N_{m0} w \delta(Z - Z_w) \delta(r - r_{10}) \Theta(\Delta_1) + N_c \left(\frac{100 \Delta_1}{q_{s1}}\right)^{K_c} \delta(r - r_{10}) \Theta(\Delta_1 - \Delta_{1w}), \quad (2)$$

$$I_{\alpha 2} = A_s e^{B_s T_s} \frac{dT}{dt} \delta(r - r_{20}) \Theta\left(-\frac{dT}{dt}\right) \Theta(\Delta_1) \Theta(T_s), \ T_s = 273.15^o K - T, \quad (3)$$

$$\Theta(x) = 1 \text{ for } x > 0, \Theta(x) = 0 \text{ for } x < 0,$$

where R_i are rates of CP's growths due to condensation (sublimation); $I_{\alpha i}$ are values that describe generation of CCN and IN; I_{fi} are values that describe freezing of droplets; δf_i describe the decrease of CP by collection; Δf_i describe turbulent transfer; w are updrafts; $\delta(x)$ is a delta-function; Δ_1 is water supersaturation; Z_w is the lowest Z with $\Delta_1 > 0$; Δ_{1w} is a value of Δ_1 at $Z=Z_w$; r_{i0} is the smallest CP size; N_{m0}, N_c, K_c, A_s, B_s are the empirical parameters (Pirnach et al., 1994, Krakovskaia et al., 1998).

Thermodynamical state (vertical profiles of T, pressure, specific humidity of the air and three projections of wind speed) of the troposphere is obtained from the 3-D mesoscale stationary model. The simulated domain is 2500x2000x10 km with spacings 50x50x0.2 km. The inputs for 3-D model are data of 10 sound ascendings made at midnights on 27.03-05.04.98 at Bellingshausen station (62°12'S, 58°58'W) that is a central point of the 3-D coordinate grid. The initial profiles of thermodynamic characteristics as inputs for 1-D model move through the 3-D domain with variable horizontal wind speeds calculated as Z-averages for the every point of the grid.

RESULTS AND DISSCUSION

Rainbands of warm (WF), cold (CF) and occluded (OF) fronts were located in the 3-D model over the Antarctic Peninsula. All them were found in areas of gravitational instability that pointed on probable embedded convection into frontal clouds. More detail results of the 3-D simulation presented in Krakovskaia (1999). Presented on Fig.1a the thermodynamical state of the troposphere obtained from 3-D LAM is used for 1-D simulation of rainbands evolution. The first track (1TR) of 1-D model starts at X=850, Y=-150 km, passes through the maximum of WF updrafts w = 38 cm/s at X=550, Y=0 km and crosses CF. The second track (2TR) starts at X=500, Y=-850 km to pass through OF with maximum w =35 cm/s and over Ukrainian Antarctic station "Akademik Vernadsky" (65°15'S, 64°16'W) (Fig.1b, 1c).

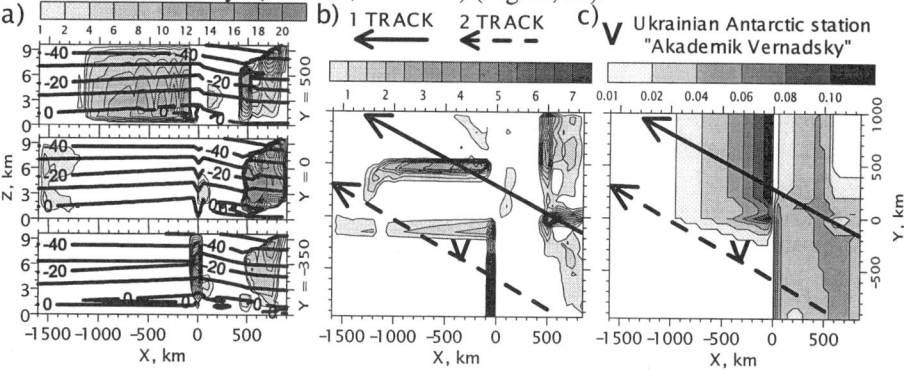

FIGURE 1. 3-D LAM outputs for 00 GMT 02.04.98: (a) vertical cross-sections of w (cm/s) (filled areas) and T (°C) (lines with numbers); (b) integral thermodynamic rate of condensation as precipitation rate (mm/h); (c) integral ice supersaturation (g/kg).

The cloud on the 1TR begins to form at t=3.8 h of the system's development within Z=5.6÷7.4 km that obviously corresponding to Ci, Cs and As cloudiness ahead WF. The diagram on Fig.2a is based on series of numerical runs for the cloud with N_c=270 (s•g)$^{-1}$ and changing K_c (ΔK_c=0.2) and A_s ($\Delta log A_s$=1) (Eq.2, 3). From the diagram the maximum of solid precipitation rates (SPR) corresponds to A_s=10^{-3} (s•g)$^{-1}$, K_c=0.2 and maximum of liquid precipitation rates (LPR) associates with almost absence of IN, that is natural, and with K_c=0.2 too. The diagram on Fig.2b is built for A_s=10^{-3} (s•g)$^{-1}$, ΔN_c=50 (s•g)$^{-1}$. The greater K_c the less N_c is needed to obtain maximum in SPR. Expectedly SPR are twice greater against on Fig.2a, while LPR are almost the same. Both maxima for SPR and LPR correspond to N_c=600÷800 (s•g)$^{-1}$ and K_c=0.2 again.

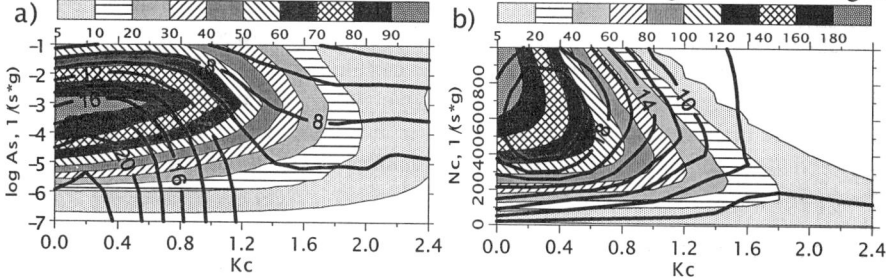

FIGURE 2. Maxima of solid (filled areas) and liquid (lines with numbers) precipitation rates (mm/h) for 24-h of cloud's development depending on parameters K_c, A_s and N_c and calculated for 1TR.

The cloud on the 2TR begins to form at t=7.6 h of the system's development within Z=4.6÷7.8 km. In difference with 1TR the similar series of runs for the cloud on 2TR show even higher SPR, but LPR are negligible for all parameters with the maximum of 2.2 mm/h for $K_c=0.4$, $N_c=270$ $(s\bullet g)^{-1}$ and $A_s=10^{-5}$ $(s\bullet g)^{-1}$. Integral microphysical characteristics of the cloud obtained in one 1-D model's run are presented on Fig.3 to show an example of cloud and precipitation time development.

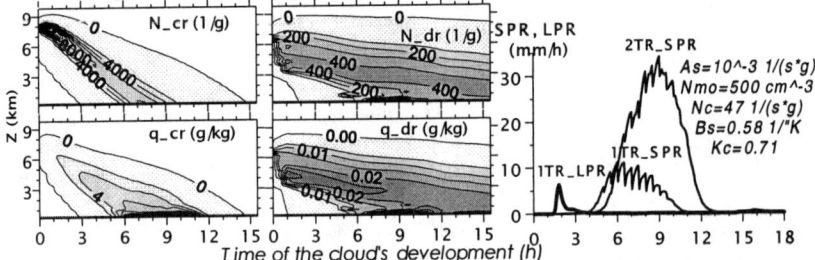

FIGURE 3. Ice crystal (N_cr) and raindrop (N_dr) concentrations, ice (q_cr) and water (q_dr) contents, SPR for 2TR, SPR and LPR for 1TR obtained in 1-D simulation with pointed parameters.

The present study does not aim at a perfect correlation between modelled and observed data, particularly in the lack of the latter. It is obvious that precipitation rates got in the simulations are far overestimated, especially for solid phase. The reason can be in the initial profiles of T, w and Δ_1. Horizontal spacings of 50 km in 3-D LAM mean that 1-D model uses the same profiles, in the present study somewhere with $w>30$ cm/s and $\Delta_1>0.1$ g/kg, about 20-30 minutes. This results in accumulation of huge reserve of water vapor in a cloud and fast growth of ice crystals. But one set of rawinsonde data per day limits decreasing of spacings in 3-D LAM. Note, obtained in Hegg et al. (1995) for the Arctic clouds mean values of $N_c=47$ cm^{-3} and $K_c=0.71$ correspond to reasonable SPR and LPR of 6-10 mm/h in the presented simulations (Figs.2b, 3 for 1TR). Moreover, precipitation of 11 and 66.9 mm have been collected at the station Bellingshausen during 16-24 GMT on 01.04.98 and 13-18, 21-22 GMT on 02.04.98 respectively. Thus, for the last day an average precipitation rate was about 11 mm/h and the peak of precipitation intensity could be far greater of this value.

It is known updrafts and supersaturation in the atmosphere determine cloud and precipitation formation processes mainly (Krakovskaia et al. 1998, etc.). The study demonstrates that at the same thermodynamic conditions the liquid phase of the precipitation formed only by coagulation processes less depends on CCN and IN concentrations, while the solid phase changes dramatically when the concentrations varying, in particular, over the Antarctic Peninsula. The study has pointed extreme values of cloud microphysical empirical parameters and showed that frontal clouds moved from the Pacific to the Antarctic under certain conditions can accumulate the huge reserve of precipitable moisture. Extensive field and numeric experiments on cloud and precipitation formation processes over the Antarctic are needed to verify the models and to obtain numerics for further cloud simulation and parameterization.

REFERENCES

1. Hegg, D.A., Ferek, R.J., and Hobbs, P.V., *J. Appl. Meteor.*, Vol.**34**, No.**9**, 2076-2082 (1995).
2. Krakovskaia, S.V., *Tr. UGMI* **247**, 17-29 (1999) (in Russian).
3. Krakovskaia, S.V., and Pirnach, A.M., *J. Atmos. Res.* **47-48**, 491-503 (1998).
4. Pirnach, A.M., and Krakovskaia, S.V., *J. Atmos. Res.* **33**, 333-365 (1994).
5. Turner, J., Lachlan-Cope, T.A., Thomas, J.P., and Colwell, S.R., *Antarc. Sci.* **7** (3), 327-337 (1995).

Laboratory Studies of Ice Nucleation in Sulfate Particles: Implications for Cirrus Clouds

Anthony J. Prenni, Matthew Wise, Sarah Brooks and Margaret A. Tolbert

Department of Chemistry & Biochemistry and CIRES, CB 216, University of Colorado, Boulder 80309

Abstract. In the laboratory, we have used FTIR spectroscopy to monitor ice nucleation from atmospherically relevant compositions of sulfate particles. Measured freezing temperatures are presented as a function of aerosol composition. We find that sulfuric acid solution aerosol exhibits greater supercooling than ammonium sulfate solution aerosol of similar weight percent. Ice saturation ratios based on these measurements are also reported. We find that ammonium sulfate solution aerosol exhibits a relatively constant ice saturation of S ~ 1.48 for ice nucleation from 232 to 222 K, while sulfuric acid solution aerosol shows an increase in ice saturation from S ~ 1.53 to S ~ 1.6 as temperature decreases from 220 K to 200 K. These high saturation ratios imply that ice nucleation from sulfate aerosols will favor the formation of a small number of large ice particles, in agreement with many observations of cirrus clouds.

INTRODUCTION

Cirrus clouds are expected to form naturally in the upper troposphere when highly dilute sulfate aerosols become supersaturated with respect to ice and homogeneously nucleate ice particles. To determine formation conditions for cirrus, a number of laboratory studies have been conducted on ice formation in H_2SO_4/H_2O aerosol (1, 3, 5, 7). The presence of ammonium ions in sulfate aerosol may influence the occurrence of cirrus clouds in the upper troposphere. A number of studies have also been conducted to determine ice formation in ammonium sulfate aerosol (2, 3, 6).

Despite the abundance of aerosol data for ice formation from sulfuric acid and ammonium sulfate solution aerosol, there remains large discrepancies in the reported freezing temperatures. In an attempt to determine the most appropriate values for ice nucleation parameters in cirrus cloud models, we have measured ice nucleation in submicrometer H_2SO_4/H_2O and $(NH_4)_2SO_4/H_2O$ aerosols using Fourier transform infrared (FTIR) spectroscopy in a temperature-controlled flow tube system.

Experimental

The experimental set-up for measuring ice nucleation has been described fully in *Prenni et al.* (8). Particles are generated by atomizing a 96 wt% sulfuric acid solution or a 27 wt% ammonium sulfate solution. The atomizer yields ~10^7 particles cm^{-3} with a mean droplet diameter of 0.3 μm and a geometric standard deviation of less than two. The aerosol passes through two temperature-controlled flow regions: the conditioning region, where aerosol composition is set, and the observation region, where the aerosol is monitored with FTIR. Prior to introduction of the aerosol, the conditioning tube walls are coated with ice. By controlling the temperature of the ice-coated conditioning region, the water vapor pressure is known. When the aerosol reaches equilibrium with temperature and water vapor in this region, it will have a well-defined composition, calculated using the Aerosol Inorganics Model (AIM) (4). For this method of determining composition to be successful, the aerosol must reach equilibrium in the conditioning region and must retain this composition in the observation region. Tests were run to ensure that these requirements were met.

After ensuring an accurate determination of aerosol composition, an infrared spectrum is taken at temperatures above the freezing temperature. The temperature of the observation region is then incrementally cooled, and aerosol phase is monitored using FTIR. Upon reaching the ice freezing point, several changes occur in the infrared for H_2SO_4/H_2O (1) and for $(NH_4)_2SO_4/H_2O$ aerosol (6). The temperature at which these spectral changes are first clearly visible is used as the freezing point, which corresponds to approximately 5 - 10% of the particle mass in our system.

Results and Discussion

Sulfuric Acid. Freezing measurements were carried out for pure water and for 15 - 24 wt% sulfuric acid, plotted as filled circles in Figure 1. Our results indicate that freezing temperature is strongly dependent on aerosol composition. This agrees well with *Koop et al.* (7), shown as a dashed line in the figure, and moderately well with the data of *Chen et al.* (3), shown as a solid line. Our results fall between those of *Bertram et al.* (1), shown as diamonds, and *Cziczo and Abbatt* (5), shown as stars, which were done under experimental conditions similar to the current study.

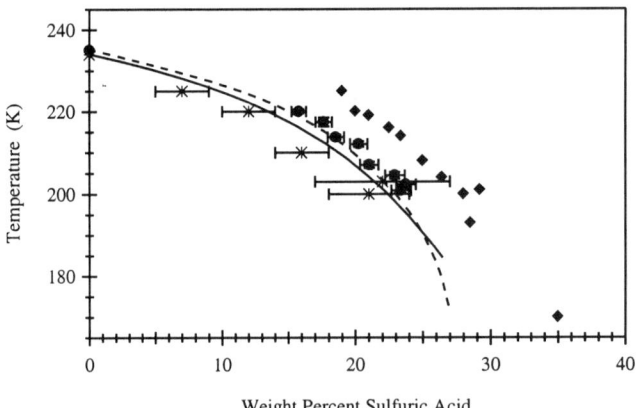

FIGURE 1. Ice formation temperatures in H_2SO_4/H_2O aerosol as a function of wt%. Included are data from this study (filled circles), data from *Bertram et al.* (1) (diamonds), data from *Cziczo and Abbatt* (5) (stars), data from *Koop et al.* (7) (dashed line), and data from *Chen et al.* (3) (solid line).

The freezing point depression is ≥ 39 °C for all of our H_2SO_4/H_2O measurements. However, we observe less supercooling than *Koop* or *Chen*. The discrepancy with *Chen* may be attributed to size differences, as we are measuring larger particles and would expect higher freezing temperatures. However, our particles are smaller than those of *Koop*. This discrepancy may be explained based on differences in freezing point determination. We report the *onset* of ice nucleation for many particles, while *Koop* measures each point independently and reports the *median* temperature. Their technique enables the measurement of a range of freezing temperatures (~4 °C), and our data appears to fall on the upper end of that temperature range.

Ammonium Sulfate. Freezing measurements were conducted for 0 - 40 wt% ammonium sulfate aerosol, shown as filled circles in Figure 2. *Cziczo and Abbatt* (6) (stars) used a technique similar to ours, but found higher freezing temperatures. Fits to the data of *Bertram et al.* (2) (dashed-dotted line) and *Chen et al.* (3) (dashed line) are also shown. *Bertram* used optical microscopy and differential scanning calorimetry, and *Chen* used a continuous flow thermal diffusion chamber. Both studies show significantly greater supercooling than the aerosol flow tube studies.

Cloud physics models often represent the degree of aerosol supercooling using:

$$T^*(wt\%) = T_{H2O} + \lambda \delta T(wt\%) \qquad (1)$$

where $T^*(wt\%)$ is the composition-dependent temperature at which ice nucleates in a solution aerosol, T_{H2O} is the supercooling of a pure water particle, δT is the composition-dependent melting point depression, and λ is the relationship between the particle nucleation point depression and the bulk melting point depression for solutions. *Chen et al.* (3) measured $\lambda = 1.75$ for ammonium sulfate solution aerosol,

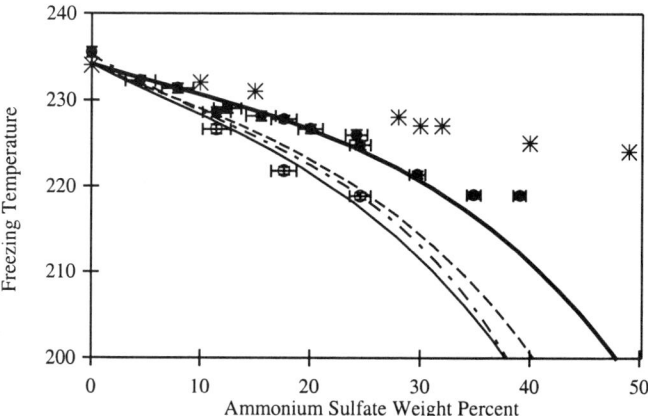

FIGURE 2. Ice formation temperatures in $(NH_4)_2SO_4/H_2O$ aerosol as a function of wt%. The onset of freezing (filled circles) and complete freezing (open circles) are shown. Fits to the data yield $\lambda = 1.2$ for freezing onset (thick solid line) and $\lambda = 2.0$ for complete freezing (thin solid line). Data from *Cziczo and Abbatt* (6) (stars), *Bertram et al.* (2) (dashed-dotted line), and *Chen et al.* (3) (dashed line) are also shown.

and *Cziczo and Abbatt* (6) determined $\lambda < 1$. Using our data, we calculate $\lambda = 1.2$ for the onset of ice formation, shown as a thick solid line in Figure 2. This corresponds to an aerosol supercooling of slightly greater than 39 °C. For several compositions, the aerosol was cooled further in 1 °C increments. The temperature at which further cooling did not yield additional ice was chosen as the temperature of complete freezing, shown in Figure 2 as open circles. These freezing temperatures were 2 - 6 °C colder than the onset freezing temperatures. Based on only three points, we determine $\lambda = 2.0$ for complete freezing of the aerosol, shown as a thin solid line fit to the data. While solution aerosol is expected to have a constant λ, this value can only be determined with size-dependent freezing measurements. We measure a range of sizes simultaneously and are thus unable to determine λ definitively. These two values for λ, 1.2 and 2.0, serve to provide a range for ammonium sulfate aerosol.

Ice Saturation. The freezing data can be used to determine the humidity conditions needed for ice formation in the atmosphere. The ice saturation ratio, S, is often used in cloud models, defined as $P_{H2O}/P(T,ice)$, where P_{H2O} is the water partial pressure at a given temperature and $P(T,ice)$ is the vapor pressure over ice at the same temperature. Although ammonium sulfate aerosol froze at higher temperatures than sulfuric acid of similar weight percent, it did not have a significantly lower saturation ratio. S was nearly constant for ammonium sulfate with a value of 1.48 ± 0.01, although the most concentrated aerosol monitored fell below 1.48. In contrast, saturation ratio increased as temperature decreased for sulfuric acid, from 1.53 at 220 K to 1.6 at 200 K.

Atmospheric Implications. Such high ice saturation ratios indicate that high humidities and/or low temperatures must be experienced for ice nucleation in the atmosphere. This requirement favors the formation of fewer ice particles. Upon freezing, these ice particles can then grow from vapor condensation. The existence of a few, large particles in cirrus clouds has been frequently observed, supporting this mechanism as a probable formation pathway in the atmosphere.

ACKNOWLEDGMENTS

This work was supported by NASA-SASS and NSF-ATM. AJP would like to acknowledge the EPA STAR Fellowship Program for funding.

REFERENCES

1. Bertram, A.K., D.D. Patterson, and J.J. Sloan, *J. Phys. Chem.*, **100**, 2376-2383 (1996).
2. Bertram, A.K., T. Koop, L.T. Molina, and M.J. Molina, *J. Phys. Chem A*, **104,** 584-588 (2000).
3. Chen, Y., P.J. DeMott, S.M. Kreidenweis, D.C. Rogers, and D.E. Sherman, *J. Atmos. Sci., accepted* (2000).
4. Clegg, S.L., P. Brimblecomb, and A. Wexler, *J. Phys. Chem A*, **102**, 2137-2154 (1998).
5. Cziczo, D.J., and J.P.D. Abbatt, *EOS Transactions, AGU Fall Meeting Suppl.*, **79**, F108 (1998).
6. Cziczo, D.J., and J.P.D. Abbatt, *J. Geophys. Res.*, **104**, 13,781-13,790 (1999).
7. Koop, T., H.P. Ng, L.T. Molina, and M.J. Molina *J. Phys. Chem A*, **102**, 8924-8931 (1998).
8. Prenni, A. J.; Wise, M. E.; Brooks, S. E.; and M.A. Tolbert, *J Geophys. Res.*, *submitted* (2000).

Laboratory Studies on the Potential of Tropospheric Insoluble Aerosol Components for Heterogeneous Ice Nucleation

Ottmar Möhler, Helmut Bunz, Aline Nink, and Ulrich Schurath

Forschungszentrum Karlsruhe, Institute of Meteorology and Climate Research, Postfach 3640, D-76021 Karlsruhe, Germany

Abstract. The large coolable and evacuable aerosol chamber AIDA is used as a cloud chamber to study processes of ice formation in tropospheric clouds. Intensity and depolarisation of forward- and back-scattered laser radiation is measured, caused by particles in a small scattering volume far from the walls. Number size distribution of interstitial aerosol and activated ice particles are measured with an optical particle counter. Droplet freezing and growth of the ice particles can thus be detected unambiguously. Particle freezing and growth is initiated by adiabatic expansion which leads to volume cooling and thus ice- and water supersaturation at constant wall temperature. Various insoluble aerosol components can be generated and added to the chamber in order to investigate their influence on ice formation processes at controlled temperatures and supersaturations.

INTRODUCTION

Ice particle formation is an important process in the troposphere and stratosphere. It can occur either by homogeneous freezing of droplets below about –35°C [1], or be heterogeneously induced by so-called ice nuclei. E.g. it is speculated that soot particles from aircraft can act as ice nuclei [2]. A quantitative description of these processes is crucial for a better understanding of the lifetime of clouds with respect to rainout, and their optical properties. Distinction between supercooled liquid and frozen aerosol particles (cloud hydrometeors; PSC particles) is essential for the investigation of these ice nucleation processes.

EXPERIMENTAL

AIDA Cloud Chamber

A schematic cross section of the AIDA cloud chamber and some analytical and technical instrumentation is shown in Figure 1. In this chamber (Volume = 84 m^3) we are able to simulate a wide range of atmospheric conditions: -90°C < T < +60°C, r.h. under static conditions near 100 %; pressures from above 1 bar to below 1 mbar; water and ice supersaturations. This covers conditions throughout the troposphere and lower stratosphere under which water clouds, mixed clouds, cirrus clouds, and even Polar Stratospheric Clouds (PSC) are formed.

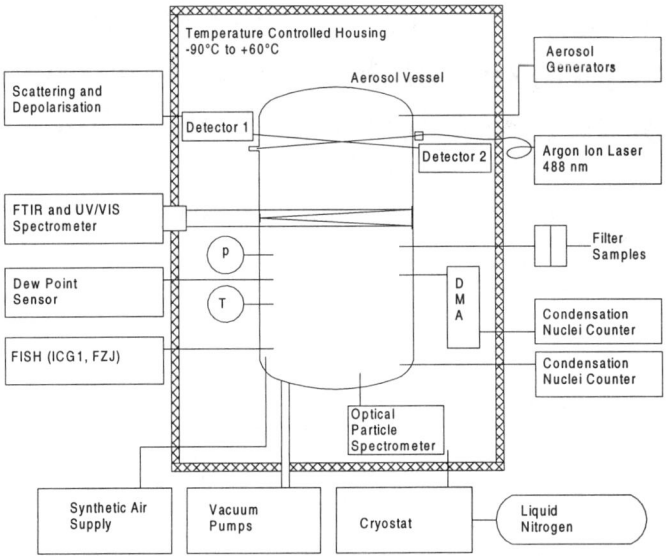

FIGURE 1. Schematic of the AIDA instrumentation for ice nucleation experiments

Ice Supersaturation

Water vapour supersaturation with respect to water or ice is achieved by controlled adiabatic expansion of the chamber air, using a large mechanical pump. This causes homogeneous volume cooling at constant wall temperature (Fig. 2), as shown by careful studies of the temperature distribution during the expansion process. The temperature difference between the gas and the walls is a result of the steady-state between adiabatic volume cooling and heat flux from the walls into the chamber.

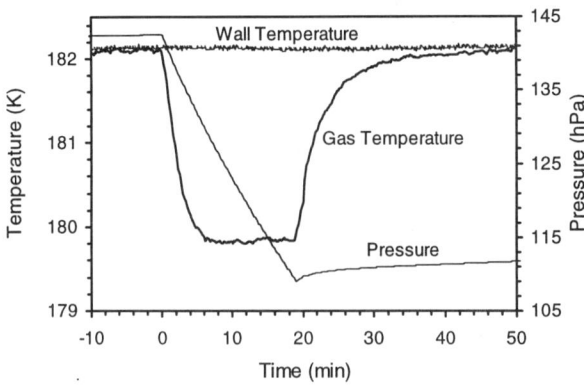

FIGURE 2. Example of an adiabatic cooling experiment started at a pressure of about 140 hPa and a temperature of 182 K.

A heat transfer model is applied to simulate the temperature changes in the aerosol vessel during pumping. Based on these temperature profiles the maximum saturation ratios with respect to ice and supercooled water were calculated assuming ice saturation in the vessel at starting the expansion and keeping the volume mixing ratios constant during pumping. The results are shown in Fig. 2 as function of the initial temperature in the vessel. With the currently available pumping system (solid lines) ice saturation ratios of up to 1.7 can be achieved. Saturation ratios up to 2.3 are reached at twice the current pumping speed that will be available in future experiments by using an additional expansion volume or an additional vacuum pump.

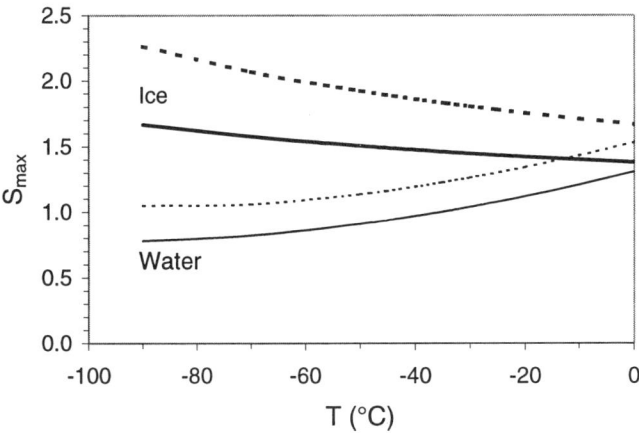

FIGURE 2. Calculated maximum super saturation ratios S_{max} with respect to ice and super cooled water that can be achieved depending on initial temperature for current pumping speeds (solid lines) and doubled pumping speed (dashed lines) that will be achieved in the future.

Optical Detection of Ice Formation

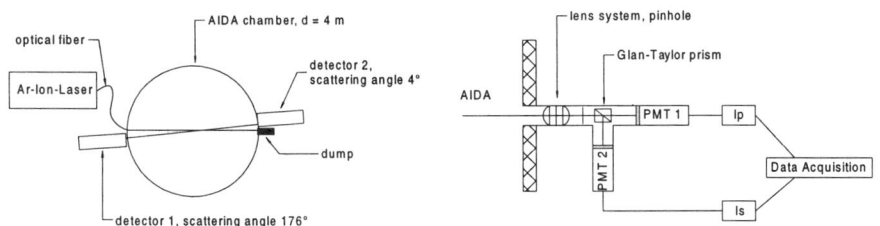

FIGURE 3. Experimental set-up of AIDA scattering and depolarisation measurements.

We have built a backward / forward scattering system consisting of a narrow Argon ion laser beam (99% polarised radiation at 488 nm) which overlaps with the line of sight of two optical detectors at scattering angles of 176° and 4°. Laser radiation is scattered by particles in the overlap volume of ~ 1.8 cm^3 in the centre of the chamber, and detected polarisation resolved. The solid angle sensed by the detector optics, at a distance of 2 m from the scattering volume, is extremely narrow. Photon counting must therefore be used for high sensitivity and time resolution. The laser source and the detectors can be attenuated by neutral density filters to avoid saturation, and to match the sensitivities of the forward and backward scattering detectors.

RESULTS

An example for the depolarisation of backward scattered laser radiation by small ice crystals, and for the increase in depolarisation upon crystal growth is depicted in Fig. 4. The pressure was 140 hPa at the beginning of the experiment, simulating lower stratospheric conditions at a temperature of 201 K. The chamber air was approximately ice saturated, because ice had previously been formed on the walls. When a very small amount of humid air was rapidly introduced into the AIDA chamber at 201 K, small ice crystals grew immediately on the pre-existing seed aerosol, and a depolarisation ratio of ~0.05 was observed. After 15 min the chamber pressure was lowered by constant pumping. While the temperature of the chamber walls remains essentially constant due to its large heat capacity, an adiabatic drop of the air temperature by $\Delta T \sim 3$ K was observed. This results in ice supersaturation which causes the ice crystals to grow, as indicated by a significant increase of the back-scattered intensity. Crystal growth also leads to an increase of the depolarisation ratio from 0.05 to a final value of about 0.08. Note that the depolarisation ratio, which is a function of the size and shape of the particles only, remains essentially constant at constant ΔT, while the back-scatter intensity drops due to ice crystal sedimentation. Further experiments are in preparation to systematically investigate the heterogeneous ice nucleation behaviour of e.g. soot and mineral aerosols.

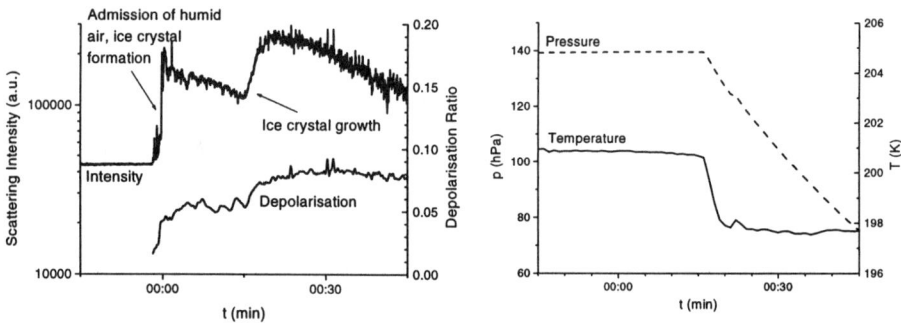

FIGURE 4. Back-scattered intensity and depolarisation ratio (left) during an expansion experiment started at 140 hPa and 201 K (right)

REFERENCES

1. Pruppacher, H. R., and Klett, J. D., *Microphysics of clouds and precipitation* The Netherlands: Kluwer Academic Publishers, 1997.
2. DeMott, P. J., Chen, Y., Kreidenweis, S. M., Rogers, D. C., and Sherman, D. E., *Geophys. Res. Lett.* **26**, 2429-2432 (1999).

Phase Transitions in Ice Phase of Clouds under Influence of Adsorbed Ions Electric Field

V. V. Klingo

Voeikov MGO RCARS, St. Petersburg, Russia

Abstract. New theoretical estimations of ice phase formation possible mechanisms in Earth's atmosphere clouds bounded with adsorbed ions have been carried out. Homogeneous mechanism of cloud droplets crystallization and heterogeneous ice nucleation by sublimation on an aerosol particle have been considered. The main principle of an electric field action on the phase transition is conserved. The electric field promotes crystallization if the difference of the electric field energy interaction with final and initial water phases in an embryo volume is negative. However, the theoretical calculations method previous published has been changed. The calculations have been performed in frame of classical theory of ice nucleation. A semispherical homogeneous embryo at a droplet surface forms at temperature slightly above -20°C. If the phase transition is heterogeneous ice nucleation by sublimation, then the ion must find oneself on the ice embryo spherical-cap. The ion provides the ice contact angle values under the characteristic cloud conditions.

Theoretical estimations of the ion electric field action on ice phase formation in clouds had been presented by Klingo (1996). Homogeneous water crystallizations and heterogeneous ice nucleation by sublimation on a substrate i.e. the aerosol particle have been considered. The estimations have been performed in an assumption that molecules of water and ice are hard dipoles with definite values of dipole moments. The ion electric field interaction with these dipoles subject to thermal motion is expressed by Langevin's function.

The main principle of the electric field action on the phase transition has been conserved but the theoretical calculations method has been changed. Mean uniform ion electric $\bar{E}$ over the volume of semispherical ice embryo V_{h_o} is set in accordance to nonuniform field in this volume. Horizontal and vertical components of this field are found. Average cosines $\overline{\cos\theta_i}$ of an angle between directions of the electric field and dipoles of water phases are determined from experimental values of dielectric constants of water $\varepsilon_2 = 90$ and ice $\varepsilon_3 = 100$, $\overline{\cos\theta_2} = \overline{\cos\theta_3} = 1$. The surface layer orientation of molecules of water and ice are not taken into account because this orientation is not grounded in the strict sense.

The phase transitions description has been performed in frame of classical nucleation theory (Fletcher, 1966). The expression of the homogeneous nucleation rate J takes in a form

$$J = e^{71.4} 10^6 \exp\left\{-\frac{G - G_E}{kT}\right\} m^{-3} s^{-1} \tag{1}$$

where G is the free energy of ice phase embryo formation in a liquid without the ion influence, G_E, is the free energy interaction difference of the ion electric field with crystal and liquid phase of water, k is Boltzmann's constant. It is assumed that a kinetics of the crystal embryo formation in the liquid phase with the electric field effect is the same as without this field.

$$G = \frac{2}{3}\pi r^3 \rho_3 L \ln(T/T_0) + \pi r^2 (\sigma_{13} - \sigma_2 + 2\sigma_{23}) \qquad (2)$$

where r is the radius of the semispherical ice embryo formed at a droplet surface, ρ_3 is the ice density, L is the specifice heat and T_0 is the temperature of an ice fusion, σ_{ik} is the surface tension coefficient of the interface between the phases i and k, index 1 corresponds an air medium, indexes 2 and 3 correspond respectively liquid and ice phases.

$$r = \frac{3\sigma_3}{\rho_3 L \ln(T_0/T)} \qquad (3)$$

$$G_E = -\frac{N}{18}\rho_3 V_{h_0} \overline{E}(p_3 \overline{\cos\theta_3} - p_2 \overline{\cos\theta_2}) \qquad (4)$$

where N is Avogadro's number, V_{h_0} is the volume of an embryo at the depth of water monomolecular layer $h_0 = 3\cdot 10^{-10}$ m, where the electric field penetrates without attenuation, the dipole moments p_i are $p_2 = 1.84D$ and $p_3 = 2.6D$.

The necessary condition for a droplet freezing during 1 sec is

$$\frac{2}{3}\pi r^3 nJ \geq 1 \qquad (5)$$

where n is the number of adsorbed ions.

The value $\overline{E}$ depends on an ion radius a. Gas ion radii of a natural atmospheric air are equal about $0.5\cdot 10^{-10}$ m.

The results of calculations according to formula (5) taking into account $a = 0.5\cdot 10^{-10}$ m and values of all necessary parameters (Dufour and Defay, 1963) show that even micron size cloud droplets are frozen during 1 sec at $T = -20°C$ under the ions action. Thus, the results coincide practically if the calculations have been conducted by two distinct methods. The value G_E (4) in exponent (1) differs in these methods approximately not more than 7% that is within the limits of the phase transition theory exactness.

If the phase transition is the heterogeneous ice nucleation by sublimation, then the ion must find oneself on the ice embryo spherical-cap. The nucleation rate calculations are performed on the basis of the expression by Fletcher (1966) for characteristic cloud values of a vapor supersaturation and a temperature. The nucleation rate depends on a contact angle of the ice on the substrate and radii ration of substrate and embryo. New method increases a contribution to the nucleation rate conditioned by the ion action roughly on 30%. However, this does not lead to considerable quantitative change of the contact angles and the radii ration which condition an ice phase appearance.

REFERENCES

Klingo, V.V. (1996), *Proceedings of 14th ICNAA*, 322-325.
Fletcher, N.H. (1966), *The Physics of Rainclouds*.
Dufour, L. and Defay, R. (1963), *Thermodynamics of Clouds*.

Influence Of Substrate Electric Charge On Heterogeneous Ice Phase Formation In Clouds

V.V. Klingo

Voeikov MGO RCARS, St. Petersburg, Russia

Abstract. The influence of substrate electric charge on heterogeneous phase transition water vapor to ice has been considered. This consideration has been conducted in the light of general investigation of the ice phase formation possible mechanisms in clouds under seeding operations for precipitation regulation purpose. The burning temperature of pyrotechnic mixtures reaches 3000 K. Reagent particles become thermoelectronic emission sources. A charges separation leads to the formation of charged particles in a cloudy medium. Water vapor molecules as multipole systems interact with the charged particle and increase their concentration at a particle surface. The action effect of the charged particle to heterogeneous crystallization is conditioned by the water vapor supersaturation increase at the particle surface. The action effect of the charged particle to heterogeneous crystallization is conditioned by the water vapor supersaturation increase at the particle surface, decrease of an ice embryo radius and accordingly the free energy of ice embryo formation. The heterogeneous ice formation at the charged particle in the cloudy medium must take place by more high temperatures and greater ice contact angles at the particle. The heterogeneous crystallization estimations have been performed on the basis of the classical nucleation theory.

The electrical charge influence of particle-substrate on heterogeneous phase transition water vapor to ice is considered in light of general investigations of ice phase formation possible mechanisms in clouds. The burning temperature of pyrotechnic mixtures, which are used under seeding operations for precipitation initiation purpose, reaches 3000 K. Hard particles with radius $10^{-6} - 10^{-8}$ form as a result of the burning. This hot particles become thermoelectronic emission sources. The positive charged particle is surrounded by an electronic cloud with decreasing density. The electrons are able to leave the attraction particle region either b them large enough thermal velocity or under the electric field action of foreign sources. A charge separation leads to emergence positive charged particles in a cloudy medium. Water vapor molecules as multipole systems interact with the charged particle, increase their own concentration at a particle surface, and decrease the free energy of ice embryo formation. Action particle electrical charge on the phase transition is only restricted water vapor concentration increase.

Space distribution of vapor concentration $n(R)$ depending on distance R from the charged particle center is described by Boltzmann's distribution

$$n(R) = n_o \exp\left\{\frac{-2}{kT}\sum_{i=1}^{4} U_i\right\}, U_1 = \frac{-q\mu \overline{\cos\beta}}{R^2}, U_4 = \frac{-\alpha q^2}{2R^4} \tag{1}$$

where n_0 is water vapor concentration in cloud medium at sufficiently large distance from charged particle. U is the interaction energy of the particle with vapor molecule moments correspondingly dipole U_1, quadrupole U_2, octople U_3, and indued (dipole) U_4; μ and α are dipole moment and polarization coefficient of water vapor molecule correspondingly, $\overline{\cos\beta}$' is the average cosine of the angle between vectors of electric field strength and of dipole moment which is expressed by Langevin's function, q is particle surface. This charge q_{max} corresponds to value of maximum superficial electric field strength E_{max} is roughly equal $3.3 \cdot 10^4$ of absolute electrostatic units.

Special analysis of Boltzmann's distribution time formation has not been conducted. However, displacement time of vapor molecules form considerable distance to charged particle surface is estimated in fractions of a second.

Vapor concentration p_v increase at charged particle surface has been found depending on the charge particle value for all the interaction forms (1) separately. It is established that total contribution in p_v has been conditioned by dipole interaction almost completely. Quadrupole and induced interactions increase vapor concentration even at q_{max} within the limits of 2.2%. The dipole interaction increases p_v and q_{max} on 120% and at $0.1q_{max}$ on 1%.

Nucleation rate calculations of heterogeneous ice nucleation by sublimation have been performed in frames of phase transition classical theory. Maximum value of ice contact angle at particle ϑ_{max}, at which charged particle forms ice embryo during 1 second, has been estimated. Dependence ϑ_{max} on charge, radius particle, and temperature of the cloud medium are presented in Table.

TABLE 1.

T,C	R,m	η 1	0.4	0.2	0.1	0
-5°	10^{-6}	55	25	20	12	12
	10^{-7}	55	25	15	10	10
20°	10^{-6}	60	30	25	20	20
	10^{-7}	55	30	25	20	20
	10^{-8}	50	20	18	15	15

Maximum value of ice contact angle at particle ϑ_{max} (in degrees) depending on charge particle in parts η of maximum charge q_{max}, radius particle R, and T of cloud medium.

PARTICLE GROWTH AND AEROSOL DYNAMICS

Investigation of the Effect of Operator Splitting on the Growth of an Aerosol Population Considering Multiphase Chemical Processes

Frank Müller

Meteorological Institute, University Hamburg, D-20146 Hamburg, Bundesstr. 55, Germany

Abstract: Since the theoretical investigation of the operator splitting error for non-linear systems is not possible at present time the paper presents a numerical investigation of this topic using a box-model framework. The considered system represents micro-physical and multiphase chemical processes relevant for an evolving poly-disperse particle population. Results show that the effect of operator splitting on the simulation results for total liquid water content, total content of chemical matter inside the particles, as well as for the total content of single aqueous phase species is increasing with increasing length of de-coupling intervals, but decreasing with increasing simulation time under the conditions specified for this study.

1. INTRODUCTION

Simultaneous numerical simulation of dynamical, micro-physical, and multiphase chemical processes is very time consuming in multi-dimensional models. Application of such numerical models with present day computer resources often requires a functional separation of the overall model and a separate sequential solution of the corresponding sub-problems. The application of the method of 'operator splitting' to solve 3D chemistry-transport-problems has been accepted for many years in the modeling of atmospheric chemical processes. Disadvantages, inherent to this method, are the creation of discontinuities of the spatial concentration distribution at the beginning of each 'chemical' time step (which lead also to an increase of the stiffness in the case of the chemical system) and splitting errors which add up to the discretization errors and other errors due to the numerical solution method.
Operator splitting is analogously applied to the complex micro-physical and multi-phase chemical processes in clouds, too. The de-coupling of the corresponding processes is expected to cause strong deviations from the solution of the completely coupled system for both the micro-physical and the chemical particle properties. This would have further consequences for the life time of clouds, their optical properties, the deposition behavior, etc.
The aim of this study is a systematic investigation of the uncertainty/variability of the simulated cloud micro-physical and chemical parameters for the de-coupled modeling of the corresponding processes in cloud models and cloud modules of air quality models as a function of the de-coupling period.

2. MODEL DESCRIPTION AND MODEL SETUP

To investigate the effect of operator splitting between micro-physical and multiphase chemical processes a simple box model is employed. A relatively simple system of processes for a size distributed particle population is considered to determine a clear cause-effect relationship. Micro-physical processes are represented by condensation/evaporation while multiphase chemistry compiles the uptake of trace gases by particles and aqueous phase reactions. Table 1 summarizes the considered gas phase species and their diffusion coefficients, $D_{g,j}$, their initial concentrations c_0, and sticking coefficients a_j. The physico-chemical aerosol parameters are arbitrarily chosen for this study. The applied maritime aerosol distribution is constructed as the superposition of three log-normal distributions according to Jaenicke (1987). The size dependent distribution of the aerosol composition was arbitrarily selected. NaCl, $(NH_4)_2SO_4$, and SiO_2 are considered as initial aerosol components. For this study the considered particle size range is arbitrarily selected according to the critical particle radius for activation, r_c, at the given relative humidity, RH. The lower limit of the considered particle size range was set to a dry particle radius of $r_c=7.4\,10^{-5}$cm. More details of the model can be found in Müller and Mauersberger (1994).

Species	O_3	HNO_3	H_2O_2	HCl	CO_2	SO_2	NH_3
$D_{g,j}$	0.15	0.13	0.18	0.18	0.16	0.13	0.25
a_j	0.52	0.11	0.13	0.08	0	0.13	0.1
C_0 (ppbv)	50	1	1	0.5	330	5	1

Table 1: Diffusion coefficients, $D_{g,j}$, sticking coefficients, a_j, and initial concentrations c_0

In the case of mutual interaction of micro-physical and multiphase chemical processes, as well as for the de-coupled treatment the Equations are solved with a Gear-solver (Hindmarsh, 1980). The de-coupling intervals are varied between 0.01s and 1000s assuming that these periods are representative for present day air quality models where dynamic and micro-physical/chemical processes are de-coupled as well. Hence, the box is regarded as 'open' with respect to the thermodynamic parameters temperature T, relative humidity RH, and pressure P. For the most part these parameters are mostly kept constant when applying the concept of operator splitting. With respect to the trace gases the box is considered to be 'closed' since trace gas concentrations are objects of the chemical processes. The thermodynamic parameters are initialized as: T=280K, P=800hPa, RH=100.1%.

3. RESULTS

The time evolution of the differences in size integrated total liquid water content (Figure 1a) increase markedly for de-coupling intervals longer than 1s. This can be explained by the characteristic times of water vapor diffusion which are in the order of a magnitude smaller than 1s. A reduction of the de-coupling interval does not notably change the total liquid water content but the size dependent water content of the particles (Figure 1c). With increasing de-coupling intervals between micro-physical and multiphase chemical processes the total water content is increasingly underestimated in the case of de-coupled treatment. The temporal evolution of the differences of total liquid water content within each single de-coupling interval reaches a maximum of about 5-8% after 300-500s simulation time. This behavior can be explained as follows: At the beginning of the growth the counteracting Kelvin effect and Raoult (or solution) effect are of special importance for the smallest particles which are characterized by the highest rates of water vapor as well as trace gas uptake, respectively. In the case of mutual interaction of dilution by condensation of water vapor and multiphase chemical processes, condensation causes enhanced fluxes of trace gases due to reduction of the trace gas specific saturation vapor pressure at the particle surface. This, in turn, regenerates the depression/reduction of the water vapor saturation ratio at the particle surface. In the case of de-coupled treatment the water vapor fluxes towards the particle surfaces decrease faster for longer de-coupling intervals due to rapid dilution since the mass of chemical substances in the individual particles is fixed during the micro-physical step. In these cases the Raoult effect rapidly loses impact on the drop specific condensation rate. This mechanism is of decreasing importance for bigger particles, and, hence of decreasing importance for increasing simulation time.

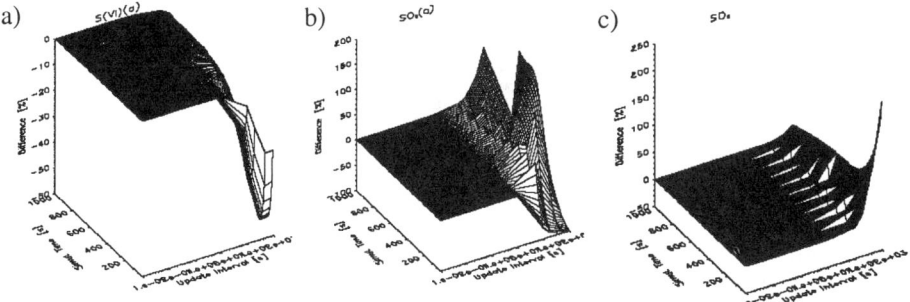

Figure 1: (a) Percent difference of the total liquid water content, (b) of the total amount of chemical matter inside the particles as a function of de-coupling interval and simulation time, and (c) of the particle size for the de-coupling period of 1000s.

The behavior of the time evolution of the differences between mutual and de-coupled treatment of the processes with respect to total mass dissolved in the particles (Figure 1b) is qualitatively different from that of total liquid water content. The total content of chemical matter inside the particles is increasingly overestimated with increasing de-coupling intervals while the overestimation decreases with increasing simulation time within each single de-coupling interval. This behavior is caused by the fact that at the beginning of each chemical operator step during one de-coupling interval the concentration gradients between the corresponding gas and aqueous phase components, respectively, are very high due to the preceding micro-physical operator step. Compared to the completely coupled case this leads to higher fluxes of gaseous precursors towards the the particle surfaces which overcompensate the opposite effect due to the inverse flux-to-particle surface relationship. Additionally, the particle volume related sulfate production in the particles is enhanced by sufficient oxidant availability in the de-coupled case through the particles are somewhat smaller (about 3-8% in radius, Figure 1c) in this case mainly during the very early simulation period. The decreasing importance of these effects with increasing simulation time is caused by the continuous depletion of the gas phase reservoir of precursor species which can also be seen in Figure 1.

The evolution of the de-coupling error with time for the case of sulfur in the corresponding phases is shown in Figure 2. It is obvious from Figure 2a that the operator splitting error of the total S(VI) concentration in the particles is strongly anti-correlated with the corresponding error of the total liquid water content (Figure 1a). This reflects the dilution relations between the completely coupled and all de-coupled cases: The maximum underestimation of the total liquid water coincides with the maximum overestimation of the sulfate concentration in the particles for the de-coupled cases.

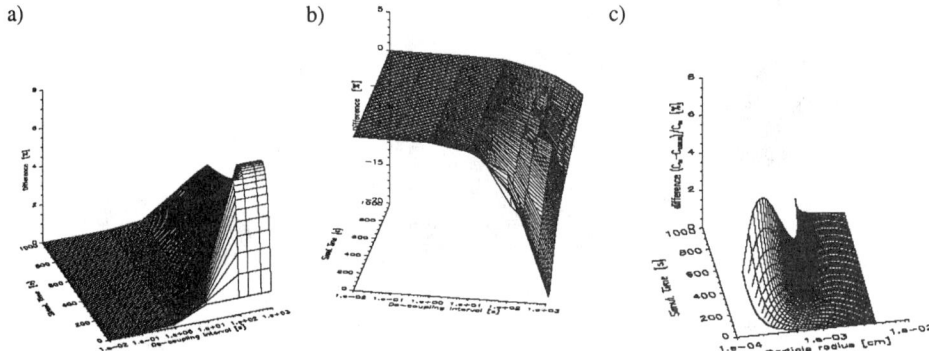

Figure 2: Percent differences of sulfur compounds as a function of de-coupling period and simulation time. a) aqueous S(VI), b) aqueous S(IV), SO_2

As expected, the evolution of the SO_2-difference (Figure 2c) as a function of simulation time and de-coupling period reflect the integral response of the gas phase concentration on the sulfur production in the particle population. The development of the S(IV)-difference expressively shows the action of operator alternation - best seen for the de-coupling intervals of 500s and 1000s. At the end of the micro-physical operator step, where no chemistry took place, the differences are negative since S(IV) was not converted by oxidation processes to S(VI) in the de-coupled cases, while in the coupled case oxidation and dilution mutually occurred. In the course of the chemical operator step, S(IV) is continuously depleted in the particles which leads to a change in sign of the corresponding operator splitting error.

4. CONCLUSIONS

The presented results indicate that de-coupling of condensation/evaporation and multiphase chemical processes for less then 10s generates no significant operating splitting errors compared to results of completely coupled simulations. For longer de-coupling intervals very strong deviations in the aqueous phase concentrations occur especially during the early phase of each de-coupling interval. Since the presented system is relatively simple further investigations are necessary to improve the understanding of non-linear interactions in more complex and, therefore, in more realistic systems.

5. REFERENCES

Hindmarsh, A.C., 1980: LSODE and LSODI, Two new initial value ordinary differential equation solvers, ACM-SIGNUM Newsl. 15, no. 4, 10-11

Jaenicke, R., 1987: Aerosolphysics and chemistry, in Landold-Börnstein Neue Serie, 46: Physical and chemical properties of the air, Springer Verlag, Berlin

Müller, F., Mauersberger, G., 1994: Case study on the interaction of size dependent multiphase chemistry and detailled microphysics, Atmos. Res., 32, 273-288

Coagulation in the Presence of Stochastic Nano-Particle Sources

A. Vrtala and P. E. Wagner

Institute for Experimental Physics, University of Vienna, Boltzmanngasse 5, A-1090 Vienna, Austria

Abstract. The behavior of atmospheric aerosols undergoing coagulation processes in the presence of aerosol particle sources with randomly distributed source strength is reported. Numerical calculations were performed for a set of different initial conditions and random source. It is shown that the interplay of aerosol formation and aging can lead to trimodal aerosols with modal radii corresponding to nucleation, the so called 'Aitken' and accumulation mode.

INTRODUCTION

It is generally accepted that aerosol effects are the most uncertain part of anthropogenic climate forcing. Human activities cause the release of large amounts of aerosols into the atmosphere. These aerosols influence the global radiation balance directly by means of scattering and absorbing solar radiation, and indirectly by acting as cloud condensation nuclei. Due to the complexity of the processes involved, numerical modeling of condensation and coagulation processes in atmospheric aerosols play a key role for a realistic interpretation of experimental findings.

Recent experiments indicate that atmospheric aerosols in the submicron range show a particle number spectrum consisting of three clearly separated modes: A nucleation mode at about 5-10 nm, a so called 'Aitken' mode at about 40-80 nm and an accumulation mode ranging from 250 to 700 nm[4-9], updating the old concept of bimodal structure introduced by Whitby[3].

In this paper, we present first results of numerical simulations of a coagulating aerosol showing an initial accumulation mode in the presence of a particle source with randomly distributed source strength.

METHOD

In order to solve the coagulation equation

$$\frac{\partial n(v,t)}{\partial t} = \frac{1}{2}\int_0^v K(v-\bar{v},\bar{v})n(v-\bar{v})n(\bar{v},t)d\bar{v} - n(v,t)\int_0^\infty K(v,\bar{v})n(\bar{v},t)d\bar{v} + s(v,t) \quad (1)$$

it is transformed into a set of coupled differential equations for the particle size distribution $n(v,t)$. A logarithmically equidistant radius-scale was chosen, details of this procedure are given elsewhere[1,2]. The coagulation function $K(v,\bar{v})$ is given by Fuchs' well known interpolation formula.

Initial aerosol particle size distributions were assumed to be of lognormal shape. According to Whitby[3] the accumulation mode is typically described by a modal radius $r_{acc} = 250 nm$, a geometric standard deviation $\sigma_{g,acc} = 2.0$, and total number concentration $N_{acc} = 3 \cdot 10^4 cm^{-3}$. The aerosol particle source $s(v,t)$ is given by a lognormal size distribution $S(v)$, the source strength is changing with time. Since daylight generally causes photochemical particle formation in the atmosphere, the particle source intensity will mainly vary according to a sinus function: $s(v,t) = S(v) \cdot (\sin(2\pi\, t/1day) + 1)$. Besides such a general trend, random bursts of particles will often be released into the atmosphere by anthropogenic sources. These events will be superimposed upon the sinus function. The source function thus takes the form: $s(v,t) = S(v) \cdot ((1-\alpha) \cdot (\sin(2\pi\, t/1day)+1) + \alpha \cdot ran\tau(t))$, where α is a weight factor in the range between zero and one characterizing the relative importance of stochastic anthropogenic influences. In the present study α is assumed to be $1/3$. $\tau(t)$ describes the speed of random changes. The random function *ran* is assumed to change its curvature at every minute (or every 10 minutes), it varies between zero and two with an average of unity. All simulations reported were carried out for a total time of 4 days with a typical accuracy of better than 2% (mass-conservation in coagulation process). The numerical integration of eq. (1) is performed by means of NAG routine D02CBF[10].

RESULTS

A representative sample of aerosol size distributions with varying source strengths is shown in figure 1. The source production rate was chosen to change its curvature every minute. The size distribution of the source is described by: $r_{nucl} = 7 nm$, $\sigma_{g,nucl} = 1.7$. It can be seen that a production term of $2 \cdot 10^5$ *particles/cc and hour* does not suffice to create a trimodal size distribution. However, a slight shoulder with a peak at about 50 *nm* radius can be observed on a log-log plot. For an average source production term of $1 \cdot 10^6$ *particles/cc and hour* an 'Aitken' mode at about 70*nm* is observed, which shifts to a radius of about 80*nm* for higher source strengths. During the coagulation process the modal radius of the nucleation mode shifts from 7*nm* to about 10 to 12 *nm*. In order to yield a modal radius of nucleation mode of about 7 *nm*, the modal radius of the source must be set to 4 *nm* at the same geometric standard deviation. This, however, enhances coagulation rates and requires even higher source production terms in order to create a trimodal aerosol size distribution. Only for a source strength of $2 \cdot 10^6$ *particles/cc and hour* a separate mode with a radius of about 40*nm* can be observed.

In order to show the variability of the aerosol size distribution after a total coagulation time of four days, several runs with varying time constants of the source term have been performed. The source curvature is chosen to change either every minute or every 10 minutes. It has been observed, that both 'Aitken' and accumulation mode are almost not influenced by the source. However, slower changes in the nucleation source term tend to exhibit more pronounced changes in the aerosol size distributions. Table 1 shows that a source with slower change rates causes a greater variability of the particle number concentration. For comparison calculations without a random source were performed, of course showing no variability. Minimum and maximum concentration values were taken for the 3rd and 4th day in order to avoid effects of initial conditions. All calculations were carried out for a random modal source radius of $7nm$.

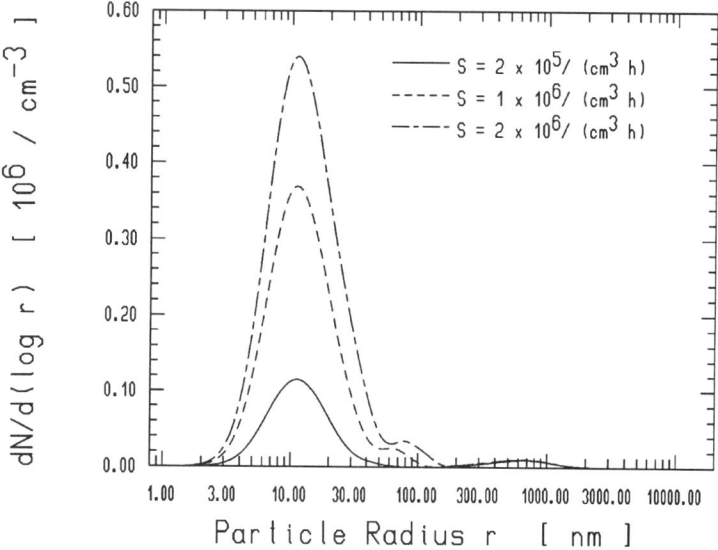

FIGURE 1. Typical aerosol size distribution after 4 days with varying source intensity. Source function changes its curvature every minute.

TABLE 1. Influence of varying nucleation mode source strengths on nucleation mode

Source strength	Sinus-type source function without random source	Random source: 1 minute	Random source: 10 minutes
2×10^5/cc h	$N_{min} = 1000$ /cc, $N_{max} = 1 \times 10^5$/cc	$N_{min} = 2 \times 10^4$/cc, $N_{max} = 9 \times 10^4$/cc, Variability: +/- 2000/cc	$N_{min} = 2 \times 10^4$/cc, $N_{max} = 2 \times 10^5$/cc, Variability: +/- 10^4/cc
2×10^6/cc h	$N_{min} = 2 \times 10^5$/cc, $N_{max} = 5 \times 10^5$/cc	$N_{min} = 2 \times 10^5$/cc, $N_{max} = 4 \times 10^5$/cc, Variability: +/- 3×10^4/cc	$N_{min} = 2.5 \times 10^5$/cc, $N_{max} = 4.5 \times 10^5$/cc, Variability: +/- 6×10^4/cc

Figure 2 shows a fit of the modal radius of the nucleation mode as function of time. While pure sinus-type source terms will permit aging and aggregation of comparatively large modal radii of about $70nm$, the presence of random sources will not allow the same amount of aging (solid curves). This is due to repeated introduction of new nucleation mode aerosol, which locks the mode at lower modal radii. A similar behavior of daily changes of nucleation mode sizes has experimentally been found by Mäkelä[5]. It should be mentioned, however that the present calculations refer rather to urban aerosols, while Mäkelä[5] reported background measurements.

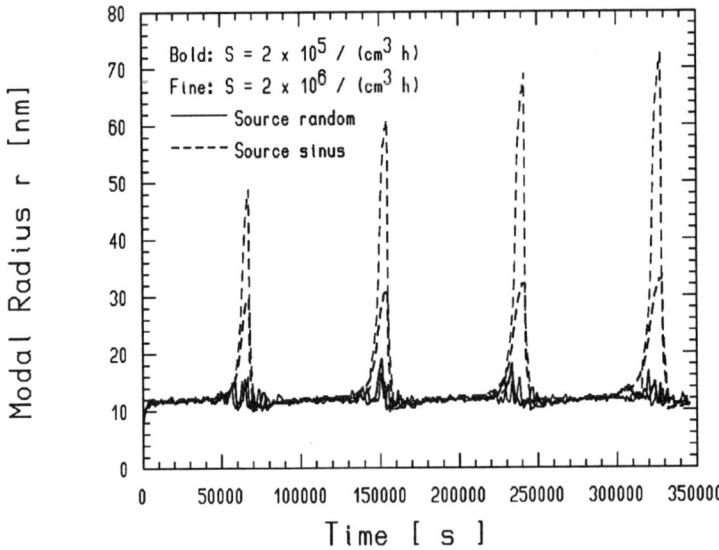

FIGURE 2. Modal radii of the nucleation mode as function of time and varying source intensities. Random source behavior is compared with pure sinus shaped source intensity time function. Modal radius of nucleation mode source is $7nm$.

ACKNOWLEDGMENTS

This study has been supported by the Hochschuljubiläumsstiftung der Stadt Wien and by the Vienna University Computer Center.

REFERENCES

1. Majerowicz, A., and Wagner, P.E., *Proc. 11th Int. Conf. On Atmospheric Aerosols Condensation and Ice Nuclei*, Budapest, Hungary, 3-8 Sept., 1984, pp. 63-67.
2. Majerowicz, A., and Wagner, P.E., *J. Aerosol Sci.* **17**, 463-465 (1986).
3. Whitby, K.T., *Atmos. Environ.* **12**, 135-159 (1978).
4. Mäkelä, J.M., et al., *Geophys. Res. Lett.* **24**, 1219-1222 (1997).
5. Mäkelä, J.M., et al., *J. Aerosol Sci.* **31**, 595-611 (2000).
6. Birmili, W., and Wiedensohler, A., *J. Aerosol Sci.* **S28**, 717-718 (1997).
7. Weber, et al., *J. Atmos Sci.* **52**, 2242-2256 (1995).
8. Temnischka, G., et al., *J. Aerosol Sci.* **20**, 1173-1176 (1989).
9. Covert, et al., *TELLUS* **48B**, 197-212 (1996).
10. NAG, Numerical Algorithms Group, *Fortran Library Manual*, Mark 14 (1990).

Growth of Homogeneously Nucleated Water Droplets: A Quantitative Comparison of Experiment and Theory

Alexander Fladerer and Reinhard Strey

Institut für Physikalische Chemie, Universität zu Köln,
Luxemburger Str. 116, D-50939 Köln, Germany

Abstract. A nucleation pulse chamber was used to homogeneously nucleate water droplets the growth of which was examined under precisely controlled conditions. Quantitative agreement between the measured experimental data and growth calculations according to the theoretical framework by *Fuchs* and *Sutugin* was found. Using calculated growth functions theoretical scattering functions were compared with experimental scattering data. Due to high data density the validity of the growth calculations could also be proved for high supersaturations (and thus high number densities of droplets) where the first *Mie*-maximum is not reached.

I. INTRODUCTION

Aerosols have a strong influence on the behavior of the atmosphere, e. g. on the heat balance and catalysis of many chemical reactions of trace gases as the production of chlorine at the surface of polar stratospheric cloud (PSC) droplets. The evolution of atmospheric aerosols through the stages of nucleation, growth and ageing is difficult to measure. To estimate their role in atmospheric models laboratory experiments under controlled conditions may help. In this paper we used the nucleation pulse chamber developed by *Strey et al.* [1] to create almost monodispersed aerosol populations by homogeneous nucleation. After the nucleation period the water clusters grow to macroscopic droplets (see Fig. 1). The growth conditions change from the kinetically controlled *free molecule regime* ($r < 10$ nm) via the *transition regime* (10 nm $< r < 200$ nm) to the diffusion controlled *continuum regime*. Because of depletion of water vapor and an increase of equilibrium vapor pressure due to condensational heat, droplet growth stops at a certain radius. For a completely monodispersed population of droplets no further development of the aerosol will take place, if coagulation is negligible. However, if there is polydispersity small droplets will evaporate and bigger droplets will continue to grow (so-called *Ostwald*-ripening).

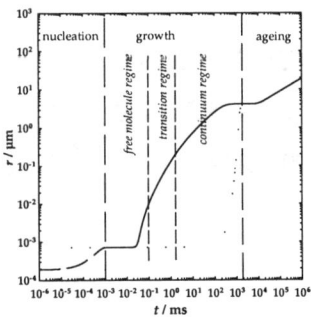

FIGURE 1. Schematic diagram of the evolution of average aerosol particle from vapor molecules to macroscopic droplets.

Up to now the growth process can only be described theoretically in the *free molecule* and in the *continuum regime*. The growth laws change continously from one borderline case to the other in the *transition regime*, for which only a numerical approximation of the temporal development of the droplet radius can be made. The solution is usually based on the laws of one borderline case and is corrected by a radius-dependent parameter. In order to describe the growth process in the *transition regime* several expressions have been formulated. In this work we measured growth curves of water droplets under varying conditions and compared them to theoretical descriptions of growth.

II. RESULTS

Using the nucleation pulse chamber it is possible to measure the growth of homogeneously nucleated droplets. To this end the *Mie*-maxima and -minima are detected as a function of time and compared to theoretical *Mie*-scattering calculations for increasing radii following the experimental approach of *Vietti* and *Schuster* [2, 3] and *Wagner* [4,5]. This procedure leads to various growth curves depending on supersaturation, number density of droplets N_d, and temperature. We calculated theoretical growth functions using the experimental parameters. We took into account the depletion of water vapor, the increase of droplet and system temperature, temperature-dependent functions of the diffusion coefficient, surface tension, liquid density and latent heat of condensation. The left plot in Fig. 2 shows theoretical growth functions in comparison with experimental data at 230 K. Quantitative agreement between *Fuchs* and *Sutugin's* theory [6] and experiment is found with no adjustable parameter. The experimental data and the growth functions agree, which indicates that the description of mass and heat flux in the *free molecule* and *transition regime* by the *Fuchs* and *Sutugin* approach is appropriately accounted for. The agreement also holds for different temperatures. Calculated growth functions and experimental data for 230 K, 240 K and 250 K are shown for comparison in the right plot in Fig. 2. We noted that the full lines in Fig. 2 differ only marginally from more recent theoretical descriptions [7].

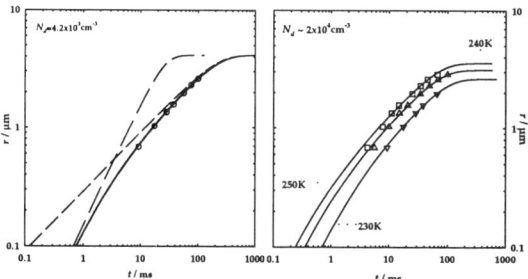

FIGURE 2. Left: Theoretical predictions for different growth regimes (dashed lines: pure *free molecule regime* and *continuum regime* calculations; solid line: *continuum regime* based calculation including *transition regime* correction according to *Fuchs* and *Sutugin*) and experimental data (circles) at 230 K. Right: Comparison of *Fuchs* and *Sutugin* growth functions (lines) and experimental data (250 K squares, 240 K triangle up, 230 K triangle down).

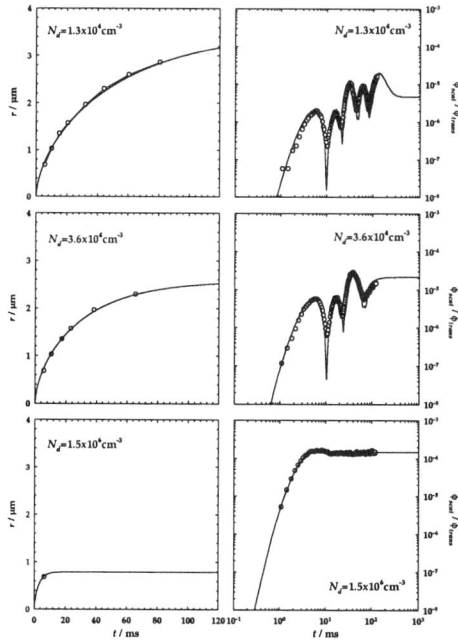

FIGURE 3. Comparison of calculated growth (left) and scattering (right) functions (lines) with experimental data (circles).

Growth of droplets in highly supersaturated vapor phases stops at small droplet radii. Thus, only one or a few *Mie*-extrema can be obtained. The theoretical growth-functions (r vs. t) permit to calculate time-dependent scattering functions ($\phi_{scat} / \phi_{trnas}$

vs. t) and thus basing the comparison on many data points. Fig. 3 shows the agreement between measured data-points and calculated scattering functions between 1.3×10^4 and 1.5×10^6 droplets / cm^3.

III. CONCLUSIONS

We found that the approach of *Fuchs* and *Sutugin* is suitable for describing growth of homogeneously nucleated water droplets. Also growth of droplet populations at the upper end of the measurable range could be compared to theoretical values by transforming growth functions to time dependent scattering functions. This seems to open a way to determine number densities of droplet populations which do not reach the size of the first *Mie*-maximum due to vapor depletion. Comparing growth experiments with theoretical results might also be used to measure unknown diffusion coefficients. However, this work has shown the validity of the *Fuchs* and *Sutugin* approach only for water droplets. We are presently working on the application to other substances, like n-pentanol and pure argon.

REFERENCES

1. Strey, R., Wagner, P. E., and Viisanen, Y., *J. Phys. Chem.* **98**, 7748 (1994).
2. Vietti, M. A., and Schuster, B. G., *J. Chem. Phys.* **58**, 434 (1973).
3. Vietti, M. A., and Schuster, B. G., *J. Chem. Phys.* **59**, 1499 (1973).
4. Wagner, P. E., *J. Colloid Interface Sci.* **44**, 181 (1973).
5. Wagner, P. E., "Aerosol Growth by Condensation" in *Topics in Current Physics* **29**: *Aerosol Microphysics II*, edited by W. H. Marlow, Springer-Verlag, Berlin, 1982, pp. 129.
6. Fuchs, N. A., and Sutugin, A. G., *Highly Dispersed Aerosols*, Ann Arbor Science Publications, Ann Arbor (1970).
7. Vesala, T., Kulmala, M., Rudolf, R., Vrtala, A., and Wagner, P. E., *J. Aerosol Sci.* **28**, 565 (1997).

Modelling Polydispersed Droplet Spectra In Nucleating Steam Flows

A.J. White[*] and M.J. Hounslow[†]

[*]*University of Durham, School of Engineering, South Road, Durham, DH1 3LE, UK.*
[†]*University of Sheffield, Chemical & Process Engineering, Mappin Street, Sheffield, S1 3JD, UK.*

Abstract. The nucleation processes occurring within steam turbine and moist air flows often lead to the formation of a polydispersion of liquid droplets for which droplet sizes encompass several orders of magnitude. This paper outlines a method for modelling such size distributions for condensing steam flow. The method uses a moment representation of the size spectra, rather than the more traditional approach of discretising the continuous range of droplet radii. Results for primary and secondary nucleating flows are presented and show very good agreement with discrete spectrum calculations, but require much less computing time. A further advantage of the method is that the equations may be cast in an Eulerian frame of reference, thereby facilitating the development of three-dimensional and unsteady computational methods for condensing flow.

INTRODUCTION

Polydispersed vapour droplet flows involving nucleation are relevant to numerous areas of scientific and engineering importance. Amongst these are high-speed aerodynamics in moist air, condensing flow in steam turbines, various chemical engineering processes and, of course, cloud formation. For the analysis of such flows it is important to model the polydispersed nature of the liquid since coupling between the phases is often strongly dependent on the size of the liquid droplets. Other two-phase behaviour may also be a function of droplet diameter (as is the case for thermodynamic "wetness" losses and droplet deposition within steam turbines).

Currently, methods for treating droplet size distributions for steam and moist air flows either employ a discretised size distribution (e.g., [1]), or use some form of averaged droplet size (e.g., [2,3]). The discretised approach provides the most comprehensive information on the shape of the size distribution, but is computationally intensive and cumbersome to implement, especially for multi-dimensional and unsteady flows. On the other hand, the use of an averaged droplet size may lead to substantially incorrect coupling between the phases, unless other information on the shape of the distribution is retained. In this respect, a number of workers (including Schnerr & Li [3]) have followed Hill [4] in using a surface-averaged droplet size and introducing the additional variables Q_1, Q_2 and Q_3, which equate to the total liquid surface, the sum of droplet radii and the total number of droplets per unit mass respectively. This is equivalent to modelling the first four moments of the size distribution whilst assuming droplet growth to be independent of radius. In this paper, a formal derivation of the moment evolution equations is outlined, and consideration given to radius-dependent growth laws. The approach follows that which has been applied to a number of areas in 'particle technology', including for example modelling the growth and aggregation of kidney stones [5].

MOMENT REPRESENTATION OF DROPLET CONSERVATION

For a compressible vapour droplet flow with zero inter-phase velocity slip, droplet number conservation may be expressed as:

$$\frac{\partial}{\partial t}(\rho f) + \nabla \cdot (\rho \mathbf{u} f) + \frac{\partial}{\partial r}(\rho G f) = \rho J \quad (1)$$

where ρ is the two-phase mixture density, $\mathbf{u}$ the velocity vector, $G=dr/dt$ is the droplet growth rate, and J is the nucleation rate per unit mass of mixture. The droplet number density function, f, is defined such that fdr is the number of droplets in the size range r to $r+dr$ (note that J is similarly defined). G, J and f are all functions of droplet radius, as well as position and time. The jth moment of the size distribution is now defined as:

$$\mu_j = \int_0^\infty f r^j dr \quad (2)$$

An expression for the evolution of this quantity is obtained by multiplying (1) by r^j and integrating over all possible radii, giving:

$$\frac{\partial}{\partial t}(\rho \mu_j) + \nabla \cdot (\rho \mathbf{u} \mu_j) = j\rho \int_0^\infty r^{j-1} fG dr + \rho J_* r_*^j \quad (3)$$

where J_* is the nucleation rate at the critical radius, r_*. (The manipulations involved in obtaining (3) are described in [6].) It is now assumed that the droplet growth rate may be expressed as the sum of a series in powers of r, i.e.,

$$G = \sum a_n r^n \quad (4)$$

The coefficients, a_n, are functions of the vapour properties only, and n may be negative if appropriate. Substituting (4) into (3) gives:

$$\frac{\partial}{\partial t}(\rho \mu_j) + \nabla \cdot (\rho \mathbf{u} \mu_j) = \rho J_* r_*^j + j\rho \sum a_n \mu_{n+j-1} \quad (5)$$

where the limits for the summation will depend on the form of the droplet growth law. Eq.(5) is the Eulerian form of the moment evolution equation. It may be recast in Lagrangian form by incorporating an expression for mass continuity.

Free Molecular Growth Regime

For free molecular growth (i.e., $Kn \gg 1$, where Kn is the Knudsen number), G rapidly adopts a value which is independent of r [4] and so may be described by a single term, a_0, in (4). It is then apparent from (5) that changes in μ_j depends only on μ_{j-1}, and hence the moment equations are closed. (Note that μ_0 does not depend on μ_{-1} because the last term in (5) is multiplied by j).

Coupling between the liquid and vapour phases is via the mass fraction of liquid, which in turn is proportional to μ_3. Thus, for most purposes, it is necessary to solve eq.(5) for $j=0,1,2,3$ only, and this is essentially the method described by Hill [4].

Continuum Growth Regime

For $Kn \ll 1$, the continuum growth model, $G = a_{-1}/r$, applies, and eq.(5) becomes:

$$\frac{\partial}{\partial t}(\rho\mu_j) + \nabla \cdot (\rho\mathbf{u}\mu_j) = \rho J_* r_*^j + j\rho a_{-1}\mu_{j-2} \qquad (6)$$

which is no longer of closed form. Closure may be achieved by approximating μ_{-1} with an appropriate extrapolation of μ_0 and μ_1, but this is likely to lead to numerical instability. Alternatively, if only even moments ($j=0,2,4$, etc.) are considered, then (6) becomes a closed equation set. A more robust approach may therefore be to solve for the first few even moments and then interpolate for μ_3 in order to couple with the gas dynamic equations. Neither of these routes has been pursued here since purely continuum regime growth is only of academic interest in the current context. (Initial growth is always in the free molecule regime for self-nucleating flows).

Droplet Growth in Low Pressure Steam

Droplet growth within steam turbines may encompass both the regimes discussed above, thus requiring a more general growth law. A modified form of Gyarmathy's expression, as discussed by Young [1], has been used as a basis here. Unfortunately, it is not possible to implement closed form moment equations for the full range of growth, but for most situations within low-pressure steam turbines, the expression is adequately approximated by:

$$G = \frac{a_{-1}}{r} + a_0 + a_1 r \qquad (7)$$

(this being accurate for $Kn > 0.6$ [6]). Eq. (7) yields moment equations that are not quite closed since the counterpart of (6) also involves a term in μ_{j-2}. However, dependence on this term is not strong, and the problem may be overcome with a simple extrapolation for μ_{-1}. (Furthermore, the first term in (7) stems from 'capillary subcooling' and is significant only in the very early stages of growth. It may therefore be neglected without serious loss of accuracy.)

Comparison With Discrete Spectrum Calculations

Fig.1 shows computations of primary and secondary nucleating flows using the current approach together with independent discrete spectrum calculations [1] (details of flow conditions and how the inter-phase coupling is achieved are given in [6]). Agreement with the full spectrum results (based on upwards of 65 droplet groups, and requiring one to two orders of magnitude more computing time) is very good for both expansions. The figure also shows results obtained using a single, averaged droplet size: for primary nucleation this gives an under-prediction of the nucleation pulse and hence a 30% over-prediction of the final droplet, whereas for secondary nucleation substantial discrepancies occur for both liquid and vapour phase quantities. These results highlight the need to model the polydispersed nature of the droplets.

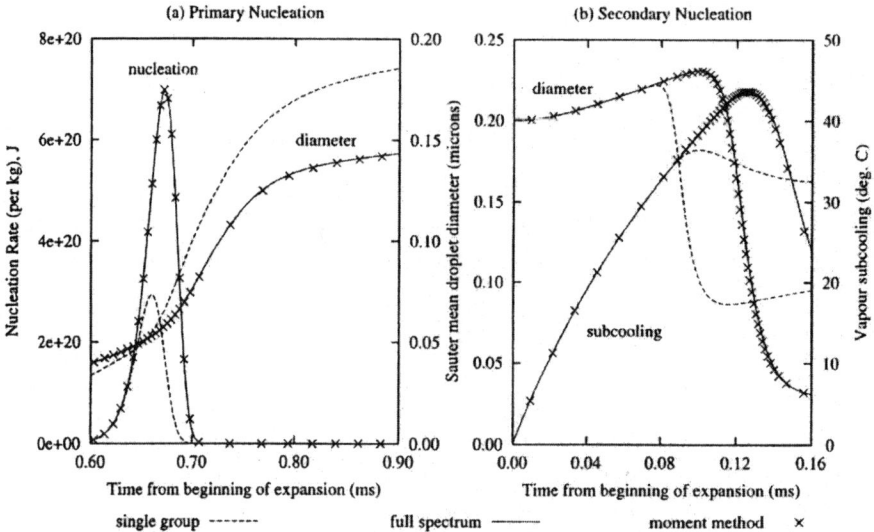

FIGURE 1. (a) Nucleation rate and average droplet diameter for a primary nucleating expansion, and (b) Average droplet diameter and vapour subcooling for a secondary nucleating expansion. Capillary subcooling has been neglected for both the moment and discrete spectrum calculations. (Calculations including capillary subcooling are presented in [6]).

CONCLUSION

A method for treating polydispersed nucleating flows of steam has been outlined. The method uses a moment representation of the droplet size distribution, and is valid for Knudsen numbers down to ~0.6. Agreement with full spectrum calculations is very good for both primary and secondary nucleating flows. Comparison with calculations using a single droplet group show that it is vital to model the full droplet polydispersion in order to maintain the correct inter-phase coupling.

REFERENCES

1. Young, J.B., *Trans. ASME – J. Turbomachinery* **114**, 569-579 (1992).
2. McCallum, M., and Hunt, R., *Int. J. for Numerical Methods* **44** (12), 1807-1821 (1999).
3. Schnerr, G.H., and Li, P., *Int. J. Multiphase Flow* **19** (5), 737-749 (1993).
4. Hill, P.G., *J. Fluid Mech.* **25** (3), 593-620 (1966).
5. Hounslow, M.J., Ryall, R.L., and Marshall, V.R., *AIChE Journal* **34** (11), 1821-1831 (1988).
6. White, A.J., and Hounslow, M.J., *Int. J. Heat & Mass Trans.* **43**, 1873-1884 (2000).

Modeling Of Homogeneous Nucleation In The Free Troposphere And Comparison With GLOBE-2 Data

Kari Klein and Owen B. Toon

Program in Atmospheric and Oceanic Sciences,
Laboratory for Atmospheric and Space Physics
University of Colorado
Boulder, CO

Abstract. During GLOBE-2, Clarke (1) observed small particles (1.5nm<r<7.5nm) in the upper troposphere, indicating a recent nucleation event. This data shows a negative correlation between the number of ultrafine particles (1.5nm<r<7.5nm) and the surface area of larger particles (.075µm < r < 3.5µm) (referred to as LOPC particles). Specifically, Clarke found that $N_{UF} \propto A_{LOPC}^{-.6}$, where N_{UF} is the number concentration of ultrafine particles (#/cm^3 @ STP) and A_{LOPC} is the surface area of the LOPC particles (µm^2/cm^3). The explanation given for this relationship (1) is that H_2SO_4 vapor is lost to the surface area of the larger particles, limiting the amount of H_2SO_4 vapor available for the production of new particles. We use a detailed microphysical model to reproduce Clarke's data (1). We found our model results to be inconsistent with the observed data, indicating deficiencies in the theory of nucleation and particle microphysics, deficiencies in our modeling techniques, and/or deficiencies in the measurement techniques used to collect the data.

INTRODUCTION

Our current understanding of aerosol nucleation is incomplete. Homogeneous binary nucleation theory is the most prevalent tool used for predicting nucleation in the troposphere. Though binary nucleation theory often predicts when nucleation will occur, the predicted nucleation rate is often insufficient to explain observations (e.g. 2). Also, there are many observations of new particle formation in conditions inconsistent with binary nucleation theory (3; 4; 5). The discrepancies between nucleation theory and observation have led us to examine observations in a more detailed way. In this paper we present a numerical study of nucleation and related aerosol physics to investigate not only if binary nucleation theory predicts when nucleation occurs, but also if it predicts an important relationship observed between new particles and pre-existing particles.

There have been several observations of ultrafine particles in the free troposphere, indicating homogeneous nucleation. Though many of the observations show a strong negative correlation between the number of new particles and pre-existing particles, these correlations have rarely been well quantified, limiting the data available for comparison with a model. One case of well-quantified data is from Clarke (1).

Clarke's data (1) has been quoted widely in the aerosol community, and is a useful data set for model comparison.

During GLOBE-2, Clarke (1) often observed many small particles (1.5nm<r<7.5nm) in the upper troposphere, indicating recent nucleation events. This data shows a negative correlation between the number of ultrafine particles (1.5nm<r<7.5nm) and the surface area of larger particles (.075μm < r < 3.5μm) (referred to as LOPC particles). Specifically, Clarke found that $N_{UF} \propto A_{LOPC}^{-0.64}$, where N_{UF} is the number concentration of ultrafine particles (#/cm^3 @ STP) and A_{LOPC} is the surface area of the LOPC particles (μm^2/cm^3). The explanation given for this relationship by Clarke (1) is that H_2SO_4 vapor is lost to the surface area of the larger particles, limiting the amount of H_2SO_4 vapor available for the production of new particles.

Our primary goal in this research is to determine if Clarke's data (1) and explanation that nucleation is controlled by loss of sulfuric acid vapor to pre-existing surface area are consistent with homogeneous binary nucleation theory.

MODEL DESCRIPTION

Ultrafine particle concentrations are controlled by a number of microphysical processes including nucleation, condensational growth, and coagulation. In considering the processes that affect the ultrafine concentration of particles, we must also consider processes that affect sulfuric acid concentration, as sulfuric acid concentration and ultrafine particle concentrations are closely related. Chemical production, namely, the oxidation of sulfur dioxide by OH controls the production of sulfuric acid. Nucleation of new particles and condensation onto existing particles controls the loss of sulfuric acid.

The model we use is an early version of the University of Colorado/ NASA Ames aerosol model. We run the model for 10 hours with a timestep of 5 seconds. Since we are interested in relatively short timescales, we are running the model as a box model. The model contains all the microphysics described above and a simple family chemistry scheme. There are 40 aerosol bins with radii ranging in size from .5 nm to 4.1 μm. We initialize the size distribution with a bimodal size distribution similar to that measured by Pueschel *et al.* (1994), representative of a typical size distribution in the upper troposphere. To vary the initial surface area we vary the number concentration of both aerosol modes by a constant. When new particles are formed via nucleation, if the nucleation scheme is one in which a critical radius is calculated, we form the new particles in the radius bin corresponding to the critical radius, otherwise we form new particles in the third radius bin, r = .794 nm.

For each study we run our model at 28 initial surface areas ranging from ~.01 μm^2/cm^3 to ~100 μm^2/cm^3. In our study we vary the temperature, the initial SO_2 mixing ratio (along with pressure), the relative humidity, and sticking coefficients (for growth and coagulation). The nucleation schemes we include in our model are: classical binary nucleation without hydration, classical binary nucleation with hydration (we use a parameterization from Kulmala et al., 1998), and kinetic nucleation.

Results And Conclusions

None of the nucleation schemes gives results similar to those observed by Clarke (1). All plots of ultrafine particle concentration versus LOPC surface area have a similar shape; at low LOPC surface areas the particle concentration is independent of surface area and at higher LOPC surface areas the ultrafine particle concentration decreases sharply with LOPC surface area. The plots of ultrafine particle concentration versus total surface area can explain this shape; at the LOPC surface areas where ultrafine particle concentrations are independent of LOPC surface area, the total surface area is constant. At lower LOPC surface areas the total surface area is controlled by the ultrafine particles. If LOPC particles are not dominating the total surface area, they are not dominating the loss of sulfuric acid due to condensation and we would not expect them to have a strong influence on the ultrafine particle concentration. At higher LOPC surface areas, the LOPC particles dominate the total surface area. In this case, sulfuric acid is quickly lost to the particles and does not build up enough for significant nucleation and subsequent growth of particles.

We cannot reproduce Clarke's data (1) with our model despite using three different nucleation schemes. At low LOPC surface area ($SA<10\mu m^2/cm^3$) ultrafine particle concentration is independent of LOPC surface area, and the total surface area is constant. At high LOPC surface area ($SA>10\mu m^2/cm^3$) ultrafine particle concentration strongly depends on LOPC surface area, and the total surface area is controlled by LOPC surface area. This system is like a switch. At high initial surface areas nucleation is effectively turned off (or at least turned very low), but at low initial surface areas nucleation occurs rapidly and the new particles dominate the total surface area. This phenomenon occurs even for the kinetic nucleation scheme, which is not very strongly dependent on sulfuric acid concentration and nucleates new particles at a more constant rate rather than in large bursts.

Our inability to reproduce this observed data may point to deficiencies in the theory of nucleation and particle microphysics, deficiencies in our modeling techniques, and/or deficiencies in the measurement techniques used to collect the data. Since our results are similar for very different nucleation mechanisms, we believe that our inability to reproduce this observed data is not likely due to deficiencies in nucleation theory. It is possible, though, that a drastically different nucleation mechanism, such as ion-ion recombination theory, may drastically change our model results. The ion-ion nucleation is limited by the production of tropospheric ions. This slow source of ions may cause the ambient surface area to exert more control on the nucleation throughout the surface area range, perhaps leading to results similar to those observed.

Another possible explanation for the discrepancy between our model results and the observed data is the limited condensing species in our model. We only treat condensation of sulfuric acid, whereas organic species and other acids may condense on ultrafine particles, contributing to their growth.

Lastly, the observed data we are comparing our model to may be anomalous. We plan to analyze more recent observed data and explore the limitations of the observed data to understand if the observations may be biased.

ACKNOWLEDGMENTS

This work was funded by NASA under grant NAG5-6504.

REFERENCES

1. Clarke, A.D., *J. Geophys. Res.* **98**, 20633-20647 (1993).
2. Raes, F., *J. Geophys. Res.* **100**, 2893-2903 (1995).
3. Weber, R.J., P.H. McMurry, F.L. Eisele, and D.J. Tanner, *J. Atm. Sci.* **52**, 2242-2257 (1995).
4. Weber, R.J., J.J. Marti, P.H. McMurry, F.L. Eisele, D.J. Tanner, and A. Jefferson, *J. Geophys. Res.* **102**, 4375-4385 (1997).
5. Clarke, A.D., D. Davis, V.N. Kapustin, F. Eisle, G. Chen, I. Paluch, D. Lenschow, A.R. Bandy, D. Thornton, K. Moore, L. Mauldin, D. Tanner, M. Litchy, M.A. Carroll, J. Collins, and G. Albercook, *Science* **282**, 89-92 (1998).

Thermodynamics and Kinetics of Condensation on Wettable Macroscopic Nuclei: New Results

Fedor M. Kuni, Alexander K. Shchekin and Alexander P. Grinin

Department of Statistical Physics, St Petersburg State University, St Petersburg, Russia 198904

Abstract. The thermodynamic and kinetic peculiarities of vapor condensation on macroscopic wettable particles are analyzed for conditions of gradual creation of the vapor metastable state. The formulas for the number of super-critical drops formed, the time of start and the duration of the stage of effective origination of super-critical drops, the width of the droplet size spectrum are considered. The dependence on parameters specifying the properties of condensing liquid and solid wettable nuclei and their interfaces, on sizes of condensation nuclei and their initial number in the vapor-gas medium, on the rate of external formation of the vapor metastable state is revealed.

INTRODUCTION

The goal of this report is to join recent results in kinetics and thermodynamics of non-steady condensation on macroscopic solid wettable nuclei. We will consider general case of a system where vapor supersaturation changes as a result of dynamic, i.e. gradual, external variation of the vapor metastable state and vapor consumption by growing super-critical drops. We will assume that the condensing drop covers the nucleus with a liquid film of uniform thickness.

A drop formed out of vapor on solid wettable condensation nucleus has two interfaces: internal (between the nucleus and enveloping liquid film) and external (between the liquid film and the vapor-gas surroundings). At the initial stage of nucleation, the internal and external interfaces overlap, and the surface forces in the thin liquid film give a significant contribution to the activation barrier of nucleation. Due to this fact the role of the surface forces becomes determinative in the whole nucleation process. It reveals through the parameters of surface tensions for interfaces and the disjoining pressure for the liquid film [1,2].

The initial stage of nucleation, on which drops overcome the activation barrier of nucleation and become of super-critical size, was considered in ref. [1]. Here we focus on the next stage of effective origination of large super-critical drops really participating in material balance within the system. This process is most sensitive to the effects of surface forces in the situation where drops should overcome a significant activation barrier. This barrier can not be too high to provide an intensive nucleation rate. In view of extremely sharp dependence of the activation barrier on vapor supersaturation in the case of condensation nuclei of macroscopic size, such a situation may take place only when the vapor metastable state is produced sufficiently slowly. Basic solution of the problem for arbitrary condensation nuclei was given in [3-5]. The results will be presented here for identical macroscopic condensation nuclei. The possible polydispersity of condensation nuclei can be incorporated into the theory [6,7].

NUCLEATION CHARACTERISTICS AT THE END OF THE STAGE OF EFFECTIVE ORIGINATION OF SUPER-CRITICAL DROPS

Let us define the ideal vapor supersaturation $\Phi \equiv (n_{tot}/n_\infty)-1$ where n_{tot} is the total number of condensing molecules per unit volume of the vapor-gas environment including the molecules in nucleating drops, n_∞ the concentration of molecules in saturated vapor. We assume that ideal supersaturation Φ grows in time t at the initial stage of effective origination of super-critical drops as

$$\Phi \equiv (t/t_\infty)^m. \qquad (1)$$

where m and t_∞ are positive parameters.

Analysis of the total number N of super-critical drops formed to the end of the stage of effective origination of the super-critical drops in the free-molecular or the diffusion regimes of drop growth gives [3-5]

$$N = \eta(-\infty)[1-\exp(-1/h)]. \qquad (2)$$

Here $\eta(-\infty)$ is the initial concentration of condensation nuclei (their number per unit volume of vapor-gas surroundings at the beginning of the stage of origination of super-critical drops). Dimensionless parameter h depends on the regime of growth of super-critical drops. As follows from (2),

$$N = \eta(-\infty)/h \qquad (h \gg 1), \qquad (3)$$

$$N = \eta(-\infty) \qquad (h \ll 1). \qquad (4)$$

According to (3) the super-critical drops capture at $h \gg 1$ only a small part of the initial number of condensation nuclei (every drop entraps one nucleus). According to (4) the super-critical drops capture at $h \ll 1$ all condensation nuclei. Therefore the magnitude of parameter h distinguishes the situations where the stage of nucleation of super-critical drops is ceased by the vapor intake into the super-critical drops or by exhaustion of the initial supply of condensation nuclei by super-critical drops originating on these nuclei.

One can find the explicit dependence of h on initial parameters m, t_∞ and $\eta(-\infty)$ of the problem. However it is more convenient to take h and the parameter $\kappa \equiv \Delta t/t_s$ (where Δt is the duration of the stage of effective origination of super-critical drops and t_s is the time lag of nucleation) as independent parameters instead of $\eta(-\infty)$ and t_∞.

Let ν be the number of condensate molecules in the whole volume of a drop including the volume of nucleus, and b_ν the chemical potential of condensate in the drop. We express b_ν in thermal units $k_B T$, where k_B is the Boltzmann constant and T the temperature of the vapor-gas environment, and count off from the value for the vapor-liquid equilibrium with the flat interface. In the case of condensation on wettable nuclei, potential b_ν as a function of ν has a maximum at $\nu = \nu_0$:

$(\partial b_v / \partial v)_{v=v_0} = 0$ [1]. The maximum determines the threshold value ζ_{th} of the vapor supersaturation. Though b_v is rather complicated function of v, the capillary contribution to b_v be dominant. Thus a good approximation for ζ_{th} is $\zeta_{th} = 2a/3v_0^{1/3}$ where $a \equiv (4\pi\gamma/k_B T)(3v_\alpha/4\pi)^{2/3}$, γ is the surface tension of condensate, v_α the molecular volume in the liquid phase.

The time t_{on} of start and the time t_{off} of termination of effective origination of super-critical drops can be determined [3-5] as

$$t_{on} \approx t_* - \Delta t/2, \qquad t_{off} \approx t_* + \Delta t/2. \tag{5}$$

Here t_* and Δt are the bearing time and the duration of the stage of effective origination of super-critical drops, respectively. Rewriting these times in terms of κ and h, one obtains

$$t_* \approx 2\kappa m\tau v_0^{4/3}/\alpha_C p_0^{1/3}, \tag{6}$$

$$\Delta t \approx (3/2^{3/2})(\kappa\tau/q\alpha_C)(v_0^{8/9}/a^{2/3}p_0^{2/9}) \tag{7}$$

where

$$q = \begin{cases} \left[(3/2^{7/2})\ln(\kappa h/6\pi \ln 2)\right]^{1/3} & (h \gg 1), \\ \left[(3/2^{7/2})\ln(\kappa/6\pi \ln 2)\right]^{1/3} & (h \ll 1), \end{cases} \tag{8}$$

α_C is the condensation coefficient, $\tau = 12/\left[(36\pi)^{1/3}v_\alpha^{2/3}n_\infty v_T\right]$ is the characteristic time in the free-molecular regime of super-critical drop growth, v_T is the thermal velocity of vapor molecules, $p_0 \equiv (3^9/2^3)v_0^7/a^3\left(\left|\partial^2 b_v/\partial v^2\right|_0\right)^3$.

Representing t_∞ and $\eta(-\infty)$ in terms of κ and h, we can write the relations for all other nucleation characteristics. We have for t_∞:

$$t_\infty \approx 2(3/2)^{1/m}(\kappa m\tau/\alpha_C)(v_0^{(4m+1)/3m}/a^{1/m}p_0^{1/3}), \tag{9}$$

while the formulas for $\eta(-\infty)$ are different in the cases of free-molecular and diffusion regimes of drop growth.

In the case of free-molecular regime of growth of super-critical drops we have for $\eta(-\infty)$, N and width $\Delta\rho$ of the final size spectrum of super-critical drops (the droplet size $\rho = v^{1/3}$):

$$\eta(-\infty) \approx (3/2)(q^2 n_\infty h p_0^{7/9}/\kappa^3 a^{2/3}v_0^{22/9}), \tag{10}$$

$$N \approx (3/2)(q^2 n_\infty p_0^{7/9}/\kappa^3 a^{2/3}v_0^{22/9}) \quad (h \gg 1), \tag{11}$$

$$\Delta\rho \approx 2^{-1/2}\kappa a^{1/3}v_0^{5/9}/qp_0^{2/9}. \tag{12}$$

This regime realizes if strong inequality $1 \ll \kappa \ll 2^{1/2} p_0^{2/9} / \alpha_C (n_\infty + n_g) v_\alpha a^{1/3} v_0^{5/9}$ fulfils (n_g is the concentration of carrier gas in the vapor-gas environment).

The diffusion regime of growth of super-critical drops requires $\kappa \gg 2^{1/2} p_0^{2/9} / \alpha_C (n_\infty + n_g) v_\alpha a^{1/3} v_0^{5/9}$. We have in this regime

$$\eta(-\infty) \approx 2^{5/4} (3\pi)^{-1/2} q^{1/2} n_\infty h (\alpha_C \tau_D / \kappa \tau)^{3/2} p_0^{4/9} / a^{1/6} v_0^{29/18}, \quad (13)$$

$$N \approx 2^{5/4} (3\pi)^{-1/2} q^{1/2} n_\infty (\alpha_C \tau_D / \kappa \tau)^{3/2} p_0^{4/9} / a^{1/6} v_0^{29/18} \quad (h \gg 1), \quad (14)$$

$$\Delta \rho \approx 2^{-1/2} \kappa \tau a^{1/3} v_0^{5/9} / q \alpha_C \tau_D p_0^{2/9}. \quad (15)$$

Here $\tau_D \approx (3/8\pi)(4\pi/3 v_\alpha)^{1/3} / D n_\infty$ is the characteristic time in the diffusion regime, D the diffusion coefficient for vapor molecules in the vapor-gas environment; the droplet size ρ is defined as $\rho = v^{2/3}$.

Specifying parameters κ and h, one can easily used the above formulas for finding the nucleation characteristics. These formulas are not restricted by a concrete choice of the dependence of the condensate chemical potential on drop size. At the same time, the role of interface properties of the wettable nuclei can be controlled in kinetics through the parameters a, p_0 and v_0.

ACKNOWLEDGMENTS

This research was supported by the Grant Center for Natural Sciences (GRACENAS), grant 97-0-14.2-31 and by Program "The Universities of Russia" (project 2145).

REFERENCES

1. Kuni, F.M., Shchekin, A.K., Rusanov, A.I., and Widom, B., *Adv. Coll. Int. Sci.*, 65, 71 (1996).
2. Shchekin A.K., Tatianenko, D.V., Kuni F.M., "Towards thermodynamics of uniform film formation on solid insoluble particles" in *Nucleation Theory and Applications*, edited by J.W.P. Schmelzer, G.Ropke and V.B. Priezzhev, Dubna, JINR, 1999, pp.320-340.
3. Grinin, A.P., Kuni, F.M., and Kurasov, V.B., *Colloid Journal of the USSR*, 52 430 (1990).
4. Grinin, A.P., Kuni, F.M., and Kurasov, V.B., *Colloid Journal of the USSR*, 52 437 (1990).
5. Kuni, F.M., Grinin, A.P., and Kurasov, V.B., "Kinetics of condensation under gradual creation of the metastable state" in *Nucleation Theory and Applications*, edited by J.W.P. Schmelzer, G.Ropke and V.B. Priezzhev, Dubna, JINR, 1999, pp.160-193.
6. Kuni, F.M., Grinin, A.P., Shchekin, A.K., and Novozhylova, T.Yu., *Colloid Journal*, **59** 183 (1997).
7. Kuni F.M., Shchekin, A.K., Grinin, A.P., "Kinetics of condensation on macroscopic solid nuclei at low dynamic vapor supersaturations" in *Nucleation Theory and Applications*, edited by J.W.P. Schmelzer, G.Ropke and V.B. Priezzhev, Dubna, JINR, 1999, pp.208-236.

The Characteristic Times of Establishing the Steady Rate of Nucleation on Soluble Aerosol Particles Containing Surfactants

Alexander K. Shchekin, Tatyana M. Yakovenko and Fedor M. Kuni

Department of Statistical Physics, St Petersburg State University, St Petersburg, Russia 198904

Abstract. The characteristic times of establishing the steady rate of nucleation on soluble aerosol particles containing surfactants have been determined in sub-equilibrium, near-equilibrium, sub-critical, near-critical and super-critical regions of droplet sizes. It has been shown that the largest is the time of establishing the steady regime in the sub-critical region of droplet sizes situated in the axis of sizes between the locations of minimum and maximum of the work of droplet formation. The role of an adsorbing monolayer of dissolved surfactant at the surface of nucleating droplet is analyzed.

1. CHARACTERISTIC REGIONS OF DROPLET SIZES

The typical dependence of the work F_ν of heterogeneous formation of a droplet on number ν molecules of condensate in the droplet is shown in Fig.1. Here ν_e, ν_{th}, ν_c are the locations of the minimum (the equilibrium droplet size), inflection and maximum (the critical droplet size) of work F_ν, $\Delta\nu_e$ and $\Delta\nu_c$ are the half-width of the well and potential hill of work F_ν, ΔF is the height of the activation barrier of nucleation. Figure 1 also shows several characteristic regions of droplet sizes. We will call region $(\nu < \nu_e - \Delta\nu_e)$ sub-equilibrium, region II $(\nu_e - \Delta\nu_e < \nu < \nu_e + \Delta\nu_e)$ near-equilibrium, region III $(\nu_e + \Delta\nu_e < \nu < \nu_c - \Delta\nu_c)$ sub-critical, region IV $(\nu_c - \Delta\nu_c < \nu < \nu_c + \Delta\nu_c)$ near-critical, and region V $(\nu > \nu_c + \Delta\nu_c)$ super-critical, correspondingly.

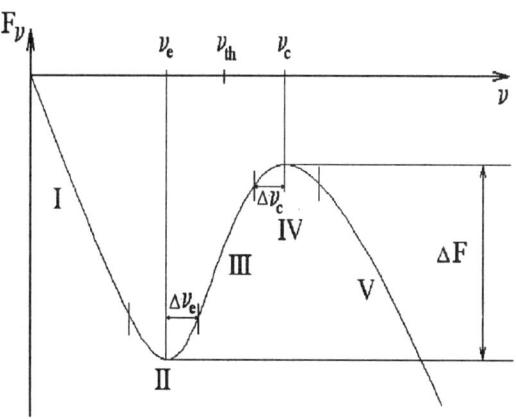

FIGURE 1. The dependence of the work F_ν of heterogeneous droplet formation on droplet size ν and the characteristic regions of droplet sizes.

The steady rate of nucleation establishes in the process of fluctuation growth of droplets over the activation barrier ΔF. One of the kinetic characteristics of nucleation, which is well defined in experiment, is the incubation time or the time lag of nucleation. This time coincides with the time of establishing the steady regime of nucleation over the entire interval of droplet sizes up to and slightly over the location of potential hill of the work of droplet formation in the axis of droplet sizes (regions I+II+III+IV in Fig.1). The goal of this communication will be finding this time in the case of heterogeneous condensation on macroscopic soluble nuclei and, in particular, on nuclei containing soluble surfactant.

Dissolution of the particle and creation of adsorbing surfactant monolayer at the vapor-droplet interface follow formation of a droplet out of supersaturated vapor onto soluble particle containing surfactant. The dissolved matter of the particle in the bulk of the droplet and the adsorbing monolayer affect the nucleation kinetics both through changing the condensate chemical potential [1-4] and the condensation coefficient for vapor molecules on the droplet surface [5,6]. For certainty, below we will assume that the condensation coefficient achieves its value for a pure condensate already in the sub-equilibrium region of droplet sizes.

2. CHARACTERISTIC TIMES OF ESTABLISHING STEADY REGIME OF NUCLEATION

In the sub-equilibrium region the rate W_v^+ of attachment of vapor molecules to droplet is much higher that the emission rate. The time of establishing the steady (in fact, equilibrium) state in this region can be determined as

$$t_I \approx \int_{v_1}^{v_e - \Delta v_e} dv / W_v^+ \qquad (1)$$

where v_1 refers to the droplet size where the condensation coefficient α_C achieves its value for pure condensate. Assuming $v_1 \ll v_e$, recognizing that for particles of macroscopic size the threshold vapor supersaturation is small compare to unity, taking into account $\Delta v_e / v_e \ll 1$, we can estimate t_I in the case of the Knudsen regime of attachment rate W_v^+ as

$$t_I \approx \tau \alpha_C^{-1} v_e^{1/3} \qquad (2)$$

where τ is the time between two successive collisions of molecule in the vapor phase.

In the near-equilibrium region the kinetic equation for droplet distribution $n_v(t)$ in sizes that determines an evolution of the droplet ensemble in time t has a form [7]

$$\partial n_v(t)/\partial t = W_{v_e}^+ \partial/\partial v \left[2(v - v_e)/(\Delta v_e)^2 + \partial/\partial v \right] n_v(t). \qquad (3)$$

Solving this equation with natural boundary conditions $n_v(t)/n_{v_e} \approx 0$ at $v < v_e - \Delta v_e$ and $v > v_e + \Delta v_e$, we find

$$t_{II} = (\Delta v_e)^2 / 2W_{v_e}^+ . \qquad (4)$$

Let us represent the distribution function $n_v(t)$ in the sub-critical region as [8]

$$n_v(t) = n_v^{(s)}(1 + \Psi_v(t)) \qquad (5)$$

where $n_v^{(s)}$ is the steady distribution function for droplets in sizes and $\Psi_v(t)$ stands for a deviation. Taking into account that $\partial F_v/\partial v \ll 1$ and $W_v^+ \approx W_{v_{th}}^+$ in the sub-critical region [1-3], we can write the governing equation for the distribution function in sizes in this region in the form

$$\partial \Psi_v(t)/\partial t = -\upsilon_v \, \partial \Psi_v(t)/\partial v + W_{v_{th}}^+ \, \partial^2 \Psi_v(t)/\partial v^2 \qquad (6)$$

where

$$\upsilon_v \equiv W_v^+ \, \partial F_v/\partial v + 2j^{(s)}/n_v^{(s)}, \qquad (7)$$

$j^{(s)}$ is the steady rate of nucleation. Analysis of eq.(6) in the pre-threshold region of vapor supersaturations where $\Delta v_e = \Delta v_c$ and $2 < \Delta F < 15$ [1-3] gives

$$t_{II} \leq (\Delta v_c)^2 / 2W_{v_{th}}^+ \left(1 + \sqrt{3\Delta F}\right). \qquad (8)$$

As was shown in [9], in the near-critical region the governing equation for droplet distribution in sizes $n_v(t)$ has a form

$$\partial n_v(t)/\partial t = W_{v_c}^+ \, \partial/\partial v \left[-2(v-v_c)/(\Delta v_c)^2 + \partial/\partial v \right] n_v(t). \qquad (9)$$

In view of the boundary conditions $(n_v(t) - n_v^{(s)})/n_v^{(e)} \approx 0$ at $v < v_c - \Delta v_c$ and $v < v_c + \Delta v_c$ where $n_v^{(e)}$ is the equilibrium distribution function for droplets in sizes, one can find here

$$t_{IV} = (\Delta v_c)^2 / 2W_{v_c}^+. \qquad (10)$$

Formula (10) can be compared with the estimates of the time lag in homogeneous nucleation obtained by several authors [10-12]. In our notation those estimates can be grouped as $t_{IV} \approx (4/9 \div 3/4)(\Delta v_c)^2/2W_{v_c}^+$ and are quite close to formula (10).

3. THE DEPENDENCE OF THE CHARACTERISTIC TIME ON SIZE AND SURFACE ACTIVITY OF SOLUBLE PARTICLES

As one can see, the largest characteristic time is t_{III}. Times t_{II} and t_{IV} are considerably smaller than t_{III}, and time t_I is the smallest one. Thus the time scale for establishing the steady rate of nucleation in heterogeneous nucleation on soluble particles can be estimated by time t_{III}.

In the case of condensation on soluble particles of surface-inactive substance [1] one can find the dependence of time $\bar{t}_{III}$ on the number v_n of molecules in the particle and on the activation barrier ΔF within the pre-threshold region of vapor supersaturations in the form $\bar{t}_{III} \propto v_n^{4/3} \left(1 + \sqrt{3\Delta F}\right)/\sqrt[3]{\Delta F}$.

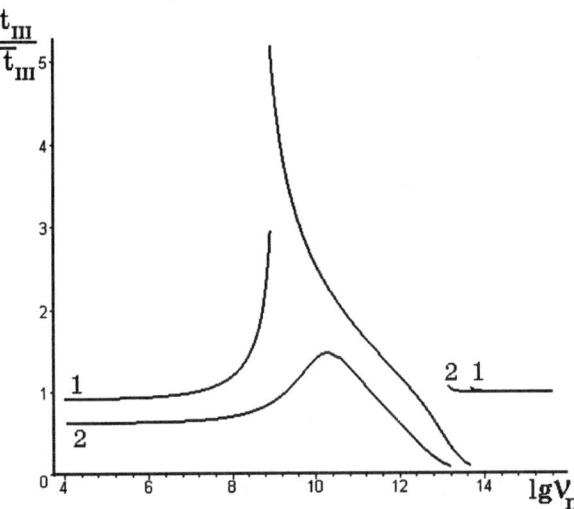

FIGURE 2. The dependence of the ratio of times $t_{III}/\bar{t}_{III}$ for surfactant and surface inactive matter of completely soluble particle on number v_n of molecules in the particle.

Figure 2 illustrates the ratio of t_{III} (determined in the case of surfactant soluble particles) and $\bar{t}_{III}$ for a fixed value of ΔF within the pre-threshold value of vapor supersaturations. We used for computations the theory and the algorithm described in [3,4]. The surface tension of pure condensate for curve 1 is larger than for curve 2. For both curves the lateral interactions in the surfactant monolayer at droplet surface were supposed significant.

ACKNOWLEDGMENTS

The work was supported by the Grant Center for Natural Sciences (GRACENAS), grant 97-0-14.2-31. Tatiana Yakovenko thanks the Nansen International Environmental and Remote Sensing Center.

REFERENCES

1. Kuni, F. M., Shchekin, A. K, and Rusanov, A. I., *Colloid J.* **55** (2), 184-192 (1993).
2. Kuni, F. M., Shchekin, A. K, and Rusanov, A. I., *Colloid J.* **55** (2), 174-183 (1993).
3. Shchekin, A. K , Kuni, F. M., Yakovenko, T. M., and Rusanov, A. I., *Colloid J.* **57 (1)**, 93-101 (1995).
4. Shchekin, A. K , Kuni, F. M., Yakovenko, T. M., and Rusanov, A. I., *Colloid J.* **57** (2), 242-248 (1995).
5. Barnes, J. T, *Adv. Coll. Int. Sci.* **26**, 89-200 (1986).
6. Shulman, M. L., Charlston, R. J., and Davis, E. J., *J. Aerosol. Sci.* **28** 737-752 (1997).
7. Kuni, F. M., Grinin, A. P, Shchekin, and A. K., Rusanov, *Colloid J.* **62 (2)** (2000)
8. Kuni, F. M., Grinin, A. P, *Colloid J. of the USSR* **46 (1)**, 23-28 (1984).
9. Melikhov, A.A., Trofimov, Yu.V., and Kuni, F.M. *Colloid J.* **56** (2), 201-204 (1994).
10. Kashchiev, D., *Surface Science.* **14**, 209-220 (1969).
11. Shi, G., and Seinfeld, J. H., *Phys. Rev. A.* **41**, 2101-2108 (1990).
12. Demo, P., and Kozisek, Z., *Phys. Rev. B.* **48,** 3620-3625 (1993); Z., *Phil. Mag. B.* **70**, 49-57 (1994).

Asymmetric Charging Of The Aerosols Including The Coagulation Mechanism

Savita Dhanorkar and A K Kamra

Indian Institute of Tropical Meteorology, Pune, India

Abstract. The charge distribution on the submicron size charged aerosols is computed from ion-aerosol balance equations including the effect of coagulation of the charged aerosols. The asymmetry in the charge distribution caused by the unequal diffusion of positive and negative ions to the aerosols and by their mutual coagulation is examined for different mean values of the aerosol radii. Our results show a decrease in the charging asymmetry for the aerosols having up to 3 elementary charges when the effect of coagulation is included.

INTRODUCTION

Charge distribution on aerosol particles mainly depends on the ion-aerosol charging mechanism and the particle-particle coagulation process. In the past, the effect of particle-particle coagulation on charge distribution has been neglected. However, due to relatively long residence time of aerosol particles in natural environment, the coagulation of charged and uncharged aerosols can significantly alter the charge distribution on the aerosol particles and thus need to be considered. Moreover, the difference in the physical properties of the positive and negative ions can cause asymmetric charging of the aerosol particles. The effect of symmetric and asymmetric bipolar charging on the coagulation of aerosol particles has been studied by Adachi et. al. (1981), Vemury et al. (1997) etc.

In this paper we include the effects of ion-aerosol attachment as well as of the particle-particle coagulation on the charge distribution of aerosol particles. The steady state charge distribution on submicron size aerosols is calculated by solving the ion-aerosol balance equations for multiply charged aerosol particles in case of asymmetric charging.

THEORY AND RESULTS

The ion-aerosol balance equations including the effect of coagulation for the monodisperse, homogeneous, chemically inactive aerosols in bipolar ion environment are given by Dhanorkar and Kamra (1997). In these equations it is assumed that the positive and negative small ions are identical except their polarity. The difference in their mobilities and masses is not considered in these equations. The ion-aerosol balance equations including the effect of coagulation of charged aerosols in case of asymmetric charging are

$$\frac{dn_1}{dt} = q - \alpha n_1 n_2 - n_1 \sum_{k=0}^{\infty} \beta_{11}^{(k)} N_1^{(k)} - n_1 \sum_{k=1}^{\infty} \beta_{12}^{(k)} N_2^{(k)} \tag{1}$$

$$\frac{dn_2}{dt} = q - \alpha n_1 n_2 - n_2 \sum_{k=0}^{\infty} \beta_{22}^{(k)} N_2^{(k)} - n_2 \sum_{k=1}^{\infty} \beta_{21}^{(k)} N_1^{(k)} \tag{2}$$

$$\frac{dN_1^{(k)}}{dt} = Q_1^{(k)} + n_1 \beta_{11}^{(k-1)} N_1^{(k-1)} + n_2 \beta_{21}^{(k+1)} N_1^{(k+1)} + \sum_{g=0}^{k} \gamma_{11}^{(g)(k-g)} N_1^{(g)} N_1^{(k-g)}$$
$$+ \sum_{g=1}^{\infty} \gamma_{12}^{(g+k)(g)} N_1^{(g+k)} N_2^{(g)} - n_1 \beta_{11}^{(k)} N_1^{(k)} - n_2 \beta_{21}^{(k)} N_1^{(k)} \qquad (3)$$
$$- N_1^{(k)} (\sum_{g=0}^{\infty} \gamma_{11}^{(k)(g)} N_1^{(g)} + \sum_{g=1}^{\infty} \gamma_{12}^{(k)(g)} N_2^{(g)})$$

$$\frac{dN_2^{(k)}}{dt} = Q_2^{(k)} + n_2 \beta_{22}^{(k-1)} N_2^{(k-1)} + n_1 \beta_{12}^{(k+1)} N_2^{(k+1)} + \sum_{g=0}^{k} \gamma_{22}^{(g)(k-g)} N_2^{(g)} N_2^{(k-g)}$$
$$+ \sum_{g=1}^{\infty} \gamma_{21}^{(g+k)(g)} N_2^{(g+k)} N_1^{(g)} - n_2 \beta_{22}^{(k)} N_2^{(k)} - n_1 \beta_{12}^{(k)} N_2^{(k)} \qquad (4)$$
$$- N_2^{(k)} (\sum_{g=0}^{\infty} \gamma_{22}^{(k)(g)} N_2^{(g)} + \sum_{g=1}^{\infty} \gamma_{21}^{(k)(g)} N_1^{(g)})$$

and

$$\frac{dN^{(0)}}{dt} = Q^{(0)} - n_1 \beta_{10} N^{(0)} - n_2 \beta_{20} N^{(0)} + n_1 \beta_{12} N_2^{(1)} + n_2 \beta_{21} N_1^{(1)}$$
$$+ \sum_{k=1}^{\infty} \gamma_{12}^{(k)(k)} [N_1^{(k)} N_2^{(k)}] + \sum_{k=1}^{\infty} \gamma_{21}^{(k)(k)} [N_1^{(k)} N_2^{(k)}] - N^{(0)} [\sum_{k=1}^{\infty} \gamma_{10} N_1^{(k)} + \sum_{k=1}^{\infty} \gamma_{20} N_2^{(k)} + \gamma_0 N^{(0)}] \qquad (5)$$

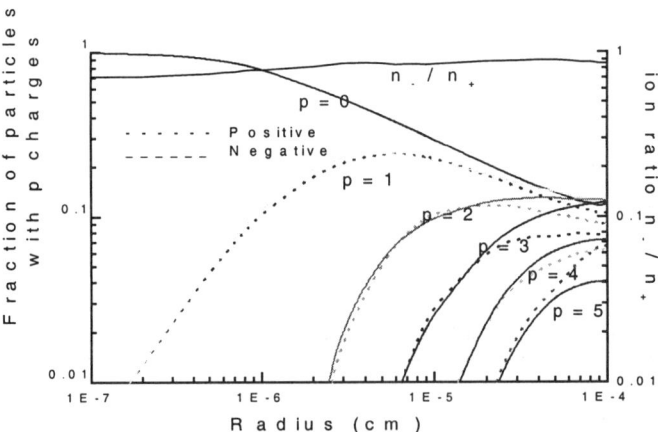

FIGURE 1. Charge distribution on the aerosols for q = 100 cm^{-3} s^{-1} and Z = 10^3 particles/cc after including the effect of coagulation where q is the rate of ion generation; α, β and γ are the recombination, attachment and coagulation coefficients and n and N are the concentrations of small and large ions, respectively.

Superscripts represent the number of elementary charges and subscripts the polarity of charges. We solve these equations for the system of aerosols having maximum of up to five charges. The change in the size of the aerosols due to coagulation of the monodisperse aerosols is neglected here. We further assume that there is no direct production of the charged aerosols.

Fig 1 shows the charge distribution of the aerosol particles when $q = 100$ cm^{-3}s^{-1} and $Z = 10^3$ particles/cc. We use the size-dependent values of the ion-aerosol attachment coefficients from Hoppel and Frick (1986). The negative and positive ions are assumed to have masses of 90 AMU and 150 AMU and mobilities 1.35 and 1.2 cm^2/Vs, respectively. The distribution on particles with mean $r \leq 0.01$ μm is nearly symmetric for the lower values of aerosol concentration. But, for particles with $r \geq 0.1$ μm the fractions of aerosols having upto 3 elementary charges of negative polarity are more than those of the positive polarity. However, the fraction of positively charged aerosols with 4 and 5 elementary charges is more than the fraction of negatively charged aerosols. The asymmetry in the particle charges increases with the number of elementary charges. Ratio of negative to positive ions first increases and then decreases slowly with the radius of aerosols and it varies between 0.65 and 0.75. It is noteworthy that the decrease in the concentration of negatively charged aerosols with 5 elementary charges coincides with the decrease in the ratio of negative to positive ions.

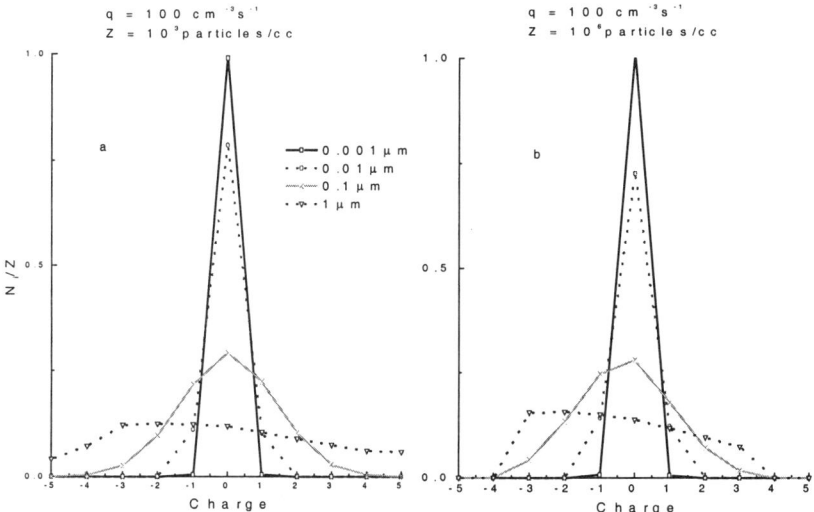

FIGURE 2. Fraction of particles carrying k elementary charges for particles of different radii

Fig 2 shows the change in the fraction of charged aerosols with the number of charges for four different mean radii where $Z = 10^3$ or 10^6 particles/cc and $q = 100$ cm^{-3}s^{-1}. For $Z = 10^3$ particles/cc the charging is almost symmetric in case of smaller particles. The charging asymmetry increases with the size of the aerosols. For the higher total aerosol concentration, the concentration of multiply charged aerosols increases with the mean radius of particles (Fig 2b).

DISCUSSION

Aerosol particles present in the atmosphere act as the centers of combination for the ions. So, an increase in the mean radius or in the concentration of aerosol particles results into the decrease in the concentration of small ions. If the ionic properties of the positive and negative ions are considered to be identical then the charge distribution on the aerosol particles can be well described by the ratio of the ion-aerosol attachment coefficients. However, the difference in the masses of positive and negative ions causes the difference in the values of the attachment coefficient. Hoppel and Frick (1986) solved the ion-aerosol balance equations and observed that the charging asymmetry depends upon the ratio of ionization rate q and the total aerosol concentration Z. They further observed that when ionization rates are very high or aerosol concentration is very low, small ion concentration $n_1 \approx n_2$ and there is a distinct difference in the number of positive and negative aerosol particles. However, if the ionization rate is low or aerosol concentration is high then the charging is nearly symmetric.

In natural environment where the aerosol concentration is very large, the effect of coagulation needs to be considered. In our calculations we include the effect of coagulation and find that coagulation of the charged aerosols causes a decrease in the charging asymmetry. For example, for the singly charged smaller particles the concentration of positive and negative particles becomes nearly equal when the effect of coagulation is included. Also, as seen in Fig 1, the charging asymmetry for the multiply charged aerosols having upto 3 charges reduces due to coagulation when compared with the results of Hoppel and Frick (1986). The aerosol particles with 4 or 5 elementary charges follow entirely different trend and show higher concentrations of positive than that of the negative particles. This decrease in the concentration of negatively charged aerosols may be due to smaller concentration of small ions of negative polarity which restricts the formation of negatively charged aerosols. Due to the difference in the respective ion-aerosol attachment coefficients the effect is more pronounced for the aerosols with 5 elementary charges than in case of the aerosols with 4 elementary charges. Hence, as shown in Fig 1 the decrease in the concentration of negative ions results into the decrease in the concentration of multiply charged aerosols of negative polarity. Further, this decrease in the multiply charged aerosol particles of negative polarity reduces the losses of the multiply charged aerosols of positive polarity due to coagulation and thus results into the larger concentration of multiply charged aerosols of positive polarity.

CONCLUSIONS

The charge distribution on the submicron size charged aerosols computed from the ion-aerosol balance equations including the effect of coagulation of the charged aerosols shows that the asymmetry in the charge distribution results when the ions of opposite polarity are considered to have different mobilities. Our results show a decrease in the charging asymmetry for the aerosols having upto 3 elementary charges when the effect of coagulation is included. However, the particles with 4 and 5 elementary charges show higher concentrations of positive than that of the negative particles.

REFERENCES

1. Adachi, M., Kousaka, Y., and Okuyama, K. J. Aerosol Sci., 16, 109-123 (1985).
2. Dhanorkar, S., and Kamra, A.K., J.Geophys. Res., 102, 30147-30159 (1997).
3. Hoppel, W.A., and Frick, G.M., Aerosol Sci. Tech., 5, 1-21 (1986).
4. Hoppel, W.A., and Frick, G.M., Aerosol Sci. Tech., 12, 471-496 (1990).
5. Vemury, S., Janzen, C., and Pratsinis, S.E., J. Aerosol Sci., 28, 599-611 (1997).

Sticking Probability and Uptake Coefficient – A Quantitative Comparison

M. Kulmala[1] and P.E. Wagner[2]

[1]*Department of Physics, University of Helsinki*
P.O. Box 9, FIN-00014 Helsinki, Finland

[2]*Insitut für Experimentalphysik, Universität Wien*
Boltzmanngasse 5, A-1090 Wien, Austria

Abstract. An analytical formula, where the uptake coefficient is given as a function of mass accommodation coefficient, has been derived. The uptake coefficient is seen always to be smaller than or equal to the mass accommodation coefficient. The independence of the uptake coefficient on mass accommodation coefficient for small Knudsen number has been shown. If the boundary conditions, particularly the vapour pressure at the droplet surface are not well defined a significant error in the value of the mass accommodation coefficient can be found. The value obtained in not well defined conditions is typically smaller than the real value.

INTRODUCTION

A unique determination of mass accommodation coefficient (i.e. sticking probability) is facilitated if [1]
1. the mass transport processes considered are significantly influenced by surface kinetics
2. additional rate determining processes (gas phase diffusion, simultaneous heat transport, liquid phase diffusion, chemical reactions,...) are quantitatively accounted for
3. thermodynamical conditions in the gas phase are well-defined
4. thermodynamic conditions in the condensed phase and particularly at the interface are well determined
5. comparatively fast mass transport processes to newly formed interfaces are considered in order to minimize pollution at the interface.

The above conditions can be fullfilled to a good approximation e.g. in experimental studies on droplet growth in supersaturated vapours [2]. In several investigations partly sticking probabilities and partly uptake coefficients have been considered experimentally in heterogeneous chemistry point of view [3,4] or condensation point of view [2] and theoretically [5-7]. However, there still exist a significant discrepancy between results. The discrepancy is mainly due to the difficulty in precise determination of mass and heat fluxes or the thermodynamical conditions at the droplet surface. In more complex systems like atmospheric trace gases or multicomponent mixtures much bigger differences of experimental values will occur.

Recently we have derived a simple formula to analyze the dependence between uptake coefficient and mass accommodation coefficient [1]. In this presentation we show numerical examples how those coefficients behave as a function of Knudsen number.

ANALYTICAL EXPRESSION

Using the determination of actual mass fluxes and maximum kinetic mass fluxes we were able to derive a simple analytical expression where uptake coefficient (γ) is given as a function of mass accommodation coefficient (α_m) and Knudsen number (Kn) [1]

$$\gamma = \frac{4}{3}\frac{p_{v,\infty} - p_{v,a}}{p_{v,\infty}} Kn \beta_m (\alpha_m)$$

(1)

Here $p_{v,a}$ is the vapour pressure at droplet surface and $p_{v,\infty}$ is the vapour pressure in gas phase. In this equation γ is given as a function of the ratio of vapour pressures, Knudsen number and the transitional correction factor (β_m). The transitional correction factor is a function of Kundsen number and sticking probability. In practise the uptake coefficient is independent of actual formula of β_m. However, in this study we have used the Fuchs-Sutugin formula [8] and

$$\gamma = \frac{4}{3}\frac{p_{v,\infty} - p_{v,a}}{p_{v,\infty}} \frac{Kn(1+Kn)}{1+\left(\frac{4}{3\alpha_m}+0.337\right)Kn + \frac{4}{3\alpha_m}Kn^2}$$

(2)

When $p_{v,a} \ll p_{v,\infty}$ the expression (2) gives correct limits for small and big particles: if $Kn \gg 1$ then $\gamma = \alpha_m$, and if $Kn \ll 1$ then γ is independent of α_m.

NUMERICAL RESULTS AND DISCUSSION

If the vapour pressure at the droplet surface as well as the vapour pressure in the gas phase are known the uptake coefficient can be calculated exactly as a function of Knudsen number and mass accomodation coefficient. In Figure 1 the uptake coefficient is given as afunction of Kn using different values of α_m. When Knudsen number is high $\gamma = \alpha_m$, and correspondingly for all α_m values the uptake coefficient approaches the mass accomodation coefficient. On the other hand for small Kn values the uptake coefficient tends to become independent of α_m. The smaller α_m is the smaller value of Kn is needed to obtain this independency. Here $p_{v,a}$ is zero.

To understand the significance of mass accommodation coefficient on uptake and condensation processes it is important to notice the difference between uptake coefficient and mass accommodation coefficient. We have recently [1] derived analytical formula between mass accommodation and uptake coefficients. Using this expression we have analysed their connection numerically. The uptake coefficient was seen to be always smaller or equal than the mass accommodation coefficient. When Knudsen number is small enough the coefficients are seen to be independent. For high Knudsen number the coefficients are equal. Undefined boundary conditions have found to be a significant source of error, and mass accommodation coefficient obtained by not well defined vapour pressure values can be significantly smaller than real values.

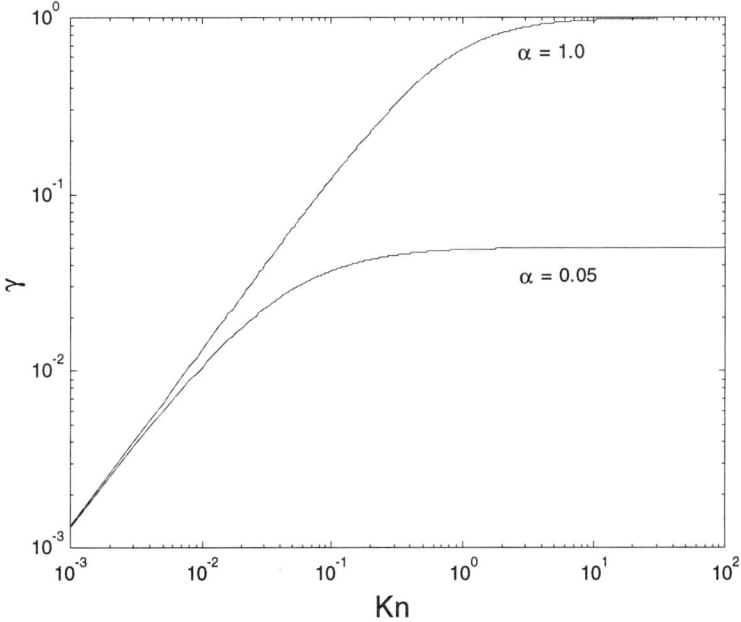

FIGURE 1. The uptake coefficient as a function of Knudsen number with two different value of the mass accommodation coefficient.

REFERENCES

1. Kulmala, M., and Wagner P.E., "Accommodation and uptake coefficients – a quantitative comparison" submitted to J. Aerosol Science (2000).
2. Wagner, P.E., "Aerosol Growth by Condensation," in Aerosol Microphysics II edited by W.H. Marlow, Springer-Verlag, Berlin, 1982, pp. 129-178
3. Davidovits, P., "Experimental Techniques and Modelling in Heterogeneous Atmospheric chemistry Studies", AAAR'99 Tacoma Washington, Notes for Tutorial, October 11, 1999.
4. Finlayson-Pitts, B.J., and Pitts J.N (2000) Chemistry of the Upper and Lower Atmosphere, Academic Press, 2000
5. Schwartz, S.E., "Mass-Transport Considerations Pertinent to Aqueous Phase Reactions of Gases in Liquid-Water Clouds", in Chemistry of Multiphase Atmospheric Systems edited by W. Jaeschke, Springer-Verlag, New York, 1986, pp. 415-471
6. Clement, C.F., Kulmala, M., and Vesala, T,. "Theoretical Consideration on Sticking Probabilities", J. Aerosol Sci., 27, 869-882 (1996)
7. Seinfeld, J.H., and Pandis, S.P,. Atmospheric Chemistry and Physics. John Wiley & Sons, 1998.
8. Fuchs, N.A., and Sutugin, A.G.,Highly dispersed aerosols, Ann Arbor Science, Ann Arbor, Michigan, 1970.

Stable Droplets in Finite Volumes

A. J. H. McGaughey and C. A. Ward

*Department of Mechanical and Industrial Engineering,
University of Toronto, Toronto, Canada M5S 3G8
416-978-4807; ward@mie.utoronto.ca*

Abstract. The stability of a single, one-component droplet has been investigated experimentally. Using such a droplet as the initial condition for an evaporation experiment, time histories of the droplet size, temperature, and temperature of the surrounding vapor are obtained. The results indicate that the liquid temperature is always less than that in the vapor.

BACKGROUND

The thermodynamic stability of a single one-component droplet in its own unbounded vapor was theoretically investigated by Gibbs (1). He predicted that such a droplet would have only one equilibrium state, and that this state would be unstable. The radius in this state is often referred to as the critical radius. With the appropriate assumptions, the Kelvin relation can predict its value.

If a system of finite volume is considered, a second equilibrium state, which is stable, has also been predicted to exist (2,3). The size of the stable droplet is a function of the system's independent variables: temperature, volume and mass. There have been no known attempts to investigate this prediction experimentally.

Continuum droplet evaporation models assume a continuous temperature profile across the liquid-vapor interface (4). Some calculations have taken into account a temperature discontinuity as predicted by Classical Kinetic Theory (CKT) (5).

Recent experimental work has questioned the validity of CKT when applied to a liquid-vapor system undergoing phase change (6-9). In both evaporation and condensation experiments, interfacial temperature discontinuities were measured that were larger than, and in the case of evaporation, in the opposite direction to those predicted by CKT. The data are in good agreement with Statistical Rate Theory (SRT), a theoretical approach based on the transition probability concept of quantum mechanics (10).

Classical nucleation theory assumes a uniform temperature in the liquid and vapor phases, and properties are evaluated at this temperature (11). To resolve discrepancies between theoretical and experimental results, corrections to the theory have been suggested (12-14). If there is a temperature discontinuity at the liquid-vapor interface during a phase change process, then the temperatures at which the properties of the two phases are to be evaluated need to be chosen carefully.

EXPERIMENTAL INVESTIGATION

An experimental apparatus has been built that can be used to investigate the stability of a pure water droplet. A simplified schematic of the test section is shown in

Figure 1. The test section is composed of two chambers: the main chamber and the auxiliary chamber. In preparation for an experiment, the test section can be evacuated to 10^{-4} Pa.

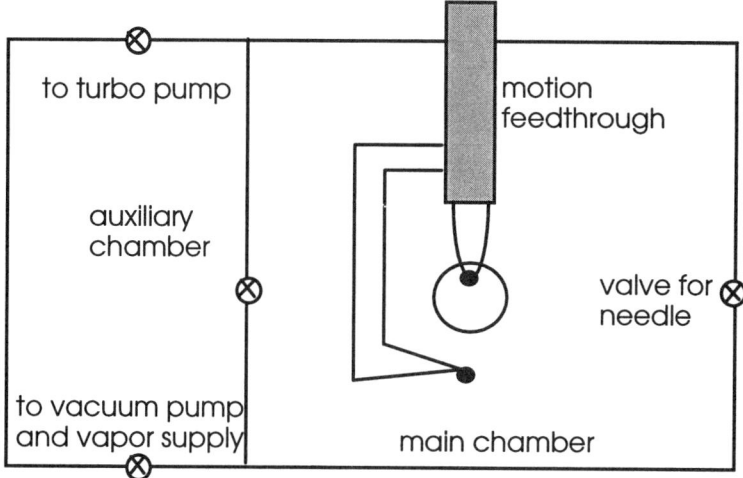

FIGURE 1. Schematic of the test section.

The two chambers are separated by a ball valve. The droplet is introduced into the main chamber through a small needle. The needle's position can be manipulated from outside the test section. The initial droplet size is controlled with a syringe pump. The droplet can have an initial diameter in the range of 1.1 to 1.8 mm. A droplet can be suspended on a thermocouple that is mounted on a linear-rotary motion feedthrough. The thermocouple has an approximately spherical bead 0.4 mm in diameter. The thermocouple wires are taken out of the system and sealed with vacuum epoxy. The needle passes though a ball valve and then a septum before leaving the system. Ideally, there is no air leakage into the chamber when the needle is removed. A telescope that can resolve to 0.01 mm is mounted on a micro-manipulator to view the droplet through a glass window. Measurements of the diameter can be taken from different perspectives by rotating the feedthrough. A second thermocouple is mounted on the feedthrough which can measure the temperature in the vapor 4 mm from the droplet thermocouple. The pressure inside the test section is measured with a mercury manometer.

Before running an experiment, the test section is pumped to a pressure below 10 Pa. Vapor at the saturation pressure corresponding to the temperature of the water supply is introduced into the test section. This pressure is measured. The main chamber is then isolated, and the auxiliary chamber evacuated to below 10 Pa. A droplet is then suspended on the thermocouple and the needle is removed from the system. The droplet diameter and the temperature of the two thermocouples are recorded at five minute intervals in order to assess the stability of the droplet. If the droplet is not stable (i.e., it is evaporating), then its size is recorded as it continues to evaporate. If the droplet is stable (i.e., its diameter is constant), the valve to the auxiliary chamber is opened and the droplet would then start to evaporate. Its diameter is recorded until it reaches a size of 0.90 mm, at which point it will no longer be spherical due to the effect of the thermocouple bead and wires. The temperatures are recorded until no liquid is present. The pressure in the final state is measured. Results from a typical experiment where a stable droplet was observed are presented in Figure 2.

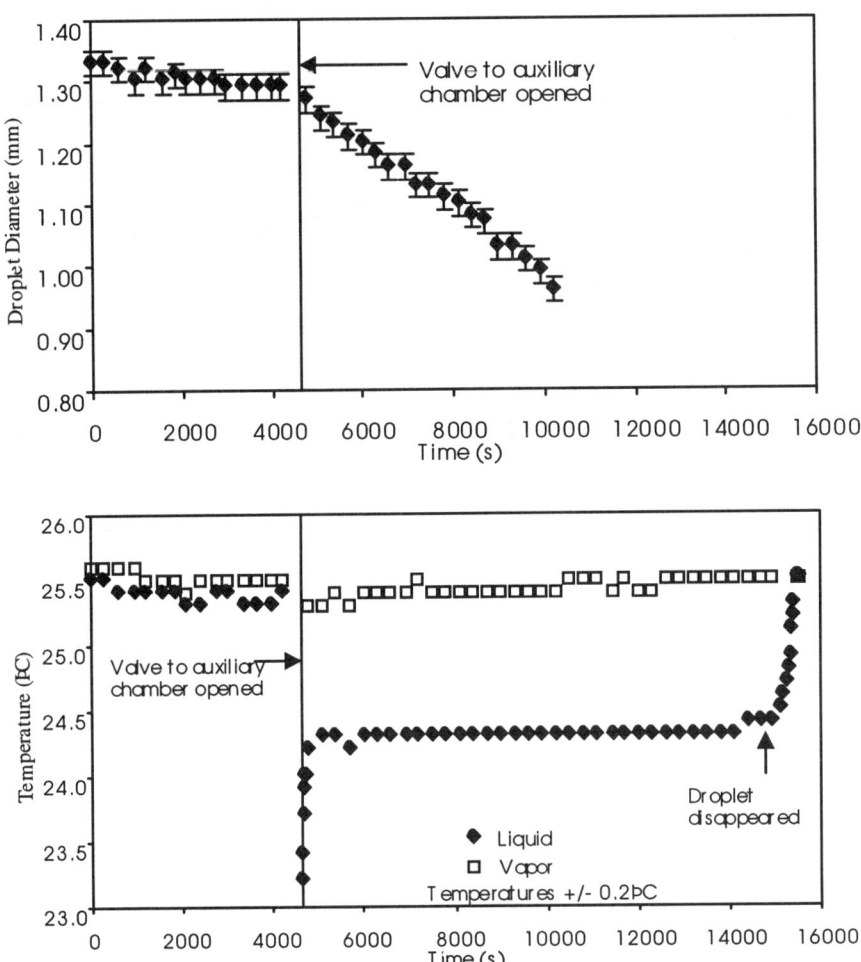

FIGURE 2. Droplet size and temperature histories.

DISCUSSION AND CONCLUSION

As seen in Figure 2, there was no observed change in the size of the droplet for a period of thirty minutes. Provided that any impurity air can be neglected, it appears that a stable, single, one-component droplet has been observed. We note from the theory that the presence of air in the system would have no noticeable effect on the stability or the predicted droplet size. During the observation period, there was no significant difference between the temperatures in the liquid and vapor phases.

When the volume was increased, the pressure went down, and evaporation started. The temperatures in the liquid suddenly decreased by 2.2°C and the vapor temperature decreased by 0.2°C. In the space of a few minutes, each of the temperatures became steady, with the temperature in the liquid being 1.1°C less than that in the vapor. Differences between 0.4 and 2.0°C have been observed in a number of experiments, with the temperature in the liquid always being less than that in the vapor. The magnitude of the difference can be attributed to the conditions of the particular experiment. It was not until the liquid had completely evaporated that the temperature read by the droplet thermocouple came back to the temperature of the vapor. This transition takes place quickly because of the thermocouple's small thermal mass (see Figure 2). The evaporation results agree well with the d^2-Law (15) with a constant slope of $-1.19 \cdot 10^{-4}$ mm^2/s.

Based on the measurements of the final system pressure and temperature, and geometrical calculations of the volumes of the two chambers, the theory predicts a stable droplet diameter of 2.12 ± 0.19 mm. The measured value is 1.29 ± 0.02 mm. The final pressure measurement is the variable most likely to cause this discrepancy. To get agreement between theory and experiment, it would need to be lower. One possible explanation is the presence of air in the system. However, based on leak tests done on the test section, the mole fraction of air in the system is less than 0.05, which would not be a sufficient amount to account for the difference. A more likely possibility is condensation on the inside walls of the test section as the droplet reaches its stable equilibrium due to spatial and temporal variations in the ambient temperature. When the valve between the chambers is opened, causing the droplet to start evaporating, the condensate would also vaporize, giving a higher final pressure reading. Before reaching any firm conclusions as to the validity of the theory or the experimental values, further work is required to refine both areas of the research.

ACKNOWLEDGMENTS

This work is supported by the Natural Sciences and Engineering Research Council of Canada and The Canadian Space Agency.

REFERENCES

1. Gibbs, J. W., *Collected Works, Volume 1*, New York: Congmans, Green and Co., 1928, pp. 237-258.
2. Rao, M., Berne, B. J., and Kalos, M. H., *Astrophysics and Space Sciences* **65**, 39-46 (1979).
3. Vogelsberger, W., *Chemical Physics Letters* **64**, 601-603 (1979).
4. Sirignano, W. A., *Journal of Fluids Engineering* **155**, 345-378 (1992).
5. Young, J. B., *International Journal of Heat and Mass Transfer* **34**, 1642-1661 (1991).
6. Fang, G., and Ward, C. A., *Physical Review E* **79**, 417-428 (1999).
7. Ward, C. A., and Fang, G., *Physical Review E* **79**, 429-440 (1999).
8. Fang, G., Ward, and C. A., *Physical Review E* **79**, 440-453 (1999).
9. Ward, C. A., and Stanga, D., submitted for publication.
10. Ward, C. A., *Journal of Chemical Physics* **67**, 229-335 (1977).
11. Abraham, F. F., *Homogeneous Nucleation Theory*, New York: Academic Press, 1974, pp. 9-30.
12. Katz, J. L., Fisk, J. A., and Rudek, M. M., "Nucleation of Single Component Supersaturated Vapors," presented at The Fourteenth International Conference on Nucleation and Aerosols, Helsinki, Finland, August 1996.
13. Granasy, L., *Journal of Chemical Physics* **104**, 5188-5198 (1996).
14. Fokin, V. M., and Zanotto, E. D., *Journal of Non-Crystalline Solids* **265**, 105-112 (2000).
15. Law, C. K., *Prog. Energy Combustion Sci.* **8**, 171-201 (1982).

Changes in the Concentration and Size Distribution of Submicron Aerosols Accompanied with the Polar Lows at Maitri, Antarctica

A K Kamra and C G Deshpande

Indian Institute of Tropical Meteorology, Pune, India

Abstract: Measurements of the submicron aerosol size distribution made during the low pressure events at the Indian Antarctic station, Maitri (70° 45'S, 11° 44'E) are reported. The total aerosol concentration increases by approximately an order of magnitude whenever a low pressure system passes over the station. The changes in the size distributions, which is generally bimodal, associated with such low pressure systems show an increase in particle concentration in nucleation mode by an order of magnitude but only marginal in accumulation mode. Observations suggest that while the increase in particle concentration in nucleation mode during these low pressure systems is due to the subsidence of midtropospheric air during the weakening of radiative inversion, the increase in the accumulation mode particles occurs either due to transportation of particles generated over sea by wave breaking or due to the erosion of exposed rocks by strong winds.

INTRODUCTION

The Antarctic aerosols have unique importance as they are free from anthropogenic pollution and thus truly represent the background aerosols which play important role in various global atmospheric processes. Earlier investigations of the properties and behavior of aerosols at Antarctica are reported by Vosrensenkii [1968], Shaw [1986, 1988], Gras and Anderson [1985], Lal and Kapoor [1989], Ito [1993], Hogan and Gow [1993], Saxena [1996]. In this paper, we report our measurements of the submicron aerosol size-distribution made at the Indian station, Maitri (70° 45' 52" S, 11° 44', 03"E, 117 m above msl), during the low pressure events in the summer of 1996-97. The results are interpreted in terms of the local meteorological and synoptic conditions.

INSTRUMENTATION AND MEASUREMENTS SITE

Measurements of size distribution of particles of diameter 13 to 1000 nm were carried out with a TSI 3030 Electrical Aerosol Analyser (EAA) system. It measures the size distribution of particles of diameter 3 to 1000 nm in 10 different size ranges. The EAA system was operated after every 3 hours to collect 5 size-distribution samples and the data was stored in a PC. In our analysis, we use the average of 5 samples collected after every 3 hours. The inlet of the EAA system was cleaned frequently to avoid accumulation of sea salt, snow or dust at the inlet.

The Indian station, Maitri is located in the Schirmacher oasis in the Dronning Maud Land, East Antarctica. The east-west trending Schirmacher oasis is exposed over an area of approximately 35 kms² with 16 kms length and a maximum width of 2.7 kms in the central part. It is at an altitude of 117 m above mean sea level. The aerosol measurements were carried out in a summer hut, namely Tirumala. The inlet of the air sampling stainless steel tube was projected out at a height of 2 m above ground through a window such that most of the time, it faces the persistent wind direction. The single storeyed building of Maitri station and other structures such as generator huts, gas plant, incinerator were about 150-200 m away in the northeast direction from the site of the measurements so that any pollutant released from them had little or no chance of reaching the site of measurements.

THE CONCENTRATION AND SIZE DISTRIBUTION OF AEROSOLS DURING THE LOW PRESSURE EVENTS

The 'wall of storms' formed by polar lows circulating around the Antarctica Convergence Zone makes a strong barrier for aerosol transport. Under the influence of depressions moving east-southeastward between latitudes 65° - 70° S, the low pressure systems sometimes penetrate upto some coastal stations such as Maitri. During our period of observations, Maitri experienced a sequence of rise and fall in the surface pressure. A dip in surface pressure at the station is most of the time preceded by several hours by a peak in aerosol concentration.

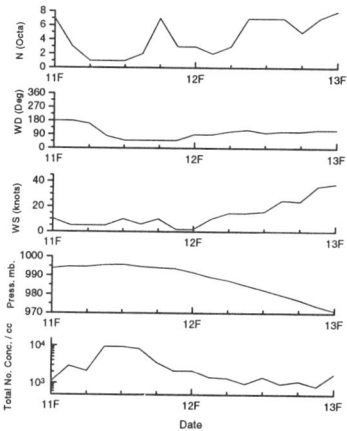

Fig.1: The total aerosol concentration alongwith the surface atmospheric pressure, wind speed, wind direction and cloud coverage during the passage of a low pressure system over Maitri on February 11-12, 1997.

The total aerosol concentration is calcualted by adding the number concentrations in each of the eight size categories from 13 to 750 nm diameter size. Our measurements in the months of January and February show some peaks of very high aerosol concentrations. Such peaks in aerosol concentrations have also been observed at other Antarctic coastal stations such as Mirny [Voskresenskii, 1968], Siple [Hogan, 1975], Syowa [Ito, 1980] and even at the South Pole [Hogan, 1979]. These peaks are associated with the passage over the station of cyclonic storms circulating around the continent of Antarctica. The 3-hourly bservations of surface pressure and cloudiness made at Maitri support such an association.

Figure 1 shows the total aerosol concentration alongwith the 3-hourly observations of surface pressure, wind speed, wind direction and total cloud coverage on February 11-12, 1997 at Maitri. The pressure starts decreasing from 1200 hours on February 11, 1997 and reaches the minimum value at 0000 hours of February 13, 1997. This was a substantial fall of 24 mb in surface pressure. The peak in total aerosol concentration showing an increase of more than an order of magnitude is observed during this period and is again followed by increase in wind speed and cloudiness. Each peak in aerosol concentration lasts for about 12 hours and it occurs before the actual surface pressure at the station reaches its minimum value. About 12 similar occurrences of the increase in total aerosol concentration associated with a frontal weather system are observed during this observational period at Maitri.

In general, the size distributions are bimodal and remain almost unchanged for several hours or days provided there is no substantial change in the meteorological conditions at the station. There is no typical diurnal pattern in the size distribution or concentration of aerosols. All size distributions are bimodal with a peak either at 75 nm or 133 nm diameter and two minima at 42 and 420 nm and are open ended at both ends.

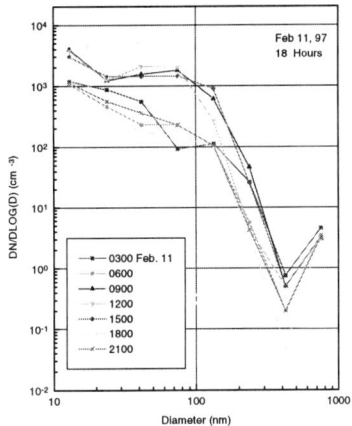

Fig.2: The aerosol size-distributions associated with the passage of low pressure system over Maitri on February 11, 1997.

The aerosol size distributions significantly change when a low pressure system passes over the station. Figure 2 shows a typical example of the changes in aerosol size distribution observed on February 11, 1997. In this case, the aerosol concentrations of particles in the nucleation mode (diameter < 100 nm) increase by about an order of magnitude with little increase, if any, in the accumulation mode (diameter > 100 nm). On some such occasions, a signficiant increase is observed only for particles in the smallest size range i.e. 13 nm diameter and other larger sizes in nuclei mode show enhanced variability. The aerosols in acccumulation mode do not show much increase during these low pressure events.

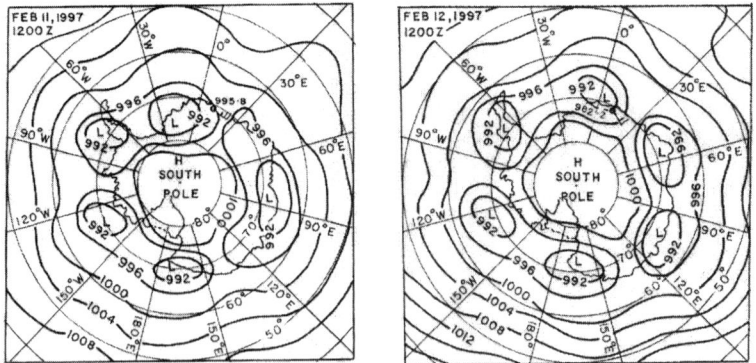

Fig.3: The six-hourly surface synoptic weather chart for 1200 hours on February 11 and 12, 1997.

In the six-hourly surface synoptic weather chart for 1200 hours on February 11, 1997, one low pressure system is located west of Maitri and another one southeast of Maitri (Figure 3). The surface weather chart for 1200 hours of February 12, 1997 shows that the system which was west of Maitri, has centered over Maitri. The peak in the total aerosol concentration at 1200 hours of February 11, 1997, is most probably associated with the low pressure system approaching from the west. Similar observations of the increase in aerosol concentration have also been reported, Voskresenskii [1968] at Mirny, by Hogan [1975] at Siple iand by Lal and Kapoor (1989) at Maitri in the Antarctic continent.

DISCUSSION

The transport, evoluation and removal of aerosol particles or trace gases over Antarctica is mainly governed by the atmospheric circulation patterns, the series of circulating cyclonic storm systems near the southern polar fronts. The observations made at the coastal stations are more difficult to interpret than those at the higher polar plateau because of greater intermixing of the effects of strong continental drainage flows due to sloped inversions and katabatic winds and the baroclinic disturbances generated over the southern ocean regions surrounding the Antarctica continent.

The process of coagulation of small particles to large particles contributes to the creation of the accumulation mode of aerosol particles. However, the contribution of coagulation process is small around Antarctica since it varies with the square of the particle concentration which is small. Other possible sources for large particles on Antarctica may be the weathering products from Earth's crust, sea-salt particles from bubble breaking at the surrounding sea surface, extraterrestrial debris, and volcanic emissions [Shaw, 1988].

The peaks in the total aerosol concentration before the arrival of a low pressure system at the stations as observed on many occasions in our measurements may be due to the subsidence of midtropospheric air when the radiative inversion weakens. The aerosols of nuclei mode produced in the upper troposphere by photochemical reactions are brought down to the surface during these events. Our observation that the aerosols during these episodes are very small particles in nucleation mode, supports such transport of aerosols. The fine crustal particles exist in the Shirmacher oasis region. Under the favourable synoptic situation for particle transport along the frontal system, these particles are carried away along with the airflow during these cyclonic storms. Also, greater vertical mixing associated with weakening of temperature inversion and moderate breeze may enhance the lifting of very fine crustal particles from the bare land area around Maitri.

ACKNOWLEDGEMENT

The authors express their gratitude to the Department of Ocean Development for supporting these studies at Antarctica. The meteorological data provided by the India Meteorological Department is thankfully acknowledged.

REFERENCES

Gras J. L. and Adriaansen,, A., J. Atmos. Chem., 3, 96-103 (1985).
Hogan A.W., J. App. Met., 14, 550-559 (1975).
Hogan, A.W., J. Appl. Meteorol., 18, 741-749 (1979).
Hogan A.W. and Gow, A. J., Tellus, 45B, 188-207 (1993).
Ito, T., Tenki Tokyo, 27, 13-24 (1980).
Ito T., Tellus, 45B, 145-159 (1993).
Lal M. and Kapoor, R.K., Atmos. Environ., 23, 803-808 (1989).
Saxena, V.K., Geophys. Res. Lett., 23, 69-72 (1996).
Shaw, G.E., J. Aerosol. Sci., 17, 937-945 (1986).
Shaw G.E., Rev. Geophys., 26, 89-112 (1988).
Voskresenskii A. I., Sov. Antark. Eksped. Trudy, 38, 149-198 (1968).

Simulation and Measurements of Dust Deposition in Macedonia

S. Nickovic[1], V. Spiridonov[2], M. Andreevska[2], and S. Music[1]

[1]*Euro-Mediterranean Centre on Insular Coastal Dynamics (IcoD), Malta, and*
[2]*Hydro-Meteorological Institute of Macedonia*

Abstract. Aerosol such as wind-blown sand and dust are transported from Sahara to Macedonia under certain atmospheric conditions. A spring case with considerable dust transport on 17 April 1999 was selected for this study. This paper presents results obtained a) by a numerical dust model, and b) by chemical analysis done in the framework of the European Monitoring Environmental Program (EMEP). A numerical simulation of dust uptake, transport and deposition has been done using the Dust Regional Atmospheric Model (DREAM), with the Eta/NCEP regional forecast model as a driving tool for dust concentration. The heavy metal elements in samples of precipitation were analyzed by chemical analysis based on an atomic absorption method. The ground measurements are used for the model validation, as well.

DUST TRANSPORT DURING APRIL 1999

Large amount of mineral dust is mobilized over Sahara and injected into the atmosphere under favorable weather conditions. Even areas far away from Sahara such as Macedonia are sometimes affected by dust transport and deposition where dust behaves as a pollutant that significantly reduces the air quality. Ambient dust concentration during such dust storms may significantly exceed international standards for allowable concentrations, and therefore cause health problems (allergy, respiratory diseases, eye infections etc.).

A strong dust storm occurred during mid April 1999 over Sahara was simulated by the dust model DREAM [1]. This model reproduces all major phases of the atmospheric dust cycle, such as dust production, transport, turbulent diffusion, and wet and dry removal.

During the considered period, a substantial downward transport of momentum existed over Sahara. At the same time, the atmospheric circulation over the Central Mediterranean was characterized by strong S-SW advection of temperature on 850 hPa (Figure 1). On 17 April, movement of the hot air tongue toward the east reduced the temperature and moderate rainfall was observed. Further eastward advancing of the cold air caused drop in temperatures over Macedonia followed by snowfall.

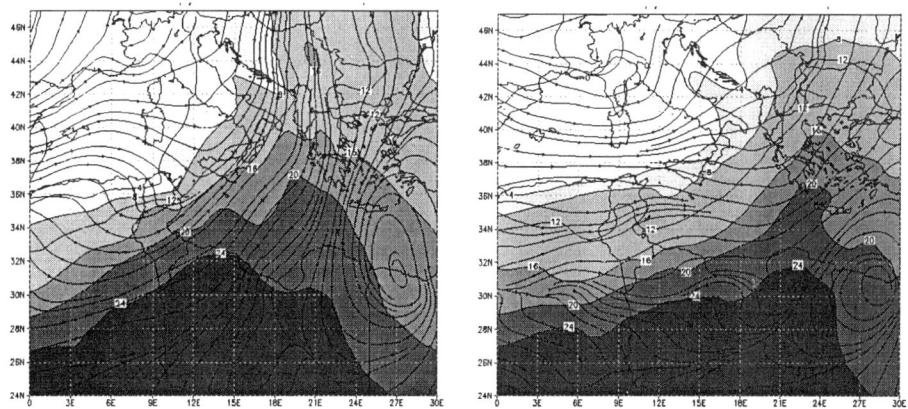

FIGURE 1. Temperature and wind streamlines at level 850 hPa valid for
18 UTC on 16 and 17 April 1999

Due to downward momentum advection, a major dust storm occurred over Sahara. The upper air weather conditions provided efficient long-range aerosol transport from dust sources towards the Balkan Peninsula. The DREAM model indicates existence of a large dust cloud in the Central Mediterranean, which extends towards Southern Europe and arrives over Macedonian region on 17 April (Figure 2a). The increased concentration is associated with total (wet and dry) aerosol deposition (Figure 2b), which covers Southern Italy and most of the Balkans. The model indicates that dust reaches rather high elevations - approximately 7 km in the vertical (Figure 3a).

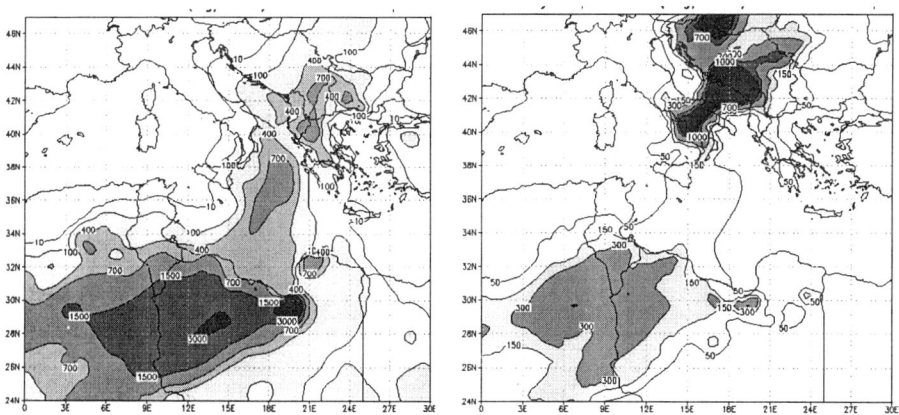

FIGURE 2. (a) Left panel: simulated surface concentration ($\mu g\ m^{-3}$); (b) Right panel: simulated total deposition ($mg\ m^{-3}$) valid for 18 UTC 17 April 1999

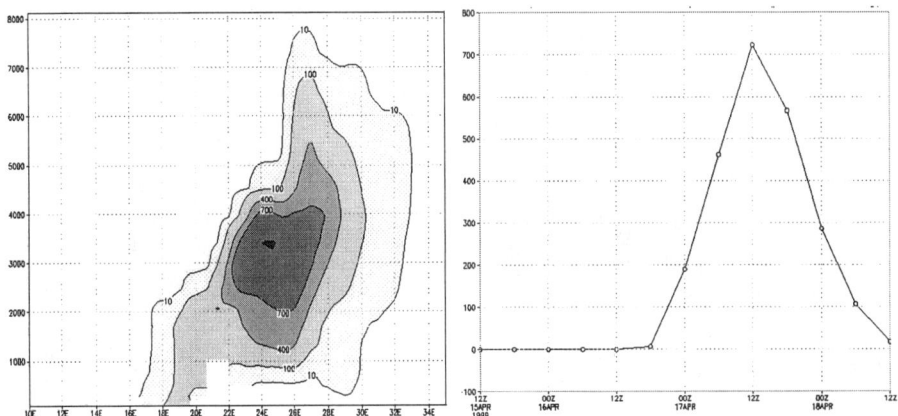

FIGURE 3. (a) Left panel: Simulated concentration (μg m^{-3}) in the vertical section crossing the latitude $\phi = 41.5°$ valid at 18 UTC 17 April 1999 (b) Right panel: Simulated time evolution of the surface concentration (μg m^{-3}) in Lazaropole, Western Macedonia ($\phi = 41.5°$, $\lambda = 20.7°$)

CHEMICAL ANALYSIS OF DUST-RELATED COMPUNDS

During 16-18 April 1999, presence of yellowish was visually evident in Macedonia. Using the atomic spectrometer VARIAN SpectAA 220, we have made chemical analysis of rainfall collected in Lazaropole, Western Macedonia ($\phi = 41.5°$, $\lambda = 20.7°$), in order to find out eventual link between model results and aerosol measurements.

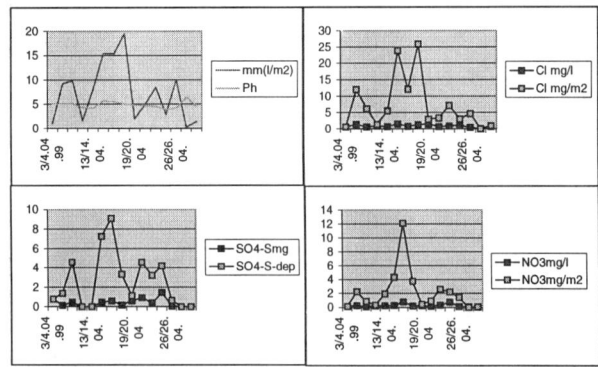

FIGURE 4. Ph-factor,, chlorides, sulfates and nitrates found in precipitation samples in Lazaropole

Figure 4 shows time evolution of Ph-factor, chlorides, sulfates and nitrates detected in the precipitation samples. One can notice existence of maximums on 17 April for all considered parameters.

Measurements of the heavy metal content in rainfall samples were performed, as well. Wet deposition of Na, K, Ca, Mg, Fe, Mn, Cu, Ni, and Zn had exceptionally high values during the critical period (a sub-set of measurements is shown in Figure 5).

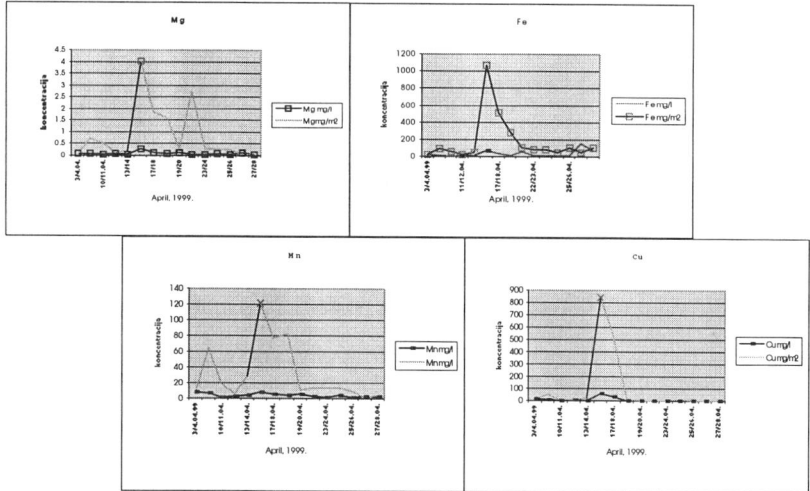

FIGURE 5. Mg, Fe, Mn and Cu in rainfall samples in Lazaropole

CONCLUSIONS

Saharan dust plum passage over Macedonia in April 1999 was traced using model simulations and ground measurements of dust. The model clearly showed that dust load and deposition affected especially the western Macedonia during the considered period. Both model and observation data sources qualitatively agree, indicating that maximum of dust activity happened during 16-18 April (compare, for example, Figure 3b with Figures 4 and 5).

This study shows that by combining in-situ measurements and model predictions, a reliable operational monitoring system can be developed, able to worn a regional community on major dust intrusions, thus reducing the risk of adverse effects caused by high natural aerosol concentrations.

ACKNOWLEDGMENT

Part of this study done at ICoD, Malta was funded by the 4-th Maltese-Italian Financial Protocol through the Project COMPASS.

REFERENCES

1. Nickovic, S., A. Papadopoulos, O. Kakaliagou, and G. Kallos: Model for prediction of desert dust cycle in the atmosphere, J. Geoph. Res., 2000, submitted.

Numerical Simulation of Frontal Rainbands over Ukraine under Different Mechanisms of Cloud and Precipitation Formation

A.M. Pirnach, S.V. Krakovskaia, A.V. Belokobylski

Ukrainian Hydrometeorological Research Institute, Kiev, Ukraine

Abstract. Three-dimension diagnostic and prognostic models were used for numerical simulation of cloud system accompanied the passage of a cyclone over Ukraine. The response of time and space distribution of cloud particle spectra, integral features and precipitation on changing of cloud and precipitation formation mechanisms (condensation, sublimation, collection by large drop and ice particles for droplets etc) was investigated. The nested and stretched grid was used to simulate the narrow band of heavy rainfall that frequently appeared ahead a cold front. The numerical experiments for different values of cloud condensation nucleus concentration were carried out for the above rainband. Comparison between spectra for different conditions was performed.

1. INTRODUCTION

The work continues the many years theoretical investigation devoted to cloud and precipitation formation that was performed in UHRI (see Buikov, 1978; Pirnach, 1979, 1998; Pirnach and Krakovskaia, 1994).

The sensitivity of cloud particle size distribution, integral cloud features and precipitation evolution to the intensity of different nucleation and coagulation mechanisms was the ultimate object of the study. The results of investigations of Twomey, 1959, Hobbs et al., 1977, Buikov, 1978, Hegg et al., 1995 etc were used in this work.

2. SHORT DESCRIPTION OF THE MODEL

The formation and development in space and time of atmospheric fronts and their cloud systems are simulated by integration of a full thermodynamic equations' set, which includes equations for air motion, water vapor content, temperature transfer, cloud droplets and ice particles distribution functions, continuity and state equations. The continuous growth of precipitation particles is accounted (see Pirnach, 1976, 1979, 1998). The parameterization of droplet and ice nucleation is used. Cloud condensation nuclei (CCN) assumed having activity at a cloud base and into a cloud by relationships which describe cloud particle (CP) generation on CCN (Buikov, 1978; Pirnach and Krakovskaya, 1994; Hobbs, 1977). The empirical parameters N_{m0}, N_c that presented the CCN concentration at a cloud base and into a cloud respectively are changed at modeling. If $N_c=0$, CCN are absent into clouds.

The analysis of cloud particle spectra into whole cloud system is impossible for this short report and herein gamma size distribution has been suggested as an appropriate form for distribution of droplets and crystals in clouds as follows

$$f(a,b,r) = A r^a e^{-br}. \qquad (1)$$

Mainly the a-parameter determines a form of gamma distribution. This parameter was used hereinafter for investigation of space and time distribution of modeled spectra and the next relationships were used to find a-parameter (see Buikov and Pirnach, 1972):

$$a = \frac{2c - 5 - \sqrt{1 + 8c}}{2(1-c)}, \qquad c = \frac{3 q_{Li}}{4\pi\rho \cdot N_i r_{mi}^3}, \qquad 1 < c < 6, \qquad (2)$$

where N_i is droplet or crystal concentration, r_{mi} is mean radius, q_{Li} is liquid water or ice content. The other particle spectra will be out from the consideration. Simulation is carried out in a co-ordinate system moving with a constant velocity. The description of model in more detail was given in (Pirnach, 1998).

3. THE SENSIVITY OF CLOUD AND PRECIPITATION EVOLUTION TO DROPLETS COLLECTION BY RAINDROPS AND ICE PARTICLES

The state of the atmosphere over Ukraine on September 30 at 00 GMT was selected as initial data for modeling. At this time the southern cyclone moved rapidly over central Ukraine and a large cloud system that included rainbands of different scale

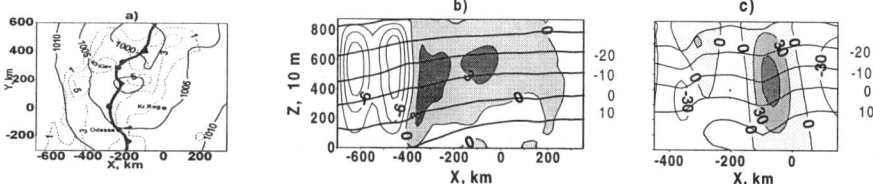

FIGURE 1. Horizontal (a) and vertical (b, c) cross-sections at time t=0 (a, b) and t=12h (c):
a) pressure, mb (solid line), temperature, °C (dashed line); front line (bold solid line) at z=0;
b) temperature, °C (bold solid line), vertical motion, cm/s (solid line) at y=400 km and t=0;
c) as b) and t=12 h.

accompanied it. Figs 1-3 and Table 1 show the time and space distribution of the temperature, pressure, vertical motion, cloud particle size distribution, integral cloud features and precipitation for two runs with and without of the coagulation processes taken into account. A regular grid with space lengths Xs= 50 km and Ys=100 km in direction normal to cold front (from west to east) and along front respectively was used. The cloud nucleation mechanism was determined by $N_{m0}=5 \cdot 10^5$ g^{-1}, $N_c=270$ g^{-1}. There is no space to locate the three-dimensional presentation of the time and spatial evolution of clouds, therefore the cross-sections with the coordinate y=400 km were selected for presentation. There was found the maximum of updraft motions reached 70 cm/s and heavy precipitation reached 10 mm/h (see Fig.2, second row).

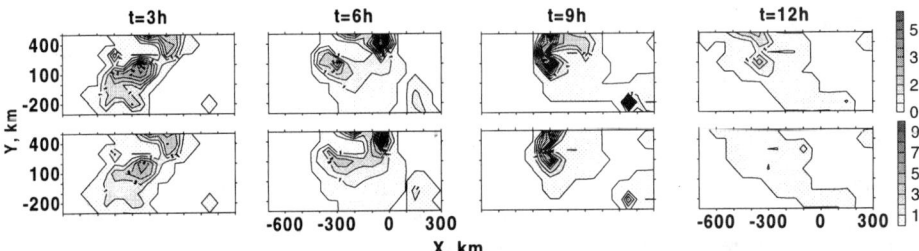

FIGURE 2. Time and space development of precipitation intensity, mm/h. First row presents a run, which excluded coagulation processes, and second row displays coagulation processes.

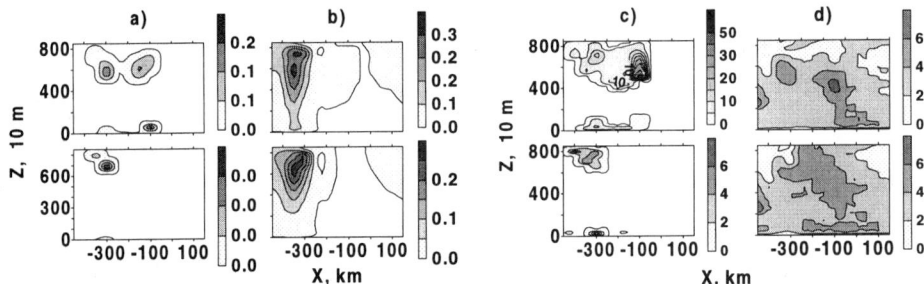

FIGURE 3. Vertical cross-sections of liquid water (a) and ice content, g/kg (b) and cloud drop (c) and ice particle spectra (d) in terms a-degree, see (2) at t=12 h. First row presents a run without coagulation processes, second row presents the process with coagulation.

The time development of precipitation presented in Fig.2 has shown that the space distribution of rainband and precipitation intensity in it has similar structure for both cases. Coagulation processes clearly increased precipitation intensity maxima, played the crucial role for small drop size spectra and sharply reduced the liquid water content. The ice particle spectra and ice content are less sensitive to precipitation formation mechanisms (see Fig.3).

4. AN IMPACT OF THE CCN-DISTRIBUTION ON PRECIPITATION CORE DEVELOPMENT

The precipitation core was found near X=-200 km and Y>50 km. The nested grid 50 km wide and with grid interval Xs=5 km was constructed for this region. The resolution for stretched grid was Xs=Ys=100 km. The coordinate Y=400 km was selected for illustration. The maximal vertical motion (of 3 m/s) and precipitation intensity exceeded 15 mm/h at t=7 h were identified in this region. But the convective cells appeared in other regions too. For example, near Y=100 km the heaviest precipitation was detected at t=2 h and its maximum reached 9 mm/h if coagulation processes have been taken into account. At t=5 h the investigated rainband was captured by the wide spread rainband. The time and space evolution for both runs was similar, but precipitation rate with coagulation clearly exceeded the rate without it.

The sensitivity of cloud particle distribution to the changing of nucleation mechanism presented by Fig.3 and Table 1. The evolution of the spatial maximum precipitation in rainband for different cloud and precipitation mechanisms has like

TABLE 1. Maxima of precipitation intensity and a-parameter maxima time development at different cloud and precipitation formation mechanisms

Time, h	Precipitation intensity maxima, mm/h					a-parameter maxima				
	1	2	3	4	5	1	2	3	4	5
1	0.5	0.3	0.8	0.2	0.9	12	31	12	12	65
2	3.7	0.8	3.9	3.8	3.2	11	12	11	11	11
3	3.1	3.5	3.7	4.0	6.3	9	10	9	9	17
4	5.5	4.5	5.4	7.4	6.6	8	24	9	9	37
5	8.6	6.4	7.8	9.9	8.7	9	11	9	9	11
6	7.0	5.5	8.6	8.6	9.6	12	13	12	11	13

1- $N_{m0}=5 \cdot 10^5 g^{-1}$, $N_c=270 g^{-1}$;   3- $N_{m0}=0$, $N_c=270 g^{-1}$;   5- As 1, coagulation is included
2- $N_{m0}=5 \cdot 10^5 g^{-1}$, $N_c=0$;   4- $N_{m0}=5 \cdot 10^5 g^{-1}$, $N_c=2700 g^{-1}$;

features. As a rule, increasing of cloud condensation nucleus concentration forced increasing of precipitation intensity. The same, the coagulation increases precipitation intensity. Only cloud condensation nucleation at a cloud base is clearly insufficient for realization the precipitable moisture in rainband.

As a rule, spectra narrowed when nucleation processes are located in limited cells that surrounded by water supersaturation region and activation processes are intensive. Spectra widened when cloud formation processes are steady. It is complicate problem that needs of more detail consideration.

5. CONCLUSIONS

The numerical simulation of a cloud system connected with the cyclone passed over Ukraine was carried out and results of the study can be summarized as follows:

If the sublimation mechanism worked successfully the time and spatial development of cloud and precipitation in both cases(with and without coagulation) is similar. The influence of coagulation is significant for the liquid water content distribution and the peak of precipitation intensity.

The time and space development of cloud drop spectra is clearly sensitive to intensity of cloud and precipitation mechanisms. They narrowed in cells with no stable processes and widened in cells with time-stable processes. The ice particle spectra are less sensitive to changing of nucleation and coagulation processes.

For inhomogeneous widespread rainbands the cloud droplet spectra have complicate structure and degree in gamma distribution changes in large limits.

6. REFERENCES

1. Buikov, M., *Numerical simulation of stratiform clouds*, Obninsc, 1978, pp. 1-68.
2. Buikov, M., Pirnach, A., *Trudy UHRI* **114**, 3-13 (1972).
3. Hegg, D. A., Ferek R. I., Hobbs, P. V., *J.Appl. Meteor.* **34**, 2076-2082 (1995).
4. Hobbs P. V., Bowdle, D., Radke, L., *Aerosol over High Plains of the United States,* Seatle, University Washington, 1977, pp. 1-144.
5. Pirnach, A., *Trudy UHRI* **144**, 20-24 (1976).
6. Pirnach, A., *Trudy UHRI* **170**, 32-43 (1979).
7. Pirnach, A., *J. Atmos. Res.*, 355-376 (1998).
8. Pirnach, A., Krakovskaia, S., *J. Atmos. Res.* **33**, 333-365 (1994).
9. Twomey, S., *Geofis. Pura Appl.* **43**, 243-249 (1959).

Investigations of Aerosol Scavenging Efficiency by Precipitation

A.A.Sinkevich, Yu.A.Dovgalyuk, M.A.Ishenko, Yu.F.Ponomarev, V.D.Stepanenko, N.E.Veremei

Voeikov Main Geophysical Observatory, Karbyshev Str.7, St.Petersburg, Russia, 194018, E-mail <sinkev@main.mgo.rssi.ru>

Abstract. Results of field experiments of aerosol scavenging by precipitation are considered. Scavenging factor was evaluated for aerosol particles within the range greater 0.3 mkm. Experiments have clearly shown that one can calculate scavenging factor only thoroughly analyzing results of each measurement because there are lot of factors (aerosol concentration variations due to synoptical processes, daily aerosol concentration trend, wind, local aerosol sorcerers, etc) that seriously complicate processing procedure. Nonstationary parameterized model of Cu cloud have been used also to calculate scavenging factor.

INTRODUCTION

Aerosol interaction with drops and precipitation is traditional problem for cloud physics scientists. Investigations have been carried out practically in all leading institutes of Rosgidromet (Russia) (see Berezinscy and Stepanov, 1989; Burtsev, 1992; Nikandrov, 1959; Petrenchuk, 1989; Sedunov, 1972).

Many scientists all over the world pay attention to different aspects of the interaction (see Castillo and Luisi, 1988; Hobbs et al, 1980; McYann and Gennings, 1988; Marahami et al,1985). Experiments had shown that aerosol particles greater than 0.3-0.5 mkm are scavenged effectively by precipitation. Most of investigations were aimed to study scavenging efficiency in dependence from aerosol dimensions. Dynamics of aerosol scavenging is the problem which have to be studied. We tried to pay attention to the mentioned problem.
Results of field experiments

The purpose of experiments were to develop method and to obtain data on the efficiency of aerosol scavenging by precipitation. Measurements were carried out in warm seasons of 1993 - 1994 and in cold period of 1997 at the meteorological station of MGO (Saint-Petersburg) and also in summers 1996-1997 in remote region from pollution sorceress-at the east of Leningrad Oblast (in the Field Experimental Base in Turgosh). Photoelectric counter PC.GTA and lidar LIVO were used to measure aerosols /Veremei et al, 1999 /.

Factors determining aerosol concentration variations: There are a lot of factors which seriously complicate possibilities to study aerosol scavenging; among them are - aerosol concentration variations due to synoptical processes, which lead to the change of air mass, daily aerosol concentration trend, wind, turbulence, unstable local aerosol sorcerers. So the first task was to study their influence on aerosol concentration in order to develop procedure to eliminate these factors from results of scavenging factor calculations.

Aerosol measurements (for particles greater than 0.3 mkm) were carried out during 39 days in 1993-1994 in St.Petersburg. Period between measurements was 1 hour. Aerosol concentration varied about 2 - 3 times a day; from day to day these variations reached 30 times. It was the result of air mass transfer, the wind near the Earth surface and local aerosol sources. The data of measurements were subdivided on 4 groups to evaluate the influence of cloudiness and precipitation regime on aerosol particle concentration. The 1-st group - the days when cloudiness was nonsignificant (related to heat convection) or there were no clouds at all. Days with significant cooling were selected in the 2-nd group from this group. Convection during the days of the 2-nd group was weak. The days of the 3-rd group are that of significant cloudiness, more often stratiform, related to the frontal situation or the warm sector of the cyclone. The great cloudiness and inversion layers in the atmosphere prevented the development of convection. The days of the 4-th group are that with precipitation in the period of observations (more often - from convective clouds). Maximum averaged aerosol concentration ($N=7.1*10^4$ l^{-1}) was registered for the 1 group. Absence of clouds lead to surface warming and so convection develops and results in aerosol transport. Cold air (2 group) prevent convection and aerosol concentration reduces significantly ($N=1.4*10^4$ l^{-1}). Clouds also prevent convection (3 group) and aerosol concentration is also rather small ($N=2.8*10^4$ l^{-1}). 4 th group include periods with precipitation and without precipitation so N ($N=5.9*10^4$ l^{-1}) is greater than in the 2 group but less than in the 1 one.

Daily aerosol trend seriously complicate results of measurements. To access this factor we have carried out measurements of aerosol concentration during several days and nights in summer 1997. To exclude antropogenic local aerosol sorceress experiments were carried out in remote region in the east of Leningrad Oblast (Field Experimental Base in Turgosh). Results of 4 experiments are presented in Fig 1.

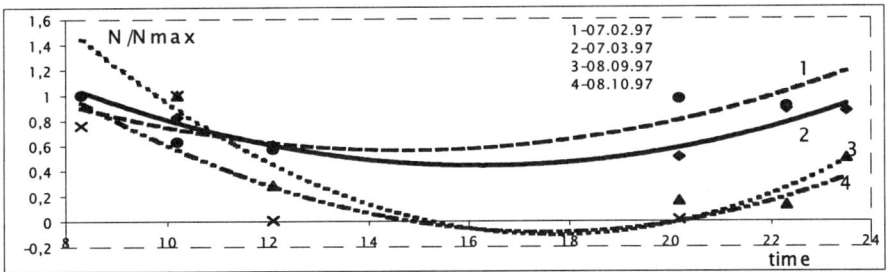

FIGURE 1. Daily aerosol trend

One can clearly see that usually aerosol concentration during morning and evening hours is greater than during day hours. It is the result of changers in solar radiation and hence - turbulence. At the same time local aerosol sorceress can significantly change difference in morning and day aerosol concentration (see curves for 09.08.and 10.08.97 - when forest fires were observed in comparison with curves for 02.07-03.07.97 when there were no powerful local aerosol souserers.

Data on aerosol scavenging by precipitation

Data on aerosol scavenging for warm periods of 1996-1997 were analyzed. Experiments were carried out in Turgosh. Only cases when measurements were carried within the periods just before precipitation, during precipitation and just after precipitation were taken into consideration. More than 200 of aerosol concentration measurements (for particles within the range 0.5-1.0 mkm) were made. Results of experiments have shown that aerosol scavenging factor for the periods just before and after precipitation take place were equal consequently to $0,07 \cdot 10^{-3}$ and $0,15 \cdot 10^{-3}$ s^{-1} (median values 0.03 and $0.02 \cdot 10^{-3}$ s^{-1}) . Aerosol scavenging factor during precipitation was equal to $0,17 \cdot 10^{-3}$ s^{-1} (median value $0.11 \cdot 10^{-3}$ s^{-1}). So one can clearly see that precipitation increase scavenging factor at the same time aerosol dry precipitation also seriously increases after finish of rain.

Lidar soundings were carried out in St.Petersburg. 17 days with precipitation were analyzed. The effect of scavenging was the most distinct in the long-time intensive precipitation. Data available show that the effect of scavenging takes place during I - 2 hours after precipitation - in the case of long-time precipitation . Significant but short-time precipitation did not result in long-time effect. In average this effect was being observed during 30 min

Examples

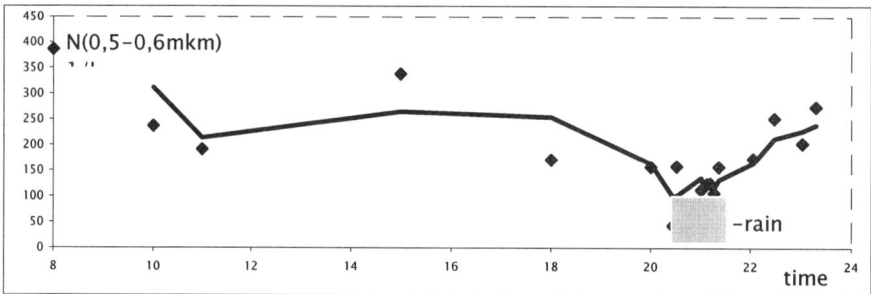

FIGURE 2. Aerosol concentration variations due to precipitation.

Results of one of experiments which was carried out in Turgosh during 07.28.97 are presented. Measurements were carried during 15 hours. Cu development take place during day hours. Rains were observed in neibohoured regions. Precipitation from Cb began at 20 hours 35 min in Turgosh (Fig.2). Rain maximum intensity was equal to 0.2 mm/min. Aerosol concentration slightly varied during day hours and it averaged value was equal to $N=247$ l^{-1}. It reduces on 40% during first 10 minutes of rain. Scavenging factor was equal to $2,09 \cdot 10^{-3}$ s^{-1}. Aerosol concentration increase take place after precipitation, it was the result of daily trend.

Numerical simulation

Nonstationary parameterized numerical model of a convective cloud was developed to study aerosol scavenging. It includes block describing the spreading of solid aerosol particles and their interaction with drops and air flows. In the subcloud layer (z=200 - 400 m) the emission of aerosol particles of a given radius (10, 20 or 30 mkm) was simulated as an instantaneous source. The variation of aerosol concentration in this layer characterized by scavenging factor was considered. Calculations showed that time variation of scavenging factor has two maxima. The first of them is a result of increasing the upward air velocity. The second maximum is caused by moving the particles downward as a result of downdrafts and falling precipitation. The value of the second maximum is greater than that of the first one by one order because the convective transfer of particles the same as their gravitational precipitating and scavenging by the rainfall is directed downward.

CONCLUSIONS

So, as a result of complex investigations of scavenging processes it is showed, that precipitation increase aerosol scavenging in comparison with dry precipitating. It obtained from field experiments that scavenging factor is greater than factor of dry aerosol precipitation in 2-2.5 times. The significant effect is reached when precipitation are intensive and have a great continuance. Available data show that the effect of scavenging the atmosphere can be observed in I - 2 hours in the cases of long-time intensive precipitation. Aerosol scavenging factor (d > I mkm), obtained from results of theoretical experiments are similar with those obtained in nature conditions. Their values are about $1,0 - 5,0 \cdot 10^{-3} \, s^{-1}$.

Acknowledgments: Investigations were supported by Rosgidromet and RFBI.

REFERENCES

1. Berezinscy N.A., Stepanov G.V. Apparatus, Method and Some Results of Aerosol Studies in Developing Cu. Trudy VGI. 1989,v.76, p.35-45.
2. Burtsev .I.I. The Results of Investigation of Conformity with Dispersion and Washout of Artificial Aerosols in Clouds and Precipitation. - International Conference on Clouds and Precipitation, Montreal, 1992, vol. II, pp. 39-41.
3. Castillo R. and De Luisi G. A Codification of Stratiform Clouds and Their Aerosol Scavenging Properties. - 10 Int. Cloud Physics Conference, Hamburg, 1988, pp. 267- 269.
4. Hobbs P. V., Smith G. L., and Radke F. Cloud-active Nuclei from Coal-fired Electric Power Plants and Their Interactions with Clouds. - J. Appl. Met., 1980, vol. 19, No. 4, pp. 439-451.
5. McYann B. T. and Gennings L. Y. Below-cloud Scavenging by Rain and Drizzle. - 10 Int. Cloud Physics Conference, Hamburg, 1988, pp. 267-269.
6. Murahami M., Kikuchi K., and Magono C. Experiments on Aerosol Scavenging by Natural Snow Crystals. - J. Met. Soc. Japan, 1985. vol. 63, pp. 119-129.
7. Nikandrov V.Ya. Clouds and Fogs Modification. Leningrad, Gidrometeoizdat, 1959, 160p.
8. Petrenchuk O.P. Atmospheric Aerosols Experimental Studies. Leningrad, Gidrometeoizdat,1989,648p.
9. Sedunov Yu.S. Physics of Drops Formation in Atmosphere. Leningrad, Gidrometeoizdat,1972, 207p.
10. Veremei N.E., Dovgaljuk Ju.A., Egorov A.D., Ishchenko M.A., Ponomarev Yu. Ph., Sinkevich A.A., Stalevich D.D., Stepanenko V.D.,Khvorostovskii K.S. A Study of Wet Scavenging of Aerosol articles by Clouds and Precipitation. Meteorology and Hydrology, 1999, N8, p.5-14.

Condensation of Supersaturated Vapor on Charged Submicrometer Particles

Chin-Cheng Chen,* Wen-Tin Tsai, and Chun-Ju Tao

Department of Chemical Engineering, National Cheng Kung University
Tainan, Taiwan, R.O.C. 701

Abstract. Condensation of supersaturated n-butanol vapor on monodisperse submicrometer D-Mannose and L-Rhamnose particles is investigated in a flow cloud chamber (FCC). The dependence of the critical supersaturation S_{cr} on particle size in the range of 30 to 90 nm is determined experimentally. The results show that the experimental S_{cr} decreases with increasing particle size, qualitatively in agreement with that predicted by Kohler theory, but quantitatively, smaller than the theoretical prediction. The condensation of supersaturated vapor on singly-positively-charged particles with diameters of 30, 60, and 90 nm is examined. A slight effect of charge on S_{cr} is observed.

INTRODUCTION

Condensation of supersaturated vapor on neutral particles is influenced by the surface properties and size of the particles. However, the dependency is not well known, and observations are reported which are both consistent and inconsistent with theory(1-3). Condensation of a vapor onto charged particles is influenced by the sign and the amount of charge and the dipole moment of the vapor molecules. Ions are believed to interact with the condensing vapor and change the energy barrier for nucleation. So far, the size of the ions tested ranges from single atoms or small molecules consisting of a few atoms(4) to macromolecules containing hundreds of atoms(5). For small ions with diameters less than 1 nm, there is no reported dependence of the nucleation rate on the size and the chemical makeup(4); but, for macromolecules with diameters of about 1 nm a decrease in the critical supersaturation with ion size is observed(5). However, only scant data are available for the condensation of vapor onto charged particles with diameters in the range of tens of nanometers.

In the present study, D-Mannose ($C_6H_{12}O_6$) and L-Rhamnose ($C_6H_{12}O_5 \cdot H_2O$) aerosols with diameters in the range of 30 nm to 90 nm were examined. The condensation of n-butanol vapor onto monodisperse particles, each carrying a single positive charge, was examined for particles with diameters of 30, 60, and 90 nm.

EXPERIMENTAL

Experimental Setup

The entire setup of experiment consists of a series of units operating as the following. Compressed air passes through an atomizer(Model 3076, TSI) to generate aerosol stream from a dilute aqueous solution of D-mannose or L-rhamnose. The aerosol stream flows through two diffusion dryers, an oven, and another two diffusion dryers in series to produce dry polydisperse aerosol, and enters an electrostatic classifier (Model 3071A, TSI) classifying the polydisperse aerosol into a single positive-charged monodisperse aerosol. The charged monodisperse particles enter an aerosol neutralizer (Model 3077, TSI) to reach a charge equilibrium. The charged particles are removed after passing through an electrostatic collector. The uncharged monodisperse aerosol is diluted with clean air, and then enters and exits the FCC. By comparing particle concentrations at the inlet and the outlet of the FCC, the removal efficiency is obtained. The particle concentration was measured by an ultrafine condensation particle counter (Model 3025A, TSI). In the measurement of charge effect, the charged particles from the classifier are diluted with clean air and then directly enter the FCC.

Measurement of Removal Efficiency

Blank runs were first carried out for each aerosol stream and then removal runs were done. The blank run results served as a base line for comparison of removal due to nucleation. In the present study, a steady flow of 2 liter/min of gas through the FCC was established, and the particle concentrations entering and exiting the FCC were recorded over a 6-min period. To minimize the uncertainty in determining the critical supersaturation, a 7 mW He-Ne laser was used to illuminate the FCC to detect any light scattering by the falling droplet produced as a result of nucleation.

The Theory of Nucleation

Since both D-mannose and L-rhamnose are soluble in n-butanol, Volmers theory(6) was used to calculate the supersaturation for neutral particles. The change in the Gibbs free energy required to form of a critical embryo, ΔG^*, is the Gibbs energy difference between the embryo with a diameter in stable equilibrium with the vapor and that in unstable equilibrium with the vapor, and an idea solution is assumed. For charged particles, they are treated as soluble ions and an addition energy term due to charge effect is included in the calculation of the change in the Gibbs free energy(7), but without sign preference effect taken into account.

RESULTS AND DISCUSSION

Comparison of The Experimental and Theoretical Critical S_{cr}

The experimental S_{cr}'s are determined from the removal efficiency curves together with observations of the laser light scattered by the falling droplets. The S_{cr} is identified as the supersaturation at which removal efficiency increases sharply and a few droplets per second per cm^2 are observed. The theoretical S_{cr}'s corresponding to a nucleation rate of one nucleation event per particle per second has been evaluated using the Volmer theory. The resultant supersaturation is very close to that calculated by Kohler theory.

Figure 1 shows a comparison of the experimental and theoretical S_{cr}'s. The results show that the nucleation ability of D-mannose and L-rhamnose aerosols is qualitatively consistent with the theoretical prediction; and that, quantitatively, the experimental S_{cr} is smaller than the value predicted by the Volmer theory for soluble particles. Note that the experimental S_{cr} for L-rhamnose is smaller than that for D-mannose. This is consistent with the fact that L-rhamnose is more soluble in n-butanol than D-mannose.

Charge Effect on The Critical Supersaturation

Figure 2 illustrates the experimental S_{cr} for D-mannose and L-rhamnose particles with diameters of 30, 60, and 90 nm, each carrying a single positive charge. The experimental values of S_{cr} for neutral particles are also shown. The charged particles seem to have lower values of S_{cr}. Although the differences are small, within experimental uncertainty there is a slight effect of charge on S_{cr}. This observation disagrees with the predictions of the classical ion-nucleation theory(6,7). For radii less than 1.0 nm, the critical supersaturation required to induced a nucleation rate of one embryo per particle per second for the charged particles is constant. For particles between 1.0 and 1.4 nm in radius, there is an obvious charge effect. As the particle size increases, the S_{cr} decreases and approaches the S_{cr} for neutral particles. The effect is negligible for particles larger than 5 nm in radius. Thus, no charge effect will be observed for particles with diameters greater than 20 nm.

The theoretical variation of S_{cr} with ion size agrees well with the experimental observation that there is no dependence of nucleation rate on size and chemical makeup for small ions containing only a few atoms(4) but there is a size dependence of S_{cr} for macromolecules containing hundreds of atoms(5). However, the slight effect on S_{cr} observed for particles larger than 30 nm is in disagreement with the theoretical predictions.

CONCLUSION

The experimental S_{cr} decreases with increasing particle size, qualitatively in agreement with that predicted by Kohler theory, but quantitatively, smaller than the theoretical prediction. For positively charged D-mannose particles with diameters of 30, 60, and 90 nm, a slight effect of charge on S_{cr} is observed, in disagreement with the theoretical prediction that no sign effect is taken into account. Further studies will be needed.

ACKNOWLEDGMENT

The support of this research by the National Science Council of the Republic of China through Grant NSC89-2214-E006-013 is gratefully acknowledged.

REFERENCES

1. Liu, B. Y. H., Pui, D. Y. H., McKenzie, R. L., Agarwal, J. K., Pohl, F. G., Preining, O., Reischl, G., Szymansky, W., and Wagner, P.E., Aerosol Sci. Technol. **3**, 107 (1984).
2. Porstendoerfer, J., Scheibel, H. G., Pohl, F. G., Preining, O., Reischl, G., and Wagner, P.E., Aerosol Sci. Technol. **4**, 65 (1985).
3. Chen, C. C., Huang, C. C., and Tao, C. J., J. Colloid Interface Sci. **211**, 193 (1999).
4. Katz, J. L., Fisk, J. A., and Chakarov, V. M., J. Chem. Phys. **101**(3), 2309 (1994).
5. Seto, T., Okuyama, K., de Juan, L., and Fernandez de la Mora, J., J. Chem. Phys. **107**(5), 1576 (1997).
6. Volmer, M., Kinetik der Phasenbildung (Theodor Steinkopff Verlag, Dresden, Germany, 1939).
7. Russell, K. C., J. Chem. Phys. **50**(4), 1809 (1969).

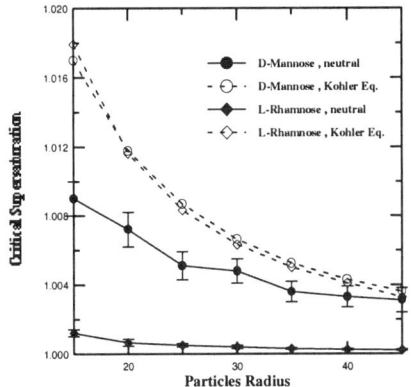

Figure 1. Variation in the S_{cr} for neutral particles as a function of particle diameter. Error bars indicate the uncertainty interval. The dashed curve indicates the theoretical value calculated by Kohler equation.

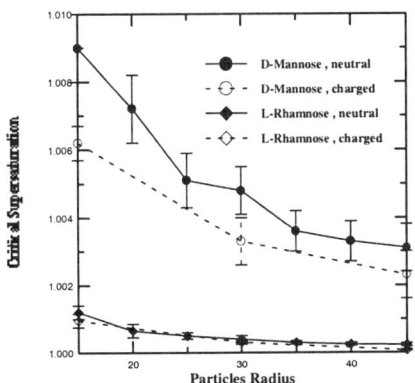

Figure 2. The effect of charge on S_{cr} for particles of D-mannose and L-rhamnose. The experimental S_{cr} for single positively charged particles of D-mannose is smaller than those for neutral particles. The filled and open data points are for neutral and charged particles, respectively. Error bars indicate the uncertainty interval.

RADIATIVE FORCING—AEROSOL CLIMATE INTERACTIONS

The Formation of Ice Clouds from Supercooled Aqueous Aerosols

Thomas Koop

Institute for Atmospheric Science
Swiss Federal Institute of Technology Zurich
ETH Hoenggerberg HPP, 8093 Zurich, Switzerland

Abstract. The formation of ice particles from liquid aqueous aerosols is of central importance for the physics and chemistry of high altitude clouds. In this paper experimental measurements and theoretical models of ice nucleation from concentrated aqueous solutions are reviewed. Both homogeneous as well as heterogeneous ice nucleation are discussed and remaining open questions are highlighted.

INTRODUCTION

Aerosols and clouds in the upper troposphere and lower stratosphere play a fundamental role in Earth's climate system and atmospheric chemistry. Cirrus ice clouds in the upper troposphere cover up to about 20-30% of the Earth and thus strongly affect its radiative properties by scattering of incoming sunlight and absorption of long-wavelength radiation from the Earth's surface.[1-3] Polar stratospheric clouds provide surfaces for important heterogeneous chlorine activation reactions occurring during ozone hole periods in the winter and spring polar stratosphere.[4]

The optical and chemical properties of ice cloud particles depend on their formation conditions which are mainly controlled by pre-existing aerosols. Ice particles can form via homogeneous ice nucleation from liquid aerosols or via heterogeneous ice nucleation on solid nuclei (see Figure 1). In this paper recent advances in both laboratory measurements as well as theoretical models on ice nucleation from aqueous aerosols in the upper troposphere and lower stratosphere will be discussed.

HOMOGENEOUS ICE NUCLEATION

The detailed microphysical description of the homogeneous nucleation of ice from pure water droplets has been an outstanding problems in the meteorological and physical science for many years. Based on significant advances in our understanding of the physics of supercooled (metastable) water,[5] improved formulations for the homogeneous ice nucleation rate coefficient have recently become available.[6,7] The nucleation of ice from supercooled aqueous solutions is an even more complicated

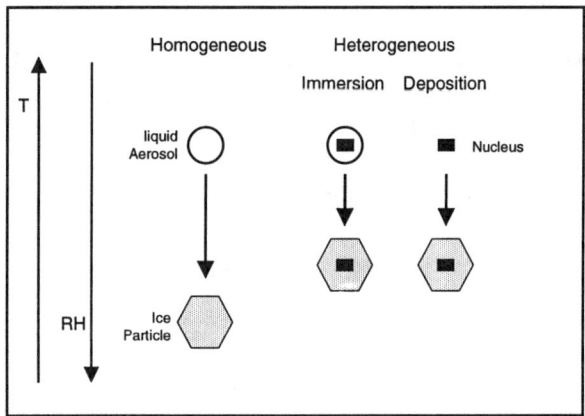

FIGURE 1. Schematic picture of different ice nucleation mechanisms. In homogeneous nucleation ice forms from purely liquid aerosols. In heterogeneous immersion freezing the nucleation of ice from liquid aerosols is facilitated by immersed heterogeneous nuclei, and thus occurs at a higher temperature or lower relative humidity (compare arrows labeled T and RH on the left hand side). In the heterogeneous deposition mode ice condenses directly from the vapor phase on dry solid nuclei.

problem than the case of pure water. Although systematic experimental studies date back to the 1960s (e.g. Ref. 8), many of these are either limited to dilute solutions at temperatures above 235 K or to solutions which are not of atmospheric interest (e.g. Ref. 9). Upper tropospheric and lower stratospheric liquid aerosols at temperatures below ~235 K predominantly consist of concentrated aqueous solution mixtures of H_2SO_4-HNO_3-NH_3-H_2O (Stratospheric aerosols: see review in Ref. 10; tropospheric aerosols: Ref. 11-15).

Since the 14th ICNAA conference in 1996 several papers have been published describing laboratory measurements of homogeneous ice nucleation from atmospherically relevant concentrated aqueous aerosols down to low temperatures (< 190 K). The experimental techniques used in these studies comprise FTIR aerosol flow tube experiments,[16,17] optical microscopy of droplets deposited on temperature-controlled quartz plates,[18,19] standard calorimetry employing water-in-oil emulsions,[19-21] and droplets stored in an electrodynamic balance.[22] Figure 2 shows a compilation of published laboratory data for H_2SO_4-H_2O and $(NH_4)_2SO_4$-H_2O droplets in the micron size range. In general there is reasonable agreement between the data obtained from the different techniques except for the fact that the ice freezing points obtained from FTIR flow tube studies (diamonds in Fig. 2) appear to be at higher temperatures than the ones obtained from other techniques (circles, squares, triangles). The reason for this disagreement is not well understood yet.

Also shown in Figure 2 are the predicted ice freezing points from several theoretical ice nucleation models. The dotted line indicates the results from a model for the binary H_2SO_4-H_2O system using classical nucleation theory.[23] This model was based on experimental data[16] which led to an improvement of an earlier approach.[24]

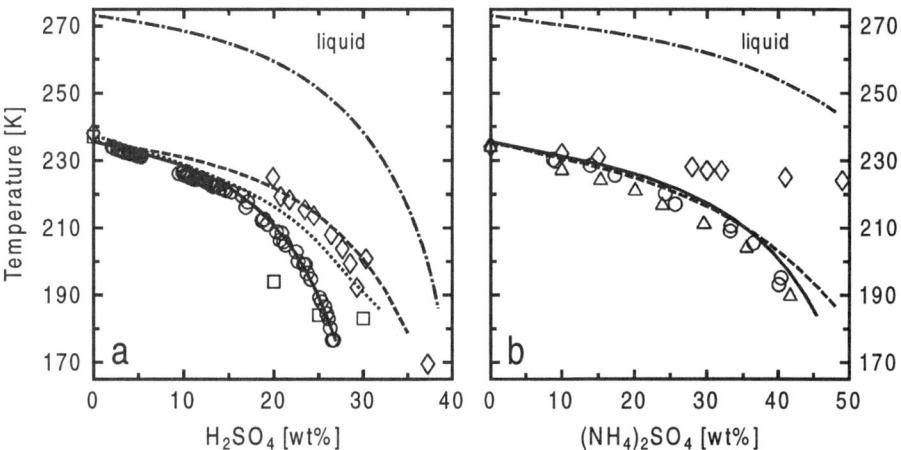

FIGURE 2. Comparison of ice nucleation measurements and models for micron sized aqueous droplets of (a) H_2SO_4 and (b) $(NH_4)_2SO_4$. Diamonds: FTIR experiments;[16,17] circles: microscope experiments;[18,19] triangles: emulsion experiments;[19-21] squares: electrodynamic balance experiments.[22] Dash-dotted line: equilibrium ice melting point curve; dotted line: classical nucleation model for H_2SO_4-H_2O;[23] dashed line: lambda model, see text;[25-27] solid line: thermodynamic model, see text.[28]

The dashed curves indicate results from a model[25-27] in which the freezing temperature of a solution droplet, T_{fr}^{Sol}, was calculated according to

$$T_{fr}^{Sol} = T_{fr}^{H2O} - \lambda \cdot \Delta T_m, \qquad (1)$$

where T_{fr}^{H2O} is the freezing temperature of a pure water droplet of the same size, ΔT_m is the equilibrium melting point depression of ice in the solution (i.e., $273.15 - T_m^{Sol}$), and λ is a constant depending on the nature of the solute. The homogenous ice nucleation rate coefficient in solution, J_{fr}^{Sol}, is predicted based on the assumption that

$$J_{fr}^{Sol}(T_{fr}^{Sol}) = J_{fr}^{H2O}(T_{fr}^{H2O}), \qquad (2)$$

where $J_{fr}^{H2O}(T_{fr}^{H2O})$ is the homogeneous ice nucleation rate coefficient for pure water, J_{fr}^{H2O}, at temperature T_{fr}^{H2O}. In the following paragraphs this model will be termed the "lambda model". The curves in Figure 2 represent a homogeneous nucleation rate of $J=10^9$ cm^{-3}s^{-1} and a value of $\lambda=1.0$ for the H_2SO_4 case, and $\lambda=1.7$ for the $(NH_4)_2SO_4$ case.[27] The formulation of Pruppacher[6] for $J_{fr}^{H2O} = f(T)$ has been used in these calculations.

The solid curves indicate the results of a semi-empirical thermodynamic ice nucleation model which is based on experimental ice freezing data of 18 different solutes in aqueous solutions.[28] This model will be termed the "thermodynamic model" in the following paragraphs. Again the curves represent a homogeneous nucleation rate of $J=10^9$ cm^{-3}s^{-1} and the $J_{fr}^{H2O} = f(T)$ formulation of Pruppacher[6] has been used.

Figure 2 a indicates that in the H$_2$SO$_4$ case the lambda model describes the FTIR data (diamonds) well, while the thermodynamic model is in better agreement with the microscope and electrodynamic balance data (circles and squares, respectively). In the (NH$_4$)$_2$SO$_4$ case both models strongly disagree with the FTIR data (diamonds), but do agree with the microscope and emulsion data (circles and triangles, respectively) reasonably well (see Fig. 2b). Note that the lambda model can be tuned to better describe the experimental data by adjusting the value of λ for each solute. This requires experimental data for this solute to be available. On the other hand, the thermodynamic model has the advantage of being solute independent, thus allowing to make predictions for solutes that have not been investigated.

To compare the results of different solutes one is often interested in the degree of supercooling below the equilibrium ice melting point before freezing occurs. In Figure 3 the supercooling in solution, ΔT^{Sol}, has been scaled to the supercooling observed in pure water, ΔT^{H2O}, by introducing the excess supercooling, ΔT^{Ex}, with:

$$\Delta T^{Ex} = \Delta T^{Sol} - \Delta T^{H2O}. \quad (3)$$

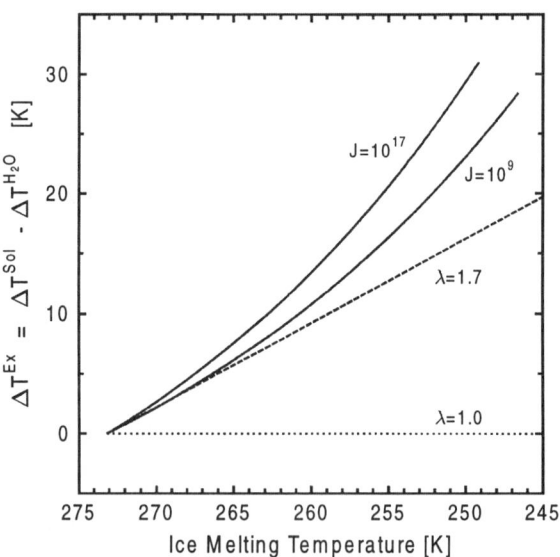

FIGURE 3. The excess supercooling, ΔT^{Ex}, as a function of the ice melting point in solution as calculated with different models. Dotted and dashed lines: calculated with the lambda model; solid lines: calculated with the thermodynamic model.

Figure 3 shows ΔT^{Ex} as a function of the ice melting point calculated using the models described above. Two cases calculated with the lambda model are shown as dotted and dashed lines. For $\lambda = 1.0$ (dotted line) the excess supercooling is zero independently of the ice melting point and, thus, the solute concentration. For $\lambda = 1.7$ (dashed line) the excess supercooling increases linearly with decreasing melting point temperature with a slope of $\lambda - 1 = 0.7$, implying that ΔT^{Ex} increases with increasing solute concentration. Note that according to the lambda-model the amount of excess supercooling is independent of the value of the nucleation rate. In contrast, in the thermodynamic model the excess supercooling depends on the nucleation rate (higher nucleation rates require a higher excess supercooling) and the excess supercooling is non-linear and generally larger than the ones predicted by the lambda model.

The same facts are illustrated in Figure 4 in a different way. The curves shown in Figures 4a and b indicate where the homogeneous ice nucleation rate is equal to 10^9 cm^{-3}s^{-1} (lower curve of each pair of same line style) and 10^{17} cm^{-3}s^{-1} (upper curves) as a function of temperature and relative humidity. Figure 4a depicts the results from the lambda model with $\lambda = 1.0$ for the H$_2$SO$_4$ case (dotted lines) and $\lambda = 1.7$ for the (NH$_4$)$_2$SO$_4$ case (dashed lines). The conversion of the solute concentration into relative humidity space was done with the help of the ion interaction model of Clegg et al.[11] As can be seen the temperature difference $\Delta T_{fr} = T_{fr}^{(J=10^9)} - T_{fr}^{(J=10^{17})}$ is constant

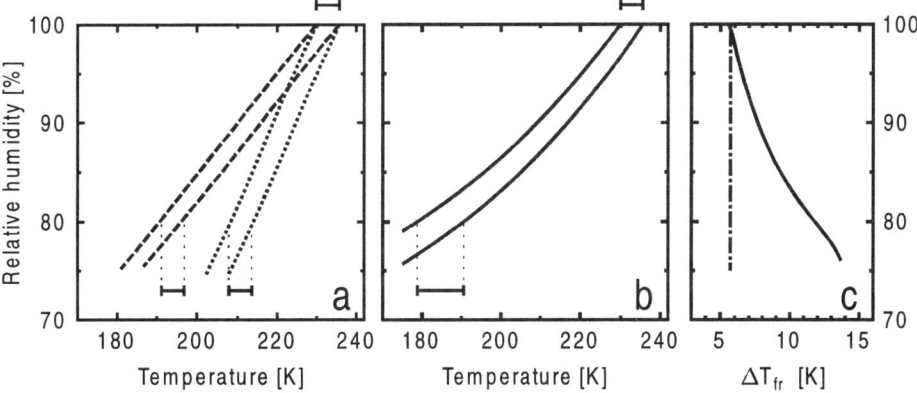

Figure 4. (a): Curves where the homogeneous ice nucleation rate coefficient, J_{fr}^{Sol}, assumes a value of 10^{17} cm^{-3}s^{-1} (left curve of each pair) and 10^9 cm^{-3}s^{-1} (right curve of each pair) based on calculations with the lambda model. Dashed lines: for (NH$_4$)$_2$SO$_4$ with $\lambda = 1.7$; dotted lines: for H$_2$SO$_4$ with $\lambda = 1.0$. The horizontal bars indicate the temperature difference between the curves, ΔT_{fr}, at 80 and 100 %RH. **(b):** Same as in (a) but calculated with the thermodynamic model. **(c):** ΔT_{fr} from panels (a) and (b) versus relative humidity. Dash-dotted line: results from the lambda model; solid line: results from the thermodynamic model.

in both the $\lambda = 1.0$ and the $\lambda = 1.7$ case (i.e. the curves are horizontally parallel). This is also indicated by the horizontal bars at 100 %RH (pure water) and 80 %RH which have the same length. Figure 4b shows results of identical calculations using the thermodynamic model. Here the two curves do not run parallel and their temperature difference increases with decreasing temperature / decreasing relative humidity. This is shown in more detail in Figure 4c. While the lambda model predicts a constant ΔT_{fr} of about 5.6 K independently of relative humidity, the thermodynamic model predicts a strong increase of ΔT_{fr} with decreasing relative humidity. Experimental data from H_2SO_4, NH_4HSO_4, and LiCl experiments indeed indicate an increase in ΔT_{fr} with decreasing relative humidity, thus providing support for the thermodynamic model.[28] However, more detailed experimental data would be helpful in determining the exact behavior of ΔT^{Ex} and ΔT_{fr}.

HETEROGENEOUS ICE NUCLEATION

In the following the thermodynamic model introduced in the last section will be used as a proxy for the homogeneous ice nucleation limit in atmospheric aerosols. The temperature and relative humidity dependence of the homogeneous nucleation of ice from micron-sized aerosols ($\approx J=10^9$ cm^{-3}s^{-1}) is indicated by the dashed line in Figure 5. In equilibrium, ice particles should always be present for conditions above the dashed curve (i.e. in the dark shaded area), which provides an upper limit for ice cloud formation. While observations of polar stratospheric clouds and lee wave-induced cirrus clouds support the predicted high supersaturations required for homogeneous ice nucleation,[28] lower maximum supersaturations are frequently observed in synoptically forced cirrus clouds,[29,30] see below. The most likely explanation for these observations is that solid nuclei induced heterogeneous ice nucleation at lower relative humidities. This is indicated in Figure 5 by two aerosol trajectories, one for a purely liquid aerosol (left) and another for an aerosol with an immersed insoluble nucleus (right). Upon increasing relative humidity the liquid aerosol becomes supersaturated with respect to ice above the ice saturation line (solid line in Figure 5). It stays in the metastable supercooled state indicated by the light shaded region until homogeneous ice nucleation occurs as it reaches a relative humidity along the homogeneous nucleation curve (dashed line). In contrast, the nucleus immersed in the other aerosol can force heterogeneous ice nucleation somewhere in the light shaded region between ice saturation (solid line) and the homogeneous nucleation limit (dashed line). The exact conditions where heterogeneous ice nucleation will occur is likely to depend on the size, morphology, and the nature of the nucleus as well as on the composition of the liquid aerosol.

There is only very limited systematic experimental data on heterogeneous ice nucleation at upper tropospheric conditions. Up to now, most attention has been given to ice nuclei effective in pure water or very dilute solutions at temperatures above 235 K [see Ref. 31 for a review of the literature]. This region of the temperature/relative humidity space is indicated by the cross-hatched area in Figure 5. However, the conditions for cirrus cloud formation in the upper troposphere are quite different with temperatures usually below 235 K and aerosols which are thought to be composed primarily of concentrated mixtures of H_2SO_4, NH_3, HNO_3 and H_2O (Ref. 11-15). The only detailed experimental study on heterogeneous ice nucleation under such conditions indicates that soot nuclei immersed in aqueous sulfuric acid may induce heterogeneous ice nucleation ice nucleation at temperatures below 220 K (Ref. 32). Other experiments indicate that gold surfaces seem to efficiently nucleate ice from aqueous sulfuric acid droplets.[33]

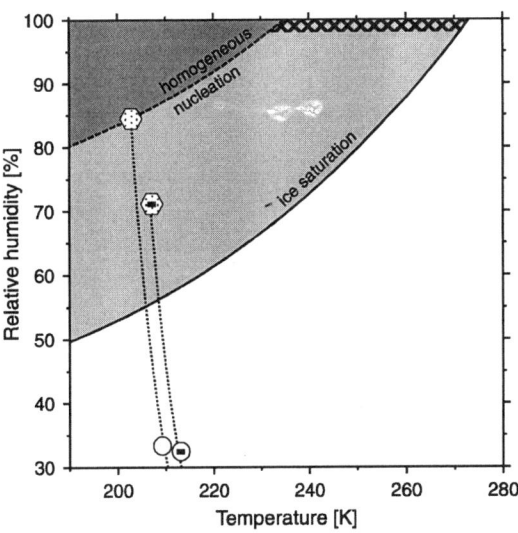

FIGURE 5. Relative humidity/temperature region of interest for atmospheric ice particle formation. The solid curve indicates the ice saturation line, i.e. above this line ice is supersaturated and liquid aerosols are metastable (light shaded region). The dashed line indicates where homogeneous ice nucleation is expected to occur in micron-sized aerosols based on laboratory experiments. Above this line liquid aerosols are unstable (dark shaded region). The cross-hatched area represents the region where almost all experiments on heterogeneous ice nuclei have been performed to date. The dotted lines depict two particle trajectories, one for a purely liquid aerosol which homogeneously nucleates ice (left), and the other for a liquid aerosol with an immersed nucleus which forces heterogeneous ice nucleation at a much lower relative humidity (right).

Heterogeneous ice nucleation is believed to be a major pathway for the formation of atmospheric ice clouds,[34] which is supported by recent field measurements. Chen et al.[35] collected ambient aerosols and ice nucleating particles in cirrus clouds and aircraft contrails and determined their elemental composition by analytical electron microscopy. They found a variety of nuclei including crustal, metallic and carbonaceous particles. The measurements showed a very pronounced enhancement of nuclei made of these materials in the ice nucleating particles when compared to the ambient aerosols. This strongly suggests that they acted as heterogeneous ice nuclei. Similar conclusions were drawn by Twohy and Gandrud[36] based on measurements of the ice nucleating particles in contrails. However, although many modeling studies have pointed out the importance of insoluble nuclei on the formation and lifetime of cirrus clouds and contrails[27,37,38] many uncertainties remain.

Figure 6 shows a comparison of predictions for the homogeneous ice nucleation from the thermodynamic model (dashed lines) together with field observations of peak relative humidities in the upper troposphere (dotted line, Ref. 29), and laboratory data for heterogeneous ice nucleation on soot particles covered by H_2SO_4 (squares, Ref. 32). The field observations show much lower relative humidities than predicted for homogeneous ice nucleation implying that either the peak relative humidities have not been sampled or that heterogeneous ice nucleation occurred at lower saturations. Interestingly, the experimental data for soot shows the same temperature / relative humidity dependence than the field observations. Whether this agreement is accidental or indicates that soot particles regularly induce ice particle formation in cirrus clouds in the upper troposphere remains to be answered. More experimental studies like the one described above[32] are certainly needed.

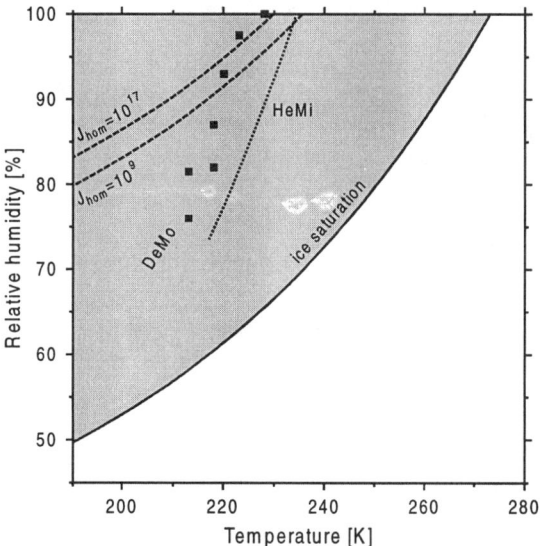

FIGURE 6. Comparison of results of the thermodynamic model for the homogeneous ice nucleation (dashed lines), laboratory experiments for ice formation on soot particles covered by H_2SO_4 (squares labeled DeMo, Ref. 32), and a parameterization of peak relative humidities observed in the upper troposphere (dotted line labeled HeMi, Ref. 29).

Recent modeling studies suggested that the degree of neutralization of upper tropospheric sulfate aerosols via NH_3 uptake may change the ice nucleation pathway of the aerosols from homogeneous nucleation to heterogeneous nucleation.[12,13] This is based on the fact that these aerosols may crystallize in the form of $(NH_4)_2SO_4$, NH_4HSO_4, $(NH_4)_3H(SO_4)_2$, or NH_4NO_3 (see Table 1). These soluble crystalline solids will deliquesce upon increasing humidities in the lower troposphere. However, at low upper tropospheric temperatures increasing relative humidity lets ice become supersaturated before the deliquescence relative humidity is reached, i.e. deliquescence can only occur in the light shaded region of Figure 5. At present there is no experimental information on whether soluble crystalline aerosols in an environment, which is supersaturated with respect to ice, will

1. deliquesce to form fully liquid aerosols (which can only freeze via homogenous ice nucleation);
2. induce heterogeneous ice nucleation directly from the vapor phase (deposition mode);[12]

3. or will induce heterogeneous ice nucleation in the immersion mode at the point of deliquescence, i.e. when the deliquescing crystal begins to be covered by a liquid.[13]

In the following the thermodynamic properties of different crystalline solids is investigated using the ion interaction solution model of Clegg et al.[11] The deliquescence relative humidity (DRH) for several crystalline phases of stratospheric and tropospheric interest are shown in Figures 7a and 7b, respectively. All DRH curves are for dry salts, i.e. it is assumed that the stoichiometry of the aerosols perfectly match the respective crystal stoichiometry. For deviating aerosol stoichiometries the DRH will always be below the values shown in Figure 7. Note that the immersion freezing becomes much more likely (and the deposition mode less likely) because a liquid solution layer will form at much lower relative humidities in the non-stoichiometric cases even when the final DRH value (where the particle becomes a fully liquid droplet) is only slightly below the value for the dry crystal.

The comparison between the DRH values of the different crystals shown in Figure 7 reveals some interesting information. The crystalline solids of stratospheric interest (Figure 7a), i.e. NAT, NAD, SAH, and SAT (see Table 1) have strongly temperature

TABLE 1. Selected crystalline solids of tropospheric and stratospheric interest.

Chemical name	Chemical formula	Abbreviation	Main area of relevance
Ammonium Sulfate	$(NH_4)_2SO_4$	ASUL	Troposphere
Ammonium Bisulfate	NH_4HSO_4	ABIS	Troposphere
Ammonium Nitrate	NH_4NO_3	ANIT	Troposphere
Letovicite	$(NH_4)_3H(SO_4)_2$	LETO	Troposphere
Sulfuric Acid Tetrahydrate	$H_2SO_4 \cdot 4H_2O$	SAT	Stratosphere
Sulfuric Acid Hemihexahydrate	$H_2SO_4 \cdot 6.5H_2O$	SAH	Stratosphere
Nitric Acid Trihydrate	$HNO_3 \cdot 3H_2O$	NAT	Stratosphere
Nitric Acid Dihydrate	$HNO_3 \cdot 2H_2O$	NAD	Stratosphere

dependent DRH curves. In contrast, the DRH curves of the tropospherically relevant crystals ASUL, ANIT, LETO, and ABIS show a much lower temperature dependence (Figure 7b). The reason for this behavior is that all the hydrates shown in Figure 7a have melting points in the temperature region shown in the figure. The DRH curves in a temperature vs. relative humidity diagram are thermodynamically identical to the melting point curve in the commonly used concentration vs. temperature phase diagrams. The melting point curve close to the maximum (i.e. close to the pure crystal stoichiometry) is usually very flat implying strong concentration changes over small temperature intervals. Since relative humidity is a strong function of solute concentration this leads to the observed behavior of strongly varying DRH values with temperature. In contrast, the melting points of the crystalline solids shown in Figure 7b are at much higher temperatures far outside the temperature region shown in the diagram. The melting point curves far from the maximum are usually very steep, i.e. only small changes in concentration are associated with large changes in temperature. Since relative humidity is not a very strong function of temperature this produces the observed weak dependence of the DRH on temperature for these crystals.

As discussed above the crystalline solids listed in Table 1 might act as heterogeneous ice nuclei in the atmosphere. They can only become effective when their DRH is larger than the ice saturation relative humidity, i.e. in the shaded region of Figure 7. This restriction excludes BISU crystals from the list of possible ice nuclei because its DRH curve is below ice saturation over the entire temperature region of atmospheric interest (see Figure 7b). For the other crystalline solids two cases have to be considered: stoichiometric aerosols which form purely crystalline ("dry") aerosols

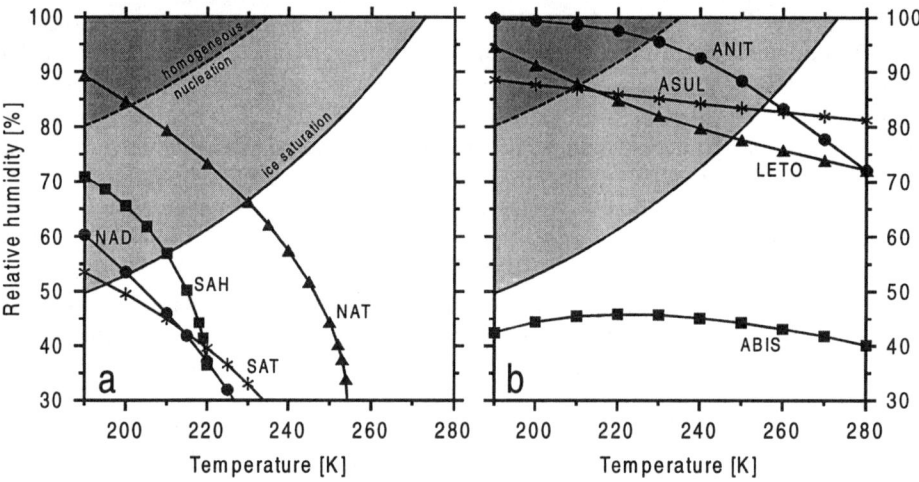

FIGURE 7. The deliquescence relative humidity curves for several crystalline solids of stratospheric and tropospheric interest (panels **a** and **b**, respectively). The solid and dashed lines as well as the shaded regions are as in Figure 5. See Table 1 for a list of the full names of the different solids.

upon efflorescence, and non-stoichiometric aerosols which form solid/liquid mixed phase aerosols. Only the purely crystalline aerosols can act as deposition nuclei. In the case of mixed phase aerosols the crystals will most likely be covered by the available liquid thus inhibiting heterogeneous ice nucleation on the crystal surface directly from the vapor phase.

In order to effectively change the formation conditions and properties of ice particles, crystalline aerosols are required to induce heterogeneous ice nucleation before homogeneous ice nucleation occurs. This defines the light shaded region above ice saturation (solid line in Figure 7) and below the homogeneous nucleation limit (dashed line) as the most important region. Hence, upper tropospheric aerosols made of ANIT, ASUL, and LETO could become ice nuclei at temperatures below ~250-260 K and above ~210-230 K.

Another interesting topic is the persistence of these crystals in upper tropospheric aerosols. BISU crystals will deliquesce as soon as the relative humidity rises above 40-50 %RH. Also ANIT, ASUL, and LETO will deliquesce at 70-80 %RH at temperatures above ~250-260 K, i.e. in the region subsaturated with respect to ice (unshaded region in Figure 7b).

In the light shaded region they can only deliquesce as long as no ice particles have formed: the presence of ice will reduce the ambient relative humidity to a value along the solid (ice saturation) line, which is below the DRH values for the three crystals.

Hence, in order to deliquesce it is required that ANIT, ASUL, or LETO do not act as ice nuclei.

In the dark shaded area the crystals cannot deliquesce, because homogeneous ice nucleation occurs along the dashed line in Figure 7b, below their DRH curves. Homogeneous ice nucleation can take place either in remaining liquid aerosols (which have not effloresced at lower relative humidity) or in the liquid part of mixed phase aerosols. This last scenario is particularly interesting because it implies that once ANIT, ASUL, or LETO have formed they will remain crystalline as long as the temperature stays below the temperature at the intersection of their DRH curve with the homogeneous nucleation curve (~230 K for ANIT, ~210 K for ASUL and LETO).

Despite the many possibilities for soluble crystalline aerosols to act as ice nuclei many uncertainties remain. Answers to the open questions can only come from concerted efforts in thermodynamic aerosol modeling (e.g. reducing the uncertainties in the DRH values at low temperatures), field missions (e.g. measurements of the chemical composition and physical state of aerosols, as well as peak relative humidity in supersaturated air), and laboratory work (e.g. experiments to determine homogeneous and heterogeneous ice nucleation rates from aerosols).

ACKNOWLEDGMENTS

The author is indebted to John Abbatt, Allan Bertram, Paul DeMott, Beiping Luo, and Thomas Peter for many discussions on aerosol thermodynamics, ice nucleation measurements, and ice nucleation models.

REFERENCES

1. Liou, K.N., *Mon. Weather Rev.* **114**, 1167-1199 (1986).
2. Baker, M.B., *Science* **276**, 1072-1078 (1997).
3. Seinfeld, J.H., and Pandis, S.N., *Atmospheric chemistry and physics*, New York: John Wiley & Sons, 1998.
4. Peter, T., *Ann. Rev. Phys. Chem.* **48**, 785-822 (1997).
5. Mishima, O., and Stanley, H. E., *Nature* **396**, 329-335 (1998).
6. Pruppacher, H.R., *J. Atmos. Sci.* **52**, 1924-1933 (1995).
7. Jeffery, C.A., and Austin, P.H., *J. Geophys. Res.* **102**, 25269-25279 (1997).
8. Pruppacher, H.R., and Neiberger, M., *J. Atmos. Sci.* **20**, 376-385 (1963).
9. Rasmussen, D. H., *J. Cryst. Growth* **56**, 56-66 (1982).
10. Carslaw, K.S., Peter, T., and Clegg, S.L., *Rev. Geophys.* **35**, 125-154 (1997).
11. Clegg, S. L., Brimblecombe, P., and Wexler, A. S., *J. Phys. Chem. A* **102**, 2137-2154 (1998).
12. Tabazadeh, A., and Toon, O.B., *Geophys. Res. Lett.* **25**, 1379-1382 (1998).
13. Martin, S.T., *Geophys. Res. Lett.*, **25**, 1657-1660 (1998).
14. Talbot, R.W., Dibb, J.E., and Loomis, M.B., *Geophys. Res. Lett.* **25**, 1367-1370 (1998).
15. Kärcher B., and Solomon, S., *J. Geophys. Res.* **104**, 27441-27459 (1999).
16. Bertram, A.K., Patterson, D.D., and Sloan, J.J., *J. Phys. Chem.* **100**, 2376-2383 (1996).
17. Cziczo D.J., and Abbatt, J.P.D., *J. Geophys. Res.* **104**, 13781-13790 (1999).
18. Koop, T., Ng, H.P., Molina, L.T., and Molina, M.J., *J. Phys. Chem. A* **102**, 8924-8931 (1998).
19. Bertram, A.K., Koop, T., Molina, L.T., and Molina, M.J., *J. Phys. Chem. A* **104**, 584-588 (2000).
20. Chang, H.Y.A., Koop, T., Molina, L.T., and Molina, M.J., *J. Phys. Chem. A* **103**, 2673-2679 (1999).

21. Koop, T., Bertram, A. K., Molina, L. T., and Molina, M. J., *J. Phys. Chem. A* **103**, 9042-9048 (1999).
22. Krämer, B., *Laboruntersuchungen zum Gefrierprozeß in polaren stratosphärischen Wolken*, Ph.D. thesis, Berlin 1998.
23. Tabazadeh, A., Jensen, E.J., and Toon, O.B., *J. Geophys. Res.* **102**, 23845-23850 (1997).
24. Jensen, E.J., Toon, O.B., and Hamill, P., *Geophys. Res. Lett.* **18**, 1857-1860 (1991).
25. Sassen, K., and Dodd, G.C., *J. Atmos. Sci.* **45**, 1357-1369 (1988).
26. DeMott, P.J., Meyers, M.P., and Cotton, W.R., *J. Atmos. Sci.* **51**, 77-90 (1994).
27. DeMott, P.J., Rogers, D.C., and Kreidenweis, S.M., *J. Geophys. Res.* **102**, 19575-19584 (1997).
28. Koop, T., Luo, B.P., Tsias, A., and Peter, T., submitted manuscript (2000).
29. Heymsfield, A.J., and Miloshevich, L.M., *J. Atmos. Sci.* **52**, 4302-4326 (1995).
30. Heymsfield, A.J., Miloshevich, L.M., Twohy, C., Sachse, G., and Oltmans, S., *Geophys. Res. Lett.* **25**, 1343-1346 (1998).
31. Pruppacher, H.R., and Klett, J.D., *Microphysics of clouds and precipitation*, Dordrecht: Kluwer, 1997.
32. DeMott, P.J., Chen, Y., Kreidenweis, S.M., Rogers, D.C., and Sherman, D.E., *Geophys. Res. Lett.* **26**, 2429-2432 (1999).
33. Martin, S.T., Salcedo, D., Molina, L.T., and Molina, M.J., *J. Phys. Chem. B.* **101**, 5307-5313 (1997).
34. Vali, G., *Ice nucleation - a review*, In: Kulmala, M. and Wagner, P.E. (Eds.), *Nucleation and atmospheric aerosols 1996*, Oxford: Elsevier, 1996.
35. Chen, Y., Kreidenweis, S.M., McInnes, L.M., Rogers, D.C., and DeMott, P.J., *Geophys. Res. Lett.* **25**, 1391-1394 (1998).
36. Twohy, C.H., and Gandrud, B.W., *Geophys. Res. Lett.* **25**, 1359-1362 (1998).
37. DeMott, P.J., Rogers, D.C., Kreidenweis, S.M., Chen, Y.L., Twohy, C.H., Baumgardner, D., Heymsfield, A.J., and Chan, K.R., *Geophys. Res. Lett.* **25**, 1387-1390 (1998).
38. Kärcher, B., Peter, T., Biermann, U.M., and Schumann, U., *J. Atmos. Sci.* **53**, 3066-3083 (1996).

What Do We Know About Phase Transition Processes Relevant to Atmospheric Aerosols?

Paul E. Wagner

Institut für Experimentalphysik, Universität Wien
Boltzmanngasse 5, A-1090 Vienna, Austria

Abstract. Various dynamical processes observed in the atmosphere are related to phase transitions from the gas phase. In this presentation selected nucleation and condensation processes are considered and their relevance to atmospheric aerosols is discussed. A few recent experimental studies are reviewed. Open questions remain particularly in the field of phase transitions in binary and multicomponent systems. The influence of miscibility and solubility of the compounds considered has not yet been clarified sufficiently. Furthermore, despite of numerous studies reported during the last few decades, the question of sticking probabilities of condensing vapor molecules remains controversial.

INTRODUCTION

Properties and dynamical changes of atmospheric aerosols have received considerable attention in recent years. A vast amount of experimental information has been reported by various research groups using different measurement techniques, a comprehensive review was presented recently [1]. It has been shown that atmospheric aerosols can often be described by superposition of three to four size distribution modes [2-4], which are frequently refered to as nucleation mode, Aitken mode, agglomeration (accumulation) mode, and coarse mode. Furthermore, the influence of various trace gases has been studied extensively [5-7].

On the other hand, several physical and chemical processes influencing formation and dynamics of atmospheric aerosols have been studied separately under controlled conditions in laboratory experiments. In this presentation particularly nucleation and condensation processes from the gas phase will be considered. Different experimental approaches and measurement methods will be mentioned. Results of selected experimental studies will be presented and their relevance for atmospheric aerosol will be discussed. Remaining open questions and need for future research will be outlined.

OBSERVATION METHODS

A crucial condition for well-defined experimental studies of nucleation or condensation processes is the preparation of mixtures of an inert gas with condensable vapors at accurately determined supersaturations (vapor phase activities, partial vapor pressures). The following techniques have been successfully applied for this purpose: Nonisothermal vapor diffusion in static [8,9] or steady state flow [10,11] systems, adiabatic expansion [12,13] and turbulent mixing [14]. As direct measurement of the temperature in supersaturated vapors is complicated, usually temperature and supersaturations need to be calculated.

Nucleation processes can generally not be detected directly. Only the particles growing subsequent to the formation of the critical clusters, are observed. The concentrations of condensing droplets can be measured by single particle counting. The Constant-Angle Mie Scattering (CAMS) method [15] allows simultaneous and independent determination of concentration and size of growing droplets during condensation. Particle growth or evaporation can also be studied using a tandem differential mobility analyzer system [16].

EXPERIMENTAL STUDIES

While homogeneous nucleation of one component systems will generally not occur in the atmosphere, homogeneous nucleation in binary or multicomponent systems is found to be quite relevant to atmospheric aerosol formation [17-19]. In fact, the formation of Cloud Condensation Nuclei (CCN) may be connected with the formation of thermodynamically stable sulphate nano-clusters by ternary homogeneous nucleation and a subsequent growth process [20].

Homogeneous nucleation has been investigated by various authors for several binary [21-25] and ternary [26] vapor mixtures. Frequently the vapor compounds water, nonane, and the homoluguous series of alcohols were considered. The choice of these compounds was motivated by the fact that for these compounds the physicochemical properties are well-known facilitating comparison to theory. However, the results are not directly applicable to atmospheric conditions. For highly nonideal mixtures the classical nucleation theory predicts unphysical behavior. It appears that for binary systems the applicability of the classical nucleation theory is quite limited. Recently it has been shown that this unphysical behavior of the binary classical nucleation theory is connected with the assumption of curvature independence of the surface tension [27]. In the case of nucleation of partially immiscible liquids a somewhat complex behavior was observed. While for alcohols the homologous trend continued, the water-nonane system showed a completely different behavior. Apparently the macroscopic properties of the compounds considered are not directly applicable to the critical nuclei.

Heterogeneous nucleation frequently occurs in the atmosphere leading to aerosol and cloud drop formation. In fact the interplay of CCN concentration and cloud liquid water content has recently been demonstrated during a measurement series in an alpine area [28]. Thus heterogeneous nucleation is relevant to atmospheric aerosols.

The few so far available quantitative experiments on heterogeneous nucleation [29-31] are resticted to unary systems. However, the atmosphere usually contains multicomponent gas mixtures. Only recently first experimental studies on binary heterogeneous nucleation were reported [32, 33]. These studies show that the macroscopic contact angle is hardly applicable for heterogeneous nucleation on nanoparticles. Unphysical predictions by the binary heterogeneous nucleation theory can occur similar as for the classical binary homogeneous nucleation theory. Furthermore, an interesting insoluble - soluble transition is observed.

It is well known that condensational growth processes in the atmosphere are influenced by the sticking probabilities for the condensing vapor molecules. Particularly the sticking probability for water molecules has received considerable attention. While early measurements indicate a value of about 0.03 [34], later a value close to 1 was determined [35]. This appears to be consistent with corresponding theoretical work [36]. However, recently an accomodation coefficient considerably below 1 was reported [37]. In this connection it should be mentioned that accurate determinations of sticking coefficients sensitively depend on well-defined boundary conditions in the experimental system [38].

CONCLUDING REMARKS

Nucleation and condensation processes in one component systems appear to be reasonably understood. However, additional studies are necessary to clarify the nucleation behavior in binary or multicomponent systems. Particularly the influence of partial miscibility and solubility of the compounds considered requires further attention. Experimental data on sticking probabilities are still controversial. Care must be taken to avoid substantial experimental errors caused by insufficient control of the experimental boundary conditions.

Finally it should be emphasized that studies of those compounds actually occurring in the atmosphere would provide a more direct applicability of laboratory experiments to atmospheric systems.

ACKNOWLEDGMENTS

This work has been supported by the Fonds zur Förderung der Wissenschaftlichen Forschung, Projekt Nr. P 9421 - GEO, and by the Hochschuljubiläumsstiftung der Stadt Wien.

REFERENCES

1. McMurry, P.H., *Atmos. Environ.* **34**, 1959 (2000).
2. Whitby, K. T., *Atmos. Environ.* **12**, 135 (1978).
3. Covert, D. S., Wiedensohler, A., Aalto, P., Heintzenberg, J., McMurry, P. H., and Leck, C., *Tellus* **48B**, 197 (1996).
4. Seinfeld, J., and Pandis, S. N., *Atmospheric Chemistry and Physics,* Wiley, New York, 1998.
5. Pandis, S. N., Paulson, S. E., Seinfeld, J. H., and Flagan, R. C., *Atmos. Environ.* **A25**, 997 (1991).
6. Zhang, S.-H., Shaw, M., Seinfeld, J. H., and Flagan, R. C., *J. Geophys. Res.* **97**, 20717 (1992).
7. Weber, R. J., McMurry, P. H., Eisele, F. L., and Tanner, D., J., *J. Atmos. Sci.* **52**, 2242 (1995).
8. Katz, J. L., and Ostermier, M., *J. Chem.Phys.* **47**, 478 (1967).
9. Hung, C., Krasnopoler, M., J., and Katz, J. L., *J. Chem.Phys.* **90**, 1856 (1989).
10. Anisimov, M. P., and Cherevko, A. G., *J. Aerosol Sci.* **16**, 97 (1985).
11. Wilck, M., Hämeri, K., Stratmann, F., and Kulmala, M., *J. Aerosol Sci.* **29**, 899 (1998).
12. Schmitt, J. L., Adams, G. W., and Zalabsky, R. A., *J. Chem.Phys.* **77**, 2089 (1982).
13. Wagner, P. E., and Strey, R., *J. Chem.Phys.* **80**, 5266 (1984).
14. Wyslouzil, B. E., Seinfeld, J. H., and Flagan, R., C., *J. Chem.Phys.* **94**, 6842 (1991).
15. Wagner, P. E., *J. Colloid Interface Sci.* **105**, 456 (1985).
16. Rader, D. J., and McMurry, P. H., *J. Aerosol Sci.* **17**, 771 (1986).
17. Raes, F., and Van Dingenen, R., *J. Geophys. Res.* **97**, 12901 (1992).
18. Arstila, H., Korhonen, P., Kulmala, M., *J. Aerosol Sci.* **30**, 131 (1999).
19. Mäkelä, J. M., Aalto, P., Jokinen, V., Pohja, T., Nissinen, A., Palmroth, S., Markkanen, T., Seitsonen, K., Lihavainen, H., and Kulmala, M., *Geophys. Res. Lett.* **24**, 1219 (1997).
20. Kulmala, M., Pirjola, L., and Mäkelä, J. M., *Nature* **404**, 67 (2000).
21. Wilemski, G., *J. Phys. Chem.* **91**, 2492 (1987).
22. Schmitt, J. L., Whitten, J., Adams, G. W., and Zalabsky, R. A., *J. Chem. Phys.* **92**, 3693 (1990).
23. Wyslouzil, B. E., Seinfeld, J. H., Flagan, R. C., and Okuyama, K., *J. Chem. Phys.* **94**, 6842 (1991).
24. Strey, R., Viisanen, Y., and Wagner, P. E., *J. Chem. Phys.* **103**, 4333 (1995).
25. Viisanen, Y., Wagner, P. E., and Strey, R., *J. Chem. Phys.* **108**, 4257 (1998).
26. Viisanen, Y., and Strey, R., *J. Chem. Phys.* **105**, 8293 (1996).
27. Laaksonen, A., McGraw, R., and Vehkamäki, H., *J. Chem. Phys.* **111**, 2019 (1999).
28. Hitzenberger, R., Giebl, H., Berner, A., Kromp, R., Reischl, G., Kasper-Giebl, A., and Puxbaum, H., "Measurements of CCN-Concentrations in the European Alpine Aerosol Using a Newly Developed Static Thermal Diffusion Chamber", this volume.
29. Porstendörfer, J., Scheibel, H., G., Pohl, F., G., Preining, O., Reischl, G., and Wagner, P. E., *Aerosol Sci. Technol.* **4**, 65 (1985).
30. Smolik, J., and Schwarz, J., *J. Colloid Interface Sci.* **185**, 382 (1997).
31. Kotzick, R., Panne, U., Niessener, R., *J. Aerosol Sci.* **28**, 725 (1997).
32. Petersen, D., Ortner, R., Vrtala, A., Laaksonen, A., Kulmala, M., and Wagner, P. E., *J. Aerosol Sci.* **30**, S35 (1999).
33. Petersen, D., Ortner, R., Vrtala, A., Laaksonen, A., Kulmala, M., and Wagner, P. E., "The Influence of Particle Solubility on Heterogeneous Nucleation in Binary Vapor Mixtures", this volume.
34. Alty, T., and Mackay, C. A., *Proc. Roy. Soc. London* **A149**, 104 (1935).
35. Wagner. P. E., in *Aerosol Microphysics II* (W. H. Marlow, Ed.) p.129. Springer, Berlin, 1982.
36. Clement, C. F., Kulmala, M., and Vesala, T., *J. Aerosol Sci.* **27**, 869 (1996).
37. Shaw, R. A., and Lamb, D., *J. Chem. Phys.* **111**, 10659 (1999).
38. Kulmala, M., and Wagner, P. E., "Sticking Probability and Uptake Coefficient - a Quantitative Comparison", this volume.

Aerosol-Cloud Interactions In Global Models Of Indirect Aerosol Radiative Forcing

Athanasios Nenes and John H. Seinfeld

Department of Chemical Engineering, California Institute of Technology, Pasadena, CA, 91125

Abstract. The sensitivity of cloud optical properties with respect to parameters that affect aerosol activation is examined. Of particular interest are the effect of volatile gases (such as HNO_3), slightly soluble and surfactant species. An adiabatic parcel model is used to simulate cloud droplet formation. Cloud optical properties are calculated from these simulations.

INTRODUCTION

Clouds are one of the key aspects of the climate system and have to be included in any climate simulation. However, computational constraints prohibit a comprehensive description of cloud formation in global climate models; instead, clouds are incorporated through simplistic parameterizations. As a result, the uncertainty introduced may be sufficiently large to rival the anthropogenic component of climate forcing. Much of this uncertainty originates in the complex relationship between cloud droplets and the aerosols they originate from. Both experimental and approximate analytical expressions that relate aerosol particles to cloud droplets have been proposed for implementation in climate models[1]. Idealized assumptions regarding the aerosol composition and cloud dynamics tend to limit the applicability of these relations. Furthermore, effects that can significantly alter the activation behavior of the aerosol, such as the presence of semi-volatile gases (e.g., HNO_3), slightly soluble[2], and surfactant species[3] are not accounted for. Errors in cloud droplet number become particularly important when cloud radiative properties are calculated, since they can be very sensitive to changes in droplet number[4].

This work attempts to determine which of the parameters affecting aerosol activation potentially has the largest effect on cloud optical properties, elucidating the parameters that should be included in a global aerosol-cloud droplet parameterization. For the purposes of this study, an adiabatic cloud parcel model with explicit aerosol microphysics is used to simulate the growth of CCN to cloud droplets within a cloud. From the simulations, cloud optical properties are calculated, and the sensitivity of the effective radius, optical thickness, and cloud albedo is computed with respect to the parameters considered.

CLOUD PARCEL AND RADIATIVE MODELS

A cloud parcel model is the simplest tool that can be used to simulate the evolution of droplet distributions throughout a non-precipitating convective cloud column. Although highly idealized, these models offer the most computationally efficient way to explore parametric space.

Cloud Parcel Model

The parcel model used is described in [5, 6]. In this model, conservation of heat and moisture are written for a rising air parcel. Moisture from the vapor phase is depleted by the aerosol population existing within the parcel. This depletion rate depends on the growth rate of each aerosol particle[6] which is primarily due to the condensation of water vapor on the particle,

$$\frac{dr}{dt} \propto \frac{(S - S_{eq})}{r} \quad (1)$$

where r is the particle size and S is the local saturation ratio. The equilibrium saturation ratio S_{eq} is equal to

$$S_{eq} = \exp\left(\frac{2\sigma}{RT\upsilon r} - \frac{i_s n_s}{n_w} - \frac{i_v n_v}{n_w} - \frac{i_{ss} n_{ss}}{n_w}\right) \quad (2)$$

where υ, σ, ρ are the molar volume, surface tension, and density of the droplet, respectively, R is the universal gas constant, T is the temperature, and n_w, n_s, n_v, n_{ss} are the number moles of water, soluble salt, volatiles, and slightly soluble species in the droplet. Finally, i_s, i_v, i_{ss} are the van't Hoff factors for the soluble salt, volatile species, and slightly soluble species dissolved within the droplet.

Equation 2 incorporates the effects considered in this study that influence aerosol activation. The first term in the exponential is the Kelvin effect, while the remaining terms express the solute, volatiles and slightly soluble effects, respectively.

In addition to water vapor, transfer of volatiles between the particles and gas phase takes place:

$$\frac{dn_v}{dt} \propto (P_v - P_{v,eq}) \quad (3)$$

where P_v is the partial pressure of the volatiles in the gas phase, and $P_{v,eq}$ is the equilibrium partial pressure of the volatiles over a given particle, and is calculated by a thermodynamic equilibrium module.

Cloud Albedo

Cloud albedo, R_c, is calculated based on the two-stream approximation of a non-absorbing, horizontally homogeneous cloud[6], $R_c = \frac{\tau}{7.7 + \tau}$, where τ is the cloud optical depth, $\tau = \int_0^H \frac{3\rho_a w_L(z)}{2\rho_w r_{eff}(z)} dz$, and $w_L(z)$ is the liquid water mixing ratio profile along the cloud column, calculated from the parcel model simulations. In addition, ρ_w is the water density, ρ_a is the air density, z is the cloud depth, and $r_{eff}(z)$ is the cloud droplet distribution effective radius.

SIMULATION PARAMETERS

Cloud optical properties depend on cloud thickness, cooling rate, volatile gas concentration, and aerosol size distribution characteristics. For each parameter, we consider values that cover the range found in experimental observations.

Cloud Parameters

Cloud thickness affects the transit time of air parcels within the cloud, and hence the time available for particle growth. We consider values ranging from 10 to 1000 m, which covers the thickness range seen for boundary layer clouds[7].

Updraft velocity influences both the transit time and the maximum supersaturation in a cloud updraft. Observed updraft velocities in boundary layer clouds vary widely, but values derived from measured vertical velocity variance are typically[7] between 30 and 50 cm s^{-1}. We explore the dependence on updraft by considering updraft velocities of 10, 30, 100, and 300 cm s^{-1}.

Finally, the volatile species considered in this study is HNO_3, and is assumed to have an initial mixing ratio ranging between 0 and 5 ppb.

Aerosol Parameters

The sensitivity of cloud optical properties with respect to aerosol distribution parameters will be explored by considering two aerosol size distributions, one representing pristine and the other polluted conditions. For this purpose, we use the marine (pristine) and urban (polluted) trimodal log-normal size distributions of [8]. The aerosol number concentration, N_i, the number mode radius, $r_{g,i}$, and geometric standard deviation, σ_i, for each aerosol type are listed in Table 1.

TABLE 1. Aerosol distribution parameters ($r_{g,i}$ in μm, N_i in cm^{-3})

Aerosol type	Nuclei mode $(r_{g,1}, \sigma_1, N_1)$	Accumulation mode $(r_{g,1}, \sigma_1, N_1)$	Coarse mode $(r_{g,1}, \sigma_1, N_1)$
Marine	0.005, 1.6, 340	0.035, 2.0, 60	0.31, 2.7, 3.1
Urban	0.007, 1.8, 106000	0.027, 2.16, 32000	0.43, 2.21, 5.4

In each of the distributions examined, the partially soluble mass fraction is allowed to vary between 0% and 50%. The solubility of the partially soluble fraction is also allowed to vary between 10^{-1} to 10^{-4} moles l^{-1}. The soluble fraction is assumed to be composed of $(NH_4)_2SO_4$, while the slightly soluble fraction is an organic with molar mass of 0.176 kg mol^{-1} and a van't Hoff factor equal to 2. When the effect of surface tension is considered, the following experimental relation[9] is used:

$$\frac{\sigma - \sigma_w}{T} = 1.87 \times 10^{-5} \ln(1 + 628.14\, C_{ss}) \quad (4)$$

where C_{ss} is the concentration of slightly soluble species (moles l^{-1}), and σ_w is the surface tension of pure water (N m^{-1}).

SIMULATION RESUTS-DISCUSSION

The "base case" simulation refers to an aerosol composed exclusively of soluble salt, without the presence of volatiles. Figure 1 presents the maximum albedo difference throughout the cloud column with respect to the base case simulation (marine aerosol), when various effects in equation (2) are included. The effect of slightly soluble species does not seem to significantly alter the cloud albedo. However, including surface tension changes can potentially play an important role for large updrafts. This effect is sensitive to the surfactant content of the aerosol, which tends to be less than 10% in pristine environments. In addition, the simulations show that solubility of the surfactant affects the maximum albedo difference between the range of 10^{-4} to 10^{-3} moles l^{-1}. Finally, volatiles can have a significant effect, being less sensitive to changes in updraft velocity.

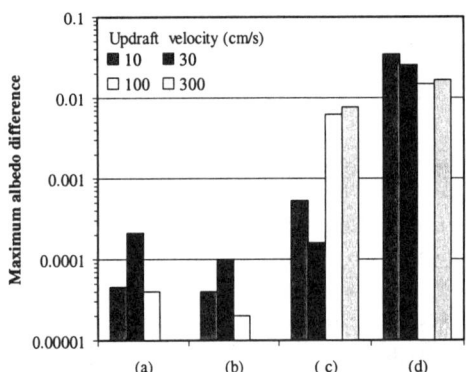

FIGURE 1. Maximum albedo difference with respect to the base case simulation, (a) 10% slightly soluble mass fraction, solubility 10^{-3} moles l^{-1} (no surface tension effects), (b) 20% slightly soluble mass fraction, solubility 10^{-3} moles l^{-1} (no surface tension effects), (c) 10% slightly soluble mass fraction, solubility 10^{-3} moles l^{-1} (with surface tension effects), (d) Initial gas-phase HNO$_3$ 5 ppb (no surface tension effects). The marine aerosol distribution is used in these simulations.

REFERENCES

1. Abdul-Razzak, H., Ghan, S. J., and Rivera-Carpio, C., *J. Geophys. Res.*, **103**, 6123-6132 (1998).
2. Laaksonen, A., Korhonen, P., Kulmala, M. and Charlson, R. J., *J.Atmos.Sci.*, **55**, 853-862 (1998).
3. Shulman, M. L., Jacobson, M. C., Charlson, R. J., Synovec, R. E., and Youngs, T. E., *Geophys.Res.Letters*, **23**, 277-280 (1996).
4. Pincus, R., and Baker, M., *Nature*, **372**, 250-252 (1994).
5. Seinfeld, J.H., and Pandis, S. N., *Atmospheric Chemistry and Physics*, New York: John Wiley (1998).
6. Nenes, A., Ghan, S., Abdul-Razzak, H., Chuang, P. Y., and Seinfeld, J.H., submitted to Tellus (2000)
7. Nichols, S., *Quart J. R. Met. Soc.*, **110**, 783-820 (1984).
8. Whitby, K.T., *Atmos. Environ.*, **12**, 135-159 (1978)
9. Facchini M. C., Mircea M., Fuzzi S., Charlson R. J., *Nature*, **401**, 257-259 (1999)

Direct And Indirect Effect Of Aerosols On The Climate Of The Southeastern U.S.

V.K. Saxena[1] and S. Menon[2]

1. Department of Marine, Earth and Atmospheric Sciences, North Carolina State University, Raleigh, NC 27695-8208
2. NASA GISS/Columbia University, 2880 Broadway, New York, NY 10025

Abstract. The effect of sulfate aerosols on cloud microphysical and optical properties has been studied using field measurements from a rural mountaintop location at the Mt. Mitchell State Park in the southeastern U.S. The sulfate concentrations were greater during 1993-97 than during 1986-89. Cloud albedo retrieved from satellite data and calculated from surface observations does not indicate a monotonic increase with increasing sulfates. The direct and cloud-mediated radiative forcing effects due to sulfate aerosols are assessed to be -4.8 and ~-4.0 W m^{-2}. These values far exceed the existing model predictions. Surface temperature records of the region during 1949-94 indicate a cooling trend tacitly supporting our assessment.

INTRODUCTION

The anthropogenic aerosols perturb the atmospheric radiation field through direct and indirect interactions with solar radiation, thereby influencing the climate. Through the direct effect, aerosols can scatter and absorb solar radiation in cloud-free air[1]. In terms of the indirect effect, aerosols composed of soluble substances such as sulfates can act as cloud condensation nuclei (CCN). Increases in CCN concentrations increase cloud droplet number (N), thereby reducing cloud droplet size, assuming the liquid water content (LWC) stays the same[2]. This enhances cloud albedo[3] and can also act to suppress drizzle production, which then increases fractional cloudiness and cloud lifetime[4]. The magnitude of the direct forcing usually relates linearly to aerosol loading for an optically "thin" atmosphere[5] and can be reliably quantified. However, the quantification of the indirect forcing is fraught with large uncertainty inherent in the relationship between aerosol number distribution and anthropogenic pollution[1]. Previous studies estimate the contribution of sulfates to indirect forcing by relating the atmospheric sulfate mass concentration to aerosol concentration (N_a) and/or CCN concentration through empirical observations and coupled chemistry-climate models[6,7]. The global-average estimates of indirect forcing from these studies range from -1.0 to -1.3 W m^{-2} with greater negative forcing for eastern U.S., southeastern Europe and eastern China[8] where sulfate emissions are greater. This could lead to changes in surface temperature and cloud reflectivity in these areas[9]. Empirical evidence of sulfate aerosol forcing on surface-air temperatures as well as on optical depth has been found[10]. Analyses of long-term surface-air temperature records indicate a cooling trend[11] for the southeastern U.S. To investigate the climatic impact of anthropogenic aerosols in the southeastern U.S., the direct and indirect effects of aerosols (primarily sulfates and to a lesser extent black carbon (BC)) were investigated using surface measurements, modeling results and remotely sensed satellite measurements. Since the observational site[13,14] intercepts relatively clean marine and highly polluted urban industrial air masses, it offers unusual and unique opportunities[15,16] to observe the impact of the air mass content on the microphysical and optical properties of the ensuing clouds.

RESULTS AND DISCUSSION

Direct Forcing

The direct radiated forcing was estimated by means of a CRM of the Climate Community Model (CCM III) that was coupled to a Mie code. The aerosol optical properties and the mass mixing ratios were used in the shortwave radiative transfer portion of the CRM to calculate forcing. The forcing due to aerosols is obtained as the difference in total solar radiation absorbed when the model is run once with and without any aerosol mass loading. During May-August of 1993-96, a total of 55 daytime cases were obtained. The average values of forcing for each of the four years are -6.36, -1.75, -8.78 and 2.42 W m^{-2}, with an overall average of ~-4.80 W m^{-2}, and values of negative -1.29, -3.60 and -10.04 W m^{-2} for marine continental and polluted continental air masses[15], respectively. Since sulfate concentrations are well correlated with BC concentrations (suggesting the same source region), the change in forcing for BC aerosols internally mixed with sulfate aerosols was calculated for ten cases with coincident sulfate and BC concentrations. The direct forcing ranges from -5.3 to +0.6 W m^{-2} for an internally mixed BC-sulfate aerosol, with BC-sulfate mass ratios ranging from 0.7 to 10.0%. The only occurrence of positive forcing was for 10% BC-sulfate mass ratio. The direct forcing for sulfate aerosols (mass ranging from 1 - 8 $\mu g\ m^{-3}$) ranges from -1.12 to -7.23 W m^{-2}, whereas that for BC aerosols (mass ranging from 0.01 to 0.2 $\mu g\ m^{-3}$) is from +0.08 to +2.12 W m^{-2}. Thus, an internally mixed BC-sulfate aerosol results in an average reduction in forcing of ~1.12 Wm^{-2}. Modeling studies indicate that the direct forcing for sulfates is reduced by 0.034 W m^{-2} for 1% of BC added to the sulfate mass mixing ratio. However, despite the reduction in forcing with the addition of BC aerosols, the values obtained for the direct forcing (minimum ~-0.07 W m^{-2} to a maximum of ~-26.4 W m^{-2}) due to sulfates are much higher than model predictions of ~-2.0 W m^{-2} for the southeastern U.S. The above estimates for forcing are uncertain by a factor of two due to the various assumptions used in the calculation of aerosol optical properties and size distribution as well as in the aerosol mass concentrations that were calculated from sulfate concentrations in cloud water.

Indirect Forcing

Since the incorporation of pollutants into clouds can affect drop-size distributions, the chemical characteristics of cloud water were investigated. Sulfates and nitrates are the two most important contributors to cloud water acidity, with average values of sulfate almost three times the average values of nitrates. The median values of cloud water sulfate were obtained for the summer months from 1986. Three intervals are chosen for cloud water pH: pH < 3.0, 3.0 < pH < 3.7 and pH > 3.7 to characterize polluted continental, continental and marine air masses, respectively. The trends in cloud water sulfate indicated that sulfate concentrations were higher during 1993-97 than during 1986-89. Previous analyses indicated a strong correlation between pH and wind direction of the air masses as well as cloud water sulfate with the source of cloud forming air masses. Thus, the long-range transport of pollutants to the Southeast, of which sulfate is predominant, can be established.

Table 1 gives regression relations between CCN, N, subcloud N_a R_{eff}, LWC and sulfates for the southeastern and northeastern U.S. A quantitative relationship that determines the increase in CCN from an increase in N_a is dependent on the chemical and dynamical characteristics of the cloud as well as on aerosol size distribution. Quantitative non-linear relationships between CCN-N were obtained from ground-based measurements for 1995-96, similar to non-linear values from other investigations on N_a and N, as shown in Table 1. The change in cloud albedo (ΔAc) resulting from an enhancement in N or CCN, when N_a increases from 100 to 1000 cm^{-3} (values are chosen to represent cleaner and polluted air masses) was calculated using a simple estimate as $\Delta Ac = 0.057\ \Delta N/N = 0.04\ \Delta CCN/CCN$ where ΔN and ΔCCN are the increases in N and CCN resulting from an increase in N_a. For the Southeast, ΔAc is ~0.33, whereas, for the North Atlantic ΔAc is ~0.25 (using equations from Table 1).

Hobbs[17] has suggested that higher pollution levels in the East Coast could account for higher sulfate production observed in clouds in the Mid-Atlantic as compared to that in the Pacific Northwest. For the Southeast for the 1995 data set, the range of sulfate mass was ~1.5 to 10 $\mu g\ m^{-3}$, with the exception of two cases that were between 20 to 30 $\mu g\ m^{-3}$ and CCN concentrations were ~130 to 1700 cm^{-3}, whereas, that for the Northeast Atlantic was 0 to 10.5 $\mu g\ m^{-3}$ for sulfates and ~50 to 250 cm^{-3} for CCN concentrations[18]. The CCN concentrations for the Southeast and for the Northeast Atlantic were reported at 1% supersaturation. Using the regression relationship between CCN and sulfate given in

Table 1, ΔCCN was calculated for an increase in sulfate mass from 2 to 5 µg m^{-3} (representative of average values of sulfate for cleaner and polluted air masses). The ΔAc calculated from the above equation is ~0.03 for the Southeast and ~0.01 for the Northeast Atlantic. The apparent sensitivity of change in N to the corresponding change in sulfate concentration was estimated from the slope of the regression relationship between logarithm of N and sulfate[19]. The higher sensitivity between N and sulfate obtained from our results compared to those at Puerto Rico for stratocumulus clouds[19] and in the Northeast[20] as indicated in Table 1, suggests that the indirect forcing is higher for the southeastern U.S.

CONCLUSIONS

Time series observations of sulfate concentrations in cloud water since 1986 indicate higher sulfate concentrations for 1993-97 as compared to that during 1986-89. The calculated change in cloud albedo from differences in CCN and N in cleaner to polluted air masses was higher for the Southeast as compared to the Northeast. Varying levels of sulfate in polluted and marine air masses, lead to changes in R_{eff} that are of sufficient magnitude to counteract warming expected due to doubling of CO_2. Higher sensitivity of N to sulfate content is obtained for the southeastern U.S. as compared to that for eastern North America and Puerto Rico. The indirect forcing effect due to sulfates for the Southeast is greater than the -4.0 W m^{-2} estimated by modeling studies[7]. A linear relationship between cloud albedo and cloud water sulfate was not found (both from satellite retrievals and calculations from in situ measurements). Variations in N and R_{eff} with varying sulfate content as well as in dynamical properties such as LWC and cloud thickness were found to be important in determining variations in cloud reflectivity. Vertical variations in LWC, N and R_{eff} must also be considered. An internal mixture of BC and sulfate reduces the sulfate forcing by ~1.12 W m^{-2}. However, despite this reduction, the combination of both direct (average ~-4.8 W m^{-2}) and indirect (greater than -4.0 W m^{-2}) radiative forcing for the sulfate aerosols for the past four years (1993-96) suggest that anthropogenic influences could balance any warming expected from the doubling of CO_2 for the southeastern U.S. On a regional scale, a cooling trend is underway for the eastern parts of U.S. Since regional climatic changes affect global climatic changes through the cloud-climate feedback mechanisms, possible consequences of the impact of reduced emissions on climate must be carefully investigated before measures are taken to control emissions and predict climate change.

ACKNOWLEDGMENTS

This research was supported by the NASA's Mission to Planet Earth (MTPE) under Contract No. NAS1-18944 from Langley Research Center, Hampton, VA and US EPA's STAR (Science to Achieve Results) grant No. R-825248.

TABLE 1. Comparisons of differences in the regression relation between cloud microphysical properties such as: cloud condensation nuclei concentration (CCN), cloud droplet number concentration (N), subcloud aerosol number concentration (N_a), cloud droplet effective radii (R_{eff}), cloud liquid water content (LWC) and sulfates for the southeastern and northeastern U.S.

Location	N and CCN/N_a	CCN and sulfate
Southeast U.S. (21)	N = 182.6 log (CCN) -333.7	CCN = 60.4(±10.6)[SO$^=_4$] + 129.3(±95.8)
Northeast U.S. (20)	NA	NA
Puerto Rico (19)	NA	NA
Northeast Atlantic (18)	NA	CCN = 15.5(±3.3)[SO$^=_4$] + 94.5(±15.8)
North Atlantic (22)	N = 215.8 log (N_a) - 382.2	NA

Location	N and sulfate	R_{eff}, N and LWC (Marine cases)
Southeast U.S. (21)	Log [N] = 0.66 (±0.05) log $[SO_4^=]_{cw}$ + 0.67 (±0.02)	1993: R_{eff} = 15.06 + 2.27 log (LWC) - 3.75 log (N) 1994: R_{eff} = 16.07 + 2.91 log (LWC) - 3.38 log (N) 1996: R_{eff} = 8.88 + 2.00 log (LWC) - 1.50 log (N)x
Northeast U.S. (20)	Log [N] = 0.26 (±0.05) log $[SO_4^=]_{cw}$ + 1.95 (±0.21)	NA
Puerto Rico (19)	Log [N] = 0.09 (±0.05) log $[SO_4^=]$ + 2.32 (±0.15)	NA
Northeast Atlantic (18)	NA	NA
North Atlantic (22)	NA	R_{eff} = 14.82 + 11.51 LWC - 4.76 log (N)

REFERENCES

1. Charlson R. J., S.E. Schwartz, J. M. Hales, R. D. Cess, J. A. Coakley, J. E. Hansen, and D. J. Hofmann, *Science* **225**, 423-430 (1992).
2. Cloud water content is a function of the dynamical processes and height above cloud base.
3. Twomey, S., *J. Atmos. Sci.* **34**, 1149-1152 (1977).
4. Albretch, B.A., *Science* **245**, 1227-1230 (1989).
5. Schwartz, S.E., *J. Aerosol Sci.* **27**, 359 (1996).
6. Jones, A., D.L. Roberts and A. Slingo, *Nature* **370**, 450-453 (1994).
7. Boucher, O. and U. Lohmann, *Tellus* **47B**, 281-300 (1995).
8. Kiehl, J.T., and B.P. Briegleb, *Science* **260**, 311-314 (1993).
9. IPCC (Intergovenmental Panel on Climate Change), Report to IPCC from the Scientific Assessment Working Group I & III, Cambridge, U.K.: Cambridge University Press (1994).
10. Karl, T.R., R,W, Knight, G. Kukla, and J. Gavin, in *Aerosol Forcing of Climate* edited by R.J. Charlson and J. Heintzberg, Chichester, U.K.: Wiley, 363-382 (1995); Saxena, V.K. and S.-C. Yu, *Geophys. Res. Lett.* **25**, 2833-2836 (1998).
11. Saxena, V.K., S.-C. Yu and J. Anderson, *Atmos. Environ.* **31**, 4211-4221 (1997).; Dettinger, M.D., M Ghil, and C.L. Keppenne, *Climatic Change* **31**, 35 (1995).
12. Charlson, R.J., J. Langer, H. Rodhe, C.B. Leovy, and S.G. Warren, *Tellus* **43AB**, 142 (1991).
13. Saxena, V.K., and N.-H. Lin, *Atmos. Environ.* **24A**, 329 (1990).
14. Saxena V.K., R.E. Stogner, A.H. Hendler, T.P. DeFelice and R.J.-Y. Yeh, *Tellus* **41B**, 92-109 (1989).
15. Ulman, J.C. and V.K. Saxena, *J. Geophys. Res.* **102**, 25451-25465 (1997).
16. Bahrmann, C.P. and V.K. Saxena, *J. Geophys. Res.* **103**, 23153-23161 (1998).
17. Hobbs, P.V., in *Aerosol-Cloud-Climate Interactions*, edited by P.V. Hobbs, San Diego: Academic Press, 33-73 (1993).
18. Hegg, D.A., *J. Geophys. Res.* **99**, 25903-25907 (1994).
19. Novakov, T., C. Rivera-Carpio, J.E. Penner, C.F. Rogers, *Tellus* **46B**, 132-141 (1994).
20. Leaitch, W.R., G.A. Isaac, J.W. Strapp, C.M. Banic, H.A. Wiebe, *J. Geophys. Res.* **97**, 2463-2474 (1992).
21. Menon, S. and V.K. Saxena, *Atmos. Res.* **47-48**, 299-315 (1998).
22. Gueltepe, I and G.A. Isaac, *Intl. J. Climatol,* **16**, 941-946 (1996).

Comparison of Simulated and Observed Aerosol Optical Depth

Nels Laulainen, Steven Ghan, Richard Easter, and Rahul Zaveri

Pacific Northwest National Laboratory, Richland, Washington, USA

Abstract. A variety of measurements have been used to evaluate the treatment of aerosol radiative properties and radiative impacts of aerosols simulated by the Model for Integrated Research on Atmospheric Global Exchanges (MIRAGE). This paper focuses on comparisons of simulated and measured aerosol optical depth (AOD). When the analyzed relative humidity is used to calculate aerosol water uptake in MIRAGE, the simulated AOD agrees with most surface measurements after cloudy conditions are filtered out and differences between model and station elevations are accounted for. Simulated AODs are low over sites in Brazil during the biomass burning season and over sites in central Canada during the wildfire season, which can be attributed to limitations in the organic and black carbon emissions data used by MIRAGE. The simulated AODs are mostly within a factor of two of satellite estimates, but MIRAGE simulates excessively high AODs off the east coast of the US and China, and too little dust off the coast of West Africa and in the Arabian Sea.

INTRODUCTION

Radiative scattering and absorption of sunlight by aerosols have a measurable impact on the surface and top-of-the-atmosphere radiation balance, also known as. the aerosol direct radiative forcing. Surface, aircraft, and satellite measurements are used to quantify this radiative impact (see *Ghan et* al. [1] for a discussion). However, such measurements cannot distinguish between the natural and anthropogenic component of the aerosol radiative forcing. Physically-based models are necessary to separate the natural and anthropogenic components of the forcing, and to consider future scenarios of emissions of aerosols and their precursor gases.

The PNNL Model for Integrated Research on Atmospheric Global Exchanges (MIRAGE) was developed for such a purpose. It consists of a detailed global tropospheric chemistry and aerosol model that predicts concentrations of oxidants, as well as aerosols and aerosol precursors, coupled to a general circulation model that predicts cloud water and cloud ice mass and cloud droplet and ice crystal number concentrations. Both number and mass of several externally-mixed log-normal aerosol size modes are predicted, with internal mixing assumed for the different aerosol components within each mode. Predicted aerosol species include sulfate, organic and black carbon, nitrate, soil dust, and sea salt. The climate model uses physically-based treatments of aerosol radiative properties (including dependence on relative humidity) and aerosol activation as cloud condensation nuclei and ice nuclei. A thorough description of MIRAGE is presented by *Easter et al.* [2].

Before models can be used to estimate the anthropogenic aerosol radiative forcing, they must be thoroughly evaluated. *Easter et al.* [3] evaluate the simulation of the concentrations of the aerosols and their precursor gases by MIRAGE. *Ghan et al.* [1] evaluate the simulation of the aerosol direct forcing. In this paper we focus on aspects of the evaluation of direct forcing related to the comparison of simulated and observed AODs not covered in *Ghan et al.* [1]. The focus of the evaluation is on the period June 1994 through May 1995.

MODEL SIMULATIONS

MIRAGE was run for the period June 1994 –May 1995 after a spin-up of three months. The horizontal resolution is T42 spectral (about $2.8°$ latitude and longitude) with 24 layers. MIRAGE simulates a mixture of aerosol that varies in space and time with emissions and with the simulated meteorology. The variations in the aerosol concentration, size distribution, and composition produce variations in the AOD, single-scattering albedo, and radiative forcing. To evaluate the simulation of AOD, we compare the MIRAGE simulation with estimates from both surface and satellite measurements. The advantage of surface measurements is high accuracy, while the advantage of satellite measurements is near-global coverage.

In comparing simulated and observed AODs it is essential to ensure that averages are formed for the same conditions. The observed AOD can only be estimated when the sun is not obscured by clouds. The simulated AOD is calculated every hour, and hence can be averaged under cloudy as well as clear conditions. It is important to filter the simulated optical depth in a manner consistent with the implicit filtering of the observed AOD. A cloud optical depth (COD) filter is applied when the simulated COD exceeds 1.0. The column maximum relative humidity filter is applied when the ECMWF analyzed column maximum relative humidity exceeds 99%. As might be expected, the filtering reduces the monthly mean simulated AOD considerably, with reductions of 15-50%. The simulated AOD with filtering agrees much better with the observed AOD than does the unfiltered AOD. The summertime maximum in AOD is simulated correctly by MIRAGE.

Further adjustments in the simulated AODs are required for grid points with surface elevation differing significantly from that of the corresponding surface station. As a first correction, we subtract from the simulated AOD the contribution from model layers below the elevation of the observing site. Such a correction can overestimate the influence of surface elevation on column optical depth, because air does not always flow around mountains. For sufficiently strong winds air can flow over mountains, carrying with it the pollutants from lower elevations. Indeed, the optical depths using the first correction are lower than observed. As a second correction, we use a Froude number parameterization to estimate the dividing streamline height (DSH) for elevated sites. Air above the DSH can pass over a topographical barrier having the same elevation as the site, while air below the DSH must move around the barrier. The Froude number corrected AOD is that of the air mass above this DSH. AODs with the Froude number correction are higher than without it and are in better agreement with the observations. All subsequent comparisons with surface station measurements of aerosol optical depth therefore use the Froude number correction to the simulated AOD.

OBSERVATIONAL

We have compiled a database of surface measurements of AOD at over 50 sites. Although the estimates of AOD from surface irradiance measurements are much more accurate than are estimates from satellite measurements, the spatial distribution of surface measurements is far from complete, with a preponderance of sites (at least in our present compilation) in the United States and rather few elsewhere. AODs have only been compiled for the period of the MIRAGE evaluation (June 1994 – May 1995). Data are not available under cloudy skies and when instruments malfunction.

Our primary source of data is from multi-filter rotating shadowband radiometer (MFRSR) observations [4]. These are mostly confined to the continental US, with a few sites distributed in Australia and Hawaii in the Pacific and in Barbados and Bermuda in the Atlantic Oceans, respectively. The total optical depth is calculated from the direct beam irradiance using the Beer-Lambert-Bouguer law. AOD is then calculated from the total optical depth by subtracting the contribution of molecular (or Rayleigh) scattering, which is a well-defined function of pressure and temperature, and an estimated contribution from ozone absorption.

Other data sources are from direct solar beam measurements, obtained with hand-held or sun-tracking sun photometers, and from stellar photometry. These data are reported either as total optical depth (TOD) or AOD, with analysis using the Beer-Lambert-Bouguer law already having been performed. The astronomical data, however, are reported as total atmospheric light extinction in units of magnitudes per airmass and so must be adjusted by a factor of about 0.92 to be consistent with AOD. The AOD data used in the model evaluation study are daily (or nightly) averaged values at a wavelength of 500 nm (or interpolated to 500 nm from about 550 nm for the astronomical values).

DISCUSSION

Figure 1 summarizes the comparison between station measurements of AOD and the MIRAGE simulation. Each point plotted represents a monthly mean (if available) for one of the 56 stations. Monthly means are formed only for times when both MIRAGE and the observations were cloud free. Although MIRAGE clearly demonstrates skill in simulating AOD, it is underestimated by up to 0.9 for some stations and months. These large differences occur primarily at stations in Brazil during the biomass burning season, stations in central Canada during the wildfire season, and stations in Asia and off the coast of Africa that are strongly influenced by soil dust. They can be attributed to limitations in the organic and black carbon emissions used by MIRAGE and difficulties of accurately simulating soil dust emissions in a global model. MIRAGE also overestimates AOD for some stations in Europe and North America by 100% or more.

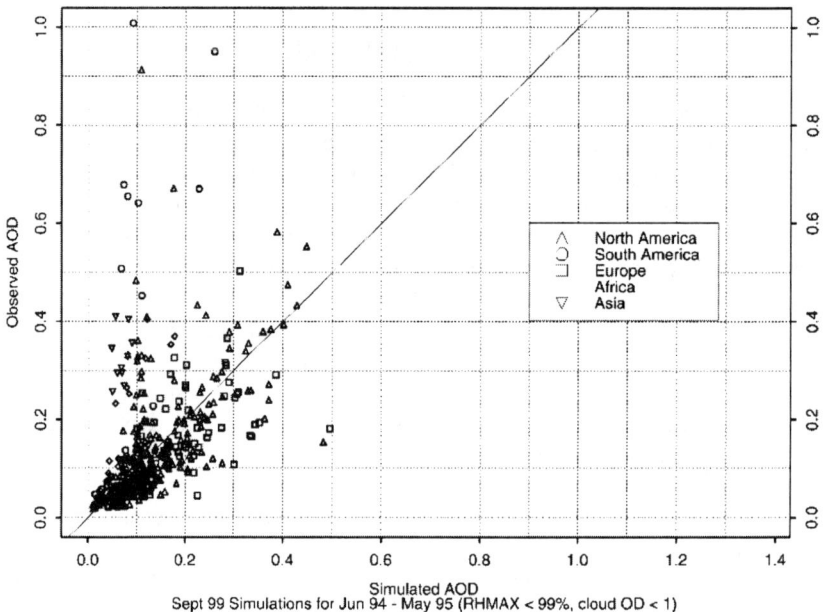

FIGURE 1. Scatterplot of the observed monthly mean aerosol optical depth for each month and station plotted versus the monthly mean aerosol optical depth simulated by MIRAGE.

ACKNOWLEDGMENTS

This study was primarily supported by the NASA Earth Science Enterprise under contract NAS5-98072. Support was also provided by the U.S. Department of Energy's Atmospheric Radiation Measurement Program and Atmospheric Chemistry Program, both part of the Office of Biological and Environmental Research. Pacific Northwest National Laboratory is operated for the DOE by Battelle Memorial Institute under contract DE-AC06-76RLO 1830. Many individuals from various institutions are gratefully acknowledged for their assistance in providing AOD data.

REFERENCES

1. Ghan, S., N. Laulainen, R. Easter, R. Wagener, S. Nemesure, E. Chapman, Y. Zhang, and R. Leung, *J. Geophys. Res.* (1999) submitted.
2. Easter, R., S. Ghan, R. Saylor, E. Chapman, H. Abdul-Razzak, R. Leung, Y. Zhang, X. Bian and R. Zaveri, *J. Geophys. Res.* (1999) submitted.
3. Easter, R. C., E. G. Chapman, S. J. Ghan, N. S. Laulainen, R. D. Saylor, and Y. Zhang, *J. Geophys. Res.* (2000) submitted.
4. Harrison, L., J. Michalsky, and J. Berndt, *Appl. Optics,* **33**, 5118-5125 (1994).

Aerosol Impact on the Earth Radiation Budget with Satellite Data

Knut W. Dammann, Rainer Hollmann, Rolf Stuhlmann

Institute for Atmospheric Physics, GKSS Research Center, D-21502 Geesthacht, Germany

Abstract. Preliminary to the launch of the first Meteosat Second Generation satellite (MSG) and within support of the Geostationary Earth Radiation Budget (GERB) instrument onboard of MSG, algorithms are tested to detect aerosol optical parameters and their possible signature on the Earth Radiation Budget (ERB) using existing data from instruments on satellites in low earth orbit like NOAA/AVHRR and ScaRaB.

INTRODUCTION

The impact of aerosols on the climate and hence the Earth Radiation Budget (ERB) can be split into a direct effect by the interaction with solar and terrestrial radiation and into an indirect effect on cloud microphysics, albedo, and precipitaion [1]. Because of the poor knowledge of global aerosol characteristics and temporal changes the magnitude of this impact is highly uncertain [2].

After the launch of Meteosat Second Generation (MSG) the Geostationary Earth Radiation Budget (GERB) instrument and the operational Spinning Enhanced Visible and InfraRed Imager (SEVIRI) will observe simultaneously regions with large tropical aerosol loadings due to both Saharan dust outbreaks and, more seasonally, biomass burning, as well as industrial pollution within their field of view. It is expected that possibly signatures of these aerosols and their impact on the ERB are detectable by a combined examination of data from the two MSG instruments. The high temporal resolution from a geostationary orbit is very useful for investigating aerosols, because their occurrence is highly variable in space and time.

In support of this future investigation and to develop and test the algorithms that will be used in it, data from satellites in low earth orbit are used to derive aerosol optical parameters and the ERB. The aerosol optical parameters are derived from measurements of the Advanced Very High Resolution Radiometer (AVHRR) onboard of the polar orbiting NOAA satellites and the ERB is measured by the Scanner for Radiation Budget (ScaRaB) on the METEOR satellite.

METEOSAT SECOND GENERATION

With the launch of the MSG Satellite, scheduled for October 2000, the GERB instrument will be put in orbit. This Instrument will be the first earth radiation budget sensor in a geostationary orbit and will measure the Earth Radiation Budget in a high temporal resolution of 15 minutes and spatial resolution of about 50km.

The instrument has a total bandpass of 0.35 to 30μm which is limited by a quartz filter to the shortwave band (0.35 to 4.9μm) and the longwave band (4.0 to 30.0μm) is provided by subtraction. GERB will be calibrated before launch and has calibration devices on board.

In addition to GERB, the operational payload of MSG, the Spinning Enhanced Visible and InfraRed Imager (SEVIRI) features 12 narrowband channels between 0.6μm and 13.4μm some of which are similar to the AVHRR instrument. The image repeat cycle of SEVIRI is 15 minutes and the spatial resolution is 3 km. These features will allow to derive aerosol optical properties with high resolution in space and time.

AEROSOL RETRIEVAL

The aerosol retrieval algorithm was developed by Higurashi and Nakajima at the Center for Climate System Research (CCSR) of the University of Tokyo [3]. The algorithm provides the aerosol optical thickness τ_a and the Ångström exponent α from measurements of channel-1 (0.63μm) and channel-2 (0.84μm) of the NOAA/AVHRR instrument for clear sky pixels over water surfaces [3]. It is based on a look-up table method from radiation transfer calculations with a combined discrete-ordinate / matrix-operator method in a coupled ocean-atmosphere system and accounts for multiple scattering and gaseous absorption in the atmosphere and rough ocean reflection [4]. To calculate the aerosol optical parameters Mie particles with a refractive index of $m = 1.5-0.005i$ are assumed and a bi-modal, log normal size distribution with a peak ratio $\gamma = c_1/c_2$ that can be translated to the Ångström exponent.

To correct for water vapor and ozone absorption and to calculate the reflection of the rough ocean surface the total water vapor amount and wind velocity are provided by ECMWF analyses and the ozone amount by TOMS measurements. To avoid the high reflection of white caps on the ocean surface wind speeds above 15 m/s are excluded from the retrieval. Also the viewing geometry and the sun zenith angle are needed for the retrieval algorithm, and areas of sun glint are also excluded.

The measured and theoretical values of the apparent reflectances in the AVHRR channels-1 and -2 are compared for first guess values of the aerosol optical thickness τ_a and the peak ratio γ. The optimal values of this two parameters are then searched by an iteration method that minimize the root mean square difference between the measured and theoretical reflectances which are reconstructed from the look-up tables [4].

SCANNER FOR RADIATION BUDGET

The Scanner for Radiation Budget (ScaRaB) has been a recent successor of the scanning broadband instruments used by Earth Radiation Budget Experiment (ERBE). ScaRaB has four channels, a visible (0.5 to 0.7μm), a shortwave (0.2 to 4.0μm), an atmospheric window (10.5 to 12.5μm) and a total channel (0.2 to 50.0μm) [5]. From March 1994 to March 1995 ScaRaB flight model one was flown on a non-sunsynchronous orbit. The height of the satellite was about 1200 km, so that the field of view of the instrument for the nadir pixel is approximately 60 × 60 km² [6][7], being to some extend larger than the Earth Radiation Budget Satellite (ERBS) field of view of 50 × 50 km².

One main emphasis during the development of ScaRaB was on calibration. The calibration error before launch was obtained to be less than 1.5% in the solar domain [8]. Reference [9] compared measurements of March 1994 ScaRaB observations with collocated ERBS nonscanner observations. Both data sets agree to 0.76 W/m² for shortwave fluxes, to 0.55 W/m² and 3.8 W/m² for longwave fluxes at night and day, respectively.

In July 1998, the 2nd flight model of ScaRaB has been brought into orbit on the Russian sunsychronous RESSURS satellite. Due to the failure of the transmitter on the RESSURS satellite at 7th of April 1999 no data could received from this time on.

DATA PROCESSING

After calibrating and navigating the AVHRR data the processing scheme APOLLO [10][11] is used to choose clear sky pixels over water surfaces. These selected data are then processed with the aerosol retrieval to get the aerosol optical parameters in AVHRR pixel resolution. To relate the aerosol information to ERB measurements, ScaRaB data from near simultaneous overpasses to that of NOAA are processed. Since ScaRaB measurements are done on a much larger spatial scale, the AVHRR aerosol information is averaged within the corresponding ScaRaB pixels by use of the ScaRaB point spread function. Finally, the average parameters and ERB measurements on the ScaRaB scale can be combined to investigate the impact of aerosols on the ERB.

RESULTS

Figure 1 shows the broadband shortwave fluxes versus the aerosol optical depth as an example of the processing. Here data over the Mediterranean Sea from overpasses of NOAA 11 on the 28th and 30th of July 1994 at 15:24 UTC and 15:00 UTC were chosen. The ScaRaB instrument observed the areas 38 and 40 minutes before AVHRR.

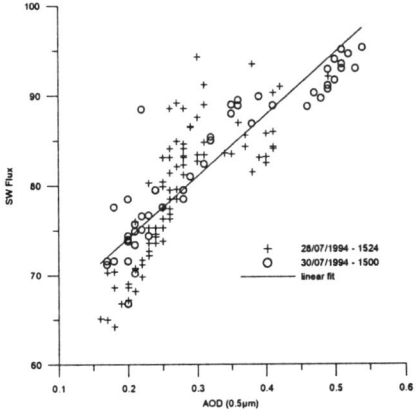

FIGURE 1. Comparison of broadband shortwave fluxes from ScaRaB and aerosol optical depth (AOD) for 28/07/1994 15:24 and 30/07/1994 - 15:00.

To exclude ScaRaB measurements with spurious cloud contamination, we apply a dynamic threshold to reject contaminated measurements. This threshold has been derived by use of the ERBE mean-directional models [12] and depends on geographic scene type and solar zenith angle.

The linear fit shows expected result, that the broadband shortwave fluxes increases with increasing aerosol optical depth.

CONCLUSIONS

In preparation of processing data in high temporal resolution from the new geostationary instruments GERB and SEVIRI onboard of MSG, data from low earth orbiting satellites are used to investigate the impact of aerosols on the ERB. First results show a resonable increase of the broadband shortwave fluxes with increasing optical depth. In future further investigation are needed to get more datasets and to compare them with ground based measurements, model results and other retrieval algorithms.

REFERENCES

1. Kaufman, Y. J., Tanré D., Gordon H. R., Nakajima T., Lenoble J., Frouin R., Grassl H., Herman B. M., King M. D. und Teillet P. M., *J. Geophys. Res.*, **102**, 16815-16830 (1997).
2. Mishchenko, M., Travis L. D., Kahn R. A. and West R. A., *J. Geophys. Res.*, **102**, 16831-16847 (1997).
3. Nakajima, T. and Higurashi A., *J. Geophys. Res.*, **102**, 16935-16946 (1997).
4. Higurashi, A. and Nakajima T., *J. Atmos. Sci.*, **56**, 905-923 (1999).
5. Viollier M., Kandel R., and Raberanto P, *Ann. Geophysicae*, **13**, 959-968 (1995).
6. Kandel R., Viollier M., Raberanto P., Duvel J. Ph., Pakomov L.A., Golovko V.A., Trishchenko A.P., Müller J., Stuhlmann R., and Raschke E., *Bull. Amer. Meteorol. Soc.*, **79**, 765-783 (1998).
7. Viollier M., Kandel R., and Raberanto P., *Ann. Geophysicae*, **13**, 959-968 (1995).
8. Müller J., Stuhlmann R., Becker R., Raschke E., Rinck H., Burkert P., Monge J.-L., Sirou F., Kandel R., Tremas T., and Pakomov L.A., *J. Atmos. Oceanic. Technol.*, **14**, 4, 802-813 (1997).
9. Bess T.D., Smith G.L., Green R.N., Rutan D.A., Kandel R.S., Raberanto P., and Viollier M., "Intercomparsion of scanning radiometer for radiation budget (ScaRaB) and earth radiation budget experiment (ERBE) results," in *Preprints of 9th conference on atmospheric radiation, Long Beach (CAL)*, Amer. Meteorol. Soc., Boston, 1997, pp. 203-207.
10. Kriebel K.T., Saunders R.W., and Gesell W, *Beitr. Phys. Atmosph.*, **63**, 165-171 (1989).
11. Saunders R.W. and Kriebel K.T, *Int. J. Remote Sensing*, **9**, 123-150 (1988).
12. Shuttles J.T., Green R.N., Minnis P., Smith G.L., Staylor W.F., Wielicki B.A., Walker I.J., Young D.F., Taylor V.R., Stowe L.L., *NASA Reference Publication 1184*, (1988).

Aerosol Effects on UV Radiation

P. Koepke, J. Reuder, H. Schwander

Meteorolog. Inst. Univ. Munich, Munich, Germany

Abstract. The reduction of erythemally weighted UV-irradiance (given as UV index, UVI) due to aerosols is analyzed by variation of the tropospheric particles in a wide, but realistic range. Varied are amount and composition of the particles and relative humidity and thickness of the mixing layer. The reduction of UVI increases with aerosol optical depth and the UV change is around 10% for a change aerosol optical depth from 0.25 to 0.1 and 0.4 respectively. Since both aerosol absorption and scattering are of relevance, the aerosol effect depends besides total aerosol amount on relative amount of soot and on relative humidity.

INTRODUCTION

Solar UV radiation is influenced by atmospheric aerosols due to scattering and absorption. This effect must be considered besides the influence of ozone and clouds, which usually are connected with UV questions, if UV is modeled or forecasted. Measurements show that UV radiation under turbid conditions can be reduced more than 25 % against clear cases [1,2,3], and that the variation of the aerosol influence for fixed turbidity is high. To get information on the reasons, the influence of tropospheric aerosol on solar UV-radiation has been analyzed by modeling, on the basis of aerosol composition and amount.

In the UV spectral range mutual effects of aerosol scattering and ozone absorption are possible. As a consequence, the aerosol effect depends not only on the aerosol amount and properties, but also on the solar zenith angle, the other atmospheric components, the receiver geometry, and of course, on the wavelength. However this paper is restricted to the UVI (UV index, a measure for the erythemal weighted global irradiance, a quantity which is proposed for public information on UV risks), to average summer conditions with a solar zenith angle of 30° and to continental aerosol types.

METHOD

The aerosol effect on the UVI is investigated by modeling the atmospheric scattering and absorption processes with a high quality radiative transfer model STAR [4] in its actual version. The model has been compared in recent years with other models and with measurements with good agreement [5,6]. The tropospheric aerosol particles are assumed to be externally mixed from four components water-soluble, insoluble, soot and sea salt, which are taken from OPAC [7]. On the basis of size distributions of the components, their change with relative humidity and their spectral complex refractive indices, the relevant radiative properties for the use in STAR are calculated under the assumption of spherical particles.

To cover the wide range of aerosol amount and properties which may occur for clear and more and more turbid continental aerosol conditions, the aerosol components are varied in the ranges which are given in Tab.1 with mass density for 50% relative humidity. To take further possible variabilities into account, the relative humidity in the mixing layer is varied with values of 50%, 70%, 80% and 90% under the assumption of fixed particle number densities. For the water-soluble and sea salt components, which are hydrophil, this results in an increase in mass density (not given in Tab. 1) and in a change in their absolute and spectral scattering properties. The range of variability which is given in Tab.1 for each of the components is not used in all the possible combinations for aerosol mixtures. Rather taken for the description of tropospheric aerosols in this study are the aerosol types "continental clean", "continental average", "continental polluted" and "urban" from OPAC [7] and additionally varied with respect to the amount of water-soluble particles. Moreover, two aerosol mixtures are added, which describe continental conditions with maritime influence. To get a further variation in optical depth, the thickness of the mixing layer, which is assumed to be filled with the aerosol type mentioned, is varied with 1 and 2 km and respectively 3 km for clear conditions. All together 112 aerosol scenarios are taken for the UV modeling. Above the mixing layer the aerosol is given with properties for stratospheric background conditions. The atmospheric properties besides the aerosol are fixed, in fact to ozone:330 DU, SO2: 1DU, surface albedo: grass, altitude: sea level, solar zenith angle 30 °

TABLE 1 Variability range of component mass density in mueg/m^3 for rel. hum. of 50%

component	water-soluble	insoluble	soot	sea salt
minimum	3.5	1.5	0.0	3.0
maximum	112.0	35.6	7.8	3.0

RESULTS

Result of the modeling are all spectral radiative properties of the aerosol mixtures which are of interest, both for comparison with measurements and to analyze the radiative influence of the aerosol, i.e. scattering and absorption coefficients, Angstrom coefficient, aerosol optical depth (AOD), single scattering albedo (SSA) and phase function. With respect to solar UV, besides UVI also UV-A and UV-B irradiances are modeled and moreover photolysis frequencies.

Shown here is the aerosol effect on UVI, given as normalized UVInorm: UVInorm is the UVI in case of the actual aerosol conditions, normalized to (divided by) the UVI for the same atmosphere but with lowest realistic aerosol amount, which is given by a visibility of 150 km and background aerosol in the upper troposphere and stratosphere. Thus the basis for the normalization is an atmosphere with an AOD at 550nm (AOD550) of 0.058 and not an atmosphere with no aerosol at all.

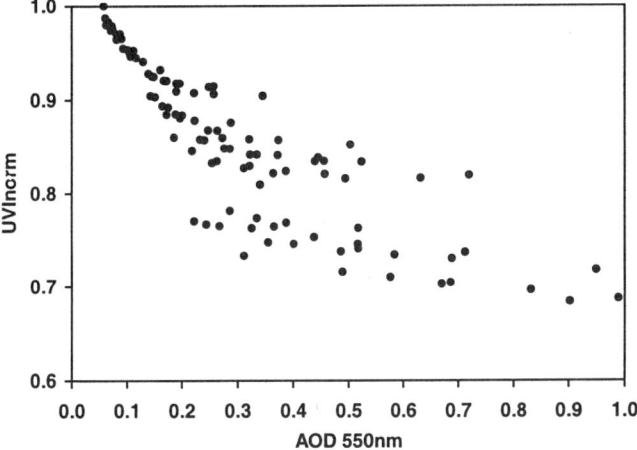

FIGURE 1. Normalized UV Index depending on aerosol optical depth at 550 nm and fixed other atmospheric parameters (see text)

Fig.1 shows UVInorm as function of AOD550. The gap around 0.8 in UVInorm is due the variation of the mixinglayer in steps of 1 km. In reality all values in the area which is stretched out by the data points will be possible and occur. The reduction of the UVI increases with optical depth, with the points in Fig.1 close together for low values of aerosol optical depth and spread over a wide range if optical depth increases. This in complete agreement with a large set of UV-B values measured in Athens [2]. In average the change in UVI is a 10 % increase or decrease for changing the AOD550 from 0.25 to 0.1 or 0.4 respectively.

Due to the strong forward scattering in the UV many of the photons from the attenuation of the direct sun become part of the diffuse radiation and thus the attenuation effect which may be described by optical depth is less pronounced for global irradiance. In case of absorption however, the optical depth results in attenuation both of direct and diffuse radiation and consequently the irradiance is reduced. Thus the aerosol influence in the UV strongly depends on the aerosol absorption and the aerosol types with high soot amount show the strongest UVI reduction. However the AOD dominantly depends on the amount of water-soluble particles and more over on relative humidity, since these particles will swell. Thus, up to an AOD550 of about 0.2, the aerosol effect can be described by the aerosol optical depth, but for higher AOD550 values, additional information is necessary.

From a sensitivity study [8] is known, that scattering and absorption coefficient together with optical depth explain more than 80 % of the aerosol effect for all UV radiation quantities and the variability in phase function, spectral behavior and height profile are of minor importance. If the UVInorm reduction is explained by a combination of aerosol scattering and absorption properties this gives good agreement, which documents that additional information on aerosol besides optical depth would be useful. However, in many cases even the aerosol optical depth is not known, and the only available information is the visibility. Under such circumstances the uncertainty of the UV effect further increases, due to the unknown aerosol height distribution [9].

CONCLUSION

The influence of aerosol on UVI may be in the same order of magnitude as the ozone variability. A variation in UVI of 20 %, as found between aerosol optical depth of 0.1 and 0.4, is similar to a variation of ozone between 300 and 360 DU. Even larger are the effects for UV-quantities where the ozone influence is reduced due to spectral weighting with relative increased contribution of the longer wavelengths to the signal. For UVI modeling, the AOD is useful only for clear conditions. For higher turbidity, information on aerosol absorption additionally must be taken into account.

REFERENCES

1. Krzyscin, J.W., and Puchalski, S., *J. Geophys. Res.* **103, D13,** 16175-16181 (1998)
2. Kylling, A., bais, A.F., Blumthaler, M., Schreder, J., Zerefos, C.S.. and Kosmidis, E., *J. Geophys. Res.* **103, D20,** 26051-26060 (1998)
3. Meleti, C. and Cappellani, F., *J. Geophys. Res.***105, D4,** 4971-4978 (2000)
4. Ruggaber, A., Dlugi, R., Nakajima, T., *J. Atmos. Chem.* **18,** 171-210 (1994)
5. Koepke, P., and 23 co-authors, *Photochem. Photobiol.* **67, 6** 657-662 (1998)
6. v. Weele, M., and 17 co-authors, *J. Geophys. Res.***105, D4,** 4915-4925 (2000)
7. Hess, M., Koepke, P., and Schult,I., *Bull. Amer. Meteor. Soc.* **79, 5,** 831-844 (1998)
8. Reuder, J. and Schwander, H. , *J. Geophys. Res.***104, D4,** 4065-4077 (1999)
9. Schwander, H., Koepke, P., and Ruggaber, A., *J. Geophys. Res.***102, D8,** 9419-9429 (1997)

Properties of Particles in the Tropopause Region

Bernd Kärcher[1] and Susan Solomon[2]

[1] *DLR Institut für Physik der Atmosphäre, D-82234 Wessling, Germany.*
[2] *NOAA, Aeronomy Laboratory R/E/AL8, Boulder, CO 80303, U.S.A.*

Abstract. Liquid aerosol particles and ice crystals in subvisible cirrus clouds in the tropopause region are investigated in terms of microphysical, chemical, and optical properties. These particle properties are examined by means of a thermodynamic equilibrium model and are related to satellite extinction measurements.

INTRODUCTION

High-altitude cirrus clouds have been identified as important components of the Earth's radiation budget. In the tropical tropopause region, subvisible cirrus clouds probably play an important role in the dehydration of air entering the upper troposphere and lower stratosphere. Besides their well-documented presence in the upper tropical troposphere, subvisible cirrus can form, albeit less frequently, near the midlatitude and polar tropopause. Aerosols and cirrus clouds may perturb chlorine chemistry and play a role in ozone chemistry at midlatitudes. Characterizing these particles in terms of chemical compositions and physical properties and understanding the link between aerosol particles and ice crystals in cirrus are key steps in assessing the impact of these particles on ozone chemistry and their effect on the radiative forcing of climate.

RESULTS AND DISCUSSION

In our presentation at the conference, we characterize particles in the tropopause region, particularly for midlatitudes, in terms of size distributions, chemical composition, and optical extinction. These particle properties are studied by means of simple models and are related to measurements taken by the Stratospheric Aerosol and Gas Experiment (SAGE). A comprehensive report on this subject has been published recently (1), and the interested reader is referred to this publication for more details.

Chemical Composition of Aerosols

Aqueous sulfate aerosols exhibit a markedly enhanced hygroscopicity and related particle growth in the presence of other condensable trace gases. We investigate aerosol growth and chemical composition by employing a thermodynamic equilibrium model in which the coupling between gas and aerosol phase is taken into account.

Under humidity conditions typical for the midlatitude tropopause, we demonstrate that nitric acid (HNO_3) can dissolve into liquid sulfuric acid (H_2SO_4/H_2O) near the ice frost point, affecting both specific surface area and extinction of the aerosol population. However, in contrast to lower stratospheric conditions, nearly pure binary HNO_3/H_2O solutions are unlikely to form, because the dew point of HNO_3 (the temperature at which the mass fraction of HNO_3 in the particles becomes equal to the mass fraction of H_2SO_4) is located well below the frost point, so that cirrus cloud formation may set in before this can happen. The presence of ammonium in sulfate solutions may affect the gas phase levels of HNO_3 and HCl, because these species are highly soluble in strongly ammoniated particles over a wide range of relative humidities.

Optical Extinction of Aerosols

The general aerosol extinction level is largely determined by the amount of particulate sulfate. Especially at cold temperatures below the ice frost point, but before cirrus clouds start forming, other condensable species may enhance the extinction. We show that aerosols in the tropopause region display a larger spread of extinction and extinction ratios than background stratospheric aerosols. Further, our results indicate that even pure H_2SO_4/H_2O aerosols can exceed the extinction values marking the transition region between aerosol and aerosol/cloud mixtures in satellite occultation data at low temperatures (< 220 K) and high water vapor mixing ratios (10-50 ppmv) that are not uncommon at the tropopause.

Physical Properties of Aerosols

We have also investigated the ranges of surface area densities of aerosol particles over a wide range of conditions (T = 200-230 K, [H_2O] = 10-100 ppmv, [H_2SO_4] = 0.02-0.5 ppbv, [HNO_3] < 0.5 ppbv). Our results suggest that specific surface areas exceeding ~ 10 μm^2 cm^{-3} are not achieved with liquid particles for [H_2SO_4] < 0.2-0.3 ppbv. Higher sulfate levels lead to high extinction, but the low extinction ratios (at 0.5 μm versus 1 μm wavelength) typical of cirrus clouds are difficult to achieve with liquid sulfate particles. However, especially during intense convective activities over populated areas, particles at altitude may contain diverse material such as ammonia, organics, minerals and metals, in addition to sulfate and nitrate. Hence, because we have only little knowledge about the concentrations of theses substances in the tropopause region, our estimated upper limit for the aerosol surface area is perhaps not entirely conclusive, but it should hold at least for remote (unpolluted) conditions.

Properties of Aerosol-Cloud Mixtures

The above findings suggest that the high average surface areas (> 10 μm^2 cm^{-3}) estimated from SAGE II data cannot be fully attributed to liquid particles. We find that only small cloud fractions detected along the line of sight of the SAGE II sensors suffice to satisfy the extinction data. Hence, midlatitude aerosol/cloud surface area densities in the tropopause region may be partly attributed to the presence of subvisible cirrus containing small (effective radius 5-10 μm) ice crystals.

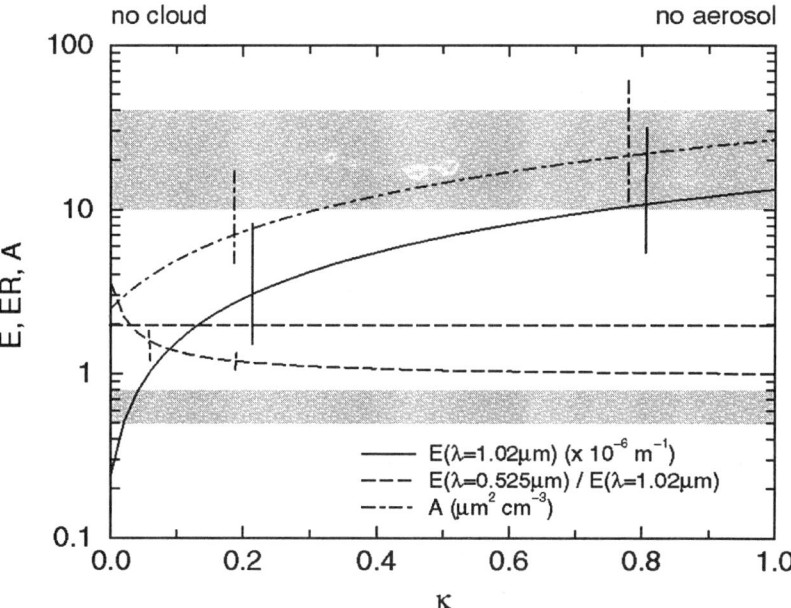

FIGURE 1. Extinction E at 1.02 μm, 0.525-vs-1.02 μm extinction ratio ER, and specific surface area A of a simulated aerosol/cloud mixture. The value of κ denotes the fraction of the SAGE II measuring volume ascribed to subvisible cirrus; the remaining fraction (1-κ) is filled with sulfate aerosol. The vertical bars indicate the ranges of E, ER, and A at selected κ values upon variation of ice water content at fixed aerosol parameters, whereby all curves have common starting points in the cloudless case. The bottom shaded area marks the transition range E(1.02 μm) = (5-8) x 10^6 m^{-1} between aerosols and aerosol/cloud mixtures. The dashed horizontal lines marks the threshold ER value of 2 below which particles are classified as cloud in conjuction with the 1-μm extinction range. The top shaded area gives the range of estimated A-values (45°N, 45°S) based on SAGE II cloud occurrence frequencies.

For the calculations shown in Figure 1, we assumed a temperature of 210 K; an ensemble of H_2SO_4/H_2O-aerosols with a mean radius of 0.08 μm and 100 pptv H_2SO_4, and 100 pptv HNO_3; a subvisible cirrus with a baseline ice water content of 0.1 mg m^{-3} (varied between 0.05 and 0.3 mg m^{-3}) and a mean crystal size of 3.8 μm. In Figure 1, we demonstrate that adding small fractions of such cloud particles to aerosols in the measuring volume of the SAGE II or other sensors dramatically changes microphysical properties of the particle mixture. Besides leading to higher surface areas, mixtures of aerosol and subvisible cirrus particles are also characterized by higher extinction efficiencies and lower extinction ratios, consistent with observations.

Implications for heterogeneous chemistry

The described partitioning of particle types in aerosol/cloud mixtures also impacts heterogeneous reaction rates. We show how the efficiency of halogen-induced heterogeneous chemistry depends on the relative abundance of liquid and ice particles in a given volume of air. For example, the reaction of $ClONO_2$ with HCl on ice particles is characterized by a high reactive uptake coefficient (~ 0.2-0.3) but is limited by gas phase diffusion when the reaction involves ice crystals in subvisible cirrus. On the other hand, similarly high uptake rates on smaller liquid particles are achieved at low temperatures (~ 205 K).

CONCLUSIONS

This work highlights the need for detailed microphysical and chemical modeling of aerosols and subvisible cirrus in order to quantify their role in atmospheric chemistry and radiative forcing, in concert with dedicated field and laboratory measurements. As discussed in more detail in (1), a better understanding of the role of ammonia and organics in particle formation and growth, ice nucleation mechanisms, the influence of small-scale waves on the temperature variability, and amplitude and frequency of occurrence of related cooling events is required before the chemical and climatic impact of particles in the tropopause region can be fully addressed.

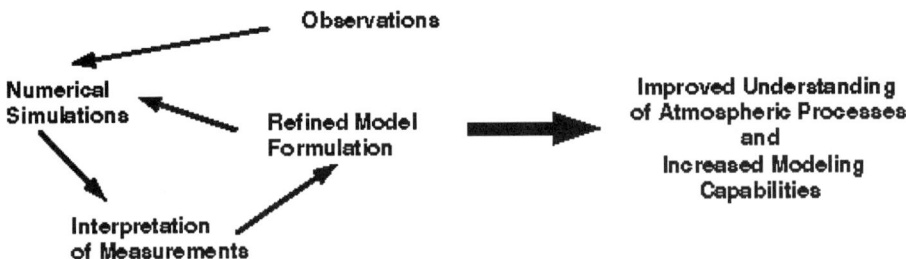

FIGURE 2. Strategy to investigate the properties of particles in the tropopause region.

Simulations of aerosol/cloud properties and comparisons with observations are interactive processes. Figure 2 depicts a possible strategy to tackles these issues using a process-related simulation model. Sequences of: numerical simulations initialized with observations — interpretation of measurements — refined model formulation, will ultimately lead to a better understanding of field data and provide means about how to correctly represent the particles in models. This forms the basis for developing a parameterization scheme of particle properties for use in large-scale models.

ACKNOWLEDGMENTS

This work was partly supported by the German Secretary of Education and Research (BMBF) within the research programme Atmospheric Aerosols (AFS).

REFERENCES

1. Kärcher, B., and Solomon, S., *J. Geophys. Res.* **104,** 27,441-27,459 (1999).

Aerosol Optical Depth over Europe: Satellite Retrieval and Modeling

C. Robles Gonzalez[*], G. de Leeuw[*], J.P. Veefkind[*,#], P.J.H. Builtjes[+],
M. van Loon[+] and M. Schaap[+]

[*]*TNO, Physics and Electronics Laboratory, P.O. Box 96864, 2509 JG The Hague, The Netherlands;*
[+]*TNO, Institute of Environmental Science. Energy Research and Process Innovation, P.O. Box 342, 7300 AH, Apeldoorn, The Netherlands.*
[#]*current address: KNMI, P.O. Box 201, 3730 AE De Bilt, The Netherlands*

Abstract. Aerosol optical depth (AOD) and Ångström coefficients over Europe retrieved from satellite data for August 1997 provide information on the spatial variations of these aerosol properties. The AOD results are compared with initial results from model calculations, showing the relative influences of sulphate and nitrate aerosol.

INTRODUCTION

Aerosols affect climate by absorbing and scattering solar radiation and changing the albedo and the lifetime of clouds. Radiative forcing by aerosols and by greenhouse gases are similar, but opposite in sign (1). However, as opposed to gases, tropospheric aerosols have a mean lifetime of up to about one week and the aerosol sources, both anthropogenic and natural, are usually rather localised. These two factors lead to a highly inhomogeneous aerosol distribution in space and time.

Satellite remote sensing provides information on the spatial distribution of column-integrated aerosol properties, such as aerosol optical depth (AOD) (2-5). However, currently no detailed information on aerosol composition or vertical distribution is available from satellite retrievals. Results are highly dependent on the information used in the retrieval algorithm, and significant errors may occur. Nevertheless, often retrieved AOD and directly measured AOD using sun photometers compare favorably (2-5). A caveat is that retrievals can only be made in cloud-free situations.

For the interpretation of aerosol retrieval results, and to fill gaps between successive retrievals, 3-D chemical transport models can be used. Data assimilation, using, e.g., satellite data, can be applied to improve the model results. Model results can be used in combination with satellite data to estimate the impact of various aerosol types on AOD, and thus on climate. Satellite retrieval can provide data with the appropriate spatial and temporal resolution to test the models and identify deficiencies.

Aerosol optical depth: retrieval

The dual view algorithm (2, 3), developed to compute the AOD for cloud-free scenes over land using ATSR-2 data (ERS-2 satellite) was applied to produce a map of the mean AOD over Europe for August 1997, see Robles-Gonzalez et al. (5) for details. Similarly, Ångström coefficients were derived, from the wavelength dependence of the AOD. Results are presented figures 1 and 2.

Aerosol optical depth: modeling

Mass concentrations of sulphate and nitrate aerosols were calculated for August 1997 using the LOTOS model (6) with an aerosol description based on the MADE model (7). The results were tested by comparison with experimental data. Both aerosol types were assumed to exist as ammonium salts (8), confined in the lower 2.5 km.

Daily maps of AOD were computed using the following equation:

$$\text{AOD}_i(\lambda) = f(\text{RH}, \lambda) \cdot \alpha_i(\lambda) \cdot B_i(\lambda) \tag{1}$$

where $\alpha_i(\lambda)$ is the mass extinction efficiency of species i (sulphate or nitrate); B_i is the column burden of species i calculated with the LOTOS model; $f(\text{RH}, \lambda)$ is a function describing the variation of the scattering coefficient with relative humidity (RH) and wavelength (λ).

The mass extinction coefficient was computed using a Mie code assuming a lognormal size distribution with a geometric mean radius of 0.05 μm, geometric standard deviation of 2.0 and a dry particle density of 1.7 g cm^{-3} applying for sulphate (9).

The size and composition of the hygroscopic sulphate and nitrate particles, and thus also their light scattering properties, change with RH. This effect was accounted for by the factor $f(\text{RH},\lambda)$. Ignoring the wavelength dependence (2), an empirical relation derived from humidity controlled nephelometry was used for f(RH) (8). The relative humidity was obtained from the LOTOS meteorological data, with different values for each of the three layers in the lower 2.5 km.

The daily AOD maps derived with the above assumptions were averaged to obtain the mean LOTOS AOD for August 1997 that are presented in figure 3.

RESULTS AND DISCUSSION

Figures 1 and 2 show the strong spatial variations of both the AOD and the Ångström coefficients over Europe. White areas indicate that no data are available; for the AOD this may be due to clouds or due to high surface reflection (e.g., over snow in the Alps); for Ångström coefficients this may also be because no fit could be made. AOD values as high as 0.6 are observed over industrialized areas. In contrast, AOD values of 0.1-0.2 were observed over relatively clean areas. Furthermore, large spatial gradients occur, with variations of a factor of 3 over a few hundreds of kilometers.

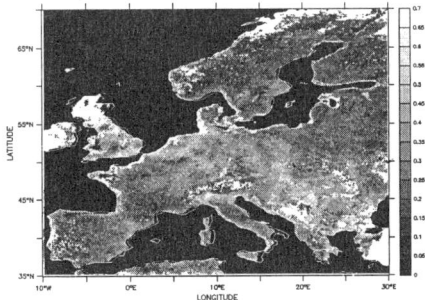

FIGURE 1. Satellite retrieved AOD at 0.55 μm.

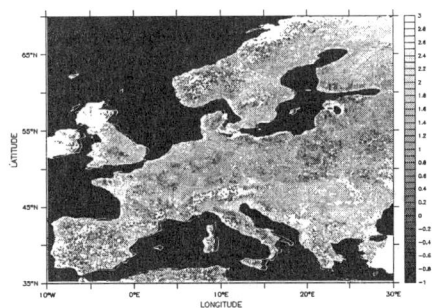

FIGURE 2. Satellite retrieved Ångström coefficients.

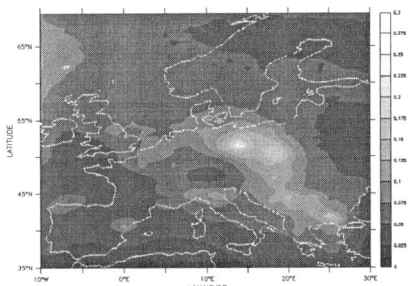

FIGURE 3. Aerosol optical depth at 0.55 μm calculated from LOTOS sulphate fields.

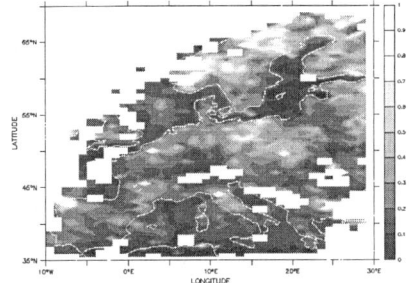

FIGURE 4. Ratio of LOTOS computed sulphate AOD and satellite retrieved AOD.

Also the Ångström coefficients appear quite variable, with relatively high values over polluted areas and relatively low values over adjacent areas. In remote areas the Ångström coefficients may have similar values as in polluted areas. The variations can in part be explained by the contribution of anthropogenic pollution to the submicron fraction, thus locally enhancing the Ångström coefficient in industrialized areas. In contrast, also variations in the natural background aerosol concentrations will occur across Europe, with different particle size distributions resulting from primary emissions and secondary aerosols produced by atmospheric processes contributing to different size fractions. The aerosol optical depth contains information on the column integrated aerosol concentrations, the Ångström coefficients contain information on the shape of the size distributions. Obviously these properties may change with height in the atmosphere, to retrieve them is a challenge.

The AOD values obtained with the LOTOS model and with the dual view algorithm show similar spatial variations but the absolute values are quite different. These differences occur because the dual view algorithm computes the total AOD, i.e. due to all aerosols present in the atmospheric column, including sulphate and nitrate but also organic compounds, dust, sea salt, etc, while the LOTOS model only considers sulphate and nitrate aerosols. The LOTOS results in figure 3 show AOD values for sulphate of up to 0.25, e.g. over Germany and Poland, while AOD values lower than 0.06 are observed over less polluted areas. The ratio of the sulphate AOD to the total AOD, Figure 4, varies from 10% to 80%. Over The Netherlands sulphate aerosol contributes between 30% and 40% to the total AOD. Higher values over, e.g., Germany and Poland suggest a mayor influence of sulphate aerosol particles to the total AOD over these areas. Lower values are observed over, e.g., France and Spain. Similar calculations show that the contribution of nitrate aerosol to the total AOD in August 1997 was relatively small for most areas (less than 10%), except for The Netherlands/Belgium and northern Italy. In these areas nitrate and sulphate have similar contributions. The results for The Netherlands, both as regards the influence of sulphate aerosol and the contribution by nitrate, are in agreement with conclusions derived from surface measurements at two different locations (8, 10).

ACKNOWLEDGEMENTS

The work described in this paper is supported by the Netherlands Space Research Foundation (SRON), contract EO-037, and the Dutch National Remote Sensing Board. The ATSR-2 data was provided by the European Space Agency (ESA).

REFERENCES

1. Charlson, R.J., Schwartz, S.E., Hales, J.M., Cess, R.D., Coackley, J.A.J., Hansen, J.E., and Hofmann, D.J. *Science*, **255**, 423-430 (1992).
2. Veefkind, J.P. *Aerosol satellite remote sensing*. PhD thesis, University of Utrecht (1999).
3. Veefkind, J.P., De Leeuw, G., Durkee, P.A., *Geophys. Res. Lett*, **25**, 3135-3138 (1998).
4. Veefkind, J.P., De Leeuw, G., Stammes, P., Koelemeijer, R.B.A., *Rem. Sens. of the Env*, in press.
5. Robles Gonzalez, C., Veefkind, J.P., De Leeuw, G., *Geophys. Res. Lett.*, in press.
6. Builtjes, P.J.H., *The LOTOS – Long Term Ozone Simulation-project. Summary report.* TNO-report, TNO-MW – R 92/240 (1992).
7. Ackermann, I.J., Memmesheimer, M., Ziegenbein, C., and Ebel, A. *Meteorol. Atmos. Phys.*, **57**, 101-114 (1995).
8. Ten Brink, H.M., Veefkind, J.P., Waijers-Ijpelaan, A., and Van der Hage, J.C., *Atm. Env.*, **30**, 4251-4261 (1996).
9. Kiehl, J.T., and Briegleb, B.P., *Science*, **260**, 311-314 (1993).
10. Diederen, H.S.M.A., Guicherit, R., and Hollander, J.C.T., *Atm. Env.*, **19**, 377-383 (1985).

Features of Aerosol Spectral Optical Depth at a Tropical Urban Environment at Pune

G.R. Aher, N. Shantikumar Singh and V. V. Agashe

Department of Physics, University of Pune, Pune 411 007, India

Abstract: The authors used a sun-tracking multiple wavelength radiometer to study characteristics of atmospheric aerosols from Pune University campus. The study shows that there is a strong influence of weather parameters like relative humidity and surface wind and atmospheric boundary layer processes such as capped inversion and upper air circulation on the temporal variation of the aerosol spectral optical depth. These are described in the paper.

INTRODUCTION

In assessing the radiative forcing due to aerosols, the parameters of importance are the columnar spectral optical depth ($\tau_{p\lambda}$), the size distribution function [$n_c(r)$] and the complex refractive index (μ). Pune is an inland tropical city, surrounded by distant low hills (100-500 m high Western Ghats) on the southern and western sides. At Pune University a sun-tracking multiple wavelength radiometer (MWR) has been in operation since 1986 for the study of atmospheric aerosols. The ground reaching, directly transmitted solar flux is measured in the spectral region 365-1020 nm at 11 narrow wavelength pass-bands from which the columnar total optical depth of the atmosphere ($\tau_{t\lambda}$) is determined following the Langley technique. Columnar aerosol optical depth ($\tau_{p\lambda}$) is deduced from $\tau_{t\lambda}$ at 11 wavelength pass-bands. Columnar spectral optical depth due to aerosols has been the focus of recent studies at this station that attempted to relate it to other variables such as aerosol size distribution and to physical parameters such as temperature inversion, relative humidity (RH), wind velocity and direction. Features of aerosol spectral optical depth, which emerge from this study, are described in this paper.

TEMPORAL VARIATION OF $\tau_{p\lambda}$

It is found that the raw data representing intensity of ground reaching flux of direct solar radiation at Pune, has a tendency to form two groups, one for the forenoon (FN) and the other for the afternoon (AN) part of the observation day. Consequently, the Langley plot of data points consists of two line segments, one for FN and the other for AN, defining the corresponding $\tau_{p\lambda}$(FN) and $\tau_{p\lambda}$(AN). This variation of $\tau_{p\lambda}$ is regarded as the diurnal variation and is related to the diurnal variation of local RH, which is high in the morning than in the afternoon. RH variation is largest in winter, when a plot of $\tau_{p\lambda}$(FN)/$\tau_{p\lambda}$(AN) vs RH defines a third degree polynomial relationship.

At times, data points in the Langley plot show small scatter from the fitted line due to unexplained fluctuations in the transmitted solar flux. On one occasion (December 13, 1999) the data scatter was also accompanied by FN-AN grouping. This case was investigated by calculating instantaneous columnar total optical depths ($\tau_{t\lambda}$) using the modified Lambert-Bouguer equation,

$$\tau_{t\lambda} = \left[\ln\left(I_{t\lambda}/I_{0\lambda}\right) - 2\ln\left(R_m/R\right)\right]/\sec Z$$

Where $I_{t\lambda}$ and $I_{0\lambda}$ represent intensities of direct solar radiation, at wavelength λ, at time t and outside the atmosphere respectively. R_m and R are respectively mean and actual sun-Earth distances at the time of observation. Z is the solar zenith angle at time t. Instantaneous columnar aerosol spectral optical depths ($\tau_{tp\lambda}$) were deduced by subtracting contributions due to molecular scattering and gaseous absorption due to O_3, NO_2 and water vapour from $\tau_{t\lambda}$.

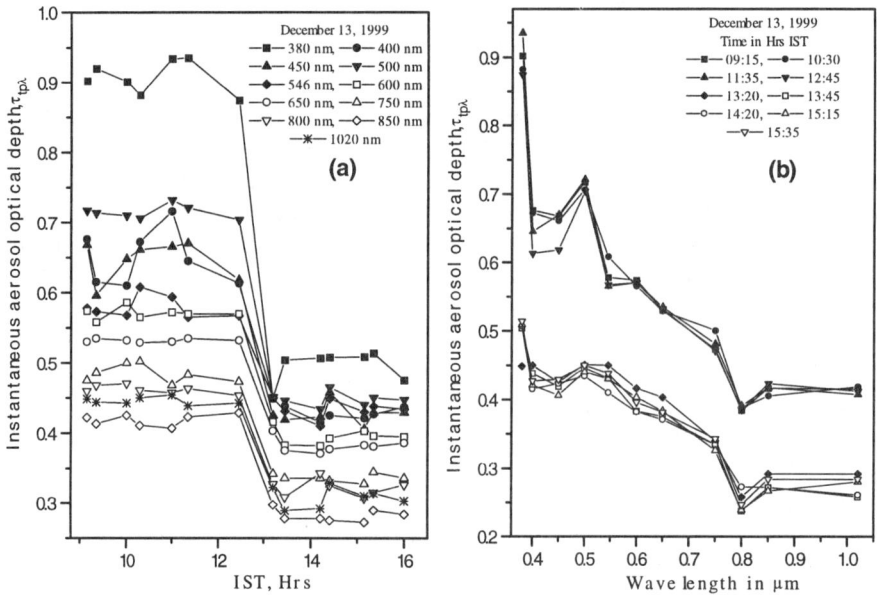

Figure. 1: Diurnal variation of (a) $\tau_{tp\lambda}$ and (b) its spectral variation

The results plotted in Fig. 1(a) for different optical filters prominently show a high value of $\tau_{tp\lambda}$ during FN which decreases steeply to a low value during AN indicating a significant reduction in the aerosol columnar content during AN. The change over from a high $\tau_{tp\lambda}$(FN) to a low $\tau_{tp\lambda}$(AN) occurs rapidly during less than an hour, between 1245 and 1320 hrs IST. Besides the steep decrease, $\tau_{tp\lambda}$ also varies during FN and also during AN. $\tau_{tp\lambda}$ also varies over wavelengths, remaining high for 380 and 500 nm wavelengths compared to other wavelengths, its magnitude decreasing as the wavelength increases. The shift from $\tau_{tp\lambda}$(FN) to $\tau_{tp\lambda}$(AN) also is large for 380 and 500 nm wavelengths and small at other wavelengths. FN-AN grouping of data is prominently seen in Fig. 1(b) showing the spectral variation of $\tau_{tp\lambda}$ at different observation times during the day. Analysis shows that whereas small variations in $\tau_{tp\lambda}$ may be produced by the surface wind and the total columnar water vapour existing in the atmosphere at different times of observation, the high value of $\tau_{tp\lambda}$ during FN is due to the confinement of aerosols due to capped inversion existing in the atmospheric boundary layer (ABL) over the observing station and the steep decrease in $\tau_{tp\lambda}$ from FN to AN is caused by the sudden dissipation of the capped inversion and dispersal of aerosols.

AVERAGE MONTHLY VARIATION

Average monthly variation of $\tau_{p\lambda}$ at different wavelengths during 1996 is shown in Fig. 2(a). The figure shows that $\tau_{p\lambda}$ is low in January, moderately high in February - March, maximum in April and decreasing in May, June and October. April maximum is the normal summer high, which is a global phenomenon attributed to increased aerosol inputs due to surface heating, transport, dust raising winds and mechanical production of aerosols (1, 2). Low value in winter (December – January), also a global feature, is attributed to the decreased aerosol input due to colder ground surface. High value in February – March, although winter months, is characteristic of this station and is due to precipitable atmospheric moisture acting as a source of haze aerosols on cold mornings in winter, which is strengthened during inversion episode (similar to the case discussed above). Occasional showers in May lower $\tau_{p\lambda}$. Decreasing trend from June to October is due to monsoon rainfall.

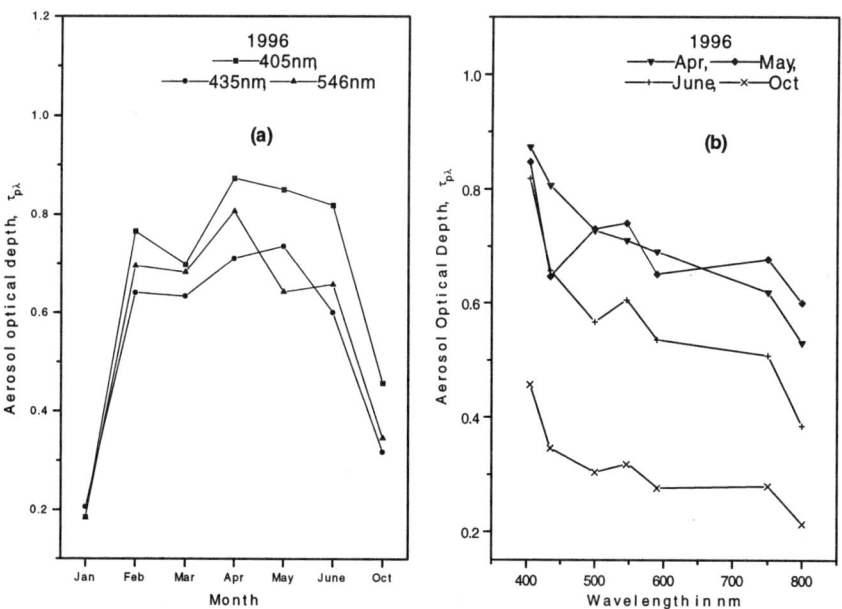

Fig. 2: (a) Monthly Variation of $\tau_{p\lambda}$, (b) Spectral Variation of $\tau_{p\lambda}$

Upper air circulation in the ABL is also found to be acting as a source (influx) or a sink (dispersal) of ambient aerosols thereby influencing the monthly variation of

$\tau_{p\lambda}$. Thus, the anticyclonic circulation in the upper air over the Arabian Sea may bring influx of marine aerosols over Pune in February – March, contributing to the moderately high value of $\tau_{p\lambda}$. In April, upper air data at 850 hPa shows marked discontinuity in the weak westerly wind indicating transition to summer season. A flat pressure gradient around the west coast and the convective activity associated with wind discontinuity affects dispersal of aerosols in the horizontal direction leading to their local confinement thereby increasing aerosol loading producing the 'summer high' in $\tau_{p\lambda}$ at Pune. Pre-monsoon upper air circulation at the top of ABL strengthens dispersal of aerosols leading to a decrease of $\tau_{p\lambda}$ in May and June. Because of low aerosol background in the post monsoon season, $\tau_{p\lambda}$ is seen to be small in October. Due to light upper wind in ABL, the possibility of influx of aerosols at Pune is small in October. This is borne out by the smaller values of aerosol mass loading (m_L) and the weighted mean radius, reff = 0.276 µm, showing dominance of ambient aerosols in October.

SPECTRAL VARIATION AND SIZE DISTRIBUTION

Fig. 2(b) shows the observed spectral variation of $\tau_{p\lambda}$ during the summer (April), and pre- and post-monsoon months (May, June and October) of 1996. The spectral variation is decay type in April. For other months the curves are similar with two superposed small peaks around 546 nm and 750 nm. The corresponding size distributions and their parameters reveal the effect of pre-monsoon weather on aerosols as described above. Size distributions (SDs) for all above months are seen to have same radius range (0.11 to 2.56 µm) for aerosol particulates. However the shape of SDs is slightly different for each month. During April and May SDs are somewhat similar to Junge power law while in June it is of modified Junge power law type and in October it is bi-modal.

CONCLUSION

The studies reveal the strong influence of surface weather parameters and ABL processes on aerosol optical depth at Pune.

ACKNOWLEDGEMENTS

The research project is supported by grants from ISRO under ISRO-GBP. The authors thank the Heads of the Departments of Physics and Space Sciences and the University authorities for their encouragement.

REFERENCES

1. Aher, G. R. and Agashe, V. V., Indian Journal of Radio & Space Physics.**27**, 53-59(1998).
2. Szymber, R.T and Sellers, W. D, J. Clim & Appl. Meterol (USA), **24**, 725-734 (1985).

Aerosol Concentrations and Scattering Coefficient at Mace Head, Ireland

C. Kleefeld[1], S. O'Reilly[1], S.G. Jennings[1], G. Kunz[2], G. de Leeuw[2],
P. Aalto[3], E. Becker[4], and C. O'Dowd[3]

[1]*Department of Experimental Physics, National University of Ireland, Galway, Ireland*
[2]*TNO Physics and Electronics Laboratory, The Hague, The Netherlands*
[3]*Department of Physics, University of Helsinki, Finland*
[4]*Centre for Marine and Atmospheric Sciences, University of Sunderland, England*

Abstract. In June 1999 an intensive measurement campaign, Particle Formation and Fate in the Coastal Environment (PARFORCE), was conducted at the coastal Global Atmospheric Watch (GAW) research station at Mace Head, Ireland. Reported are the results of the aerosol scattering measurements. Aerosol scattering is dominated by the local sea-salt source and this is supported by closure study results.

INTRODUCTION

Scattering of solar radiation by atmospheric aerosol particles affects the Earth's energy balance and contributes to the negative radiative forcing of climate. Since aerosol concentrations are highly variable in space and time, an assessment of the overall impact of aerosols on the climate system requires a characterisation of aerosol properties and their controlling factors on a regional scale. In recent years the importance of sea-salt aerosols for radiative processes in the marine boundary layer has been established (Murphy *et al.*, 1998; Latham and Smith, 1990).

A unique platform for studying marine aerosols is the Global Atmospheric Watch (GAW) research station at Mace Head, located on the west coast of Ireland (53° 19' N, 9° 54' W). A wind direction sector ranging from 180° to 300° opens to the Atlantic Ocean and so the research station can be designated as clean marine in character. A distortion of the wind field is caused by two islands off the coast of Mace Head, situated out about 4 km in a westerly direction and about 3 km in a south-westerly direction, respectively.

The objective of this paper is to identify the relative contributions of local primary and secondary aerosol sources to the observed dynamics of the aerosol scattering coefficient (σ_{sp}). In a closure study the measured σ_{sp} values are subject to a comparison with calculated scattering coefficients derived from aerosol size distribution measurements.

Scattering measurements in Mace Head started in early 1997 and form an important component of the ongoing long-term measurement program. In this contribution, we report scattering measurements performed in June 1999 during the PARFORCE '99 campaign (New Particle Formation and Fate in the Coastal Environment). The intensive campaign provides a broad range of meteorological and aerosol physical parameters which are essential to the interpretation of the σ_{sp} data. Secondary aerosol production and particle growth under low tide conditions is one focus of the PARFORCE project.

MEASUREMENTS

Measurements of total aerosol scattering are conducted at Mace Head by means of nephelometry using a single-wavelength (λ = 550 nm) integrating nephelometer TSI Model 3551. The detection limit of the nephelometer was defined by a signal-to-noise ratio of 2 with a noise level assessed as the standard deviation of the zero baseline measurements over a time period of 24 hours. Thus, the mean detection limit of the scattering measurements calculates to 0.3 Mm^{-1}. Scattering values are reported for relative humidities < 50%. No size segregated sampling of the aerosol was conducted.

Aerosol particle size distributions have been measured during the PARFORCE '99 campaign using a Differential Mobility Particle Sizer (DMPS, size range from 3 to 800 nm, University of Helsinki) and a Forward Scattering Spectrometer Probe (FSSP, size range from 0.5 to 47 µm, CMAS).

Total particulate mass concentrations for particle diameter d < 2.5 µm ($PM_{2.5}$) were measured by means of a Tapered Element Oscillating Microbalance (TEOM).

CALCULATIONS

The closure study was based on the method used by Quinn et al. (1995) with supplements from Sloane (1984) and Quinn and Coffman (1998). A Mie model (Bohren and Huffman, 1983) is used to calculate scattering coefficients from the aerosol particle size distribution measurements. The calculations are performed at a wavelength of 550 nm using a model adapted to account for non-ideal integration and illumination of the nephelometer. The aerosol particles are approximated as spherical and assumed to consist of sea salt, sulphate and black carbon (soot). Both external and internal mixtures of aerosols are considered. The internally mixed aerosols are approximated by concentric sphere scatters with a core of carbon and a water-soluble outer layer. The effect of relative humidity variations on particle size and scattering coefficient is considered.

CONCLUSIONS

In June 1999, the hourly averages of σ_{sp} range from 2.8 to 82.9 Mm^{-1} (see Fig. 1), resulting in a mean value of 15.8 ± 10.0 Mm^{-1}. These scattering values compare very well with similar measurements conducted in the marine boundary layer by Quinn et al. (1998).

A signature of the local sea-salt source can be identified by correlating the scattering coefficients with wind speed. An increase in wind speed triggers an exponential increase of the minimal baseline aerosol scattering level. Increasing the wind speed from 2 to 15 ms^{-1} results in a σ_{sp} enhancement of one order of magnitude.

An investigation of the distribution of σ_{sp} with wind direction reveals that maximum scattering coefficients are confined to a small sector ranging from 160° to 200°. The sector does not represent the dominant wind direction of the measurement period. In this sector, 64% of the σ_{sp} variability can be attributed to $PM_{2.5}$ concentrations, thus indicating that particles < 2.5 µm diameter dominate the scattering. Considering all wind direction sectors, $PM_{2.5}$ concentrations explain about 36% of the variability in the scattering coefficients.

The maximum σ_{sp} values are believed to reflect a geographical effect due to the islands located within the 160° to 200°-sector. LIDAR observations (TNO, The Netherlands) indicate particle plumes coming occasionally off the islands. As an example, Fig. 2 shows a horizontal LIDAR scan of the atmospheric volume backscatter coefficient, coded in false color. The image shows aerosol plumes coming off the islands and extending into easterly directions.

A relation between an increase in σ_{sp} and secondary aerosol production due to particle nucleation and growth at low tide cannot be unequivocally established. Four cases of an increase in aerosol scattering of about 25% could be identified and will be a subject of further analyses.

A closure study performed for a particle size range from 3 nm to 2.5 µm diameter reveals that the measured and calculated values of scattering coefficients agree within 6%, which is within experimental uncertainty. This indicates that the assumption of a sea-salt dominated aerosol used in the closure study is justified.

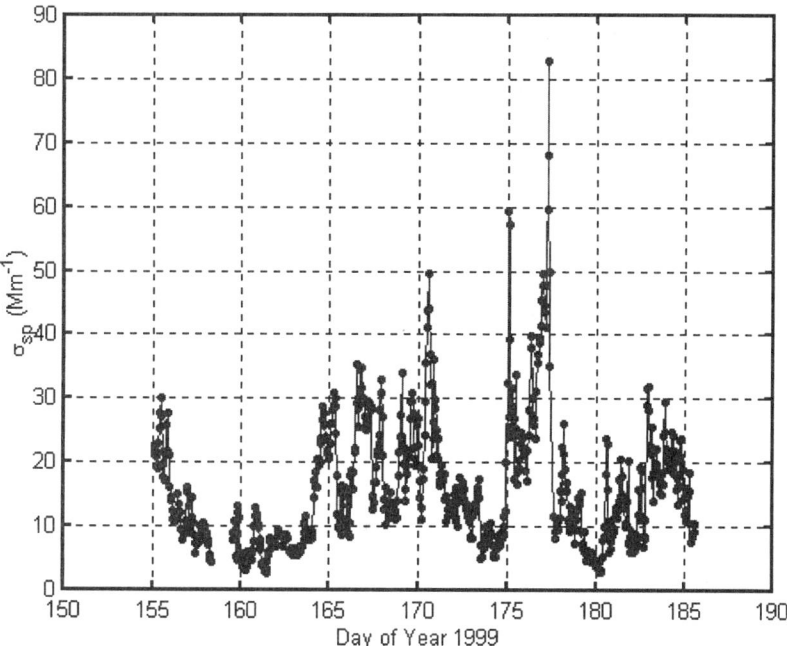

FIGURE 1. Hourly averages of aerosol scattering coefficients measured during the period from 4 June (DoY 155) to 4 July 1999 (DoY 185).

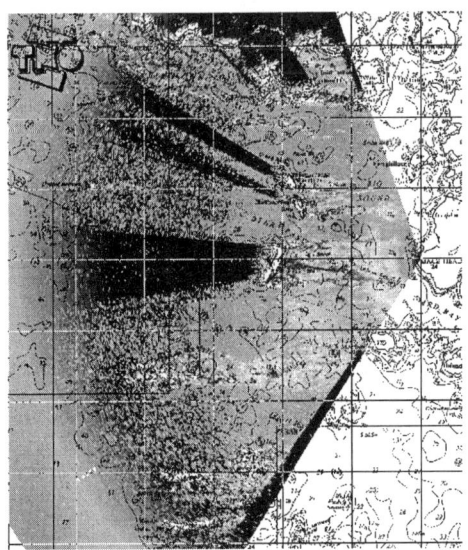

FIGURE 2. Horizontal LIDAR scan, showing the atmospheric volume backscatter coefficient coded in false color (June 26, 1999).

REFERENCES

1. Bohren, C.F. and Huffman, D.R., *Absorption and scattering of light by small particles*, New York: John Wiley, New York, 1983, pp. 477-489.
2. Latham, J. and Smith, M.H., *Nature* **347**, 372-373 (1990).
3. Murphy, D.M. *et al, Nature* **392**, 62-65 (1998).
4. Quinn, P.K. and D. J. Coffman *J. Geophys. Res.* **103**, 16575-16596 (1998).
5. Quinn, P.K., Marshall, S.F., Bates, T.S., Covert, D.S. and Kappustin, V.N., *J. Geophys. Res.* **100**, 8977-8991 (1995).
6. Sloane, C.S., *Atmos. Environ.* **18**, 871-878 (1984)

Parameterization of the Optical Properties of Sulfate Aerosols

J. Li, S. Schmitt
Canadian Center For Climate, Atmospheric Environment Service,
University of Victoria, Victoria, British Columbia, Canada

J.G.D. Wong
Canadian Center For remote Sensing,
Natural Resources Canada, Ottawa, Ontario, Canada

J.S. Dobbie
Atmospheric Science Program, Department of Physics,
Dalhousie University, Halifax, Nova Scotia, Canada

P. Chylek
Atmospheric Science Program, Departments of Physics and Oceanography
Dalhousie University, Halifax, Nova Scotia, Canada

INTRODUCTION

In this work, a general parameterization of aerosol optical properties as functions of relative humidity is presented, including the single scattering albedo, asymmetry, and backscatter optical properties. We offer a novel, rigorous formulation of the optical properties in terms of the dry size distribution. Our parameterization of the optical properties exhibits a simple dependence on relative humidity, which makes it suitable for use in climate models.

GROWTH RATE AND SIZE DISTRIBUTION

The study focuses on sulphuric acid and ammonium sulphate as these compounds are believed to be the dominant sulphate compounds influencing radiative transfer studies.

A useful term for describing the growth of aerosol particles as a function of the ambient relative humidity is the growth factor, η. It is defined as the ratio of the aerosol particle radius, r_d at the specified relative humidity, H to the radius of the corresponding dry aerosol particle r_d. The growth factor depends on a number of things, in particular, the size of the particle and relative humidity,

$$\frac{r}{r_d} = \eta_d(r, H) \quad (1)$$

A value of η corresponds to the growth of a particle from a radius r to a larger radius of $r = \eta r_d$. In view of this, the wet size distribution, n', is related to the dry size distribution, n, in the following way

$$n(r_d) \rightarrow n'(r) = n(r/\eta)\frac{d(r\eta)}{dr} \quad (2)$$

This prescription constrains the number of particles in the interval of r_d to r_d+dr_d to equal the number of wet, growing particles in the interval from r_d to r_d+dr_d. That is, $n(r_d)dr_d=n'(r)dr$.

As a consequence of Eq.(2) the total number of particles are constrained to remain constant during growth. The manner in which the wet size distribution is related to the dry size distribution, shown in Eq.(2), is crucial to the aerosol optical property calculation.

The average of a physical quantity $F(r)$ weighted by a size distribution $n'(r)$ is given by

$$F(r) = \int_0^\infty F(r)n'(r)dr / N_0 \tag{3}$$

To perform two stream radiative transfer calculations, the extinction coefficient, single scattering albedo, and asymmetry factor are required. However, for growing sulfate-water particles the dielectric properties are a function of the dry aerosol size and the amount of growth that has taken place. Fortunately, however, the optical properties can be formulated in terms of the dry radius and the growth factor η.

PARAMETERIZATION OF THE AEROSOL OPTICAL PROPERTIES

For both ammonium sulfate and sulfuric acid, a four-band scheme is presented.

Measurements show that the aerosol distributions usually have effective radii in the range from 0.1 to 1 µm (Lacis and Mishchenko, 1995) and effective variance of 0.2 for a gamma distribution or equivalently an effective variance of 0.693 in log-normal distribution. Compared to the influence of effective radius, the effective variance is believed to have less importance for aerosol radiative properties, especially for a distribution with large effective radius. Chylek and Wong (1995) showed that for the log-normal distribution, different values of effective variance but constant effective radius yield similar results for the back scattered fraction.

To parameterize the optical properties of wet aerosols based on the initial dry aerosol properties, a relationship between WAC and DAC must be established. It will certainly depend on the relative humidity, H, and, at first glance, we might also expect it to depend on the effective radius of the dry aerosol size distribution. We find that is strongly dependent on the relative humidity, H, but largely insensitive to changes in the effective radius of the dry aerosol size distribution. We found differences to be usually less than 2% for different effective radii in the range between 0.1 - 1 µm and relative humidities up to 95%. Therefore, the parameterization of R is taken as independent of effective radius of the original dry particle distribution. We based the parameterization on the averaged value of R for three effective radii used in the parameterization. For $NH_4 2SO_4$, R is parameterized simply as;

$$R = \exp(k_1 H + k_2 H^6 + k_3 H^{12}) \tag{4}$$

This is valid for $H>0.35$, since below the crystallization point the growth rate is set to unity. For H_2SO_4, the growth rate rises very rapidly for very low R. We are not sure of the growth rate below $H<0.05$, but relative humidity in this range seldom occurs in the atmosphere. Therefore, we consider growth of H_2SO_4 for $H>0.05$ and parameterize R as;

$$R = R_5 \exp(k_1 H' + k_2 H'^6 + k_3 H'^{12}) \tag{5}$$

Where R_5 is the value of R at $H=0.05$ ($R_5 = 1.443$) for $HSSO_4$), $H'=H-0.05$. In Eqs.(4) and (5), the terms proportional to H^6 and H^{12} are used because R increases very rapidly when H is close to 100 %.

This parameterization of R can be incorporated in chemical transport models where the transformation from dry source particles to wet particle loading is being considered. This will influence aerosol transport magnitudes since sedimentation for dry aerosols and wet aerosols is different. To date, chemical models ignore this factor.

The specific extinction coefficient, asymmetry factor, and single scattering albedo for each band are parameterized in the following way:

$$\Psi_i = a_1 + a_2 H + a_3 H^8 \qquad (6)$$

$$1 - \omega_i = b_1 + b_2 + b_3 H^8 \qquad (7)$$

and,

$$g_i = c_1 + c_2 H + c_3 H^8 \qquad (8)$$

Terms involving H^8 account for the rapid changes in optical properties for relative humidity close to 100%.

It should be noted that for different effective radius size, the specific extinction can show completely different behavior in response to changes in H.

For increases in H, the specific extinction generally increases for an effective radius of r_e= 0.166 µm, for both bands 1 and 2; whereas, for the larger effective radius, r_e=0.5 µm, the specific extinction for band 1 now decreases in response to increases in H and band 2 shows a significant reduction in slope relative to the smaller effective radius results. Now, the extinction coefficient is the product of WAC and the specific extinction. By comparing the dependence of WAC and specific extinction to increases in relative humidity, we see that changes in WAC will dominate over changes in specific extinction for a variation in H. That is, the extinction will always tend to increase, regardless of any decreases in specific extinction. However, with an increase of H, the extinction coefficient for particles with a smaller effective radius will increase more rapidly compared to particles with a larger effective radius, keeping in mind that WAC was insensitive to changes in effective radius. Therefore, the aerosol radiative forcing is more sensitive to distributions with smaller effective radius for changes of H.

In addition to these trends, it is found that variation rates of the single scattering albedo for band 4 and asymmetry factor for all bands are larger for distribution with smaller effective radius than for the distribution with larger effective radius.

For aerosol radiative forcing calculations, a commonly used radiative transfer scheme is the thin optical depth approximation. In this approximation an important factor is the back-scattered fraction, β. β specifies the fraction of radiation scattered into the backward hemisphere relative to the direction of the incident electromagnetic radiation. Because of its importance, we have parameterized the backscatter fraction, β, in the following way for each band interval,

$$\beta = f_1 + f_2 H + f_3 H^8 \qquad (9)$$

A formulation of the thin optical depth approximation that is commonly used in calculating the aerosol radiative forcing is;

$$\Delta F = -\frac{S_0}{4} T_1^2 (1 - A_c)(1 - R_s)^2 \beta \tau \qquad (10)$$

Where S_0 is the solar constant, T_1 is the transmission above the aerosol layer, A_c is the cloud coverage, R_s is the surface albedo, β is the back scattered fraction, and τ is the optical depth of aerosol layer. As we showed in a previous section, β has a tendency to decrease as H increases, especially for H greater than about 30%, for all cases of dry effective radius size. This result is consistent with larger particles scattering more radiation towards the forward direction than smaller particles do. So, ignoring the variation of will lead to an over estimation of the aerosol radiative forcing.

Therefore, we suggest that the sensitivity of radiative forcing to variations in relative humidity, under the thin atmospheric approximation, should be considered as dependent on a combined factor: βτ, β(H) τ_H -> $f(H)\beta\tau$. From our parameterization, this combined parameter can easily be obtained. We note that for a geometrical aerosol thickness of z the dry aerosol product is $\beta\tau=DACf_{1a1z}$. Therefore, from Eqs. (9) and (10), our $f(H)$ is given by;

$$f(H) = R(1+\frac{a_2}{a_1}H+\frac{a_3}{a_1}H^8)(1+\frac{f_2}{f_1}H+\frac{f_3}{f_1}H^8) \quad (11)$$

We have noticed that there is a stronger radiative forcing dependence on H for sulfuric acid than for ammonium sulfate. It is also found that the radiative forcing is more sensitive to size distributions with a smaller effective radius and longer wavelengths (near infrared region as opposed to visible). For example, at an effective radius of 0.166 μm, the rate of change of radiative forcing to an increase of H for the near infrared band is nearly double that of the same increase in H in the visible band, for large values of H. In the previous studies of aerosol radiative forcing more attention has been given to the visible region; whereas, our work shows that in some circumstances the near infrared is more important. By including the relative humidity variations of β in $f(H)$, the radiative forcing in the visible is decreased. β decreases more rapidly in the visible than for the near infrared because Mie calculations are a function of the size parameter $X=2\pi r\lambda$ and backscatter is most pronounced for small values of X. Thus β for a wet distribution will decrease more rapidly in the visible than in the infrared for the same dry aerosol distribution.

REFERENCES

1. Chylek, J. and J. G. D. Wong, 1998: Erroneous use of the modified Kohler equation in cloud and aerosol physics applications. *J. Atmos. Sci.*, **53**, 1473-1476.
2. Lacis, A. A. and M. I. Mishchenko, (1995): Climate forcing, climate sensitivity, and climate response: a radiative modeling perspective on atmospheric aerosols. In: Aerosol Forcing of Climate, ed. R. J. Charlson and J. Heintzenberg, pp. 11-42. John Wiley & Son Press.
3. Langner, J. H., and H. Rodhe, (1991): A global three-dimensional model of the tropospheric sulfur cycle. *J. Atmos. Chem.*, **13**, 255-263.
4. Li, J. and V. Ramaswamy, (1996): Four stream spherical harmonic expansion approximation for solar radiative transfer. *J. Atmos. Sci.*, **53**, 1174-1186.
5. Li, J. and G. J. Boer, 1999: The continuity equation for the stratospheric aerosol and its characteristic curves, *J. Atmos. Sci.*, **57**, 442-451.
6. Tang, I. N., (1996): Chemical and size effects of hygroscopic aerosols on light scattering coefficients. *J. Geophys. Res.*, **101**, 19245-19250.
7. Toon, O. B. and J. B. Pollack and B. N. Khare, (1976). The optical constants of several atmospheric aerosol species: ammonium sulfate, ammonium oxide and sodium chloride. *J. Geophys. Res.*, **81**, 5733-5748.

Aerosol Pollution of the Atmosphere and Its Influence on the Direct Solar Radiation in Some Regions of Georgia

A. Amiranashvili[1], V. Amiranashvili[1], K. Tavartkiladze[2]

[1] *Institute of Geophysics, Georgian Academy of Sciences, 1. M. Aleksidze Str., 380093 Tbilisi, Georgia*
[2] *Institute of Geography, Georgian Academy of Sciences, 8., M. Aleksidze Str., 380093 Tbilisi, Georgia*

Abstract. The impact of the atmospheric pollution on the direct solar radiation regime in Georgia was studied. As a quantitative parameter for the atmospheric pollution level the optical depth of atmospheric aerosols (τ_a) was used. Computer calculations of the direct Solar radiation using the data on (τ_a) for Georgia showed that the increasing level of the atmospheric pollution causes a reduction in the direct incident Solar radiation irradiance.

INTRODUCTION

Due to the global climate warming investigations of the radiative effects of the aerosol pollution of the atrmosphere are drawing particular attention. In Georgia regular actinometric observations were carried out at six stations till 1990. In Tbilisi (403 m above the sea level) since 1928. At the other 5 stations since the middle of the fifties (Telavi - 568, Tsalka 1457, Anaseuli - 158, Senaki - 40, Sokhumi 116 m above the sea level). In this work the results of the investigations of the long term variations of the mean annual values of atmospheric aerosol optical depth τ_a at the mentioned stations and calculations of the direct solar radiation attenuation caused by the increase of the air pollution by aerosols are presented.

METHODS

τ_a values were calculated by the method (Tavartkiladze, 1989). Using the data of Tbilisi, by means of the method (Obukhov, 1960) the data from the other stations were extrapolated to the period 1928-1990. Finally, by the method (Amiranashvili, 1997) the background (τ_B), anthropogenic (τ_A) and random (τ_R) components of τa were determined. The same work gives a method for the calculation of the direct Solar radiation attenuation.

RESULTS

The variations of the anthropogenic and background components of atmospheric aerosol optical depth τ_{A+B} can be described by the following empiric expression:

$$\tau_{A+B} = \begin{cases} \tau_B & \text{when } t \leq T \\ \dfrac{a}{(58-|1985-t|)^{\frac{1}{2}}} \exp\left(b|1985-t|^{\frac{4}{3}} \right) & \text{when } t \leq T \end{cases} \quad (1)$$

where a,b,T are empiric coefficients, whose values are presented in Table 1, t - years. The values of the anthropogenic (τ_A) and random (τ_R) components of τ_a were easily determined from the following formulas:

$$\tau_A = \tau_{A+B} - \tau_B; \quad \tau_R = \tau_a - \tau_{A+B} \quad (2)$$

Table 2 presents estimated values of atmospheric aerosol optical depth and its anthropogenic (τ_A) and random (τ_R) components in various regions of Georgia in two twenty year periods. According to this table the increase in the aerosol optical depth in 1971-1990 at all six stations was caused mainly by the increase of the anthropogenic component.

Table 3 presents the results of the calculations of the direct Solar radiation attenuation in the ultraviolet (UV), visible (Vis.) and infrared (IR) spectrum ranges of the Solar readiation in various regions of Georgia in 1928-1950 and 1971-1990.

TABLE 1. Values of a,b,T and τ_B

Parameter	Tbilisi	Telavi	Tsalka	Anaseuli	Senaki	Sokhumi
a	1.029	0.890	0.518	0.713	0.739	0.760
b	-0.013	-0.013	-0.010	-0.012	-0.011	-0.011
T	1935	1935	1937	1936	1936	1936
τ_B	0.023	0.015	0.016	0.021	0.021	0.016

TABLE 2. Values of τ_a, τ_A and τ_R for the wavelength $\lambda=1$ mcm in various regions of Georgia in 1928-1950 and 1971-1990

Period	Station	Tbilisi	Telavi	Tsalka	Anaseuli	Senaki	Sokhumi
1928-1950	τ_a	0.046±0.013	0.041±0.012	0.038±0.005	0.041±0.009	0.044±0.008	0.040±0.009
	τ_A	0.004±0.005	0.003±0.004	0.001±0.002	0.003±0.003	0.002±0.003	0.002±0.003
	τ_R	0.019±0.012	0.023±0.011	0.020±0.005	0.017±0.009	0.021±0.007	0.022±0.008
1971-1990	τ_a	0.145±0.023	0.134±0.025	0.079±0.015	0.111±0.025	0.102±0.028	0.106±0.025
	τ_A	0.090±0.013	0.077±0.011	0.035±0.004	0.058±0.007	0.056±0.007	0.060±0.008
	τ_R	0.032±0.018	0.042±0.017	0.028±0.014	0.032±0.021	0.025±0.021	0.030±0.021

TABLE 3. Values of the direct Solar radiation attenuation (Wt/m^2) in various regions of Georgia in two twenty year periods

Period	Station	Tbilisi	Telavi	Tsalka	Anaseuli	Senaki	Sokhumi
1928-1950	UV	65±4	63±3	57±3	65±2	67±2	66±3
	Vis.	137±17	124±12	91±10	140±10	153±11	146±14
	IR	293±8	285±6	241±28	323±6	331±18	331±6
	Total	495±28	472±19	389±36	528±18	551±28	543±21
1971-1990	UV	73±2	70±3	60±3	72±3	73±3	72±3
	Vis.	217±19	193±22	110±8	213±22	217±29	213±24
	IR	327±9	313±10	229±12	351±10	358±13	350±10
	Total	617±29	576±34	399±17	636±34	648±44	635±36

Finally in Table 4 the results of the calculations of the ratio of the aerosol and total atmospheric attenuation of the direct Solar radiation averaged out in the period 1928-1990 to the values in 1928 is presented.

TABLE 4. Increase of the mean attenuation of the direct Solar radiation in 1928-1990 in comparison to 1928 (%)

Attenuation type	Spectrum range	Tbilisi	Telavi	Tsalka	Anaseuli	Senaki	Sokhumi
Aerosol attenuation	UV	268	260	150	226	174	189
	Vis.	276	286	152	231	178	205
	IR	289	302	169	242	188	221
Total atmospheric attenuation	UV	17	18	8	17	14	15
	Vis.	87	90	33	74	61	71
	IR	18	16	4	13	12	9

Thus, according to Tables 3 and 4 in Georgia the increase of the atmospheric pollution in 1928-1990 caused a considerable decrease in the direct Solar radiation irradiance. This effect manifested itself most intensively in Tbilisi and Telavi. The least attenuation of the direct Solar radiation was observed at station Tsalka.

REFERENCES

1. Tavartkiladze, K.A., *Modelling of the Aerosol Attenuation of Radiation and Atmospheric Pollution Control Methods*, Tbilisi: Metsniereba, 1989, p. 204.
2. Obukhov, A. M., *Izvestia AN SSSR, Ser. Geophys.* 3, 432-435 (1960).
3. Amiranashvili, V.A., and Tavartkiladze, K., *Bulletin of the Geogian Acad.Sci.*, **155 N3**, 360-362 (1997).

AN EVALUATION OF CHEMICAL AND SIZE EFFECTS ON RADIATIVE PROPERTIES OF MULTI-COMPONENT AEROSOLS

Shaocai Yu[1], V. K. Saxena[2] and B.N. Wenny[2]

1. Nicholas School of the Environment, Duke University, Durham, NC 27708
2. Department of Marine, Earth and Atmospheric Sciences, North Carolina State University, Raleigh, NC 27695-8208, USA

Abstract. The sensitivity of aerosol radiative properties and radiation transmission to aerosol composition, size distribution, relative humidity (RH) is examined in this paper. Mie calculation and a tropospheric visible radiation model are used. The partial molar refraction method and the volume-average method were used to calculate the real and imaginary components of the refractive index of real aerosols respectively. The sensitivity test shows that both the extinction coefficient and asymmetry factor increase by ~48% with the real component varying from 1.40 to 1.65., and the single scattering albedo decreases 24% with the imaginary component varying from −0.005 to −0.1. Both aerosol radiative properties and radiation transmission are very sensitive to the change in geometric mean radius and geometric standard deviation.

INTRODUCTION

Evaluation of aerosol direct radiative forcing is complicated by the fact that aerosols are highly non-uniformly distributed over the Earth and comprise a variety of chemical species, and their abundance varies with particle size, location and time.[1,2,3] As mentioned by Penner et al. (1994), one of the central scientific questions related to the direct radiative influence of aerosols is how the aerosol composition and size distribution affect the optical depth and radiative properties of aerosols, including the dependence on relative humidity.[4] Up to today the sensitivity of direct aerosol forcing to chemical composition, size distribution and relative humidity on a global scale has been tested with a "reference box model" and a GCM model.[1,2] All of these model studies on direct aerosol forcing only focused on anthropogenic sulfate aerosols. In this study the sensitivity of aerosol radiative properties to size distribution is tested on the basis of the calculation of the particle radiative properties for the accumulation mode only.

RESULTS AND DISCUSSION

Atmospheric aerosol particles are composed of complex mixtures of natural and anthropogenic chemical species that include (1) water-soluble inorganic and organic compounds such as sulfate, nitrate, formate and acetate, (2) water-insoluble inorganic and organic compounds such as elemental carbon, Al_2O_3 and n-alkanes, and (3) water itself. Table 1 lists the chemical components in atmospheric aerosol.

One of the central questions for prediction of radiative properties of aerosol particles is to accurately calculate their refractive index. It has been shown that the partial molar refraction approach is applicable to calculate refractive index for an ionic solid-aqueous electrolyte mixture.[5,6] Table 1 lists the ratios of the partial molar refraction to molecular weight for different components. The average value (1.86 g cm^{-3}) is used as the mean density for water-soluble parts in aerosol particles. The density (1.40 g cm^{-3}) and refractive index (1.55) is used for other organic compounds. The imaginary part of multi-component aerosols was calculated using the volume average of the imaginary parts of refractive index of the individual species. It was found that the complex refractive indices for urban, continental and marine aerosols were $1.575 - 0.027i$, $1.557 - 0.016i$, $1.479 - 0.0027i$ respectively. Hanel (1976) measured the real part of the refractive index for urban aerosols in the city of Mainz, Germany, to be 1.57 ± 0.04.[7] It was found that the growth factors from 30% to 80% RH range from 1.57 to 1.92 (average 1.77 ± 0.12) and from 1.55 to 1.90 (average 1.71 ± 0.11) for scattering and extinction coefficients, respectively. This is in agreement with the criterion value of the hydroscopic growth factor (1.7 ± 0.3) utilized to date as a first estimate in climate change modeling studies.[1]

The parameters used for three types of aerosols are assumed as follows: (1) urban aerosols of Meszaros (1978)[8], N=560 cm^{-3}, D_g=0.100 μm, σ_g =2.0, m=1.590-0.027i, RH=80%; (2) continental aerosols of Hoppel et al. (1984)[9], N=3000 cm^{-3}, D_g=0.080 μm, σ_g =2.0, m=1.564-0.016i, RH=80%; (3) marine aerosols of Gathman (1983)[10], N=67 cm^{-3}, D_g=0.266 μm, σ_g =1.622, m=1.479-0.003i, RH=80%. RH is set to be 80% in the Mie calculation. It was found that the scattering and extinction coefficients increase by about 48% and asymmetry factor decrease by 6% with the real component increasing from 1.4 to 1.65. But the single scattering albedo is insensitive to the change in the real component. The scattering coefficient and single scattering albedo decrease by 18% and 24% with imaginary component varying from –0.005 to 0.10, respectively. The extinction coefficient and asymmetry factor are insensitive to the change in imaginary component. It was found that the scattering and extinction coefficients and asymmetry factor are very sensitive to the change in geometric standard deviation. The scattering and extinction coefficients increase by 389.3 times and 334 times respectively with geometric standard deviation varying from 1.2 to 3.0.

It is interesting to note that the wavelength dependence of aerosol radiative properties is very sensitive to geometric mean radius, and rather complicated. The radiation transmission was calculated for the assumed 2 kilometer aerosol layer by Madronich's Tropospheric Ultraviolet-Visible Radiation Transfer Model (TUVRTM).[11] The sensitivity of aerosol-induced radiation transmission changes at 580 nm is tested under the following constant conditions: date=7/01/1995, O_3=278

DU, ground albedo=0.15, air pressure=940 mb, Latitude=35.63, longitude=82.33, UT=17.90, zenith angle=13.31, the aerosol layer=2km. Three aerosol radiative properties (optical depth, asymmetry factor and single scattering albedo) are needed in TUVRTM model to calculate the aerosol-induced radiation transmission change. Figure 1 shows the sensitivity of radiation transmission to the imaginary component, number concentration and size distribution for three types of aerosols. Note that the radiation transmission at 580 nm is 0.911 without the aerosol layer under the assumed atmospheric conditions. It is interesting to note that the radiation transmission is not sensitive to the changes in the above parameters if total number concentration is small such as that in the case of maritime aerosols of Gathman[10] (total number concentration is only 67 cm^{-3}). The radiation transmission is sensitive to the change in imaginary component and number concentration with the decrease of visible radiation transmission by 2.7% and 4.2% when the imaginary component varies from -0.005 to -0.1 and number concentration from 50 to 4000 cm^{-3}, respectively. The radiation transmission is very sensitive to the changes in geometric mean radius and geometric standard deviation.

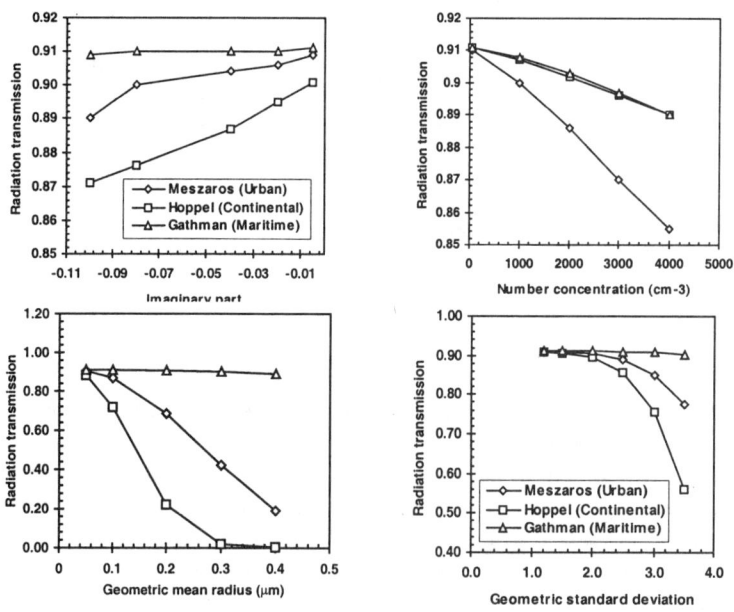

FIGURE 1. The radiation transmission at 580 nm across a 2-km aerosol layer as a function of RH, real and imaginary parts, number concentration and size distribution for three types of aerosols.

TABLE 1. The partial molar refraction ratio of aerosol chemical components. MH (1961) is Moelwyn-Hughes (1961). R_i/M_i is ratio of partial molar refraction and molar molecular weight.

Species	Ri/Mi	Reference	Species	Ri/Mi	Reference
Soluble			$(OOCCOO)^{2-}$	0.165	This study
H^+	0.000	5	$CH_3S(O)_2OH$	0.175	This study
OH^-	0.261	5	H_2O	0.206	6
F^-	0.114	5	**Insoluble**		
Br^-	0.148	6	C	0.176	This study
Cl^-	0.237	6	SiO_2	0.124	6
NO_3^-	0.164	5	Al_2O_3	0.104	6
SO_4^{2-}	0.140	6	Fe_2O_3	0.139	6
Na^+	0.040	6	PbO	0.082	6
NH_4^+	0.271	6	Pb	0.045	6
K^+	0.078	6	n-Hexadecanoic acid	0.305	This study
Ca^{2+}	0.048	6	n-Octadecanoic acid	0.307	This study
Mg^{2+}	0.001	6	Malonic acid	0.166	This study
$HCOO^-$	0.161	This study	Succinic acid	0.205	This study
CH_3COO^-	0.219	This study	Glutaric acid	0.216	This study
$CH_3CH_2COO^-$	0.241	This study	$C_6H_4(COOH)_2$	0.241	This study
$CH_3(CO)COO^-$	0.203	This study	other organics	0.240	This study

ACKNOWLEDGMENTS

This research was supported by the NASA's Mission to Planet Earth (MTPE) under Contract No. NAS1-18944 from Langley Research Center, Hampton, VA and US EPA's STAR (Science to Achieve Results) grant No. R-825248.

REFERENCES

1. Charlson R. J., S.E. Schwartz, J. M. Hales, R. D. Cess, J. A. Coakley, J. E. Hansen, and D. J. Hofmann, *Science* **225,** 423-430 (1992).
2. Kiehl, J.T., and B.P. Briegleb, *Science* **260,** 311-314 (1993).
3. Yu, S., *Atmos. Res.,* in press (2000).
4. Penner J.E., R. J. Charlson, J. M. Hales, N. S. Laulainen, R. Leifer, T. Novakov, J. Ogren, L. F. Radke, S. E. Schwartz, and L. Travis, *Bull. Amer. Meteorol. Soc.* **75,** 375-400 (1994).
5. Moelwyn-Hughes, E.A., *Physical Chemistry,* New York: Pergamon Press, 1961.
6. Stelson, A.W., *Environ. Sci. Technol.* **24,** 1676-1679 (1990).
7. Hanel, G., *Adv. Geophys.* **19,** 73-188 (1976).
8. Maszaros, A., *Atmos. Environ.* **12,** 2425-2428 (1978).
9. Hoppel, W.A., R. Larson, and M.A. Vietti, *Atmos. Environ* **18,** 1613-1621 (1984).
10. Gathman, S.C., *Opt. Engineer.* **22,** 56-62 (1983).
11. Madronich, S., "The Atmosphere and UV-B Radiation at Ground Level," in *Environmental UV Photobiology,* edited by A.R. Young, New York: Plenum Press, 1993.

AEROSOLS AND CLOUD PROPERTIES

Organic Aerosols As Cloud Condensation Nuclei

Timothy M. Raymond and Spyros N. Pandis

Department of Chemical Engineering
Carnegie Mellon University, Pittsburgh, PA 15213

Abstract. The CCN activities of both pure organic aerosol species as well as secondary organic particles created in a smog chamber were investigated. Activation diameters were experimentally determined using a Tandem Differential Mobility Analyzer (TDMA) and a thermal diffusion Cloud Condensation Nucleus Counter (CCNC). Laboratory experiments were performed at 0.3% and 1.0% supersaturations with sodium chloride, ammonium sulfate, glutaric acid, adipic acid, hexadecane, hexadecanol, myristic acid, palmitic acid, and stearic acid. Of the organic species, only glutaric and adipic acid were found to activate at supersaturations less than 1% and at diameters smaller than 250 nm. The secondary organic aerosol produced in the smog chamber by oxidizing α-pinene vapor with ozone was found to have activation diameters of 33 nm and 136 nm at 1.0% and 0.3% supersaturations respectively. Experiments confirmed that Köhler Theory works well in predicting activation of soluble inorganic species while an extension of the theory is necessary to account for the low solubility organic species studied.

INTRODUCTION

Organic matter makes up a large percentage of atmospheric particulate matter. Atmospheric aerosols usually have a mixed chemical composition including a variety of inorganic and organic species. Organic carbon can represent 20-50% of the fine aerosol mass and exists as a complex mixture of hundreds of organic compounds.[1,2] Water condensation on atmospheric particles is strongly affected by the composition, structure, and surface properties of a particle.[3]

The properties of these organic aerosols are not well understood. As much as 60% of the organic particulate matter may be water soluble.[4] Our overall understanding of the system is currently limited by our inability to deal with the complexity of the organic aerosol fraction, and to quantify the interactions between inorganic and organic aerosol components.[5] Some organic species have been shown to activate to cloud droplets while others may or may not hinder activation of otherwise active inorganic particles by means of coating them with an organic layer.[6,7,8]

The first step in studying activation properties of atmospheric aerosols is to understand the individual properties of the pure chemical species that compose them. In the present work, the CCN activity of a number of pure organic aerosol components known to exist in typical atmospheric aerosols has been studied. The substances studied cover a range of biogenic and anthropogenic aerosol components.[9,10]

EXPERIMENTAL PROCEDURE

The basic experimental setup used is shown in Figure 1. The system consists of an aerosol generation system, aerosol conditioning, particle size selection, a system for size and distribution measurement of the final aerosol, and the cloud condensation nucleus counter for assessing activation. For smog chamber studies, the aerosol generation system was replaced with the smog chamber and the particles produced *in-situ* by chemical reaction were fed to the system.

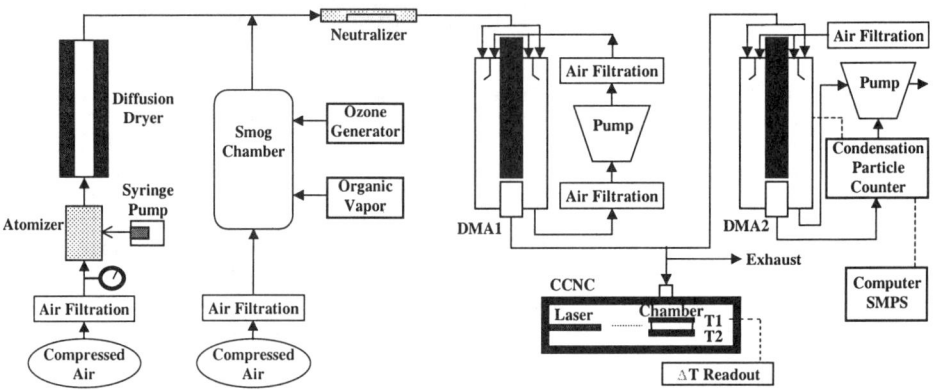

FIGURE 1. Experimental apparatus.

The aerosol generation system atomizes a solution of the desired substance with a high velocity air jet (Constant Output Atomizer 3076, TSI). The solution droplets enter a diffusion dryer and exit as smaller, dry particles with a wide size distribution. The polydisperse dry particles are then charged in a Kr-85 bipolar aerosol neutralizer (Aerosol Neutralizer 3077, TSI) and the aerosol reaches a nearly Boltzmann equilibrium distribution of charges. The aerosol passes into a differential mobility analyzer (DMA 1) where particles of a particular size are 'selected' (3071A, TSI). The particles have a narrow size distribution centering on the selected size. The particles can then be measured by a scanning mobility particle sizer system (SMPS 3934, TSI) and their ability to become CCN by a cloud condensation nucleus counter, or CCNC (M1 model, DH Associates). The CCNC is a thermal diffusion chamber-type counter and works by providing a selected water vapor supersaturation between 0.2% and 2.0% by maintaining a temperature gradient between two moistened plates. A laser is used to illuminate the droplets, and a camera and internal counting mechanism measure the droplet concentration. The SMPS works similarly to the DMA but instead of 'selecting' a particle size, the particles are scanned through the entire size range and counted with a condensation particle counter (Condensation Particle Counter 3010, TSI). A computer controls the particle scanning and counting using software developed for the system (SMPS software version 3.0, TSI). The particles are simultaneously measured in the CCNC for activation.

Comparison of the SMPS and CCNC data results in the percent activation at a given particle size and supersaturation. A sample experimental activation curve is shown below in Figure 2. The activation diameter is indicated as D_{50}, the basis of this method being that at 50% activation, half the particles have a diameter greater than D_{50} corresponding to the actual diameter.[6]

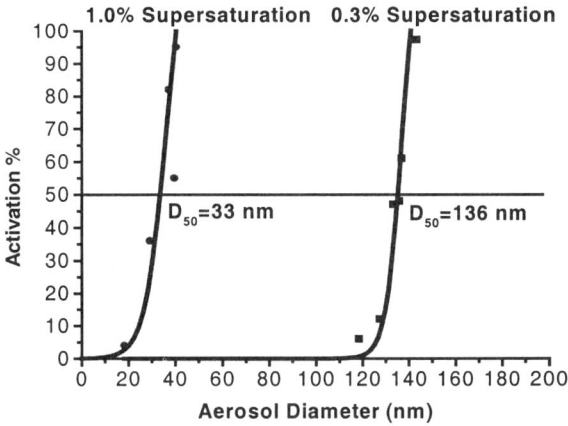

FIGURE 2. Activation results for smog chamber oxidation of α-pinene.

RESULTS AND DISCUSSION

Köhler theory, based on the thermodynamic balance of the competing Raoult and Kelvin effects, can be used to predict the CCN activation of inorganic salts.[11,12] The Köhler theory is limited by its assumption of a constant mass of solute to either soluble inorganic aerosol particles or insoluble matter. For the present work, a modified Köhler equation is proposed taking into account the solubility of the solute, non-ideality of the solution, and the surface tension effect.[13] The equation is given below.

$$S(D_p) = \gamma_w \left\{ \frac{M_s\left[(D_p)^3 \rho_{drop} - (d_s)^3 \rho_s\right]}{M_s(D_p)^3 \rho_{drop} - M_s(d_s)^3 \rho_s + M_w(d_s)^3 \rho_s v \varepsilon_m} \right\} \exp\left(\frac{4M_w \sigma_{sol}}{RT\rho_w D_p}\right) \quad (1)$$

where for any ε_m less than unity

$$\varepsilon_m = \left[\frac{(D_p)^3 C_{sat} \rho_{drop}}{(d_s)^3 \rho_s}\right] \quad (2)$$

and where S is the water vapor saturation relative to a flat pure water surface required for droplet equilibrium, R is the gas constant, T is temperature, σ_w is the air-liquid surface tension, M_w is the molecular weight of water, M_s is the molecular weight of the solute, ρ_w is the density of water, ρ_s is the density of the solute, d_s is the dry particle diameter, D_p is the droplet diameter, and v is the average number of ions into which a solute molecule dissociates. With complete dissolution, ε_m is equal to unity. ε_m is the dissolved mass fraction of the solute particle, γ_w is the water activity coefficient of the solution, and C_{sat} is the saturation concentration of the solute. In the above equation, the properties of the droplet, containing an insoluble part and the aqueous solution, and the solution alone are designated by the subscripts 'drop' and 'sol' respectively. The resulting supersaturation curve has a maximum at the D_p corresponding to the critical droplet diameter and critical supersaturation. Experimentally determined and theoretically calculated activation diameters for two typical atmospheric supersaturations are given in Table 1.

TABLE 1. Experimental and Theoretical Activation Diameters in Nanometers.

Chemical Species	0.3% Supersaturation		1.0% Supersaturation	
	Experiment	Theory	Experiment	Theory
Sodium Chloride	50	49	17	22
Ammonium Sulfate	69	60	26	27
Glutaric Acid	81	72	41	32
Adipic Acid	165	175	98	95
Oxidized α-Pinene	136	-	33	-
Hexadecane Hexadecanol Myristic Acid Palmitic Acid Stearic Acid	No activation up to 1.0% Supersaturation and 250 nm diameter particles in both experiment and theory			

ACKNOWLEDGMENTS

We gratefully acknowledge the National Science Foundation for funding of this work (ATM-9814116).

REFERENCES

1. Saxena, P., and Hildemann, L. M., *Journal of Atmospheric Chemistry* **24**, 57-109 (1996).
2. Andrews, E., Kreidenweis, S. M., Penner, J. E., and Larson, S. M., *Journal of Geophysical Research* **102**, 21,997-22,012 (1997).
3. Kotzick, R., Panne, U., and Niessner, R., *J. Aerosol Sci.* **28**, 725-735 (1997).
4. Kerminen, V., *J. Aerosol Sci.* **28**, 121-132 (1997).
5. Ji, Q., Shaw, G. E., and Cantrell, W., *J. of Geophysical Res.* **103**, 28,013-28,019 (1998).
6. Cruz, C. N., and Pandis, S. N., *J. of Geophysical Res.* **103**, 13,111-13,123 (1998).
7. Corrigan, C. E., and Novakov, T., *Atmospheric Environment* **33**, 2661-2668 (1999).
8. Gill, P. S., Graedel, T. E., and Weschler, C. J., *Rev. of Geophys. and Space Phys.* **21**, 903-920 (1983).
9. Shulman, M. L., Charlson, R. J., and Davis, E. J., *J. Aerosol Sci.* **28**, 737-752 (1997).
10. Limbeck, A., and Puxbaum, H., *Atmospheric Environment* **33**, 1847-1852 (1999).
11. Köhler, H., *Trans. Faraday Soc.* **32**, 1152-1161 (1936).
12. Pruppacher, H. R., and Klett, J. D., *Microphysics of Clouds and Precipitation,* Boston: D. Reidel Pub. Co., 1980.
13. Shulman, M. L., Jacobson, M. C., Carlson, R. J., Synovec, R. E., and Young, T. E., *Geophysical Research Letters* **23**, 277-280 (1996).

Ultrathin Subvisible Cirrus Clouds at the Tropical Tropopause

Th. Peter[1], B.P. Luo[2], Ch. Kiemle[3], H. Flentje[3], M. Wirth[3], S. Borrmann[4],
A. Thomas[5], A. Adriani[6], F. Cairo[6], G. Di Donfrancesco[6], L. Stefanutti[7],
V. Santacesaria[7], K.S. Carslaw[8], A.R. MacKenzie[9]

[1] *Laboratorium für Atmosphärenphysik (LAPETH), ETH, Zurich, Switzerland*
[2] *Max-Planck-Institut für Chemie. P.O. Box 3060, 55020 Mainz, Germany*
[3] *DLR Oberpfaffenhofen, 82234 Wessling, Germany*
[4] *ICG-1, Forschungszentrum Juelich, 52425 Jülich, Germany*
[5] *Institut für Physik der Atmosphäre, Gutenberg Universität, 55099 Mainz, Germany*
[6] *CNR-IFA, Via del Fosso del Cavaliere, 100, 00133 Roma, Italy*
[7] *IROE-CNR "Nello Carrara", Via Panciatichi 64, 50127 Firenze, Italy*
[8] *School of the Environment, University of Leeds, Leeds LS2 9JT, United Kingdom*
[9] *Environmental Science Dept., Lancaster University, Lancaster LA1 4YQ, United Kingdom*

Abstract. Subvisible cirrus clouds with a vertical thickness of only 100-300 m but horizontal extent of thousands of square kilometers have been detected at the tropical tropopause around 17 km altitude during the European-Union-funded APE-THESEO campaign. The cloud layers have been characterized by measurements on board of two aircraft: the Russian high-flying research aircraft Geophysica, which performed *in situ* measurements of the cloud layers; and a German Falcon research aircraft flying up to 13 km altitude and directing the Geophysica into these clouds, which remained invisible for the Geophysica pilot even during level flight within the layer. Both *in situ* and remote measurements suggest that the condensed phase volume ranges between 1 and 5 $\mu m^3 cm^{-3}$. If the particles consisted of water ice, this would correspond to 10-40 ppbv condensed water. Concerning the condensed mixing ratio this would correspond to the thinnest ice cloud ever observed. As a matter of fact, to condense only 10-40 ppbv of H_2O in equilibrium requires temperatures to remain in an interval of 20-70 mK below the frost point. As the cloud layers have an extensive horizontal size this requirement can hardly be satisfied, leading to the conclusion that the clouds are not composed of water ice.

INTRODUCTION

Cirrus clouds in the tropical upper troposphere are receiving increasing attention within several international research projects [1]. Issues of central interest are the clouds' radiative forcing, the mechanism by which they promote the dehydration of upper tropospheric / lower stratospheric air, their global climatology, and their microphysics, in particular their formation process. During APE-THESEO in February and March 1999 on the Seychelles Islands in the western Indian ocean (5°S, 55°E), we have performed *in-situ* aircraft and airborne lidar measurements in the upper troposphere and lower stratosphere at tropical latitudes. The OLEX lidar system (Ozone Lidar EXperiment) on the board of the Falcon aircraft has directed the high flying Geophysica into the regions of interest. *In situ* measurements include besides all the meteorological data (temperature, wind speed and wind direction), the total and gas phase H_2O, the amount of H_2O and HNO_3 in the condensed phase, the particle size distribution, the backscatter ratio of particles within a few hundred meters of the aircraft in the parallel and depolarized channels.

CLOUD OBSERVATIONS AT THE TROPICAL TOPOPAUSE

The extensive OLEX lidar measurements obtained during APE-THESEO deliver a unique overview of cirrus clouds in the tropics. A large variety of cirrus clouds with different spatial distribution and backscatter and depolarizations ratios have been observed. Besides the opaque (visible) and the subvisible cirrus clouds, a hitherto not yet observed ultra-thin class of cirrus clouds was often found as shown in Fig.1. The ultra-thin clouds are characterized by
1. a horizontal extent of several thousands of square kilometers, intermitted by similarly large regions without such layers;
2. a vertical thickness of only a few hundred meters;
3. high horizontal homogeneity as evidenced by the lidar backscatter signal;
4. moderate depolarization ratios in the sharp layers (as the upper layer in Fig.1a) indicating non-spherical, crystalline particles; or vanishing depolarization in the diffuse layers (as the lower layer in Fig.1a) indicating spherical droplets;
5. extremely small amounts of condensed phase material as evidenced by very small backscatter ratios as well as by *in situ* measurements;
6. a vertical positioning of the layers just at or somewhat below the local tropopause (17 to 18 km), i.e. close to the temperature minimum.

Ultra-thin cloud layers have been observed during about half of all occasions, while during the others the tropopause region was either cloud-free or filled with thicker cirrus or other clouds. The aerosol backscatter ratio ($R - 1$) of the layers was always less than 6 at wavelength 1064 nm and less than 0.2 at 532 nm. For comparison, this is smaller than that of typical type 1b polar stratospheric clouds (with $R > 2$ at 532 nm [2]), which consist of ternary $HNO_3/H_2SO_4/H_2O$ droplets and belong to the thinnest, haze-like clouds known at all.

Figure 1a also shows the flight path of Geophysica (thin white line). Comparison with in-situ measurements is made in Fig.1b by showing the backscatter ratio measured by MAS at 532 nm. Geophysica crossed the upper ultra-thin layer several times and the MAS signal shows excellent correlation along the flight route. In Fig.2 the size distributions measured by the FSSP on board Geophysica are shown. The radius of the cloud particles is 3-6 µm with a mean value of ~ 4 µm. The number concentration of particles with $r \geq 2$ µm is $8.3 \cdot 10^{-3}$ cm^{-3}, i.e. about 1:6000 of all particles are in the cloud mode. The cloud mode corresponds to a specific particle volume of 2.3 µm^{3}cm^{-3}. These numbers have large uncertainties as we have only a couple of counts (i.e. bad statistics) and we have to group several size bins together. Irrespective of these uncertainties an important point clearly emerges: the amount of material condensed in the cloud particles is extremely small, corresponding to only about 10 to 40 ppbv of water if the particles consisted of ice.

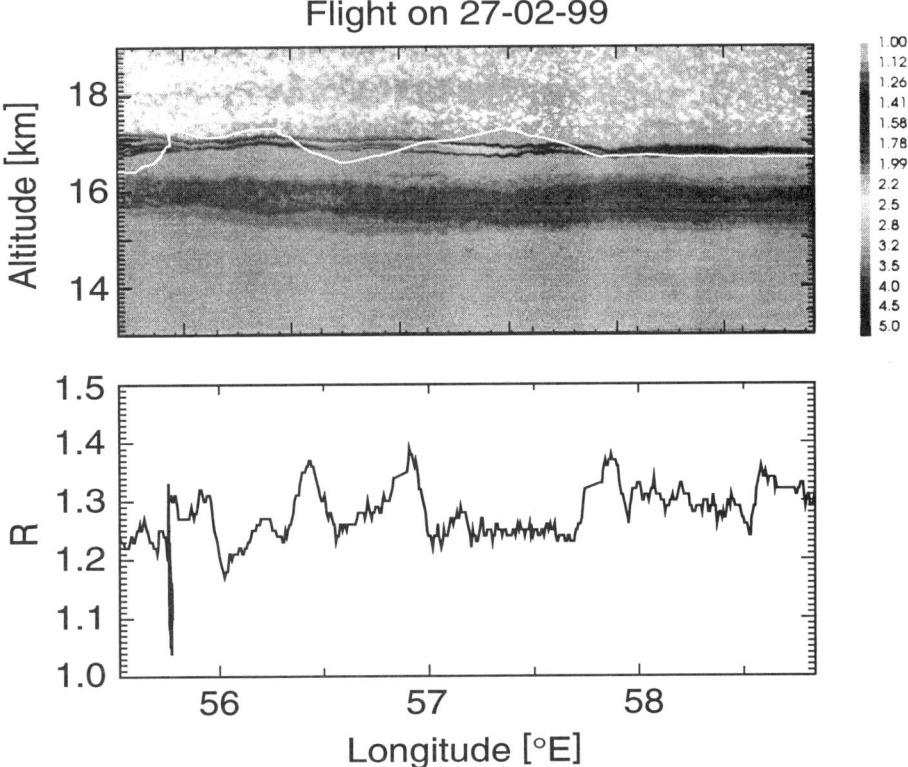

FIGURE 1. (a) Lidar backscatter ratio at 1064 nm wavelength measured by OLEX on 27 February 1999. Two subvisible cirrus layers. Upper layer with sharp contours at local tropopause, depolarizing. Thin white curve marks flight track of Geophysica. Lower diffuse layer without depolarization. (b) Measurement of backscatter spectrometer (MAS) at 532 nm on board Geophysica (not yet corrected for constant offset).

We simulated the lidar backscatter signal at 1064 nm for the upper sharp thin layer in Fig.1a using the T-matrix method [3]. For the total water mixing ratio we may assume 2-3 ppmv H_2O as was measured around the local tropopause by a hygrometer on board Geophysica. Various microphysical assumptions for the particle distribution can are made. The particles in this thin cloud cannot be liquid particles, as depolarization signals has been detected by both the OLEX lidar and the MAS instrument. Thus, the particles must be aspherical, i.e. solid particles. The first obvious guess is that they could be water ice particles. However, the volume of the ice particles and therefore also their backscatter ratio shows a very steep increase when the temperature drops by only a fraction of a degree below the frost point. To be compatible with the observed volume of ~ 2.3 $\mu m^3 cm^{-3}$ and backscatter ratio of ~ 4 at wavelength 1064 nm, the temperature would have to stay in the extremely narrow interval of 0.07 K just below the frost point. When $T = T_{frost} - 0.14$ K, the particles would grow and develop a volume of ~ 5 $\mu m^3 cm^{-3}$ and corresponding backscatter ratio of R ~ 8 within minutes, whereas the slightest increase of the temperature above T_{frost} will make the particles disappear within minutes. Both cases are incompatible with the observations. On the other hand, the 70-mK interval is highly unlikely in view of the vast horizontal extent

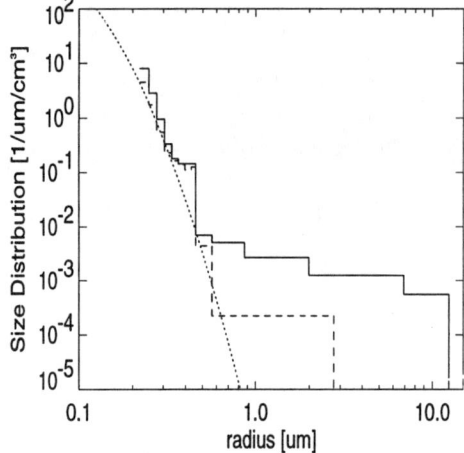

FIGURE 2. The particle size distribution measured by the Forward Scattering Spectrometer Probe (FSSP) on 24 February 1999. Solid line: FSSP particle size distribution while Geophysica was crossing ultrathin cloud at 17 km. Dashed line: the size distribution of the background aerosol measured by FSSP. Dotted line: log-normal distribution with mode radius $r_m = 0.062$ μm, standard deviation $\sigma = 1.58$ and the total particle concentration $n = 50$ cm^{-3}. Calculation of scattering cross sections by means of T-matrix method approximating the particles by spheroids with aspect ratio 0.75. Refraction index assumed to be 1.48 corresponding to nitric acid trihydrate (NAT), see text.

of the cloud layers. The altitude of these clouds changes smoothly by up to 100 m (Fig.1a). This alone corresponds to a temperature change of nearly 1 K if the layer is adiabatically transported. The clouds would show a dramatic response to such temperature changes if they were composed of water ice. In reality, the backscatter ratio does not show a correlation with changes in the layer altitude. Even larger temperature perturbations have to be expected if there are cumulonimbus clouds or the outflowing cirrus decks below the thin layer at the tropopause. Therefore, the total amount of condensed material in combination with the homogeneity of their optical properties are strong arguments against the assumption that the clouds are made of ice particles.

In its presentation at the ICNAA this paper will discuss will examine the possibility that the solid particles are composed of other material than water ice. It is evident that the layer must be related to the special conditions at the local tropopause, most likely to the prevailing extremely low temperatures (- 90°C) and the upwelling motion of the inner tropical convergence zone (ITCZ). Volcanic ashes (silicate particles) or meteoritic material (silicates or metallic particles) and volatile particles like nitric acid hydrates will be considered. Implications for stratospheric chemistry and climate will be discussed briefly.

REFERENCES

1. Wang, P.H., D. Rind, C.R. Trepte, G.S. Kent, G.K. Yue, K.M. Skeens, An empirical model study of the tropospheric meridional circulation based on SAGE II observations, *J. Geophys. Res.* **103**, 13801-13818 (1998).
2. Tsias, A., M. Wirth, K.S. Carslaw, J. Biele, H. Mehrtens, J. Reichardt, C. Wedekind, V. Weiss, W. Renger, R. Neuber, U. von Zahn, B. Stein, V. Santacesaria, L. Stefanutti, F. Fierli, J. Bacmeister, Th. Peter, Aircraft lidar observations of an enhanced type Ia polar stratospheric clouds during APE-POLECAT, *J. Geophys. Res.* **104**, 23961-23969 (1999).
3. Mishchenko M.I., Light scattering by randomly oriented axially symmetric particles, *J. Opt. Soc. Am. A*, **8**, 871-882 (1991).

Large Stratospheric Particles Observed by Lidar During the SOLVE Mission

Beiping Luo[1], Rong-Ming Hu[2], Thomas Peter[3], Kenneth S. Carslaw[2], Chris A. Hostetler[4], Lamont Poole[4], Thomas J. McGee[5], John F. Burris[5]

[1] *Max-Planck-Institut für Chemie. P.O. Box 3060, 55020 Mainz, Germany*
[2] *School of the Environment, University of Leeds, Leeds LS2 9JT, United Kingdom*
[3] *Laboratorium für Atmosphärenphysik (LAPETH), ETH, Zurich, Switzerland*
[4] *NASA Langley Research Center (LaRC), Mail Stop 417, Hampton, VA 23681-2199*
[5] *Laboratory for Atmospheres, NASA Goddard Space Flight Center, Greenbelt, MD 20771, USA*

Abstract. Arctic stratospheric aerosols and clouds have been observed during the SOLVE/THESEO-2000 campaign in the winter 1999/2000 with the newly developed LaRC aerosol lidar with 532 nm parallel and perpendicular and 1064 nm parallel backscattering channels on board the NASA DC-8 research aircraft. The lidar data have been evaluated by means of the T-matrix method for the backscatter by non-spherical particles in the micron-size range. Similar to previous winters several polar stratospheric cloud (PSC) types have been observed: (1) type 2 PSCs which consist of water ice particles (typical particle radii $r = 1$-2 μm); (2) type 1b PSCs which are consistent with ternary solution droplets ($r \approx 0.3$ μm); and (3) enhanced type 1a clouds which are most likely composed of nitric acid trihydrate (NAT) particles ($r = 1.5$-0.3 μm). Besides these well-known PSC types other very thin type 1a clouds have been found, that can be explained only in terms of very large particles ($r > 3$ μm). This finding corroborates *in situ* measurements, which have first detected these particles during the same campaign. The lidar images reveal the full extent of these clouds, typically a 2-3 km thick layer extending over thousands of square kilometers. As these particles sediment with $v > 0.5$ km/day and are most probably composed of NAT, they might lead to substantial stratospheric denitrification with important implications for Arctic ozone loss.

INTRODUCTION

Polar stratospheric clouds (PSC) may affect ozone in two important ways: first, cloud particle surfaces host heterogeneous reactions which convert chemically inactive chlorine species into active ozone-depleting ClO_x. Second, the particles absorb nitric acid and thus reduce NO_x in the gas phase. This is considered an important prerequisite for severe ozone loss due to the suppression of $NO_x + ClO_x$ deactivation reactions. The NO_x reduction may become permanent *via* removal of HNO_3 due to the particles' gravitational settling (denitrification). Denitrification has been observed regularly in the Antarctic, while Arctic stratospheric winters remain often too warm to allow for the formation of sufficiently large HNO_3-containing particles with fall velocities above 100 m/day. Here we describe and interpret the first measurements of a new aerosol lidar system developed by NASA Langley Research Center and operated during the NASA-sponsored SAGE III Ozone Loss and Validation Experiment (SOLVE) and European-Union-sponsored Third European Stratospheric Experiment on Ozone (THESEO-2000) in the winter 1999/2000 on board the NASA DC-8. This lidar measures profiles of the aerosol and/or cloud backscatter ratio at 532 and 1064 nm and the aerosol/cloud depolarization ratio at 532 nm. For the first time lidar measurements indicate the existence of horizontally and vertically extensive regions filled with particles sufficiently large to cause denitrification in the Arctic.

THE MISSION ON 23 JANUARY 2000

The variety of polar stratospheric aerosols and clouds and the existence of extensive air masses filled with particularly large particles is revealed by the lidar measurements carried out during several of the SOLVE DC-8 flights. On 23 January 2000 the DC-8 performed an 8.5-hr flight, see Fig.1, with the first half over Scandinavia (7:50-12:30 UTC) and the second half crossing the northern Atlantic (12:30-17:30 UTC) and the east coast of Greenland (15:30 UTC). Fig.1 displays the aerosol backscatter ratio at 1064 nm (S_{1064}). Not shown are the backscatter at 532 nm and the aerosol depolarization ratio at 532 nm (δ_{532}). Several types of PSCs can be clearly distinguished:

Type 2 PSCs. Clouds with $S_{1064} > 100$ are prominent over the east coast of Greenland (around 15:30 UTC) indicating the formation of ice clouds. As often observed during the entire SOLVE campaign, the east coast of Greenland is the origin of gravity wave activity, which may lead to strong mesoscale cooling to temperatures below the frost point and subsequent ice formation. The aspherical ice particles strongly depolarize the backscattered light ($\delta_{532} > 30$ %).

Type 1b PSCs. Over Scandinavia between 19 and 22 km altitude (< 9:00 UTC) and over the east coast of Greenland in between 17 and 23 km (around 16:00 UTC) clouds with moderate backscatter and very low depolarization ($\delta_{532} < 4$ %). These are the characteristics of liquid clouds, which previously have been identified as ternary solution droplets ($HNO_3/H_2SO_4/H_2O$).

Enhanced Type 1a PSCs. Regions in Figure 1 with similar backscatter ratio S_{1064} as

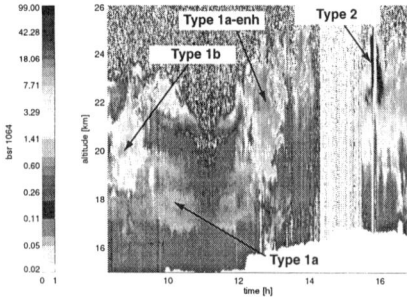

FIGURE 1. Aerosol backscatter ratio at 1064 nm of polar stratospheric clouds during a flight extending from southern Finland to the east coast of central Greenland and back on 23 January 2000 (no data around 15:00 UTC).

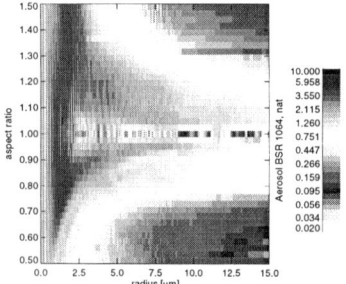

FIGURE 2. T-matrix calculation of backscatter ratio at 1064 nm for spheroidal nitric acid trihydrate particles with 1 ppbv condensed HNO_3. Backscatter scales linearly with condensed HNO_3 mixing ratio.

type 1b clouds but relatively high depolarization ratio ($\delta_{532} > 20$ %) have been encountered above 19 km downstream of the Greenland ice clouds (> 16:00 UTC) as well as over Scandinavia (8:00-9:00 and 12:00-13:30 UTC). Most likely these clouds contain HNO_3 hydrate particles, probably NAT. The type 1a clouds encountered here show a relatively high backscatter ratio and have been termed type 1a-enh [1].

Thin type 1a PSCs. There are extensive regions over Scandinavia between 8:00 and 13:00 UTC (with a gap around 11:00 UTC) at 17-19 km altitude filled by depolarizing clouds with small backscatter ratio ($S_{1064} < 1$). Apart of their widespread occurrence these clouds do not appear to be very dramatic. However, it is shown below that these clouds can be explained only in terms of very large particles.

Background aerosol. On 23 January 2000 there are only a few regions filled merely with background aerosols, e.g. at high altitudes (> 24 km) or at low altitudes over Scandinavia (< 17 km) early in the flight (< 9:00 UTC). Interestingly, the background aerosols at low altitude show a non-negligible backscatter ratio which indicates enhanced H_2SO_4 levels. These might be related to the eruptions of volcanoes (Shishaldin, Aleutian Islands, April 1999; or Gua Gau Pichincha, Ecuador, October 1999).

T-MATRIX METHOD

For spherical particles, the backscatter ratio can be calculated by means of analytical expressions [2]. However, the inversion becomes much more complicated, when the particles are aspherical. First, the shape of the observed particles is usually not known. Second, while the backscatter cross sections of sufficiently large arbitrarily shaped particles can be calculated by means of ray tracing techniques, for micron-sized particles (radius-to-wavelength ratio close to unity) the available techniques apply only to rotationally symmetric particles [3,4].

For simple, rotationally symmetric dielectric bodies the T-matrix method [3] allows to compute the particle backscatter cross sections by rigorously solving the Maxwell equations. Previously we have shown that the optical properties of PSC particles may be successfully approximated by applying the T-matrix method to prolate and oblate spheroids [5]. Figure 2 shows the backscatter ratio at 1064 nm and the depolarization ratio at 532 nm of spheroids of various equivalent-sphere radii and asphericities ranging from aspect ratio 0.5 (prolate, i.e. cigar-like particle, long half-axis twice as long as short half-axis) to 1.5 (oblate, i.e. plate-like particle, long half-axis 1.5-times longer than short half-axis). The results in Fig.2 apply to the index of refraction of nitric acid trihydrate (averaged over anisotropies) and assume that 1 ppbv of HNO_3 have condensed as NAT. The backscatter ratio scales linearly with the mixing ratio of HNO_3 in the condensed phase, while the depolarization ratio is independent of it. Figure 2 can be used directly to interpret the measurements in Fig.1. As an evident example, based on Fig.2 the low depolarization ratios of type 1b clouds observed on 23 January 2000 may be directly identified as spherical particles (white stripe in Fig.2, lower panel).

The size of the cloud particles observed during SOLVE forces us to extend the existing T-matrix method to particles with 15 µm radius, see Fig.2. T-matrix-based integrations of the Maxwell equations are computer time intensive. In particular when the scatterering particles are much larger than the wavelength of the scattered light,

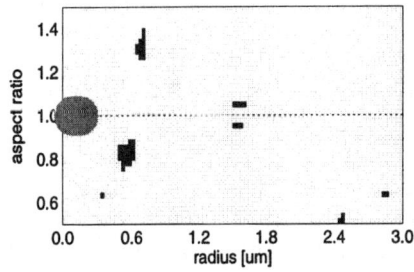 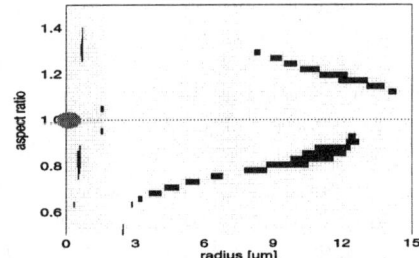

FIGURE 3. Radii (*r*) and shapes of PSC types derived from T-matrix-based evaluation of DC-8 lidar measurements on 23 January 2000 (see Fig.1). Left panel: background aerosol (H_2SO_4/H_2O droplets, $r \sim 0.15$ μm), type 1b ($HNO_3/H_2SO_4/H_2O$ droplets, $r \sim 0.25$ μm), type 1a-enh (NAT, $r \sim 0.5$ μm), type 2 (ice, $r \sim 1.5$ μm). Right panel: extension to larger radii featuring thin type 1a clouds (NAT, $r > 3$ μm).

more precisely when the Mie parameter $x = 2r\pi/\lambda$ is much larger than unity, there is slow convergence of the series of special functions involved in the computations, with simultaneous high-precision requirements. In Fig.2 we encounter Mie parameters up to 90, involving computations of several weeks of CPU time on a modern workstation.

T-MATRIX-BASED DATA EVALUATION

In Fig.3 we display the results of the T-matrix calculations in size/shape phase space for the PSCs observed over the east coast of Greenland on 23 January 2000. The identification of size and shape of cloud particles requires to search for matches between the observed backscatter and depolarization ratios in Fig.1 and those generated in Figure 2. All calculations have been performed assuming lognormal size and (where applicable) shape with lognormal widths 1.1-1.6. This leads to some smoothing of the results, otherwise they are relatively independent of the particular choice of the distribution widths. The most striking result displayed in Fig.3 is the large size of the particles in the optically thin type 1a clouds ($r > 3$ μm). On the other hand, the lidar measurements constrain the condensed mass only by requiring that more than 1 ppbv HNO_3 should be condensed, which in combination with radii in the wide range 3-13 μm yields a poor constraint to particle number densities. The large radii corroborate *in situ* measurements [Fahey, personal communication], which suggest the presence of large HNO_3-containing particles. It is unclear at present whether in previous years similar particles have been observed but not properly identified.

REFERENCES

1. Tsias, A. *et al.*, *J. Geophys. Res.* **104**, 23961-23969 (1999).
2. Born M., E. Wolf, Principles of Optics, Cambridge Uni. Press, 1997.
3. Mishchenko M.I., *J. Opt. Soc. Am. A*, **8**, 871-882 (1991).
4. Macke, A., M.I. Mishchenko, K., Muinonen, B.E. Carlson, *Optics Letters* **20** (19), 1934-1936 (1995).
5. Carslaw, K.S. *et al.*, *Nature* **391**, 675-678 (1998); Carslaw, K.S. *et al.*, *J. Geophys. Res.* **100**, 5785-5796 (1998).

Lack of Closure Between Dry and Wet Aerosol Measurements: Results from ACE-2

J.R.Snider[1], S.Guibert[2], J.L. Brenguier[2]

[1]*University of Wyoming, Laramie, Wyoming;* [2]*Meteo France, CNRM/GMEI, Toulouse, France;*

Abstract. Using data from the ACE-2 field experiment we examine observed and predicted particle size spectra at relative humidities that are representative of the region immediately below the base of marine stratocumulus clouds. The comparison consists of airborne particle size distribution measurements made with an FSSP-300, at ambient relative humidities and north of Tenerife, Spain, and inferred wet aerosol spectra derived using Kohler theory. The theory was initialized with both aerosol size distribution (measured at RH=20%) and aerosol composition measurements made at the Punto del Hidalgo (PDH) site, located on the north coast of Tenerife. The comparison shows that the FSSP-300 reported substantially lower concentrations in the tail of the wet size distribution (wet diameter greater than 0.38 μm) compared to that inferred using the PDH data and Kohler theory.

Introduction

The Cloudy-Column portion of the second aerosol characterization experiment (ACE-2) was an investigation of indirect effect of aerosols on marine stratocumulus microphysical, radiative, and morphological properties [2]. The aerosol indirect effect refers to the response of cloud properties to changing physical and chemical characteristics of the atmospheric aerosol. Cloud droplet number concentration is an important indicator of cloud property changes and is influenced by a subset of the aerosol that is called the cloud condensation nuclei (CCN). Changes in CCN, and therefore droplet concentration, affect both cloud albedo [9] and cloud precipitation efficiency [1]. The transformation of aerosols into cloud droplets is referred to as the activation process. An intermediate step is the deliquescence and growth of the sub-cloud aerosol with increasing relative humidity through the sub-cloud layer. Here we examine observations of marine boundary layer wet aerosol size spectra below cloud base and the degree to which these observations conform to predictions based on ground-based measurements of the dry aerosol size spectra and chemical composition. This particular study is one of the several ways to test the ACE-2 field measurements for consistency. It also facilitates an evaluation of the factors that control the apportionment of water into haze aerosol and the theory that describes that process.

The ACE-2 Measurements

Aerosol physical and chemical properties (APP) were measured at the Punta del Hidalgo (PDH) site located on the north coast of Tenerife [5]. Measurements made at PDH consist of size spectra (6 to 500 nm), measured at 20% relative humidity (RH) using a differential mobility analyser (DMA), chemical composition of particles smaller than 1 µm, and aerosol hygroscopicity [7]. A brief summary of this data set is shown in the first five rows of Tables 1a and 1b. Because the ammonium to sulfate mole ratio is ~2, because sulfate and ammonium dominate the composition, and because the results of Swietlicki et al. [7] show that the PDH aerosol typically exists as an internal mixture of water soluble and insoluble components, we have assumed that individual particles can be modelled as a mixture of ammonium sulfate and insoluble mass. The mixing state was described using the values of the soluble fraction (ε) shown in Tables 1a-1b. These were derived using the PDH data [6]. Water uptake as a function of RH was predicted using the thermochemical data presented in Tang and Munkelwitz [8].

Table 1a – Measurements from PDH (rows 4-6) and concentration comparisons (7-10)					
Date	June 21	June 24	June 25	June 26	July 4
Airmass	Interm.	Clean	Clean	Clean	Clean
nss-SO4, µg m^{-3}	0.9	0.2	0.2	0.2	0.3
NH_4/SO_4 mole ratio	1.8	2.0	1.9	2.0	1.0
ε, soluble fraction	0.5	0.7	0.8	0.5	0.7
N_{PCASP}/N_{PDH} (D>0.10 µm)	1.1	NA	NA	NA	NA
N_{FSSP3}/N_{PDH} (D>0.38 µm)	0.3	0.3	0.3	0.4	0.3
N_{FSSP3}/N_{PDH} (D>0.43 µm)	0.2	0.2	0.3	0.4	0.2
N_{FSSP3}/N_{PDH} (D>0.48 µm)	0.1	0.2	0.3	0.3	0.3

Table 1b – As in Table 1a but for polluted and intermediate airmass conditions at the end of ACE-2					
Date	July 7	July 9	July 17	July 18	July 19	July 21
Airmass	Polluted	Polluted	Polluted	Polluted	Polluted	Interm.
nss-SO4, µg m^{-3}	5.3	2.8	1.6	3.1	1.4	0.6
NH_4/SO_4 mole ratio	1.7	1.7	1.7	1.7	1.1	0.6
ε, soluble fraction	0.8	0.8	0.6	0.7	0.9	0.8
N_{PCASP}/N_{PDH} (D>0.10 µm)	NA	NA	0.6	0.6	0.4	2.2
N_{FSSP3}/N_{PDH} (D>0.38 µm)	0.2	0.2	0.3	0.2	0.1	0.2
N_{FSSP3}/N_{PDH} (D>0.43 µm)	0.1	0.2	0.3	0.2	0.1	0.2
N_{FSSP3}/N_{PDH} (D>0.48 µm)	0.1	0.2	0.2	0.2	0.1	0.2

Airborne measurements of the sub-cloud aerosol during ACE-2 were obtained from the Meteo France Merlin. With the exception of July 21, all of the Merlin measurements were made north of PDH. We consider data obtained from two optical particle counters operated on the Merlin: 1) the FSSP-300 for measurements of the sub-cloud aerosol (0.38<D<21 µm) at ambient RH, and 2) the PCASP for measurement of dried aerosol (0.1<D<3 µm) at a temperature at least 15 °C warmer than ambient [6]. The FSSP-300 data was conditionally sampled to exclude measurements made above cloud base, at less than 87% RH, and at RH larger than 97%. We utilized FSSP-300 sizing thresholds that correspond to the refractive index of pure water (n=1.33).

Data Analysis

The PDH APP data are the most complete for the ACE-2. However, those measurement systems sampled at a height of 55 m, while the cloud base altitudes observed by the Merlin were between 650 m and 1400 m. In addition, most of the flights were performed at a distance of 60 to 120 km (up to 250 km on 7 July) north of PDH. We addressed these issues by comparing dry aerosol size distributions measured by both Merlin PCASP and the PDH DMA. An example is shown in Figure 1. In Tables 1a and 1b we report the concentration ratios (Merlin PCASP divided by PDH). For 17, 18, and 19 July the PCASP concentration is ~50% of that measured at PDH. The large concentration ratio observed on July 21 is attributed to island pollution, and on 21 June the PCASP was configured differently so that the possibility of incomplete particle drying can not be excluded. The results presented here suggest that dry particle concentrations at PDH exceeded those observed by the Merlin by about a factor of two. Factors contributing to this disparity will be discussed.

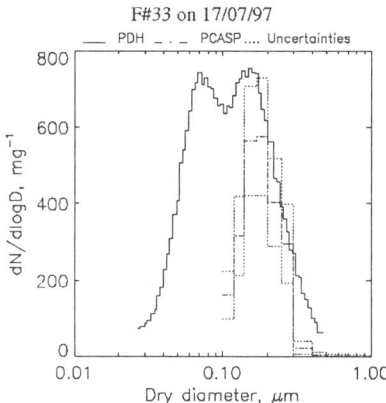

Figure 1: Size distributions of dry aerosol measured at PDH (thick line) and on the Merlin with the PCASP (dot-dashed line). Variability in the PCASP measurement is indicated by ± one standard deviation.

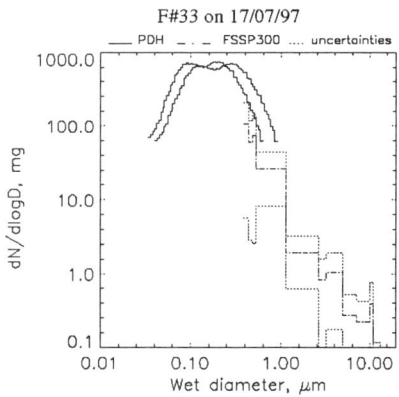

Figure 2 – Predicted (87 and 97% RH; thick lines) and observed size spectra (87 to 97 % RH, dot-dashed line). Variability in the FSSP-300 measurement is indicated by ± one standard deviation.

The size spectra and composition measured at PDH were used with Kohler theory to calculate the wet particle PDH spectra. The calculations are compared to the FSSP-300 measurements in Figure 2 and in the final three rows of Tables 1a-1b. The concentration ratios (N_{FSSP3}/N_{PDH}) were calculated by integrating the FSSP-300 spectra from 1 μm up to the specified limits (Tables 1a-1b) and then normalizing by the PDH result evaluated at RH=92%. There is a clear disparity between the two results with the FSSP-300 reporting lower concentrations. Since there is evidence that the Merlin sampled lower accumulation mode aerosol concentrations than the PDH (cf. Figure 1), we cannot exclude the possibility that the disparity shown in the final three rows of Tables 1a-1b may have resulted from the fact that the two measurement platforms were not collocated. However, after correcting for that difference the N_{FSSP3}/N_{PDH} ratios still remain substantially below unity. Hence, the possibility of either measurement bias or inaccuracies in Kohler theory must be addressed.

Conclusions

The results presented here suggest a disparity between measured and inferred particle concentrations at sizes larger than 0.38 µm. Plausible explanations for this disparity include measurement error and invalid assumptions associated with the theory used to transform spectra, expressed as a function of dry size, to spectra expressed as a function of wet particle size. The latter possibility is consistent with the findings of McInnes et al. [4] who show that an unusually small diameter growth factor (1.1 at RH=85%) is needed to reconcile dry size spectra and relative increases in scattering extinction with increasing RH. If this possibility holds true for the ACE-2 data set, it would corroborate the biases seen in attempted CCN closure studies [3][6][10], and would also help to guide continued investigations of these related phenomena.

References

1. Albrecht, B. A., 1989: Aerosols, cloud microphysics, and fractional cloudiness. *Science*, 245, 1227-1230.
2. Brenguier, J.L., P.Y.Chuang, Y.Fouquart, D.W.Johnson, F.Parol, H.Pawlowska, J.Pelon, L.Schüller, F.Schröder, and J.R.Snider, An overview of the ACE-2 CLOUDY COLUMN closure experiment, in press *Tellus*, 2000
3. Chuang, P.Y., D.R.Collins, H.Pawlowska, J.R.Snider, H.H.Jonsson, J.L.Brenguier, R.C.Flagan, and J.H.Seinfeld, CCN measurements during ACE-2 and their relationship to cloud microphysical properties, in press *Tellus*, 2000
4. McInnes L., M. Bergin, J. Ogren, and S. Schwartz, Apportionment of lightscattering and hygroscopic growth to aerosol composition, Geophys. Res. Lett., 513-516, 1998.
5. Putaud, J.P., R.Van Dingenen, M.Mangoni, A.Virkkula, and F.Raes, Chemical mass closure and origin assessment of the submicron aerosol in the marine boundary layer and the free troposphere at Tenerife during ACE-2, *Tellus*, in press, 2000
6. Snider, J.R. and J.-L. Brenguier, A comparison of cloud condensation nuclei and cloud droplet measurements obtained during ACE-2, in press *Tellus*, 2000
7. Swietlicki, E., et al., Hygroscopic properties of aerosol particles in the north-eastern Atlantic during ACE-2, in press *Tellus*, 2000
8. Tang, I.N., and H.R.Munkelwitz, Water activities, densities, and refractive indicies of aqueous sulfates and sodium nitrate droplets of atmospheric importance, J.Geophys.Res., 99, 18801-18808, 1994
9. Twomey S., 1977: The influence of pollution on the shortwave albedo of clouds. J. Atmos. Sci., 34, 1149-1152.
10. Wood, R., et al., Boundary layer and aerosol evolution during the third lagrangian experiment in ACE-2, in press *Tellus*, 2000

Observations of the Evolution of Particulate in an Urban Plume and Following Interaction With Cloud

Keith N. Bower[1], M.J.Flynn[1], T.W.Choularton[1], R.A.Burgess[1], H.Coe[1], E.Swietlicki[2], B.Martinsson[2], J.Zhou[2], A.Wiedensohler[3], W.Birmilli[3], K.Müller[3], and A.Berner[4].

[1] *The Physics Department, UMIST, PO Box 88, Manchester M60 1QD, UK.*
[2] *Division of Nuclear Physics, Lund University, PO Box 118, S-22100, Lund, Sweden.*
[3] *Institute for Tropospheric Research, Permoserstr.15, D-04318 Leipzig, Germany.*
[4] *Institut fur Experimentalphysik, Universitat Wien, , Strudlhofgasse 4, A-1090, Wien, Austria.*

Abstract. An investigation into the characteristics of aerosol produced in city center Manchester, and the evolution of such aerosol as they advect out of the city and pass over the Pennines. Measurements were made of aerosol properties within the city and at three sites to the east of the city, including a site at the top of Holme Moss which is often in cloud. These measurements showed strong correlation between particle number and traffic flow within the city. Outside the city as the airmass advected over the Pennines it was observed that the particles were a mixture of sulphates, nitrates, chlorides, and organics, and that most were only weakly hygroscopic. Hence despite the high number of aerosol present, few were effective as CCN and the number of cloud droplets formed was limited in most cases to about 600cm^{-3}. This was in contrast to the ACE-2 experiment where a much more aged plume from Western Europe was sampled on Tenerife, and droplet numbers of upto 2800cm^{-3} were observed. More work is needed to establish the details of the aging process responsible for these changes in hygroscopicity.

INVESTIGATIONS CARRIED OUT IN THE MANCHESTER URBAN PLUME

Introduction

A number of experiments have been conducted to investigate the formation of particles in the atmosphere in the centre of Manchester, their evolution as the plume in which they reside moves out from the city, and the subsequent interaction of these particles and gases in the urban plume with cloud. It is to be expected that as the plume is advected away from the city that the new particles will evolve by coagulation, deposition of gases onto their surfaces, and by chemical reactions at the surface or within the body of the particles. If they are (or become) sufficiently soluble they may subsequently act as cloud condensation nuclei (CCN).

In Spring 1997 and 1999 experiments were carried out at the UMIST field site at Holme Moss (a hilltop in the Southern Pennines, to the east of the city of Manchester), to investigate the interaction of the Manchester urban plume with the hill cap cloud which often forms at the site (as moisture laden air carried in the prevailing westerly wind, is forced to rise as it passes over the elevated terrain). The 1997 experiment was a pilot study, but in 1999 a collaborative field project was carried out by four international groups at and around Holme Moss. This multi national experiment was conducted under the auspices of the EUROTRAC II subproject PROCLOUD. Collaborating institutes included the University of Lund, Sweden (LUND), the University of Vienna, Austria (IEP), and the Institute for Tropospheric Research in Leipzig, Germany (IFT).

Aims and Methodology

The aims of the 1999 PROCLOUD experiment were to examine the effects of the properties of the aerosol and trace gases in the urban airmass flowing into cloud on the microphysics and chemistry of the cloud and to investigate the role of the cloud in modifying the properties of the aerosol and trace gases emerging downwind of the cloud system.

To achieve the stated aims, measurements were made at 4 ground based sites. Site 1 was located on a rooftop at UMIST in central Manchester while site 2 was located approximately 17 km to the east of the city, in the outer suburbs away from major local pollution sources. Site 3, which served as the in-cloud site, was on the top of Holme Moss (located approximately 25 km ENE of the city center). The fourth site was a further 3 km downwind (in the prevailing SW winds), to the east of the ridge, allowing study of material emerging from the cloud. At each site measurements were made of the aerosol size distribution. Additionally at site 2, aerosol hygroscopic growth factors as a function of size were measured. Measurements of the size resolved aerosol chemistry were made at sites 2 and 4 while at site 3, measurements of cloud microphysics and cloud water chemistry were made in addition to those of the interstitial aerosol. Gas phase species pertinent to the evolution of the aerosol population were measured at the relevant sites.

Results

The city of Manchester itself was a major source of aerosol. Figure 1 presents aerosol and gases as a function of time of day averaged over a 1 month period in late autumn and shows that aerosol in the accumulation mode rises sharply with NOx and CO (not shown on figure) during the morning peak period. This trend is also followed by the recently formed particles. Most of these particles are probably associated with the combustion in motor vehicle engines.

SO_2 concentrations were typically 2 ppbv to 5 ppbv and it is evident that ultra-fine particles are produced at a maximum rate around the middle of the day when photochemical activity is a maximum. Ammonia concentrations in central Manchester were high, typically 5 to 10 ppbv with a maximum of 17 ppbv. It is likely, therefore, that nucleation of particles through the sulphuric acid ammonia route is significant, especially around the middle of the day although organic compounds may also have been involved. It is also evident that some ultra-fine particle production occurs throughout the 24 hour period although the concentrations of such particles are much lower at night than during the day. The mechanism of formation of these particles is not known.

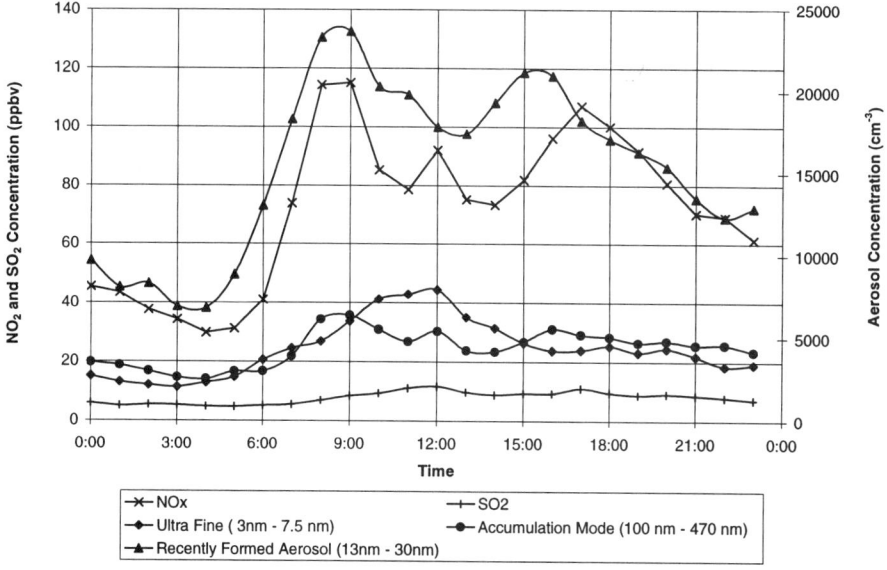

FIGURE 1. Plot of Monthly Averages of Gas Phase and Aerosol Data as a Function of Time of Day, for City Center Manchester.

The particles advected out of the city, which are formed in the urban area consist of a mixture of soluble and insoluble material. The soluble portion was found to consist of sulphates, nitrates and chlorides, while it was expected that the insoluble material would consist of organics. The insoluble organic compounds dominate the aerosol mass and the hygroscopic growth factor of the vast majority of the particles in the urban plume (90% by number) is very small. This result was obtained from the operation of a hygroscopic tandem Differential Mobility sizing Instrument operated by the group from the University of Lund.

Figure 2 shows the relationship between cloud droplet number concentrations measured in the cloud at the summit of Holme Moss during the 1999 campaign and the number of particles greater than $0.1\mu m$ diameter entering the cloud. It can be seen that during periods with low aerosol number the droplet number closely tracked the aerosol number. These were periods with low NOx and CO and were periods when highly polluted air was not reaching the site. These were mostly at night when the urban emissions were a minimum. During periods when the urban plume with its high aerosol loading entered the cloud (clearly visible in the figure) only a small increase in the droplet number was observed for most of the time. This result is consistent with the measurements of the hygroscopic properties of the aerosol discussed above.

It is to be noted, however, that this is very different to the results obtained from a similar experiment conducted on Tenerife as part of ACE-2. In this experiment a very aged aerosol plume from Europe was studied and it was found that most of the particles had become highly hygroscopic and that the droplet number concentration in the cloud was nearly equal to the number of aerosol particles larger than 0.042μm diameter with a maximum number concentration of 2800cm^{-3}. This is also shown in figure 2.

During the Holme Moss project a few cases were observed where more aged urban aerosol distributions were reaching the measurement sites, possibly from Ireland, it was found that these were more soluble, and that the soluble fraction of these particles contained more ammonium sulphate than those originating from Manchester. However the full details of the mechanisms responsible for this plume ageing need further investigation.

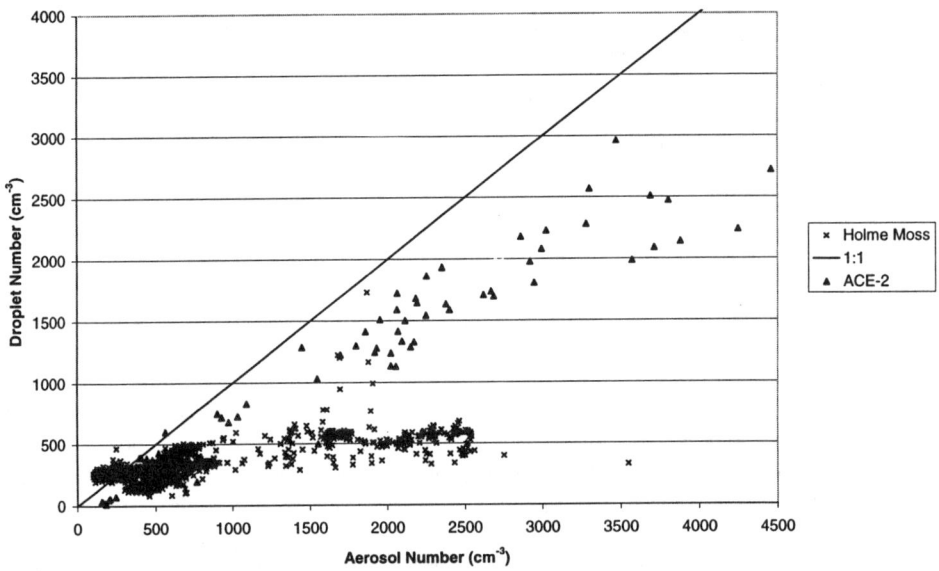

Figure 2. Relationship Between Cloud Droplet Number and Aerosol Number for Holme Moss, and ACE-2. Holme Moss data is FSSP droplet number plotted against downwind ASASP-X aerosol number >0.1μm, ACE-2 data is DAA droplet number plotted against upwind DMPS aerosol number >0.042μm.

ACKNOWLEDGMENTS

This work was supported by the United Kingdom Natural Environment Research Council.

Aerosol Fluxes Over an Urban Canopy

J.R. Dorsey, E.G. Nemitz, M. Theobold, P.I. Williams, D. Fowler, M.W. Gallagher

Physics Dept., UMIST, UK

Abstract. Total aerosol number fluxes in the size range $11nm < D_p < 3000nm$ have been measured over a large UK city. Measurements were made using the eddy covariance technique, with a condensation particle counter as the aerosol detector. Size resolved fluxes and concentrations were also taken for aerosol larger than 100nm. Supporting measurements of micrometeorology, aerosol size spectra and trace gas concentrations were also made. Results are discussed in terms of meteorological conditions, aerosol dynamics and anthropogenic activity.

URBAN AEROSOL MEASUREMENTS

Aerosols produced within cities have a variety of effects upon both the urban environment and downwind areas. Because they appear at high concentrations at all sizes in cities, aerosols are thought to be linked to high asthma incidences. Thus the most important immediate effect of urban aerosols is on human health. Particulate pollution is transported out of cities in the urban plume, and can be deposited at great distances downwind into potentially sensitive ecosystems.

In this work, aerosol fluxes and concentrations over a city were measured directly for the first time, using the eddy covariance technique. Measurements were made from the top of a tower in the east end of Edinburgh city centre. The work was conducted as part of SASUA, an ongoing campaign based project aimed at quantifying the Sources and Sinks of Urban Aerosol.

MEASUREMENT TECHNIQUES

Total aerosol number flux was determined using a condensation particle counter (CPC) adapted at UMIST for high frequency use. The greater size range of the CPC ($11nm < D_p < 3000nm$) meant more particles were detected by this system than by conventional instruments. This both helped with the statistical requirements of the eddy covariance method, and allowed the fluxes of fine mode aerosol to be investigated. A similar system has been successfully used before over a Scots Pine forest (Buzorius et. al., 1998).

Size segregated fluxes of aerosols larger than 100nm were measured using conventional optical particle counters (PMS model ASASP-X), and the behavior of fine mode aerosols could be studied by subtracting this accumulation mode flux from the total flux. Size spectra for aerosol between 5nm and 450nm at ten-minute time resolution were also measured. Trace gas data from a nearby long term monitoring station and traffic census data were made available by the city council. The results have been interpreted using these supporting measurements.

RESULTS

Data were taken over a wide range of meteorological conditions during May and October 1999, and over different surface types depending upon the wind direction. These surfaces included the city centre to the west, suburbs to the north-east, and an area of parkland to the south-east.

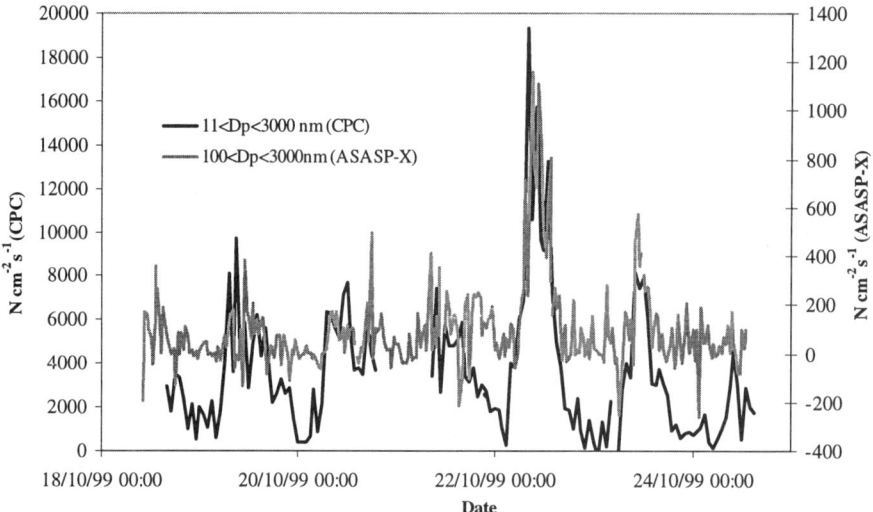

FIGURE 1. Total and accumulation mode aerosol number fluxes during October 1999.

Figure 1 shows total and accumulation mode aerosol fluxes over part of the October 1999 field campaign. The total number flux is roughly an order of magnitude higher than the accumulation mode flux. This reflects the fact that the majority of urban aerosol are smaller than 100nm. The much higher variability of the total flux may be due to the fact that new particles produced within the city are in the fine mode, and possibly do not grow into the accumulation mode until they have been advected out of the city. For example, a diurnal cycle can clearly be seen in the total flux trace. This is mainly caused by the sharp increase in aerosol concentration associated with the morning and evening rush hours. Another factor is the increase in emission velocity which occurs during the day.

Figure 2 shows the aerosol emission velocity and concentration on an average day. Data used to produce this plot were taken from the May and October 1999 field campaigns. Note that, on average, the city emits aerosol all the time. Net deposition events are rare, and never very strong. The emission velocity peaks during the day, as solar heating makes the boundary layer less stable.

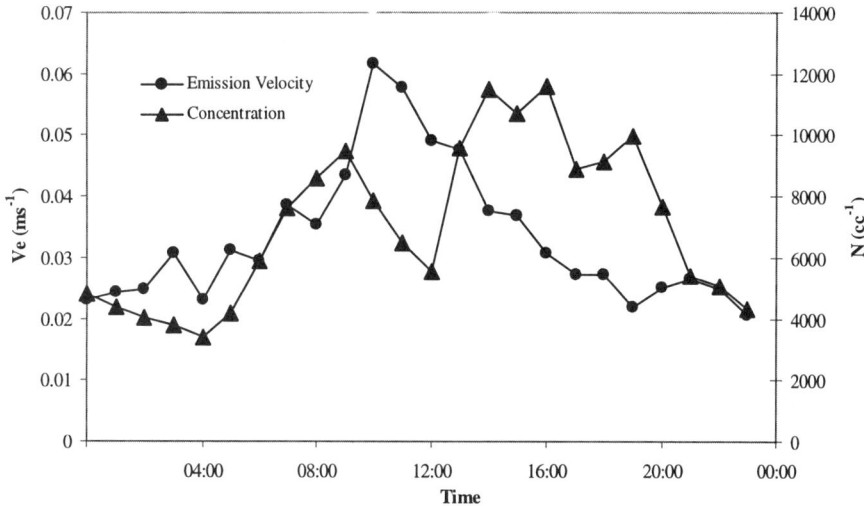

FIGURE 2. Average diurnal emission velocity and concentration.

The concentration trace clearly shows the effect of the morning and evening rush hours. The reason for the second peak in the evenings at around 19:00 is less clear, as the traffic flow reliably peaks before that time. However, it is noted that NO_x concentration often does show a peak at that time of day. This may indicate secondary particle production from gases produced during the rush hour.

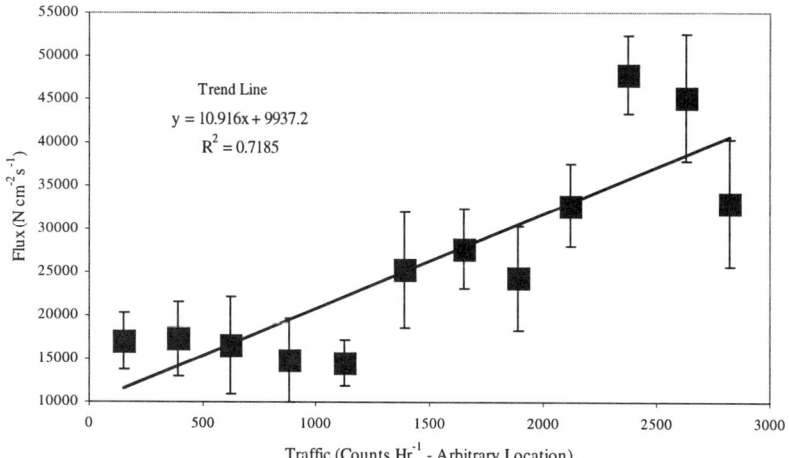

FIGURE 3. Total aerosol flux ($11nm < D_p < 3000nm$) against traffic flow in the city.

The concentration trace clearly shows the effect of the morning and evening rush hours.

Figure 3 shows the relationship between the aerosol flux and traffic flow in the city. Again, the data used in this plot were taken from the May and October 1999 campaigns. The traffic activity was measured at only one point in the city, but a linear relation between census points across the city has been found. Although the data shows a lot of variability, a trend of increasing flux with increasing traffic flow is visible. It is hoped that analysis of a larger data set, stratified by stability and wind direction will allow a better parameterization of the relationship.

CONCLUSIONS

Aerosol fluxes over a variety of sizes have been directly measured over a city for the first time. A relationship between total aerosol number flux and traffic activity in the city has been demonstrated. The mechanisms behind aerosol concentration and fluxes have been investigated. It is hoped, in future, to use this data and further measurements to parameterize the net flux from the city in terms of quantities such as trace gas concentrations and anthropogenic activity. These parameterizations will be used in a model and tested against observations.

ACKNOWLEDGMENTS

SASUA is supported by NERC under the URGENT thematic programme (Grant 022244). The help of Edinburgh City Council in providing additional data and infrastructure is gratefully acknowledged.

REFERENCES

1. Buzorius, G., Rannick, J. Mäkelä, J. M., Vesala, T. and Kulmala, M. (1998). *J. Aerosol Science* **29**, 1/2, 157-171.

Cloud Liquid Water Content Responses to Hygroscopic Seeding of Warm Clouds

S. S. Kandalgaonkar, G. K. Manohar and M. I. R. Tinmaker
Indian Institute of Tropical Meteorology, Pune 411 008

Abstract. The cloud liquid water content (CLWC) data in time and space from a total of 96 pairs of target (T) and control (C) experiments were analyzed in this study to compare the responses of CLWC to hygroscopic seeding of warm clouds. Our results of various approaches taken for this analysis have indicated significant modifications in the CLWC for the T clouds as against C clouds. Analysis of changes in CLWC in the T clouds after and before seeding have pointed out their increasing trend of values with increment in the number of seeded traverse in most cases. These results have shown that CLWC in the T clouds increases following the seeding treatment in the range 9-26%. Similar comparisons in the C clouds have indicated obvious diminution in CLWC that lies in the range 5-11 %. These results are the clear indications of influence on microphysical growth and decay of such clouds that arises from hygroscopic seeding and not seeding respectively of warm clouds. Analysis of spatial responses of CLWC to seeding has shown that the optimum effect of seeding may be achieved for a suitable cloud in the altitude range 5750-6250 ft. (a.s.l.) in the Pune area. It is believed that this study has provided adequate support in favour of the hypothesis of hygroscopic seeding of warm clouds.

INTRODUCTION

A well designed randomized warm cloud modification experiment was carried out in Maharashtra State during 11 summer monsoon season (1973-74, 1976, 1979-1986). Some results in the early years of these experiments have appeared in the literature (Kapoor et al., 1975; Ramachandra Murty et al., 1975; Parasnis et al., 1982). The objective of this study is to examine the response of CLWC (cloud liquid water content) to seeding during target and control experiments of six year period (1981-1986). The physical process involved in the initiation and development of rain in warm clouds are condensation collision-coalescence and breakup. The concept of warm cloud modification to increase rainfall, is based on the modification of rain processes through seeding the clouds with either hygroscopic material thereby tapping the potential precipitation efficiency of the cloud systems. The measurements of CLWC were carried out in not seeded (control) and seeded clouds were used for documenting the warm cloud responses to hygroscopic seeding. The results of this study are presented in this paper.

AREA, DESIGN, SEEDABLE DAYS AND SEEDING TECHNIQUE

The experimental area is located in the western ghats in the Deccan plateau region at about an altitude of 550 m. It is about 40 km east of Pune. The freezing level of the clouds in this area during summer months is about 6 km and majority of clouds do not reach higher than 5 km. Hence the dominant rain forming process in these clouds is the collision-coalescence and break-up. The cross over design having two sectors with a buffer between has been adopted. The area of each sector is 1600 km^2.

The classification of seedable days has been based on forecast of amount of low clouds, wind, current weather conditions and radiosonde observations carried out at Pune a few hours before the commencement of actual seeding.

The aircraft used for seeding was a Dakota (DC-3) which was fitted with seeding equipment and several instruments. CLWC was measured by employing a Johnson Williams (J-W) hot wire liquid water content meter. It measures the CLWC that is contributed by cloud droplets of diameters $\leq$ 30 µm. During an experiment CLWC meter reading were continuously monitored.

DATA

The experiment was performed during the summer monsoon months (June-September) for six year period (1981-1986). During this period actual experimental days were 96 (see Table 1) but the Target / Control operations were performed only on 88 / 82 days as CLWC data on remaining days were not available due to some technical difficulty.

RESULTS AND DISCUSSION

Temporal Variation of CLWC

Table 1 gives the year-wise (1981-1986) mean values of CLWC and their standard deviation for traverses 1-5 of Target / Control experiments. This data have been analyzed in four different ways to study the response of CLWC due to seeding.

(i)   % change in CLWC in Tr. IV with respect to Tr. III
(ii) % change in CLWC in Tr. V with respect to Tr. III
(iii) % change in CLWC in Tr. V with respect to Tr. I
(iv) % change in CLWC in Tr.(3+4+5) with respect to Tr.(1+2+3)

The results of the above four categories suggests that on an average CLWC in Target cloud showed increase in the range 9 to 26%, while in control cloud decrease in CLWC was in the range 5 to 12% respectively. Thus the net response of CLWC to hygroscopic cloud seeding was positive. This particular result is also verified by examining the positive or negative response of CLWC on individual experimental day on which target and control operations was carried out. This examination showed that out of 88 target clouds cases on 55 target cloud showed 29% increase in CLWC while 33 target clouds showed 14% decrease in CLWC. The result of control cloud cases was almost in the opposite way compared to the target clouds.

TABLE 1: Year-wise (1981-1986) mean values and their standard deviation for the traverse 1-5 of CLWC for the target and control experiments.

Year of Experi-ment	No. of Experimental Days	Average cloud liquid water content (gm m^{-3}) in Target and Control Traverses				
		Target Traverse No.				
		1	2	3	4	5
1981	10	.26 (0.09)	.31 (.14)	.36 (.20)	.31 (.13)	.35 (.18)
1982	17	.45 (.17)	.44 (.18)	.43 (.21)	.48 (.25)	.56 (.29)
1983	15	.45 (.14)	.41 (.16)	.43 (.15)	.46 (.14)	.53 (.21)
1984	23	.48 (.17)	.48 (.19)	.47 (.23)	.53 (.24)	.64 (.42)
1985	18	.47 (.11)	.53 (.13)	.48 (.11)	.55 (.17)	.53 (.13)
1986	13	.42 (.13)	.46 (.15)	.40 (.14)	.49 (.20)	.56 (.25)

Year of Experi-ment	No. of Experimental Days	Average cloud liquid water content (gm m^{-3}) in Target and Control Traverses				
		Control Traverse No.				
		1	2	3	4	5
1981	10	.36 (.24)	.35 (.12)	.32 (.24)	.30 (.09)	.28 (.14)
1982	17	.44 (.21)	.42 (.22)	.40 (.15)	.39 (.15)	.33 (.16)
1983	15	.56 (.34)	.54 (.21)	.48 (.23)	.46 (.17)	.44 (.18)
1984	23	.35 (.16)	.42 (.22)	.43 (.17)	.40 (.17)	.41 (.23)
1985	18	.50 (.20)	.53 (.21)	.52 (.18)	.49 (.20)	.45 (23)
1986	13	.39 (.13)	.39 (.10)	.34 (.13)	.36 (.11)	.38 (.10)

Spatial Variation of CLWC

Figure 1 gives the cloud number density for each altitude thickness of 250 ft. a.s.l and % response of CLWC for target and control clouds cases. The CLWC responses for each T and C cloud at these altitude interval was worked out by considering the % change in traverse V with respect to traverse III and shown in the Figure 1. A curve of best fit is drawn through these data points to describe the average altitudinal distribution of clouds in the Pune region. The curve shows three peaks of which peak around 6500 ft is more pronounced. The result of the altitudinal response of CLWC shown in this figure is self clear. The target clouds in the altitude range 5750 ft to 6250 ft (asl) offer optimum results of cloud seeding but not so far in control ones.

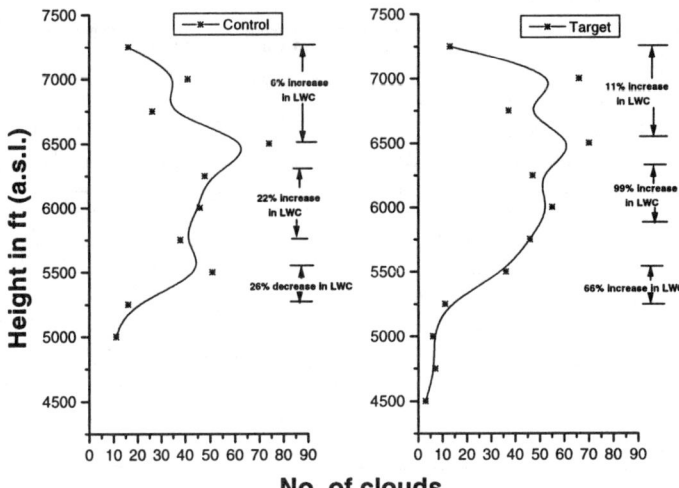

Figure 1. Cloud number density in altitude thickness of 250 ft above 4250 ft a.s.l. up to 7500 ft a.s.l. and CLWC during target and control clouds.

CONCLUSIONS

1. Analysis of changes in CLWC in target clouds after and before seeding suggested the increasing trend of values with respect to increment in the number of seeded traverses in most cases. The increase in the target clouds is about 9-26%. Similar comparison for control clouds indicated obvious diminution in CLWC that lies in the range 5-11%.

2. Spatial variation of CLWC has shown that optimum effect of seeding may be achieved for a suitable cloud in the altitude range 5750-6250 ft (asl) in the Pune area.

ACKNOWLEDGEMENTS

The authors are thankful to the Director, IITM, Pune for conducting this experiments. The authors are also thankful to Dr. A.S.R. Murty for his constant encouragement during the period of this study.

REFERENCES

1. Kapoor, R.K., Paul, S.K., Ramachandra Murty, A.S., Krishna, K., Ramana Murty, Bh.V., 1975. Study of drop size distribution in warm clouds subject to repeated seeding. *J. Weather Modification*, **7**, 116-126.
2. Ramachandra Murty, A.S., Selvam, A.M. and Ramana Murty, Bh.V., 1975. Dynamic effect of salt seeding in warm cumulus clouds. *J. Weather Modification*, **7**, 31-43.
3. Parasnis, S.S., Selvam, A.M., Ramachandra Murty, A.S., Ramana Murty, Bh.V., 1982. Dynamics responses of warm monsoon clouds to salt seeding. *J. Weather Modification*, **14**, 35-37.

Variations of Atmospheric Aerosols Inside and Outside Cloud Air

Suvarna Kandalgaonkar and M.I.R. Tinmaker
Indian Institute of Tropical Meteorology, Pune 411 008, India
e-mail : sskandal@tropmet.ernet.in

Abstract. Aitken Nuclei are present in the size range of 0.001 to 0.1 μm in appreciable concentrations in the continental and maritime environments. The main sources of these nuclei are mainly trace gases and they form out of gas to particle conversion which can occur through the nucleation of aerosol from the supersaturated gases and by the photochemical reactions associated with the absorption of solar radiation by molecules. Gas to particle conversions may be enhanced by high relative humidity and the presence of liquid water. During the summer monsoon months of 1980-1982 warm cloud modification experiment was conducted by IITM at Pune. During this experiment Aitken nuclei observations were made inside and outside the stratocumulus and cumulus clouds at the same altitude during three monsoon seasons. The results of the study suggested that observed higher concentration of Aitken nuclei inside the cloud may be due to more active gas-to-particle conversion process than in the cloud free air.

INTRODUCTION

Atmospheric aerosols are a complex mixture of solid and / or liquid particles suspended in the gaseous medium (which is generally air). The size of the aerosols vary in the range about 10^{-4} μm to tens of micrometers and it depends upon the location of measurements. Atmospheric aerosols play an important role in the control of many atmospheric processes like visibility, radiation balance, atmospheric electricity, air pollution etc (Junge, 1963). Many cloud modification experiments were performed to study the effect of cloud seeding on cloud electrical, dynamical and microphysical parameters. These experiments showed that initial spectrum of cloud droplet sizes is of fundamental importance for the further growth of cloud and it ultimately depends upon the number concentration of Aitken nuclei. The main sources of these nuclei are mainly trace gases and they form out of gas-to-particle conversion which can occur through the nucleation of aerosol from the supersaturated gases and by the photochemical reactions associated with the absorption of solar radiation by molecules. Gas-to-particle conversions may be enhanced by high relative humidity and the presence of liquid water. The presence of AN acts as the primary source for cloud nucleation which is very critical and a vital factor for the further development of cloud. With this view in mind, the observation of AN were carried out inside and outside the cloud air during the warm cloud modification experiment conducted by IITM over the Pune region.

DATA AND MEASUREMENTS

During the summer monsoon months of 1980-1982 warm cloud modification experiment was conducted at Pune. During this experiment a portable expansion counter is used (Khemani et al., 1985) to measure the AN concentration in the size range of 0.001 to 0.1 µm. The nuclei counter is volume controlled and therefore can be used for aircraft measurements in the lower atmosphere. With this counter the observations of AN were taken inside and outside the stratocumulus and cumulus clouds at the same altitude during the three monsoon season (1980-1982).

RESULTS

Table 1 gives the average concentration of AN inside the clouds and cloud free air during the three monsoon seasons. From this table it is seen that inside the cloud the concentration is observed to be higher (6.3×10^4 cm^{-3}) than the air outside the clouds (3.3×10^4 cm^{-3}). Observed higher concentration of AN inside the cloud suggests that the gas-to-particle conversion process are more active inside the cloud due to the presence of liquid water and high relative humidity (Tinmaker, 1994).

TABLE 1. Average concentrations of Aitken Nuclei ($N \times 10^4$ cm^{-3}) inside clouds and cloud free air during three monsoon seasons at 1.5 km level

Year	No. of Observations	Inside clouds	Cloud-free air
1980	16	7.4	4.2
1981	28	7.1	3.9
1982	25	4.4	1.9

CONCLUSIONS

1. The average concentration of AN has been found to be significantly higher inside the clouds (6.3×10^4 cm^{-3}) than in the air outside the clouds (3.3×10^4 cm^{-3}).

2. The higher concentration of AN observed inside the clouds suggest that gas-to-particle conversion processes are more active inside the cloud than in the cloud-free air.

ACKNOWLEDGEMENTS

The authors are thankful to the Director, IITM, Pune for conducting this experiments. The authors are also thankful to Dr. A.S.R. Murty for his constant encouragement during the period of this study.

REFERENCES

1. Junge, 1963 : *Air chemistry and radioactivity*, New York and London, Academic Press, 382 pp.
2. Khemani, L.T., Momin, G.A., Naik, Medha S., Kumar, R. and Murty, Bh.V.R., 1985 : Observations of Aitken nuclei and trace gases in different environments in India, *Water, Air and Soil Pollution,* **24**, 131.
3. Tinmaker, M.I.R., 1994 : *Physical aspects of the precipitation processes in monsoon cloud*, M.Sc. Thesis, University of Pune, Pune, 179.

Tower Based Measurements of Micrometeorological Exchange Parameters and Heat Fluxes Above a City

J.R. Dorsey, E.G. Nemitz, M.W. Gallagher, M. Theobold, D. Fowler

Physics Dept., UMIST, UK

Abstract. Measurements of micrometeorological parameters over the City of Edinburgh have been made using the eddy covariance technique. Preliminary values for urban heat fluxes are presented, and vertical momentum transport is discussed. Certain patterns are visible in the behaviour of friction velocity and sensible heat flux, but more data will be required to properly establish values for the key parameters.

MEASUREMENTS

The micrometeorology of urban areas has not been extensively studied in contrast to other surface types. However, an understanding of the urban boundary layer is important in connection with the transport of air pollution within and out of cities. In the current work, measurements were made from a tower 70m above street level in the city of Edinburgh in the UK. The work was carried out as part of a project seeking to quantify the sources and sinks of urban aerosol by the use of micrometeorological techniques (SASUA).

The instruments were mounted on top of the Nelson monument, a 35m tower on Calton Hill in the east end of the city. This made the measurement height 70m above street level. A Gill Solent HS sonic anemometer was used in conjunction with a Campbell Scientific KH2O Krypton hygrometer to make the measurements. The high frequency (> 10Hz) data were used to calculate parameters by eddy covariance. Data were taken during two field campaigns, in May and October 1999.

RESULTS

Figure 1 shows the frequency of occurrence of each wind direction during the May and October 1999 field campaigns. Data are averaged over one hour, and sorted into bins of 5°. The three prevailing directions are approximately south-west, South-East and North-East. To the South-West of the measurement site is the centre of the city of Edinburgh. To the North-East there is a small park and several suburbs. To the South-East is predominantly parkland.

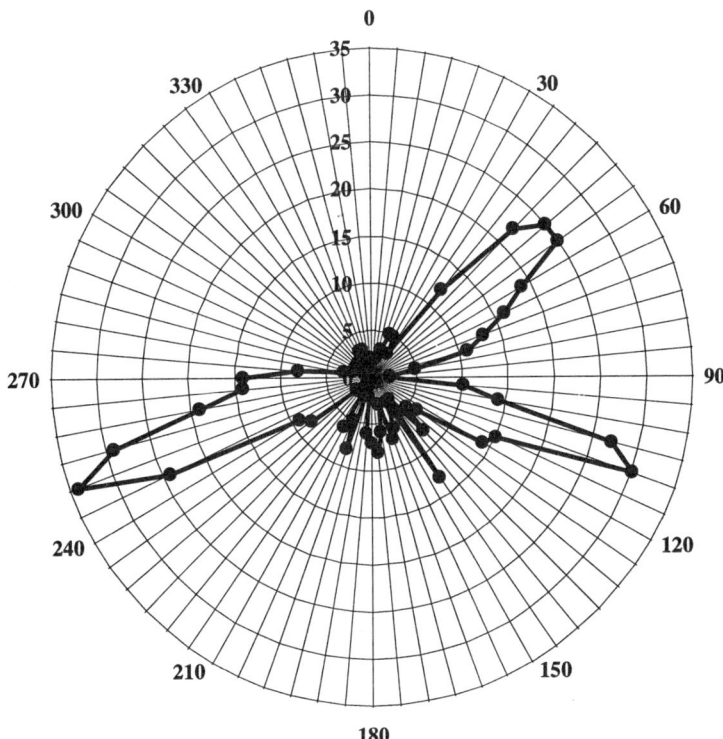

FIGURE 1. Histogram of hourly averaged wind direction over the May and October 1999 campaigns.

The directions shown in figure 1 put the measurement footprint over different surface types. However these are all representative of some common aspect of city surfaces. In this analysis the data are treated together to give preliminary values averaged over all urban surface subtypes.

Figure 2 shows the average diurnal heat fluxes over the city for the same period. The latent heat flux exhibits a clear peak during the day. The low value of this peak may be explained by the overcast conditions often encountered in Edinburgh. There is no obvious clear pattern to the latent heat flux. The pattern for latent heat flux is controlled by more variable factors, such as precipitation, rather than solar forcing.

It may be possible to gain further insight into the energy balance of the city when more data are available. Separation of 'wet' and 'dry' days could help refine the estimate for latent heat flux. Separation of data into different wind directions could also clarify the sensible heat flux from different surface types.

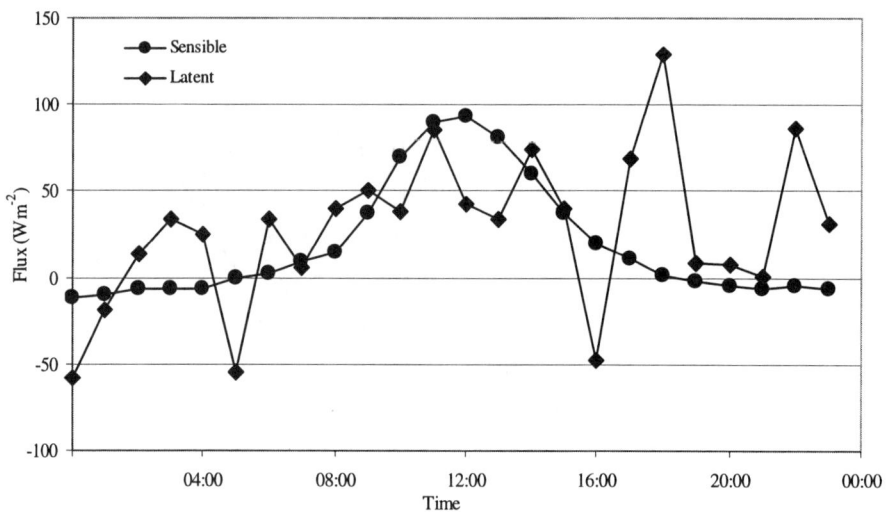

FIGURE 2. Average diurnal heat fluxes.

Figure 3 shows the relationship between mean wind speed and friction velocity. The correlation is clearly linear, with the scatter caused by inclusion of data from different stability regimens.

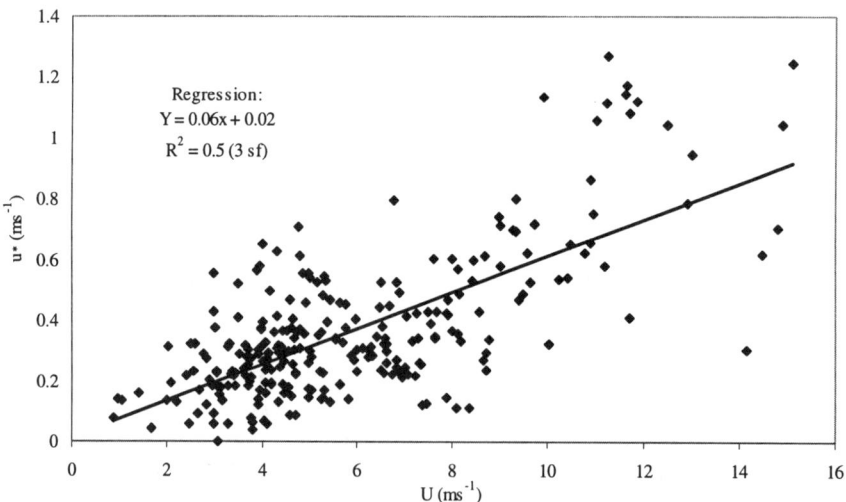

FIGURE 3. Friction velocity vs. mean wind speed.

Stratification of these data into different wind directions and stability cases could provide estimates of roughness length in the city. Again, more data will be required to draw firm conclusions.

CONCLUSIONS

Measurements of heat fluxes and friction velocity have been made over a city. A clear diurnal pattern for sensible heat flux has been found. Latent heat flux proves to be more variable, with individual precipitation events being a major factor in its value. A linear relation between friction velocity and mean wind speed has been found, averaged over the surface types included in this dataset. Further work is required to characterize the city accurately, and to deduce the parameters applicable to various urban surface subtypes.

ACKNOWLEDGMENTS

SASUA is supported by NERC under the URGENT thematic programme (Grant 022244). The help of Edinburgh City Council in providing additional data and infrastructure is gratefully acknowledged.

Physics and Chemistry of Atmospheric Aerosol Particles at Zedang and Jinghong, China

Yang Jun, Li Zihua, and Zhu Bin

Nanjing Institute of Meteorology, Nanjing 210044, P.R.China

Abstract. Atmospheric aerosol particles were measured at the Jinghong and Zedang meteorological observatory in the winters of 1997/98 and 1998/99, and their physicochemical properties, such as mass concentration, size distribution, visible light-absorption coefficient and chemical composition, were analyzed. Results show that aerosol particles at the two sites have significant physical and chemical difference.

INTRODUCTION

Regional field experiments for atmospheric aerosol are needed as the only way of measuring such parameters as the size distribution or chemical composition at different geographical locations. A series of measurements of atmospheric aerosol particles was undertaken at Zedang (91°46'E, 29°15'N, 3560m A.S.L.), Tibet by the use of a nine-stage cascade impactor (Andersen 1 ACFM non-viable ambient particle sizing sampler) and an optical particle counter (OPC). This site is located in the middle valley of Tibet Plateau and 2 km off the coast of Brahmaputra. Other site of Jinghong (104°04'E, 21°52'N, 554m A.S.L.), Yunnan located in the southern part of the valleys of the Hengduan Mountains and surrounded by primitive tropical rain forests and rubber trees. The same property of our two sites is the ecological environments have been affected by human activities only to a slight extent and can be seem as typical region of China even on the earth.

RESULTS AND DISCUSSION

Aerosol Particle Concentration

Mean number concentrations N with diameter ≥ 0.3 µm at Zedang and Jinghong are 28.7 and 98.3 cm^{-3}, respectively. The former is the smallest as compared with other Chinese observations and an order of magnitude smaller than concentrations found in some big city such as Beijing [1] and Shanghai [2]. Concentration at Jinghong agrees quite well with results from a few other clean locations, such as Xianghe and Xinglong [3]. N reached a maximum in the midmorning (Fig.1) and then exhibit sharp decreases through the noon to the minimums before the evening. Due to ground inversion layer develops, N increases after sunset and reaches its second maximum about 20:00.

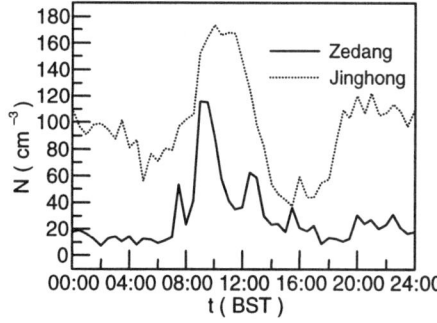
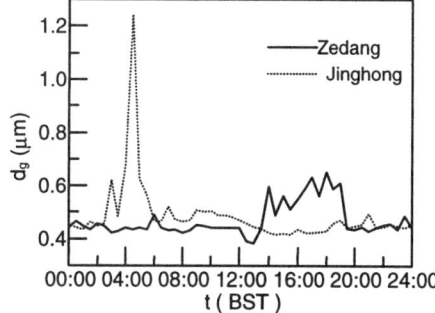

FIGURE 1. Diurnal variation of aerosol concentration in Zedang and Jinghong.

FIGURE 2. Diurnal variation of the geometric mean diameter of aerosol particles.

Stokes settling contributes mainly to the decrease of N in the night and another minimum occurred before sunrise. Weak wind lasts at Zedang in the night for relative constant N. Diurnal variations of geometric mean diameter d_g display a drastic difference with a maximum of 1.24 μm at 4:30 in Jinghong and a high value period from 14:00 to 19:00 in Zedang (Fig.2). We attribute the peak of d_g in Jinghong to fog droplets. Jinghong fog layer reaches ground about 04:00 with decreased N due to small contribution of coarse mode particles and the increase of deposition velocity. Variations of coarse particle numbers and d_g at Zedang agree well with wind velocity.

Increase of coarse particle numbers and d_g, decrease of N correspond to higher wind velocity (Table 1), which indicates that the coarse particle is maintained by mechanical lifting. Critical diameter for coefficient changed from negative to positive is 1.0μm. N at Jinghong is significantly correlated with relative humidity RH because more tiny particles grow considerably with increasing humidity for OPC detected. Negative correlation between N and visibility L is also significant, which due to that more particles can attenuate more radiation and fine particles have more efficient.

Generally, total mass concentrations of Zeang and Jinghong are 261.1 and 111.6 μg m^{-3}, respectively. Mass concentrations both of fine and coarse particles at Zedang are higher than Jinghong, which is different to the comparison of N. The accumulation mode masses of 93.8 and 65.4 μg m^{-3} contribute 36 and 59% of the total mass for Zedang and Jinghong, respectively.

Linear absorption coefficients of aerosols to visible light were determined by blackness of filter [4]. The absorption coefficient of 2.12×10^{-6} m^{-1} at Jinghong is the smallest compared with New York, Beijing and Nanjing [5], and the value of 3.55×10^{-4} m^{-1} at Zedang is similar with Beijing and two orders of magnitude larger than Jinghong, which probably due to the higher mass concentration of Zedang.

TABLE 1. Correlation between Aerosol Particle Concentration and Meteorological Parameters for Different Size Range.

d (μm)		>=0.3	0.3-2.0	>2.0	0.4-0.5	0.5-0.6	0.8-1.0	1.2-1.5
Zedang	V(m/s)	-0.2537	-0.2681	**0.41046**	-0.343	NS	NS	0.33151
Jinghong	L(km)	-0.6242	-0.6241	-0.50177	NS	**-0.7045**	**-0.7887**	NS
	RH(%)	0.52965	0.53018	0.282028	NS	0.55879	0.51907	NS

NS, not significant; underline and bold are significant at the threshold of 5% and 1% respectively.

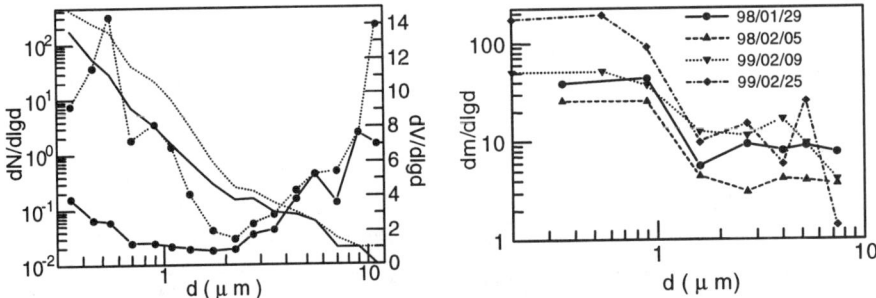

FIGURE 3. Average number (no symbol) and volume (circle) spectra at Zedang (solid), Jinghong (dot).

FIGURE 4. Mass concentration spectra at zedang indicate with beginning data of each measurement.

Aerosol Size Distribution

The number spectra (Fig.3) present distinctly difference for the two sites with obvious low values at Zedang fine particles and similar coarse mode distribution, also suggesting that the anthropogenic origin has a small influence at Zedang. The more significant difference in accumulation mode is present on volume concentration size distribution. In Jinghong, accumulation mode and coarse mode peak in the vicinity of 0.55 μm and above 10μm, respectively. The corresponding values are about 0.3 and 10μm. in Zedang. We can see from Figure 4 that mass concentrations of coarse particles at Zedang fluctuate with size more significant and two lines of 1999 are all above those of 1998. To compare the two curves of 1998, which were taken for successive periods, mass concentrations from 5 February decrease entirely by a factor of 40 to 50% under the curve from 29 January. With analogous surface wind velocity through each period, the main cause for this difference is snow with a precipitation of

TABLE 2. Element Abundance (ppm), Concentrations (ng m-3) and Enrichment Factors of Atmospheric Aerosol in Zedang and Jinghong.

Site Element	Zedang abundance	concentration	EF	Jinghong abundance	concentration	EF
Na	11548.580	3898.749	0.658	7270.194	810.52225	0.584
Mg	10796.460	3476.139	0.832	7784.107	824.495	0.810
Al	90006.720	28900.450	1.784	41731.720	4580.9135	1.152
Si	229230.000	71602.980	1.332	101626.260	11131.745	0.821
P	661.607	195.740	1.015	1102.630	120.45815	2.346
S	1751.313	518.135	10.853	61596.230	6704.6925	527.629
K	24230.840	10789.680	1.507	28015.820	3091.6165	2.438
Ca	28648.920	13206.640	1.272	68068.930	7592.1925	4.265
Ti	4268.224	1271.337	1.496	4357.681	477.8823	2.220
V	78.625	32.716	0.836	154.174	17.8332	2.681
Mn	921.038	272.218	1.452	1251.503	136.8721	2.946
Fe	32663.100	9241.550	1.000	22324.680	2445.845	1.000
Ni	16.848	8.477	0.314	--*	--	--
Zn	116.754	50.859	2.687	659.349	74.23115	21.600
As	124.538	36.845	111.479	327.564	36.98285	418.337
Pb	132.322	39.148	16.400	97.020	9.5718	15.277

* concentration below the blank level.

0.1 mm on 5 February. The result gives some indication of the wet remove course of snow can significantly decrease the aerosol mass concentrations.

Aerosol Chemistry

Aerosol elements (Table 2) were analyzed by X-ray fluorescence analysis in Center of Materials Analysis in Nanjing University. Crustal elements (Si, Al, Ca, Fe, Mg, K, Ti) have higher levels than pollutant elements (S, As, Pb, Zn, V) in the both locations. It is believed that soil derived components are mainly responsible for these aerosol components and crustal materials are more abundant in aerosols at Zedang than Jinghong. Ca and K have higher abundance at Jinghong probably due to construction and agricultural fertilizer aerosol source in Jinghong. Concentration of S is a lot bigger in Jinghong than Zedang and also is the third bigger one in comparison to other elements in Jinghong. This result shows the influence of coal burning on air quality even in Jinghong as a relatively clean region. Absolute concentrations of chromium and lead in Zedang are higher than in Jinghong because of a higher soil chromium concentration and a relatively important traffic circulation around the site of Zedang.

CONCLUSIONS

Mean number concentrations ≥ 0.3 μm at Zedang and Jinghong are 28.7 and 98.3 cm^{-3} and influenced by wind velocity and humidity, respectively. The former is the lowest value in China detected until now. Mean mass concentrations at the two sites are 261.12 and 111.63 μg/m^3, respectively. Significant lower size distribution at Zedang during fine mode size range and the similar of coarse mode indicates human origin only has a small contribution to aerosol particles at Zedang.

The lowest absorption coefficient of aerosol detected in Jinghong with a value of 2.12×10^{-6} m^{-1}, and this value at Zedang is two orders of magnitude higher due to higher mass concentration.

Crustal elements are the main components of aerosol particles at the two sites with pollutant elements have a higher enrichment at Jinghong compared to Zedang.

ACKNOWLEDGMENTS

Professor Huang Shihong from Nanjing University provided Anderson cascade impactor and filter for this paper. Financial support was provided by meteorological fund for youth from CMA and Chinese nature science fund grant 49675250.

REFERENCES

1. You, R. G. et al., *Scientia Atmos. Sinica* **7**, 88-94 (1983).
2. Zhang, W. et al., *Scientia Atmos. Sinica* **14**, 225-231 (1990).
3. Chen, J. R., Zhou, W. X., and An, Q., *J. Nanjing Inst. Meteorol.* **19**, 374-378 (1996).
4. Hogan, A. W., *J. Aerosol Sci.* **15**, 1-12 (1984).
5. Huang, S. H., and Hogan, A. W., *J. Nanjing Univ. (Natural sciences edition)* **24**, 130-135 (1988).

Model Studies on the Effect of Nitric Acid Vapour on Cirrus Cloud Formation

Jukka Hienola, Markku Kulmala and Ari Laaksonen*

Department of Physics, P.O. Box 9, FIN-00014 University of Helsinki, Finland
*Department of Applied Physics, University of Kuopio, P.O. Box 1627, FIN-70211 Kuopio, Finland

Abstract. The effect of elevated nitric acid concentrations on mixed phase cirrus cloud (containing both supercooled liquid and frozen ice particles) cloud particle size and number concentration changes are studied using a detailed multicomponent condensation model. Our model calculations suggest that high nitric acid volume mixing ratios, which are measured in the upper troposphere, can decrease the geometric mean radius of activated cirrus cloud particle distributions and increase the activated cloud particle number concentrations. These changes can also have effects on cirrus cloud optical properties, such as cloud shortwave albedo.

INTRODUCTION

Cirrus clouds play a significant role in determining the Earth's radiation balance, and therefore influence the climate. It has been estimated that these clouds regularly cover 20% of the Earth's total area [1]. The optical properties of a cirrus cloud depend on its ability to scatter and absorb incoming solar and outgoing terrestrial radiation, and therefore cloud microphysical properties, such as cloud particle size distribution, cloud particle number concentration and particle shape, have major effects on cloud's optical properties.

Nitrogen oxide emissions from fossil fuel burning are converted to more stable reservoir species, such as nitric acid, in the atmosphere. In the upper troposphere, sources of hygroscopic nitric acid vapour are large polluted air masses that are elevated from planetary boundary layer, and nitrogen oxides emitted by subsonic aircraft.

Nitric acid vapour can condense on small supercooled droplets and reduce the vapour pressure of water above small solution droplets, resulting in an increase in water vapour condensation on droplets. During condensational growth of a supercooled droplet population, the smaller droplets aquire higher nitric acid mass fractions in comparison with larger ones and this solute effect decreases the Köhler curve maxima for small droplets. As a result, more of the smaller liquid droplets can grow to cloud drop size and the resulting cloud contains more numerous but smaller supercooled cloud droplets [2].

However, many studies indicate that upper tropospheric clouds contain both liquid and solid phase particles. The formation of an ice phase can take place by three different ways: 1) via water vapour adsorption onto an ice nucleus surface, 2) via transformation of supercooled droplet to an ice particle, or 3) via collision of supercooled liquid droplet with an ice nucleus and initiation of ice formation. In this study, we consider the second case, where the probability to have an ice nucleation event inside a droplet volume depends on the droplet size, temperature and chemical composition. If the liquid aerosols have already undergone activation to cloud droplets, they can be considered to be more or less pure water, because the absolute amount of water vapour in the gas phase is much higher than the amount of e.g. nitric acid vapour, and therefore water is major condensing substance. On the other hand, if the initial temperature is low enough, a major fraction of initial liquid phase droplets may never activate before freezing, even if they contain some dissolved species which can decrease particle freezing temperature below 235 K (Schaefer point), and the microphysical properties of the resulting mixed phase cloud can be totally different than in the case, where some fraction of droplets is activated before freezing.

We have studied how condensable nitric acid vapour can change supercooled and frozen cirrus cloud microphysical properties. To estimate these effects, we have developed a numerical cloud model, which will be described in the next section. We will also discuss the effect of nitric acid vapour on cirrus cloud formation.

THE MODEL

We have developed a cirrus cloud model (CCM) which can be used to study the formation and growth of mixed phase clouds of supercooled and frozen cloud particles. The CCM is based on the liquid phase cloud model presented by Kulmala *et al.* [3] with a detailed multicomponent condensation model by Mattila *et al.* [4]. The model describes a simple adiabatic air parcel, which contains some initial condensation nucleus (CN) distribution and condensable gases, such as water and nitric acid vapours. The initial CN distribution can be varied from monodisperse to multimodal log-normal distributions of H_2SO_4/H_2O particles. Some fraction of the particles can be also insoluble. Partial vapour pressures above droplet surfaces are calculated using expressions given by Luo *et al.* [5].

The ice nucleation rate inside a droplet volume is calculated using a classical homogenous binary nucleation theory for water/sulphuric acid-solutions in the form presented by MacKenzie *et al.* [6]. Nucleation of solid phase in liquid volume is assumed to be a stochastic process, i.e. different nucleation events are independent of each other [7].

The model contains a stiff set of ordinary differential equations for ambient temperature, total pressure, vapour pressures of condensing species and mass fluxes of different species for different size classes and phases. Heat fluxes are approximated using a second order series expansion of the Clausius-Clapeyron equation, as described by Lehtinen *et al.* [8], and enthalphy changes related to phase transitions are calculated as presented by Hienola *et al.* [9]. The set of ordinary differential equations is solved using our own Fortran 90 version of FORTRAN 77 routine DDASSL by Brenan *et al.* [10]. DDASSL solves a first order differential-algebraic system of $G(X,Y,Y') = 0$ using the Petzold-Gear BDF method.

RESULTS AND DISCUSSION

We have studied the effect of gaseous nitric acid vapour on mixed phase cirrus cloud formation and microphysical properties in upper tropospheric conditions. We have used initial H_2SO_4/H_2O (80%/20% w/w) aerosol distribution, which is based on upper tropospheric measurements [11,12]. The initial log-normal distribution contains 1000 droplets per cm^3 with initial geometric mean radius of 15 nm and standard deviation of 1.5. The initial distribution is divided into 33 different size bins. The model runs were performed starting from the altitude of 7600 m, corresponding to a initial temperature of 239 K. The air parcel is lifted with a constant updraft velocity of 10 cm per second.

The initial gas phase concentrations of nitric acid applied in different model runs were 10 pptv and 100 pptv representing clean or remote upper tropospheric conditions, and 900 pptv and 4 ppbv representing polluted air. Volume mixing ratios as high as 4 ppbv have been reported in the literature [13].

The initial water vapour saturation ratio was set to 0.99. Owing to air parcel updraft motion, air temperature starts to decrease and saturation ratio of water starts to increase, while water vapour depletion by condensational growth of activated cloud droplets tends to decrease it. The saturation ratio of water vapour exceeds the value of unity, but after some time, the depletion of water vapour by condensational growth becomes more dominant than the effect of air parcel updraft motion, and the saturation ratio starts to decrease to unity. Also the latent heat released in condensation increases the air temperature somewhat, tending to decrease the saturation ratio.

Cloud droplet freezing probability depends on droplet size and temperature, and after some time, the largest droplets start to freeze. High initial water vapour saturation ratio results in very rapid growth of subsequently frozen ice particles due to fact that saturation vapour pressure of supercooled water is higher than saturation vapour pressure of ice (i.e. $S_i > S_w$). Therefore, if air is saturated with respect to ice, it is subsaturated with respect to liquid water and ice particles and supercooled droplets cannot coexist in thermodynamical equilibrium with each other in same air parcel. After aqueous droplets start to freeze, saturation ratio of water starts to decrease rapidly resulting in evaporation of supercooled droplets. The water vapour condensation on ice particles continues until the saturation ratio with respect to ice has decreased to unity.

At elevated nitric acid mixing ratios, significant amounts of nitric acid is taken up by the droplets during the activation process. The extra nitric acid in small droplets decreases water vapour pressure above droplet surfaces. As a result, the ambient saturation ratio, which is needed for droplet activation decreases, and more droplets can activate and grow to cloud droplet sizes. However, if the amount of cloud droplets increases, there is more droplets scavenging water from the gas phase, and thus the geometric mean radius (GMR) of activated cloud droplet population decreases due to elevated nitric acid concentrations in the gas phase [2]. After the droplets start to freeze, ice particles grow very fast depleting all available water vapour from the gas phase. Because of increased nitric acid volume mixing ratios, the GMR of activated droplet population decreases, and therefore also resulting GMR of ice particle distribution decrease (Fig. 1a). With nitric acid volume mixing ratio increase from 10 pptv to 4 ppbv, the GMR of solid phase particle distribution decreases from approximately 48 microns to 25 microns.

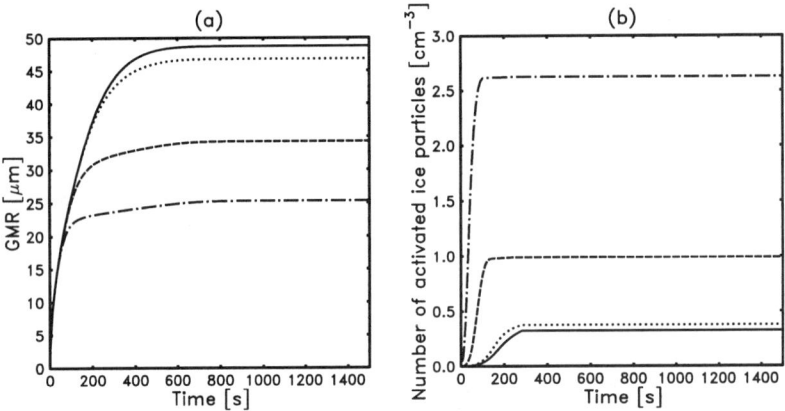

FIGURE 1. The effect of gaseous nitric acid on (a) geometric mean radius of activated ice particle distribution [μm] and (b) the number of activated ice particles [cm^{-3}] as a function of time [s]. Initial gas phase nitric acid volume mixing ratio is 10 pptv (solid line), 100 pptv (dotted line), 900 pptv (dashed line) and 4000 pptv (dash-dotted line).

Elevated nitric acid volume mixing ratios have also effects on cirrus cloud particle number concentration, as noted above. Laaksonen et al. [2] estimated the effect of nitric acid on supercooled clouds, which contain only liquid phase droplets. They found that 1.5 - 4.0 times more droplets activated depending on initial nitric acid concentrations and updraft velocities. This effect can be seen also in case of mixed phase clouds with liquid droplets and solid ice particles, although most of the frozen particles are already activated. Therefore, the effect of nitric acid on solid phase particle activation results mainly from the effect of nitric acid on liquid phase activation. With nitric acid volume mixing ratio increase from 10 pptv to 4 ppbv, the number concentration of ice particles increase from 0.3 particles per cm^3 to 2.7 particles per cm^3 (Fig. 1b).

The effect of initial values, such as initial temperature, updraft velocity, initial aerosol particle distribution and water vapour saturation ratio have strong influence on the resulting droplet and ice particle population. For example, if the initial temperature is low enough (e.g. near 235 K), initial aerosol droplets have no time to grow to cloud droplet sizes before they freeze, although for small (nonactivated) droplets also the droplet chemical composition can change the freezing behaviour of the droplet population. On the other hand, if the initial temperature is relatively high (e.g. above 243 K), a supercooled cloud droplet population can exist before cloud droplets start to freeze. Therefore the history of air parcel has also effects on mixed phase cloud formation and furthermore on cloud microphysical properties.

ACKNOWLEDGMENTS

J. Hienola acknowledge financial support by the Vilho, Yrjö and Kalle Väisälä Foundation.

REFERENCES

1. Liou, K.-N., Mon. Wea. Rev. 114, 1167-1199 (1986).
2. Laaksonen, A., Hienola, J., Kulmala M., and Arnold, F., Geophys. Res. Lett. 24, 3009-3012 (1997).
3. Kulmala, M., Laaksonen, A., Korhonen, P., Vesala, T., Ahonen, T. and Barrett, J.C., J. Geophys. Res. 98, 22949-22958 (1993).
4. Mattila, T., Kulmala, M. and Vesala, T., J. Aerosol Sci. 28, 553-564 (1997).
5. Luo, P.B., Carslaw, K.S., Peter, T. and Glegg, S.L., Geophys. Res. Lett. 22, 247-250 (1995).
6. MacKenzie, A.R., Laaksonen, A., Batris, E. and Kulmala, M., J. Geophys. Res. 103, 10875-10884 (1998).
7. Koop, T., Luo, B.P., Biermann, U.M. And Peter, T.H., Nucleation and Atmospheric Aerosols - Proceedings of the 14th ICNAA conference, 318-321 (1996).
8. Lehtinen, K.E.J., Kulmala, M., Vesala, T. and Jokiniemi, J.K., J. Aerosol Sci. 29, 1035-1044 (1998).
9. Hienola, J., Kulmala, M., and Laaksonen, A., "Condensation and Evaporation of Water Vapor in Mixed Aerosols of Liquid Droplets and Ice: Numerical Comparison of Growth Rate Expressions," submitted to J. Aerosol Sci. (2000).
10. Brenan, K.E, Campbell, S.L. And Petzold L.R., Numerical Solution of Initial-Value Problems in Differential-Algebraic Equations, 2nd ed., SIAM (1996).
11. Pueschel R.F., Livingston, J.M., Ferry, G.F. and DeFelice, T.E., Atmos. Environ. 28, 951-960 (1994).
12. Jensen, E.J., Toon, O.B., Westphal, D.L., Kinne, S. and Heymsfield, A.J., J. Geophys. Res. 99, 10421-10442 (1994).
13. Schneider, J., Arnold, F., Burger, V., Droste-Franke, B., Grimm, F., Kirchner, G., Klemm, M., Stilp, T., Wohlfrom, K.-H., Siegmund, P. and van Velthoven P.F.J., J. Geophys. Res. 103, 25337-25343 (1998).

GLOBAL ATMOSPHERIC AEROSOLS

Nucleation Properties of Aerosols in the Atmospheres of Mars and Titan

David L. Glandorf[1], Daniel B. Curtis[1], Tony Colaprete[2,3], Owen B. Toon[3,4], and Margaret A. Tolbert[1]

[1]*Department of Chemistry and Biochemistry and CIRES,* [2]*Department of Astrophysical and Planetary Sciences,* [3]*LASP, and* [4]*Department of Atmospheric and Oceanic Sciences, University of Colorado, Campus Box 216 Boulder 80309*

Abstract. Mars, with a rich CO_2 atmosphere and Titan, with an abundance of hydrocarbons, have both been proposed as models for the atmosphere of early Earth. In the current atmospheres of Mars and Titan, CO_2 and hydrocarbons, respectively, may condense to form clouds. In this paper, we explore cloud formation processes in these two very different planetary atmospheres. Under early Martian conditions, infrared scattering by CO_2 clouds could have warmed the planet's surface above freezing. However, the radiative effect of the clouds depends strongly on the nucleation and growth kinetics of the cloud particles. We experimentally examine the nucleation and growth of CO_2 on water ice under Martian conditions. We find that a critical saturation of S=1.3 is required for nucleation, corresponding to a contact parameter of m=0.95. After nucleation, growth of CO_2 proceeds rapidly without a surface kinetic barrier. Using a microphysical cloud model, our data suggest that CO_2 clouds are best described as "snow", having a small number of very large particles. Titan's atmosphere may have clouds of ethane and methane and may even support a cycle analogous to the hydrologic cycle on Earth. Alternatively, Titan's atmosphere may be highly supersaturated in organics if no suitable particles are available for nucleation. Titan's organic haze particles, dubbed tholins, may provide a suitable nucleation surface. We have prepared a laboratory sample of tholins and will be performing nucleation experiments of ethane on the tholins.

INTRODUCTION

Clouds are ubiquitous in planetary atmospheres; indeed Earth is one of the few planets that is not completely cloud covered. Like Earth, Mars has numerous water-ice clouds. Clouds are important to the Martian thermal budget; they cause deep self-sustaining thermal inversions. Cloud formation is also a major avenue for removing the omnipresent dust from the atmosphere, partly by nucleation scavenging. In the current Martian atmosphere, there have been several observations of CO_2 clouds. About 25% of the CO_2 atmosphere condenses out during the polar night, much of it falling to the ground as dry-ice snow to form the seasonal polar caps. CO_2 clouds may have been more common in the geologic past if Mars had a more substantial atmosphere. Infrared scattering by large CO_2 cloud particles may have once created a significant greenhouse effect on Mars (1), perhaps warming enough to create rivers, whose remnant valleys are seen today. However, the radiative effect of CO_2 clouds depends strongly on the size and vertical distribution of the particles, as well as the total cloud optical depth. These parameters in turn depend on the nucleation mechanism. Homogeneous nucleation is unlikely in the Martian atmosphere due to the very high supersaturations required to overcome the energy barrier to form a critical germ. Heterogeneous nucleation on preexisting aerosols may provide a pathway for nucleation at lower saturation ratios. Possible substrates in the Martian atmosphere include dust, water ice particles, and water-covered dust particles (2). Here we present laboratory data for CO_2 nucleation on water ice and use a microphysical cloud model to apply this data to the Martian atmosphere.

Titan is a vastly different planet than Mars or Earth. Though its atmosphere, like Earth's, is mainly composed of nitrogen, it lacks oxygen and instead is rich in hydrocarbons. From space Titan appears as a brownish-red, featureless sphere due to the hydrocarbon haze which extends 200 km above the planet's surface. Ethane may condense near 50 km altitude, just above Titan's tropopause, and methane clouds may float in the lower atmosphere above seas or lakes of hydrocarbon mixtures. The Cassini/Huygens spacecraft should provide a better understanding of these clouds in a few years. Earth-based and Voyager data suggest that Titan's lower atmosphere, which contains condensable meters of methane, is relatively cloud free and highly supersaturated. Possibly, nucleation of methane is difficult on Titan. Early Earth may have had an atmosphere like Titan's. Hence a better understanding of the hydrocarbon haze, which now shields Titan's surface from ultraviolet light and heats the upper atmosphere, might shed light on our own distant past. Here we discuss laboratory studies of ethane nucleation on tholins as a step in understanding aerosols on Titan.

Experimental

Substrates (water-ice for Mars and tholins for Titan) were prepared on a silicon surface. The surface was placed in a vacuum chamber and maintained at a constant low temperature (3). The surface was continually probed with FTIR transmission spectra collected every 2 sec. The temperature of the silicon wafer was measured using three thermocouples, calibrated as described below. After preparation of the substrate, the gas pressure (CO_2 or ethane) was then incrementally increased until the gas nucleated. Nucleation was observed by changes in the infrared spectrum upon condensation. After nucleation, the pressure was varied to perform a frost point calibration. The pressure at which no condensation or evaporation is observed is used with literature vapor pressures to calibrate the surface temperature (4). The saturation ratio is then the ratio of the nucleation pressure to the frost point pressure.

RESULTS AND DISCUSSION FOR MARTIAN CO_2 CLOUDS

Figure 1a shows a typical experiment for CO_2 nucleation on water ice at 133 K. The CO_2 pressure was incrementally increased until nucleation occurred, after which the pressure of CO_2 dropped toward its equilibrium value. The dashed line is the saturation ratio, while the solid line is the integrated FTIR absorbance of the 3700 cm^{-1} band of condensed CO_2. The nucleation event can be identified at approximately 7 minutes by the rise in absorbance and the subsequent decrease in saturation ratio. We find that a critical saturation of S=1.3 is required for nucleation of CO_2 on water ice for temperatures in the range of 130 to 140 K. Using nucleation theory we can extract a contact parameter between ice and CO_2 of m=0.952 (5).

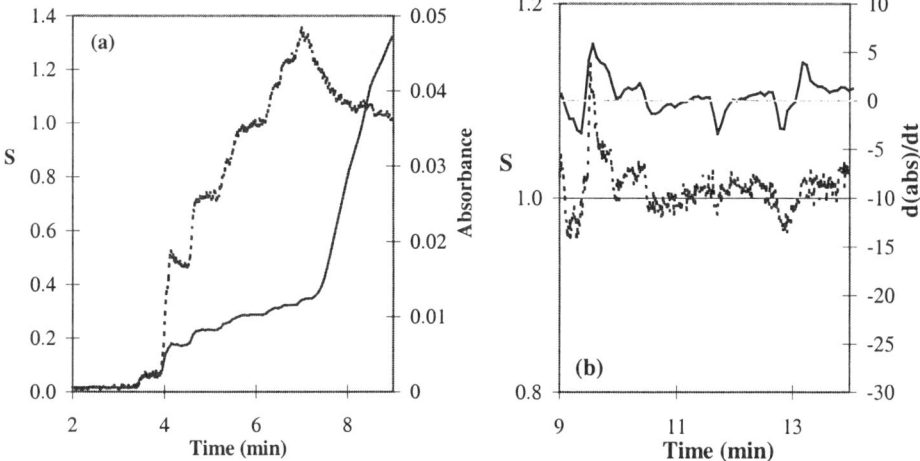

FIGURE 1. Typical experiment. The dashed line is the saturation ratio (S). The solid line is the integrated absorbance of the CO_2 3700 cm^{-1} overtone band in 1a, and its time derivative in 1b. Figure 1a shows the nucleation event, here occurring at around 7 minutes. Figure 1b shows the growth and loss of the film after nucleation, depending on the saturation ratio.

After nucleation, growth occurred whenever S>1 and evaporation whenever S<1 (Figure 1b). To quantify the growth, the infrared absorbance of the 3700 cm^{-1} band of condensed CO_2 was used with literature optical constants (6) and an optical model (5) to determine the film thickness. A comparison of our film growth rates to theoretical models assuming no activation barrier for growth (7) gives good agreement. Thus, after nucleation, CO_2 growth is rapid and proceeds without a surface kinetic barrier.

We use a time dependent ice cloud model (8, 9) to simulate CO_2 condensation in the Martian atmosphere. For cooling rates from 1 to 100 K/day, the results indicate that a very small number of large particles will form. Essentially, the high saturation ratio needed for nucleation results in the rapid growth of the nucleated particles, leading to vapor condensation onto a small number of large particles. Additionally, since supersaturations > 30% are required to form particles, at least 30% of the atmosphere mass will condense when these clouds form. The end result would be a "snow" flurry of large CO_2 particles. Although these particles are likely to create an IR scattering greenhouse effect, the clouds would probably be patchy and short lived, limiting the amount of warming.

Results and Discussion for Titan Ethane Clouds

The haze particles found high in Titan's atmosphere are thought to be solid organic particles formed photochemically. These solid materials, produced in laboratory simulations, were dubbed "tholins" by *Sagan and Khare* (10). The confirmation that laboratory tholin material matches the optical properties of Titan's haze material was achieved in 1984 by *Khare et al.* (11). We will be studying this tholin material as a nucleation surface for hydrocarbon clouds in Titan's atmosphere.

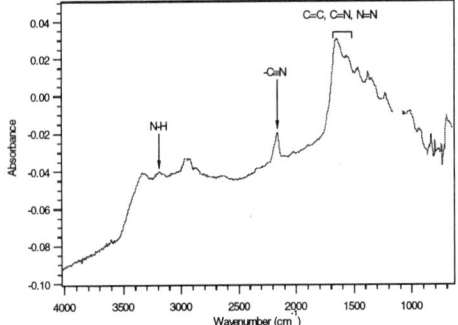

FIGURE 2. FTIR spectrum of laboratory-produced Titan tholin material. The negative absorbances are due to the fact that tholin material has a lower refractive index than silicon.

We have produced tholins in the laboratory by electrical discharge of 2 Torr of a gas mixture of 90% N_2 and 10% CH_4. A silicon wafer is placed inside the glass chamber, and the electrical discharge is applied for 48 hours. A tannish-brown solid is observed to form on the chamber walls and the wafer. The FTIR spectrum of the material (Figure 2) closely matches previously published spectra (12) Experiments are currently underway to probe the nucleation of ethane on this tholin sample. These results should provide insight into the amount of hydrocarbon supersaturation expected in Titan's atmosphere.

ACKNOWLEDGMENTS

This work supported by the NASA Astrobiology Institute under grant NCC2-5300.

REFERENCES

1. Forget, F., and Pierrehumbert, R. *Science*, **278**, 1273-1276 (1997).
2. Gooding, J. *Icarus*, **66**, 56-74 (1986).
3. Tisdale, R., D. Glandorf, M. Tolbert, and O. Toon. *J. Geophy. Res.*, **99**, 25631-25654 (1994).
4. Brown, G., and W. Ziegler. *Adv. Cryogenic Eng.*, **25**, 662-670 (1979).
5. Pruppacher, H., and J. Klett. *Microphysics of Clouds and Precipitation* Dordrech: Kluwer Acad. Publ., 1997.
6. Warren, S. *Appl. Opt.* **25**, 2650-2674 (1986).
7. Young, J. *Int. J. Heat. Mass Transfer*, **36**, 2941-2956 (1993).
8. Michelangi, D., Toon, O., Haberle, R., and Pollack, J. *Icarus*, **102**, 261-285 (1993).
9. Colaprete, A., O. Toon, and J. Magalhaes. *J. Geophy. Res.*, **104**, 9043-9053 (1999).
10. Sagan, C., and B. Khare. *Nature*, **277**, 102-107 (1979).
11. Khare, B., C. Sagan, E. Arakawa, F. Suits, T. Callcott, and M. Williams. *Icarus* **60**, 127-137 (1984).
12. McDonald, G., W. Thompson, M. Heinrich, B. Khare, and C. Sagan. *Icarus,* **108**, 137-145 (1994).

Ion and Nano-particle Measurement in Ion-induced Nucleation Process

Kikuo Okuyama, Manabu Shimada and Chan S. Kim

Department of Chemical Engineering, Hiroshima University, 4-1, Kagamiyama 1-chome, Higashi-Hiroshima 739-8527, Japan

Abstract. Ion-induced nucleation is the process leading supersaturated vapors to condense on ions and has become an attractive research subject to understand the gas-to-particle conversion process involving generation of ions and nanometer sized particles (nano-particles). In order to measure the mobility distribution of ions and nano-particles in the ion-induced nucleation process, the current aerosol measurement technology has been recently improved. The performance of these methods was confirmed by measuring the electrical mobility distribution of water ions generated by various methods. Ion-induced nucleation experiments using these improved aerosol measurement capabilities were performed for the following two cases: (1) condensation of supersaturated vapors onto monovalent and divalent ions of varying size, and (2) ion and nano-particle formation from SO_2, NO_2, H_2O, NH_3 and air mixture by ionizing irradiation. The results clearly show that ion-induced nucleation depends greatly on the sign of ion, ion size, type of vapor and supersaturation ratio.

INTRODUCTION

In any partially ionized and highly supersaturated environment, nucleation can occur on ions in one of two ways. One route is by ordinary homogeneous nucleation that occurs in the absence of ions as long as a sufficient degree of saturation of the vapor exists. The second is nucleation resulting from ion-molecule reactions, so called ion-induced nucleation. It is well known that ion-induced nucleation of vapor species occurs preferentially as compared to purely homogeneous nucleation. A well-designed experimental method that enables a quantitative comparison of ion-induced nucleation to homogeneous nucleation can provide data to evaluate theories of both processes. Various aerosol measurement equipment including a Differential Mobility Analyzer (DMA), a Mixing-type Condensation Nucleation Counter (CNC), a Particle Size Magnifier (PSM), and a Faraday Cup Electrometer (FCE) have been improved to permit the simultaneous measurement of ions and nanometer-sized particles (nano-particles).

In this lecture, after reviewing the electrical mobility distributions of ions measured by this improved aerosol measurement equipment, recent investigations concerning ion-induced nucleation will be presented. These experiments, the condensation of supersaturated vapors on monovalent and divalent ions of various size generated by electrospray, and ion-induced and binary homogeneous nucleation in $SO_2/NO_2/NH_3/H_2O/N_2$ gas mixtures using α-ray irradiation, were performed to show the dependency of the ion-induced nucleation process on the sign and size of ions, the type of vapor and the supersaturation ratio.

MEASUREMENT AND SIZING OF NANO-PARTICLES AND IONS USING AEROSOL TECHNOLOGY

When performing measurements using a DMA, bipolar ions are generated by α-ray irradiation, and the aerosol particles acquire a bipolar charge. These charged particles are then electrically classified by differences in their electrical mobility, and the concentration of the classified particles are measured by one or a combination of a FCE, a CNC or a PSM to obtain the electrical mobility distribution of ions and nano-particles in the gas stream. Subsequently, the particles size distribution can be obtained by the proper data reduction method.

Figure 1(a) shows a schematic diagram of a modified type of DMA, combined with a FCE and mass spectrometer to measure ions, molecular clusters and nano-particles [1]. This measurement system also has significant potential for the in-situ, simultaneous measurement of nano-particles and ions at low pressures down to about 1 Torr by using a low-pressure type DMA and FCE as shown in Fig. 1(b) [2,3,4].

A number of types of CNCs have been conventionally used for measurement of the number concentration of nano-particles due to the difficulty involved in the detection of such small particles by light scattering methods. The working principle of the CNC is based on the deposition of supersaturated vapors on very small particles, which is both opposed by their large curvature and promoted by their charge. However, the detection limit of these instruments is about 2-3nm. Below these sizes condensational growth becomes difficult due to the Kelvin effect. Moreover, deposition losses within the apparatus become remarkably large due to Brownian diffusion.

Figure 2(a) shows a mixing-type CNC where ethylene glycol is used as the condensable vapor [5]. This CNC can measure particles larger than a few nm in

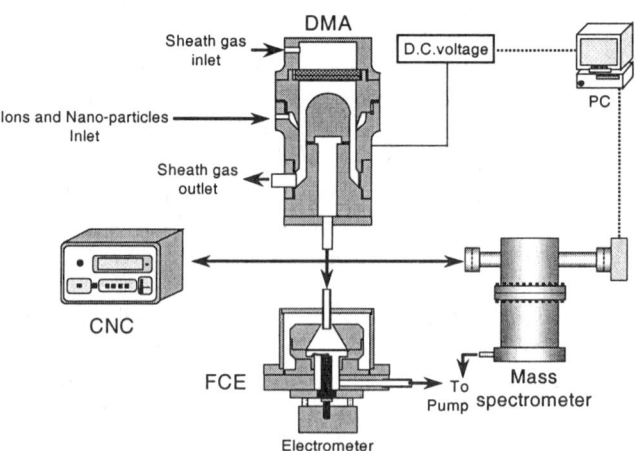

FIGURE 1(A). Schematic diagram of experimental setup to measure ions and nano-particles.

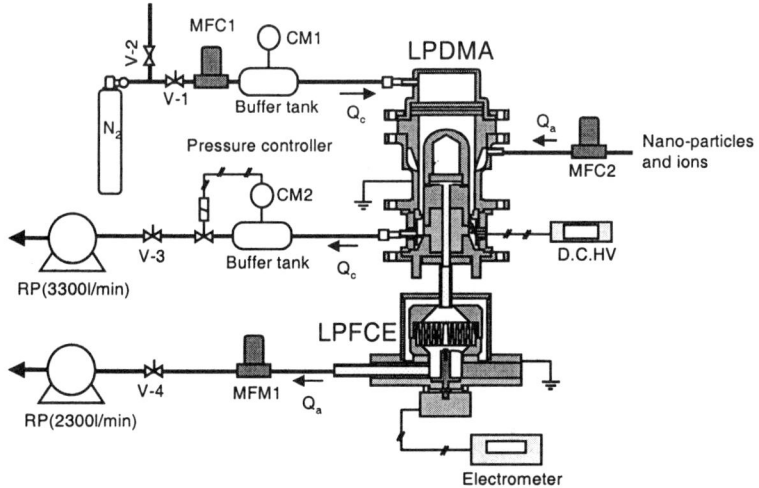

FIGURE 1(B). Experimental setup of low-pressure DMA (LPDMA) system.

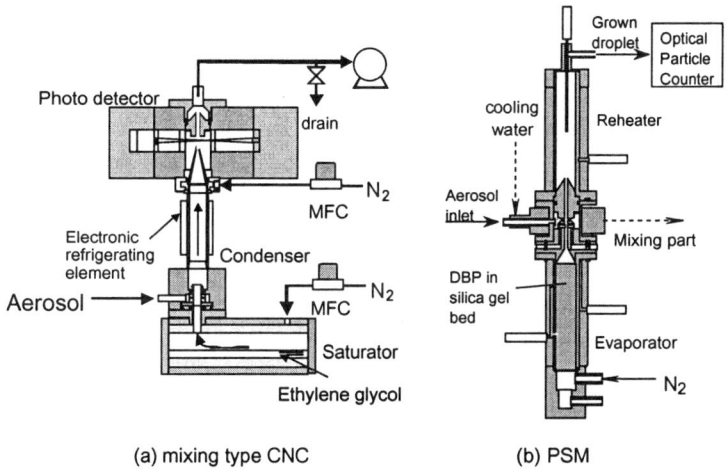

FIGURE 2. Diagram of mixing type CNC and PSM

diameter by using a cooling system after the mixing chamber. Figure 2(b) shows a PSM, which uses dibutyl phthalate (DBP) as the condensable vapor [6]. This PSM can also become a mixing-type CNC by adding particle detection equipment at the outlet. In both the mixing-type CNC and the PSM, a hot stream of nitrogen saturated with vapors after passing through the evaporator is mixed with the cool aerosol stream. The subsequent temperature drop creates a supersaturated vapor mixture. In this supersaturated condition, nano-particles can grow into large droplets that can be detected using an optical counter. A FCE is used to measure the quantity of electrical charge on the impinging particles. Although the lowest detection limit of commercially available FCEs is 10^{-15}A, we improved the electrometer to provide detection limits down to 10^{-16}A [1]. It is inevitable that in these ion-induced nucleation experiments, the use of mass spectrometer will be necessary to measure the ion species and molecular clusters in order to understand the chemical reaction that occur before nano-particle formation.

MEASUREMENT OF MOBILITY OF IONS

A number of different ions can be generated by either the electrolytic dissociation of gases by ionizing radiation (α-rays, β-rays, etc.), corona discharge, or electrospraying. These ions consist of positively and/or negatively charged molecular clusters. In the atmosphere, most of the positive ions are in the form of $H_3O^+(H_2O)n$ (n = 4-6), which have electrical mobilities ranging from 2.6×10^{-5} to 1.12×10^{-3} m^2/V/s. The composition of negative ions in the atmosphere is more complex because they are formed from various gas species such as CO_2, NO, O_3. Examples of some negative ions are $O_2^-(H_2O)n$, $CO_4^-(H_2O)n$, $NO_3^-(H_2O)n$, (n = 2-4), and they have electrical mobilities of 7.0×10^{-5} to 1.19×10^{-3} m^2/V/s. An ion of about 1 nm in size consists of 3-5 water-clusters and an ion of about 2 nm consists of 30-50 water-clusters.

Ions generated by radiolysis of gas

A modified DMA with an Integrated Electrometer (IE) was used to measure the sizes and relative numbers of positive and negative ions produced by a 6.07 MBq 241-Am sources as shown in Fig. 3(a). Figure 3(b) shows the measured electrical mobility distribution of ions in N_2, air and Ar, respectively [1]. From this figure, it is clear that negative ions have larger electrical mobility and lower peak number concentration than positive ions.

Ions generated by corona discharge

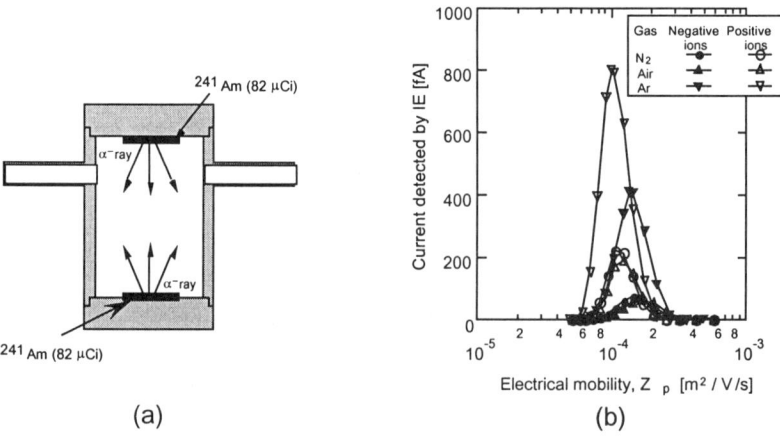

(a)   (b)

FIGURE 3. Schematic illustration of ion generation chamber equipped with two [241]Am α-ray sources, and measured mobility distributions of ions generated by [241]Am α-ray source in N_2, air and Ar; Aerosol and sheath flow rates of DMA are 3.0 l/min and 42 l/min, respectively.

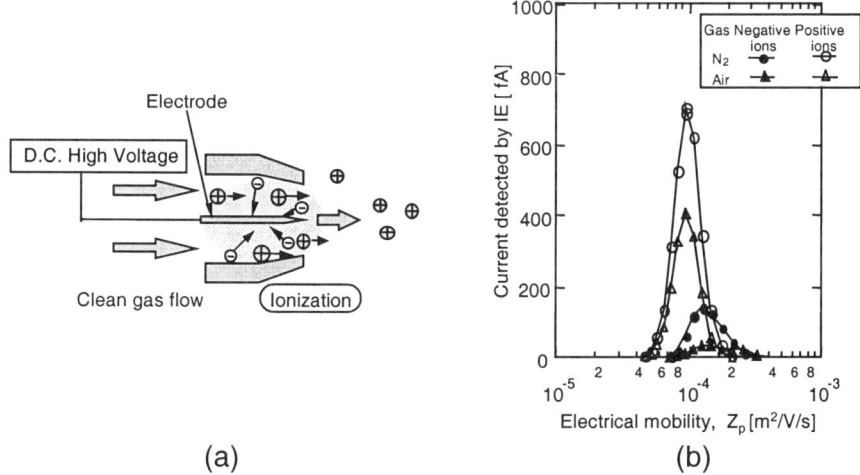

FIGURE 4. Schematic illustration of corona discharging ionizer, and measured mobility distributions of ions generated by corona discharging in N_2 and air; Aerosol and sheath flow rates of DMA are 4.0 l/min and 42 l/min, respectively.

Electrical mobility distribution of ions produced by positive and negative corona discharge was also measured [1]. An Ultraclean Ionizer (RI-4C, Harada Sangyo, Japan) shown in Fig. 4(a) was used as a corona discharge source.

In the ionizer, a clean gas flow surrounds a silica electrode to suppress particle deposition at the tip of electrode. The ionizer can generate either positive or negative ions, depending on the polarity of the voltage. Figure 4(b) shows the measured electrical mobility distribution of ions in N_2 and air. The peak values of the electrical mobility are the same for both gases. The ion concentration for N_2 is larger than air because the ionization potential for N_2 is lower than that for the O_2 that is contained in air.

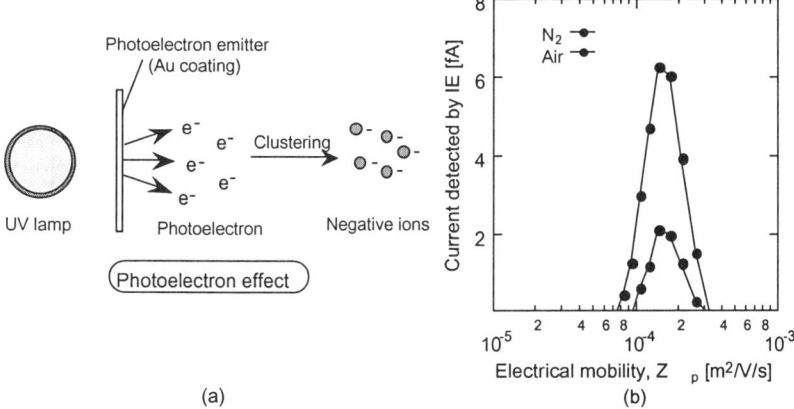

FIGURE 5. Schematic illustration of UV/Photoelectron method, and measured mobility distributions of negative ions generated by UV/photoelectron method in N_2 and air; Aerosol and sheath flow rates of DMA are 4.0 l/min and 19.8 l/min, respectively.

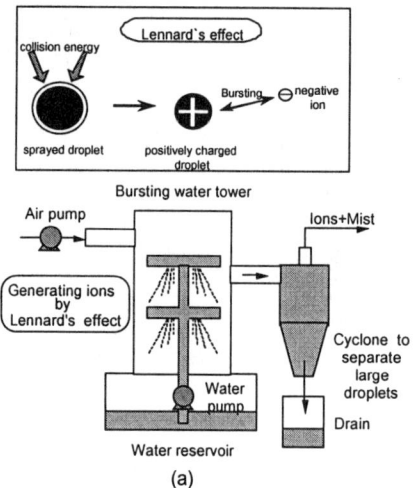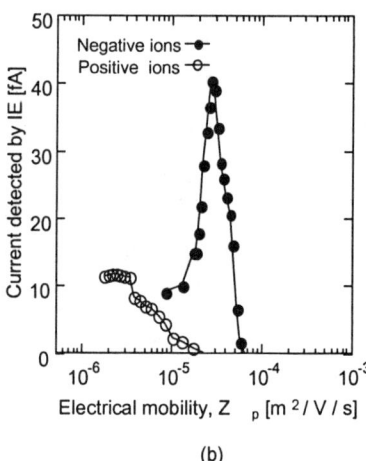

FIGURE 6. Schematic illustration of apparatus for generation of ions by the Lennard's effect, and measured mobility distributions of ions generated by the Lennard's effect; Aerosol and sheath flow rates of DMA are 4.0 l/min and 40 l/min, respectively.

Ions generated by UV/photoelectron method

When ultraviolet light with photon energy exceeding a certain threshold value is irradiated onto a metal surface, photoelectrons are emitted from the surface. In the experiments, we used a combination of a lamp with a short wavelength (about 254 nm) and an Au thin film surface (Fig. 5(a)). The resulting photoelectrons produce negative ion clusters such as $O_2^-(H_2O)n$, $O^-(H_2O)n$, $OH^-(H_2O)n$, $NO^-(H_2O)n$, $NO_2^-(H_2O)n$, etc. by clustering with water vapor or oxygen atoms. The measured mobility distributions of the ions produced in N_2 and air are shown in Fig. 5(b) [1]. As observed for other methods for producing ions, the number concentration of ions for N_2 is larger than air. However, the peak mobility is found to be same for both gases.

Ions generated by the Lennard Effect

Negative ions and positively charged droplets are generated when water droplets burst in air (Lennard's effect). Figure 6(a) shows the experimental apparatus based on this ion generation principle. In this apparatus, water is continuously sprayed into

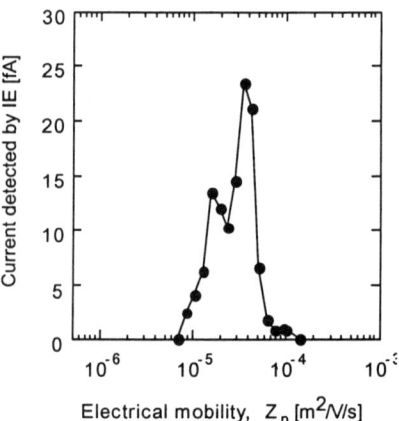

FIGURE 7. Measured mobility distributions of ions generated by a waterfall; Aerosol and sheath flow rates of DMA are 2.0 l/min and 20 l/min, respectively.

air with large droplets separated from the gas using a cyclone. Ions and small droplets suspended in air exit the apparatus. As seen from Fig. 6(b) [1], the negative ions have a peak mobility at 2.7×10^{-5} m^2/V/s. The positive species have a broader distribution and a lower concentration than the negative ions. Figure 7 shows the electrical mobility distribution of negative ions generated by a waterfall. The results show that the peak in the electrical mobility distribution is close to that seen in Fig. 6 (b).

CONDENSATION OF SUPERSATURATED VAPOR ON MONOVALENT AND DIVALENT IONS OF VARYING SIZE

Ion-induced nucleation by seed ions of controlled size and charge was examined for several monovalent positive and negative ions as well as divalent cations generated by electrospray ionization [5]. Electrospray makes it possible to generate a mixture of a few different types of ions by the ionization of solutions. As shown in Fig. 8, the resulting ions are classified by the modified DMA to provide a know composition of ion-clusters and to select the number of charges on the ions. This mobility-selected fraction of ions is then introduced into the PSM, where ions are mixed with a DBP vapor at a well-characterized supersaturation, S. This experiment provides the following interesting results: None of anions tested is activated but singly charged cations with mobilities from 0.48 up to 0.93 cm^2/V/s are activated at values of ln S 30% smaller than predicted by Thomson's model. As the mass of the ion-cluster increases, the Kelvin effect can be quantitatively used to predict activation, however when the mass of decreases, the Kelvin effect cannot be used in this manner.

Recently, Gamero-Castano and Fernandenz de la Mora modified the mixing part of the PSM so that higher supersaturation ratios could be attained without particle formation by homogeneous nucleation, which enabled the activation of singly charged, sub-nanometer particles using the PSM [5,7].

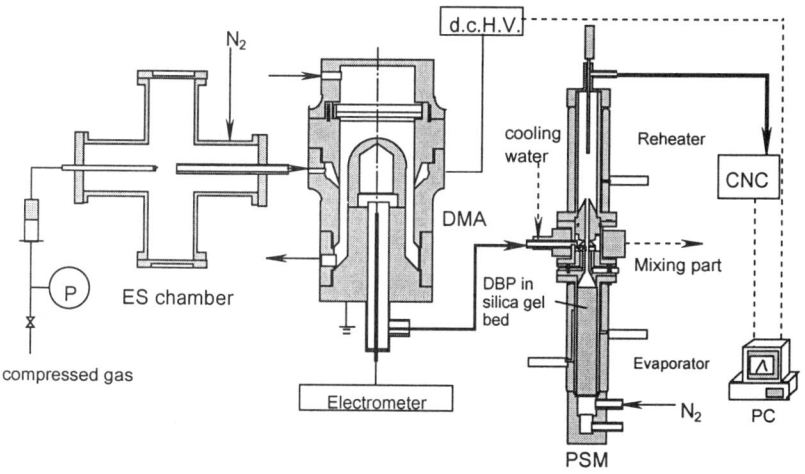

FIGURE 8. Schematic diagram of experimental setup for studying ion-induced nucleation by seed ions.

NANO PARTICLE FORMATION FROM NH₃/SO₂/NO₂/H₂O/AIR MIXTURE BY ION-INDUCED NUCLEATION

Gas-to-particle conversion in a humid atmosphere containing small amounts of gaseous impurities, such as SO_2, NO_2 and NH_3, by ionizing radiation is an important particle formation mechanism in the atmosphere [8,9,10]. Most of the ammonia exists in soil and is emitted in atmosphere. However, if NH_3 exists together with SO_2 resulting from anthropogenic sources, it plays a significant role in particle formation by generating, at high enough concentrations, a solid phase $(NH_4)_2SO_2$ particle.

Figure 9 shows a schematic diagram of the experimental system designed to investigate the particle formation dynamics in a mixture of SO_2 and NH_3. The experimental apparatus consists of a continuous flow gas-generation system, an α-ray ionization chamber, an ultrafine aerosol CNC (TSI model 3025), and a DMA/FCE system. Nano-particles generated in the ionization chamber were sent to the CNC to determine the particle number concentration, and nano-particles and ions were directed to the DMA/FCE particle sizing system to measure their electrical mobility distributions. A small ionization chamber of dimension 71×30×20 mm was used. The upper and lower walls of the chamber act as electrodes, and two radioactive sources of 3.03 MBq (82 µCi) were placed on the lower wall. The upper electrode is connected to an electrometer to measure the ion current. The lower electrode is connected to the ground when the total particle number concentration and the electrical mobility distribution are measured, and a dc voltage of +300V is applied to the lower electrode in order to measure the charged-particle fraction.

The effect of NH_3 gas concentration in the $NH_3/SO_2/H_2O/Air$ mixture on the number concentration of the generated particles is shown in Fig. 10(a). The particle number concentrations increase in the presence of NH_3 by about a factor of 2-4,

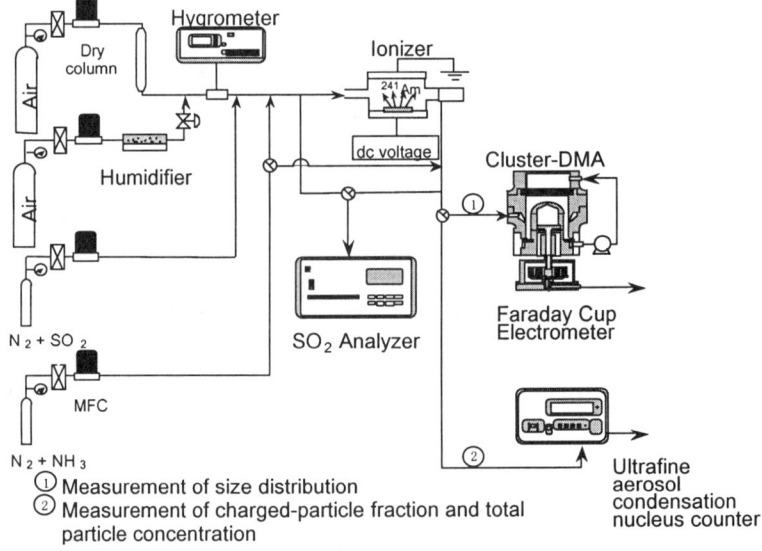

FIGURE 9. Schematic diagram of the experimental system ($NH_3/SO_2/H_2O/Air$ mixture).

depending on the SO_2 concentration, but are independent of the level of NH_3 in the range of 0.71-3.56 ppm SO_2.

These results indicate that the presence of NH_3 leads to a product with a vapor pressure lower than that of H_2SO_4, which is produced in the SO_2/H_2O/Air system. If NH_3 gas reacts strictly with the liquid particles of H_2SO_4 or H_2SO_4/H_2O generated from the oxidation of SO_2, an increase in particle number concentration should not be observed when NH_3 is present. Furthermore, since particles are not generated with the $NH_3/SO_2/H_2O$/Air mixture in the absence of ^{241}Am α-ray radioactive source in the ionization chamber, NH_3 does not react directly with SO_2 to produce particles under the current experimental conditions. While particle number concentrations at NH_3 mixing ratios of 0.71-3.6 ppm are 2-4 times higher than those in the absence of NH_3, it is noteworthy that the addition of NH_3 does not result in an enormous increase in number of new particles.

Figure 10(b) shows effect of the NH_3 concentration on the charged-particle fraction of nano-particles generated at various SO_2 mixing ratios. In the absence of SO_2, all particles are charged, although their number concentration is very low. These few particles must be formed by ion-induced nucleation in H_2O /Air and NH_3/ H_2O /Air mixtures. In the absence of NH_3, the charged-particle fraction decreases linearly from 1.0 to 0.5 over the range of SO_2 concentration studied. This is caused by an increase in the binary homogeneous nucleation of H_2O vapor and H_2SO_4 produced from the oxidation of SO_2. When NH_3 is present, the charged-particle fraction decreases steeply from 1.0 to 0.3 when the SO_2 concentration increases from 0 to 1 ppm, and it reaches a constant value of 0.3 for SO_2 concentrations exceeding 1 ppm. The steep reduction in charged-particle fraction in the presence of NH_3 suggests that NH_3 vapor leads primarily to neutral particles by homogeneous nucleation.

Figure 11(a) and 11(b) show the effects of NH_3 gas concentration on the electrical mobility distributions for negative ions and negatively charged particles, and positive

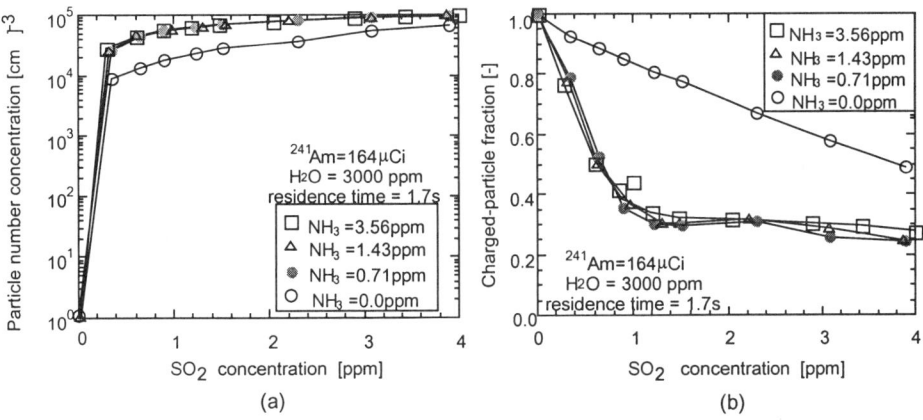

FIGURE 10. (a) Number concentration of particles generated at SO_2 levels of 0-4 ppm, NH_3 levels of 0-3.56 ppm, ^{241}Am radioactive source of 6.06 MBq, H_2O concentration of 3000 ppm, and residence time of 1.7 s. (b) Charged-particle fraction of particles generated at SO_2 levels of 0-4 ppm, NH_3 levels of 0-3.56 ppm, ^{241}Am radioactive source of 6.06 MBq, H_2O concentration of 3000 ppm, and residence time of 1.7 s.

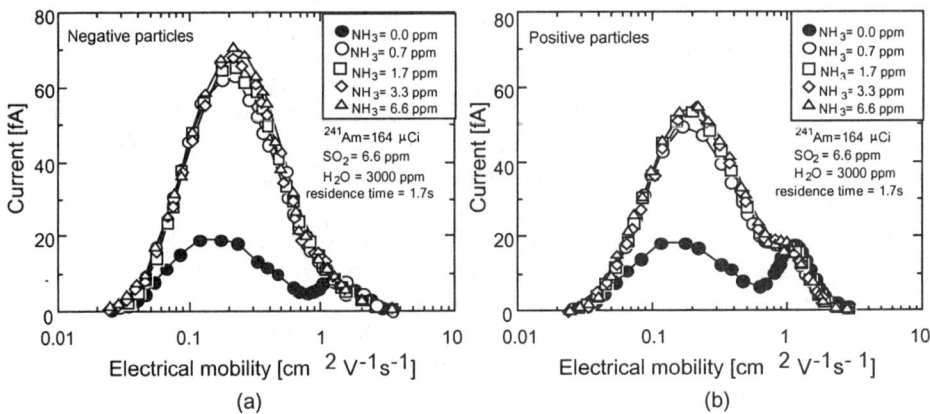

FIGURE 11. (a) Electrical mobility distributions of negative ion and negatively charged particles at NH_3 levels of 0-6.6 ppm, ^{241}Am radioactive source of 6.06 MBq, SO_2 levels of 6.6 ppm, H_2O concentration of 3000 ppm, and residence time of 1.7 s. (b) Electrical mobility distributions of positive ion and positively charged particles at NH_3 levels of 0-6.6 ppm, ^{241}Am radioactive source of 6.06 MBq, SO_2 levels of 6.6 ppm, H_2O concentration of 3000 ppm, and residence time of 1.7 s.

ions and positively charged particles, respectively. In the absence of NH_3, that is, an $SO_2/H_2O/Air$ mixture, the electrical mobility distributions exhibit bimodal peaks from ions and charged nano-particles. A peak of around 1.2-1.4 $cm^2/V/s$ is associated with the ions generated by the α-ray irradiation, and one of about 0.12-0.15 $cm^2/V/s$ with nano-particles formed from H_2SO_4 molecules and H_2O vapor. When comparing negatively and positively charged particles, the peak height of the positively charged particles is slightly smaller than that of negatively charged particles, although the peak height of positive ions is larger than that of negative ions. In the presence of NH_3, the electrical mobility distributions become unimodal with a shoulder remaining for the ion peak. The peak height for the $NH_3/SO_2/H_2O/Air$ mixture significantly exceeds that for the $SO_2/H_2O/Air$ mixture. The electrical mobility at the peak of the $NH_3/SO_2/H_2O/Air$ mixture is 0.2 $cm^2/V/s$, somewhat exceeding that of the $SO_2/H_2O/Air$ mixture. In the presence of NH_3, the peak height and electrical mobility at the peak are independent of the level of NH_3 in the measured range of 0.7-6.6 ppm.

Figure 12(a), (b) and (c) show the particle formation mechanisms clarified in this study [8,9,10]. In the H_2O/Air mixture, that is, when there is no SO_2 present in the mixture, particles with very low concentrations of 0.1-1 cm^{-3} are formed only by ion-induced nucleation of H_2O vapor (Fig. 12(a)). Although OH radicals are produced by the ion-molecule reaction, new products with low vapor pressure are not produced because there are no molecules that react with the OH radicals. In the $NH_3/H_2O/Air$ mixture, particles are also generated only by ion-induced nucleation. Although the OH radical reacts slowly with NH_3 and produces molecules such as NH_2, NH_2O, HNO, and NO by chain reactions, these products are gaseous at room temperature and thus do not form particles. In the $SO_2/H_2O/Air$ mixture, particles are formed by competition between ion-induced nucleation and binary homogeneous nucleation (Fig. 12(b)). The hydroxyl radicals react with SO_2 and produce SO_3, which ultimately produces H_2SO_4 by reaction with H_2O. The H_2SO_4 vapor forms particles not only by

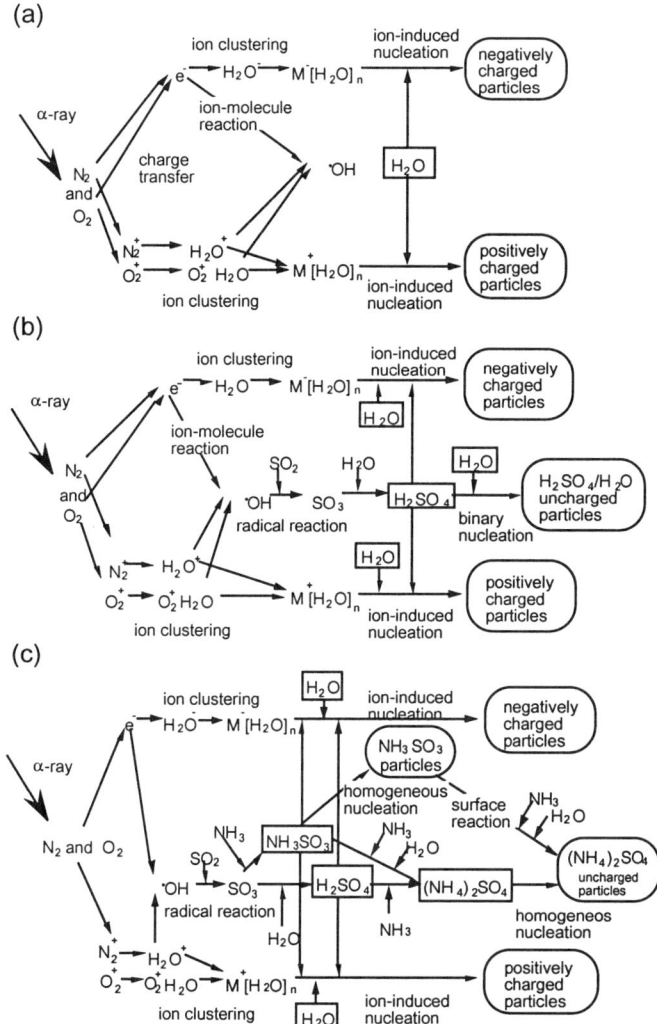

FIGURE 12. (a) Particle formation mechanism in H_2O/Air mixture. (b) Particle formation mechanism in SO_2/H_2O/Air mixture. (c) Particle formation mechanism in $NH_3/SO_2/H_2O$/Air mixture.

binary homogeneous nucleation with H_2O vapor but also by ion-induced nucleation. In this case, SO_2 and H_2O activate primarily by binary homogeneous nucleation, and particle formation by ion-induced nucleation is faster than that by binary homogeneous nucleation.

In the $NH_3/SO_2/H_2O$/Air mixture, particles are also formed by ion-induced nucleation and homogeneous nucleation (Fig. 12(c)). The addition of NH_3 complicates the homogeneous nucleation process because is presence forms new products with low vapor pressures. SO_3 molecules produced by the reaction of OH radicals and SO_2 react with both of NH_3 and H_2O and produce NH_3SO_3 and H_2SO_4. Moreover, the H_2SO_4 reacts with NH_3 and produces $(NH_4)_2SO_4$. H_2SO_4 is a liquid and $(NH_4)_2SO_4$ and NH_3SO_3 are solid at room temperature. Therefore, homogeneous nucleation is induced by $(NH_4)_2SO_4$ and/or NH_3SO_3. Ion-induced nucleation is thought to be caused by H_2SO_4 and/or NH_3SO_3 because H_2SO_4 and NH_3SO_3 are produced at an earlier stage than $(NH_4)_2SO_4$.

The effects of NO_2 gas on the particle formation in the SO_2/H_2O/Air mixture were also studied experimentally using the same system shown in Fig. 9. No particles were produced in the NO_2/H_2O/Air mixture. Particle generation in the $SO_2/NO_2/H_2O$/Air mixture was enhanced by NO_2 levels of 0.04-1.99 ppm, and depended on the NO_2 concentration.

CONCLUSIONS

(1) Improved aerosol measurement capabilities using a modified DMA, a FCE, a mixing–type CNC and a PSM were capable of measuring nano-particles down to a few nm as well as high concentrations of ions at pressures down to 1 Torr.

(2) The mobilities of positive and negative ions produced by the various methods based on the different formation mechanisms were measured. Positive ions in the atmospheric have smaller electrical mobility than negative ones for those generated by α-ray irradiation, corona discharge, the UV/photoelectron method and the Lennard's effect.

(3) Vapor nucleation induced by seed ions of controlled size and charge generated by electrospray and classified by a DMA was studied using a PSM. This experimental system is available to evaluate the ion-induced nucleation phenomena for various ions in different vapors.

(4) In order to evaluate the competition between ion-induced nucleation and binary homogeneous nucleation, the formation of nano-particles and ions in $NH_3/SO_2/NO_2/H_2O$/Air mixture by α-ray irradiation was studied. The results clearly show that the ion-induced nucleation greatly depends on the composition of the gas and the supersaturation ratio.

REFERENCES

1. Okuyama, K., Shimada, M., Okita, A., Otani, Y., and Cho, S. J., *J. Aerosol Res. Jpn.* **13**, 83-93 (1998).
2. Seto, T., Nakamoto, T., Okuyama, K., Adachi, M., Kuga, Y., and Takeuchi, K., *J. Aerosol Sci.* **28**, 193-206 (1997).
3. Camata, R. P., Hirosawa, M., Okuyama, K. and Takeuchi, K., *J. Aerosol Sci.* **31**, 391-401 (2000).
4. Seol, K. S., Tsutatani, Y., Camata, R. P., Yabumoto, J., Isomura, S., Okada, Y., Okuyama, K., and Takeuchi, K., *J.Aerosol Sci.*, in press (2000).
5. Seto, T., Okuyama, K., de Juan, L., and Fernandez de la Mora, J., *J. Chem. Phys.* **107**, 1576-1585 (1997).
6. Okuyama, K., Kousaka, Y., and Motouchi, T., *Aerosol Sci. Technol.* **3**, 353-366 (1984).
7. Gamero-Castano, M., and Fernandez de la Mora, J., *J. Aerosol Sci.*, in press (2000).
8. Adachi, M., Okuyama, K., and Seinfeld, J. H., *J. Aerosol Sci.* **23**, 327-337 (1992)
9. Kim, T. O., Adachi, M., Okuyama, K., and Seinfeld, J. H., *Aerosol Sci. Technol.* **26**, 527-543 (1997)
10. Kim, T. O., Ishida, T., Adachi, M., Okuyama, K., and Seinfeld, J. H., *Aerosol Sci. Technol.* **29**, 111-125 (1998)

Modeling of Global Sulfate Aerosol Number Concentrations

Michael Herzog[1], Joyce E. Penner[1], John J. Walton[1],
Sonia M. Kreidenweis[2], Debra Y. Harrington[2]

[1] *University of Michigan, Ann Arbor, MI*
[2] *Colorado State University, Fort Collins, CO*

Abstract. A two-moment, two-mode model of sulfate aerosol dynamics has been added to the University of Michigan three-dimensional chemical transport, transformation and deposition model, GRANTOUR. The two-moment model predicts both aerosol number and mass concentrations, and was chosen based on its computational efficiency and the small number of prognostic variables, which reduce storage requirements in the large-scale model. Simulations were performed to investigate the processes that control predicted number concentrations, and comparisons with observations were used to suggest additional features that should be added to the coupled gas-phase chemistry / aerosol model to bring the predictions more in line with observations.

INTRODUCTION

There has been increased recognition of importance of aerosols to climate, and a corresponding interest in including the indirect and direct effects of particles in models for climate and climate change prediction. Sulfate compounds have been of particular interest for a number of reasons: they generally constitute a major mass fraction of particulate matter; sulfur has both natural and anthropogenic sources; and sulfate particle formation is important in determining aerosol number concentrations, which are needed to compute indirect effects.

Aerosol number concentrations should respond to variations in sources and sinks of particles and account for the observed significant temporal and spatial variability in aerosol distributions (d'Almeida et al., 1991). Thus, in further developing an understanding of how the atmospheric aerosol will respond to changes in source strengths, a method is needed for predicting not only particulate mass production and loss rates, but also particulate number production and loss.

MODEL DESCRIPTION

GRANTOUR Transport and Chemistry

GRANTOUR is a global transport, transformation and deposition model which has been applied to three-dimensional tropospheric chemistry studies (Walton et al., 1988; Penner et al., 1994; Chuang et al., 1997). It is formulated as a Lagrangian parcel model, typically run with 50,000 air parcels which represent constant air mass parcels in the atmosphere. Each parcel carries trace constituents represented by mass mixing ratios. Advection of the parcels is carried out on an Eulerian grid. Parcel species concentrations are mapped to the Eulerian grid in order to calculate changes due to diffusional mixing

(Walton et al., 1988); these changes are then applied to the parcels. The model runs off-line by using wind and precipitation data from a GCM.

The species that have been included in the global sulfur cycle model are sulfur dioxiode (SO_2), dimethylsulfide (DMS), and sulfuric acid (H_2SO_4), or sulfate. The gaseous emissions are treated in three families divided by source type: anthropogenic, oceanic and vegetative, and volcanic. All sources are imposed at the center of grid boxes and, except for the volcanoes, are assumed to be well mixed in the lowest 100 mbar. The volcanic sources are assumed to be well mixed below their height of injection, which was determined as in Spiro et al. (1992).

To account for fast conversion of SO_2 to $SO4^{2-}$ in combustion plumes, based on observations of power plant plumes (Meagher et al., 1978), 3% of the mass of anthropogenic and biomass burning SO_2 emissions was removed from the gaseous source file and input as primary emissions of particles. In accordance with the methodology used by Binkowski and Shankar (1995) in their regional model, 20% and 80% of the mass of the direct particle source was input to modes 1 and 2, respectively, in number concentrations determined by assuming lognormal modes with geometric number mean diameters of 0.01 and 0.07 µm and standard deviations of 1.6 and 2.0 µm, respectively.

The scavenging coefficients were set to reproduce measured washout ratios (Penner et al., 1994); we applied the scavenging coefficients to the sulfuric acid vapor and to mode 2 particles, which are more likely to serve as CCN.

Sulfate Aerosol Dynamics

The model of sulfate aerosol dynamics is essentially that used by Kreidenweis and Seinfeld (1988) and Kreidenweis et al. (1991), and is described fully in Harrington and Kreidenweis (1998). Two modes that represent Aitken and accumulation mode particles in the fine particle fraction of the aerosol are modeled; a representation of coarse mode particulate matter is not yet included. The model is driven by a chemical source rate of $H_2SO_4(g)$, supplied from the chemistry model, and includes parameterized equations for binary nucleation of sulfuric acid particles from the vapor phase, condensational growth of each mode, and inter- and intra-modal coagulation (Youngblood and Kreidenweis, 1994; Harrington and Kreidenweis, 1998; Kreidenweis and Harrington, 1998). The aerosol modes are characterized by the time-varying mass mean particle size in each mode, determined from mass and number concentration of each mode, rather than by fixed mean diameters or bin boundaries for the two modes, as has been done in other studies (e.g. Raes and Van Dingenen, 1992; Russel et al., 1994).

The addition of particulate mass via the aqueous transformation of SO_2 is assumed to increase only the mass of mode 2 particles over the chemistry-aerosol coupling timestep. The rationale behind this treatment is that the larger mode 2 particles are the likely cloud condensation nuclei (CCN), and thus would receive this additional sulfate mass during a cloud cycle.

RESULTS

Results from the final, converged, annual cycle are reported below.

Aerosol Number Concentrations

Although the modeled particulate sulfate mass exhibits an annual cycle because of seasonal differences in transport and in the magnitude of oxidation and removal pathways, the simulated surface number concentrations over most of the globe exhibit very little month-to-month variability.

This is because they are dominated by the direct emissions from the anthropogenic SO_2 sources, which have no seasonal dependence in the source files used here. Annual averages of the predicted total number concentrations (N_{tot}) at the surface and at 355mbar are shown in Figures 1a and 1b, respectively.

The results shown in Figure 1a indicate that the model reproduces expected gradients in N_{tot} at the surface, with highest values over industrialized source regions and decreasing concentrations with increasing distance from the sources (Pruppacher and Klett, 1997). Peaks in modeled annual average

N_{tot} occur over the eastern U.S., eastern Europe, and China, and exceed 10,000 cm^{-3}, consistent with observations of rural and urban aerosols. The maximum concentration for particles aged an hour, the coupling timestep used in our model, is about 10^6 cm^{-3} (Hinds, 1982); considering this, and taking into account the large grid volumes and thus large spatial averaging used in the global model, we do not expect to reproduce localized very high N_{tot} such as those observed near city centers and in plumes.

Modeled estimates of 1,000-5,000 cm^{-3} over remote land regions (e.g., Alaska, northeastern Asia, N. Africa, and Australia) are similar to the range reported in the literature. In polar regions, the model reproduces expected lower Antartic particle number concentrations (100-200 cm^{-3}) and higher Artic N_{tot} (500-several thousand cm^{-3}). Observed and simulated particle concentrations generally decrease with altitude. In the upper troposphere, modeled N_{tot} have a stronger seasonal dependence; annual averages range from 200-1,200 cm^{-3}, again in good agreement with the literature estimates.

A number of observations and model predictions at the surface and at other levels have been compiled, data are displayed in Figure 2. The best agreement between simulations and observations was in marine regions and in the free and upper troposphere. The number concentrations in the accumulation mode over the central U.S. and Canada were simulated better at upper levels than they were in the boundary layer, where they were underpredicted. The largest differences between observations and simulations occur in polar regions, particularly in the annual cycle. Peak number concentrations are observed in late summer in Antarctica, but while modeled number concentrations were similar to observations for the winter and fall, they did not have a pronounced annual cycle and did not reproduce the observed maxima. In the Artic, the model again failed to produce an annual cycle, but in that case overpredicted number concentrations significantly.

DISCUSSIONS AND CONCLUSIONS

Results from sensitivity tests demonstrate the importance of proper treatment of the surface sources of particulate matter, and of the competition between new particle formation and condensational growth for condensable vapor. First, other sources of particulate mass and number concentration that interact with the sulfur emissions should be added to the model, including sea salt, dust, and carbonaceous particles, and the hygroscopic growth of mixed particles estimated. Second, processes that occur on sub-grid scales must be properly parameterized. The sulfur emissions occur on spatial scales much smaller than the grid sizes used in the global model, and injecting the gaseous emissions immediately into the surface grid boxes dilutes surface species concentrations artificially. It is known that some conversion of sulfur gases to particles occurs in plume, probably due to both homogeneous nucleation and to conversion on particles emitted with the gases; by accounting for this with the assumed 3% rate, a more realistic simulation of surface number concentrations is achieved. However, the most appropriate mass conversion percentages, their variation with location and time, and the input size distribution of these particles, should be developed using data and models with higher resolution.

Our studies indicated that number concentrations in the upper troposphere were not strongly affected by the treatment of surface sources, and that nucleation played a large role in determining number concentrations at higher altitudes. The nonlinearity of the assumed nucleation rate expression, with respect to relative humidity, temperature and $H_2SO_4(g)$ concentration, was particularly influential, and this sensitivity emphasizes the need for a better understanding of the factors controlling new particle formation in the atmosphere.

The results presented here are not intended to represent a fully realized formulation of the complex atmospheric aerosol system, or even the sulfate aerosol fraction. They are, however, a first step towards adding prognostic aerosol equations to a large-scale model.

ACKNOWLEDGEMENTS

This material is based upon work supported by the NASA AEAP Program.

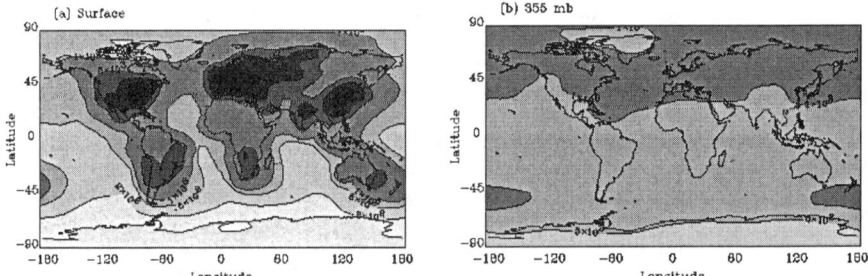

FIGURE 1. Annual average total particle number concentrations [cm^{-3}], base case simulation: a) surface, b) 355 mbar.

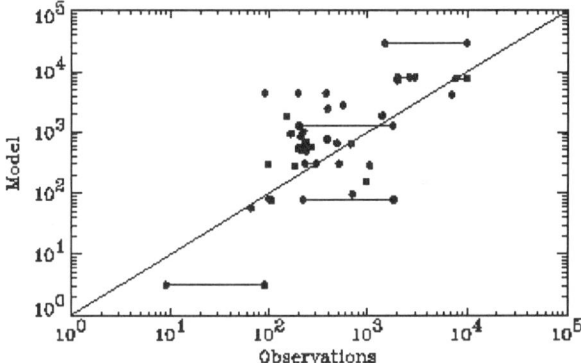

FIGURE 2. Comparison of model predictions of particle number concentrations [cm^{-3}] with observations.

REFERENCES

1. Binkowski, F. S., and Shankar, U., J. Geophys. Res. 100, 26191-26209 (1995).
2. Chuang C. C., Penner, J. E., Taylor, K. E., Grossman, A. S., and Walton, J. J., J. Geophys. Res. 102, 3761-3778 (1997).
3. Harrington, D. Y., and Kreidenweis, S. M., Atmos. Environ. 32, 1691-1700 (1998).
4. Kreidenweis, S. M., and Harrington, D. Y., Atmos. Environ. 32, 1701-1709 (1998).
5. Hinds, W. C., Aerosol technology: Properties, behavior, and measurement of airborne particles, John Wiley & Sons, New York, 1982.
6. Kreidenweis, S. M., and Seinfeld, J. H., Atmos. Environ. 2, 283-296 (1988).
7. Kreidenweis, S. M., Yin, F. D., Wang, S. C., Grosjean, D., Flagan, R. C., and Seinfeld J. H., Atmos. Environ. 25A, 2491-2500 (1991).
8. Meagher, J. F., Stockburger, L., Bailey, E. M., and Huff, O., Atmos. Environ. 12, 2197-2203 (1978).
9. Penner, J. E., Atherron, C. S., and Graedel, T. E., "Global emissions and models of photochemically active compounds", in Global Atmospheric-Biospheric Chemistry, edited by R. G. Prinn, Plenum Press, New York, 1994, pp. 223-247.
10. Pruppacher, H. R., and Klett, J. D., Microphysics of clouds and precipitation, Kluwer Academic Publishers, London, 1997.
11. Raes, F., and Van Dingenen, R., J. Geophys. Res. 97, 12901-12912 (1992).
12. Russel, L. M., Pandis, S. N., and Seinfeld J. H., J. Geophys. Res. 99, 20989-21003 (1994).
13. Spiro, P.A., Jacob, D. J., and Logan, J. A., J. Geophys. Res. 97, 6023-6036 (1992)
14. Walton, J. J., MacCracken, M. C., and Ghan, S. J., J. Geophys. Res. 93, 8339-8354 (1988).
15. Youngblood, D. A., and Kreidenweis, S. M., Further development and testing of a bimodal aerosol dynamic model, Colorado State University, Department of Atmospheric Science, Report No. 550, 1994.

Modelling and Observations of Aerosol Properties in the Clean and Polluted Marine Boundary Layer and Free Troposphere

Elisabetta Vignati, Frank Raes, Rita Van Dingenen, and Jean-Philippe Putaud

European Commission, Joint Research Centre, Environment Institute, TP 280, Via E. Fermi, 21020 Ispra (VA), Italy

Abstract. The characteristic time of many of the microphysical aerosol processes is days up to several weeks, hence longer than the residence time of the aerosol within a typical atmospheric compartment such as the marine boundary layer, the free troposphere etc. To understand aerosol properties, one cannot confine the discussion to such compartments. This paper presents simulations using a box model that describes aerosol microphysics within the context of atmospheric dynamics that connects those compartments. The model results for a Clean Marine and a Polluted Continental air mass are compared to observations.

INTRODUCTION

In the atmosphere, particles are produced from anthropogenic activities, such as fuel combustion and also from natural sources, such as dust, sea spray, and due to volcano activities. They are also formed in the atmosphere by gas-to-particle conversion processes. Particles evolve in size and composition through condensation of vapour species or by evaporation, coagulating with other particles, by chemical reaction, or by activation in the presence of supersaturated water. Aerosols are critical players in the hydrological cycle and climate system. It is therefore needed to understand their cycling in the atmosphere, and to be able to predict their characteristics. This presentation is focused on modelling the microphysics of particle formation and evolution through the Hadley/Walker cells, and comparison with observations.

MODELLING OF AEROSOL MICROPHYSICS AND COMPARISON WITH OBSERVATIONS

The Hadley/Walker cells are characterised by upward motions due to convection in the tropics, or along frontal surfaces in the mid-latitudes, compensated by slow and large scale subsidence in the subtropical and polar regions. A box model in which the aerosol dynamics processes are implemented is "moved" through the cells. The processes are related to the H_2SO_4-SO_2 aerosol system resulting also from biogenic DMS emissions over the oceans. This simulation is done with the AERO3 model, which considers the primary emissions of insoluble soot particles as well (1). AERO3 allows for the internal mixing of the particles by coagulation, condensation, nucleation and in-cloud SO_2 oxidation. In the Clean Marine (CM) case the box is "moved" through an oceanic Hadley/Walker cell. The Polluted Continental (PC) scenario studies aerosol cycling in a Hadley/Walker cell that has its convective updraft over the polluted continent. In both cases the simulation starts with the aerosol that enters a convective cloud, where the fraction of SO_2 is oxidised in the cloud droplets and the fraction of the aerosol that is activated at a supersaturation of 2 % is completely removed by precipitation. DMS and the remainder of the SO_2 and the inactivated aerosol is injected into the Free Troposphere (FT), where an immediate dilution by a factor of 4 is assumed to account for the turbulent mixing at the exit of the cloud. Subsequently the aerosol plume travels for 15 days in the FT, where it is slowly dispersed due to atmospheric turbulence. In the plume DMS is oxidised by OH to SO_2 and further to H_2SO_4. If the conditions allow, nucleation can occur, in which case and the aerosol further develops by coagulation and condensation only. As a rough simulation of the conditions expected in the free troposphere, the plume experiences 90 % relative humidity during the first day after the cloud injection into the FT. Afterwards the relative humidity is kept at 40%. During the first 8 days the temperature is maintained constant at 239 K. During the last 7 days the plume subsides, and the temperature increases linearly to the MBL value. After 15 days, the aerosol entrains into the MBL, where it travels for another 7 days. There, the SO_2 concentration responds to the oxidation of locally emitted DMS with OH radicals, and the MBL aerosol size distribution is governed by the aerosol dynamical processes, by cycling through MBL clouds and by entrainment of the free tropospheric aerosol. The latter is the one resulting after the initial 15 days of transport in the FT.

Clean Marine case

The model is initialised with a typical clean marine aerosol size distribution with log-normal parameters given in Table 1. The particles are assumed pure H_2SO_4-H_2O droplets. Initially the total particle number sharply decreases, as all particles larger than a critical diameter of 20 nm are activated and eventually removed in the convective cloud.

TABLE 1 Parameters for the initial log-normal distributions

		Clean Marine	Polluted Continental
Aitken mode (dry)	N (/cm^3)	300	2800
	D_p (nm)	36	76
	σ	1.42	1.52
Accumulation mode (dry)	N (/cm^3)	70	350
	D_p (nm)	150	230
	σ	1.47	1.36

The remaining surface area, which is small, together with the initial high relative humidity and the low temperatures, leads to a burst of nucleation of new particles. After the first day the particle number steadily decreases due to coagulation, while condensation of H_2SO_4 increases the aerosol mass. The distribution after 15 days of transport in the FT is mono-modal, in agreement with measurements at the free tropospheric station of Izana only in the cleanest conditions (2). When after 15 days this aerosol entrains into the MBL, in-situ MBL nucleation is quenched by the surface area of the entrained aerosol. Entrainment and MBL cloud processing determine the final aerosol size distribution. The latter exhibits the same features as the initial size distribution (Figure 1a), which supports the sequence of microphysical and dynamical processes used to explain these size distributions. The mode at $D_{p,dry}$ = 40 nm is due to the shape of the FT size distribution, while the second mode is formed by the cloud processing in the MBL stratiform clouds.

According to this model, formation by nucleation in the upper troposphere in cloud outflow and subsequent coagulation during subsidence would explain the observed increase in total number concentration with increasing altitude over the remote oceans. Condensation increases the aerosol mass and shifts the aerosol size distribution towards larger sizes, but due to the decreasing availability of SO_2 the importance of the process diminishes during transport in the FT. Entrainment of free tropospheric particles in the MBL is the process determining the final MBL particle number, whereas cloud processing by marine stratus clouds is the main contributor of the aerosol mass increase.

Polluted Continental case

The difference with the previous simulation is that the model is initialized with an aerosol size distribution and SO_2 concentration, typical of continental polluted conditions (3). The aerosol is an external/internal mixture consisting of water-soluble H_2SO_4 and insoluble soot. The average water-soluble mass fraction is 28%. Only the insoluble and part of the mixed particles survive wet deposition in the convective cloud. Although this offers a larger surface area for condensation than in the clean case, nucleation still does occur because of the larger concentration of SO_2 that survives through the cloud.

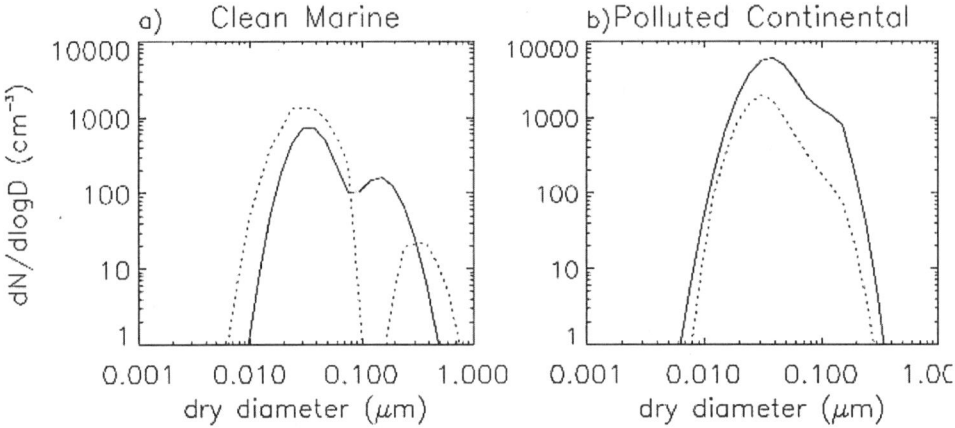

FIGURE 1 Initial (solid line) and final (dotted line) size distributions for the Clean Marine (CM) (a) and Polluted Continental (PC) (b) case.

Twelve hours after the injection in the FT, condensation of H_2SO_4 and coagulation with the nucleated soluble particles have transformed the insoluble particles into mixed particles. During the FT transport the total number of particles, which initially are mainly soluble, decreases due to coagulation. The mixed particle number changes very little in the FT, as there are no insoluble particles available to create new mixed particles. Figure 1b shows the size distribution after 15 days of transport in the FT. The distribution is bimodal, with the largest mode being the boundary layer aerosol, modified by convective transport and transport in the FT, and the smallest mode the new aerosol nucleated near the convective cloud. The bimodal distribution is in qualitative agreement with most measurements at the free tropospheric site of Izana (4).

REFERENCES

1. Vignati E., *Modelling interaction between aerosols and gaseous compounds in the polluted marine atmosphere*. PhD Thesis, Risø National Laboratory, Report No. Risø-R-1163(EN), pp 133 (1999).
2. Raes, F., R. Van Dingenen, E. Cuevas, P.F.J. Van Veithoven, and J.M. Prospero, Observations of aerosols in the free troposphere and marine boundary layer of the subtropical Northeast Atlantic: Discussion of processes determining their size distribution, *J. Geophys. Res.*, 102, 21315-21328 (1997).
3. Bates T.S., P. K. Quinn, D.S. Covert, D.J. Caffman, J.E. Johnson, and A. Wiedensohler, Aerosol physical properties and processes in the lower marine boundary layer: a comparison of shipboard sub-micron data from ACE 1 and ACE 2, *Tellus*, In press (2000).
4. Raes F., R. Van Dingenen, E. Vignati, J. Wilson, J.P. Putaud, J.H. Seinfeld, and P. Adams, Formation and cycling of aerosols in the global troposphere. *Atmospheric Environment,* Accept. for publication (2000).

The Free Tropospheric Aerosol, Origin and Properties

H.C. Hansson

Stockholm University, 106 91 Stockholm, Sweden

Abstract. FREETROPE, a subproject within the ACE-2 experiment, focused on the origin and the properties of the free tropospheric (FT) aerosol, was conducted during the summer 1997 on and close to Tenerife. Major findings discussed are chemical and physical properties and how they relate to what was found for the marine boundary layer (MBL) in simultaneous measurements. The observations indicate that the FT has a strong influence of aerosol originating from combustion. Newly formed particles were only observed once in the FT. In all indicating advection of particles from ground based sources contribute to a considerable fraction of the aerosol mass.

1. INTRODUCTION

One of the subprojects of Aerosol Characterization Experiment 2 (ACE-2) with in the IGAC (International Global Atmospheric Chemistry) project was the "Free tropospheric aerosols and their mixing with the Marine Boundary layer (FREETROPE). This subproject focused on the properties of the atmospheric aerosol in the free troposphere, how it develops during transport and if possibly it significantly influences the content and development of the boundary layer aerosol. For an overview of ACE-2 see Raes et al., 2000 [1].

FREETROPE was conducted during the summer1997 at the island Tenerife in conjunction with other ACE-2 subprojects. Two groundbased stations, one at a light house (50 m asl) at the north western shore of Tenerife and one at the observatory Izana (2500 m asl) situated at the mountain ridge dividing the island in south – north direction, were used for detailed chemical and physical characterization of the aerosol. By restricting the sampling such that local influence was minimized, the data collected are considered as representative of the marine boundary layer and the free troposphere [2, 3]. From daily meteorological soundings was seen that Izana at midday often was situated close to a slight shift in the potential temperature that is interpreted as the boundary between an intermediate layer, the buffer layer, and the FT. It is assumed that Izana gives a representative signal of the FT, as the measurements are limited to nighttime when up slope air motions has stopped.

One aircraft, the Dutch Cessna Citation, operated by the University of Delft and University of Utrecht, was equipped with gas and aerosol instrumentation focused on tracing incidents of nucleation and particle growth in the free troposphere [4].

2. RESULTS AND DISCUSSION

The chemical composition was found to differ between the FT and the MBL concerning background conditions, i.e. originating from the Mid Atlantic or Arctic [2]. Putaud et al. made a major effort in assuring the best possible accuracy in their measurements and attempted error estimate to address significant variations. The major difference found was a generally higher content of organic and black carbon in the FT. When the FT aerosol was traced back to the North American Continent the OC/BC fraction increased with more than a factor 2. However the stated accuracy of the OC/BC measurements was strongly questioned by Huebert and Charlson [5], resulting in a debate whether the FT aerosol really is enriched in OC/BC compared with the MBL. It is very interesting whether a general difference between the FT and the MBL were found as it reflects how major mixing processes affect the chemical and physical characteristics of the aerosol.

The particle size distributions in the FT were found to vary most drastically, which was interpreted as a result of a stable layering in the FT, where each layer had its own history concerning what processes the particles gone through. The MBL measurements showed generally a very stable size distribution, with a clearly discernable Aitken and accumulation mode. In both FT and MBL no particle modes were found below 20 nm, indicating the absence of recent nucleation events. A clear dependence on volume concentration were found in the MBL as with increasing concentrations, i.e. higher anthropogenic influence, the particle mode diameters grew and the highest concentrations observed the Aitken was nearly not separable from the accumulation mode [3]. Another finding was a small number of nuclei mode particles was observed during almost the whole measurement period, mostly in the range 10 – 50 particles/cc and at a few occasions reaching to 100.

Another major difference found comparing the FT and the MBL was the hygroscopic growth factors (HGF), that shows the change in particle diameter when the particles go from low RH (< 20%) to 90% RH. The MBL aerosol showed consistently a unimodal behavior with a mean growth factor of 1.65 to 1.75 when increasing particle diameter from 30 to 450 nm [6]. At some few occasions a mode with a HGF of about 1.4 is observed. In the FT, the measurements were only made during a shorter time period of about 10 days and only at one particle diameter, 50 nm. Two hygroscopic modes were then observed with HGF at about 1.4 and 1.6, respectively, with the lower one dominating in occurrence.

The chemistry strongly influences the uptake of water at different relative humidity. The response on the HGF as measured with the instruments used in ACE2, the Tandem Differential Mobility Analyzer (TDMA) has been carefully investigated in laboratory experiments using pure inorganic as well as with added organic substances. [7, 8, 9]. They find excellent agreement with theory when assuming the organic components as an insoluble fraction of the particle. One ambient study, Zhang et al., [10], indicates the organic fraction in an urban aerosol inhibits water uptake while it enhances the water uptake in an aged aerosol. The passive role of organic components is confirmed in a closure study on polluted and marine background aerosol [11]. However the errors in that comparison allow deviations from the assumption. The database of the ACE-2 measurements is quite limited but the HGF's found indicate a higher non-soluble fraction in the FT that might be attributed to a higher fraction of OC/BC.

Aircraft measurements with not only the Cessna Citation, but also aircraft involved in the ACE-2, did see very few occasions of nucleation, i.e. detection of a particle mode in the size range below 10 nm. The Cessna detected at the only early morning hour flight the only substantial increase in particle number below 10 nm. Returning to the same air parcel twice during the four-hour flight, they observed the nucleation mode disappear [4]. They interpreted this nucleation event as a result of strong advection of the air parcel in a frontal system close to the North American East Coast, cleansing the air from larger particles and thus lowering the particle surface and the air temperature, thus in total facilitating nucleation. Concentrations of acetylene, propane and ethane were in this and other flights found to be substantially increased at 6 to 11 km altitude compared to lower levels [12].

In conclusion, there is a set of indications concerning the gas composition, chemical composition and physical properties that the FT aerosol contains more pollution derived compounds than found in the boundary layer. However it should be noted that the measurements on ground cover a limited time on only one island and there have been raised questions concerning the accuracy of the OC/BC measurements. Consequently much more data is needed, not only over longer time periods but also from many other sites, to draw any general conclusions. Further the circumstances in both layers seem also to be such that not any nucleation of significance is taking place.

These results can be considered as logical in the view of likely characteristics of anthropogenic influenced continental aerosol and how those can be modified by different processes occurring in an advection through clouds and further transport in the FT. When going to higher relative humidities and especially into supersaturated environments as clouds several theoretical investigations have shown that the organic components can have a crucial influence on the water uptake (e.g. [13]). Martinsson et al., [14], made a closure study allowing a test of whether organic components in background air can be assumed to be hygroscopically inactive in a orographic cloud. However one case of four disproved the hypothesis. Still the inorganic compounds are much more soluble than the known organic compounds and as well concentration of e.g. sulfates is usually higher than the water-soluble organic compounds and thus dominating the droplet formation. This means that it is mainly the particles with sufficient high amount of inorganic compounds that will form cloud droplets. Consequently these particles/droplets will encounter a higher risk of being scavenged by precipitation.

Small particles with low amounts of inorganic salts will have the largest possibility to advect through a cloud to high altitude, i. e. to the upper FT without being scavenged. Consequently the aerosol should be enriched on small particles and on insoluble compounds as BC, the non water soluble OC, and perhaps also slightly soluble OC. Accordingly if measurements where made the size distribution should be shifted towards a domination of fine Aitken mode particles and the HGF should be dominated by particles with a low HGF. Well into the upper FT the aerosol mainly change due to nucleation, coagulation and condensation. Liquid phase processes are most likely insignificant due to low cloud activity and low relative humidity.

Accordingly to the measurements during ACE2, nucleation occurs rarely and thus particles grow by condensation while subsiding into the lower troposphere.

For the general composition of the FT aerosol and its properties it is most important to identify the main paths for aerosols to enter the upper FT as well as understand how these effects the properties of the particles. FREETROPE contribute to this showing that the FT aerosol is not only sulfates formed at high altitude but also contain OC/BC coming most likely from particles advected from ground. Clouds clearly do not scavenge all particles.

REFERENCES

1. Raes, F., Bates, T., McGovern, F., and Van Liedekerke, M., The Second Aerosol Characterization Experiment (ACE-2): General overview and main results, *Tellus* **52B**, 119-110 (2000)
2. Putaud, J.-P., VanDingenen, R., Mangoni, M., Virkkula, Maring, H., Prospero., J. M., Swietlicki, E., Berg, O. H., Hillamo, R., and J. Mäkelä, Chemical mass closure and assessment of the origin of the submicron aerosol in the marine boundary layer and the free troposphere at Tenerife during ACE-2, *Tellus* **52B**, 141-168 (2000)
3. VanDingenen, R., 2000, personal communication
4. De Reus, M., Ström, J., Curtius, J., Pirjola, L., Vignatti, E., Arnold, F., Hansson, H-C, Kulmala M., Lelieveld, J., and Raes., F., Aerosol production and growth in the upper free troposphere, 2000, submitted to *J. Geophys. Res.*
5. Huebert, B. J. and Charlson, R. J., Uncertainties in Data on Organic Aerosols, A short communication to Tellus, *Tellus* **52B** (2000)
6. Swietlicki, E., Zhou, J., Covert, D.S., Hämeri, K., Busch, B., Väkevä, M., Dusek, U., Berg, O. H., Wiedensohler, A., Aalto, P., Mäkelä, J., Martinsson, B.G., Papaspiropoulos, G., Mentes, B., Frank, G. And Stratmann, F., Hygroscopic properties of aerosol particles in the north-eastern Atlantic during ACE-2, *Tellus* **52B**, 201-227 (2000)
7. Svenningsson, B., Ph. D. Thesis, Hygroscopic Growth of Atmospheric Aerosol Particles and Its Relation to Nucleation Scavenging in Clouds, Depart. of Nucl. Physics, Univ. of Lund, Sweden, ISBN 91-628-2764-2 (1997)
8. Hansson, H-C, Rood, M. J., Koloutsou-Vakakis, S., Hämeri, K., Orisini, D., and Wiedensohler, A., NaCl Aerosol Particle Hygroscopicity Dependence on Mixing with Organic Compounds, *J. of Atm. Chem.* **31**, 321-346 (1999)
9. Virkkula, A., VanDingenen, R., Raes, F. and Hjort, J., Hygroscopic properties of aerosol fromed by oxidation of limonene, α-pinene, and β-pinene, *J. of Geophys. Res.*, **104, D3**, 3569-3579 (1999)
10. Zhang, X. Q., McMurry, P. H., Hering, S. V. And Casuccio, G. S., Mixing characteristic and water content of submicron aerosols measured in Los Angeles and at the Grand Canyon, *Atmos. Environ., Part. A*, **27**, 1593-1607 (1993)
11. Swietlicki, E., Zhou, J., Berg, O. H., Martinsson, B. G., Frank, G., Cederfält, S.-I., Dusek, U., Berner, A., Birmili, W., Wiedensohler, A., Yuskiewicz, B., Bowers, K. N., A closure study of sub-micrometer aerosol particle hygroscopic behavior, *Atmospheric Res.*, **50**, 205-240 (1999)
12. Scheeren, B., 2000, personal communication.
13. Kulmala, M., Laaksonen, A., and Charlson, R. J., Clouds without supersaturation, *Nature* **388**, 336-337
14. Martinsson, B. G., Frank, G., Cederfält, S.-I., Swietlicki, E., Berg, O. H., Zhou, J., Bower, K. N., Bradbury, C., Birmili, W., Stratmann, F., Wendisch, M., Wiedensohler, A., Yuskiewicz, B., Droplet nucleation and growth in orographic clouds in relation to the aerosol population, *Atmospheric Res.*, **50**, 289-315 (1999).

Mineral Aerosol Production, Transport, and Removal During ACE-2: Comparisons of an Event Model to Satellite

Peter R. Colarco and Owen B. Toon

Laboratory for Atmospheric and Space Physics
Program in Atmospheric and Oceanic Sciences
University of Colorado, Boulder, Colorado 80309

Abstract. The long-range transport of mineral aerosols has climatic implications because of the aerosols' radiative properties. We have incorporated an aerosol microphysical model into the NCAR Model for Atmospheric Transport and Chemistry (MATCH) in order to carry out three-dimensional simulations of dust emission and transport from the Sahara Desert. A detailed source scheme has been implemented which takes into account the surface soil and roughness properties in determining the spatial and temporal variation in dust emissions. The calculated dust fluxes are partitioned across a large number of size bins, and we treat the transport and removal processes on each size bin independently to examine how the aerosol size distribution evolves. We account for dust particle removal by sedimentation, deposition, and wet scavenging processes. The combined model is driven by assimilated meteorological fields and has been applied to a case study of the ACE-2 experiment (June and July, 1997). We constrain dust fields predicted by our model against the remotely sensed and in-situ data available from the experiment time frame. The model is able to reproduce dust optical depths and a plume shape in good agreement with observations made by the AVHRR and TOMS satellite instruments. We present detailed radiative transfer calculations based on our modeled dust fields in order to simulate radiances observed by satellite.

MODEL DESCRIPTION

Mineral aerosols scatter and absorb solar and infrared radiation, but their radiatively important properties are not well known. In particular, the spatial and temporal variations of the dust particle size distribution are poorly understood. In order to address this issue we have developed a coupled transport and microphysical model capable of simulating the dust lifecycle. This model predicts the three-dimensional spatial distribution of dust in the atmosphere, and includes an explicit representation of the aerosol size distribution and detailed parameterizations of production and removal mechanisms. Since the model is driven by observed wind fields it is useful for simulating real events which can be validated against empirical data sets.

The model is built from the NCAR Model for Atmospheric Transport and Chemistry (MATCH) (1). MATCH is an off-line chemical tracer model which includes the same basic physics package as the NCAR CCM2 and can be driven by assimilated meteorological fields. We run MATCH at T63 horizontal resolution (approximately 1.8° × 1.8°) and 28 vertical layers, driving it with NCEP reanalyses.

The dynamical fields MATCH computes are fed into our aerosol microphysical model, a version of the NASA Ames Aerosol and Radiation Model (2), where dust production, transport, and removal is calculated.

Dust production is calculated using the source parameterization of Marticorena and Bergametti (3), which accounts for surface soil and roughness properties in determining the mass flux of dust emitted as a function of the 10-m wind speeds. We partition the dust flux across 30 radius bins (0.05 – 30.0 µm) using a size distribution representative of observed background desert aerosol size distributions, and we treat the transport and removal processes affecting each size bin independently in order to examine the evolution of the particle size distribution. We account for particle removal by gravitational sedimentation, turbulent deposition, and wet scavenging.

The model is applied to a case study of dust emission and transport from the Sahara Desert during the ACE-2 experiment time frame (June and July, 1997). In order to speed up our aerosol model calculations we reduce the global dynamical fields predicted by MATCH to a regional grid of the same resolution (Figure 1). A detailed cartography of Saharan surface characteristics is implemented in our source scheme (4). Dust emissions and transport are calculated for the period June 18 – July 18, 1997. In the following sections we describe two approaches to model validation, first by direct comparison to observations made during ACE-2 and then through radiative transfer calculations that can be validated against satellite observations.

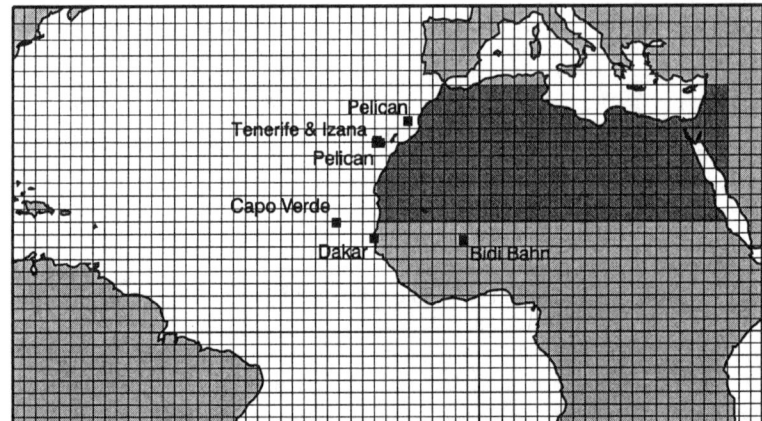

FIGURE 1. The above figure shows the model domain for our simulations, where the grid lines indicate the horizontal resolution of the model. The shaded area over northern Africa is our Saharan source region. The shaded boxes indicate ACE-2 and associated measurement sites.

COMPARISON TO ACE-2 CORRELATED OBSERVATIONS

While a number of observations of Saharan dust were made during ACE-2, we restrict our attention to two sets of measurements, one from aircraft vertical profiles and the other from surface based sunphotometer measurements. We consider vertical profiles flown through the dust plume near Tenerife by the Cirpas Pelican remotely piloted aircraft on July 8. The aircraft had several particle counters on board that were able to resolve the aerosol size distribution as a function of altitude. Comparison of retrieved volume size distributions to those modeled shows good qualitative agreement for points above 2000 m in altitude and somewhat poorer agreement for lower points. The modeled atmospheric column used for this comparison shows low-level precipitation, which is consistent with satellite imagery of this region, though the aircraft flew through a cloud free environment while taking profiles. This suggests that the horizontal resolution of boundary layer precipitation is a possible explanation for the discrepancy between the modeled and measured vertical distribution of aerosols at lower altitudes.

We also compare our model results to retrievals made by the AERONET network of ground-based sunphotometers. By assuming the index of refraction of the dust, we compute the temporal evolution of the column aerosol optical depth. In general there is a very good agreement in both the temporal variation and magnitudes of optical depth (Figure 2). Additionally, there is a good agreement between the modeled column integrated size distribution and that retrieved at the free troposphere AERONET site at Izana (altitude 2500 m).

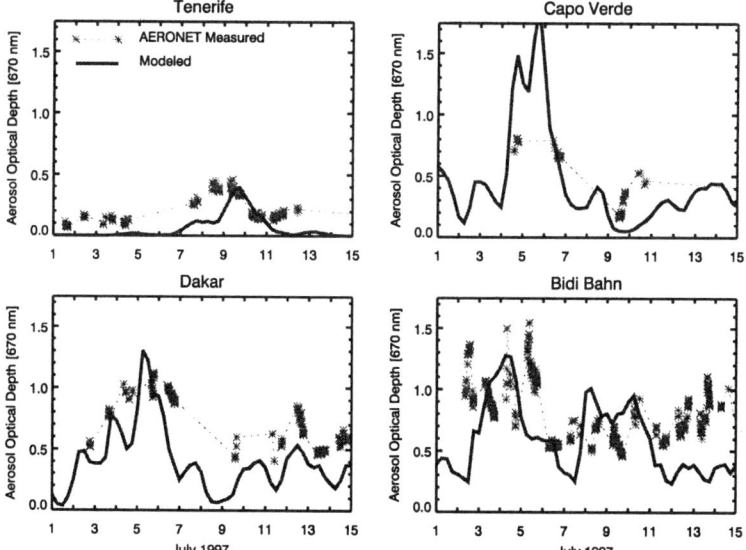

FIGURE 2. Comparison of the modeled to AERONET retrieved 670 nm optical depth at four sites: Tenerife, Capo Verde, Dakar, and Bidi Bahn.

COMPARISON TO SATELLITE OBSERVATIONS

We apply Mie scattering theory to compute the phase functions and scattering and extinction coefficients associated with our predicted dust fields. Integrating over the modeled size distributions we calculate column optical depths at 0.63 μm for direct comparison to the retrieved AVHRR aerosol optical thickness product and qualitative comparison to the TOMS aerosol index product. These comparisons show a good agreement between the model and the satellite observations in the placement and extent of the dust plume as well as the magnitude of the optical depth. A further series of calculations are carried out using the modeled aerosol phase functions and a multi-scattering radiative transfer code in order to compute top-of-atmosphere backscatter radiances for more direct comparison against the TOMS measurements.

ACKNOWLEDGMENTS

We would like to acknowledge the assistance of Phil Rasch with the MATCH model and Beatrice Marticorena for help with the dust lifting parameterization. Thanks also to Phil Russell for assistance with the ACE-2 data sets. Additional thanks to Brent Holben and Didier Tanré for the availability of the AERONET data. One author (PRC) is supported by the NASA Earth System Science Fellowship.

REFERENCES

1. Rasch, P.J., N.M. Mahowald, and B.E. Eaton, *Journal of Geophysical Research* **102**, 28127-28138 (1997).
2. Toon, O.B., R.P. Turco, D. Westphal, R. Malone, and M.S. Liu, *Journal of the Atmospheric Sciences* **45**, 2123-2143 (1988).
3. Marticorena, B., and G. Bergametti, *Journal of Geophysical Research* **100**, 16415-16430 (1995).
4. Marticorena, B., G. Bergametti, B. Aumont, Y. Callot, C. N'Doumé, and M. Legrand, *Journal of Geophysical Research* **102**, 4387-4404 (1997).

PLUME PROCESSING OF JET ENGINE EXHAUST AEROSOLS INJECTED INTO THE UPPER TROPOSPHERE AND LOWER STRATOSPHERE

D.E. Hagen, P.D. Whitefield, J. Paladino, and O. Schmid

Cloud and Aerosol Science Laboratory, Department of Physics, University of Missouri-Rolla, Rolla, MO 65401 U.S.A.

Abstract. This paper describes measurement results obtained from jet engine exhaust emission studies with a focus on particulate emissions. The measurement venues include both ground based studies with no plume processing and airborne cases with variable amounts of plume processing.

INTRODUCTION

Aviation has experienced rapid expansion as the world economy has grown. Over the last 10 years passenger traffic on scheduled airlines has increased by 60%, and over the next 10 to 15 years demand for air travel is expected to grow by about 5% per year (Brasseur et al. [1]). These aircraft emit gases and particles into the atmosphere directly. Most of these emissions occur in the upper troposphere and lower stratosphere while the aircraft expend most of their fuel during any given flight plan. These gases and particles alter the concentration of atmospheric greenhouse gases and trigger the formation of condensation trails (contrails) and may increase cirrus cloudiness (Penner et al. [2]). This paper will discuss evidence for plume processing, and in particular gas to particle conversion, of jet engine exhaust aerosols injected into the upper troposphere and lower stratosphere. In particular, the paper will focus on measurements and their interpretation made by the Cloud and Aerosol Sciences Laboratory at the University of Missouri – Rolla (UMR).

DISCUSSION

These measurements, both ground based and in flight, have been made with the Mobile Aerosol Sampling system (MASS) developed at UMR during the early 1990's. A brief description of the MASS facility will be given including discussion of methodological and configurational differences for ground-based versus airborne measurements (Hagen et al. [3], Whitefield et al. [4]). The same suite of instrumentation was used to sample exhaust emission particulates within one meter of the engine exhaust plane in ground measurements, and in flight to sample the emissions at distances from 100m to 10's of kilometers behind the engine. In some cases the same aircraft and engines were studied in ground test and airborne campaigns.

The study of atmospheric effects of aircraft emissions has been of considerable interest to the international aerospace community during the last decade. As a result, several ground-based and airborne measurement campaigns have been funded by NASA, the European Economic Community, the United States Air Force and members of the Aerospace Industry, in which UMR has had the opportunity participate (see table 1).

This paper will review and intercompare the particulate data from these campaigns. Evidence for gas to particle conversion, as plumes evolve, will be presented. Table 1 presents a list of the field campaigns from which this data is taken.

TABLE 1

Recent Field Campaigns associated with Plume Processing of particulates

Campaign Name	Reference or Campaign Description and supporting agencies
POLINAT 1 (airborne)	Pollution from Aircraft Emissions in the North Atlantic Flight Corridor, Contract no. EV5V-CT93-0310 (DG 12 DTEE) Ed. U. Schumann, Commission of European Communities Final Report October 1996. (EEC, NASA)
AEDC (ground-based)	Experimental Characterization of Gas Turbine Emissions at Simulated Flight Altitude Conditions. Ed. Robert Howard, AEDC TR-96-3 June 1996. (NASA)
SUCCESS (airborne)	Hagen, D., P. Whitefield, J. Paladino, M. Trueblood, and H. Lilenfeld, "Particulate sizing and emission indices for a jet engine exhaust sampled at cruise", Geophys. Res. Letts. 25, 1681-1684 (1998). (NASA) Paladino, J., P. Whitefield, D. Hagen, A. Hopkins, and M. Trueblood, "Particle concentration characterization for jet engine emissions under cruise conditions", Geophys. Res. Letts. 25, 1697-1700 (1998). (NASA)
SNIFF (airborne	Workshop on Aerosols and Particulates from Aircraft Gas

and ground-based)	Turbine Engines, NASA/CP-1999-208918, June 1999. (NASA)
POLINAT 2 SONEX (airborne)	Pollution from Aircraft Emissions in the North Atlantic Flight Corridor (POLINAT 2), E.E.C. Air Pollution Research Report 68, Luxembourg, December 1998.(EEC,NASA)
F100 (ground-based)	Engine Gaseous, Aerosol Precursor and Particulate at Simulated Flight Altitude Conditions, NASA/TM-1998-208509, October 1998. (NASA)

ACKNOWLEDGMENTS

The authors would like to acknowledge the support and sponsorship of the research described in this paper by the University of Missouri Research Board, the DLR Oberpfaffenhofen Institute of Atmospheric Physics, National Aeronautics and Space Administration, and the U.S. Air Force.

REFERENCES

1. Brasseur, G.P., R.A. Cox, D. Hauglustaine, I. Isaksen, J. Lelieveld, D.H. Lister, R. Saussen, U. Schumann, A. Wahner, and P. Wiesen., European scientific assessment of the atmospheric effects of aircraft emissions, Atmospheric Environment, 32, 2327-2422 (1998).
2. Penner, J.E. and D.H. Lister, The Intergovernmental Panel on Climate Control (IPCC) Special Report on Aviation and the Global Atmosphere, published in May 1999..
3. Hagen, D., P. Whitefield, J. Paladino, M. Trueblood, and H. Lilenfeld, "Particulate sizing and emission indices for a jet engine exhaust sampled at cruise", Geophys. Res. Letts. 25, 1681-1684 (1998.
4. Whitefield, P.D., D.E. Hagen, J. Paladino and H.V. Lilenfeld, Particulate Characterization in the Near Field of Commercial Transport Aircraft Exhaust Plumes Using the UMR-MASS, Part 1. J. Geophys. Res. Atmospheres, 101, 19551.

MICROSCOPIC PARTICLE PROPERTIES

Novel Measurements of Atmospheric Aerosol Properties

Peter H. McMurry, William D. Dick, Xin Wang

Department of Mechanical Engineering
University of Minnesota
Minneapolis, MN 55455

Abstract. Methods that were recently developed in our laboratory to measure fundamental aerosol properties including density, water content, refractive index and shape are discussed. All of these techniques involve measurements on particles that have first been classified according to electrical mobility with a differential mobility analyzer (DMA). Density is measured by using Ehara's aerosol particle mass analyzer (APM) to measure the mass distribution of mobility-classified particles, while relative humidity-dependent water uptake involves use of the tandem differential mobility analyzer (TDMA). Refractive index and shape are measured by means of multiangle light scattering (MALS). Concurrent measurements of size-resolved chemical composition enabled us to determine the dependence of water uptake, refractive index and shape on the major submicron species including sulfates, organic carbon, ammonium, and soil dust.

INTRODUCTION

Atmospheric aerosol particles are highly diverse. Particles of a given size can include several particle types having distinct compositions, and particles of a given type often include multiple species. Furthermore, the average particle composition varies with time, size and location. This complexity makes it difficult to obtain accurate aerosol measurements since the response of aerosol measuring instruments depends on properties which are composition-dependent and are often not well known. For example, impactors classify particles according to aerodynamic size, which depends on particle size, shape, and density; optical particle counters classify particles according to optical equivalent size, which depends on size, shape, and refractive index; and electrical mobility analyzers classify particles according to electrical mobility equivalent size, which depends on size and shape. Determining the relationship between "sizes" measured with these instruments requires knowledge of the pertinent aerosol properties. Similarly, aerosol effects including light scattering and deposition in the lungs or on surfaces depend on particle properties.

In previous work, information on aerosol properties has typically been calculated from measured aerosol composition. Problems with this approach include uncertainties in properties of major aerosol species, especially organic carbon, and the impossibility of obtaining information on particle-to-particle variations from filter or impactor samples typically used to obtain information on composition.

In this paper we summarize several new approaches that our research group has developed to measure aerosol properties.

Water Content: The Tandem Differential Mobility Analyzer (TDMA)

The TDMA [1] has been widely used to measure the relative humidity (RH)-dependent water uptake of atmospheric aerosols [2,3,4]. TDMA data has recently been used to show that organic aerosols contribute to water uptake [5,6]. In this paper we summarize our most recent discoveries on the contribution of particulate organic carbon to water uptake during measurements in the Great Smoky Mountain National Park during the summer of 1995.

Figure 1a shows a typical example of the sensitivity of particle volume to relative humidity. In these measurements, 0.2 µm particles selected at 7 to 8% RH with the first differential mobility analyzer (DMA1) were humidified to the value shown on the abscissa of Figure 1a and classified according to size with DMA2. Note that particles grew monotonically with RH, even at very low values of RH.

The Zdanovskii-Stokes-Robinson thermodynamic model (ZSR, [7]) was used to calculate water uptake for the inorganic constituents of this aerosol (sulfate and ammonium were the only inorganics present in significant amounts). We assumed that the organic carbon (OC) and sulfate ($SO_4^=$) were internally mixed. The result of these calculations is shown in the curve labeled "deliquescent sulfate." Because the aerosol was dried to 8%, which is below the crystallization point for aqueous letovicite and

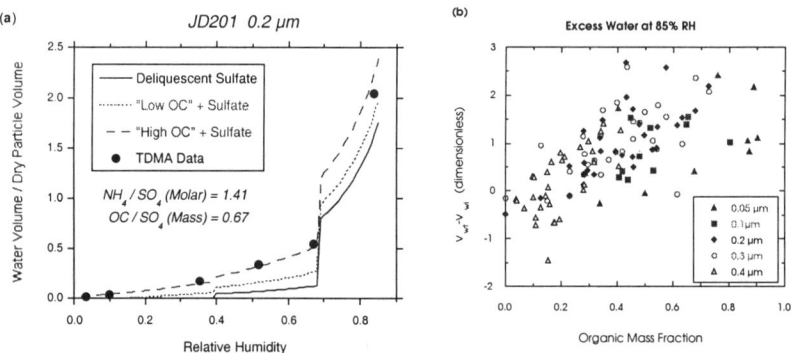

FIGURE 1. (a) Measured wet-to-dry volume ratios for 0.2 µm particles on July 20, 1995 in the Great Smoky Mountain National Park. Note that measured growth on this occasion significantly exceeds that predicted for internally mixed sulfate/organic carbon particles if the OC is assumed to be nonhygroscopic (solid line). Agreement between model and measurement is improved when the water uptake by OC is also included, as shown by the two dashed lines. (b) Dependence of "excess" water on measured OC mass fraction. Excess water is defined as the difference between water content measured by the TDMA, and water content that can be explained for sulfates using the ZSR thermodynamic model. (Journal of Geophysical Research Atmospheres 105(D1)1471-1479.)

ammonium sulfate solutions, we assume that these compounds are in their crystalline states at the inlet to the relative humidity conditioner downstream of DMA1. Note that the measured water uptake exceeds the value obtained with the ZSR model. We refer to the difference between the measured water uptake and the calculated water uptake for the inorganics as "excess water." Figure 1b shows the relationship between this excess water and the OC mass fraction for all data obtained in this study. Note that excess water tends to increase with increasing $OC/SO_4^=$ ratio. From this relationship we infer that OC is, to some extent, hygroscopic. Based on information similar to that shown in Figure 1b, estimates for the range of RH-dependent OC water uptake during this study were obtained. The dashed curves in Figure 1a compare the calculated water uptake including that taken up by OC and by the inorganics with TDMA measurements. Two curves that cover the range of OC water uptake observed during the study are shown. Note that agreement between measured and calculated water uptake improves significantly when the water uptake by OC is taken into account. Note also that OC appears to be responsible for most of the water uptake for RH < 50 to 70%. At elevated relative humidities the amount of water taken up by OC is typically small to that taken up by the inorganics, but it is nevertheless significant. Measurements from other studies in other locations suggest that OC water uptake depends very much on the OC composition. In urban areas OC tends to be less hygroscopic than in rural areas. We believe this may be because OC in aged rural aerosols may contain more highly oxygenated organic compounds. Additional work is needed to quantify the relationship between OC hygroscopicitiy and OC composition.

Particle Shape: Multiangle Light Scattering (MALS)

In his doctoral research, William Dick [8] investigated the use of multiangle light scattering from individual submicron particles to measure particle shape and refractive index. In this section we briefly describe the instrument used for this work [9] and discuss its use for measuring particle shape.

A photograph of the MALS instrument is shown in Figure 2. A circularly polarized argon ion laser beam (λ=488 nm) is focused and projected horizontally through the chamber, intersecting a vertically flowing stream of sheathed particle-laden air at the center of the chamber. Light scattered by an individual particle within the viewing volume is collected by 14 optical fibers with lenses that subtend collection half-angles of 1.25°. The fibers guide the light to 14 mini-photomultipliers for conversion to detectable current levels. Amplified Gaussian-shaped signals are recorded with a 16-channel transient waveform digitizer set to a sampling frequency of ~3 MHz per channel. For every particle detected, 16 sets of 512 digitized signal values are transferred from the digitizer to a PC. Post-measurement digital integration of amplitude is performed on all recorded pulses to minimize the effects of signal noise [10].

FIGURE 2. DAWN-A Multiangle light scattering instrument used to measure particle shape and refractive index [9]. Monodisperse particles that have been classified according to size with a DMA enter the hollow, spherical scattering chamber from the top. The particles pass through a beam of focused, circularly polarized argon ion light at the center of the chamber. The scattered light is simultaneously collected at each of fourteen detectors located over a range of polar (θ) and azimuthal angles (φ).

Light scattering measurements are made for particles of known size, thereby eliminating one of the three unknowns (size, shape, and refractive index) from the inverse problem. Monodispersions of laboratory or atmospheric aerosols are selected with a DMA and then are drawn into the DAWN-A scattering chamber. Shape information is obtained by placing detectors at all eight available azimuthal angular positions at a fixed polar angle of $\theta=55°$. Refractive index is inferred from measurements made with the remaining six detectors positioned at polar angles of

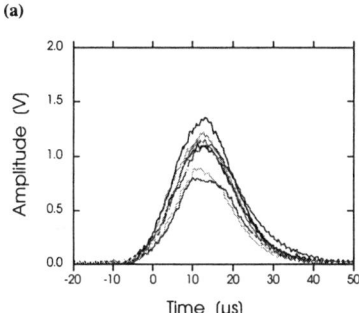
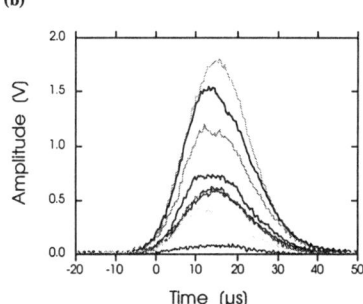

FIGURE 3. Aximuthal scattering signatures for (a) spherical aqueous 0.8 μm NaCl droplets at 79% RH and (b) 0.8 μm mobility-equivalent diameter NaCl crystalline cubes at 6% RH. (Measurement Sci. Technol. 9:183-196. Reproduced by permission of IOP Publishing Limited.)

θ=40, 75, 90, 125, and 140° and from the average responses of the eight 55° shape detectors.

Data showing azimuthal variabilities for spherical and nonspherical particles is shown in Figures 3a and b. Note that responses for the spherical NaCl droplets are more uniform than for the NaCl cubes. We define the "sphericity index," SPX, as:

$$SPX = 1 - \frac{\sqrt{\sum_{i=1}^{8}(x_i - \overline{x})^2}}{7\overline{x}} \quad (1)$$

where x_i is the peak response of the detector at azimuthal position φ_i and $\overline{x}$ is the mean response for the eight azimuthal detectors.

Data showing SPX distributions for monodisperse laboratory-generated DOS spheres and for NaCl cubes of 0.2 μm and 0.5 μm are shown in Figure 4. Note that

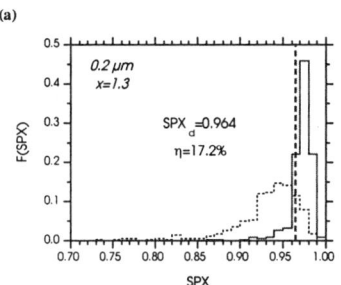
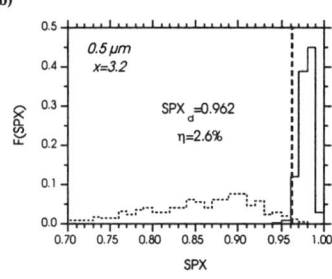

FIGURE 4. SPX distributions for monodisperse laboratory-generated aerosols of 0.2 μm (a) and 0.5 μm (b). The solid line is for spherical DOS oil droplets, and the dashed line is for NaCl cubes. The vertical dashed line nominally distinguishes spherical from nonspherical particles. In each case, SPX distributions were obtained by measuring the instruments' response to 512 particles. (Measurement Sci. Technol. 9:183-196. Reproduced by permission of IOP Publishing Limited.)

MALS provides reasonable distinction between cubes and spheres for particles as small as 0.2 μm

Figure 5 shows the results of atmospheric measurements for 0.4 and 0.5 μm particles. The open squares show the fraction of particles found to be nonspherical using MALS. The aerosol was dried to less than 10% RH before the MALS measurements were carried out. The solid triangles show the fraction of the aerosol that was soil dust, as determined by chemical analyses of impactor samples. Note that except for the period from July 24-26, there is a reasonably good correlation between the fraction of the aerosol that is nonspherical and the fraction that is soil dust. This suggests that many of the nonspherical particles were crustal in origin. An independent analyses has shown that Saharan dust impacted our sampling site during the July 24-26 period. We hypothesize that because these crustal particles had undergone atmospheric processing during transport from Africa, they may have acquired a coating of sulfate or other compounds that made them appear more spherical. We have not confirmed this experimentally.

Also shown in Figure 5a is the fraction of the aerosol that was "less" hygroscopic, as determined with a TDMA. Again, there appears to be a correlation between the less hygroscopic fraction and the nonspherical fraction, suggesting that the same particles that were nonspherical were also nonhygroscopic. On average, about 10 to 15% of the submicron particles were nonspherical, although this fraction varied considerably from day-to-day.

Particle Refractive Index: Multiangle Light Scattering (MALS)

Mie theory provides the angular scattering characteristics for homogeneous spherical particles of known refractive index. In this work we assume that Mie theory accurately characterizes scattering from the particles that were found to be spherical using the technique described in the previous section. We then use Mie theory to find the best value of the refractive index to match the measured dependence of scattering

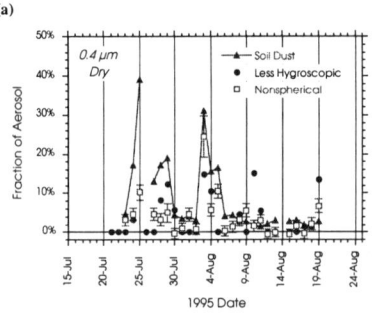
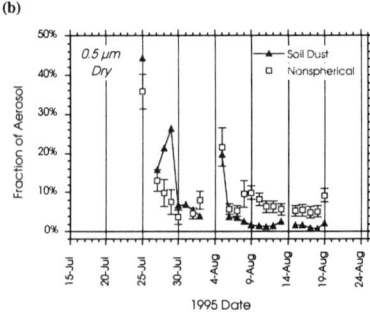

FIGURE 5. Nonspherical particle fractions measured by MALS in the Great Smoky Mountain National Park during the summer of 1995 are shown with the open squares for 0.4 μm (a) and 0.5 μm particles. Also shown is the fraction of the aerosol that was soil dust (solid triangles) and (in Figure 5a) the fraction of the aerosol in "less" hygroscopic particles, as determined with the TDMA. (Measurement Sci. Technol. 9:183-196. Reproduced by permission of IOP Publishing Limited.)

intensity on polar angle (θ). We did not have adequate sensitivity to recover both the real and imaginary component of refractive indices. Therefore, the imaginary component was calculated from elemental carbon concentration measured with an impactor.

Examples of the dependence of scattering intensity on polar angle for 0.2 and 0.8 μm atmospheric particles measured in the Great Smoky Mountain National Park are shown in Figure 6. The expected range of angular response for each size is represented by an upper curve (broken line) that has been calculated with the refractive index of solid ammonium sulfate (n=1.531 at λ=488 nm) and a lower curve (dashed curve) calculated with the index of water (n=1.337 at λ=488 nm). The solid line shows the best fit of Mie theory to the data.

A model was developed to calculate refractive indices based on the measured size-resolved aerosol chemical composition and water content. The water content included both water associated with inorganics and with organic carbon, as described above. Details of the model are described by Dick [8]. The measured and calculated refractive indices are compared in Figures 7a and b. Note that day-to-day variabilities in aerosol composition explain much of the measured variability in measured refractive index. The measured and modeled refractive indices agree to within about 0.02 refractive index units. The large differences between the "wet" and "dry" refractive indices reflect the dependence of refractive index on particle water content. Similar results were obtained for particles of 0.2, 0.4, 0.5, 0.6, 0.7 and 0.8 μm.

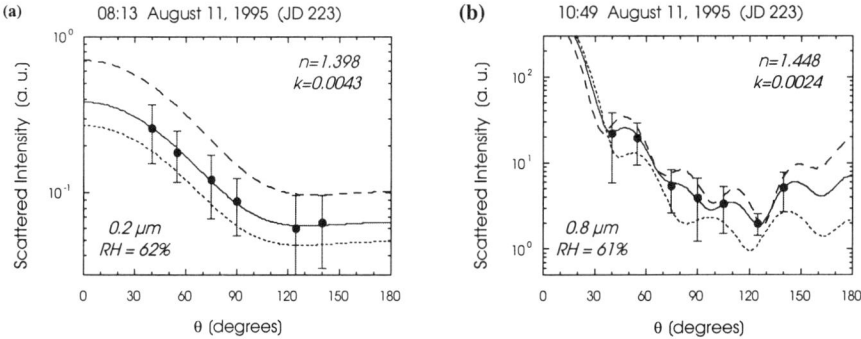

FIGURE 6. Angular dependence of light scattering for 0.2 and 0.8 μm particles. The solid line in the center shows the best fit of Mie theory to our MALS data. Upper and lower bounds for scattering are shown by the upper and lower curves, which apply to water and solid ammonium respectively.

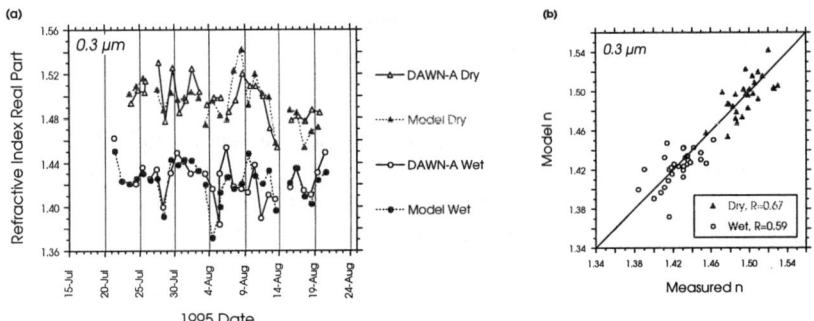

FIGURE 7. (a) Time series of refractive indices for dry (4-10% RH) and wet (44-76% RH) 0.3 μm particles. Both measured and calculated refractive indices are shown. (b) Correlation of calculated and measured refractive indices for all 0.3 μm atmospheric particles measured in the Great Smoky Mountain National Park during July and August 1995.

Particle Density: The Aerosol Particle Mass Analyzer (APM)

The aerosol particle mass analyzer described by Ehara and coworkers [12] was used during measurements in Atlanta, GA during the summer of 1999 to measure aerosol density. A schematic of this instrument is shown in Figure 8. The instrument consists of two cylindrical electrodes separated by a narrow annular gap. The entire assembly rotates about its axis at a known speed typically in the range of 1,000 to 7,000 rpm. The outer electrode is maintained at ground and while a voltage is applied to the inner electrode. As charged particles flow through the annular space separating the two electrodes they are subjected to a centrifugal force, which pushes them radially outwards, and an electrical force which draws them inwards. The voltage to the center electrode is adjusted until these forces balance, at which point particles penetrate through the annular region and can be detected downstream with a condensation particle counter (CPC).

We used the APM for the first time in studies of urban Atlanta, GA, aerosols. In these experiments, the TDMA was placed upstream of the APM to enable us to select particles of known electrical mobility equivalent size at a known relative humidity. In all cases, particles were dried to below 10% RH before they flowed into the APM. The experimental procedure involved first measuring the penetration of polystyrene latex spheres (PSL) through the APM as a function of the APM classifying voltage. The TDMA classifying voltage was adjusted to provide maximum penetration of the PSL spheres prior to carrying out the APM penetration measurements. After the PSL

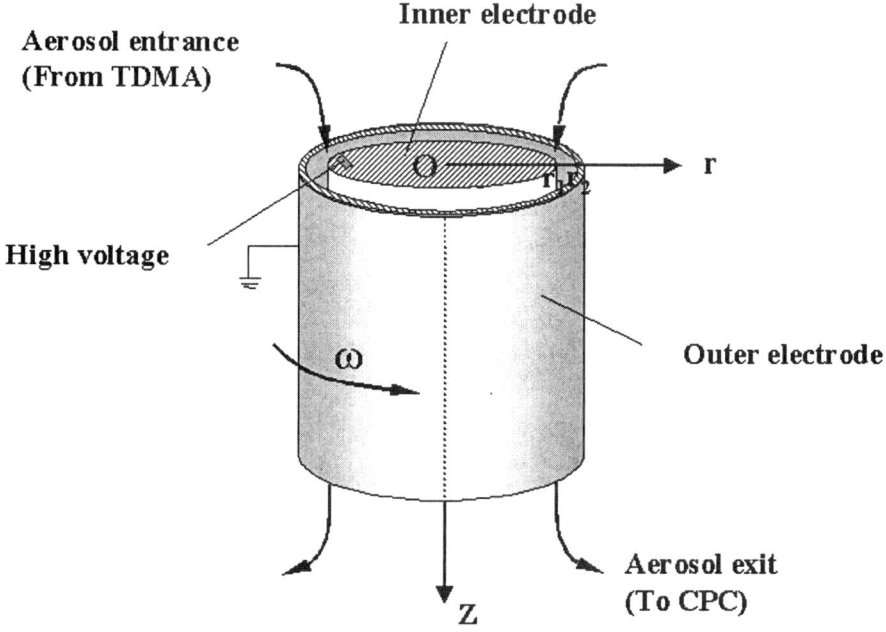

FIGURE 8. Schematic of the aerosol particle mass analyzer (APM).

penetration through the APM was completed, the penetration of atmospheric particles of precisely the same electrical mobility equivalent size was measured. The electrical mobility equivalent size was kept constant by operating the TDMA at exactly the same flow rates and classifying voltages for the atmospheric particles and the PSL.

Examples of our data from Atlanta, GA is shown in Figures 9a and b. These plots show the concentration downstream of the APM as a function of the APM classifying voltage for 0.107 μm and 0.309 μm mobility equivalent diameter particles. Data for both PSL and atmospheric particles are shown. Note that the 0.107 μm atmospheric particle data in Figure 9a show a single peak that occurs at a higher classifying voltage than was measured for PSL. In contrast, the 0.309 μm atmospheric particle data in Figure 9b show as many as four peaks, suggesting that particles of several distinct masses were present at the entrance to the APM. We used the APM to measure the mass distributions of 0.107 and 0.309 μm particles on ten days during August 1999 and we found that the mass distributions varied significantly from day to day, although the peak to the right of the PSL peak seen in Figures 9a and b was usually the dominant peak.

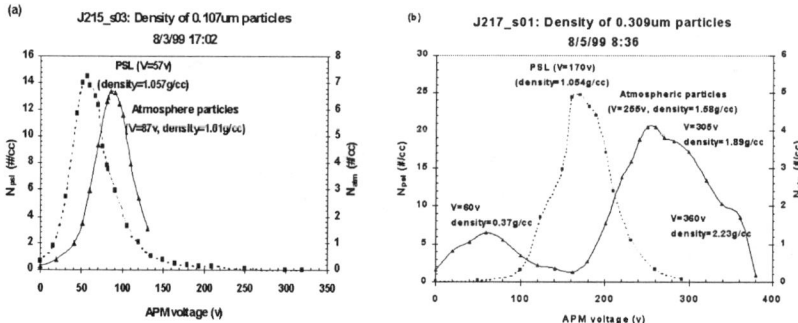

FIGURE 9. Concentration downstream of the APM as a function of APM classifying voltage for (a) 0.107 µm and (b) 0.309 µm particles

If particles are assumed to be spherical, then the density of atmospheric particles can be inferred from these data. The force balance on particles that penetrate through the APM can be expressed as:

$$qE = m\omega^2 r = \rho v \omega^2 r \qquad (2)$$

where the charge, q, equals a single elementary charge under our operating conditions, the electric field E varies in proportion to the APM classifying voltage, m is particle mass, ω is the rotational speed of the APM, r is the radial distance of the narrow annular gap from the axis of rotation, ρ is particle material density and v is particle volume. For the PSL and atmospheric particles of the same mobility equivalent size we operate at fixed rotational speed and vary the voltage. It follows that the density of the atmospheric particles is:

$$\rho_{atm} = \rho_{psl} \frac{V_{atm}}{V_{psl}} \qquad (3)$$

In deriving Equation (3) we assumed that the volume of the PSL and atmospheric particles is equal. Because both particle types passed through the TDMA before they were classified by the APM, this is precisely true if the atmospheric particles are spherical. However, if they are nonspherical then this will not be true. A proper analysis would require that we account for the effect of shape on aerodynamic drag as the nonspherical particles pass through the DMA (because transport through the APM does not involve drag, the APM measurements are unaffected by shape). In this paper we assume that all particles are spherical, although this is unlikely to be true. Thus, the densities shown in Figures 9 must be interpreted in light of this caveat. We believe the major peak to the right of the PSL peak applies to the most common type of atmospheric particle. Based on TDMA data and on impactor data, we believe these are hygroscopic compounds that contain primarily organic carbon and sulfates. Because the sulfates are hygroscopic, we believe these particles are spherical. Although we did not carry out MALS measurements in Atlanta, our MALS observations at the nearby Great Smoky Mountain National Park in the summer of 1995 support this hypothesis. These measurements showed that particles that apparently consisted of sulfates and organics appeared to be spherical even at relative humidities below 10%. If so, then the densities indicated for these particles (1.60±0.05 gcm^{-3} at ~7%RH) would be true material densities. We think it likely that the low mass particles shown in Figure 9b (peak at 60 V, density = 0.37 g/cm^3) probably consist of chain agglomerate carbon particles, similar to those emitted by diesel engines [13]. The true material density of such particles would, therefore, exceed 0.37 g/cm^3. The shoulders to the right of the 1.58 gcm^{-3} peak on Figure 9b were apparently produced by particles having higher densities. This diversity of densities for particles of a give mobility equivalent size is a new observation. Further work to identify the composition of these different particle types is needed.

CONCLUSIONS

We have developed techniques to measure aerosol properties including water uptake, shape, refractive index and density. Because such measurements require specialized equipment and personnel, we believe a high priority should be placed on developing experimentally verified models that can be used to infer properties from measured composition. We have made progress at developing models that are consistent with our measurements of water uptake and refractive index. This modeling work has shown that the uptake of water by organic carbon must be taken into account to obtain agreement between calculated and measured values.

ACKNOWLEDGMENTS

This research was supported by the Electric Power Research Institute (grant EPRI W09116-08/W404105-01) and by the Environmental Protection Agency (Georgia Tech Subgrant G-35-W62-G1; EPA Prime Grant Number R 826372-01-0). Although the research described in this article has been funded in part by the United States Environmental Protection Agency, it has not been subjected to the Agency's required peer and policy review and therefore does not necessarily reflect the views of the Agency and no official endorsement should be inferred.

REFERENCES

1. Rader, D. J., and McMurry, P. H., *J. Aerosol Sci.* **17**, 771-787 (1986).
2. McMurry, P. H., and Stolzenburg, M. R., *Atmos. Environ.* **23**, 497-507 (1989).
3. Svenningsson, B., Hansson, H. C., Wiedensohler, A., Noone, K., Ogren, J., Hallberg, A., Colvile, R., *Tellus* **44B**, 556-569 (1992).
4. Berg, O. H., Swietlicki, E., and Krejci, R., *J. Geophys. Res.* **103(D13)**, 16535-16545 (1999).

5. Saxena, P., Hildemann, L. M., McMurry, P. H., and Seinfeld, J. H., *J. Geophys. Res.* **100(D9)**, 18755-18770 (1995)
6. Dick, W. D., Saxena, P., and McMurry, P. H., *J. Geophys Res.* **105(D1)**, 1471-1479 (2000).
7. Stokes, R. H. and Robinson, R. A., *J. Phys. Chem.* **70(7)**, 2126-2131 (1966).
8. Dick, W. D., "Multiangle Light Scateering Techniques for Measuring Shape and Refractive Index of Submicron Atmospheric Particles," Ph.D. Thesis, University of Minnesota (1998).
9. Wyatt, P. J., Schehrer, K. L., Phillips, S. D., Jackson, C., Chang, Y. J., Parker, R. G., Phillips, D. T., Bottiger, J. R. *Appl. Opt,.* **27(2)**, 217-221 (1988).
10. Sachweh, B. A., Dick, W. D., and McMurry, P. H., Aerosol Sci. Technol. **23**, 373-391 (1995).
11. Dick, W. D., Ziemann, P. J., Huang, P.-F., McMurry, P. H., *Measurement Sci. Technol.* **9**, 183-196 (1998)
12. Ehara, K., Hagwood, C., Coakley, K. J., *J. Aerosol Sci.* **27**, 217-234 (1996).
13. McMurry, P. H., Litchy, M., Huang, P.F., Cai, X.P., Turpin, B.J., Dick, W.D., Hanson, A., *Atmos. Environ.* **30**, 101-108 (1996).

Application of Nucleation Theories to Atmospheric Aerosol Formation

Ari Laaksonen

Department of Applied Physics, University of Kuopio, Finland

Abstract. One-component and multicomponent nucleation theories are considered from the viewpoint of their predictive success. Special attention is given to the classical predictions regarding nucleation of sulfuric acid–water and sulfuric acid-water-ammonia. The application of the classical theory to atmospheric aerosol formation is discussed together with the possibility of explaining the observed correlation between sulfuric acid vapor concentration and new particle formation rate with the aid of ion-induced nucleation, or with the ternary nucleation mechanism involving ammonia.

INTRODUCTION

During the recent years it has become obvious that homogeneous nucleation events of fresh aerosol particles take frequently place in the atmosphere. This rather surprising fact is a result of a number of experimental studies of the evolution of ultrafine aerosol size distributions in different environments such as the marine boundary layer [1]-[5], forest areas [6], [7], plumes of pollution sources [8] and the upper troposphere [9]-[11]. Earlier, it was thought that atmospheric nucleation events probably require very special environmental conditions, e.g. very low temperatures (lower stratosphere), very high humidity (vicinity of evaporating clouds), or very low pre-existing aerosol concentrations. The new observations, which were made possible by advances in aerosol measurement techniques, have clearly changed this view.

What are the molecular species that nucleate in the atmosphere? At the moment, no clear-cut answer can be given. The reason for this is that the aerosol instruments measure concentrations of particles larger than about 3 nm (i.e. critical nuclei are not directly observed), and furthermore, there are no chemical characterization techniques available for particles in the ultrafine size range. From the theoretical viewpoint this is a somewhat frustrating situation: Even if the theory predicts that nucleation can take place at conditions in which a nucleation event was actually observed, one does not really know whether this is a valid prediction because the molecular species that cause the particle formation could be completely different from those that were assumed to be nucleating! The situation can only be remedied with experimental information concerning the nucleus compositions, and in the meantime the best we can do is to make good guesses as to what the particle forming species are, formulate theories on their nucleation properties, and try to validate the theories against laboratory experiments and computer simulations. This seems like a straightforward program, but it is of course complicated by several factors. First, the atmospheric aerosol forming molecules are presumably fairly non-ideal, and the only theory that has so far been applied to describe the nucleation of such species is the classical nucleation theory (including certain variants), which has several shortcomings. Secondly, the nucleating molecules necessarily have to have very low vapor pressures, which makes quantitative nucleation experiments very difficult to perform. Thirdly, the intermolecular potentials of the molecules presumed to be nucleating at atmospheric conditions are complicated, decreasing the reliability of the computer simulations.

Despite of the difficulties associated with the theoretical, experimental, and numerical studies of nucleation of atmospherically interesting molecules, some progress has been made in the recent years. Below, I will first survey the current status of one-component and multicomponent nucleation theories, and then focus on the more specific theoretical descriptions of sulfuric acid-water and sulfuric acid-water-ammonia nucleation and the reliability of the theoretical predictions. Finally, I will consider the atmospheric observations from the point of view of available theories.

ONE-COMPONENT NUCLEATION THEORIES

Nucleation theories can be categorized into purely molecular theories, phenomenological theories, and intermediate theories. The purely molecular theories (see e.g. [12]) are only applicable to simple molecules, and therefore I will not consider them further (although they are expected to yield valuable information for construction of more reliable phenomenological theories.) The phenomenological theories such as the classical nucleation theory (CNT) predict nucleation rates using as input various macroscopic, measurable fluid properties. Beside CNT, different scaling approaches have gained popularity in the recent years. Finally, the intermediate theories contain molecular-level information but also some degree of averaging. For example, the density functional theory (DFT) [13], which treats the fluid density as a continuos variable, uses intermolecular potentials and is therefore not applicable to real molecules. The diffuse interface theory (DIT) [14], on the other hand, gives the work of nucleus formation in principle in a similar manner as the DFT. However, the DIT relates the work of formation to a characteristic interface thickness, which is obtained from macroscopic fluid properties, and thus the theory is applicable to real molecules.

The CNT is based on two assumptions, namely the incompressibility assumption and the capillarity approximation. The first one is used in deriving the Kelvin equation

$$r^* = 2\sigma v / kT \ln(S) \qquad (1)$$

where r^* is the critical radius of the cluster, S denotes the saturation ratio of the vapor, T is temperature, k is the Boltzmann constant, v is the liquid-phase molecular volume and σ is the surface tension. The capillarity approximation equates the nucleus surface tension with that of macroscopic fluid with flat interface. The work of nucleus formation is

$$W^* = (1/3)A\sigma \tag{3}$$

where A denotes the surface area of the nucleus (assumed spherical). The nucleation rate is given by

$$J = K\exp(-W^*/kT) \tag{2}$$

where K is a kinetic pre-factor.

With the development of experimental techniques for measurement of nucleation rates at constant temperature it has been realized that the CNT predicts the saturation ratio dependence of J rather well for several substances whereas the temperature dependence is generally poorer. The Nucleation Theorem [15] relates the saturation ratio dependence of W^* and the number of molecules in the critical cluster as

$$dW^*/d(kT\ln S)_T = -g^*. \tag{4}$$

Furthermore, it can be shown that $d\ln(J)/(kT\ln S)$ is approximately equal to $(-g^*+1)$. A correct S-dependence of J therefore translates into a correct prediction of the molecular content of the critical nucleus. In general, the CNT seems to predict the nucleus size quite nicely for non-polar and weakly polar (water, alcohols) substances. However, for strongly polar substances the predictions are not that good, which may be due to the chainlike nature of the nuclei of such molecules [16].

Based on the success of CNT in predicting the S-dependence of J (and W^*), a theory [17] can be formulated in which the molecular content of the nucleus is the same as in CNT ($g^* = g^*_{CNT}$) and the work of nucleus formation is given by

$$W^* = W^*_{CNT} - D(T) \tag{5}$$

with $D(T)$ a function that depends on temperature only. Talanquer [18] proposed a theory in which the function $D(T)$ is obtained by requiring W^* to vanish at the spinodal, which is of course physically a reasonable requirement, and, as shown by Talanquer, this does improve the temperature dependence of nucleation rates for certain molecules. However, it should be noted that making W^* go to zero at some finite S can only lower W^* relative to W^*_{CNT} (which approaches zero as S approaches infinity), and therefore the predicted nucleation rate is closer to experimental rates only if the CNT prediction is too low – if CNT predicts a higher nucleation rate than what is observed experimentally, this kind of correction can only worsen the prediction (which is the case with e.g. water).

Other theories that have had some success in predicting one-component nucleation rates include the scaled model of Hale [19] and the DIT. Hale's scaled model is based on CNT, but the experimental surface tension is replaced with a correlation giving the surface tension at a given temperature as a function of the critical temperature, liquid density, and the Eotvos constant. The scaled model seems to work reasonably well even in some cases in which the classical predictions are not satisfactory. The DIT gives the work of nucleus formation as an integral over the interfacial enthalpy and entropy profiles, and the integral is related to a characteristic interface thickness. This leads to easily solvable nucleation formulas, which in principle have a level of approximation lower than the CNT equations. The input parameters are similar as with CNT (liquid density, surface tension; however, the heat of vaporization is also needed). For many substances the DIT predictions are better than CNT predictions, but there are exceptions, notably water and some alcohols.

MULTICOMPONENT NUCLEATION

The CNT for multicomponent nucleation [20] is a rather straightforward generalization of the one-component theory. The nucleus composition for a n-component nucleus is found by solving the simultaneous equations

$$\ln(P_1/P_{sol,1})/v_1 = \ln(P_2/P_{sol,2})/v_2 = \ldots = \ln(P_n/P_{sol,n})/v_n \quad (6)$$

where Pi denotes the partial pressure of component i in the vapor phase, Pi,sol is the equilibrium vapor pressure of i with respect to the liquid solution, and v_i is the partial molecular volume of component i. When the nucleus mole fractions have been found, the radius of the nucleus can be calculated from

$$r^* = 2\sigma v_i /[kT\ln(P_i/P_{i,sol})]. \quad (7)$$

The work of nucleus formation is obtained from Eq. (3) (σ being now the solution surface tension).

The multicomponent CNT has been tested experimentally only with binary vapor mixtures. It has been found that when the nucleating species form fairly ideal bulk liquid solutions, the predictions of the theory are fairly reasonable [21]. However, when one of the components tends to concentrate on the solution surface, the theory can yield unphysical results. For example, with water-alcohol mixtures (where the alcohol is a surface active agent lowering the nucleus surface tension strongly) the theory predicts at certain vapor compositions a decreasing nucleation rate with increasing water vapor density as the alcohol vapor density is fixed. Such behavior violates the multicomponent Nucleation Theorem [16], because is corresponds to a negative number of water molecules in the critical nucleus. We have recently shown

[22] that the unphysical predictions follow directly from the capillarity approximation, i.e. the assumption that the nucleus surface tension is independent of cluster size and can be equated with the bulk solution surface tension.

Although the CNT can yield unphysical predictions of the nucleus *overall* composition, the nucleus "interior" composition obtained by solving equations (6) may not be too much different from reality. The reason for this is that Eq. (6) can be shown to follow directly from the incompressibility assumption, and there is no need to evoke the surface tension in deriving it [22]. For example, in the ethanol-water system the CNT predicts that the nucleus interior is almost pure water up to quite high normalized vapor phase ethanol mole fractions [23]. This is in accord with molecular dynamics simulations of ethanol-water clusters [24], which showed that in nucleus-sized clusters with overall ethanol mole fraction of 0.12, all ethanol molecules repartitioned to the surface. Similar cluster compositions were found in a DFT study of a model molecular system [25], in which the components enhance each other's nucleation ability in all proportions although they exhibit partial immiscibility in the bulk liquid (such behavior has been seen in experiments of nucleation of water and higher alcohols [26]). The model system consisted of spherical Lennard-Jones atoms (monomers) and dumbbell molecules of two Lennard-Jones atoms (dimers). The interaction potentials between the unlike species were chosen so that one end of the dimer was "monophilic" (liked the monomer) while the other end was "monophobic". Figure 1 shows density profiles of the monomer and both dimer ends in a critical nucleus found in a monomer-rich vapor. It is seen that all the dimers are concentrated on the nucleus surface, the "monophobic" ends pointing toward the vapor. The nucleus contains 159 monomers and 28 dimers, while a nucleus in a pure monomer vapor (with same vapor density as in the case of Fig. 1) contains 302 atoms. The enhancement in nucleation rate caused by the dimers is in this case more than 20 orders of magnitude, which shows the important role that adsorption of molecules capable of reducing the surface energy can play in homogeneous nucleation.

SULFURIC ACID

Sulfuric acid is the molecular species most often associated with atmospheric nucleation phenomena. The simplest scheme is that sulfuric acid and water vapors undergo binary nucleation. In more complicated schemes the nucleation threshold is further lowered by the participation of ions or a third molecular species such as ammonia. The conclusions of whether or not sulfuric acid causes nucleation at given atmospheric conditions are usually based on CNT predictions, and I will therefore consider shortly the CNT for sulfuric acid-water and sulfuric acid-water-ammonia in the light of experimental results and computer simulations.

Hydration. The heat of mixing of sulfuric acid with water is very high, which indicates that gas-phase sulfuric acid molecules may form hydrates, i.e. bind with one

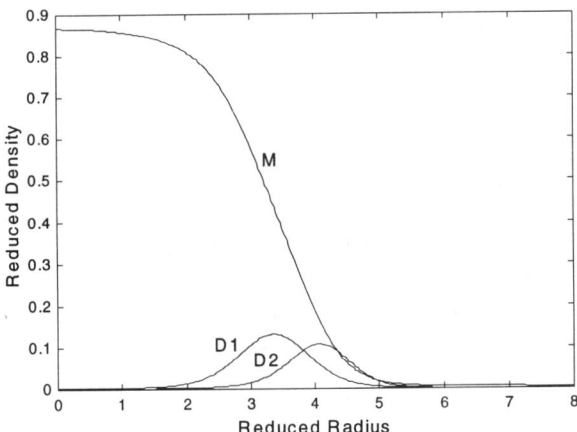

FIGURE 1. Density profiles for a critical nucleus in a surface active binary system. The line marked with "M" indicates monomer density and the lines marked with "D1" and "D2" densities of the two atoms constituting the dimer molecule. The "monophobic" ends (D2) of the dimers are oriented toward the vapor.

or more water molecules. This effectively reduces the gas-phase activity of sulfuric acid and thereby makes nucleation less favorable. The classical hydrate interaction model [27] accounts for the reduction in the sulfuric acid activity by treating the hydrates as small liquid droplets which possess thermodynamic properties (densities, vapor pressures, surface tensions) that can be related with those of macroscopic liquids. This is of course even more questionable in the case of hydrates than in the case of critical nuclei. However, there is some evidence from experimental studies and computer simulations that the classical predictions of the extent of hydration in sulfuric acid vapor may not be complete nonsense. McGraw and Weber [28] calculated the total concentration of hydrates in H_2SO_4 vapor by subtracting the free acid concentration (obtained from bulk acid activities) from total acid concentration (obtained from gas phase mass spectrometer measurements [29]). They concluded that the classical model overestimates hydration somewhat, but a significant fraction of the acid is still bound in hydrates. Hanson and Eisele [30] obtained similar results at low relative humidities (below 20%) from a study of H_2SO_4 diffusion coefficient as a function of RH; at higher humidities they saw rather good agreement with the classical model.

The results of H_2SO_4 hydration from computer simulations have been somewhat mixed. Several *ab initio* studies [31]-[34] indicate that sulfuric acid would not form stable hydrates in the gas-phase while a Monte Carlo study [35] applying a sulfuric acid potential model based on Hartree-Fock calculations suggested that essentially all gas-phase H_2SO_4 molecules would be hydrated. As pointed out by Hanson and Eisele [30], the molecular level theories would need to describe the hydration energies to accuracies better than 1 kcal/mol to correctly predict the hydrate distributions. In view of the finding of Meijer and Sprik [36] that the energy of hydration of gas-phase SO_3 is underestimated in molecular level calculations by about 15 kcal/mol when the commonly used BLYP functional is applied, it is quite possible that the *ab initio* calculations have underestimated the H_2SO_4 hydration energies as well. However, the *ab initio* studies have yielded valuable information about the proton transfer in sulfuric acid-water clusters. It takes at least three water molecules to initialize the proton transfer [32], and proton transferred and neutral structures can coexist for hydrates with up to five water molecules [34]. Presumably, the dissociation of the second proton requires the presence of several additional waters. This indicates that the bulk properties, notably the activity coefficients, of sulfuric acid solutions differ from those of small clusters, and if the classical liquid drop model predicts the amount of hydration more or less correctly, it may be due to cancellation of errors.

Nucleation rates. Only very recently have sulfuric acid-water nucleation rate measurements been carried out in which the sulfuric acid vapor concentration was measured directly [37], [38]. The measurement of $[H_2SO_4]$ is important, because the absolute concentration of sulfuric acid in these experiments is very low (on the order of 10^{10} molecules per cc), and some of the vapor may be lost on chamber walls before entering the nucleation region (the experiments are typically carried out in flow tubes). We found earlier [39] that the CNT predicts reasonably well the outcome of sulfuric acid-water experiments at 298 K when the relative humidity is in the range of 40-50% and nucleation rates between about 1-1000 $cm^{-3}s^{-1}$. However, there was a discrepancy between our experimental results and those of Wyslouzil et al. [40]. The new experiments seem to corroborate the conclusion that the CNT predicts sulfuric acid-water nucleation rates that are within one or two orders of magnitude of the measured values both at 295 K [37] and at 240 K [38] (in the former study the nucleation rates were between about 0.01-1000 $cm^{-3}s^{-1}$ and in the latter study in the range of 10^6 $cm^{-3}s^{-1}$). Furthermore, Ball et al. [37] concluded that the $[H_2SO_4]$ dependence of the predicted nucleation rate was more or less correct, while the RH dependence was too steep. According to the multicomponent Nucleation Theorem this would mean that the CNT predicts the number of acid molecules in the critical nucleus quite well but overestimates the number of water molecules. Interestingly, this implies that the CNT predicts both the critical nucleus mole fraction and the radius of the nucleus incorrectly (I assume here that the "interior" mole fraction given by Eq. (6) equals the overall mole fraction, which should be a reasonable approximation because the system is not surface active). However, as noted above, the only assumption behind Eq. (6) is incompressibility of the liquid phase, and one would therefore expect that the predicted cluster composition corresponds more closely to reality than the radius of the cluster and the molecular numbers. The incorrect prediction of the composition may reflect the finding from the molecular simulations [32]-[34] that the bulk thermodynamic quantities, especially activity coefficients, are not applicable to small sulfuric acid-water clusters.

Does the CNT really perform as well for sulfuric acid-water as the experiments indicate? Caution must be exercised here: There are various formulations of the thermodynamic quantities of H_2SO_4-H_2O solutions available in the literature, which makes the matter complicated. Specifically, it seems that the differences in various CNT calculations that we have earlier [39], [41] attributed to application of different versions of the theory itself are in reality caused by application of different formulas for activity coefficients [42]. Also, there still exists a large uncertainty in the vapor pressure of pure sulfuric acid [43]. Taken together, it should be carefully checked how the different formulations of activity coefficients, surface tension and solution densitiy affect the both the absolute nucleation rates and their dependences on temperature and vapor concentrations. Of course, the experimental data is still scarce, and more measurements are needed before we can reliably compare theory and experiments.

The effect of ammonia. The effect of ammonia on sulfuric acid-water nucleation was first studied theoretically by Coffman and Hegg [44] who used the CNT, but had to apply activity coefficients limited to dilute solutions. A more complete thermodynamic model was developed recently [45] and applied in calculation of ternary nucleation rates. Both studies suggest that levels of around 10 ppt of ammonia could boost the sulfuric-acid water nucleation substantially. Furthermore, at typical ambient ammonia levels the sulfuric acid concentration required for substantial nucleation to take place is as low as one thousandth of that causing binary nucleation. Preliminary experiments [37] indicate that at least ammonia levels of several tens of ppt promote the nucleation rate. Two molecular level computer simulation studies [46], [47] have been performed on the effect of one ammonia molecule on sulfuric acid hydrates containing up to 5 waters. Although the proton transfer from acid to ammonia can be initiated by a single water molecule, only the presence of 4 waters made it energetically favorable [47]. Furthermore, it was concluded that a single NH_3 molecule is not sufficient to stabilize the hydrates enough to cause nucleation [47].

NUCLEATION IN THE ATMOSPHERE

Does sulfuric acid/water nucleation explain atmospheric aerosol formation? Results from CNT calculations indicate that in certain cases it does. We conducted recently a modeling study on nucleation events observed in the Finnish arctic [9]. Some of the observed events were associated with clean marine air masses, while others were caused by pollution plumes originating from smelters located in the Kola peninsula. These smelters are the largest sulfur dioxide sources in Northern Europe, and therefore it is quite possible that sulfuric acid-water nucleation takes place in the plumes. Indeed, a combined atmospheric chemistry/aerosol dynamics model predicted that nucleation would take place in the selected cases – the model actually underpredicted the measured concentrations of fresh particles. In contrast, we were not able to reproduce the marine nucleation events even by setting the concentration of preexisting aerosols (which act as a sink for sulfuric acid vapor) to zero. We believe that the results of the study gave fairly strong indication that the marine related nucleation events are produced by a different mechanism than the pollution events, which are possibly due to sulfuric acid-water.

Other studies have indicated that the CNT for sulfuric acid/water can predict observed nucleation events especially in cold and humid environments with low aerosol surface area (remote troposphere, vicinity of evaporating clouds) [48], [6]. However, nucleation events observed in the marine boundary layer and in a clean continental environment could not be explained, and it was suggested that ammonia [48] and in some cases atmospheric ions [6] may accelerate the nucleation at sulfuric acid concentrations too low for binary nucleation to take place.

An interesting feature discovered by Weber et al. [49], [50] in connection with the boundary layer nucleation events is a correlation between new particle production rate and H_2SO_4 concentration. The formation rate is proportional to the first to second power of $[H_2SO_4]$, which, according to the Nucleation Theorem indicates (assuming that temperature and relative humidity are fairly constant) that the critical nuclei contain only one to two sulfuric acids. Weber et al. showed that such a correlation cannot be produced by the binary CNT for sulfuric acid and water. Is it possible that ion-induced nucleation could produce the correlation? To be able to compare predictions with observations, we must first calculate the formation rate of 3 nm particles at given nucleation rate. This can be done if the particle growth rate and the concentration of preexisting aerosols (which act as a coagulational sink for the nucleated clusters are known). In a simplified scheme we can make the following assumptions: (1) Critical clusters are produced at nucleation rate J; (2) clusters grow by addition of monomers only; (3) Clusters do not evaporate; (4) Clusters coagulate with preexisting particles but not with other clusters; (5) The system is in steady-state; (6) The growth of the clusters is controlled by collisions of acid molecules, and after the addition of an acid, equilibrium with respect to ambient humidity is immediately restored by addition of water molecules; (7) The preexisting particle size distribution can be described as monodisperse. Following usual nucleation kinetics, we can now write the following equations that describe the time dependence of concentrations of clusters containing 1,2,...n acid molecules more than the critical nucleus:

$$dN_1/dt = J - K_1N_1N_p - C_1N_1 = 0$$
$$dN_2/dt = C_1N_1 - K_2N_2N_p - C_2N_2 = 0$$
$$...$$
$$dN_n/dt = C_{n-1}N_{n-1} - K_nN_nN_p - C_nN_n = 0 \quad (8)$$

Here N_i is the concentration of clusters containing (i-1) acids more than the critical nucleus, N_p is the preexisting particle concentration, K_i is the coagulation coefficient between the i-clusters and preexisting particles, and C_i is the condensation rate of acid on i-clusters. The rate of formation of n-clusters is now

$$C_{n-1}N_{n-1} = JC_1 C_2 ... C_{n-1} / [(C_1 + K_1N_p)(C_2 + K_2N_p)... (C_{n-1} + K_{n-1}N_p)] \quad (9)$$

This expression gives results in quite reasonable agreement with less approximate treatments [49].

Consider now ion-induced nucleation. According to the classical description [51], ion-induced sulfuric acid-water nucleation takes place at somewhat lower H_2SO_4 concentration than the homogeneous nucleation, but the slopes of the rates vs. $[H_2SO_4]$

are almost the same until a threshold value is reached, after which the ion-induced rate is constant (equaling the ion production rate). Thus, below the threshold ion-induced nucleation cannot produce the observed correlation. Above the threshold [H_2SO_4] the rate is constant, and Eq. (9) indicates that the slope of observed 3-nm particle formation rate is a result from competition between condensational growth and coagulation of the clusters. If the coagulation term is large enough, the 3-nm formation rate is directly proportional to the product of the C_i in the numerator of Eq. (9), which in turn (assuming free molecular condensation) is proportional to [H_2SO_4]$^{(n-1)}$. A 3-nm particle contains roughly 50 acid molecules, and therefore the observed rate has a very steep dependence on sulfuric acid vapor concentration if the preexisting aerosol concentration is high. Figure 2 shows 3-nm particle formation rates (solid lines) when the ion-induced nucleation rate is 1 cm^{-3}s^{-1}, and the preexisting aerosol surface areas correspond to observed values [50]. The calculated lines are clearly different from the observations (dashed lines).

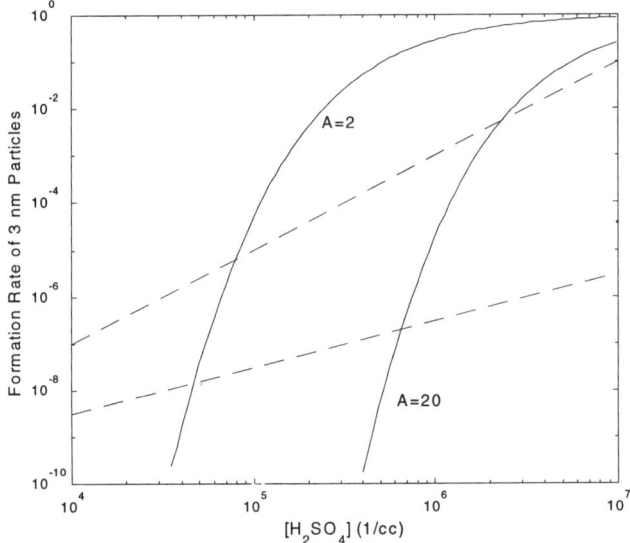

FIGURE 2. The formation rate of 3 nm particles as a function of sulfuric acid vapor concentration. The dashed lines indicate the lower and upper bounds of the correlation observed at Mauna Loa, Hawaii [49], [50]. The solid lines are calculated using Eq. (9), assuming an ion-induced nucleation rate of 1 cm^{-3}s^{-1} and preexisting aerosol surface areas of 2 and 20 µm^2/cc.

Can the observed correlation shown in Fig. 2 be explained by sulfuric acid-water nucleation boosted by ammonia? Equation (9) indicates, that the slope of the observed 3-nm particle formation rate vs. [H_2SO_4] can in practice only become steeper than the slope of the nucleation rate, not gentler. Kulmala et al. [52] have recently suggested, based on CNT calculations, that ternary nucleation of sulfuric acid, water and ammonia actually takes place in the atmosphere very often, but in most cases the freshly formed stable clusters grow so slowly that they have been scavenged by coagulation before reaching the size of 3 nm. Only if there is some other supersaturated vapor present that can make the clusters grow at a higher rate will they survive to detectable size range. This is a plausible scheme, but can only explain the observed correlation if the critical nuclei contain only one or two acids and the growth is fast enough that very little coagulation takes place. However, the atmospheric concentrations of sulfuric acid and ammonia are so low, at least in remote areas, that the CNT predicts the critical nuclei to contain more than two acids even in ternary critical nuclei [45]. One possibility is, that the species that is responsible for the fast growth of the molecules takes also part in the nucleation. Marti et al. [53] have suggested that organic molecules may have been condensing on the freshly nucleated particles observed at a continental site. Organic molecules are often surface active, and adsorption of organics could lower the surface energy of critical nuclei substantially (see Fig. 1), making them very small. If an organic vapor responsible for the subsequent growth of the particles also promotes their nucleation, the atmospheric nucleation events may not be occurring more often than long-term observations [7] indicate.

CONCLUSION

Atmospheric nucleation events take place quite frequently. Some of the events can be predicted using the binary CNT for sulfuric acid-water, and others possibly by a ternary mechanism involving ammonia. Whether the predictions can be used to *explain* the observations remains unclear. More experimental data is needed in order to really understand the predictive success of the CNT. Monte Carlo simulations will probably be very useful in exploring the nucleation ability of various molecular species, but reliable intermolecular potential models are needed. For sulfuric acid-water such models could be developed based on molecular level calculations and experimental information on the hydration of acid vapor. However, no matter how sophisticated the theoretical methods are, their predictions on atmospheric nucleation can ultimately be verified only when the nucleating molecular species can be experimentally identified.

REFERENCES

1. Covert D.S., Kapustin V.N., Quinn P.K. and Bates T.S., New Particle Formation in the Marine Boundary Layer. J. Geophys. Res., 97, 20581-20589 (1992).
2. Hoppel W.A., Frick G.M. and Fitzgerald J.W., Marine boundary layer measurements of new particle formation and the effects nonprecipating clouds have on aerosol size distribution. J. Geophys. Res., 99, 14443-14495 (1994).
3. Weber R.J., McMurry P.H., Eisele F.L. and Tanner D.J. Measurement of expected nucleation precursor species and 3-500 nm diameter particles at Mauna Loa observatory, Hawaii. J. Atmos. Sci., 52, 2242-2257 (1995).

4. Weber R.J. et al. A study of new particle formation and growth involving biogenic and trace gas species measured during ACE 1. J. Geophys. Res. 103, 16385-16396 (1998).
5. Clarke, A.D., Kapustin, V.N., Eisele, F.L., Weber R.J. and McMurry, P.H., Particle production near marine clouds: Sulfuric acid and predictions from classical binary nucleation. Geophys. Res. Lett. 26, 2425-2428 (1999).
6. Mäkelä J.M., Aalto P., Jokinen V., Pohja T., Nissinen A., Palmroth S., Markkanen T., Seitsonen K., Lihavainen H., Kulmala M., Observations of ultrafine aerosol particle formation and growth in boreal forest. Geophys. Res. Lett., 1219-1222 (1997).
7. Leaitch W.R., Bottenheim J. W., Biesenthal T.A.,., Li S.-M., Liu P.S.K., Asalian K., Dryfhout-Clark H, Hopper F. and Brechtel F. A case study of gas-to-particle conversion in an eastern Canadian forest. J. Geophys. Res. 104, 8095-8112 (1999).
8. Pirjola L., Laaksonen A., Aalto P. and Kulmala M. Sulfate aerosol formation in the Arctic boundary layer. J. Geophys. Res. 103, 8309-8321 (1998).
9. Clarke A.D., Atmospheric nuclei in the remote troposphere. J. Atmos. Chem. 14, 479-488 (1992).
10. Brock C.A., Hamill P., Wilson J.C., Jonsson H.H. and Chan K.R., Particle formation in the upper troposphere: A source of nuclei for the stratospheric aerosol. Science, 270, 1650-1653 (1995).
11. Clarke A.D. et al., Nucleation in the equatorial free troposphere: Favorable environments during PEM-Tropics. J. Geophys. Res. 104, 5735-5744 (1999).
12. Senger B., Schaaf P., Corti D.S., Bowles R, Voegel J.-C., and Reiss H., A molecular theory of the homogeneous nucleation rate. I. Formulation and fundamental issues.
13. Oxtoby D.W. and Evans R., Nonclassical nucleation theory for the gas-liquid transition. J. Chem. Phys. 89, 7521-7530 (1988).
14. Granasy L., Diffuse interface theory for homogeneous vapor condensation. J. Chem. Phys. 104, 5188-5198 (1996).
15. Oxtoby D.W. and Kashchiev D., A general relation between the nucleation work and the size of the nucleus in multicomponent nucleation. J. Chem. Phys. 100, 7665-7671 (1994).
16. ten Wolde P.R., Oxtoby D.W. and Frenkel D., Coil-globule transition in gas-liquid nucleation of polar fluids. Phys. Rev. Lett. 81, 3695-3698 (1998).
17. McGraw R. and Laaksonen A., Scaling properties of the critical nucleus in classical and molecular-based theories of vapor-liquid nucleation. Phys. Rev. Lett. 76, 2754-2757 (1996).
18. Talanquer V, A new phenomenological approach to gas-liquid nucleation based on the scaling properties of the critical nucleus. J. Chem. Phys. 106, 9957-9960 (1997).
19. Hale B.N., Application of a scaled homogeneous nucleation rate formalism to experimental data at $T<<T_c$. Phys. Rev. A 33, 4156-4163 (1986).
20. Wilemski G., Composition of the critical nucleus in multicomponent vapor nucleation. J. Chem. Phys. 80, 1370-1372 (1984).
21. Strey R. and Viisanen Y., Measurement of the molecular content of binary nuclei. Use of the nucleation rate surface for ethanol-hexanol. J. Chem. Phys. 99, 4693-4704 (1993).
22. Laaksonen A., McGraw R and Vehkamäki H, Liquid drop formalism and free-energy surfaces in binary homogeneous nucleation theory. J. Chem. Phys. 111, 2019-2027 (1999).
23. Viisanen Y, Strey R., Laaksonen A and Kulmala M., Measurement of the molecular content of binary nuclei II. Use of the nucleation rate surface for water-ethanol. J. Chem. Phys 100, 6062-6072 (1994).
24. Tarek M. and Klein M.L, Molecular dynamics study of two-component systems: The shape and surface structure of water/ethanol droplets. J. Phys. Chem. A 101, 8639-8642 (1997).
25. Napari I. and Laaksonen A., Surfactant effects and an order-disorder transition in binary gas-liquid nucleation. Phys. Rev. Lett. 84, 2184-2187 (2000).
26. Strey R., Viisanen Y. and Wagner P.E. Measurement of the molecular content of binary nuclei III. Use of the nucleation rate surfaces for water-n-acohol systems. J. Chem. Phys. 103, 4333-4345 (1995).
27. Jaecker-Voirol A., Mirabel P. and Reiss H., Hydrates in supersaturated binary sulfuric acid-water vapor: A reexamination. J. Chem. Phys. 87, 4849-4852 (1987)
28. McGraw R. and Weber R. J., Hydrates in binary sulfuric acid-water vapor: Comparison of CIMS measurements with the liquid-drop model. Geophys. Res. Lett. 25, 3143-3146 (1998).

29. Marti J.J., Jefferson A, Cai X.P., Richert C., McMurry P.H. and Eisele F, H_2SO_4 vapor pressure of sulfuric acid and ammonium sulfate solutions. J. Geophys. Res. 102, 3725-3735 (1997).
30. Hanson D.R. and Eisele F., Diffusion in humidified nitrogen: Hydrated H_2SO_4. J. Phys. Chem. A 104, 1715-1719 (2000)
31. Morokuma K. and Mugurama C., Ab initio molecular orbital study of the mechanism of the gas phase reaction $SO_3 + H_2O$: Importance of the second water molecule. J. Am. Che. Soc. 116, 10316-10317 (1994)
32. Arstila H., Laasonen K. and Laaksonen A, Ab initio study of gas-phase sulphuric acid hydrates containing 1 to 3 water molecules. J. Chem. Phys., 108, 1031-1039 (1998).
33. Bandy A.R. and Ianni J.C., Study of the hydrates of H_2SO_4 using density functional theory. J. Phys. Chem. A 102, 6533-6539 (1998).
34. Re S., Osamura Y., and Morokuma R, Coexistence of neutral and ion-pair clusters of hydrated sulfuric acid $H_2SO_4(H_2O)_n$ ($n = 1-5$) – A molecular orbital study. J. Phys. Chem. A 103, 3535-3547 (1999).
35. Kusaka I., Wang Z.-G. and Seinfeld J.H., Binary nucleation of sulfuric acid-water: Monte Carlo simulation. J. Chem. Phys. 108, 6829-6848 (1998).
36. Meijer E.J. and Sprik M., A density functional study of the addition of water to SO_3 in the gas phase and in aqueous solution. J. Phys. Chem. A 102, 2893-2898 (1998).
37. Ball S.M., Hanson D.R., Eisele F.L. and McMurry P.H., Laboratory studies of particle nucleation: Initial results for H_2SO_4, H_2O, and NH_3 vapors. J. Geophys. Res. 104, 23709-23718 (1999).
38. Eisele F.L. and Hanson D.R., First measurement of prenucleation molecular clusters. J. Phys. Chem. A 104, 830-836 (2000).
39. Viisanen Y, Kulmala M. and Laaksonen A., Experiments on gas-liquid nucleation of sulfuric acid and water. J. Chem. Phys. 107, 920-926 (1997).
40. Wyslouzil B.E., Seinfeld J.H., Flagan R.C. and Okuyama K., Binary nucleation in acid-water systems. II. Sulfuric acid-water and comparison with methanesulfonic acid-water. J. Chem. Phys. 94, 6842-6850 (1991).
41. Kulmala M., Laaksonen A. and Pirjola L., Parametrizations for sulfuric acid/water nucleation rates. J. Geophys. Res. 103, 8301-8307 (1998).
42. Kulmala M., personal communication (1999).
43. Ayers G.P., Gillett R.W. and Gras J.L., On the vapor pressure of sulfuric acid. Geophys. Res. Lett. 7, 433-436 (1980).
44. Coffman D.J. and Hegg D.A. A preliminary study of the effect of ammonia on particle nucleation in the marine boundary layer. J. Geophys. Res. 100, 7147-7160 (1995)
45. KorhonenP., Kulmala M., Laaksonen A., Viisanen Y., McGraw R., Seinfeld J.H., Ternary nucleation of H_2SO_4, NH_3 and H_2O in the atmosphere. J. Geophys. Res., 104, 26349-26353 (1999).
46. Larson L.J., Largent A. and Tao F.-M. Structure of the sulfuric acid-ammonia system and the effect of water molecules in the gas phase. J. Phys. Chem A. 103, 6786-6792 (1999).
47. Ianni J.C. and Bandy A.R. A density functional study of the hydrates of $NH_3 \cdot H_2SO_4$ and its implications for the formation of new atmospheric particles. J. Phys. Chem A. 103, 2801-2811 (1999).
48. Weber R.J. et al. New particle formation in the remote troposphere: A comparison of observations at various sites. Geophys. Res. Lett. 26, 307-310m (1999)
49. Weber R.J. Studies of new particle formation in the remote troposphere. Ph.D. thesis, University of Minnesota (1995)
50. Weber R.J., Marti J.J., McMurry P.H., Eisele F.L. and Tanner D.J. Measured atmospheric new particle formation rates: Implications for nucleation mechanisms. Chem. Eng. Comm. 151, 53.64 (1996).
51. Raes F. and Janssens A. Ion-induced aerosol formation in a H_2O- H_2SO_4 system –I. Extension of the classical theory and search for experimental evidence. J. Aerosol Sci. 16, 217-227 (1985).
52. Kulmala M., Pirjola L. and Mäkelä J.M., Stable sulphate clusters as a source of new atmsopheric particles. Nature, 404, 66-69 (2000).
53. Marti J.J., Weber R.J. and McMurry P.H. New particle formation at a remote continental site: Assessing the contributions of SO_2 and organic precursors. J. Geophys. Res. 102, 6331-6339 (1997).

Aerosol SANS: A new method to probe the structure of nanodroplets

Barbara E. Wyslouzil[*], Gerald Wilemski[†], and Reinhard Strey[‡]

[*]*Department of Chemical Engineering, Worcester Polytechnic Institute, Worcester MA 01609, USA,*
[†]*Department of Physics, University of Missouri, Rolla, MO 65409, USA,*
[‡]*Institute for Physical Chemistry, University of Cologne, D-50939 Cologne, Germany*

Abstract. Small angle neutron scattering (SANS) is proving to be an invaluable tool for probing the structure of nanodroplets. By fitting the SANS scattering spectrum one can derive the key properties of the aerosol, i.e. the number density, the average particle size, and the polydispersity, under conditions where conventional aerosol measurements are not possible. Furthermore, in binary nanodroplets it appears feasible to identify scattering from an enriched surface layer of the droplet, a capability that is relevant for understanding the aerosol formation process as well as the heterogeneous chemistry of the multi-component nanodroplets.

INTRODUCTION

Nanodroplet aerosols form readily in the supersonic expansions that occur, for example, in turbomachinery, jet exhausts and volcanic eruptions. Thus understanding particle formation and growth when cooling rates approach 1K/µs is of broad scientific interest. From a fundamental point of view, particles with radii <10 nm are important because they lie in the critical transition zone between large molecular clusters and bulk material. Furthermore, in multi-component nanodroplets, there is strong theoretical and indirect experimental evidence that the surface and interior compositions of the droplets can differ significantly. Since surface enrichment affects nucleation, growth and evaporation kinetics, and the heterogeneous chemistry of aerosol droplets, direct evidence for this phenomenon can deepen our fundamental understanding of aerosol formation and growth processes.

SANS is a well established method for examining the properties of microemulsions.[1-4]. The experimental challenges inherent in applying SANS to aerosols are best understood by comparing the characteristics of these two systems. Microemulsions are almost ideal samples: they are stable, the droplet number density N is usually >10^{16} cm^{-3} leading to a volume fraction ϕ of the disperse phase in the range 0.01-0.9, and, finally, less than 1 cm^3 of material is required per sample. In contrast, an aerosol is not stable and must be produced continuously in the neutron beam line. Finite nucleation rates and rapid coagulation combine to keep N below ~10^{14} cm^{-3} and ϕ below ~10^{-5}. Thus the scattering signals from aerosols are many orders of magnitude smaller than from microemulsions. Fortunately, background scattering in the aerosol SANS experiments is primarily instrument related and is also several orders of magnitude lower than in a typical microemulsion experiment.

EXPERIMENTAL

Aerosols with $N \sim 10^{12}$ cm^{-3} and a mean particle radius $\langle r \rangle \sim 10$ nm, are produced in a supersonic nozzle by rapidly expanding a dilute vapor mixture of D_2O (or other condensible liquid) and N_2. As illustrated in Fig. 1, our experiments[5] clearly prove the feasibility of applying SANS to liquid droplet aerosols in the 5-10 nm size range even though N is many orders of magnitude lower than in a typical microemulsion. Unlike earlier light scattering experiments[6], the structure in the SANS spectrum let us determine simultaneously for the first time $\langle r \rangle$, N, and the average width of the distribution σ using only the simple assumption that the aerosol has a log-normal or Gaussian distribution. In our first experiments[5] we found that $\langle r \rangle$ increased with the partial pressure of the condensible vapor p_v but did not depend strongly on the total pressure at the inlet to the nozzle p_0, while N decreased with p_v thereby confirming our model predictions of droplet nucleation and growth in the supersonic nozzle.

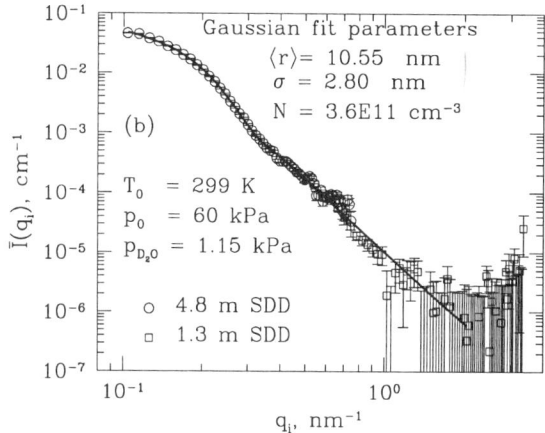

FIGURE 1. The circularly averaged scattered intensity from a pure D_2O aerosol. The solid line shows the fit to the data for the parameters given in the figure.

We also clearly observe a Doppler shift in the momentum of the scattered neutrons that arises from the directed motion of the high velocity (400-450 m/s) particles and the geometry of our experiment. The presence of a Doppler shift is unique to our SANS experiments. Because the size distribution parameters derived from the experimental data depend critically on correctly interpreting the 2-d scattering patterns, in our second set of experiments[7] we designed a sequence of runs to carefully explore the phenomenon. Figure 2 illustrates the 2-d scattering patterns measured for constant inlet conditions and three different neutron wavelengths. As the neutron wavelength changes from 0.5 to 1.5 nm, corresponding to a change in the neutron velocity from 800 to 267 m/s, we see a clear change from a nearly circular to a highly elliptical scattering pattern. Two-body scattering theory[8] quantitatively predicts the changes that occur as the relative velocity of the neutrons and particles is varied. The contour lines superimposed on the experimental data are the predicted scattering intensities based on the parameters of the aerosol size distribution derived from fitting the data at λ =1 nm. As a potentially interesting application we can use the Doppler shift as a simultaneous flow diagnostic to directly measure the velocity of the particles[7,8]. For a test case, the average velocity derived from the Doppler shift was within 2% of the mean flow velocity, 440 m/s, expected in the viewing volume from the pressure trace data under these conditions.

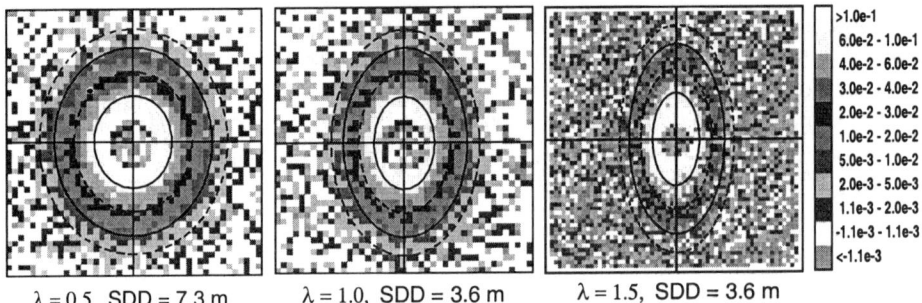

FIGURE 2. The observed 2-d scattering patterns as a function of neutron wavelength and sample-to-detector distance (SDD). The solid contour closest to the center of the detector corresponds to an absolute scattering intensity of 0.08 cm^{-1}. The remaining contours correspond to absolute intensities of 0.03, 0.008 and 0.003 cm^{-1} respectively.

Finally, our binary aerosol experiments have included aerosols formed from H_2O, D_2O, and n-butanol or its fully deuterated analogue d-butanol. Our preliminary results for D_2O-H_2O aerosols are presented elsewhere in these proceedings. In contrast to the complete miscibility found in D_2O-H_2O mixtures, bulk samples of H_2O and n-butanol have a wide miscibility gap near room temperature. No one has measured a lower critical solution temperature for the system. Thus, our working assumption is that binary nanodroplets containing H_2O (or D_2O) and d-butanol (or n-butanol) will exhibit a water-rich core and an alcohol-rich shell. Since the scattering length of H is negative and about half the magnitude of that of D, scattering signals are usually dominated by the scattering intensity from the deuterated compound. In experiments with D_2O-n-butanol droplets (not shown here) the signal intensity decreases as q^{-4} in the high q region, which is consistent with the picture that a D_2O-rich core contributes to most of the signal. In contrast, Fig. 3 illustrates that the scattering from the H_2O-d-butanol aerosol is much weaker and, although quite noisy, the signal appears to fall off as q^{-2} in the high q region, which is consistent with the picture of shell-scattering.

In summary, aerosol SANS provides us with a powerful new way to study the properties of nanometer sized liquid droplets in the environment in which they form. Combined with pressure trace measurements and modeling, SANS provides additional information critical to our understanding of droplet formation and droplet growth in the nanometer size regime.

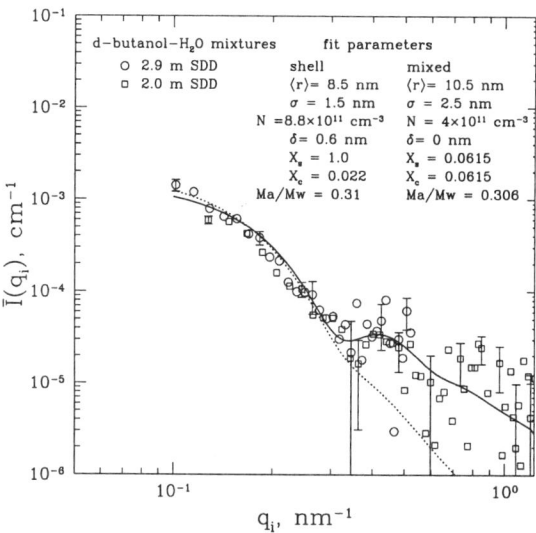

FIGURE 3. The observed SANS spectra is compared to a diffuse shell and a well mixed droplet model. The parameters of the fit are given in the legend. The data are more consistent with the shell model than with the well-mixed droplet model. For clarity only every 5th error bar is displayed.

ACKNOWLEDGMENTS

We thank our NIST contacts Dr. J. Barker and Dr. C. Glinka for their help and for useful discussions. This work was supported by the National Science Foundation, the Donors of the Petroleum Research Fund of the ACS, the Engineering Research Program of the US DOE, the University of Missouri Research Board, and by a NATO Travel Grant. The work was based on activities supported by the National Science Foundation under agreement No. DMR-9423101.

REFERENCES

1. Gradzielski, M, Langevin, D., Magid L., and Strey, R., *J. Phys. Chem* **99**, 13232 (1995).
2. Kotlarchyk, M., Chen, S.H., Huang, J.S., and Kim, M.V., *Phys. Rev. A* **29**, 2054 (1984).
3. Mortensen, K., *J. Phys. Condens. Matter* **8**, A103 (1996).
4. Strey, R., Winkler, J., and Magid, L., *J. Phys. Chem.* **95**, 7502 (1991).
5. Wyslouzil, B.E., Cheung, J.L., Wilemski, G., and Strey, R., *Phys. Rev. Letters* **79**, 431 (1997).
6. Moses, C.A. and Stein, G.D., *J. Fluids Eng.* **100**, 311 (1978).
7. Wyslouzil, B.E., Wilemski, G., Cheung, J.L., Strey, R., and Barker, J., *Phys. Rev. E* **60**, 4330 (1999).
8. Wilemski, G., *Phys. Rev. E* **61**, 557 (2000).

Phase Changes in Internally Mixed Organic/Sulfate Aerosols

Sarah D. Brooks, Anthony J. Prenni, Matthew E. Wise and Margaret A. Tolbert

Department of Chemistry & Biochemistry and CIRES, CB 216, University of Colorado, Boulder 80309

Abstract. Using a temperature controlled flow tube system equipped with FTIR detection of particle phase and relative humidity, we have measured the deliquescence (uptake of water) and efflorescence (loss of water) of internally mixed ammonium sulfate/maleic acid particles. Our results indicate that crystalline ammonium sulfate particles remain dry until reaching a deliquescence relative humidity of approximately 81%. In contrast, internally mixed particles deliquesce at significantly lower relative humidities. Results are presented for the deliquescence and efflorescence phase changes of mixed ammonium sulfate/maleic acid aerosols as a function of maleic acid wt%. The results suggest that the presence of water-soluble organics in tropospheric aerosol could alter the conditions under which phase changes occur in the atmosphere.

INTRODUCTION

While the deliquescence and efflorescence of pure ammonium sulfate is well established, recent field data shows evidence that aerosols may be composed of up to 50% or more organic material (3, 5, 6). An understanding of the phase behavior of these mixed particles is important since the probability of heterogeneous reactions can depend strongly on the particle phase. For example, the hydrolysis of N_2O_5 to form nitric acid is much more likely to occur on a liquid than on a solid with reaction probabilities of 0.02-0.06 and <0.003, for liquid and solid ammonium sulfate, respectively (1, 4).

A significant portion of the total organic content of the tropospheric aerosol has been identified as water soluble organics (WSO), including diacids. While insoluble organics may inhibit water uptake by atmospheric aerosols, WSOs may enhance water uptake. We have chosen maleic acid as a representative diacid based on its high solubility and presence in the atmospheric (2, 6). Pure ammonium sulfate makes the transition from a solid to a liquid at a characteristic deliquescence relative humidity (RH). However, the liquid sulfate does not immediately solidify upon exposure to RH's below the deliquescence point, but instead remains liquid until recrystallizing at a significantly lower efflorescence RH. Quantifying the effect of maleic acid on the deliquescence and efflorescence points of ammonium sulfate aerosol will improve our understanding of the phase of mixed organic/sulfate aerosols in the troposphere.

Experimental

A temperature-controlled flow tube system equipped with an Nicolet 730 FTIR spectrometer was used to monitor the uptake and loss of water by aerosols. A complementary study of the deliquescence properties of bulk ammonium sulfate, maleic acid, and a eutectic mixture was conducted to compare deliquescence properties of aerosols versus bulk materials.

For flow tube experiments, aerosol particles are generated using an atomizer (TSI Model 3076) and a syringe pump (Harvard Apparatus 22). A maleic acid/ammonium sulfate solution is prepared with the desired acid to sulfate weight ratio. The atomizer output is approximately 10^7 particles/cm^3, with a mean diameter of ~0.35 microns. For deliquescence experiments, aerosol liquid water content is reduced as the aerosol passes through a diffusion dryer, ensuring crystalline aerosol prior to entering the flow tubes. Upon leaving the dryer, the aerosol flow is mixed with a dilution flow of humidified air from a temperature controlled bubbler. During the experiment, the relative humidity is incrementally increased by increasing the humidified flow from the bubbler.

Two 80 cm pretubes allow the aerosol to mix and equilibrate with its surrounding environment before reaching the 80 cm single pass FTIR observation tube. All flow tubes are double jacketed allowing methanol/water coolant to circulate through the outer tubes. The coolant is controlled by a Neslab ULT-95 refrigerator, and temperature is monitored with 5 thermistors.

An EdgeTech chilled mirror hygrometer measures the total water in the system. In addition, gas phase water concentrations are measured using FTIR spectroscopy calibrated with the hygrometer. The FTIR spectrometer also monitors condensed phase composition of the aerosol particles.

For efflorescence experiments, deliquesced aerosol from the atomizer is sent directly into the first pretube. The pretube has an additional inlet for a dry nitrogen dilution line to enter the flow and mix with the aerosol at the temperature of interest. The relative humidity is decreased by incrementally increasing this dilution flow.

The bulk property measurements were conducted using a sealed vessel and an Extech RH meter. At 273 K, a saturated solution of ammonium sulfate was placed in the vessel, and the solution was stirred using a magnetic stir bar. After the system was given several hours to equilibrate, the RH above the solution was recorded. Similar measurements were taken for a saturated maleic acid solution and for a eutectic solution. The composition of the eutectic solution was determined by adding alternating aliquots of maleic acid and ammonium sulfate to a stirred solution in an ice bath, until saturation with respect to each component was reached.

Results and Discussion

Figure 1 shows results of the deliquescence experiments as a function of wt% maleic acid. The deliquescence points in the figure represent the RH of the solid to liquid phase transition. The diamonds represent experiments conducted in a flow tube, and the circles represent the bulk measurements of RH at saturation. For pure ammonium sulfate, no condensed phase water is apparent in the IR until the deliquescence point. In contrast, in the mixed aerosol experiments, a small amount of liquid water is taken up at RH's below the deliquescence point. We have found that the amount of water required to deliquesce the aerosol is not taken up until well above the eutectic point for most mixed aerosols, and that the deliquescence point depends of the ratio of the components. Predictions derived from the Gibbs-Duhem equation indicate that the deliquescence point of a multicomponent system will always be lower than the deliquescence points of the single components (7, 8). As the weight percent maleic acid is increased, the deliquescence phase transition occurs at progressively lower relative humidities over the current range of experiments. Pure maleic acid deliquesces at a 73% RH, a higher relative humidity than many of the mixed maleic acid/ammonium sulfate aerosols.

The aerosol experiments agree well with bulk measurements for pure ammonium sulfate and pure maleic acid, indicating that the deliquescence phase change is driven by the thermodynamic properties of the system. The lowest deliquescence point measured, 56% RH, was the bulk measurement for the eutectic composition, containing 47 wt% maleic acid.

Figure 2 shows preliminary efflorescence results at 288 K from the aerosol flow tube experiments. Pure ammonium sulfate effloresces at 34% RH at 288 K. Within experimental error, a 33 wt% maleic acid effloresces at the same point as the pure ammonium sulfate. In contrast, a mixed aerosol of 50 wt % remains a liquid down to 15% RH or lower. Further experiments are being conducted to determine whether these aerosols with high maleic acid loadings ever dry out at >0 % RH.

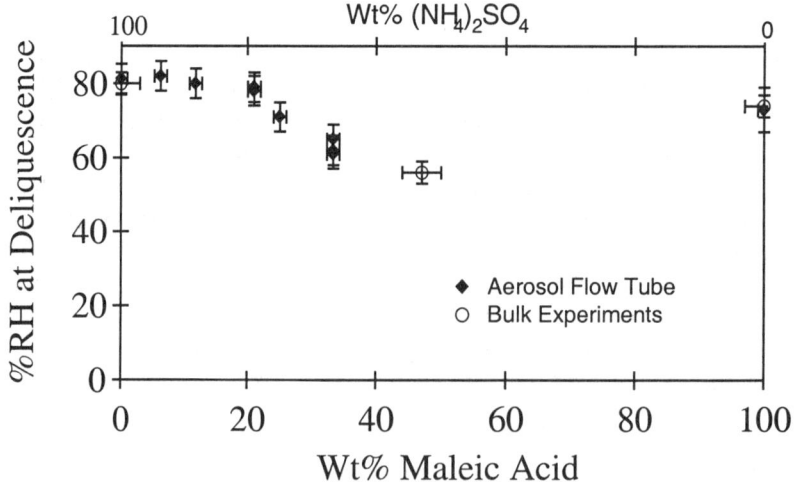

Figure 1. Summary of the deliquescence experiments at 273 K.

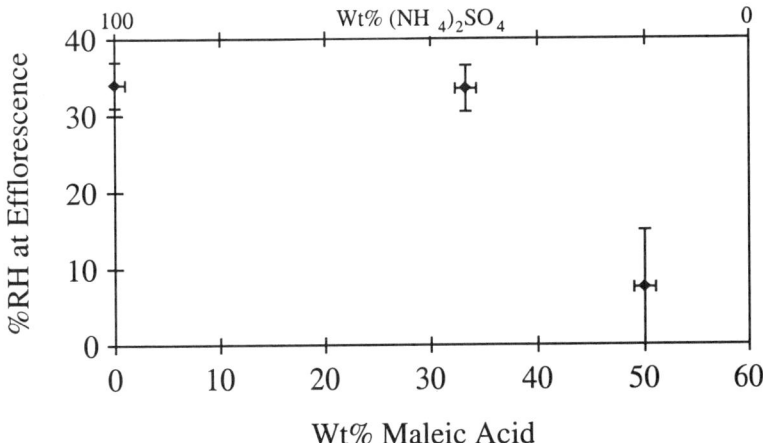

Figure 2. Summary of the preliminary efflorescence experiments at 288 K.

Atmospheric Implications

To the extent that maleic acid is representative of the WSOs, our results indicate that under tropospheric conditions, the greater the WSO content, the more likely an aerosol is to be liquid. Futhermore, once atmospheric aerosols deliquesce, they may remain liquids for the duration of their tropospheric lifetimes. The lower deliquescence RH's and efflorescence RH's for the mixed maleic acid/ammonium sulfate aerosols suggest that atmospheric aerosols are likely to be liquid under a greater range of atmospheric conditions than previously expected if they contain WSOs with similar properties to maleic acid.

ACKNOWLEDGMENTS

This work was supported by NASA-SASS and NSF-ATM.

REFERENCES

1. Hu, J.H., and J.P.D. Abbatt, *J. Phys. Chem A*, **101**, 871-878 (1996).
2. Kawamura, K., and K. Ikushima, *Environ. Sci. Technol.*, **27**, 2227-2235 (1993).
3. Murphy, D.M., D.S. Thomas, and M.J. Mahoney, *Science,* **282**, 1664-1669 (1998).
4. Mozurkewich, M. and J.G. Calvert, *J. Geophys. Res.,* **93**, 15,889-15,896 (1988).
5. Onasch, T. B., R.L. Siefert, S.D. Brooks, A.J. Prenni, B. Murray, M.A. Wilson, and M.A. Tolbert, *J. Geophys. Res.,* **104**, 21,317-21,326 (1999).
6. Saxena, P., and L.M. Hildemann, *J. Atmos. Chem.,* **24**, 57-109 (1998).
7. Tang, I. N., and H. R. Munkelwitz, *J. Appl. Meteorol,* **33**, 791-796 (1994).
8. Wexler, A.S., and J.H. Seinfeld, *Atmos. Environ.*, **25A**, 2731-2746 (1991).

Characterization of Aerosol Particles by Their Heterogeneous Nucleation: Activity at Low Supersaturations with Respect to Water and Various Organic Vapors

W. Holländer, W. Dunkhorst and H. Windt

Fraunhofer Institute of Toxicology and Aerosol Research
D-30625 Hannover, Germany
Email: hollaender@ita.fhg.de

Abstract. An expansion type cloud condensation nucleus counter / Kelvin spectrometer has been operated with water and various organic liquids. Heterogeneous nucleation is markedly different depending on the particle / vapor material combination.

INTRODUCTION

Assessing the anthropogenic impact on cloud nucleation is an important issue in the global change context. Normally, this impact is determined by simultaneously measuring cloud droplet spectra, total and interstitial aerosol Kelvin spectra and chemical aerosol composition accompanied by air trajectory calculations. This is exactly what is done in the current project 'Physical and chemical characterization of the interstitial aerosol' funded by the German Ministry of Research and Technology under grant # 07AF101/0. For this field project we had developed a Kelvin spectrometer based on rapid expansion and constant angle Mie scattering (CAMS, Wagner, 1985) of the growing monodisperse droplets as well as their extinction. This device can be operated with virtually all noncorrosive and nontoxic liquids with large enough vapor pressure for droplets to grow well into the Mie region. It was the objective of this preliminary laboratory study to determine the heterogeneous nucleation behavior of various organic liquids with respect to graphite.

KELVIN SPECTROMETER CHARACTERISTICS

Expansion type counters do not suffer from concentration artefacts due to vapor depletion; details of the present device shown schematically in Fig.1 are given elsewhere (Holländer et al.; in preparation).

cloud condensation nucleus counter

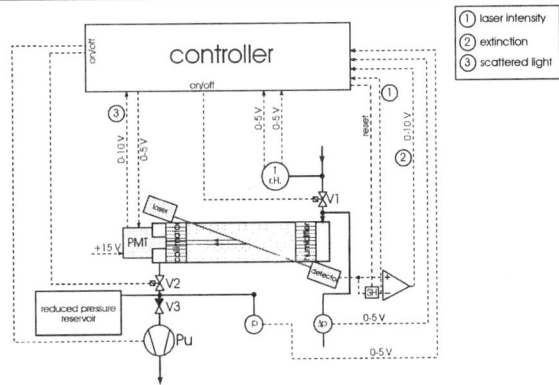

FIGURE 1. : Scheme of the cloud condensation nucleus counter / Kelvin spectrometer. After rapid expansion within tens of milliseconds, particles nucleate and grow to a cloud of monodisperse droplets. Mie structures allow sizing at any time, and from extinction, the droplet concentration can be derived with good accuracy using Mie theory.

The experimental scattering intensity $J_{30} = S_{30} \cdot \Delta\Omega \cdot N \cdot V$ under a certain angle (30° in our case) is given by the product of Mie scattering intensity in that direction S_{30}, droplet concentration N, solid aperture angle $\Delta\Omega$ and an amplification factor V. The extinction coefficient $E = -\ln(T) = \beta \cdot DECS \cdot N \cdot L$ is according to Lambert-Beer's law with the transmission T over length L, and the dimensionless extinction cross section $DECS = \alpha_{Mie}^2 \cdot Q_e$ with $\beta = \lambda_{light}^2 / (4\pi)$. Q_e is the extinction efficiency. The experimental ratio E/J_{30} does not depend on the concentration and is proportional to the ratio $DECS/S_{30}$ obtained from theory, which uniquely characterizes droplet size.

Therefore, by comparing experimental E/J_{30} with theoretical $DECS/S_{30}$ values (e.g. Fig. 2) we obtain the size as a function of time which must fit growth theory as described e.g. in Wagner (1982).

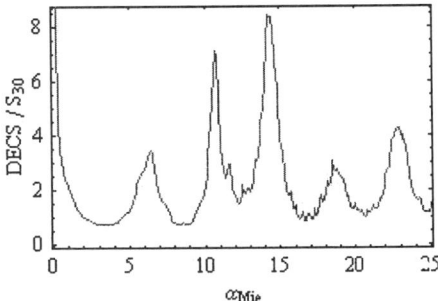

FIGURE 2 Calculated Mie - ratio of DECS and unpolarized scattering intensity under 30° S_{30} for butanol (n = 1.40). Although in principle the extinction to scattering ratios look similar for all liquids, increasing refractive index compresses the structures which also may have more or less pronounced extrema.

LABORATORY METHODS AND RESULTS

Graphite aerosol was produced by a Palas GFG1000 generator. Two fixed distance graphite electrodes are spark eroded in argon atmosphere. We operated the generator at 10 Hz with a mass production rate of 200 µg/h, a median diameter of about 50 nm and a geometric standard deviation of about 2. (Palas, 1999).

An example for the E/J_{30} ratio experimentally determined as a function of time is shown in the next figure.

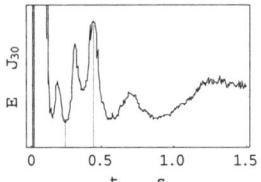 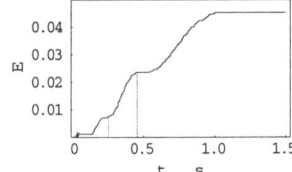

FIGURE 3. Left: the experimental ratio E/J_{30} for butanol favorably compares with Fig 2, the theoretical ratio $DECS/S_{30}$. Right: reference points for the extinction coefficient E are chosen preferably close to plateaus for determination of droplet concentration. For butanol they are at α_{Mie} = 8.5 and 14.3.

Of course, the present device is capable of using other operating liquids as well. For a given saturation ratio S, the Kelvin diameter $d_k = 4M\sigma / \left(RT \cdot \rho_{liq} \cdot \ln(S)\right)$ depends mostly on the molecular weight M, the surface tension σ, and the density ρ_{liq} of the liquid.

The above formula assumes that the condensable liquid wets the surface; if the contact angle is larger than zero degrees and therefore a spherical cap with contact angle ϕ forms on the insoluble particle, following Twomey (1977), corrections must be introduced which is not easy in praxi, however, since contact angles sensitively depend on surface conditions like smoothness or contaminations.

From a practical point of view, this might be employed for selective detection of particles non-wettable by one liquid but wettable by another. In the practical context of atmospheric research, any easily feasible differentiation between wettable-insoluble, soluble and hydrophobic particles would be useful. Therefore, we looked into the potential applicability of n- butanol $C_4H_{10}O$ (which is already widely used for CNCs at high supersaturations), trichloromethane $CHCl_3$, tetrachloromethane, CCl_4, and tetrachloroethylene, C_2Cl_4. It turned out that $CHCl_3$ and CCl_4 did not work for the above aerosol at our standard expansion ratios.

As a conclusion from the results shown in Fig. 4 we can state that butanol and tetrachloroethylene seem more efficient in detecting laboratory graphite aerosol than water which is plausible considering the contact angles of these liquids with graphite.

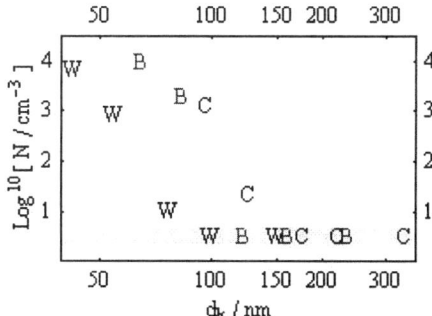

FIGURE 4. Response of water (W), butanol (B) and tetrachloroethylene (C) to polydisperse graphite aerosol.

In the atmosphere, all particles are exposed to water and it remains to be checked whether under such conditions selectivity will be similarly good. If this should be the case, parallel operation with different liquids of Kelvin spectrometers set at equal Kelvin detection sizes should allow to quickly characterize the atmospheric particles capable of condensational growth at low supersaturations.

CONCLUSIONS

We have described an instrument with the following characteristics:
- Supersaturation is achieved by rapid adiabatic expansion
- Droplet growth is monodisperse allowing application of Mie theory
- Actual droplet concentration is experimentally determined by scattering and extinction signals
- Kelvin spectrum is measured sequentially using various supersaturations.

We have also shown by laboratory experiments with various organic liquids that the instrument is capable of indicating different wetting properties of the vapor-particle combination. Further experiments will have to show whether this technique can be applied to in situ characterization of atmospheric particles.

REFERENCES

1. Wagner, P. E., "A Constant-Angle Mie Scattering Method (CAMS) for Investigation of Particle Formation Processes", J. Colloid Interface Sci. 105, 456 (1985).
2. Holländer, W., W. Dunkhorst & H. Windt, "Design and characterization of an expansion type CCN"; in preparation
3. Palas GFG1000 manual; see also http://www.palas.de/trock.htm
4. Twomey S., Atmospheric aerosols; Elsevier 1977
5. Wagner, P. E.,"Aerosol growth by condensation"; pp 129 - 178 in: Topics in Current Physics - Aerosol Microphysics; Ed. W. H. Marlow; Springer 1982.

Fixation and Chemical Analysis of Single Liquid Particle

M. Kasahara, S. Akashi, C.-J. Ma and S. Tohno

Graduate School of Energy Science, Kyoto University
Uji, Kyoto, 611-0011 JAPAN

Abstract. The sampling method and treatment procedures to fix liquid droplet as a solid particle were investigated and the elemental analysis of the fixed single particle was also tried applying PIXE and micro-PIXE analyses. Small liquid particles like fog droplet could be easily fixed by exposure to cyanoacrylate vapor within several minutes. Although large liquid particles like raindrops were also fixed successively, some of them were not perfect. Raindrops were easily fixed by freezing method. They existed in stable by exposure to cyanoacrylate vapor after freezing. The elemental concentration of single raindrop separated into 5 size ranges was determined using PIXE and micro-PIXE analysis. The concentration was dependent upon the raindrop size.

INTRODUCTION

Last several years, the importance of global environmental problems has been recognized at the worldwide. Atmospheric aerosols play an important role in such a problem. The physical and chemical properties of aerosols are fundamental to understand the effects of aerosols on atmospheric environment and behaviors of aerosols in the atmosphere. The characteristics of aerosol particles are described by a number of physical and chemical factors. In general, the most important factors are the concentration, particle size and chemical composition.

The properties of aerosols are usually evaluated as the average of a large number of particles, namely as bulk sample. Although the grasp of properties of the single particles is important rather than the average properties of a number of particles, there is a limitation in the chemical analysis of single particle, especially single liquid particle because of the technical problems.

In this study, the sampling method and treatment procedures to fix liquid droplet as the solid particle were investigated and the elemental analysis of the fixed single particle was tried applying PIXE and micro-PIXE analyses.

FIXATION OF LIQUID PARTICLES

Chemical analysis of liquid aerosol particles is quite difficult in the form of a single particle using a common analytical technique. In order to characterize individually liquid particles such as fog droplets and raindrops, two fixation techniques shown in

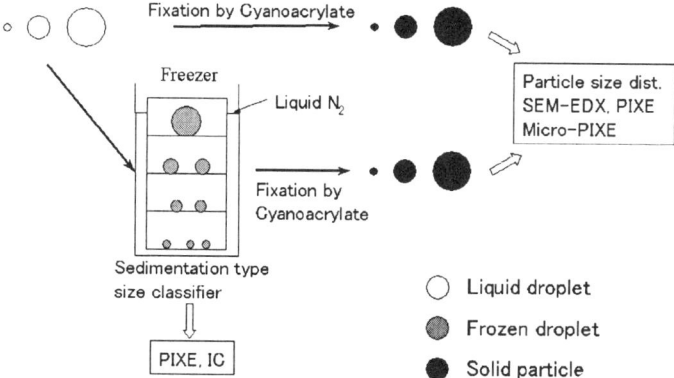

FIGURE 1. Fixation technique of liquid particles.

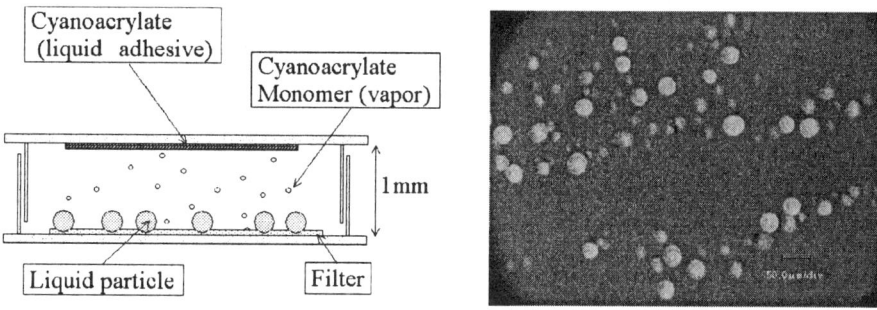

FIGURE 2a. Reaction chamber with Cyanoacrylate.

FIGURE 2b. Picture of fixed liquid particle.

Figure 1. were tested. One is a chemical reaction method with α-cyanoacrylate vapor and another is a freezing method with liquid nitrogen.

Fixation of Liquid Particles with Cyanoacrylate Vapor

Small liquid particles like fog droplets were directly collected onto the Fluoropore filter which has water repellent and exposed to ethyl α-cyanoacrylate vapor in a small exposure chamber shown in Figure 2a for fixation. Cyanoacrylate vapor was generated by evaporating instantaneous adhesive. Liquid particles were easily fixed to solid particles within several minutes depending upon the droplet size. This technique allows direct visualization and chemical analysis of single liquid particles by digital and optical microscope or SEM-EDX. The fixed particles showed spherical shape and were stable even under vacuum condition. Figure 2b shows an example of photograph of digital microscope of the fixed liquid particles. The particle size of the fixed droplets shown in the figure was ranged from about 10 to 50 μm.

Large liquid particles like raindrops were also tried to fix by exposure to α-cyanoacrylate vapor. Although they were successfully fixed, some of them were not perfectly fixed and broken by stabbing with the needle. Particle size was confirmed not to change during fixation by comparing the particle size distribution of fixed particles with that of original liquid droplets.

Fixation of Liquid Particles by Freezing Method

Raindrops were sampled in the liquid nitrogen as shown in Figure 1. Sampling apparatus consists of a dewar vessel and 5-stage stainless steel sieves with the different mesh size from 0.21 to 1.0mm. Raindrops fallen directly into the liquid nitrogen were frozen and sank to the lower stage owing to their high density. The frozen raindrops were treated in two different ways. One is that the frozen raindrops were exposed to cyanoacrylate vapor to fix stably. Another is that one frozen raindrop on the each sieve was taken out using a vacuum pipette and was dried after placing onto the polycarbonate film to prepare a sample for PIXE analysis. Handling was performed in a clean air system filled with the cooling nitrogen gas. Particle size distribution of raindrops was determined by counting the fixed particles on each stage. The elemental concentrations of single raindrops separated into 5 size ranges were determined using PIXE and micro-PIXE analysis.

CHEMICAL ANALYSIS OF SINGLE LIQUID PARTICLES

Particle Induced X-ray Emission (PIXE) is one of the most powerful analytical methods. The analysis of atmospheric aerosols is one of the most suitable fields for PIXE analysis. In most cases, PIXE is used for the analysis of bulk aerosol samples. In this study PIXE was, however, used for the elemental analysis of single liquid particles. Further, micro-PIXE was also used to obtain the elemental distribution map in a single particle.

PIXE analysis was performed with a 2.0 MeV proton beam from Tandem accelerator at Department of Nuclear Energy, Kyoto University. The ion beam size was 6mm diameter and it could cover the entire single raindrop sample. The concentrations of 15 elements (Si, S, Cl, K, Ti, V, Cr, Mn, Fe, Ni, Cu, Zn, Br, Pb) were determined quantitatively. The elemental masses in one raindrop are shown as a function of raindrop diameter in Figure 3. Raindrops were sampled at soon (0mm) and after

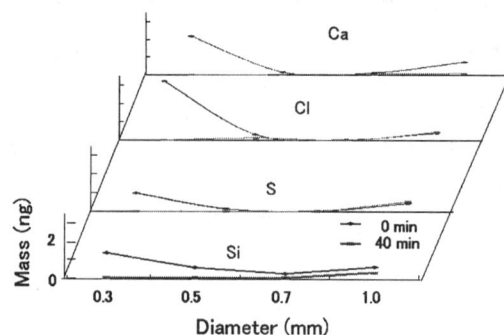

FIGURE 3. Elemental mass in a raindrop as a function of raindrop diameter.

40 mm rainfall from the beginning of rain. The elemental concentrations in raindrops were extremely high at the beginning of rain, especially in the smaller raindrops. It means that the wet scavenging is very effective for the smaller liquid particles and at the start of the rain.

Micro-PIXE analysis was carried out using the micro-beam system at the Advance Radiation Technology Center, Japan Atomic Energy Research Institute. It consists of a 3 MV single-ended electrostatic accelerator, a beam control system and a data taking system from X-ray detector. Micro-PIXE measurements were performed with a scanning 2.5 MeV H^+ beam having the beam diameter of 1-2 μm and the beam current of >100pA. Beam scanned 128x128 pixels corresponding the beam area, X-ray emitted from each pixel stored in 1024 channel according to its X-ray energy, namely the kind of element. Figure 4 shows the digital microphotograph of trace of raindrop, PIXE spectrum and elemental distribution map of Ca and S in a part of one raindrop. Elemental analysis of single raindrop could be performed by using PIXE and micro-PIXE techniques.

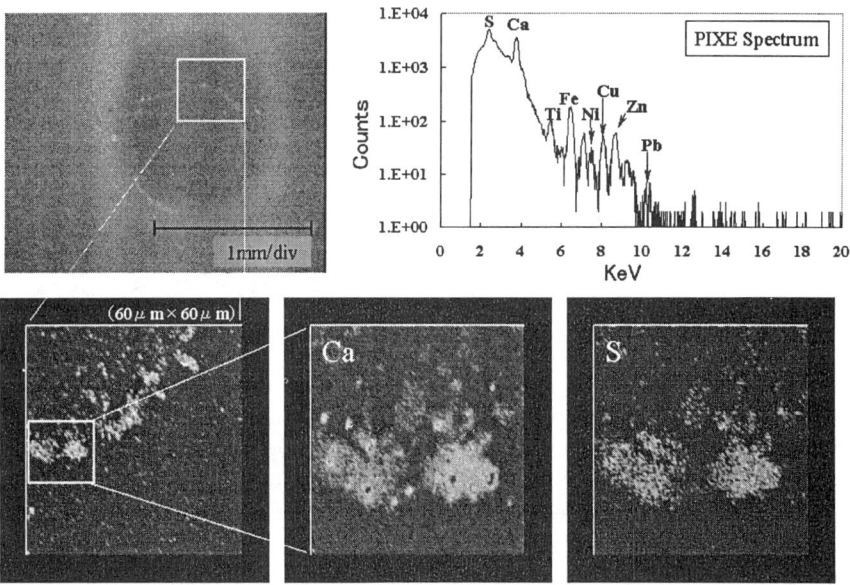

FIGURE 4. PIXE spectrum and elemental distribution map obtained from analysis of single raindrops by micro-PIXE

ACKNOWLEDGMENTS

This work was supported partly by funds from the program of the Research for the Future (RFTF) of the Japan Society for the Promotion of Science (JSPS-RFTF 97P01002) and Grant-in-Aid for Scientific Research (B) under Grant No.09044161 from Ministry of Education, Science, Sports and Culture, Japan.

Atmospheric Aerosols in the Asian Part of the Former Soviet Union

R.Van Grieken[1], R. Jaenicke[2], K. P. Koutsenogii[3],
T.V. Khodzher[4], and G.N. Kulipanov[5]

[1] *Universitaire Instelling Antwerpen, Belgium*
[2] *Johannes Gutenberg University, Mainz, Germany*
[3] *Iinstitute of Chemical Kinetics and Combustion SB RAS, Novosibirsk, Russia*
[4] *Limnological Institute SB RAS, Irkutsk, Russia*
[5] *Budker Institute of Nuclear Physics SB RAS, Novosibirsk, Russia*

The paper presents the results of the study of atmospheric aerosols in the Siberian region. This research was focused on studying the chemical and biological characteristics of both natural and pollution aerosols in this vast territory of the Asian part of the former Soviet Union (FSU). The data received within the frame of this project formed the basis for the evaluation of the impact of these aerosols on the Arctic region and the global climate. The Asian part of FSU is of scientific interest in the context of environmental chemistry, air pollution and study of atmospheric aerosols.

The specific relevance of aerosol research in Siberia is as follows. There are areas that are very remote from industrial or densely populated centres. In these circumstances, aerosol characterization contributes to the definition of global continental «background» or «baseline» aerosol, i.e. the aerosol which should occur in natural circumstances and on which all pollution is superimposed. Recently, in the context of global and long-term climatic changes, baseline aerosols and long-range transport of particulate air pollutants have become a topic of much debate. This is because it has been thought that aerosols might compensate for the well-known greenhouse effect to a significant. In this context, more and more attention is being paid to long-range effects of continental aerosols on the polar regions.

Research in Norway and Alaska showed that Western or Central Siberia may be a very important source of pollutants affecting the air composition in the Arctic region. Cities and regions in South Siberia are enormously polluted by heavy metals, the level of pollution being many times higher than in the Western world, thus the health of local population is seriously affected. In most cases, gigantic point sources are in the area, which is simple from the viewpoint of environmental chemistry and unambiguous results should be obtained easily.

Atmospheric deposition is a possible cause of the existence of heavy metals in the southern part of Lake Baikal. In the northern part of the lake, the environment is unpolluted. However in the southern part of the lake is considerably threatened by pollution from Baikalsk wood-pulp mill and industrial enterprises of Irkutsk, Angarsk,

Shelekhovo and Ulan-Ude. The analysis of aerosols in this part of the lake allows one to quantify the deposition flux of heavy metals, and to compare it to the amount of pollutants brought by various discharges and by rivers.

In the characterization of aerosol particles, normally a large amount of soil dust and marine salt particles are found. All other natural and pollution particles are usually numerically insignificant in atmospheric aerosols. Central Siberia is remote for several thousand kilometers from all soil dust and in winter seas are frozen for a long period. Every characterization of aerosol particles is therefore, limited to the most interesting particles, namely those which are transported over long distances and are pollution particles.

The project was divided into the following phases:
- specific sampling and monitoring of aerosols in Siberia
    Sampling stations operated in high pollution, background pollution and remote sites. For practical reasons, these sites were based at geophysical and biological stations of the Siberian Branch of the Russian Academy of Sciences located in different climatic and landscape regions of Western and Eastern Siberia.
- measurement of the particle size distribution and chemical composition of aerosol particles depending on geographical and seasonal variations.

Fig.1 shows the relative concentration (X_{Fe}) of the different elements, measurement in AA samples both in Southwest Siberia (a, b, c) and North Siberia (d).

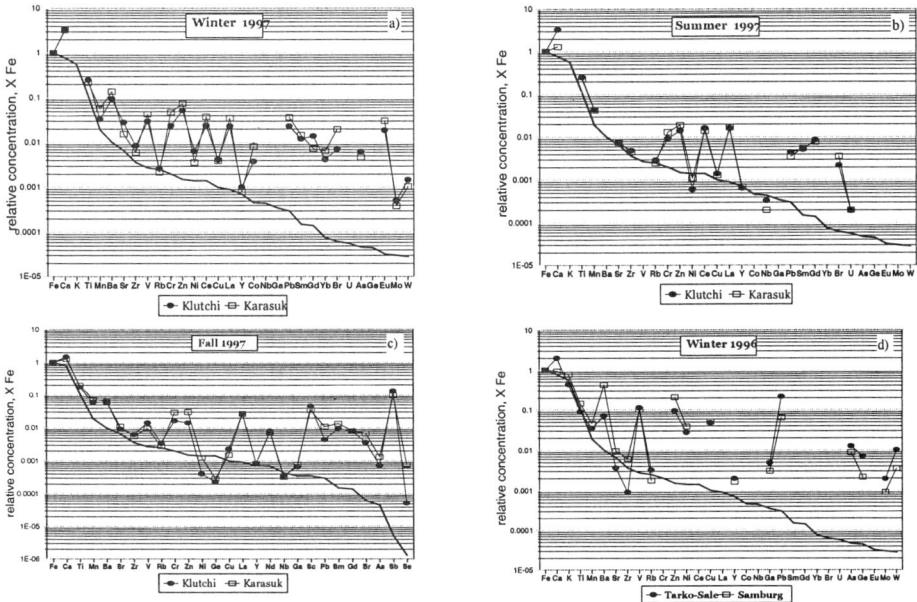

FIGURE 1. Multielemental composition of the atmospheric aerosol in West Siberia.

The dots are the geometrical mean for the measurements in Kluchi (forest - steppe zone), the squares for Karasuk (steppe zone). On Fig 1d represent the measurement in Tarko-Sale (forest-tundra zone) and Samburg (tundra zone).

$$X_{Fe} = \frac{C_i}{C_{Fe}}$$

with C_i the geometric mean mass concentration of element i and C_{Fe} the geometric mean mass concentration for Fe. From these figures it can be seen that there is a good agreement between the dots and the squares. Since the distance between Kljuchi and Karasuk is about 450 km and between Tarko-Sale and Samburg nearly 250 km, this means that the multielemental composition of the atmospheric aerosol in West Siberia is spatially independent for each of the different seasons.

To understand the source of the atmospheric aerosol one should look at the thick curve in Fig 1a, b, c, d. This curve represents the content of the different elements in the earth core (crust). From these figures is can be seen that the number of elements which coincide with the curve is dependent on the season. This group of elements belongs to the soil-erosion type. If the X_{Fe} value for specific element exceeds more ten times the earth content, this element is attributed to anthropogenic sources. Since the ratio $X_{Fe}/(X_{Fe})_{crust}$ by definition is co-called enrichment factor (EF), all elements can be divided into two classes. The first class, with EF<10, originates from natural sources (soil erosion). The second one, with EF>10, comes from anthropogenic sources. From Fig.1 on can easily see that the ratio between the number of natural and anthropogenic elements changes for the different seasons. In winter most of the elements seem to originate from anthropogenic sources. In the summer on the other hand, most elements can be attributed to soil-erosion sources.

Table 1 represents the mass concentrations on two classes in winter. On can see that the mass concentration of the elements belonging to the soil-erosion sources in the southern region exceeds the corresponding values from the northern region by a factor of 6 (Ca, Y) to 60 (Zr).

TABLE 1. Mass concentration of the atmospheric aerosol different elements from North and South West Siberia region in winter.

Region	Site	Anthropogenic sources, ng/m³						
		V	Zn	Ni	Cu	Pb	As	Ge
North	Tarko-Sale	16	18	5	9	40	6	1
	Samburg	13	27	4	7	9	1	0,3
South	Kluchi	21	38	2	7	15	1	0,7
		Soil-erosion sources, ng/m³						
		Ca	Ti	Mn	Sr	Zr	Rb	Y	Fe
North	Tarko-Sale	380	17	8	0.7	0.2	0.7	0.2	190
	Samburg	120	18	6	1,2	0,8	0,3	0,2	130
South	Kluchi	2400	300	90	21	12	6	1,2	1800

For the elements belonging to the anthropogenic sources the mass concentration from the northern region are the similar to or exceed the values for southern region. This fact means that the second class has regional scale while the first one has a global scale.

The total chemical composition of AA in Siberia and seasonal change one shows on

Fig.2. It is easy to see that in summer season the soil erosion process in the important source of AA in Siberia.

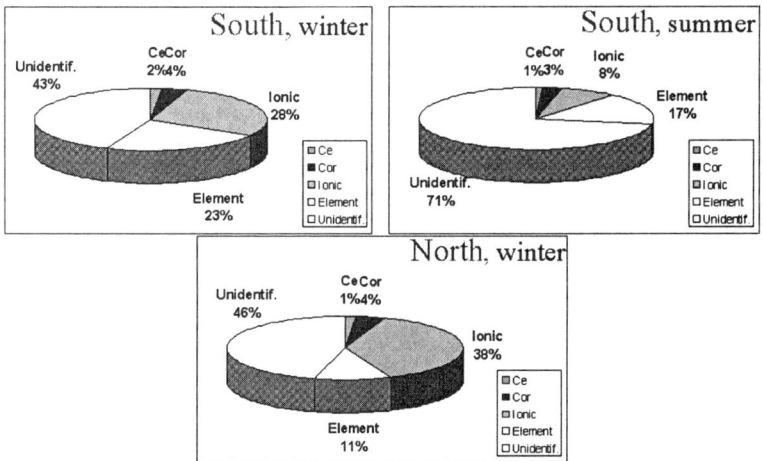

FIGURE 2. The atmospheric aerosol chemical composition in West Siberia.

From late spring till early autumn, forest fires occupy the areas of several million hectares in Siberia.

They emit several million tons of aerosols into the atmosphere. Smoke tails from large forest fires can be followed at a distance of up to several hundred and even thousand kilometres while the height of smoke column reaches several kilometres [1]. The studies were carried out under the support of INTAS (grant No 93-0182).

The more detailed results received in frame of this project are presented also in [2 - 6].

REFERENCES

1. E. Valendik. *Sib. Ecol J.*, 1996, V.3, No 1.
2. Final report to INTAS. Project 93 - 0182, 1.4.1995 - 31.3.1996, 74p.
3. Periodic report to INTAS. Project 93 - 0!82 - ext. 1.6.1996 - 31.5.1997, 47p.
4. Final report to INTAS. Project 93 - 0182 - ext. 1.6.1997 - 31.5.1998, 104 p.
5. K.P. Koutsenogii, P.K. Koutsenogii. Monitoring of Chemical and Disperse Composition of Atmospheric Aerosols in Siberia. *Chemistry for Sustainable Development*, 1997, V. 5, p. 429-442.
6. *Atmospheric and Oceanic Optics*. Special Issues on Siberian Aerosol, 1994, V.7, No8, pp. 541-628, 1996, V 9, No 6, pp. 445-564, 1997, V.10, No 6, pp. 353-436, 1998, V.11, No 6, pp.481-586.

AEROSOL CHARACTERIZATION AND PROPERTIES

Vertical Distribution Characteristics of Atmospheric Aerosols in Liaoning, NE China

Yang Jun[a], Zhou Deping[b], Gong Fujiu[b], Gao Jianchun[c], and Li Zihua[a]

[a]*Nanjing Institute of Meteorology, Nanjing 210044, P.R.China*
[b]*Liaoning Rain-Making Office, Shenyang 110015, P.R.China*
[c]*Research Institute of Applied Meteorology of Beijing, Beijing 100025, P.R.China*

Abstract. 16 aircraft missions were conducted for the measurement of atmospheric aerosols in separate days of late spring and early summer of 1996 and 1997. The paper deals with detailed analysis of the variation in vertical distributions of the concentration of the particles and their size distribution at 0~5 km above ground, with the relations to temperature and relative humidity documented in general. Evidence suggests that the concentrations show differing distribution feature in vertical above and below the cap of the mixed layer; the particle size distribution is subject to a range of forming mechanisms, displaying a multi-modal pattern; the horizontal concentration experiences remarkable variation; temperature and relative humidity stratifications have conspicuous influence on the concentration and size distribution of aerosols.

INTRODUCTION

The concentration and size distribution of atmospheric aerosol and their spatial patterns represent key factors affecting atmospheric radiative transfer, and their observations are, however, quite limited so that further field observations over different regions are necessary. Aircraft observation made with the city of Shenyang as the center over much of Liaoning province except the Peninsula was undertaken in April, May and June with aircraft speed at 42~70, averaging 59 m/s, sampling in level flight, descent and ascent (mean ascending at 3.5 m/s) for the day length (0607~1752 BST) at the height of 18.3~5156 m covering the entire mixed layer to the mid-lower troposphere except the immediately near-surface layer. The PMS airborne PCASP-100X (powerless cavity aerosol size prober) and FSSP-100 (forward scattering aerosol size prober) were utilized for our measurement.

RESULTS AND DISCUSSION

Based on the study of optimal estimates E from the ordered samplings x(n) of multiple distribution models, Hoaglin et al. [1] suggested using trimmed mean with 25% discarded from each side. This method of resistance analysis could reduce adverse effects of outliers in data and to differentiate the dominant from anomalous behaviors in the sequence in order to explore the causes of the outliers in a separate manner. Thus, we are allowed to get a reliable vertical distribution feature of aerosols and its relation to atmospheric conditions.

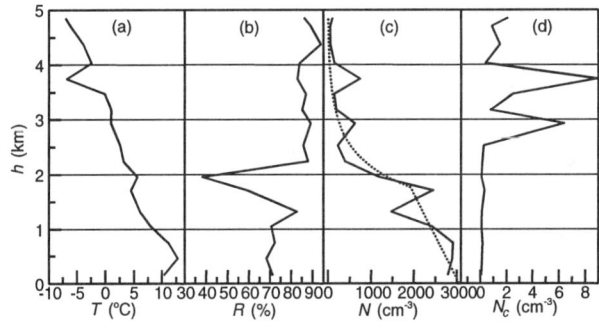

FIGURE 1. Bulk profiles of the concentration of total particles N (c), of coarse particles N_c (d), temperature T (a) and relative humidity (b).

Number Concentration of Aerosols

Data were separated at 300-m intervals before calculating for the estimates E, resulting in the height-dependent variations of concentration, temperature and relative humidity (Fig.1). The mean height of mixed layer top is around 1700 m above ground with a maximum concentration emerges there. The concentration below 1000 m does not satisfy the general exponential decay law and changes insignificantly. This is due to our observation was conducted in the daytime when near-surface turbulence mixing was so robust that the height-dependent distribution of aerosols was homogeneous. In addition, observation began normally from the very morning in which a temperature inversion was not disintegrated entirely, leading to the growth of the concentration as a function of altitude below 743.2 m. It is only above 1700 m that the concentration reduces with altitude in good conformity with the exponential law. Coarse particles make ignorable contribution to total concentration, forming 0.02%, on average, and maximizing at 2.50% in the height of 4372 m. The cause of these high values is clouds therein during flight mission with relative humidity in excess of 90%.

Thus at least two functions are needed for investigating the cases in the mixed layer and that in free atmosphere. The estimates of the concentration at a 100-m interval

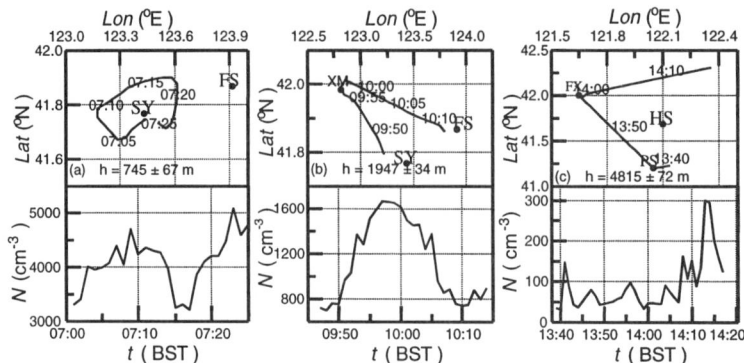

FIGURE 2. Variation of the concentration of total particles measured in the level flight (FS: Fushun, FX: Fuxin, HS: Heishan, PS: Panshan, SY: Shenyang, XM: Xinming) on April 30, 1996 (a), April 7, 1997 (b) and May 14, 1997 (c).

were calculated and by use of a linear and an exponential decay law, respectively, we obtain the fitting (Fig.1c, dotted line) of concentration N(h):

$$N(h) = \begin{cases} 3055 - 0.64h & ; \quad h < 1760 \text{m} \\ 1930 \exp\left(\dfrac{-(h-1760)}{580}\right) & ; \quad h \geq 1760 \text{m} \end{cases} \quad (1)$$

As seen from Fig.1c, fitting from (1) is considerably superior to that from a usual scheme of exponential decay rates. Eq.(2) can be put into a general form as

$$N(h) = \begin{cases} N_b - kh & ; \quad h < h_m \\ N_t \exp\left(\dfrac{h_m - h}{H_p'}\right) & ; \quad h \geq h_m \end{cases} \quad (2)$$

where N_b and N_t are the concentration at ground and the cap of the mixed layer, respectively; k the altitude-dependent lapse rate in the mixed layer; h_m the height of the cap; H_p' the scale height of aerosols in free atmosphere which differs from the usual one [2] and it represents the depth of homogeneous distribution with N_t.

Kendall's rank correlation coefficients between N and T, R, arriving at 0.735 and -0.567 (significance level < 0.01) mean a remarkable correlation between them. The significant positive correlation between N and T is determined largely by their height-dependent reduction, on the whole. From the vertical distribution of Fig.1, due to the checking by inversion, fine particles are conspicuously accumulated below. On the other hand, altitude-varying R in the troposphere is far more complicated than that of temperature such that the sign is variable for N-R correlation.

Fig.2 presents the variation of the concentration measured from level flights in which temperature and relative humidity experienced no great change. We notice that the concentrations are distributed inhomogeneously in horizontal. Also, the concentration is relatively bigger in the vicinity of the city of Shenyang in a pattern with higher (lower) values in the E-W (N-S) orientation, which is due to the fact that industries are located mostly in the western part of the city, with mountains on its east and west sides to keep pollutants from dispersal in contrast to the N-S orientation.

The Size Distribution of aerosol Concentration

Except the height of >4000, the main peak are located between 0.1~0.7 μm and maximal diameter is 0.13 μm (Fig.3). In addition to the peak, a secondary peak is between 1.5~3.0 μm. With increasing height, the size distribution is close to the background pattern and the maximum diameter shifts to smaller sizes. Beyond 4000 m the maximal diameter ≤ 0.1 μm. Size distribution range beyond 1000 m is widened because that during observation, the sky was so cloudy that the FSSP-100 detector could not distinguish between cloud and non-cloud particles.

The volume concentration distributions indicate clearly a bi-modal and even a multi-modal, of which a secondary peak around 10 μm is marked by a maximum of marine coarse-particle aerosols mode in the size range. This is because about half of the observation was affected, to varying degree, by marine atmospheric from the Bay of Liaodong and northern Huanghai Sea under the effect of southerly flow in addition to the observation over the coastal belt. Also, the volume distribution undergoes small

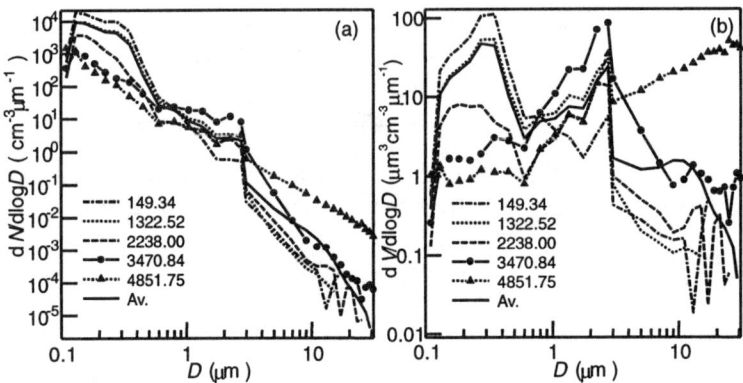

FIGURE 3. Averaged concentration distribution (a) and mean volume distribution (b) of aerosols at a range of altitude (m).

variation as a function of mixing layer altitude and, when entering the free atmosphere, the accumulative mode volumetric concentration reduces quickly.

CONCLUDING REMARKS

The particles of aerosol are piled up considerably below an inversion layer. The vertical concentration distribution is obtained for the mixed layer and free atmosphere for which the fitting from a linear and an exponential decay law, respectively, is superior to that from a single exponential decay law, leading to the lapse rate of ~0.64 particle per 100 m rise for the concentration in the mixed layer over the study region.

The size distribution shows a maximum in the accumulative and coarse-particle modes and a multi-modal form in relation to a range of aerosol sources, with the maximum diameter of 0.13 μm below 4000 m altitude. With increased height the maximum of the accumulative mode shifts towards smaller sizes and above 4000 m the size distribution approaches to the atmospheric background counterpart.

Great difference in the concentration may occur when horizontal temperature and relative humidity are steady, and the concentration is under the significant influence of topography and the distribution of industrial regions.

ACKNOWLEDGMENTS

This work is supported by the National Natural Science Foundation of China under Grant 49675250.

REFERENCES

1. Hoaglin, D. C., F. Mostteller and J. W. Tukey, *Understanding Robust and Exploratory Data Analysis*, John Wiley and Sons, New York, 1983, Translated by Chen Zhonglian et al, Beijing, China Statistic Press, 1996, pp.311-349 (in Chinese).
2. Jaenicke, R., "Vertical Distribution of Atmospheric Aerosols" in *Nucleation and Atmospheric Aerosols*, edited by N. Fukata et al., Proceedings of 13th ICNAA, Hampton, 1992, pp. 417-425.

Mass Balance Of Aerosol Particles As A Function Of Their Size

Molnár, A.[1], Mészáros, E.[2], Feczkó, T.[2] and Temesi, D.[1]

[1] *Air Chemistry Group of Hungarian Academy of Sciences, H-8201 Veszprém, PO Box 158, Hungary*
[2] *University of Veszprém, H-8201 Veszprém, PO Box 158, Hungary*

Abstract. The mass concentration of different chemical species of fine aerosol particles was measured as a function of their size. The obtained values are compared to total mass concentrations. The results are mainly discussed concerning the ratio of organic carbon to inorganic sulfate.

INTRODUCTION

Atmospheric aerosol particles play an important part in the control of the state of the atmosphere. Since the effects of the particles depend on their size and chemical composition, the determination of the mass balance as a function of the size is of particular importance. The chemical mass balance of coarse and fine particles has been separately study in several works[1]. However, more detailed information about fine particles is still needed. This is true in particular concerning the relative importance of inorganic sulfate and organic compounds since a considerable body of evidences suggest that fine particles in the atmosphere consist mainly of these two kinds of species. Recent measurements also show that an important part of organic carbon can be found in water soluble species[2]. Thus, it is evident that organic particles also takes place in the control of atmospheric processes like cloud formation and visibility degradation. For comparing the possible role of sulfate and organic particles, the parallel determination of their physical and chemical properties is obviously needed. For this reason a program was initiated in Hungary to determine the necessary particle properties in rural continental air. Similar investigation was previously carried out by Neusüss et al.[3].

METHODS OF MEASUREMENTS

In the summer of 1999 a sampling campaign was performed at the Hungarian background air pollution monitoring station, K-puszta. This sampling site is on a forest clearing, approximately in the middle of the country, on the Great Hungarian Plain. The average height of this region is 100-200 m a.s.l. Because of the lack of closed large emission sources, aerosol characteristics measured at K-puszta represent the Central-Eastern European regional air.

The program included the measurements of the number and mass size distributions as well as the chemical composition of aerosol particles. Beside, other aerosol parameters (scattering and absorption coefficients) were also monitored during the sampling period. Briefly, the air sampled at a height of 10 m above the surface was driven under ambient temperature and humidity conditions through an electrical low pressure impactor (ELPI) consisting of 13 stages[4]. Before entering the impactor the particles are electrically charged by a corona charger. In the sampler the charged particles are deposited owing to their inertia on impactor stages with cut sizes from 0.031 to 10.8 μm. The charged particles impacting on aluminum foils create electric signals which are converted into number size distribution. By assuming spherical particles of unit density, these size spectra are recalculated as surface or mass size distributions. The resolution time in this study is five minutes. In this way detailed number and mass size distributions can be obtained mainly for fine-mode particles (8 of 12 stages belong to this size fraction).

The aerosol samples for different size ranges can be subsequently chemically analyzed. For obtaining sufficient amount of material for chemical analyses a sampling period of at least half a day is necessary.

The sampling was carried out between July 12 and August 8, 1999 always in daylight. The sampling started at sunrise and ended at sunset. Accordingly, ELPI have generally been operated for 15-16 hours a day, with a 24.6 l min^{-1} flow rate. The samples on Al-foils were stored in refrigerator until analysis.

The chemical analysis were made in the laboratories of University of Veszprém. Each impactor stages were cut in half. One half was used in the analysis of inorganic ions (sulfate, nitrate, chloride, ammonium, potassium, sodium and calcium) and low molecular weight organic acids (only oxalic acid was determined occasionally) by capillary electrophoresis. From the other halves of the samples total carbon concentration was measured as carbon dioxide by catalytic combustion method with oxygen at 680°C.

RESULTS OF THE MEASUREMENTS

The average results of the sampling program are summarized in Table 1.

TABLE 1. Chemical Composition of Fine Aerosol Particles as a Function of Their Size (Note That Other Species Analyzed Constitute a Negligible Part of The Total Mass).

Stage	1	2	3	4	5	6	7	8	Total
Cut off diameter (μm)	0.03	0.06	0.11	0.18	0.27	0.42	0.68	1.05	
Total mass (μgm^{-3})	1.38*	0.49*	0.60*	1.64	3.60	7.63	2.88	1.09	19.31
Carbon (μgm^{-3})	1.21	0.41	0.45	0.61	0.94	1.40	0.65	0.27	6.35
Sulfate (μgm^{-3})	0.09	0.02	0.07	0.31	0.72	1.32	0.37	0.04	2.94
Nitrate (μgm^{-3})	d	d	d	0.00	0.01	0.03	0.02	0.03	<0.1
Ammonium (μgm^{-3})	0.03	0.01	0.03	0.11	0.26	0.51	0.13	0.02	1.10
Potassium (μgm^{-3})	0.04	0.05	0.05	0.05	0.01	0.01	0.02	0.02	0.25
Missing mass	?	?	?	0.56	1.66	4.36	1.69	0.71	8.57
Hypothetical carbon factor	~1	~1	~1	1.9	2.8	4.1	3.6	2.6	2.3

*Mass is estimated by adding up the different components analyzed, d: below detection limit.

The first thing to be mentioned is that for the smallest three size ranges the total aerosol mass is the value of the sum of the mass of the species analyzed. This is explained by the fact that on these impactor stages ELPI gave much smaller values that those followed from chemical analyses. The reason for this discrepancy is an open question, which needs further considerations.

The data tabulated show that the majority of the mass of fine aerosol particles can be found in the 0.27-0.78 µm size as expected. In this size range the particles consist mostly of sulfate and carbon compounds. One has to note in this respect that the ratio of elemental carbon to organic carbon is very low at this sampling station which means that the figures given refer practically to organic carbon. Data also indicate that sulfate ions in the size range where the majority of the mass of fine particles is detected are neutralized by ammonium ions in agreement with many previous studies. Sulfate and ammonium give 20 % and 7 % of the total mass in this interval, respectively. The concentration of nitrate and potassium is not too important, while other species analyzed constitute an even more negligible fraction of the total mass.

The most important result of this program is the fact that a considerable fraction of the carbon mass is found in the size range with particle diameter below about 0.1 µm. The major part of this carbon can be found in particles with sizes between 0.03 and 0.06 µm. Since in this size range even a relatively important fraction of sulfate is detected, one can speculate that these particles, organic carbon in particular, give a predominant part of the total number concentration of aerosol particles. This finding is in good agreement with that of Novakov and Penner[5] measured in Puerto Rico.

It can be seen from data tabulated that the sum of the mass of species analyzed, mostly that of sulfate and carbon, is smaller than the total mass in the size range above 0.18 µm (for particles with smaller diameters this is questionable). In an earlier work the concentration of different metals was also determined for the same station[6]. Since metal concentrations of fine particles reported are very small with those tabulated, it is assumed with some caution that missing masses in Table 1 can be explained by the fact that carbon as an element and not organic carbon compounds were detected. In the last row of Table 1 hypothetical carbon factors are given, which are the multiplication factors of carbon concentrations to obtain the total mass measured by ELPI. Results make it possible that this factor increases with increasing particle size and they are higher that those generally found in the literature. However, the results of Zappoli et al.[2] indicate that in Italy and Hungary an important part of organic substances is composed of macromolecular species, which explains at east partly the high carbon factors received.

The sulfate to carbon ratios as a function of particle size are represented in Table 2.

TABLE 2. The carbon-sulfate ratio as a function of the size distribution.

Stage	1	2	3	4	5	6	7	8	Total
Cut off diameter (µm)	0.03	0.06	0.11	0.18	0.27	0.42	0.68	1.05	
C/SO4	13.4	20.5	6.4	2.0	1.3	1.1	1.8	6.8	5.1

The figures given indicate that the ratio of sulfate mass to that of organic carbon has a minimum in the size range where the majority of fine aerosol mass is found. The ratio is high in particular below 0.1 µm when it is above ten and even reaches twenty. It is assumed that this is caused by the fact that small organic carbon particles were continuously formed in the air during the sampling period (mainly sunny weather), while the major part of sulfate particles was already in the accumulation mode. The increase of carbon to sulfate ratio above 0.68 µm in diameter involves that in this size range there is an other carbon source, probably the release of primary bio-aerosols by the vegetation.

CONCLUSIONS

On the basis of the results of this program one can conclude that organic carbon even in elemental form has a higher mass concentration than sulfate ions under regional continental conditions. On the other hand, the majority of particles in the diameter range below 0.1µm consists of organic carbon at least in daylight during summertime. To be able for estimating the role of organic particles in cloud formation, the determination of water solubility of these very small organic particles would be of crucial importance. The research will continue in this direction.

ACKNOWLEDGMENTS

The authors are thankful to Hungarian Scientific Research Fund (OTKA) for funding this work (Project number T030226).

REFERENCES

1. Heintzenberg, J., Fine particles in the global troposphere. A review. *Tellus*, **41B**, 149-160 (1989).
2. Zappoli, S., Andracchio, A., Fuzzi, S., Facchini, M.C., Gelencsér, A., Kiss, G., Krivácsy, Z., Molnár, A., Mészáros, E., Hansson, H.-C., Rossman, K., and Zebühr, Y., 1999: Inorganic, organic and macromolecular components of fine aerosol in different areas in Europe in relation to their water solubility. *Atmospheric Environment*, **33**, 2733-2743 (1999).
3. Neusüss, C., Brueggemann, E., Gnauk, T., Wex, H., Herrmann, H., and Wiedensohler, A., Chemical composition and mass closure of the size-segregated atmospheric aerosol in Falkenberg during LACE. *J. Aerosol Sci.*, **30**, (Suppl. 1) 913-914 (1999).
4. Laitinen, A., Hautanen, J., Keslkinen, J., Moision, N., Marjamäki, M., Elsilä, A., Real time measurement of the size distribution of urban air aerosols with electric low pressure impactor. *J. Aerosol Sci.*, **27**, (Suppl. 1) 299-300 (1996).
5. Novakov, T., and Penner, J., 1993: Large contribution of organic aerosol to cloud–condensation nuclei concentration. *Nature*, **365**, 823-826 (1993).
6. Molnár, A., Mészáros, E., Bozó, L., Borbély-Kiss, I., Koltay, E. and Szabó, Gy., 1993: Elemental composition of atmospheric aerosol particles under different conditions in Hungary. *Atmospheric Environment*, **27A**, 2457-2461 (1993).

Method For Volatility Measurements on Polydisperse Aerosol

Otmar Schmid, Donald E. Hagen, Philip D. Whitefield,
Alfred R. Hopkins, Ben Eimer

Cloud and Aerosol Science Laboratory of the University of Missouri-Rolla

Abstract. We describe a method for measuring the amount of volatile material in the aerosol phase using a thermal discriminator. This method, which requires the measurement of the particle size distributions of the heated (through discriminator) and non-heated (bypassing discriminator) sample aerosol, includes the effects due to both particle loss and partially volatile aerosols. Tests with polydisperse internally mixed, i.e. partially volatile, aerosol (not shown here) indicate a high degree of accuracy of this method even for ultrafine particles.

INTRODUCTION

The thermal discriminator is a device, which exploits differences in aerosol volatility to discriminate between particles of different chemical composition. In recent years, the discriminator technique has been widely applied [1, 2, 3, 4] to examine the possible formation of sulfuric acid aerosols in aircraft emissions due to the sulfur impurities contained in aviation kerosene [5]. For accurate measurements of the amount of sulfuric acid in the aerosol phase, the systematic effects due to the presence of partially volatile aerosol and particle loss have to be considered [6]. The method for volatility measurements, presented here, includes these effects.

THERMAL DISCRIMINATOR

Principle And Technical Description

The thermal discriminator, which essentially consists of a heated tube section followed by a cooling section, exposes the sample aerosol to a relatively high temperature T_{dc}. As sample aerosol passes through the discriminator, particles evaporate partially or completely, depending on the relative amount of compounds, which are volatile at T_{dc}. The resulting size change is a measure of the amount of volatile material on the aerosol.

In addition to evaporation, the sample aerosol will experience wall losses due to various mechanisms such as diffusion, thermophoresis, electrophoresis, and impaction. Since most of these effects depend on particle size, we may expect that particle loss is a complex function of particle chemistry as well as initial and final particle diameter. This aspect will be examined below, based on the experimentally determined penetration coefficient of a discriminator, which was designed at the University of Missouri-Rolla (UMR).

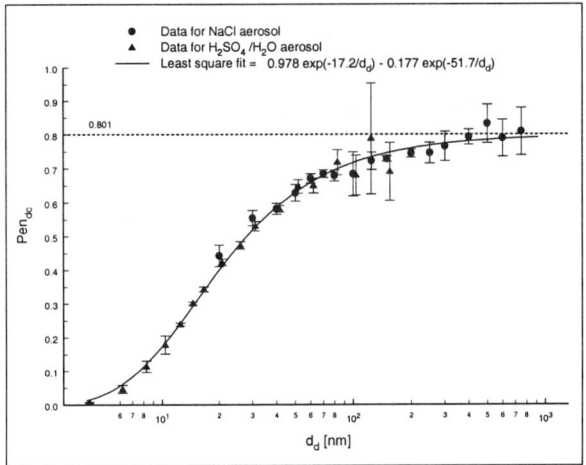

FIGURE 1. Penetration coefficient of Pen_{dc} the UMR discriminator for singly charged NaCl (non-volatile) and H_2SO_4/H_2O (98 % volatile) aerosol as a function of downstream diameter d_d.

The heated section of this UMR discriminator is a 1.44 m long stainless steel tube with an inner diameter of 0.187 inch (4.750 mm). The tubing is tightly wound to a coil with six turns and enclosed in a thermally insulated box. For a typical flow rate of 1.4 liter/min, the residence time in the heated section, which is kept at $T_{dc} = 300\,°C$, is about 0.56 s. Downstream of the heated section is a 1.3 m long cooling section in which the aerosol is allowed to cool back to laboratory temperature (~20°C).

Penetration Coefficient Of The UMR Discriminator

The penetration coefficient Pen_{dc} of the UMR discriminator was measured using a tandem Differential Mobility Analyzer (DMA). For details, please refer to [7]. Figure 1 depicts Pen_{dc} for singly charged NaCl and H_2SO_4/H_2O aerosol as a function of downstream diameter d_d. NaCl particles are completely non-volatile at $T_{dc} = 300\,°C$. The ideally completely volatile H_2SO_4/H_2O aerosol used here was in fact only about 98 % (by volume) volatile, due to the presence of some non-volatile impurities.

It is evident from Figure 1 that, for large d_d (200 nm < d_d < 750 nm), Pen_{dc} is almost constant and approaches an asymptotic value of 0.80. On the other hand, Pen_{dc} decreases monotonically to zero with decreasing d_d indicating increased particle loss for smaller particles. However, the most important result of Figure 1 is that Pen_{dc} is a function of d_d only, i.e. the initial particle size and the particle chemistry do not affect Pen_{dc}. We will show below that this feature is the basis for a relatively simple algorithm for data reduction. One of the most likely explanations for this feature is that the sulfuric acid evaporates almost instantaneously (evaporation time is much smaller than residence time in discriminator) leaving a non-volatile residual particle of diameter d_d behind. Due to the rapid evaporation of all volatile material, the effective size of the particles in the discriminator for both non-volatile and partially volatile particles is identical to their final size d_d. For a more detailed discussion of this and other aspects of the performance of the UMR discriminator, please refer to [7].

Acquisition And Reduction Of Data

Figure 2 depicts the experimental setup for volatility measurements using a thermal discriminator. The basic idea is to determine the total particle size distribution and the size distribution of the non-volatile fraction of the sample aerosol by measuring the non-heated (bypassing the discriminator) and the heated (through discriminator) distributions. These distributions and the experimentally determined Pen_{dc} can be used to calculate the amount of volatile material in the aerosol phase.

Appropriate setting of valves S1 and S2 (see Figure 2) allows the measurement of the non-heated (total) and the heated (non-volatile) mobility spectra $\{N_{tot,i}, Z_i\}$ and $\{N_{nv,i}, Z_i\}$, where $N_{tot,i}$ and $N_{nv,i}$ are the respective particle number concentrations measured by the Condensation Nucleus Counter (CNC) at the electrical mobilities Z_i, with $i = 1, 2,..., i_{max}$, which are stepwise selected by the DMA. According to Figure 2, these spectra can be converted into the differential size distributions of the total and the non-volatile fraction of the sample aerosol, n_{tot} and n_{nv}, using

$$N_{tot,i} = \sum_n n_{tot,n,i} Pen_{1,n,i} P_{n,i} \eta_{n,i} Pen_{0,n,i}, \qquad (1)$$

and
$$N_{nv,i} = \sum_n n_{nv,n,i} Pen_{dc,n,i} Pen_{2,n,i} P_{n,i} \eta_{n,i} Pen_{0,n,i}, \qquad (2)$$

and employing a standard DMA inversion algorithm [e.g. 8]. $Pen_{1,n,i}$ and $Pen_{2,n,i}$ are the penetration coefficients of the connecting tubing for bypassing and passing through the discriminator, respectively. $P_{n,i}$, $\eta_{n,i}$, $n_{tot,n,i}$, and $n_{nv,n,i}$ represent the charging probability, the CNC detection efficiency, n_{tot}, and n_{nv}, respectively, each of them evaluated at the electrical mobility Z_i. $Pen_{0,i}$ is the penetration coefficient of the DMA which is given by the area underneath the transfer function of the DMA. The effects of multiple charging are included by summing over all charge states $n = 1, 2,...$

Integration over all particle sizes yields the total and non-volatile aerosol volume densities (aerosol volume per unit volume of carrier gas)

$$V_{tot} = \frac{\pi}{6} \int_{-\infty}^{\infty} d_v^3 n_{tot} d(\log d_v) \qquad (3)$$

and
$$V_{nv} = \frac{\pi}{6} \int_{-\infty}^{\infty} d_v^3 n_{nv} d(\log d_v), \qquad (4)$$

respectively. And finally we find the volatile volume ratio R_v, which is a measure for the amount of volatile material in the aerosol phase, from

$$R_v \equiv \frac{V_{tot} - V_{nv}}{V_{tot}} = 1 - \frac{V_{nv}}{V_{tot}}. \tag{5}$$

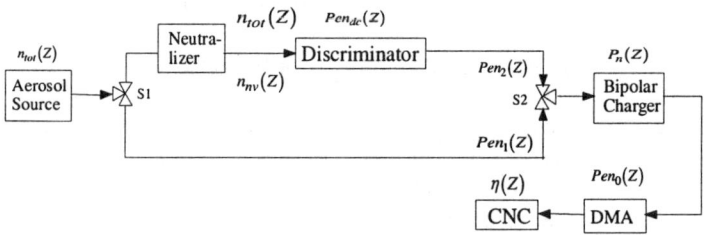

FIGURE 2. Experimental setup for volatility measurements.

CONCLUDING REMARKS

The accuracy of the algorithm for data reduction, proposed here, hinges on three main assumptions: (1) complete removal of volatile compounds from the aerosol phase in the discriminator, (2) no recondensation in the cooling section of the discriminator, and (3) Pen_{dc} is a function of d_d only. Assumptions (1) and (2) guarantee that the particles exiting the discriminator contain only non-volatile material, while assumption (3) is a necessary requirement for equation (2). The validity of all three assumptions has been shown experimentally for the UMR discriminator using sulfuric acid as volatile material [7].

Tests with polydisperse internally mixed aerosol (not shown here) indicate a high degree of accuracy of this method even for ultrafine particles.

ACKNOWLEDGMENTS

The authors would like to acknowledge NASA, USAF, and UMR for partially sponsoring the research described in this paper.

REFERENCES

1. Anderson, B.E., Cofer, W.R., Barrick, J.D., Bagwell, D.R., and Hudgins, C.H., *Geophys. Res. Let.* **25**, 1693-1696 (1998).
2. Schröder, F.P., Kärcher, B., Petzold, A., Baumann, R., Busen, R., Hoell, C., and Schumann, U., *Geophys. Res. Lett.* **25**, 2789-2792 (1998).
3. Paladino, J., Whitefield, P., Hagen, D., Hopkins, A.R., Trueblood, M., *Geophys. Res. Lett.* **25**, 1697-1700 (1998).
4. Ferry, G.V., Pueschel, R.F., Strawa, A.W., Kondo, Y., Howard, S.D., Verma, S., Mahoney, M.J., Bui, T.P., Hannan, J.R., and Fuelberg, H.E. *Geophys. Res. Lett.* **26**, 2399-2402 (1999).
5. Brown, R.C., Miake-Lye, R.C., Anderson, M.R., Kolb, C.E., and Resch, T.J., *Geophys. Res.* **101**, 22,939-22,953 (1996).
6. Kreidenweis, S.M., McInnes, L.M., and Brechtel, F.J., *J. Geophys. Res.*, **103**, 16,511-16,524 (1996).
7. Schmid, O., Ph.D. Thesis submitted at the University of Missouri-Rolla, 2000.
8. Hagen, D.E., and Alofs, D.J., *Aerosol Sci. Technol.* **2**, 465-475 (1983).

OBSERVATIONAL RESEARCHES ON SAND AEROSOL SIZE DISTRIBUTION IN HELANSHAN AREA[*]

Niu Shengjie[1] and Zhang Chengchang[2]

[1]*NingXia Institute of Meteorological Science*
Yin Chuan 750002, P. R. China
[2]*Beijing Institute of Meteorology*
Bei jing, l0008l, P. R. China

1. INTRODUCTION

Dust and sand phenomena (dust, sand blowing and sand storm) frequently occur in Helanshan area, northwest China. It is one of the natural disasters in northwest China. It can make a great economic loss and kill people sometimes, especially when black storm occurs, it also pollutes atmospheric environment, so many scientists concerns on its formation mechanism. We made comprehensive measurements about dust and sand phenomena on April and May during 1996-1999. We only focus on sand aerosol size distribution in this paper.

2. INSTRUMENT AND EXPERIMENT

Sand aerosol particles are sampled by APS--3310 (made in U.S.A.), its range is 0.504 to 30. 5 µm, and the range divided into 58 intervals. It includes $D_{(partical\ diamentes)} >$ 0. 486 µm and D > 32.8 µm , so its total intervals are 60. This instrument can display particle number spectral, surface spectral and volume spectral (mass spectral) automatically. Sample height is usually at 4 to 8 meters. We measured three sandstorm events at Yanchi, Ningxia (April 22, 23 and 27 in 1998), one sandstorm event was at Jilantai (April, 29 in 1996), and the another one was at Alanshan Zouqi, Inner Mongolia (April 10 in 1999).

3. SAND AEROSOL SIZE DISTRIBUTION

3.1. Background

This proiect is supported by National Natural Science Foundation of China.Sand particle size distribution in Yanchi (south Maowusu desert) is two peaks spectral (see Fig. l), its peak diameter (d_p) is located on 0.90 to 1.32 µm and 1.50 to 2.30 µm

---

[*] This project is supported by National Natural Science Foundation of China

respectively. d_p of mass spectral is located on 3.28 to 4.07 µm. Sand particle spectral in Alashan Zuoqi (east of Tegeli dessert) presents single peak feature, its d_p is located on 0.90 to 1.04 µm, its mass spectral also presents single peak feature, d_p is 2.84 to 4.70 µm. The ratio of number concentration which diameter is greater than 7.0 µm (N_7) to total number concentration (N_T) is 99.6%, and the ratio of mass concentration which diameter is greater than 7.0 µm(M_7)to total mass concentration (M_T) is 90.0%.

3.2. Particle Spectral Features under Dust and Sandblowing weather

Spectral pattern changes with wind direction (wd) under dust and sand blowing weather in Yanchi, there is two peak spectral when wd is E or S, d_p is 0.78 to 1.6µm and 2.31 to 2.29 µm, the diameter for all particles is smaller than 7.0 µm (see Fig. 2). There is single peak spectral when wd is W or N. d_p is 0.84 to 0.96 µm, N_7/N_T is 99.8%,mass spectral are single peak, and can be fitted by log-normal distribution, d_p is 3.28 to 3.79 µm, M_7/M_T is 88.2% to 91.6%.Number spectral in Alashan Zuoqi is single peak spectral. d_p is 0.84 to 0.90 µm, N_7/N_T is 99.6%, mass spectral are also single peak, and d_p is 3.0 to 4.7 µm, M_7/M_T is 78.9% to 80.9%.

3.3 Particle Spectral under Sandstorm Weather

Fig.3a, 3b, 3c present the changes of total number concentration (CN),total mass concentration (CM) and wind velocity (V) with time in the former, middle and last stage of sandstorm in Yanchi on April 27, 1998, respectively. CM reaches its maximum (11.3 mg/m^3) in 20 minutes, and then, decreasing gradually, wind velocity changes waved, CN, CM and V does not exist correlation at this stage, and particle number spectral has two peaks, d_p is 0.91 to 1.72 µm, and 1.72 to 2.00 µm respectively.
In the middle stage of sandstorm, the changes of CN, CM and V with time exist a good correlation, particle spectral has single peak. n the last stage, the changes of CN and CM have the same trend.

4. CONCLUSION

4.1. Sand aerosol spectral of background has different features in different geographic locations. It expresses a two peak feature in Yanchi, but one peak in Alashan Zuoqi and Jilantai, mass spectral are all single peak, and can be fitted by log-normal distribution, spectral width has a big difference.

4.2. Under dust and sandblowing weather, particle size distribution presents two peaks feature when wind is E or S in Yanchi only. Mass spectral can be fitted by log-normal distribution for all places.

4.3. When sandstorm occurs, the changes of CN and CM with time have a good correlation at last two stages.

Chemical Characterization Of Water Soluble Organic Compounds In Tropospheric Fine Aerosol

G. Kiss[a], A. Gelencsér[a], A. Hoffer[b], Z. Krivácsy[a], E. Mészáros[b], Á. Molnár[a] and B. Varga[b],

[a]*Air Chemistry Group of the Hungarian Academy of Sciences, University of Veszprém*
[b]*Dept. of Earth and Environmental Sciences, University of Veszprém*
8201 Veszprém, POB 158, Hungary, e-mail:kissgy@almos.vein.hu

Abstract. Water soluble organic compounds were investigated in aerosol samples collected in a rural station in Hungary. The share of water soluble organic carbon (WSOC) reached 58% of the total carbon and 40% of the water soluble species (both organic and inorganic). Most of the WSOC was characterized by polyconjugated (e.g. aromatic) structure, acidic functional groups and a continuous distribution in the mass to charge ratio between 100 and 700. Individual compounds with molecular weight of several hundreds have also been detected.

INTRODUCTION

The influence of tropospheric aerosol particles on climate highly depends on their chemical composition. In the past few years numerous indications were found that organic compounds alter the hygroscopic behaviour of aerosol particles and play an important role in atmospheric processes (Novakov and Penner, 1993; Saxena et al., 1995). Despite its importance our present knowledge on the organic fraction of the aerosol is rather incomplete. This is particularly so for the water soluble organic compounds which accounted for 20% to 67% of the total organic carbon present in aerosol samples collected at different locations in the world (Saxena et al., 1995; Zappoli et al., 1999) However, this is the fraction which is likely to increase the hygroscopicity of aerosol particles and play a role in cloud formation. In this paper water soluble organic compounds are characterized by total organic carbon analysis, spectroscopic techniques and liquid chromatography-mass spectrometry.

EXPERIMENTAL

Aerosol samples (n=28) were collected at K-puszta (φ=46°58', λ=19°33', h=136 m a.s.l.) about 70 km southeast of Budapest, Hungary in August 1998. The ambient air was sampled with a flow rate of 30 L/min in a forest clearance on the Great Hungarian Plain. Coarse aerosol particles (d>1 µm) were removed by a cascade impactor of two stages and the fine aerosol particles were collected on a quartz back-up filter.

Black carbon was monitored by a soot photometer at 565 nm. Total carbon was measured by a total organic carbon analyzer (Astro 2100, Zellweger Analytics) from the spots of the original samples. Water soluble organic carbon was determined by soaking spots in Milli-Q water for 12 hours and analyzing the remaining carbon on the spot after drying. Liquid chromatographic separation was performed with a Waters 2690 Separations Module on a LiChrospher 100 RP-18 column connected to a MicroMass Quattro II mass spectrometer equipped with an atmospheric pressure ionisation source used in the electrospray mode. UV and fluorescence spectra were recorded with a Waters 490M UV detector and a Waters 470 fluorescence detector, respectively.

RESULTS AND DISCUSSION

Share Of The Water Soluble Organic Compounds

In all the fine aerosol samples the concentrations of inorganic ions, total carbon (TC), water soluble organic carbon (WSOC) and black carbon (BC) were determined. Since carbonate was not found in the samples TC was regarded as the sum of total organic carbon (TOC) and black carbon. Water insoluble organic carbon (WINSOC) was calculated as TC-WSOC-BC. On average the total carbon concentration was 8.3 µg/m^3 (SD = 2.2 µg/m^3) which was distributed among the carbon forms as shown in Fig.1A. Water soluble organic carbon accounted for 58% and 64% of TC and TOC, respectively. The share of WSOC among the water soluble species was also significant as shown in Fig.1B. The mass concentration of WSOC (4.8 µg/m^3, SD = 1.6 µg/m^3) was equal to that of sulfate (4.8 µg/m^3, SD = 2.2 µg/m^3) which was the most abundant inorganic ion.

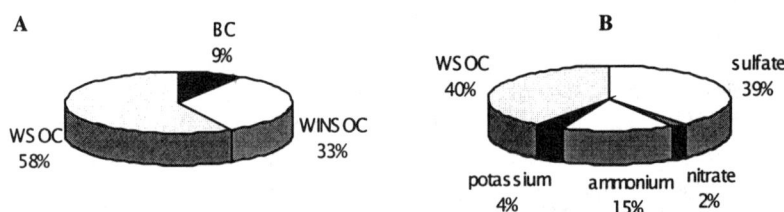

FIGURE 1. Average distribution of carbon (a) and water soluble species (b) in the aerosol samples.

The share of water soluble organic compounds is even more significant if their carbon content is converted to mass. Since the water soluble organic compounds carry polar functional groups (e.g. hydroxyl, carboxyl, carbonyl) we think that the conversion factor can be 1.5 or even greater. By applying 1.5 we obtained that the mass concentration of the water soluble organic substances in the fine aerosol samples was about the same as that of the sum of all inorganic species.

UV And Fluorescence Spectroscopy

To characterize the water soluble organic compounds UV spectroscopy and fluorescence spectrometry were applied. The UV spectrum of an aqueous aerosol extract is shown in Fig. 2A. The absorption of the dissolved substances decreased sharply between 200 nm and 240 nm and slowly above 240 nm. Since nitrate ion has strong absorption below 240 nm we assumed that its spectrum was superimposed on that of the water soluble organic compounds. Therefore the inorganic ions were separated from the organic species by solid phase extraction. The aerosol extract was acidified to pH=2 and passed through an extraction cartridge filled with non-polar packing. Then the retained organic components were eluted from the cartridge with methanol, evaporated to dryness then taken up with Milli-Q water. The TOC analysis of the effluent revealed that 74% of the WSOC was retained on the cartridge and 26% was present in the effluent. The UV spectra of the effluent and the eluate are shown in Fig. 2A, too. It can be seen that the effluent had strong UV absorption mainly below 240 nm (caused by the nitrate ion and some organic constituents) while the spectrum of the eluate had absorption up to 400 nm due to the presence of polyconjugated organic compounds. The presence of such structures in the eluate was confirmed by fluorescence emission spectra, too. There was an emission band present in the spectrum of both the aqueous aerosol extract and the eluate but there was hardly any fluorescence observed in the effluent (Fig. 2B).

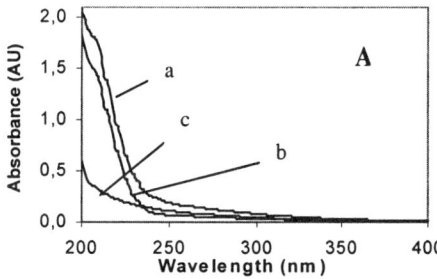
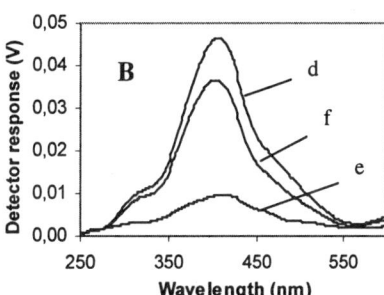

FIGURE 2. UV and fluorescence emission spectra of an aerosol extract (a,d), effluent (b,e) and eluate (c,f) from the solid phase extraction column.

These results indicate that one fourth of the WSOC content of the aerosol samples is represented by very hydrophylic compounds without polyconjugated structure while there is another fraction which can be characterized by aromatic and possibly other polyconjugated components and may account for three fourth of the WSOC.

Liquid Chromatography-Mass Spectrometry

The aim of the LC-MS analysis was to obtain information on the ionic character, polarity and molecular mass of the water soluble organic compounds. Most of the components eluted close to the dead volume of the chromatographic column indicating that these compounds were ionic or neutral possessing polar functional groups. Since these compounds gave the most intense signal in the negative ion mode we concluded that most of the water soluble organic carbon was represented by compounds carrying easily ionizable most probably acidic functional groups. The mass spectra of these early eluting components revealed that the water soluble carbon content was shared among a great number of compounds producing a practically continuous distribution of ions in the m/z (mass to charge ratio) =100 to 700 range (Fig. 3A). Several individual ions from m/z=45 to m/z=373 were superimposed on the "ion hump". These results indicated that water soluble organic compounds with molecular weight of several hundreds were present in atmospheric aerosol. Since humic like substances had been found in aerosol samples earlier (Zappoli et al., 1999) the mass spectrum of a fulvic acid reference material was recorded under the same conditions (Fig 3B). Although the two spectra were not identical high degree of similarity could be observed in terms of complexity and the shape of the ion distribution.

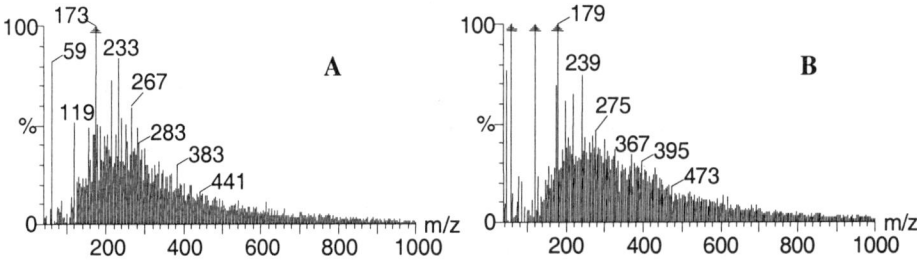

FIGURE 3. Mass spectrum of the water soluble organic compounds present in aerosol samples (A) and the spectrum of the aqueous solution of a Nordic reference fulvic acid under the same conditions (B).

Conclusions

Water soluble organic compounds accounted for 58% of total carbon and approximately half of the water soluble mass in tropospheric fine aerosol collected at K-puszta, Hungary. Spectroscopic analyses revealed the presence of polyconjugated (e.g. aromatic) parts in these components having molecular weight probably in the 100 Da to 700 Da range.

ACKNOWLEDGMENTS

This work was supported by the Hungarian National Scientific Research Fund (OTKA Project No. T030186, T030226, F029610, F029607). The courtesy of the Institue of Applied Environmental Research of University of Stockholm for the possibility to perform LC-MS analysis is gratefully acknowledged.

REFERENCES

1. Novakov, T. and Penner, J. E., *Nature*, **365**, 823-826 (1993)
2. Saxena P., Hildemann, L.M., McMurry, P.H., Seinfeld, J.H., *J. Geophys. Res.*, **100**, 18755-70 (1995)
3. Zappoli, S, Andracchio, A., Fuzzi, S., Facchini, M.C., Gelencsér, A., Kiss, G., Krivácsy, Z., Molnár, Á., Mészáros, E., Hansson, H-C., Rosman, K. and Zebühr, Y., *Atmos. Environ.*, **33**, 2733-2743 (1999)

The Electric Charging Of Aerosols In High Ionized Atmosphere

F. Gensdarmes [1,2], D. Boulaud[1], A. Renoux[2]

[1] Institut de Protection et de Sûreté Nucléaire, Département de Prévention et d'Etude des Accidents, Service d'Etudes et de Recherches en Aérocontamination et en Confinement. CEA/Saclay, Bat 389, 91191 Gif-sur-Yvette cedex, France.

[2] Université Paris XII, Laboratoire de Physique des Aérosols et de Transfert des Contaminations. Av. du Général de Gaulle, 94010 Créteil cedex, France.

Abstract. Our interest is the knowledge of electrical properties of aerosols in high ionized atmosphere, which could arise from hypothetical accident in nuclear power plant. Ionic properties are important in order to model correctly the aerosol charge distribution in such atmosphere. Consequently we have developed an experimental device which allows us to produce high ionic bipolar concentrations, up to 10^{14} m^{-3} with ^{60}Co sources, and to measure both aerosol charge distribution and ions properties. Experimental results show that the aerosol charge distribution is negative, according to ions properties measurements and Clement and Harrison expression. For 0.3 µm latex particles the mean charge is -0.6e for ionic concentrations around 10^{12} m^{-3}. The mean charge is reduce to -0.2e in very high ionized atmosphere, when ions concentrations reach 10^{14} m^{-3}. In such case ions mobilities are lower and tend toward equal value, which leads to a reduction of charge asymmetry.

INTRODUCTION

In case of accident within a nuclear power plant, gamma irradiation could involve a high ionized atmosphere where the dose rate can reach 10 kGy/h. In such case the knowledge of the aerosol charge distribution is usefull for modelling aerosol transport and deposition in a close vessel.

In a bipolar ionic atmosphere, Bricard and Pradel (1966) [1] showed that the electrical charge distribution of an aerosol does not necessarily reach the Boltzmann equilibrium. Asymmetric charging could arise from the different diffusive properties between positive and negative ions. This asymmetry is represented by a parameter which is the positive to negative electrical conductivities ratio, defined by the expression (1).

$$X = \frac{\mu_+ \cdot n_+}{\mu_- \cdot n_-}, \qquad (1)$$

where μ_+ and μ_- are respectively the positive and negative ion mobilities ($m^2 \cdot V^{-1} \cdot s^{-1}$), n_+ and n_- respectively the positive and negative ion concentrations (m^{-3}).

Several analytical expressions exist to determine the aerosol charge distribution according to ionic properties in the continuum regime [2]. However there is very few experiments where both aerosol charge distribution and ionic properties are measured together.

The aim of this work is to measure the asymmetry parameter and the resulting charge distribution on aerosols for differents ionization rate.

EXPERIMENTAL DEVICE

Our experimental device represented figure 1 allows us to produce high ionic bipolar concentrations, upper than 10^{14} m^{-3} with ^{60}Co sources. Such concentrations are obtained for dose rates of about 7 kGy/h. We can vary the ionization rate, which is equivalent to the dose rate, by changing the distance between the ^{60}Co sources and the sampling volume. The positive and negative electrical conductivities are measured with a cylindrical capacitor by using the Gerdien method, described in detail by Tammet (1967) [3]. The charge distribution of monodisperse aerosol is determined with a radial flow mobility analyzer of order 1 [4]. In our experiments, the relative humidity of air is between 11 and 17 %, the temperature is 20°C.

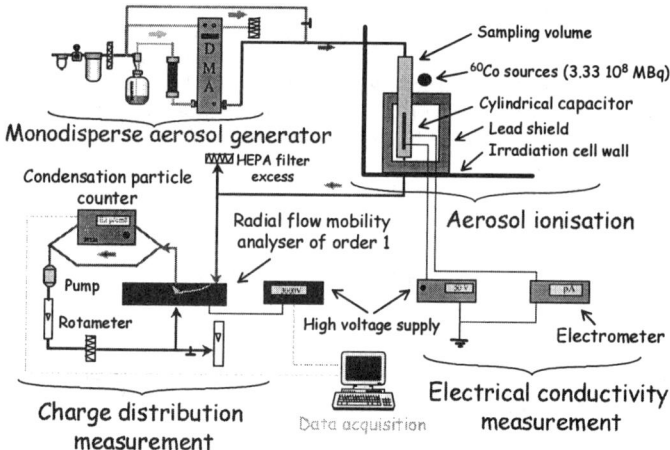

FIGURE 1. Experimental Device.

RESULTS

Figure 2, shows the two slopes of the curves relating to the small positive and negative ions, which are proportional respectively to conductivities due to positive and negative ions. Thus, the asymmetry parameter is given by the ratio of the two slopes. For a dose rate equal to 100 Gy/h the value obtained is X=0.88. The positive and negative ions concentrations are calculated from the respective saturations currents and the flow rate in the capacitor, we obtained respectively n_+=1.1E^{13} m^{-3} and n_-=9.8E^{12} m^{-3}. Figure 3 shows the experimental charge distribution for monodisperse aerosol of 0.3 µm in diameter, obtained with a dose rate equal to 100 Gy/h and total concentration of 490 p/cm^3. We also represent the Boltzmann equilibrium as well as the distribution of Clement and Harrison (1992) [2] calculated according to the given value in experiments of the asymmetry parameter. The mean charge calculated from the experimental distribution (figure 3) is J=-0.39 e, the standard deviation associated is equal to 0.10 e. The mean charge calculated from Clement and Harrison equation is J=-0.35 e.

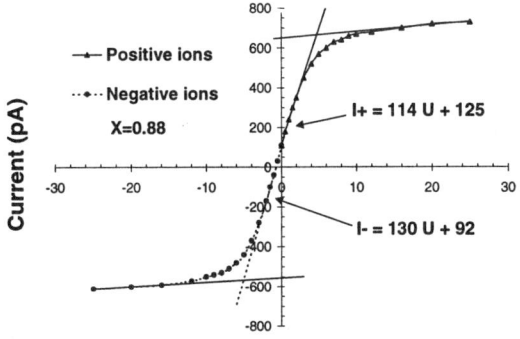

FIGURE 2. Measurement of Asymmetry Ratio between Positive and Negative Ion Conductivities for a Dose Rate equal to 100 Gy/h.

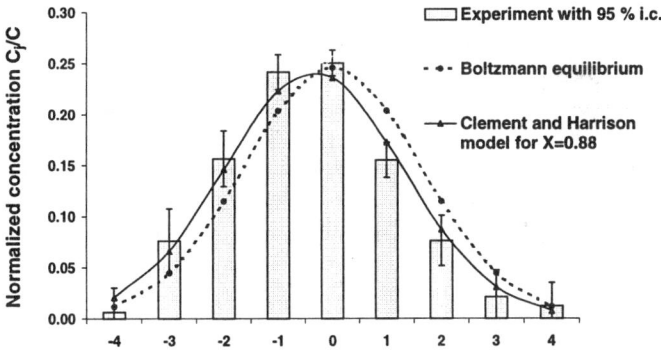

FIGURE 3. Charge Distribution for 0.3 µm Monodisperse Aerosol for a Dose Rate equal to 100 Gy/h.

Figure 4 represents the aerosol mean charge as a function of a charge parameter on the assumption that the medium is electrically neutral [1]. The charge parameter is defined by Q/C^2, where Q represent the number of pair of small ions fixed per $m^3 \cdot s$ by the aerosol particles and C the total aerosol concentration (m^{-3}). The Bricard expression [1] shows that the mean charge reach a value independent of Q/C^2 for a ionization rate sufficiently high. This was checked experimentaly by Pourprix (1973) [5] for a charge parameter smaller than $8E^{-8}$ $m^3 \cdot s^{-1}$. Our experimental results are in good agreement with the theoretical law for a charge parameter up to $2E^{-7}$ $m^3 \cdot s^{-1}$. For higher charge parameter our results show that the mean charge decrease. This result is explained by a reduction of the asymmetry ratio, which was also measured in our experiments. This shows that for a constant sampling time, the ratio between the mobility of the positive and negative ions is not independent of the ionization rate. Moreover, that affects the aerosol charge distribution.

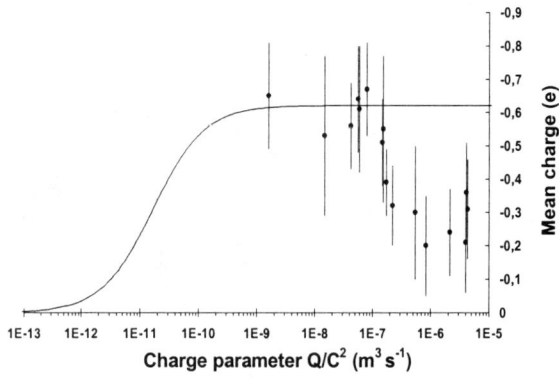

FIGURE 4. Aerosol Mean Charge Related to the Intensity of Ionization.

CONCLUSIONS

Our experimental results show that the analytical expression of Clement and Harrison describes well the charge distribution of an aerosol in a bipolar medium, if one takes into account the ion asymmetry ratio of the medium. Moreover, our results show that the asymmetry parameter is not constant, it depends on the ionization rate. This parameter approaches the unity in high ionized atmosphere. The whole of these results shows that it is necessary to know the ionic properties of the medium in order to correctly calculate the charge distribution of an aerosol.

REFERENCES

1. Bricard, J. and Pradel, J., "Electric charge and radioactivity of naturally occurring aerosols" in *Aerosol Science*. edited by C.N. Davies, Academic Press, 1966, pp. 87-109.
2. Clement, C. F. and Harrison, R. G., *J. Aerosol Science* **23**, 481-504 (1992).
3. Tammet, H.F., *The aspiration method for the determination of atmospheric-ion spectra*. edited by B. Benny, Israel program for scientific translations, 1967.
4. Mesbah, B., Le S.M.E.C. (Spectromètre de Mobilité Electrique Circulaire). Théorie, Performances et Applications. Thèse de l'Université Paris XII, 1994. Rapport CEA-R-5693.
5. Pourprix, M., Un nouveau précipitateur électrostatique : Application à l'étude de la charge des aérosols par diffusion d'ions bipolaires. Thèse de docteur ingénieur. Université Paris VI, 1973.

Comments on the Soluble Particle Identification by the Spot Test Technique

Josef Podzimek

Cloud and Aerosol Sciences Laboratory
University of Missouri-Rolla, Rolla, MO 65409-0640, USA

Abstract. The results of old and new measurements of salt nuclei are reviewed in order to estimate the effects of particle size and of environmental humidity or temperature on the size distribution of potential cloud condensation nuclei. Several possible improvements of the sampling and evaluation technique are discussed.

INTRODUCTION

More than one hundred years ago, R.E. Liesegang described how a crystal of silver nitrate placed on a glass slide covered by a thin gelatinous layer containing dissolved potassium dichromate formed concentric rings of silver dichromate precipitate around the original crystal. Several years later, Wilhelm Ostwald attempted to explain this phenomenon by the so-called supersaturation theory. Since that time, several other theories were developed, such as adsorption, coagulation, and diffusion wave theories, which were then described in a survey article by Stern (1954). Common to all theories was the critical quantity of reaction products governing the formation of precipitation zones, such as solubility product, supersaturation, etc.

The laboratory "spot test technique" (or Liesegang ring method) was applied and adapted to routine aerosol sampling by Winckelmann (1931), among others. Later, other investigators described in nearly one hundred articles the results of the sampling of atmospheric aerosol particles containing specific ions (chlorides or sulphates, for example) and the advantages and drawbacks related to the sample preparation and evaluation. The main results of these studies were recently summarized by J. Podzimek and M. Podzimek (1999).

Some of the subjects discussed most frequently were the slide (film) coating by a sensitized gelatin layer, the storing of slides before their exposure, the treatment of exposed slides, and the effects of environmental humidity, temperature, and sampled particle properties on the determination of the magnification factor. This factor enables the conversion of the circular spots in the sensitized gelatin into real sizes of "spherical" salt particles. Our contribution will focus on several of the aforementioned subjects and on the results obtained during laboratory measurements with sale particles and salt solution droplets.

PARAMETERS IMPORTANT FOR THE DETERMINATION OF THE PARTICLE MAGNIFICATION FACTORS

a) Thickness and composition of the sensitized gelatin layer.

In spite of the application of different substrates for specific particle (droplet) deposition (see J. Podzimek and M. Podzimek, 1999), we will limit our discussion to the case when salt nuclei (NaCl) collide with a sensitized gelatin layer. This layer usually was prepared from 8% gelatin (b.w.) and had a thickness between 20 µm and 100 µm. The slides covered by a gelatin layer were then dipped for 20 seconds into a 5% (b.w.) solution of silver nitrate and stored afterwards in the dark. Exposing the slide to solar or UV radiation made the spots visible around the deposited salt particles containing mainly silver chloride and silver nitrate and thus suitable for microscopic evaluation. Example of such an evaluation of salt solution spots is presented in Fig. 1.

The production of a layer of uniform thickness and the knowledge of its thickness are essential for the correct determination of the salt particle (nuclei) size distribution. This holds true especially in the case of a broad salt particle size distribution, when submicron nuclei might coexist with activated particles larger than 5.0 µm. The real particle magnification factors of large nuclei depend on the whether the diffusion spheres will reach the slide substrate and continue to grow in two dimensions. This might explain the deviations from the simple relationship between salt solution droplet and Liesegang circle diameter found e.g. by Ueno and Sano (1971).

b) Effects of humidity and temperature.

Both parameters can affect the formation of spots in gelatin layers during sample preparation, slide exposure and storage. Measurements taken recently confirmed the results of older investigations by Yue and Podzimek (1980), which uncovered the considerable effects of environmental humidity and of the humidity maintained during the sampling and the initial period of sample storage on the calculated magnification factors. The samples taken and stored at environmental humidity above 80% exhibited magnification factors between 1.5 and 2.0 times larger on average than samples taken at humidities at about 72%. Some hysteresis of magnification factor curves for salt nuclei was usually observed while increasing and decreasing humidity in the container with exposed slides.

The temperature effect is closely related to the environmental humidity and affects the solution droplet evaporation and ion diffusion in the gelatin layer. However, the increase of atmospheric temperature from +2°C to +21°C caused an increase of the salt particle magnification factor of less than 20% (Podzimek, 1959).

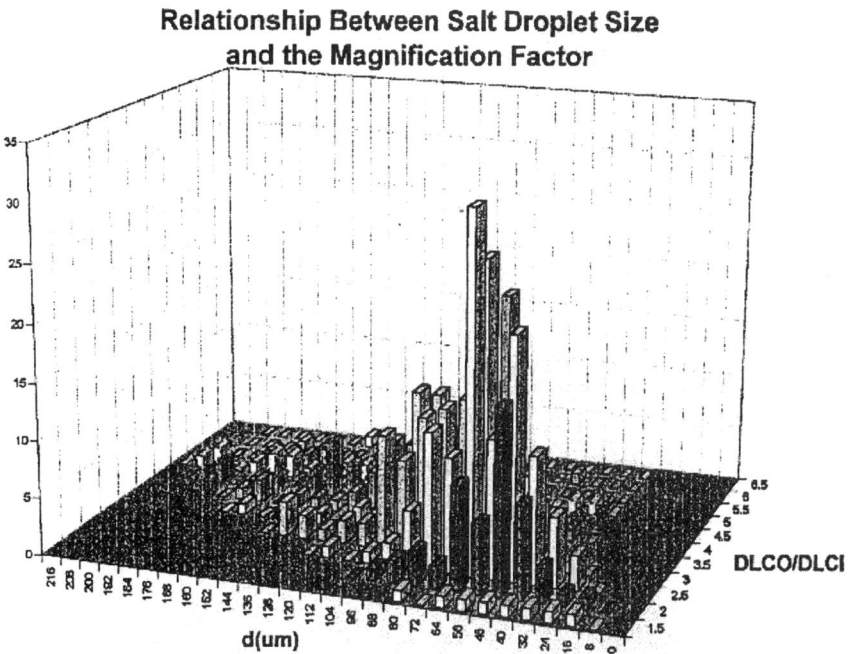

Fig. 1 Frequency of occurrence of salt solution (5% b.w.) droplet sizes (d) and corresponding spot magnification factors (M = DLCO/DLCI). DLCO and DLCI are diameters of outer and inner Liesegang rings. Gelatin layer thickness was about 50 µm.

c) Effects of the concentrations of inner and outer electrolytes and of the sample evaluation procedure on magnification factors.

In general, there was a strong indication during our experiments that the diffusing ring solution was supersaturated with the precipitate once the ring appeared. Lower salt (such as NaCl) concentrations in sprayed "wet" aerosol lead to higher magnification factors (for example, for 1% NaCl solution droplets, M. F. = 2.58 and for 6% NaCl solution droplets, M. F. = 2.08 at humidities to 72%). This certainly affects the calculation of dry salt particle magnification factors, assuming that the diffusion process of the outer electrolytes starts around the salt crystals at higher relative humidity in the container.

The circle (spot) formation can be slowed down or inhibited by gelatin concentrations higher than 10% (b.w.) or by the addition of the soluble product of the reaction ($NaNO_3$). In order for find very small circles in an optical or scanning electron microscope, all samples with deposited NaCl particles were stored in darkness until being exposed to a mercury lamp for 10 to 20 minutes. This yielded brown-reddish circular spots on the gelatin's yellowish background. Different procedures have been described for the identification of Cl and other ions (such as

SO_4, NH_4, NO_3) on different deposition substrates (see reference such as J. Podzimek and M. Podzimek, 1999).

POTENTIAL IMPROVEMENTS OF THE SPOT TEST TECHNIQUE

There are still several questions which must be satisfactorily answered before widely applying the spot test technique for the detection of atmospheric aerosol particles (see for example Gerard et al. 1989 a,b). For instance, more advanced three-dimensional models would better describe the complex diffusion processes during the spot formation (which could include the counterdiffusion of inner electrolyte ions). Also, the spot test requires strict control of environmental conditions (mainly humidity) during sampling and processing of aerosol samples and for determining the particle (droplet) magnification factor and impactor collection efficiency for specific situations.

The main problem is still the time-consuming sample evaluation, which can be facilitated only for a very few precipitates (such as salt particle spots having a substantially different color than the gelatin layer). The other difficulty of finding a suitable spot test technique is related to particles in polluted air which often have a mixed microstructure (solid and liquid, or different chemical composition).

REFERENCES

1. Gerard, R., Villard, A., and Serpolay, R., Atmos. Res., **22**, 335-350 (1989a).
2. Gerard, R., Villard, A., and Serpolay, R., Atmos. Res., **22**, 351-371 (1989b).
3. Podzimek, J. Studia Geophys. Geod., **3**, 256-280 (1959).
4. Podzimek, J. and Podzimek, M., "Liesegang Ring Technique Applied to the Chemical Idenfication of Atmospheric Aerosol Particles" in Analytical Chemistry of Aerosols (ed. By K. R. Spurny), Lewis Publ., Boca Raton, 1999, pp. 231-242.
5. Stern, K.H., Chem. Revs., **54**, 79-99 (1954).
6. Ueno, Y., and Sano, I., Bull. Chem. Soc. Japan, **44**, 637-641 (1971).
7. Winckelmann, J., Mikrochemie, **12**, 437-439 (1931).
8. Yue, P.C., and Podzimek, J., Ind. Eng. Chem. Prod. Res. Dev., **19**, 42-46 (1980).

Development of DMA-Faraday Cup Electrometer System for Measurement of Submicron Aerosol Particles

Manabu Shimada, Ferry Iskandar, and Kikuo Okuyama

Department of Chemical Engineering, Hiroshima University
1-4-1 Kagamiyama, Higashi Hiroshima 739-8527, Japan

Abstract. A measurement system consisting of a Long Differential Mobility Analyzer (LDMA) and a Faraday Cup Electrometer (FCE) was developed to investigate the performance of the system for measuring the size distribution of submicron airborne particles. The LDMA was designed to have an extended classification zone in order to classify particles up to 1 μm. A new data reduction program was prepared to convert the electrical mobility distribution data obtained with the FCE into a particle size distribution. By using the program, the size distribution of TiO_2 particles determined by the present system agreed with that measured by a combination of the LDMA and a condensation nucleus counter. The system was also evaluated using polydispersed, submicron SiO_2 particles generated by a spray drying method. The results show that the present system performs well and is a useful tool available for measuring particles size over a wide range.

INTRODUCTION

Theoretical and experimental investigations of the classification of airborne particles by Differential Mobility Analyzers (DMA) have been performed for many years using a number of variable factors that influence classification accuracy. Although the measurement of submicron size particles with diameters up to 1 μm are important for cases such as atmospheric dust or cigarette smoke, most currently existing DMA types are only able to classify aerosol particles with diameters up to 300 nm in size. Condensation Particle Counters (CPC) are widely used to count particles classified by the DMA. However, CPC's suffer from a limited range of operating environments and a high cost per unit. Therefore, a cheaper particle counting instrument feasible in many operating environments is needed.

In this study, we developed a combination LDMA-FCE (Long DMA-Faraday Cup Electrometer) system to measure particles with sizes up to 1 μm in diameter. We also evaluated the performance of this system experimentally. The measured aerosol size distribution of the new system was compared with the results from a LDMA-CPC system. Furthermore, this system was used to measure the size distribution of spherical, submicron particles generated by spray drying.

EXPERIMENTAL SETUP AND EVALUATION PROCEDURES

Figure 1 shows a schematic diagram of the LDMA-FCE system. In order to extend the classification size to particles up to 1 μm, the length of the annular classification region of the LDMA was lengthened to approximately 1 meter, with inner and outer rod diameter of 15 and 20 mm, respectively. The sheath flow is introduced from 2 mm-hole at the top of inner rod and is smoothed by passing through an annular nylon mesh filter. Charged aerosol is introduced through a 1mm small circular slit and joins with the sheath airflow in the classifying region. After classification, the aerosol is introduced into the FCE and is measured as an electric current value by a High-Sensitivity-Electrometer. In a preliminary experiment, the LDMA was examined using the Tandem-DMA method, which combines the LDMA with another standard DMA (Model 3071, TSI) using a particle standard of PSL (polystyrene latex) as a calibration aerosol. The result confirmed that the LDMA produced a well-characterized, monodisperse aerosol distribution.

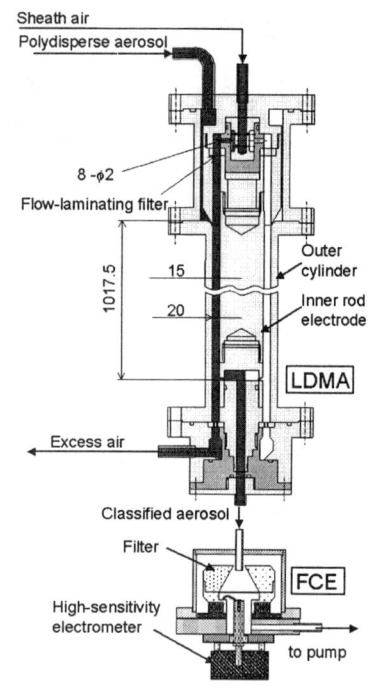

FIGURE 1. LDMA-FCE measurement system.

Considering the fact that large particles also have multiple charges, a new data reduction technique becomes necessary to obtain the actual size distribution from the raw measurements, particularly for aerosols that contain large particles. In addition, the data reduction technique used when the number concentration of particles is obtained by CPC [1] is different than when measured by the FCE. Therefore, a new data reduction technique was created to convert the measured amount of charge to a particle number concentration.

The experimental apparatus is shown in Fig. 2. The apparatus consisted of an aerosol generator, a particle collector, and a particle measurement system. To confirm the accuracy of the new data reduction technique, the size distribution of TiO_2 particles generated by a CVD method, with a mean diameter of 100 nm, was measured by the LDMA-FCE system and compared with measurements using an LDMA-CPC system. Afterward, a sample of polydispersed submicron SiO_2 particles was also measured using the LDMA-FCE system. The SiO_2 particles were generated using a spray drying method from a nanosized colloidal SiO_2 suspension (Nissan Chemical Ind. Co. Ltd., primary particle size 10-20 nm), which was nebulized by an ultrasonic oscillator (Omron Co. Ltd., NE-U11B), and dried at temperature of 600 °C in an electric tubular furnace. The morphology and size distribution of the SiO_2 particles was also observed and measured by SEM photographs on particles collected in an electrostatic precipitator.

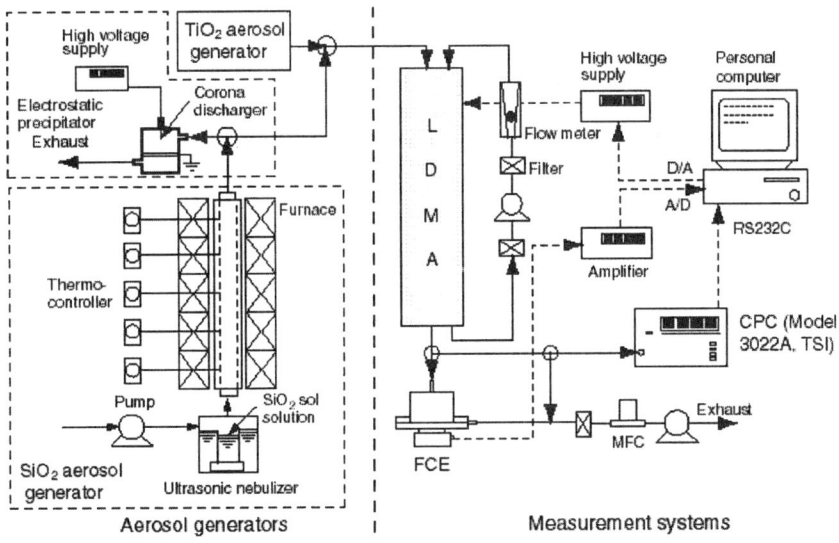

FIGURE 2. Experimental apparatus.

EXPERIMENTAL RESULT AND DISCUSSION

Size Distribution of TiO_2 Particles

Figure 3a shows the electrical mobility distributions of TiO_2 particles that have been measured by both the LDMA-FCE and LDMA-CPC before inversion with the data reduction program. Both measurements show that the equivalent electrical mobility values (around 100 nm in diameter size) correspond at the peak of the distributions. However, the peak value of the elementary electric charge measured by FCE is higher than the peak value of the number concentration measured by CPC, because the LDMA also classified particles having multiple unit charges. Figure 3b shows the particle size distribution after inversion. Good agreement is obtained between the particle size distributions measured by the CPC and the FCE. Therefore, we conclude that the LDMA-FCE system is capable of correctly measuring aerosol size distributions.

Size Distribution of SiO_2 Particles

Figure 4 shows a Scanning Electron Microscope (SEM) photograph of SiO_2 particles collected by the electrostatic precipitator. The SiO_2 particles were generated by spray drying resulting in a spherical particle morphology. Figure 5 shows the size distribution of the SiO_2 particles measured from LDMA-FCE system. As a result, the size distribution measured by LDMA-FCE system has a main peak value around 200 nm, which is close to that obtained by observation of the SEM photograph.

CONCLUSION

An LDMA-FCE particle measurement system was developed and found suitable to measure size distributions of submicron particles, after the application of a data reduction program capable of converting the measured electric current distribution to a size distribution.

REFERENCE

1. Adachi, M., Okuyama K., Kousaka Y., Moon S.W., and Seinfield J. H., *Aerosol Sci. Tech.*, **12**, 225-239 (1990).

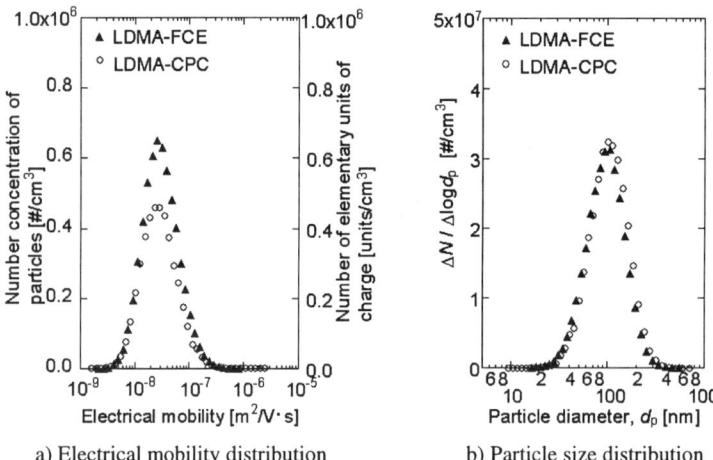

a) Electrical mobility distribution   b) Particle size distribution

FIGURE 3. Electrical mobility and particle size distributions obtained with LDMA-FCE and LDMA-CPC systems.

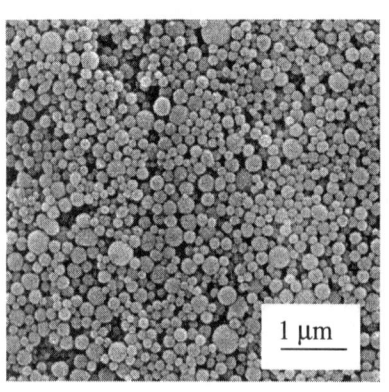

FIGURE 4. SEM picture of SiO_2 particles prepared by spray drying.

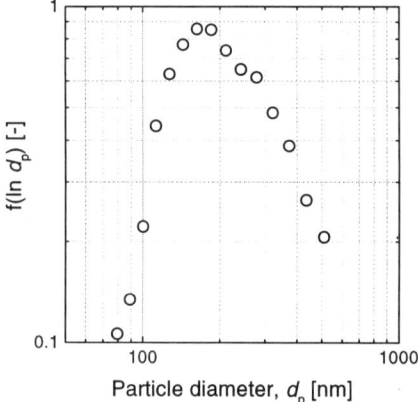

FIGURE 5. Size distribution of spray-dried SiO_2 particles measured with LDMA-FCE system.

Hygroscopic Properties of Ultrafine Particles in Coastal and Forest Environments

Hämeri, K. and Väkevä, M.

Department of Physics
FIN-00014 University of Helsinki
Finland

Abstract. The hygroscopic properties of recently produced ultrafine particles were measured in forestal and coastal environments using an Ultrafine Tandem Differential Mobility Analyser. The size range of the ambient aerosol particles that was investigated ranged from 8 nm to 30 nm in dry diameter. The measurements in a boreal forest site (SMEAR II station, Hyytiälä, Finland) showed that the growth factors during the days with production of ultrafine particles have a clear diurnal behaviour. The minimum growth factor was ca. 1.12-1.13 and was obtained during nighttime. The maximum values were obtained between 13:00 and 16:00. The median values for the maximum growth factor for 10 nm dry diameter particles was 1.26 and for 20 nm dry size 1.28. The measurements in a coastal site (Mace Head station, Carna, Ireland) showed that the new particles produced during nucleation bursts were non- or slightly hygroscopic, with growth factor ca. 1.0-1.1 for 8 nm and 10 nm particles. The background ultrafine particles that were not connected with nucleation bursts were found to be more hygroscopic.

INTRODUCTION

The origin of submicrometer particles in ambient air is widely studied phenomenon. Several field studies in a variety of environments have been devoted to this subject (e.g. Aalto et al. (1995); Hämeri et al. (1996); Mäkelä et al. (1997); O'Dowd et al. (1998); Schröder & Ström (1997); Väkevä et al. (2000); Weber et al. (1997) and Wiedensohler et al. (1993)). New particle formation events have been detected in continental and marine conditions as well as in free troposphere. These studies indicate that natural processes lead to significant production of ultrafine particles.

Despite of the intense research activity, the detailed particle production mechanisms are still unclear. One possible way to obtain information on the particle production is to determine the chemical composition of newly formed particles. This is difficult due to small mass of the particles in ultrafine size range and the relatively short sampling times required in order obtaining representative sample of the particle burst. Widely used technique to gain indirect information on chemical composition of individual particles is the determination of hygroscopic properties with a Tandem Differential Mobility Analyser. In this study we report investigations that were done using a recently developed Ultrafine Tandem Differential Mobility Analyser (UF-TDMA, Hämeri et al., 2000). This instrument is designed to measure hygroscopic diameter growth factors of individual aerosol particles in the dry particle diameter range 8-30 nm when taken from dry state (relative humidity RH<5%) to high relative humidity (typically RH=90%).

The measurements presented here were performed under two EU funded projects: BIOFOR and PARFORCE. These projects focus on forestal and coastal environments respectively. These both are regions where systematic particle production bursts were observed to take place (O'Dowd et al., 1998; Mäkelä et al., 1997). The ultrafine particle properties reported here differ significantly in different environments. Nevertheless, the results give clear information on the solubility of the particles and consequently define the characteristics of components responsible for the growth to detectable sizes.

EXPERIMENTAL

The BIOFOR field campaign took place at a boreal forest site (SMEAR II) in Southern Finland. Sample for the UF-TDMA measurements was taken above the forest canopy at 67 m height. Measurements were performed during three periods: April-May 1998, August 1998 and March-April 1999. The focus here will be on the campaign spring 1999, because that was the period when most of the nucleation bursts were observed. During the intensive field study 10 nm, 15 nm and 20 nm dry particles were measured continuously starting every 10 min; i.e. one size was measured every half an hour. This was thought to be the most efficient way of determining both the growth factor of each of the sizes as a function of time and the growth factor development of the nucleation mode particle population as the particles grew to larger sizes due to condensation and coagulation. During the 1999 campaign also bigger dry sizes (20, 35, 50, 73, 109,166, 264 nm) were monitored with an H-TDMA (see e.g. Swietlicki et al., 1999). These measurements were performed inside the canopy some hundred meters from the UF-TDMA measurements.

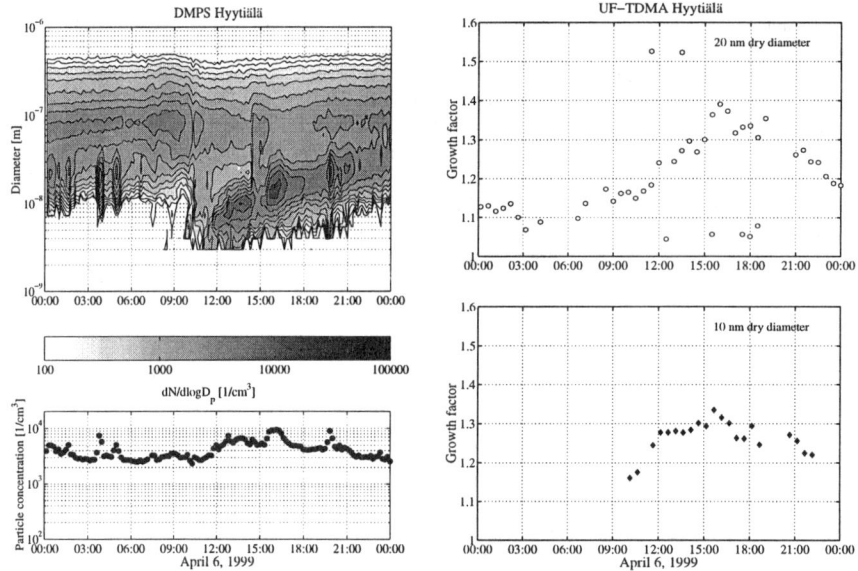

FIGURE 1. Typical nucleation event as detected in Hyytiälä forest site during April 6, 1999. The evolution of the size distribution during the day is shown (left) together with the temporal variation of the hygroscopic growth factors for ultrafine particles (right).

The PARFORCE field campaign in September 1998 and June 1999 was undertaken at the Mace Head Atmospheric Monitoring Station on the west coast of Ireland. In Mace Head the sampling for UF-TDMA was done by the sea at the height of 10 m. During the summer 1999 sizes of 8 nm, 10 nm and 20 nm were monitored. These sizes were selected because the size distribution measurements during 1998 campaign showed that the recently formed particles were typically smaller than 10 nm, and 8 nm dry size would better represent the newly formed particles while 20 nm was considered to represent the background ultrafine particles.

RESULTS

Hyytiälä, Finland, BIOFOR campaign

At the boreal forest field station diurnal pattern of growth factors for all particles with sizes between 10 and 30 nm was observed during nucleation days. Figure 1 shows an example of a nucleation day during the BIOFOR campaign. The nucleation event starts ca. 10:00 and the particles grow towards the Aitken mode during the course of the day. The lowest values for growth factors were detected in early morning and late at night, the maximum values were obtained during afternoon. The growth factors of the ultrafine particles start to increase typically some hours before the nucleation burst is observed. This behaviour of growth factors was, however, observed also on several days when no clear nucleation was detected indicating similar condensation processes onto pre-existing particle surfaces. The H-TDMA measurements showed that the diurnal behaviour of Aitken mode particles was consistent with nucleation mode particles.

Mace Head, Ireland, PARFORCE campaign

In the coastal environment of Mace Head the recently formed particles were found to be non- or slightly hygroscopic while the background ultrafine particles were clearly more hygroscopic. The hygroscopic properties of newly formed 8 nm particles were found to change drastically as the nucleation burst started. This behaviour is clearly seen in Figure 2, where a typical nucleation burst during a period of clean air masses originating from south-west is shown. The summertime monitoring revealed a temporal pattern in growth factors of 8 and 10 nm (sometimes also 20 nm) resembling that observed in BIOFOR campaigns. During these days also the magnitude of the growth factors was similar to observations above a boreal forest.

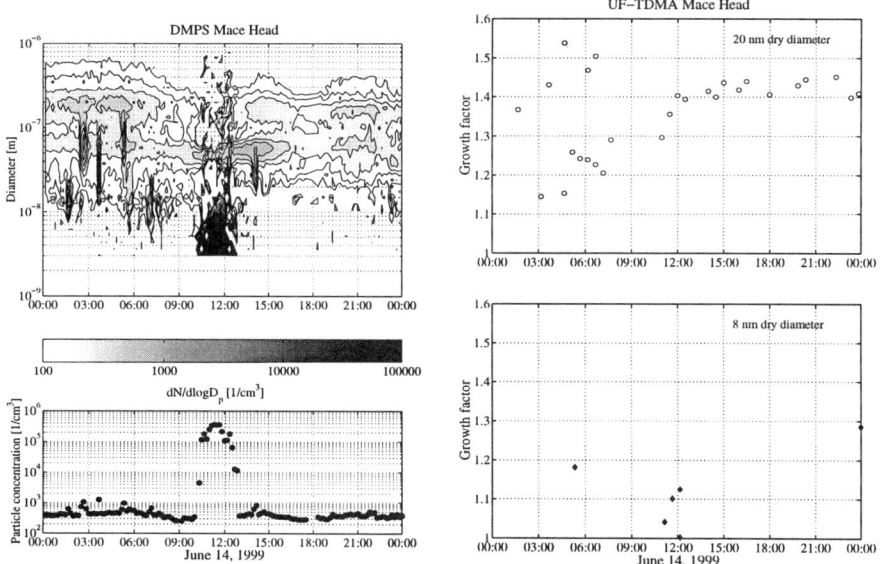

FIGURE 2. Typical nucleation burst with the source region close to the sampling point as detected in Mace•Head coastal site during June 14, 1999. The evolution of the size distribution during the day is shown (left) together with the temporal variation of the hygroscopic growth factors for ultrafine particles (right).

CONCLUSIONS

In forestal environment clear diurnal pattern of growth factors for particles with sizes between 10 nm and 30 nm were observed during the nucleation days. Lowest values were detected during late evening and early morning, the maximum of the growth factor was observed during midday. The highest values of the growth factors during the nucleation events were similar to those of ammonium sulphate.

In clean marine coastal environment the size of the recently formed particles remained typically below 10 nm. This observation is also visible when comparing the growth factors of 8 nm and 20 nm particles. The pre-existing background aerosol (20 nm) have clearly a different composition than the newly formed ones (8 nm). Low growth factors of 8 nm particles indicate that they consist of organic or other non- or weakly soluble species, possibly accompanied by a small soluble part. Simultaneous observations of 20 nm particles, on the other hand, reveal significant soluble fractions typical for sulphate aerosol in various degrees of neutralisation.

ACKNOWLEDGMENTS

The authors would like to thank Pasi Aalto and Ismo Koponen for their help in conducting the measurements. The staffs in SMEAR II and Hyytiälä field stations and Mace Head Atmospheric Monitoring Station are acknowledged. This study was funded by the European Commission Contracts No. ENV4-CT97-0526 (PARFORCE) and ENV4-CT97-0405 (BIOFOR).

REFERENCES

1. Aalto, P., Kulmala, M. and Nilsson, E.D. Nucleation events on the Värriö environmental measurement station. J. Aerosol Sci. 26, S411-S412 (1995).
2. Hämeri, K., Kulmala, M., Aalto, P., Leszynski, K., Visuri, R. and Hämekoski, K. The investigations of aerosol particle formation in urban background area of Helsinki. Atmos. Res. 41, 281-298 (1996).
3. Hämeri, K., Väkevä, M., Hansson, H.-C. and Laaksonen, A. Hygroscopic Growth of Ultrafine Ammonium Sulphate Aerosol Measured Using an Ultrafine Tandem Differential Mobility Analyser. J. Geophys. Res. Submitted (2000).
4. Mäkelä, J.M., Aalto, P., Jokinen, V., Pohja, T., Nissinen, A., Palmroth, S., Markkanen, T., Seitsonen, K., Lihavainen, H. and Kulmala, M. Observations of ultrafine aerosol particle formation and growth in boreal forest, Geophys. Res. Lett., 24, 1219-1222 (1997).
5. O'Dowd, C.D., Geever, M., Hill, M.K., Smith, M.H. and Jennings, S.G. New particle formation: Nucleation rates and spatial scales in the clean marine coastal environment. Geophys. Res. Lett., 25, 1661-1664 (1998).
6. Schröder, F. and Ström, J. Aircraft measurements of sub micrometer aerosol particles (> 7 nm) in the midlatitude free troposphere and tropopause region. Atmos. Res. 44, 333-356 (1997).
7. Swietlicki, E., Zhou, J., Berg, O.H., Martinsson, B.G., Frank, G., Cederfelt, S.I., Dusek, U., Berner, A., Birmili, W., Wiedensohler, A., Yuskiewicz, B. and Bower, K.N. A closure study of sub-micrometer aerosol particle hygroscopic behaviour, Atm. Res., 50, 205-240 (1999).
8. Väkevä, M., Hämeri, K., Puhakka, T., Nilsson, D., Hohti, H. and Mäkelä, J.M. Effects of meteorological processes on aerosol particle size distribution in an urban background area. J. Geophys. Res. In press (2000).
9. Weber, R.J., Marti, J.J., McMurry, P.H., Eisele, F.L., Tanner, D.J. and Jefferson, A. Measurements of new particle formation and ultrafine particle growth rates at a clean continental site. J. Geophys. Res. 102, 4376-4385 (1997).
10. Wiedensohler, A., Aalto, P., Covert, D., Heintzenberg, J. and McMurry, P. Intercomparison of tree methods to determine size distributions of ultrafine aerosols with low number concentrations. J. Aerosol Sci. 24, 551-554 (1993).

Study of the Transport of Contaminants from Norilsk Integrated Mining Plant to the North of Western Siberia

Vladimir F. Raputa[1], Antonina I. Smirnova[2], Konstantin P. Koutzenogii[2], Boris S. Smolyakov[3], and Tatjana V. Yaroslavtseva[2]

[1] *Institute of Chemical Kinetics and Combustion SB RAS, 630090, Novosibirsk, Russia.*
[2] *Institute of Computing Mathematics and Mathematical Geophysics SB RAS, 630090, Novosibirsk, Russia*
[3] *Institute of Inorganic Chemical SB RAS, 630090, Novosibirsk, Russia*

Abstract. In this paper an attempt has been done to develop this method. We have performed a joint analysis of the data of the daily measurements of the chemical composition of atmospheric aerosols and wind direction. To this end, we used the autumn series of observations made in Samburg in 1998 with the winds of the north-east sector. The joint analysis of the surface wind fields and the data on the chemical composition of atmospheric aerosols allowed us to reveal a number of cases where the exhausts of the contaminants of the Norilsk plant were carried out to the north of Western Siberia, to establish relations between the observed concentrations and the initial exhausts of dominating substances and elements.

INTRODUCTION

There are no great sources of anthropogenic pollution of atmospheric air on the north of Western Siberia. The nearest large sources are situated at distances of hundreds and even thousands of kilometers. They include industrial enterprises of the South and Middle Urals, Kolskii peninsula, Norilsk, the South of Western Siberia.

Systematic lasting observations of the chemical composition of atmospheric aerosols in a number of settlements to the north of Western Siberia make it possible to separately study the above sources during regional pollution of the territories mentioned [1]. The particular interest are the cases of prolonged winds of certain directions. A joint analysis of the data of observations of the daily dynamics of a change in the chemical composition of atmospheric aerosols and wind regime allows one to identify in some cases the source of pollution, establish a specific aerosol composition, the characteristic trace substances and chemical elements.

This paper studies atmospheric aerosols arriving from Norilsk region. The main sources of atmospheric pollution of the city and its neighborhood are the enterprises of non-ferrous metallurgy [2]. The exhausts of sulfur dioxide of these enterprises are more than 2 mln. ton/year. Sulfur dioxide makes up 96% of the exhausts, dust, and carbon oxide and nitrogen oxide amount to 1-1.5%.

Among heavy metals, predominant are copper, nickel, cobalt, zinc, lead, arsenic, selenium, and antimony. The emissions are performed by enterprises at great heights and are carried out over great distances by strong winds.

Atmospheric Transport and Chemical Composition of Aerosols

The long continuous observations of both the composition of atmospheric aerosols and wind direction and velocity at various heights allow one to reveal the qualitative and quantitative relations to the sources of emission. When taking into account the mutual arrangement of Norilsk plant and observation sites (settlements Samburg and Tarko-Sale), of most interest are the observation periods corresponding to the stable directions of winds of the north-east sector. In this case, with respect to the distance between sources and receivers, one can estimate the dynamics of the arrival of aerosol contaminants.

It is noteworthy, that at relatively short times of transport (of order of several days), the main amount of exhausts will be located within the boundary atmospheric layer and distributed vertically. In this case, with reasonable input parameters, the concentration of contaminants can be calculated from the equation below:

$$\frac{\partial S}{\partial t} + U \frac{\partial S}{\partial x} = K \frac{\partial^2 S}{\partial y^2} - \sigma S + F(x,y,t), \qquad (1)$$

where S is the concentration of contaminants averaged over the thickness of the boundary atmospheric layer, U is the wind velocity along the axis, K is the coefficient of turbulent diffusion across the wind direction, σ is the coefficient taking into account the dry deposition of pollutants, F(x,y,t) is a function describing the position and emission regime of the sources of pollution.

Equation (1) is suitable for describing the main processes of contaminant transport and transformation on regional scale and performing scenario calculations. Numerical simulation of pollution processes under current conditions causes noticeable difficulties and needs a definite fitting of eq. (1) to the observation data using the corresponding statements of reverse problems.

Fig. 1 shows the daily dynamics of a change in the content of substances typical of the exhausts of the Norilsk plant. These include SO_4^{2-}, Cu, Ni, Se, Pb [1]. The wind direction and velocity measured by a weather bureau of Tarko-Sale are shown at a weathercock level. To gain a notion of wind direction at heights of about 500 or 1000 m it is necessary to introduce a correction by rotating velocity vectors (Fig. 1) by 20-40° clockwise. As follows from Fig. 1, the fairly lasting winds of the north-east sector were observed from 18 to 22, September and from October, 27 to November, 4, 1998. In these periods, there were the mutually consistent jumps in the concentrations of the substances under study to the relatively high values which unambiguously indicates the sources of their origin.

A staggered change in concentrations points to both a relatively limited width of the contaminant torch carried from Norilsk and the side crossing of the point of observation by the torch. During these observation periods, the concentration of nitrates decreases and can reach the values 100-200 times smaller than those of sulfates. The same ratio is between the initial exhausts of sulfur dioxide and nitric oxides [1].

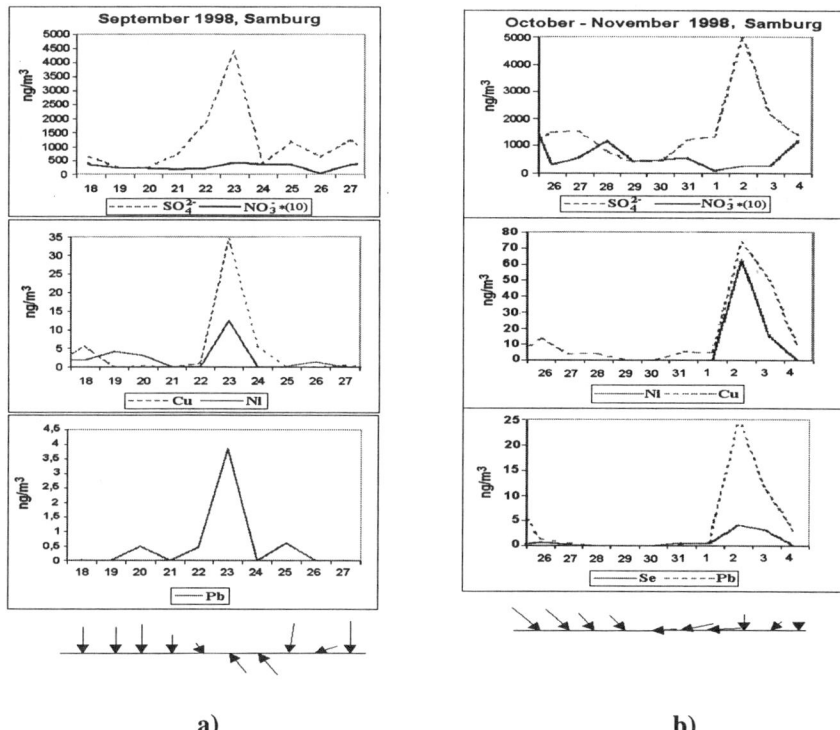

FIGURE 1. Daily dynamics of a change in concentrations of SO_4^{2-}, NO_3^-, Cu, Ni, Se, Pb and wind directions from September 18 to September 27 (a) and from October 26 to November 4 (b), 1998.

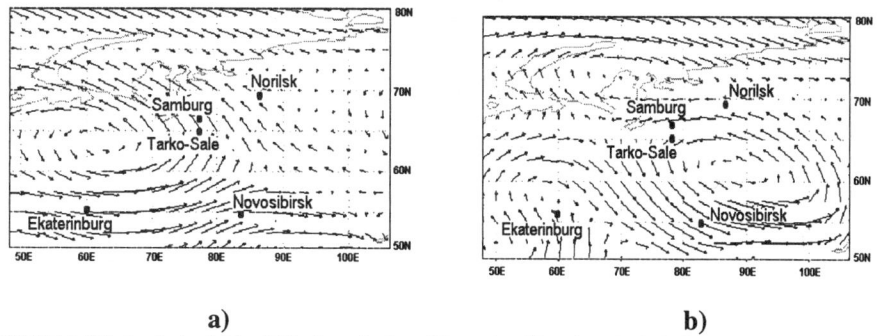

FIGURE 2. Wind velocity field at 850 mb on the 1st of November (a) and on the 2nd of November (b) of 1998.

The relatively high concentrations of copper, nickel, selenium, lead, sulfates and the accordingly low concentrations of nitrates observed from 9 to 13, October, indicate their relation to the exhausts of Norilsk. The short periods of aerosol emission from Norilsk are October, 19 and 26.

The data from Fig.1 on the surface wind velocities are rather fragmental and give no complete notion of the directions of the average transfer of aerosol contaminants in the mixing layer. They can be used in the preliminary stage of investigations. The potentialities of analysis can be intensified by using the maps of the spatial distribution of wind fields during certain time periods of observation of the territories under study which is illustrated in Figs. 2a and b. The figures show the fields of wind velocities at 850 mb for November 1 and 2, 1998 taken from the data base Reanalysis (NCEP/NCAR - National Centers of environmental forecast, National Center of atmospheric investigations, USA).

At that time, at the observation site (Samburg) we recorded a considerable increase in the concentrations of SO_4^{2-}, Cu, Ni, Pb, Se, etc., typical of the exhausts of Norilsk. Figure 2a shows that on the 1st of November 1998, the wind over the territory between Samburg and Norilsk was of the north-west direction which excluded the direct ingress of the exhausts of the Norilsk plant into the measurement sites. As follows from Fig. 2b, on the 2nd of November 1998, the pattern of wind field in this region became more favorable with respect to the transport of aerosols and was in agreement with the recorded high level of the concentrations of specific atmospheric contaminants. A deeper analysis of the conditions for the passage of the city torch needs additional data on the temperature, pressure and other meteorological characteristics.

CONCLUSIONS

The north-east sector winds do not necessarily guarantee the direct transport of Norilsk exhausts to observation sites, which is due to the insufficient width of the city torch of contaminants. More often, this happens through the side drift of the torch stream. Using more complete meteorological information, theoretical concepts of the basic physico-chemical mechanisms of contaminant transport, the source structure and the data of observations of the chemical composition of atmospheric aerosols makes it possible to construct the quantitative models of regional contamination of the territories to the north of Western Siberia on a basis of reverse problems.

REFERENCES

1. Koutzenogii, K.P., G.A.Kovalskaya, A.I.Smirnova, et al. *Atmospheric and oceanic optics*, v.11, No. 6, 625-631 (1998).
2. Bezuglaya, E.Yu., Rastorgueva, G.P. and Smirnova I.V. *How does the industrial city breathe?* Hydrometeoizdat, Leningrad, 1991, 256 p.

FROM AEROSOL MICROPHYSICS TO GEOPHYSICS USING THE METHOD OF MOMENTS

R. McGraw, D. L. Wright, C. M. Benkovitz, and S. E. Schwartz

Atmospheric Sciences Division, Department of Environmental Science
Brookhaven National Laboratory, Upton NY 11973

Abstract. We describe new developments in the application of the Quadrature Method of Moments (QMOM) [1]. These include the first application of the QMOM in a 3-D chemical transformation and transport model on the sub-hemispheric scale [2]. The QMOM simultaneously tracks an arbitrary (even) number of moments of a particle size distribution directly in space and time without the need for explicitly representing the distribution itself. The present implementation evolves the six lowest-order radial moments for each of several externally-mixed aerosol populations. From these moments we report modeled geographic distributions of several aerosol properties, including a shortwave radiative forcing obtained using the Multiple Isomomental Distribution Aerosol Surrogate (MIDAS) technique [3]. These results demonstrate the capabilities of these moment-based techniques to simultaneously represent aerosol nucleation, condensation, coagulation, dry deposition, wet removal, cloud activation, and transport processes in a large-scale model, and to yield aerosol optical properties and radiative influence from the modeled aerosol moments. We report on recent extensions of the method for simulation of internal mixtures and generally-mixed aerosols, and on a bivariate extension of the QMOM for modeling simultaneous coagulation and sintering of particle populations [4].

BACKGROUND

The method of moments (MOM) provides an alternative to bin-sectional and modal methods for representation of aerosol microphysical processes in complex particle population models. Unlike the sectional approach, the MOM makes no attempt to track the full particle size distribution (PSD) during a simulation – only its lower order moments. Advantages of the method include great computational efficiency, as only a few variables need to be integrated to accurately obtain important integral properties of the underlying PSD, and freedom from numerical diffusion. The MOM was developed by chemical engineers as a useful way to model the performance of batch fed crystallizers [5]. Later the method was brought to the attention of the aerosol community by Friedlander [6], and soon after applied to an investigation of oscillatory nucleation and growth [7]. There it was found that nonlinearity inherent in condensation feedback - as particles grow they consume condensable vapor and thus regulate their own formation - can lead to interesting aerosol dynamic behavior, including oscillatory nucleation and growth rates. Analysis of the compact, closed-form moment evolution equations for free-molecular growth led to a full determination of the stability properties of aerosol reactor operation, including effects due to nucleation nonlinearity and background aerosol loading [7].

Despite its potential advantages in aerosol science, the MOM has yet to gain wide acceptance. Problems with the method included the closure requirement, a severe restriction on the kind of particle growth laws that could be handled and, of course, the perceived limitation that the method yields only moments of the PSD rather than the PSD itself. Through recent developments these limitations have been largely overcome, and new versions of the MOM have become powerful tools for representation of aerosol microphysical processes under rather general conditions. Closure limitations have largely been overcome through the use of interpolation schemes for obtaining moments not explicitly tracked [8], and through the introduction of quadrature-based moment methods [1,9]. In particular, the quadrature method of moments (QMOM) permits closure for arbitrarily complex growth laws [1], and the method is free from the potentially severe restriction of requiring an assumed PSD shape (as in modal methods). Furthermore, although the PSD itself remains unknown, integrals of known kernel functions over the PSD can nonetheless be accurately estimated using moment-based and quadrature methods. In recent applications, reliable estimates for aerosol physical and optical properties (generally within an accuracy of a few percent) were obtained using only the tracked lower-order moments of the PSD [10,1,3]. Perhaps the most extensive application of moment methods to date has been to the representation of aerosol microphysical processes in regional-to-global scale atmospheric transport/tranformation models. This is illustrated in the next section.

REPRESENTATION OF AEROSOL MICROPHYSICAL PROCESSES IN A SUB-HEMISPHERIC MODEL

Some general features of our current QMOM-based aerosol module [2] include: Integration of the moment evolution equations for the 6 lowest-order radial moments for each of several externally-mixed aerosol populations. These populations include sulfate, sea salt, and continental background aerosol (2 modes). Collectively they are represented by 24 moments, plus sulfate mass per particle for each of the 3 non-sulfate populations, for a total of 27 variables. The module includes binary sulfuric acid-water nucleation, with no 'tuning' factor used – as none was needed to obtain copious new particle formation, and other physical processes (see Abstract). The sulfate aerosol is allowed to undergo a size-resolved cloud activation. The Global Chemistry Model driven by Observation-derived meteorological data (GChM-O) [11] serves as the platform into which we implemented the new aerosol module for the current model runs. For aerosol transport and mixing, the model employs a 4th-order, positive definite advection scheme and observation-derived (ECMWF) meteorological data. The horizontal resolution is 1.125 degrees, and there are 15 vertical levels extending to about 100 hPa.

Figure 1 shows a sampling of results from a simulation period that began October 1, 1986, with no aerosol initially present in the model domain. No fluxes from regions outside the model domain were included in the simulation.

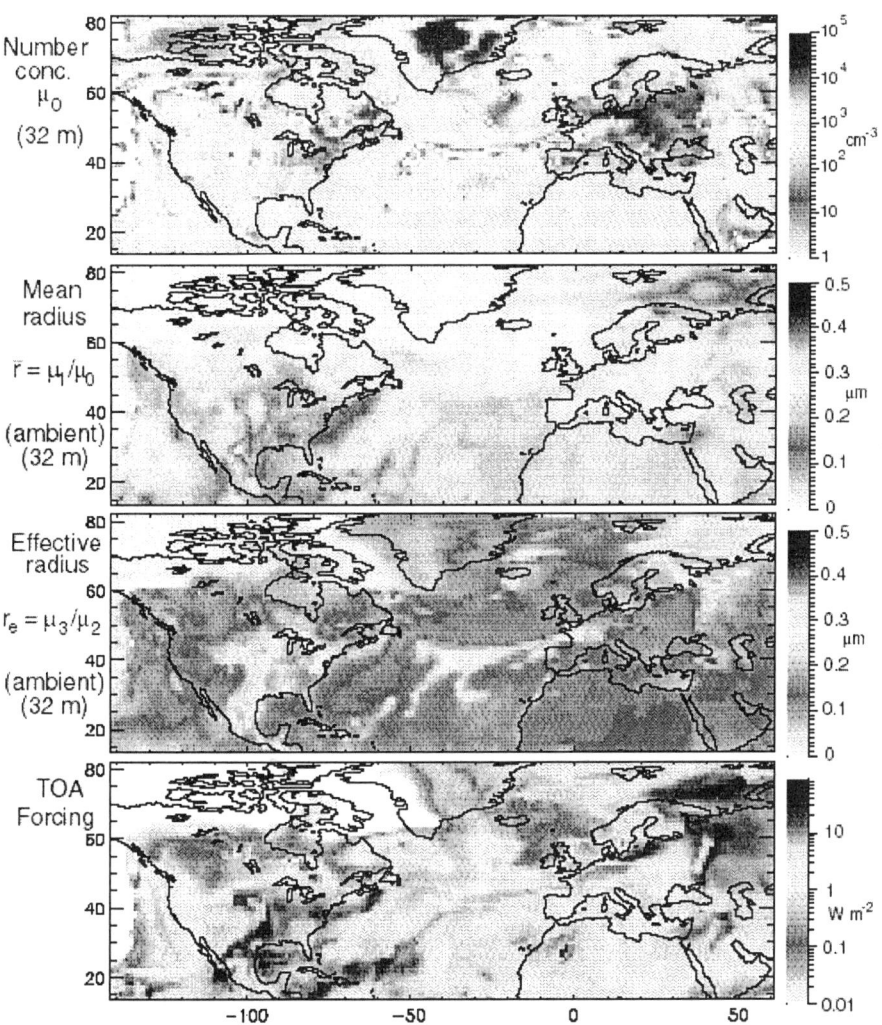

FIGURE 1. Modeled sulfate aerosol on October 22, 1986 0600UT. Top to bottom: number concentration, number mean radius, effective radius, and magnitude of the clear-sky shortwave forcing at the top of the atmosphere (ΔF) at a wavelength of 550nm. White regions indicate values below the plotting scale. From Wright et al. (2000) [2].

FUTURE DIRECTIONS

Currently, most atmospheric models are capable of treating only the two extreme mixing states of the generally-mixed particle population: internal and external mixtures. Generally-mixed, multicomponent particle populations need to be described by a fully multivariate distribution $f(m_1, m_2, ...)$ where m_i is the mass of component i in the particle. Most sectional models of the atmospheric aerosol are 1-dimensional in that they classify the particle population using only a single volume or mass coordinate. This representation generally forces the internal mixture approximation whereby $f(m_1, m_2,...) \cong f(m)$ where m is the total particle mass. That is, it is assumed that all particles having the same total mass have the same composition. We have recently succeeded, in a collaboration with Daniel Rosner of Yale University, in developing a fully bivariate extension of the QMOM for modeling simultaneous coagulation and sintering of particle populations in flames [4]. The resulting bivariate model, whose variables are particle volume and particle surface area (variables chosen to provide a parameterized description of complex particle shape [12]), can easily be adapted to provide a representation for a generally-mixed aerosol population having two components and the fully bivariate distribution $f(m_1, m_2)$. The extension of moment methods, perhaps in combination with different methodologies such as Monte Carlo [12], to higher-order multivariate representations would have immediate and revolutionary implications for aerosol modeling.

ACKNOWLEDGMENTS

This research was supported in part by NASA through interagency agreement number W-18,429 as part of its interdisciplinary research program on tropospheric aerosols, and in part by the U.S. Department of Energy (DOE) as part of the Atmospheric Chemistry Program.

REFERENCES

1. McGraw, R. "Description of aerosol dynamics by the quadrature method of moments", Aerosol Sci. Technol., 27, 255-265 (1997).
2. Wright, D. L., McGraw, R., Benkovitz, C. M., and Schwartz, S. E. "Six moment representation of multiple aerosol populations in a sub-hemispheric chemical transport model", Geophys. Res. Lett, 27, 967-970 (2000).
3. Wright D. L. "Retrieval of optical properties of atmospheric aerosols from moments of the particle size distribution", J. Aerosol Sci. 31, 1-18 (2000).
4. Wright D. L., McGraw R., and Rosner D. E. "Bivariate extension of the quadrature method of moments for modeling simultaneous coagulation and sintering of particle populations", submitted for publication (2000).
5. Hulburt, H. M. and Katz, S. "Some problems in particle technology: A statistical mechanical formulation", Chem. Eng. Sci. 19, 555-574 (1964).
6. Friedlander S. K. "Dynamics of aerosol formation by chemical reaction", Ann. New York Acad. Sci. 404, 354-364 (1983).
7. McGraw, R. and Saunders J. H. "A condensation feedback mechanism for oscillatory nucleation and growth", Aerosol Sci. Technol. 3, 367-380 (1984).
8. Frenklach, M. and Harris, S. J. "Aerosol dynamics modeling using the method of moments", J. Coll. Interface Sci. 118, 252-261 (1987).
9. Barrett, J. C. and Webb, N. A. "A comparison of some approximate methods for solving the aerosol general dynamic equation", J. Aerosol Sci. 29, 31-39 (1998).
10. McGraw, R., Huang, P. I., and Schwartz, S. E. "Optical properties of atmospheric aerosols from moments of the particle size distribution", Geophys. Res. Lett. 22, 2929-2932 (1995).
11. Benkovitz, C. M. and Schwartz S. E. "Evaluation of modeled sulfate and SO_2 over North America and Europe for four seasonal months in 1986-1987", J. Geophys. Res. 102, 25305-25338 (1997).
12. Tandon, P. and Rosner, D. E. "Monte Carlo simulation of particle aggregation and simultaneous restructuring", J. Coll. Interface Sci. 213, 273-286 (1999).

Coating Ambient Particles for Enhanced Detection by Mass Spectrometry

David B. Kane, Berk Oktem, and Murray V. Johnston

Department of Chemistry and Biochemistry
University of Delaware
Newark, DE 19716

Abstract. Ultraviolet absorbing coatings are used to enhance the detection of aerosol particles by aerosol mass spectrometry (AMS). The coating is accomplished in-line with the aerosol sampling by passing the aerosol through a flow diffusion cloud chamber. It is observed that coating the particles with a 2.5 nm layer of benzoic acid increases the particle hit rate by 2 times when ablated by 266 nm radiation. The coating causes a greater enhancement of the hit rate closer to the ablation threshold, and thicker coatings cause greater increases the hit rate.

INTRODUCTION

Aerosol mass spectrometers (AMS) typically use ultraviolet laser desorption ionization (LDI) to produce ions from particles in the source region of the mass spectrometer. However, the ablation threshold and ablation efficiency are well known to be dependent on the composition of the material being ionized.[1,2] This dependence arises from the absorption cross-section and ionization potential of the material undergoing LDI. The UV absorption of many constituents of atmospheric aerosol particles are limited at the typical wavelengths (193 nm, 248 nm, 266 nm) used in single particle mass spectrometers. The low absorption of some species leads to low ion yields from these species that may cause them to go undetected. To overcome this limitation high laser fluences are often used. While this increases the ability to detect a variety of atomic species in the particles, much of the chemical information of the particle may be lost. In particular for organic compounds in the particles are extensively fragmented at the high photon flux required to ablate most particles. This limits the ability to determine the molecular content of the particles, making it possible only to determine if the particle is organic in nature.

In this report we discuss the use of a thin coating of a nonvolatile aromatic compound to enhance the detection of nonabsorbing and weakly absorbing species with an AMS. The coating is achieved by mixing the aerosol with a supersaturated vapor of an aromatic compound. The vapor condenses in a thin layer on the surface of the particles. Here it is shown that coating particles with benzoic acid increases the hit rate when the particles are ablated at 266 nm.

EXPERIMENTAL METHODS:

A schematic of the experimental setup is shown in Figure 1. The aerosol mass spectrometer consists of an aerosol inlet, an ablation laser, and a reflectron time of flight mass analyzer. Details of the design and operation of the aerosol mass spectrometer are available in the literature.[1] The 4th harmonic of a Nd:YAG at 266 nm was used as the ablation laser. The polydisperse aerosols used in these experiments were produced with an atomizer. The size distributions of the coated and uncoated aerosols were measured with a differential mobility analyzer (DMA). The thickness of the coating was determined by the change in the diameter of the average particle.

Particle Coating

Aerosol particles were coated with a thin layer of benzoic acid by mixing them with a supersaturated benzoic acid vapor in a flow diffusion cloud chamber (FCC).[3] A schematic of the FCC is shown in Figure 1. It consists of a saturator, a preheating region, and a cooled condensation region. When operating air flows through the saturator, filled with the coating material (benzoic acid in the experiments described here), and into a preheating region kept at a slightly higher temperature than the saturator. The aerosol is mixed with the vapor in the preheating region. From the preheating region of the FCC the air-vapor-aerosol mixture flows into the cooled condensation region. The rapid change in the temperature from the preheated region to the condensation region causes the vapor to become supersaturated.[3] In the condensation region the supersaturated vapor will condense on the aerosol particles by heterogeneous nucleation. The benzoic acid supersaturation and the residence time of the aerosol particles in the supersaturated vapor control the thickness of the benzoic acid coating on the particles.

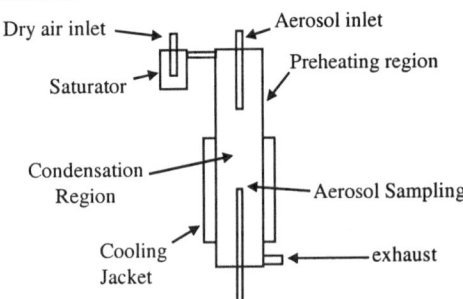

FIGURE 1. Schematic drawing of the flow cloud chamber used for coating aerosol particles.

Figure 2 shows the changes in the size distribution of oleic acid particles after coating at two different conditions. For the milder coating conditions, labeled coating 1, the maximum of the size distribution does not shift within the resolution of the DMA used to size the particles. This coating is considered to be <2.5 nm thick. For coating 1 an increase in the number density of smaller particles is observed due to the fast growth of smaller particles. For coating 2, the diameter of the maximum aerosol density in the particle size distribution shifts by ~5 nm indicating a ~2.5 nm thick coating on the particles. With this method controlled coatings have been grown on particles of oleic acid, NaCl, KCl-NaCl, and ammonium sulfate.

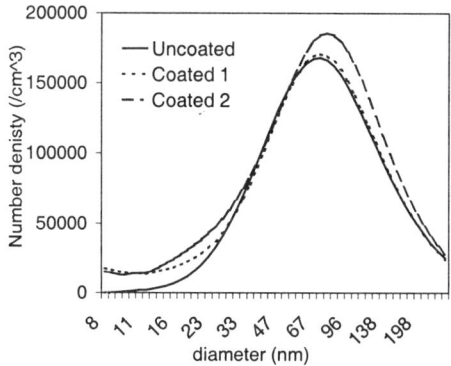

FIGURE 2. Size distributions of the uncoated and benzoic acid coated particles of oleic acid.

EXPERIMENTAL RESULTS

The detection efficiency of an AMS is proportional to the hit rate, the number of particles detected in given time period. The hit rate is then normalized to the number density of the aerosol, to account for changes in the number density of the different aerosols. Note, since the volume flow rate into the inlet is constant, the hit rate reported in hits/(kiloparticle/cm^3) is dependent on only the transmission efficiency of the inlet and the ablation efficiency of the particles.

It has been observed that coating particles with benzoic acid enhances the detection of NaCl, oleic acid and ammonium sulfate particles. Benzoic acid is an aromatic molecule with a high absorbance at 266 nm the wavelength of the ablation laser. Figure 3 shows the effect of coating benzoic acid on the hit rate of oleic acid particles as a function of the ablation laser energy. A coating of benzoic acid on polydisperse oleic acid particles with an average diameter of 150 nm enhances the particle detection efficiency from 50% to 3 times depending on the laser energy and the coating thickness. The relative increase in the hit rate due to the coating increases as the laser energy decreases, suggesting that the coating has a greater effect at low laser fluence.

From other experimental data is evident that the thickness of the benzoic acid coating affects the hit rate. For example a 10 nm coating on oleic acid particles increased the hit rate by 150% at a laser energy of 5.5 mj, compared to the 50% increase in the hit rate of particles coated with 2.5 nm of benzoic acid at the same laser laser energy, 5.5 mj.

In order to get an accurate understanding of how the particle coating enhances the detection of particles it is necessary to account for changes in the particle distribution. Coating the particles affects the distribution of the particles in two ways. First mixing the aerosol with the benzoic acid vapor dilutes the aerosol particles, typically from 20 to 100% depending on the coating conditions. Second changing the particle size changes the detection efficiency of the AMS. Increasing the diameter of the particles increases the hit rate at constant aerosol densities. [1]

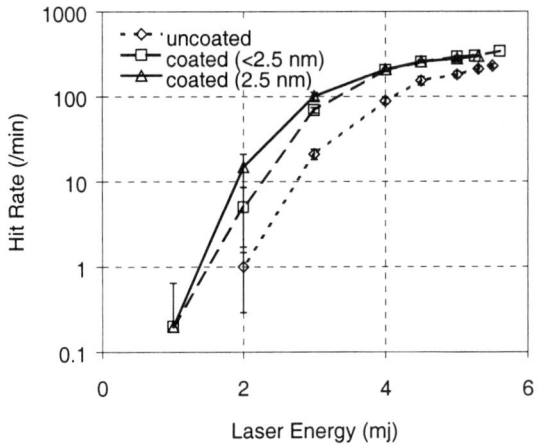

FIGURE 3. Effect of coating on the hit rate of oleic acid particles.

It is also important to examine the effects that the coating has on the mass spectra. In particular for chemical analysis of ambient aerosols when the composition of the particles is not known, it is essential that the coating not change the mass spectra in an unpredictable fashion. The benzoic acid coating does not significantly change the mass spectra of any particles studied to date (oleic acid, NaCl, NaCl-KCl and ammonium sulfate).

This work continues with testing a variety of coating materials on different particles, and different ablation wavelengths. The ultimate goal is to find coating materials which both enhance the ablation efficiency as well as facilitate the determination of the particle's chemical composition from the mass spectra

ACKNOWLEDGMENTS

This work was supported by the National Science Foundation, Grant No. CH9629672.

REFERENCES

1. David B. Kane, Berk Oktem, and Murray Johnston, Aerosol Science and Technology (2000) in press.
2. David S. Thomson, Ann M. Middlebrook, and Daniel M. Murphy, Aerosol Science and Technology **26**, 544 (1997).
3. Vivek Vohra and Richard H. Heist, Journal of Chemical Physics **104**, 382 (1991).

Variations of the Weight Concentrations of Dust, Nitrogen Oxides, Sulphur Dioxide and Ozone in the Surface Air in Tbilisi

A.Amiranashvili[1], V.Amiranashvili[1], T.Gzirishvili[1], G.Gunia[2], L.Intskirveli[2], J. Kharchilava[1]

[1] *Institute of Geophysics, Georgian Academy of Sciences, 1. M. Aleksidze Str., 380093 Tbilisi, Georgia*
[2] *Institute of Hydrometeorology, Georgian Academy of Sciences, 150., D. Agmashenebeli Ave., 380012 Tbilisi, Georgia*

Abstract. The data of the Georgian air pollution monitoring network were analysed in order to establish any trends in the variations of the near-ground concentrations of various substances. It was shown that despite a considerable fall-down in the Georgian economy in the recent years, the overall atmospheric pollution level did not change significantly.

INTRODUCTION

The City of Tbilisi represents the largest industrial center of Georgia. It is located on both banks of the River Mtkvari at 380-750 m above the sea level extending to more than 30 km. Its area is about 350 km^2. The main sources of the atmospheric pollution here are industry, power plants, transport, etc. Systematic observations on the content of dust, nitrogen and sulphur oxides in the near-earth air began in 1975 (Gunia, 1985), on the content of ozone - in 1981 (Kharchilava et al., 1988). The work presents the data on the variations of the mentioned minor atmospheric components in Tbilisi in 1981-1997.

METHODS

The method of the measurement of the weight concentration of solid aerosol admixtures, nitrogen and sulphur oxides in the near-earth atmosphere is presented in (Gunia, 1985). The measurements of the surface ozone concentration (SOC) were carried out constantly using electrochemical ozonometers made in the former German Democratic Republic, not less than three weeks in a month (Kharchilava et al., 1988). In this work data on the surface ozone concentration for 15 o'clock (the time of actual maximum of SOC) are presented.

RESULTS

After the collapse of the former Soviet Union Georgia underwent a strong fall-down of the industrial activity. Correspondingly the emissions of the pollutants in the atmosphere decreased significantly. For example in 1991-1996 in comparison to 1985-1990 the emissions of SO_2 decreased approximately four times, NO_X - three times (Beritashvili et al., 1998), solid admixtures - five times. Thus, after 1991 among the polluting admixtures the share of NO_X emissions (mainly due to the transport) has increased. Correspondingly the surface concentrations of the mentioned admixtures have also changed (Table 1). As this table shows, after 1991 the surface concentration of solid aerosols and sulphur dioxide have decreased considerably. The NO_X content has changed insignificantly (in the last 3-4 years the amount of transport and fuel used by it have been already comparable to their amounts in the eighties).

The increase in the atmospheric NO_X share entailed a considerable elevation of the concentration of surface ozone of the photochemical origine. In 1991-1997 in comparison to 1981-1990 the mean annual SOC in Tbilisi increased almost by 50%. Its biggest increase in the second period (more than twice) was detected in the cold season from November to March. Thus, the abrupt industrial fall-down and decreasing of dust and sulphur dioxide emissions in the atmosphere have not entailed any considerable improvements in the environmental state of the atmospheric air over Tbilisi.

TABLE 1. Mean annual concentrations of some atmospheric admixtures in Tbilisi (mcg/m^3)

Year	Dust	SO_2	NO_2	NO	O_3
1981	600	180	120	-*	34
1982	400	150	70	-	45
1983	500	90	50	-	-
1984	400	90	50	-	37
1985	400	90	50	-	29
1986	400	110	50	-	40
1987	500	100	40	30	44
1988	400	110	40	30	39
1989	500	113	70	40	38
1990	300	61	-	-	40
1991	400	6	50	40	46
1992	400	7	50	30	56
1993	300	13	40	30	57
1994	300	7	40	30	52
1995	300	-	40	30	56
1996	300	9	40	30	56
1997	-	-	-	-	60

*- data are available not for all months

TABLE 2. Intraannual variations of mean monthly SOC in Tbilisi in 1981-1990 and 1991-1997 (mcg/m3) (σ - standard deviation, C_V - variation coefficient)

Period	Param.	I	II	III	IV	V	VI	VII	VIII	IX	X	XI	XII	I-XII
I) 1981-1990	Mean	23	27	36	52	54	54	49	43	45	35	21	13	38
	σ	14	10	10	10	8	6	14	12	9	10	9	6	8
	C_V	61	37	28	19	15	11	29	28	20	29	43	46	21
II) 1991-1997	Mean	49	55	82	70	66	61	57	55	53	46	43	30	56
	σ	13	9	14	18	14	9	9	8	4	10	14	7	17
	C_V	27	16	17	26	21	15	16	15	8	22	33	23	30
Ratio II/I in %-s		213	204	228	135	122	113	116	128	118	131	205	231	147
α significance level according to Student's criterium		0.001	0.001	0.001	0.05	0.1	0.2	0.2	0.1	0.1	0.01	0.001	0.01	

REFERENCES

1. Gunia, G.S., *Issues of Atmospheric Pollution Monitoring in Georgian SSR*, Leningrad: Gidrometeoizdat, 1985, p. 84.
2. Kharchilava, D.F., and Amiranashvili, A., *Studies of Atmospheric Ozone Variations in Soviet Georgia. Results of Research on the International Geophysical Projects*, Moscow, 1988, p. 114.
3. Beritashvili, B.Sh., et al., *Bulletin of the National Climate Research Centre of Georgia* **7(E)**, 55-62 (1998).

Temporal Variation Of Atmospheric Aerosol Size Distribution At A Tropical Station, Mysore (12 N) From Ground Based Sunphotometer Studies

B. S. N. Prasad, B Narasimhamurthy and N. V. Raju

Dept of Physics, University of Mysore, Manasagangothri, Mysore - 570 006, INDIA

INTRODUCTION

Atmospheric aerosols are important in studies involving atmospheric visibility, radiative transfer, cloud physics, climate change studies etc. Spectral attenuation of solar radiation in the visible and infrared region depends on the atmospheric aerosol number density and size distribution. These two quantities are related to the aerosol extinction or aerosol optical depth (AOD) $\tau_{p\lambda}$ by the Angstrom's equation $\tau_{p\lambda} = \beta \lambda^{-\alpha}$, where α and β are related to size and number density of aerosols respectively and λ is in µm. Since the mechanism for the generation and removal of atmospheric aerosols is controlled both by natural (meteorological) and anthropogenic (pollution) processes, the parameters $\tau_{p\lambda}$, α and β would exhibit short term and long term trends. Atmospheric aerosol optical depth ($\tau_{p\lambda}$) in the VIS and near IR can be determined from ground based sunphotometers - Multiwavelength Radiometer (MWR) and following Langley technique. The Angstrom turbidity parameters α and β can be obtained from a plot of AODs versus wavelength (µm) on a log-log scale. The integrated columnar precipitable water vapour W can also be evaluated from the MWR operating at suitable wavelengths. Extracting the information on aerosol size distribution from the AOD data is an important aspect in atmospheric aerosol studies. Inversion of the AOD data ($\tau_{p\lambda}$) involving Fredholm integral equations (first kind) yield results on aerosol size and number density. From these, the aerosol columnar content, mass loading, mean radius, effective radius, etc., can be computed and examined for seasonal and annual variations. Laboratory measurements and analytical studies of the effects of changes in ambient relative humidity (RH) on aerosol characteristics/properties for different aerosol types and models are reported. The studies by Shettle and Fenn (1979) and by d' Almeida (1991) have indicated a decrease in the refractive index of aerosols and significant increase in particle size (resulting from the condensation growth of aerosols) with increase in RH. Such effects lead to increase in aerosol extinction coefficient (Shettle and Fenn, 1979) depending on the type of aerosols. Studies of aerosol characteristics with changes in humidity levels in natural atmospheric environment are limited. From simultaneous estimates of aerosol spectral optical depth ($\tau_{p\lambda}$) and precipitable water vapour content

(W) for very clear and stable atmospheric conditions, Tomasi (1979) has shown a linear relation between $\tau_{p\lambda}$ and W for small values W. Increased atmospheric turbidity associated with increase in precipitable water vapour content for a tropical location has been reported by Mohamed and Frangi (1983). Several of these aerosol characteristics are examined from our studies on aerosol size distribution derived from MWR data on AODs.

INSTRUMENTATION AND DATA ANALYSIS

Aerosol optical depth $\tau_{p\lambda}$ and integrated (columnar) precipitable water vapour W are evaluated from solar extinction measurements using a multiwavelength radiometer based on the principles of filter wheel sunphotometer (Moorthy et al., 1993, 1994). MWR measurements at ten narrow wavelength bands (in the VIS and near IR, 400-1025 nm) give the total atmospheric optical depths from which the aerosol optical depth are derived after applying corrections due to molecular scattering, and ozone and water vapour absorption. The precipitable water vapour 'W' is obtained from the MWR data at 850, 940 and 1025 nm following the procedure adopted by the Indian MWR network stations (Moorthy et al., 1994).

Mysore (12° N, 76° E) is a continental station, located on the Deccan plateau of Indian sub-continent. The geographic features of the station are; altitude 767 m (msl), the Arabian sea at a distance of 200 km on west, the Bay of Bengal 400 km away on the east, the Indian ocean 500km away in the south and a vast land mass of several thousand kilometers in the north. It lies almost in between the equator and tropic of cancer. It has the meteorological features of tropical zone. Mysore receives rainfall during south-west and north-east monsoons (June to November) averaging to 800-1000 mm annually. A few showers also occur during March and April months. Data collection at Mysore is generally not possible during May/June-September/October because of the prevailing monsoon conditions (rains and extensive cloud cover). Thus the data coverage is generally for the period from November of any year to May of the following year. MWR is operated on days with clear sky conditions when no visible clouds are present across or near the solar disc. The data availability however is not uniform for all the years of MWR observations at Mysore (see table 1). Regular data for all the years (1988-1999) are available for winter (December, January and February) and summer (March, April and May) seasons and only for a few years for the monsoon (NE) season (September, October and November).

Using the aerosol optical depths at eight wavelengths (excluding 940 nm) the columnar (height integrated) size distribution functions of aerosols $n_c(r)dr$ have been determined by numerical inversion of the integral equation (Fredholm integral) relating $\tau_{p\lambda}$ to $n_c(r)dr$, aerosol radius r and the Mie extinction efficiency parameter Q_{ext}. The solutions $n_c(r)dr$ have been obtained following the iterative inversion method described by King (1982). From the derived size distribution profiles, the total columnar content N_T, weighted mean radius R, effective radius R_{eff} and mass loading m_l are computed for the aerosols in the radius range from 0.1 µm to 5 µm. The expressions used are :

$$N_T = \int n_c(r)dr \ ; \ R = (\int n_c(r)rdr) / (\int n_c(r)dr)$$

$$R_{eff} = (\int n_c(r)r^3 dr) / (\int n_c(r)r^2 dr) \ ; \ m_l = (4\pi d_p / 3)\int n_c(r) \ r^3 dr$$

where d_p is the density of the aerosols.

RESULTS AND DISCUSSION

The observed AODs for Mysore show a seasonal variation with larger values for summer months than for the winter months. These general features and any interannual variations in these are explained in terms of the source and sinks for the atmospheric aerosols such as primary production of aerosols (e.g. wind generated dust) and secondary production (e.g. photochemical), and wet scavenging during prolonged monsoon rains (S-W and N-E) and occasional heavy showers during March - April month in any year. The aerosol size distribution show predominant bimodal (BM) type with two distinct modes one at the accumulation mode (small radius, 0.1-0.5 μm) and other in the coarse particle mode (large particles, 0.5-5 μm). The power law unimode (PU) type of distribution is also prevalent where a power law distribution is evident at small particle range (r < 0.8 μm) and a distinct mode at a higher radius around 1 μm. A modified Junge (MJ) type of distribution is seen very rarely and is characterized by a near flat region at lower particle sizes with the number density decreasing slowly with increase in particle radius followed by a nearly monatomic decrease (nearly linear on a log-log scale). Unimodal (UM) distribution is also seen occasionally. The following table summarizes the monthly aerosol size distribution seen at Mysore.

Table 1.

Years	Months									
	Sep	Oct	Nov	Dec	Jan	Feb	Mar	Apr	May	Jun
1988/89	—	BM	PU	MJ	BM	BM	PU	PU	PU	—
1989/90	BM	PU	—	BM	BM	PU	PU	BM	BM	—
1990/91	BM	PU	UM	BM	BM	BM	BM	PU	BM	—
1991/92	BM	BM	BM	BM	BM	BM	BM	BM	MJ	—
1992/93	—	—	UM	UM	BM	BM	BM	BM	—	—
1993/94	—	BM	BM	PU	BM	BM	BM	BM	PU	—
1994/95	—	—	UM	BM	PU	BM	BM	BM	—	—
1995/96	—	—	—	PU	PU	PU	PU	PU	PU	—
1996/97	—	—	—	PU	PU	PU	PU	PU	PU	PU
1997/98	—	—	—	PU	PU	PU	PU	PU	PU	PU
1998/99	—	—	—	PU	PU	PU	PU	PU	—	—
MJ : Modified Junge; UM : Unimodal ; BM: Bimodal; PU: Power law - Unimodal										

The total columnar aerosol mass loading (m_l) and the effective radius (R_{eff}) is larger for summer months than for the winter months. The mass loading is sensitive to the large particle end of the size distribution. Changes in the small particle end of the distribution does not affect the mass loading significantly but the effective radius is affected. The total aerosol concentration (columnar) is considered in two size regimes, the accumulation regime N_a (r = 0.1 to 0.5 μm) and the coarse particle regime N_c (r = 0.5 to 5 μm). The ratio (N_c / N_a) show patterns similar to that seen for R_{eff}. These results from aerosol size distribution data are also analyzed to examine the effect of precipitable water vapour W on aerosol characteristics.

REFERENCES

d' Almeida G A, Koepke P and Shettle E P, Atmospheric Aerosols : Global climatology and radiative characteristics, A Deepak publishing, Virginia, USA (1991).

King M D, J Atmos Sci, **39**, (1982) 1356.

Mohamed A B and Frangi J P, J Clim Appl Meteorol, **22**, (1983) 1820.

Moorthy K K, Nair P R, Prasad B S N, Muralikrishnan N, Gayathri H B, Narasimhamurthy B, Niranjan K, Ramesh Babu V, Sathyanarayana G V, Agashe V V, Aher G R, Risal Singh and Srivastava B N, Indian J Radio & Space Phys, **22** (1993), 243.

Moorthy K K, Prasad B S N, Chakravarty S C and Chandrasekaran S, ISRO-IMAP-SR-43-94, Indian Space Research Organisation, Bangalore (1994).

Shettle G E and Fenn R W, AFGL-TR-79-0214, Air Force Geophysics Laboratory, MA, USA (1979).

Tomasi C, Quart J Roy Meteorol Soc, **105** (1979) 1027.

Classification of Aerosol Substances by Use of Specific Vapour Pressures

Jürgen Müller

Umweltbundesamt
Paul-Ehrlich-Str. 29, 63225 Langen/Germany

Abstract. Atmospheric particulate and semi-volatile substances are classified by use of specific vapour pressure (p_s). Airborne particulate matter consists of many different species which all have individual mass size distributions. To all substances of $p_s < 1$ Pa a Mass Median Diameter (MMD) may be assigned. Substances of $p_s < 10^{-11}$ Pa totally are bound in particulate phase. Substances of p_s between 10^0 and 10^{-11} Pa, which comprises almost all organic compounds, partly are in gaseous phase and partly in particulate phase. Airborne substances of $p_s > 1$ Pa entirely are in gaseous phase.

Substances of p_s between 10^0 and 10^{-24} Pa belong to the fine particle mode (particle diameter below 2,5 μm). Substances of $p_s < 10^{-24}$ Pa belong to the coarse particle mode (particle diameter above 2,5 μm).

By knowledge of p_s of a substance it's MMD can be calculated. Thus all airborne particulate substances can be classified in respect to particle size.

In addition the amount of the gaseous and particulate fraction of a semi-volatile substance may be related to p_s or alternatively to MMD. Semi-volatile substances are bound in the particle diameter range between 0,1 and 1 μm, where the maximum of the fine particle mode is located.

By knowledge of the MMD the calculation of the tropospheric residence time of particulate and semi-volatile substances becomes simplified.

INTRODUCTION

Particulate matter (PM) of ambient air consists of particles in a size range between about 0,01 and 30 μm. Whitby (1972) classified the mass modalities and Jaenicke (1978) quantified the residence time of airborne PM.

By modern chemical analysis within the last decades many inorganic and organic substances of PM were detected. By impactor measurements it was found that in urban and rural air away of specific sources the fine particle mode (diameter below 2,5 μm) and coarse particle mode (diameter above 2,5 μm) arrange themselves in roughly equal masses. It was observed that every particulate substance has a specific mass size distribution which do not change significantly between urban and rural air (Müller 1986). The absolute concentration between urban and rural air may be very different but the relative mass size distributions are quite stable.

The Mass Median Diameter (MMD) is known as a key parameter in order to predict the dry and wet removal processes of a particulate substance. By knowledge of the MMD we also can assess inhalability and the amounts of deposition in our respiratory tract. By use of the MMD's particulate substances may also be categorized into local, regional and long-range pollutants.

MEASUREMENTS AND CLASSIFICATION

By impactor sampling carried out at many sites and chemical analysis of the mass fractions in the different size ranges many organic and inorganic substances were detected.

Empirically it was found that the MMD of an ambient air particulate substance quantitatively can be related to it's specific vapour pressure (p_s).

$$(1) \quad MMD = -d_o \left(\frac{1}{5} \lg \frac{p_s}{p_o} + 1\right)$$

$d_o = 0{,}5 \ \mu m, \quad p_o = 10^6 \ Pa, \quad p_s < 1 \ Pa$

Organic compounds usually have boiling points below 1000 °C and p_s-data are listed in handbooks.

Many metals and metal oxides, however, have boiling points up to 4000 °C which induce very low non-measurable vapour pressures. In this case the MMD may be assessed by use of the melting point (T_F) and boiling point (T_B). Empirically it was found:

$$(2) \quad MMD = d_o \left(\frac{2 T_m}{T_o + T} - 1\right)$$

$d_o = 0{,}5 \ \mu m \ ; \quad T_m = \dfrac{T_F + T_B}{2}, \quad T_m > 150 \ °C$

$T_o = 273{,}15 \ °C, \quad T = $ Ambient air temperature °C

For many inorganic substances only melting points (T_F) but no boiling points (T_B) are available. By knowledge of T_F using the approximate empirical equation $T_B = 1{,}75 \ T_F$ also fictive T_B's can be calculated. Thus, also for very involatile inorganic substances MMD's can be assessed.

Due to the fact that airborne PM only fractionally is water soluble it has to be dissolved in strong acids in order to obtain aqueous solutions for analysis by AAS, ICP-MS or ICP-OES. By this procedure inorganic species of PM are destroyed and only analysis of elements are feasible.

However, the modality of a measured elemental mass size distribution also is a fingerprint of the different elemental species (Müller 1998). To each mode of the distribution by use of equation (1) or (2) species of different specific vapour pressure can be assigned. Thus, also heavy-volatile inorganic species may be identified.

By sampling with a glass fiber filter backed by polyurethane (PU)-foam it was found that substances in an p_s-range between 1 and 10^{-11} Pa are semi-volatile. Airborne substances with p_s below 10^{-11} Pa totally are in particulate phase and p_s above 1 Pa totally are in gaseous phase. Semi-volatile substances correspond to MMD's between 0,1 and 1 µm. Within the diameter-range of 0,3 and 0,7 µm around the maximum of the fine mode the particulate fraction is proportional to the MMD.

By use of p_s or the corresponding T_m the chemical composition of airborne PM can be put in order in respect to particle size.

By use of p_s also the chemical structure of a single aerosol particle can be put in order (Müller 1988). Compact particles consist of an involatile core surrounded by shells of rising volatility. The core mostly consists of metal oxides and the shells of salts, acids, and organic compounds. By aging of the aerosol polar organics become dominant as was found in samples collected in remote areas.

RESIDENCE TIME

In order to predict the atmospheric fate of particulate substances and semi-volatile substances the residence time in dependence of physico-chemical and meteorological parameters have to be known. The MMD of a substance plays a key role in the calculation of total residence time. The atmospheric elimination processes predominantly are a sum of degradation by chemical gas-phase reactions (τ_{gr}), gaseous dry deposition (τ_{gdry}), particulate dry deposition (τ_{pdry}) and wet deposition ($\tau_{(g+p)wet}$).

The residence times of the different processes are reciprocally added:

$$(3) \quad \frac{1}{\tau_{(g+p)}} = \frac{1}{\tau_{gr}} + \frac{1}{\tau_{gdry}} + \frac{1}{\tau_{pdry}} + \frac{1}{\tau_{(g+p)wet}}$$

In the case of semi-volatile substances τ_{gr} is dominated by photochemical reactions of the gas-phase fraction x_g with tropospheric OH-radicals (reaction constant K_{OH}):

$$(4) \quad \frac{1}{\tau_{gr}} = x_g \cdot K_{OH} \cdot [OH] \; ; \quad x_g = \frac{m_g}{m_g + m_p}$$

$[OH] = 10^6$ molec./cm³ (diurnal average)

The residence time of the gaseous fraction induced by dry deposition (τ_{gdry}) up to date is based on the measurement of the deposition velocities (V_{gdry}) of the different substances.

$$(5) \quad \frac{1}{\tau_{gdry}} = \frac{V_{gdry}}{H} \; ; \quad V_{gdry} = \text{deposition velocity of the gaseous fraction}$$

H = 1000 m (yearly average of mixing height)

By use of a slightly modified formula of Jaenicke (1978) the dry residence time of an aerosol substance may be related to it's MMD:

$$(6) \quad \frac{1}{\tau_{pdry}} = \frac{V_o}{2H} \left[\left(\frac{MMD}{d_o}\right)^2 + \left(\frac{d_o}{MMD}\right)^2 \right]$$

Under Middle European conditions the constants have the following yearly average values: $V_o = 0{,}1$ cm/s, $H = 1000$ m, $d_o = 0{,}5$ µm, $\tau_{(g+p)wet} = 6$ d

For $\tau_{(g+p)\,wet}$ under Middle European conditions 6 d is chosen which means that by rain-out and wash-out processes the lower troposphere totally is cleaned one time within 6 days. However episodes of longer and shorter dryness may also occur.

For particle-bound substances it is assumed that water solubility plays only a minor role in wet removal processes. Particles are incorporated by rain or cloud droplets without considering the solubility of the substances bound inside the particles. Therefore also particulate substances of low water solubility are observed in rainwater.

If the gaseous fraction of a semi-volatile substance ascends in the troposphere, by temperature decrease it is transformed into particulate phase and removed as a particle. Due to it's relatively low vapour pressure a semi-volatile substance is not able to reach the stratosphere.

REFERENCES

1. Jaenicke, R., *Bericht Bunsengesellschaft Physikalische Chemie* **82**, 1198-1202 (1978)
2. Whitby, J.K., Lin, B.Y., Husar, R. and N. Barsic, *Journal of Colloid and Interface Science* **39** (1972)
3. Müller, J. *Journal of Aerosol Science* **17**, 277-282 (1986)
4. Müller, J., *Journal of Aerosol Science* **19, No. 7**, 1161-1164 (1988)
5. Müller, J., *Journal of Aerosol Science* **29, Suppl. 1**, 219-220 (1998)

Aerodynamic Particle Size Registration by TV-Methods

S.M. Kolomiets

Typhoon Research and Production Association, 82 Lenin Avenue, Obninsk, Kaluga Oblast 249020, Russia, e-mail: kolomiets@mail.ru

Abstract. A possibility is shown to simultaneously determine the aerodynamic and geometrical sizes of suspended particles at some modification of the known television methods. Under certain conditions it is possible to estimate particle matter density A relative error in the determination of the density is minimal for particles close in form to spheres. In this case it may be about 20 %. In the opposite case this error may be significantly higher.

INTRODUCTION

For a description of aerosols besides their geometrical sizes an aerodynamic size is widely used; i.e. a certain particle is given the size of a spherical particle of a known matter density having the same time of the rate relaxation (defined by the Stokes law) as the particle chosen [1]. This size is one of the basic characteristics governing the particle motion dynamics. It depends on aerosol matter density; in case of non-spherical particles it also depends on the particle form and orientation in space. Simultaneous estimation both of aerodynamic and geometrical sizes is important for many problems. In particular, such measurements can be realized by a Doppler laser rate-meter [2]. But in the latter case it is implicitly assumed that the particles are closed in form to spherical ones so that the single parameter measured characterizes completely a particle form.

The given paper is a consideration of a possibility to estimate the aerodynamic size with the use of television (TV) methods being more informative in particle form estimation.

ESSENCE

The essence is in the following. An acoustic wave is excited in the particle flow. The wave propagates in the direction normal to the video camera optical axis. Some light pulses in each frame illuminate the particle flow; and the pulses are synchronized with the acoustic wave frequency. Particle images are registered in time moments corresponding to the pulses; the image coordinates are determined which estimates the time of particle relaxation. Simultaneously particle geometrical sizes and its orientation in space are determined. The results of a comparison make it possible to draw certain conclusions of the dependence of the aerodynamic size on the particle parameters. In particular, under certain condition the particle matter density can be determined.

The choice of the acoustic wave propagation direction ensures the same particle image sharpness in different time moments. Usually the flow is directed transversely to the receiving objective optical axis so that the acoustic wave can propagate either along with the flow or in the direction normal to the flow and the objective optical axis. Let us consider the first case.

Assume that the particle flow moves in the 0Z direction and the acoustic wave propagates in the same direction with some frequency Ω and phase F. In this case the coordinate z of the particle at an arbitrary time moment t is determined by the relationship:

$$z(t) = z^* + Vt + A\cos(\Omega t + F - \varphi), \qquad (1)$$

where z^* is the particle coordinate at time moment $t = 0$ corresponding to its purely progressive motion; V is the flow rate; A is the particle displacement amplitude in the acoustic wave; φ is the phase shift between acoustic oscillations and particle oscillations. Here $A = A_0 (1 + \Omega^2 \tau^2)^{-1/2}$; $\tg \varphi = \Omega \tau$; where $A_0 = A(\tau = 0) = (2J/\rho_m C_m)^{1/2}/\Omega$; J is the acoustic wave intensity; ρ_m, C_m are the density of the medium and sound velocity in it; τ is the relaxation time, $\tau = (\rho - \rho_m)D^2/(18\eta)$, where ρ is the particle matter density; η is the medium viscosity; D is the particle aerodynamic size sought [3].

It should be noted that the "electrical" parameters Ω, F can be rather stable with time and their measurement is easily made by traditional means. The parameters V, A_0 depend significantly on the particular device construction and can vary with time due to different causes. Therefore a constant control of these parameters would be desirable. From this viewpoint (1) comprises, generally speaking, four unknown values: z^*; V; A_0; τ. We shall assume that in each frame four light pulses illuminate the particle flow. Then the values $z_i = z(t_i)$ at time moments t_i will be known (i = 0; 1; 2; 3). Considering that z^* is of no practical interest we shall further on use the values: $s_1 = z_1 - z_0$; $s_2 = z_2 - z_0$; $s_3 = z_3 - z_0$. The analysis shows that it is suitable to choose $\Omega t_i + F = k \pi/2$; $k = 0; \pm 1; \pm 2; \ldots$. Then we have:

$$\begin{aligned} s_1 &= VT/4 - A_0 \cos\varphi \, (\cos\varphi - \sin\varphi); \\ s_2 &= VT/2 - 2 A_0 \cos^2\varphi; \\ s_3 &= 3VT/4 - A_0 \cos\varphi \, (\cos\varphi + \sin\varphi); \end{aligned} \qquad (2)$$

where $T = 2\pi/\Omega$, i.e. the light pulses repetition period is equal to T/4.

From three equations (2) the relaxation time is easily found:

$$\Omega \tau = \tg\varphi = (2s_1 - s_2)/(s_1 + s_3 - 2s_2), \qquad (3)$$

when necessary other parameters V, A_0 can be found.

It is obvious that if one of the parameters V, A_0 is known three light pulses will be enough, and when both parameters V, A_0 are known only two pulses will be sufficient.

Relationships (1)-(3) describe the case when the acoustic wave propagates in the direction of the flow. This case is governed, in particular, by the optical scheme of the device "ARFA" [4]. In other cases the wave may propagate also in the direction perpendicular to that of the flow. Then in (1)-(3) we may put V = 0. Here τ is likely to be determined only by two values, for example, s_1 and s_2.

MEASUREMENT RANGES AND ERRORS

Let us estimate now possible measurement ranges and errors. Assume that absolute errors in s_1, s_2, s_3 are statistically independent and equal to the same value δz, then from (3) the absolute error $\delta \tau$ in the definition of τ can be easily found:

$$\Omega \delta \tau = \delta (tg\varphi) = (1 + tg^2\varphi)(5 + 6 tg^2\varphi - 8tg\varphi)^{1/2}(\delta z/A_0)/2. \qquad (4)$$

In particular, if $tg\varphi \ll 1$, then $\delta(tg\varphi) \approx (5^{1/2}/2)(\delta z/A_0)$; if $tg\varphi \approx 1$, then $\delta(tg\varphi) \approx 3^{1/2}(\delta z/A_0)$; and if $tg\varphi \gg 1$, then $\delta(tg\varphi) \approx (6^{1/2}/2) tg^3\varphi (\delta z/A_0)$.

Assume that for the lower and upper measurement limits τ_{min} and τ_{max} the corresponding relative errors do not exceed 50 %. Let $\delta z/A_0 \approx 2.10^{-2}$. Then $\tau_{max}/\tau_{min} = 106$. If the wave propagates in the direction perpendicular to that of the flow then $\tau_{max}/\tau_{min} = 159$. Knowing τ, one can find the aerodynamic size as $D^2 = 18\eta \tau/(\rho - \rho_m)$. If ρ is unknown, then when calculating D it is assumed that $\rho = \rho^*$, where ρ^* is some conditional value (usually used for the device calibration). So as far as $\tau \sim D^2$, then $D_{max}/D_{min} \approx 10^1$.

For spherical particles D is non-other then its geometrical diameter. For non-spherical particles τ and consequently D depend both on particle form and particle orientation relatively to the direction of the acoustic wave propagation. At the same time the above-mentioned device "ARFA" [4] gives a possibility to define geometrical sizes of each particle projections onto two mutually perpendicular planes. For particles of an ellipsoid of rotation form the definition is possible both of basic geometric sizes and particle orientation in space [5].

Let us estimate possible measurement ranges and errors for particles in air ($\rho \gg \rho_m$) using (4). For TV-methods the size measurement lower limit D_{min} is determined by the optical resolution and usually is equal to 1...2 µm. Let us take for certainty that $D_{min} = 2$ µm. Then at $\rho = 1$ g/cm^3 we have: $\tau_{min} = 1.2.10^{-5}$ s. For the above value $\delta z/A_0 \approx 2.10^{-2}$ it is obvious that $\tau_{max} = 1.3.10^{-3}$ s, $D_{max} = 20$ µm. For the parameters given above the acoustic frequency is rather low, $\Omega = 3725$ s^{-1}; $f = \Omega/2\pi \approx 600$ Hz. If A_0 is assumed to be equal to 0.05 mm then the acoustic wave intensity J is low as well: $J = 1$ mW/sm^2. For particles close in form to spherical ones at comparing of τ and D their matter density can be found. It may be shown that for particles close in form to spheres, $\delta \rho/\rho \approx 0.22$. Here $\delta \rho$ is absolute measurement error of the corresponding value. It is clear that this error will grow with an increasing difference of particle form from a spherical one.

In the foregoing aerosol particles have been considered. For hydrosols the situation is somewhat different and the size range is another than that of aerosols. For example, $D_{min} = 10$ µm, $D_{max} = 100$ µm. It should be noted that the estimation of matter density of particles suspended in water could be used in different technological processes. But their consideration is beyond the scope of the given paper.

REQUIREMENTS TO FLOW RATE

Let us now assess the requirements to the flow rate. Usually in TV-methods the permissible rate range is rather wide - from 10^0 to 10^2 cm/s: $L/T_0 < V < \delta D/t_p$. Here L is the video camera field of vision size along the flow, T_0 is the period of frame repetition (usually $T_0 = 20$ ms), t_p is the duration of light pulses. But when defining the aerodynamic size additional limitations put.

Consider now the case when the acoustic wave propagates in the direction of the flow. Then for unambiguity of the results the following condition should be met: $s_1 < s_2 < s_3$. For (3) it means that $VT > 4.84 A_0$. From the other side $(s_3)_{max} = 3VT/4$ should be less then L. For practical purposes it may be shown that if $A_0 = 0.05$ mm; $L = 0.45$ mm; $1/T = 600$ Hz; then 15 cm/s $\leq V \leq 18$ cm/s.

Consider now the case of the acoustic wave propagation normally to the flow. It is clearly seen that there is one additional limitation connected with the field of vision size L along the flow: $VT/2 < L$. The second one connected with the field of vision size L' in perpendicular direction: $2A_0 < L'$. It may be shown that for the same values of the parameters and $L' = 0.6$ mm the range of rates is considerably wider: 2.2 cm/s $\leq V \leq 27$ cm/s. But there are some "problems" in this case.

CONCLUSIONS

Thus a possibility is shown to simultaneously determine the aerodynamic and geometrical sizes of suspended particles at some modification of the known television methods. Under certain conditions it is possible to estimate particle matter density. The lower size limit of an aerosol particle under control is units of micrometers and that of hydrosols is tens of micrometers. The upper limit decreases with increasing density of particle matter. In particular, at the difference of particle and environment densities of 1 g/cm^3 it makes tens of micrometers for aerosols and hundreds of micrometers for hydrosols. The acoustic wave intensity is rather low for aerosols and high for hydrosols. A relative error in the determination of the density is minimal for particles close in form to spheres. In this case it may be about 20 %. In the opposite case this error may be significantly higher.

ACKNOWLEDGMENTS

The work was supported by the Russian Foundation for Basic Research (Grant 98-05-64096).

REFERENCES

1. Reist, P.C., *Introduction to aerosol science,* Moscow: Publishing House "Mir", 1967, pp. 60 - 70.
2. Kolomiets, S.M., *Journal of Optical Technology* **64** (11), 45 - 47 (1997).
3. Mednikov, E.A., *Acoustic coagulation and aerosol precipitation,* Moscow: Publishing House of the USSR Academy of Sciences, 1963, pp. 35 - 40.
4. Kolomiets, G.A., and Kolomiets, S.M., *Atmosphere and Ocean Optics* **12** (6), 534 - 536 (1999).
5. Kolomiets, S.M., *Atmosphere and Ocean Optics* **12** (8), 670 - 671 (1999).

Particle Image Formation by TV-Analyzer "ARFA"

S.M. Kolomiets

Typhoon Research and Production Association, 82 Lenin Avenue, Obninsk, Kaluga Oblast 249020, Russia, e-mail: kolomiets@mail.ru

Abstract. The peculiarities of suspended particle images construction in the television aerosol size and shape analyzer "ARFA" are considered. It is shown that under certain conditions images positions relative to some certain point in the registration plane determine unambiguously the particle position in the plane perpendicular to the flow direction. For some aerosol particles of different shapes given are the experimental results - two images of each particle corresponding to its projections onto two mutually perpendicular planes.

INTRODUCTION

Among the various methods of determining the sizes and shape of suspended particles most informative are the television (TV) methods. Usually one image of every particle corresponding to its projection onto the plane perpendicular to the light beam optical axis is registered [1]. But known are the schemes allowing to form in one registering plane (in the photocathode screen of video camera) simultaneously two images corresponding to its projection onto two mutually perpendicular planes [2]. This significantly increases the potentialities of the methods. Besides the images contain in principal the information on a particle location in the device count volume that is rather important from the viewpoint of bounding the count volume along the axis of the light beam of the acceptable depth of field.

The formation of images is rather dependent on a device optical scheme. In the given paper peculiarities of suspended particle image formation as applied to the scheme of analyzer "ARFA" [2] are considered.

OPTICAL SCHEME

The scheme of the analyzer is given in Fig. 1a. The mirrors 5 and 7 are positioned and oriented so that the light-beam axis at the exit of objective 8 (the OY axis) is perpendicular to the light-beam axis at the exit of objective 2 (the OX axis). Objectives 4 and 8 are positioned so that the object plane axial point of objective 4 coincides with the image plane axial point of objective 8 at some point O (Fig. 1b). Objective 9, coaxial with objective 8, optically conjugated this point O with some point O* in the recording plane (Fig. 1c) i.e. the photocathode screen of video camera 10. The light-beam scheme (objectives 4 and 8, collecting lens 6 and mirrors 5 and 7) constructs in particle flow 3, perpendicular to the plane of figure, an intermediate (aerial) real image 13 of a particle 12 (Fig. 1b). This image corresponds (for illustration) to a case when various points of a particle do not shade one another. Objective 9 constructs in the registration plane (Fig. 1c) real image 15 of intermediate image 13. Image15

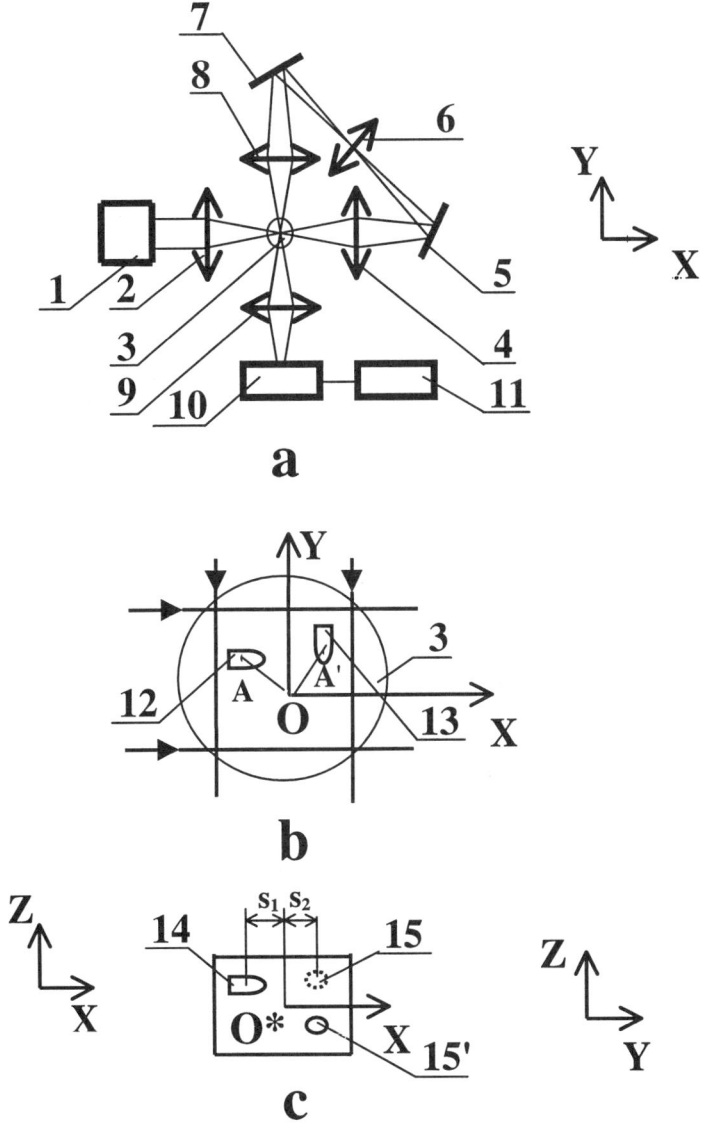

FIGURE 1. AEROSOL PARTICLE SIZES AND SHAPE ANALYZER "ARFA":
(a) overall diagram; (b) view of the beam intersection region; (c) view of recording plane:

(1) illuminator; (2, 4, 8, 9) objectives; (3) particle flow; (5, 7) mirrors; (6) collecting lens; (10) video camera; (11) processing block; (12) particle; (13) intermediate particle image; (14) image of the particle itself; (15) or (15') image of the intermediate particle image.

corresponds to the particle projection onto the Y0Z plane since, when constructing this image, the beam from the objective 2 propagating along the 0X axis illuminates the particle. Simultaneously objective 9 constructs in the same plane a real image 14 just of the same particle 12. However image 14 corresponds to the particle projection onto the X0Z plane since, when constructing this image, the beam from the objective 8 propagating along the 0Y axis illuminates the particle.

PARTICLE COORDINATES DEFINITION

Let us consider (Fig. 1b) some arbitrary point "A" of an object (particle 12) with the coordinates $\{x_b, y_b\}$. Let $\{x_b', y_b'\}$ be the coordinates of the point A', i.e. image of point A constructed by the optical scheme (4, 5, 6, 7, 8). For the real parameters of the above scheme as it shown [2], the lateral and longitudinal magnifications are practically equal to unity (can differ from unity no more then some percent). It is easily seen that $x_b' = y_b$, $y_b' = -x_b$. If $\angle AOX = \varphi$; $\angle A'OX = \varphi'$, then $\operatorname{tg}\varphi = y_b/x_b$; $\operatorname{tg}\varphi' = 1/\operatorname{tg}\varphi$. It means that AO $\perp$ A'O. Then one can see that $AO^2 = A'O^2 = x^2 + y^2 = x'^2 + y'^2$. That means that the point A' is rotated around the axis origin by $\pi/2$ relative to the point A. The rotation is in a clockwise direction. Inasmuch as this is valid for all the points of the "object", image 13 is rotated as the whole by $\pi/2$ in a clockwise direction relative to particle 12. Later on for simplicity we shall take that the point A corresponds to the center of particle 12, and the point A' corresponds to the center of its image 13. It should be mentioned that the point O* position (in the horizontal direction) is controlled by the instrument alignment with the help of corresponding devices.

The registration plane (Fig. 1c) is optically conjugated with the X0Z plane by objective 9 (the magnification of which is V). Therefore, later on we shall consider only the coordinates x_b, y_b in the plane X0Y and corresponding coordinates x_c, y_c in the registration plane. It is easily seen that $s_1 = Vx_b$; $s_2 = Vy_b$. Then the distance R of the point A from the origin of the coordinates has the form: $R^2 = (s_1^2 + s_2^2)/V^2$.

The optical scheme 4, 5, 6, 7, 8 constructs a direct (non-reversed) image. If the axes of objectives 4 and 8 are in one plane (the plane X0Y) the coordinates z_b of the points A, A' are equal. Then the points of image 14 have the same coordinates z_c as the corresponding points of image 15. Above it has been mentioned that image 13 is rotated by $\pi/2$ in a clockwise direction relative to the object - particle 12. That is if particle 12 is in the II or IV quadrants of the coordinate plane X0Y then its image is correspondingly in the I or III quadrants. Here s_1 and s_2 have different signs. If the particle is in the I or III quadrants its image is correspondingly in the II or IV quadrants. Here s_1 and s_2 have the same signs. Generally speaking it is difficult to say according to images 14, 15 which of them is the image of the particle itself and which is the image of the intermediate image. Therefore $\operatorname{tg}\varphi$ can also have two meanings: $(\operatorname{tg}\varphi)_1 = s_1/s_2$; $(\operatorname{tg}\varphi)_2 = s_2/s_1$. In particular, if $x_b = y_b$ (i.e. the point A is on the bisectrix of the angle between the axes 0X and 0Y) the points A and A' (with the same coordinates z_c) will be superimposed. There also exists a certain "dead" zone of particle positions where images 14 and 15 are superimposed. But the possibility of such superposition can be easily avoided by a proper alignment of the optical scheme. For example, mirror 7 can shift the intermediate image in the direction 0Z (in Fig. 1.c "shifted" image 15' corresponds to this case). It is significant that in this case we know deliberately which of images 14, 15 corresponds to the particle itself and which corresponds to its intermediate image. Therefore the particle position is determined unambiguously.

For the images not to be superimposed (even partially) at any particle spatial position the images separation should be above the upper limit of the size range. From the other hand it is clear that the shifted image should be completely in the video camera field of view. Both images will be registered only for the particles being distant from the field of view boundary towards the flow (for Fig. 1 it is the lower boundary) by the value no less then the separation chosen. That is the separation of the images decreases the "operating" field of view towards the flow. Therefore it would be desirable that the separation of images be equal to the upper limit of the size range.

Thus in the case under consideration if the positions of the images s_1, s_2 in the registration plane are known the position of a particle can be determined unambiguously in the plane X0Y: $R^2 = (s_1^2 + s_2^2)/V^2$, $\text{tg}\varphi = s_2/s_1$. Such information is important both for excluding particles being outside the limits of the depth of field and for some applied problems.

The images are put into a PC with the help of "miroVideo DC 10" board. At a proper image processing it is not difficult to find both each image area and the length of its projections onto mutually perpendicular coordinate axes 0X, 0Y, 0Z (Fig. 1.c). For particles close in shape to an ellipsoid of rotation one can calculate both geometric sizes of the particle at its arbitrary orientation in space [3]. For some aerosol particles of different shapes given are the experimental results - two image of each particle corresponding to its projections onto two mutually perpendicular planes.

CONCLUSIONS

Thus, the peculiarities of suspended particle images construction in the television aerosol size and shape analyzer "ARFA" are considered. It is shown that under certain conditions images positions relative to some certain point in the registration plane determine unambiguously the particle position in the plane perpendicular to the flow direction. For some aerosol particles of different shapes given are the experimental results - two images of each particle corresponding to its projections onto two mutually perpendicular planes.

ACKNOWLEDGMENTS

The work was supported by the Russian Foundation for Basic Research (Grant 98-05-64096).

REFERENCES

1. Belyaev, S.P., Nikiforova, N.K., Smirnov, V.V., et al. *Optoelectronic Methods of Studying Aerosols,* Moscow: Publishing House "Energoizdat", 1981, pp. 126 - 134.
2. Kolomiets, G.A., and Kolomiets, S.M., *Atmosphere and Ocean Optics* 12 (6), 534 - 536 (1999).
3. Kolomiets, S.M., *Atmosphere and Ocean Optics* 12 (8), 670 - 671 (1999).

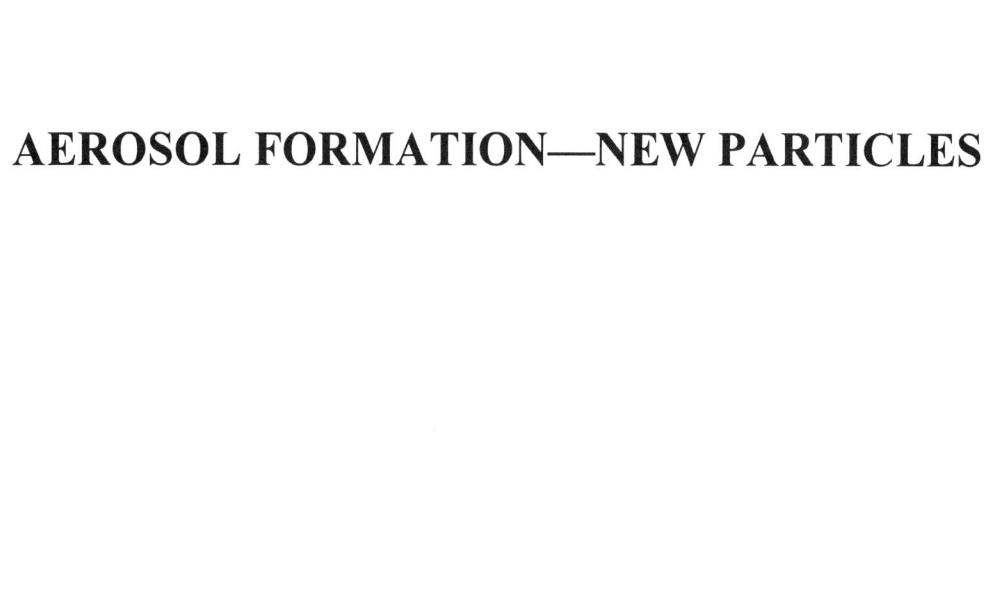
AEROSOL FORMATION—NEW PARTICLES

Analysis of the Formation and Growth of Atmospheric Aerosols Using DMPS Data

M. Dal Maso, M. Kulmala, J.M. Mäkelä, P. Aalto and C.D. O'Dowd

Department of Physics, University of Helsinki
P.O. Box 9, Helsinki, Finland

Abstract. Formation and growth of aerosol particles during two international field campaigns have been analyzed. Aerosol particle size distributions measured using DMPS data are used together with five simple equations to obtain concentrations of condensable vapor, 3nm and 1 nm size particles as well as to obtain source rate of condensable vapor. For this analysis the measured growth rates, condensational and coagulational sinks are needed.

INTRODUCTION

Nucleation, i.e. formation of ultrafine particles that are detected at sizes of a few nanometers, has been observed frequently at a boreal forest site in Hyytiälä, Southern Finland [1], and in the coastal boundary layer [2]. Here, a method involving only the size distribution data and the concepts of condensational sink and coagulational sink [3] is used to evaluate data obtained in international field campaigns BIOFOR and PARFORCE. The BIOFOR campaign took place at the Hyytiälä forest station and PARFORCE at the Mace Head Atmospheric Research Station, Ireland. The BIOFOR data used here was collected using two Differential Mobility Particle Sizer (DMPS) instruments, the first measuring 3-10 nm particles and the second 10-500 nm [4]. The data obtained from both instruments was combined to one size spectrum.

METHODS TO ESTIMATE AEROSOL AND VAPOUR PROPERTIES FROM EXPERIMENTAL DATA

The observed growth of nucleation mode particles, the source rate of condensable material and the changes of the nucleation mode concentration during the particle formation events are analyzed using the following five equations for the vapor concentration, particle number concentration and particle growth [3]. For a condensable vapor X, the time dependence of the vapor concentration C can be presented as [3,5]

$$\frac{dC}{dt} = Q - CS \cdot C \qquad (1)$$

Here Q is the source rate and CS is the condensational sink. The condensational sink is based on the concept presented by Pirjola et al. [6]. It gives the value of how rapidly condensable vapor molecules will condense on the aerosol, and it is strongly dependent of the aerosol size distribution.

The growth rate of a particle with a radius r can be expressed as [7]

$$\frac{dr}{dt} = \frac{m_v \beta_m DC}{r\rho} \qquad (2)$$

In this r is the particle radius, m_v is the molecular mass of the condensable vapor X, B_m is the transitional correction factor for the mass flux, D is the diffusion coefficient and p is the particle density. The time evolution of the aerosol number concentration N in a size class i can be given as

$$\frac{dN_i}{dt} = J_i - CoagS_i \cdot N_i \qquad (3)$$

where J_i is the formation rate of particles of size i. $CoagS$ is the coagulation sink. The coagulation sink gives the rate of removal of particles of size i by coagulation to bigger particles. The coagulational sink can be determined from

$$CoagS_i = \sum_j K_{ij} N_j \qquad (4)$$

where K_{ij} is the coagulation coefficient.

Using these equations with the measured DMPS data, we are able to calculate values for the nucleation rate of 1 nm particles, as well as the concentration of 1 nm particles. The equation for J_1 becomes [3]

$$J_1 = \left[\frac{dN_3}{dt} + K_3 N_3\right] e^{Kt} \qquad (5)$$

Here $dN3/dt$ is the observed formation rate of 3 nm particles, K_3 is the coagulation sink for 3nm particles, K is the coagulation sink for particles between 1 and 3 nm and t is the time that a particle needs to grow from 1nm 3 nm.

DATA ANALYSIS

The analyzed days are event days during the BIOFOR 3 campaign and PARFORCE campaign. The DMPS data for these days shows a clear increase of small particle concentration. We used the DMPS data plots to visually determine the start and end times of the particle formation period. We also determined the maximum size that the particles reached by the end of the formation period. Using these numbers, we were able to estimate the number of newly formed particles.

We got an estimate for dN3 / dt, the apparent nucleation rate, by taking the observed concentration of newly formed particles and dividing it by the formation time. Typical values for the apparent nucleation rate range from 0.1 to 1 $cm^{-3}s^{-1}$ [8].

The coagulation sink was determined from the size distributions given by the DMPS measurements. We calculated the coagulation coefficients for 1 nm, 2 nm and 3 nm particles with every other size class and made the summation for every measured DMPS spectrum.

The expression for J_1 depends exponentially of both the growth time of a particle to the 3 nm size and the coagulation sink used. The growth time was estimated from the growth rate of the newly formed particles. For BIOFOR, we estimated the growth rates to be from 2 to 4 nm/h in these cases, and calculated upper and lower estimates for J_1 and N1. For the PARFORCE data we used an estimate for the growth rate that was determined visually from the DMPS data plots. The coagulation sink K representing the removal of particles smaller than 3 nm was taken as the average of the coagulation sink for 1 and 2 nm particles, because most of the coagulation occurs in this size range.

RESULTS AND DISCUSSION

In Figure 1 the daily behavior of the coagulational sink is presented. The coagulational sink is calculated for 1nm, 2 nm and 3 nm particles. One can see that the coagulational sink behaves the same way as the condensational sink, because they are both very dependent on the aerosol size distribution. The condensational sink is calculated using the mass accommodation coefficient to be unity. The numerical value of the condensational sink is bigger than the values of coagulational sinks. From a coagulation point of view the sink is biggest for 1 nm particles.

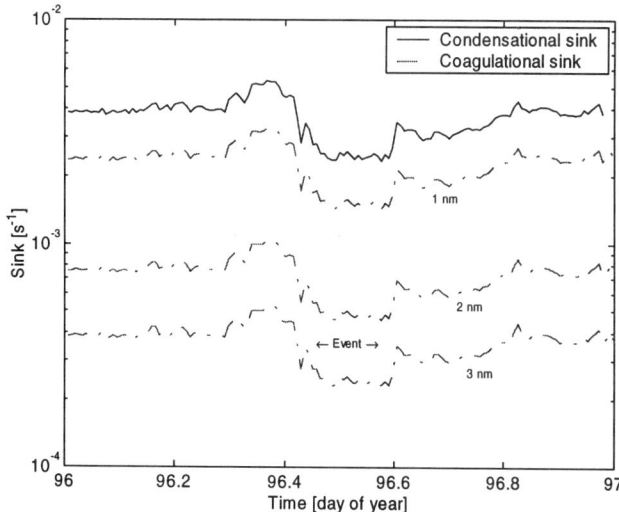

FIGURE 1. The behavior of the condensational and coagulational sink during a typical event day. The coagulational sink for 1 nm particles is much bigger than the sink for 2 and 3 nm particles.

Table 1 shows the results of the calculations made. For BIOFOR, there are two values given for J_1 and N_1 each day, representing the values calculated with the upper and lower estimate for the growth rates of small particles. A higher estimate gives a lower result for J_1, because the 1-3 nm particles have more time to be removed by coagulation. The range of the values also includes the variation of the coagulation sink during the particle formation period. The same is true also for the PARFORCE data, except that the growth rates are estimated visually from data. The estimates for the PARFORCE J_3 are probably too small, as they are taken from the DMPS data only, not taking into account the CPC measurements. Typically J_1 is significantly bigger at PARFORCE conditions than at BIOFOR conditions.

TABLE 1. Formation and growth properties of nucleation mode aerosols

Measurement	$N3$ [cm^{-3}]	$J3$ [s^{-1}cm^{-3}]	$J1$ [cm^{-3}]	$N1$ [s^{-1}cm^{-3}]	Growth rate (nm/h)
05.04.1999, Hyytiälä, BIOFOR 3	$7.1 \cdot 10^3$	$2.3 \cdot 10^{-1}$	10-250	$3 \cdot 10^3 - 1 \cdot 10^5$	2-4
06.04.1999, Hyytiälä, BIOFOR 3	$2.5 \cdot 10^3$	$1.6 \cdot 10^{-1}$	10-160	$3 \cdot 10^3 - 1 \cdot 10^5$	2-4
14.04.1999, Hyytiälä, BIOFOR 3	$3.4 \cdot 10^3$	$1.8 \cdot 10^{-1}$	5-90	$1 \cdot 10^3 - 1 \cdot 10^4$	2-4
14.06.1999, Mace Head, PARFORCE	$2.0 \cdot 10^5$	ca. 35	200-1000	ca. $3 \cdot 10^5$	ca. 6
24.06.1999, Mace Head, PARFORCE	$1.8 \cdot 10^4$	ca. 1	$10^3 - 10^5$	ca. $5 \cdot 10^6$	ca. 3

REFERENCES

1. Mäkelä J.M., Aalto P., Jokinen V., Pohja T., Nissinen A., Palmroth S., Markkanen T., Seitsonen K., Lihavainen H., Kulmala M., Observations of ultrafine aerosol particle formation and growth in boreal forest. Geophys. Res. Lett., 1219-1222 (1997).
2. O'Dowd C., McFiggins G., Creasey D.J., Pirjola L., Hoell C., Smith M.H., Allan B.J., Plane J.M.C., Heard D.E., Lee J.D., Pilling M.J. and Kulmala M. On the photochemical production of new particles in the coastal boundary layer. Geophys. Res.Letters. 26, 1707-1710 (1999)
3. M. Kulmala, M. Dal Maso, J.M. Mäkelä, L. Pirjola, M. Väkevä, P. Aalto, P. Miikkulainen, K. Hämeri and C.D. O'Dowd Growth and composition of nucleation mode particles, submitted to Tellus. (2000)
4. P.Aalto, K.Hämeri, E.Becker, R.Weber, J.Salm, J.M.Mäkelä, M.Väkevä, I.Koponen, H.Karlsoon, C.Hoell, Colin O'Dowd, Physical properties of aerosols during nucleation events, submitted to Tellus. (2000)
5. Kulmala, M., A. Toivonen, J. M. Mäkelä and A. Laaksonen Analysis of the growth of nucleation mode particles observed in Boreal forest. *Tellus* 50B, 449-462. (1998).
6. Pirjola, L., Kulmala, M., Wilck, M., Bischoff, A., Stratmann, F., and Otto, E. Effects of aerosol dynamics on the formation of sulphuric acid aerosols and cloud condensation nuclei. J. Aerosol Sci., 30, 1079-1094. (1999)
7. Kulmala, M. Nucleation as an aerosol physical problem. University of Helsinki, Department of Physics. Ph.D. thesis (1988)
8. J.M.Mäkelä, M.Dal Maso, A. Laaksonen, L. Pirjola, P. Keronen, L.Laakso and M.Kulmala, Characteristics of the aerosol particle formation events observed at a boreal forest site in southern Finland, submitted to *Boreal Environment Research,* (2000)

Long-term Measurements of Events of New Particle Formation at Hohenpeissenberg: Methods of Analysis and Climatology

W. Birmili[#], A. Wiedensohler[#], C. Plass-Dülmer[+] and H. Berresheim[+]

[#]*Institute for Tropospheric Research, Permoserstrasse 15, 04318 Leipzig, Germany*
email: wolfi@tropos.de
[+]*German Weather Service, MOHp, Albin-Schwaiger-Weg 10, 82832 Hohenpeissenberg Germany*

Abstract. Atmospheric new particle formation is investigated, based on long-term observations of the aerosol number size distributions, relevant gaseous components including H_2SO_4 and OH, and meteorological parameters. Measurements were conducted at Meteorological Observatory Hohenpeissenberg, a rural continental mountain site in southern Germany. A rough classification yielded approximately 50 significant events of new particle formation between April 1998 and February 2000. Midday peak concentrations of ultrafine particle concetation (3-11 nm) ranged between 3000 and 40,000 cm^{-3} with H_2SO_4 concentrations being mostly in the order of 10^7 cm^{-3}. A case study from April 20, 1998 provided insight into the aerosol dynamic processes associated with an event of new particle formation: The event was observed under intense solar radiation, with total particle number concentrations increasing from 6000 to 25000 cm^{-3} within one hour, and ultrafine particles (3-11 nm) accounting for more than 50 % of total number. A lower limit of the particle nucleation rate was estimated to be 3 cm^{-3} s^{-1}, being consistent with present models of ternary nucleation involving the H_2SO_4 / H_2O / NH_3 system. Roughly 80 % of the subsequent drop in ultrafine mode particle number concentration could be explained by coagulation. The observed particle growth rate of 2.1+/- 0.1 nm/h was largely attributed to the condensation of measured H_2SO_4, assuming neutralization by ammonia.

Introduction

Atmospheric particles affect the earth's radiation budget due to their light-scattering and cloud-forming properties (Charlson and Heintzenberg, 1995), and also potentially the atmospheric photo-oxidant budget (Ravishankara, 1997). An important source process of particles is atmospheric nucleation, initiated by supersaturated vapours that form molecule clusters. Newly formed particles can grow by further vapour condensation but also self-coagulation. Particles and precursor gases can also be scavenged by pre-existing aerosol surfaces, leading to a loss in number concentrations. Based on data from a continental measurement site, aerosol dynamic processes are investigated that might have caused the observations of new particle formation, subsequent particle growth, and number concentration decrease.

Experimental

Measurements were conducted at Meteorological Observatory Hohenpeissenberg, a mountain research station in Southern Germany (Birmili et al., 2000). Dry submicrometer particle size distributions (3-700 nm) were measured using a Differential Mobility Particle Sizer. Ambient conditions were accounted for by using a hygroscopic particle growth factor of 1.15 (40 % r.H.). Gas phase H_2SO_4 and OH concentrations were determined by atmospheric pressure chemical ionization mass spectrometry (AP/CIMS) (Berresheim et al. 2000a) Other gas phase data were taken from the WMO Global Atmospheric Watch program (GAW). A preliminary analysis of the dataset yielded roughly 50 events of significant particle formation between April 1998 and February 2000, corresponding to a frequency of ~10 %. On event days, peak ultrafine particle concentrations were between 3000 and 40,000 cm^{-3}, H_2SO_4 concentrations being mostly on the order of 10^7 cm^{-3}.

Case study on April 20, 1998

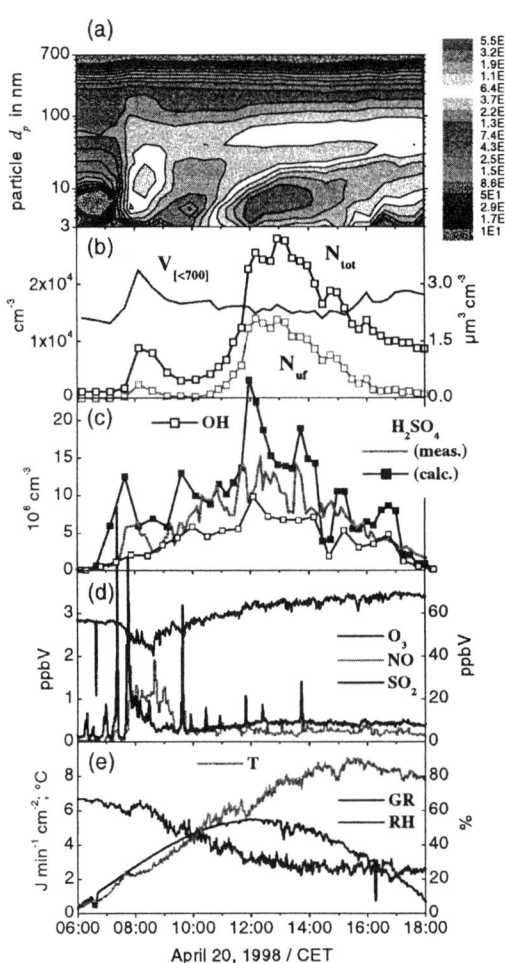

The figure on the left shows the evolution of the particle size distribution ($dN/dlog$ d_p in cm^{-3}), ultrafine particle concentrations N_{uf} (size range 3-11 nm), total particle concentrations N_{tot}, H_2SO_4, OH, and other parameters on April 20, 1998. This day featured intense solar radiation ("GR"), and a strong temperature ("T") increase and relative humidity ("RH") decrease during daytime, occurring in connection with a synoptic high pressure ridge and subsiding air. The wind direction was mainly south-westerly (Föhn conditions), at low wind speed (< 3 m s^{-1}). Radiosonde data from Munich (13:00) indicated a neutrally stratified boundary layer with a capping inversion at a height of 2000 m. A well-mixed boundary layer can therefore be assumed. Between 11:00 and 12:00 both N_{uf} and N_{tot} increased strongly, N_{tot} up to around 25000 cm^{-3}, with at least 50 % contributed by particles < 10 nm. The H_2SO_4 concentration reached $1.5*10^7$ cm^{-3}, after having gradually increased since 09:00. OH showed a similar midday maximum around 10^7 cm^{-3}. From 13:30 to 18:00, N_{tot} decreased to about 6000 cm^{-3}, and N_{uf} to less than 1000 cm^{-3}, respectively. H_2SO_4 and OH concentrations decreased with solar radiation approaching night-time levels after 18:00. Following the onset of new particle formation, the particle size distribution also gradually shifted to larger sizes (Fig. on previous page, top graph). Lognormal functions were fitted to the particle size distributions, yielding a particle growth rate of 2.1 +/- 0.1 nm/h. This can be seen in the time evolution of mode diameters in the figure below. After 14:45, a second ultrafine particle mode was identified.

H_2SO_4 concentrations were calculated from a simple chemical mass balance model considering production via the reaction of OH and SO_2, and the loss onto pre-existing aerosol. Details are described in a second paper (Berresheim et al., 2000b, this issue). The differences between modelled and measured H_2SO_4 were significant, corresponding to an unexplained loss rate of H_2SO_4 in the range of $0-6*10^7$ cm^{-3} s^{-1} (maximum at 12:00). This loss might correspond to a removal of H_2SO_4 due to new particle formation. In view of all the uncertainties involved in the present calculations we conclude that, on April 20, this rate was at least on the order of a few 10^4 molec. cm^{-3} s^{-1} during the main particle formation event. Assuming 3 nm particles of pure H_2SO_4 this rate would be equivalent to a production rate of a few 10^2 p. cm^{-3} s^{-1}.

For the presented case, a crude lower limit of the particle nucleation rate was estimated by dividing the number increase of particles > 3 nm by its corresponding duration. From Fig. 1, we assumed an increase by 20000 cm^{-3} over 2 hours (10:30-12:30), yielding a rate of roughly 3 cm^{-3} s^{-1}. Following the parametrization of Kulmala et al. (1998), a binary sulphuric acid/water nucleation rate of 1 cm^{-3} s^{-1} would require $3*10^9$ cm^{-3} H_2SO_4, i.e. a few hundred times more than actually observed. At Hohenpeissenberg, an agricultural area, one can safely assume that ambient ammonia levels usually exceed 20 pptv significantly. Using this and recent results of ternary H_2SO_4 / H_2O / NH_3 nucleation theory (Korhonen et al., 1999), only 1 to $2*10^7$ cm^{-3} H_2SO_4 would be necessary, which is very close to the observed values around 10^7 cm^{-3}. It appears thus likely that a ternary nucleation process was responsible for the particle formation event observed here. The empirical rate 3 cm^{-3} s^{-1} is by two orders of magnitude lower that the equivalent rate derived from the deficiency in H_2SO_4 balance in the previous section. An explanation for this could be (1) overestimation of the "time of nucleation" when determining the empirical rate, or (2) particle number concentration

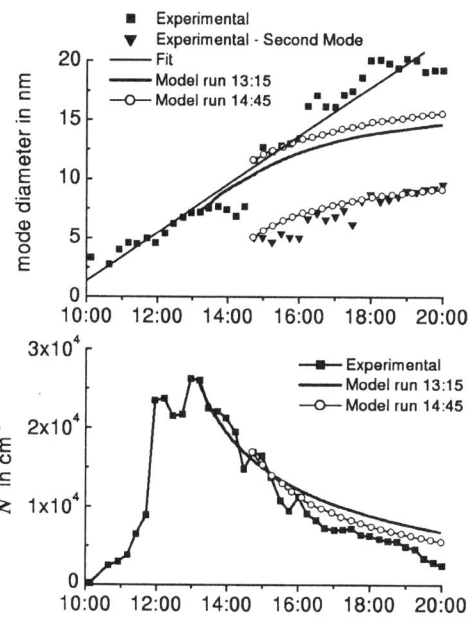

loss of produced particles < 3 nm due to coagulation.

To investigate the observed effects of, both particle number decrease, and the shift in the particle size distribution following the main nucleation event, we used a simple sectional aerosol model (38 sections) to simulate vapour condensation ($\alpha=1$), and coagulation. Each H_2SO_4 molecule was assumed to be neutralized by two molecules of NH_3. Calculations were performed for 283 K, 900 mbar, and 40 % RH. The figure on the previous page shows the time evolution of modelled and experimental mode parameters. One can see that shortly after model initialization (13:15: one ultrafine mode; 14:45: two ultrafine modes), the modelled mode diameters evolve growth rates very similar to the experimental data. After 16:00, however, the modelled growth slows down remarkably, as a direct effect of the decreasing H_2SO_4 concentration, but in contrast to the observed, rather linear growth. Concerning the decrease in ultrafine modal number concentration, we found that, for both model runs, the calculated particle concentrations closely followed the measured values (previous page, bottom graph) until larger deviations also occurred after 16:00. Referring to the 18:00 values, the particle number decrease was still described to 80 % by the model.

Conclusions and Outlook

The analysis of a selected event of new particle formation suggested that (1) instead of binary H_2SO_4 / H_2O nucleation, rather a ternary mechanism might have applied on the day of concern, (2) the growth of particles was likely a result of H_2SO_4 condensation including neutralization by ammonia, and (3) coagulation was a process efficient enough to account for 80 % the observed decrease in particle number concentration. The presented methods will be applied to more case studies, and specific data will be presented especially on the long-term behaviour of observed particle formation, and the climatological aspects.

Acknowledgments

We thank T. Elste for providing H_2SO_4 and OH data, S. Gilge and U. Kaminski for providing GAW data, R. Ruf, P. Settele, and R. Wilhelm for DMPS maintenance, and the German ministry for education and research for support by grant AFS 07AF201.

References

Berresheim, H., Elste, T., Plass-Dülmer, C., Eisele, F. L., and Tanner, D. J. (2000a) Chemical ionization mass spectrometer for long-term measurements of atmospheric OH and H_2SO_4. Int. J. Mass Spectrom., in press.

Berresheim, H., Elste, T., Plass-Dülmer, C, Birmili, W., Wiedensohler, A., O'Dowd, C. D., Hansson, H.-C., and J. M. Mäkelä (2000b), Observed H_2SO_4 and OH concentrations and their relation to particle nucleation in marine and continental rural air, this issue

Birmili, W. and Wiedensohler, A., and Plass-Dülmer, C., and Berresheim, H. (2000) Evolution of newly formed Aerosol Particles in the continental Boundary Layer: A Case Study including OH and H_2SO_4 Measurements. Geophys. Res. Letters, conditionally accepted.

Charlson, R. J. and Heintzenberg, J. (1995) Aerosol Forcing of Climate. John Wiley & Sons Ltd.

Korhonen, P., Kulmala, M., Laaksonen, A., Viisanen, Y., McGraw, R., and Seinfeld, J. H. (1999) Ternary nucleation of H_2SO_4, NH_3, and H_2O in the atmosphere. J. Geophys. Res., 104, 26349-26354.

Kulmala, M., Laaksonen, A., and Pirjola, L. (1998) Parametrizations for sulphuric acid/water nucleation rates. J. Geophys. Res., 103, 8301-8307.

Ravishankara, A. R. (1997) Heterogeneous and multiphase chemistry in the troposphere. Science, 276, 1058-1065.

Observed H_2SO_4 And OH Concentrations And Their Relation To Particle Nucleation Events In Marine And Rural Continental Air

H. Berresheim[1], T. Elste[1], C. Plass-Dülmer[1], W. Birmili[2], A. Wiedensohler[2], C.D. O'Dowd[3], H.C. Hansson[4], and J.M. Mäkelä[5]

[1]*German Weather Service, Meteorological Observatory, Albin-Schwaiger-Weg 10, 82383 Hohenpeissenberg, Germany.*
[2]*Institute for Tropospheric Research, Leipzig, Germany.*
[3]*Centre for Marine and Atmospheric Sciences, School of the Environment, University of Sunderland, Sunderland, United Kingdom.*
[4]*Air Pollution Laboratory, Stockholm University, Institute of Applied Environmental Research, Stockholm, Sweden.*
[5]*Department of Physics, University of Helsinki, Helsinki, Finland.*

Abstract. High-time resolution measurements of ambient H_2SO_4 and OH concentrations were made in conjunction with particle nucleation events observed at a coastal marine (Mace Head, Ireland) and a rural continental site (Hohenpeissenberg, Germany). Strong nucleation events occurred on a daily basis at Mace Head but much less frequently at Hohenpeissenberg. The potential flux of H_2SO_4 molecules contributing to new particle formation can be reasonably estimated for the rural event conditions (on the order of a few 10^4 molec. cm^{-3} s^{-1} or higher) by comparing calculated and measured H_2SO_4 levels. However, for the coastal marine site simple balance calculations assuming the SO_2+OH reaction to be the only source of gaseous H_2SO_4 resulted in H_2SO_4 concentrations much lower than observed levels suggesting the presence of an unaccounted major source of H_2SO_4 in this environment. Other observations at Mace Head indicated a significant suppression of ambient OH concentrations, in particular during a midday particle burst at low tide. However, no general correlations were observed between tidal cycles and H_2SO_4 and OH diel variations. It is concluded that nucleation events observed in both environments are related to quite different aerosol gaseous precursor chemistry.

1. INTRODUCTION

To better understand the mechanisms of new particle formation in different natural environments concurrent measurements of relevant aerosol, gaseous, and meteorological parameters have been conducted since April 1998 at Hohenpeissenberg, a rural mountain site in southern Germany (see also Birmili et al., this issue, and Birmili et al. (1)), and – on a field campaign basis - at Mace Head on the west coast of Ireland (see also O'Dowd et al., this issue). The present work highlights results obtained from corresponding measurements of gaseous sulfuric acid (H_2SO_4), i.e., the potentially dominant gaseous precursor for new particles, and of hydroxyl radical (OH) concentrations and potential relationships between both species and particle nucleation events observed in both environments.

2. METHODS

Atmospheric H_2SO_4 and OH concentrations were measured by chemical ionization mass spectrometry (CIMS) involving titration of OH with added SO_2, formation of HSO_4^- product ions by reaction of NO_3^- ions with H_2SO_4, and subsequent mass spectrometric analysis of the reactant and product ions. Detection limits for OH and H_2SO_4 at 5 min signal integration are estimated to be 5×10^5 cm^{-3} and 3×10^4 cm^{-3}, respectively. The instrument and the measurement techniques have been described in detail by Berresheim et al. (2).

3. RESULTS

Figure 1 shows an example of total particle number and H_2SO_4 and OH concentrations measured at Hohenpeissenberg on May 15, 1998. A strong particle nucleation event was observed on this day starting with a rapid increase of ultrafine particle concentrations to approximately 20,000 cm^{-3} between about 1000-1200 CET and declining after about 1330-1400 CET. Most of the days with similar nucleation events observed at this rural background site were characterized by intensive solar radiation conditions with maximum new particle numbers occurring around midday hours (in the 1000-1400 CET period) suggesting a major influence of photochemical processes on the particle nucleation process. Corresponding concentrations of H_2SO_4 were relatively high peaking at levels around 10^7 cm^{-3} or higher. On the other hand, it can be deduced from other days showing similar high H_2SO_4 concentrations and intensive radiation but no major particle bursts that these two criteria alone are insufficient to trigger a nucleation event. A further indication that H_2SO_4 may be involved in new particle formation in rural background air - besides reaching relatively high maximum concentrations on event days - is the time lag often observed between the increasing slopes in the H_2SO_4 and aerosol particle number profiles on event days (Figure 1) which can be on the order of 0-2 hours.

Assuming that the only source of H_2SO_4 is oxidation of SO_2 by OH and its sinks are a) deposition to the existing total aerosol particle surface and b) condensation to form new particles (homogeneous nucleation) a simple steady-state balance calculation can be made to assess the contribution of b), i.e., the "nucleation flux" $J(H_2SO_4)$ of H_2SO_4 molecules:

$$k\,[SO_2]\,[OH] - CS\,[H_2SO_4] = J(H_2SO_4)$$

Here k is the rate constant of the SO_2+OH reaction (8.5×10^{-13} cm^3 molecule^{-1} s^{-1}, (3)), CS is the condensational sink which is calculated assuming wet spherical particles, a mass accommodation coefficient α of unity, and Dahneke's interpolation formula for vapor diffusion in the transition regime (4). Figure 2 shows the results of this calculation for May 15, 1998, based on actual concentrations of SO_2, OH, and H_2SO_4 measured at the Hohenpeissenberg site. It can be clearly seen that the peak of the observed new particle production nearly coincides with a significant maximum in the H_2SO_4 "nucleation flux" (1.5×10^5 molecules cm^{-3} s^{-1}). A pollution episode around

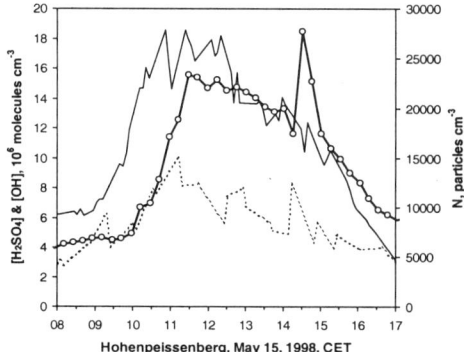

FIGURE 1. Concentrations of total particle number N (thick line with circles indicating 15 min averages), H_2SO_4 (continuous line) and OH (broken line) measured at Hohenpeissenberg on a day with a strong particle nucleation event occurring around midday.

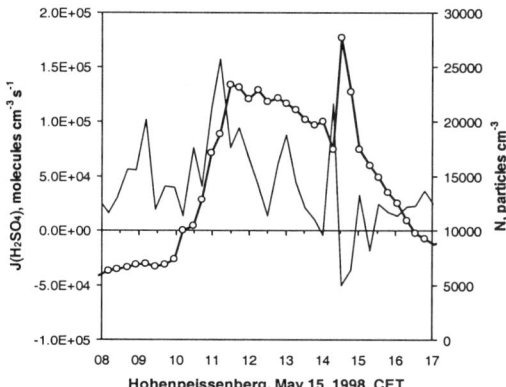

FIGURE 2. Calculated H_2SO_4 "nucleation flux" $J(H_2SO_4)$ contributing to particle nucleation on May 15, 1998 (see Figure 1) and corresponding measured total particle number concentration N; see text for details.

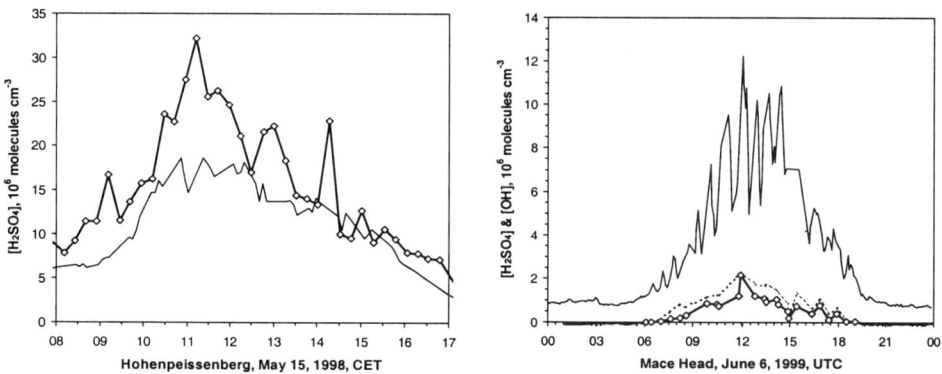

FIGURE 3. Measured (lines) vs. calculated (thick lines with circles) H_2SO_4 concentrations for selected days with particle nucleation events at Hohenpeissenberg and Mace Head, respectively (OH measurement data: broken line).

1430 CET disturbed the steady-state balance by producing highly variable SO_2, OH and aerosol surface concentrations.

The H_2SO_4 concentration may be calculated in a first approach by setting $J(H_2SO_4) = 0$ in the above equation and using the remaining expression $k[SO_2][OH]/CS$. As expected the resulting maximum values were significantly higher than corresponding measured H_2SO_4 levels for days with particle nucleation events in rural continental air. However, as shown by comparison in Figure 3, preliminary results obtained for the coastal marine site (Mace Head) showed a significant underestimation of observed H_2SO_4 calculations independent of the occurrence of nucleation events. Therefore, in contrast to background continental site a major source of H_2SO_4 in addition to the SO_2+OH reaction may exist in coastal marine air which has not been accounted for in the above equation.

During the June 1999 field campaign at Mace Head the OH/H_2SO_4 concentration ratios were relatively low compared to ratios typically observed at Hohenpeissenberg in the summer. These general observations suggested that not only H_2SO_4 but also OH steady-state concentrations may be strongly affected by unknown photochemical processes particular to this coastal marine region. Supporting this hypothesis, results from measurements at Mace Head on June 16, 1999, are presented when a strong nucleation event occurred around noontime during low tide. Steady westerly winds advected relatively clean marine air to the site. As can be seen, significant deviations between the OH concentration and the ozone photolysis frequency $j(O^1D)$ were observed concurrently with this event suggesting a rapid short-term increase in unknown OH sink processes.

ACKNOWLEDGMENTS

We would like to thank P. Settele and R. Ruf for assistance in the measurements at Hohenpeissenberg and Mace Head, and J. Strom and S. Gilge for providing the SO_2 data. This work was supported financially through grants 07AF201A/8 (German BMBF aerosol research program, AFS) and ENV4-CT97-0526 (European Commission, project PARFORCE).

REFERENCES

1. Birmili, W., Wiedensohler, A., Plass-Dülmer, C., and Berresheim, H., Evolution of newly formed aerosol particles in the continental boundary layer: A case study including OH and H_2SO_4 measurements, *Geophys. Res. Lett.*, in press, 2000.
2. Berresheim, H., Elste, T., Plass-Dülmer, C., Eisele, F.L., and Tanner, D.J., Chemical ionization mass spectrometer for long-term measurements of atmospheric OH and H_2SO_4, *Int. J. Mass Spectrom.*, in press, 2000.
3. De More, W.B., Sander, S.P., Golden, D.M., Hampson, R.F., Kurylo, M.J., Howard, C.J., Ravishankara, A.R., Kolb, C.E., and Molina, M.J., JPL Publ. 97-4, Pasadena, 1997.
4. Seinfeld, J.H., and Pandis, S.P., *Atmospheric Chemistry and Physics*, 2nd edition, John Wiley, New York, 1998.

PARFORCE: Objectives and Achievements

C.D. O'Dowd[1], J.M. Mäkelä[1], P. Korhonen[7], K. Hämeri[1], M.Väkevä[1], L. Pirjola[1], H-C Hansson[2], S.G. Jennings[3], G. de Leeuw[4], G. Kunz[4], H. Berresheim[5], R.M Harrison[6], A.G. Allen[6], Y. Viisanen[7] and M. Kulmala[1].

[1]*Department of Physics, PO Box 9, FIN-00014, University of Helsinki, Finland,*
[2]*Stockholm University, 10691 Stockholm, Sweden,*
[3]*Physics Department, NUI, Galway, Ireland,*
[4]*TNO Physics & Electronics Laboratory, PO Box 96864, 2509JG, The Hague, The Netherlands,*
[5] *Deutscher Wetterdienst, 82383 Hohenpeissenberg, Germany,*
[6]*University of Birmingham, Institute of Public Health, Birmingham, B15 2TT, England,*
[7]*Finish Meteorological Institute, Helsinki, FIN-00810, Finland*

Abstract. Understanding the formation of natural particles in the atmosphere, and their growth to radiatively active sizes, is critical to quantifying the role of anthropogenic emissions on cloud formation, climate change and public health. Only a few regions have been identified as strong natural sources of aerosols in the boundary layer: in particular, the coastal region seems to be the strongest natural source of these new particles. The PARFORCE programme was designed to elucidate and understand the underlying processes leading to observed coastal nucleation and to quantify the factors promoting coastal nucleation. Initial results indicate that nucleation rates in the coastal environment are of the order of 10^7 cm^{-3} s^{-1} and can be explained by ternary nucleation of sulphuric acid, water vapour and ammonia; however, growth to detectable sizes can only be explained by additional condensation of, probably, organic vapour – otherwise these new stable embryos are lost due to coagulation. The primary biogenic condensing species leading to the observed particle concentrations is thought to be a halocarbon derivative. Peak concentration of particles at sizes >3 nm can reach 1,000,000 cm^{-3} after a coastal nucleation event and these events occur almost on a daily basis over considerable spatial scales.

INTRODUCTION

Elevated aerosol particle concentrations have been observed for some time in the coastal environment and have been associated with the formation of new particles from the gas phase. O'Dowd et al., (1998, 1999) illustrated that these events occurred regularly in the coastal atmosphere and that they were associated with the occurrence of low tide and solar irradiation. Additionally, they highlighted a possible link with the OH radical, suggesting that the photochemical production of one or more of the aerosol precursor species. Following these initial reports, the PARFORCE programme (New Particle Formation and Fate in the Coastal Environment), funded by the EU, was enabled. The objectives of this investigation were to elucidate and understand the processes and conditions which promote and control homogeneous heteromolecular nucleation in the coastal boundary layer. Primary objectives were:

(1) Determine the conditions and the rates under which homogeneous heteromolecular nucleation occurs in the coastal boundary layer (i.e. pre-existing aerosol surface area; precursor gas concentration; micro- and macro-scale meteorology).

(2) Examine whether these nucleation events can be explained by binary or ternary heteromolecular nucleation of the following schemes: H_2SO_4-H_2O or NH_3-H_2SO_4-H_2O or NH_3-H_2O-MSA/HCl/HNO_3, or whether alternative nucleation schemes are likely to explain the observed events i.e. organic embryo formation followed by organic and/or sulphate growth.

(3) Examine the influence of anthropogenic/continental air parcel mixing on coastal nucleation.

(4) Explore the growth rates of newly-formed particles.

(5) Examine the long-term frequency of occurrence of coastal nucleation bursts and their duration. Two intensive field campaigns were conducted at Mace Head during 1998 and 1999 supported by continuous measurements of ultra-fine particle concentration for a two year period.

RESULTS

Two very successful intensive field campaigns were conducted in 1998 and 1999, supported by a continuous longterm measurements programme at Mace Head. The main results of PARFORCE may be summarised as follows:

- Biogenic coastal aerosol formation events occur approximately 90% of days of the year (Figure 1).

- Peak aerosol concentrations can exceed 1,000,000 cm^{-3} during nucleation bursts which can last from 2-8 hours (Figure 2).

- Coastal nucleation events occur during daylight hours coinciding with low tide. The source of aerosol precursors is the exposed tidal zone.

- Coastal nucleation occurs simultaneously all along the coastline suggesting that coastal regions provide a very significant source of natural aerosol particles.

- Formation rates of detectable-sized particle (i.e. d>3nm) are of the order of 10^4-10^5 cm^{-3} s^{-1}.

- Nucleation rates of new particles are of the order of 10^7-10^8 cm^{-3} s^{-1}.

- Nucleation does not appear to be inhibited by aerosol condensation sinks due to the strength of the precursor gas emissions.

- Enhanced drag, and thus turbulence, is observed during low tide resulting in enhanced mixing and water vapour flux, thus helping to promote nucleation.

- H_2SO_4 and OH concentrations followed an expected diurnal profile with maximum concentrations occurring around the middle of the day. Acid concentrations reached levels in excess of 10^7 molecules cm^{-3} on 73% of the measurement days. Acid concentration on polluted days exceeded 10^8 molecules cm^{-3}.

- OH concentrations had peak values of 2-4 x 10^6 molecules cm^{-3}. MSA concentrations exhibited peak values in marine air masses with concentrations between 2-4 x 10^6 molecules cm^{-3}.

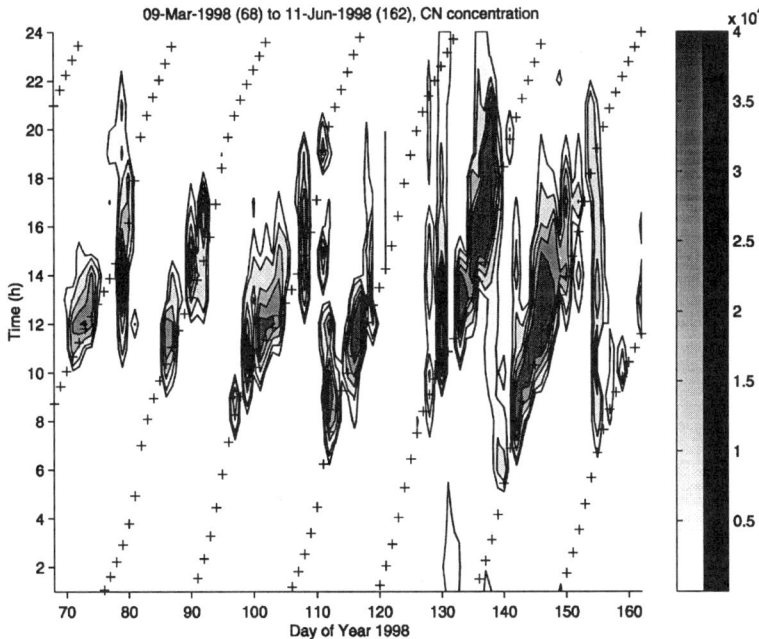

FIGURE 1: Long term record of elevated particle concentrations and low tide occurrence at Mace Head. + marks the occurrence of low tide. Grey scale bar is particle concentration in cm^{-3}.

- No correlation existed between peak H_2SO_4 concentrations and low tide occurrence or nucleation events.
- Nucleation was observed for H_2SO_4 concentrations >2 x 10^6 molecules cm^{-3}.
- A significant source of night-time H_2SO_4 was observed, suggesting the presence of another important SO_2 oxidant.
- Dibromomethene, chloriodomethane, tribrobomethane and diiodomethane were observed to be the most dominant VOCs emitted from various marine biota in the coastal zone.
- Modelling studies indicate that ternary nucleation of H_2SO_4, water and ammonia is the most likely particle formation mechanisms for given acid and ammonia conditions, however, an additional condensable species is required to

grow these new stable cluster particles to detectable sizes (>3nm). This additional condensable species is thought to be emitted during low tide.

- Growth-factor analysis of recently formed 8 nm particles indicate that these particles comprise some species significantly less soluble than sulphate aerosol, suggesting VOC derivatives as the likely condensable species.

- Transmission Electron Microscopy (TEM) samples indicated a significant presence of Iodine in recently formed ultra-fine particles.

- Airborne measurements indicate that these coastal nucleation events are not unique to Mace Head and occur all along the coastline. Given off shore flow, these coastal plumes can extend out over the ocean for several 100s of km. Modelling predictions indicated that after 2-3 days, these coastal plumes can lead to enhancements of radiatively-active aerosol and CCN by 2-10 times above that of the pre-existing population.

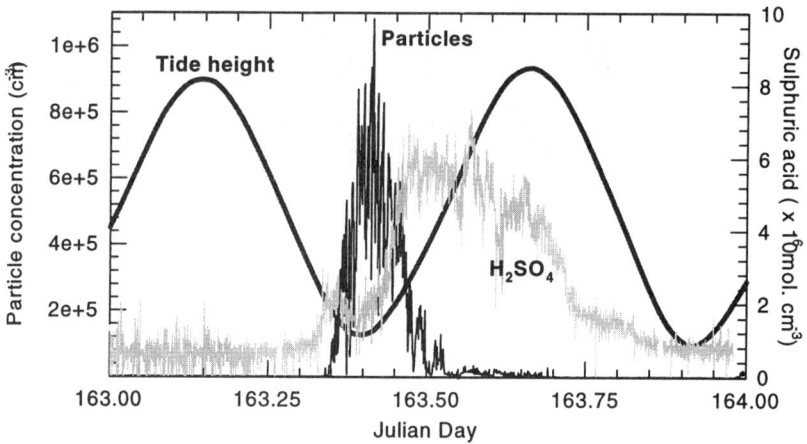

FIGURE 2: Tide height, H_2SO_4 and particle concentration (>3) for a typical event.

ACKNOWLEDGMENTS

This work was supported by the European Commission contract ENV4-CT97-0526.

REFERENCES

1. O'Dowd, C.D., M. Geever, M.K. Hill, S.G. Jennings, M.H. Smith, *Geophys. Res. Letters.*, **25**, 1661-1664 (1998).

2. O'Dowd, C.D., G. McFiggens, L. Pirjola, D. J. Creasey, C. Hoell, M.H. Smith B. Allen, J.M.C. Plane, D. E. Heard, J. D. Lee, M. J. Pilling, and M. Kulmala, *Geophys. Res. Letters.*, **26**, 1707-1711 (1999).

Observations And Models Of Particle Nucleation Near Cloud Outflows

Charles Clement[1], Ian Ford[2] and Cynthia Twohy[3]

[1]*15 Witan Way, Wantage, Oxfordshire OX12 9EU, United Kingdom*
[2]*Department of Physics and Astronomy, University College London, Gower Street, London WC1E 6BT, United Kingdom*
[3]*College of Oceanic and Atmospheric Sciences, Oregon State University, Corvallis, Oregon 97331, U.S.A.*

Abstract. Aerosol has been detected in considerable concentrations in the outflow of an anvil cloud during a major storm over the midwestern United States. We make rough calculations to try to interpret the measurements using simple models of particle nucleation and coagulation. We find it difficult to account for the observations, particularly in understanding the origin of the observed number density.

INTRODUCTION

On May 8, 1996, as part of the SUCCESS campaign (1), a NASA DC8 aircraft flew through and near the cirrus outflow of a large storm over the midwestern United States. Measurements were made at 1 s intervals of various gas concentrations, condensation nuclei, cloud surface area and meteorological variables. The aim of this work is to interpret the nature of the particles observed.

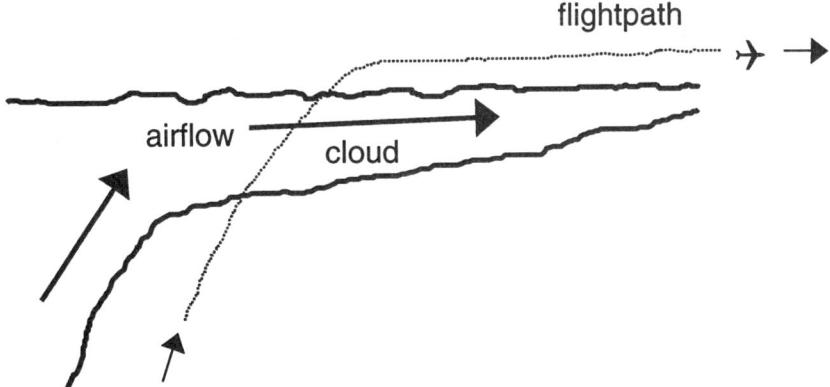

FIGURE 1. Sketch of the flightpath of the aircraft in relation to the anvil structure of the cloud and the presumed airflow through it.

MEASUREMENTS

Particles were found with radii in the range 10 - 25 nm, and were observed with fluctuating concentrations up to a maximum of 1.25×10^4 cm^{-3} at the ambient temperature, $T \approx 215$ K, and pressure, p = 200 mb (in standard temperature and pressure the concentration would be a factor of four larger).

In the rising region of the flight, most gas concentrations and the aerosol concentration were strongly correlated. The almost perfect correlation found between the concentrations of CO and CH$_4$ in this region has been interpreted as due to the recent mixing of two air masses with internally uniform concentrations (2). The correlation with aerosol concentration then indicates that one of the air masses, presumably that from the outflow of the storm, contained an almost uniform aerosol concentration which had been formed before the mixing took place. An examination of the behaviour of the correlations at different times and over what are effectively different length scales reveals differences in gas and aerosol mixing properties.

The high volatility of the observed aerosol indicates it could consist of sulphuric acid droplets nucleated in the relatively aerosol free outflow of the storm, where the acid is being formed from reactions involving SO$_2$ carried up from the surface. We have attempted a quantitative analysis to verify this hypothesis, based on a molecular production rate in the region of 10^{11}m^{-3}s^{-1} which is consistent with estimates of the likely SO$_2$ concentration.

MODELLING

The nucleation phase was examined using an analytic model for nucleation bursts (3) and a parametrised rate model for sulphuric acid/water nucleation (4). At the very low ambient temperatures, the model predicts a dependence of the nucleation rate on acid concentration, c, close to c^2 which is characteristic of barrierless nucleation (where the dimer is at least as likely to grow as it is to decay).

For T=215 K and a water vapour concentration of 50 ppm, the nucleation pulse is predicted to last about 5 minutes and to produce an initial particle number density of about $N = 7 \times 10^6$ cm^{-3}, of tiny droplets containing only 3 or 4 acid molecules. This concentration is much lower in size and total mass, and much higher in number density, than the observed concentration. The model predictions are illustrated in Figure 2. P is the source rate of sulphuric acid vapour, which is uncertain, and two predictions of N are given, one most appropriate for low P, and one for high P. There is only slight difference between the two predictions. In Figure 3, the size of the nucleated particles over the same range of P is shown, together with the rate of growth of these particles due to acid and water condensation following the pulse.

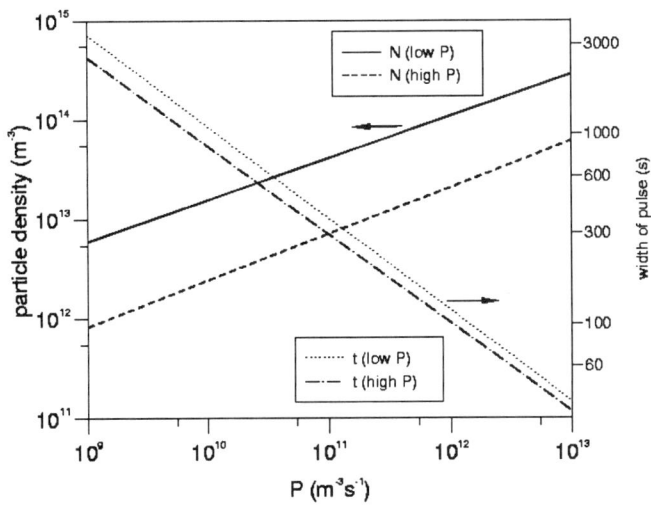

FIGURE 2. Particle yield and timescale for nucleation pulse at $T=215$K and 50 ppm water, for a sulphuric acid molecular production rate P.

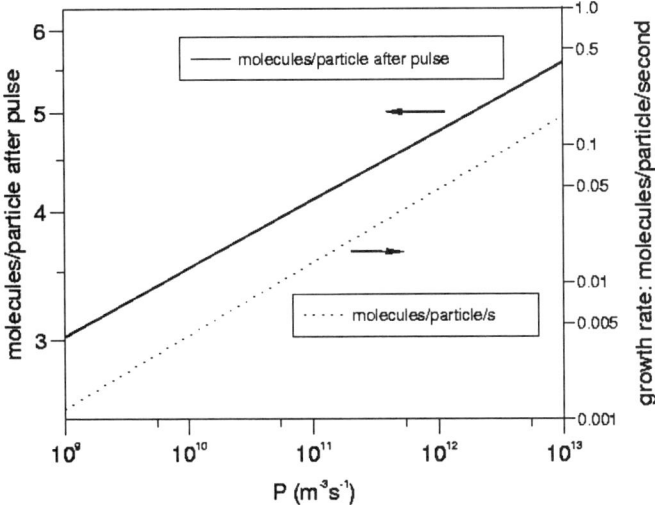

FIGURE 3. Size of nucleated particles in terms of sulphuric acid molecules, and growth rate of each particle after nucleation.

Of course, such a distribution of particles will coagulate, and further material can condense on it. We allow the aerosol to grow assuming that the additional mass is sulphuric acid and water, driven by the production of acid vapour. Not all the actual aerosol was likely to have been observed, and allowing for a range of uncertainty due to this missing mass, the growth time needed to reproduce the observed mass concentration with the assumed molecular production rate $P=10^{11} m^{-3} s^{-1}$ would be between 48 minutes and 7 hours. This agrees satisfactorily with the timescale for the storm, which had certainly lasted for several hours.

To complete the interpretation, we need to reproduce the observed particle number concentration, for which a reduction by a factor of over 500 from the initial value predicted by our model is required. Coagulation in the molecular regime is appropriate for the range of sizes of these particles. We therefore employ a collision probability that increases as the square root of the radius for equal sizes, but which can be much larger between unequal sized droplets. Intriguingly, in making preliminary estimates of the coagulation of the model nucleated aerosol, we have encountered great difficulties in accounting for the data. However, it is clear that at least several hours of aging are required to account for the observed particle distribution.

CONCLUSIONS

The nucleation burst observed was very large in spatial extent (greater than 200 km across), and such upper troposphere bursts following major storms are significant producers of the atmospheric aerosol which act as CCN and play a role in affecting the climate. Our results indicate that the burst was probably due to sulphuric acid nucleation from SO_2 of a largely anthropogenic origin, but that the quantitative features of the burst would repay further observations and study.

REFERENCES

1. Toon, O.B. and Miake-Lye, R.C., Subsonic aircraft: contrail and cloud effects special study (SUCCESS). *Geophys. Res. Lett.* **25**, 1109 (1998).
2. Clement, C. F., Ford, I. J., and Twohy, C. H., Mixing of atmospheric gas concentrations, to appear in *Phys. Rev. Lett.* (2000).
3. Clement, C. F. and Ford, I. J., Gas-to-particle conversion in the atmosphere: II. analytical models of nucleation bursts. *Atmos. Environment* **33**, 489 (1999).
4. Kulmala, M., Laaksonen, A. and Pirjola, L., Parametrization for sulfuric acid / water nucleation rates. *J. Geophys. Res.* **103**, 88301 (1998).

New Particle Formation and Hygroscopical Growth in the Lithuanian Coastal Environment

V. Ulevicius, D. Sopauskiene, G. Mordas, S. Stapcinskaite

Institute of Physics, A. Goštauto 12, LT-2600 Vilnius, Lithuania

Abstract - The events with well-developed temperature inversions were analysed. During one type of events increasing number concentration of Aitken mode particles caused increase in accumulation mode particles growth rate. Possibly, the oxidation of dimethylsulfide (DMS) occurred mostly via abstraction way and predominantly H_2SO_4 particles were generated. Also during the another type of events growth rate of accumulation mode particles increased with increasing Aitken mode particle number concentration. This result supports the hypothesis that nighttime chemistry was involved in new particle formation and growth.
Experimental results showed that the formation and growth of new particles were clearly observed in the coastal zone.

INTRODUCTION

Coastal zones create regions, where particles are formed together from sea and land (natural and/or anthropogenic) sources and from gas-to-particle conversion. Specific meteorological conditions, which predominant in coastal zone can influence aerosol particles characteristics [1,2]. Relative humidity changes in atmosphere control aerosol formation and change, as well efficiency of heterogeneous reactions [3].

The goal of this study was to consider the possible mechanisms that could be responsible for new particles formation and growth in the Lithuanian coastal boundary layer. Information about particle growth was obtained analysing aerosol particle size distributions and estimating particle diameter growth rates. These experiments were performed at Preila located on the Curonian Spit, which separates the Curonian Bay and the Baltic Sea.

METHODS

The information about mechanisms of chemical reaction predominant in the gas-to-particle conversion was obtained from calculations of particle diameter growth rates. Particle diameter growth rates, $\Delta D_p / \Delta t$, were calculated from the rate at which the concentration, $N(D_p, t)$, of particles with diameter larger than D_p increased with time. Calculations were performed using the following expression [4]:

$$\frac{\Delta D_p}{\Delta t} = \frac{N(D_p, t_2) - N(D_p, t_1)}{(t_2 - t_1) n_{av}(D_p)} \quad (1)$$

where

$$n_{av}(D_p) = [n(D_p,t_1) + n(D_p,t_2)]/2 \qquad (2)$$

The aerosol particle number concentrations and size distributions in the 10 to 200 nm size range were measured using an electrostatic aerosol size spectrometer (ELAS 5Mc) developed by the Lithuanian Institute of Physics [5]. The temperature was measured at 2 and 8 m above ground with temperature sensor CAMPBELL (model PT 100/3), and relative humidity was measured at height of 8 m with relative humidity sensor ROTRONIC (model MP-100A).

RESULTS AND DISCUSSION

Experiments described in this study were performed during June – August 1997. It was chosen few events for investigation of aerosol particles growth regularity. Predominantly particles formation and growth was observed in the morning and evening hours. Therefore morning and evening events were analysed.

Typical morning event was chosen on August 11. Temperature inversion reached 3,7 ^{0}C (Fig. 1A) that morning. Relative humidity was > 96 %. Intensity of UV radiation started to increase at 5:30 a.m.; temperature inversion at this time was still not broken up. At these meteorological parameters concentration of Aitken mode particles (Fig. 1B), because of the start of photochemical reactions, increased from 3880 cm^{-3} to 8810 cm^{-3} at 6:00 a.m. After that it started to decrease because solar

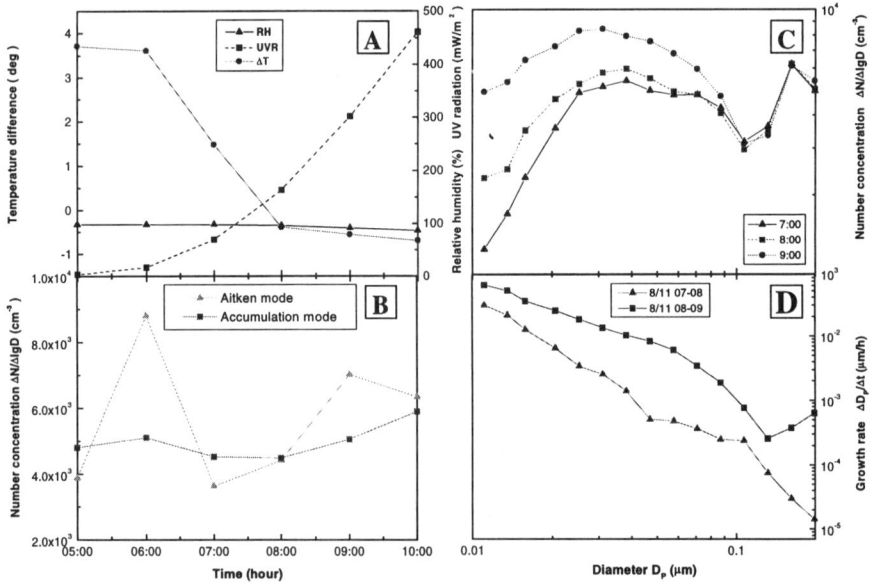

FIGURE 1. Morning event on August 11, Preila. **A.** Variations of meteorological parameters in time. **B.** Accumulation and Aitken modes particles concentration variations in time. **C.** Aerosol particles size distribution. **D.** Aerosol growth rate dependence on diameter.

radiation eroded inversion and stagnant air was mixed with fresh air from upper layers of the atmosphere. Aitken mode particles concentration increased to 7020 cm^{-3} at about 9:00 a.m. This increase in concentration is concerned with particles transference from the higher atmosphere layers. Accumulation mode particles concentration changes were not so significant as it were for Aitken mode. Aerosol particles size distribution on August 11 during morning event is shown in Fig. 1C. Maximum concentration was at 9:00 a.m. Temperature inversion at this time was already broken up, good mixing occurred, intensity of UV radiation reached more than 300 mW/m^2, therefore it might be that particles were brought from higher layers. Aerosol particles growth rate dependence on its initial diameter is shown in Fig 1D. Growth rate at 9:00 a.m. was much higher than at 8:00 a.m. Aitken mode particles concentration reached second maximum at 9:00 a.m. This change in the particle diameter growth rate possibly can be explained by the ratio of methanesulfonic acid (MSA) to non-sea-salt sulphate in morning hours. If the oxidation of dimethylsulfide (DMS) occurred mostly via abstraction pathway, dominant H_2SO_4 particles were generated from DMS. These particles due to coagulation with the larger particles increased hygroscopicity of accumulation mode particle. During this event particles growth rate was decreasing when its initial diameter increased. Shape of these curves shows that there took place low-pressure vapours, that were formed during reactions in gaseous phase condensation. Growth rate at 9:00 a.m. for particles larger 0,13 µm increases with initial particle diameter. Consequently, growth occurred because of the chemical reaction in droplets.

Evening event was chosen in August 8. Temperature inversion this evening

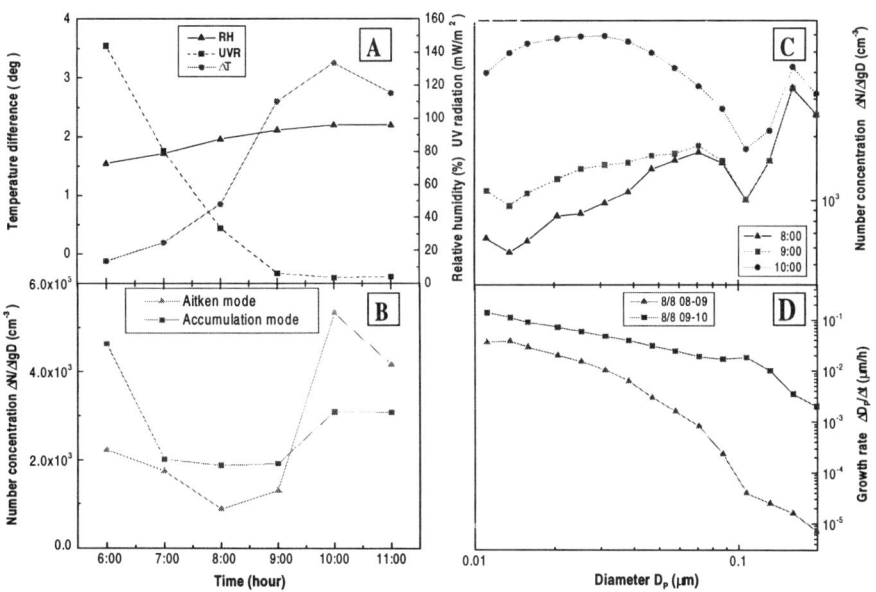

FIGURE 2. Evening event on August 8, 1997, Preila. **A.** Variations of meteorological parameters in time. **B.** Accumulation and Aitken modes particles concentration variations in time. **C.** Aerosol particles size distribution. **D.** Aerosol growth rate dependence on diameter.

reached 3,2 ^{0}C (Fig. 2A). Intensity of UV radiation from 6:00 p.m. decreased and at the 9:00 p.m. it was only about 6 mW/m^2. Relative humidity increased from 72 % at 6:00 p.m. to 96 % at 11:00 p.m. Aerosol particles concentration decreased at 7:00 p.m. (Fig. 2B). Later concentration started to increase and at 10:00 p.m. concentration of Aitken mode particles was 5320 cm^{-3}. This case might be concerned with change in wind direction. Wind changed direction in almost 200 degrees. Wind was blowing from the land in the beginning of the event. Later it turned direction and started to blow from the sea, at the same time bringing aerosol particles. Aerosol particles size distribution is shown in Fig. 2C. High concentration might be concerned with change in wind direction. Maximum concentration was at 10:00 p.m. Solar radiation was already low, therefore increase in concentration of particles was not related with chemical reactions. Temperature inversion at this time was highest (3,2 ^{0}C) and relative humidity reached 96 %. The increase in concentration might be influenced by specific nighttime chemical reactions that occur over the sea. The central species in nighttime chemistry is the nitrate free radical NO_3. Conversion of NO_X to peroxyacetyl nitrate (PAN) and HNO_3 is rapid. HNO_3 is formed by homogeneous hydrolysis of N_2O_5 and by nitrate radical reactions with organic gases [6]. Aerosol particles growth rate dependence on initial diameter is shown in Fig. 2D. Growth rate at 10:00 p.m. was higher than at 9:00 p.m. It was caused by increase of relative humidity. Particles growth rate was decreasing when initial particle diameter was increasing. Shape of these curves show that there took place condensation of low-pressure vapours, that were formed during reactions in gaseous phase, i.e. diffusive growth occurred.

CONCLUSIONS

Experimental results showed that the new particle formation and growth were clearly observed at coastal zone. The observed aerosol production and growth can be explained by the photochemical and nighttime chemical reactions. On the other hand there is a need to continue measurement of the submicron aerosols to investigate new particle formation process in more detail.

REFERENCES

1. O'Dowd C., McFiggans G., Creasey D.J., Pirjola L., Hoell C., Smith M.H., Allan B.J., Plane J.M.C., Heard D.E., Lee J.D., Pilling M.J. Kulmala M., *Geophysical Research Letters*, 26, 12, 1707-1710 (1999).
2. Despiau S., Cougnenc S. and Resch F., *Journal Aerosol Science*, 27, 403-415 (1996).
3. Ulevicius V., Trakumas S. and GirgzdysA., *Atmospheric Environment*, 28, 795-800 (1994).
4. McMurry, P.H. and Wilson, J.C., *Atmospheric Environment*, 16:121-134 (1982).
5. Ulevicius, V.A., Juozaitis, A.A., Balsys, A.S. and Girgzdys, A.J., *Fizika Atmosfery*, 12:177-184 (1988).
6. Russel A.G., Cass G.R., and Seinfeld J.H., *Environmental Science & Technology*, 20, 1167-1172 (1986).

Aerosol and Trace Gas Measurements over the Birmingham Conurbation during PUMA

Bower,K.N., Beswick,K.M., Burgess,R.A., Stromberg,I.M. and Gallagher,M.W.

The Physics Department, UMIST, PO Box 88, Manchester, M60 1QD

Abstract. The UMIST instrumented light aircraft (Cessna 182) was used during summer 1999 and winter 2000 to make measurements of aerosol and trace gases in the Boundary Layer (BL) and Free Troposphere (FT) both upwind and downwind of the city of Birmingham in the UK. These measurements were made as part of the NERC URGENT/ASURE program PUMA (Pollution of the Urban Midlands Atmosphere). Gradients in the concentrations of fine and accumulation mode aerosol and of the trace gases NOx and ozone were observed both in the vertical and horizontally across the conurbation. Examples are presented. These, and thermodynamic data have been made available to initialise and validate output from a high spatial resolution meso-scale model (coupled to dispersion and atmospheric chemistry modules) developed elsewhere in PUMA (in order to predict primary and secondary air pollutant concentrations at urban background locations across the West Midlands region of England).

THE PUMA PROGRAM

Introduction.

The principal objective of the PUMA (Pollution of the Urban Midlands Atmosphere) research program is to apply a high spatial resolution meso-scale model to the West Midlands region of England, to validate it in three dimensions; and to incorporate a coupled dispersion and atmospheric chemistry module. The aim is to develop the model to be able to predict primary and secondary air pollutant concentrations at urban background locations across the conurbation (with a 2km horizontal and 25m vertical resolution), to validate it against measurements of CO, SO_2, NO_X, O_3, inorganic particulate matter etc. and to compare it with the output from existing models. In order to gain an insight into the chemical processes controlling the composition of the urban atmosphere, the aim has been to measure a wide range of transient and long-lived chemical species (including hydrocarbons, carbonyl compounds, oxyacids of nitrogen and the free radical species OH, HO_2, RO_2 and NO_3) at a number of ground based sites within the region during intensive measurement campaigns. It is intended that a management model, applicable to national and local governments will be produced, in order to assess the impact on air quality of specific control strategies for a wide range of criteria pollutants and on a range of timescales from minutes to years.

The specific contribution made by UMIST to the PUMA consortium is to use its instrumented light aircraft (a Cessna C–182J, call sign G–AVCV) to provide information on the vertical structure of the atmosphere within the PUMA model domain (e.g. profiles of temperature, dewpoint pressure etc), as well as measurements of the gradients in the concentrations of trace gases and aerosol properties in both the vertical and horizontal under a variety of atmospheric and thermal conditions (i.e. to support the PUMA model development program by providing a database of information to aid in the initialisation and validation of their 3–D models).

Methodology.

Two intensive field measurement campaigns have been carried out within PUMA, in June-July 1999 and January-February 2000. In both summer and winter campaigns, the UMIST C–182 aircraft was used to make measurements of aerosol and trace gases in the Boundary Layer (BL) and Free Troposphere (FT) both upwind and downwind of the conurbation surrounding the city of Birmingham. Horizontal legs were generally flown at a fixed altitude (around 580m in 1999) in a flight pattern around the conurbation designed to enable straight and level legs to be flown both perpendicular to and along the prevailing wind direction. In addition, profiles were flown from near ground level up to an altitude of around 2000m in an area just to the west of the box, so as to characterise the vertical structure of the atmosphere from ground level up to the FT. Figure 1 shows a map of the UK, the computed (ECMWF) airflow back trajectories for June 16th 1999, and the flight track of the aircraft on that day.

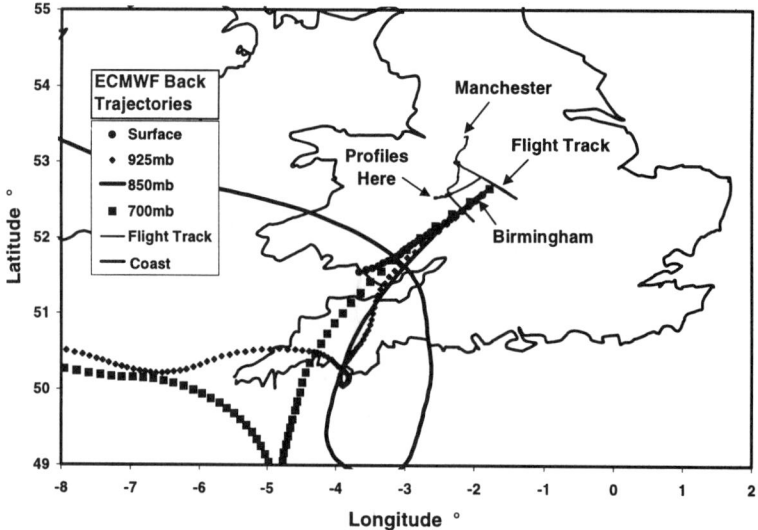

Figure 1: The PUMA experimental area, the UMIST C-182 Flight Track on June 16th 1999, and the computed ECMWF 5 day back trajectories at indicated levels.

In the summer 1999 campaign, 9 flights were carried out on 8 days in a variety of atmospheric conditions and wind directions (e.g. from clean south westerlies on June 16th – Flight 1, to more polluted easterlies on July 9th – Flight 9). In 2000, 7 flights were carried out on 4 separate winter days.

Measurements performed on each of the flights included many of the following: total condensation nuclei (CN) concentrations using a TSI 3010 particle counter (>50 nm); accumulation mode aerosol size spectra in the range 0.1-3.0 µm (diameter) and number concentration (N) using a PMS ASASP-X laser scattering probe; NO_2 or NO_X using a Scintrex LMA-4; O_3 using a GFAS OS-G-2 ozone sonde; plus standard fit measurements of temperature (T), dewpoint temperature (Td), pressure (p), GPS position etc. Turbulence was measured (on some 1999 flights) using a five-hole differential pressure device. Wind trajectory analysis was also carried out at a number of pressure levels to determine the bulk dynamical properties at each level

Results

The profiles carried out to the west of the region (generally upwind of the Birmingham conurbation) often reveal significant variations in concentrations of CN, N, NO_2 and O_3 with height. Profiles of T and Td were used to help determine the depth of the BL. A typical feature of the daytime polluted urban BL measurements (under unstable conditions) is the presence of significant concentrations of fine and accumulation mode aerosol in the lower part of the BL. Towards the top of the BL, N usually declines, whereas CN concentrations can be significantly enhanced, probably due to a mixture of gas to particle conversion processes and entrainment.

Figure 2 shows an ascent profile of aerosol concentration for flight 8 (8th July

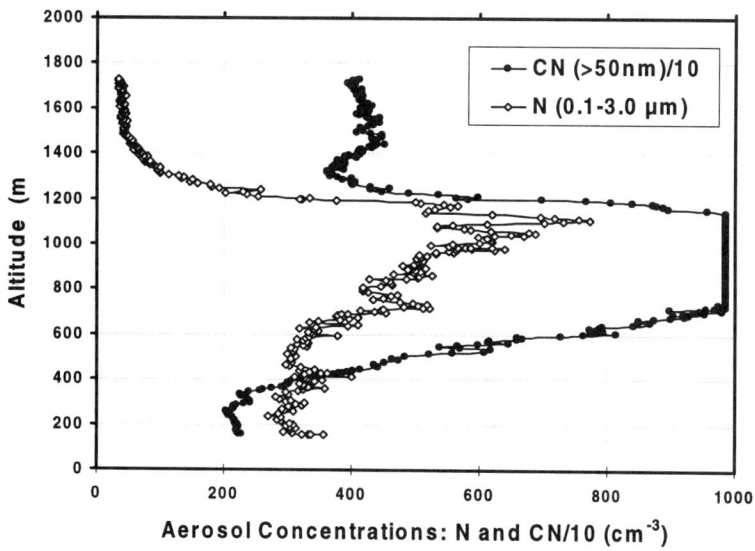

Figure 2: Ascent profile of aerosol concentration on flight 8, 8th July 1999.

1999). Here both CN and N increase towards the BL top (CN actually saturates the 3010 counter at 10^5 cm^{-3} in this case). Profiles of NO_x and O_3 often show evidence of an anti-correlation between the two gases with NO_x declining and O_3 generally increasing with altitude.

Significant concentration gradients were also observed between horizontal flight legs upwind and downwind of the city, in both summer and wintertime experiments. In June-July 1999, concentration gradients in CN were typically around 10^3 particles cm^{-3}. Concentrations of accumulation mode aerosols were also elevated downwind. Gradients of around 6 ppbv for NO_X and 10 ppbv for ozone were typical across the conurbation. Figure 3 shows the increase in concentration downwind of the city for NOx during Flight 2 on June 18th. Cross-conurbation gradients of NO_2 and CN were similar during winter 2000. However, wintertime ozone concentrations and gradients were smaller, the upwind to downwind variation being no more than 4 ppbv. Significant temperature gradients of between 0.5 and 1 °C were also observed.

Small sections of flights intercepting plumes from nearby small towns or point sources such as power stations were also often encountered and readily identifiable.

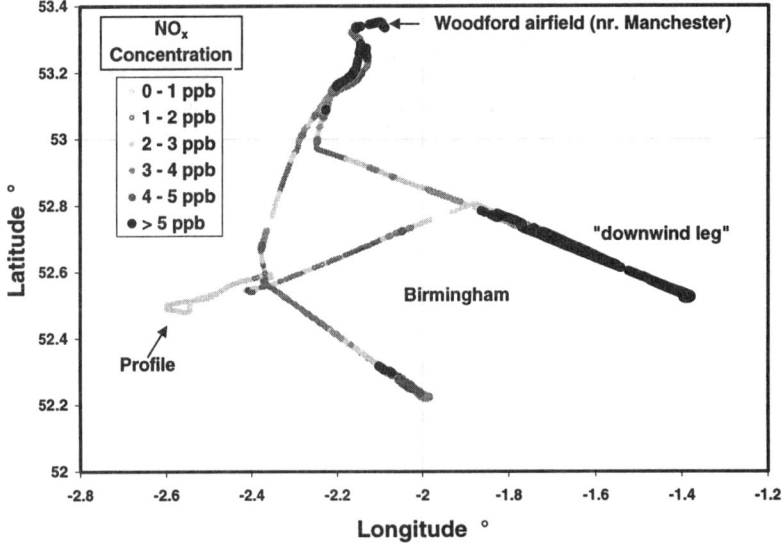

Figure 3: NOx concentrations on the Flight Track around the Birmingham conurbation on June 18th 1999.

ACKNOWLEDGMENTS

This work was supported by the UK Natural Environment Research Council (NERC) grant GST/02/2244 through the URGENT program. We should also like to thank the British Atmospheric Data Centre (BADC) and ECMWF for providing access to the PUMA back trajectory service.

Global transport of gaseous pollutants in the atmosphere and aerosol particle formation through kinetic processes

A.E.Aloyan, V.O.Arutyunyan

Institute for Numerical Mathematics, RAS, Gubkin str., 8, Moscow, Russia

Abstract. A mathematical model for the transport of multicomponent gaseous pollutants accounting for the photochemical transformations and aerosol formation in the Northern hemisphere is considered. Nucleation, condensation, and coagulation processes are taken into account. Aerosol size spectrum is divided into 25 bins (starting from 10^{-5} μm) by doubling the particle masses. Oxidation of sulfur dioxide into sulfuric acid vapor is considered. At some levels, this brings to supersaturation of H_2SO_4 - H_2O with further deposition of the water solution of sulfuric acid onto the surfaces of condensation nuclei. Main principles of the model development and numerical realization are discussed and the results of numerical experiments are given based on calculations of sulfuric acid transformation and contrail formation in the troposphere of the Northern hemisphere.

I INTRODUCTION

Formation and removal of aerosol particles in the atmosphere are described by a number of processes: nucleation, condensation, coagulation, chemical transformation in the gas and aqueous phases, interphase exchange, hydrodynamics, dry and wet deposition, etc. The chemical composition of the particles depends essentially on size, which is critical to atmospheric physics and chemistry. Sulfate aerosol particles play an important role in the atmospheric disperse media. Mostly, these particles are formed through homogeneous nucleation of water and sulfuric acid.

II STATEMENT OF THE PROBLEM

The numerical model of global transport of multicomponent gaseous pollutants and aerosols in the Northern hemisphere (accounting for the photochemical transformations, nucleation, condensation, and coagulation) is formulated in the spherical system of coordinates (λ, ψ, z), where λ is the longitude, ψ is the complement of latitude and z is the altitude measured from the underlying surface. The main equations for the dynamics of pollutant concentration are written down as

$$\frac{\partial C_i}{\partial t} + \frac{u}{a\sin\psi}\frac{\partial C_i}{\partial \lambda} + \frac{v}{a}\frac{\partial C_i}{\partial \psi} + w\frac{\partial C_i}{\partial z} = F_{\text{gas}} - P_{\text{nucl}} - P_{\text{cond}} + P_{\text{phot}}$$

$$+\frac{\partial}{\partial z}\nu\frac{\partial C_i}{\partial z} + \frac{1}{a^2\sin^2\psi}\frac{\partial}{\partial \lambda}\mu\frac{\partial C_i}{\partial \lambda} + \frac{1}{a^2\sin\psi}\frac{\partial}{\partial \psi}\mu\sin\psi\frac{\partial C_i}{\partial \psi}, \qquad (1)$$

$$\frac{\partial \varphi_k}{\partial t} + \frac{u}{a\sin\psi}\frac{\partial \varphi_k}{\partial \lambda} + \frac{v}{a}\frac{\partial \varphi_k}{\partial \psi} + (w-w_g)\frac{\partial \varphi_k}{\partial z} = F_{\text{aer}} + P_{\text{cond}} + P_{\text{coag}} + P_{\text{nucl}}$$

$$+\frac{\partial}{\partial z}\nu\frac{\partial \varphi_k}{\partial z} + \frac{1}{a^2\sin^2\psi}\frac{\partial}{\partial \lambda}\mu\frac{\partial \varphi_k}{\partial \lambda} + \frac{1}{a^2\sin\psi}\frac{\partial}{\partial \psi}\mu\sin\psi\frac{\partial \varphi_k}{\partial \psi}. \qquad (2)$$

Here $C_i, (i = 1, N_g)$, $\varphi_k, (k = 1, N_a)$ are concentrations of gaseous pollutants and aerosols, respectively; N_g and N_a are the numbers of gaseous and aerosol components, respectively; $\mathbf{u} = (u, v, w)$ is the wind velocity vector along λ, ψ, z directions, respectively; w_g is the gravitational settling; μ, ν are the horizontal and vertical turbulence exchange coefficients, respectively; a is the average radius of the Earth; F_{gas} and F_{aer} are gas and aerosol emission source terms; P_{nucl}, P_{cond}, P_{coag}, and P_{phot} are operators describing nucleation, condensation, coagulation, and photochemical transformation, respectively.

Equations (1) and (2) are solved under boundary conditions of periodicity at the horizontal direction; the fluxes on the upper boundary ($z=9.8$km) are taken to be zero; on the lower boundary $z = h$ the boundary conditions are set according to the Monin-Obuchov theory. The deposition is accounted in the model through the dependence on the Henry constant (K_H).

The model consists of the following main blocks.

Photochemistry. The photochemical transformations of S-C-N species are considered in the model. The gas-phase pollutant sources are global emission data of SO_2, NO_x and CH_4 from the earth's surface.

Nucleation. New particle formation from vapors occurs mainly via binary homogeneous nucleation of sulfuric acid and water drops. The nucleation rate J depends on the mass concentration in the vapor phase (c), relative humidity of air (Rh), and temperature (T). The critical radius and nucleation rate are computed based on [6].

Condensation. The kinetic equation for the spatially inhomogeneous case has the form:

$$\frac{\partial f_g}{\partial t} + \frac{\partial (v_g f_g)}{\partial g} + \text{div}\mathbf{u}f_g = J(t)\delta[g - \gamma g_*(t)] + F_T \qquad (3)$$

where F_T describes the turbulent diffusion both in vertical and horizontal directions; $f_g(t)$ is nonequilibrium distribution function (size spectrum) and J is determined by the method described in [6].

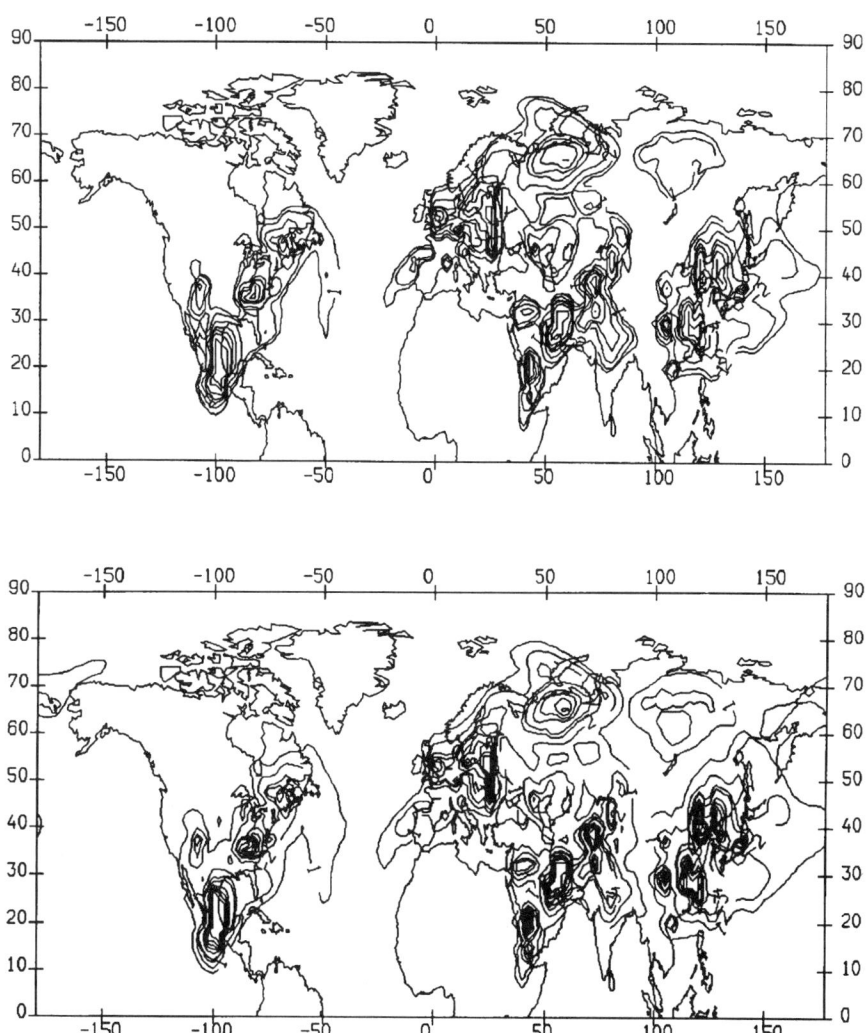

The growth rate v_g accounts for any exchange regime between vapor and particle surface [2].

Coagulation. The particle concentration and size spectrum evolution is described by the Smoluchowski equation as done in the model [5].

The main principles of constructing the numerical methods for the above-listed models can be found in [1,4,5,3].

III NUMERICAL RESULTS

Using the model described above, numerical experiments were carried out to study the spatial and temporal variability of the gaseous pollutants and aerosols in the Northern hemisphere for the period of January to December, 1992. The computational domain has the following sizes: 144 (number of grid points by longitude) × 73 (that by latitude) × 15 (that by vertical).

The numerical experiments were performed for different seasons of year (January, April, July, and October) and time intervals comprising 10 and 30 days. Calculated were concentrations of sulfur dioxide and sulfuric acid vapors as well as the nonequilibrium size distribution functions in every point of the grid domain.

Figure 1 demonstrates the concentration fields in the plain (ψ, λ) (January; 30 days period) for two distinct aerosol particle size fractions.

Depending on the type of circulation and transformation processes, the acid aerosol particle concentration peaks vary with continents. Precipitation over the reference calculation period bring to scavenging of acid particles and enhanced acidity of rainwater. Finally, it should be noted that the account of kinetic processes of transformation in modeling the pollutant global transport in the atmosphere allow the local areas of abundance of acid particles be separated out and contrail formation be traced.

REFERENCES

1. Marchuk, G.I, Aloyan, A.E., *Izv. AN: FAO* **31**, 5, 597-606 (1995) (In Russian).
2. Sutuguin, A.G., Tokar, Ya.I., *Colloid Journal* **47**, 2, 341-347, (1985) (In Russian).
3. Aloyan, A.E., Egorov, V.A., Marchuk, G.I., and Piskunov, V.N. *Russ. J. Num. Anal. Math. Modell.* **7**, 7, 457-471 (1992).
4. Aloyan, A.E., Arutyunyan, V.O. and Marchuk, G.I., *Russ. J. Num. Anal. Math. Modelling* **10**, 2, 93-114 (1995).
5. Aloyan, A.E., Arutyunyan, V.O., Lushnikov, A.A., Zagainov, V.A., *J. Aeros. Sci.* **28**, 1, 67-85 (1997).
6. Kulmala, M., Laaksonen, A., Pirjola, L., *J. Geophys. Res.* **103**, 8301-8307 (1998).
7. Wexler, A.S., J.H. Lurmann, and Seinfeld, J.H., *Atmos. Environ.* **28**, 531-546 (1994).

Particle Formation From The Oxidation Of Alpha-Pinene By Ozone

James Fitzgerald[1], William Hoppel[1], Glendon Frick[1], Peter Caffrey[1], Louise Pasternack[1], Dean Hegg[2], Song Gao[2], John Ambrusko[3], William Sullivan[3], Richard Leaitch[4] and Christopher Cantrell[5]

[1] *Naval Research Laboratory, 4555 Overlook Ave. SW, Washington D.C. 20375-5000*
[2] *University of Washington, Atmos. Sci. & Geophys. Bldg., mail code 351640, Seattle, WA 98195-1640*
[3] *Calspan-University of Buffalo Research Center, 4455 Genesee St., Buffalo, NY 14225*
[4] *Atmospheric Environmental Service, 4905 Dufferin St., Downsview, Ontario M3H 5T4*
[5] *National Center for Atmospheric Research, 1850 Table Mesa Dr., Boulder, CO 80303*

Abstract. Nucleation and growth of SOA were observed from the reaction products of α-pinene and ozone, utilizing relatively low concentrations of α-pinene (16 ppb) and ozone (95 ppb). Good agreement between observed and modeled results was obtained when the yield was about 5.2% and the SVP of the condensing species was about 0.008 ppb. Our results suggest that the nucleation rate in the α-pinene/ozone system may be limited by the initial nucleation steps.

INTRODUCTION

Secondary organic aerosol (SOA) is formed when VOC gas-phase oxidation products have sufficiently low vapor pressure to condense, causing an increase in the mass loading of particulate matter in the atmosphere. α-pinene is particularly interesting in this regard, because observations in the laboratory (Hatakeyama et al., 1989, Hoffmann et al., 1998) suggest that some reaction product(s) of α-pinene and ozone have a sufficiently low vapor pressure to nucleate new particles, even for very small amounts of reacted α-pinene, provided the preexisting aerosol concentration is low. The results of experiments which describe particle formation and growth in the α-pinene/ozone system are described in this paper.

EXPERIMENTAL RESULTS AND MODEL SIMULATION

The experiments described here were carried out in the Calspan – Univ. of Buffalo 590 m^3 environmental chamber with teflon coated interior. Overnight filtering reduced concentrations of measured gases to sub ppb levels (below detection limits of monitoring instruments) and particle concentrations to less than 0.1 cm^{-3}. A more detailed description of the chamber can be found in a report on characterization of the chamber (Hoppel et al.,1999).

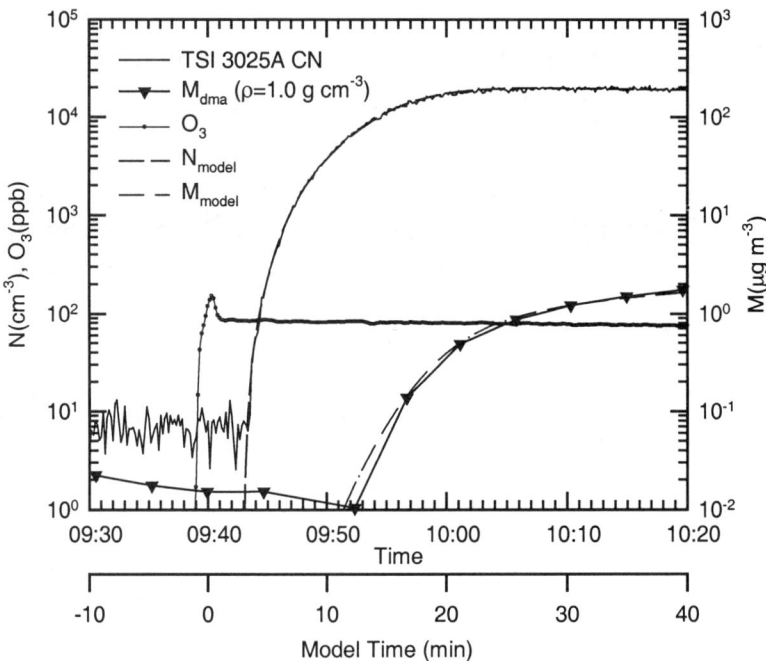

FIGURE 1. Temporal variation of particle number and mass concentrations resulting from oxidation of 16ppb α-pinene by 95 ppb of ozone.

Figure 1 shows the temporal evolution of ozone, ultrafine particle concentration, and aerosol mass for the α-pinene/ozone oxidation experiment of 9 November 1998. First α-pinene was introduced into the chamber and about 15 minutes later ozone was injected over a period of about 90 seconds. The initial concentrations of α-pinene and ozone were 16 ppb and 95 ppb respectively. After four minutes, the concentration of ultrafine particles, as measured with the TSI 3025 CN counter, increased dramatically and after 15 minutes had reached a concentration of 20K cm^{-3}, as shown in Figure 1. The evolution of the particle size distribution as measured with a differential mobility analyzer (DMA) is shown in Figure 2 where the increasing numbers of particles is a result of nucleation and the growth to larger radii is the result of condensation of SOA material. The mass concentrations calculated from the measured size distributions are shown in Figure 1 by inverted triangles. Because of the rapid growth of the particles, the nucleation rate (NR) as a function of time can be obtained quite accurately by taking the time derivative of the ultrafine particle concentration. The experimentally determined nucleation rates are shown in Figure 3. The curve labeled NR$_{fit}$ is a fit to the nucleation rates that was used for the model simulations.

A sectional aerosol model (Fitzgerald et al., 1998) using 32 size sections spanning the range from 0.6 nm to 85 nm, was employed to simulate the experiments. Aerosol processes determining the evolution of the particle size distribution are condensation, nucleation, coagulation, and wall deposition. A single (effective total) molar yield is used to describe the yield of the condensing product. Inputs to the model are the nucleation rate, and initial concentrations of reactants. Unknowns are the molar yield and saturation vapor pressure (SVP), which are constrained to unique sets of values by conservation of mass and are determined by matching the modeled results to the observed results. This gave a molar yield of 5.2% and SVP of about 0.008 ppb. Model results (total mass and number) are shown by the dashed lines in Figure 1. The observed and modeled number concentration should agree since the NR was an experimental input into the model. Conservation of mass was applied at 30 minutes so the modeled mass and measured mass agree at that point in time. The aerosol mass yield (defined as the aerosol mass divided by the mass of reacted α-pinene) is shown in Figure 5. The goodness of the fit at other times is a measure of how well the assumption of the single component representation fits the data.

NUCLEATION

Figure 3 shows the (total effective) vapor concentration of the condensing product(s) as predicted by the model. The maximum vapor concentration ($\sim 2.7 \times 10^8$) is reached at the same time the NR peaks. A plot of NR vs. vapor concentration is shown in Figure 4. It is not known whether a single product is responsible for the nucleation or possibly several reaction products are participating in the observed nucleation and condensation. The dashed lines shown in Figure 4 are the nucleation lines generated by classical homogeneous nucleation theory for reasonable values of molecular weight and density and several values of surface tension. It is seen that the slope of the observed nucleation line is less than that which can be accounted for by classical theory. Smaller slopes may well be a sign that the rate determining step in nucleation is the formation of dimers and trimers rather than conditions at the critical radius determined by classical nucleation theory.

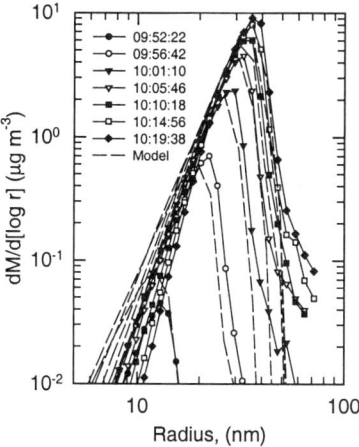
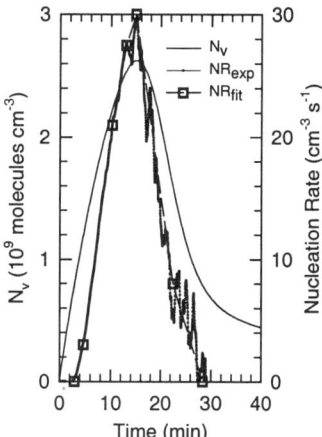

FIGURE 2. Evolution of the measured and predicted aerosol mass distribution.

FIGURE 3. Temporal plot of nucleation rate and vapor concentration of condensing species.

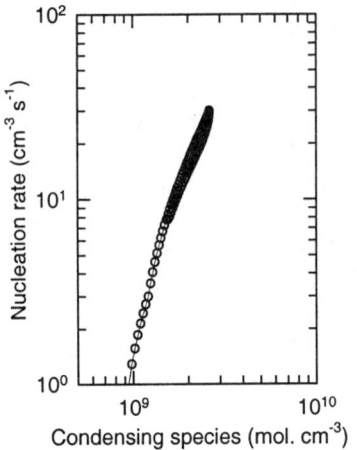 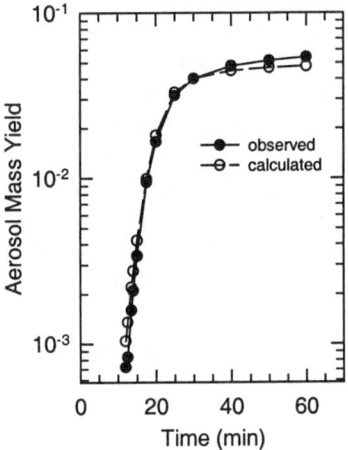

FIGURE 4. Nucleation rate as a function of vapor concentration of condensing species.

FIGURE 5. Measured and predicted aerosol mass yields as a function of time.

CONCLUSIONS

Nucleation and growth of SOA were observed from the reaction of α-pinene and ozone, utilizing relatively low concentrations of α-pinene (15 ppb) and ozone (95 ppb). Good agreement between observed and modeled results was obtained when the yield was about 5.2% and the SVP of the condensing species was about 0.008 ppb. The SVP must be viewed as an upper limit since lower values would give equally good results. While only a single condensing species was needed to obtain agreement with observations, the possibility that there was more than one low SVP species contributing to the observed growth cannot be ruled out, in which case the single component yield and SVP represent an "effective" single component yield and SVP.

We are unable to explain the observed nucleation in the α-pinene/ozone system in terms of classical nucleation theory. The slope of the nucleation rate versus vapor pressure of the nucleating species, suggests that the nucleation rate may be limited by the initial nucleation steps (i.e., dimer, trimer, or adduct formation).

REFERENCES

1. Fitzgerald, J. W., Hoppel, W. A., and Gelbard, F., *J. Geophys. Res.*, **103**, 16,085-16,102, (1998).
2. Hatakeyama, S., Katsuyuki, I., Fukuyama, T., and Akimoto, H., *J. Geophys. Res.*, **94**, 13,013-13,024, (1989).
3. Hoffmann, T., Odum, J. R., Bowman, F., Collins, D., Klockow, D., Flagan, R. C., and Seinfeld, J. H., *J. Atmos. Chem.*, **26**, 189-222, (1997).
4. Hoppel, W., Frick, G., Caffrey, P., Pasternack, L., Albrechcinski, T., Ambrusko, J. R., Sullivan, W., Hegg, D., and Gao, S., Rep. MR 6110-99-8370, 58 pp, Naval Research Lab, Washington, D.C., (1999).

CONDENSATION AND CLOUD CONDENSATION NUCLEI

Cloud Condensation Nuclei Spectral Climatology

James G. Hudson, Seong Soo Yum, and Yonghong Xie

Desert Research Institute,
University of Nevada,
2215 Raggio Pkwy.,
Reno, NV, 89512-1095, USA

Abstract. Airborne measurements of cloud condensation nuclei spectra at several locations around the world are presented. Vertical distributions show gradients that appear to represent cloud scavenging and continental (anthropogenic) particle production.

INTRODUCTION

Aircraft measurements were made with the Desert Research Institute (DRI) instantaneous cloud condensation nucleus (CCN) spectrometer [1]. All of the reported measurements were made outside of clouds to avoid splashing artifacts and reductions of concentrations that are within droplets that can not be sampled. This instrument deduces CCN spectra from the droplet size distribution out of the cloud chamber, which exposes the sample to an increasing field of supersaturations. Particles with lower critical supersaturations (S_c) produce larger droplets and the relationship between droplet size (really just the channel number of the output of the white light optical particle counter) and S_c is determined by inputting monodisperse aerosol of known composition. Figure 1 shows a typical calibration with monodisperse ammonium sulfate particles. CCN spectral information is possible because this relationship is monotonic and by assuming that all particles with the same S_c produce similar responses within the instrument.

RESULTS

Figure 2 shows the apparent response of cloud microphysics to CCN. Figure 3 relates various boundary layer CCN spectra. Similarly high concentrations abounded in continental air at all S. There are many similarities among the maritime spectra in both Hemispheres except for the only wintertime project (SOCEX1), which shows much lower concentrations, especially at the lowest S (largest particles). Measurements made in cloudier conditions were lower, usually especially the lowest S. Florida maritime apparently had considerable continental influence especially at the lowest S.

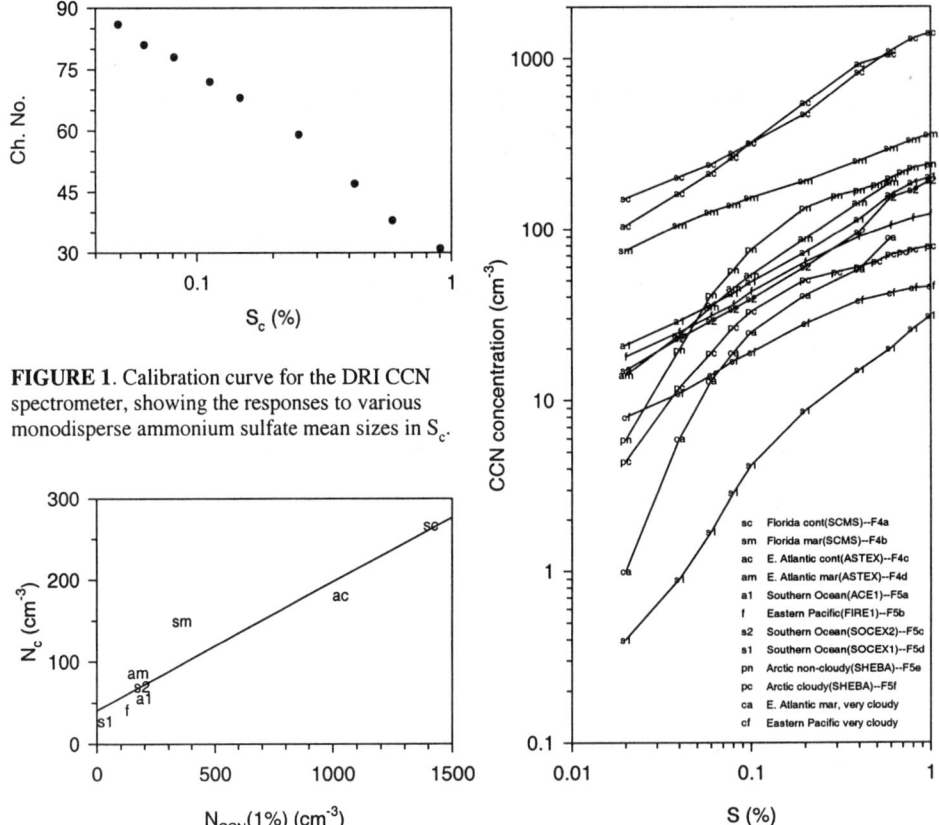

FIGURE 1. Calibration curve for the DRI CCN spectrometer, showing the responses to various monodisperse ammonium sulfate mean sizes in S_c.

FIGURE 2. Average cloud droplet concentrations versus CCN(1% S) concentrations from various projects. Project symbols are shown in Fig. 3.

FIGURE 3. Boundary layer average CCN spectra from various projects. These are traditional cumulative CCN spectra as a function of S.

Figures 4 and 5 show average vertical distributions in various projects and air masses. Variability is displayed by the standard deviations (error bars) of the concentrations at highest S. Figures 4 a and b show little vertical variability. Figure 4c displays a continental low level layer transported 1200 km from Europe. At higher altitudes 4c shows similarity to the corresponding maritime air mass (Fig. 4d). Figs. 4 b and d seem to show the effects of cloud scavenging at low altitudes. Figures 4d, 5 a-c show similarity among maritime air masses in both Hemispheres. Figs. 5 e and f in the arctic are also similar except for scavenging by low level stratus (5f). Figure 5d (the only wintertime project) shows seasonality compared to 5 a and c [2].

ACKNOWLEDGMENTS

NOAA/NASA Climate Change and Data Detection Program, NA67RJ0146.

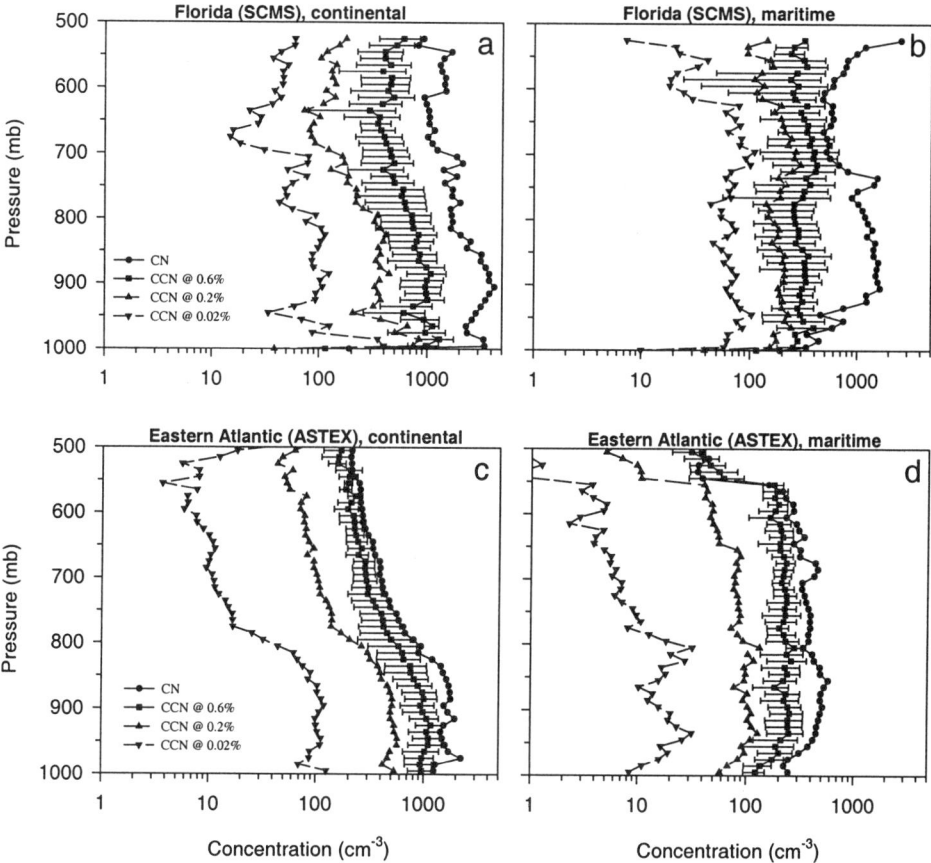

FIGURE 4. Total Particles (CN) and CCN concentrations at 3 supersaturations as a function of pressure altitude. These are averages within 10 mb pressure altitude bins. Panels a and b are from the Small Cumulus Microphysics Study (SCMS) in summer 1995 on the east coast of Florida [3]. Panel c and d are from the Atlantic Stratocumulus Transition Experiment (ASTEX) in summer 1992 [4]. This experiment was centered at the Azores Islands.

REFERENCES

1. Hudson, J. G., *J. Atmos. & Ocean. Techn.* **6**, 1055-1065 (1989).
2. Ayers, G.P., and Gras, J. L., *Nature* **353**, 834-835 (1991).
3. Hudson, J.G., and Yum, S.S., J. Atmos. Sci. (in review).
4. Hudson, J.G., and Xie, Y., *J. Geophys. Res.* **104**, 30219-30229 (1999).
5. Hudson, J.G., Xie, Y., and Yum, S.S., *J. Geophys. Res.* **103**, 16609-16624 (1998)
6. Yum, S.S., and J.G. Hudson, J. Geophys. Res., (in review).

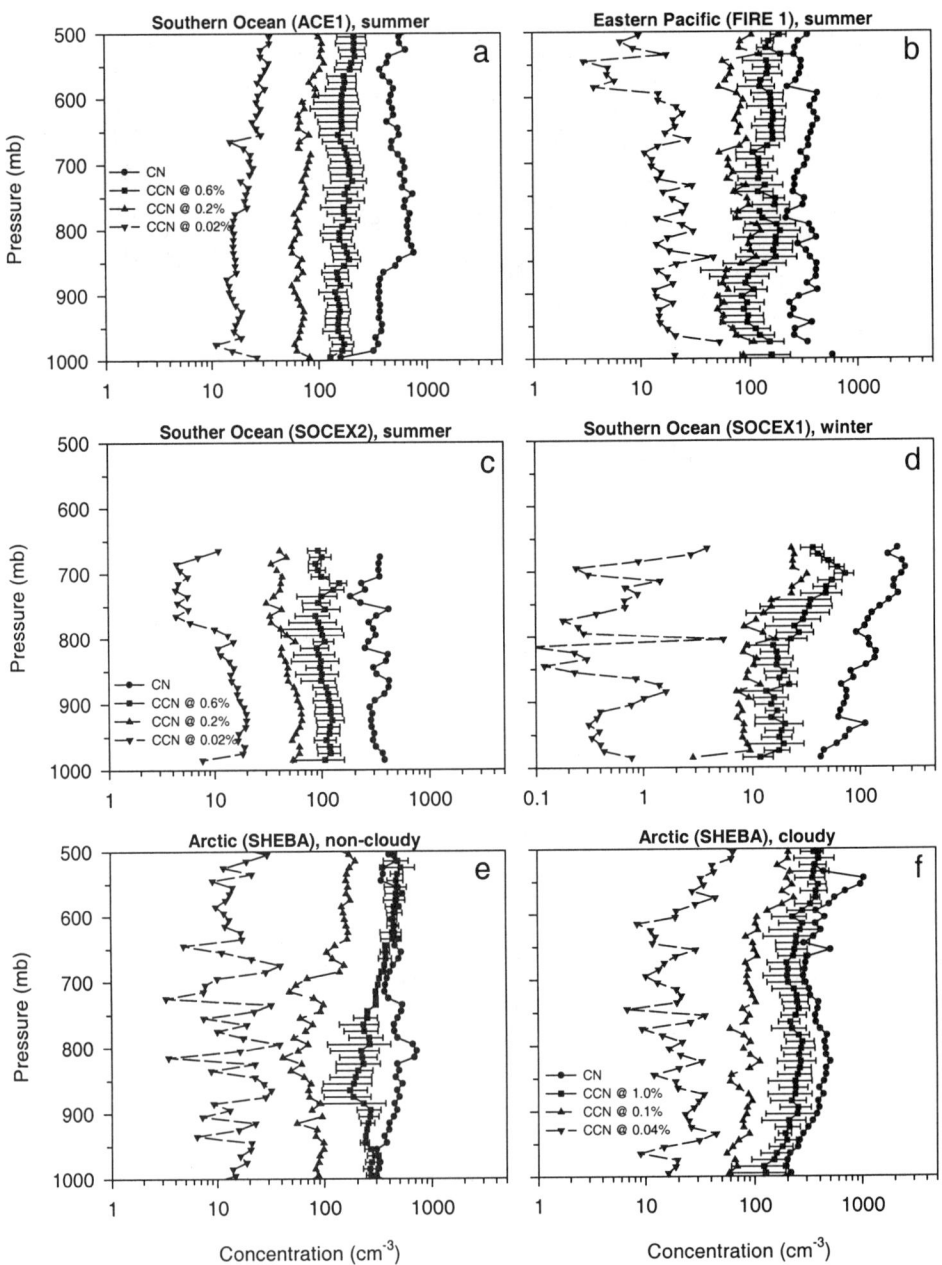

FIGURE 5. As Fig. 4 for other projects. Panels a, c, and d show data from the Southern Ocean: a--ACE1, spring 1995 [5]; c--SOCEX2, summer 1995; and d--SOCEX1, winter 1993. Panel b is from FIRE1 in the eastern Pacific summer 1987. Panels e and f are from the Surface Heat Budget of the Arctic Ocean (SHEBA) spring 1998 north of Alaska [6]. Panel e shows measurements with no low clouds while panel f shows measurements with low clouds.

On the Distribution of Condensation Nuclei (CN) in the Upper Troposphere / Lower Stratosphere and the Nature of CN Sources and Sinks

Andrew G. Detwiler and L. Ronald Johnson

Institute of Atmospheric Sciences
South Dakota School of Mines and Technology
Rapid City, SD, 57701-3995
USA
Andrew.Detwiler@sdsmt.edu

Abstract. Condensation nuclei (CN) concentrations in the upper troposphere / lower stratosphere were measured for 15 months in the late 1970's during the Global Atmospheric Sampling Program. Automated instrument packages on 4 commercial transport aircraft recorded CN concentrations several times per hour during routine passenger-carrying operations above 6 KM MSL over much of the Northern Hemisphere. These data show clear evidence that CN are well-mixed in cloud-free tropospheric regions, that the dominant sources of CN are in the troposphere or at the earth's surface, and that the cold trap at the tropopause is a region of primary CN nucleation or a region where otherwise undetectable CN are caused to grow to sizes detectable by the GASP CN instrumentation.

INTRODUCTION

The troposphere has long been understood as the main source of condensation nuclei (CN) for the lower stratosphere. See, e.g. [1]; [2] and [3]. Hofmann [4] discusses observations showing a consistent peak in CN number mixing ratio just below the tropopause over Laramie, Wyoming, using a series of 11 years of balloon-borne CN counter observations from 1980-1990. He notes that this region exhibits a relative minimum in larger aerosol particles, and hypothesizes that the lack of competition for condensable vapors in this region leads to an environment favorable for nucleation of new particles. Detwiler et al [5] present observations that, within their limits, do not support this near-tropopause maximum in CN mixing ratio, but rather show CN to be well-mixed vertically within the cloud-free troposphere, with mixing ratios decreasing with greater distance above the tropopause. We present here additional observations that aid in understanding the balance of processes regulating CN concentration in the tropopause region.

OBSERVATIONS

The Global Atmospheric Sampling Program (GASP) was sponsored by the U. S. National Aeronautics and Space Administration during the 1970's. The goal of the program was to gain a better understanding of atmospheric chemistry and physics in the cross-tropopause region in which the world's commercial subsonic aircraft fleet does most of its flying. Five Boeing 747 aircraft were equipped with automated instrumentation for recording concentrations of various atmospheric gases and counters to monitor aerosol particle concentrations in various size ranges, during normal revenue operation. A more complete review of GASP and this instrumentation is given in [5] and [6].

We focus here on only two constituents, ozone and CN. Ozone was monitored by modified Dasibi UV absorption instruments. Condensation nuclei were monitored using modified Environment One condensation nucleus counters. The Environment One counters operate using a cloud chamber in which cyclic expansions cause nucleation of water droplets on aerosol particles, and particle concentration is inferred by measuring optical extinction of the resulting cloud.

During GASP sampling, ambient air was brought inside the aircraft hull, then pressurized to cabin pressure using a special pressurization system, before entering the CN counter. Comparisons are given in [5] between GASP CN concentrations and CN concentrations measured using the Rosen balloon-borne CN counters launched from Laramie, Wyoming, on days when GASP aircraft flew over or near Laramie. The GASP CN sampling system characteristically reported from 1/3 to 1/6 the CN concentration reported by the Rosen CN balloon-borne counter (from which the measurements discussed in [4] were obtained) at the same altitude. Detwiler et al [5] hypothesize that the sampling/pressurization/cloud-forming process in the GASP system resulted in the loss of a substantial fraction of the smaller and/or more volatile CN that were counted by the Rosen counter.

Fifteen months of observations from the period 1978-1979, when CN counters were deployed on 4 GASP aircraft operating over the Northern Hemisphere, were analysed. Within this period, there were 4788 condensation nuclei observations in cloud-free air above 6 km MSL. The majority of these observations were obtained over North America, with smaller numbers of observations acquired over Europe, non-Communist Asia, and the Atlantic and Pacific Oceans. See [5] and [6] for more details.

RESULTS

We show in Figure 1 the relationship between CN number concentration and ozone mixing ratios from this set of clear-air observations. The automated GASP sampling system recorded observations and/or calibration data at 5-minute intervals. The ozone mixing ratios in Figure 1 are averages over a 16-second period just prior to recording by the data system. The CN concentrations are the minimum values observed over the 5-minutes elapsed since the previous time data were recorded. A clear region exists in

Figure 1 for a range of ozone values around ~100 ppbv where minimum CN concentrations are rarely less than a few 10's per cm^3. Smaller minimum concentrations are often reported at higher and lower ozone values. An ozone mixing ratio of ~100 ppbv is often taken to indicate the borderline between lower values characteristic of the troposphere and higher values characteristic of the stratosphere. Wozniak [6] shows that ozone mixing ratios in the GASP data set increase from a few 10's of ppbv to hundreds of ppbv, going from below to above the tropopause height interpolated to the aircraft position from the archived global U.S. National Meteorological Center analyses nearest the time of the observation. A value of 100 ppbv typically occurs anywhere from 2 km below to 2 km above the interpolated tropopause height. A plot similar in format to Figure 1, but not shown, in which minimum CN concentration is plotted relative to distance from the tropopause, shows the region containing relatively few low-CN observations to extend from ~2 km above to ~2 km below the interpolated tropopause height.

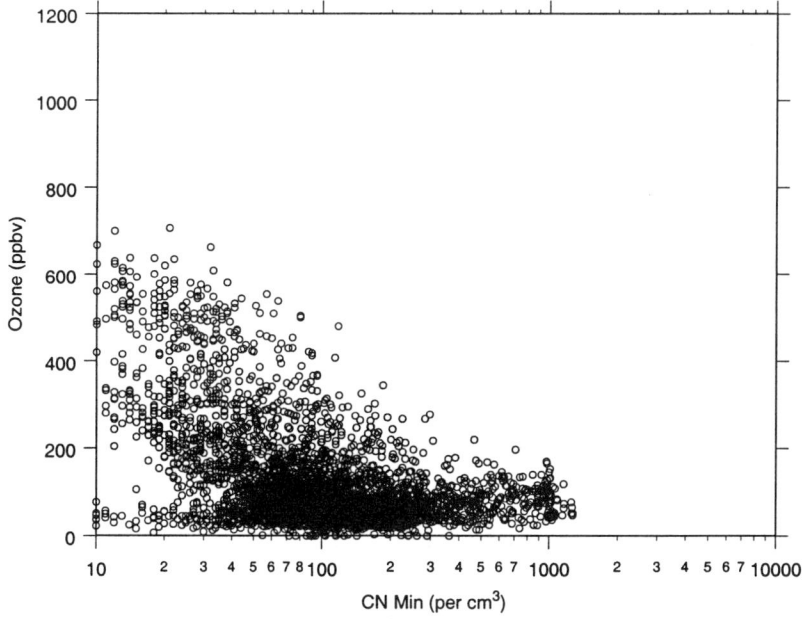

FIGURE 1. Ozone mixing ratio is plotted versus the minimum CN concentration observed over the 5-minute period preceding the recording of the observations. Data are taken from 4788 clear-air observations from commercial aircraft flights over the Northern Hemisphere during the period January, 1978 through March, 1979.

We interpret the lack of observations of low concentrations of CN in the tropopause region as an indication that this is a region in which CN are being nucleated, or relatively small and volatile CN are growing to sizes at which the GASP sampling system detects them. Although the GASP instrument system also included an optical particle counter to sample concentrations of particles tenths of micrometers and larger in diameter, these data have several characteristics that render them suspect. It is therefore impossible to verify whether the anti-correlation between higher concentrations of CN and lower concentrations of larger particles reported in [4] is also characteristic of regions sampled by the GASP aircraft. Although the CN data in Figure 1 are shown as concentrations, in [5] these data are plotted in the form of mixing ratios, and a consistent relative maximum in CN mixing ratio just below the tropopause is not observed. The tropopause data in [5] are interpolated from twice-daily relatively coarse global analyses, while Hofmann [4] determines tropopause height directly from thermodynamic data obtained precisely at the location and time of the CN sounding. This difference in spatial and temporal resolution may in part account for the lack of sub-tropopause maximum in CN mixing ratio in GASP. It also may be that the particles comprising the sub-tropopause maximum seen in [4] are of a type that are not detected by the GASP instrumentation.

CONCLUSIONS

We find that there is a relative dearth of low CN concentrations in the tropopause region sampled during GASP, indicating that this region is one which either new particles are nucleated, or very fine particles characteristically grow to sizes at which they were detectable by the GASP instrumentation. Overall particle number mixing ratios, however, do not peak in this region, suggesting that these newly detectable particles coagulate with other particles on relatively short time scales.

ACKNOWLEDGMENTS

This work was supported in part by NASA grant NAG 5-2711. We thank Allison (Wozniak) Schauer for the initial exploratory analysis of these data, and the NASA GASP program and its manager, J. D. Holdeman, for its diligent efforts in acquiring, performing quality control, and archiving these unique observations.

REFERENCES

1. Cadle, R. D., and C. S. Kiang, *Rev. Geophys. Space Phys.*, **15**, 195-202 (1977).
2. Turco, R. P., P. Hamill, O. B. Toon, R. C. Whitten, and C. S. Kiang, *J. Atmos. Sci.*, **36**, 699-717 (1979).
3. Toon, O. B., R. P. Turco, P. Hamill, C. S. Kiang, and R. C. Whitten, *J. Atmos. Sci.*, **36**, 718-736 (1979)
4. Hofmann, D. J., *J. Geophys. Res.*, **98**, 12753-12766 (1993).
5. Detwiler, A. G., L. R. Johnson, and A. G. Wozniak, *J. Geophys. Res.*, (in press).
6. Wozniak, A. G., 1997. *Exploratory Analysis of Global Atmospheric Sampling Program Data*. M. S. Thesis, South Dakota School of Mines and Technology. Rapid City, SD, 173 pp.

Measurements of CCN-Concentrations in the European Alpine Aerosol Using A Newly Developed Static Thermal Diffusion Counter

R. Hitzenberger*, H. Giebl*, A. Berner*, R. Kromp*, G. Reischl*, A. Kasper-Giebl**, H. Puxbaum**

University of Vienna, Institute for Experimental Physics, Boltzmanng. 5, A-1090 Vienna, Austria
**University of Technology of Vienna, Institute of Analytical Chemistry, Getreidemarkt 9, A-1060 Vienna, Austria*

Abstract. The CCN counter developed at the University of Vienna operates on the principle of a static thermal diffusion chamber. Since 1997, it was used to obtain CCN concentrations in the European alpine background aerosol during intensive measurement campaigns. The 1997 campaign was performed on Mt. Sonnblick (3104 m a.s.l.), while in 1999 and 2000, intensive campaigns were performed on Mt. Rax (1644 m a.s.l.). CCN concentrations at 0.5% supersaturation were found to be comparable ar both sites and also comparable to earlier measurements performed with a commercial CCN counter (DH Associates) on Mt. Sonnblick. Simultaneous measurements of CCN concentration, aerosol number size distribution (measured with a differential mobility particle spectrometer) and cloud liquid water content provided insights into the aerosol/cloud dynamics on Mt. Rax

INTRODUCTION

In the debate on the indirect aerosol effect on the global radiative balance[1], knowledge of the cloud nucleating properties of the aerosol is crucial. The data base on CCN concentrations is still not large, even though several large cloud studies were conducted in the last decade. Most CCN measurements were performed in the marine background aerosol (e.g. during ACE I[2] and ACE II) or at coastal stations under conditions of maritime flow[3]. In the continental aerosol, few CCN data exist despite the intensive studies of continental clouds (e.g. in the GCE campaigns in the Po valley, Italy[4], at Mt. Kleiner Feldberg[5], Germany, and at Great Dun Fell, Great Britain[6]). The only published data of CCN concentrations measured with a CCN counter in Europe are still those collected during the field campaigns on Mt. Sonnblick[7]. In this paper, we present first results of CCN measurements performed at alpine stations in Austria with the new CCN counter as well as first insights into the aerosol/cloud dynamics.

EXPERIMENTAL DETAILS

Site Description

All field campaigns described here were parts of intensive cloud studies, where the physical and chemical properties of the aerosol were investigated also under no cloud conditions. Mt. Sonnblick is a single peak of 3104m altitude in the Central Range of the Austrian Alps (Province of Salzburg), while the station on Mt. Rax was an alpine shelter located at 1644 m altitude at the eastern edge of a mid-level mountain plateau at the eastern rim of the Alps (Province of Lower Austria). Under cloud conditions, both stations receive boundary layer air, because of convective processes, while under no-cloud conditions, free tropospheric air can be sometimes found at Mt. Sonnblick. During the measurement campaigns, the meteorological situation was mostly characterized by a succession of frontal passages with good mixing even under no-cloud conditions.

The 1997 Mt. Sonnblick campaign was conducted in spring (April and May). CCN measurements were possible only under no-cloud conditions restricting the number of sampling days to 7. The Rax campaigns were conducted during 3 weeks in April 1999 and 4 weeks in March 2000.

Instruments

The University of Vienna CCN counter operates on the principle of a static thermal diffusion chamber (Twomey-type counter). CCN are measured automatically in cycles lasting 5 minutes using a computer controlled CCD-camera / image analysis system. Nominal supersaturations can be set between 0.2 and 1.5% by adjusting the temperature difference in the chamber. Details of the instrument and its calibration procedure are described elsewhere[8].

During the Rax campaigns, the CCN counter was equipped with a 1µm pre-precipitator to enable also sampling of non-activated CCN left in the interstitial aerosol. Additional information on the aerosol dynamics is available. The number size distribution of the aerosol below 1µm diameter was measured with a differential mobility particle spectrometer[9] (DMPS) on a time scale of 6 minutes. In parallel to the aerosol measurements, a Gerber PVM-100 (particle volume monitor) probe was used to detect the presence of a cloud and to record its liquid water content on a time basis of 3 seconds.

RESULTS

CCN Concentrations

CCN concentrations were measured at both stations at nominal supersaturations of 0.5%. Table 1 shows average CCN concentrations as well as minima and maxima for the 1997 Sonnblick and 1999 Rax campaigns measured under no-cloud conditions. The results of the 2000 Rax campaign were not yet evaluated at press time.

TABLE 1. CCN concentrations at 0.5% supersaturation, out of cloud

	Mt. Sonnblick, 1997	Mt. Rax, 1999
Average concentration [cm^{-3}]	384	362
5-minute maximum	3816	615
5-minute minimum	36	54

These values are well comparable to those measured earlier at Mt. Sonnblick with a commercial instrument[7], where average CCN concentrations ranged from 29 to 786 cm^{-3} with averages around 240 (fall 1995) and 400 (summer 1996). Minima are in the range given for maritime background aerosols[10,11], while maxima are comparable to concentrations measured in the urban area of Vienna in spring[7] (average 1353cm^{-3}, range 170 – 2630 cm^{-3}). The Sonnblick 1997 maximum occurred during a 20 minute period with extremely high concentrations. If these are removed, the average drops tp 1311cm^{-3}.

Aerosol/Cloud Dynamics

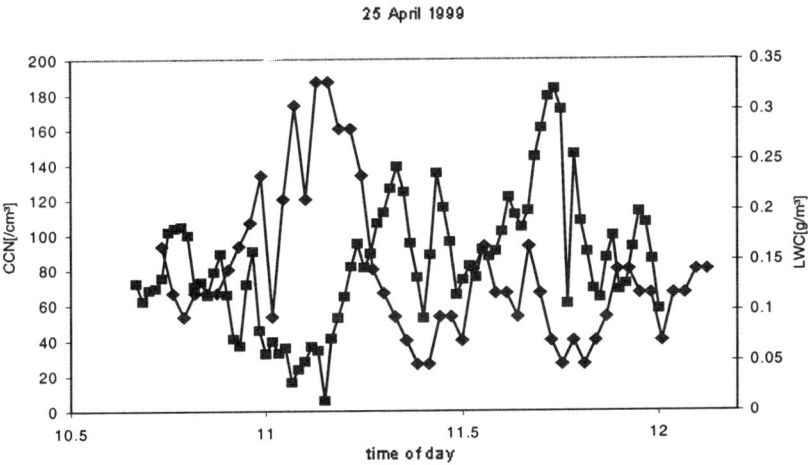

FIGURE 1. CCN concentration (diamonds) and cloud LWC (squares) at Mt. Rax, April 25, 1999 during a patchy cloud.

An interesting feature was found at Mt. Rax under conditions of patchy clouds. Figure 1 shows the variation of both CCN concentration and cloud liquid water content (LWC) measured on April 25, 1999. CCN concentrations drop to a minimum of 13 cm^{-3} whenever the LWC increases, and recover when LWC drops to the no-cloud value of 0.02 g/cm³. Figure 2 shows examples of the DMPS size distribution on April 16. In the early morning, no cloud was present. Cloudy patches started around noon, and during the afternoon, the station was within a cloud. A distinct change in size distribution was observed, with decreasing total numbers and scavenging of particles larger than about 30nm.

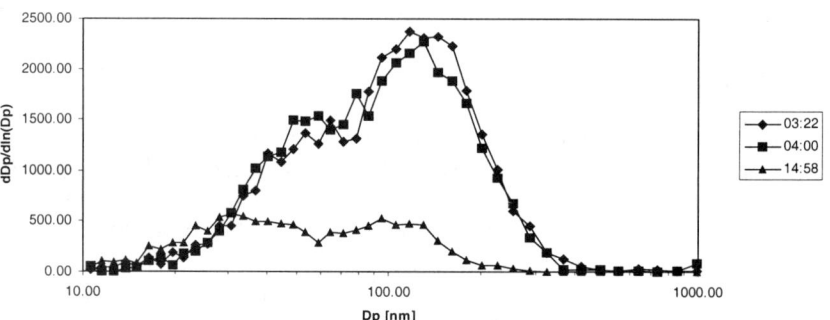

FIGURE 2. Aerosol number size distribution, April 16, 1999. The upper curves are size distributions measured in pre-cloud conditions, while the lower curve is representative of the interstitial aerosol size distribution under cloud conditions.

ACKNOWLEDGEMENTS

The work was funded by the Austrian Science Fund, grants P 103 28-CHE and P 131 43-CHE. The authors wish to acknowledge the assistance of the staff of both the Sonnblick Observatory and the Ottohaus (Rax), as well as the Zentralanstalt für Meteorologie und Geodynamik and the Sonnblickverein for the use of the facilities at the Sonnblick Observatory and the Österreichischer Alpenverein for the use of the facilities of the Ottohaus, Rax. We express our sincerest gratitude to all participants of the Sonnblick and Rax field campaigns for their valuable assistance.

REFERENCES

1. Hudson, J. G., *J. Appl. Meteorol.* **32**, 596 –607 (1993).
2. Bates, T. S. et al., *J. Geophys. Res.* **103**D, 16297-16318 (1998).
3. Jennings, S. G. et al., "Surface CCN measurements at Mace Head, on the west coast of Ireland", in Nucleation and atmospheric aerosols 1996, edited by M. Kulmala and P. E. Wagner, Oxford, Pergamon press, 1996, 800-803.
4. Fuzzi, S. et al., *Contr. Atmos. Phys.* **71**, 3-19 (1998).
5. Wobrock, W. et al., *J. Atmos. Chem.* **19**, 3-35 (1994).
6. Choularton, T. et al., *Atmos. Environ.* **31**, 2393-2405 (1997).
7. Hitzenberger, R. et al., *Atmos. Environ.* **33**, 2647-2659 (1999).
8. Giebl. H. et al., manuscript in work (2000).
9. Reischl, G., *Aerosol Sci. Technol.* **14**, 5-24 (1991).
10. Charlson, R. J. et al., *Nature* **326**, 655-661 (1987).
11. Hegg, D. A. et al., *J. Geophys. Res.* **96**D, 18727 – 18733 (1991).

Cloud Condensation Nuclei Measurement Uncertainties: Implications for Cloud Models

J.R.Snider[1], W.Cantrell[2], G.Shaw[3], and D.Delene[4]

[1]*University of Wyoming, Laramie, Wyoming*
[2]*Indiana University, Bloomington, Indiana*
[3]*University of Alaska, Fairbanks, Alaska*
[4]*University of Colorado, Boulder, Colorado*

Abstract. Cloud condensation nuclei (CCN) spectra can be either measured directly or predicted using Kohler theory initialized with aerosol physicochemical property (APP) data. APP (aerosol composition and size spectra), the CCN spectrum, cloud droplet number concentration (*CDNC*), and vertical velocity (*w*) at the base of a convective cloud updraft are linked by what is commonly called the cloud droplet activation process. Using data sets obtained in cloud regions unaffected by entrainment and drizzle scavenging a comparison of predicted and observed droplet activation is possible. Assessments of the measurement uncertainties associated with APP, CCN, *CDNC*, and *w* are necessary for these closure studies. Here we summarize laboratory comparisons of CCN counter response to aerosols of known size and composition. Uncertainties associated with the CCN measurement are quantified. We also present field measurements from ACE-2. The latter show agreement between measured *CDNC* and predictions based on CCN and *w*. Also using ACE-2 data, we document a systematic bias between *CDNC* derived from APP, and *CDNC* derived from CCN. The implied disparity between CCN and APP is most pronounced when north-eastern Atlantic airmasses are influenced by continental pollution.

Introduction

Cloud condensation nuclei (CCN) spectra are measured by exposing aerosol to a controlled supersaturation while monitoring droplet nucleation and growth. The measurement represents an assessment of the evolution of aerosols to cloud droplets in an environment which is not perturbed by the additional processes that influence droplet size and concentration within natural clouds (i.e., entrainment and drizzle scavenging). The measurement of the CCN is therefore one aspect of the current attempts to quantify aerosol/cloud interactions and to also parameterize them in climate and atmospheric chemistry models. These ideas are summarized in Figure 1. The atmospheric fields of cloud droplet number concentration (*CDNC*), aerosol physicochemical properties (APP), and CCN are connected by two subprocesses. These are referred to as Kohler (I) and parcel (II) theory. Cloud model simulations often combine both subprocesses (I and II) and these can be validated independently by measuring the CCN spectra. These validation studies, or closure experiments, require assessments of the measurement uncertainties associated with each of the atmospheric fields. In this report we present uncertainties associated with three CCN instruments. We also examine the degree of closure between measurements of *CDNC*, CCN and *w* and between measurements of APP and CCN. Data sets used in these closure studies consist of CCN, from the University of Wyoming static diffusion chamber [4], vertical velocity and cloud microphysical data from the Meteo France Merlin [4], and measurements of submicrometric aerosol size distribution and composition from the Punto del Hidalgo site (PDH) [3].

Figure 1 – APP (i.e., aerosol size and composition spectra) are related to CCN via Kohler theory (I). CCN are related to *CDNC* by parcel models (II), either analytic [6] or numerical.

The Alaska Comparison

In November 1997 investigators from the University of Wyoming, the Desert Research Institute, and the University of Alaska met at the University of Alaska to intercompare CCN measurement systems [5]. We challenged two continuous flow and two static diffusion CCN measurement systems with quasi-monodisperse ammonium sulfate aerosols. We also measured the aerosol size distribution. This was transformed to a CCN activation spectrum via Kohler theory and the derived CCN spectrum was taken to be a reference. A total of seven tests were conducted, and within each of these CCN measurements at a range of applied supersaturation were obtained. Results from three of the four instruments are summarized in Table 1, where the symbols WYO1, WYO2 and CCNR represent the Wyoming aircraft CCN instrument [4], the Wyoming balloon CCN [1], and the CCN Remover [2], respectively. The bias relative to the reference is characterized by mean values, *rerr*, that range between -38 and +19%, and its standard deviation, σ_{rerr}. For the instrument used in the closure studies presented below (WYO1), the results indicate that departures between the direct and derived CCN measurement should seldom exceed ± 50%. Also, data from the WYO1 is relatively unbiased.

Table 1 – Results of the CCN Comparison for Applied Supersaturations > 0.3%											
WYO1				WYO2				CCNR			
n_{tot}	n_{test}	rerr %	σ_{rerr} %	n_{tot}	n_{test}	rerr %	σ_{rerr} %	n_{tot}	n_{test}	rerr %	σ_{rerr} %
255	5	2	45	73	3	-38	29	32	3	19	39

n_{tot} is the total number of CCN measurements conducted, n_{test} is the number of tests involving that particular instrument, *rerr* is the relative average disparity, expressed as the average of the ratio of the departure between the CCN and reference concentrations divided by the reference concentration, and σ_{rerr} is the *rerr* standard deviation.

Closure between *CDNC* and CCN Measurements

The comparisons shown in Figure 2 are based on airborne measurements of *w*, CCN (from WYO1), and *CDNC* obtained by the Merlin during the ACE-2 Cloudy Column experiments. On the ordinate we present predicted values of *CDNC*. These were evaluated by inputting measurements of *w* and CCN into the Twomey equation [6]. The set of *w* values used in the calculation was sampled during straight- and level-flight traverses of the cloud layer. Data points plotted in Figure 2 are averages of the probability density functions (pdf) both for the observed $CDNC$ (pdf($CDNC$)) and w (pdf(w)), with the latter transformed to a pdf($CDNC$) via the Twomey equation. This calculation is constrained to values of $w > 0$ and the constraint omits some of the cloud regions that have been affected by entrainment and drizzle formation. CCN measurements input into the calculation were not obtained simultaneous with the pdf($CDNC$) and pdf(w), but were sampled within 60 km of the measurements of cloud microphysical properties. Details of this approach are discussed in reference 4. Figure 2 indicates a degree of closure that is consistent with experimental uncertainties associated with the WYO1 CCN measurement (i.e., a $\pm 50\%$ relative error in concentration).

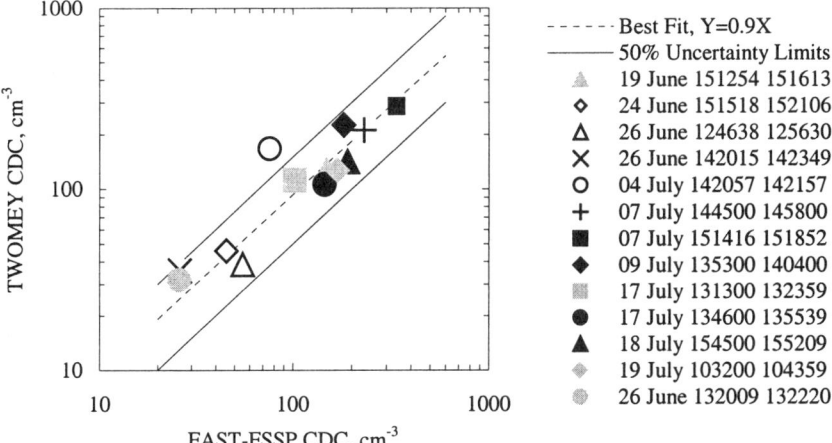

Figure 2 – A successful closure result based on ACE-2 (1997) field measurements. The compared values of $CDNC$ correspond to measurements from the Meteo France Fast FSSP and calculations based on the Twomey equation initialized with measurements of w and CCN. The $CDNC$ and w data were sampled during the straight- and level-flight cloud traverse intervals indicated in the caption. The standard error in the slope of the best-fit line (dashed) is ± 0.09. The solid lines indicate $\pm 50\%$ relative uncertainty compared to a line of perfect agreement (i.e., y=x).

Predicting *CDNC* from ACE-2 CCN and APP Data

We initialized a parcel model with both the WYO1 CCN measurements (see above) and the APP measurements obtained from the PDH monitoring site [3]. We find that $CDNC$ values derived from the PDH data exceed those based on the CCN and by amounts that exceed the maximum probable uncertainty associated with the WYO1 CCN measurement. We also found a positive correlation between the ratio of the two derived $CDNC$ values (i.e., $CDNC_{PDH}/CDNC_{CCN}$) and non-seasalt sulfate concentration. Finally, we note that non-seasalt sulfate is commonly used as an indicator of continental aerosol transport to the north-eastern Atlantic. In light of the closure shown in the previous section, and because of the disparity discussed in this section, we surmise that APP data may substantially overpredict $CDNC$ relative to both in-cloud observations of $CDNC$ and calculations based on CCN. The implication is that APP data may overpredict the extent of the stratocumulus albedo increase attributable to the anthropogenic aerosols.

Discussion and Conclusions

We have presented uncertainties associated with CCN measurements obtained from three CCN instruments. For the CCN instrument used in ACE-2 (WYO1) these uncertainties are unbiased, but associated with a statistical error ($\pm 50\%$) that reflects a nonsystematic departure between the direct and derived CCN concentration measurements. The latter was obtained by inputting measurements of an ammonium sulfate aerosol spectrum into Kohler theory. Further, we show the results of a successful closure experiment which involves comparisons between measurements of *CDNC* and predicted values of *CDNC*. The latter was based on measurements of both CCN and w. In contrast, the Punto del Hidalgo (PDH) APP data was found to consistently overpredict *CDNC* relative to CCN, and the degree of this overprediction increased with increasing levels continental pollution.

Several phenomena, both atmospheric and instrumental, may explain the implied disparity between the CCN and APP. First, since we are comparing data from two different measurement platforms (the Merlin and PDH) it is important to ascertain that there were no gradients in the measured fields. Second, since the WYO1 data is limited to applied supersaturation (S) values larger than 0.2% those measurements do not capture the slope of the activation spectrum at the low values of S that are important for droplet activation in stratocumulus. And finally, selection of *CDNC* and w from the cloud insitu data involve the implicit assumption that cloud regions of CCN activation can be accurately identified. Each of these issues must be addressed, and the accuracy and precision of all relevant measurement systems must be scrutinized before definite conclusions can be drawn from the field measurements

References

1. Delene, D., T.Deshler, P.Wechsler, and G.Vali, A balloon-borne cloud condensation nuclei counter, *J. Geophys. Res.*, 103, 8927-8934, 1998.
2. Ji, Q., G.E. Shaw, and W. Cantrell, A new instrument for measuring cloud condensation nuclei: Cloud condensation nucleus "remover", *J. Geophys. Res.*, 103, 28013-28019, 1998
3. Putaud, J.P., R.Van Dingenen, M.Mangoni, A.Virkkula, and F.Raes, Chemical mass closure and origin assessment of the submicron aerosol in the marine boundary layer and the free troposphere at Tenerife during ACE-2, *Tellus*, in press, 2000
4. Snider, J.R. and J.-L. Brenguier, A comparison of cloud condensation nuclei and cloud droplet measurements obtained during ACE-2, in press *Tellus*, 2000
5. Snider, J.R., W.Cantrell, G.Shaw, J.Hudson, D.Delene, and S.S.Yum, Comparisons of four cloud condensation nucleus measurement systems, manuscript submitted to *J. Atmos. Oceanic Technol.*, 2000
6. Twomey, S., The nuclei of natural cloud formation. Part II: The supersaturation in natural clouds and the variation of cloud droplet concentration, *Geophy. Pura. Appl.*, 43, 243-249, 19

Maritime CCN Measurement and Delayed Droplet Growth

John Gras

CSIRO Atmospheric Research
Aspendale 3195, VIC., Australia

Abstract. Measurements at Cape Grim, Tasmania, during the recent Southern Ocean Photochemistry Experiment (SOAPEX-2), using a static Cloud Condensation Nucleus (CCN) counter, show delayed growth in development of the peak in light scattering when sampling marine air. In polluted air the time taken to reach the peak in scattering is similar to that for the inorganic salts used to calibrate the counter.

INTRODUCTION

Uncertainty in the magnitude of indirect forcing of climate, through modification of cloud properties by aerosol particles, is now recognised as a major contributor to the current uncertainty in climate prediction. In part this arises because of an incomplete understanding or description of global aerosol chemistry and a lack of strong evidence that, in general, cloud properties can be accurately predicted from the pre-cloud aerosol. Instrumental CCN measurements provide a relatively controlled environment in which the chemical and microphysical interaction in water vapour nucleation at low supersaturation can be tested, independent of the wider set of variables that control clouds in the real atmosphere. Closure studies such as those by Bigg (1), Covert et al. (2) and Chuang et al. (3) that attempt to reconcile observed and calculated CCN concentrations have not, in general, been as successful as might be hoped. This indicates a less-than-optimum representation of some physico-chemical factors in modelled CCN concentration or unrecognised instrumental problems.

Instrumental measurements of CCN concentration aim to determine the number of particles that nucleate water droplets at specified supersaturation levels, typically from a tenth of one percent to around 1 percent (RH =100.1% to 101%). This differs from real clouds in a few aspects. Cloud peak-supersaturation, whilst generally in this range, is not a constant but is set by a balance between the growing supersaturation in an ascending air parcel and the relief by condensation of water vapour in the cloud. Cloud droplet numbers are controlled largely by the number of particles that activate at peak supersaturation, which is achieved a small distance above cloud base. Other potentially important factors may or may not be simulated well in CCN chambers. These include kinetic limitations to particle growth, particularly of large or giant particles, entrainment of air into the cloud, deactivation of some previously activated drops due to competition for water vapour and the influence of soluble gases on the ionic composition of the growing droplet. The actual importance of these and other effects will depend strongly on the environment in which the cloud develops. Instrumental measurements, in principle, afford a useful means for the determination of the number of particles that have the potential to form cloud droplets at a given supersaturation. Such measurements intrinsically combine the aerosol chemical and microphysical factors relevant to cloud droplet nucleation in real clouds.

This work describes observations on delayed peak scattering in a static cloud chamber and their association with particle number concentration in marine and continentally-modified marine air at Cape Grim, Tasmania. Measurements were made as part of the second Southern Ocean Photochemical Experiment, SOAPEX-2, during January and February 1999.

CCN INSTRUMENTATION, OBSERVATIONS & DISCUSSION

CCN concentrations were determined using an automated static thermal diffusion cloud chamber (STG CCN) with photometric detection. Details of the cloud chamber, which is based on the design of Lala and Jiusto (4), are given in Gras et al. (5). In this design, CCN concentration is derived using white light scattered at 45° to the forward direction. Calibration of the amplitude of the scattering peak in terms of particle concentration is by monodisperse soluble particles, which are counted using a CN counter operated in parallel with the CCN counter. The procedure yields a linear response over the range of concentrations normally encountered at Cape Grim.

The hourly median concentration of CCN active at 0.5% supersaturation, obtained during SOAPEX-2, is shown in Fig. 1. The 25th and 75th percentiles of the hourly distribution of the time taken to reach a peak in scattered light amplitude are also plotted. These data show a systematic relationship between CCN concentration and the occurrence of delayed scattering peaks. At low CCN concentrations significantly more clouds show later peaks in scattering and a much greater variability in the delay. In the present study no obvious delay in the onset of the scattering peak was evident. For high CCN concentrations median time to peak and inter-quartile range correspond closely to those for soluble calibration salts, e.g. $(NH_4)_2 SO_4$.

At Cape Grim there is a seasonal cycle in CCN concentration in clean air and the concentration also changes strongly with changes in air mass origin. Elevated particle concentrations, such as those in Fig. 1, typically result from recent transit (less than two days) of sample air over industrial/urban source areas in Tasmania or south-eastern Australia. Delays in peak scattering development during SOAPEX-2 were associated mainly with clean air masses with low CCN concentration. This is consistent with a series of measurements at Cape Grim using organic solvents in place of water in a STG CCN counter, reported by Bigg (6). Over a 7-day period (with similar season and concentration range to the SOAPEX-2 observations) organic condensates (ethanol and cyclo-hexane) produced comparable droplet growth to that with water during clean maritime (low CCN) conditions. In modified air masses with local sources, condensation and growth were minimal using organic condensates, and organic condensation decreased with increasing particle concentration (6).

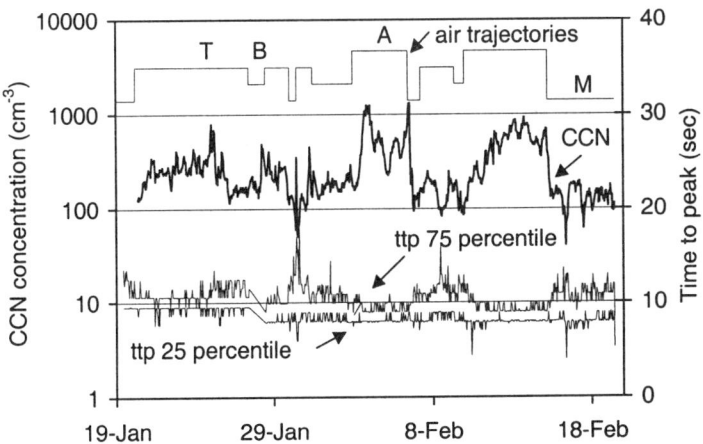

FIGURE 1. Variation of hourly median CCN concentration with corresponding 25th and 75th percentiles of the time to peak distributions. Trajectory sources shown are (A) S. E. Australia, (B) Bass Strait, (T) Tasmania and (M) clean marine – Southern Ocean.

Taken together, Bigg's observations and the present measurements imply more active organic and hydrophobic components in aged, clean maritime air, compared with aerosol from relatively fresh emissions. This appears contrary to the sequence of change in surface activity proposed by Ellison et al. (7).

ORIGIN OF DELAYS & CONSEQUENCES

"Normal" time to peak scattering is taken from calibration of the STG CCN counter with monodisperse particles of soluble salts. For the range of calibration particle size used (up to 250 nm diameter) time to peak scattering does not depend on particle size. Several possible mechanisms may control, or contribute to, differential delays in the time taken to reach a scattering peak for ambient particles. Rate of growth of a droplet depends, in part, on the condensation or mass accommodation coefficient (8), and this dependence has been used, for example, for determination of the condensation coefficient (9). One explanation of the present observations is that the condensation coefficient depends on aerosol type or air mass. Clean marine air masses at Cape Grim will have a history of cloud processing and particles with a smaller condensation coefficient should be more resistant to removal. The underlying cause of a small condensation coefficient appears to be chemical. In the absence of detailed chemical/structural data for the particles in the few tens to few hundred nanometre size range, it remains unclear if this results from an organic surface film as proposed by Gill et al. (10), Bigg (1) and others, or mixtures of soluble and slightly soluble (organic) fractions (11). In the present context it appears unlikely to result from inorganic slightly soluble species.

From a practical perspective slowly developing droplets represent a problem for static CCN counters using photometric detection. Calibration is carried out with a salt where all particles grow at a similar rate. For extended cloud-particle growth times, sedimentation of faster-growing droplets from the sensing volume may become appreciable before slower-growing droplets can contribute fully to the scattering peak giving lower apparent CCN concentrations. In part this may explain some differences in measured and observed CCN concentrations in earlier closure studies. Other types of CCN counter may also be affected, depending on the available growth time and the threshold for inclusion of activated droplets.

An important conclusion is that information additional to CCN concentration can be derived from the time evolution of the droplet cloud in CCN chambers. This could potentially be useful for determining the variation in condensation coefficient and at a minimum it provides an indicator for possible departure from calibration conditions of the CCN counter.

ACKNOWLEDGMENTS

This work was carried out as part of the Cape Grim Baseline Air Pollution Program. Assistance of the Cape Grim staff is gratefully acknowledged.

REFERENCES

1. Bigg, E. K., Discrepancy between observation and prediction of concentration of cloud condensation nuclei, *Atmos. Res.* **20**, 82-86 (1986).
2. Covert, D. S., Gras, J.L, Wiedensohler, A. and Stratmann, F., Comparison of directly measured CCN with CCN modelled from the number size distribution in the marine boundary layer during ACE-1 at Cape Grim Tasmania, *J. Geophys. Res.* **103**, 16597-16608 (1998).
3. Chuang, P.Y., Collins, D.R., Pawlowska, H., Snider, J.R., Johnsson, H.H. Brenguier, J-L., Flagan, R.C. and Seinfeld, J.H., CCN measurements during ACE-2 and their relationship to cloud microphysical properties, *Tellus* in press.
4. Lala, G.G. and Jiusto, J.E., An automated light scattering CCN counter, *J. Appl. Meteor.* **16**, 413-418 (1977).
5. Gras, J.L., Jennings, S.G. and Geever, M., CCN determination, comparing counters with single-drop-counting and photometric detectors at Mace Head Ireland, *Időjárás* **100**, 171-181 (1996).
6. Bigg, E. K. Technique for studying the chemistry of cloud condensation nuclei, *Atmos. Res.* **20**, 75-80 (1986).
7. Ellison, G.B., Tuck, A.F. and Vaida, V., Atmospheric processing of organic aerosols, *J. Geophys. Res.* **104**, 11633-11641 (1999).
8. Fukuta, N. and Walter, L., Kinetics of hydrometeor growth from a vapor spherical model, *J. Atmos. Sci.* **27**, 1160-1172 (1974).
9. Chodes, N., Warner, J. and Gagin, A., A determination of the condensation coefficient of water from the growth rate of small droplets, *J. Atmos. Sci.* **31**, 1351-1357 (1974).
10. Gill, P.S., Graedel, T.E. and Weschler, C.J., Organic films on atmospheric aerosol particles, fog droplets, cloud droplets, raindrops and snowflakes. *Rev. Geophys.* **21**, 903-920 (1983).
11. Shulman, M.L., Jacobsen, M.C., Charlson, R.J., Synovec, R.E. and Young, T.E., Dissolution behavior and surface tension effects of organic compounds in nucleating cloud droplets, *Geophys. Res. Lett.* **23**, 277-280 (1996).
12. Laaksonen, A., Korhoen, P., Kulmala, M., and Charlson, R.J., Modification of the Köhler equation to include soluble trace gases and slightly soluble substances, *J. Atmos.Sci.* **55**, 853-862 (1998).

Size Distribution and Critical Supersaturation Spectrum of The Aerosol from an Electrically Heated Nichrome Wire

Max B. Trueblood[1,5], Millard A. Carter[2], Donald E. Hagen[1], Philip D. Whitefield[3], and Josef Podzimek[4]

1-Department of Physics and Cloud and Aerosol Sciences Lab, University of Missouri – Rolla 65401
2 - Shaw Services, Inc, Baton Rouge, LA 70801
3 – Department of Chemistry, University of Missouri – Rolla, 65401
4 – Department of Mechanical Engineering, University of Missouri – Rolla 65401

Abstract. *An aerosol generator was constructed that produces particles from the Joule heating of a nichrome wire. This novel generator produced a stable output (+/- 20%) over a 7 hour period. The approximate mean diameter was 24 nm. The critical supersaturation spectrum was measured and found to be quite high (5.4% for 24 nm particles), showing that this nichrome aerosol was highly insoluble. Energy dispersive spectroscopy showed that the percentage of Cr far exceeded that of Ni in the particles, whereas in the bulk wire, Cr has only one third the concentration of Ni.*

INTRODUCTION

The use of nichrome wire aerosol for the calibration of condensation nuclei counters was described by, e.g., Liu et al. (1975) who used, besides other test aerosols, nichrome wire particles of 25 and 32 nm for the calibration of the Pollak counter. Podzimek and Kassner (1976) compared the so called UMR-AANC (University of Missouri – Rolla Absolute Aitken Nuclei Counter) with the SANDS (Stratospheric Aitken Nuclei Detection System) counter. At very low nuclei counts (< 700/cc) the two counters displayed very similar concentrations, if nichrome particles were used. They disagreed considerably for larger sodium chloride nuclei. Comparison of different condensation nuclei counters with the UMR-AANC described by Podzimek et al. (1982) using five different aerosols at concentrations varying from 300/cc to 20,000/cc, showed, however, the largest differences if nichrome wire aerosol were used. One possible explanation of these discrepancies was the difference in the particle's residence time in the different counters, which might discriminate between nuclei with a simple vs. complex composition and surface structure. Carter (1989) found that the detection efficiency of the Nolan Pollak counter was a strong function of the concentration, dwelling time and chemical properties of the aerosol. Carter found differences of as much as 33 % by varying the residence time.

The present study describes the construction / performance of a simple nichrome wire aerosol generator and investigates some of the physico–chemical properties of the generated ultrafine particles.

5 – Author to whom correspondence should be addressed.

NICHROME AEROSOL GENERATOR DESCRIPTION

The nichrome aerosol generator (Fig. 1) consists of a Pyrex tube (34 mm ID X 900 mm L) housing the nichrome wire.

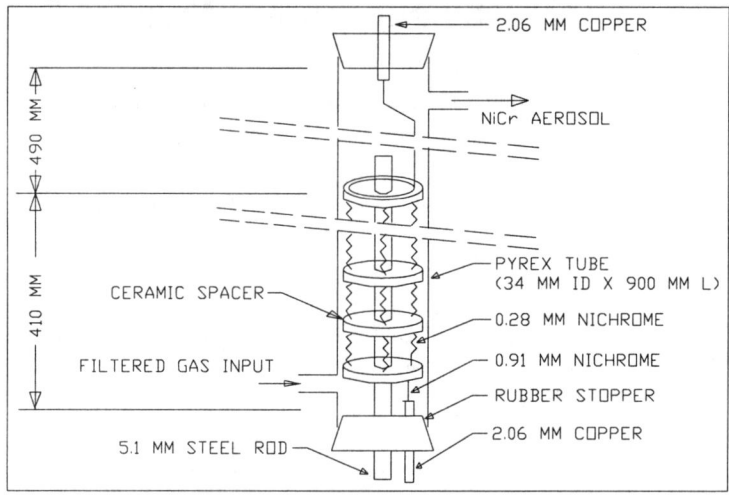

FIGURE 1. Schematic diagram of the nichrome wire aerosol generator.

All nichrome–nichrome joints were spot welded. The four long coils of small diameter nichrome wire were each made from a single piece of 29 gauge nichrome wire (0.287 mm dia X 2000 mm L) (60% Ni, 20% Cr, 20% Fe). Using a lathe running at a low speed, each piece was wound (under about 20 pounds of tension) onto a 6 mm diameter mandrel so as to form a uniform coil. To ensure that all four coils conducted the same electric current, all four of these original straight pieces of nichrome wire were within a few millimeters of the same length. Effectively all the Joule heating occurred in these coils.

Two electric heating elements (designed for a residential electric heating furnace and purchased from the local hardware store) were disassembled and some of their parts used for the present generator. These two heating elements provided twelve ceramic discs that served as spacers for the four nichrome coils, keeping them in place even when they were heated to a dull red during the generating session (the same role they played in the heating element for the residential furnace). The ceramic discs (44.4 mm OD X 10.4 mm L) had one 5.6 mm hole in the center and four 8.6 mm holes equally spaced on a 27.2 mm bolt circle. The discs were spaced about every 38 mm and maintained in that place by a 5 mm dia steel rod. Unless the nichrome wire coils were held in place by the ceramic discs, they would expand, squirm around, and touch one another causing a short circuit when heated to dull red, thus causing unpredictable results. A flow of 4.0 to 6.0 L/m of dry nitrogen or filtered room air was forced through the generator.

The nichrome wire aerosol generators this laboratory has used in the past have been composed of only one small diameter coil. These generators were quite unstable, and would burn out in a few tens of minutes. In the present generator, if one of the four legs becomes hotter than the others (and its resistance increases), then the current load is shifted to the other three, preventing that one leg from undergoing thermal runaway.

STABILITY AND SIZE DISTRIBUTION MEASUREMENTS

The system to measure the output stability of the generator consisted of the generator and a condensation nucleus counter (CNC) (MN: 112L428G1, SN: 3264196, General Electric, Schenectady, NY 12345) Fig. 2 shows that after the initial 100 minute burn in period, the aerosol concentration was constant to within +/- 20% over a 400 minute interval. The generator consumed approximately 600W.

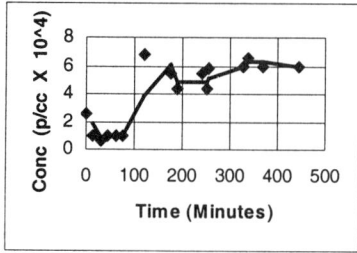

FIGURE 2. Concentration vs. time for the nichrome wire aerosol generator.

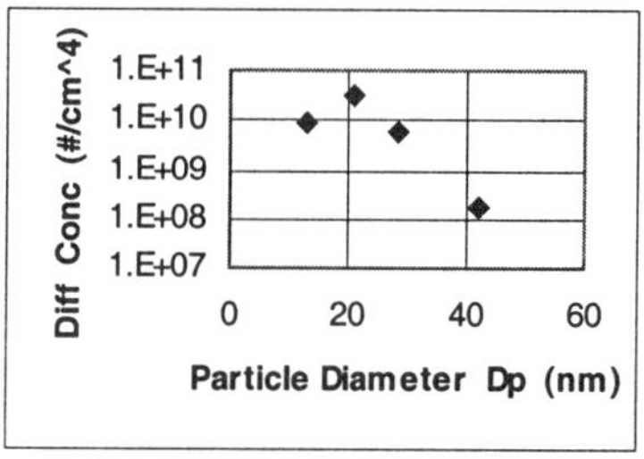

FIGURE 3. Differential concentration dN/dDp vs. particle diameter Dp at the outlet of the aerosol generator.

To measure the size distribution, the output of the generator was fed into the differential mobility analyzer (DMA) (MN: 3071, TSI, PO Box 64394, St Paul, MN 55164 - 0394) (Knutson and Whitby, 1975a; Knutson and Whitby, 1975b; Kousaka et al., 1981; Fissan et al., 1996) while the concentration in the DMA monodisperse output was measured by the alternating gradient chamber (ALGR) (Hoppel et al., 1979; Hoppel et al., 1980; Hagen et al., 1990). The readings from the ALGR were used as the input to an inversion program (Hagen & Alofs, 1983), which yielded the differential concentration dN/dDp vs. dDp at the outlet of the aerosol generator (Fig. 3). Note that the peak in the dN/dDp vs. Dp was at approximately 24 nm.

SUPERSATURATION SPECTRUM MEASUREMENTS

The critical supersaturation of the nichrome aerosol was measured by simultaneously sending a portion of the monodisperse output of the DMA into both the continuous flow diffusion chamber (CFD) (Alofs, 1978; Alofs and Trueblood, 1981) and the alternating gradient chamber (ALGR). The CFD is a continuous flow thermal diffusion cloud chamber that imposes a known supersaturation on the test aerosol. (The ALGR was used so that any time variations in the generator output were normalized away.) If the supersaturation applied by the CFD (SScfd) is greater than the nuclei's critical supersaturation (SSc), then freely growing water drops form on the nuclei and are hence detected. Fig. 4 is a plot of the normalized concentration vs. the SScfd for Dp = 24nm. It is clear that the SSc(Dp=24nm) = 5.40%. (The other data point shown is SSc(Dp=28nm) = 4.80%). Fig. 5 is a plot of the critical supersaturation SSc vs. particle diameter Dp for the nichrome aerosol, and for comparison NaCl and disodium fluorescein (uranine), a laser dye. Of all those shown, the nichrome aerosol has the highest SSc for a given Dp. Such a high SSc may make this aerosol useful for certain tests, such as a challenge aerosol for condensation nucleus counters.

Although a plot is not shown, energy dispersive spectroscopy (EDS) performed on the nichrome particles showed that the amount of chromium in the particles far outweighed the amount of nickel. Although phase diagrams for nichrome show that Cr is still a solid at dull red, it is well know in the metallurgical industry that an alloy containing Cr will lose (by sublimation) a few percent of Cr when melted. Of the three constituents of nichrome (Ni 60%, Cr 20%, Fe 20%), only one oxide (CrO_3) is soluble. If the aerosol studied here were composed completely of CrO_3, then the SSc of the 24 nm particles should be 0.8 to 1.6%, depending on the number of ions CrO_3 forms in water. Thus the aerosol is probably a mix of the oxides of Cr, Ni, and Fe.

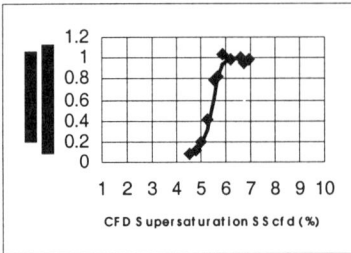 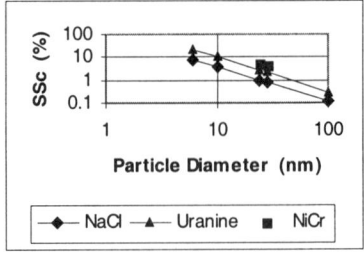

FIGURE 4. Normalized concentration vs. CFD supersaturation for 24 nm nichrome particles.

FIGURE 5. Critical super-saturation vs. dry particle diameter Dp for NaCl, disodium fluorescein, and nichrome.

REFERENCES

1. Alofs, D. J. and Trueblood, M. B., *J. Rech. Atmos.*, **15**, 219 - 223 (1981).
2. Alofs, D. J. , *J. Applied Meteorology,* **17,** 1286 - 1297 (1978).
3. Carter, M. A., M. S. Thesis, Mech Engr, Univ of Missouri – Rolla, (1989)
4. Fissan, H., Hummes, D., Stratmann, F., Buscher, P., Neumann, S., Pui, D. Y. H., and Chen, D., *Aerosol Sci. and Technol.*, **24**, 1 - 13 (1996).
5. Hagen, D. E. and Alofs, D. J., *Aerosol Science and Technology* **2**, 465 - 475, (1983).
6. Hagen, D. E., Trueblood, M. B., and Alofs, D. J., *Aerosol Science & Technol*, **12,** 547 - 560 (1990).
7. Hoppel, W. A, Twomey, S., and Wojciechowski, T. A., *J. Aerosol Sci*, **10**, 369 - 373 (1979).
8. Hoppel, W. A, Twomey, S., and Wojciechowski, T. A., *J. Aerosol Sci*, **11,** 21 - 422 (1980).
9. Knutson, E. O., Whitby, K. T., *J. Aerosol Sci.* **6**, 443 - 451 (1975a).
10. .Knutson, E. O., Whitby, K. T., *J. Aerosol Sci.*, **6**, 453 - 460 (1975b).
11. Kousaka, Y., Okuyama, K., and Endo, Y., *J. Aerosol Sci.*, **12,** 339 - 348 (1981).
12. Liu, B. Y. H., Pui, D. Y. H., Hogan, A. W. and Rich, T. A., J. Appl. Meteor., **14,** 46 - 51 (1975)
13. Podzimek, J. and Kassner, J. L. Jr., J. Appl. Meteor., **15,** 1333 - 1336 (1976).
14. Podzimek, J. Carstens, J. C., and Yue, P. C., Atmos. Environment, **16** , 1 - 11 (1982)

Performance Evaluation of Mixing Type CNC at Low Pressure

Kikuo Okuyama[*], Manabu Shimada[*], Chan S. Kim[*], Yoshifumi Itoh[*], and Melissa M. Lunden[†]

[*]*Department of Chemical Engineering, Hiroshima University, 1-4-1 Kagamiyama, Higashi-Hiroshima 739-8527, Japan*
[†]*Environmental Energy Technologies Division, Ernest Orando Lawrence Berkeley National Laboratory, One Cyclotron Road, Berkeley, CA 94720, USA*

Abstract. A mixing type condensation nucleus counter (MTCNC) was used to detect particles suspended under low-pressure conditions for the first time. The experimental results for the counting efficiency of particles down to 2 nm in diameter show that the MTCNC can measure particles at low pressures above 80 Torr by adjusting the aerosol flow rate, amount of vapor and temperature depending on the desired operation pressure. At pressures above 80 Torr, the optimal operation conditions for the MTCNC for variables such as the saturator temperature, condenser temperature and mixing ratio were close to those under atmospheric pressure. Since the condensational growth of particles depends greatly upon the residence time, the current MTCNC needs improvement to increase the growth time under even lower-pressure conditions.

INTRODUCTION

A number of different condensation nucleus counters (CNCs) have been developed and used to measure the number concentration of aerosol particles at standard atmospheric pressure conditions. Recently, it has become important to be able to use the CNC to detect aerosol particles under the low-pressure conditions found in semiconductor preparation processes and in the stratosphere. This study reports on the first application of a mixing type condensation nucleus counter (MTCNC) to operation under such low-pressure conditions. The effectiveness of the low-pressure mixing type CNC (LPMTCNC) was evaluated by measuring the counting efficiency for various mixing conditions, vapor temperatures and operating pressures.

MIXING TYPE CONDENSATION NUCLEUS COUNTER

Figure 1 shows the MTCNC tested, which was developed by Kousaka et al. [1,2] and commercialized by Kanomax Japan, Inc., Japan. Particle-free air flows through the heated, thermostatically controlled saturator where it becomes saturated with vapor from a liquid reservoir. The aerosol is drawn into the MTCNC and is mixed with the warm saturated vapor-air mixture. Turbulent mixing of the aerosol flow with the saturated air is achieved by passing the aerosol through very small holes in the mixing zone. Condensational growth of particles takes place primarily in the mixing zone and the ensuing pipe, whose temperature is controlled by cooling with an electric refrigeration element. Changing the saturator temperature while keeping the mixing ratio constant can control the supersaturation in the mixing zone. In these experiments, ethylene glycol was used as the condensable vapor. The supersaturated ethylene glycol vapor condenses onto the surface of the aerosol particles by heterogeneous condensation and the particles grow to optically detectable droplets. A diode laser irradiates the grown droplets, and the frequency of pulses produced by the scattered light is counted to obtain the number of particles flowing per unit time. In this counting region of the MTCNC, the aerosol droplets are focused along to centerline by surrounding the droplet flow with a sheath gas stream.

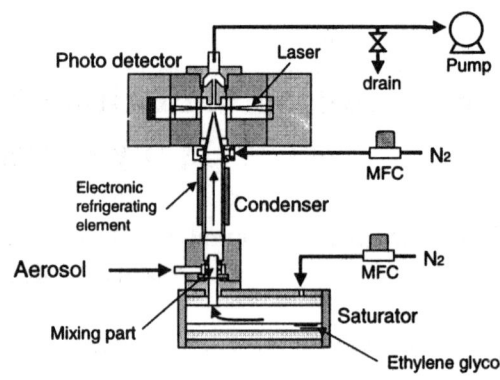

FIGURE 1. Diagram of mixing type CNC (MTCNC).

EXPERIMENTAL APPARATUS AND METHOD

The experimental setup is shown in Fig. 2. NaCl aerosol particles are generated by an evaporation-condensation type generator and acquire a bipolar charge in an Am-241 α-ray source. Then particles are classified to a specific particle diameter by a differential mobility analyzer (DMA). The number concentration of the particles is adjusted to 10^2-10^3#/cm^3 using a diluter, and the particle stream is directed to a Faraday cup electrometer (FCE) to determine the absolute particle number

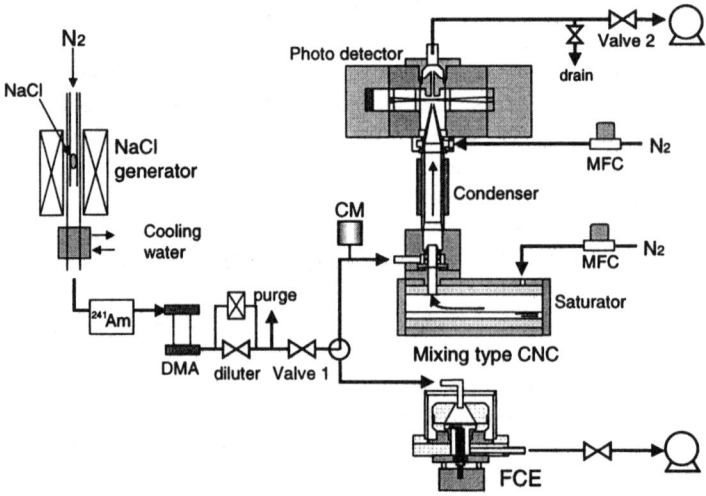

FIGURE 2. Schematic diagram of experimental setup.

concentration. This number concentration is compared to that obtained by the MTCNC to determine the counting efficiency of the instrument. The aerosol flow rate and pressure are controlled by valve 1 and valve 2 shown in Fig. 2. To characterize instrument performance, we measured how the counting efficiency changes as a function of the operating conditions of the MTCNC by adjusting the saturator temperature, the condenser temperature, the aerosol flow rate, and the mixing ratio.

EXPERIMENTAL RESULTS AND DISCUSSION

The experiment results of the counting efficiency under various operating pressures and aerosol flow rates are shown in Fig. 3. The aerosol flow rate was equal to the mass flow rate of aerosol supplied to the MTCNC at atmospheric pressure. The counting efficiency is defined as the ratio of the number concentration measured by the MTCNC to that measured by the FCE. When aerosol flow rate was 300 scm^3/min, the counting efficiency gradually falls off with decreasing pressure. Below 150 Torr, the counting efficiency falls off much more suddenly. This result means that sufficient condensational growth did not occur due to the reduction of the residence time inside the MTCNC caused by the decreased pressure. For an aerosol flow rate of 100 and 200 scm^3/min, the counting efficiency goes through a maximum as a function of pressure. This might be explained by poor mixing of the vapor with the aerosol due to the low aerosol flow rate and the deposition of particles onto the wall by Brownian diffusion. However, it can be concluded that the LPMTCNC can measure aerosol particles at low pressures down to about 80 Torr by changing the aerosol flow rate to the proper value for the desired pressure range.

Figure 4 shows the change of counting efficiency as a function of particle diameter at two low-pressure conditions. The counting efficiency was equal to unity for particles larger than 5 nm when the aerosol flow rate was 200 scm^3/min and the operating pressure was 200 Torr. However, when the aerosol flow rate was 300 scm^3/min and operating pressure was 133 Torr, the counting efficiency decreased from unity due to a low supersaturation ratio and an insufficient residence time for particle growth.

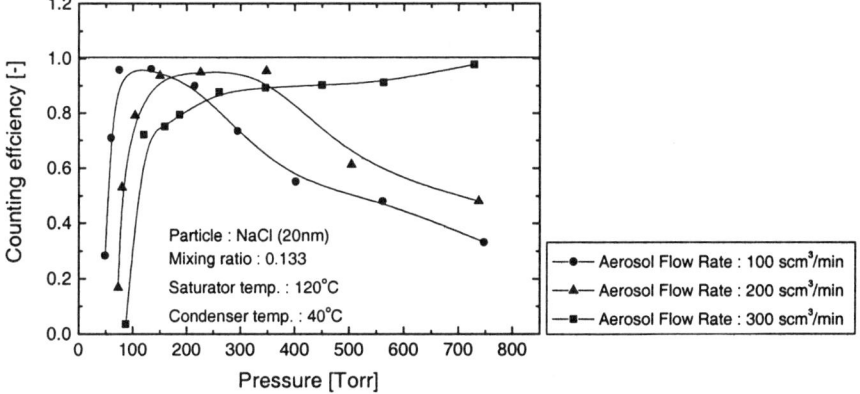

FIGURE 3. Change of counting efficiency with pressure and aerosol flow rate.

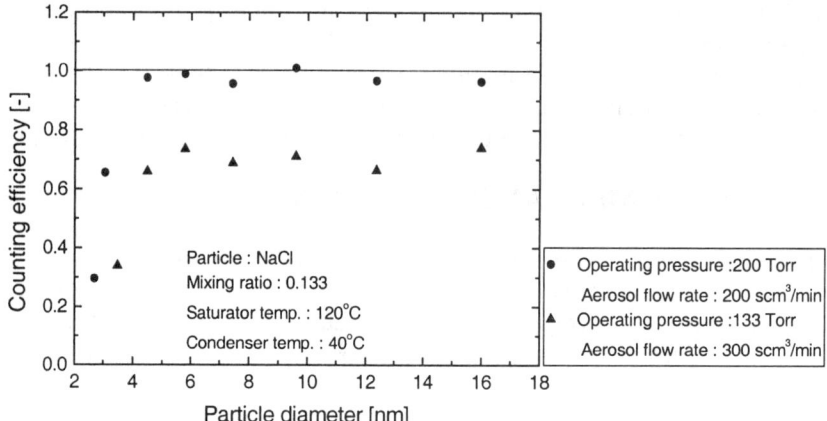

FIGURE 4. Change of counting efficiency with particle diameter.

SUMMARY

This work presents the first application of a mixing type condensation nucleus counter (MTCNC) to low-pressure conditions. At low pressure, the counting efficiency tends to drop because there is insufficient residence time for particle growth by condensation. However, the current MTCNC can measure particles larger than 5nm at pressures above 80 Torr by choosing the optimal aerosol flow rate for each operation pressure. Table 1 shows the experimentally determined optimal conditions for the use of the LPMTCNC.

TABLE 1. Optimal condition of mixing type CNC.

	Operation pressure	Optimal condition
Saturator Temp.	760-80 Torr	120 °C
Condenser Temp.	760-80 Torr	40 °C
Mixing Ratio	760-80 Torr	0.133
	760-350 Torr	300 scm^3/min
Aerosol Flow Rate	350-180 Torr	200 scm^3/min
	180- 80 Torr	100 scm^3/min

ACKNOWLEDGMENT

Part of this study was supported by a Grant-in-Aid from Semiconductor Technology Academic Research Center (STARC).

REFERENCES

1. Kousaka, Y., Okuyama, K., Niida, T., Hosokawa, T., and Mimura, T., *J. Aerosol Sci.* **13**, 231-240 (1982).
2. Okuyama, K., Kousaka, Y., and Motouchi, T., *Aerosol Sci. Technol.* **7**, 353-366 (1984).

Is There Aerosol/Cloud Layer Near Global Tropopause?

G. C. Asnani and M. K. Rama Varma Raja

Indian Institute of Tropical Meteorology, Dr. Homi Bhabha Road, Pashan, Pune-411008, Ph: 91-020-5880347, Fax:91-020-5890347,

Abstract. During recent years, observational evidence has become available to show that just below the tropopause, within a couple of kilometers inside the troposphere, there is abundance of water vapour as well as of aerosols. These constitute material, which is favourable for the formation of visible/subvisible cloud layers or haze layers in the presence of very low temperatures which prevail in this region. Evidence obtained from Indian MST Radar data and from published literature is produced to support the hypothesis that throughout the world, the troposphere is capped by haze /cloud layer.

Important Evidence for global Aerosol Layer near Tropopause:

1. MST Radar is a powerful tool for probing the atmosphere upto very great heights including the whole troposphere, stratosphere and even mesosphere. One important feature of reflectivity pattern in MST Radar data has been the layered structure of reflectivity at various levels in the troposphere and even at higher levels. In the troposphere, there is a bunch of layers of high MST Radar reflectivity. One of these layers of high reflectivity is always near the tropopause which is near 17 km in the tropics, decreasing to 10 km or less in the extra-tropics and in polar regions. Fig.1 is an example of such high reflectivity layers.

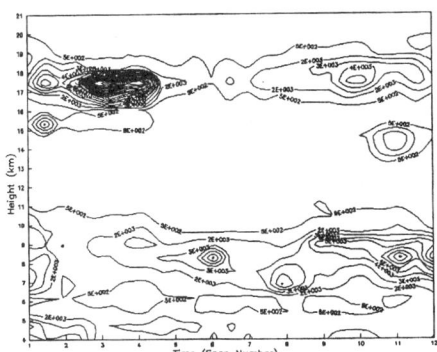

2. It has been customary to dispose of these layers of high MST Radar reflectivity as layers of "mechanical turbulence", even clear air turbulence or thermo-dynamically

"stable stratification". This customary concept has recently encountered observational and theoretical objections (Cho 1998). If these are thermo-dynamically stable layers, then mechanical turbulence, is unlikely in any layer as a whole. When the layers of high MST Radar reflectivity are 1-2 km thick, as is observed almost on all occasions, then we have to look for a physical mechanism other than mechanical turbulence, which creates smaller-scale patches a few meters thick, and having alternatively high and low static stability, with strong temperature gradients in the vertical, alternately positive and negative. Looking at the available observational data, we have come to the conclusion that we have to give up the traditional idea that the high reflectivity layers seen in the MST Radar data are due to mechanical turbulence.

3. Using 2,50,000 MST Radar observations taken at Gadanki (13.47° N, 79.18°E), India, throughout 14 months, September 95 - November 96, we plotted scatter diagrams of reflectivity versus vertical wind shear. The very unconventional result is that reflectivity decreases almost exponentially with increase in vertical wind shear; it is verified that Richardson number will be less than 1/4 when vertical wind shear is more than 25 ms-1/km (Rama Varma Raja, 1999). These scatter diagrams leave us in no doubt that mechanical turbulence is not the main cause of high MST Radar reflectivity.

4. If mechanical turbulence is ruled out, then what is the cause of high reflectivity? We have come to the following conclusions:
(i) 1-2 km thick layers of high reflectivity are the layers of visible/subvisible clouds or haze layers. They have relatively high humidity and high aerosol content. Most of the conventional humidity sensors are not sufficiently sensitive to detect patches of high and low relative humidity inside these 1-2 km thick layers.
(ii) MST Radar observations show alternating layers of vertical upward and downward motion along the vertical. The vertical wave-length of this vertical motion is generally of the order of 5 km, i.e. 2.5 km of upward vertical motion and 2.5 km of downward motion. Upward vertical motion causes higher vertical lapse rate (decrease of static stability) and higher relative humidity. These vertical motions with alternating signs in the vertical create vertical convergence and divergence of moisture, aerosol content, momentum and kinetic energy. Lower temperatures, higher humidity and higher aerosol content are favourable for formation of visible/subvisible cloud /haze layers. These alternating vertical motions are attributed to inertio-gravity waves in the atmosphere. We call these 1-2 km thick layers of high reflectivity as "Mother Cloud
Layers"(Rama Varma Raja et al. 1999).
(iii) Inside these Mother Cloud Layers, there are microphysical and microdynamical processes associated with visible/subvisible hydrometeors or aerosol ensembles. Hydrometeors being heavier than air, tend to descend inside the Mother Cloud layer. In their vertical descent, these hydrometeors exchange heat and water vapour with their environment, causing sharp vertical gradients of temperature, humidity and density including these due to hydrometeor loading. Brunt-Vaisala oscillations and Kelvin-Helmholtz waves also get

generated which also create vertical gradients of temperature and humidity. These vertical gradients and discontinuities in temperature, humidity and density, and also possibly electrical field associated with change of phase of water substance create large vertical gradients in refractive index of cloud air. These vertical gradients and discontinuities in refractive index cause high reflectivity with respect to the vertical beam of MST Radar from the ground. The reflection/scatter of the radar beam is partly due to Rayleigh type scattering but is considerably due to the Bragg type scattering. It is this Bragg type scatter which is the source of high reflectivity observed in the MST Radar data.

5. Aircraft observations with very sensitive sensors (strom et al 1997), lidar observations (Nee et al. 1998), satellite observations (Newell et al. 1997), and MST Radar observations elsewhere in the world (Ram Varma Raja et al. 1999, Nastrom et al. 1998, Cifelli et al. 1994) confirm that there are inertio-gravity waves; there are layers of high and low relative humidity and vertical temperature lapse rates which tend to create Mother Layers 1-2 km thick, with vertical spacing of the order of 5 km, favourable for formation of visible/subvisibe clouds. Within these mother layers, there are thinner patches of very steep vertical gradients of temperature, humidity and hydrometeor content, which are clearly very favourable for giving high MST Radar reflectivity of Bragg type; that 1-2 km thick layer just below the tropopause is always favourable for this type of reflection/scatter of MST Radar beam; the micro-physical processes create vertical and horizontal gradients of refractive index, having a large spectrum of horizontal and vertical wave-lengths to such an extent that one can mistake it as due to clear air turbulence. Then, we would be missing the micro-physical and micro-dynamical processes connected with hydrometeors inside the Mother layers.

Large-scale stable vertical lapse rates of temperature near the tropopause by themselves are not capable of creating mechanical turbulence; but high water vapour content, high aerosol content and very low temperatures are favourable for formation of visible or sub-visible hydrometeors or haze particles creating Bragg-type reflection/scatter of MST Radar beam.

Analytical Prediction of Homogeneous Nucleation in Rapidly Expanding Pure Vapours

J.B.Young[1] and L.Huang[2]

1. Cambridge University Engineering Department, Cambridge, CB2 1PZ, UK.
jby@eng.cam.ac.uk
2. Dept. of Mech Eng., Hong Kong Polytechnic University, Hunghom, Kowloon, Hong Kong.
mm1huang@polyu.edu.hk

Abstract. The paper describes an analytical method for predicting the maximum vapour subcooling and final droplet number distribution in a rapidly expanding pure vapour undergoing homogeneous nucleation. This type of flow is characterised by very high nucleation rates which are quenched by the release of latent heat as the droplets grow rapidly to macroscopic sizes. Numerical calculation methods are normally used to predict the conditions at maximum subcooling but these are computationally time consuming and difficult to implement in a multi-dimensional flow calculation. By introducing certain assumptions capitalising on the short time-scale of the nucleation process, the paper shows that it is possible to obtain an elegant analytical solution for any prescribed pressure distribution. The analysis provides physical insight into flow condensation processes and incorporation into CFD procedures should remove the need for very fine meshes and excessive computing resources.

Introduction

Condensation by homogeneous nucleation in rapidly expanding pure vapours is of practical importance in engineering devices such as nozzles and steam turbines. The process is characterised by high subcoolings and nucleation rates which are quenched by the latent heat released by the nucleated droplets as they grow to macroscopic size. Once the nucleation process is complete, the droplet number distribution associated with a particular fluid particle (in the Lagrangian sense) remains constant (in the absence of agglomeration) and hence defines the size of the droplets present in the downstream flow. The accurate prediction of this size distribution is of the greatest importance. In a steam turbine, for example, the droplet deposition rate onto the blades (leading to the formation of coarse water and, ultimately, blade erosion) is a strong function of the size of the fog droplets formed earlier by homogeneous nucleation.

Conventional Calculation Procedures

The calculation of the droplet number distribution is normally performed numerically. Quite apart from uncertainties in the expression used for the nucleation rate, accurate predictions require detailed tracking of the growth of large numbers of droplet groups (typically 50) through the short period of intense nucleation. Such calculations are feasible for one-dimensional nozzle flows but are computationally cumbersome and inelegant for multi-dimensional flow calculations.

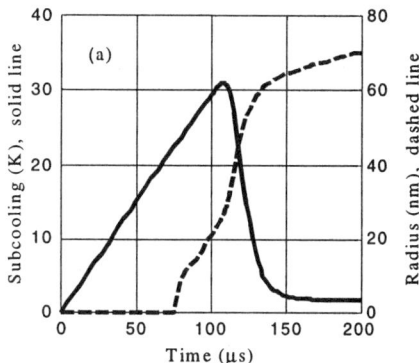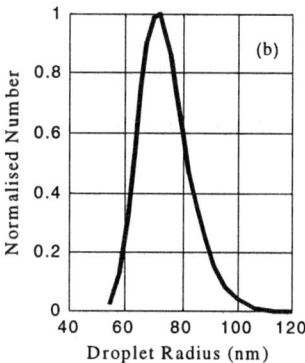

FIGURE 1. Numerical calculation of homogeneous condensation of steam expanding at a constant rate of $k_p = 5000$ s^{-1} with Wilson point pressure of 0.45 bar. (a) Subcooling (K) and Sauter mean droplet radius (nm). (b) Normalised droplet number distribution at end of expansion.

Figure 1 shows the results of a typical calculation of the condensation of pure steam expanding from a dry saturated state at 1 bar pressure with a constant expansion rate of $k_p = 5000$ s^{-1}. [$k_p = -d(\ln p)/dt$ with p being the pressure at Lagrangian time t.] The period of intense nucleation near maximum subcooling is divided into a large number of time increments. Within each increment, the nucleation rate I (which can be calculated from any prescription deemed suitable) is assumed to remain constant. Droplets formed in this increment then grow (subject to a prescribed growth law) at a rate dependent on the local subcooling. The total wetness and latent heat release is obtained by summing over all droplet groups. The heat release eventually checks the increase of the subcooling, the flow returns towards equilibrium and nucleation ceases. The calculation scheme was invented by K. Oswatitsch in 1941 and has been used in numerous studies in much the same form ever since. It is the purpose of this paper to provide an analytical theory to make such numerical calculations redundant.

Outline of the Theory

Full details of the mathematical analysis can be found in Huang and Young [1]. In that paper it is shown that the nub of the problem is the prediction of the maximum subcooling at the so-called 'Wilson point'. Once conditions at the Wilson point are known, all other physical quantities follow, including the droplet number distribution.

The vapour subcooling is defined by $\Delta T = T_s - T$, where T_s is the saturation temperature at the local pressure p and T is the actual vapour temperature. ΔT is expanded as a Taylor series about the Wilson point, truncated at the second-order term. The first derivative of ΔT vanishes by definition and hence,

$$\Delta T(t) \cong \Delta T_w + \frac{1}{2}\Delta \ddot{T}_w (t-t_w)^2 \qquad (1)$$

where the dot notation denotes differentiation with respect to Lagrangian time. The fluid dynamic equations for the conservation of momentum and energy are now combined with the Clapeyron equation to give an equation for the time rate of change of subcooling in terms of the local liquid mass fraction Y and expansion rate k_p,

$$\frac{\Delta \dot{T}}{T_s} + L\dot{Y} - \frac{\gamma-1}{\gamma}\left(\frac{L-1}{L}\right)k_p \cong 0 \qquad (2)$$

where $L = h_{fg}/c_p T_s$ and $\gamma = c_p/c_v$. Equation (2) involves only minor approximations and, being in Lagrangian form, is quite general. Writing B for the coefficient of k_p, differentiating with respect to time (assuming L and B are constant through the nucleation period) and specialising to the Wilson point results in the two equations,

$$\dot{S}_w = \frac{\Delta \dot{T}_w}{T_{sw}} = B_w k_{pw} - L_w \dot{Y}_w = 0 \qquad (3)$$

$$\ddot{S}_w = \frac{\Delta \ddot{T}_w}{T_{sw}} = B_w \dot{k}_{pw} - L_w \ddot{Y}_w \qquad (4)$$

Expressions for the Y derivatives are found from the nucleation and droplet growth rate equations. These are expressed in the forms,

$$I = I_0 \exp\left(-\frac{\Theta^2}{S^2}\right) \qquad \dot{r} = C_r S \qquad (5)$$

where r is the radius of a droplet. Irrespective of the prescription adopted, I_0, Θ and C_r are all weak functions of T and remain almost constant during the intense nucleation phase. The equation for the droplet growth rate neglects the effect of high surface curvature and this is probably the most serious approximation in the whole theory. Substituting equation (1) into the expression for I gives,

$$I = I_w \exp\left(-\frac{(t_w-t)^2}{\tau_n^2}\right) \quad \text{with} \quad \tau_n^2 = \frac{-S_w^3}{\Theta_w \ddot{S}_w} \qquad (6)$$

τ_n is the time during which I changes by a factor of e from its value at the Wilson point. The wetness fraction at any time t is the sum of contributions from all droplets nucleated upstream (starting from $t = t_s$ when the saturation line is crossed). Thus,

$$Y(t) = \frac{4}{3}\pi \rho_f \int_{t_s}^{t} I(t_n) r^3(t,t_n) dt_n \qquad (7)$$

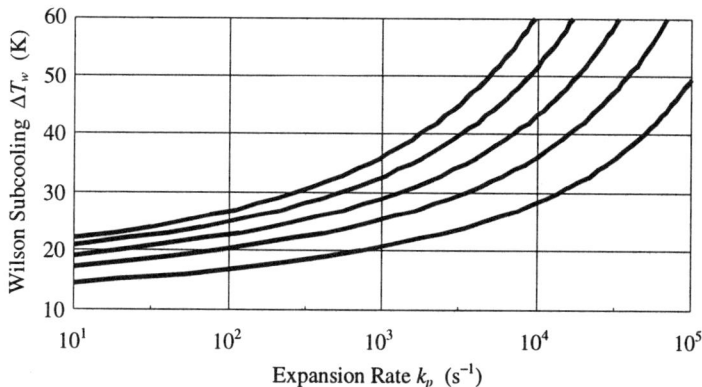

FIGURE 2. Variation of maximum subcooling ΔT_w with expansion rate k_p for steam with Wilson point pressures $p_w = 0.1$ bar (top curve), 0.3 bar, 1.0 bar, 3.0 bar, 10.0 bar (bottom curve).

where ρ_f is the liquid density and $r(t,t_n)$ is the radius, at time t, of a droplet nucleated at time t_n. Introducing equations (5) and (6) into (7) and performing the integrations gives, after some manipulation,

$$\dot{Y}_w = \pi^{3/2} \rho_f I_w (C_r S_w)^3 \tau_n^3 \tag{8}$$

$$\ddot{Y}_w = 4\pi \rho_f I_w (C_r S_w)^3 \tau_n^2 \tag{9}$$

If k_p and its time derivative are specified, equations (3), (4), (8) and (9) can be solved algebraically to give $S_w = \Delta T_w/T_{sw}$ the dimensionless subcooling at the Wilson point. All other variables can then be determined, including the droplet number distribution. For the special case when k_p is constant, a closed form solution is obtained,

$$\left(\frac{S_w^2}{\Theta_w}\right)^6 \exp\left(-\frac{\Theta_w}{S_w^2}\right) = \frac{64 B_w^4}{\pi^3 L_w \rho_f I_0 C_r^3 \Theta_w^3} (-k_p)^4 \tag{10}$$

Figure 2 shows the variation of maximum subcooling with expansion rate for steam at various Wilson point pressures, calculated from equation (10). Equivalent numerical calculations indicate that the accuracy is good. Not only does the analytical method provide physical insight into the flow condensation process but it is also suitable for inclusion in CFD procedures where the pressure field is automatically generated as part of the solution.

REFERENCES

1. Huang, L. and Young, J.B., *Proc. R. Soc. Lond.* A, **452**, 1459-1473 (1996).

Homogeneous Nucleation and Shock Wave Interaction in Condensing Steam Flows

A.J.White[1] and J.B.Young[2]

1. University of Durham, School of Engineering, South Road, Durham, DH1 3LE, UK.
   Alexander.White@durham.ac.uk
2. Cambridge University Engineering Department, Trumpington St., Cambridge, CB2 1PZ, UK.
   jby@eng.cam.ac.uk

Abstract. It is well known that the release of latent heat by a homogeneously nucleating steam flow can have a strong effect on the flow dynamics and this is particularly true if the flow is transonic or supersonic with shock waves present. Most research has been aimed at calculating the overall flow behaviour but equally important is the effect of the shocks on the nucleation process itself and the resulting droplet size distribution. This paper explores the shock-nucleation interaction process in a pure steam flow and shows how the downstream droplet size distribution can be affected by the presence of shocks in both steady and unsteady flow.

Introduction

There have been many studies of homogeneous flow condensation both in pure steam [1] and moist air [2], and the influence of the condensation process on the flow dynamics in nozzles and turbine cascades is now well documented. The effects are particularly marked in transonic flows where the interaction between a shock wave and a nucleation zone can give rise to various periodically unsteady flow patterns, both symmetrical and asymmetrical [2]. Most research has been directed at predicting the overall flow behaviour but the influence of a shock on droplet nucleation is also important because it can dramatically affect the downstream droplet size distribution. One practical application is in condensing steam turbines where the entropy creation due to condensation, the deposition rate of droplets on the blades, and the predisposition for secondary nucleations all depend on the size distribution of droplets formed upstream by homogeneous nucleation.

Shock-Nucleation Interaction in Steady Flow

When pure steam, initially slightly superheated, expands in a nozzle it becomes subcooled. As the subcooling increases, the homogeneous nucleation rate rises resulting in the production of droplets which grow by vapour condensation to macroscopic dimensions. The latent heat released by the growing droplets eventually checks the increase of subcooling, the two-phase flow returns towards thermodynamic equilibrium and nucleation ceases. The mean droplet radius downstream of the nucleation zone depends on the local steam pressure and rate of expansion and lies in the range 0.05 – 0.5 µm in most practical applications.

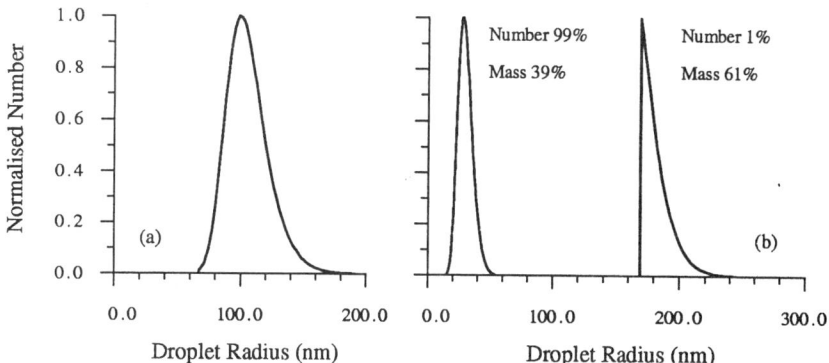

FIGURE 1. Droplet size distributions produced by homogeneous nucleation in a rapidly expanding steam flow. (a) Downstream of a 'naturally terminated' nucleation ($p_w = 0.5$ bar, $k_p = 1000$ s^{-1}). (b) Bimodal distribution resulting from a 'shock terminated' primary nucleation followed by a secondary 'naturally terminated' nucleation. (Each mode of the distribution is individually normalised for clarity.)

The standard numerical method for calculating the droplet size distribution resulting from such a 'naturally terminated' nucleation is described in [3]. Figure 1(a) shows a typical droplet number distribution calculated in this way using classical nucleation theory with the self-consistent and non-isothermal corrections. The pressure p_w at maximum subcooling (the Wilson point) was 0.5 bar and the rate of expansion ($k_p = -d(\ln p)/dt$) was 1000 s^{-1}. The mean droplet diameter produced in a nozzle flow of this type can be measured optically and the few measurements available give reasonable agreement with the calculations. In principle, the size distribution can also be obtained optically but, with such small droplets, the accuracy is poor and the theoretical results have not yet been validated in this respect.

Under certain flow conditions, a shock wave may intersect the region of intense nucleation and terminate the production of droplets by its temperature increase. (Shock thicknesses are of the order of a few mean free paths and droplets traverse the wave in a few nanoseconds. This is much shorter than the time required for significant heat release by condensation which is of the order of microseconds.) Downstream of the wave, there are a number of possibilities depending on the shock strength and position in the nucleation zone, and the rate of expansion in the downstream region.

If the shock is sufficiently strong that the vapour becomes superheated, all the droplets evaporate in the downstream region. For weaker shocks, however, the vapour may remain subcooled. If 'shock termination' of nucleation is such as to leave a substantial population of droplets in the flow and the subsequent rate of expansion is comparatively low, droplet growth by condensation may be sufficient to inhibit further increase in subcooling with the result that the flow reverts towards thermodynamic equilibrium without further droplet production. However, although the mass fraction of liquid is then approximately the same as without the shock present, there will be fewer droplets per unit mass of steam than if nucleation had terminated naturally and the mean droplet diameter may be substantially larger.

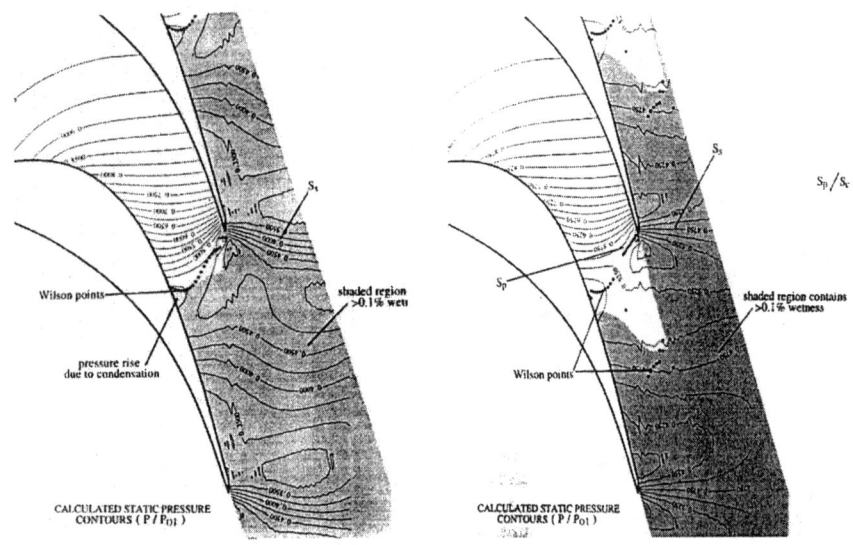

FIGURE 2. Calculations of homogeneous nucleation in a steam turbine cascade. Left diagram shows 'naturally terminated' nucleation. Right diagram shows 'shock terminated' nucleation in the upper section of the flow passage with completion further downstream.

For a given liquid mass fraction, the mean droplet diameter is inversely proportional to the cube root of the total droplet number. Thus, if termination occurs at a point where the nucleation rate is, say, 1000 times smaller than the maximum rate achieved with natural termination, the mean droplet diameter will be larger by a factor of about 10. Also, the distribution will be sharply truncated at the cut-off point and will be strongly skewed like the primary group shown in Fig. 1(b).

Alternatively, if shock termination depletes the droplet population significantly and the rate of expansion in the downstream region is high, condensation on existing droplets may be insufficient to prevent an increase of subcooling with the result that a secondary homogeneous nucleation occurs. The overall droplet number distribution then becomes bimodal as shown by the example in Fig. 1(b).

Shock termination of nucleation in steady flow has been observed experimentally in a steam turbine cascade and is illustrated in Fig. 2. In the left diagram, the steam nucleates naturally, the computed Wilson points being shown by the line of dots just upstream of an oblique shock which originates at the trailing edge of the upper blade and extends across the passage. The path of the shock can be traced by careful study of the constant pressure contours but is much easier to see in the schlieren photograph of the original experiment, Fig. 10 of [1]. In the right diagram of Fig. 2, the pressure ratio across the cascade is (very) slightly reduced. In the upper section of the passage, shock termination of nucleation occurs and the condensation process is not completed until much further downstream. In the lower section nucleation is terminated naturally. This behaviour is confirmed in the experimental schlieren photograph, Fig. 11 of [1].

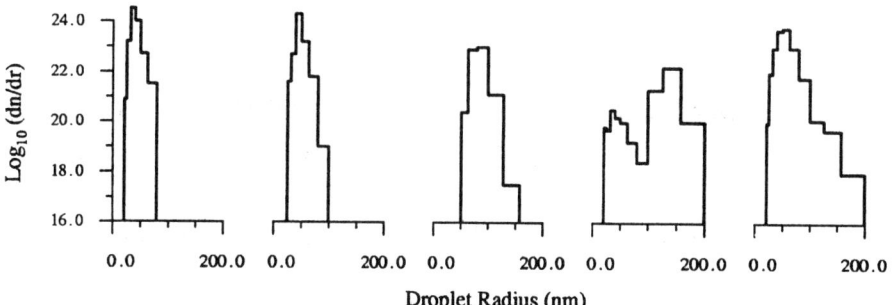

FIGURE 3. Computed instantaneous droplet number histograms downstream of a periodically unsteady condensation process (frequency 400 Hz) occurring in the throat region of a converging-diverging nozzle. The five distributions correspond to time $0T$, $0.2T$, $0.4T$, $0.6T$ and $0.8T$, where T is the period. The distribution becomes bimodal at $0.6T$, the larger droplets having nucleated upstream of the moving shock and the smaller ones resulting from a secondary nucleation downstream.

Unsteady Flow

The droplet size distributions associated with unsteady condensation processes with shock waves are even more complex. Consider, for example the periodically unsteady shock motion occurring in the throat region of a converging-diverging nozzle when the subcooling attains its maximum value at about the same position. A number of different types of unsteady shock patterns have been observed [2] but the simplest involves a single shock forming just downstream of the throat, growing in strength as it moves upstream and then decaying as it enters the converging section. This process is repeated periodically. During its upstream travel, the shock completely traverses the nucleation zone with the result that all conditions ranging from natural termination to complete extinction of nucleation are realised. An observer stationed downstream at the nozzle outlet would experience a time varying droplet number distribution which would be repeated periodically at the same frequency as the shock motion. The results of a calculation for a typical periodically-unsteady nozzle condensation process are shown in Fig. 3. The time-averaged droplet size distribution (of which only the mean value can currently be measured) is clearly much broader than the distribution produced under steady flow conditions. Further details can be found in [4].

REFERENCES

1. White, A.J. and Young, J.B., *Phil. Trans. R. Soc. Lond.* A, **354**, 59-88 (1996).
2. Adam, S. and Schnerr, G.H., *J. Fluid Mech.*, **348**, 1-28 (1997).
3. Huang, L. and Young, J.B., "Analytical Prediction of Homogeneous Nucleation in Rapidly Expanding Pure Vapours" in *Nucleation and Atmospheric Aerosols 2000*, edited by B.Hale and M.Kulmala, AIP Conference Proceedings, New York: American Institute of Physics, 2000.
4. White, A.J. and Young, J.B., *AIAA J. Propulsion and Power*, **9**, 579-587 (1993).

Stable Sulfate Clusters as a Source of New Atmospheric Particles

M. Kulmala, L. Pirjola, J.M. Mäkelä and C.D. O'Dowd

Department of Physics, University of Helsinki
P.O. Box 9, Helsinki, Finland

Abstract. The importance of atmsopheric aerosols to the global radiation balance, to cloud formation, and to alleged human health effects has motivated us to investigate the formation and growth of atmospheric aerosol particles. Formation of new atmospheric particles (diameter between 3 and 10 nm) has been observed in a wide variety of low and high altitude locations. These aerosol particles are a source of Aitken mode nuclei and, after growing, one of the major sources of atmospheric cloud condensation nuclei (CCN) thus affecting cloud formation and global radiation balance. While some studies show that the measured new particle formation rates could be explained by binary sulphuric acid – water nucleation, others, particularly those in the marine boundary layer, or over the continental sites, indicate that observed ambient nucleation rates exceed those predicted by the binary scheme. The observed ambient H_2SO_4 concentrations are much less than needed for binary nucleation but appropriate for ternary nucleation (water – sulphuric acid – ammonia). Using a ternary nucleation modeling scheme with aerosol dynamics, nucleation is predicted to be ubiquitous in the troposphere yielding a reservoir of thermodynamically stable 1-3 nm clusters (TSC). The most likely pathway for CCN formation thus involves growth of these clusters up to 30-100 nm sizes via condensation of condensable vapours

INTRODUCTION

Formation and growth of atmospheric aerosols have been studied experimentally and theoretically [1]. Bursts of recently formed particles have been observed in several regions around the world, for example, in the free troposphere [2], in the marine boundary layer [3], at coastal sites [4], in the vicinity of evaporating clouds [5], in Arctic areas [6], in urban areas and in stack plumes [7], and in boreal forests [8,9]. Several nucleation mechanisms have been proposed to explain this particle production, along with meteorological-related nucleation enhancement processes such as turbulent fluctuations, waves and mixing [10,11].

Typically, the formation of atmospheric aerosols is attributed to binary nucleation of water and sulphuric acid [12], however, the binary theory is able to predict the nucleation rates only at some extreme conditions of low temperatures, high relative humidities, small pre-existing aerosol concentrations and at high sulphuric acid concentrations. The recently developed ternary nucleation model [13,14] of water – sulphuric acid – ammonia gives significantly higher nucleation rates and thus predicts nucleation under typical tropospheric sulphuric acid ($1 \cdot 10^5$-$1 \cdot 10^7$ cm^{-3}) [2] and ammonia (some ppt) concentrations. Although using different thermodynamical data (vapour pressures, surface tension) both models [13,14] are in reasonable agreement showing e.g. the key role of sulphuric acid molecules in the nucleation process and the effect of ammonia. Our recent model study shows how ternary nucleation together with subsequent condensation of other vapours is able to give one possible explanation on atmospheric particle formation [1]. Model calculations illustrate that ternary nucleation happens very easily in daytime, thus, raising the question: Why do we not observe new particle production as often in the atmosphere?

THE HYPOTHESIS

We have recently presented the following hypothesis [1], which – in principle – enables us to explain all observed particle production in the atmosphere. (1) In the atmosphere nucleation is occurring almost everywhere at least in daytime. The conditions in the free troposphere (cooler temperature, less pre-existing aerosols) will even favour nucleation more. (2) Nucleation is maintaining a reservoir of thermodynamically stable clusters (TSC) which are too small to be detected. (3) Under certain conditions TSCs grow to detectable sizes and further to CCN. The phenomena incolved are described in schematig picture (Figure 1).

Theoretically predictable pathways of atmospheric nucleation are at least ion induced [15] and ternary $H_2SO_4 - NH_3 - H_2O$ [13,14] mechanisms. The newly formed particles, whose sizes are greater than a critical cluster (about 1 nm) but smaller than 3 nm, are called thermodynamically stable clusters (TSC). There are two possibilities for TSCs growth to detectable size particles (>3 nm in diameter). First, if the concentration of pre-existing aerosols is low due to precipitation or some other reason, self-coagulation of TSCs becomes significant and the growth to 3 nm size only takes in the order of an hour. Secondly, if there is a high source of available condensable vapours (such as organics, inorganic acids and ammonia) growth through condensation to detectable sizes and even to Aitken mode size occurs over a time scale of about 1–2 hour [1].

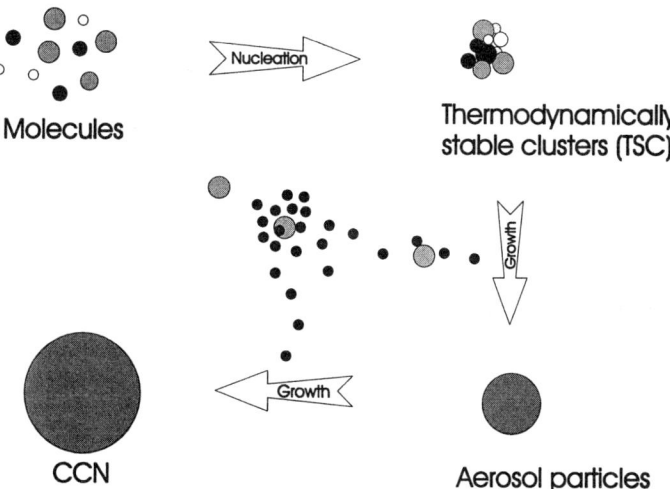

FIGURE 1. The schematic diagram. Thermodynamically stable clusters (TSC) are formed by homogeneous or ion induced multicomponent nucleation. Then they grow to detectable aerosol particles and further to CCN size by coagulation and condensation of available vapours.

RESULTS AND DISCUSSION

The hypothesis is based on theoretical calculations and tested by model simulations [1]. More than 100 model runs are performed using a sectional aerosol dynamic model AEROFOR [16] under clear sky conditions, and the modeled results are compared with several experimental ones [4,9]. We have simulated [1] the particle number size distribution, total particle number concentration (Ntot) and the concentration of particles with diameters greater than 3 nm (N3) and greater than 10 nm (N10) during 7 days period using a) constant nucleation rate (like ion induced nucleation with any gases), and b) ternary nucleation. Our simulations show the existence of TSCs and a steady state will occur after some days. In the first case, the total number concentration increases significantly and the steady state value is achieved within 20 minutes. However, the amount of detectable aerosols (N_3) remains small [1]. This means that in practise only few of the newly formed particles (TSCs) grow and therefore, a reservoir of 1×10^5 TSCs per cm^3 exists with the nucleation rate 100 cm^{-3} s^{-1}. However, the reservoir depends on the nucleation rate (see Figure 2).

We have tested how well the hypothesis is able to explain experimental observations of atmospheric particle formation and growth under forest conditions at the Hyytiälä measurement station, Finland [8,9] and under coastal conditions at Mace Head, Ireland [4]. In both cases the agreement between theoretical values and experimental observations has been found to be good.

As a conclusion [1] we can say that the TSCs are able to explain observed aerosol formation bursts in the atmosphere. The occurrence of bursts is connected to the activation of TSCs with extra condensable vapours (e.g. after advection over sources or mixing). On the other hand, a small pre-existing aerosol concentration favours TSC production and subsequent self-coagulation of TSCs is able to grow TSCs to be detected. To explain the observed number concentrations the nucleation rate of 50-100 cm^{-3} s^{-1} is needed. Since the ionization rate under typical tropospheric conditions is in the order of 1-5 ion pairs/(cm^3s) [15] we can rule out ion induced nucleation as a probable aerosol formation route. If there exists pathways other than ternary nucleation of water-ammonia-sulphuric acid, they should occur as easily as the ternary one. However, additional condensable vapours will still be required to activate TSCs to detectable sizes and further to cloud condensation nuclei.

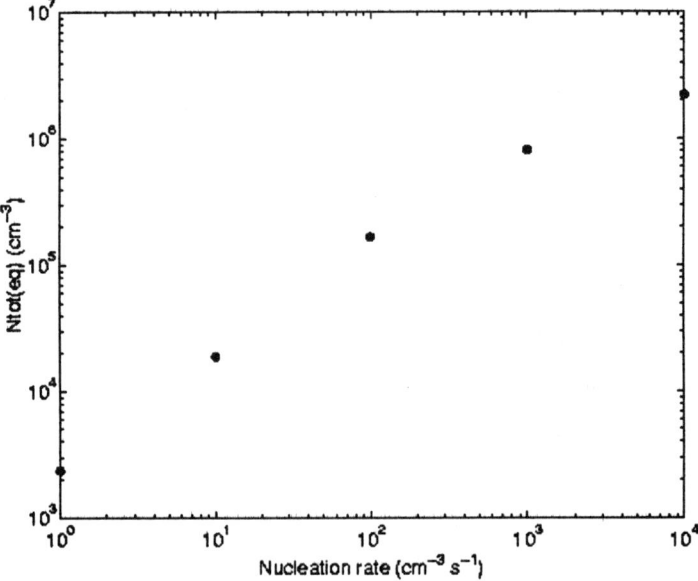

FIGURE 2. The steady state number concentration of thermodynamically stable clusters as a function of constant nucleation rate.

REFERENCES

1. Kulmala M., Pirjola L., Mäkelä J.M., Stable sulphate clusters as a source of new atmsopheric particles. Nature, 404, 66-69 (2000).
2. Weber R.J.. et al., (1999) New particle formation in the remote troposphere: A comparison of observations at various sites. *Geophys. Res. Lett.*, 26, 307-310 (1999).
3. Covert D.S., Kapustin V.N., Quinn P.K. and Bates T.S., New Particle Formation in the Marine Boundary Layer. J. Geophys. Res., 97, 20581-20589 (1992).
4. O'Dowd C., McFiggins G., Creasey D.J., Pirjola L., Hoell C., Smith M.H., Allan B.J., Plane J.M.C., Heard D.E., Lee J.D., Pilling M.J. and Kulmala M. On the photochemical production of new particles in the coastal boundary layer. Geophys. Res.Letters. 26, 1707-1710 (1999)
5. Clarke, A.D., Varner, J.L. *et al.* Particle production in the remote marine atmosphere: Cloud outflow and subsidence during ACE 1. *J. Geophys. Res.*, 103, 16397-16409 (1998).
6. Wiedensohler, A., Covert, D.S., Swietlicki, E., Aalto, P., Heintzenberg, J., and Leck, C. Occurrence of an ultrafine particle mode less than 20 nm in diameter in the marine boundary layer during Arctic summer and autumn. *Tellus*, 48B, 213-222. (1996)
7. Kerminen, V.-M., and Wexler, A.S. Post-fog nucleation of H_2SO_4-H_2O particles in smog, *Atmos. Environ.*, 28, 2399-2406 (1994).
8. Kulmala, M., A. Toivonen, J. M. Mäkelä and A. Laaksonen Analysis of the growth of nucleation mode particles observed in Boreal forest. *Tellus* 50B, 449-462. (1998).
9. Mäkelä J.M., Aalto P., Jokinen V., Pohja T., Nissinen A., Palmroth S., Markkanen T., Seitsonen K., Lihavainen H., Kulmala M., Observations of ultrafine aerosol particle formation and growth in boreal forest. Geophys. Res. Lett., 1219-1222 (1997).
10. Easter, R.C., and Peters, L.K. Binary homogeneous nucleation: Temperature and relative humidity fluctuations, nonlinearity, and aspects of new particle production in the atmosphere. *J. Applied Met.*, 33, 775-784 (1994).
11. Nilsson, E.D. and Kulmala, M. The potential for atmospheric mixing processes to enhance the binary nucleation rate. *J. Geophys. Res.*, 103, 1381-1389 (1998).
12. Kulmala, M., Laaksonen, A., and Pirjola, L. Parameterizations for sulphuric acid/water nucleation rates. *J. Geophys. Res.*, 103, 8301-8308 (1998).
13. KorhonenP., Kulmala M., Laaksonen A., Viisanen Y., McGraw R., Seinfeld J.H., Ternary nucleation of H_2SO_4, NH_3 and H_2O in the atmosphere. J. Geophys. Res., 104, 26349-26353 (1999).
14. Coffmann D.J. and Hegg. D.A., A Preliminary study of the effect of ammonia on particle nucleation in the marine boundary layer. J. Geophys. Res., 100, 7147-7160 (1995).
15. Raes, F. and Janssens, A. Ion-induced aerosol formation in a H_2O-H_2SO_4 system - 1. Extension of the classical theory and search for experimental evidence. *J. Aerosol Sci.*, 16, 217-227 (1985).
16. Pirjola, L. Effects of the increased UV radiation and biogenic VOC emissions on ultrafine aerosol formation, *J. Aerosol. Sci.*, 30, 355-367 (1999).

Characteristics of the Three Years Continuous Data on New Particle Formation Events Observed at a Boreal Forest Site

J.M.Mäkelä, M.Dal Maso, A.Laaksonen*, L.Pirjola, P.Keronen and M.Kulmala

Department of Physics, P.O. Box 9, Siltavuorenpenger 20 D, FIN-00014 University of Helsinki, Finland
* University of Kuopio, Department of Applied Physics POB 1627, 70211 Kuopio, Finland

Abstract. We have analysed 184 formation events of new atmospheric aerosol particles, observed at a boreal forest site in Hyytiälä, Southern Finland. Recognition, selection and classification of the formation events is based on continuous experimental size distribution data of submicron particles from a period of Jan 31st 1996 - Sept 18th 1999 (1327 days). The apparent particle formation rates vary in the range of 0.001-1 particles/(cm^3 s). The ultrafine particle growth rates vary in the range of 1-17 nm/h.

INTRODUCTION

Ultrafine particle formation has been observed in the atmosphere already for a few decades [1],[2]. Recently, high time resolution size distribution data have also been obtained from the new particle formation bursts [3]-[6]. Continuous aerosol size distribution measurements with a twin-DMPS instrument were started in Southern Finland at the Hyytiälä measurement site (61 51'N, 24 17'E) in January 1996 [3]. Since then, about 50-60 particle formation events have been observed annually. The fundamental micro-physical nucleation process occurring in these events remains unknown. Most often the occurrence of these events correlate with high levels of visible and UV-light as well as with vertical mixing of lower tropospheric air (height ≈ 0-1000 m). It is not yet clear whether the origin of the precursors is **i)** photochemistry of atmospheric gaseous compounds, **ii)** thermodynamics favouring nucleation and condensation processes due to mixing of air parcels with different temperature and humidity, **iii)** mixing of two or more species from different layers of air, **iv)** some of i-iii) combined [7],[8]. Moreover, the identity of the actual chemical compounds involved in nucleation and condensation growth of the particles is not known due to the small prevailing concentration of these species. A full interpretation of the data is difficult, since the compounds involved have not yet been experimentally identified, the quantities of matter being small. The compounds a) initiating the nucleation and b) providing the condensational growth may well be different, and their chemical identity may as well vary (horizontally and vertically) between different locations in the troposphere.

ANALYSIS

The appearance of a typical particle formation event recorded by a DMPS-measurement is shown in Figure 1. At around midday the newly formed particles enter the measurement range with initial sizes of 4-10 nm, and grow larger with a growth rate of few nanometers per hour, reaching 20-50 nm by the evening. This pattern is very distinctive, and can be seen almost throughout the year, most often on sunny days [3]. Most often the particles start to be observed at around 3-5 nm, which means that they have already grown for some time starting from their initial size. Note, that the evolution towards higher particle sizes seen in the particle size spectra during the particle formation process is interpreted as a particle growth process, assuming that the aerosol is quite homogeneous in a larger scale air mass. Based on the particle size distribution data we can categorise the measurement period into days during which some new particle formation occurred and into days with no significant particle formation (non-event days). From the particle formation days we can furthermore select the ones in which the duration of particle growth was sufficiently long to permit calculation of certain growth-related quantities. These days are termed event days. The events were divided into three separate classes with a quality number ranging from 1 to 3. The best ones, which showed a clear nucleation mode that was easily distinguishable until it had grown to the Aitken mode size, were classified as class 1. This classification is somewhat subjective, but usable. From the sets of DMPS-distributions we have estimated several features of the events such as 'apparent' particle formation rates (particles/cm^3/s) calculated from observed nucleation mode particle concentrations in the starts and in the end of the event and particle growth rate (nm/h) from the subsequent spectra in DMPS data. To describe the effective ability of the pre-existing aerosol particles to reduce the gas phase concentration of the condensable species, we have calculated the condensational sink CS, which is introduced by the equation for the condensable species molecular number concentration, N_a

$$\frac{dN_a}{dt} = -4\pi \cdot CS \cdot D \cdot (N_a - N_{a,a}) \text{ where } CS = \int_0^\infty r\beta_M n(r)dr = \sum_i \beta_i r_i N_i \quad (1) \& (2)$$

where D is the diffusion coefficient of the condensable species in the gas phase, and $N_{a,a}$ is the the molecular number concentration of the condensable species at particle surface and CS is the condensational sink. Here r_i is the radius of the i:th size class, N_i is the respective number concentration and β_i is given by Fuchs and Sutugin [9].

RESULTS

The number of observed events per year during the measurement period is shown in Table 1. The overall coverage of the DMPS-data is on the order of 97-98 %, except for 1998 being only 93 %.

Table 1. Occurrence of the events

Year	Class 1	Class 2	Class 3	Total
1996	14	18	11	43
1997	12	17	14	43
1998	13	11	14	38
1999	18	19	23	60
Total	57	65	62	184

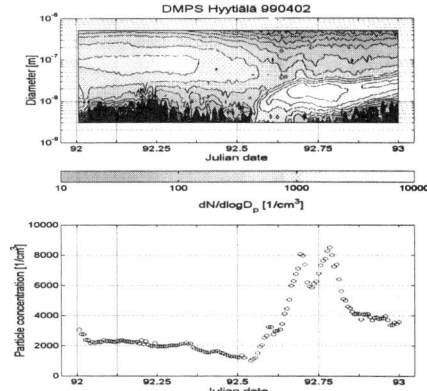

Figure 1. Typical particle formation event as a surface plot, particle size distribution in Hyytiälä April 2nd 1999.

A clear annual bimodal pattern is observed for the whole period, showing a larger peak in event number frequency in the spring (March-April-May) and a smaller one in the autumn (September-October). In December no events have been observed to have taken place. The particle formation bursts always take place during daytime. No more than a single event has been observed to occur per day. The events always start at least 2 hours after sunrise, and on the average 3-4 hours from sunrise (Figure 2). The particle formation period usually lasts for several hours. The duration has two annual peaks coinciding with the spring and autumn frequency maxima of the events. The average yield of new particles within one day's formation event has also two annual maxima. From the two quantities, the yield of new particles and duration of the particle formation period, one can derive the apparent new particle formation rate vs. day of the year, as shown in Figure 3. Particle growth during the formation period has been estimated by determining the growth rate (dD_p/dt in nm/h) from the evolution of maximum size of the event mode. Subsequent growth has been estimated from particle size at 8 hours after start of event. The maximum size of the event particles at 8 hours after the start of event shows a clear dependence on temperature. A positive correlation with temperature may be explained by temperature dependence of biogenic vapour emissions [10]. However, it is not known with certainty that the precursors of the condensing species originate from vegetation. The wind direction analysis reveals that largest sizes for 8 hour growth are achieved with wind directions arriving from the sector WD=280° - 300. This, again may be explained by the low values of condensational sinks in air masses arriving from these directions (North Atlantic).

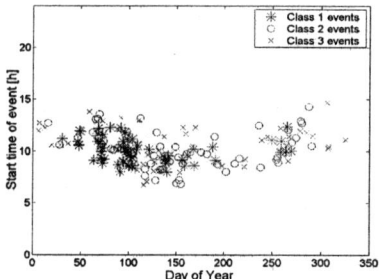

Fig. 2. Starting time of particle formation events vs day of year.

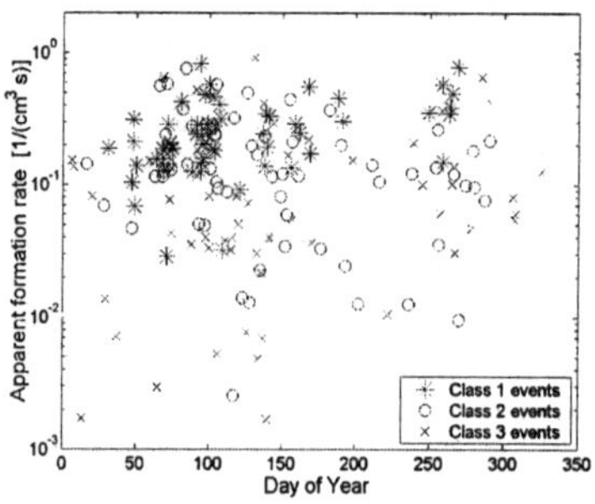

Figure 3. Apparent particle formation rate, (new particle yield)/(duration of particle formation).

The maximum size of the event particles at eight hours after event start plotted vs. day of year (Figure 4) reveals that the particles tend to grow larger in the summer. However, it remains unclear why particle formation is not observed to take place in the mid-summer, when one would expect e.g. the highest monoterpene concentrations. If the particle growth is very fast in the summertime and, moreover, if the particles are assumedly formed slightly higher up in the summertime, then, it is possible that all the particles grow in few minutes close to Aitken size already before getting to the ground level. When measuring down at the ground level, it is possible that we actually miss a fraction of the particle formation events classifying the ground level data only. The nucleation mode maximum size growth rate in the beginning of the event derived from the maximum size of event particles, is shown in Figure 5. The value of the estimated particle growth rate (dD_p/dt) is on the average 6-8 nm/h. But, the growth rate during summer time may be as high as 14-17 nm/h. The growth rate and the apparent particle formation rate do correlate, but the large variety in the data points suggests of a very weak connection between growth and formation.

The amount of visible light as well as UV-radiation was seen to correlate with the growth of the particles. However, all types of radiation are also directly linked with total amount of sunlight, and therefore the extent of vertical mixing also correlates with both UVA and UVB. The set of event data suggests a mean value of -0.015 K/m for the typical vertical potential temperature gradient $\Delta\Theta/\Delta z$ during the particle formation event. The reference data from non-event days gives a mean value of -0.005 K/m for the gradient In addition to the data presented, we also searched for a correlation for the particle formation rate and particle growth with several other quantities such as NO_x, SO_2, O_3 and RH, but no clear correlation was found.

I

Ibrahim, Y., 27
Imre, D., 428
Intskirveli, L., 793
Ishenko, M. A., 538
Iskandar, F., 773
Itkin, A. L., 37, 307
Itoh, M., 225
Itoh, Y., 877

J

Jackson, K. A., 406
Jaenicke, R., 740
Jayaraman, D., 178, 323
Jennings, S. G., 597, 827
Jianchun, G., 747
Johnson, B. A., 284
Johnson, L. R., 857
Johnston, M. V., 789
Jun, Y., 650, 747
Jurek, Z., 245

K

Kamra, A. K., 515, 526
Kandalgaonkar, S. S., 639, 643
Kane, D. B., 103, 789
Kärcher, B., 585
Kasahara, M., 736
Kashchiev, D., 147
Kasper-Giebl, A., 861
Kathmann, S. M., 197, 201
Katz, J. L., 35, 284
Kaupužs, J., 221
Kaverin, A. M., 420
Kelton, K. F., 386
Keronen, P., 896
Kevrekidis, P. G., 217
Kharchilava, J., 793
Khodzher, T. V., 740
Kholmurodov, K. T., 253
Kiefer, J., 260
Kiemle, C., 619
Kim, C. S., 665
Kiss, G., 761
Kissane, M. P., 362

Kleefeld, C., 597
Klein, K., 503
Klingo, V. V., 347, 351, 479, 482
Km, C. S., 877
Knott, M., 213
Koepke, P., 581
Kole, T. P., 35
Kolomiets, S. M., 804, 808
Koop, T., 549
Korhonen, P., 111, 827
Koropchak, J. A., 99
Koutzenogii, K. P., 740, 781
Krakovskaia, S. V., 467, 534
Kreidenweis, S. M., 443, 447, 451, 677
Krivácsy, Z., 761
Kromp, R., 861
Kshevetskiy, M. S., 299
Kulipanov, G. N., 740
Kulmala, M., 84, 111, 135, 366, 519, 654, 815, 827, 892, 896
Kuni, F. M., 131, 507, 511
Kunz, G., 597, 827
Kurasov, V., 257
Kusaka, I., 402

L

Laakso, L., 366
Laaksonen, A., 111, 135, 654, 711, 896
Lamanna, G., 287
Laulainen, N., 573
Laulainen, N. S., 55, 71
Lazaridis, M., 217
Leaitch, R., 847
Lee, D. W., 139
Li, J., 601
Li, J.-S., 15
Lihavainen, H., 84, 284
Löbbus, M., 291
Lovejoy, E. R., 327
Lunden, M. M., 877
Luo, B. P., 619, 623

M

Ma, C.-J., 736
MacKenzie, A. R., 619
Mahnke, R., 221, 229

Mäkelä, J. M., 366, 815, 823, 827, 892, 896
Maksimov, I. L., 15, 272
Manohar, G. K., 639
Maršík, F., 237
Martinez, D., 280
Martinsson, B., 631
Matas, J., 311
Matsumoto, M., 123, 151
Mavliev, R., 139
McClurg, R. B., 88
McFiggans, G., 331
McGaughey, A. J. H., 522
McGee, T. J., 623
McGraw, R., 3, 111, 373, 428, 785
Mclean, M., 241
McMurry, P. H., 699
Menon, S., 569
Mészáros, E., 751, 761
Mikheev, V. B., 55, 71
Miloshev, G., 455
Mirabel, P., 41
Möhler, O., 475
Molnár, A., 751, 761
Mordas, G., 835
Müller, F., 487
Müller, J., 800
Müller, K., 631
Music, S., 530

N

Narasimhamurthy, B., 796
Nasibulin, A. G., 99
Nemitz, E. G., 635, 646
Nenes, A., 565
Nickovic, S., 530
Nink, A., 475
Noppel, M., 339
Nuth, III, J. A., 75, 264

O

O'Dowd, C., 597
O'Dowd, C. D., 815, 823, 827, 892
Oh, K. J., 11, 107
Ohguchi, K., 123, 151
Oktem, B., 789

Okuyama, K., 665, 773, 877
Onasch, T. B., 428
O'Reilly, S., 597
Ortner, R., 135
Oxtoby, D., 402
Oxtoby, D. W., 245, 398

P

Padtberg, K., 432
Pal, P., 167
Paladino, J., 693
Pandis, S. N., 615
Pasternack, L., 847
Peeters, P., 67, 143
Penner, J. E., 677
Pervukhin, V. V., 71
Peter, T., 619, 623
Peters, F., 79
Petersen, D., 135
Pinaev, V., 205, 315, 358
Pirjola, L., 827, 892, 896
Pirnach, A. M., 467, 534
Plass-Dülmer, C., 819, 823
Plaude, N. O., 463
Podzimek, J., 769, 873
Poltavtsev, V. I., 276
Polygalov, Y. I., 354
Ponomarev, Y. F., 538
Poole, L., 623
Prasad, B. S. N., 796
Prenni, A. J., 428, 471, 728
Putaud, J.-P., 681
Puxbaum, H., 861
Puzyrewski, R., 303

R

Rabeony, M., 27
Raes, F., 681
Raju, N. V., 796
Ramasamy, P., 319, 323
Rama Varma Raja, M. K., 881
Ramløv, H., 410
Ranasinghe, K. S., 418
Raputa, V. F., 781
Rathke, B., 414
Ray, C. S., 418

Raymond, T. M., 615
Reischl, G., 861
Reischl, G. P., 335
Reiss, H., 3, 23, 41, 181
Renoux, A., 765
Reuder, J., 581
Robles Gonzalez, C., 589
Rogers, D. C., 443, 447, 451
Rudek, M. M., 284
Rusyniak, M., 27, 171

S

Sakiyama, K., 225
Salter, B. C., 284
Sanada, M., 272
Santacesaria, V., 619
Saxena, V. K., 569, 608
Schaaf, P., 3, 23
Schaap, M., 589
Schenter, G. K., 197, 201
Schmelzer, J. W. P., 155
Schmid, O., 693, 755
Schmitt, J. L., 51
Schmitt, S., 601
Schurath, U., 475
Schwander, H., 581
Schwartz, S. E., 785
Seinfeld, J. H., 111, 565
Senger, B., 23
Shandakov, S., 315, 358
Shandakov, S. D., 276
Shandakova, G. V., 276
Shantikumar Singh, N., 593
Shaw, G., 865
Shaymordanov, I. N., 354
Shchekin, A. K., 299, 507, 511
Shengjie, N., 759
Shimada, M., 665, 773, 877
Shneidman, V. A., 406
Shvets, I., 315
Sinkevich, A. A., 538
Smirnova, A. I., 781
Smolík, J., 311
Smolyakov, B. S., 781
Snider, J. R., 627, 865
Solomon, S., 585
Sopauskiene, D., 835
Spiridonov, V., 530

Stapcinskaite, S., 835
Stefanutti, L., 619
Stepanenko, V. D., 538
Streletzky, K. A., 59, 63
Strey, R., 7, 59, 95, 119, 249, 414, 495, 724
Stromberg, I. M., 839
Stuhlmann, R., 577
Subramanian, C., 178, 323
Sullivan, W., 847
Swietlicki, E., 631

T

Takano, H., 225
Talanquer, V., 398
Tanimura, S., 268
Tao, C.-J., 542
Tauer, K., 432
Tavartkiladze, K., 605
Temesi, D., 751
Theobold, M., 635, 646
Thomas, A., 619
Timoshenko, S. A., 354
Timoshina, L. V., 99
Tinmaker, M. I. R., 639, 643
Tinsley, B. A., 459
Tohno, S., 736
Tolbert, M. A., 428, 471, 661, 728
Tomida, H., 225
Toon, O. B., 503, 661, 689
Trueblood, M. B., 873
Tsai, W.-T., 542
Twohy, C., 831

U

Ulevicius, V., 835

V

Väkevä, M., 777, 827
Van Brunt, K., 51
Vandike, J. L., 424
Van Dingenen, R., 681
van Dongen, M. E. H., 67, 143, 287
Van Grieken, R., 740

van Loon, M., 589
Varadarajan, E., 319
Varga, B., 761
Veefkind, J. P., 589
Vehkamäki, H., 91, 213
Veremei, N. E., 538
Vignati, E., 681
Viisanen, Y., 7, 84, 111, 827
Voegel, J.-C., 23
Vogelsberger, W., 291
Vrtala, A., 135, 335, 491
Vychuzhanina, M. V., 463

Wölk, J., 7, 59
Wonczak, S., 95
Wong, J. G. D., 601
Wright, D. L., 785
Wyslouzil, B. E., 59, 63, 724

X

Xie, Y., 853

W

Wagner, P. E., 135, 335, 491, 519, 561
Walton, J. J., 677
Wang, H.-C., 139
Wang, X., 699
Ward, C. A., 522
Warshavsky, V. B., 299
Wenny, B. N., 608
White, A. J., 499, 888
Whitefield, P. D., 693, 755, 873
Wiedensohler, A., 631, 819, 823
Wilcox, C. F., 159
Wilemski, G., 15, 63, 343, 724
Williams, P. I., 331, 635
Wilson, P. W., 410
Windt, H., 732
Wirth, M., 619
Wise, M., 471
Wise, M. E., 728

Y

Yakovenko, T. M., 511
Yaroslavtseva, T. V., 781
Yasuoka, K., 123, 151, 253, 268
Ye, P., 19
Yoo, S., 107
Young, J. B., 884, 888
Yu, S., 608
Yum, S. S., 853

Z

Zach, S., 335
Zachariassen, K. E., 410
Zaveri, R., 573
Ždímal, V., 311
Zeng, X. C., 11, 107, 253
Zhang, Y.-X., 159
Zhou, J., 631
Zihua, L., 650, 747